Student's Solutions Manual
to accompany Jon Rogawski's

Single Variable
CALCULUS
EARLY TRANSCENDENTALS

SECOND EDITION

BRIAN BRADIE
Christopher Newport University

ROGER LIPSETT

W. H. FREEMAN AND COMPANY
NEW YORK

ISBN-13: 978-1-4292-5500-4
ISBN-10: 1-4292-5500-5

Printed in the United States of America

First Printing

W. H. Freeman and Company, 41 Madison Avenue, New York, NY 10010
Houndmills, Basingstoke RG21 6XS, England
www.whfreeman.com

CONTENTS

1 PRECALCULUS REVIEW

1.1 Real Numbers, Functions, and Graphs

Preliminary Questions

1. Give an example of numbers a and b such that $a < b$ and $|a| > |b|$.

SOLUTION Take $a = -3$ and $b = 1$. Then $a < b$ but $|a| = 3 > 1 = |b|$.

2. Which numbers satisfy $|a| = a$? Which satisfy $|a| = -a$? What about $|-a| = a$?

SOLUTION The numbers $a \geq 0$ satisfy $|a| = a$ and $|-a| = a$. The numbers $a \leq 0$ satisfy $|a| = -a$.

3. Give an example of numbers a and b such that $|a + b| < |a| + |b|$.

SOLUTION Take $a = -3$ and $b = 1$. Then

$$|a + b| = |-3 + 1| = |-2| = 2, \qquad \text{but} \qquad |a| + |b| = |-3| + |1| = 3 + 1 = 4.$$

Thus, $|a + b| < |a| + |b|$.

4. What are the coordinates of the point lying at the intersection of the lines $x = 9$ and $y = -4$?

SOLUTION The point $(9, -4)$ lies at the intersection of the lines $x = 9$ and $y = -4$.

5. In which quadrant do the following points lie?

(a) $(1, 4)$ **(b)** $(-3, 2)$ **(c)** $(4, -3)$ **(d)** $(-4, -1)$

SOLUTION

(a) Because both the x- and y-coordinates of the point $(1, 4)$ are positive, the point $(1, 4)$ lies in the first quadrant.

(b) Because the x-coordinate of the point $(-3, 2)$ is negative but the y-coordinate is positive, the point $(-3, 2)$ lies in the second quadrant.

(c) Because the x-coordinate of the point $(4, -3)$ is positive but the y-coordinate is negative, the point $(4, -3)$ lies in the fourth quadrant.

(d) Because both the x- and y-coordinates of the point $(-4, -1)$ are negative, the point $(-4, -1)$ lies in the third quadrant.

6. What is the radius of the circle with equation $(x - 9)^2 + (y - 9)^2 = 9$?

SOLUTION The circle with equation $(x - 9)^2 + (y - 9)^2 = 9$ has radius 3.

7. The equation $f(x) = 5$ has a solution if (choose one):

(a) 5 belongs to the domain of f.

(b) 5 belongs to the range of f.

SOLUTION The correct response is **(b)**: the equation $f(x) = 5$ has a solution if 5 belongs to the range of f.

8. What kind of symmetry does the graph have if $f(-x) = -f(x)$?

SOLUTION If $f(-x) = -f(x)$, then the graph of f is symmetric with respect to the origin.

Exercises

1. Use a calculator to find a rational number r such that $|r - \pi^2| < 10^{-4}$.

SOLUTION r must satisfy $\pi^2 - 10^{-4} < r < \pi^2 + 10^{-4}$, or $9.869504 < r < 9.869705$. $r = 9.8696 = \frac{12337}{1250}$ would be one such number.

In Exercises 3–8, express the interval in terms of an inequality involving absolute value.

3. $[-2, 2]$

SOLUTION $|x| \leq 2$

5. $(0, 4)$

SOLUTION The midpoint of the interval is $c = (0 + 4)/2 = 2$, and the radius is $r = (4 - 0)/2 = 2$; therefore, $(0, 4)$ can be expressed as $|x - 2| < 2$.

7. $[1, 5]$

SOLUTION The midpoint of the interval is $c = (1 + 5)/2 = 3$, and the radius is $r = (5 - 1)/2 = 2$; therefore, the interval $[1, 5]$ can be expressed as $|x - 3| \leq 2$.

In Exercises 9–12, write the inequality in the form $a < x < b$.

9. $|x| < 8$

SOLUTION $-8 < x < 8$

11. $|2x + 1| < 5$

SOLUTION $-5 < 2x + 1 < 5$ so $-6 < 2x < 4$ and $-3 < x < 2$

In Exercises 13–18, express the set of numbers x satisfying the given condition as an interval.

13. $|x| < 4$

SOLUTION $(-4, 4)$

15. $|x - 4| < 2$

SOLUTION The expression $|x - 4| < 2$ is equivalent to $-2 < x - 4 < 2$. Therefore, $2 < x < 6$, which represents the interval $(2, 6)$.

17. $|4x - 1| \leq 8$

SOLUTION The expression $|4x - 1| \leq 8$ is equivalent to $-8 \leq 4x - 1 \leq 8$ or $-7 \leq 4x \leq 9$. Therefore, $-\frac{7}{4} \leq x \leq \frac{9}{4}$, which represents the interval $[-\frac{7}{4}, \frac{9}{4}]$.

In Exercises 19–22, describe the set as a union of finite or infinite intervals.

19. $\{x : |x - 4| > 2\}$

SOLUTION $x - 4 > 2$ or $x - 4 < -2 \Rightarrow x > 6$ or $x < 2 \Rightarrow (-\infty, 2) \cup (6, \infty)$

21. $\{x : |x^2 - 1| > 2\}$

SOLUTION $x^2 - 1 > 2$ or $x^2 - 1 < -2 \Rightarrow x^2 > 3$ or $x^2 < -1$ (this will never happen) $\Rightarrow x > \sqrt{3}$ or $x < -\sqrt{3} \Rightarrow$ $(-\infty, -\sqrt{3}) \cup (\sqrt{3}, \infty)$.

23. Match (a)–(f) with (i)–(vi).

(a) $a > 3$

(b) $|a - 5| < \dfrac{1}{3}$

(c) $\left|a - \dfrac{1}{3}\right| < 5$

(d) $|a| > 5$

(e) $|a - 4| < 3$

(f) $1 \leq a \leq 5$

(i) a lies to the right of 3.
(ii) a lies between 1 and 7.
(iii) The distance from a to 5 is less than $\frac{1}{3}$.
(iv) The distance from a to 3 is at most 2.
(v) a is less than 5 units from $\frac{1}{3}$.
(vi) a lies either to the left of -5 or to the right of 5.

SOLUTION

(a) On the number line, numbers greater than 3 appear to the right; hence, $a > 3$ is equivalent to the numbers to the right of 3: **(i)**.

(b) $|a - 5|$ measures the distance from a to 5; hence, $|a - 5| < \frac{1}{3}$ is satisfied by those numbers less than $\frac{1}{3}$ of a unit from 5: **(iii)**.

(c) $|a - \frac{1}{3}|$ measures the distance from a to $\frac{1}{3}$; hence, $|a - \frac{1}{3}| < 5$ is satisfied by those numbers less than 5 units from $\frac{1}{3}$: **(v)**.

(d) The inequality $|a| > 5$ is equivalent to $a > 5$ or $a < -5$; that is, either a lies to the right of 5 or to the left of -5: **(vi)**.

(e) The interval described by the inequality $|a - 4| < 3$ has a center at 4 and a radius of 3; that is, the interval consists of those numbers between 1 and 7: **(ii)**.

(f) The interval described by the inequality $1 < x < 5$ has a center at 3 and a radius of 2; that is, the interval consists of those numbers less than 2 units from 3: **(iv)**.

25. Describe $\{x : x^2 + 2x < 3\}$ as an interval. *Hint:* Plot $y = x^2 + 2x - 3$.

SOLUTION The inequality $x^2 + 2x < 3$ is equivalent to $x^2 + 2x - 3 < 0$. In the figure below, we see that the graph of $y = x^2 + 2x - 3$ falls below the x-axis for $-3 < x < 1$. Thus, the set $\{x : x^2 + 2x < 3\}$ corresponds to the interval $-3 < x < 1$.

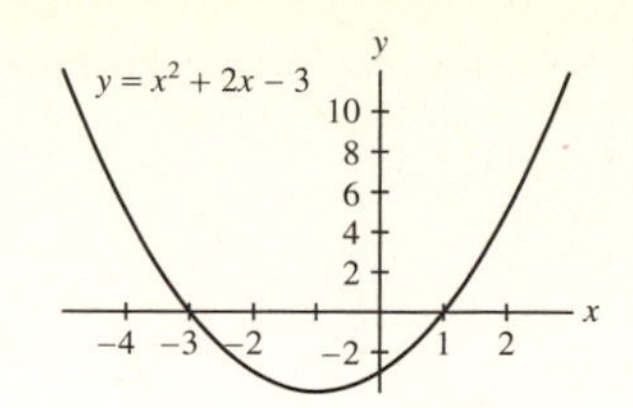

27. Show that if $a > b$, then $b^{-1} > a^{-1}$, provided that a and b have the same sign. What happens if $a > 0$ and $b < 0$?

SOLUTION Case 1a: If a and b are both positive, then $a > b \Rightarrow 1 > \frac{b}{a} \Rightarrow \frac{1}{b} > \frac{1}{a}$.

Case 1b: If a and b are both negative, then $a > b \Rightarrow 1 < \frac{b}{a}$ (since a is negative) $\Rightarrow \frac{1}{b} > \frac{1}{a}$ (again, since b is negative).

Case 2: If $a > 0$ and $b < 0$, then $\frac{1}{a} > 0$ and $\frac{1}{b} < 0$ so $\frac{1}{b} < \frac{1}{a}$. (See Exercise 2f for an example of this).

29. Show that if $|a - 5| < \frac{1}{2}$ and $|b - 8| < \frac{1}{2}$, then $|(a + b) - 13| < 1$. *Hint:* Use the triangle inequality.

SOLUTION

$$\begin{aligned}|a + b - 13| &= |(a - 5) + (b - 8)| \\ &\le |a - 5| + |b - 8| \quad \text{(by the triangle inequality)} \\ &< \frac{1}{2} + \frac{1}{2} = 1.\end{aligned}$$

31. Suppose that $|a - 6| \le 2$ and $|b| \le 3$.

(a) What is the largest possible value of $|a + b|$?

(b) What is the smallest possible value of $|a + b|$?

SOLUTION $|a - 6| \le 2$ guarantees that $4 \le a \le 8$, while $|b| \le 3$ guarantees that $-3 \le b \le 3$. Therefore $1 \le a + b \le 11$. It follows that

(a) the largest possible value of $|a + b|$ is 11; and

(b) the smallest possible value of $|a + b|$ is 1.

33. Express $r_1 = 0.\overline{27}$ as a fraction. *Hint:* $100r_1 - r_1$ is an integer. Then express $r_2 = 0.2666\ldots$ as a fraction.

SOLUTION Let $r_1 = .\overline{27}$. We observe that $100r_1 = 27.\overline{27}$. Therefore, $100r_1 - r_1 = 27.\overline{27} - .\overline{27} = 27$ and

$$r_1 = \frac{27}{99} = \frac{3}{11}.$$

Now, let $r_2 = 0.2\overline{666}$. Then $10r_2 = 2.\overline{666}$ and $100r_2 = 26.\overline{666}$. Therefore, $100r_2 - 10r_2 = 26.\overline{666} - 2.\overline{666} = 24$ and

$$r_2 = \frac{24}{90} = \frac{4}{15}.$$

35. The text states: *If the decimal expansions of numbers a and b agree to k places, then* $|a - b| \le 10^{-k}$. Show that the converse is false: For all k there are numbers a and b whose decimal expansions *do not agree at all* but $|a - b| \le 10^{-k}$.

SOLUTION Let $a = 1$ and $b = 0.\overline{9}$ (see the discussion before Example 1). The decimal expansions of a and b do not agree, but $|1 - 0.\overline{9}| < 10^{-k}$ for all k.

37. Find the equation of the circle with center $(2, 4)$:

(a) with radius $r = 3$.

(b) that passes through $(1, -1)$.

SOLUTION

(a) The equation of the indicated circle is $(x - 2)^2 + (y - 4)^2 = 3^2 = 9$.

(b) First determine the radius as the distance from the center to the indicated point on the circle:

$$r = \sqrt{(2 - 1)^2 + (4 - (-1))^2} = \sqrt{26}.$$

Thus, the equation of the circle is $(x - 2)^2 + (y - 4)^2 = 26$.

39. Determine the domain and range of the function

$$f : \{r, s, t, u\} \to \{A, B, C, D, E\}$$

defined by $f(r) = A$, $f(s) = B$, $f(t) = B$, $f(u) = E$.

SOLUTION The domain is the set $D = \{r, s, t, u\}$; the range is the set $R = \{A, B, E\}$.

In Exercises 41–48, find the domain and range of the function.

41. $f(x) = -x$

SOLUTION D : all reals; R : all reals

43. $f(x) = x^3$

SOLUTION D : all reals; R : all reals

45. $f(x) = |x|$

SOLUTION D : all reals; $R : \{y : y \geq 0\}$

47. $f(x) = \dfrac{1}{x^2}$

SOLUTION $D : \{x : x \neq 0\}$; $R : \{y : y > 0\}$

In Exercises 49–52, determine where $f(x)$ is increasing.

49. $f(x) = |x + 1|$

SOLUTION A graph of the function $y = |x + 1|$ is shown below. From the graph, we see that the function is increasing on the interval $(-1, \infty)$.

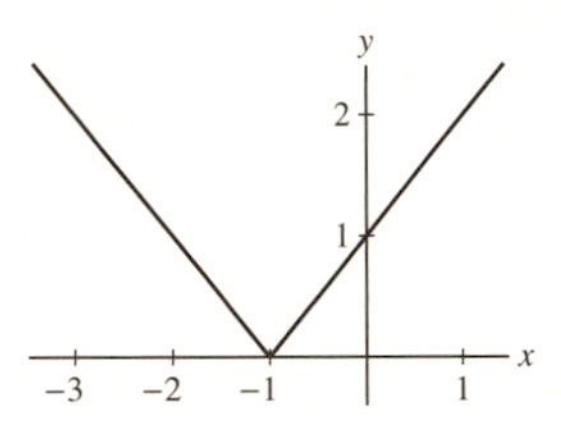

51. $f(x) = x^4$

SOLUTION A graph of the function $y = x^4$ is shown below. From the graph, we see that the function is increasing on the interval $(0, \infty)$.

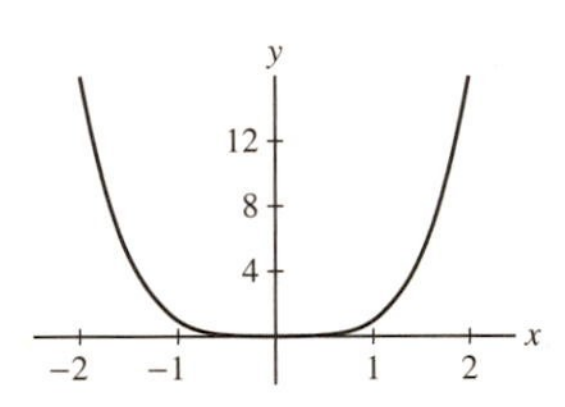

In Exercises 53–58, find the zeros of $f(x)$ and sketch its graph by plotting points. Use symmetry and increase/decrease information where appropriate.

53. $f(x) = x^2 - 4$

SOLUTION Zeros: ± 2
Increasing: $x > 0$
Decreasing: $x < 0$
Symmetry: $f(-x) = f(x)$ (even function). So, y-axis symmetry.

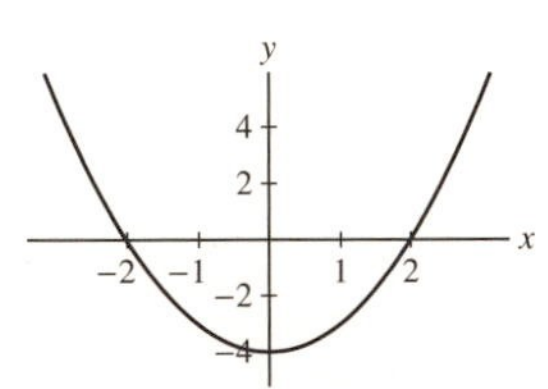

55. $f(x) = x^3 - 4x$

SOLUTION Zeros: $0, \pm 2$; Symmetry: $f(-x) = -f(x)$ (odd function). So origin symmetry.

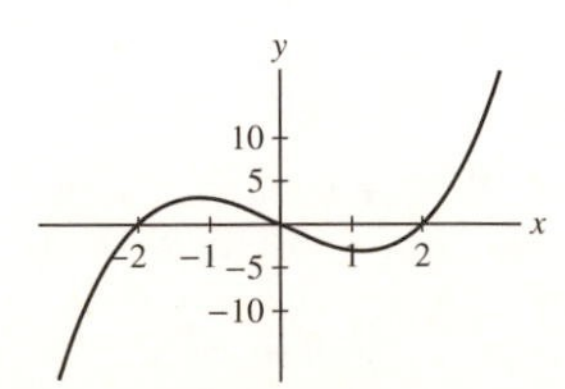

57. $f(x) = 2 - x^3$

SOLUTION This is an x-axis reflection of x^3 translated up 2 units. There is one zero at $x = \sqrt[3]{2}$.

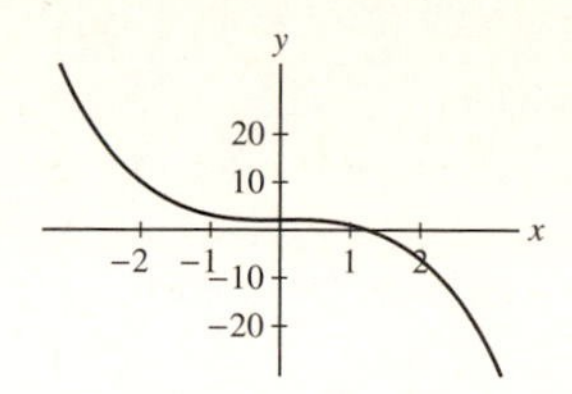

59. Which of the curves in Figure 26 is the graph of a function?

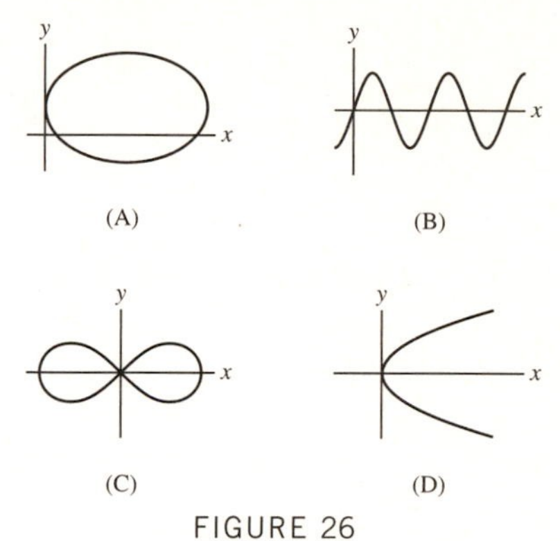

FIGURE 26

SOLUTION (B) is the graph of a function. (A), (C), and (D) all fail the vertical line test.

61. Determine whether the function is even, odd, or neither.

(a) $f(t) = \dfrac{1}{t^4 + t + 1} - \dfrac{1}{t^4 - t + 1}$

(b) $g(t) = 2^t - 2^{-t}$

(c) $G(\theta) = \sin\theta + \cos\theta$

(d) $H(\theta) = \sin(\theta^2)$

SOLUTION

(a) This function is odd because

$$f(-t) = \frac{1}{(-t)^4 + (-t) + 1} - \frac{1}{(-t)^4 - (-t) + 1}$$
$$= \frac{1}{t^4 - t + 1} - \frac{1}{t^4 + t + 1} = -f(t).$$

(b) $g(-t) = 2^{-t} - 2^{-(-t)} = 2^{-t} - 2^t = -g(t)$, so this function is odd.

(c) $G(-\theta) = \sin(-\theta) + \cos(-\theta) = -\sin\theta + \cos\theta$ which is equal to neither $G(\theta)$ nor $-G(\theta)$, so this function is neither odd nor even.

(d) $H(-\theta) = \sin((-\theta)^2) = \sin(\theta^2) = H(\theta)$, so this function is even.

63. Show that $f(x) = \ln\left(\dfrac{1-x}{1+x}\right)$ is an odd function.

SOLUTION

$$f(-x) = \ln\left(\frac{1-(-x)}{1+(-x)}\right)$$
$$= \ln\left(\frac{1+x}{1-x}\right) = -\ln\left(\frac{1-x}{1+x}\right) = -f(x),$$

so this is an odd function.

In Exercises 65–70, let $f(x)$ be the function shown in Figure 27.

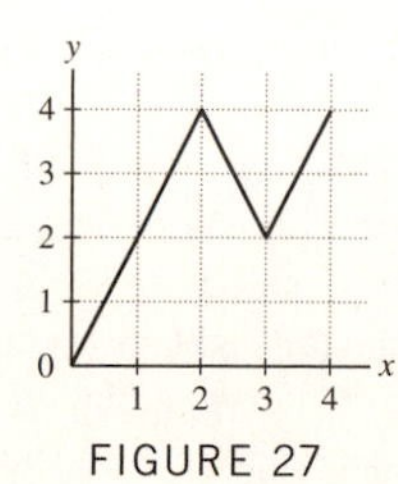

FIGURE 27

65. Find the domain and range of $f(x)$?

SOLUTION $D: [0, 4]$; $R: [0, 4]$

67. Sketch the graphs of $f(2x)$, $f\left(\frac{1}{2}x\right)$, and $2f(x)$.

SOLUTION The graph of $y = f(2x)$ is obtained by compressing the graph of $y = f(x)$ horizontally by a factor of 2 (see the graph below on the left). The graph of $y = f(\frac{1}{2}x)$ is obtained by stretching the graph of $y = f(x)$ horizontally by a factor of 2 (see the graph below in the middle). The graph of $y = 2f(x)$ is obtained by stretching the graph of $y = f(x)$ vertically by a factor of 2 (see the graph below on the right).

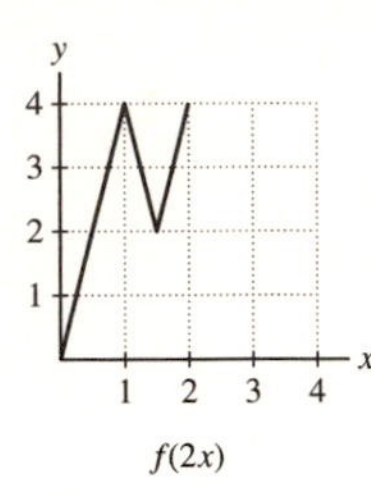

f(2x)

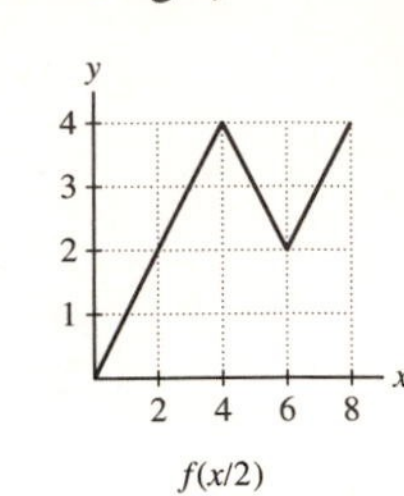

f(x/2)

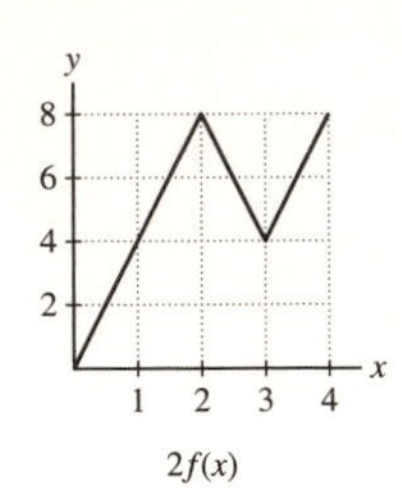

2f(x)

69. Extend the graph of $f(x)$ to $[-4, 4]$ so that it is an even function.

SOLUTION To continue the graph of $f(x)$ to the interval $[-4, 4]$ as an even function, reflect the graph of $f(x)$ across the y-axis (see the graph below).

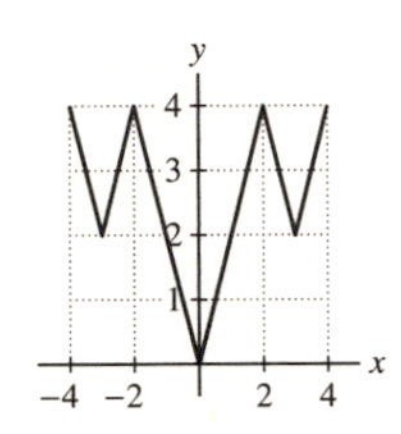

71. Suppose that $f(x)$ has domain $[4, 8]$ and range $[2, 6]$. Find the domain and range of:

(a) $f(x) + 3$
(b) $f(x + 3)$
(c) $f(3x)$
(d) $3f(x)$

SOLUTION

(a) $f(x) + 3$ is obtained by shifting $f(x)$ upward three units. Therefore, the domain remains $[4, 8]$, while the range becomes $[5, 9]$.

(b) $f(x + 3)$ is obtained by shifting $f(x)$ left three units. Therefore, the domain becomes $[1, 5]$, while the range remains $[2, 6]$.

(c) $f(3x)$ is obtained by compressing $f(x)$ horizontally by a factor of three. Therefore, the domain becomes $[\frac{4}{3}, \frac{8}{3}]$, while the range remains $[2, 6]$.

(d) $3f(x)$ is obtained by stretching $f(x)$ vertically by a factor of three. Therefore, the domain remains $[4, 8]$, while the range becomes $[6, 18]$.

73. Suppose that the graph of $f(x) = \sin x$ is compressed horizontally by a factor of 2 and then shifted 5 units to the right.

(a) What is the equation for the new graph?

(b) What is the equation if you first shift by 5 and then compress by 2?

(c) GU Verify your answers by plotting your equations.

SOLUTION

(a) Let $f(x) = \sin x$. After compressing the graph of f horizontally by a factor of 2, we obtain the function $g(x) = f(2x) = \sin 2x$. Shifting the graph 5 units to the right then yields

$$h(x) = g(x - 5) = \sin 2(x - 5) = \sin(2x - 10).$$

(b) Let $f(x) = \sin x$. After shifting the graph 5 units to the right, we obtain the function $g(x) = f(x - 5) = \sin(x - 5)$. Compressing the graph horizontally by a factor of 2 then yields

$$h(x) = g(2x) = \sin(2x - 5).$$

(c) The figure below at the top left shows the graphs of $y = \sin x$ (the dashed curve), the sine graph compressed horizontally by a factor of 2 (the dash, double dot curve) and then shifted right 5 units (the solid curve). Compare this last graph with the graph of $y = \sin(2x - 10)$ shown at the bottom left.

The figure below at the top right shows the graphs of $y = \sin x$ (the dashed curve), the sine graph shifted to the right 5 units (the dash, double dot curve) and then compressed horizontally by a factor of 2 (the solid curve). Compare this last graph with the graph of $y = \sin(2x - 5)$ shown at the bottom right.

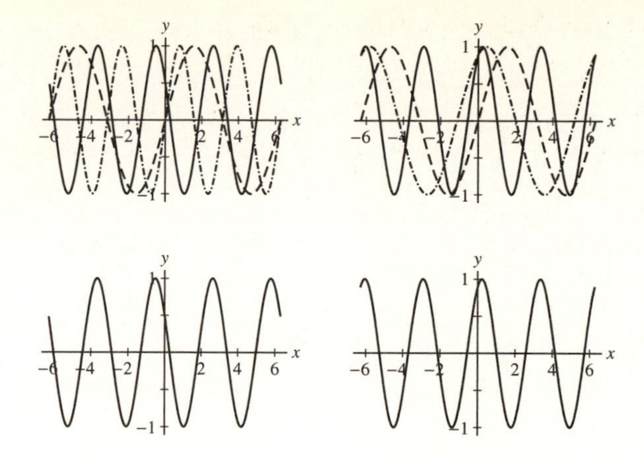

75. Sketch the graph of $f(2x)$ and $f\left(\frac{1}{2}x\right)$, where $f(x) = |x| + 1$ (Figure 28).

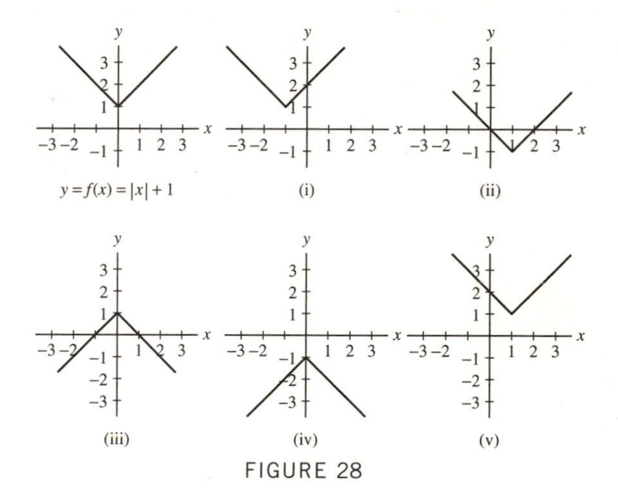

FIGURE 28

SOLUTION The graph of $y = f(2x)$ is obtained by compressing the graph of $y = f(x)$ horizontally by a factor of 2 (see the graph below on the left). The graph of $y = f(\frac{1}{2}x)$ is obtained by stretching the graph of $y = f(x)$ horizontally by a factor of 2 (see the graph below on the right).

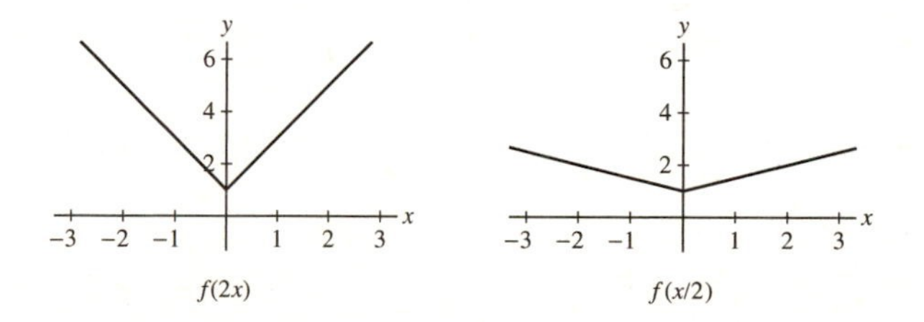

77. Define $f(x)$ to be the larger of x and $2 - x$. Sketch the graph of $f(x)$. What are its domain and range? Express $f(x)$ in terms of the absolute value function.

SOLUTION

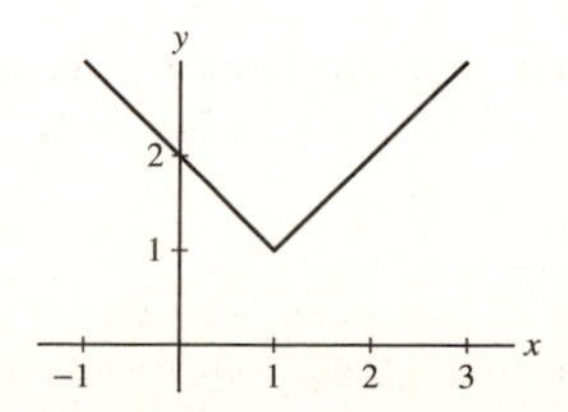

The graph of $y = f(x)$ is shown above. Clearly, the domain of f is the set of all real numbers while the range is $\{y \mid y \geq 1\}$. Notice the graph has the standard V-shape associated with the absolute value function, but the base of the V has been translated to the point $(1, 1)$. Thus, $f(x) = |x - 1| + 1$.

79. Show that the sum of two even functions is even and the sum of two odd functions is odd.

SOLUTION Even: $(f+g)(-x) = f(-x) + g(-x) \stackrel{\text{even}}{=} f(x) + g(x) = (f+g)(x)$

Odd: $(f+g)(-x) = f(-x) + g(-x) \stackrel{\text{odd}}{=} -f(x) + -g(x) = -(f+g)(x)$

81. Prove that the only function whose graph is symmetric with respect to both the y-axis and the origin is the function $f(x) = 0$.

SOLUTION Suppose f is symmetric with respect to the y-axis. Then $f(-x) = f(x)$. If f is also symmetric with respect to the origin, then $f(-x) = -f(x)$. Thus $f(x) = -f(x)$ or $2f(x) = 0$. Finally, $f(x) = 0$.

Further Insights and Challenges

83. Show that a fraction $r = a/b$ in lowest terms has a *finite* decimal expansion if and only if

$$b = 2^n 5^m \quad \text{for some } n, m \geq 0.$$

Hint: Observe that r has a finite decimal expansion when $10^N r$ is an integer for some $N \geq 0$ (and hence b divides 10^N).

SOLUTION Suppose r has a finite decimal expansion. Then there exists an integer $N \geq 0$ such that $10^N r$ is an integer, call it k. Thus, $r = k/10^N$. Because the only prime factors of 10 are 2 and 5, it follows that when r is written in lowest terms, its denominator must be of the form $2^n 5^m$ for some integers $n, m \geq 0$.

Conversely, suppose $r = a/b$ in lowest with $b = 2^n 5^m$ for some integers $n, m \geq 0$. Then $r = \frac{a}{b} = \frac{a}{2^n 5^m}$ or $2^n 5^m r = a$. If $m \geq n$, then $2^m 5^m r = a 2^{m-n}$ or $r = \frac{a 2^{m-n}}{10^m}$ and thus r has a finite decimal expansion (less than or equal to m terms, to be precise). On the other hand, if $n > m$, then $2^n 5^n r = a 5^{n-m}$ or $r = \frac{a 5^{n-m}}{10^n}$ and once again r has a finite decimal expansion.

85. A function $f(x)$ is symmetric with respect to the vertical line $x = a$ if $f(a - x) = f(a + x)$.

(a) Draw the graph of a function that is symmetric with respect to $x = 2$.

(b) Show that if $f(x)$ is symmetric with respect to $x = a$, then $g(x) = f(x + a)$ is even.

SOLUTION

(a) There are many possibilities, one of which is

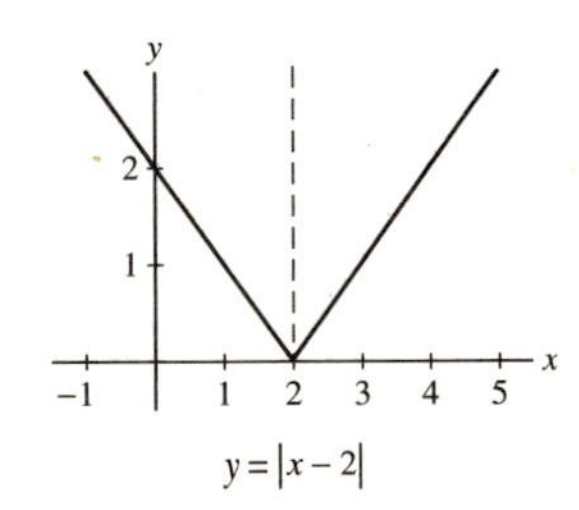

(b) Let $g(x) = f(x + a)$. Then

$$\begin{aligned} g(-x) = f(-x + a) &= f(a - x) \\ &= f(a + x) \qquad \text{symmetry with respect to } x = a \\ &= g(x) \end{aligned}$$

Thus, $g(x)$ is even.

1.2 Linear and Quadratic Functions

Preliminary Questions

1. What is the slope of the line $y = -4x - 9$?

SOLUTION The slope of the line $y = -4x - 9$ is -4, given by the coefficient of x.

2. Are the lines $y = 2x + 1$ and $y = -2x - 4$ perpendicular?

SOLUTION The slopes of perpendicular lines are negative reciprocals of one another. Because the slope of $y = 2x + 1$ is 2 and the slope of $y = -2x - 4$ is -2, these two lines are *not* perpendicular.

3. When is the line $ax + by = c$ parallel to the y-axis? To the x-axis?

SOLUTION The line $ax + by = c$ will be parallel to the y-axis when $b = 0$ and parallel to the x-axis when $a = 0$.

4. Suppose $y = 3x + 2$. What is Δy if x increases by 3?

SOLUTION Because $y = 3x + 2$ is a linear function with slope 3, increasing x by 3 will lead to $\Delta y = 3(3) = 9$.

5. What is the minimum of $f(x) = (x + 3)^2 - 4$?

SOLUTION Because $(x + 3)^2 \geq 0$, it follows that $(x + 3)^2 - 4 \geq -4$. Thus, the minimum value of $(x + 3)^2 - 4$ is -4.

6. What is the result of completing the square for $f(x) = x^2 + 1$?

SOLUTION Because there is no x term in $x^2 + 1$, completing the square on this expression leads to $(x - 0)^2 + 1$.

Exercises

In Exercises 1–4, find the slope, the y-intercept, and the x-intercept of the line with the given equation.

1. $y = 3x + 12$

SOLUTION Because the equation of the line is given in slope-intercept form, the slope is the coefficient of x and the y-intercept is the constant term: that is, $m = 3$ and the y-intercept is 12. To determine the x-intercept, substitute $y = 0$ and then solve for x: $0 = 3x + 12$ or $x = -4$.

3. $4x + 9y = 3$

SOLUTION To determine the slope and y-intercept, we first solve the equation for y to obtain the slope-intercept form. This yields $y = -\frac{4}{9}x + \frac{1}{3}$. From here, we see that the slope is $m = -\frac{4}{9}$ and the y-intercept is $\frac{1}{3}$. To determine the x-intercept, substitute $y = 0$ and solve for x: $4x = 3$ or $x = \frac{3}{4}$.

In Exercises 5–8, find the slope of the line.

5. $y = 3x + 2$

SOLUTION $m = 3$

7. $3x + 4y = 12$

SOLUTION First solve the equation for y to obtain the slope-intercept form. This yields $y = -\frac{3}{4}x + 3$. The slope of the line is therefore $m = -\frac{3}{4}$.

In Exercises 9–20, find the equation of the line with the given description.

9. Slope 3, y-intercept 8

SOLUTION Using the slope-intercept form for the equation of a line, we have $y = 3x + 8$.

11. Slope 3, passes through $(7, 9)$

SOLUTION Using the point-slope form for the equation of a line, we have $y - 9 = 3(x - 7)$ or $y = 3x - 12$.

13. Horizontal, passes through $(0, -2)$

SOLUTION A horizontal line has a slope of 0. Using the point-slope form for the equation of a line, we have $y - (-2) = 0(x - 0)$ or $y = -2$.

15. Parallel to $y = 3x - 4$, passes through $(1, 1)$

SOLUTION Because the equation $y = 3x - 4$ is in slope-intercept form, we can readily identify that it has a slope of 3. Parallel lines have the same slope, so the slope of the requested line is also 3. Using the point-slope form for the equation of a line, we have $y - 1 = 3(x - 1)$ or $y = 3x - 2$.

17. Perpendicular to $3x + 5y = 9$, passes through $(2, 3)$

SOLUTION We start by solving the equation $3x + 5y = 9$ for y to obtain the slope-intercept form for the equation of a line. This yields

$$y = -\frac{3}{5}x + \frac{9}{5},$$

from which we identify the slope as $-\frac{3}{5}$. Perpendicular lines have slopes that are negative reciprocals of one another, so the slope of the desired line is $m_{\perp} = \frac{5}{3}$. Using the point-slope form for the equation of a line, we have $y - 3 = \frac{5}{3}(x - 2)$ or $y = \frac{5}{3}x - \frac{1}{3}$.

19. Horizontal, passes through $(8, 4)$

SOLUTION A horizontal line has slope 0. Using the point slope form for the equation of a line, we have $y - 4 = 0(x - 8)$ or $y = 4$.

21. Find the equation of the perpendicular bisector of the segment joining (1, 2) and (5, 4) (Figure 11). *Hint:* The midpoint Q of the segment joining (a, b) and (c, d) is $\left(\frac{a+c}{2}, \frac{b+d}{2}\right)$.

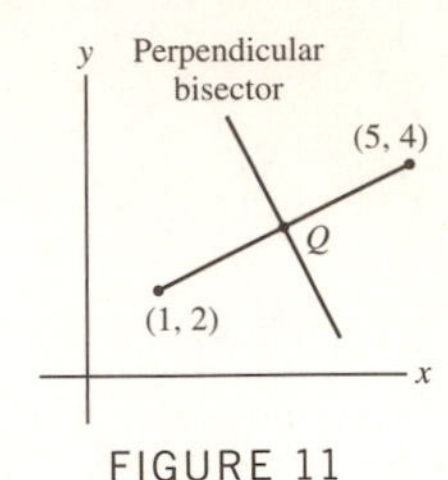

FIGURE 11

SOLUTION The slope of the segment joining (1, 2) and (5, 4) is

$$m = \frac{4-2}{5-1} = \frac{1}{2}$$

and the midpoint of the segment (Figure 11) is

$$\text{midpoint} = \left(\frac{1+5}{2}, \frac{2+4}{2}\right) = (3, 3)$$

The perpendicular bisector has slope $-1/m = -2$ and passes through (3, 3), so its equation is: $y - 3 = -2(x - 3)$ or $y = -2x + 9$.

23. Find an equation of the line with x-intercept $x = 4$ and y-intercept $y = 3$.

SOLUTION From Exercise 22, $\frac{x}{4} + \frac{y}{3} = 1$ or $3x + 4y = 12$.

25. Determine whether there exists a constant c such that the line $x + cy = 1$:

(a) Has slope 4
(b) Passes through (3, 1)
(c) Is horizontal
(d) Is vertical

SOLUTION

(a) Rewriting the equation of the line in slope-intercept form gives $y = -\frac{x}{c} + \frac{1}{c}$. To have slope 4 requires $-\frac{1}{c} = 4$ or $c = -\frac{1}{4}$.
(b) Substituting $x = 3$ and $y = 1$ into the equation of the line gives $3 + c = 1$ or $c = -2$.
(c) From (a), we know the slope of the line is $-\frac{1}{c}$. There is no value for c that will make this slope equal to 0.
(d) With $c = 0$, the equation becomes $x = 1$. This is the equation of a vertical line.

27. Materials expand when heated. Consider a metal rod of length L_0 at temperature T_0. If the temperature is changed by an amount ΔT, then the rod's length changes by $\Delta L = \alpha L_0 \Delta T$, where α is the thermal expansion coefficient. For steel, $\alpha = 1.24 \times 10^{-5}\,^\circ\text{C}^{-1}$.
(a) A steel rod has length $L_0 = 40$ cm at $T_0 = 40°$C. Find its length at $T = 90°$C.
(b) Find its length at $T = 50°$C if its length at $T_0 = 100°$C is 65 cm.
(c) Express length L as a function of T if $L_0 = 65$ cm at $T_0 = 100°$C.

SOLUTION

(a) With $T = 90°$C and $T_0 = 40°$C, $\Delta T = 50°$C. Therefore,

$$\Delta L = \alpha L_0 \Delta T = (1.24 \times 10^{-5})(40)(50) = .0248 \quad \text{and} \quad L = L_0 + \Delta L = 40.0248 \text{ cm}.$$

(b) With $T = 50°$C and $T_0 = 100°$C, $\Delta T = -50°$C. Therefore,

$$\Delta L = \alpha L_0 \Delta T = (1.24 \times 10^{-5})(65)(-50) = -.0403 \quad \text{and} \quad L = L_0 + \Delta L = 64.9597 \text{ cm}.$$

(c) $L = L_0 + \Delta L = L_0 + \alpha L_0 \Delta T = L_0(1 + \alpha \Delta T) = 65(1 + \alpha(T - 100))$

29. Find b such that $(2, -1)$, $(3, 2)$, and $(b, 5)$ lie on a line.

SOLUTION The slope of the line determined by the points $(2, -1)$ and $(3, 2)$ is

$$\frac{2-(-1)}{3-2} = 3.$$

To lie on the same line, the slope between (3, 2) and $(b, 5)$ must also be 3. Thus, we require

$$\frac{5-2}{b-3} = \frac{3}{b-3} = 3,$$

or $b = 4$.

31. The period T of a pendulum is measured for pendulums of several different lengths L. Based on the following data, does T appear to be a linear function of L?

L (cm)	20	30	40	50
T (s)	0.9	1.1	1.27	1.42

SOLUTION Examine the slope between consecutive data points. The first pair of data points yields a slope of

$$\frac{1.1-0.9}{30-20}=0.02,$$

while the second pair of data points yields a slope of

$$\frac{1.27-1.1}{40-30}=0.017,$$

and the last pair of data points yields a slope of

$$\frac{1.42-1.27}{50-40}=0.015$$

Because the three slopes are not equal, T does not appear to be a linear function of L.

33. Find the roots of the quadratic polynomials:

(a) $4x^2-3x-1$

(b) x^2-2x-1

SOLUTION

(a) $x=\dfrac{3\pm\sqrt{9-4(4)(-1)}}{2(4)}=\dfrac{3\pm\sqrt{25}}{8}=1 \text{ or } -\dfrac{1}{4}$

(b) $x=\dfrac{2\pm\sqrt{4-(4)(1)(-1)}}{2}=\dfrac{2\pm\sqrt{8}}{2}=1\pm\sqrt{2}$

In Exercises 34–41, complete the square and find the minimum or maximum value of the quadratic function.

35. $y=x^2-6x+9$

SOLUTION $y=(x-3)^2$; therefore, the minimum value of the quadratic polynomial is 0, and this occurs at $x=3$.

37. $y=x^2+6x+2$

SOLUTION $y=x^2+6x+9-9+2=(x+3)^2-7$; therefore, the minimum value of the quadratic polynomial is -7, and this occurs at $x=-3$.

39. $y=-4x^2+3x+8$

SOLUTION $y=-4x^2+3x+8=-4(x^2-\frac{3}{4}x+\frac{9}{64})+8+\frac{9}{16}=-4(x-\frac{3}{8})^2+\frac{137}{16}$; therefore, the maximum value of the quadratic polynomial is $\frac{137}{16}$, and this occurs at $x=\frac{3}{8}$.

41. $y=4x-12x^2$

SOLUTION $y=-12(x^2-\frac{x}{3})=-12(x^2-\frac{x}{3}+\frac{1}{36})+\frac{1}{3}=-12(x-\frac{1}{6})^2+\frac{1}{3}$; therefore, the maximum value of the quadratic polynomial is $\frac{1}{3}$, and this occurs at $x=\frac{1}{6}$.

43. Sketch the graph of $y=x^2+4x+6$ by plotting the minimum point, the y-intercept, and one other point.

SOLUTION $y=x^2+4x+4-4+6=(x+2)^2+2$ so the minimum occurs at $(-2,2)$. If $x=0$, then $y=6$ and if $x=-4$, $y=6$. This is the graph of x^2 moved left 2 units and up 2 units.

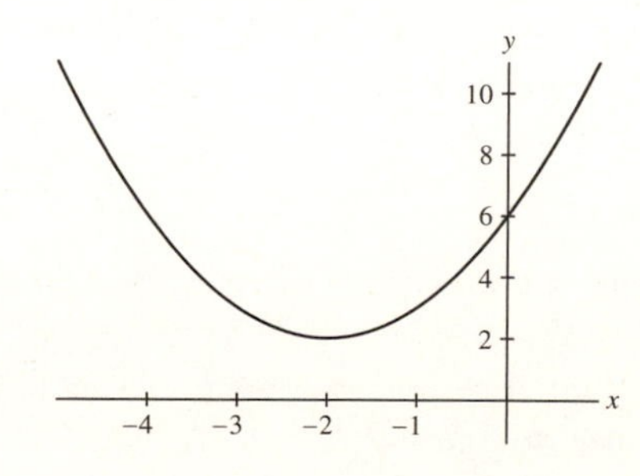

45. For which values of c does $f(x) = x^2 + cx + 1$ have a double root? No real roots?

SOLUTION A double root occurs when $c^2 - 4(1)(1) = 0$ or $c^2 = 4$. Thus, $c = \pm 2$.

There are no real roots when $c^2 - 4(1)(1) < 0$ or $c^2 < 4$. Thus, $-2 < c < 2$.

47. Prove that $x + \frac{1}{x} \geq 2$ for all $x > 0$. *Hint:* Consider $(x^{1/2} - x^{-1/2})^2$.

SOLUTION Let $x > 0$. Then

$$\left(x^{1/2} - x^{-1/2}\right)^2 = x - 2 + \frac{1}{x}.$$

Because $(x^{1/2} - x^{-1/2})^2 \geq 0$, it follows that

$$x - 2 + \frac{1}{x} \geq 0 \qquad \text{or} \qquad x + \frac{1}{x} \geq 2.$$

49. If objects of weights x and w_1 are suspended from the balance in Figure 13(A), the cross-beam is horizontal if $bx = aw_1$. If the lengths a and b are known, we may use this equation to determine an unknown weight x by selecting w_1 such that the cross-beam is horizontal. If a and b are not known precisely, we might proceed as follows. First balance x by w_1 on the left as in (A). Then switch places and balance x by w_2 on the right as in (B). The average $\bar{x} = \frac{1}{2}(w_1 + w_2)$ gives an estimate for x. Show that $\bar{x}$ is greater than or equal to the true weight x.

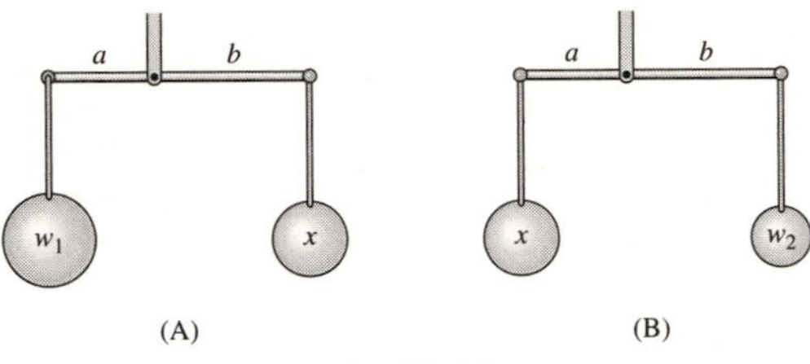

FIGURE 13

SOLUTION First note $bx = aw_1$ and $ax = bw_2$. Thus,

$$\begin{aligned}
\bar{x} &= \frac{1}{2}(w_1 + w_2) \\
&= \frac{1}{2}\left(\frac{bx}{a} + \frac{ax}{b}\right) \\
&= \frac{x}{2}\left(\frac{b}{a} + \frac{a}{b}\right) \\
&\geq \frac{x}{2}(2) \qquad \text{by Exercise 47} \\
&= x
\end{aligned}$$

51. Find a pair of numbers whose sum and product are both equal to 8.

SOLUTION Let x and y be numbers whose sum and product are both equal to 8. Then $x + y = 8$ and $xy = 8$. From the second equation, $y = \frac{8}{x}$. Substituting this expression for y in the first equation gives $x + \frac{8}{x} = 8$ or $x^2 - 8x + 8 = 0$. By the quadratic formula,

$$x = \frac{8 \pm \sqrt{64 - 32}}{2} = 4 \pm 2\sqrt{2}.$$

If $x = 4 + 2\sqrt{2}$, then

$$y = \frac{8}{4 + 2\sqrt{2}} = \frac{8}{4 + 2\sqrt{2}} \cdot \frac{4 - 2\sqrt{2}}{4 - 2\sqrt{2}} = 4 - 2\sqrt{2}.$$

On the other hand, if $x = 4 - 2\sqrt{2}$, then

$$y = \frac{8}{4 - 2\sqrt{2}} = \frac{8}{4 - 2\sqrt{2}} \cdot \frac{4 + 2\sqrt{2}}{4 + 2\sqrt{2}} = 4 + 2\sqrt{2}.$$

Thus, the two numbers are $4 + 2\sqrt{2}$ and $4 - 2\sqrt{2}$.

Further Insights and Challenges

53. Show that if $f(x)$ and $g(x)$ are linear, then so is $f(x) + g(x)$. Is the same true of $f(x)g(x)$?

SOLUTION If $f(x) = mx + b$ and $g(x) = nx + d$, then

$$f(x) + g(x) = mx + b + nx + d = (m + n)x + (b + d),$$

which is linear. $f(x)g(x)$ is not generally linear. Take, for example, $f(x) = g(x) = x$. Then $f(x)g(x) = x^2$.

55. Show that $\Delta y/\Delta x$ for the function $f(x) = x^2$ over the interval $[x_1, x_2]$ is not a constant, but depends on the interval. Determine the exact dependence of $\Delta y/\Delta x$ on x_1 and x_2.

SOLUTION For x^2, $\dfrac{\Delta y}{\Delta x} = \dfrac{x_2^2 - x_1^2}{x_2 - x_1} = x_2 + x_1$.

57. Let $a, c \neq 0$. Show that the roots of

$$ax^2 + bx + c = 0 \quad \text{and} \quad cx^2 + bx + a = 0$$

are reciprocals of each other.

SOLUTION Let r_1 and r_2 be the roots of $ax^2 + bx + c$ and r_3 and r_4 be the roots of $cx^2 + bx + a$. Without loss of generality, let

$$r_1 = \frac{-b + \sqrt{b^2 - 4ac}}{2a} \Rightarrow \frac{1}{r_1} = \frac{2a}{-b + \sqrt{b^2 - 4ac}} \cdot \frac{-b - \sqrt{b^2 - 4ac}}{-b - \sqrt{b^2 - 4ac}}$$

$$= \frac{2a(-b - \sqrt{b^2 - 4ac})}{b^2 - b^2 + 4ac} = \frac{-b - \sqrt{b^2 - 4ac}}{2c} = r_4.$$

Similarly, you can show $\dfrac{1}{r_2} = r_3$.

59. Prove **Viète's Formulas**: The quadratic polynomial with α and β as roots is $x^2 + bx + c$, where $b = -\alpha - \beta$ and $c = \alpha\beta$.

SOLUTION If a quadratic polynomial has roots α and β, then the polynomial is

$$(x - \alpha)(x - \beta) = x^2 - \alpha x - \beta x + \alpha\beta = x^2 + (-\alpha - \beta)x + \alpha\beta.$$

Thus, $b = -\alpha - \beta$ and $c = \alpha\beta$.

1.3 The Basic Classes of Functions

Preliminary Questions

1. Give an example of a rational function.

SOLUTION One example is $\dfrac{3x^2 - 2}{7x^3 + x - 1}$.

2. Is $|x|$ a polynomial function? What about $|x^2 + 1|$?

SOLUTION $|x|$ is not a polynomial; however, because $x^2 + 1 > 0$ for all x, it follows that $|x^2 + 1| = x^2 + 1$, which is a polynomial.

3. What is unusual about the domain of the composite function $f \circ g$ for the functions $f(x) = x^{1/2}$ and $g(x) = -1 - |x|$?

SOLUTION Recall that $(f \circ g)(x) = f(g(x))$. Now, for any real number x, $g(x) = -1 - |x| \leq -1 < 0$. Because we cannot take the square root of a negative number, it follows that $f(g(x))$ is not defined for any real number. In other words, the domain of $f(g(x))$ is the empty set.

4. Is $f(x) = \left(\frac{1}{2}\right)^x$ increasing or decreasing?

SOLUTION The function $f(x) = (\frac{1}{2})^x$ is an exponential function with base $b = \frac{1}{2} < 1$. Therefore, f is a decreasing function.

5. Give an example of a transcendental function.

SOLUTION One possibility is $f(x) = e^x - \sin x$.

Exercises

In Exercises 1–12, determine the domain of the function.

1. $f(x) = x^{1/4}$

SOLUTION $x \geq 0$

3. $f(x) = x^3 + 3x - 4$

SOLUTION All reals

5. $g(t) = \dfrac{1}{t+2}$

SOLUTION $t \neq -2$

7. $G(u) = \dfrac{1}{u^2 - 4}$

SOLUTION $u \neq \pm 2$

9. $f(x) = x^{-4} + (x-1)^{-3}$

SOLUTION $x \neq 0, 1$

11. $g(y) = 10^{\sqrt{y} + y^{-1}}$

SOLUTION $y > 0$

In Exercises 13–24, identify each of the following functions as polynomial, rational, algebraic, or transcendental.

13. $f(x) = 4x^3 + 9x^2 - 8$

SOLUTION Polynomial

15. $f(x) = \sqrt{x}$

SOLUTION Algebraic

17. $f(x) = \dfrac{x^2}{x + \sin x}$

SOLUTION Transcendental

19. $f(x) = \dfrac{2x^3 + 3x}{9 - 7x^2}$

SOLUTION Rational

21. $f(x) = \sin(x^2)$

SOLUTION Transcendental

23. $f(x) = x^2 + 3x^{-1}$

SOLUTION Rational

25. Is $f(x) = 2^{x^2}$ a transcendental function?

SOLUTION Yes.

In Exercises 27–34, calculate the composite functions $f \circ g$ and $g \circ f$, and determine their domains.

27. $f(x) = \sqrt{x}, \quad g(x) = x + 1$

SOLUTION $f(g(x)) = \sqrt{x+1}$; $D: x \geq -1$, $\quad g(f(x)) = \sqrt{x} + 1$; $D: x \geq 0$

29. $f(x) = 2^x, \quad g(x) = x^2$

SOLUTION $f(g(x)) = 2^{x^2}$; D: $\mathbf{R}$, $\quad g(f(x)) = (2^x)^2 = 2^{2x}$; D: $\mathbf{R}$

31. $f(\theta) = \cos\theta, \quad g(x) = x^3 + x^2$

SOLUTION $f(g(x)) = \cos(x^3 + x^2)$; D: $\mathbf{R}$, $\quad g(f(\theta)) = \cos^3\theta + \cos^2\theta$; D: $\mathbf{R}$

33. $f(t) = \dfrac{1}{\sqrt{t}}, \quad g(t) = -t^2$

SOLUTION $f(g(t)) = \dfrac{1}{\sqrt{-t^2}}$; D: Not valid for any t, $\quad g(f(t)) = -\left(\frac{1}{\sqrt{t}}\right)^2 = -\frac{1}{t}$; $D: t > 0$

35. The population (in millions) of a country as a function of time t (years) is $P(t) = 30.2^{0.1t}$. Show that the population doubles every 10 years. Show more generally that for any positive constants a and k, the function $g(t) = a2^{kt}$ doubles after $1/k$ years.

SOLUTION Let $P(t) = 30 \cdot 2^{0.1t}$. Then

$$P(t+10) = 30 \cdot 2^{0.1(t+10)} = 30 \cdot 2^{0.1t+1} = 2(30 \cdot 2^{0.1t}) = 2P(t).$$

Hence, the population doubles in size every 10 years. In the more general case, let $g(t) = a2^{kt}$. Then

$$g\left(t + \frac{1}{k}\right) = a2^{k(t+1/k)} = a2^{kt+1} = 2a2^{kt} = 2g(t).$$

Hence, the function g doubles after $1/k$ years.

Further Insights and Challenges

In Exercises 37–43, we define the first difference δf *of a function* $f(x)$ *by* $\delta f(x) = f(x+1) - f(x)$.

37. Show that if $f(x) = x^2$, then $\delta f(x) = 2x + 1$. Calculate δf for $f(x) = x$ and $f(x) = x^3$.

SOLUTION $f(x) = x^2$: $\delta f(x) = f(x+1) - f(x) = (x+1)^2 - x^2 = 2x + 1$
$f(x) = x$: $\delta f(x) = x + 1 - x = 1$
$f(x) = x^3$: $\delta f(x) = (x+1)^3 - x^3 = 3x^2 + 3x + 1$

39. Show that for any two functions f and g, $\delta(f+g) = \delta f + \delta g$ and $\delta(cf) = c\delta(f)$, where c is any constant.

SOLUTION $\delta(f+g) = (f(x+1) + g(x+1)) - (f(x) - g(x))$

$$= (f(x+1) - f(x)) + (g(x+1) - g(x)) = \delta f(x) + \delta g(x)$$

$$\delta(cf) = cf(x+1) - cf(x) = c(f(x+1) - f(x)) = c\delta f(x).$$

41. First show that

$$P(x) = \frac{x(x+1)}{2}$$

satisfies $\delta P = (x+1)$. Then apply Exercise 40 to conclude that

$$1 + 2 + 3 + \cdots + n = \frac{n(n+1)}{2}$$

SOLUTION Let $P(x) = x(x+1)/2$. Then

$$\delta P(x) = P(x+1) - P(x) = \frac{(x+1)(x+2)}{2} - \frac{x(x+1)}{2} = \frac{(x+1)(x+2-x)}{2} = x + 1.$$

Also, note that $P(0) = 0$. Thus, by Exercise 40, with $k = 1$, it follows that

$$P(n) = \frac{n(n+1)}{2} = 1 + 2 + 3 + \cdots + n.$$

43. This exercise combined with Exercise 40 shows that for all whole numbers k, there exists a polynomial $P(x)$ satisfying Eq. (1). The solution requires the Binomial Theorem and proof by induction (see Appendix C).

(a) Show that $\delta(x^{k+1}) = (k+1)x^k + \cdots$, where the dots indicate terms involving smaller powers of x.

(b) Show by induction that there exists a polynomial of degree $k+1$ with leading coefficient $1/(k+1)$:

$$P(x) = \frac{1}{k+1}x^{k+1} + \cdots$$

such that $\delta P = (x+1)^k$ and $P(0) = 0$.

SOLUTION

(a) By the Binomial Theorem:

$$\delta(x^{n+1}) = (x+1)^{n+1} - x^{n+1} = \left(x^{n+1} + \binom{n+1}{1}x^n + \binom{n+1}{2}x^{n-1} + \cdots + 1\right) - x^{n+1}$$

$$= \binom{n+1}{1}x^n + \binom{n+1}{2}x^{n-1} + \cdots + 1$$

Thus,

$$\delta(x^{n+1}) = (n+1)x^n + \cdots$$

where the dots indicate terms involving smaller powers of x.

(b) For $k = 0$, note that $P(x) = x$ satisfies $\delta P = (x+1)^0 = 1$ and $P(0) = 0$.
Now suppose the polynomial

$$P(x) = \frac{1}{k}x^k + p_{k-1}x^{k-1} + \cdots + p_1 x$$

which clearly satisfies $P(0) = 0$ also satisfies $\delta P = (x+1)^{k-1}$. We try to prove the existence of

$$Q(x) = \frac{1}{k+1}x^{k+1} + q_k x^k + \cdots + q_1 x$$

such that $\delta Q = (x+1)^k$. Observe that $Q(0) = 0$.
If $\delta Q = (x+1)^k$ and $\delta P = (x+1)^{k-1}$, then

$$\delta Q = (x+1)^k = (x+1)\delta P = x\delta P(x) + \delta P$$

By the linearity of δ (Exercise 39), we find $\delta Q - \delta P = x\delta P$ or $\delta(Q - P) = x\delta P$. By definition,

$$Q - P = \frac{1}{k+1}x^{k+1} + \left(q_k - \frac{1}{k}\right)x^k + \cdots + (q_1 - p_1)x,$$

so, by the linearity of δ,

$$\delta(Q - P) = \frac{1}{k+1}\delta(x^{k+1}) + \left(q_k - \frac{1}{k}\right)\delta(x^k) + \cdots + (q_1 - p_1) = x(x+1)^{k-1} \qquad \boxed{1}$$

By part (a),

$$\delta(x^{k+1}) = (k+1)x^k + L_{k-1,k-1}x^{k-1} + \ldots + L_{k-1,1}x + 1$$
$$\delta(x^k) = kx^{k-1} + L_{k-2,k-2}x^{k-2} + \ldots + L_{k-2,1}x + 1$$
$$\vdots$$
$$\delta(x^2) = 2x + 1$$

where the $L_{i,j}$ are real numbers for each i, j.
To construct Q, we have to group like powers of x on both sides of Eq. (1). This yields the system of equations

$$\frac{1}{k+1}\left((k+1)x^k\right) = x^k$$
$$\frac{1}{k+1}L_{k-1,k-1}x^{k-1} + \left(q_k - \frac{1}{k}\right)kx^{k-1} = (k-1)x^{k-1}$$
$$\vdots$$
$$\frac{1}{k+1} + \left(q_k - \frac{1}{k}\right) + (q_{k-1} - p_{k-1}) + \cdots + (q_1 - p_1) = 0.$$

The first equation is identically true, and the second equation can be solved immediately for q_k. Substituting the value of q_k into the third equation of the system, we can then solve for q_{k-1}. We continue this process until we substitute the values of $q_k, q_{k-1}, \ldots q_2$ into the last equation, and then solve for q_1.

1.4 Trigonometric Functions

Preliminary Questions

1. How is it possible for two different rotations to define the same angle?

SOLUTION Working from the same initial radius, two rotations that differ by a whole number of full revolutions will have the same ending radius; consequently, the two rotations will define the same angle even though the measures of the rotations will be different.

2. Give two different positive rotations that define the angle $\pi/4$.

SOLUTION The angle $\pi/4$ is defined by any rotation of the form $\frac{\pi}{4} + 2\pi k$ where k is an integer. Thus, two different positive rotations that define the angle $\pi/4$ are

$$\frac{\pi}{4} + 2\pi(1) = \frac{9\pi}{4} \qquad \text{and} \qquad \frac{\pi}{4} + 2\pi(5) = \frac{41\pi}{4}.$$

3. Give a negative rotation that defines the angle $\pi/3$.

SOLUTION The angle $\pi/3$ is defined by any rotation of the form $\frac{\pi}{3} + 2\pi k$ where k is an integer. Thus, a negative rotation that defines the angle $\pi/3$ is

$$\frac{\pi}{3} + 2\pi(-1) = -\frac{5\pi}{3}.$$

4. The definition of $\cos\theta$ using right triangles applies when (choose the correct answer):

(a) $0 < \theta < \frac{\pi}{2}$ **(b)** $0 < \theta < \pi$ **(c)** $0 < \theta < 2\pi$

SOLUTION The correct response is **(a)**: $0 < \theta < \frac{\pi}{2}$.

5. What is the unit circle definition of $\sin\theta$?

SOLUTION Let O denote the center of the unit circle, and let P be a point on the unit circle such that the radius $\overline{OP}$ makes an angle θ with the positive x-axis. Then, $\sin\theta$ is the y-coordinate of the point P.

6. How does the periodicity of $\sin\theta$ and $\cos\theta$ follow from the unit circle definition?

SOLUTION Let O denote the center of the unit circle, and let P be a point on the unit circle such that the radius $\overline{OP}$ makes an angle θ with the positive x-axis. Then, $\cos\theta$ and $\sin\theta$ are the x- and y-coordinates, respectively, of the point P. The angle $\theta + 2\pi$ is obtained from the angle θ by making one full revolution around the circle. The angle $\theta + 2\pi$ will therefore have the radius $\overline{OP}$ as its terminal side. Thus

$$\cos(\theta + 2\pi) = \cos\theta \quad \text{and} \quad \sin(\theta + 2\pi) = \sin\theta.$$

In other words, $\sin\theta$ and $\cos\theta$ are periodic functions.

Exercises

1. Find the angle between 0 and 2π equivalent to $13\pi/4$.

SOLUTION Because $13\pi/4 > 2\pi$, we repeatedly subtract 2π until we arrive at a radian measure that is between 0 and 2π. After one subtraction, we have $13\pi/4 - 2\pi = 5\pi/4$. Because $0 < 5\pi/4 < 2\pi$, $5\pi/4$ is the angle measure between 0 and 2π that is equivalent to $13\pi/4$.

3. Convert from radians to degrees:

(a) 1 **(b)** $\frac{\pi}{3}$ **(c)** $\frac{5}{12}$ **(d)** $-\frac{3\pi}{4}$

SOLUTION

(a) $1\left(\frac{180^\circ}{\pi}\right) = \frac{180^\circ}{\pi} \approx 57.3^\circ$

(b) $\frac{\pi}{3}\left(\frac{180^\circ}{\pi}\right) = 60^\circ$

(c) $\frac{5}{12}\left(\frac{180^\circ}{\pi}\right) = \frac{75^\circ}{\pi} \approx 23.87^\circ$

(d) $-\frac{3\pi}{4}\left(\frac{180^\circ}{\pi}\right) = -135^\circ$

5. Find the lengths of the arcs subtended by the angles θ and ϕ radians in Figure 20.

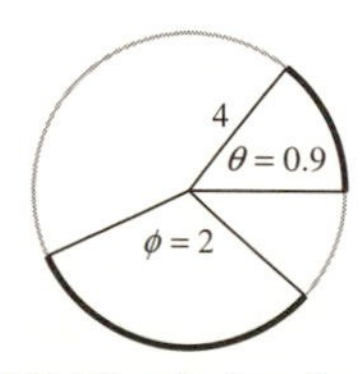

FIGURE 20 Circle of radius 4.

SOLUTION $s = r\theta = 4(.9) = 3.6$; $s = r\phi = 4(2) = 8$

7. Fill in the remaining values of $(\cos\theta, \sin\theta)$ for the points in Figure 22.

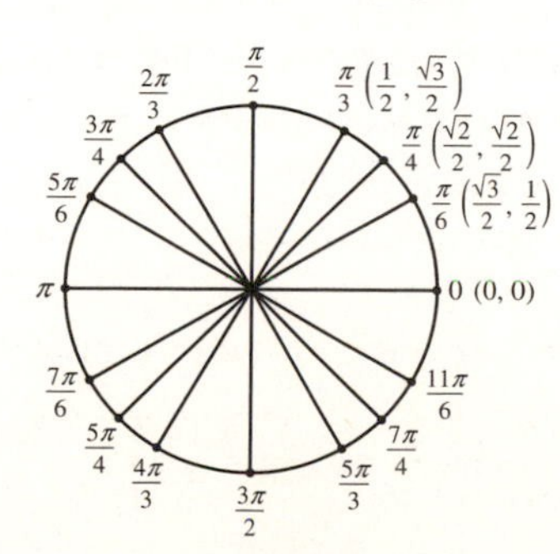

FIGURE 22

SOLUTION

θ	$\frac{\pi}{2}$	$\frac{2\pi}{3}$	$\frac{3\pi}{4}$	$\frac{5\pi}{6}$	π	$\frac{7\pi}{6}$
$(\cos\theta, \sin\theta)$	$(0, 1)$	$\left(\frac{-1}{2}, \frac{\sqrt{3}}{2}\right)$	$\left(\frac{-\sqrt{2}}{2}, \frac{\sqrt{2}}{2}\right)$	$\left(\frac{-\sqrt{3}}{2}, \frac{1}{2}\right)$	$(-1, 0)$	$\left(\frac{-\sqrt{3}}{2}, \frac{-1}{2}\right)$
θ	$\frac{5\pi}{4}$	$\frac{4\pi}{3}$	$\frac{3\pi}{2}$	$\frac{5\pi}{3}$	$\frac{7\pi}{4}$	$\frac{11\pi}{6}$
$(\cos\theta, \sin\theta)$	$\left(\frac{-\sqrt{2}}{2}, \frac{-\sqrt{2}}{2}\right)$	$\left(\frac{-1}{2}, \frac{-\sqrt{3}}{2}\right)$	$(0, -1)$	$\left(\frac{1}{2}, \frac{-\sqrt{3}}{2}\right)$	$\left(\frac{\sqrt{2}}{2}, \frac{-\sqrt{2}}{2}\right)$	$\left(\frac{\sqrt{3}}{2}, \frac{-1}{2}\right)$

In Exercises 9–14, use Figure 22 to find all angles between 0 *and* 2π *satisfying the given condition.*

9. $\cos\theta = \frac{1}{2}$

SOLUTION $\theta = \frac{\pi}{3}, \frac{5\pi}{3}$

11. $\tan\theta = -1$

SOLUTION $\theta = \frac{3\pi}{4}, \frac{7\pi}{4}$

13. $\sin x = \frac{\sqrt{3}}{2}$

SOLUTION $x = \frac{\pi}{3}, \frac{2\pi}{3}$

15. Fill in the following table of values:

θ	$\frac{\pi}{6}$	$\frac{\pi}{4}$	$\frac{\pi}{3}$	$\frac{\pi}{2}$	$\frac{2\pi}{3}$	$\frac{3\pi}{4}$	$\frac{5\pi}{6}$
$\tan\theta$							
$\sec\theta$							

SOLUTION

θ	$\frac{\pi}{6}$	$\frac{\pi}{4}$	$\frac{\pi}{3}$	$\frac{\pi}{2}$	$\frac{2\pi}{3}$	$\frac{3\pi}{4}$	$\frac{5\pi}{6}$
$\tan\theta$	$\frac{1}{\sqrt{3}}$	1	$\sqrt{3}$	und	$-\sqrt{3}$	-1	$-\frac{1}{\sqrt{3}}$
$\sec\theta$	$\frac{2}{\sqrt{3}}$	$\sqrt{2}$	2	und	-2	$-\sqrt{2}$	$-\frac{2}{\sqrt{3}}$

17. Show that if $\tan\theta = c$ and $0 \le \theta < \pi/2$, then $\cos\theta = 1/\sqrt{1+c^2}$. *Hint:* Draw a right triangle whose opposite and adjacent sides have lengths c and 1.

SOLUTION Because $0 \le \theta < \pi/2$, we can use the definition of the trigonometric functions in terms of right triangles. $\tan\theta$ is the ratio of the length of the side opposite the angle θ to the length of the adjacent side. With $c = \frac{c}{1}$, we label the length of the opposite side as c and the length of the adjacent side as 1 (see the diagram below). By the Pythagorean theorem, the length of the hypotenuse is $\sqrt{1+c^2}$. Finally, we use the fact that $\cos\theta$ is the ratio of the length of the adjacent side to the length of the hypotenuse to obtain

$$\cos\theta = \frac{1}{\sqrt{1+c^2}}.$$

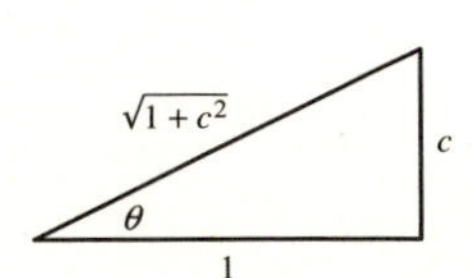

In Exercises 19–24, assume that $0 \le \theta < \pi/2$.

19. Find $\sin\theta$ and $\tan\theta$ if $\cos\theta = \frac{5}{13}$.

SOLUTION Consider the triangle below. The lengths of the side adjacent to the angle θ and the hypotenuse have been labeled so that $\cos\theta = \frac{5}{13}$. The length of the side opposite the angle θ has been calculated using the Pythagorean theorem: $\sqrt{13^2 - 5^2} = 12$. From the triangle, we see that

$$\sin\theta = \frac{12}{13} \quad \text{and} \quad \tan\theta = \frac{12}{5}.$$

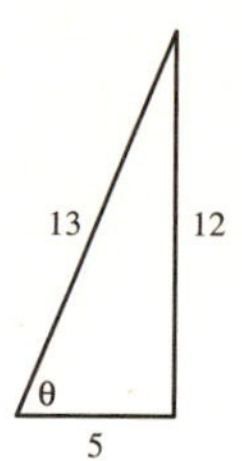

21. Find $\sin\theta$, $\sec\theta$, and $\cot\theta$ if $\tan\theta = \frac{2}{7}$.

SOLUTION If $\tan\theta = \frac{2}{7}$, then $\cot\theta = \frac{7}{2}$. For the remaining trigonometric functions, consider the triangle below. The lengths of the sides opposite and adjacent to the angle θ have been labeled so that $\tan\theta = \frac{2}{7}$. The length of the hypotenuse has been calculated using the Pythagorean theorem: $\sqrt{2^2 + 7^2} = \sqrt{53}$. From the triangle, we see that

$$\sin\theta = \frac{2}{\sqrt{53}} = \frac{2\sqrt{53}}{53} \quad \text{and} \quad \sec\theta = \frac{\sqrt{53}}{7}.$$

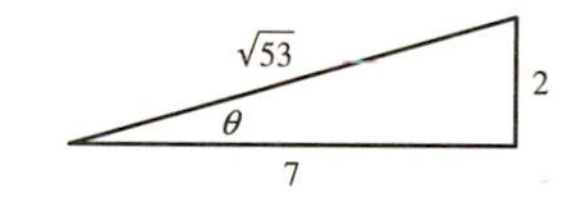

23. Find $\cos 2\theta$ if $\sin\theta = \frac{1}{5}$.

SOLUTION Using the double angle formula $\cos 2\theta = \cos^2\theta - \sin^2\theta$ and the fundamental identity $\sin^2\theta + \cos^2\theta = 1$, we find that $\cos 2\theta = 1 - 2\sin^2\theta$. Thus, $\cos 2\theta = 1 - 2(1/25) = 23/25$.

25. Find $\cos\theta$ and $\tan\theta$ if $\sin\theta = 0.4$ and $\pi/2 \le \theta < \pi$.

SOLUTION We can determine the "magnitude" of $\cos\theta$ and $\tan\theta$ using the triangle shown below. The lengths of the side opposite the angle θ and the hypotenuse have been labeled so that $\sin\theta = 0.4 = \frac{2}{5}$. The length of the side adjacent to the angle θ was calculated using the Pythagorean theorem: $\sqrt{5^2 - 2^2} = \sqrt{21}$. From the triangle, we see that

$$|\cos\theta| = \frac{\sqrt{21}}{5} \quad \text{and} \quad |\tan\theta| = \frac{2}{\sqrt{21}} = \frac{2\sqrt{21}}{21}.$$

Because $\pi/2 \le \theta < \pi$, both $\cos\theta$ and $\tan\theta$ are negative; consequently,

$$\cos\theta = -\frac{\sqrt{21}}{5} \quad \text{and} \quad \tan\theta = -\frac{2\sqrt{21}}{21}.$$

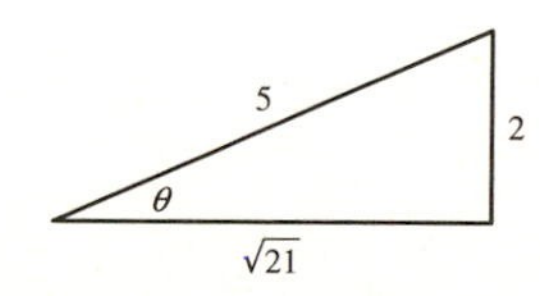

27. Find $\cos\theta$ if $\cot\theta = \frac{4}{3}$ and $\sin\theta < 0$.

SOLUTION We can determine the "magnitude" of $\cos\theta$ using the triangle shown below. The lengths of the sides opposite and adjacent to the angle θ have been labeled so that $\cot\theta = \frac{4}{3}$. The length of the hypotenuse was calculated using the Pythagorean theorem: $\sqrt{3^2 + 4^2} = 5$. From the triangle, we see that

$$|\cos\theta| = \frac{4}{5}.$$

Because $\cot\theta = \frac{4}{3} > 0$ and $\sin\theta < 0$, the angle θ must be in the third quadrant; consequently, $\cos\theta$ will be negative and

$$\cos\theta = -\frac{4}{5}.$$

29. Find the values of $\sin\theta$, $\cos\theta$, and $\tan\theta$ for the angles corresponding to the eight points in Figure 23(A) and (B).

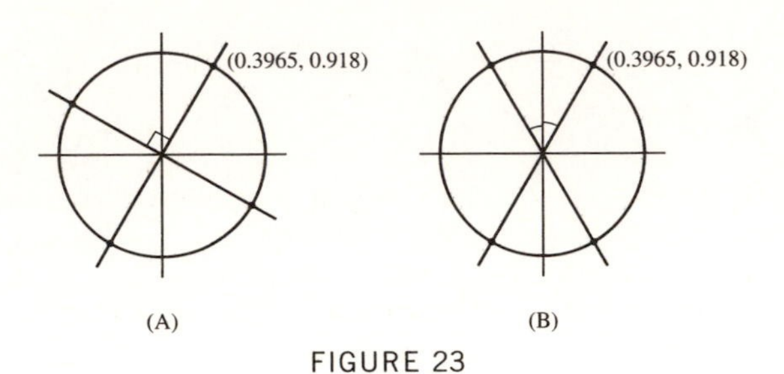

FIGURE 23

SOLUTION Let's start with the four points in Figure 23(A).

- The point in the first quadrant has coordinates $(0.3965, 0.918)$. Therefore,

$$\sin\theta = 0.918, \quad \cos\theta = 0.3965, \quad \text{and} \quad \tan\theta = \frac{0.918}{0.3965} = 2.3153.$$

- The coordinates of the point in the second quadrant are $(-0.918, 0.3965)$. Therefore,

$$\sin\theta = 0.3965, \quad \cos\theta = -0.918, \quad \text{and} \quad \tan\theta = \frac{0.3965}{-0.918} = -0.4319.$$

- Because the point in the third quadrant is symmetric to the point in the first quadrant with respect to the origin, its coordinates are $(-0.3965, -0.918)$. Therefore,

$$\sin\theta = -0.918, \quad \cos\theta = -0.3965, \quad \text{and} \quad \tan\theta = \frac{-0.918}{-0.3965} = 2.3153.$$

- Because the point in the fourth quadrant is symmetric to the point in the second quadrant with respect to the origin, its coordinates are $(0.918, -0.3965)$. Therefore,

$$\sin\theta = -0.3965, \quad \cos\theta = 0.918, \quad \text{and} \quad \tan\theta = \frac{-0.3965}{0.918} = -0.4319.$$

Now consider the four points in Figure 23(B).

- The point in the first quadrant has coordinates $(0.3965, 0.918)$. Therefore,

$$\sin\theta = 0.918, \quad \cos\theta = 0.3965, \quad \text{and} \quad \tan\theta = \frac{0.918}{0.3965} = 2.3153.$$

- The point in the second quadrant is a reflection through the y-axis of the point in the first quadrant. Its coordinates are therefore $(-0.3965, 0.918)$ and

$$\sin\theta = 0.918, \quad \cos\theta = -0.3965, \quad \text{and} \quad \tan\theta = \frac{0.918}{0.3965} = -2.3153.$$

- Because the point in the third quadrant is symmetric to the point in the first quadrant with respect to the origin, its coordinates are $(-0.3965, -0.918)$. Therefore,

$$\sin\theta = -0.918, \quad \cos\theta = -0.3965, \quad \text{and} \quad \tan\theta = \frac{-0.918}{-0.3965} = 2.3153.$$

- Because the point in the fourth quadrant is symmetric to the point in the second quadrant with respect to the origin, its coordinates are $(0.3965, -0.918)$. Therefore,

$$\sin\theta = -0.918, \quad \cos\theta = 0.3965, \quad \text{and} \quad \tan\theta = \frac{-0.918}{0.3965} = -2.3153.$$

31. Refer to Figure 24(B). Compute $\cos\psi$, $\sin\psi$, $\cot\psi$, and $\csc\psi$.

(A) (B)

FIGURE 24

SOLUTION By the Pythagorean theorem, the length of the side opposite the angle ψ in Figure 24(B) is $\sqrt{1-0.3^2} = \sqrt{0.91}$. Consequently,

$$\cos\psi = \frac{0.3}{1} = 0.3, \quad \sin\psi = \frac{\sqrt{0.91}}{1} = \sqrt{0.91}, \quad \cot\psi = \frac{0.3}{\sqrt{0.91}} \quad \text{and} \quad \csc\psi = \frac{1}{\sqrt{0.91}}.$$

33. Use the addition formula to compute $\cos\left(\frac{\pi}{3} + \frac{\pi}{4}\right)$ exactly.

SOLUTION

$$\cos\left(\frac{\pi}{3} + \frac{\pi}{4}\right) = \cos\frac{\pi}{3}\cos\frac{\pi}{4} - \sin\frac{\pi}{3}\sin\frac{\pi}{4}$$
$$= \frac{1}{2}\cdot\frac{\sqrt{2}}{2} - \frac{\sqrt{3}}{2}\cdot\frac{\sqrt{2}}{2} = \frac{\sqrt{2}-\sqrt{6}}{4}.$$

In Exercises 35–38, sketch the graph over $[0, 2\pi]$.

35. $2\sin 4\theta$

SOLUTION

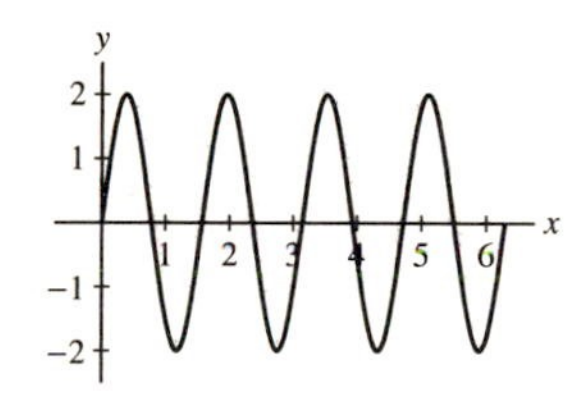

37. $\cos\left(2\theta - \frac{\pi}{2}\right)$

SOLUTION

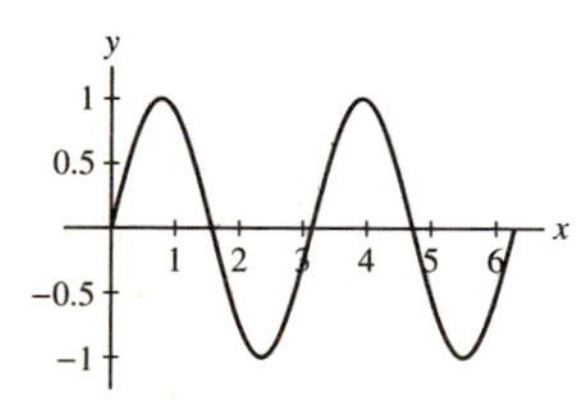

39. How many points lie on the intersection of the horizontal line $y = c$ and the graph of $y = \sin x$ for $0 \le x < 2\pi$? *Hint:* The answer depends on c.

SOLUTION Recall that for any x, $-1 \le \sin x \le 1$. Thus, if $|c| > 1$, the horizontal line $y = c$ and the graph of $y = \sin x$ never intersect. If $c = +1$, then $y = c$ and $y = \sin x$ intersect at the peak of the sine curve; that is, they intersect at $x = \frac{\pi}{2}$. On the other hand, if $c = -1$, then $y = c$ and $y = \sin x$ intersect at the bottom of the sine curve; that is, they intersect at $x = \frac{3\pi}{2}$. Finally, if $|c| < 1$, the graphs of $y = c$ and $y = \sin x$ intersect twice.

In Exercises 41–44, solve for $0 \le \theta < 2\pi$ *(see Example 4).*

41. $\sin 2\theta + \sin 3\theta = 0$

SOLUTION $\sin\alpha = -\sin\beta$ when $\alpha = -\beta + 2\pi k$ or $\alpha = \pi + \beta + 2\pi k$. Substituting $\alpha = 2\theta$ and $\beta = 3\theta$, we have either $2\theta = -3\theta + 2\pi k$ or $2\theta = \pi + 3\theta + 2\pi k$. Solving each of these equations for θ yields $\theta = \frac{2}{5}\pi k$ or $\theta = -\pi - 2\pi k$. The solutions on the interval $0 \le \theta < 2\pi$ are then

$$\theta = 0, \frac{2\pi}{5}, \frac{4\pi}{5}, \pi, \frac{6\pi}{5}, \frac{8\pi}{5}.$$

43. $\cos 4\theta + \cos 2\theta = 0$

SOLUTION $\cos\alpha = -\cos\beta$ when $\alpha + \beta = \pi + 2\pi k$ or $\alpha = \beta + \pi + 2\pi k$. Substituting $\alpha = 4\theta$ and $\beta = 2\theta$, we have either $6\theta = \pi + 2\pi k$ or $4\theta = 2\theta + \pi + 2\pi k$. Solving each of these equations for θ yields $\theta = \frac{\pi}{6} + \frac{\pi}{3}k$ or $\theta = \frac{\pi}{2} + \pi k$. The solutions on the interval $0 \le \theta < 2\pi$ are then

$$\theta = \frac{\pi}{6}, \frac{\pi}{2}, \frac{5\pi}{6}, \frac{7\pi}{6}, \frac{3\pi}{2}, \frac{11\pi}{6}.$$

In Exercises 45–54, derive the identity using the identities listed in this section.

45. $\cos 2\theta = 2\cos^2\theta - 1$

SOLUTION Starting from the double angle formula for cosine, $\cos^2\theta = \frac{1}{2}(1 + \cos 2\theta)$, we solve for $\cos 2\theta$. This gives $2\cos^2\theta = 1 + \cos 2\theta$ and then $\cos 2\theta = 2\cos^2\theta - 1$.

47. $\sin\frac{\theta}{2} = \sqrt{\frac{1 - \cos\theta}{2}}$

SOLUTION Substitute $x = \theta/2$ into the double angle formula for sine, $\sin^2 x = \frac{1}{2}(1 - \cos 2x)$ to obtain $\sin^2\left(\frac{\theta}{2}\right) = \frac{1 - \cos\theta}{2}$. Taking the square root of both sides yields $\sin\left(\frac{\theta}{2}\right) = \sqrt{\frac{1 - \cos\theta}{2}}$.

49. $\cos(\theta + \pi) = -\cos\theta$

SOLUTION From the addition formula for the cosine function, we have

$$\cos(\theta + \pi) = \cos\theta\cos\pi - \sin\theta\sin\pi = \cos\theta(-1) = -\cos\theta$$

51. $\tan(\pi - \theta) = -\tan\theta$

SOLUTION Using Exercises 48 and 49,

$$\tan(\pi - \theta) = \frac{\sin(\pi - \theta)}{\cos(\pi - \theta)} = \frac{\sin(\pi + (-\theta))}{\cos(\pi + (-\theta))} = \frac{-\sin(-\theta)}{-\cos(-\theta)} = \frac{\sin\theta}{-\cos\theta} = -\tan\theta.$$

The second to last equality occurs because $\sin x$ is an odd function and $\cos x$ is an even function.

53. $\tan x = \frac{\sin 2x}{1 + \cos 2x}$

SOLUTION Using the addition formula for the sine function, we find

$$\sin 2x = \sin(x + x) = \sin x\cos x + \cos x\sin x = 2\sin x\cos x.$$

By Exercise 45, we know that $\cos 2x = 2\cos^2 x - 1$. Therefore,

$$\frac{\sin 2x}{1 + \cos 2x} = \frac{2\sin x\cos x}{1 + 2\cos^2 x - 1} = \frac{2\sin x\cos x}{2\cos^2 x} = \frac{\sin x}{\cos x} = \tan x.$$

55. Use Exercises 48 and 49 to show that $\tan\theta$ and $\cot\theta$ are periodic with period π.

SOLUTION By Exercises 48 and 49,

$$\tan(\theta + \pi) = \frac{\sin(\theta + \pi)}{\cos(\theta + \pi)} = \frac{-\sin\theta}{-\cos\theta} = \tan\theta,$$

and

$$\cot(\theta + \pi) = \frac{\cos(\theta + \pi)}{\sin(\theta + \pi)} = \frac{-\cos\theta}{-\sin\theta} = \cot\theta.$$

Thus, both $\tan\theta$ and $\cot\theta$ are periodic with period π.

57. Use the Law of Cosines to find the distance from P to Q in Figure 26.

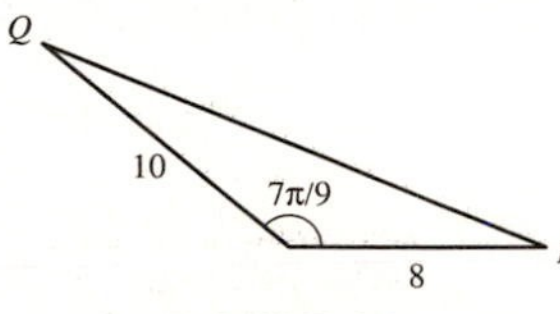

FIGURE 26

SOLUTION By the Law of Cosines, the distance from P to Q is

$$\sqrt{10^2 + 8^2 - 2(10)(8)\cos\frac{7\pi}{9}} = 16.928.$$

Further Insights and Challenges

59. Use the addition formula to prove

$$\cos 3\theta = 4\cos^3\theta - 3\cos\theta$$

SOLUTION

$$\begin{aligned}\cos 3\theta &= \cos(2\theta + \theta) = \cos 2\theta \cos\theta - \sin 2\theta \sin\theta = (2\cos^2\theta - 1)\cos\theta - (2\sin\theta\cos\theta)\sin\theta \\ &= \cos\theta(2\cos^2\theta - 1 - 2\sin^2\theta) = \cos\theta(2\cos^2\theta - 1 - 2(1 - \cos^2\theta)) \\ &= \cos\theta(2\cos^2\theta - 1 - 2 + 2\cos^2\theta) = 4\cos^3\theta - 3\cos\theta\end{aligned}$$

61. Let θ be the angle between the line $y = mx + b$ and the x-axis [Figure 28(A)]. Prove that $m = \tan\theta$.

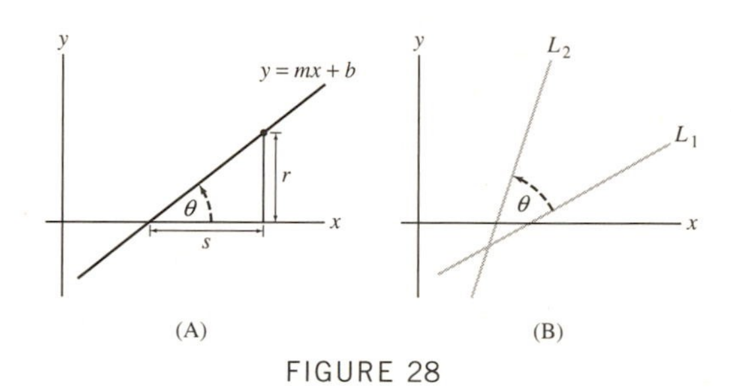

(A) (B)

FIGURE 28

SOLUTION Using the distances labeled in Figure 28(A), we see that the slope of the line is given by the ratio r/s. The tangent of the angle θ is given by the same ratio. Therefore, $m = \tan\theta$.

63. Perpendicular Lines Use Exercise 62 to prove that two lines with nonzero slopes m_1 and m_2 are perpendicular if and only if $m_2 = -1/m_1$.

SOLUTION If lines are perpendicular, then the angle between them is $\theta = \pi/2 \Rightarrow$

$$\cot(\pi/2) = \frac{1 + m_1 m_2}{m_1 - m_2}$$

$$0 = \frac{1 + m_1 m_2}{m_1 - m_2}$$

$$\Rightarrow m_1 m_2 = -1 \Rightarrow m_1 = -\frac{1}{m_2}$$

1.5 Inverse Functions

Preliminary Questions

1. Which of the following satisfy $f^{-1}(x) = f(x)$?

(a) $f(x) = x$

(b) $f(x) = 1 - x$

(c) $f(x) = 1$

(d) $f(x) = \sqrt{x}$

(e) $f(x) = |x|$

(f) $f(x) = x^{-1}$

SOLUTION The functions **(a)** $f(x) = x$, **(b)** $f(x) = 1 - x$ and **(f)** $f(x) = x^{-1}$ satisfy $f^{-1}(x) = f(x)$.

2. The graph of a function looks like the track of a roller coaster. Is the function one-to-one?

SOLUTION Because the graph looks like the track of a roller coaster, there will be several locations at which the graph has the same height. The graph will therefore fail the horizontal line test, meaning that the function is *not* one-to-one.

3. The function f maps teenagers in the United States to their last names. Explain why the inverse function f^{-1} does not exist.

SOLUTION Many different teenagers will have the same last name, so this function will not be one-to-one. Consequently, the function does not have an inverse.

4. The following fragment of a train schedule for the New Jersey Transit System defines a function f from towns to times. Is f one-to-one? What is $f^{-1}(6{:}27)$?

Trenton	6:21
Hamilton Township	6:27
Princeton Junction	6:34
New Brunswick	6:38

SOLUTION This function is one-to-one, and $f^{-1}(6{:}27) = \text{Hamilton Township}$.

5. A homework problem asks for a sketch of the graph of the *inverse* of $f(x) = x + \cos x$. Frank, after trying but failing to find a formula for $f^{-1}(x)$, says it's impossible to graph the inverse. Bianca hands in an accurate sketch without solving for f^{-1}. How did Bianca complete the problem?

SOLUTION The graph of the inverse function is the reflection of the graph of $y = f(x)$ through the line $y = x$.

6. Which of the following quantities is undefined?

(a) $\sin^{-1}\left(-\frac{1}{2}\right)$
(b) $\cos^{-1}(2)$
(c) $\csc^{-1}\left(\frac{1}{2}\right)$
(d) $\csc^{-1}(2)$

SOLUTION **(b)** and **(c)** are undefined. $\sin^{-1}\left(-\frac{1}{2}\right) = -\frac{\pi}{6}$ and $\csc^{-1}(2) = \frac{\pi}{6}$.

7. Give an example of an angle θ such that $\cos^{-1}(\cos\theta) \neq \theta$. Does this contradict the definition of inverse function?

SOLUTION Any angle $\theta < 0$ or $\theta > \pi$ will work. No, this does not contradict the definition of inverse function.

Exercises

1. Show that $f(x) = 7x - 4$ is invertible and find its inverse.

SOLUTION Solving $y = 7x - 4$ for x yields $x = \dfrac{y+4}{7}$. Thus, $f^{-1}(x) = \dfrac{x+4}{7}$.

3. What is the largest interval containing zero on which $f(x) = \sin x$ is one-to-one?

SOLUTION Looking at the graph of $\sin x$, the function is one-to-one on the interval $[-\pi/2, \pi/2]$.

5. Verify that $f(x) = x^3 + 3$ and $g(x) = (x-3)^{1/3}$ are inverses by showing that $f(g(x)) = x$ and $g(f(x)) = x$.

SOLUTION

- $f(g(x)) = \left((x-3)^{1/3}\right)^3 + 3 = x - 3 + 3 = x.$
- $g(f(x)) = \left(x^3 + 3 - 3\right)^{1/3} = \left(x^3\right)^{1/3} = x.$

7. The escape velocity from a planet of radius R is $v(R) = \sqrt{\dfrac{2GM}{R}}$, where G is the universal gravitational constant and M is the mass. Find the inverse of $v(R)$ expressing R in terms of v.

SOLUTION To find the inverse, we solve

$$y = \sqrt{\frac{2GM}{R}}$$

for R. This yields

$$R = \frac{2GM}{y^2}.$$

Therefore,

$$v^{-1}(R) = \frac{2GM}{R^2}.$$

In Exercises 8–15, find a domain on which f is one-to-one and a formula for the inverse of f restricted to this domain. Sketch the graphs of f and f^{-1}.

9. $f(x) = 4 - x$

SOLUTION The linear function $f(x) = 4 - x$ is one-to-one for all real numbers. Solving $y = x - 4$ for x gives $x = 4 - y$. Thus, $f^{-1}(x) = 4 - x$.

11. $f(x) = \dfrac{1}{7x - 3}$

SOLUTION The graph of $f(x) = 1/(7x - 3)$ given below shows that f passes the horizontal line test, and is therefore one-to-one, on its entire domain $\{x : x \neq \frac{3}{7}\}$. Solving $y = 1/(7x - 3)$ for x gives

$$x = \frac{1}{7y} + \frac{3}{7}; \quad \text{thus,} \quad f^{-1}(x) = \frac{1}{7x} + \frac{3}{7}.$$

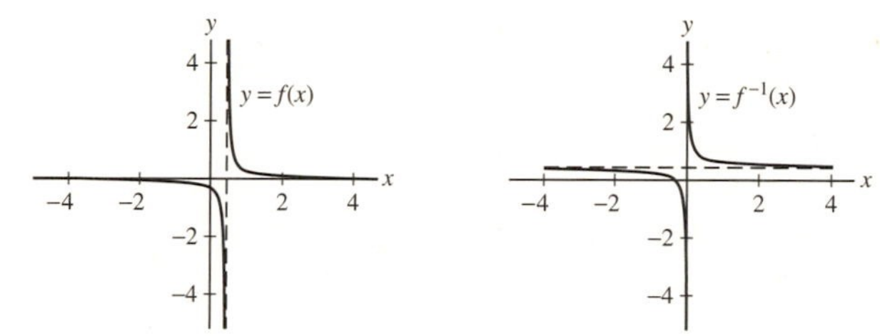

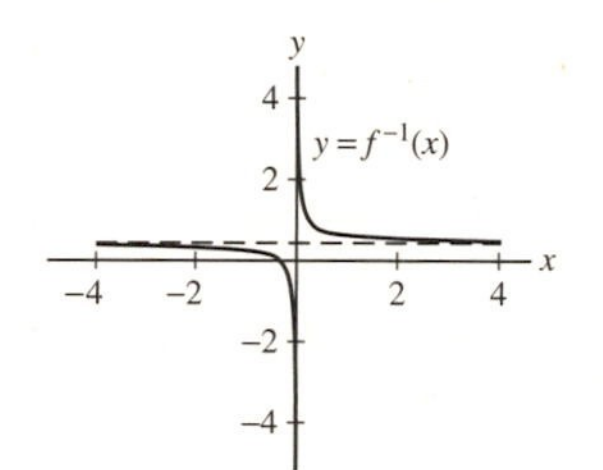

13. $f(x) = \dfrac{1}{\sqrt{x^2 + 1}}$

SOLUTION To make the function $f(x) = \dfrac{1}{\sqrt{x^2 + 1}}$ one-to-one, we must restrict the domain to either $\{x : x \geq 0\}$ or $\{x : x \leq 0\}$. If we choose the domain $\{x : x \geq 0\}$, then solving $y = \dfrac{1}{\sqrt{x^2 + 1}}$ for x yields

$$x = \frac{\sqrt{1 - y^2}}{y}; \quad \text{hence,} \quad f^{-1}(x) = \frac{\sqrt{1 - x^2}}{x}.$$

Had we chosen the domain $\{x : x \leq 0\}$, the inverse would have been

$$f^{-1}(x) = -\frac{\sqrt{1 - x^2}}{x}.$$

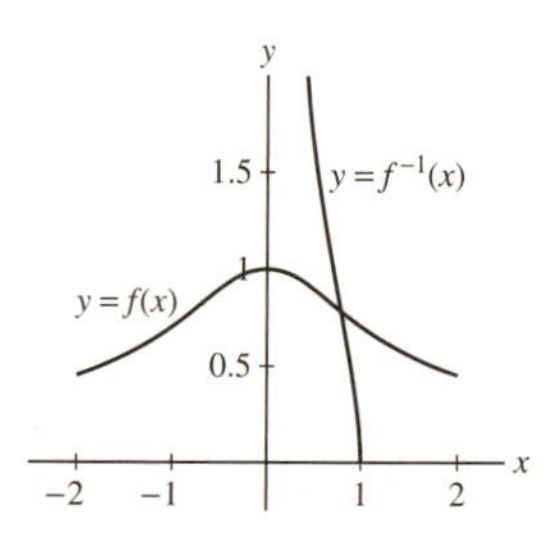

15. $f(x) = \sqrt{x^3 + 9}$

SOLUTION The graph of $f(x) = \sqrt{x^3 + 9}$ given below shows that f passes the horizontal line test, and therefore is one-to-one, on its entire domain $\{x : x \geq -9^{1/3}\}$. Solving $y = \sqrt{x^3 + 9}$ for x yields $x = (y^2 - 9)^{1/3}$. Thus, $f^{-1}(x) = (x^2 - 9)^{1/3}$.

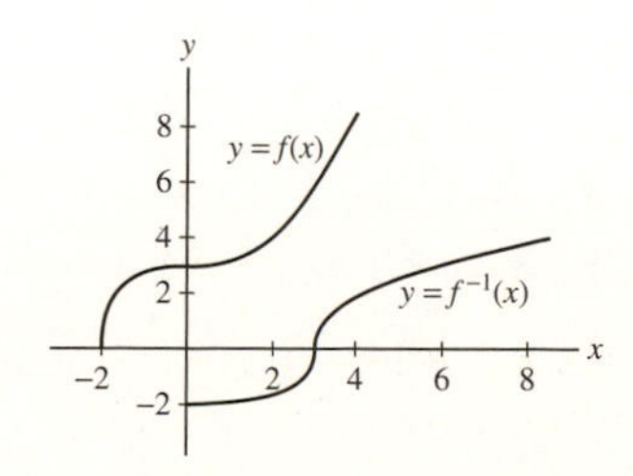

17. Which of the graphs in Figure 20 is the graph of a function satisfying $f^{-1} = f$?

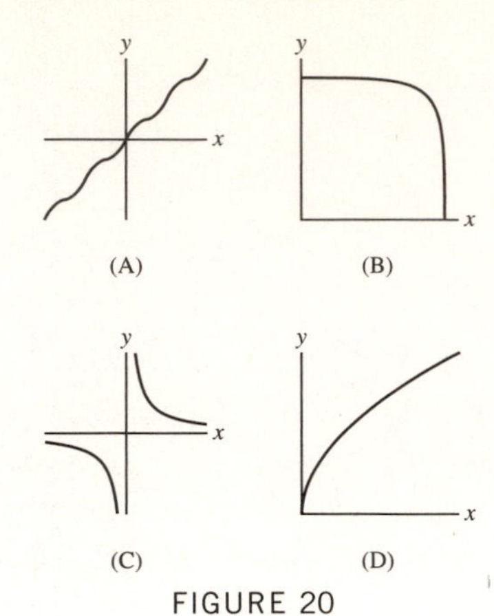

FIGURE 20

SOLUTION Figures (B) and (C) would not change when reflected around the line $y = x$. Therefore, these two satisfy $f^{-1} = f$.

19. Let $f(x) = x^7 + x + 1$.

(a) Show that f^{-1} exists (but do not attempt to find it). *Hint:* Show that f is increasing.

(b) What is the domain of f^{-1}?

(c) Find $f^{-1}(3)$.

SOLUTION

(a) The graph of $f(x) = x^7 + x + 1$ is shown below. From this graph, we see that $f(x)$ is a strictly increasing function; by Example 3, it is therefore one-to-one. Because f is one-to-one, by Theorem 3, f^{-1} exists.

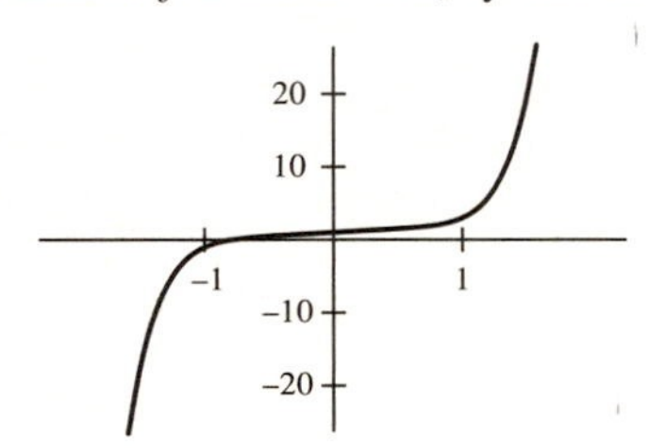

(b) The domain of $f^{-1}(x)$ is the range of $f(x) : (-\infty, \infty)$.

(c) Note that $f(1) = 1^7 + 1 + 1 = 3$; therefore, $f^{-1}(3) = 1$.

21. Let $f(x) = x^2 - 2x$. Determine a domain on which f^{-1} exists, and find a formula for f^{-1} for this domain of f.

SOLUTION From the graph of $y = x^2 - 2x$ shown below, we see that if the domain of f is restricted to either $x \leq 1$ or $x \geq 1$, then f is one-to-one and f^{-1} exists. To find a formula for f^{-1}, we solve $y = x^2 - 2x$ for x as follows:

$$y + 1 = x^2 - 2x + 1 = (x - 1)^2$$

$$x - 1 = \pm\sqrt{y + 1}$$

$$x = 1 \pm \sqrt{y + 1}$$

If the domain of f is restricted to $x \leq 1$, then we choose the negative sign in front of the radical and $f^{-1}(x) = 1 - \sqrt{x + 1}$. If the domain of f is restricted to $x \geq 1$, we choose the positive sign in front of the radical and $f^{-1}(x) = 1 + \sqrt{x + 1}$.

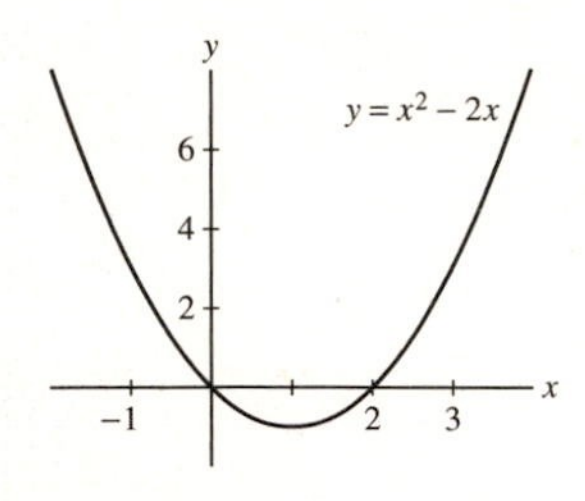

In Exercises 23–28, evaluate without using a calculator.

23. $\cos^{-1} 1$

SOLUTION $\cos^{-1} 1 = 0$.

25. $\cot^{-1} 1$

SOLUTION $\cot^{-1} 1 = \frac{\pi}{4}$.

27. $\tan^{-1} \sqrt{3}$

SOLUTION $\tan^{-1} \sqrt{3} = \tan^{-1}\left(\frac{\sqrt{3}/2}{1/2}\right) = \frac{\pi}{3}$.

In Exercises 29–38, compute without using a calculator.

29. $\sin^{-1}\left(\sin \dfrac{\pi}{3}\right)$

SOLUTION $\sin^{-1}(\sin \frac{\pi}{3}) = \frac{\pi}{3}$.

31. $\cos^{-1}\left(\cos \dfrac{3\pi}{2}\right)$

SOLUTION $\cos^{-1}(\cos \frac{3\pi}{2}) = \cos^{-1}(0) = \frac{\pi}{2}$. The answer is not $\frac{3\pi}{2}$ because $\frac{3\pi}{2}$ is not in the range of the inverse cosine function.

33. $\tan^{-1}\left(\tan \dfrac{3\pi}{4}\right)$

SOLUTION $\tan^{-1}(\tan \frac{3\pi}{4}) = \tan^{-1}(-1) = -\frac{\pi}{4}$. The answer is not $\frac{3\pi}{4}$ because $\frac{3\pi}{4}$ is not in the range of the inverse tangent function.

35. $\sec^{-1}(\sec 3\pi)$

SOLUTION $\sec^{-1}(\sec 3\pi) = \sec^{-1}(-1) = \pi$. The answer is not 3π because 3π is not in the range of the inverse secant function.

37. $\csc^{-1}(\csc(-\pi))$

SOLUTION No inverse since $\csc(-\pi) = \frac{1}{\sin(-\pi)} = \frac{1}{0} \longrightarrow \infty$.

In Exercises 39–42, simplify by referring to the appropriate triangle or trigonometric identity.

39. $\tan(\cos^{-1} x)$

SOLUTION Let $\theta = \cos^{-1} x$. Then $\cos\theta = x$ and we generate the triangle shown below. From the triangle,

$$\tan(\cos^{-1} x) = \tan\theta = \frac{\sqrt{1-x^2}}{x}.$$

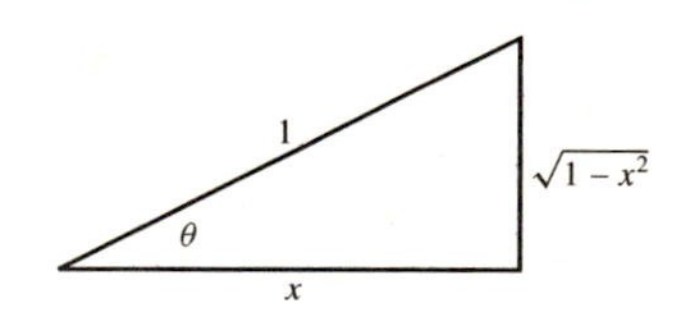

41. $\cot(\sec^{-1} x)$

SOLUTION Let $\theta = \sec^{-1} x$. Then $\sec\theta = x$ and we generate the triangle shown below. From the triangle,

$$\cot(\sec^{-1} x) = \cot\theta = \frac{1}{\sqrt{x^2-1}}.$$

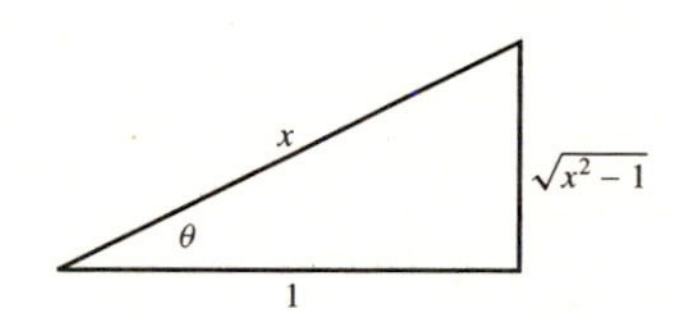

In Exercises 43–50, refer to the appropriate triangle or trigonometric identity to compute the given value.

43. $\cos(\sin^{-1} \frac{2}{3})$

SOLUTION Let $\theta = \sin^{-1} \frac{2}{3}$. Then $\sin\theta = \frac{2}{3}$ and we generate the triangle shown below. From the triangle,

$$\cos\left(\sin^{-1} \frac{2}{3}\right) = \cos\theta = \frac{\sqrt{5}}{3}.$$

45. $\tan\left(\sin^{-1} 0.8\right)$

SOLUTION Let $\theta = \sin^{-1} 0.8$. Then $\sin\theta = 0.8 = \frac{4}{5}$ and we generate the triangle shown below. From the triangle,

$$\tan(\sin^{-1} 0.8) = \tan\theta = \frac{4}{3}.$$

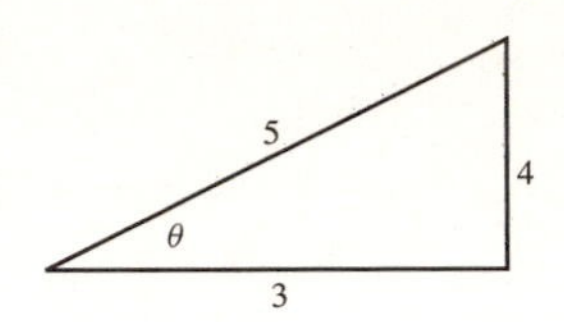

47. $\cot\left(\csc^{-1} 2\right)$

SOLUTION $\csc^{-1} 2 = \frac{\pi}{6}$. Hence, $\cot(\csc^{-1} 2) = \cot\frac{\pi}{6} = \sqrt{3}$.

49. $\cot\left(\tan^{-1} 20\right)$

SOLUTION Let $\theta = \tan^{-1} 20$. Then $\tan\theta = 20$, so $\cot(\tan^{-1} 20) = \cot\theta = \frac{1}{\tan\theta} = \frac{1}{20}$.

Further Insights and Challenges

51. Show that if $f(x)$ is odd and $f^{-1}(x)$ exists, then $f^{-1}(x)$ is odd. Show, on the other hand, that an even function does not have an inverse.

SOLUTION Suppose $f(x)$ is odd and $f^{-1}(x)$ exists. Because $f(x)$ is odd, $f(-x) = -f(x)$. Let $y = f^{-1}(x)$, then $f(y) = x$. Since $f(x)$ is odd, $f(-y) = -f(y) = -x$. Thus $f^{-1}(-x) = -y = -f^{-1}(x)$. Hence, f^{-1} is odd.

On the other hand, if $f(x)$ is even, then $f(-x) = f(x)$. Hence, f is not one-to-one and f^{-1} does not exist.

1.6 Exponential and Logarithmic Functions

Preliminary Questions

1. Which of the following equations is incorrect?

(a) $3^2 \cdot 3^5 = 3^7$

(b) $(\sqrt{5})^{4/3} = 5^{2/3}$

(c) $3^2 \cdot 2^3 = 1$

(d) $(2^{-2})^{-2} = 16$

SOLUTION

(a) This equation is correct: $3^2 \cdot 3^5 = 3^{2+5} = 3^7$.
(b) This equation is correct: $(\sqrt{5})^{4/3} = (5^{1/2})^{4/3} = 5^{(1/2)\cdot(4/3)} = 5^{2/3}$.
(c) This equation is incorrect: $3^2 \cdot 2^3 = 9 \cdot 8 = 72 \neq 1$.
(d) this equation is correct: $(2^{-2})^{-2} = 2^{(-2)\cdot(-2)} = 2^4 = 16$.

2. Compute $\log_{b^2}(b^4)$.

SOLUTION Because $b^4 = (b^2)^2$, $\log_{b^2}(b^4) = 2$.

3. When is $\ln x$ negative?

SOLUTION $\ln x$ is negative for $0 < x < 1$.

4. What is $\ln(-3)$? Explain.

SOLUTION $\ln(-3)$ is not defined.

5. Explain the phrase "The logarithm converts multiplication into addition."

SOLUTION This phrase is a verbal description of the general property of logarithms that states

$$\log(ab) = \log a + \log b.$$

6. What are the domain and range of $\ln x$?

SOLUTION The domain of $\ln x$ is $x > 0$ and the range is all real numbers.

7. Which hyperbolic functions take on only positive values?

SOLUTION $\cosh x$ and $\operatorname{sech} x$ take on only positive values.

8. Which hyperbolic functions are increasing on their domains?

SOLUTION $\sinh x$ and $\tanh x$ are increasing on their domains.

9. Describe three properties of hyperbolic functions that have trigonometric analogs.

SOLUTION Hyperbolic functions have the following analogs with trigonometric functions: parity, identities and derivative formulas.

Exercises

1. Rewrite as a whole number (without using a calculator):

(a) 7^0

(b) $10^2(2^{-2}+5^{-2})$

(c) $\dfrac{(4^3)^5}{(4^5)^3}$

(d) $27^{4/3}$

(e) $8^{-1/3}\cdot 8^{5/3}$

(f) $3\cdot 4^{1/4}-12\cdot 2^{-3/2}$

SOLUTION

(a) $7^0=1$.
(b) $10^2(2^{-2}+5^{-2})=100(1/4+1/25)=25+4=29$.
(c) $(4^3)^5/(4^5)^3=4^{15}/4^{15}=1$.
(d) $(27)^{4/3}=(27^{1/3})^4=3^4=81$.
(e) $8^{-1/3}\cdot 8^{5/3}=(8^{1/3})^5/8^{1/3}=2^5/2=2^4=16$.
(f) $3\cdot 4^{1/4}-12\cdot 2^{-3/2}=3\cdot 2^{1/2}-3\cdot 2^2\cdot 2^{-3/2}=0$.

In Exercises 2–10, solve for the unknown variable.

3. $e^{2x}=e^{x+1}$

SOLUTION If $e^{2x}=e^{x+1}$ then $2x=x+1$, and $x=1$.

5. $3^x=\left(\frac{1}{3}\right)^{x+1}$

SOLUTION Rewrite $(\frac{1}{3})^{x+1}$ as $(3^{-1})^{x+1}=3^{-x-1}$. Then $3^x=3^{-x-1}$, which requires $x=-x-1$. Thus, $x=-1/2$.

7. $4^{-x}=2^{x+1}$

SOLUTION Rewrite 4^{-x} as $(2^2)^{-x}=2^{-2x}$. Then $2^{-2x}=2^{x+1}$, which requires $-2x=x+1$. Solving for x gives $x=-1/3$.

9. $k^{3/2}=27$

SOLUTION Raise both sides of the equation to the two-thirds power. This gives $k=(27)^{2/3}=(27^{1/3})^2=3^2=9$.

In Exercises 11–26, calculate without using a calculator.

11. $\log_3 27$

SOLUTION $\log_3 27=\log_3 3^3=3\log_3 3=3$.

13. $\ln 1$

SOLUTION $\ln 1=0$.

15. $\log_2(2^{5/3})$

SOLUTION $\log_2 2^{5/3}=\dfrac{5}{3}\log_2 2=\dfrac{5}{3}$.

17. $\log_{64} 4$

SOLUTION $\log_{64} 4=\log_{64} 64^{1/3}=\dfrac{1}{3}\log_{64} 64=\dfrac{1}{3}$.

19. $\log_8 2+\log_4 2$

SOLUTION $\log_8 2+\log_4 2=\log_8 8^{1/3}+\log_4 4^{1/2}=\dfrac{1}{3}+\dfrac{1}{2}=\dfrac{5}{6}$.

21. $\log_4 48-\log_4 12$

SOLUTION $\log_4 48-\log_4 12=\log_4\dfrac{48}{12}=\log_4 4=1$.

23. $\ln(e^3) + \ln(e^4)$

SOLUTION $\ln(e^3) + \ln(e^4) = 3 + 4 = 7$.

25. $7^{\log_7(29)}$

SOLUTION $7^{\log_7(29)} = 29$.

27. Write as the natural log of a single expression:

(a) $2\ln 5 + 3\ln 4$ **(b)** $5\ln(x^{1/2}) + \ln(9x)$

SOLUTION

(a) $2\ln 5 + 3\ln 4 = \ln 5^2 + \ln 4^3 = \ln 25 + \ln 64 = \ln(25 \cdot 64) = \ln 1600$.
(b) $5\ln x^{1/2} + \ln 9x = \ln x^{5/2} + \ln 9x = \ln(x^{5/2} \cdot 9x) = \ln(9x^{7/2})$.

In Exercises 29–34, solve for the unknown.

29. $7e^{5t} = 100$

SOLUTION Divide the equation by 7 and then take the natural logarithm of both sides. This gives

$$5t = \ln\left(\frac{100}{7}\right) \quad \text{or} \quad t = \frac{1}{5}\ln\left(\frac{100}{7}\right).$$

31. $2^{x^2-2x} = 8$

SOLUTION Since $8 = 2^3$, we have $x^2 - 2x - 3 = 0$ or $(x-3)(x+1) = 0$. Thus, $x = -1$ or $x = 3$.

33. $\ln(x^4) - \ln(x^2) = 2$

SOLUTION $\ln(x^4) - \ln(x^2) = \ln\left(\dfrac{x^4}{x^2}\right) = \ln(x^2) = 2\ln x$. Thus, $2\ln x = 2$ or $\ln x = 1$. Hence, $x = e$.

35. Use a calculator to compute $\sinh x$ and $\cosh x$ for $x = -3, 0, 5$.

SOLUTION

x	-3	0	5
$\sinh x = \dfrac{e^x - e^{-x}}{2}$	$\dfrac{e^{-3} - e^3}{2} = -10.0179$	$\dfrac{e^0 - e^0}{2} = 0$	$\dfrac{e^5 - e^{-5}}{2} = 74.203$
$\cosh x = \dfrac{e^x + e^{-x}}{2}$	$\dfrac{e^{-3} + e^3}{2} = 10.0677$	$\dfrac{e^0 + e^0}{2} = 1$	$\dfrac{e^5 + e^{-5}}{2} = 74.210$

37. Show, by producing a counterexample, that $\ln(ab)$ is not equal to $(\ln a)(\ln b)$.

SOLUTION Let $a = e^2$ and $b = e^3$. Then $ab = e^5$ and $\ln(ab) = \ln(e^5) = 5$; however,

$$(\ln a)(\ln b) = (\ln e^2)(\ln e^3) = 2(3) = 6.$$

39. Show that $y = \tanh x$ is an odd function.

SOLUTION $\tanh(-x) = \dfrac{e^{-x} - e^{-(-x)}}{e^{-x} + e^{-(-x)}} = \dfrac{e^{-x} - e^x}{e^{-x} + e^x} = -\dfrac{e^x - e^{-x}}{e^x + e^{-x}} = -\tanh x$.

41. The **Gutenberg–Richter Law** states that the number N of earthquakes per year worldwide of Richter magnitude at least M satisfies an approximate relation $\log_{10} N = a - M$ for some constant a. Find a, assuming that there is one earthquake of magnitude $M \ge 8$ per year. How many earthquakes of magnitude $M \ge 5$ occur per year?

SOLUTION Substituting $N = 1$ and $M = 8$ into the Gutenberg–Richter law and solving for a yields

$$a = 8 + \log_{10} 1 = 8.$$

The number N of earthquakes of Richter magnitude $M \ge 5$ then satisfies

$$\log_{10} N = 8 - 5 = 3.$$

Finally, $N = 10^3 = 1000$ earthquakes.

43. Refer to the graphs to explain why the equation $\sinh x = t$ has a unique solution for every t and why $\cosh x = t$ has two solutions for every $t > 1$.

SOLUTION From its graph we see that $\sinh x$ is a one-to-one function with $\lim_{x\to-\infty} \sinh x = -\infty$ and $\lim_{x\to\infty} \sinh x = \infty$. Thus, for every real number t, the equation $\sinh x = t$ has a unique solution. On the other hand, from its graph, we see that $\cosh x$ is not one-to-one. Rather, it is an even function with a minimum value of $\cosh 0 = 1$. Thus, for every $t > 1$, the equation $\cosh x = t$ has two solutions: one positive, the other negative.

45. Prove the addition formula for $\cosh x$.

SOLUTION

$$\cosh(x+y) = \frac{e^{x+y}+e^{-(x+y)}}{2} = \frac{2e^{x+y}+2e^{-(x+y)}}{4}$$
$$= \frac{e^{x+y}+e^{-x+y}+e^{x-y}+e^{-(x+y)}}{4} + \frac{e^{x+y}-e^{-x+y}-e^{x-y}+e^{-(x+y)}}{4}$$
$$= \left(\frac{e^x+e^{-x}}{2}\right)\left(\frac{e^y+e^{-y}}{2}\right) + \left(\frac{e^x-e^{-x}}{2}\right)\left(\frac{e^y-e^{-y}}{2}\right)$$
$$= \cosh x \cosh y + \sinh x \sinh y.$$

47. An (imaginary) train moves along a track at velocity v. Bionica walks down the aisle of the train with velocity u in the direction of the train's motion. Compute the velocity w of Bionica relative to the ground using the laws of both Galileo and Einstein in the following cases.
(a) $v = 500$ m/s and $u = 10$ m/s. Is your calculator accurate enough to detect the difference between the two laws?
(b) $v = 10^7$ m/s and $u = 10^6$ m/s.

SOLUTION Recall that the speed of light is $c \approx 3 \times 10^8$ m/s.
(a) By Galileo's law, $w = 500 + 10 = 510$ m/s. Using Einstein's law and a calculator,

$$\tanh^{-1}\frac{w}{c} = \tanh^{-1}\frac{500}{c} + \tanh^{-1}\frac{10}{c} = 1.7 \times 10^{-6};$$

so $w = c \cdot \tanh(1.7 \times 10^{-6}) \approx 510$ m/s. No, the calculator was not accurate enough to detect the difference between the two laws.
(b) By Galileo's law, $u + v = 10^7 + 10^6 = 1.1 \times 10^7$ m/s. By Einstein's law,

$$\tanh^{-1}\frac{w}{c} = \tanh^{-1}\frac{10^7}{3 \times 10^8} + \tanh^{-1}\frac{10^6}{3 \times 10^8} \approx 0.036679,$$

so $w \approx c \cdot \tanh(0.036679) \approx 1.09988 \times 10^7$ m/s.

Further Insights and Challenges

49. Verify the formula $\log_b x = \dfrac{\log_a x}{\log_a b}$ for $a, b > 0$ such that $a \neq 1, b \neq 1$.

SOLUTION Let $y = \log_b x$. Then $x = b^y$ and $\log_a x = \log_a b^y = y \log_a b$. Thus, $y = \dfrac{\log_a x}{\log_a b}$.

51. Prove that every function $f(x)$ can be written as a sum $f(x) = f_+(x) + f_-(x)$ of an even function $f_+(x)$ and an odd function $f_-(x)$. Express $f(x) = 5e^x + 8e^{-x}$ in terms of $\cosh x$ and $\sinh x$.

SOLUTION Let $f_+(x) = \frac{f(x)+f(-x)}{2}$ and $f_-(x) = \frac{f(x)-f(-x)}{2}$. Then $f_+ + f_- = \frac{2f(x)}{2} = f(x)$. Moreover,

$$f_+(-x) = \frac{f(-x)+f(-(-x))}{2} = \frac{f(-x)+f(x)}{2} = f_+(x),$$

so $f_+(x)$ is an even function, while

$$f_-(-x) = \frac{f(-x)-f(-(-x))}{2}$$
$$= \frac{f(-x)-f(x)}{2} = -\frac{(f(x)-f(-x))}{2} = -f_-(x),$$

so $f_-(x)$ is an odd function.
For $f(x) = 5e^x + 8e^{-x}$, we have

$$f_+(x) = \frac{5e^x + 8e^{-x} + 5e^{-x} + 8e^x}{2} = 8\cosh x + 5\cosh x = 13\cosh x$$

and

$$f_-(x) = \frac{5e^x + 8e^{-x} - 5e^{-x} - 8e^x}{2} = 5\sinh x - 8\sinh x = -3\sinh x.$$

Therefore, $f(x) = f_+(x) + f_-(x) = 13\cosh x - 3\sinh x$.

1.7 Technology: Calculators and Computers

Preliminary Questions

1. Is there a definite way of choosing the optimal viewing rectangle, or is it best to experiment until you find a viewing rectangle appropriate to the problem at hand?

SOLUTION It is best to experiment with the window size until one is found that is appropriate for the problem at hand.

2. Describe the calculator screen produced when the function $y = 3 + x^2$ is plotted with viewing rectangle:

(a) $[-1, 1] \times [0, 2]$ **(b)** $[0, 1] \times [0, 4]$

SOLUTION

(a) Using the viewing rectangle $[-1, 1]$ by $[0, 2]$, the screen will display nothing as the minimum value of $y = 3 + x^2$ is $y = 3$.

(b) Using the viewing rectangle $[0, 1]$ by $[0, 4]$, the screen will display the portion of the parabola between the points $(0, 3)$ and $(1, 4)$.

3. According to the evidence in Example 4, it appears that $f(n) = (1 + 1/n)^n$ never takes on a value greater than 3 for $n > 0$. Does this evidence *prove* that $f(n) \le 3$ for $n > 0$?

SOLUTION No, this evidence does not constitute a proof that $f(n) \le 3$ for $n \ge 0$.

4. How can a graphing calculator be used to find the minimum value of a function?

SOLUTION Experiment with the viewing window to zoom in on the lowest point on the graph of the function. The y-coordinate of the lowest point on the graph is the minimum value of the function.

Exercises

The exercises in this section should be done using a graphing calculator or computer algebra system.

1. Plot $f(x) = 2x^4 + 3x^3 - 14x^2 - 9x + 18$ in the appropriate viewing rectangles and determine its roots.

SOLUTION Using a viewing rectangle of $[-4, 3]$ by $[-20, 20]$, we obtain the plot below.

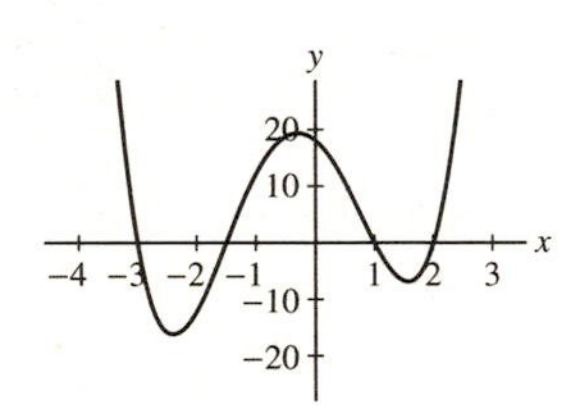

Now, the roots of $f(x)$ are the x-intercepts of the graph of $y = f(x)$. From the plot, we can identify the x-intercepts as -3, -1.5, 1, and 2. The roots of $f(x)$ are therefore $x = -3$, $x = -1.5$, $x = 1$, and $x = 2$.

3. How many *positive* solutions does $x^3 - 12x + 8 = 0$ have?

SOLUTION The graph of $y = x^3 - 12x + 8$ shown below has two x-intercepts to the right of the origin; therefore the equation $x^3 - 12x + 8 = 0$ has two positive solutions.

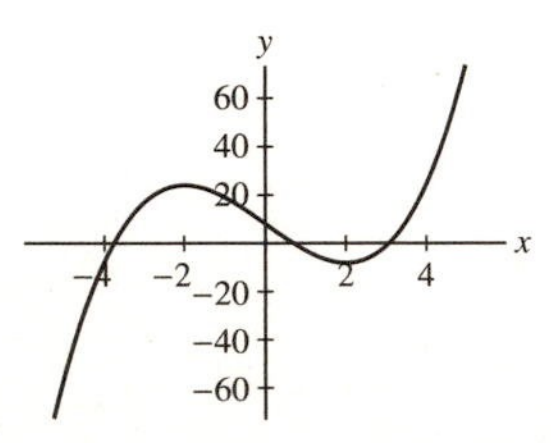

5. Find all the solutions of $\sin x = \sqrt{x}$ for $x > 0$.

SOLUTION Solutions to the equation $\sin x = \sqrt{x}$ correspond to points of intersection between the graphs of $y = \sin x$ and $y = \sqrt{x}$. The two graphs are shown below; the only point of intersection is at $x = 0$. Therefore, there are no solutions of $\sin x = \sqrt{x}$ for $x > 0$.

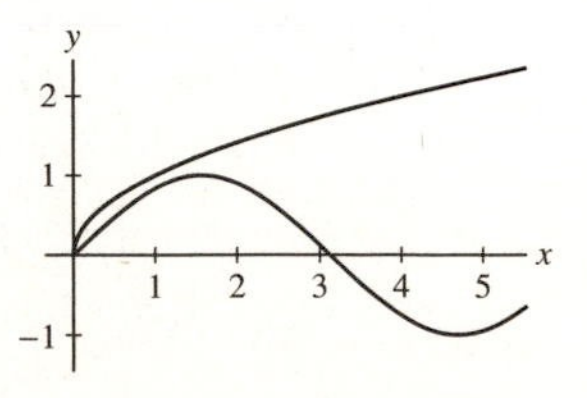

7. Let $f(x) = (x - 100)^2 + 1000$. What will the display show if you graph $f(x)$ in the viewing rectangle $[-10, 10]$ by $[-10, 10]$? Find an appropriate viewing rectangle.

SOLUTION Because $(x - 100)^2 \geq 0$ for all x, it follows that $f(x) = (x - 100)^2 + 1000 \geq 1000$ for all x. Thus, using a viewing rectangle of $[-10, 10]$ by $[-10, 10]$ will display nothing. The minimum value of the function occurs when $x = 100$, so an appropriate viewing rectangle would be $[50, 150]$ by $[1000, 2000]$.

9. Plot the graph of $f(x) = x/(4 - x)$ in a viewing rectangle that clearly displays the vertical and horizontal asymptotes.

SOLUTION From the graph of $y = \dfrac{x}{4 - x}$ shown below, we see that the vertical asymptote is $x = 4$ and the horizontal asymptote is $y = -1$.

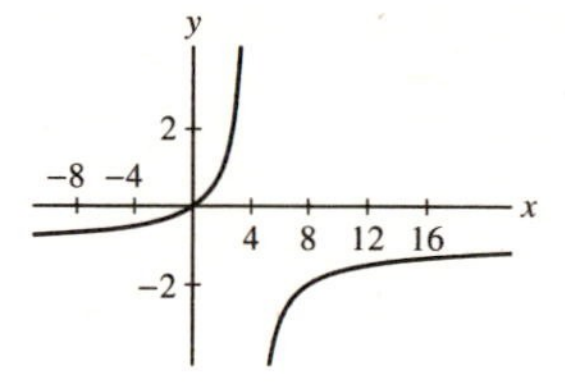

11. Plot $f(x) = \cos(x^2)\sin x$ for $0 \leq x \leq 2\pi$. Then illustrate local linearity at $x = 3.8$ by choosing appropriate viewing rectangles.

SOLUTION The following three graphs display $f(x) = \cos(x^2)\sin x$ over the intervals $[0, 2\pi]$, $[3.5, 4.1]$ and $[3.75, 3.85]$. The final graph looks like a straight line.

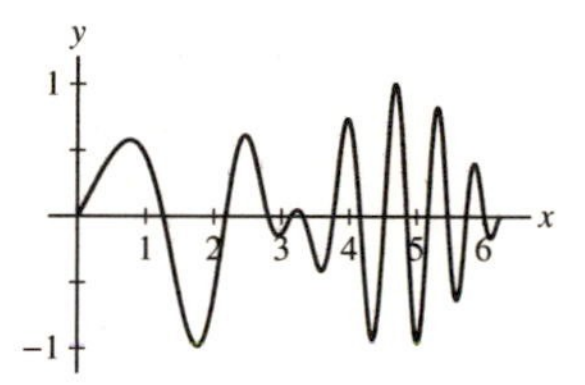

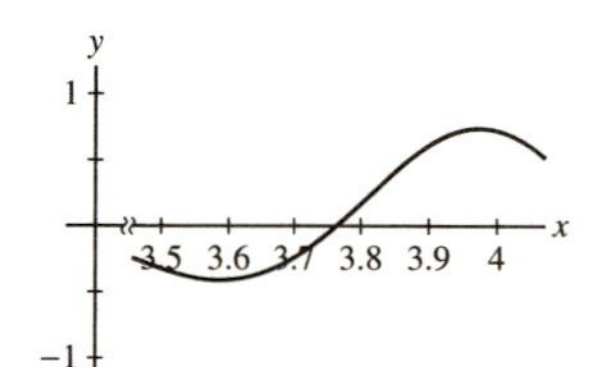

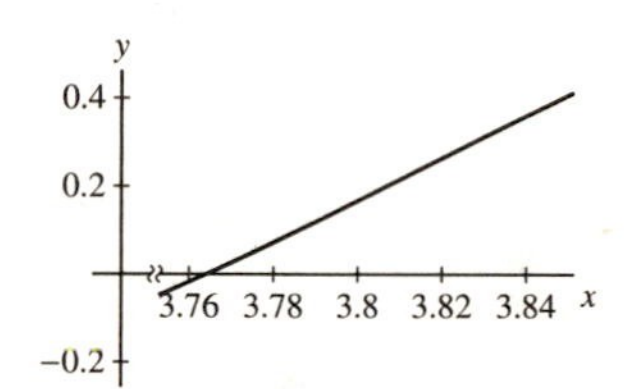

In Exercises 13–18, investigate the behavior of the function as n or x grows large by making a table of function values and plotting a graph (see Example 4). Describe the behavior in words.

13. $f(n) = n^{1/n}$

SOLUTION The table and graphs below suggest that as n gets large, $n^{1/n}$ approaches 1.

n	$n^{1/n}$
10	1.258925412
10^2	1.047128548
10^3	1.006931669
10^4	1.000921458
10^5	1.000115136
10^6	1.000013816

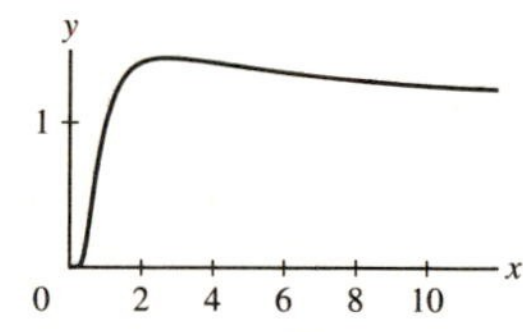

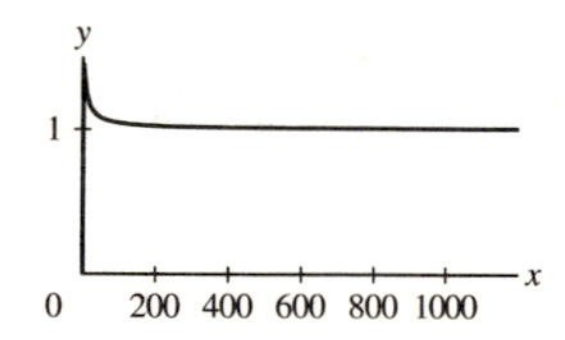

15. $f(n) = \left(1 + \dfrac{1}{n}\right)^{n^2}$

SOLUTION The table and graphs below suggest that as n gets large, $f(n)$ tends toward ∞.

n	$\left(1 + \frac{1}{n}\right)^{n^2}$
10	13780.61234
10^2	$1.635828711 \times 10^{43}$
10^3	$1.195306603 \times 10^{434}$
10^4	$5.341783312 \times 10^{4342}$
10^5	$1.702333054 \times 10^{43429}$
10^6	$1.839738749 \times 10^{434294}$

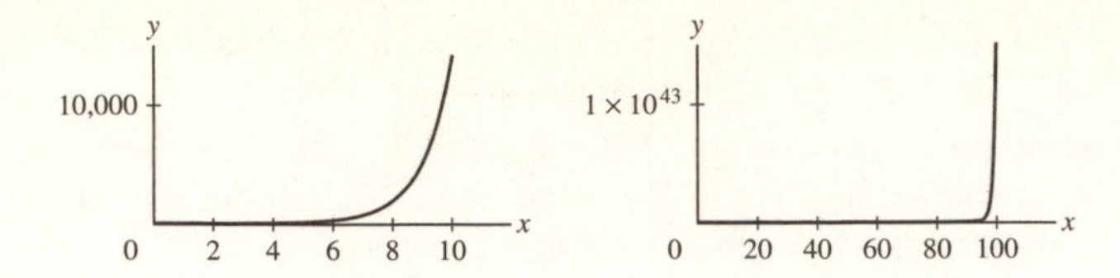

17. $f(x) = \left(x \tan \frac{1}{x}\right)^x$

SOLUTION The table and graphs below suggest that as x gets large, $f(x)$ approaches 1.

x	$\left(x \tan \frac{1}{x}\right)^x$
10	1.033975759
10^2	1.003338973
10^3	1.000333389
10^4	1.000033334
10^5	1.000003333
10^6	1.000000333

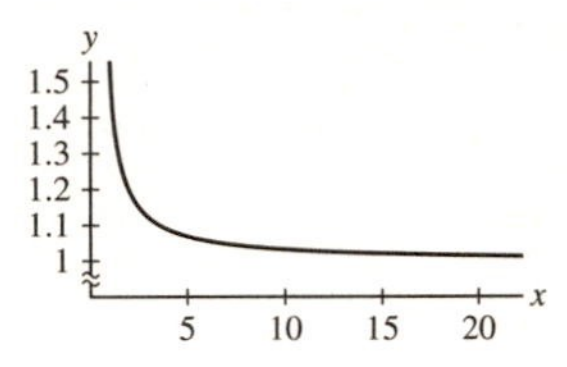

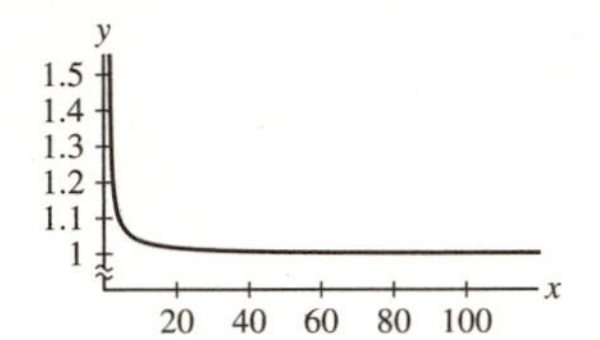

19. The graph of $f(\theta) = A\cos\theta + B\sin\theta$ is a sinusoidal wave for any constants A and B. Confirm this for $(A, B) = (1, 1)$, $(1, 2)$, and $(3, 4)$ by plotting $f(\theta)$.

SOLUTION The graphs of $f(\theta) = \cos\theta + \sin\theta$, $f(\theta) = \cos\theta + 2\sin\theta$ and $f(\theta) = 3\cos\theta + 4\sin\theta$ are shown below.

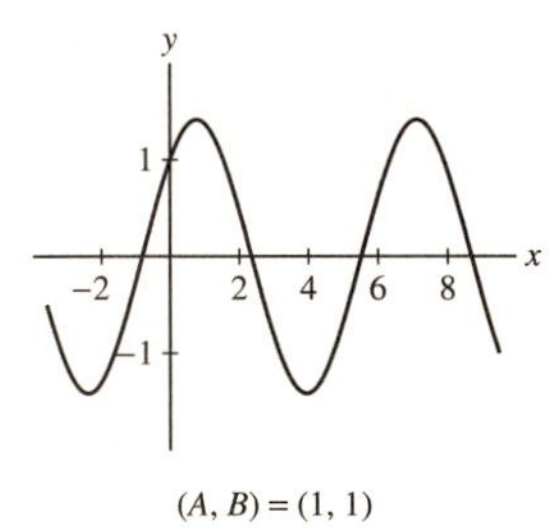

$(A, B) = (1, 1)$

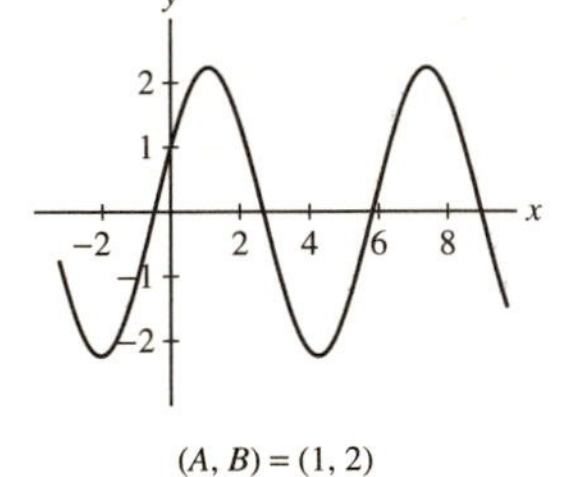

$(A, B) = (1, 2)$

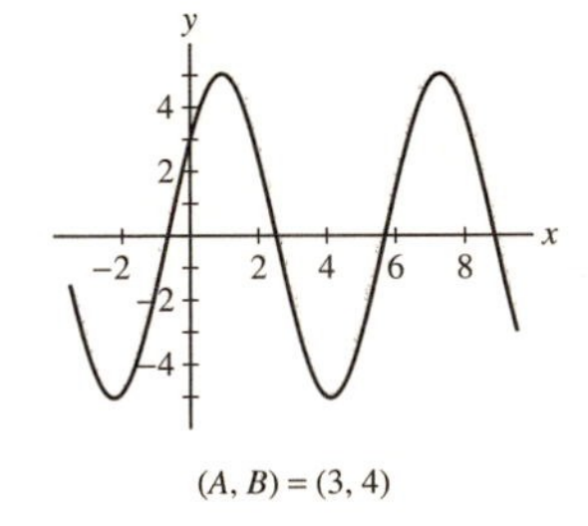

$(A, B) = (3, 4)$

21. Find the intervals on which $f(x) = x(x+2)(x-3)$ is positive by plotting a graph.

SOLUTION The function $f(x) = x(x+2)(x-3)$ is positive when the graph of $y = x(x+2)(x-3)$ lies above the x-axis. The graph of $y = x(x+2)(x-3)$ is shown below. Clearly, the graph lies above the x-axis and the function is positive for $x \in (-2, 0) \cup (3, \infty)$.

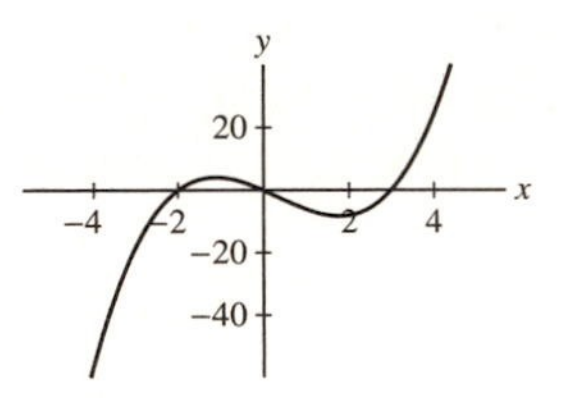

Further Insights and Challenges

23. CAS Let $f_1(x) = x$ and define a sequence of functions by $f_{n+1}(x) = \frac{1}{2}(f_n(x) + x/f_n(x))$. For example, $f_2(x) = \frac{1}{2}(x+1)$. Use a computer algebra system to compute $f_n(x)$ for $n = 3, 4, 5$ and plot $f_n(x)$ together with $\sqrt{x}$ for $x \geq 0$. What do you notice?

SOLUTION With $f_1(x) = x$ and $f_2(x) = \frac{1}{2}(x+1)$, we calculate

$$f_3(x) = \frac{1}{2}\left(\frac{1}{2}(x+1) + \frac{x}{\frac{1}{2}(x+1)}\right) = \frac{x^2+6x+1}{4(x+1)}$$

$$f_4(x) = \frac{1}{2}\left(\frac{x^2+6x+1}{4(x+1)} + \frac{x}{\frac{x^2+6x+1}{4(x+1)}}\right) = \frac{x^4+28x^3+70x^2+28x+1}{8(1+x)(1+6x+x^2)}$$

and

$$f_5(x) = \frac{1+120x+1820x^2+8008x^3+12870x^4+8008x^5+1820x^6+120x^7+x^8}{16(1+x)(1+6x+x^2)(1+28x+70x^2+28x^3+x^4)}.$$

A plot of $f_1(x)$, $f_2(x)$, $f_3(x)$, $f_4(x)$, $f_5(x)$ and $\sqrt{x}$ is shown below, with the graph of $\sqrt{x}$ shown as a dashed curve. It seems as if the f_n are asymptotic to $\sqrt{x}$.

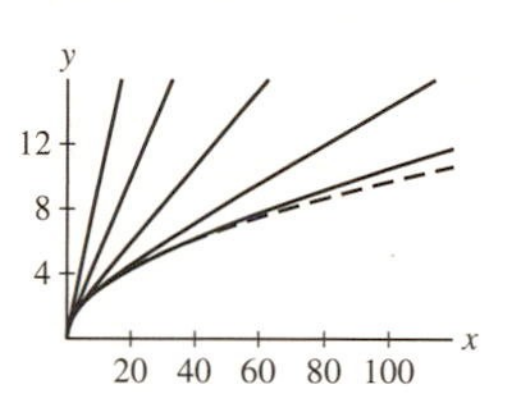

CHAPTER REVIEW EXERCISES

1. Express $(4, 10)$ as a set $\{x : |x - a| < c\}$ for suitable a and c.

SOLUTION The center of the interval $(4, 10)$ is $\frac{4+10}{2} = 7$ and the radius is $\frac{10-4}{2} = 3$. Therefore, the interval $(4, 10)$ is equivalent to the set $\{x : |x - 7| < 3\}$.

3. Express $\{x : 2 \le |x - 1| \le 6\}$ as a union of two intervals.

SOLUTION The set $\{x : 2 \le |x - 1| \le 6\}$ consists of those numbers that are at least 2 but at most 6 units from 1. The numbers larger than 1 that satisfy these conditions are $3 \le x \le 7$, while the numbers smaller than 1 that satisfy these conditions are $-5 \le x \le -1$. Therefore $\{x : 2 \le |x - 1| \le 6\} = [-5, -1] \cup [3, 7]$.

5. Describe the pairs of numbers x, y such that $|x + y| = x - y$.

SOLUTION First consider the case when $x + y \ge 0$. Then $|x + y| = x + y$ and we obtain the equation $x + y = x - y$. The solution of this equation is $y = 0$. Thus, the pairs $(x, 0)$ with $x \ge 0$ satisfy $|x + y| = x - y$. Next, consider the case when $x + y < 0$. Then $|x + y| = -(x + y) = -x - y$ and we obtain the equation $-x - y = x - y$. The solution of this equation is $x = 0$. Thus, the pairs $(0, y)$ with $y < 0$ also satisfy $|x + y| = x - y$.

In Exercises 7–10, let $f(x)$ be the function shown in Figure 1.

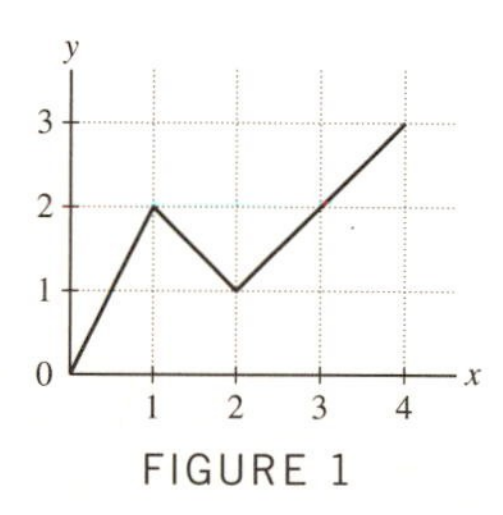

FIGURE 1

7. Sketch the graphs of $y = f(x) + 2$ and $y = f(x + 2)$.

SOLUTION The graph of $y = f(x) + 2$ is obtained by shifting the graph of $y = f(x)$ up 2 units (see the graph below at the left). The graph of $y = f(x + 2)$ is obtained by shifting the graph of $y = f(x)$ to the left 2 units (see the graph below at the right).

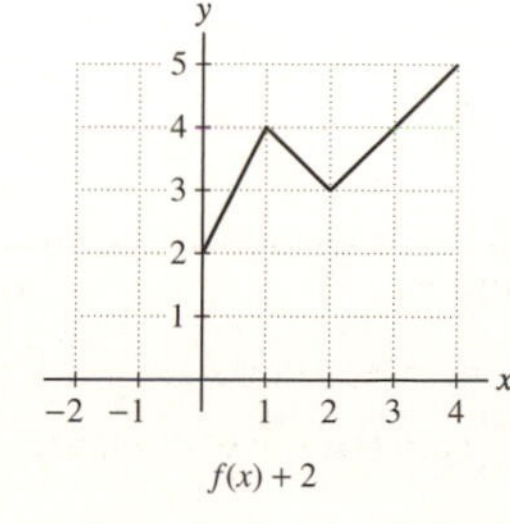

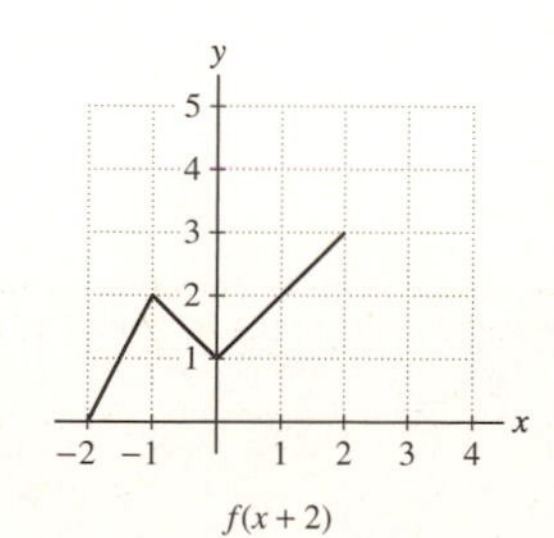

9. Continue the graph of $f(x)$ to the interval $[-4, 4]$ as an even function.

SOLUTION To continue the graph of $f(x)$ to the interval $[-4, 4]$ as an even function, reflect the graph of $f(x)$ across the y-axis (see the graph below).

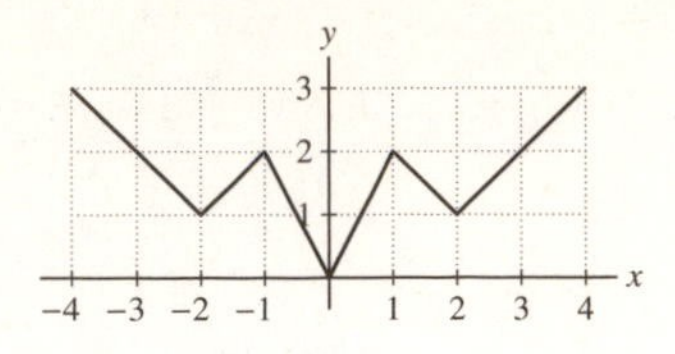

In Exercises 11–14, find the domain and range of the function.

11. $f(x) = \sqrt{x+1}$

SOLUTION The domain of the function $f(x) = \sqrt{x+1}$ is $\{x : x \geq -1\}$ and the range is $\{y : y \geq 0\}$.

13. $f(x) = \dfrac{2}{3-x}$

SOLUTION The domain of the function $f(x) = \dfrac{2}{3-x}$ is $\{x : x \neq 3\}$ and the range is $\{y : y \neq 0\}$.

15. Determine whether the function is increasing, decreasing, or neither:

(a) $f(x) = 3^{-x}$

(b) $f(x) = \dfrac{1}{x^2+1}$

(c) $g(t) = t^2 + t$

(d) $g(t) = t^3 + t$

SOLUTION

(a) The function $f(x) = 3^{-x}$ can be rewritten as $f(x) = (\frac{1}{3})^x$. This is an exponential function with a base less than 1; therefore, this is a decreasing function.

(b) From the graph of $y = 1/(x^2+1)$ shown below, we see that this function is neither increasing nor decreasing for all x (though it is increasing for $x < 0$ and decreasing for $x > 0$).

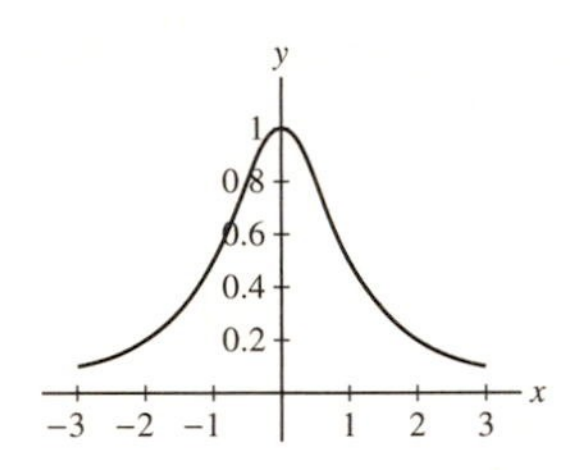

(c) The graph of $y = t^2 + t$ is an upward opening parabola; therefore, this function is neither increasing nor decreasing for all t. By completing the square we find $y = (t + \frac{1}{2})^2 - \frac{1}{4}$. The vertex of this parabola is then at $t = -\frac{1}{2}$, so the function is decreasing for $t < -\frac{1}{2}$ and increasing for $t > -\frac{1}{2}$.

(d) From the graph of $y = t^3 + t$ shown below, we see that this is an increasing function.

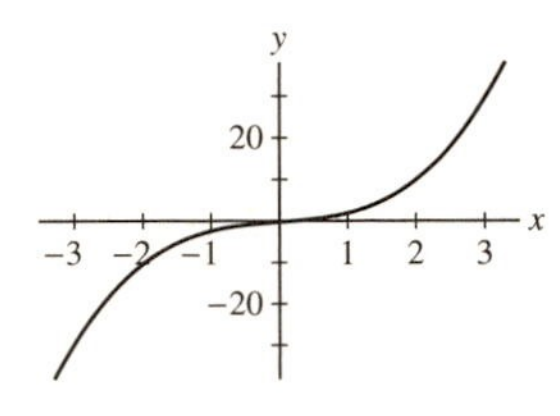

In Exercises 17–22, find the equation of the line.

17. Line passing through $(-1, 4)$ and $(2, 6)$

SOLUTION The slope of the line passing through $(-1, 4)$ and $(2, 6)$ is

$$m = \frac{6-4}{2-(-1)} = \frac{2}{3}.$$

The equation of the line passing through $(-1, 4)$ and $(2, 6)$ is therefore $y - 4 = \frac{2}{3}(x+1)$ or $2x - 3y = -14$.

19. Line of slope 6 through $(9, 1)$

SOLUTION Using the point-slope form for the equation of a line, the equation of the line of slope 6 and passing through $(9, 1)$ is $y - 1 = 6(x - 9)$ or $6x - y = 53$.

21. Line through (2, 3) parallel to $y = 4 - x$

SOLUTION The equation $y = 4 - x$ is in slope-intercept form; it follows that the slope of this line is -1. Any line parallel to $y = 4 - x$ will have the same slope, so we are looking for the equation of the line of slope -1 and passing through (2, 3). The equation of this line is $y - 3 = -(x - 2)$ or $x + y = 5$.

23. Does the following table of market data suggest a linear relationship between price and number of homes sold during a one-year period? Explain.

Price (thousands of \$)	180	195	220	240
No. of homes sold	127	118	103	91

SOLUTION Examine the slope between consecutive data points. The first pair of data points yields a slope of

$$\frac{118 - 127}{195 - 180} = -\frac{9}{15} = -\frac{3}{5},$$

while the second pair of data points yields a slope of

$$\frac{103 - 118}{220 - 195} = -\frac{15}{25} = -\frac{3}{5}$$

and the last pair of data points yields a slope of

$$\frac{91 - 103}{240 - 220} = -\frac{12}{20} = -\frac{3}{5}.$$

Because all three slopes are equal, the data does suggest a linear relationship between price and the number of homes sold.

25. Find the roots of $f(x) = x^4 - 4x^2$ and sketch its graph. On which intervals is $f(x)$ decreasing?

SOLUTION The roots of $f(x) = x^4 - 4x^2$ are obtained by solving the equation $x^4 - 4x^2 = x^2(x - 2)(x + 2) = 0$, which yields $x = -2$, $x = 0$ and $x = 2$. The graph of $y = f(x)$ is shown below. From this graph we see that $f(x)$ is decreasing for x less than approximately -1.4 and for x between 0 and approximately 1.4.

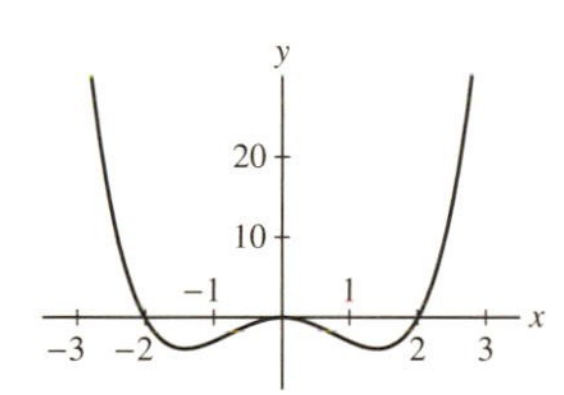

27. Let $f(x)$ be the square of the distance from the point (2, 1) to a point $(x, 3x + 2)$ on the line $y = 3x + 2$. Show that $f(x)$ is a quadratic function, and find its minimum value by completing the square.

SOLUTION Let $f(x)$ denote the square of the distance from the point (2, 1) to a point $(x, 3x + 2)$ on the line $y = 3x + 2$. Then

$$f(x) = (x - 2)^2 + (3x + 2 - 1)^2 = x^2 - 4x + 4 + 9x^2 + 6x + 1 = 10x^2 + 2x + 5,$$

which is a quadratic function. Completing the square, we find

$$f(x) = 10\left(x^2 + \frac{1}{5}x + \frac{1}{100}\right) + 5 - \frac{1}{10} = 10\left(x + \frac{1}{10}\right)^2 + \frac{49}{10}.$$

Because $(x + \frac{1}{10})^2 \geq 0$ for all x, it follows that $f(x) \geq \frac{49}{10}$ for all x. Hence, the minimum value of $f(x)$ is $\frac{49}{10}$.

In Exercises 29–34, sketch the graph by hand.

29. $y = t^4$

SOLUTION

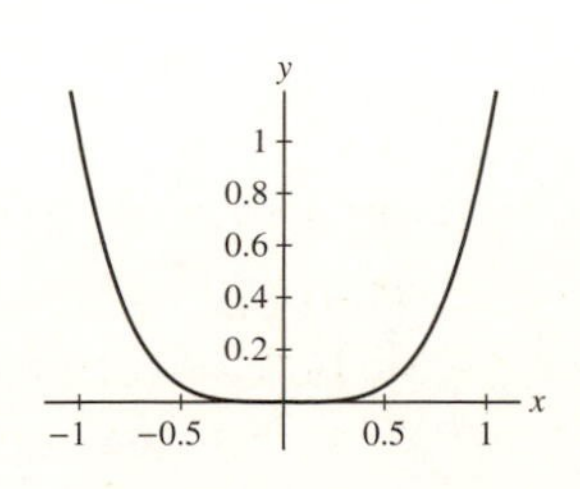

31. $y = \sin\dfrac{\theta}{2}$

SOLUTION

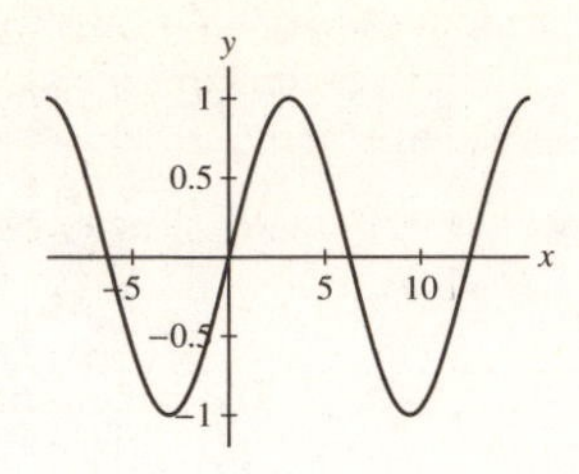

33. $y = x^{1/3}$

SOLUTION

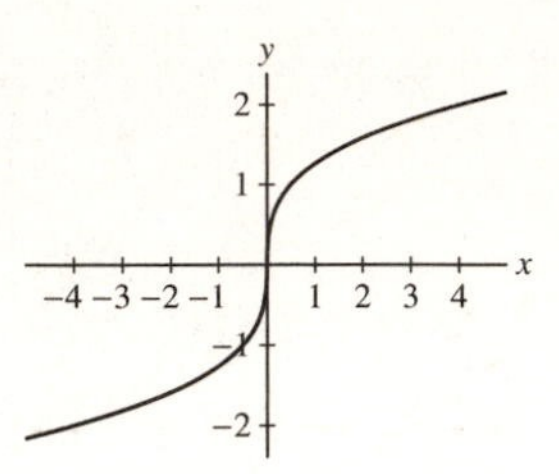

35. Show that the graph of $y = f\left(\frac{1}{3}x - b\right)$ is obtained by shifting the graph of $y = f\left(\frac{1}{3}x\right)$ to the right $3b$ units. Use this observation to sketch the graph of $y = \left|\frac{1}{3}x - 4\right|$.

SOLUTION Let $g(x) = f(\frac{1}{3}x)$. Then

$$g(x - 3b) = f\left(\frac{1}{3}(x - 3b)\right) = f\left(\frac{1}{3}x - b\right).$$

Thus, the graph of $y = f(\frac{1}{3}x - b)$ is obtained by shifting the graph of $y = f(\frac{1}{3}x)$ to the right $3b$ units.

The graph of $y = |\frac{1}{3}x - 4|$ is the graph of $y = |\frac{1}{3}x|$ shifted right 12 units (see the graph below).

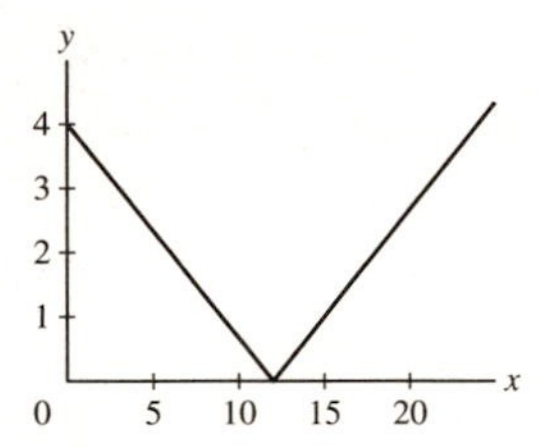

37. Find functions f and g such that the function

$$f(g(t)) = (12t + 9)^4$$

SOLUTION One possible choice is $f(t) = t^4$ and $g(t) = 12t + 9$. Then

$$f(g(t)) = f(12t + 9) = (12t + 9)^4$$

as desired.

39. What is the period of the function $g(\theta) = \sin 2\theta + \sin\frac{\theta}{2}$?

SOLUTION The function $\sin 2\theta$ has a period of π, and the function $\sin(\theta/2)$ has a period of 4π. Because 4π is a multiple of π, the period of the function $g(\theta) = \sin 2\theta + \sin\theta/2$ is 4π.

41. Give an example of values a, b such that

(a) $\cos(a + b) \neq \cos a + \cos b$

(b) $\cos\dfrac{a}{2} \neq \dfrac{\cos a}{2}$

SOLUTION

(a) Take $a = b = \pi/2$. Then $\cos(a + b) = \cos\pi = -1$ but

$$\cos a + \cos b = \cos\frac{\pi}{2} + \cos\frac{\pi}{2} = 0 + 0 = 0.$$

(b) Take $a = \pi$. Then

$$\cos\left(\frac{a}{2}\right) = \cos\left(\frac{\pi}{2}\right) = 0$$

but

$$\frac{\cos a}{2} = \frac{\cos\pi}{2} = \frac{-1}{2} = -\frac{1}{2}.$$

43. Solve $\sin 2x + \cos x = 0$ for $0 \le x < 2\pi$.

SOLUTION Using the double angle formula for the sine function, we rewrite the equation as $2 \sin x \cos x + \cos x = \cos x(2 \sin x + 1) = 0$. Thus, either $\cos x = 0$ or $\sin x = -1/2$. From here we see that the solutions are $x = \pi/2$, $x = 7\pi/6$, $x = 3\pi/2$ and $x = 11\pi/6$.

45. GU Use a graphing calculator to determine whether the equation $\cos x = 5x^2 - 8x^4$ has any solutions.

SOLUTION The graphs of $y = \cos x$ and $y = 5x^2 - 8x^4$ are shown below. Because the graphs do not intersect, there are no solutions to the equation $\cos x = 5x^2 - 8x^4$.

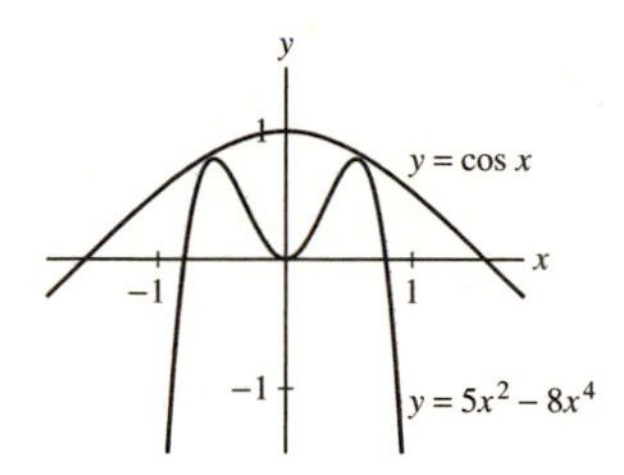

47. Match each quantity (a)–(d) with (i), (ii), or (iii) if possible, or state that no match exists.

(a) $2^a 3^b$

(b) $\dfrac{2^a}{3^b}$

(c) $(2^a)^b$

(d) $2^{a-b}3^{b-a}$

(i) 2^{ab}

(ii) 6^{a+b}

(iii) $\left(\frac{2}{3}\right)^{a-b}$

SOLUTION

(a) No match.

(b) No match.

(c) **(i)**: $(2^a)^b = 2^{ab}$.

(d) **(iii)**: $2^{a-b}3^{b-a} = 2^{a-b}\left(\dfrac{1}{3}\right)^{a-b} = \left(\dfrac{2}{3}\right)^{a-b}$.

49. Find the inverse of $f(x) = \sqrt{x^3 - 8}$ and determine its domain and range.

SOLUTION To find the inverse of $f(x) = \sqrt{x^3 - 8}$, we solve $y = \sqrt{x^3 - 8}$ for x as follows:

$$y^2 = x^3 - 8$$
$$x^3 = y^2 + 8$$
$$x = \sqrt[3]{y^2 + 8}.$$

Therefore, $f^{-1}(x) = \sqrt[3]{x^2 + 8}$. The domain of f^{-1} is the range of f, namely $\{x : x \ge 0\}$; the range of f^{-1} is the domain of f, namely $\{y : y \ge 2\}$.

51. Find a domain on which $h(t) = (t - 3)^2$ is one-to-one and determine the inverse on this domain.

SOLUTION From the graph of $h(t) = (t - 3)^2$ shown below, we see that h is one-to-one on each of the intervals $t \ge 3$ and $t \le 3$.

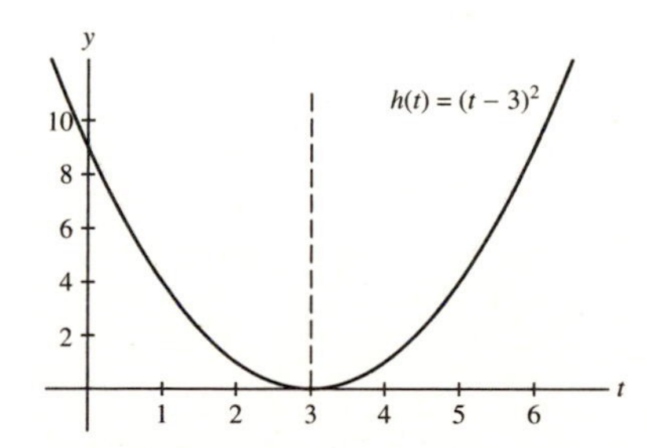

We find the inverse of $h(t) = (t - 3)^2$ on the domain $\{t : t \le 3\}$ by solving $y = (t - 3)^2$ for t. First, we find

$$\sqrt{y} = \sqrt{(t - 3)^2} = |t - 3|.$$

Having restricted the domain to $\{t : t \le 3\}$, $|t - 3| = -(t - 3) = 3 - t$. Thus,

$$\sqrt{y} = 3 - t$$
$$t = 3 - \sqrt{y}.$$

The inverse function is $h^{-1}(t) = 3 - \sqrt{t}$. For $t \ge 3$, $h^{-1}(t) = 3 + \sqrt{t}$.

53. Suppose that $g(x)$ is the inverse of $f(x)$. Match the functions (a)–(d) with their inverses (i)–(iv).

(a) $f(x)+1$ **(b)** $f(x+1)$ **(c)** $4f(x)$ **(d)** $f(4x)$

(i) $g(x)/4$ **(ii)** $g(x/4)$ **(iii)** $g(x-1)$ **(iv)** $g(x)-1$

SOLUTION

(a) (iii): $f(x)+1$ and $g(x-1)$ are inverse functions:

$$f(g(x-1))+1=(x-1)+1=x;$$
$$g(f(x)+1-1)=g(f(x))=x.$$

(b) (iv): $f(x+1)$ and $g(x)-1$ are inverse functions:

$$f(g(x)-1+1)=f(g(x))=x;$$
$$g(f(x+1))-1=(x+1)-1=x.$$

(c) (ii): $4f(x)$ and $g(x/4)$ are inverse functions:

$$4f(g(x/4))=4(x/4)=x;$$
$$g(4f(x)/4)=g(f(x))=x.$$

(d) (i): $f(4x)$ and $g(x)/4$ are inverse functions:

$$f(4\cdot g(x)/4)=f(g(x))=x;$$
$$\frac{1}{4}g(f(4x))=\frac{1}{4}(4x)=x.$$

2 LIMITS

2.1 Limits, Rates of Change, and Tangent Lines

Preliminary Questions

1. Average velocity is equal to the slope of a secant line through two points on a graph. Which graph?

SOLUTION Average velocity is the slope of a secant line through two points on the graph of position as a function of time.

2. Can instantaneous velocity be defined as a ratio? If not, how is instantaneous velocity computed?

SOLUTION Instantaneous velocity cannot be defined as a ratio. It is defined as the limit of average velocity as time elapsed shrinks to zero.

3. What is the graphical interpretation of instantaneous velocity at a moment $t = t_0$?

SOLUTION Instantaneous velocity at time $t = t_0$ is the slope of the line tangent to the graph of position as a function of time at $t = t_0$.

4. What is the graphical interpretation of the following statement? The average rate of change approaches the instantaneous rate of change as the interval $[x_0, x_1]$ shrinks to x_0.

SOLUTION The slope of the secant line over the interval $[x_0, x_1]$ approaches the slope of the tangent line at $x = x_0$.

5. The rate of change of atmospheric temperature with respect to altitude is equal to the slope of the tangent line to a graph. Which graph? What are possible units for this rate?

SOLUTION The rate of change of atmospheric temperature with respect to altitude is the slope of the line tangent to the graph of atmospheric temperature as a function of altitude. Possible units for this rate of change are °F/ft or °C/m.

Exercises

1. A ball dropped from a state of rest at time $t = 0$ travels a distance $s(t) = 4.9t^2$ m in t seconds.

(a) How far does the ball travel during the time interval [2, 2.5]?

(b) Compute the average velocity over [2, 2.5].

(c) Compute the average velocity for the time intervals in the table and estimate the ball's instantaneous velocity at $t = 2$.

Interval	[2, 2.01]	[2, 2.005]	[2, 2.001]	[2, 2.00001]
Average velocity				

SOLUTION

(a) During the time interval [2, 2.5], the ball travels $\Delta s = s(2.5) - s(2) = 4.9(2.5)^2 - 4.9(2)^2 = 11.025$ m.

(b) The average velocity over [2, 2.5] is

$$\frac{\Delta s}{\Delta t} = \frac{s(2.5) - s(2)}{2.5 - 2} = \frac{11.025}{0.5} = 22.05 \text{ m/s}.$$

(c)

time interval	[2, 2.01]	[2, 2.005]	[2, 2.001]	[2, 2.00001]
average velocity	19.649	19.6245	19.6049	19.600049

The instantaneous velocity at $t = 2$ is 19.6 m/s.

3. Let $v = 20\sqrt{T}$ as in Example 2. Estimate the instantaneous rate of change of v with respect to T when $T = 300$ K.

SOLUTION

T interval	[300, 300.01]	[300, 300.005]
average rate of change	0.577345	0.577348
T interval	[300, 300.001]	[300, 300.00001]
average rate of change	0.57735	0.57735

The instantaneous rate of change is approximately 0.57735 m/(s · K).

In Exercises 5 and 6, a stone is tossed vertically into the air from ground level with an initial velocity of 15 m/s. *Its height at time t is* $h(t) = 15t - 4.9t^2$ m.

5. Compute the stone's average velocity over the time interval [0.5, 2.5] and indicate the corresponding secant line on a sketch of the graph of $h(t)$.

SOLUTION The average velocity is equal to

$$\frac{h(2.5) - h(0.5)}{2} = 0.3.$$

The secant line is plotted with $h(t)$ below.

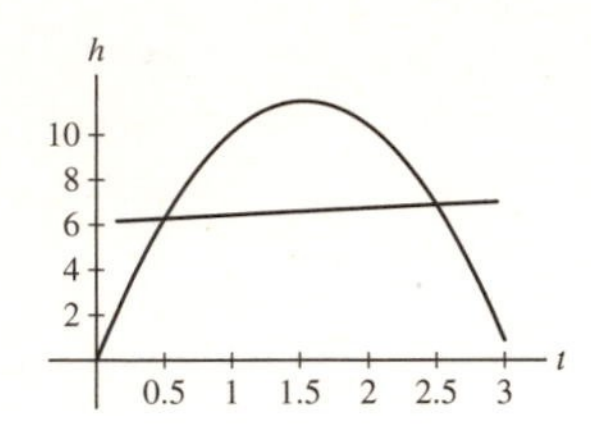

7. With an initial deposit of \$100, the balance in a bank account after t years is $f(t) = 100(1.08)^t$ dollars.

(a) What are the units of the rate of change of $f(t)$?

(b) Find the average rate of change over [0, 0.5] and [0, 1].

(c) Estimate the instantaneous rate of change at $t = 0.5$ by computing the average rate of change over intervals to the left and right of $t = 0.5$.

SOLUTION

(a) The units of the rate of change of $f(t)$ are dollars/year or \$/yr.

(b) The average rate of change of $f(t) = 100(1.08)^t$ over the time interval $[t_1, t_2]$ is given by

$$\frac{\Delta f}{\Delta t} = \frac{f(t_2) - f(t_1)}{t_2 - t_1}.$$

time interval	[0, .5]	[0, 1]
average rate of change	7.8461	8

(c)

time interval	[0.5, 0.51]	[0.5, 0.501]	[0.5, 0.5001]
average rate of change	8.0011	7.9983	7.9981
time interval	[0.49, 0.5]	[0.499, 0.5]	[0.4999, 0.5]
average rate of change	7.9949	7.9977	7.998

The rate of change at $t = 0.5$ is approximately \$8/yr.

9. Figure 8 shows the estimated number N of Internet users in Chile, based on data from the United Nations Statistics Division.

(a) Estimate the rate of change of N at $t = 2003.5$.

(b) Does the rate of change increase or decrease as t increases? Explain graphically.

(c) Let R be the average rate of change over [2001, 2005]. Compute R.

(d) Is the rate of change at $t = 2002$ greater than or less than the average rate R? Explain graphically.

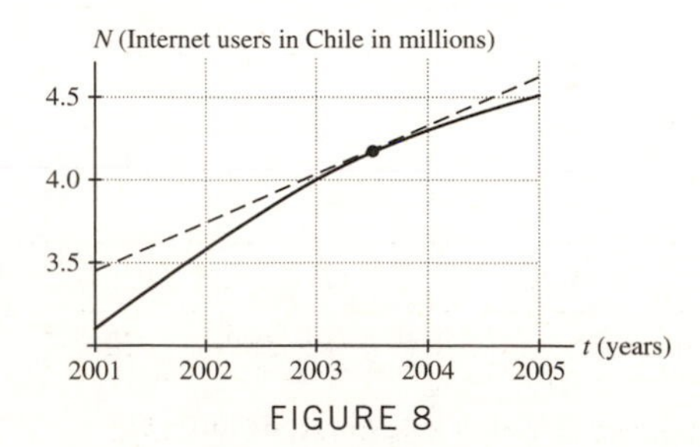

FIGURE 8

SOLUTION

(a) The tangent line shown in Figure 8 appears to pass through the points (2002, 3.75) and (2005, 4.6). Thus, the rate of change of N at $t = 2003.5$ is approximately

$$\frac{4.6 - 3.75}{2005 - 2002} = 0.283$$

million Internet users per year.

(b) As t increases, we move from left to right along the graph in Figure 8. Moreover, as we move from left to right along the graph, the slope of the tangent line decreases. Thus, the rate of change decreases as t increases.

(c) The graph of $N(t)$ appear to pass through the points (2001, 3.1) and (2005, 4.5). Thus, the average rate of change over [2001, 2005] is approximately

$$R = \frac{4.5 - 3.1}{2005 - 2001} = 0.35$$

million Internet users per year.

(d) For the figure below, we see that the slope of the tangent line at $t = 2002$ is larger than the slope of the secant line through the endpoints of the graph of $N(t)$. Thus, the rate of change at $t = 2002$ is greater than the average rate of change R.

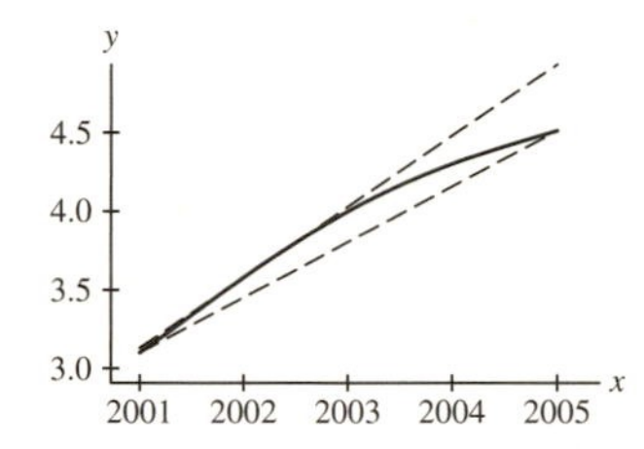

In Exercises 11–18, estimate the instantaneous rate of change at the point indicated.

11. $P(x) = 3x^2 - 5;\quad x = 2$

SOLUTION

x interval	[2, 2.01]	[2, 2.001]	[2, 2.0001]	[1.99, 2]	[1.999, 2]	[1.9999, 2]
average rate of change	12.03	12.003	12.0003	11.97	11.997	11.9997

The rate of change at $x = 2$ is approximately 12.

13. $y(x) = \dfrac{1}{x+2};\quad x = 2$

SOLUTION

x interval	[2, 2.01]	[2, 2.001]	[2, 2.0001]	[1.99, 2]	[1.999, 2]	[1.9999, 2]
average rate of change	−0.0623	−0.0625	−0.0625	−0.0627	−0.0625	−0.0625

The rate of change at $x = 2$ is approximately −0.06.

15. $f(x) = e^x;\quad x = 0$

SOLUTION

x interval	[−0.01, 0]	[−0.001, 0]	[−0.0001, 0]	[0, 0.01]	[0, 0.001]	[0, 0.0001]
average rate of change	0.9950	0.9995	0.99995	1.0050	1.0005	1.00005

The rate of change at $x = 0$ is approximately 1.00.

17. $f(x) = \ln x;\quad x = 3$

SOLUTION

x interval	[2.99, 3]	[2.999, 3]	[2.9999, 3]	[3, 3.01]	[3, 3.001]	[3, 3.0001]
average rate of change	0.33389	0.33339	0.33334	0.33278	0.33328	0.33333

The rate of change at $x = 3$ is approximately 0.333.

19. The height (in centimeters) at time t (in seconds) of a small mass oscillating at the end of a spring is $h(t) = 8\cos(12\pi t)$.
(a) Calculate the mass's average velocity over the time intervals [0, 0.1] and [3, 3.5].
(b) Estimate its instantaneous velocity at $t = 3$.

SOLUTION

(a) The average velocity over the time interval $[t_1, t_2]$ is given by $\dfrac{\Delta h}{\Delta t} = \dfrac{h(t_2) - h(t_1)}{t_2 - t_1}$.

time interval	[0, 0.1]	[3, 3.5]
average velocity	−144.721 cm/s	0 cm/s

(b)

time interval	[3, 3.0001]	[3, 3.00001]	[3, 3.000001]	[2.9999, 3]	[2.99999, 3]	[2.999999, 3]
average velocity	−0.5685	−0.05685	−0.005685	0.5685	0.05685	0.005685

The instantaneous velocity at $t = 3$ seconds is approximately 0 cm/s.

21. Assume that the period T (in seconds) of a pendulum (the time required for a complete back-and-forth cycle) is $T = \frac{3}{2}\sqrt{L}$, where L is the pendulum's length (in meters).
(a) What are the units for the rate of change of T with respect to L? Explain what this rate measures.
(b) Which quantities are represented by the slopes of lines A and B in Figure 10?
(c) Estimate the instantaneous rate of change of T with respect to L when $L = 3$ m.

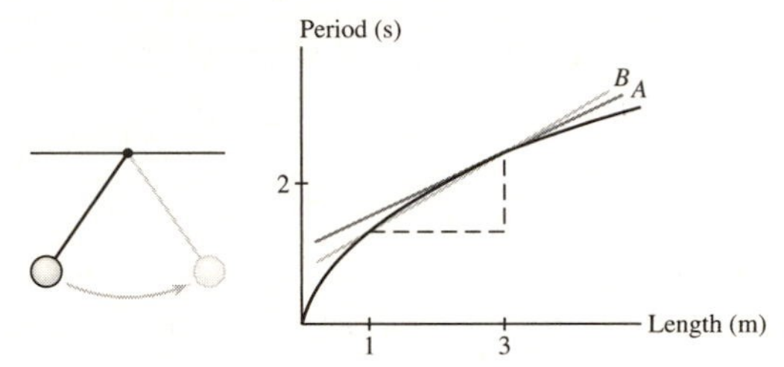

FIGURE 10 The period T is the time required for a pendulum to swing back and forth.

SOLUTION
(a) The units for the rate of change of T with respect to L are seconds per meter. This rate measures the sensitivity of the period of the pendulum to a change in the length of the pendulum.
(b) The slope of the line B represents the average rate of change in T from $L = 1$ m to $L = 3$ m. The slope of the line A represents the instantaneous rate of change of T at $L = 3$ m.
(c)

time interval	[3, 3.01]	[3, 3.001]	[3, 3.0001]	[2.99, 3]	[2.999, 3]	[2.9999, 3]
average velocity	0.4327	0.4330	0.4330	0.4334	0.4330	0.4330

The instantaneous rate of change at $L = 1$ m is approximately 0.4330 s/m.

23. GU An advertising campaign boosted sales of Crunchy Crust frozen pizza to a peak level of S_0 dollars per month. A marketing study showed that after t months, monthly sales declined to

$$S(t) = S_0 g(t), \quad \text{where } g(t) = \frac{1}{\sqrt{1+t}}.$$

Do sales decline more slowly or more rapidly as time increases? Answer by referring to a sketch of the graph of $g(t)$ together with several tangent lines.

SOLUTION We notice from the figure below that, as time increases, the slopes of the tangent lines to the graph of $g(t)$ become less negative. Thus, sales decline more slowly as time increases.

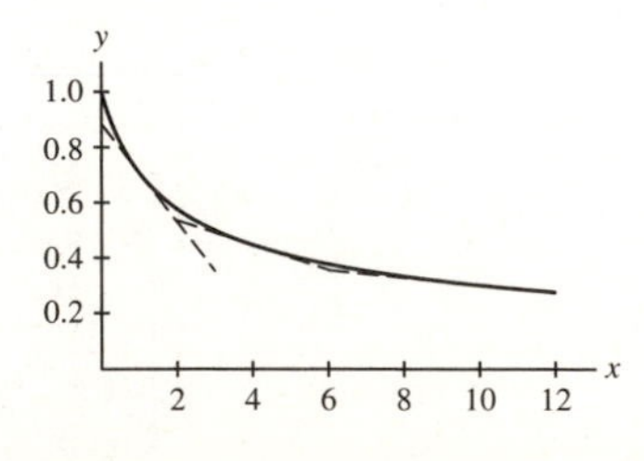

25. The graphs in Figure 13 represent the positions s of moving particles as functions of time t. Match each graph with a description:

(a) Speeding up

(b) Speeding up and then slowing down

(c) Slowing down

(d) Slowing down and then speeding up

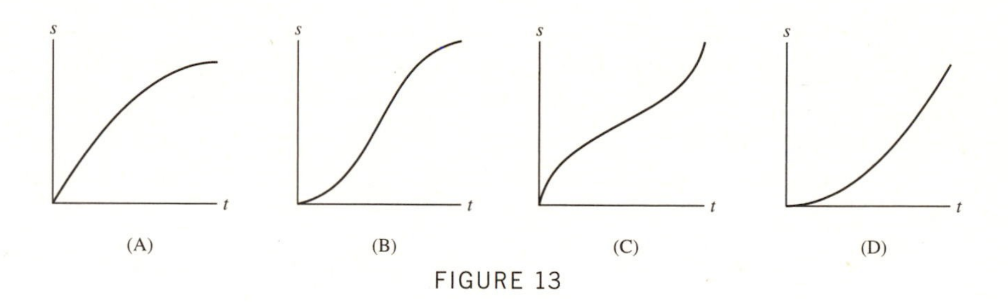

FIGURE 13

SOLUTION When a particle is speeding up over a time interval, its graph is bent upward over that interval. When a particle is slowing down, its graph is bent downward over that interval. Accordingly,

- In graph (A), the particle is (c) slowing down.
- In graph (B), the particle is (b) speeding up and then slowing down.
- In graph (C), the particle is (d) slowing down and then speeding up.
- In graph (D), the particle is (a) speeding up.

27. The fungus *Fusarium exosporium* infects a field of flax plants through the roots and causes the plants to wilt. Eventually, the entire field is infected. The percentage $f(t)$ of infected plants as a function of time t (in days) since planting is shown in Figure 15.

(a) What are the units of the rate of change of $f(t)$ with respect to t? What does this rate measure?

(b) Use the graph to rank (from smallest to largest) the average infection rates over the intervals $[0, 12]$, $[20, 32]$, and $[40, 52]$.

(c) Use the following table to compute the average rates of infection over the intervals $[30, 40]$, $[40, 50]$, $[30, 50]$.

Days	0	10	20	30	40	50	60
Percent infected	0	18	56	82	91	96	98

(d) Draw the tangent line at $t = 40$ and estimate its slope.

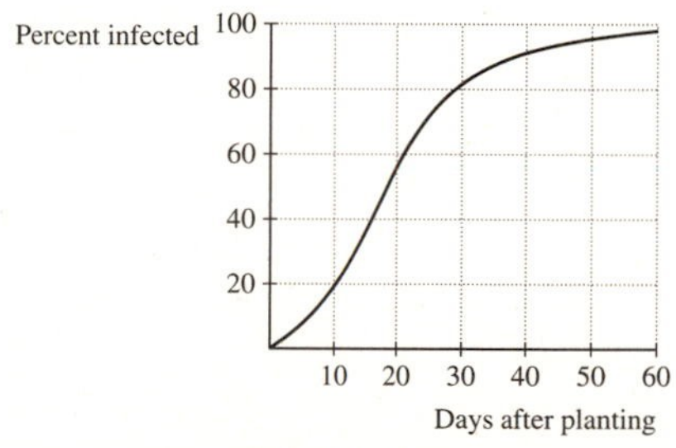

FIGURE 15

SOLUTION

(a) The units of the rate of change of $f(t)$ with respect to t are percent /day or %/d. This rate measures how quickly the population of flax plants is becoming infected.

(b) From smallest to largest, the average rates of infection are those over the intervals $[40, 52]$, $[0, 12]$, $[20, 32]$. This is because the slopes of the secant lines over these intervals are arranged from smallest to largest.

(c) The average rates of infection over the intervals $[30, 40]$, $[40, 50]$, $[30, 50]$ are 0.9, 0.5, 0.7 %/d, respectively.

(d) The tangent line sketched in the graph below appears to pass through the points $(20, 80)$ and $(40, 91)$. The estimate of the instantaneous rate of infection at $t = 40$ days is therefore

$$\frac{91 - 80}{40 - 20} = \frac{11}{20} = 0.55\%/\text{d}.$$

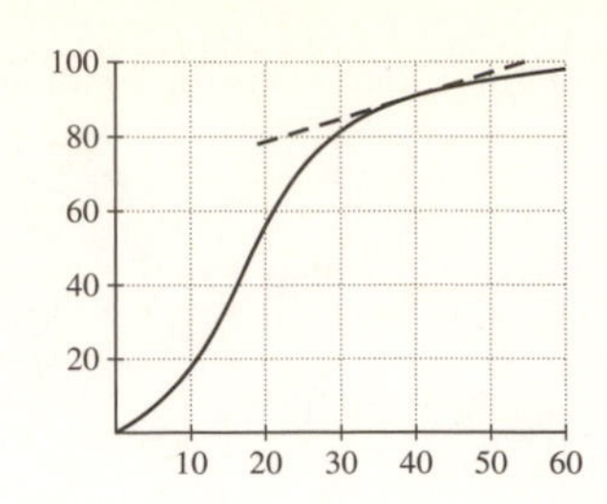

29. If an object in linear motion (but with changing velocity) covers Δs meters in Δt seconds, then its average velocity is $v_0 = \Delta s/\Delta t$ m/s. Show that it would cover the same distance if it traveled at constant velocity v_0 over the same time interval. This justifies our calling $\Delta s/\Delta t$ the *average velocity*.

SOLUTION At constant velocity, the distance traveled is equal to velocity times time, so an object moving at constant velocity v_0 for Δt seconds travels $v_0 \delta t$ meters. Since $v_0 = \Delta s/\Delta t$, we find

$$\text{distance traveled} = v_0 \delta t = \left(\frac{\Delta s}{\Delta t}\right) \Delta t = \Delta s$$

So the object covers the same distance Δs by traveling at constant velocity v_0.

31. Which graph in Figure 16 has the following property: For all x, the average rate of change over $[0, x]$ is greater than the instantaneous rate of change at x. Explain.

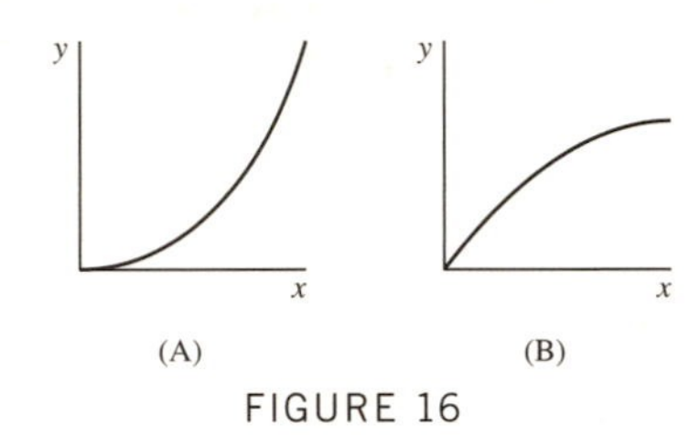

FIGURE 16

SOLUTION

(a) The average rate of change over $[0, x]$ is greater than the instantaneous rate of change at x: (B).

(b) The average rate of change over $[0, x]$ is less than the instantaneous rate of change at x: (A)

The graph in (B) bends downward, so the slope of the secant line through $(0, 0)$ and $(x, f(x))$ is larger than the slope of the tangent line at $(x, f(x))$. On the other hand, the graph in (A) bends upward, so the slope of the tangent line at $(x, f(x))$ is larger than the slope of the secant line through $(0, 0)$ and $(x, f(x))$.

Further Insights and Challenges

33. Let $Q(t) = t^2$. As in the previous exercise, find a formula for the average rate of change of Q over the interval $[1, t]$ and use it to estimate the instantaneous rate of change at $t = 1$. Repeat for the interval $[2, t]$ and estimate the rate of change at $t = 2$.

SOLUTION The average rate of change is

$$\frac{Q(t) - Q(1)}{t - 1} = \frac{t^2 - 1}{t - 1}.$$

Applying the difference of squares formula gives that the average rate of change is $((t + 1)(t - 1))/(t - 1) = (t + 1)$ for $t \neq 1$. As t gets closer to 1, this gets closer to $1 + 1 = 2$. The instantaneous rate of change is 2.

For $t_0 = 2$, the average rate of change is

$$\frac{Q(t) - Q(2)}{t - 2} = \frac{t^2 - 4}{t - 2},$$

which simplifies to $t + 2$ for $t \neq 2$. As t approaches 2, the average rate of change approaches 4. The instantaneous rate of change is therefore 4.

35. Find a formula for the average rate of change of $f(x) = x^3$ over $[2, x]$ and use it to estimate the instantaneous rate of change at $x = 2$.

SOLUTION The average rate of change is

$$\frac{f(x) - f(2)}{x - 2} = \frac{x^3 - 8}{x - 2}.$$

Applying the difference of cubes formula to the numerator, we find that the average rate of change is

$$\frac{(x^2 + 2x + 4)(x - 2)}{x - 2} = x^2 + 2x + 4$$

for $x \neq 2$. The closer x gets to 2, the closer the average rate of change gets to $2^2 + 2(2) + 4 = 12$.

2.2 Limits: A Numerical and Graphical Approach

Preliminary Questions

1. What is the limit of $f(x) = 1$ as $x \to \pi$?

SOLUTION $\lim_{x\to\pi} 1 = 1$.

2. What is the limit of $g(t) = t$ as $t \to \pi$?

SOLUTION $\lim_{t\to\pi} t = \pi$.

3. Is $\lim_{x\to 10} 20$ equal to 10 or 20?

SOLUTION $\lim_{x\to 10} 20 = 20$.

4. Can $f(x)$ approach a limit as $x \to c$ if $f(c)$ is undefined? If so, give an example.

SOLUTION Yes. The limit of a function f as $x \to c$ does not depend on what happens *at* $x = c$, only on the behavior of f as $x \to c$. As an example, consider the function

$$f(x) = \frac{x^2 - 1}{x - 1}.$$

The function is clearly not defined at $x = 1$ but

$$\lim_{x\to 1} f(x) = \lim_{x\to 1} \frac{x^2 - 1}{x - 1} = \lim_{x\to 1} (x + 1) = 2.$$

5. What does the following table suggest about $\lim_{x\to 1-} f(x)$ and $\lim_{x\to 1+} f(x)$?

x	0.9	0.99	0.999	1.1	1.01	1.001
$f(x)$	7	25	4317	3.0126	3.0047	3.00011

SOLUTION The values in the table suggest that $\lim_{x\to 1-} f(x) = \infty$ and $\lim_{x\to 1+} f(x) = 3$.

6. Can you tell whether $\lim_{x\to 5} f(x)$ exists from a plot of $f(x)$ for $x > 5$? Explain.

SOLUTION No. By examining values of $f(x)$ for x close to but greater than 5, we can determine whether the one-sided limit $\lim_{x\to 5+} f(x)$ exists. To determine whether $\lim_{x\to 5} f(x)$ exists, we must examine value of $f(x)$ on both sides of $x = 5$.

7. If you know in advance that $\lim_{x\to 5} f(x)$ exists, can you determine its value from a plot of $f(x)$ for all $x > 5$?

SOLUTION Yes. If $\lim_{x\to 5} f(x)$ exists, then both one-sided limits must exist and be equal.

Exercises

In Exercises 1–4, fill in the tables and guess the value of the limit.

1. $\lim_{x\to 1} f(x)$, where $f(x) = \dfrac{x^3 - 1}{x^2 - 1}$.

x	$f(x)$	x	$f(x)$
1.002		0.998	
1.001		0.999	
1.0005		0.9995	
1.00001		0.99999	

SOLUTION

x	0.998	0.999	0.9995	0.99999	1.00001	1.0005	1.001	1.002
$f(x)$	1.498501	1.499250	1.499625	1.499993	1.500008	1.500375	1.500750	1.501500

The limit as $x \to 1$ is $\frac{3}{2}$.

3. $\lim_{y\to 2} f(y)$, where $f(y) = \dfrac{y^2 - y - 2}{y^2 + y - 6}$.

y	$f(y)$	y	$f(y)$
2.002		1.998	
2.001		1.999	
2.0001		1.9999	

SOLUTION

y	1.998	1.999	1.9999	2.0001	2.001	2.02
$f(y)$	0.59984	0.59992	0.599992	0.600008	0.60008	0.601594

The limit as $y \to 2$ is $\frac{3}{5}$.

5. Determine $\lim_{x\to 0.5} f(x)$ for $f(x)$ as in Figure 9.

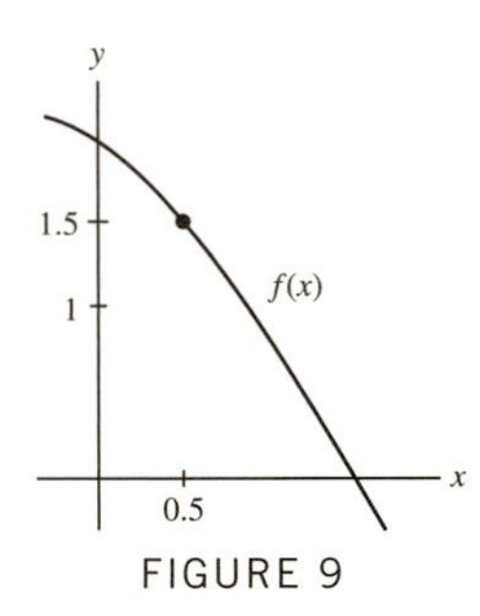

FIGURE 9

SOLUTION The graph suggests that $f(x) \to 1.5$ as $x \to 0.5$.

In Exercises 7 and 8, evaluate the limit.

7. $\lim_{x\to 21} x$

SOLUTION As $x \to 21$, $f(x) = x \to 21$. You can see this, for example, on the graph of $f(x) = x$.

In Exercises 9–16, verify each limit using the limit definition. For example, in Exercise 9, show that $|3x - 12|$ can be made as small as desired by taking x close to 4.

9. $\lim_{x\to 4} 3x = 12$

SOLUTION $|3x - 12| = 3|x - 4|$. $|3x - 12|$ can be made arbitrarily small by making x close enough to 4, thus making $|x - 4|$ small.

11. $\lim_{x\to 3} (5x + 2) = 17$

SOLUTION $|(5x + 2) - 17| = |5x - 15| = 5|x - 3|$. Therefore, if you make $|x - 3|$ small enough, you can make $|(5x + 2) - 17|$ as small as desired.

13. $\lim_{x\to 0} x^2 = 0$

SOLUTION As $x \to 0$, we have $|x^2 - 0| = |x + 0||x - 0|$. To simplify things, suppose that $|x| < 1$, so that $|x + 0||x - 0| = |x||x| < |x|$. By making $|x|$ sufficiently small, so that $|x + 0||x - 0| = x^2$ is even smaller, you can make $|x^2 - 0|$ as small as desired.

15. $\lim_{x\to 0} (4x^2 + 2x + 5) = 5$

SOLUTION As $x \to 0$, we have $|4x^2 + 2x + 5 - 5| = |4x^2 + 2x| = |x||4x + 2|$. If $|x| < 1$, $|4x + 2|$ can be no bigger than 6, so $|x||4x + 2| < 6|x|$. Therefore, by making $|x - 0| = |x|$ sufficiently small, you can make $|4x^2 + 2x + 5 - 5| = |x||4x + 2|$ as small as desired.

In Exercises 17–36, estimate the limit numerically or state that the limit does not exist. If infinite, state whether the one-sided limits are ∞ or $-\infty$.

17. $\lim_{x\to 1} \frac{\sqrt{x}-1}{x-1}$

SOLUTION

x	0.9995	0.99999	1.00001	1.0005
$f(x)$	0.500063	0.500001	0.49999	0.499938

The limit as $x \to 1$ is $\frac{1}{2}$.

19. $\lim_{x\to 2} \frac{x^2+x-6}{x^2-x-2}$

SOLUTION

x	1.999	1.99999	2.00001	2.001
$f(x)$	1.666889	1.666669	1.666664	1.666445

The limit as $x \to 2$ is $\frac{5}{3}$.

21. $\lim_{x\to 0} \frac{\sin 2x}{x}$

SOLUTION

x	−0.01	−0.005	0.005	0.01
$f(x)$	1.999867	1.999967	1.999967	1.999867

The limit as $x \to 0$ is 2.

23. $\lim_{\theta\to 0} \frac{\cos\theta - 1}{\theta}$

SOLUTION

x	−0.05	−0.001	0.001	0.05
$f(x)$	0.0249948	0.0005	−0.0005	−0.0249948

The limit as $x \to 0$ is 0.

25. $\lim_{x\to 4} \frac{1}{(x-4)^3}$

SOLUTION

x	3.99	3.999	3.9999	4.0001	4.001	4.01
$f(x)$	-10^6	-10^9	-10^{12}	10^{12}	10^9	10^6

The limit does not exist. As $x \to 4-$, $f(x) \to -\infty$; similarly, as $x \to 4+$, $f(x) \to \infty$.

27. $\lim_{x\to 3+} \frac{x-4}{x^2-9}$

SOLUTION

x	3.01	3.001	3.0001	3.00001
$f(x)$	−16.473	−166.473	−1666.473	−16666.473

As $x \to 3+$, $f(x) \to -\infty$.

29. $\lim_{h \to 0} \sin h \cos \frac{1}{h}$

SOLUTION

h	-0.01	-0.001	-0.0001	0.0001	0.001	0.01
$f(h)$	-0.008623	-0.000562	0.000095	-0.000095	0.000562	0.008623

The limit as $x \to 0$ is 0.

31. $\lim_{x \to 0} |x|^x$

SOLUTION

x	-0.05	-0.001	-0.00001	0.00001	0.001	0.05
$f(x)$	1.161586	1.006932	1.000115	0.999885	0.993116	0.860892

The limit as $x \to 0$ is 1.

33. $\lim_{t \to e} \frac{t - e}{\ln t - 1}$

SOLUTION

r	$e - 0.01$	$e - 0.001$	$e - 0.0001$	$e + 0.0001$	$e + 0.001$	$e + 0.01$
$f(t)$	2.713279	2.717782	2.718232	2.718332	2.718782	2.723279

The limit as $t \to 0$ is approximately 2.718. (The exact answer is e.)

35. $\lim_{x \to 1-} \frac{\tan^{-1} x}{\cos^{-1} x}$

SOLUTION

x	0.999	0.9999	0.99999	0.999999	0.9999999
$f(x)$	17.549	55.532	175.619	555.360	1756.204

The limit as $x \to 1-$ does not exist.

37. The **greatest integer function** is defined by $[x] = n$, where n is the unique integer such that $n \leq x < n + 1$. Sketch the graph of $y = [x]$. Calculate, for c an integer:

(a) $\lim_{x \to c-} [x]$ **(b)** $\lim_{x \to c+} [x]$

SOLUTION Here is a graph of the greatest integer function:

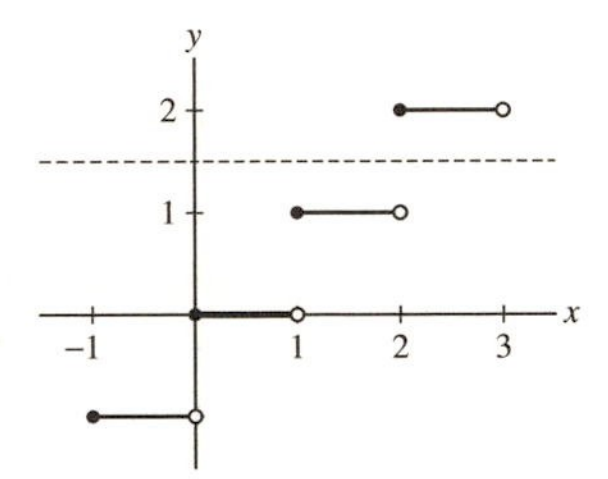

(a) From the graph, we see that, for c an integer,

$$\lim_{x \to c-} [x] = c - 1.$$

(b) From the graph, we see that, for c an integer,

$$\lim_{x \to c+} [x] = c.$$

In Exercises 39–46, determine the one-sided limits numerically or graphically. If infinite, state whether the one-sided limits are ∞ or $-\infty$, and describe the corresponding vertical asymptote. In Exercise 46, $[x]$ is the greatest integer function defined in Exercise 37.

39. $\lim\limits_{x\to 0\pm} \dfrac{\sin x}{|x|}$

SOLUTION

x	-0.2	-0.02	0.02	0.2
$f(x)$	-0.993347	-0.999933	0.999933	0.993347

The left-hand limit is $\lim\limits_{x\to 0-} f(x) = -1$, whereas the right-hand limit is $\lim\limits_{x\to 0+} f(x) = 1$.

41. $\lim\limits_{x\to 0\pm} \dfrac{x - \sin|x|}{x^3}$

SOLUTION

x	-0.1	-0.01	0.01	0.1
$f(x)$	199.853	19999.8	0.166666	0.166583

The left-hand limit is $\lim\limits_{x\to 0-} f(x) = \infty$, whereas the right-hand limit is $\lim\limits_{x\to 0+} f(x) = \frac{1}{6}$. Thus, the line $x = 0$ is a vertical asymptote from the left for the graph of $y = \frac{x-\sin|x|}{x^3}$.

43. $\lim\limits_{x\to -2\pm} \dfrac{4x^2+7}{x^3+8}$

SOLUTION The graph of $y = \frac{4x^2+7}{x^3+8}$ for x near -2 is shown below. From this graph, we see that

$$\lim_{x\to -2-} \frac{4x^2+7}{x^3+8} = -\infty \quad \text{while} \quad \lim_{x\to -2+} \frac{4x^2+7}{x^3+8} = \infty.$$

Thus, the line $x = -2$ is a vertical asymptote for the graph of $y = \frac{4x^2+7}{x^3+8}$.

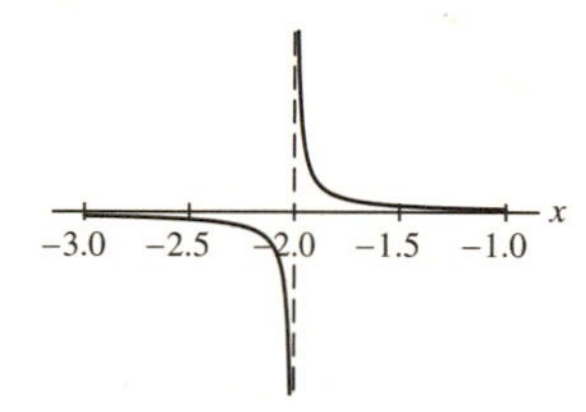

45. $\lim\limits_{x\to 1\pm} \dfrac{x^5+x-2}{x^2+x-2}$

SOLUTION The graph of $y = \frac{x^5+x-2}{x^2+x-2}$ for x near 1 is shown below. From this graph, we see that

$$\lim_{x\to 1\pm} \frac{x^5+x-2}{x^2+x-2} = 2.$$

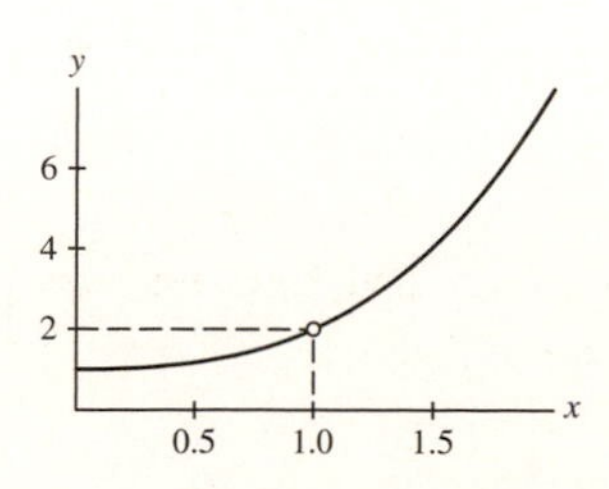

47. Determine the one-sided limits at $c = 2, 4$ of the function $f(x)$ in Figure 12. What are the vertical asymptotes of $f(x)$?

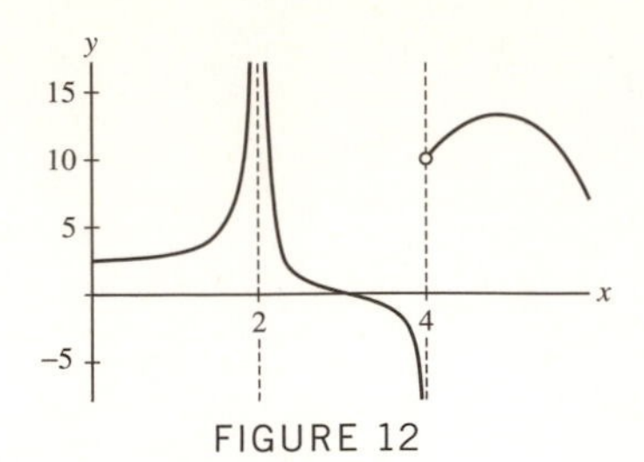

FIGURE 12

SOLUTION

- For $c = 2$, we have $\lim_{x\to 2-} f(x) = \infty$ and $\lim_{x\to 2+} f(x) = \infty$.
- For $c = 4$, we have $\lim_{x\to 4-} f(x) = -\infty$ and $\lim_{x\to 4+} f(x) = 10$.

The vertical asymptotes are the vertical lines $x = 2$ and $x = 4$.

In Exercises 49–52, sketch the graph of a function with the given limits.

49. $\lim_{x\to 1} f(x) = 2, \quad \lim_{x\to 3-} f(x) = 0, \quad \lim_{x\to 3+} f(x) = 4$

SOLUTION

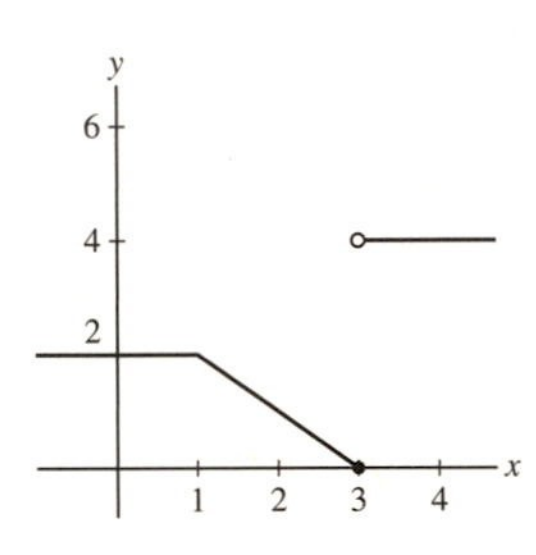

51. $\lim_{x\to 2+} f(x) = f(2) = 3, \quad \lim_{x\to 2-} f(x) = -1, \; \lim_{x\to 4} f(x) = 2 \neq f(4)$

SOLUTION

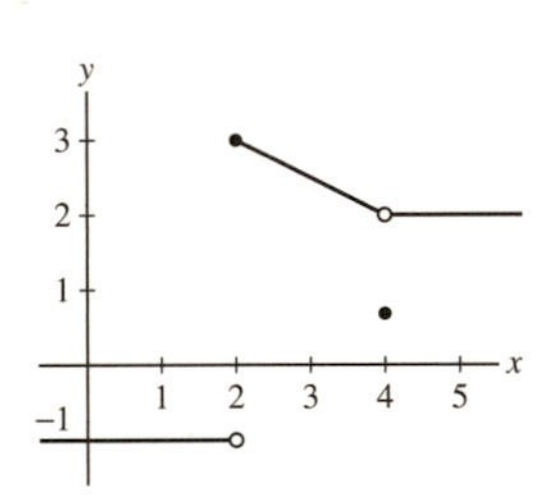

53. Determine the one-sided limits of the function $f(x)$ in Figure 14, at the points $c = 1, 3, 5, 6$.

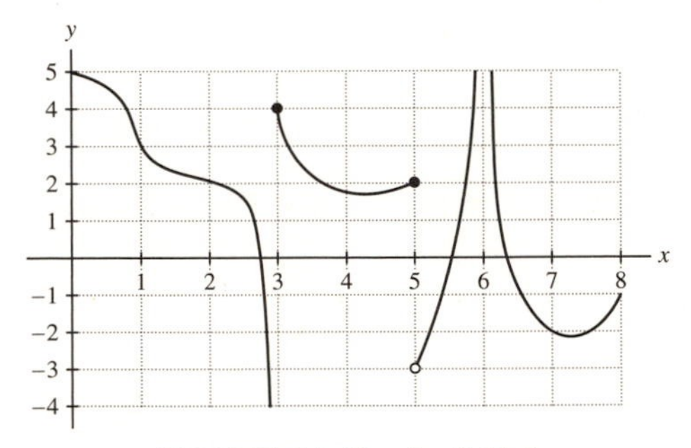

FIGURE 14 Graph of $f(x)$

SOLUTION

- $\lim_{x\to 1-} f(x) = \lim_{x\to 1+} f(x) = 3$
- $\lim_{x\to 3-} f(x) = -\infty$

- $\lim_{x \to 3+} f(x) = 4$
- $\lim_{x \to 5-} f(x) = 2$
- $\lim_{x \to 5+} f(x) = -3$
- $\lim_{x \to 6-} f(x) = \lim_{x \to 6+} f(x) = \infty$

GU *In Exercises 55–60, plot the function and use the graph to estimate the value of the limit.*

55. $\lim_{\theta \to 0} \dfrac{\sin 5\theta}{\sin 2\theta}$

SOLUTION

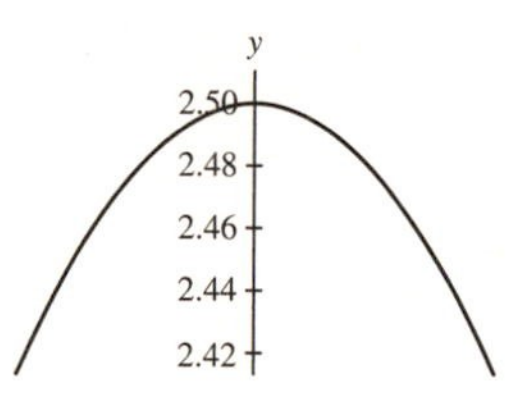

From the graph of $y = \dfrac{\sin 5\theta}{\sin 2\theta}$ shown above, we see that the limit as $\theta \to 0$ is $\frac{5}{2}$.

57. $\lim_{x \to 0} \dfrac{2^x - \cos x}{x}$

SOLUTION

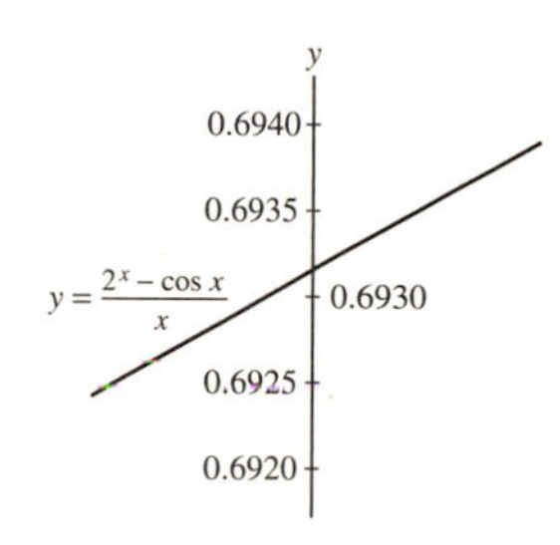

The limit as $x \to 0$ is approximately 0.693. (The exact answer is $\ln 2$.)

59. $\lim_{\theta \to 0} \dfrac{\cos 7\theta - \cos 5\theta}{\theta^2}$

SOLUTION

From the graph of $y = \dfrac{\cos 7\theta - \cos 5\theta}{\theta^2}$ shown above, we see that the limit as $\theta \to 0$ is -12.

61. Let n be a positive integer. For which n are the two infinite one-sided limits $\lim_{x \to 0\pm} 1/x^n$ equal?

SOLUTION First, suppose that n is even. Then $x^n \geq 0$ for all x, and $\frac{1}{x^n} > 0$ for all $x \neq 0$. Hence,

$$\lim_{x \to 0-} \frac{1}{x^n} = \lim_{x \to 0+} \frac{1}{x^n} = \infty.$$

Next, suppose that n is odd. Then $\frac{1}{x^n} > 0$ for all $x > 0$ but $\frac{1}{x^n} < 0$ for all $x < 0$. Thus,

$$\lim_{x \to 0-} \frac{1}{x^n} = -\infty \quad \text{but} \quad \lim_{x \to 0+} \frac{1}{x^n} = \infty.$$

Finally, the two infinite one-sided limits are equal whenever n is even.

63. GU In some cases, numerical investigations can be misleading. Plot $f(x) = \cos\frac{\pi}{x}$.

(a) Does $\lim_{x\to 0} f(x)$ exist?

(b) Show, by evaluating $f(x)$ at $x = \pm\frac{1}{2}, \pm\frac{1}{4}, \pm\frac{1}{6}, \ldots$, that you might be able to trick your friends into believing that the limit exists and is equal to $L = 1$.

(c) Which sequence of evaluations might trick them into believing that the limit is $L = -1$.

SOLUTION Here is the graph of $f(x)$.

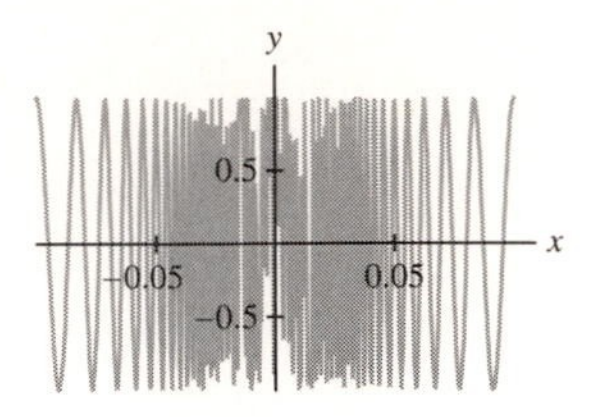

(a) From the graph of $f(x)$, which shows that the value of $f(x)$ oscillates more and more rapidly as $x \to 0$, it follows that $\lim_{x\to 0} f(x)$ does not exist.

(b) Notice that

$$f\left(\pm\frac{1}{2}\right) = \cos\frac{\pi}{\pm 1/2} = \cos \pm 2\pi = 1;$$
$$f\left(\pm\frac{1}{4}\right) = \cos\frac{\pi}{\pm 1/4} = \cos \pm 4\pi = 1;$$
$$f\left(\pm\frac{1}{6}\right) = \cos\frac{\pi}{\pm 1/6} = \cos \pm 6\pi = 1;$$

and, in general, $f(\pm\frac{1}{2n}) = 1$ for all integers n.

(c) At $x = \pm 1, \pm\frac{1}{3}, \pm\frac{1}{5}, \ldots$, the value of $f(x)$ is always -1.

Further Insights and Challenges

65. Investigate $\lim_{\theta\to 0} \dfrac{\sin n\theta}{\theta}$ numerically for several values of n. Then guess the value in general.

SOLUTION

- For $n = 3$, we have

θ	-0.1	-0.01	-0.001	0.001	0.01	0.1
$\dfrac{\sin n\theta}{\theta}$	2.955202	2.999550	2.999996	2.999996	2.999550	2.955202

The limit as $\theta \to 0$ is 3.

- For $n = -5$, we have

θ	-0.1	-0.01	-0.001	0.001	0.01	0.1
$\dfrac{\sin n\theta}{\theta}$	-4.794255	-4.997917	-4.999979	-4.999979	-4.997917	-4.794255

The limit as $\theta \to 0$ is -5.

- We surmise that, in general, $\lim_{\theta\to 0} \dfrac{\sin n\theta}{\theta} = n$.

67. Investigate $\lim_{x\to 1} \frac{x^n - 1}{x^m - 1}$ for (m, n) equal to (2, 1), (1, 2), (2, 3), and (3, 2). Then guess the value of the limit in general and check your guess for two additional pairs.

SOLUTION

-

x	0.99	0.9999	1.0001	1.01
$\frac{x-1}{x^2-1}$	0.502513	0.500025	0.499975	0.497512

The limit as $x \to 1$ is $\frac{1}{2}$.

x	0.99	0.9999	1.0001	1.01
$\frac{x^2-1}{x-1}$	1.99	1.9999	2.0001	2.01

The limit as $x \to 1$ is 2.

x	0.99	0.9999	1.0001	1.01
$\frac{x^2-1}{x^3-1}$	0.670011	0.666700	0.666633	0.663344

The limit as $x \to 1$ is $\frac{2}{3}$.

x	0.99	0.9999	1.0001	1.01
$\frac{x^3-1}{x^2-1}$	1.492513	1.499925	1.500075	1.507512

The limit as $x \to 1$ is $\frac{3}{2}$.

- For general m and n, we have $\lim_{x\to 1} \frac{x^n - 1}{x^m - 1} = \frac{n}{m}$.
-

x	0.99	0.9999	1.0001	1.01
$\frac{x-1}{x^3-1}$	0.336689	0.333367	0.333300	0.330022

The limit as $x \to 1$ is $\frac{1}{3}$.

x	0.99	0.9999	1.0001	1.01
$\frac{x^3-1}{x-1}$	2.9701	2.9997	3.0003	3.0301

The limit as $x \to 1$ is 3.

x	0.99	0.9999	1.0001	1.01
$\frac{x^3-1}{x^7-1}$	0.437200	0.428657	0.428486	0.420058

The limit as $x \to 1$ is $\frac{3}{7} \approx 0.428571$.

69. GU Plot the graph of $f(x) = \dfrac{2^x - 8}{x - 3}$.

(a) Zoom in on the graph to estimate $L = \lim_{x\to 3} f(x)$.

(b) Explain why

$$f(2.99999) \le L \le f(3.00001)$$

Use this to determine L to three decimal places.

SOLUTION

(a)

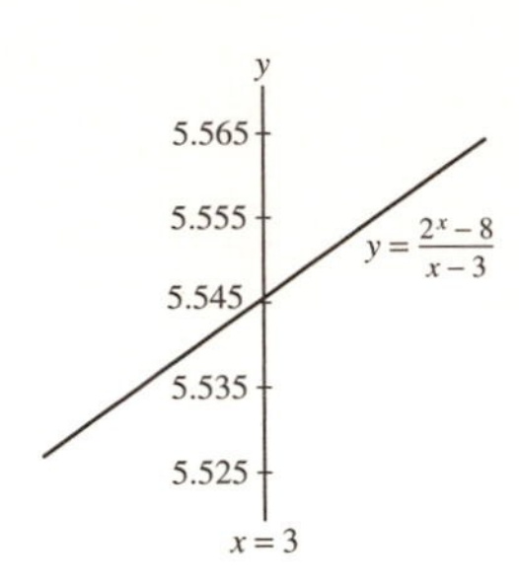

(b) It is clear that the graph of f rises as we move to the right. Mathematically, we may express this observation as: whenever $u < v$, $f(u) < f(v)$. Because

$$2.99999 < 3 = \lim_{x\to 3} f(x) < 3.00001,$$

it follows that

$$f(2.99999) < L = \lim_{x\to 3} f(x) < f(3.00001).$$

With $f(2.99999) \approx 5.54516$ and $f(3.00001) \approx 5.545195$, the above inequality becomes $5.54516 < L < 5.545195$; hence, to three decimal places, $L = 5.545$.

2.3 Basic Limit Laws

Preliminary Questions

1. State the Sum Law and Quotient Law.

SOLUTION Suppose $\lim_{x\to c} f(x)$ and $\lim_{x\to c} g(x)$ both exist. The Sum Law states that

$$\lim_{x\to c} (f(x) + g(x)) = \lim_{x\to c} f(x) + \lim_{x\to c} g(x).$$

Provided $\lim_{x\to c} g(x) \neq 0$, the Quotient Law states that

$$\lim_{x\to c} \frac{f(x)}{g(x)} = \frac{\lim_{x\to c} f(x)}{\lim_{x\to c} g(x)}.$$

2. Which of the following is a verbal version of the Product Law (assuming the limits exist)?

(a) The product of two functions has a limit.

(b) The limit of the product is the product of the limits.

(c) The product of a limit is a product of functions.

(d) A limit produces a product of functions.

SOLUTION The verbal version of the Product Law is **(b)**: The limit of the product is the product of the limits.

3. Which statement is correct? The Quotient Law does not hold if:

(a) The limit of the denominator is zero.

(b) The limit of the numerator is zero.

SOLUTION Statements **(a)** is correct. The Quotient Law does not hold if the limit of the denominator is zero.

Exercises

In Exercises 1–24, evaluate the limit using the Basic Limit Laws and the limits $\lim_{x\to c} x^{p/q} = c^{p/q}$ *and* $\lim_{x\to c} k = k$.

1. $\lim_{x\to 9} x$

SOLUTION $\lim_{x\to 9} x = 9$.

3. $\lim_{x\to\frac{1}{2}} x^4$

SOLUTION $\lim_{x\to\frac{1}{2}} x^4 = \left(\frac{1}{2}\right)^4 = \frac{1}{16}.$

5. $\lim_{t\to 2} t^{-1}$

SOLUTION $\lim_{t\to 2} t^{-1} = 2^{-1} = \frac{1}{2}.$

7. $\lim_{x\to 0.2} (3x+4)$

SOLUTION Using the Sum Law and the Constant Multiple Law:

$$\lim_{x\to 0.2} (3x+4) = \lim_{x\to 0.2} 3x + \lim_{x\to 0.2} 4$$

$$= 3\lim_{x\to 0.2} x + \lim_{x\to 0.2} 4 = 3(0.2) + 4 = 4.6.$$

9. $\lim_{x\to -1} (3x^4 - 2x^3 + 4x)$

SOLUTION Using the Sum Law, the Constant Multiple Law and the Powers Law:

$$\lim_{x\to -1} (3x^4 - 2x^3 + 4x) = \lim_{x\to -1} 3x^4 - \lim_{x\to -1} 2x^3 + \lim_{x\to -1} 4x$$

$$= 3\lim_{x\to -1} x^4 - 2\lim_{x\to -1} x^3 + 4\lim_{x\to -1} x$$

$$= 3(-1)^4 - 2(-1)^3 + 4(-1) = 3 + 2 - 4 = 1.$$

11. $\lim_{x\to 2} (x+1)(3x^2-9)$

SOLUTION Using the Product Law, the Sum Law and the Constant Multiple Law:

$$\lim_{x\to 2} (x+1)\left(3x^2-9\right) = \left(\lim_{x\to 2} x + \lim_{x\to 2} 1\right)\left(\lim_{x\to 2} 3x^2 - \lim_{x\to 2} 9\right)$$

$$= (2+1)\left(3\lim_{x\to 2} x^2 - 9\right)$$

$$= 3(3(2)^2 - 9) = 9.$$

13. $\lim_{t\to 4} \frac{3t-14}{t+1}$

SOLUTION Using the Quotient Law, the Sum Law and the Constant Multiple Law:

$$\lim_{t\to 4} \frac{3t-14}{t+1} = \frac{\lim_{t\to 4}(3t-14)}{\lim_{t\to 4}(t+1)} = \frac{3\lim_{t\to 4} t - \lim_{t\to 4} 14}{\lim_{t\to 4} t + \lim_{t\to 4} 1} = \frac{3\cdot 4 - 14}{4+1} = -\frac{2}{5}.$$

15. $\lim_{y\to\frac{1}{4}} (16y+1)(2y^{1/2}+1)$

SOLUTION Using the Product Law, the Sum Law, the Constant Multiple Law and the Powers Law:

$$\lim_{y\to\frac{1}{4}} (16y+1)(2y^{1/2}+1) = \left(\lim_{y\to\frac{1}{4}} (16y+1)\right)\left(\lim_{y\to\frac{1}{4}} (2y^{1/2}+1)\right)$$

$$= \left(16\lim_{y\to\frac{1}{4}} y + \lim_{y\to\frac{1}{4}} 1\right)\left(2\lim_{y\to\frac{1}{4}} y^{1/2} + \lim_{y\to\frac{1}{4}} 1\right)$$

$$= \left(16\left(\frac{1}{4}\right)+1\right)\left(2\left(\frac{1}{2}\right)+1\right) = 10.$$

17. $\lim_{y\to 4} \dfrac{1}{\sqrt{6y+1}}$

SOLUTION Using the Quotient Law, the Powers Law, the Sum Law and the Constant Multiple Law:

$$\lim_{y\to 4} \frac{1}{\sqrt{6y+1}} = \frac{1}{\lim_{y\to 4} \sqrt{6y+1}} = \frac{1}{\sqrt{6 \lim_{y\to 4} y + 1}}$$

$$= \frac{1}{\sqrt{6(4)+1}} = \frac{1}{5}.$$

19. $\lim_{x\to -1} \dfrac{x}{x^3+4x}$

SOLUTION Using the Quotient Law, the Sum Law, the Powers Law and the Constant Multiple Law:

$$\lim_{x\to -1} \frac{x}{x^3+4x} = \frac{\lim_{x\to -1} x}{\lim_{x\to -1} x^3 + 4 \lim_{x\to -1} x} = \frac{-1}{(-1)^3 + 4(-1)} = \frac{1}{5}.$$

21. $\lim_{t\to 25} \dfrac{3\sqrt{t} - \frac{1}{5}t}{(t-20)^2}$

SOLUTION Using the Quotient Law, the Sum Law, the Constant Multiple Law and the Powers Law:

$$\lim_{t\to 25} \frac{3\sqrt{t} - \frac{1}{5}t}{(t-20)^2} = \frac{3\sqrt{\lim_{t\to 25} t} - \frac{1}{5}\lim_{t\to 25} t}{\left(\lim_{t\to 25} t - 20\right)^2} = \frac{3(5) - \frac{1}{5}(25)}{5^2} = \frac{2}{5}.$$

23. $\lim_{t\to \frac{3}{2}} (4t^2 + 8t - 5)^{3/2}$

SOLUTION Using the Powers Law, the Sum Law and the Constant Multiple Law:

$$\lim_{t\to \frac{3}{2}} (4t^2 + 8t - 5)^{3/2} = \left(4 \lim_{t\to \frac{3}{2}} t^2 + 8 \lim_{t\to \frac{3}{2}} t - 5\right)^{3/2} = (9 + 12 - 5)^{3/2} = 64.$$

25. Use the Quotient Law to prove that if $\lim_{x\to c} f(x)$ exists and is nonzero, then

$$\lim_{x\to c} \frac{1}{f(x)} = \frac{1}{\lim_{x\to c} f(x)}$$

SOLUTION Since $\lim_{x\to c} f(x)$ is nonzero, we can apply the Quotient Law:

$$\lim_{x\to c} \left(\frac{1}{f(x)}\right) = \frac{\left(\lim_{x\to c} 1\right)}{\left(\lim_{x\to c} f(x)\right)} = \frac{1}{\lim_{x\to c} f(x)}.$$

In Exercises 27–30, evaluate the limit assuming that $\lim_{x\to -4} f(x) = 3$ *and* $\lim_{x\to -4} g(x) = 1$.

27. $\lim_{x\to -4} f(x)g(x)$

SOLUTION $\lim_{x\to -4} f(x)g(x) = \lim_{x\to -4} f(x) \lim_{x\to -4} g(x) = 3 \cdot 1 = 3.$

29. $\lim_{x\to -4} \dfrac{g(x)}{x^2}$

SOLUTION Since $\lim_{x\to -4} x^2 \neq 0$, we may apply the Quotient Law, then applying the Powers Law:

$$\lim_{x\to -4} \frac{g(x)}{x^2} = \frac{\lim_{x\to -4} g(x)}{\lim_{x\to -4} x^2} = \frac{1}{\left(\lim_{x\to -4} x\right)^2} = \frac{1}{16}.$$

31. Can the Quotient Law be applied to evaluate $\lim_{x\to 0}\frac{\sin x}{x}$? Explain.

SOLUTION The limit Quotient Law *cannot* be applied to evaluate $\lim_{x\to 0}\frac{\sin x}{x}$ since $\lim_{x\to 0} x = 0$. This violates a condition of the Quotient Law. Accordingly, the rule *cannot* be employed.

33. Give an example where $\lim_{x\to 0}(f(x)+g(x))$ exists but neither $\lim_{x\to 0} f(x)$ nor $\lim_{x\to 0} g(x)$ exists.

SOLUTION Let $f(x) = 1/x$ and $g(x) = -1/x$. Then $\lim_{x\to 0}(f(x)+g(x)) = \lim_{x\to 0} 0 = 0$ However, $\lim_{x\to 0} f(x) = \lim_{x\to 0} 1/x$ and $\lim_{x\to 0} g(x) = \lim_{x\to 0} -1/x$ do not exist.

Further Insights and Challenges

35. Suppose that $\lim_{t\to 3} tg(t) = 12$. Show that $\lim_{t\to 3} g(t)$ exists and equals 4.

SOLUTION We are given that $\lim_{t\to 3} tg(t) = 12$. Since $\lim_{t\to 3} t = 3 \neq 0$, we may apply the Quotient Law:

$$\lim_{t\to 3} g(t) = \lim_{t\to 3}\frac{tg(t)}{t} = \frac{\lim_{t\to 3} tg(t)}{\lim_{t\to 3} t} = \frac{12}{3} = 4.$$

37. Assuming that $\lim_{x\to 0}\frac{f(x)}{x} = 1$, which of the following statements is necessarily true? Why?

(a) $f(0) = 0$

(b) $\lim_{x\to 0} f(x) = 0$

SOLUTION

(a) Given that $\lim_{x\to 0}\frac{f(x)}{x} = 1$, it is not necessarily true that $f(0) = 0$. A counterexample is provided by $f(x) = \begin{cases} x, & x \neq 0 \\ 5, & x = 0 \end{cases}$.

(b) Given that $\lim_{x\to 0}\frac{f(x)}{x} = 1$, it is necessarily true that $\lim_{x\to 0} f(x) = 0$. For note that $\lim_{x\to 0} x = 0$, whence

$$\lim_{x\to 0} f(x) = \lim_{x\to 0} x\frac{f(x)}{x} = \left(\lim_{x\to 0} x\right)\left(\lim_{x\to 0}\frac{f(x)}{x}\right) = 0\cdot 1 = 0.$$

39. Suppose that $\lim_{h\to 0} g(h) = L$.

(a) Explain why $\lim_{h\to 0} g(ah) = L$ for any constant $a \neq 0$.

(b) If we assume instead that $\lim_{h\to 1} g(h) = L$, is it still necessarily true that $\lim_{h\to 1} g(ah) = L$?

(c) Illustrate (a) and (b) with the function $f(x) = x^2$.

SOLUTION

(a) As $h \to 0$, $ah \to 0$ as well; hence, if we make the change of variable $w = ah$, then

$$\lim_{h\to 0} g(ah) = \lim_{w\to 0} g(w) = L.$$

(b) No. As $h \to 1$, $ah \to a$, so we should not expect $\lim_{h\to 1} g(ah) = \lim_{h\to 1} g(h)$.

(c) Let $g(x) = x^2$. Then

$$\lim_{h\to 0} g(h) = 0 \quad \text{and} \quad \lim_{h\to 0} g(ah) = \lim_{h\to 0}(ah)^2 = 0.$$

On the other hand,

$$\lim_{h\to 1} g(h) = 1 \quad \text{while} \quad \lim_{h\to 1} g(ah) = \lim_{h\to 1}(ah)^2 = a^2,$$

which is equal to the previous limit if and only if $a = \pm 1$.

2.4 Limits and Continuity

Preliminary Questions

1. Which property of $f(x) = x^3$ allows us to conclude that $\lim_{x\to 2} x^3 = 8$?

SOLUTION We can conclude that $\lim_{x\to 2} x^3 = 8$ because the function x^3 is continuous at $x = 2$.

2. What can be said about $f(3)$ if f is continuous and $\lim_{x\to 3} f(x) = \frac{1}{2}$?

SOLUTION If f is continuous and $\lim_{x\to 3} f(x) = \frac{1}{2}$, then $f(3) = \frac{1}{2}$.

3. Suppose that $f(x) < 0$ if x is positive and $f(x) > 1$ if x is negative. Can f be continuous at $x = 0$?

SOLUTION Since $f(x) < 0$ when x is positive and $f(x) > 1$ when x is negative, it follows that

$$\lim_{x\to 0+} f(x) \le 0 \quad \text{and} \quad \lim_{x\to 0-} f(x) \ge 1.$$

Thus, $\lim_{x\to 0} f(x)$ does not exist, so f cannot be continuous at $x = 0$.

4. Is it possible to determine $f(7)$ if $f(x) = 3$ for all $x < 7$ and f is right-continuous at $x = 7$? What if f is left-continuous?

SOLUTION No. To determine $f(7)$, we need to combine either knowledge of the values of $f(x)$ for $x < 7$ with *left*-continuity or knowledge of the values of $f(x)$ for $x > 7$ with right-continuity.

5. Are the following true or false? If false, state a correct version.

(a) $f(x)$ is continuous at $x = a$ if the left- and right-hand limits of $f(x)$ as $x \to a$ exist and are equal.

(b) $f(x)$ is continuous at $x = a$ if the left- and right-hand limits of $f(x)$ as $x \to a$ exist and equal $f(a)$.

(c) If the left- and right-hand limits of $f(x)$ as $x \to a$ exist, then f has a removable discontinuity at $x = a$.

(d) If $f(x)$ and $g(x)$ are continuous at $x = a$, then $f(x) + g(x)$ is continuous at $x = a$.

(e) If $f(x)$ and $g(x)$ are continuous at $x = a$, then $f(x)/g(x)$ is continuous at $x = a$.

SOLUTION

(a) False. The correct statement is "$f(x)$ is continuous at $x = a$ if the left- and right-hand limits of $f(x)$ as $x \to a$ exist and equal $f(a)$."

(b) True.

(c) False. The correct statement is "If the left- and right-hand limits of $f(x)$ as $x \to a$ are equal but not equal to $f(a)$, then f has a removable discontinuity at $x = a$."

(d) True.

(e) False. The correct statement is "If $f(x)$ and $g(x)$ are continuous at $x = a$ and $g(a) \neq 0$, then $f(x)/g(x)$ is continuous at $x = a$."

Exercises

1. Referring to Figure 14, state whether $f(x)$ is left- or right-continuous (or neither) at each point of discontinuity. Does $f(x)$ have any removable discontinuities?

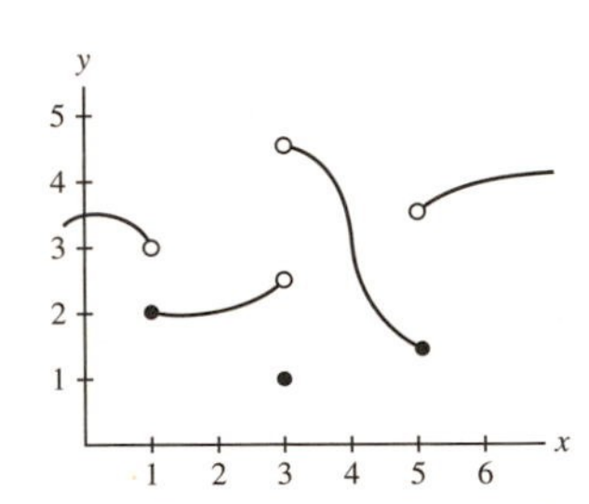

FIGURE 14 Graph of $y = f(x)$

SOLUTION

- The function f is discontinuous at $x = 1$; it is right-continuous there.
- The function f is discontinuous at $x = 3$; it is neither left-continuous nor right-continuous there.
- The function f is discontinuous at $x = 5$; it is left-continuous there.

However, these discontinuities are not removable.

Exercises 2–4 refer to the function $g(x)$ in Figure 15.

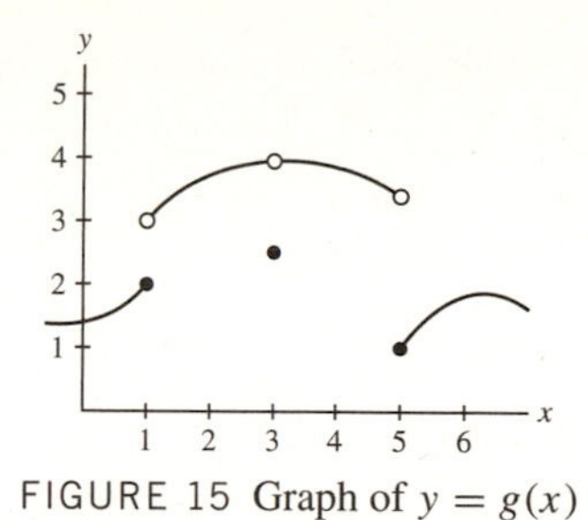

FIGURE 15 Graph of $y = g(x)$

3. At which point c does $g(x)$ have a removable discontinuity? How should $g(c)$ be redefined to make g continuous at $x = c$?

SOLUTION Because $\lim_{x\to 3} g(x)$ exists, the function g has a removable discontinuity at $x = 3$. Assigning $g(3) = 4$ makes g continuous at $x = 3$.

5. In Figure 16, determine the one-sided limits at the points of discontinuity. Which discontinuity is removable and how should f be redefined to make it continuous at this point?

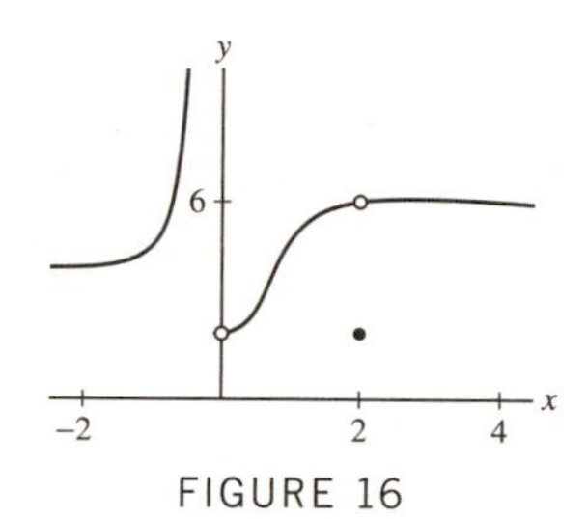

FIGURE 16

SOLUTION The function f is discontinuous at $x = 0$, at which $\lim_{x\to 0-} f(x) = \infty$ and $\lim_{x\to 0+} f(x) = 2$. The function f is also discontinuous at $x = 2$, at which $\lim_{x\to 2-} f(x) = 6$ and $\lim_{x\to 2+} f(x) = 6$. Because the two one-sided limits exist and are equal at $x = 2$, the discontinuity at $x = 2$ is removable. Assigning $f(2) = 6$ makes f continuous at $x = 2$.

In Exercises 7–16, use the Laws of Continuity and Theorems 2 and 3 to show that the function is continuous.

7. $f(x) = x + \sin x$

SOLUTION Since x and $\sin x$ are continuous, so is $x + \sin x$ by Continuity Law (i).

9. $f(x) = 3x + 4\sin x$

SOLUTION Since x and $\sin x$ are continuous, so are $3x$ and $4\sin x$ by Continuity Law (ii). Thus $3x + 4\sin x$ is continuous by Continuity Law (i).

11. $f(x) = \dfrac{1}{x^2 + 1}$

SOLUTION

- Since x is continuous, so is x^2 by Continuity Law (iii).
- Recall that constant functions, such as 1, are continuous. Thus $x^2 + 1$ is continuous.
- Finally, $\dfrac{1}{x^2+1}$ is continuous by Continuity Law (iv) because $x^2 + 1$ is never 0.

13. $f(x) = \cos(x^2)$

SOLUTION The function $f(x)$ is a composite of two continuous functions: $\cos x$ and x^2, so $f(x)$ is continuous by Theorem 5, which states that a composite of continuous functions is continuous.

15. $f(x) = e^x \cos 3x$

SOLUTION e^x and $\cos 3x$ are continuous, so $e^x \cos 3x$ is continuous by Continuity Law (iii).

In Exercises 17–34, determine the points of discontinuity. State the type of discontinuity (removable, jump, infinite, or none of these) and whether the function is left- or right-continuous.

17. $f(x) = \dfrac{1}{x}$

SOLUTION The function $1/x$ is discontinuous at $x = 0$, at which there is an infinite discontinuity. The function is neither left- nor right-continuous at $x = 0$.

19. $f(x) = \dfrac{x-2}{|x-1|}$

SOLUTION The function $\dfrac{x-2}{|x-1|}$ is discontinuous at $x = 1$, at which there is an infinite discontinuity. The function is neither left- nor right-continuous at $x = 1$.

21. $f(x) = \left[\frac{1}{2}x\right]$

SOLUTION The function $\left[\frac{1}{2}x\right]$ is discontinuous at even integers, at which there are jump discontinuities. Because

$$\lim_{x\to 2n+}\left[\frac{1}{2}x\right] = n$$

but

$$\lim_{x\to 2n-}\left[\frac{1}{2}x\right] = n-1,$$

it follows that this function is right-continuous at the even integers but not left-continuous.

23. $f(x) = \dfrac{x+1}{4x-2}$

SOLUTION The function $f(x) = \dfrac{x+1}{4x-2}$ is discontinuous at $x = \frac{1}{2}$, at which there is an infinite discontinuity. The function is neither left- nor right-continuous at $x = \frac{1}{2}$.

25. $f(x) = 3x^{2/3} - 9x^3$

SOLUTION The function $f(x) = 3x^{2/3} - 9x^3$ is defined and continuous for all x.

27. $f(x) = \begin{cases} \dfrac{x-2}{|x-2|} & x \neq 2 \\ -1 & x = 2 \end{cases}$

SOLUTION For $x > 2$, $f(x) = \dfrac{x-2}{(x-2)} = 1$. For $x < 2$, $f(x) = \dfrac{(x-2)}{(2-x)} = -1$. The function has a jump discontinuity at $x = 2$. Because

$$\lim_{x\to 2-} f(x) = -1 = f(2)$$

but

$$\lim_{x\to 2+} f(x) = 1 \neq f(2),$$

it follows that this function is left-continuous at $x = 2$ but not right-continuous.

29. $g(t) = \tan 2t$

SOLUTION The function $g(t) = \tan 2t = \dfrac{\sin 2t}{\cos 2t}$ is discontinuous whenever $\cos 2t = 0$; i.e., whenever

$$2t = \frac{(2n+1)\pi}{2} \qquad \text{or} \qquad t = \frac{(2n+1)\pi}{4},$$

where n is an integer. At every such value of t there is an infinite discontinuity. The function is neither left- nor right-continuous at any of these points of discontinuity.

31. $f(x) = \tan(\sin x)$

SOLUTION The function $f(x) = \tan(\sin x)$ is continuous everywhere. Reason: $\sin x$ is continuous everywhere and $\tan u$ is continuous on $\left(-\frac{\pi}{2}, \frac{\pi}{2}\right)$—and in particular on $-1 \leq u = \sin x \leq 1$. Continuity of $\tan(\sin x)$ follows by the continuity of composite functions.

33. $f(x) = \dfrac{1}{e^x - e^{-x}}$

SOLUTION The function $f(x) = \dfrac{1}{e^x - e^{-x}}$ is discontinuous at $x = 0$, at which there is an infinite discontinuity. The function is neither left- nor right-continuous at $x = 0$.

In Exercises 35–48, determine the domain of the function and prove that it is continuous on its domain using the Laws of Continuity and the facts quoted in this section.

35. $f(x) = 2\sin x + 3\cos x$

SOLUTION The domain of $2\sin x + 3\cos x$ is all real numbers. Both $\sin x$ and $\cos x$ are continuous on this domain, so $2\sin x + 3\cos x$ is continuous by Continuity Laws (i) and (ii).

37. $f(x) = \sqrt{x}\sin x$

SOLUTION This function is defined as long as $x \geq 0$. Since $\sqrt{x}$ and $\sin x$ are continuous, so is $\sqrt{x}\sin x$ by Continuity Law (iii).

39. $f(x) = x^{2/3}2^x$

SOLUTION The domain of $x^{2/3}2^x$ is all real numbers as the denominator of the rational exponent is odd. Both $x^{2/3}$ and 2^x are continuous on this domain, so $x^{2/3}2^x$ is continuous by Continuity Law (iii).

41. $f(x) = x^{-4/3}$

SOLUTION This function is defined for all $x \neq 0$. Because the function $x^{4/3}$ is continuous and not equal to zero for $x \neq 0$, it follows that

$$x^{-4/3} = \frac{1}{x^{4/3}}$$

is continuous for $x \neq 0$ by Continuity Law (iv).

43. $f(x) = \tan^2 x$

SOLUTION The domain of $\tan^2 x$ is all $x \neq \pm(2n-1)\pi/2$ where n is a positive integer. Because $\tan x$ is continuous on this domain, it follows from Continuity Law (iii) that $\tan^2 x$ is also continuous on this domain.

45. $f(x) = (x^4+1)^{3/2}$

SOLUTION The domain of $(x^4+1)^{3/2}$ is all real numbers as $x^4+1 > 0$ for all x. Because $x^{3/2}$ and the polynomial x^4+1 are both continuous, so is the composite function $(x^4+1)^{3/2}$.

47. $f(x) = \dfrac{\cos(x^2)}{x^2-1}$

SOLUTION The domain for this function is all $x \neq \pm 1$. Because the functions $\cos x$ and x^2 are continuous on this domain, so is the composite function $\cos(x^2)$. Finally, because the polynomial x^2-1 is continuous and not equal to zero for $x \neq \pm 1$, the function $\dfrac{\cos(x^2)}{x^2-1}$ is continuous by Continuity Law (iv).

49. Show that the function

$$f(x) = \begin{cases} x^2+3 & \text{for } x < 1 \\ 10-x & \text{for } 1 \leq x \leq 2 \\ 6x-x^2 & \text{for } x > 2 \end{cases}$$

is continuous for $x \neq 1, 2$. Then compute the right- and left-hand limits at $x = 1, 2$, and determine whether $f(x)$ is left-continuous, right-continuous, or continuous at these points (Figure 17).

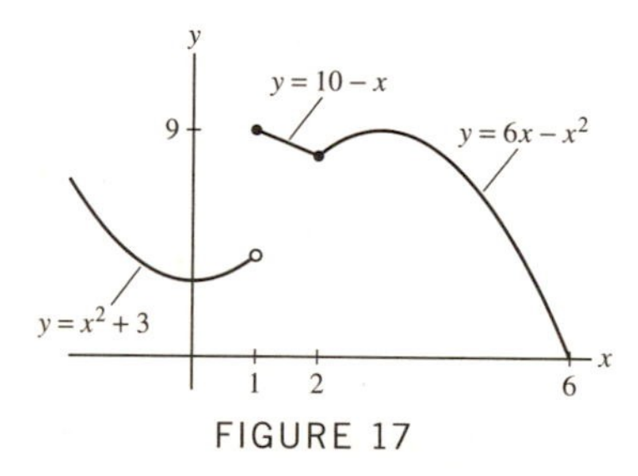

FIGURE 17

SOLUTION Let's start with $x \neq 1, 2$.

- Because x is continuous, so is x^2 by Continuity Law (iii). The constant function 3 is also continuous, so x^2+3 is continuous by Continuity Law (i). Therefore, $f(x)$ is continuous for $x < 1$.
- Because x and the constant function 10 are continuous, the function $10-x$ is continuous by Continuity Law (i). Therefore, $f(x)$ is continuous for $1 < x < 2$.
- Because x is continuous, x^2 is continuous by Continuity Law (iii) and $6x$ is continuous by Continuity Law (ii). Therefore, $6x-x^2$ is continuous by Continuity Law (i), so $f(x)$ is continuous for $x > 2$.

At $x = 1$, $f(x)$ has a jump discontinuity because the one-sided limits exist but are not equal:

$$\lim_{x\to 1-} f(x) = \lim_{x\to 1-} (x^2+3) = 4, \qquad \lim_{x\to 1+} f(x) = \lim_{x\to 1+} (10-x) = 9.$$

Furthermore, the right-hand limit equals the function value $f(1) = 9$, so $f(x)$ is right-continuous at $x = 1$. At $x = 2$,

$$\lim_{x\to 2-} f(x) = \lim_{x\to 2-} (10-x) = 8, \qquad \lim_{x\to 2+} f(x) = \lim_{x\to 2+} (6x-x^2) = 8.$$

The left- and right-hand limits exist and are equal to $f(2)$, so $f(x)$ is continuous at $x = 2$.

In Exercises 51–54, sketch the graph of $f(x)$. At each point of discontinuity, state whether f is left- or right-continuous.

51. $f(x) = \begin{cases} x^2 & \text{for } x \le 1 \\ 2 - x & \text{for } x > 1 \end{cases}$

SOLUTION

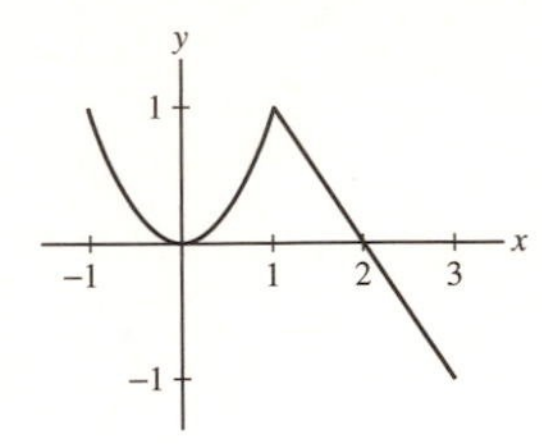

The function f is continuous everywhere.

53. $f(x) = \begin{cases} \dfrac{x^2 - 3x + 2}{|x - 2|} & x \neq 2 \\ 0 & x = 2 \end{cases}$

SOLUTION

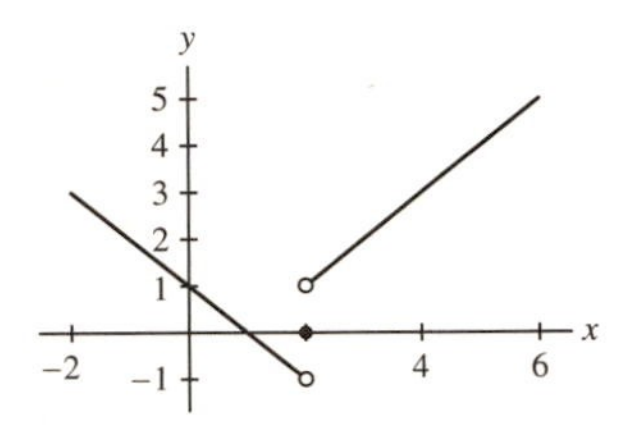

The function f is neither left- nor right-continuous at $x = 2$.

55. Show that the function

$$f(x) = \begin{cases} \dfrac{x^2 - 16}{x - 4} & x \neq 4 \\ 10 & x = 4 \end{cases}$$

has a removable discontinuity at $x = 4$.

SOLUTION To show that $f(x)$ has a removable discontinuity at $x = 4$, we must establish that

$$\lim_{x \to 4} f(x)$$

exists but does not equal $f(4)$. Now,

$$\lim_{x \to 4} \frac{x^2 - 16}{x - 4} = \lim_{x \to 4} (x + 4) = 8 \neq 10 = f(4);$$

thus, $f(x)$ has a removable discontinuity at $x = 4$. To remove the discontinuity, we must redefine $f(4) = 8$.

In Exercises 57–59, find the value of the constant (a, b, or c) that makes the function continuous.

57. $f(x) = \begin{cases} x^2 - c & \text{for } x < 5 \\ 4x + 2c & \text{for } x \ge 5 \end{cases}$

SOLUTION As $x \to 5-$, we have $x^2 - c \to 25 - c = L$. As $x \to 5+$, we have $4x + 2c \to 20 + 2c = R$. Match the limits: $L = R$ or $25 - c = 20 + 2c$ implies $c = \frac{5}{3}$.

59. $f(x) = \begin{cases} x^{-1} & \text{for } x < -1 \\ ax + b & \text{for } -1 \le x \le \frac{1}{2} \\ x^{-1} & \text{for } x > \frac{1}{2} \end{cases}$

SOLUTION As $x \to -1-$, $x^{-1} \to -1$ while as $x \to -1+$, $ax + b \to b - a$. For f to be continuous at $x = -1$, we must therefore have $b - a = -1$. Now, as $x \to \frac{1}{2}-$, $ax + b \to \frac{1}{2}a + b$ while as $x \to \frac{1}{2}+$, $x^{-1} \to 2$. For f to be continuous at $x = \frac{1}{2}$, we must therefore have $\frac{1}{2}a + b = 2$. Solving these two equations for a and b yields $a = 2$ and $b = 1$.

61. Define $g(t) = \tan^{-1}\left(\frac{1}{t-1}\right)$ for $t \neq 1$. Answer the following questions, using a plot if necessary.

(a) Can $g(1)$ be defined so that $g(t)$ is continuous at $t = 1$?

(b) How should $g(1)$ be defined so that $g(t)$ is left-continuous at $t = 1$?

SOLUTION

(a) From the graph of $g(t)$ shown below, we see that g has a jump discontinuity at $t = 1$; therefore, $g(a)$ cannot be defined so that g is continuous at $t = 1$.

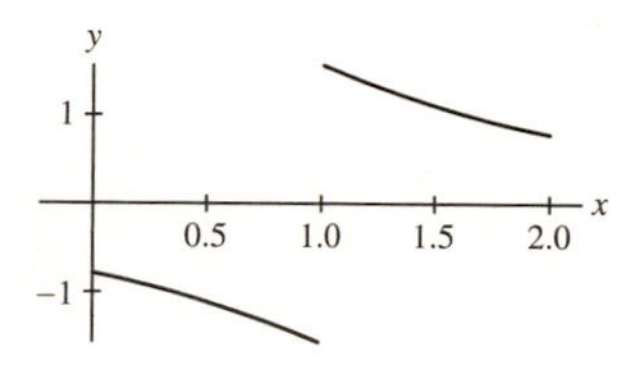

(b) To make g left-continuous at $t = 1$, we should define

$$g(1) = \lim_{t\to 1-} \tan^{-1}\left(\tfrac{1}{t-1}\right) = -\frac{\pi}{2}.$$

In Exercises 63–66, draw the graph of a function on $[0, 5]$ *with the given properties.*

63. $f(x)$ is not continuous at $x = 1$, but $\lim_{x\to 1+} f(x)$ and $\lim_{x\to 1-} f(x)$ exist and are equal.

SOLUTION

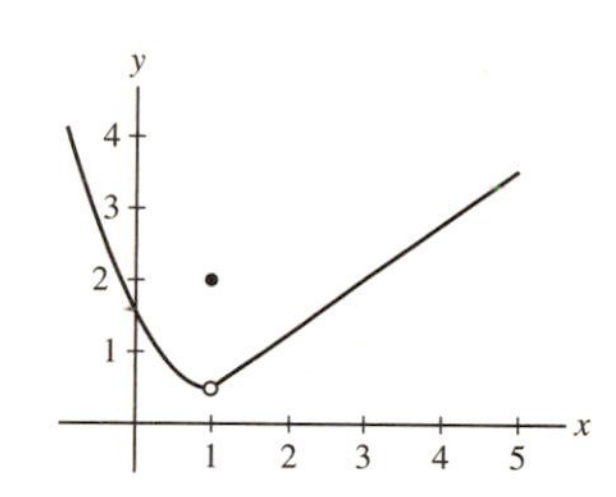

65. $f(x)$ has a removable discontinuity at $x = 1$, a jump discontinuity at $x = 2$, and

$$\lim_{x\to 3-} f(x) = -\infty, \qquad \lim_{x\to 3+} f(x) = 2$$

SOLUTION

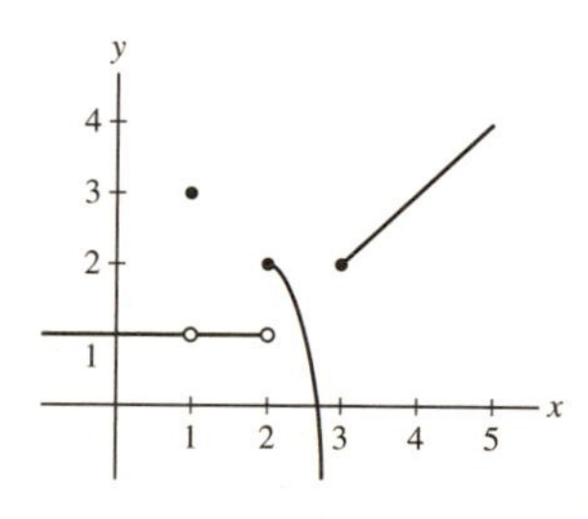

In Exercises 67–80, evaluate using substitution.

67. $\lim_{x\to -1} (2x^3 - 4)$

SOLUTION $\lim_{x\to -1} (2x^3 - 4) = 2(-1)^3 - 4 = -6.$

69. $\lim_{x\to 3} \frac{x+2}{x^2+2x}$

SOLUTION $\lim_{x\to 3} \frac{x+2}{x^2+2x} = \frac{3+2}{3^2+2\cdot 3} = \frac{5}{15} = \frac{1}{3}$

71. $\lim_{x\to \frac{\pi}{4}} \tan(3x)$

SOLUTION $\lim_{x\to \frac{\pi}{4}} \tan(3x) = \tan(3\cdot\frac{\pi}{4}) = \tan(\frac{3\pi}{4}) = -1$

73. $\lim_{x \to 4} x^{-5/2}$

SOLUTION $\lim_{x \to 4} x^{-5/2} = 4^{-5/2} = \frac{1}{32}$.

75. $\lim_{x \to -1} (1 - 8x^3)^{3/2}$

SOLUTION $\lim_{x \to -1} (1 - 8x^3)^{3/2} = (1 - 8(-1)^3)^{3/2} = 27$.

77. $\lim_{x \to 3} 10^{x^2 - 2x}$

SOLUTION $\lim_{x \to 3} 10^{x^2 - 2x} = 10^{3^2 - 2(3)} = 1000$.

79. $\lim_{x \to 4} \sin^{-1}\left(\frac{x}{4}\right)$

SOLUTION $\lim_{x \to 4} \sin^{-1}\left(\frac{x}{4}\right) = \sin^{-1}\left(\lim_{x \to 4} \frac{x}{4}\right) = \sin^{-1}\left(\frac{4}{4}\right) = \frac{\pi}{2}$

81. Suppose that $f(x)$ and $g(x)$ are discontinuous at $x = c$. Does it follow that $f(x) + g(x)$ is discontinuous at $x = c$? If not, give a counterexample. Does this contradict Theorem 1 (i)?

SOLUTION Even if $f(x)$ and $g(x)$ are discontinuous at $x = c$, it is *not* necessarily true that $f(x) + g(x)$ is discontinuous at $x = c$. For example, suppose $f(x) = -x^{-1}$ and $g(x) = x^{-1}$. Both $f(x)$ and $g(x)$ are discontinuous at $x = 0$; however, the function $f(x) + g(x) = 0$, which is continuous everywhere, including $x = 0$. This does not contradict Theorem 1 (i), which deals only with continuous functions.

83. Use the result of Exercise 82 to prove that if $g(x)$ is continuous, then $f(x) = |g(x)|$ is also continuous.

SOLUTION Recall that the composition of two continuous functions is continuous. Now, $f(x) = |g(x)|$ is a composition of the continuous functions $g(x)$ and $|x|$, so is also continuous.

85. In 2009, the federal income tax $T(x)$ on income of x dollars (up to \$82,250) was determined by the formula

$$T(x) = \begin{cases} 0.10x & \text{for } 0 \le x < 8350 \\ 0.15x - 417.50 & \text{for } 8350 \le x < 33{,}950 \\ 0.25x - 3812.50 & \text{for } 33{,}950 \le x < 82{,}250 \end{cases}$$

Sketch the graph of $T(x)$. Does $T(x)$ have any discontinuities? Explain why, if $T(x)$ had a jump discontinuity, it might be advantageous in some situations to earn *less* money.

SOLUTION $T(x)$, the amount of federal income tax owed on an income of x dollars in 2009, might be a discontinuous function depending upon how the tax tables are constructed (as determined by that year's regulations). Here is a graph of $T(x)$ for that particular year.

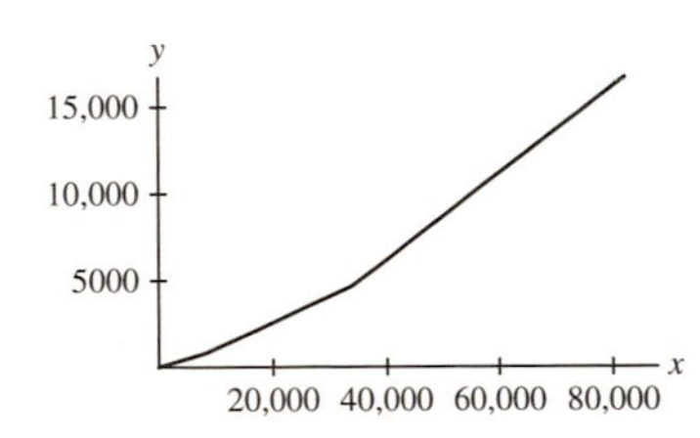

If $T(x)$ had a jump discontinuity (say at $x = c$), it might be advantageous to earn slightly less income than c (say $c - \epsilon$) and be taxed at a lower rate than to earn c or more and be taxed at a higher rate. Your net earnings may actually be more in the former case than in the latter one.

Further Insights and Challenges

87. Give an example of functions $f(x)$ and $g(x)$ such that $f(g(x))$ is continuous but $g(x)$ has at least one discontinuity.

SOLUTION Answers may vary. The simplest examples are the functions $f(g(x))$ where $f(x) = C$ is a constant function, and $g(x)$ is defined for all x. In these cases, $f(g(x)) = C$. For example, if $f(x) = 3$ and $g(x) = [x]$, g is discontinuous at all integer values $x = n$, but $f(g(x)) = 3$ is continuous.

89. Show that $f(x)$ is a discontinuous function for all x where $f(x)$ is defined as follows:

$$f(x) = \begin{cases} 1 & \text{for } x \text{ rational} \\ -1 & \text{for } x \text{ irrational} \end{cases}$$

Show that $f(x)^2$ is continuous for all x.

SOLUTION $\lim_{x\to c} f(x)$ does not exist for any c. If c is irrational, then there is always a rational number r arbitrarily close to c so that $|f(c) - f(r)| = 2$. If, on the other hand, c is rational, there is always an *irrational* number z arbitrarily close to c so that $|f(c) - f(z)| = 2$.

On the other hand, $f(x)^2$ is a constant function that always has value 1, which is obviously continuous.

2.5 Evaluating Limits Algebraically

Preliminary Questions

1. Which of the following is indeterminate at $x = 1$?

$$\frac{x^2+1}{x-1}, \qquad \frac{x^2-1}{x+2}, \qquad \frac{x^2-1}{\sqrt{x+3}-2}, \qquad \frac{x^2+1}{\sqrt{x+3}-2}$$

SOLUTION At $x = 1$, $\frac{x^2-1}{\sqrt{x+3}-2}$ is of the form $\frac{0}{0}$; hence, this function is indeterminate. None of the remaining functions is indeterminate at $x = 1$: $\frac{x^2+1}{x-1}$ and $\frac{x^2+1}{\sqrt{x+3}-2}$ are undefined because the denominator is zero but the numerator is not, while $\frac{x^2-1}{x+2}$ is equal to 0.

2. Give counterexamples to show that these statements are false:

(a) If $f(c)$ is indeterminate, then the right- and left-hand limits as $x \to c$ are not equal.

(b) If $\lim_{x\to c} f(x)$ exists, then $f(c)$ is not indeterminate.

(c) If $f(x)$ is undefined at $x = c$, then $f(x)$ has an indeterminate form at $x = c$.

SOLUTION

(a) Let $f(x) = \frac{x^2-1}{x-1}$. At $x = 1$, f is indeterminate of the form $\frac{0}{0}$ but

$$\lim_{x\to 1-} \frac{x^2-1}{x-1} = \lim_{x\to 1-} (x+1) = 2 = \lim_{x\to 1+} (x+1) = \lim_{x\to 1+} \frac{x^2-1}{x-1}.$$

(b) Again, let $f(x) = \frac{x^2-1}{x-1}$. Then

$$\lim_{x\to 1} f(x) = \lim_{x\to 1} \frac{x^2-1}{x-1} = \lim_{x\to 1} (x+1) = 2$$

but $f(1)$ is indeterminate of the form $\frac{0}{0}$.

(c) Let $f(x) = \frac{1}{x}$. Then f is undefined at $x = 0$ but does not have an indeterminate form at $x = 0$.

3. The method for evaluating limits discussed in this section is sometimes called "simplify and plug in." Explain how it actually relies on the property of continuity.

SOLUTION If f is continuous at $x = c$, then, by definition, $\lim_{x\to c} f(x) = f(c)$; in other words, the limit of a continuous function at $x = c$ is the value of the function at $x = c$. The "simplify and plug-in" strategy is based on simplifying a function which is indeterminate to a continuous function. Once the simplification has been made, the limit of the remaining continuous function is obtained by evaluation.

Exercises

In Exercises 1–4, show that the limit leads to an indeterminate form. Then carry out the two-step procedure: Transform the function algebraically and evaluate using continuity.

1. $\lim_{x\to 6} \frac{x^2-36}{x-6}$

SOLUTION When we substitute $x = 6$ into $\frac{x^2-36}{x-6}$, we obtain the indeterminate form $\frac{0}{0}$. Upon factoring the numerator and simplifying, we find

$$\lim_{x\to 6} \frac{x^2-36}{x-6} = \lim_{x\to 6} \frac{(x-6)(x+6)}{x-6} = \lim_{x\to 6} (x+6) = 12.$$

3. $\displaystyle\lim_{x\to-1} \frac{x^2+2x+1}{x+1}$

SOLUTION When we substitute $x = -1$ into $\frac{x^2+2x+1}{x+1}$, we obtain the indeterminate form $\frac{0}{0}$. Upon factoring the numerator and simplifying, we find

$$\lim_{x\to-1} \frac{x^2+2x+1}{x+1} = \lim_{x\to-1} \frac{(x+1)^2}{x+1} = \lim_{x\to-1} (x+1) = 0.$$

In Exercises 5–34, evaluate the limit, if it exists. If not, determine whether the one-sided limits exist (finite or infinite).

5. $\displaystyle\lim_{x\to7} \frac{x-7}{x^2-49}$

SOLUTION $\displaystyle\lim_{x\to7} \frac{x-7}{x^2-49} = \lim_{x\to7} \frac{x-7}{(x-7)(x+7)} = \lim_{x\to7} \frac{1}{x+7} = \frac{1}{14}.$

7. $\displaystyle\lim_{x\to-2} \frac{x^2+3x+2}{x+2}$

SOLUTION $\displaystyle\lim_{x\to-2} \frac{x^2+3x+2}{x+2} = \lim_{x\to-2} \frac{(x+1)(x+2)}{x+2} = \lim_{x\to-2} (x+1) = -1.$

9. $\displaystyle\lim_{x\to5} \frac{2x^2-9x-5}{x^2-25}$

SOLUTION $\displaystyle\lim_{x\to5} \frac{2x^2-9x-5}{x^2-25} = \lim_{x\to5} \frac{(x-5)(2x+1)}{(x-5)(x+5)} = \lim_{x\to5} \frac{2x+1}{x+5} = \frac{11}{10}.$

11. $\displaystyle\lim_{x\to-\frac{1}{2}} \frac{2x+1}{2x^2+3x+1}$

SOLUTION $\displaystyle\lim_{x\to-\frac{1}{2}} \frac{2x+1}{2x^2+3x+1} = \lim_{x\to-\frac{1}{2}} \frac{2x+1}{(2x+1)(x+1)} = \lim_{x\to-\frac{1}{2}} \frac{1}{x+1} = 2.$

13. $\displaystyle\lim_{x\to2} \frac{3x^2-4x-4}{2x^2-8}$

SOLUTION $\displaystyle\lim_{x\to2} \frac{3x^2-4x-4}{2x^2-8} = \lim_{x\to2} \frac{(3x+2)(x-2)}{2(x-2)(x+2)} = \lim_{x\to2} \frac{3x+2}{2(x+2)} = \frac{8}{8} = 1.$

15. $\displaystyle\lim_{t\to0} \frac{4^{2t}-1}{4^t-1}$

SOLUTION $\displaystyle\lim_{t\ to0} \frac{4^{2t}-1}{4^t-1} = \lim_{t\ to0} \frac{(4^t-1)(4^t+1)}{4^t-1} = \lim_{t\to0} (4^t+1) = 2.$

17. $\displaystyle\lim_{x\to16} \frac{\sqrt{x}-4}{x-16}$

SOLUTION $\displaystyle\lim_{x\to16} \frac{\sqrt{x}-4}{x-16} = \lim_{x\to16} \frac{\sqrt{x}-4}{(\sqrt{x}+4)(\sqrt{x}-4)} = \lim_{x\to16} \frac{1}{\sqrt{x}+4} = \frac{1}{8}.$

19. $\displaystyle\lim_{y\to3} \frac{y^2+y-12}{y^3-10y+3}$

SOLUTION $\displaystyle\lim_{y\to3} \frac{y^2+y-12}{y^3-10y+3} = \lim_{y\to3} \frac{(y-3)(y+4)}{(y-3)(y^2+3y-1)} = \lim_{y\to3} \frac{(y+4)}{(y^2+3y-1)} = \frac{7}{17}.$

21. $\displaystyle\lim_{h\to0} \frac{\sqrt{2+h}-2}{h}$

SOLUTION $\displaystyle\lim_{h\to0} \frac{\sqrt{h+2}-2}{h}$ does not exist.

- As $h \to 0+$, we have $\displaystyle\frac{\sqrt{h+2}-2}{h} = \frac{(\sqrt{h+2}-2)(\sqrt{h+2}+2)}{h(\sqrt{h+2}+2)} = \frac{h-2}{h(\sqrt{h+2}+2)} \to -\infty.$
- As $h \to 0-$, we have $\displaystyle\frac{\sqrt{h+2}-2}{h} = \frac{(\sqrt{h+2}-2)(\sqrt{h+2}+2)}{h(\sqrt{h+2}+2)} = \frac{h-2}{h(\sqrt{h+2}+2)} \to \infty.$

23. $\lim_{x\to 4} \frac{x-4}{\sqrt{x}-\sqrt{8-x}}$

SOLUTION

$$\lim_{x\to 4} \frac{x-4}{\sqrt{x}-\sqrt{8-x}} = \lim_{x\to 4} \frac{(x-4)(\sqrt{x}+\sqrt{8-x})}{(\sqrt{x}-\sqrt{8-x})(\sqrt{x}+\sqrt{8-x})} = \lim_{x\to 4} \frac{(x-4)(\sqrt{x}+\sqrt{8-x})}{x-(8-x)}$$

$$= \lim_{x\to 4} \frac{(x-4)(\sqrt{x}+\sqrt{8-x})}{2x-8} = \lim_{x\to 4} \frac{(x-4)(\sqrt{x}+\sqrt{8-x})}{2(x-4)}$$

$$= \lim_{x\to 4} \frac{(\sqrt{x}+\sqrt{8-x})}{2} = \frac{\sqrt{4}+\sqrt{4}}{2} = 2.$$

25. $\lim_{x\to 4} \left(\frac{1}{\sqrt{x}-2} - \frac{4}{x-4}\right)$

SOLUTION $\lim_{x\to 4} \left(\frac{1}{\sqrt{x}-2} - \frac{4}{x-4}\right) = \lim_{x\to 4} \frac{\sqrt{x}+2-4}{(\sqrt{x}-2)(\sqrt{x}+2)} = \lim_{x\to 4} \frac{\sqrt{x}-2}{(\sqrt{x}-2)(\sqrt{x}+2)} = \frac{1}{4}.$

27. $\lim_{x\to 0} \frac{\cot x}{\csc x}$

SOLUTION $\lim_{x\to 0} \frac{\cot x}{\csc x} = \lim_{x\to 0} \frac{\cos x}{\sin x} \cdot \sin x = \cos 0 = 1.$

29. $\lim_{t\to 2} \frac{2^{2t}+2^t-20}{2^t-4}$

SOLUTION $\lim_{t\to 2} \frac{2^{2t}+2^t-20}{2^t-4} = \lim_{t\to 2} \frac{(2^t+5)(2^t-4)}{2^t-4} = \lim_{t\to 2} (2^t+5) = 9.$

31. $\lim_{x\to \frac{\pi}{4}} \frac{\sin x - \cos x}{\tan x - 1}$

SOLUTION $\lim_{x\to \frac{\pi}{4}} \frac{\sin x - \cos x}{\tan x - 1} \cdot \frac{\cos x}{\cos x} = \lim_{x\to \frac{\pi}{4}} \frac{(\sin x - \cos x)\cos x}{\sin x - \cos x} = \cos\frac{\pi}{4} = \frac{\sqrt{2}}{2}.$

33. $\lim_{\theta\to \frac{\pi}{4}} \left(\frac{1}{\tan\theta - 1} - \frac{2}{\tan^2\theta - 1}\right)$

SOLUTION $\lim_{\theta\to \frac{\pi}{4}} \left(\frac{1}{\tan\theta - 1} - \frac{2}{\tan^2\theta - 1}\right) = \lim_{\theta\to \frac{\pi}{4}} \frac{(\tan\theta+1)-2}{(\tan\theta+1)(\tan\theta-1)} = \lim_{\theta\to \frac{\pi}{4}} \frac{1}{\tan\theta+1} = \frac{1}{2}.$

35. GU Use a plot of $f(x) = \frac{x-4}{\sqrt{x}-\sqrt{8-x}}$ to estimate $\lim_{x\to 4} f(x)$ to two decimal places. Compare with the answer obtained algebraically in Exercise 23.

SOLUTION Let $f(x) = \frac{x-4}{\sqrt{x}-\sqrt{8-x}}$. From the plot of $f(x)$ shown below, we estimate $\lim_{x\to 4} f(x) \approx 2.00$; to two decimal places, this matches the value of 2 obtained in Exercise 23.

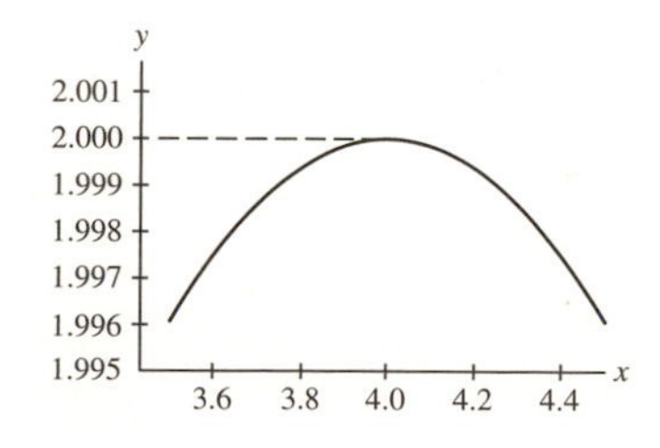

In Exercises 37–42, evaluate using the identity

$$a^3 - b^3 = (a-b)(a^2+ab+b^2)$$

37. $\lim_{x\to 2} \frac{x^3-8}{x-2}$

SOLUTION $\lim_{x\to 2} \frac{x^3-8}{x-2} = \lim_{x\to 2} \frac{(x-2)\left(x^2+2x+4\right)}{x-2} = \lim_{x\to 2} \left(x^2+2x+4\right) = 12.$

39. $\lim_{x\to 1} \frac{x^2-5x+4}{x^3-1}$

SOLUTION $\lim_{x\to 1} \frac{x^2-5x+4}{x^3-1} = \lim_{x\to 1} \frac{(x-1)(x-4)}{(x-1)\left(x^2+x+1\right)} = \lim_{x\to 1} \frac{x-4}{x^2+x+1} = -1.$

41. $\lim_{x\to 1} \frac{x^4-1}{x^3-1}$

SOLUTION

$$\lim_{x\to 1} \frac{x^4-1}{x^3-1} = \lim_{x\to 1} \frac{(x^2-1)(x^2+1)}{(x-1)(x^2+x+1)} = \lim_{x\to 1} \frac{(x-1)(x+1)(x^2+1)}{(x-1)(x^2+x+1)} = \lim_{x\to 1} \frac{(x+1)(x^2+1)}{(x^2+x+1)} = \frac{4}{3}.$$

43. Evaluate $\lim_{h\to 0} \frac{\sqrt[4]{1+h}-1}{h}$. *Hint:* Set $x = \sqrt[4]{1+h}$ and rewrite as a limit as $x \to 1$.

SOLUTION Let $x = \sqrt[4]{1+h}$. Then $h = x^4 - 1 = (x-1)(x+1)(x^2+1)$, $x \to 1$ as $h \to 0$ and

$$\lim_{h\to 0} \frac{\sqrt[4]{1+h}-1}{h} = \lim_{x\to 1} \frac{x-1}{(x-1)(x+1)(x^2+1)} = \lim_{x\to 1} \frac{1}{(x+1)(x^2+1)} = \frac{1}{4}.$$

In Exercises 45–54, evaluate in terms of the constant a.

45. $\lim_{x\to 0} (2a+x)$

SOLUTION $\lim_{x\to 0} (2a+x) = 2a.$

47. $\lim_{t\to -1} (4t-2at+3a)$

SOLUTION $\lim_{t\to -1} (4t-2at+3a) = -4+5a.$

49. $\lim_{h\to 0} \frac{2(a+h)^2-2a^2}{h}$

SOLUTION $\lim_{h\to 0} \frac{2(a+h)^2-2a^2}{h} = \lim_{h\to 0} \frac{4ha+2h^2}{h} = \lim_{h\to 0} (4a+2h) = 4a.$

51. $\lim_{x\to a} \frac{\sqrt{x}-\sqrt{a}}{x-a}$

SOLUTION $\lim_{x\to a} \frac{\sqrt{x}-\sqrt{a}}{x-a} = \lim_{x\to a} \frac{\sqrt{x}-\sqrt{a}}{\left(\sqrt{x}-\sqrt{a}\right)\left(\sqrt{x}+\sqrt{a}\right)} = \lim_{x\to a} \frac{1}{\sqrt{x}+\sqrt{a}} = \frac{1}{2\sqrt{a}}.$

53. $\lim_{x\to 0} \frac{(x+a)^3-a^3}{x}$

SOLUTION $\lim_{x\to 0} \frac{(x+a)^3-a^3}{x} = \lim_{x\to 0} \frac{x^3+3x^2a+3xa^2+a^3-a^3}{x} = \lim_{x\to 0} (x^2+3xa+3a^2) = 3a^2.$

Further Insights and Challenges

In Exercises 55–58, find all values of c such that the limit exists.

55. $\lim_{x\to c} \frac{x^2-5x-6}{x-c}$

SOLUTION $\lim_{x\to c} \frac{x^2-5x-6}{x-c}$ will exist provided that $x-c$ is a factor of the numerator. (Otherwise there will be an infinite discontinuity at $x=c$.) Since $x^2-5x-6=(x+1)(x-6)$, this occurs for $c=-1$ and $c=6$.

57. $\lim_{x\to 1} \left(\frac{1}{x-1} - \frac{c}{x^3-1}\right)$

SOLUTION Simplifying, we find

$$\frac{1}{x-1} - \frac{c}{x^3-1} = \frac{x^2+x+1-c}{(x-1)(x^2+x+1)}.$$

In order for the limit to exist as $x \to 1$, the numerator must evaluate to 0 at $x=1$. Thus, we must have $3-c=0$, which implies $c=3$.

59. For which sign $\pm$ does the following limit exist?

$$\lim_{x\to 0}\left(\frac{1}{x} \pm \frac{1}{x(x-1)}\right)$$

SOLUTION

- The limit $\lim_{x\to 0}\left(\frac{1}{x} + \frac{1}{x(x-1)}\right) = \lim_{x\to 0}\frac{(x-1)+1}{x(x-1)} = \lim_{x\to 0}\frac{1}{x-1} = -1.$
- The limit $\lim_{x\to 0}\left(\frac{1}{x} - \frac{1}{x(x-1)}\right)$ does not exist.
  - As $x \to 0+$, we have $\frac{1}{x} - \frac{1}{x(x-1)} = \frac{(x-1)-1}{x(x-1)} = \frac{x-2}{x(x-1)} \to \infty.$
  - As $x \to 0-$, we have $\frac{1}{x} - \frac{1}{x(x-1)} = \frac{(x-1)-1}{x(x-1)} = \frac{x-2}{x(x-1)} \to -\infty.$

2.6 Trigonometric Limits

Preliminary Questions

1. Assume that $-x^4 \le f(x) \le x^2$. What is $\lim_{x\to 0} f(x)$? Is there enough information to evaluate $\lim_{x\to \frac{1}{2}} f(x)$? Explain.

SOLUTION Since $\lim_{x\to 0} -x^4 = \lim_{x\to 0} x^2 = 0$, the squeeze theorem guarantees that $\lim_{x\to 0} f(x) = 0$. Since $\lim_{x\to\frac{1}{2}} -x^4 = -\frac{1}{16} \neq \frac{1}{4} = \lim_{x\to\frac{1}{2}} x^2$, we do not have enough information to determine $\lim_{x\to\frac{1}{2}} f(x)$.

2. State the Squeeze Theorem carefully.

SOLUTION Assume that for $x \neq c$ (in some open interval containing c),

$$l(x) \le f(x) \le u(x)$$

and that $\lim_{x\to c} l(x) = \lim_{x\to c} u(x) = L$. Then $\lim_{x\to c} f(x)$ exists and

$$\lim_{x\to c} f(x) = L.$$

3. If you want to evaluate $\lim_{h\to 0}\frac{\sin 5h}{3h}$, it is a good idea to rewrite the limit in terms of the variable (choose one):

(a) $\theta = 5h$ **(b)** $\theta = 3h$ **(c)** $\theta = \frac{5h}{3}$

SOLUTION To match the given limit to the pattern of

$$\lim_{\theta\to 0}\frac{\sin\theta}{\theta},$$

it is best to substitute for the argument of the sine function; thus, rewrite the limit in terms of **(a)**: $\theta = 5h$.

Exercises

1. State precisely the hypothesis and conclusions of the Squeeze Theorem for the situation in Figure 6.

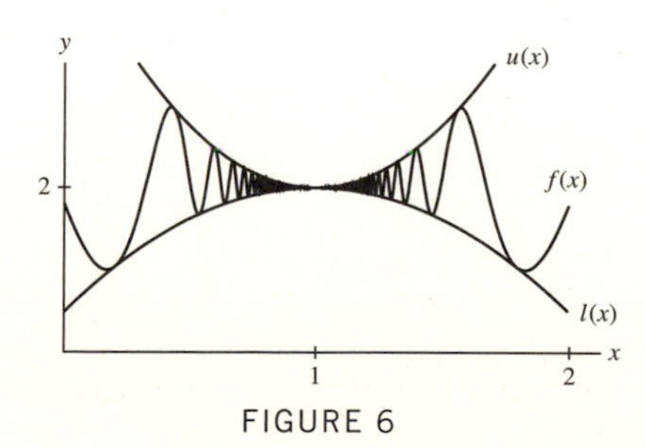

FIGURE 6

SOLUTION For all $x \neq 1$ on the open interval (0, 2) containing $x = 1$, $\ell(x) \leq f(x) \leq u(x)$. Moreover,

$$\lim_{x\to 1} \ell(x) = \lim_{x\to 1} u(x) = 2.$$

Therefore, by the Squeeze Theorem,

$$\lim_{x\to 1} f(x) = 2.$$

3. What does the Squeeze Theorem say about $\lim_{x\to 7} f(x)$ if $\lim_{x\to 7} l(x) = \lim_{x\to 7} u(x) = 6$ and $f(x), u(x)$, and $l(x)$ are related as in Figure 8? The inequality $f(x) \leq u(x)$ is not satisfied for all x. Does this affect the validity of your conclusion?

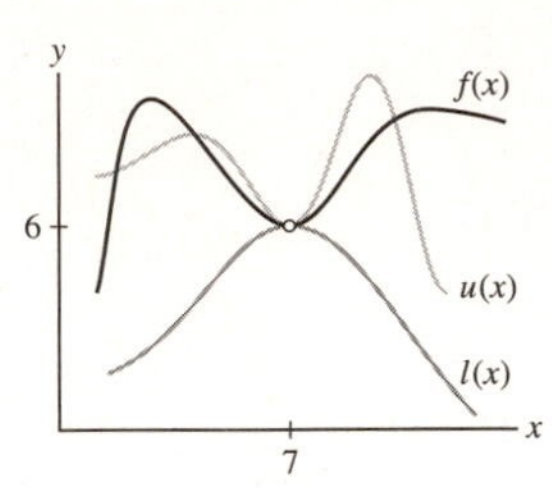

FIGURE 8

SOLUTION The Squeeze Theorem does not require that the inequalities $l(x) \leq f(x) \leq u(x)$ hold for all x, only that the inequalities hold on some open interval containing $x = c$. In Figure 8, it is clear that $l(x) \leq f(x) \leq u(x)$ on some open interval containing $x = 7$. Because $\lim_{x\to 7} u(x) = \lim_{x\to 7} l(x) = 6$, the Squeeze Theorem guarantees that $\lim_{x\to 7} f(x) = 6$.

5. State whether the inequality provides sufficient information to determine $\lim_{x\to 1} f(x)$, and if so, find the limit.

(a) $4x - 5 \leq f(x) \leq x^2$
(b) $2x - 1 \leq f(x) \leq x^2$
(c) $4x - x^2 \leq f(x) \leq x^2 + 2$

SOLUTION

(a) Because $\lim_{x\to 1} (4x - 5) = -1 \neq 1 = \lim_{x\to 1} x^2$, the given inequality does *not* provide sufficient information to determine $\lim_{x\to 1} f(x)$.
(b) Because $\lim_{x\to 1} (2x - 1) = 1 = \lim_{x\to 1} x^2$, it follows from the Squeeze Theorem that $\lim_{x\to 1} f(x) = 1$.
(c) Because $\lim_{x\to 1} (4x - x^2) = 3 = \lim_{x\to 1} (x^2 + 2)$, it follows from the Squeeze Theorem that $\lim_{x\to 1} f(x) = 3$.

In Exercises 7–16, evaluate using the Squeeze Theorem.

7. $\lim_{x\to 0} x^2 \cos \frac{1}{x}$

SOLUTION Multiplying the inequality $-1 \leq \cos \frac{1}{x} \leq 1$, which holds for all $x \neq 0$, by x^2 yields $-x^2 \leq x^2 \cos \frac{1}{x} \leq x^2$. Because

$$\lim_{x\to 0} -x^2 = \lim_{x\to 0} x^2 = 0,$$

it follows by the Squeeze Theorem that

$$\lim_{x\to 0} x^2 \cos \frac{1}{x} = 0.$$

9. $\lim_{x\to 1} (x - 1) \sin \frac{\pi}{x - 1}$

SOLUTION Multiplying the inequality $\left|\sin \frac{\pi}{x-1}\right| \leq 1$, which holds for $x \neq 1$, by $|x - 1|$ yields $\left|(x - 1) \sin \frac{\pi}{x-1}\right| \leq |x - 1|$ or $-|x - 1| \leq (x - 1) \sin \frac{\pi}{x-1} \leq |x - 1|$. Because

$$\lim_{x\to 1} -|x - 1| = \lim_{x\to 1} |x - 1| = 0,$$

it follows by the Squeeze Theorem that

$$\lim_{x\to 1} (x - 1) \sin \frac{\pi}{x - 1} = 0.$$

11. $\lim_{t\to 0} (2^t - 1)\cos\frac{1}{t}$

SOLUTION Multiplying the inequality $\left|\cos\frac{1}{t}\right| \le 1$, which holds for $t \ne 0$, by $|2^t - 1|$ yields $\left|(2^t - 1)\cos\frac{1}{t}\right| \le |2^t - 1|$ or $-|2^t - 1| \le (2^t - 1)\cos\frac{1}{t} \le |2^t - 1|$. Because

$$\lim_{t\to 0} -|2^t - 1| = \lim_{t\to 0} |2^t - 1| = 0,$$

it follows by the Squeeze Theorem that

$$\lim_{t\to 0} (2^t - 1)\cos\frac{1}{t} = 0.$$

13. $\lim_{t\to 2} (t^2 - 4)\cos\frac{1}{t-2}$

SOLUTION Multiplying the inequality $\left|\cos\frac{1}{t-2}\right| \le 1$, which holds for $t \ne 2$, by $|t^2 - 4|$ yields $\left|(t^2 - 4)\cos\frac{1}{t-2}\right| \le |t^2 - 4|$ or $-|t^2 - 4| \le (t^2 - 4)\cos\frac{1}{t-2} \le |t^2 - 4|$. Because

$$\lim_{t\to 2} -|t^2 - 4| = \lim_{t\to 2} |t^2 - 4| = 0,$$

it follows by the Squeeze Theorem that

$$\lim_{t\to 2} (t^2 - 4)\cos\frac{1}{t-2} = 0.$$

15. $\lim_{\theta\to\frac{\pi}{2}} \cos\theta\cos(\tan\theta)$

SOLUTION Multiplying the inequality $|\cos(\tan\theta)| \le 1$, which holds for all θ near $\frac{\pi}{2}$ but not equal to $\frac{\pi}{2}$, by $|\cos\theta|$ yields $|\cos\theta\cos(\tan\theta)| \le |\cos\theta|$ or $-|\cos\theta| \le \cos\theta\cos(\tan\theta) \le |\cos\theta|$. Because

$$\lim_{\theta\to\frac{\pi}{2}} -|\cos\theta| = \lim_{\theta\to\frac{\pi}{2}} |\cos\theta| = 0,$$

it follows from the Squeeze Theorem that

$$\lim_{\theta\to\frac{\pi}{2}} \cos\theta\cos(\tan\theta) = 0.$$

In Exercises 17–26, evaluate using Theorem 2 as necessary.

17. $\lim_{x\to 0} \frac{\tan x}{x}$

SOLUTION $\lim_{x\to 0} \frac{\tan x}{x} = \lim_{x\to 0} \frac{\sin x}{x}\frac{1}{\cos x} = \lim_{x\to 0} \frac{\sin x}{x} \cdot \lim_{x\to 0} \frac{1}{\cos x} = 1 \cdot 1 = 1.$

19. $\lim_{t\to 0} \frac{\sqrt{t^3+9}\,\sin t}{t}$

SOLUTION $\lim_{t\to 0} \frac{\sqrt{t^3+9}\,\sin t}{t} = \lim_{t\to 0} \sqrt{t^3+9} \cdot \lim_{t\to 0} \frac{\sin t}{t} = \sqrt{9} \cdot 1 = 3.$

21. $\lim_{x\to 0} \frac{x^2}{\sin^2 x}$

SOLUTION $\lim_{x\to 0} \frac{x^2}{\sin^2 x} = \lim_{x\to 0} \frac{1}{\frac{\sin x}{x}\,\frac{\sin x}{x}} = \lim_{x\to 0} \frac{1}{\frac{\sin x}{x}} \cdot \lim_{x\to 0} \frac{1}{\frac{\sin x}{x}} = \frac{1}{1} \cdot \frac{1}{1} = 1.$

23. $\lim_{\theta\to 0} \frac{\sec\theta - 1}{\theta}$

SOLUTION $\lim_{\theta\to 0} \frac{\sec\theta - 1}{\theta} = \lim_{\theta\to 0} \frac{1 - \cos\theta}{\theta\cos\theta} = \lim_{\theta\to 0} \frac{1 - \cos\theta}{\theta} \cdot \lim_{\theta\to 0} \frac{1}{\cos\theta} = 0 \cdot 1 = 0.$

25. $\lim_{t\to\frac{\pi}{4}} \frac{\sin t}{t}$

SOLUTION $\frac{\sin t}{t}$ is continuous at $t = \frac{\pi}{4}$. Hence, by substitution

$$\lim_{t\to\frac{\pi}{4}} \frac{\sin t}{t} = \frac{\frac{\sqrt{2}}{2}}{\frac{\pi}{4}} = \frac{2\sqrt{2}}{\pi}.$$

27. Let $L = \lim_{x\to 0} \dfrac{\sin 14x}{x}$.

(a) Show, by letting $\theta = 14x$, that $L = \lim_{\theta\to 0} 14\dfrac{\sin\theta}{\theta}$.

(b) Compute L.

SOLUTION

(a) Let $\theta = 14x$. Then $x = \frac{\theta}{14}$ and $\theta \to 0$ as $x \to 0$, so

$$L = \lim_{x\to 0} \frac{\sin 14x}{x} = \lim_{\theta\to 0} \frac{\sin\theta}{(\theta/14)} = \lim_{\theta\to 0} 14\frac{\sin\theta}{\theta}.$$

(b) Based on part (a),

$$L = 14 \lim_{\theta\to 0} \cdot \frac{\sin\theta}{\theta} = 14.$$

In Exercises 29–48, evaluate the limit.

29. $\lim_{h\to 0} \dfrac{\sin 9h}{h}$

SOLUTION $\lim_{h\to 0} \dfrac{\sin 9h}{h} = \lim_{h\to 0} 9\,\dfrac{\sin 9h}{9h} = 9.$

31. $\lim_{h\to 0} \dfrac{\sin h}{5h}$

SOLUTION $\lim_{h\to 0} \dfrac{\sin h}{5h} = \lim_{h\to 0} \dfrac{1}{5}\dfrac{\sin h}{h} = \dfrac{1}{5}.$

33. $\lim_{\theta\to 0} \dfrac{\sin 7\theta}{\sin 3\theta}$

SOLUTION We have

$$\frac{\sin 7\theta}{\sin 3\theta} = \frac{7}{3}\left(\frac{\sin 7\theta}{7\theta}\right)\left(\frac{3\theta}{\sin 3\theta}\right)$$

Therefore,

$$\lim_{\theta\to 0} \frac{\sin 7\theta}{3\theta} = \frac{7}{3}\left(\lim_{\theta\to 0} \frac{\sin 7\theta}{7\theta}\right)\left(\lim_{\theta\to 0} \frac{3\theta}{\sin 3\theta}\right) = \frac{7}{3}(1)(1) = \frac{7}{3}$$

35. $\lim_{x\to 0} x \csc 25x$

SOLUTION Let $h = 25x$. Then

$$\lim_{x\to 0} x \csc 25x = \lim_{h\to 0} \frac{h}{25} \csc h = \frac{1}{25} \lim_{h\to 0} \frac{h}{\sin h} = \frac{1}{25}.$$

37. $\lim_{h\to 0} \dfrac{\sin 2h \sin 3h}{h^2}$

SOLUTION

$$\lim_{h\to 0} \frac{\sin 2h \sin 3h}{h^2} = \lim_{h\to 0} \frac{\sin 2h \sin 3h}{h \cdot h} = \lim_{h\to 0} \frac{\sin 2h}{h} \frac{\sin 3h}{h}$$
$$= \lim_{h\to 0} 2\frac{\sin 2h}{2h} 3\frac{\sin 3h}{3h} = \lim_{h\to 0} 2\frac{\sin 2h}{2h} \lim_{h\to 0} 3\frac{\sin 3h}{3h} = 2 \cdot 3 = 6.$$

39. $\lim_{\theta\to 0} \dfrac{\sin(-3\theta)}{\sin(4\theta)}$

SOLUTION $\lim_{\theta\to 0} \dfrac{\sin(-3\theta)}{\sin(4\theta)} = \lim_{\theta\to 0} \dfrac{-\sin(3\theta)}{3\theta} \cdot \dfrac{3}{4} \cdot \dfrac{4\theta}{\sin(4\theta)} = -\dfrac{3}{4}.$

41. $\lim_{t\to 0} \dfrac{\csc 8t}{\csc 4t}$

SOLUTION $\lim_{t\to 0} \dfrac{\csc 8t}{\csc 4t} = \lim_{t\to 0} \dfrac{\sin 4t}{\sin 8t} \cdot \dfrac{8t}{4t} \cdot \dfrac{1}{2} = \dfrac{1}{2}.$

43. $\lim_{x\to 0} \frac{\sin 3x \sin 2x}{x \sin 5x}$

SOLUTION $\lim_{x\to 0} \frac{\sin 3x \sin 2x}{x \sin 5x} = \lim_{x\to 0} \left(3 \frac{\sin 3x}{3x} \cdot \frac{2}{5} \frac{(\sin 2x)\,/\,(2x)}{(\sin 5x)\,/\,(5x)}\right) = \frac{6}{5}.$

45. $\lim_{h\to 0} \frac{\sin(2h)(1-\cos h)}{h^2}$

SOLUTION $\lim_{h\to 0} \frac{\sin(2h)(1-\cos h)}{h^2} = \lim_{h\to 0} \frac{\sin(2h)}{h} \lim_{h\to 0} \frac{1-\cos h}{h} = 1 \cdot 0 = 0.$

47. $\lim_{\theta\to 0} \frac{\cos 2\theta - \cos\theta}{\theta}$

SOLUTION

$$\lim_{\theta\to 0} \frac{\cos 2\theta - \cos\theta}{\theta} = \lim_{\theta\to 0} \frac{(\cos 2\theta - 1) + (1-\cos\theta)}{\theta} = \lim_{\theta\to 0} \frac{\cos 2\theta - 1}{\theta} + \lim_{\theta\to 0} \frac{1-\cos\theta}{\theta}$$
$$= -2 \lim_{\theta\to 0} \frac{1-\cos 2\theta}{2\theta} + \lim_{\theta\to 0} \frac{1-\cos\theta}{\theta} = -2 \cdot 0 + 0 = 0.$$

49. Calculate $\lim_{x\to 0-} \frac{\sin x}{|x|}$.

SOLUTION

$$\lim_{x\to 0-} \frac{\sin x}{|x|} = \lim_{x\to 0-} \frac{\sin x}{-x} = -1$$

51. Prove the following result stated in Theorem 2:

$$\lim_{\theta\to 0} \frac{1-\cos\theta}{\theta} = 0 \qquad \boxed{7}$$

Hint: $\frac{1-\cos\theta}{\theta} = \frac{1}{1+\cos\theta} \cdot \frac{1-\cos^2\theta}{\theta}$.

SOLUTION

$$\lim_{\theta\to 0} \frac{1-\cos\theta}{\theta} = \lim_{\theta\to 0} \frac{1}{1+\cos\theta} \cdot \frac{1-\cos^2\theta}{\theta} = \lim_{\theta\to 0} \frac{1}{1+\cos\theta} \cdot \frac{\sin^2\theta}{\theta}$$
$$= \lim_{\theta\to 0} \frac{1}{1+\cos\theta} \cdot \lim_{\theta\to 0} \frac{\sin^2\theta}{\theta} = \lim_{\theta\to 0} \frac{1}{1+\cos\theta} \cdot \lim_{\theta\to 0} \sin\theta \frac{\sin\theta}{\theta}$$
$$= \lim_{\theta\to 0} \frac{1}{1+\cos\theta} \cdot \lim_{\theta\to 0} \sin\theta \cdot \lim_{\theta\to 0} \frac{\sin\theta}{\theta} = \frac{1}{2} \cdot 0 \cdot 1 = 0.$$

In Exercises 53–55, evaluate using the result of Exercise 52.

53. $\lim_{h\to 0} \frac{\cos 3h - 1}{h^2}$

SOLUTION We make the substitution $\theta = 3h$. Then $h = \theta/3$, and

$$\lim_{h\to 0} \frac{\cos 3h - 1}{h^2} = \lim_{\theta\to 0} \frac{\cos\theta - 1}{(\theta/3)^2} = -9 \lim_{\theta\to 0} \frac{1-\cos\theta}{\theta^2} = -\frac{9}{2}.$$

55. $\lim_{t\to 0} \frac{\sqrt{1-\cos t}}{t}$

SOLUTION $\lim_{t\to 0+} \frac{\sqrt{1-\cos t}}{t} = \sqrt{\lim_{t\to 0+} \frac{1-\cos t}{t^2}} = \sqrt{\frac{1}{2}} = \frac{\sqrt{2}}{2}$; on the other hand, $\lim_{t\to 0-} \frac{\sqrt{1-\cos t}}{t} = -\sqrt{\lim_{t\to 0-} \frac{1-\cos t}{t^2}} = -\sqrt{\frac{1}{2}} = -\frac{\sqrt{2}}{2}.$

Further Insights and Challenges

57. Use the result of Exercise 52 to prove that for $m \neq 0$,

$$\lim_{x\to 0} \frac{\cos mx - 1}{x^2} = -\frac{m^2}{2}$$

SOLUTION Substitute $u = mx$ into $\dfrac{\cos mx - 1}{x^2}$. We obtain $x = \frac{u}{m}$. As $x \to 0$, $u \to 0$; therefore,

$$\lim_{x\to 0} \frac{\cos mx - 1}{x^2} = \lim_{u\to 0} \frac{\cos u - 1}{(u/m)^2} = \lim_{u\to 0} m^2 \frac{\cos u - 1}{u^2} = m^2\left(-\frac{1}{2}\right) = -\frac{m^2}{2}.$$

59. **(a)** Investigate $\displaystyle\lim_{x\to c} \frac{\sin x - \sin c}{x - c}$ numerically for the five values $c = 0, \frac{\pi}{6}, \frac{\pi}{4}, \frac{\pi}{3}, \frac{\pi}{2}$.

(b) Can you guess the answer for general c?

(c) Check that your answer to (b) works for two other values of c.

SOLUTION

(a)

x	$c - 0.01$	$c - 0.001$	$c + 0.001$	$c + 0.01$
$\dfrac{\sin x - \sin c}{x - c}$	0.999983	0.99999983	0.99999983	0.999983

Here $c = 0$ and $\cos c = 1$.

x	$c - 0.01$	$c - 0.001$	$c + 0.001$	$c + 0.01$
$\dfrac{\sin x - \sin c}{x - c}$	0.868511	0.866275	0.865775	0.863511

Here $c = \frac{\pi}{6}$ and $\cos c = \frac{\sqrt{3}}{2} \approx 0.866025$.

x	$c - 0.01$	$c - 0.001$	$c + 0.001$	$c + 0.01$
$\dfrac{\sin x - \sin c}{x - c}$	0.504322	0.500433	0.499567	0.495662

Here $c = \frac{\pi}{3}$ and $\cos c = \frac{1}{2}$.

x	$c - 0.01$	$c - 0.001$	$c + 0.001$	$c + 0.01$
$\dfrac{\sin x - \sin c}{x - c}$	0.710631	0.707460	0.706753	0.703559

Here $c = \frac{\pi}{4}$ and $\cos c = \frac{\sqrt{2}}{2} \approx 0.707107$.

x	$c - 0.01$	$c - 0.001$	$c + 0.001$	$c + 0.01$
$\dfrac{\sin x - \sin c}{x - c}$	0.005000	0.000500	−0.000500	−0.005000

Here $c = \frac{\pi}{2}$ and $\cos c = 0$.

(b) $\displaystyle\lim_{x\to c} \frac{\sin x - \sin c}{x - c} = \cos c.$

(c)

x	$c - 0.01$	$c - 0.001$	$c + 0.001$	$c + 0.01$
$\dfrac{\sin x - \sin c}{x - c}$	−0.411593	−0.415692	−0.416601	−0.420686

Here $c = 2$ and $\cos c = \cos 2 \approx -0.416147$.

x	$c-0.01$	$c-0.001$	$c+0.001$	$c+0.01$
$\dfrac{\sin x - \sin c}{x - c}$	0.863511	0.865775	0.866275	0.868511

Here $c = -\frac{\pi}{6}$ and $\cos c = \frac{\sqrt{3}}{2} \approx 0.866025$.

2.7 Limits at Infinity

Preliminary Questions

1. Assume that

$$\lim_{x\to\infty} f(x) = L \quad \text{and} \quad \lim_{x\to L} g(x) = \infty$$

Which of the following statements are correct?

(a) $x = L$ is a vertical asymptote of $g(x)$.

(b) $y = L$ is a horizontal asymptote of $g(x)$.

(c) $x = L$ is a vertical asymptote of $f(x)$.

(d) $y = L$ is a horizontal asymptote of $f(x)$.

SOLUTION

(a) Because $\lim_{x\to L} g(x) = \infty$, $x = L$ is a vertical asymptote of $g(x)$. This statement is correct.

(b) This statement is not correct.

(c) This statement is not correct.

(d) Because $\lim_{x\to\infty} f(x) = L$, $y = L$ is a horizontal asymptote of $f(x)$. This statement is correct.

2. What are the following limits?

(a) $\lim_{x\to\infty} x^3$ **(b)** $\lim_{x\to-\infty} x^3$ **(c)** $\lim_{x\to-\infty} x^4$

SOLUTION

(a) $\lim_{x\to\infty} x^3 = \infty$

(b) $\lim_{x\to-\infty} x^3 = -\infty$

(c) $\lim_{x\to-\infty} x^4 = \infty$

3. Sketch the graph of a function that approaches a limit as $x \to \infty$ but does not approach a limit (either finite or infinite) as $x \to -\infty$.

SOLUTION

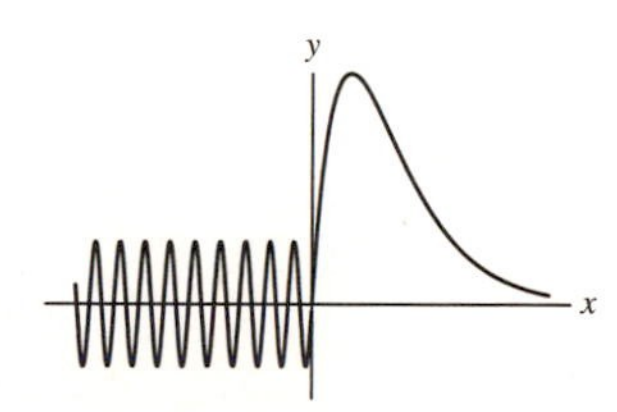

4. What is the sign of a if $f(x) = ax^3 + x + 1$ satisfies $\lim_{x\to-\infty} f(x) = \infty$?

SOLUTION Because $\lim_{x\to-\infty} x^3 = -\infty$, a must be negative to have $\lim_{x\to-\infty} f(x) = \infty$.

5. What is the sign of the leading coefficient a_7 if $f(x)$ is a polynomial of degree 7 such that $\lim_{x\to-\infty} f(x) = \infty$?

SOLUTION The behavior of $f(x)$ as $x \to -\infty$ is controlled by the leading term; that is, $\lim_{x\to-\infty} f(x) = \lim_{x\to-\infty} a_7x^7$. Because $x^7 \to -\infty$ as $x \to -\infty$, a_7 must be negative to have $\lim_{x\to-\infty} f(x) = \infty$.

6. Explain why $\lim_{x\to\infty} \sin\frac{1}{x}$ exists but $\lim_{x\to 0} \sin\frac{1}{x}$ does not exist. What is $\lim_{x\to\infty} \sin\frac{1}{x}$?

SOLUTION As $x \to \infty$, $\frac{1}{x} \to 0$, so

$$\lim_{x\to\infty} \sin\frac{1}{x} = \sin 0 = 0.$$

On the other hand, $\frac{1}{x} \to \pm\infty$ as $x \to 0$, and as $\frac{1}{x} \to \pm\infty$, $\sin\frac{1}{x}$ oscillates infinitely often. Thus

$$\lim_{x\to 0} \sin\frac{1}{x}$$

does not exist.

Exercises

1. What are the horizontal asymptotes of the function in Figure 6?

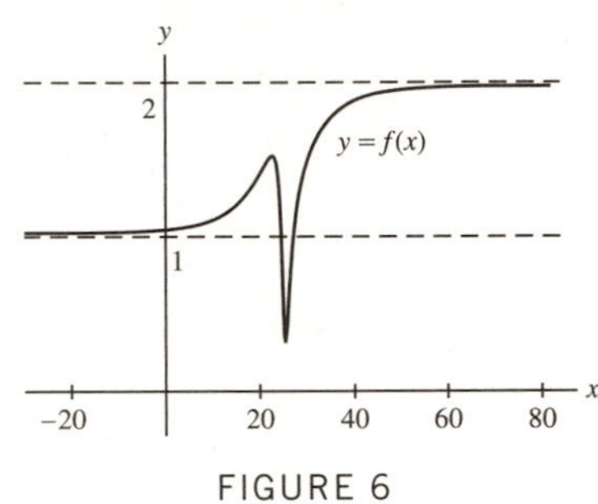

FIGURE 6

SOLUTION Because

$$\lim_{x\to-\infty} f(x) = 1 \quad \text{and} \quad \lim_{x\to\infty} f(x) = 2,$$

the function $f(x)$ has horizontal asymptotes of $y = 1$ and $y = 2$.

3. Sketch the graph of a function $f(x)$ with a single horizontal asymptote $y = 3$.

SOLUTION

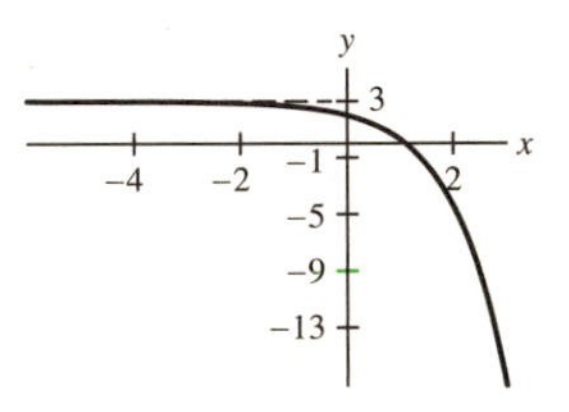

5. GU Investigate the asymptotic behavior of $f(x) = \dfrac{x^3}{x^3 + x}$ numerically and graphically:

(a) Make a table of values of $f(x)$ for $x = \pm 50, \pm 100, \pm 500, \pm 1000$.

(b) Plot the graph of $f(x)$.

(c) What are the horizontal asymptotes of $f(x)$?

SOLUTION

(a) From the table below, it appears that

$$\lim_{x\to\pm\infty} \frac{x^3}{x^3 + x} = 1.$$

x	± 50	± 100	± 500	± 1000
$f(x)$	0.999600	0.999900	0.999996	0.999999

(b) From the graph below, it also appears that

$$\lim_{x\to\pm\infty} \frac{x^3}{x^3 + x} = 1.$$

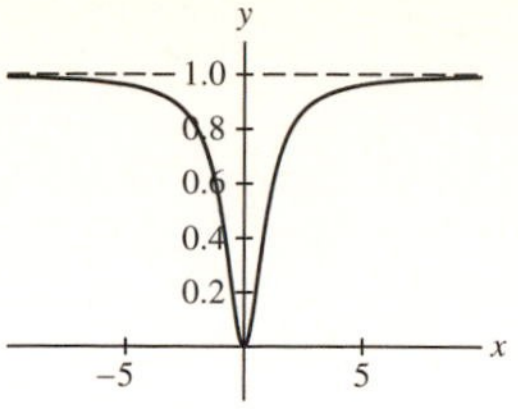

(c) The horizontal asymptote of $f(x)$ is $y = 1$.

In Exercises 7–16, evaluate the limit.

7. $\lim_{x\to\infty} \frac{x}{x+9}$

SOLUTION

$$\lim_{x\to\infty} \frac{x}{x+9} = \lim_{x\to\infty} \frac{x^{-1}(x)}{x^{-1}(x+9)} = \lim_{x\to\infty} \frac{1}{1+\frac{9}{x}} = \frac{1}{1+0} = 1.$$

9. $\lim_{x\to\infty} \frac{3x^2+20x}{2x^4+3x^3-29}$

SOLUTION

$$\lim_{x\to\infty} \frac{3x^2+20x}{2x^4+3x^3-29} = \lim_{x\to\infty} \frac{x^{-4}(3x^2+20x)}{x^{-4}(2x^4+3x^3-29)} = \lim_{x\to\infty} \frac{\frac{3}{x^2}+\frac{20}{x^3}}{2+\frac{3}{x}-\frac{29}{x^4}} = \frac{0}{2} = 0.$$

11. $\lim_{x\to\infty} \frac{7x-9}{4x+3}$

SOLUTION

$$\lim_{x\to\infty} \frac{7x-9}{4x+3} = \lim_{x\to\infty} \frac{x^{-1}(7x-9)}{x^{-1}(4x+3)} = \lim_{x\to\infty} \frac{7-\frac{9}{x}}{4+\frac{3}{x}} = \frac{7}{4}.$$

13. $\lim_{x\to-\infty} \frac{7x^2-9}{4x+3}$

SOLUTION

$$\lim_{x\to-\infty} \frac{7x^2-9}{4x+3} = \lim_{x\to-\infty} \frac{x^{-1}(7x^2-9)}{x^{-1}(4x+3)} = \lim_{x\to-\infty} \frac{7x-\frac{9}{x}}{4+\frac{3}{x}} = -\infty.$$

15. $\lim_{x\to-\infty} \frac{3x^3-10}{x+4}$

SOLUTION

$$\lim_{x\to-\infty} \frac{3x^3-10}{x+4} = \lim_{x\to-\infty} \frac{x^{-1}(3x^3-10)}{x^{-1}(x+4)} = \lim_{x\to-\infty} \frac{3x^2-\frac{10}{x}}{1+\frac{4}{x}} = \frac{\infty}{1} = \infty.$$

In Exercises 17–22, find the horizontal asymptotes.

17. $f(x) = \frac{2x^2-3x}{8x^2+8}$

SOLUTION First calculate the limits as $x \to \pm\infty$. For $x \to \infty$,

$$\lim_{x\to\infty} \frac{2x^2-3x}{8x^2+8} = \lim_{x\to\infty} \frac{2-\frac{3}{x}}{8+\frac{8}{x^2}} = \frac{2}{8} = \frac{1}{4}.$$

Similarly,

$$\lim_{x\to-\infty} \frac{2x^2-3x}{8x^2+8} = \lim_{x\to-\infty} \frac{2-\frac{3}{x}}{8+\frac{8}{x^2}} = \frac{2}{8} = \frac{1}{4}.$$

Thus, the horizontal asymptote of $f(x)$ is $y = \frac{1}{4}$.

19. $f(x) = \dfrac{\sqrt{36x^2+7}}{9x+4}$

SOLUTION For $x > 0$, $x^{-1} = |x^{-1}| = \sqrt{x^{-2}}$, so

$$\lim_{x\to\infty} \frac{\sqrt{36x^2+7}}{9x+4} = \lim_{x\to\infty} \frac{\sqrt{36+\frac{7}{x^2}}}{9+\frac{4}{x}} = \frac{\sqrt{36}}{9} = \frac{2}{3}.$$

On the other hand, for $x < 0$, $x^{-1} = -|x^{-1}| = -\sqrt{x^{-2}}$, so

$$\lim_{x\to-\infty} \frac{\sqrt{36x^2+7}}{9x+4} = \lim_{x\to-\infty} \frac{-\sqrt{36+\frac{7}{x^2}}}{9+\frac{4}{x}} = \frac{-\sqrt{36}}{9} = -\frac{2}{3}.$$

Thus, the horizontal asymptotes of $f(x)$ are $y = \frac{2}{3}$ and $y = -\frac{2}{3}$.

21. $f(t) = \dfrac{e^t}{1+e^{-t}}$

SOLUTION With

$$\lim_{t\to\infty} \frac{e^t}{1+e^{-t}} = \frac{\infty}{1} = \infty$$

and

$$\lim_{t\to-\infty} \frac{e^t}{1+e^{-t}} = 0,$$

the function $f(t)$ has one horizontal asymptote, $y = 0$.

In Exercises 23–30, evaluate the limit.

23. $\displaystyle\lim_{x\to\infty} \frac{\sqrt{9x^4+3x+2}}{4x^3+1}$

SOLUTION For $x > 0$, $x^{-3} = |x^{-3}| = \sqrt{x^{-6}}$, so

$$\lim_{x\to\infty} \frac{\sqrt{9x^4+3x+2}}{4x^3+1} = \lim_{x\to\infty} \frac{\sqrt{\frac{9}{x^2}+\frac{3}{x^5}+\frac{2}{x^6}}}{4+\frac{1}{x^3}} = 0.$$

25. $\displaystyle\lim_{x\to-\infty} \frac{8x^2+7x^{1/3}}{\sqrt{16x^4+6}}$

SOLUTION For $x < 0$, $x^{-2} = |x^{-2}| = \sqrt{x^{-4}}$, so

$$\lim_{x\to-\infty} \frac{8x^2+7x^{1/3}}{\sqrt{16x^4+6}} = \lim_{x\to-\infty} \frac{8+\frac{7}{x^{5/3}}}{\sqrt{16+\frac{6}{x^4}}} = \frac{8}{\sqrt{16}} = 2.$$

27. $\displaystyle\lim_{t\to\infty} \frac{t^{4/3}+t^{1/3}}{(4t^{2/3}+1)^2}$

SOLUTION $\displaystyle\lim_{t\to\infty} \frac{t^{4/3}+t^{1/3}}{(4t^{2/3}+1)^2} = \lim_{t\to\infty} \frac{1+\frac{1}{t}}{(4+\frac{1}{t^{2/3}})^2} = \frac{1}{16}.$

29. $\displaystyle\lim_{x\to-\infty} \frac{|x|+x}{x+1}$

SOLUTION For $x < 0$, $|x| = -x$. Therefore, for all $x < 0$,

$$\frac{|x|+x}{x+1} = \frac{-x+x}{x+1} = 0;$$

consequently,

$$\lim_{x\to-\infty} \frac{|x|+x}{x+1} = 0.$$

31. Determine $\lim_{x\to\infty} \tan^{-1} x$. Explain geometrically.

SOLUTION As an angle θ increases from 0 to $\frac{\pi}{2}$, its tangent $x = \tan\theta$ approaches ∞. Therefore,

$$\lim_{x\to\infty} \tan^{-1} x = \frac{\pi}{2}.$$

Geometrically, this means that the graph of $y = \tan^{-1} x$ has a horizontal asymptote at $y = \frac{\pi}{2}$.

33. According to the **Michaelis–Menten equation** (Figure 7), when an enzyme is combined with a substrate of concentration s (in millimolars), the reaction rate (in micromolars/min) is

$$R(s) = \frac{As}{K+s} \qquad (A, K \text{ constants})$$

(a) Show, by computing $\lim_{s\to\infty} R(s)$, that A is the limiting reaction rate as the concentration s approaches ∞.

(b) Show that the reaction rate $R(s)$ attains one-half of the limiting value A when $s = K$.

(c) For a certain reaction, $K = 1.25$ mM and $A = 0.1$. For which concentration s is $R(s)$ equal to 75% of its limiting value?

Leonor Michaelis
1875–1949

Maud Menten
1879–1960

FIGURE 7 Canadian-born biochemist Maud Menten is best known for her fundamental work on enzyme kinetics with German scientist Leonor Michaelis. She was also an accomplished painter, clarinetist, mountain climber, and master of numerous languages.

SOLUTION

(a) $\lim_{s\to\infty} R(s) = \lim_{s\to\infty} \frac{As}{K+s} = \lim_{s\to\infty} \frac{A}{1+\frac{K}{s}} = A.$

(b) Observe that

$$R(K) = \frac{AK}{K+K} = \frac{AK}{2K} = \frac{A}{2},$$

have of the limiting value.

(c) By part (a), the limiting value is 0.1, so we need to determine the value of s that satisfies

$$R(s) = \frac{0.1s}{1.25+s} = 0.075.$$

Solving this equation for s yields

$$s = \frac{(1.25)(0.075)}{0.025} = 3.75 \text{ mM}.$$

In Exercises 35–42, calculate the limit.

35. $\lim_{x\to\infty} \left(\sqrt{4x^4+9x} - 2x^2\right)$

SOLUTION Write

$$\sqrt{4x^4+9x} - 2x^2 = \left(\sqrt{4x^4+9x} - 2x^2\right)\frac{\sqrt{4x^4+9x}+2x^2}{\sqrt{4x^4+9x}+2x^2}$$

$$= \frac{(4x^4+9x) - 4x^4}{\sqrt{4x^4+9x}+2x^2} = \frac{9x}{\sqrt{4x^4+9x}+2x^2}.$$

Thus,

$$\lim_{x\to\infty} (\sqrt{4x^4+9x} - 2x^2) = \lim_{x\to\infty} \frac{9x}{\sqrt{4x^4+9x}+2x^2} = 0.$$

37. $\lim_{x\to\infty} (2\sqrt{x} - \sqrt{x+2})$

SOLUTION Write

$$2\sqrt{x} - \sqrt{x+2} = (2\sqrt{x} - \sqrt{x+2})\frac{2\sqrt{x}+\sqrt{x+2}}{2\sqrt{x}+\sqrt{x+2}}$$
$$= \frac{4x-(x+2)}{2\sqrt{x}+\sqrt{x+2}} = \frac{3x-2}{2\sqrt{x}+\sqrt{x+2}}.$$

Thus,

$$\lim_{x\to\infty}(2\sqrt{x} - \sqrt{x+2}) = \lim_{x\to\infty}\frac{3x-2}{2\sqrt{x}+\sqrt{x+2}} = \infty.$$

39. $\lim_{x\to\infty} (\ln(3x+1) - \ln(2x+1))$

SOLUTION Because

$$\ln(3x+1) - \ln(2x+1) = \ln\frac{3x+1}{2x+1}$$

and

$$\lim_{x\to\infty}\frac{3x+1}{2x+1} = \frac{3}{2},$$

it follows that

$$\lim_{x\to\infty}(\ln(3x+1) - \ln(2x+1)) = \ln\frac{3}{2}.$$

41. $\lim_{x\to\infty} \tan^{-1}\left(\frac{x^2+9}{9-x}\right)$

SOLUTION Because

$$\lim_{x\to\infty}\frac{x^2+9}{9-x} = \lim_{x\to\infty}\frac{x+\frac{9}{x}}{\frac{9}{x}-1} = \frac{\infty}{-1} = -\infty,$$

it follows that

$$\lim_{x\to\infty}\tan^{-1}\left(\frac{x^2+9}{9-x}\right) = -\frac{\pi}{2}.$$

43. Let $P(n)$ be the perimeter of an n-gon inscribed in a unit circle (Figure 8).

(a) Explain, intuitively, why $P(n)$ approaches 2π as $n\to\infty$.

(b) Show that $P(n) = 2n\sin\left(\frac{\pi}{n}\right)$.

(c) Combine (a) and (b) to conclude that $\lim_{n\to\infty}\frac{n}{\pi}\sin\left(\frac{\pi}{n}\right) = 1$.

(d) Use this to give another argument that $\lim_{\theta\to 0}\frac{\sin\theta}{\theta} = 1$.

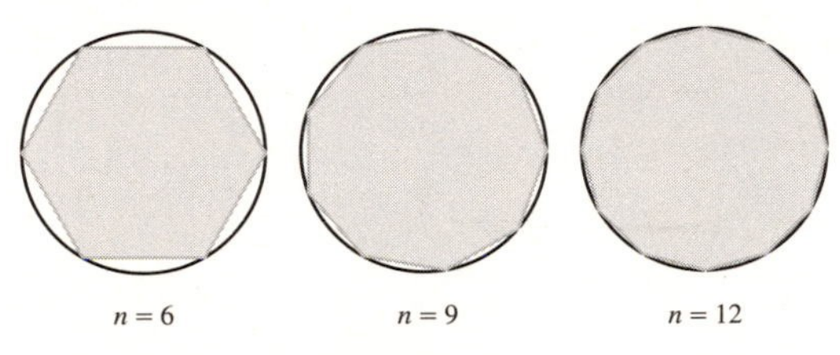

FIGURE 8

SOLUTION

(a) As $n\to\infty$, the n-gon approaches a circle of radius 1. Therefore, the perimeter of the n-gon approaches the circumference of the unit circle as $n\to\infty$. That is, $P(n)\to 2\pi$ as $n\to\infty$.

(b) Each side of the n-gon is the third side of an isosceles triangle with equal length sides of length 1 and angle $\theta = \frac{2\pi}{n}$ between the equal length sides. The length of each side of the n-gon is therefore

$$\sqrt{1^2 + 1^2 - 2\cos\frac{2\pi}{n}} = \sqrt{2(1 - \cos\frac{2\pi}{n})} = \sqrt{4\sin^2\frac{\pi}{n}} = 2\sin\frac{\pi}{n}.$$

Finally,

$$P(n) = 2n\sin\frac{\pi}{n}.$$

(c) Combining parts (a) and (b),

$$\lim_{n\to\infty} P(n) = \lim_{n\to\infty} 2n\sin\frac{\pi}{n} = 2\pi.$$

Dividing both sides of this last expression by 2π yields

$$\lim_{n\to\infty} \frac{n}{\pi}\sin\frac{\pi}{n} = 1.$$

(d) Let $\theta = \frac{\pi}{n}$. Then $\theta \to 0$ as $n \to \infty$,

$$\frac{n}{\pi}\sin\frac{\pi}{n} = \frac{1}{\theta}\sin\theta = \frac{\sin\theta}{\theta},$$

and

$$\lim_{n\to\infty} \frac{n}{\pi}\sin\frac{\pi}{n} = \lim_{\theta\to 0} \frac{\sin\theta}{\theta} = 1.$$

Further Insights and Challenges

45. Every limit as $x \to \infty$ can be rewritten as a one-sided limit as $t \to 0+$, where $t = x^{-1}$. Setting $g(t) = f(t^{-1})$, we have

$$\lim_{x\to\infty} f(x) = \lim_{t\to 0+} g(t)$$

Show that $\lim_{x\to\infty} \frac{3x^2 - x}{2x^2 + 5} = \lim_{t\to 0+} \frac{3 - t}{2 + 5t^2}$, and evaluate using the Quotient Law.

SOLUTION Let $t = x^{-1}$. Then $x = t^{-1}$, $t \to 0+$ as $x \to \infty$, and

$$\frac{3x^2 - x}{2x^2 + 5} = \frac{3t^{-2} - t^{-1}}{2t^{-2} + 5} = \frac{3 - t}{2 + 5t^2}.$$

Thus,

$$\lim_{x\to\infty} \frac{3x^2 - x}{2x^2 + 5} = \lim_{t\to 0+} \frac{3 - t}{2 + 5t^2} = \frac{3}{2}.$$

47. Let $G(b) = \lim_{x\to\infty} (1 + b^x)^{1/x}$ for $b \geq 0$. Investigate $G(b)$ numerically and graphically for $b = 0.2, 0.8, 2, 3, 5$ (and additional values if necessary). Then make a conjecture for the value of $G(b)$ as a function of b. Draw a graph of $y = G(b)$. Does $G(b)$ appear to be continuous? We will evaluate $G(b)$ using L'Hôpital's Rule in Section 4.5 (see Exercise 69 in Section 4.5).

SOLUTION

- $b = 0.2$:

x	5	10	50	100
$f(x)$	1.000064	1.000000	1.000000	1.000000

It appears that $G(0.2) = 1$.

- $b = 0.8$:

x	5	10	50	100
$f(x)$	1.058324	1.010251	1.000000	1.000000

It appears that $G(0.8) = 1$.

- $b = 2$:

x	5	10	50	100
$f(x)$	2.012347	2.000195	2.000000	2.000000

It appears that $G(2) = 2$.

- $b = 3$:

x	5	10	50	100
$f(x)$	3.002465	3.000005	3.000000	3.000000

It appears that $G(3) = 3$.

- $b = 5$:

x	5	10	50	100
$f(x)$	5.000320	5.000000	5.000000	5.000000

It appears that $G(5) = 5$.

Based on these observations we conjecture that $G(b) = 1$ if $0 \le b \le 1$ and $G(b) = b$ for $b > 1$. The graph of $y = G(b)$ is shown below; the graph does appear to be continuous.

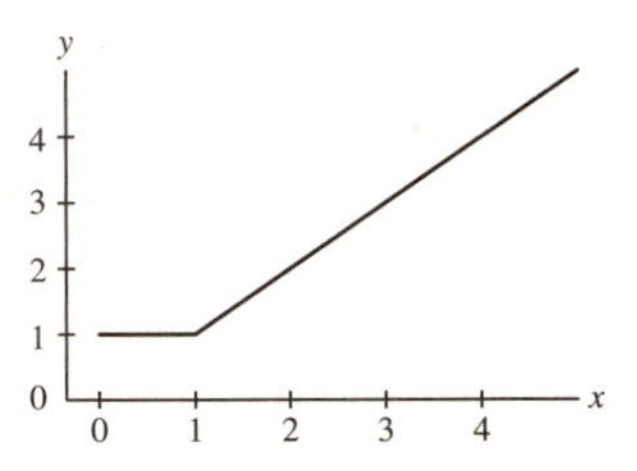

2.8 Intermediate Value Theorem

Preliminary Questions

1. Prove that $f(x) = x^2$ takes on the value 0.5 in the interval $[0, 1]$.

SOLUTION Observe that $f(x) = x^2$ is continuous on $[0, 1]$ with $f(0) = 0$ and $f(1) = 1$. Because $f(0) < 0.5 < f(1)$, the Intermediate Value Theorem guarantees there is a $c \in [0, 1]$ such that $f(c) = 0.5$.

2. The temperature in Vancouver was 8°C at 6 AM and rose to 20°C at noon. Which assumption about temperature allows us to conclude that the temperature was 15°C at some moment of time between 6 AM and noon?

SOLUTION We must assume that temperature is a continuous function of time.

3. What is the graphical interpretation of the IVT?

SOLUTION If f is continuous on $[a, b]$, then the horizontal line $y = k$ for every k between $f(a)$ and $f(b)$ intersects the graph of $y = f(x)$ at least once.

4. Show that the following statement is false by drawing a graph that provides a counterexample:

If $f(x)$ is continuous and has a root in $[a, b]$, then $f(a)$ and $f(b)$ have opposite signs.

SOLUTION

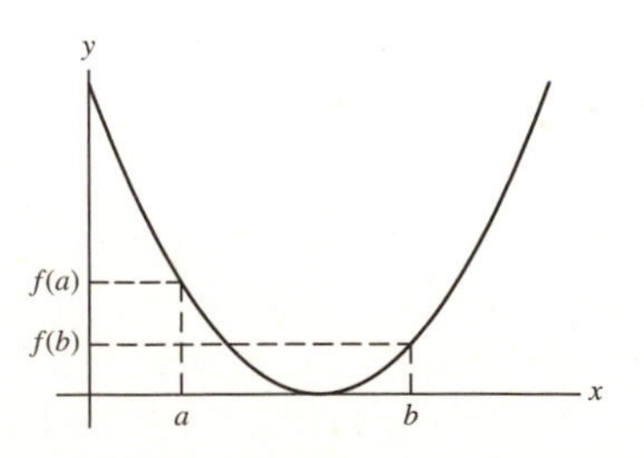

5. Assume that $f(t)$ is continuous on $[1, 5]$ and that $f(1) = 20$, $f(5) = 100$. Determine whether each of the following statements is always true, never true, or sometimes true.

(a) $f(c) = 3$ has a solution with $c \in [1, 5]$.

(b) $f(c) = 75$ has a solution with $c \in [1, 5]$.

(c) $f(c) = 50$ has no solution with $c \in [1, 5]$.

(d) $f(c) = 30$ has exactly one solution with $c \in [1, 5]$.

SOLUTION

(a) This statement is sometimes true.

(b) This statement is always true.

(c) This statement is never true.

(d) This statement is sometimes true.

Exercises

1. Use the IVT to show that $f(x) = x^3 + x$ takes on the value 9 for some x in $[1, 2]$.

SOLUTION Observe that $f(1) = 2$ and $f(2) = 10$. Since f is a polynomial, it is continuous everywhere; in particular on $[1, 2]$. Therefore, by the IVT there is a $c \in [1, 2]$ such that $f(c) = 9$.

3. Show that $g(t) = t^2 \tan t$ takes on the value $\frac{1}{2}$ for some t in $\left[0, \frac{\pi}{4}\right]$.

SOLUTION $g(0) = 0$ and $g(\frac{\pi}{4}) = \frac{\pi^2}{16}$. $g(t)$ is continuous for all t between 0 and $\frac{\pi}{4}$, and $0 < \frac{1}{2} < \frac{\pi^2}{16}$; therefore, by the IVT, there is a $c \in [0, \frac{\pi}{4}]$ such that $g(c) = \frac{1}{2}$.

5. Show that $\cos x = x$ has a solution in the interval $[0, 1]$. *Hint:* Show that $f(x) = x - \cos x$ has a zero in $[0, 1]$.

SOLUTION Let $f(x) = x - \cos x$. Observe that f is continuous with $f(0) = -1$ and $f(1) = 1 - \cos 1 \approx 0.46$. Therefore, by the IVT there is a $c \in [0, 1]$ such that $f(c) = c - \cos c = 0$. Thus $c = \cos c$ and hence the equation $\cos x = x$ has a solution c in $[0, 1]$.

In Exercises 7–16, prove using the IVT.

7. $\sqrt{c} + \sqrt{c+2} = 3$ has a solution.

SOLUTION Let $f(x) = \sqrt{x} + \sqrt{x+2} - 3$. Note that f is continuous on $\left[\frac{1}{4}, 2\right]$ with $f(\frac{1}{4}) = \sqrt{\frac{1}{4}} + \sqrt{\frac{9}{4}} - 3 = -1$ and $f(2) = \sqrt{2} - 1 \approx 0.41$. Therefore, by the IVT there is a $c \in \left[\frac{1}{4}, 2\right]$ such that $f(c) = \sqrt{c} + \sqrt{c+2} - 3 = 0$. Thus $\sqrt{c} + \sqrt{c+2} = 3$ and hence the equation $\sqrt{x} + \sqrt{x+2} = 3$ has a solution c in $\left[\frac{1}{4}, 2\right]$.

9. $\sqrt{2}$ exists. *Hint:* Consider $f(x) = x^2$.

SOLUTION Let $f(x) = x^2$. Observe that f is continuous with $f(1) = 1$ and $f(2) = 4$. Therefore, by the IVT there is a $c \in [1, 2]$ such that $f(c) = c^2 = 2$. This proves the existence of $\sqrt{2}$, a number whose square is 2.

11. For all positive integers k, $\cos x = x^k$ has a solution.

SOLUTION For each positive integer k, let $f(x) = x^k - \cos x$. Observe that f is continuous on $\left[0, \frac{\pi}{2}\right]$ with $f(0) = -1$ and $f(\frac{\pi}{2}) = \left(\frac{\pi}{2}\right)^k > 0$. Therefore, by the IVT there is a $c \in \left[0, \frac{\pi}{2}\right]$ such that $f(c) = c^k - \cos(c) = 0$. Thus $\cos c = c^k$ and hence the equation $\cos x = x^k$ has a solution c in the interval $\left[0, \frac{\pi}{2}\right]$.

13. $2^x + 3^x = 4^x$ has a solution.

SOLUTION Let $f(x) = 2^x + 3^x - 4^x$. Observe that f is continuous on $[0, 2]$ with $f(0) = 1 > 0$ and $f(2) = -3 < 0$. Therefore, by the IVT, there is a $c \in (0, 2)$ such that $f(c) = 2^c + 3^c - 4^c = 0$.

15. $e^x + \ln x = 0$ has a solution.

SOLUTION Let $f(x) = e^x + \ln x$. Observe that f is continuous on $[e^{-2}, 1]$ with $f(e^{-2}) = e^{e^{-2}} - 2 < 0$ and $f(1) = e > 0$. Therefore, by the IVT, there is a $c \in (e^{-2}, 1) \subset (0, 1)$ such that $f(c) = e^c + \ln c = 0$.

17. Carry out three steps of the Bisection Method for $f(x) = 2^x - x^3$ as follows:

(a) Show that $f(x)$ has a zero in $[1, 1.5]$.

(b) Show that $f(x)$ has a zero in $[1.25, 1.5]$.

(c) Determine whether $[1.25, 1.375]$ or $[1.375, 1.5]$ contains a zero.

SOLUTION Note that $f(x)$ is continuous for all x.

(a) $f(1) = 1$, $f(1.5) = 2^{1.5} - (1.5)^3 < 3 - 3.375 < 0$. Hence, $f(x) = 0$ for some x between 1 and 1.5.

(b) $f(1.25) \approx 0.4253 > 0$ and $f(1.5) < 0$. Hence, $f(x) = 0$ for some x between 1.25 and 1.5.

(c) $f(1.375) \approx -0.0059$. Hence, $f(x) = 0$ for some x between 1.25 and 1.375.

19. Find an interval of length $\frac{1}{4}$ in $[1, 2]$ containing a root of the equation $x^7 + 3x - 10 = 0$.

SOLUTION Let $f(x) = x^7 + 3x - 10$. Observe that f is continuous with $f(1) = -6$ and $f(2) = 124$. Therefore, by the IVT there is a $c \in [1, 2]$ such that $f(c) = 0$. $f(1.5) \approx 11.59 > 0$, so $f(c) = 0$ for some $c \in [1, 1.5]$. $f(1.25) \approx -1.48 < 0$, and so $f(c) = 0$ for some $c \in [1.25, 1.5]$. This means that $[1.25, 1.5]$ is an interval of length 0.25 containing a root of $f(x)$.

In Exercises 21–24, draw the graph of a function $f(x)$ on $[0, 4]$ with the given property.

21. Jump discontinuity at $x = 2$ and does not satisfy the conclusion of the IVT.

SOLUTION The function graphed below has a jump discontinuity at $x = 2$. Note that while $f(0) = 2$ and $f(4) = 4$, there is no point c in the interval $[0, 4]$ such that $f(c) = 3$. Accordingly, the conclusion of the IVT is *not* satisfied.

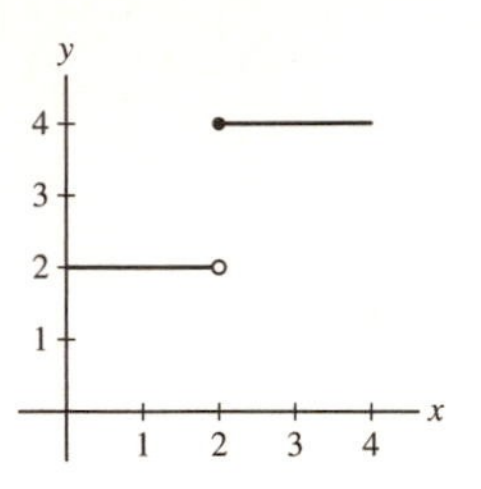

23. Infinite one-sided limits at $x = 2$ and does not satisfy the conclusion of the IVT.

SOLUTION The function graphed below has infinite one-sided limits at $x = 2$. Note that while $f(0) = 2$ and $f(4) = 4$, there is no point c in the interval $[0, 4]$ such that $f(c) = 3$. Accordingly, the conclusion of the IVT is *not* satisfied.

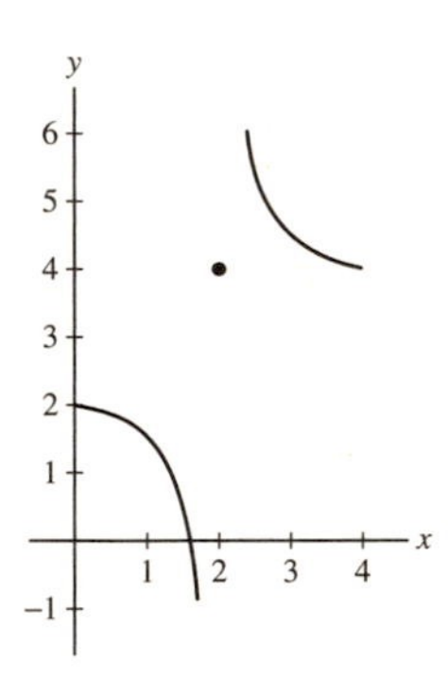

25. Can Corollary 2 be applied to $f(x) = x^{-1}$ on $[-1, 1]$? Does $f(x)$ have any roots?

SOLUTION No, because $f(x) = x^{-1}$ is not continuous on $[-1, 1]$. Even though $f(-1) = -1 < 0$ and $f(1) = 1 > 0$, the function has no roots between $x = -1$ and $x = 1$. In fact, this function has no roots at all.

Further Insights and Challenges

27. Show that if $f(x)$ is continuous and $0 \le f(x) \le 1$ for $0 \le x \le 1$, then $f(c) = c$ for some c in $[0, 1]$ (Figure 6).

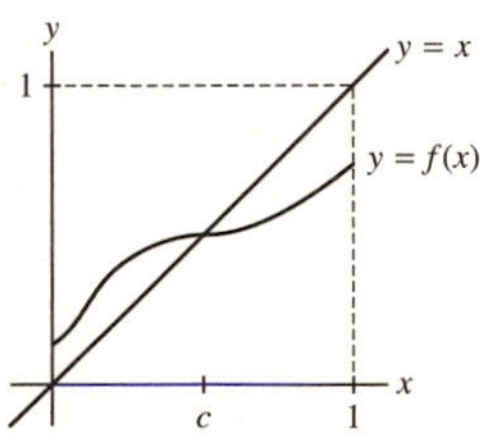

FIGURE 6 A function satisfying $0 \le f(x) \le 1$ for $0 \le x \le 1$.

SOLUTION If $f(0) = 0$, the proof is done with $c = 0$. We may assume that $f(0) > 0$. Let $g(x) = f(x) - x$. $g(0) = f(0) - 0 = f(0) > 0$. Since $f(x)$ is continuous, the Rule of Differences dictates that $g(x)$ is continuous. We need to prove that $g(c) = 0$ for some $c \in [0, 1]$. Since $f(1) \le 1$, $g(1) = f(1) - 1 \le 0$. If $g(1) = 0$, the proof is done with $c = 1$, so let's assume that $g(1) < 0$.

We now have a continuous function $g(x)$ on the interval $[0, 1]$ such that $g(0) > 0$ and $g(1) < 0$. From the IVT, there must be some $c \in [0, 1]$ so that $g(c) = 0$, so $f(c) - c = 0$ and so $f(c) = c$.

This is a simple case of a very general, useful, and beautiful theorem called the **Brouwer fixed point theorem**.

29. **Ham Sandwich Theorem** Figure 7(A) shows a slice of ham. Prove that for any angle θ $(0 \le \theta \le \pi)$, it is possible to cut the slice in half with a cut of incline θ. *Hint:* The lines of inclination θ are given by the equations $y = (\tan\theta)x + b$, where b varies from $-\infty$ to ∞. Each such line divides the slice into two pieces (one of which may be empty). Let $A(b)$ be the amount of ham to the left of the line minus the amount to the right, and let A be the total area of the ham. Show that $A(b) = -A$ if b is sufficiently large and $A(b) = A$ if b is sufficiently negative. Then use the IVT. This works if $\theta \ne 0$ or $\frac{\pi}{2}$. If $\theta = 0$, define $A(b)$ as the amount of ham above the line $y = b$ minus the amount below. How can you modify the argument to work when $\theta = \frac{\pi}{2}$ (in which case $\tan\theta = \infty$)?

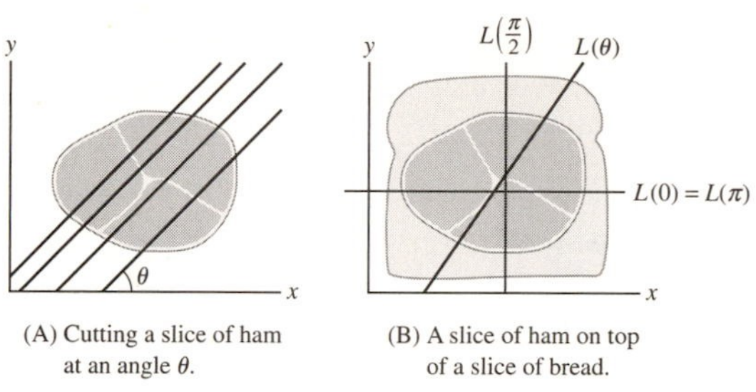

(A) Cutting a slice of ham at an angle θ.

(B) A slice of ham on top of a slice of bread.

FIGURE 7

SOLUTION Let θ be such that $\theta \ne \frac{\pi}{2}$. For any b, consider the line $L(\theta)$ drawn at angle θ to the x axis starting at $(0, b)$. This line has formula $y = (\tan\theta)x + b$. Let $A(b)$ be the amount of ham above the line minus that below the line.

Let $A > 0$ be the area of the ham. We have to accept the following (reasonable) assumptions:

- For low enough $b = b_0$, the line $L(\theta)$ lies entirely below the ham, so that $A(b_0) = A - 0 = A$.
- For high enough b_1, the line $L(\theta)$ lies entirely above the ham, so that $A(b_1) = 0 - A = -A$.
- $A(b)$ is continuous as a function of b.

Under these assumptions, we see $A(b)$ is a continuous function satisfying $A(b_0) > 0$ and $A(b_1) < 0$ for some $b_0 < b_1$. By the IVT, $A(b) = 0$ for some $b \in [b_0, b_1]$.

Suppose that $\theta = \frac{\pi}{2}$. Let the line $L(c)$ be the vertical line through $(c, 0)$ $(x = c)$. Let $A(c)$ be the area of ham to the left of $L(c)$ minus that to the right of $L(c)$. Since $L(0)$ lies entirely to the left of the ham, $A(0) = 0 - A = -A$. For some $c = c_1$ sufficiently large, $L(c)$ lies entirely to the right of the ham, so that $A(c_1) = A - 0 = A$. Hence $A(c)$ is a continuous function of c such that $A(0) < 0$ and $A(c_1) > 0$. By the IVT, there is some $c \in [0, c_1]$ such that $A(c) = 0$.

2.9 The Formal Definition of a Limit

Preliminary Questions

1. Given that $\lim_{x\to 0} \cos x = 1$, which of the following statements is true?

(a) If $|\cos x - 1|$ is very small, then x is close to 0.

(b) There is an $\epsilon > 0$ such that $|x| < 10^{-5}$ if $0 < |\cos x - 1| < \epsilon$.

(c) There is a $\delta > 0$ such that $|\cos x - 1| < 10^{-5}$ if $0 < |x| < \delta$.

(d) There is a $\delta > 0$ such that $|\cos x| < 10^{-5}$ if $0 < |x - 1| < \delta$.

SOLUTION The true statement is **(c)**: There is a $\delta > 0$ such that $|\cos x - 1| < 10^{-5}$ if $0 < |x| < \delta$.

2. Suppose it is known that for a given ϵ and δ, $|f(x) - 2| < \epsilon$ if $0 < |x - 3| < \delta$. Which of the following statements must also be true?

(a) $|f(x) - 2| < \epsilon$ if $0 < |x - 3| < 2\delta$

(b) $|f(x) - 2| < 2\epsilon$ if $0 < |x - 3| < \delta$

(c) $|f(x) - 2| < \frac{\epsilon}{2}$ if $0 < |x - 3| < \frac{\delta}{2}$

(d) $|f(x) - 2| < \epsilon$ if $0 < |x - 3| < \frac{\delta}{2}$

SOLUTION Statements **(b)** and **(d)** are true.

Exercises

1. Based on the information conveyed in Figure 5(A), find values of L, ϵ, and $\delta > 0$ such that the following statement holds: $|f(x) - L| < \epsilon$ if $|x| < \delta$.

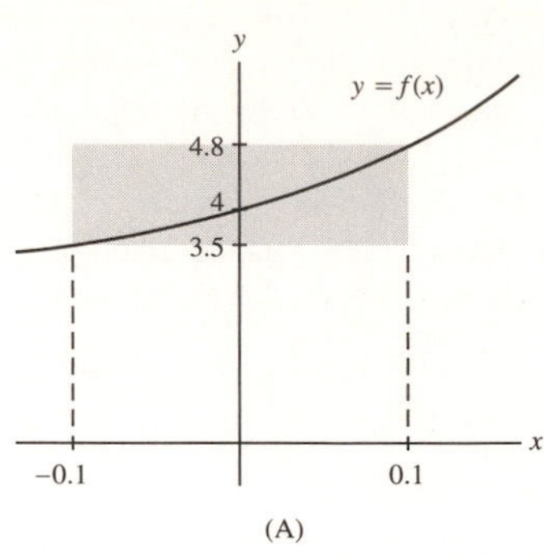

(A) (B)

FIGURE 5

SOLUTION We see $-0.1 < x < 0.1$ forces $3.5 < f(x) < 4.8$. Rewritten, this means that $|x - 0| < 0.1$ implies that $|f(x) - 4| < 0.8$. Replacing numbers where appropriate in the definition of the limit $|x - c| < \delta$ implies $|f(x) - L| < \epsilon$, we get $L = 4$, $\epsilon = 0.8$, $c = 0$, and $\delta = 0.1$.

3. Consider $\lim_{x\to 4} f(x)$, where $f(x) = 8x + 3$.

(a) Show that $|f(x) - 35| = 8|x - 4|$.

(b) Show that for any $\epsilon > 0$, $|f(x) - 35| < \epsilon$ if $|x - 4| < \delta$, where $\delta = \frac{\epsilon}{8}$. Explain how this proves rigorously that $\lim_{x\to 4} f(x) = 35$.

SOLUTION

(a) $|f(x) - 35| = |8x + 3 - 35| = |8x - 32| = |8(x - 4)| = 8\,|x - 4|$. (Remember that the last step is justified because $8 > 0$).

(b) Let $\epsilon > 0$. Let $\delta = \epsilon/8$ and suppose $|x - 4| < \delta$. By part **(a)**, $|f(x) - 35| = 8|x - 4| < 8\delta$. Substituting $\delta = \epsilon/8$, we see $|f(x) - 35| < 8\epsilon/8 = \epsilon$. We see that, for any $\epsilon > 0$, we found an appropriate δ so that $|x - 4| < \delta$ implies $|f(x) - 35| < \epsilon$. Hence $\lim_{x\to 4} f(x) = 35$.

5. Consider $\lim_{x\to 2} x^2 = 4$ (refer to Example 2).

(a) Show that $|x^2 - 4| < 0.05$ if $0 < |x - 2| < 0.01$.

(b) Show that $|x^2 - 4| < 0.0009$ if $0 < |x - 2| < 0.0002$.

(c) Find a value of δ such that $|x^2 - 4|$ is less than 10^{-4} if $0 < |x - 2| < \delta$.

SOLUTION

(a) If $0 < |x - 2| < \delta = 0.01$, then $|x| < 3$ and $\left|x^2 - 4\right| = |x - 2||x + 2| \le |x - 2|\,(|x| + 2) < 5|x - 2| < 0.05$.

(b) If $0 < |x - 2| < \delta = 0.0002$, then $|x| < 2.0002$ and

$$\left|x^2 - 4\right| = |x - 2||x + 2| \le |x - 2|\,(|x| + 2) < 4.0002|x - 2| < 0.00080004 < 0.0009.$$

(c) Note that $\left|x^2 - 4\right| = |(x + 2)(x - 2)| \le |x + 2|\,|x - 2|$. Since $|x - 2|$ can get arbitrarily small, we can require $|x - 2| < 1$ so that $1 < x < 3$. This ensures that $|x + 2|$ is at most 5. Now we know that $\left|x^2 - 4\right| \le 5|x - 2|$. Let $\delta = 10^{-5}$. Then, if $|x - 2| < \delta$, we get $\left|x^2 - 4\right| \le 5|x - 2| < 5 \times 10^{-5} < 10^{-4}$ as desired.

7. Refer to Example 3 to find a value of $\delta > 0$ such that

$$\left|\frac{1}{x} - \frac{1}{3}\right| < 10^{-4} \qquad \text{if} \qquad 0 < |x - 3| < \delta$$

SOLUTION The Example shows that for any $\epsilon > 0$ we have

$$\left|\frac{1}{x} - \frac{1}{3}\right| \le \epsilon \quad \text{if } |x - 3| < \delta$$

where δ is the smaller of the numbers 6ϵ and 1. In our case, we may take $\delta = 6 \times 10^{-4}$.

9. GU Plot $f(x) = \sqrt{2x-1}$ together with the horizontal lines $y = 2.9$ and $y = 3.1$. Use this plot to find a value of $\delta > 0$ such that $|\sqrt{2x-1} - 3| < 0.1$ if $|x - 5| < \delta$.

SOLUTION From the plot below, we see that $\delta = 0.25$ will guarantee that $|\sqrt{2x-1} - 3| < 0.1$ whenever $|x - 5| \le \delta$.

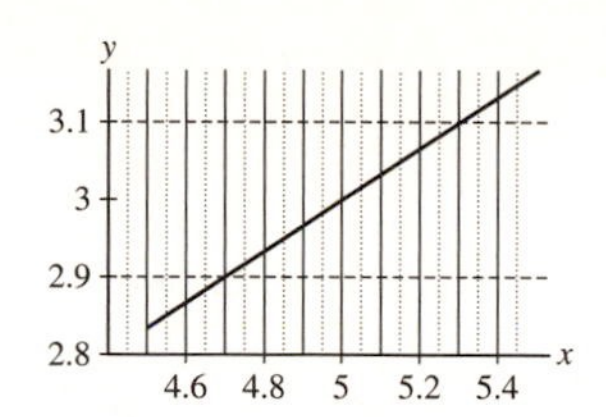

11. GU The number e has the following property: $\lim_{x\to 0} \frac{e^x - 1}{x} = 1$. Use a plot of $f(x) = \frac{e^x - 1}{x}$ to find a value of $\delta > 0$ such that $|f(x) - 1| < 0.01$ if $|x - 1| < \delta$.

SOLUTION From the plot below, we see that $\delta = 0.02$ will guarantee that

$$\left|\frac{e^x - 1}{x} - 1\right| < 0.01$$

whenever $|x| < \delta$.

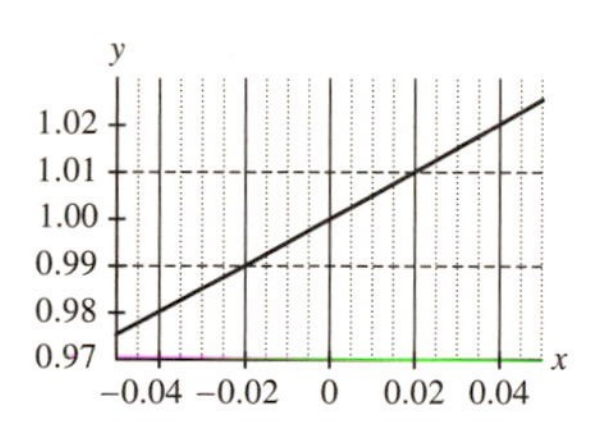

13. Consider $\lim_{x\to 2} \frac{1}{x}$.

(a) Show that if $|x - 2| < 1$, then

$$\left|\frac{1}{x} - \frac{1}{2}\right| < \frac{1}{2}|x - 2|$$

(b) Let δ be the smaller of 1 and 2ϵ. Prove:

$$\left|\frac{1}{x} - \frac{1}{2}\right| < \epsilon \quad \text{if} \quad 0 < |x - 2| < \delta$$

(c) Find a $\delta > 0$ such that $\left|\frac{1}{x} - \frac{1}{2}\right| < 0.01$ if $|x - 2| < \delta$.

(d) Prove rigorously that $\lim_{x\to 2} \frac{1}{x} = \frac{1}{2}$.

SOLUTION

(a) Since $|x - 2| < 1$, it follows that $1 < x < 3$, in particular that $x > 1$. Because $x > 1$, then $\frac{1}{x} < 1$ and

$$\left|\frac{1}{x} - \frac{1}{2}\right| = \left|\frac{2-x}{2x}\right| = \frac{|x-2|}{2x} < \frac{1}{2}|x - 2|.$$

(b) Let $\delta = \min\{1, 2\epsilon\}$ and suppose that $|x - 2| < \delta$. Then by part (a) we have

$$\left|\frac{1}{x} - \frac{1}{2}\right| < \frac{1}{2}|x - 2| < \frac{1}{2}\delta < \frac{1}{2} \cdot 2\epsilon = \epsilon.$$

(c) Choose $\delta = 0.02$. Then $\left|\frac{1}{x} - \frac{1}{2}\right| < \frac{1}{2}\delta = 0.01$ by part (b).

(d) Let $\epsilon > 0$ be given. Then whenever $0 < |x - 2| < \delta = \min\{1, 2\epsilon\}$, we have

$$\left|\frac{1}{x} - \frac{1}{2}\right| < \frac{1}{2}\delta \le \epsilon.$$

Since ϵ was arbitrary, we conclude that $\lim_{x\to 2} \frac{1}{x} = \frac{1}{2}$.

15. Let $f(x) = \sin x$. Using a calculator, we find:

$$f\left(\frac{\pi}{4} - 0.1\right) \approx 0.633, \quad f\left(\frac{\pi}{4}\right) \approx 0.707, \quad f\left(\frac{\pi}{4} + 0.1\right) \approx 0.774$$

Use these values and the fact that $f(x)$ is increasing on $\left[0, \frac{\pi}{2}\right]$ to justify the statement

$$\left|f(x) - f\left(\frac{\pi}{4}\right)\right| < 0.08 \quad \text{if} \quad \left|x - \frac{\pi}{4}\right| < 0.1$$

Then draw a figure like Figure 3 to illustrate this statement.

SOLUTION Since $f(x)$ is increasing on the interval, the three $f(x)$ values tell us that $0.633 \le f(x) \le 0.774$ for all x between $\frac{\pi}{4} - 0.1$ and $\frac{\pi}{4} + 0.1$. We may subtract $f(\frac{\pi}{4})$ from the inequality for $f(x)$. This show that, for $\frac{\pi}{4} - 0.1 < x < \frac{\pi}{4} + 0.1$, $0.633 - f(\frac{\pi}{4}) \le f(x) - f(\frac{\pi}{4}) \le 0.774 - f(\frac{\pi}{4})$. This means that, if $|x - \frac{\pi}{4}| < 0.1$, then $0.633 - 0.707 \le f(x) - f(\frac{\pi}{4}) \le 0.774 - 0.707$, so $-0.074 \le f(x) - f(\frac{\pi}{4}) \le 0.067$. Then $-0.08 < f(x) - f(\frac{\pi}{4}) < 0.08$ follows from this, so $|x - \frac{\pi}{4}| < 0.1$ implies $|f(x) - f(\frac{\pi}{4})| < 0.08$. The figure below illustrates this.

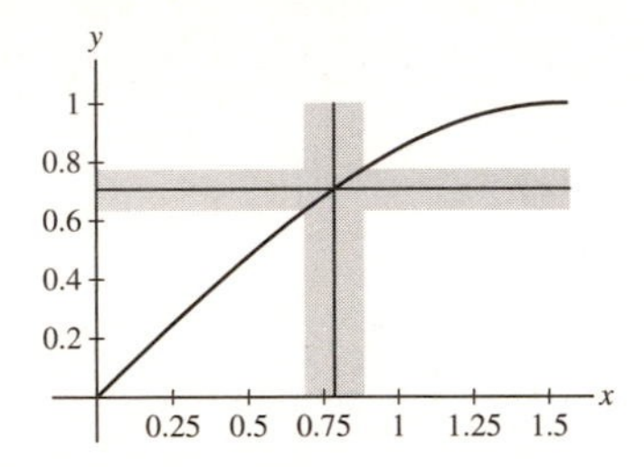

17. Adapt the argument in Example 2 to prove rigorously that $\lim_{x \to c} x^2 = c^2$ for all c.

SOLUTION To relate the gap to $|x - c|$, we take

$$\left|x^2 - c^2\right| = |(x + c)(x - c)| = |x + c|\,|x - c|\,.$$

We choose δ in two steps. First, since we are requiring $|x - c|$ to be small, we require $\delta < |c|$, so that x lies between 0 and $2c$. This means that $|x + c| < 3|c|$, so $|x - c||x + c| < 3|c|\delta$. Next, we require that $\delta < \dfrac{\epsilon}{3|c|}$, so

$$|x - c||x + c| < \frac{\epsilon}{3|c|} 3|c| = \epsilon,$$

and we are done.

Therefore, given $\epsilon > 0$, we let

$$\delta = \min\left\{|c|, \frac{\epsilon}{3|c|}\right\}.$$

Then, for $|x - c| < \delta$, we have

$$|x^2 - c^2| = |x - c|\,|x + c| < 3|c|\delta < 3|c|\frac{\epsilon}{3|c|} = \epsilon.$$

In Exercises 19–24, use the formal definition of the limit to prove the statement rigorously.

19. $\lim_{x \to 4} \sqrt{x} = 2$

SOLUTION Let $\epsilon > 0$ be given. We bound $|\sqrt{x} - 2|$ by multiplying $\dfrac{\sqrt{x} + 2}{\sqrt{x} + 2}$.

$$|\sqrt{x} - 2| = \left|\sqrt{x} - 2\left(\frac{\sqrt{x} + 2}{\sqrt{x} + 2}\right)\right| = \left|\frac{x - 4}{\sqrt{x} + 2}\right| = |x - 4|\left|\frac{1}{\sqrt{x} + 2}\right|.$$

We can assume $\delta < 1$, so that $|x - 4| < 1$, and hence $\sqrt{x} + 2 > \sqrt{3} + 2 > 3$. This gives us

$$|\sqrt{x} - 2| = |x - 4|\left|\frac{1}{\sqrt{x} + 2}\right| < |x - 4|\frac{1}{3}.$$

Let $\delta = \min(1, 3\epsilon)$. If $|x - 4| < \delta$,

$$|\sqrt{x} - 2| = |x - 4|\left|\frac{1}{\sqrt{x} + 2}\right| < |x - 4|\frac{1}{3} < \delta\frac{1}{3} < 3\epsilon\frac{1}{3} = \epsilon,$$

thus proving the limit rigorously.

21. $\lim_{x\to 1} x^3 = 1$

SOLUTION Let $\epsilon > 0$ be given. We bound $\left|x^3 - 1\right|$ by factoring the difference of cubes:

$$\left|x^3 - 1\right| = \left|(x^2 + x + 1)(x - 1)\right| = |x - 1|\left|x^2 + x + 1\right|.$$

Let $\delta = \min(1, \frac{\epsilon}{7})$, and assume $|x - 1| < \delta$. Since $\delta < 1$, $0 < x < 2$. Since $x^2 + x + 1$ increases as x increases for $x > 0$, $x^2 + x + 1 < 7$ for $0 < x < 2$, and so

$$\left|x^3 - 1\right| = |x - 1|\left|x^2 + x + 1\right| < 7|x - 1| < 7\frac{\epsilon}{7} = \epsilon$$

and the limit is rigorously proven.

23. $\lim_{x\to 2} x^{-2} = \frac{1}{4}$

SOLUTION Let $\epsilon > 0$ be given. First, we bound $x^{-2} - \frac{1}{4}$:

$$\left|x^{-2} - \frac{1}{4}\right| = \left|\frac{4 - x^2}{4x^2}\right| = |2 - x|\left|\frac{2 + x}{4x^2}\right|.$$

Let $\delta = \min(1, \frac{4}{5}\epsilon)$, and suppose $|x - 2| < \delta$. Since $\delta < 1$, $|x - 2| < 1$, so $1 < x < 3$. This means that $4x^2 > 4$ and $|2 + x| < 5$, so that $\frac{2 + x}{4x^2} < \frac{5}{4}$. We get:

$$\left|x^{-2} - \frac{1}{4}\right| = |2 - x|\left|\frac{2 + x}{4x^2}\right| < \frac{5}{4}|x - 2| < \frac{5}{4}\cdot\frac{4}{5}\epsilon = \epsilon.$$

and the limit is rigorously proven.

25. Let $f(x) = \frac{x}{|x|}$. Prove rigorously that $\lim_{x\to 0} f(x)$ does not exist. *Hint:* Show that for any L, there always exists some x such that $|x| < \delta$ but $|f(x) - L| \geq \frac{1}{2}$, no matter how small δ is taken.

SOLUTION Let L be any real number. Let $\delta > 0$ be any small positive number. Let $x = \frac{\delta}{2}$, which satisfies $|x| < \delta$, and $f(x) = 1$. We consider two cases:

- $(|f(x) - L| \geq \frac{1}{2})$: we are done.
- $(|f(x) - L| < \frac{1}{2})$: This means $\frac{1}{2} < L < \frac{3}{2}$. In this case, let $x = -\frac{\delta}{2}$. $f(x) = -1$, and so $\frac{3}{2} < L - f(x)$.

In either case, there exists an x such that $|x| < \frac{\delta}{2}$, but $|f(x) - L| \geq \frac{1}{2}$.

27. Let $f(x) = \min(x, x^2)$, where $\min(a, b)$ is the minimum of a and b. Prove rigorously that $\lim_{x\to 1} f(x) = 1$.

SOLUTION Let $\epsilon > 0$ and let $\delta = \min(1, \frac{\epsilon}{2})$. Then, whenever $|x - 1| < \delta$, it follows that $0 < x < 2$. If $1 < x < 2$, then $\min(x, x^2) = x$ and

$$|f(x) - 1| = |x - 1| < \delta < \frac{\epsilon}{2} < \epsilon.$$

On the other hand, if $0 < x < 1$, then $\min(x, x^2) = x^2$, $|x + 1| < 2$ and

$$|f(x) - 1| = |x^2 - 1| = |x - 1|\,|x + 1| < 2\delta < \epsilon.$$

Thus, whenever $|x - 1| < \delta$, $|f(x) - 1| < \epsilon$.

29. First, use the identity

$$\sin x + \sin y = 2\sin\left(\frac{x + y}{2}\right)\cos\left(\frac{x - y}{2}\right)$$

to verify the relation

$$\sin(a + h) - \sin a = h\frac{\sin(h/2)}{h/2}\cos\left(a + \frac{h}{2}\right) \qquad \boxed{6}$$

Then use the inequality $\left|\frac{\sin x}{x}\right| \leq 1$ for $x \neq 0$ to show that $|\sin(a + h) - \sin a| < |h|$ for all a. Finally, prove rigorously that $\lim_{x\to a} \sin x = \sin a$.

SOLUTION We first write

$$\sin(a+h) - \sin a = \sin(a+h) + \sin(-a).$$

Applying the identity with $x = a + h$, $y = -a$, yields:

$$\sin(a+h) - \sin a = \sin(a+h) + \sin(-a) = 2\sin\left(\frac{a+h-a}{2}\right)\cos\left(\frac{2a+h}{2}\right)$$

$$= 2\sin\left(\frac{h}{2}\right)\cos\left(a+\frac{h}{2}\right) = 2\left(\frac{h}{h}\right)\sin\left(\frac{h}{2}\right)\cos\left(a+\frac{h}{2}\right) = h\frac{\sin(h/2)}{h/2}\cos\left(a+\frac{h}{2}\right).$$

Therefore,

$$|\sin(a+h) - \sin a| = |h|\left|\frac{\sin(h/2)}{h/2}\right|\left|\cos\left(a+\frac{h}{2}\right)\right|.$$

Using the fact that $\left|\frac{\sin\theta}{\theta}\right| < 1$ and that $|\cos\theta| \le 1$, and making the substitution $h = x - a$, we see that this last relation is equivalent to

$$|\sin x - \sin a| < |x - a|.$$

Now, to prove the desired limit, let $\epsilon > 0$, and take $\delta = \epsilon$. If $|x - a| < \delta$, then

$$|\sin x - \sin a| < |x - a| < \delta = \epsilon,$$

Therefore, a δ was found for arbitrary ϵ, and the proof is complete.

Further Insights and Challenges

In Exercises 31–33, prove the statement using the formal limit definition.

31. The Constant Multiple Law [Theorem 1, part (ii) in Section 2.3, p. 77]

SOLUTION Suppose that $\lim_{x\to c} f(x) = L$. We wish to prove that $\lim_{x\to c} af(x) = aL$.

Let $\epsilon > 0$ be given. $\epsilon/|a|$ is also a positive number. Since $\lim_{x\to c} f(x) = L$, we know there is a $\delta > 0$ such that $|x - c| < \delta$ forces $|f(x) - L| < \epsilon/|a|$. Suppose $|x - c| < \delta$. $|af(x) - aL| = |a||f(x) - aL| < |a|(\epsilon/|a|) = \epsilon$, so the rule is proven.

33. The Product Law [Theorem 1, part (iii) in Section 2.3, p. 77]. *Hint:* Use the identity

$$f(x)g(x) - LM = (f(x) - L)\,g(x) + L(g(x) - M)$$

SOLUTION Before we can prove the Product Law, we need to establish one preliminary result. We are given that $\lim_{x\to c} g(x) = M$. Consequently, if we set $\epsilon = 1$, then the definition of a limit guarantees the existence of a $\delta_1 > 0$ such that whenever $0 < |x - c| < \delta_1$, $|g(x) - M| < 1$. Applying the inequality $|g(x)| - |M| \le |g(x) - M|$, it follows that $|g(x)| < 1 + |M|$. In other words, because $\lim_{x\to c} g(x) = M$, there exists a $\delta_1 > 0$ such that $|g(x)| < 1 + |M|$ whenever $0 < |x - c| < \delta_1$.

We can now prove the Product Law. Let $\epsilon > 0$. As proven above, because $\lim_{x\to c} g(x) = M$, there exists a $\delta_1 > 0$ such that $|g(x)| < 1 + |M|$ whenever $0 < |x - c| < \delta_1$. Furthermore, by the definition of a limit, $\lim_{x\to c} g(x) = M$ implies there exists a $\delta_2 > 0$ such that $|g(x) - M| < \frac{\epsilon}{2(1+|L|)}$ whenever $0 < |x - c| < \delta_2$. We have included the "1+" in the denominator to avoid division by zero in case $L = 0$. The reason for including the factor of 2 in the denominator will become clear shortly. Finally, because $\lim_{x\to c} f(x) = L$, there exists a $\delta_3 > 0$ such that $|f(x) - L| < \frac{\epsilon}{2(1+|M|)}$ whenever $0 < |x - c| < \delta_3$. Now, let $\delta = \min(\delta_1, \delta_2, \delta_3)$. Then, for all x satisfying $0 < |x - c| < \delta$, we have

$$\begin{aligned} |f(x)g(x) - LM| &= |(f(x) - L)g(x) + L(g(x) - M)| \\ &\le |f(x) - L|\,|g(x)| + |L|\,|g(x) - M| \\ &< \frac{\epsilon}{2(1+|M|)}(1+|M|) + |L|\frac{\epsilon}{2(1+|L|)} \\ &< \frac{\epsilon}{2} + \frac{\epsilon}{2} = \epsilon. \end{aligned}$$

Hence,

$$\lim_{x\to c} f(x)g(x) = LM = \lim_{x\to c} f(x) \cdot \lim_{x\to c} g(x).$$

35. Here is a function with strange continuity properties:

$$f(x) = \begin{cases} \dfrac{1}{q} & \text{if } x \text{ is the rational number } p/q \text{ in lowest terms} \\ 0 & \text{if } x \text{ is an irrational number} \end{cases}$$

(a) Show that $f(x)$ is discontinuous at c if c is rational. *Hint:* There exist irrational numbers arbitrarily close to c.

(b) Show that $f(x)$ is continuous at c if c is irrational. *Hint:* Let I be the interval $\{x : |x - c| < 1\}$. Show that for any $Q > 0$, I contains at most finitely many fractions p/q with $q < Q$. Conclude that there is a δ such that all fractions in $\{x : |x - c| < \delta\}$ have a denominator larger than Q.

SOLUTION

(a) Let c be any rational number and suppose that, in lowest terms, $c = p/q$, where p and q are integers. To prove the discontinuity of f at c, we must show there is an $\epsilon > 0$ such that for any $\delta > 0$ there is an x for which $|x - c| < \delta$, but that $|f(x) - f(c)| > \epsilon$. Let $\epsilon = \frac{1}{2q}$ and $\delta > 0$. Since there is at least one irrational number between any two distinct real numbers, there is some irrational x between c and $c + \delta$. Hence, $|x - c| < \delta$, but $|f(x) - f(c)| = |0 - \frac{1}{q}| = \frac{1}{q} > \frac{1}{2q} = \epsilon$.

(b) Let c be irrational, let $\epsilon > 0$ be given, and let $N > 0$ be a prime integer sufficiently large so that $\frac{1}{N} < \epsilon$. Let $\frac{p_1}{q_1}, \dots, \frac{p_m}{q_m}$ be all rational numbers $\frac{p}{q}$ in lowest terms such that $|\frac{p}{q} - c| < 1$ and $q < N$. Since N is finite, this is a finite list; hence, one number $\frac{p_i}{q_i}$ in the list must be closest to c. Let $\delta = \frac{1}{2}|\frac{p_i}{q_i} - c|$. By construction, $|\frac{p_i}{q_i} - c| > \delta$ for all $i = 1 \dots m$. Therefore, for any rational number $\frac{p}{q}$ such that $|\frac{p}{q} - c| < \delta$, $q > N$, so $\frac{1}{q} < \frac{1}{N} < \epsilon$.

Therefore, for any *rational* number x such that $|x - c| < \delta$, $|f(x) - f(c)| < \epsilon$. $|f(x) - f(c)| = 0$ for any irrational number x, so $|x - c| < \delta$ implies that $|f(x) - f(c)| < \epsilon$ for any number x.

CHAPTER REVIEW EXERCISES

1. The position of a particle at time t (s) is $s(t) = \sqrt{t^2 + 1}$ m. Compute its average velocity over $[2, 5]$ and estimate its instantaneous velocity at $t = 2$.

SOLUTION Let $s(t) = \sqrt{t^2 + 1}$. The average velocity over $[2, 5]$ is

$$\frac{s(5) - s(2)}{5 - 2} = \frac{\sqrt{26} - \sqrt{5}}{3} \approx 0.954 \text{ m/s}.$$

From the data in the table below, we estimate that the instantaneous velocity at $t = 2$ is approximately 0.894 m/s.

interval	[1.9, 2]	[1.99, 2]	[1.999, 2]	[2, 2.001]	[2, 2.01]	[2, 2.1]
average ROC	0.889769	0.893978	0.894382	0.894472	0.894873	0.898727

3. For a whole number n, let $P(n)$ be the number of *partitions* of n, that is, the number of ways of writing n as a sum of one or more whole numbers. For example, $P(4) = 5$ since the number 4 can be partitioned in five different ways: 4, $3 + 1$, $2 + 2$, $2 + 1 + 1$, and $1 + 1 + 1 + 1$. Treating $P(n)$ as a continuous function, use Figure 1 to estimate the rate of change of $P(n)$ at $n = 12$.

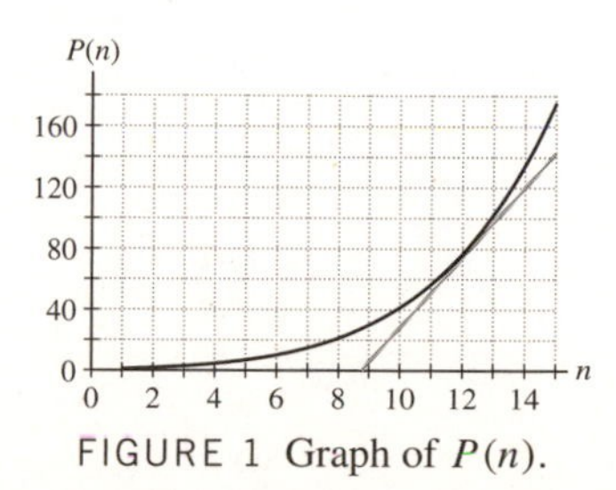

FIGURE 1 Graph of $P(n)$.

SOLUTION The tangent line drawn in the figure appears to pass through the points $(15, 140)$ and $(10.5, 40)$. We therefore estimate that the rate of change of $P(n)$ at $n = 12$ is

$$\frac{140 - 40}{15 - 10.5} = \frac{100}{4.5} = \frac{200}{9}.$$

In Exercises 5–10, estimate the limit numerically to two decimal places or state that the limit does not exist.

5. $\lim_{x\to 0} \dfrac{1-\cos^3(x)}{x^2}$

SOLUTION Let $f(x) = \frac{1-\cos^3 x}{x^2}$. The data in the table below suggests that

$$\lim_{x\to 0} \frac{1-\cos^3 x}{x^2} \approx 1.50.$$

In constructing the table, we take advantage of the fact that f is an even function.

x	± 0.001	± 0.01	± 0.1
$f(x)$	1.500000	1.499912	1.491275

(The exact value is $\frac{3}{2}$.)

7. $\lim_{x\to 2} \dfrac{x^x-4}{x^2-4}$

SOLUTION Let $f(x) = \frac{x^x-4}{x^2-4}$. The data in the table below suggests that

$$\lim_{x\to 2} \frac{x^x-4}{x^2-4} \approx 1.69.$$

x	1.9	1.99	1.999	2.001	2.01	2.1
$f(x)$	1.575461	1.680633	1.691888	1.694408	1.705836	1.828386

(The exact value is $1 + \ln 2$.)

9. $\lim_{x\to 1} \left(\dfrac{7}{1-x^7} - \dfrac{3}{1-x^3}\right)$

SOLUTION Let $f(x) = \left(\frac{7}{1-x^7} - \frac{3}{1-x^3}\right)$. The data in the table below suggests that

$$\lim_{x\to 1} \left(\frac{7}{1-x^7} - \frac{3}{1-x^3}\right) \approx 2.00.$$

x	0.9	0.99	0.999	1.001	1.01	1.1
$f(x)$	2.347483	2.033498	2.003335	1.996668	1.966835	1.685059

(The exact value is 2.)

In Exercises 11–50, evaluate the limit if it exists. If not, determine whether the one-sided limits exist (finite or infinite).

11. $\lim_{x\to 4} (3 + x^{1/2})$

SOLUTION $\lim_{x\to 4} (3 + x^{1/2}) = 3 + \sqrt{4} = 5.$

13. $\lim_{x\to -2} \dfrac{4}{x^3}$

SOLUTION $\lim_{x\to -2} \dfrac{4}{x^3} = \dfrac{4}{(-2)^3} = -\dfrac{1}{2}.$

15. $\lim_{t\to 9} \dfrac{\sqrt{t}-3}{t-9}$

SOLUTION $\lim_{t\to 9} \dfrac{\sqrt{t}-3}{t-9} = \lim_{t\to 9} \dfrac{\sqrt{t}-3}{(\sqrt{t}-3)(\sqrt{t}+3)} = \lim_{t\to 9} \dfrac{1}{\sqrt{t}+3} = \dfrac{1}{\sqrt{9}+3} = \dfrac{1}{6}.$

17. $\lim_{x\to 1} \dfrac{x^3-x}{x-1}$

SOLUTION $\lim_{x\to 1} \dfrac{x^3-x}{x-1} = \lim_{x\to 1} \dfrac{x(x-1)(x+1)}{x-1} = \lim_{x\to 1} x(x+1) = 1(1+1) = 2.$

19. $\lim_{t\to 9} \frac{t-6}{\sqrt{t}-3}$

SOLUTION Because the one-sided limits

$$\lim_{t\to 9-} \frac{t-6}{\sqrt{t}-3} = -\infty \qquad \text{and} \qquad \lim_{t\to 9+} \frac{t-6}{\sqrt{t}-3} = \infty,$$

are not equal, the two-sided limit

$$\lim_{t\to 9} \frac{t-6}{\sqrt{t}-3} \qquad \text{does not exist.}$$

21. $\lim_{x\to -1+} \frac{1}{x+1}$

SOLUTION For $x > -1$, $x+1 > 0$. Therefore,

$$\lim_{x\to -1+} \frac{1}{x+1} = \infty.$$

23. $\lim_{x\to 1} \frac{x^3-2x}{x-1}$

SOLUTION Because the one-sided limits

$$\lim_{x\to 1-} \frac{x^3-2x}{x-1} = \infty \qquad \text{and} \qquad \lim_{x\to 1+} \frac{x^3-2x}{x-1} = -\infty,$$

are not equal, the two-sided limit

$$\lim_{x\to 1} \frac{x^3-2x}{x-1} \qquad \text{does not exist.}$$

25. $\lim_{x\to 0} \frac{e^{3x}-e^x}{e^x-1}$

SOLUTION

$$\lim_{x\to 0} \frac{e^{3x}-e^x}{e^x-1} = \lim_{x\to 0} \frac{e^x(e^x-1)(e^x+1)}{e^x-1} = \lim_{x\to 0} e^x(e^x+1) = 1\cdot 2 = 2.$$

27. $\lim_{x\to 1.5} \frac{[x]}{x}$

SOLUTION $\lim_{x\to 1.5} \frac{[x]}{x} = \frac{[1.5]}{1.5} = \frac{1}{1.5} = \frac{2}{3}.$

29. $\lim_{z\to -3} \frac{z+3}{z^2+4z+3}$

SOLUTION

$$\lim_{z\to -3} \frac{z+3}{z^2+4z+3} = \lim_{z\to -3} \frac{z+3}{(z+3)(z+1)} = \lim_{z\to -3} \frac{1}{z+1} = -\frac{1}{2}.$$

31. $\lim_{x\to b} \frac{x^3-b^3}{x-b}$

SOLUTION $\lim_{x\to b} \frac{x^3-b^3}{x-b} = \lim_{x\to b} \frac{(x-b)(x^2+xb+b^2)}{x-b} = \lim_{x\to b} (x^2+xb+b^2) = b^2+b(b)+b^2 = 3b^2.$

33. $\lim_{x\to 0} \left(\frac{1}{3x} - \frac{1}{x(x+3)}\right)$

SOLUTION $\lim_{x\to 0} \left(\frac{1}{3x} - \frac{1}{x(x+3)}\right) = \lim_{x\to 0} \frac{(x+3)-3}{3x(x+3)} = \lim_{x\to 0} \frac{1}{3(x+3)} = \frac{1}{3(0+3)} = \frac{1}{9}.$

35. $\lim_{x\to 0-} \frac{[x]}{x}$

SOLUTION For x sufficiently close to zero but negative, $[x] = -1$. Therefore,

$$\lim_{x\to 0-} \frac{[x]}{x} = \lim_{x\to 0-} \frac{-1}{x} = \infty.$$

37. $\lim_{\theta\to\frac{\pi}{2}} \theta\sec\theta$

SOLUTION Because the one-sided limits

$$\lim_{\theta\to\frac{\pi}{2}-} \theta\sec\theta = \infty \qquad \text{and} \qquad \lim_{\theta\to\frac{\pi}{2}+} \theta\sec\theta = -\infty$$

are not equal, the two-sided limit

$$\lim_{\theta\to\frac{\pi}{2}} \theta\sec\theta \qquad \text{does not exist.}$$

39. $\lim_{\theta\to 0} \dfrac{\cos\theta - 2}{\theta}$

SOLUTION Because the one-sided limits

$$\lim_{\theta\to 0-} \frac{\cos\theta - 2}{\theta} = \infty \qquad \text{and} \qquad \lim_{\theta\to 0+} \frac{\cos\theta - 2}{\theta} = -\infty$$

are not equal, the two-sided limit

$$\lim_{\theta\to 0} \frac{\cos\theta - 2}{\theta} \qquad \text{does not exist.}$$

41. $\lim_{x\to 2-} \dfrac{x-3}{x-2}$

SOLUTION For x close to 2 but less than 2, $x - 3 < 0$ and $x - 2 < 0$. Therefore,

$$\lim_{x\to 2-} \frac{x-3}{x-2} = \infty.$$

43. $\lim_{x\to 1+} \left(\dfrac{1}{\sqrt{x-1}} - \dfrac{1}{\sqrt{x^2-1}} \right)$

SOLUTION $\lim_{x\to 1+} \left(\dfrac{1}{\sqrt{x-1}} - \dfrac{1}{\sqrt{x^2-1}} \right) = \lim_{x\to 1+} \dfrac{\sqrt{x+1} - 1}{\sqrt{x^2-1}} = \infty.$

45. $\lim_{x\to\frac{\pi}{2}} \tan x$

SOLUTION Because the one-sided limits

$$\lim_{x\to\frac{\pi}{2}-} \tan x = \infty \qquad \text{and} \qquad \lim_{x\to\frac{\pi}{2}+} \tan x = -\infty$$

are not equal, the two-sided limit

$$\lim_{x\to\frac{\pi}{2}} \tan x \qquad \text{does not exist.}$$

47. $\lim_{t\to 0+} \sqrt{t}\cos\dfrac{1}{t}$

SOLUTION For $t > 0$,

$$-1 \le \cos\left(\frac{1}{t}\right) \le 1,$$

so

$$-\sqrt{t} \le \sqrt{t}\cos\left(\frac{1}{t}\right) \le \sqrt{t}.$$

Because

$$\lim_{t\to 0+} -\sqrt{t} = \lim_{t\to 0+} \sqrt{t} = 0,$$

it follows from the Squeeze Theorem that

$$\lim_{t\to 0+} \sqrt{t}\cos\left(\frac{1}{t}\right) = 0.$$

49. $\lim_{x\to 0} \dfrac{\cos x - 1}{\sin x}$

SOLUTION

$$\lim_{x\to 0} \frac{\cos x - 1}{\sin x} = \lim_{x\to 0} \frac{\cos x - 1}{\sin x} \cdot \frac{\cos x + 1}{\cos x + 1} = \lim_{x\to 0} \frac{-\sin^2 x}{\sin x(\cos x + 1)} = -\lim_{x\to 0} \frac{\sin x}{\cos x + 1} = -\frac{0}{1+1} = 0.$$

51. Find the left- and right-hand limits of the function $f(x)$ in Figure 2 at $x = 0, 2, 4$. State whether $f(x)$ is left- or right-continuous (or both) at these points.

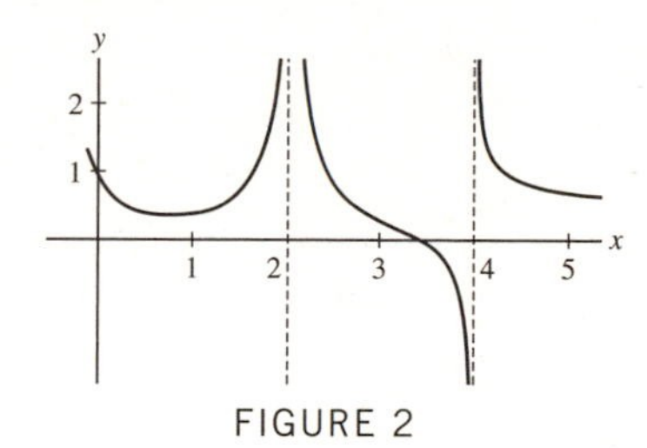

FIGURE 2

SOLUTION According to the graph of $f(x)$,

$$\lim_{x\to 0-} f(x) = \lim_{x\to 0+} f(x) = 1$$

$$\lim_{x\to 2-} f(x) = \lim_{x\to 2+} f(x) = \infty$$

$$\lim_{x\to 4-} f(x) = -\infty$$

$$\lim_{x\to 4+} f(x) = \infty.$$

The function is both left- and right-continuous at $x = 0$ and neither left- nor right-continuous at $x = 2$ and $x = 4$.

53. Graph $h(x)$ and describe the discontinuity:

$$h(x) = \begin{cases} e^x & \text{for } x \le 0 \\ \ln x & \text{for } x > 0 \end{cases}$$

Is $h(x)$ left- or right-continuous?

SOLUTION The graph of $h(x)$ is shown below. At $x = 0$, the function has an infinite discontinuity but is left-continuous.

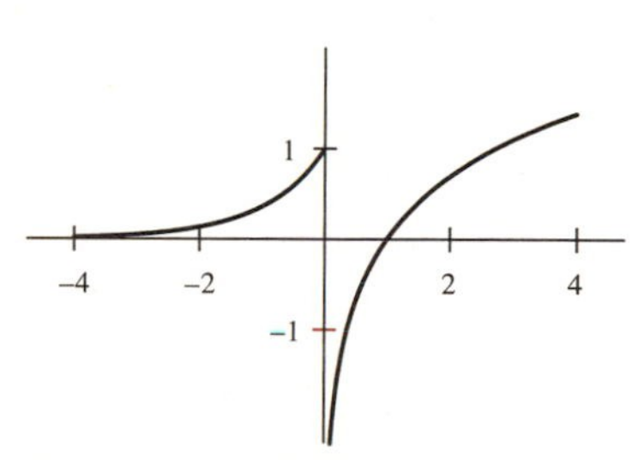

55. Find the points of discontinuity of

$$g(x) = \begin{cases} \cos\left(\dfrac{\pi x}{2}\right) & \text{for } |x| < 1 \\ |x - 1| & \text{for } |x| \ge 1 \end{cases}$$

Determine the type of discontinuity and whether $g(x)$ is left- or right-continuous.

SOLUTION First note that $\cos\left(\frac{\pi x}{2}\right)$ is continuous for $-1 < x < 1$ and that $|x - 1|$ is continuous for $x \le -1$ and for $x \ge 1$. Thus, the only points at which $g(x)$ might be discontinuous are $x = \pm 1$. At $x = 1$, we have

$$\lim_{x\to 1-} g(x) = \lim_{x\to 1-} \cos\left(\frac{\pi x}{2}\right) = \cos\left(\frac{\pi}{2}\right) = 0$$

and

$$\lim_{x\to 1+} g(x) = \lim_{x\to 1+} |x - 1| = |1 - 1| = 0,$$

so $g(x)$ is continuous at $x = 1$. On the other hand, at $x = -1$,

$$\lim_{x\to -1+} g(x) = \lim_{x\to -1+} \cos\left(\frac{\pi x}{2}\right) = \cos\left(-\frac{\pi}{2}\right) = 0$$

and

$$\lim_{x\to -1-} g(x) = \lim_{x\to -1-} |x-1| = |-1-1| = 2,$$

so $g(x)$ has a jump discontinuity at $x = -1$. Since $g(-1) = 2$, $g(x)$ is left-continuous at $x = -1$.

57. Find a constant b such that $h(x)$ is continuous at $x = 2$, where

$$h(x) = \begin{cases} x+1 & \text{for } |x| < 2 \\ b - x^2 & \text{for } |x| \ge 2 \end{cases}$$

With this choice of b, find all points of discontinuity.

SOLUTION To make $h(x)$ continuous at $x = 2$, we must have the two one-sided limits as x approaches 2 be equal. With

$$\lim_{x\to 2-} h(x) = \lim_{x\to 2-} (x+1) = 2+1 = 3$$

and

$$\lim_{x\to 2+} h(x) = \lim_{x\to 2+} (b - x^2) = b - 4,$$

it follows that we must choose $b = 7$. Because $x + 1$ is continuous for $-2 < x < 2$ and $7 - x^2$ is continuous for $x \le -2$ and for $x \ge 2$, the only possible point of discontinuity is $x = -2$. At $x = -2$,

$$\lim_{x\to -2+} h(x) = \lim_{x\to -2+} (x+1) = -2+1 = -1$$

and

$$\lim_{x\to -2-} h(x) = \lim_{x\to -2-} (7 - x^2) = 7 - (-2)^2 = 3,$$

so $h(x)$ has a jump discontinuity at $x = -2$.

In Exercises 58–63, find the horizontal asymptotes of the function by computing the limits at infinity.

59. $f(x) = \dfrac{x^2 - 3x^4}{x - 1}$

SOLUTION Because

$$\lim_{x\to\infty} \frac{x^2 - 3x^4}{x-1} = \lim_{x\to\infty} \frac{1/x^2 - 3}{1/x^3 - 1/x^4} = -\infty$$

and

$$\lim_{x\to -\infty} \frac{x^2 - 3x^4}{x-1} = \lim_{x\to -\infty} \frac{1/x^2 - 3}{1/x^3 - 1/x^4} = \infty,$$

it follows that the graph of $y = \dfrac{x^2 - 3x^4}{x - 1}$ does not have any horizontal asymptotes.

61. $f(u) = \dfrac{2u^2 - 1}{\sqrt{6 + u^4}}$

SOLUTION Because

$$\lim_{u\to\infty} \frac{2u^2 - 1}{\sqrt{6+u^4}} = \lim_{u\to\infty} \frac{2 - 1/u^2}{\sqrt{6/u^4 + 1}} = \frac{2}{\sqrt{1}} = 2$$

and

$$\lim_{u\to -\infty} \frac{2u^2 - 1}{\sqrt{6+u^4}} = \lim_{u\to -\infty} \frac{2 - 1/u^2}{\sqrt{6/u^4 + 1}} = \frac{2}{\sqrt{1}} = 2,$$

it follows that the graph of $y = \dfrac{2u^2 - 1}{\sqrt{6 + u^4}}$ has a horizontal asymptote of $y = 2$.

63. $f(t) = \dfrac{t^{1/3} - t^{-1/3}}{(t - t^{-1})^{1/3}}$

SOLUTION Because

$$\lim_{t\to\infty} \frac{t^{1/3} - t^{-1/3}}{(t - t^{-1})^{1/3}} = \lim_{t\to\infty} \frac{1 - t^{-2/3}}{(1 - t^{-2})^{1/3}} = \frac{1}{1^{1/3}} = 1$$

and

$$\lim_{t\to-\infty} \frac{t^{1/3} - t^{-1/3}}{(t - t^{-1})^{1/3}} = \lim_{t\to-\infty} \frac{1 - t^{-2/3}}{(1 - t^{-2})^{1/3}} = \frac{1}{1^{1/3}} = 1,$$

it follows that the graph of $y = \dfrac{t^{1/3} - t^{-1/3}}{(t - t^{-1})^{1/3}}$ has a horizontal asymptote of $y = 1$.

65. Assume that the following limits exist:

$$A = \lim_{x\to a} f(x), \qquad B = \lim_{x\to a} g(x), \qquad L = \lim_{x\to a} \frac{f(x)}{g(x)}$$

Prove that if $L = 1$, then $A = B$. *Hint:* You cannot use the Quotient Law if $B = 0$, so apply the Product Law to L and B instead.

SOLUTION Suppose the limits A, B, and L all exist and $L = 1$. Then

$$B = B \cdot 1 = B \cdot L = \lim_{x\to a} g(x) \cdot \lim_{x\to a} \frac{f(x)}{g(x)} = \lim_{x\to a} g(x)\frac{f(x)}{g(x)} = \lim_{x\to a} f(x) = A.$$

67. In the notation of Exercise 65, give an example where L exists but neither A nor B exists.

SOLUTION Suppose

$$f(x) = \frac{1}{(x-a)^3} \qquad \text{and} \qquad g(x) = \frac{1}{(x-a)^5}.$$

Then, neither A nor B exists, but

$$L = \lim_{x\to a} \frac{(x-a)^{-3}}{(x-a)^{-5}} = \lim_{x\to a} (x-a)^2 = 0.$$

69. Let $f(x) = x\left[\frac{1}{x}\right]$, where $[x]$ is the greatest integer function. Show that for $x \neq 0$,

$$\frac{1}{x} - 1 < \left[\frac{1}{x}\right] \leq \frac{1}{x}$$

Then use the Squeeze Theorem to prove that

$$\lim_{x\to 0} x\left[\frac{1}{x}\right] = 1$$

Hint: Treat the one-sided limits separately.

SOLUTION Let y be any real number. From the definition of the greatest integer function, it follows that $y - 1 < [y] \leq y$, with equality holding if and only if y is an integer. If $x \neq 0$, then $\frac{1}{x}$ is a real number, so

$$\frac{1}{x} - 1 < \left[\frac{1}{x}\right] \leq \frac{1}{x}.$$

Upon multiplying this inequality through by x, we find

$$1 - x < x\left[\frac{1}{x}\right] \leq 1.$$

Because

$$\lim_{x\to 0} (1 - x) = \lim_{x\to 0} 1 = 1,$$

it follows from the Squeeze Theorem that

$$\lim_{x\to 0} x\left[\frac{1}{x}\right] = 1.$$

71. Use the IVT to prove that the curves $y = x^2$ and $y = \cos x$ intersect.

SOLUTION Let $f(x) = x^2 - \cos x$. Note that any root of $f(x)$ corresponds to a point of intersection between the curves $y = x^2$ and $y = \cos x$. Now, $f(x)$ is continuous over the interval $[0, \frac{\pi}{2}]$, $f(0) = -1 < 0$ and $f(\frac{\pi}{2}) = \frac{\pi^2}{4} > 0$. Therefore, by the Intermediate Value Theorem, there exists a $c \in (0, \frac{\pi}{2})$ such that $f(c) = 0$; consequently, the curves $y = x^2$ and $y = \cos x$ intersect.

73. Use the IVT to show that $e^{-x^2} = x$ has a solution on $(0, 1)$.

SOLUTION Let $f(x) = e^{-x^2} - x$. Observe that f is continuous on $[0, 1]$ with $f(0) = e^0 - 0 = 1 > 0$ and $f(1) = e^{-1} - 1 < 0$. Therefore, the IVT guarantees there exists a $c \in (0, 1)$ such that $f(c) = e^{-c^2} - c = 0$.

75. Give an example of a (discontinuous) function that does not satisfy the conclusion of the IVT on $[-1, 1]$. Then show that the function

$$f(x) = \begin{cases} \sin \dfrac{1}{x} & x \neq 0 \\ 0 & x = 0 \end{cases}$$

satisfies the conclusion of the IVT on every interval $[-a, a]$, even though f is discontinuous at $x = 0$.

SOLUTION Let $g(x) = [x]$. This function is discontinuous on $[-1, 1]$ with $g(-1) = -1$ and $g(1) = 1$. For all $c \neq 0$, there is no x such that $g(x) = c$; thus, $g(x)$ does not satisfy the conclusion of the Intermediate Value Theorem on $[-1, 1]$.

Now, let

$$f(x) = \begin{cases} \sin\left(\frac{1}{x}\right) & \text{for } x \neq 0 \\ 0 & \text{for } x = 0 \end{cases}$$

and let $a > 0$. On the interval

$$x \in \left[\frac{a}{2 + 2\pi a}, \frac{a}{2}\right] \subset [-a, a],$$

$\frac{1}{x}$ runs from $\frac{2}{a}$ to $\frac{2}{a} + 2\pi$, so the sine function covers one full period and clearly takes on every value from $-\sin a$ through $\sin a$.

77. GU Plot the function $f(x) = x^{1/3}$. Use the zoom feature to find a $\delta > 0$ such that $|x^{1/3} - 2| < 0.05$ for $|x - 8| < \delta$.

SOLUTION The graphs of $y = f(x) = x^{1/3}$ and the horizontal lines $y = 1.95$ and $y = 2.05$ are shown below. From this plot, we see that $\delta = 0.55$ guarantees that $|x^{1/3} - 2| < 0.05$ whenever $|x - 8| < \delta$.

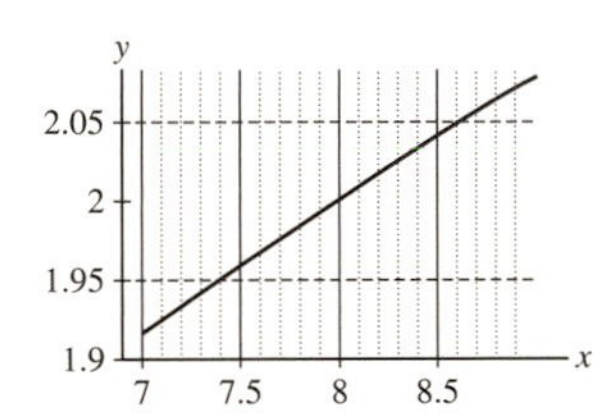

79. Prove rigorously that $\lim_{x \to -1} (4 + 8x) = -4$.

SOLUTION Let $\epsilon > 0$ and take $\delta = \epsilon/8$. Then, whenever $|x - (-1)| = |x + 1| < \delta$,

$$|f(x) - (-4)| = |4 + 8x + 4| = 8|x + 1| < 8\delta = \epsilon.$$

3 DIFFERENTIATION

3.1 Definition of the Derivative

Preliminary Questions

1. Which of the lines in Figure 10 are tangent to the curve?

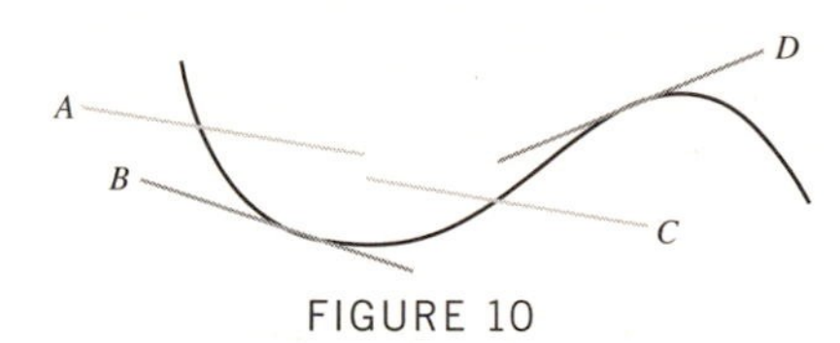

FIGURE 10

SOLUTION Lines B and D are tangent to the curve.

2. What are the two ways of writing the difference quotient?

SOLUTION The difference quotient may be written either as

$$\frac{f(x) - f(a)}{x - a}$$

or as

$$\frac{f(a+h) - f(a)}{h}.$$

3. Find a and h such that $\dfrac{f(a+h) - f(a)}{h}$ is equal to the slope of the secant line between $(3, f(3))$ and $(5, f(5))$.

SOLUTION With $a = 3$ and $h = 2$, $\dfrac{f(a+h) - f(a)}{h}$ is equal to the slope of the secant line between the points $(3, f(3))$ and $(5, f(5))$ on the graph of $f(x)$.

4. Which derivative is approximated by $\dfrac{\tan\left(\frac{\pi}{4} + 0.0001\right) - 1}{0.0001}$?

SOLUTION $\dfrac{\tan\left(\frac{\pi}{4} + 0.0001\right) - 1}{0.0001}$ is a good approximation to the derivative of the function $f(x) = \tan x$ at $x = \frac{\pi}{4}$.

5. What do the following quantities represent in terms of the graph of $f(x) = \sin x$?

(a) $\sin 1.3 - \sin 0.9$ **(b)** $\dfrac{\sin 1.3 - \sin 0.9}{0.4}$ **(c)** $f'(0.9)$

SOLUTION Consider the graph of $y = \sin x$.

(a) The quantity $\sin 1.3 - \sin 0.9$ represents the difference in height between the points $(0.9, \sin 0.9)$ and $(1.3, \sin 1.3)$.

(b) The quantity $\dfrac{\sin 1.3 - \sin 0.9}{0.4}$ represents the slope of the secant line between the points $(0.9, \sin 0.9)$ and $(1.3, \sin 1.3)$ on the graph.

(c) The quantity $f'(0.9)$ represents the slope of the tangent line to the graph at $x = 0.9$.

Exercises

1. Let $f(x) = 5x^2$. Show that $f(3+h) = 5h^2 + 30h + 45$. Then show that

$$\frac{f(3+h) - f(3)}{h} = 5h + 30$$

and compute $f'(3)$ by taking the limit as $h \to 0$.

SOLUTION With $f(x) = 5x^2$, it follows that

$$f(3+h) = 5(3+h)^2 = 5(9+6h+h^2) = 45+30h+5h^2.$$

Using this result, we find

$$\frac{f(3+h)-f(3)}{h} = \frac{45+30h+5h^2-5\cdot 9}{h} = \frac{45+30h+5h^2-45}{h} = \frac{30h+5h^2}{h} = 30+5h.$$

As $h \to 0$, $30+5h \to 30$, so $f'(3) = 30$.

In Exercises 3–6, compute $f'(a)$ in two ways, using Eq. (1) and Eq. (2).

3. $f(x) = x^2 + 9x, \quad a = 0$

SOLUTION Let $f(x) = x^2 + 9x$. Then

$$f'(0) = \lim_{h\to 0} \frac{f(0+h)-f(0)}{h} = \lim_{h\to 0} \frac{(0+h)^2+9(0+h)-0}{h} = \lim_{h\to 0} \frac{9h+h^2}{h} = \lim_{h\to 0}(9+h) = 9.$$

Alternately,

$$f'(0) = \lim_{x\to 0} \frac{f(x)-f(0)}{x-0} = \lim_{x\to 0} \frac{x^2+9x-0}{x} = \lim_{x\to 0}(x+9) = 9.$$

5. $f(x) = 3x^2 + 4x + 2, \quad a = -1$

SOLUTION Let $f(x) = 3x^2 + 4x + 2$. Then

$$f'(-1) = \lim_{h\to 0} \frac{f(-1+h)-f(-1)}{h} = \lim_{h\to 0} \frac{3(-1+h)^2+4(-1+h)+2-1}{h}$$

$$= \lim_{h\to 0} \frac{3h^2-2h}{h} = \lim_{h\to 0}(3h-2) = -2.$$

Alternately,

$$f'(-1) = \lim_{x\to -1} \frac{f(x)-f(-1)}{x-(-1)} = \lim_{x\to -1} \frac{3x^2+4x+2-1}{x+1}$$

$$= \lim_{x\to -1} \frac{(3x+1)(x+1)}{x+1} = \lim_{x\to -1}(3x+1) = -2.$$

In Exercises 7–10, refer to Figure 11.

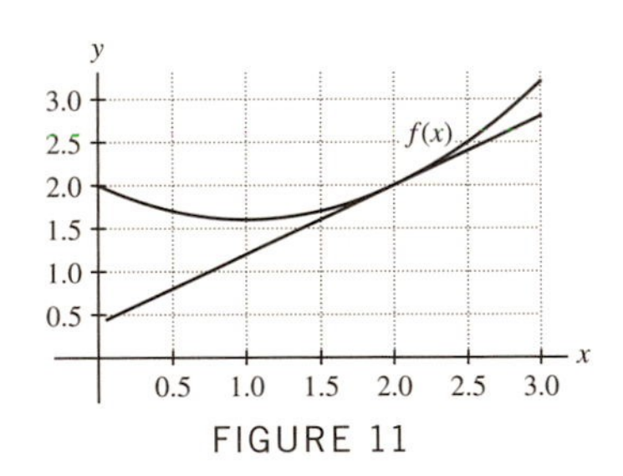

FIGURE 11

7. Find the slope of the secant line through $(2, f(2))$ and $(2.5, f(2.5))$. Is it larger or smaller than $f'(2)$? Explain.

SOLUTION From the graph, it appears that $f(2.5) = 2.5$ and $f(2) = 2$. Thus, the slope of the secant line through $(2, f(2))$ and $(2.5, f(2.5))$ is

$$\frac{f(2.5)-f(2)}{2.5-2} = \frac{2.5-2}{2.5-2} = 1.$$

From the graph, it is also clear that the secant line through $(2, f(2))$ and $(2.5, f(2.5))$ has a larger slope than the tangent line at $x = 2$. In other words, the slope of the secant line through $(2, f(2))$ and $(2.5, f(2.5))$ is larger than $f'(2)$.

9. Estimate $f'(1)$ and $f'(2)$.

SOLUTION From the graph, it appears that the tangent line at $x = 1$ would be horizontal. Thus, $f'(1) \approx 0$. The tangent line at $x = 2$ appears to pass through the points $(0.5, 0.8)$ and $(2, 2)$. Thus

$$f'(2) \approx \frac{2-0.8}{2-0.5} = 0.8.$$

In Exercises 11–14, refer to Figure 12.

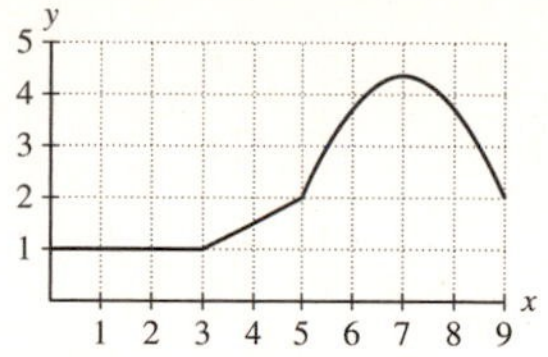

FIGURE 12 Graph of $f(x)$.

11. Determine $f'(a)$ for $a = 1, 2, 4, 7$.

SOLUTION Remember that the value of the derivative of f at $x = a$ can be interpreted as the slope of the line tangent to the graph of $y = f(x)$ at $x = a$. From Figure 12, we see that the graph of $y = f(x)$ is a horizontal line (that is, a line with zero slope) on the interval $0 \le x \le 3$. Accordingly, $f'(1) = f'(2) = 0$. On the interval $3 \le x \le 5$, the graph of $y = f(x)$ is a line of slope $\frac{1}{2}$; thus, $f'(4) = \frac{1}{2}$. Finally, the line tangent to the graph of $y = f(x)$ at $x = 7$ is horizontal, so $f'(7) = 0$.

13. Which is larger, $f'(5.5)$ or $f'(6.5)$?

SOLUTION The line tangent to the graph of $y = f(x)$ at $x = 5.5$ has a larger slope than the line tangent to the graph of $y = f(x)$ at $x = 6.5$. Therefore, $f'(5.5)$ is larger than $f'(6.5)$.

In Exercises 15–18, use the limit definition to calculate the derivative of the linear function.

15. $f(x) = 7x - 9$

SOLUTION

$$\lim_{h\to 0} \frac{f(a+h) - f(a)}{h} = \lim_{h\to 0} \frac{7(a+h) - 9 - (7a - 9)}{h} = \lim_{h\to 0} 7 = 7.$$

17. $g(t) = 8 - 3t$

SOLUTION

$$\lim_{h\to 0} \frac{g(a+h) - g(a)}{h} = \lim_{h\to 0} \frac{8 - 3(a+h) - (8 - 3a)}{h} = \lim_{h\to 0} \frac{-3h}{h} = \lim_{h\to 0} (-3) = -3.$$

19. Find an equation of the tangent line at $x = 3$, assuming that $f(3) = 5$ and $f'(3) = 2$?

SOLUTION By definition, the equation of the tangent line to the graph of $f(x)$ at $x = 3$ is $y = f(3) + f'(3)(x - 3) = 5 + 2(x - 3) = 2x - 1$.

21. Describe the tangent line at an arbitrary point on the "curve" $y = 2x + 8$.

SOLUTION Since $y = 2x + 8$ represents a straight line, the tangent line at any point is the line itself, $y = 2x + 8$.

23. Let $f(x) = \frac{1}{x}$. Does $f(-2 + h)$ equal $\frac{1}{-2 + h}$ or $\frac{1}{-2} + \frac{1}{h}$? Compute the difference quotient at $a = -2$ with $h = 0.5$.

SOLUTION Let $f(x) = \frac{1}{x}$. Then

$$f(-2 + h) = \frac{1}{-2 + h}.$$

With $a = -2$ and $h = 0.5$, the difference quotient is

$$\frac{f(a+h) - f(a)}{h} = \frac{f(-1.5) - f(-2)}{0.5} = \frac{\frac{1}{-1.5} - \frac{1}{-2}}{0.5} = -\frac{1}{3}.$$

25. Let $f(x) = 1/\sqrt{x}$. Compute $f'(5)$ by showing that

$$\frac{f(5+h) - f(5)}{h} = -\frac{1}{\sqrt{5}\sqrt{5+h}(\sqrt{5+h} + \sqrt{5})}$$

SOLUTION Let $f(x) = 1/\sqrt{x}$. Then

$$\frac{f(5+h)-f(5)}{h} = \frac{\frac{1}{\sqrt{5+h}} - \frac{1}{\sqrt{5}}}{h} = \frac{\sqrt{5}-\sqrt{5+h}}{h\sqrt{5}\sqrt{5+h}}$$

$$= \frac{\sqrt{5}-\sqrt{5+h}}{h\sqrt{5}\sqrt{5+h}}\left(\frac{\sqrt{5}+\sqrt{5+h}}{\sqrt{5}+\sqrt{5+h}}\right)$$

$$= \frac{5-(5+h)}{h\sqrt{5}\sqrt{5+h}(\sqrt{5+h}+\sqrt{5})} = -\frac{1}{\sqrt{5}\sqrt{5+h}(\sqrt{5+h}+\sqrt{5})}.$$

Thus,

$$f'(5) = \lim_{h\to 0}\frac{f(5+h)-f(5)}{h} = \lim_{h\to 0} -\frac{1}{\sqrt{5}\sqrt{5+h}(\sqrt{5+h}+\sqrt{5})}$$

$$= -\frac{1}{\sqrt{5}\,\sqrt{5}(\sqrt{5}+\sqrt{5})} = -\frac{1}{10\sqrt{5}}.$$

In Exercises 27–44, use the limit definition to compute $f'(a)$ and find an equation of the tangent line.

27. $f(x) = 2x^2 + 10x, \quad a = 3$

SOLUTION Let $f(x) = 2x^2 + 10x$. Then

$$f'(3) = \lim_{h\to 0}\frac{f(3+h)-f(3)}{h} = \lim_{h\to 0}\frac{2(3+h)^2 + 10(3+h) - 48}{h}$$

$$= \lim_{h\to 0}\frac{18 + 12h + 2h^2 + 30 + 10h - 48}{h} = \lim_{h\to 0}(22 + 2h) = 22.$$

At $a = 3$, the tangent line is

$$y = f'(3)(x-3) + f(3) = 22(x-3) + 48 = 22x - 18.$$

29. $f(t) = t - 2t^2, \quad a = 3$

SOLUTION Let $f(t) = t - 2t^2$. Then

$$f'(3) = \lim_{h\to 0}\frac{f(3+h)-f(3)}{h} = \lim_{h\to 0}\frac{(3+h) - 2(3+h)^2 - (-15)}{h}$$

$$= \lim_{h\to 0}\frac{3 + h - 18 - 12h - 2h^2 + 15}{h}$$

$$= \lim_{h\to 0}(-11 - 2h) = -11.$$

At $a = 3$, the tangent line is

$$y = f'(3)(t-3) + f(3) = -11(t-3) - 15 = -11t + 18.$$

31. $f(x) = x^3 + x, \quad a = 0$

SOLUTION Let $f(x) = x^3 + x$. Then

$$f'(0) = \lim_{h\to 0}\frac{f(h)-f(0)}{h} = \lim_{h\to 0}\frac{h^3 + h - 0}{h}$$

$$= \lim_{h\to 0}(h^2 + 1) = 1.$$

At $a = 0$, the tangent line is

$$y = f'(0)(x-0) + f(0) = x.$$

33. $f(x) = x^{-1}, \quad a = 8$

SOLUTION Let $f(x) = x^{-1}$. Then

$$f'(8) = \lim_{h\to 0}\frac{f(8+h)-f(8)}{h} = \lim_{h\to 0}\frac{\frac{1}{8+h} - \left(\frac{1}{8}\right)}{h} = \lim_{h\to 0}\frac{\frac{8-8-h}{8(8+h)}}{h} = \lim_{h\to 0}\frac{-h}{(64+8h)h} = -\frac{1}{64}$$

The tangent at $a = 8$ is

$$y = f'(8)(x-8) + f(8) = -\frac{1}{64}(x-8) + \frac{1}{8} = -\frac{1}{64}x + \frac{1}{4}.$$

35. $f(x) = \dfrac{1}{x+3}, \quad a = -2$

SOLUTION Let $f(x) = \frac{1}{x+3}$. Then

$$f'(-2) = \lim_{h\to 0} \frac{f(-2+h) - f(-2)}{h} = \lim_{h\to 0} \frac{\frac{1}{-2+h+3} - 1}{h} = \lim_{h\to 0} \frac{\frac{1}{1+h} - 1}{h} = \lim_{h\to 0} \frac{-h}{h(1+h)} = \lim_{h\to 0} \frac{-1}{1+h} = -1.$$

The tangent line at $a = -2$ is

$$y = f'(-2)(x+2) + f(-2) = -1(x+2) + 1 = -x - 1.$$

37. $f(x) = \sqrt{x+4}, \quad a = 1$

SOLUTION Let $f(x) = \sqrt{x+4}$. Then

$$f'(1) = \lim_{h\to 0} \frac{f(1+h) - f(1)}{h} = \lim_{h\to 0} \frac{\sqrt{h+5} - \sqrt{5}}{h} = \lim_{h\to 0} \frac{\sqrt{h+5} - \sqrt{5}}{h} \cdot \frac{\sqrt{h+5} + \sqrt{5}}{\sqrt{h+5} + \sqrt{5}}$$

$$= \lim_{h\to 0} \frac{h}{h(\sqrt{h+5} + \sqrt{5})} = \lim_{h\to 0} \frac{1}{\sqrt{h+5} + \sqrt{5}} = \frac{1}{2\sqrt{5}}.$$

The tangent line at $a = 1$ is

$$y = f'(1)(x-1) + f(1) = \frac{1}{2\sqrt{5}}(x-1) + \sqrt{5} = \frac{1}{2\sqrt{5}}x + \frac{9}{2\sqrt{5}}.$$

39. $f(x) = \dfrac{1}{\sqrt{x}}, \quad a = 4$

SOLUTION Let $f(x) = \dfrac{1}{\sqrt{x}}$. Then

$$f'(4) = \lim_{h\to 0} \frac{f(4+h) - f(4)}{h} = \lim_{h\to 0} \frac{\frac{1}{\sqrt{4+h}} - \frac{1}{2}}{h} = \lim_{h\to 0} \frac{\frac{2-\sqrt{4+h}}{2\sqrt{4+h}} \cdot \frac{2+\sqrt{4+h}}{2+\sqrt{4+h}}}{h} = \lim_{h\to 0} \frac{\frac{4-4-h}{4\sqrt{4+h}+2(4+h)}}{h}$$

$$= \lim_{h\to 0} \frac{-1}{4\sqrt{4+h} + 2(4+h)} = -\frac{1}{16}.$$

At $a = 4$ the tangent line is

$$y = f'(4)(x-4) + f(4) = -\frac{1}{16}(x-4) + \frac{1}{2} = -\frac{1}{16}x + \frac{3}{4}.$$

41. $f(t) = \sqrt{t^2+1}, \quad a = 3$

SOLUTION Let $f(t) = \sqrt{t^2+1}$. Then

$$f'(3) = \lim_{h\to 0} \frac{f(3+h) - f(3)}{h} = \lim_{h\to 0} \frac{\sqrt{10+6h+h^2} - \sqrt{10}}{h}$$

$$= \lim_{h\to 0} \frac{\sqrt{10+6h+h^2} - \sqrt{10}}{h} \cdot \frac{\sqrt{10+6h+h^2} + \sqrt{10}}{\sqrt{10+6h+h^2} + \sqrt{10}}$$

$$= \lim_{h\to 0} \frac{6h+h^2}{h(\sqrt{10+6h+h^2} + \sqrt{10})} = \lim_{h\to 0} \frac{6+h}{\sqrt{10+6h+h^2} + \sqrt{10}} = \frac{3}{\sqrt{10}}.$$

The tangent line at $a = 3$ is

$$y = f'(3)(t-3) + f(3) = \frac{3}{\sqrt{10}}(t-3) + \sqrt{10} = \frac{3}{\sqrt{10}}t + \frac{1}{\sqrt{10}}.$$

43. $f(x) = \dfrac{1}{x^2+1}, \quad a = 0$

SOLUTION Let $f(x) = \dfrac{1}{x^2+1}$. Then

$$f'(0) = \lim_{h\to 0} \frac{f(0+h) - f(0)}{h} = \lim_{h\to 0} \frac{\frac{1}{(0+h)^2+1} - 1}{h} = \lim_{h\to 0} \frac{\frac{-h^2}{h^2+1}}{h} = \lim_{h\to 0} \frac{-h}{h^2+1} = 0.$$

The tangent line at $a = 0$ is

$$y = f(0) + f'(0)(x - 0) = 1 + 0(x - 1) = 1.$$

45. Figure 13 displays data collected by the biologist Julian Huxley (1887–1975) on the average antler weight W of male red deer as a function of age t. Estimate the derivative at $t = 4$. For which values of t is the slope of the tangent line equal to zero? For which values is it negative?

FIGURE 13

SOLUTION Let $W(t)$ denote the antler weight as a function of age. The "tangent line" sketched in the figure below passes through the points $(1, 1)$ and $(6, 5.5)$. Therefore

$$W'(4) \approx \frac{5.5 - 1}{6 - 1} = 0.9 \text{ kg/year}.$$

If the slope of the tangent is zero, the tangent line is horizontal. This appears to happen at roughly $t = 10$ and at $t = 11.6$. The slope of the tangent line is negative when the height of the graph decreases as we move to the right. For the graph in Figure 13, this occurs for $10 < t < 11.6$.

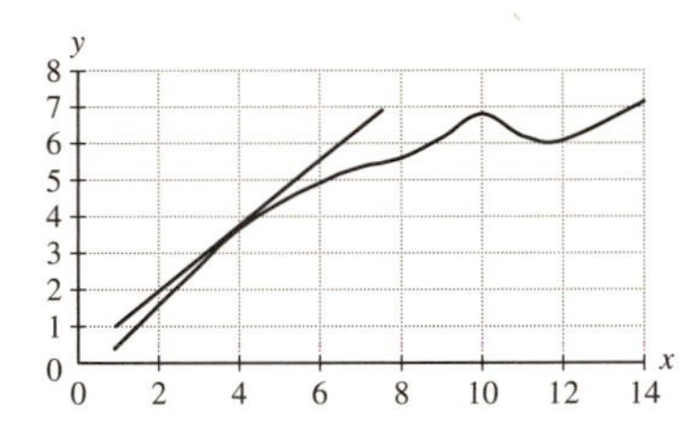

47. GU Let $f(x) = \dfrac{4}{1+2^x}$.

(a) Plot $f(x)$ over $[-2, 2]$. Then zoom in near $x = 0$ until the graph appears straight, and estimate the slope $f'(0)$.

(b) Use (a) to find an approximate equation to the tangent line at $x = 0$. Plot this line and $f(x)$ on the same set of axes.

SOLUTION

(a) The figure below at the left shows the graph of $f(x) = \frac{4}{1+2^x}$ over $[-2, 2]$. The figure below at the right is a close-up near $x = 0$. From the close-up, we see that the graph is nearly straight and passes through the points $(-0.22, 2.15)$ and $(0.22, 1.85)$. We therefore estimate

$$f'(0) \approx \frac{1.85 - 2.15}{0.22 - (-0.22)} = \frac{-0.3}{0.44} = -0.68$$

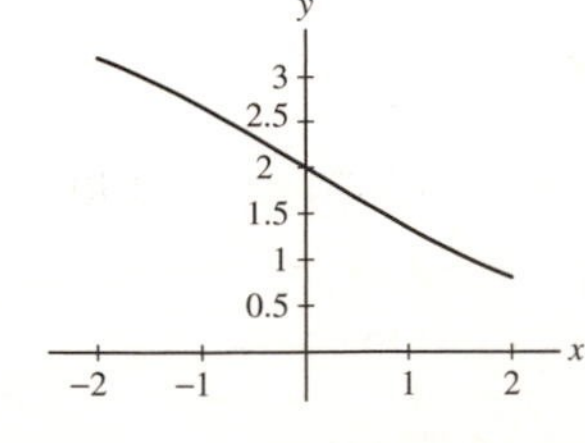

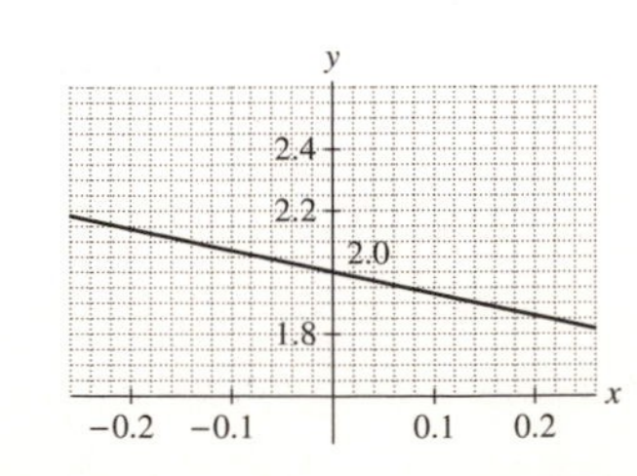

(b) Using the estimate for $f'(0)$ obtained in part (a), the approximate equation of the tangent line is

$$y = f'(0)(x - 0) + f(0) = -0.68x + 2.$$

The figure below shows the graph of $f(x)$ and the approximate tangent line.

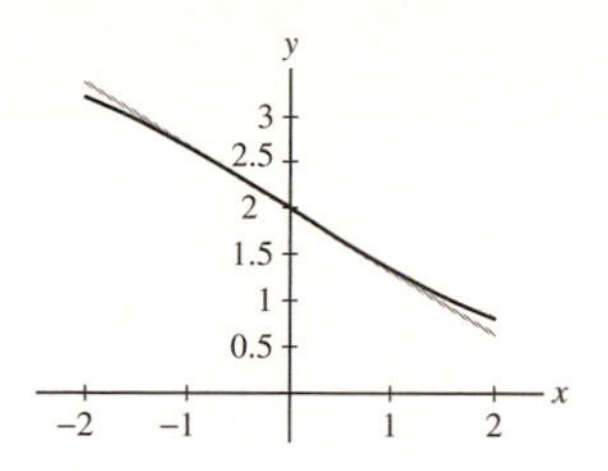

49. Determine the intervals along the x-axis on which the derivative in Figure 15 is positive.

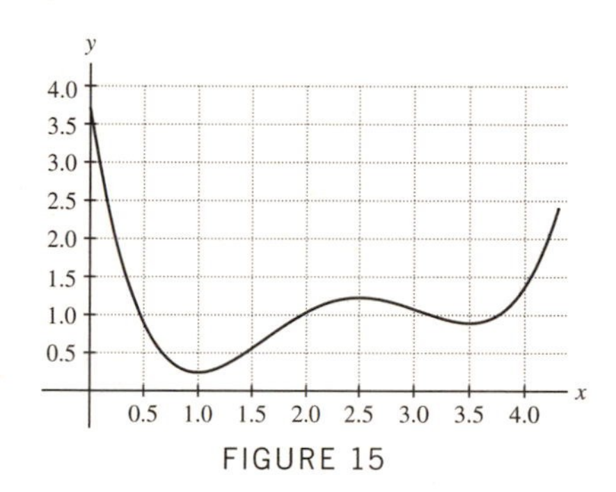

FIGURE 15

SOLUTION The derivative (that is, the slope of the tangent line) is positive when the height of the graph increases as we move to the right. From Figure 15, this appears to be true for $1 < x < 2.5$ and for $x > 3.5$.

In Exercises 51–56, each limit represents a derivative $f'(a)$. Find $f(x)$ and a.

51. $\displaystyle\lim_{h\to 0} \frac{(5+h)^3 - 125}{h}$

SOLUTION The difference quotient $\dfrac{(5+h)^3 - 125}{h}$ has the form $\dfrac{f(a+h) - f(a)}{h}$ where $f(x) = x^3$ and $a = 5$.

53. $\displaystyle\lim_{h\to 0} \frac{\sin\left(\frac{\pi}{6} + h\right) - 0.5}{h}$

SOLUTION The difference quotient $\dfrac{\sin(\frac{\pi}{6} + h) - .5}{h}$ has the form $\dfrac{f(a+h) - f(a)}{h}$ where $f(x) = \sin x$ and $a = \frac{\pi}{6}$.

55. $\displaystyle\lim_{h\to 0} \frac{5^{2+h} - 25}{h}$

SOLUTION The difference quotient $\dfrac{5^{(2+h)} - 25}{h}$ has the form $\dfrac{f(a+h) - f(a)}{h}$ where $f(x) = 5^x$ and $a = 2$.

57. Apply the method of Example 6 to $f(x) = \sin x$ to determine $f'\left(\frac{\pi}{4}\right)$ accurately to four decimal places.

SOLUTION We know that

$$f'(\pi/4) = \lim_{h\to 0} \frac{f(\pi/4 + h) - f(\pi/4)}{h} = \lim_{h\to 0} \frac{\sin(\pi/4 + h) - \sqrt{2}/2}{h}.$$

Creating a table of values of h close to zero:

h	-0.001	-0.0001	-0.00001	0.00001	0.0001	0.001
$\dfrac{\sin(\frac{\pi}{4} + h) - (\sqrt{2}/2)}{h}$	0.7074602	0.7071421	0.7071103	0.7071033	0.7070714	0.7067531

Accurate up to four decimal places, $f'(\frac{\pi}{4}) \approx 0.7071$.

59. For each graph in Figure 16, determine whether $f'(1)$ is larger or smaller than the slope of the secant line between $x = 1$ and $x = 1 + h$ for $h > 0$. Explain.

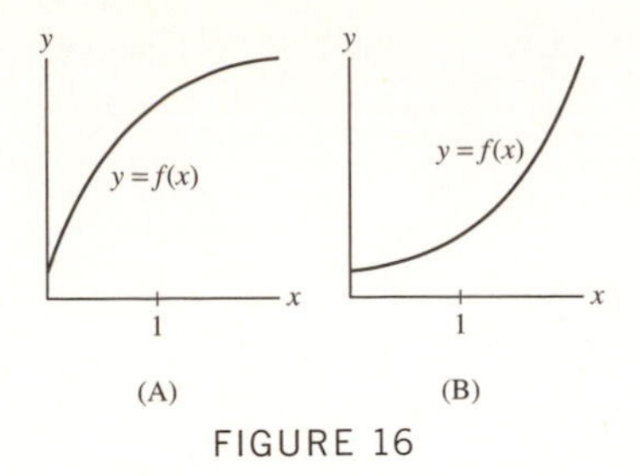

FIGURE 16

SOLUTION

- On curve (A), $f'(1)$ is larger than

$$\frac{f(1+h) - f(1)}{h};$$

the curve is bending downwards, so that the secant line to the right is at a lower angle than the tangent line. We say such a curve is **concave down**, and that its derivative is *decreasing*.

- On curve (B), $f'(1)$ is smaller than

$$\frac{f(1+h) - f(1)}{h};$$

the curve is bending upwards, so that the secant line to the right is at a steeper angle than the tangent line. We say such a curve is **concave up**, and that its derivative is *increasing*.

61. GU Sketch the graph of $f(x) = x^{5/2}$ on [0, 6].

(a) Use the sketch to justify the inequalities for $h > 0$:

$$\frac{f(4) - f(4-h)}{h} \le f'(4) \le \frac{f(4+h) - f(4)}{h}$$

(b) Use (a) to compute $f'(4)$ to four decimal places.

(c) Use a graphing utility to plot $f(x)$ and the tangent line at $x = 4$, using your estimate for $f'(4)$.

SOLUTION

(a) The slope of the secant line between points $(4, f(4))$ and $(4 + h, f(4 + h))$ is

$$\frac{f(4+h) - f(4)}{h}.$$

$x^{5/2}$ is a smooth curve increasing at a faster rate as $x \to \infty$. Therefore, if $h > 0$, then the slope of the secant line is greater than the slope of the tangent line at $f(4)$, which happens to be $f'(4)$. Likewise, if $h < 0$, the slope of the secant line is less than the slope of the tangent line at $f(4)$, which happens to be $f'(4)$.

(b) We know that

$$f'(4) = \lim_{h \to 0} \frac{f(4+h) - f(4)}{h} = \lim_{h \to 0} \frac{(4+h)^{5/2} - 32}{h}.$$

Creating a table with values of h close to zero:

h	−0.0001	−0.00001	0.00001	0.0001
$\frac{(4+h)^{5/2} - 32}{h}$	19.999625	19.99999	20.0000	20.0000375

Thus, $f'(4) \approx 20.0000$.

(c) Using the estimate for $f'(4)$ obtained in part (b), the equation of the line tangent to $f(x) = x^{5/2}$ at $x = 4$ is

$$y = f'(4)(x - 4) + f(4) = 20(x - 4) + 32 = 20x - 48.$$

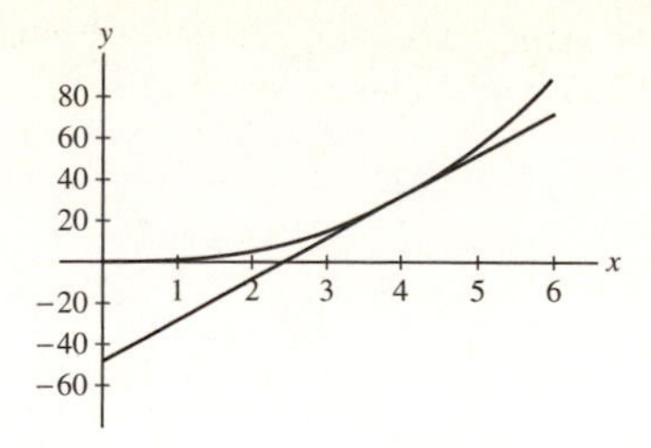

63. GU Use a plot of $f(x) = x^x$ to estimate the value c such that $f'(c) = 0$. Find c to sufficient accuracy so that

$$\left|\frac{f(c+h) - f(c)}{h}\right| \leq 0.006 \qquad \text{for} \quad h = \pm 0.001$$

SOLUTION Here is a graph of $f(x) = x^x$ over the interval [0, 1.5].

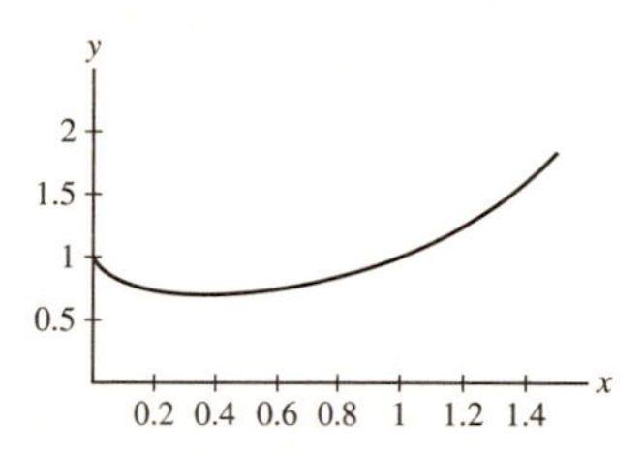

The graph shows one location with a horizontal tangent line. The figure below at the left shows the graph of $f(x)$ together with the horizontal lines $y = 0.6$, $y = 0.7$ and $y = 0.8$. The line $y = 0.7$ is very close to being tangent to the graph of $f(x)$. The figure below at the right refines this estimate by graphing $f(x)$ and $y = 0.69$ on the same set of axes. The point of tangency has an x-coordinate of roughly 0.37, so $c \approx 0.37$.

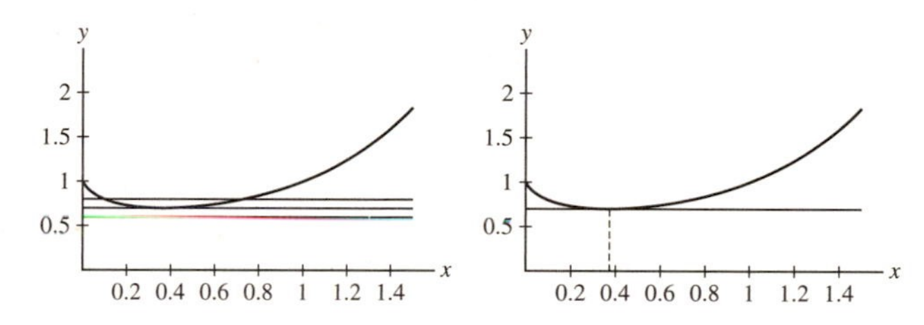

We note that

$$\left|\frac{f(0.37 + 0.001) - f(0.37)}{0.001}\right| \approx 0.00491 < 0.006$$

and

$$\left|\frac{f(0.37 - 0.001) - f(0.37)}{0.001}\right| \approx 0.00304 < 0.006,$$

so we have determined c to the desired accuracy.

In Exercises 65–71, estimate derivatives using the ***symmetric difference quotient*** *(SDQ), defined as the average of the difference quotients at h and $-h$:*

$$\frac{1}{2}\left(\frac{f(a+h) - f(a)}{h} + \frac{f(a-h) - f(a)}{-h}\right) = \frac{f(a+h) - f(a-h)}{2h} \qquad \boxed{4}$$

The SDQ usually gives a better approximation to the derivative than the difference quotient.

65. The vapor pressure of water at temperature T (in kelvins) is the atmospheric pressure P at which no net evaporation takes place. Use the following table to estimate $P'(T)$ for $T = 303, 313, 323, 333, 343$ by computing the SDQ given by Eq. (4) with $h = 10$.

T (K)	293	303	313	323	333	343	353
P (atm)	0.0278	0.0482	0.0808	0.1311	0.2067	0.3173	0.4754

SOLUTION Using equation (4),

$$P'(303) \approx \frac{P(313) - P(293)}{20} = \frac{0.0808 - 0.0278}{20} = 0.00265 \text{ atm/K};$$

$$P'(313) \approx \frac{P(323) - P(303)}{20} = \frac{0.1311 - 0.0482}{20} = 0.004145 \text{ atm/K};$$

$$P'(323) \approx \frac{P(333) - P(313)}{20} = \frac{0.2067 - 0.0808}{20} = 0.006295 \text{ atm/K};$$

$$P'(333) \approx \frac{P(343) - P(323)}{20} = \frac{0.3173 - 0.1311}{20} = 0.00931 \text{ atm/K};$$

$$P'(343) \approx \frac{P(353) - P(333)}{20} = \frac{0.4754 - 0.2067}{20} = 0.013435 \text{ atm/K}$$

In Exercises 67 and 68, traffic speed S along a certain road (in km/h) varies as a function of traffic density q (number of cars per km of road). Use the following data to answer the questions:

q (density)	60	70	80	90	100
S (speed)	72.5	67.5	63.5	60	56

67. Estimate $S'(80)$.

SOLUTION Let $S(q)$ be the function determining S given q. Using equation (4) with $h = 10$,

$$S'(80) \approx \frac{S(90) - S(70)}{20} = \frac{60 - 67.5}{20} = -0.375;$$

with $h = 20$,

$$S'(80) \approx \frac{S(100) - S(60)}{40} = \frac{56 - 72.5}{40} = -0.4125;$$

The mean of these two symmetric difference quotients is -0.39375 kph·km/car.

Exercises 69–71: The current (in amperes) at time t (in seconds) flowing in the circuit in Figure 19 is given by Kirchhoff's Law:

$$i(t) = Cv'(t) + R^{-1}v(t)$$

where $v(t)$ is the voltage (in volts), C the capacitance (in farads), and R the resistance (in ohms, Ω).

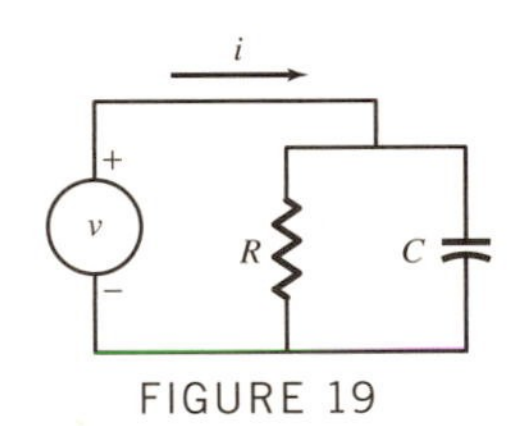

FIGURE 19

69. Calculate the current at $t = 3$ if

$$v(t) = 0.5t + 4 \text{ V}$$

where $C = 0.01$ F and $R = 100\ \Omega$.

SOLUTION Since $v(t)$ is a line with slope 0.5, $v'(t) = 0.5$ volts/s for all t. From the formula, $i(3) = Cv'(3) + (1/R)v(3) = 0.01(0.5) + (1/100)(5.5) = 0.005 + 0.055 = 0.06$ amperes.

71. Assume that $R = 200\ \Omega$ but C is unknown. Use the following data to estimate $v'(4)$ (by an SDQ) and deduce an approximate value for the capacitance C.

t	3.8	3.9	4	4.1	4.2
$v(t)$	388.8	404.2	420	436.2	452.8
$i(t)$	32.34	33.22	34.1	34.98	35.86

SOLUTION Solving $i(4) = Cv'(4) + (1/R)v(4)$ for C yields

$$C = \frac{i(4) - (1/R)v(4)}{v'(4)} = \frac{34.1 - \frac{420}{200}}{v'(4)}.$$

To compute C, we first approximate $v'(4)$. Taking $h = 0.1$, we find

$$v'(4) \approx \frac{v(4.1) - v(3.9)}{0.2} = \frac{436.2 - 404.2}{0.2} = 160.$$

Plugging this in to the equation above yields

$$C \approx \frac{34.1 - 2.1}{160} = 0.2 \text{ farads.}$$

Further Insights and Challenges

73. Explain how the symmetric difference quotient defined by Eq. (4) can be interpreted as the slope of a secant line.

SOLUTION The symmetric difference quotient

$$\frac{f(a+h) - f(a-h)}{2h}$$

is the slope of the secant line connecting the points $(a-h, f(a-h))$ and $(a+h, f(a+h))$ on the graph of f; the difference in the function values is divided by the difference in the x-values.

75. Show that if $f(x)$ is a quadratic polynomial, then the SDQ at $x = a$ (for any $h \neq 0$) is *equal* to $f'(a)$. Explain the graphical meaning of this result.

SOLUTION Let $f(x) = px^2 + qx + r$ be a quadratic polynomial. We compute the SDQ at $x = a$.

$$\begin{aligned}\frac{f(a+h) - f(a-h)}{2h} &= \frac{p(a+h)^2 + q(a+h) + r - (p(a-h)^2 + q(a-h) + r)}{2h} \\ &= \frac{pa^2 + 2pah + ph^2 + qa + qh + r - pa^2 + 2pah - ph^2 - qa + qh - r}{2h} \\ &= \frac{4pah + 2qh}{2h} = \frac{2h(2pa+q)}{2h} = 2pa + q\end{aligned}$$

Since this doesn't depend on h, the limit, which is equal to $f'(a)$, is also $2pa + q$. Graphically, this result tells us that the secant line to a parabola passing through points chosen symmetrically about $x = a$ is always parallel to the tangent line at $x = a$.

3.2 The Derivative as a Function

Preliminary Questions

1. What is the slope of the tangent line through the point $(2, f(2))$ if $f'(x) = x^3$?

SOLUTION The slope of the tangent line through the point $(2, f(2))$ is given by $f'(2)$. Since $f'(x) = x^3$, it follows that $f'(2) = 2^3 = 8$.

2. Evaluate $(f - g)'(1)$ and $(3f + 2g)'(1)$ assuming that $f'(1) = 3$ and $g'(1) = 5$.

SOLUTION $(f - g)'(1) = f'(1) - g'(1) = 3 - 5 = -2$ and $(3f + 2g)'(1) = 3f'(1) + 2g'(1) = 3(3) + 2(5) = 19$.

3. To which of the following does the Power Rule apply?

(a) $f(x) = x^2$ **(b)** $f(x) = 2^e$
(c) $f(x) = x^e$ **(d)** $f(x) = e^x$
(e) $f(x) = x^x$ **(f)** $f(x) = x^{-4/5}$

SOLUTION

(a) Yes. x^2 is a power function, so the Power Rule can be applied.
(b) Yes. 2^e is a constant function, so the Power Rule can be applied.
(c) Yes. x^e is a power function, so the Power Rule can be applied.
(d) No. e^x is an exponential function (the base is constant while the exponent is a variable), so the Power Rule does not apply.
(e) No. x^x is not a power function because both the base and the exponent are variable, so the Power Rule does not apply.
(f) Yes. $x^{-4/5}$ is a power function, so the Power Rule can be applied.

4. Choose (a) or (b). The derivative does not exist if the tangent line is: (a) horizontal (b) vertical.

SOLUTION The derivative does not exist when: (b) the tangent line is vertical. At a horizontal tangent, the derivative is zero.

5. Which property distinguishes $f(x) = e^x$ from all other exponential functions $g(x) = b^x$?

SOLUTION The line tangent to $f(x) = e^x$ at $x = 0$ has slope equal to 1.

Exercises

In Exercises 1–6, compute $f'(x)$ using the limit definition.

1. $f(x) = 3x - 7$

SOLUTION Let $f(x) = 3x - 7$. Then,

$$f'(x) = \lim_{h\to 0} \frac{f(x+h) - f(x)}{h} = \lim_{h\to 0} \frac{3(x+h) - 7 - (3x-7)}{h} = \lim_{h\to 0} \frac{3h}{h} = 3.$$

3. $f(x) = x^3$

SOLUTION Let $f(x) = x^3$. Then,

$$f'(x) = \lim_{h\to 0} \frac{f(x+h) - f(x)}{h} = \lim_{h\to 0} \frac{(x+h)^3 - x^3}{h} = \lim_{h\to 0} \frac{x^3 + 3x^2h + 3xh^2 + h^3 - x^3}{h}$$

$$= \lim_{h\to 0} \frac{3x^2h + 3xh^2 + h^3}{h} = \lim_{h\to 0} (3x^2 + 3xh + h^2) = 3x^2.$$

5. $f(x) = x - \sqrt{x}$

SOLUTION Let $f(x) = x - \sqrt{x}$. Then,

$$f'(x) = \lim_{h\to 0} \frac{f(x+h) - f(x)}{h} = \lim_{h\to 0} \frac{x + h - \sqrt{x+h} - (x - \sqrt{x})}{h} = 1 - \lim_{h\to 0} \frac{\sqrt{x+h} - \sqrt{x}}{h} \cdot \left(\frac{\sqrt{x+h} + \sqrt{x}}{\sqrt{x+h} + \sqrt{x}}\right)$$

$$= 1 - \lim_{h\to 0} \frac{(x+h) - x}{h(\sqrt{x+h} + \sqrt{x})} = 1 - \lim_{h\to 0} \frac{1}{\sqrt{x+h} + \sqrt{x}} = 1 - \frac{1}{2\sqrt{x}}.$$

In Exercises 7–14, use the Power Rule to compute the derivative.

7. $\left.\frac{d}{dx} x^4\right|_{x=-2}$

SOLUTION $\frac{d}{dx}\left(x^4\right) = 4x^3$ so $\left.\frac{d}{dx}x^4\right|_{x=-2} = 4(-2)^3 = -32.$

9. $\left.\frac{d}{dt} t^{2/3}\right|_{t=8}$

SOLUTION $\frac{d}{dt}\left(t^{2/3}\right) = \frac{2}{3}t^{-1/3}$ so $\left.\frac{d}{dt}t^{2/3}\right|_{t=8} = \frac{2}{3}(8)^{-1/3} = \frac{1}{3}.$

11. $\frac{d}{dx} x^{0.35}$

SOLUTION $\frac{d}{dx}\left(x^{0.35}\right) = 0.35(x^{0.35-1}) = 0.35x^{-0.65}.$

13. $\frac{d}{dt} t^{\sqrt{17}}$

SOLUTION $\frac{d}{dt}(t^{\sqrt{17}}) = \sqrt{17}t^{\sqrt{17}-1}$

In Exercises 15–18, compute $f'(x)$ and find an equation of the tangent line to the graph at $x = a$.

15. $f(x) = x^4, \quad a = 2$

SOLUTION Let $f(x) = x^4$. Then, by the Power Rule, $f'(x) = 4x^3$. The equation of the tangent line to the graph of $f(x)$ at $x = 2$ is

$$y = f'(2)(x-2) + f(2) = 32(x-2) + 16 = 32x - 48.$$

17. $f(x) = 5x - 32\sqrt{x}, \quad a = 4$

SOLUTION Let $f(x) = 5x - 32x^{1/2}$. Then $f'(x) = 5 - 16x^{-1/2}$. In particular, $f'(4) = -3$. The tangent line at $x = 4$ is

$$y = f'(4)(x-4) + f(4) = -3(x-4) - 44 = -3x - 32.$$

19. Calculate:

(a) $\frac{d}{dx}12e^x$ **(b)** $\frac{d}{dt}(25t - 8e^t)$ **(c)** $\frac{d}{dt}e^{t-3}$

Hint for (c): Write e^{t-3} as $e^{-3}e^t$.

SOLUTION

(a) $\frac{d}{dx}12e^x = 12\frac{d}{dx}e^x = 12e^x$.

(b) $\frac{d}{dt}(25t - 8e^t) = 25\frac{d}{dt}t - 8\frac{d}{dt}e^t = 25 - 8e^t$.

(c) $\frac{d}{dt}e^{t-3} = e^{-3}\frac{d}{dt}e^t = e^{-3} \cdot e^t = e^{t-3}$.

In Exercises 21–32, calculate the derivative.

21. $f(x) = 2x^3 - 3x^2 + 5$

SOLUTION $\frac{d}{dx}\left(2x^3 - 3x^2 + 5\right) = 6x^2 - 6x$.

23. $f(x) = 4x^{5/3} - 3x^{-2} - 12$

SOLUTION $\frac{d}{dx}\left(4x^{5/3} - 3x^{-2} - 12\right) = \frac{20}{3}x^{2/3} + 6x^{-3}$.

25. $g(z) = 7z^{-5/14} + z^{-5} + 9$

SOLUTION $\frac{d}{dz}\left(7z^{-5/14} + z^{-5} + 9\right) = -\frac{5}{2}z^{-19/14} - 5z^{-6}$.

27. $f(s) = \sqrt[4]{s} + \sqrt[3]{s}$

SOLUTION $f(s) = \sqrt[4]{s} + \sqrt[3]{s} = s^{1/4} + s^{1/3}$. In this form, we can apply the Sum and Power Rules.

$$\frac{d}{ds}\left(s^{1/4} + s^{1/3}\right) = \frac{1}{4}(s^{(1/4)-1}) + \frac{1}{3}(s^{(1/3)-1}) = \frac{1}{4}s^{-3/4} + \frac{1}{3}s^{-2/3}.$$

29. $g(x) = e^2$

SOLUTION Because e^2 is a constant, $\frac{d}{dx}e^2 = 0$.

31. $h(t) = 5e^{t-3}$

SOLUTION $\frac{d}{dt}5e^{t-3} = 5e^{-3}\frac{d}{dt}e^t = 5e^{-3}e^t = 5e^{t-3}$.

In Exercises 33–36, calculate the derivative by expanding or simplifying the function.

33. $P(s) = (4s - 3)^2$

SOLUTION $P(s) = (4s - 3)^2 = 16s^2 - 24s + 9$. Thus,

$$\frac{dP}{ds} = 32s - 24.$$

35. $g(x) = \dfrac{x^2 + 4x^{1/2}}{x^2}$

SOLUTION $g(x) = \dfrac{x^2 + 4x^{1/2}}{x^2} = 1 + 4x^{-3/2}$. Thus,

$$\frac{dg}{dx} = -6x^{-5/2}.$$

In Exercises 37–42, calculate the derivative indicated.

37. $\left.\frac{dT}{dC}\right|_{C=8}$, $T = 3C^{2/3}$

SOLUTION With $T(C) = 3C^{2/3}$, we have $\frac{dT}{dC} = 2C^{-1/3}$. Therefore,

$$\left.\frac{dT}{dC}\right|_{C=8} = 2(8)^{-1/3} = 1.$$

39. $\left.\dfrac{ds}{dz}\right|_{z=2}$, $s = 4z - 16z^2$

SOLUTION With $s = 4z - 16z^2$, we have $\dfrac{ds}{dz} = 4 - 32z$. Therefore,

$$\left.\frac{ds}{dz}\right|_{z=2} = 4 - 32(2) = -60.$$

41. $\left.\dfrac{dr}{dt}\right|_{t=4}$, $r = t - e^t$

SOLUTION With $r = t - e^t$, we have $\dfrac{dr}{dt} = 1 - e^t$. Therefore,

$$\left.\frac{dr}{dt}\right|_{t=4} = 1 - e^4.$$

43. Match the functions in graphs (A)–(D) with their derivatives (I)–(III) in Figure 13. Note that two of the functions have the same derivative. Explain why.

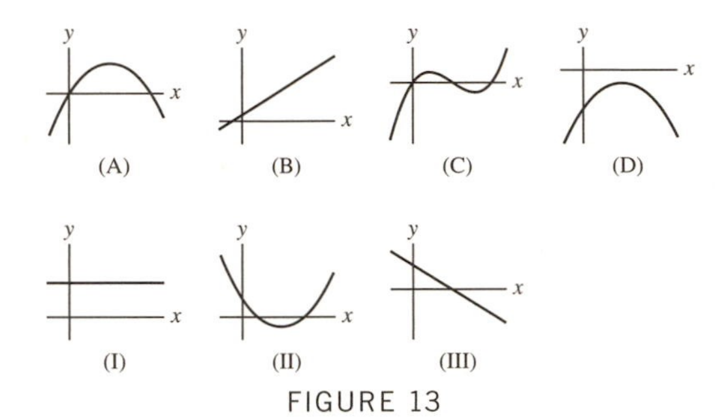

FIGURE 13

SOLUTION

- Consider the graph in (A). On the left side of the graph, the slope of the tangent line is positive but on the right side the slope of the tangent line is negative. Thus the derivative should transition from positive to negative with increasing x. This matches the graph in (III).
- Consider the graph in (B). This is a linear function, so its slope is constant. Thus the derivative is constant, which matches the graph in (I).
- Consider the graph in (C). Moving from left to right, the slope of the tangent line transitions from positive to negative then back to positive. The derivative should therefore be negative in the middle and positive to either side. This matches the graph in (II).
- Consider the graph in (D). On the left side of the graph, the slope of the tangent line is positive but on the right side the slope of the tangent line is negative. Thus the derivative should transition from positive to negative with increasing x. This matches the graph in (III).

Note that the functions whose graphs are shown in (A) and (D) have the same derivative. This happens because the graph in (D) is just a vertical translation of the graph in (A), which means the two functions differ by a constant. The derivative of a constant is zero, so the two functions end up with the same derivative.

45. Assign the labels $f(x)$, $g(x)$, and $h(x)$ to the graphs in Figure 15 in such a way that $f'(x) = g(x)$ and $g'(x) = h(x)$.

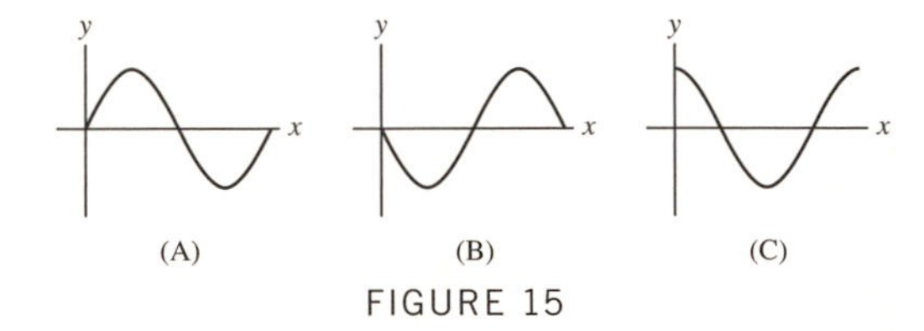

FIGURE 15

SOLUTION Consider the graph in (A). Moving from left to right, the slope of the tangent line is positive over the first quarter of the graph, negative in the middle half and positive again over the final quarter. The derivative of this function must therefore be negative in the middle and positive on either side. This matches the graph in (C).

Now focus on the graph in (C). The slope of the tangent line is negative over the left half and positive on the right half. The derivative of this function therefore needs to be negative on the left and positive on the right. This description matches the graph in (B).

We should therefore label the graph in (A) as $f(x)$, the graph in (B) as $h(x)$, and the graph in (C) as $g(x)$. Then $f'(x) = g(x)$ and $g'(x) = h(x)$.

47. Use the table of values of $f(x)$ to determine which of (A) or (B) in Figure 17 is the graph of $f'(x)$. Explain.

x	0	0.5	1	1.5	2	2.5	3	3.5	4
$f(x)$	10	55	98	139	177	210	237	257	268

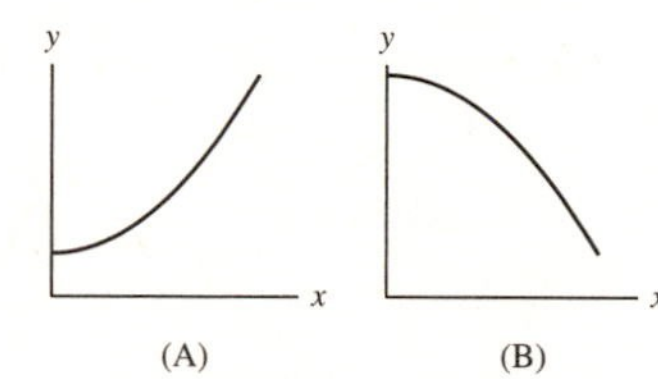

FIGURE 17 Which is the graph of $f'(x)$?

SOLUTION The increment between successive x values in the table is a constant 0.5 but the increment between successive $f(x)$ values decreases from 45 to 43 to 41 to 38 and so on. Thus the difference quotients decrease with increasing x, suggesting that $f'(x)$ decreases as a function of x. Because the graph in (B) depicts a decreasing function, (B) might be the graph of the derivative of $f(x)$.

49. Compute the derivatives, where c is a constant.

(a) $\dfrac{d}{dt}ct^3$ **(b)** $\dfrac{d}{dy}(9c^2y^3 - 24c)$ **(c)** $\dfrac{d}{dz}(5z + 4cz^2)$

SOLUTION

(a) $\dfrac{d}{dt}ct^3 = 3ct^2$.

(b) $\dfrac{d}{dz}(5z + 4cz^2) = 5 + 8cz$.

(c) $\dfrac{d}{dy}(9c^2y^3 - 24c) = 27c^2y^2$.

51. Find the points on the graph of $y = x^2 + 3x - 7$ at which the slope of the tangent line is equal to 4.

SOLUTION Let $y = x^2 + 3x - 7$. Solving $dy/dx = 2x + 3 = 4$ yields $x = \frac{1}{2}$.

53. Determine a and b such that $p(x) = x^2 + ax + b$ satisfies $p(1) = 0$ and $p'(1) = 4$.

SOLUTION Let $p(x) = x^2 + ax + b$ satisfy $p(1) = 0$ and $p'(1) = 4$. Now, $p'(x) = 2x + a$. Therefore $0 = p(1) = 1 + a + b$ and $4 = p'(1) = 2 + a$; i.e., $a = 2$ and $b = -3$.

55. Let $f(x) = x^3 - 3x + 1$. Show that $f'(x) \geq -3$ for all x and that, for every $m > -3$, there are precisely two points where $f'(x) = m$. Indicate the position of these points and the corresponding tangent lines for one value of m in a sketch of the graph of $f(x)$.

SOLUTION Let $P = (a, b)$ be a point on the graph of $f(x) = x^3 - 3x + 1$.

- The derivative satisfies $f'(x) = 3x^2 - 3 \geq -3$ since $3x^2$ is nonnegative.
- Suppose the slope m of the tangent line is greater than -3. Then $f'(a) = 3a^2 - 3 = m$, whence

$$a^2 = \frac{m+3}{3} > 0 \quad \text{and thus} \quad a = \pm\sqrt{\frac{m+3}{3}}.$$

- The two parallel tangent lines with slope 2 are shown with the graph of $f(x)$ here.

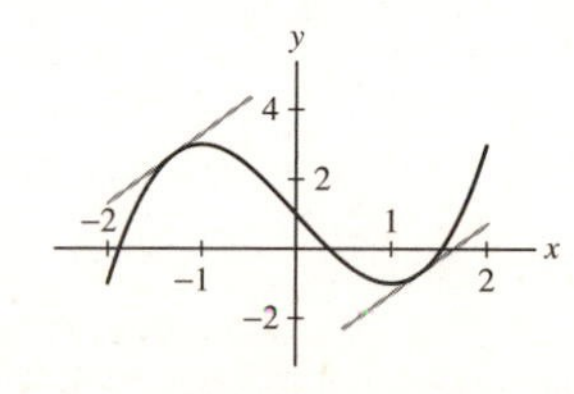

57. Compute the derivative of $f(x) = x^{3/2}$ using the limit definition. *Hint:* Show that

$$\frac{f(x+h) - f(x)}{h} = \frac{(x+h)^3 - x^3}{h}\left(\frac{1}{\sqrt{(x+h)^3} + \sqrt{x^3}}\right)$$

SOLUTION Once we have the difference of square roots, we multiply by the conjugate to solve the problem.

$$f'(x) = \lim_{h \to 0} \frac{(x+h)^{3/2} - x^{3/2}}{h} = \lim_{h \to 0} \frac{\sqrt{(x+h)^3} - \sqrt{x^3}}{h} \left(\frac{\sqrt{(x+h)^3} + \sqrt{x^3}}{\sqrt{(x+h)^3} + \sqrt{x^3}} \right)$$

$$= \lim_{h \to 0} \frac{(x+h)^3 - x^3}{h} \left(\frac{1}{\sqrt{(x+h)^3} + \sqrt{x^3}} \right).$$

The first factor of the expression in the last line is clearly the limit definition of the derivative of x^3, which is $3x^2$. The second factor can be evaluated, so

$$\frac{d}{dx} x^{3/2} = 3x^2 \frac{1}{2\sqrt{x^3}} = \frac{3}{2} x^{1/2}.$$

59. Let $f(x) = xe^x$. Use the limit definition to compute $f'(0)$, and find the equation of the tangent line at $x = 0$.

SOLUTION Let $f(x) = xe^x$. Then $f(0) = 0$, and

$$f'(0) = \lim_{h \to 0} \frac{f(0+h) - f(0)}{h} = \lim_{h \to 0} \frac{he^h - 0}{h} = \lim_{h \to 0} e^h = 1.$$

The equation of the tangent line is

$$y = f'(0)(x - 0) + f(0) = 1(x - 0) + 0 = x.$$

61. Biologists have observed that the pulse rate P (in beats per minute) in animals is related to body mass (in kilograms) by the approximate formula $P = 200m^{-1/4}$. This is one of many *allometric scaling laws* prevalent in biology. Is $|dP/dm|$ an increasing or decreasing function of m? Find an equation of the tangent line at the points on the graph in Figure 18 that represent goat ($m = 33$) and man ($m = 68$).

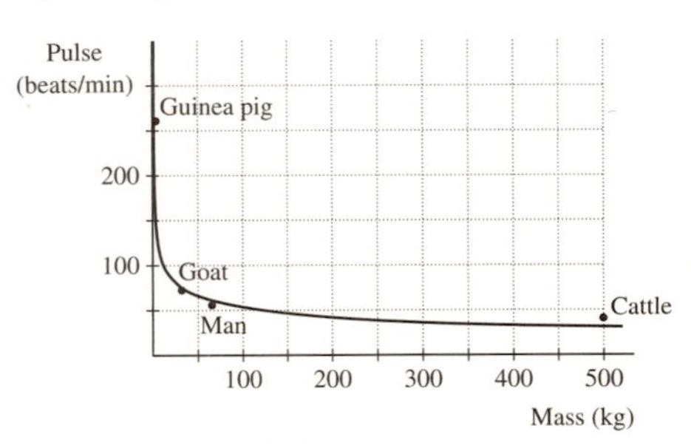

FIGURE 18

SOLUTION $dP/dm = -50m^{-5/4}$. For $m > 0$, $|dP/dm| = |50m^{-5/4}|$. $|dP/dm| \to 0$ as m gets larger; $|dP/dm|$ gets smaller as m gets bigger.

For each $m = c$, the equation of the tangent line to the graph of P at m is

$$y = P'(c)(m - c) + P(c).$$

For a goat ($m = 33$ kg), $P(33) = 83.445$ beats per minute (bpm) and

$$\frac{dP}{dm} = -50(33)^{-5/4} \approx -0.63216 \text{ bpm/kg}.$$

Hence, $y = -0.63216(m - 33) + 83.445$.

For a man ($m = 68$ kg), we have $P(68) = 69.647$ bpm and

$$\frac{dP}{dm} = -50(68)^{-5/4} \approx -0.25606 \text{ bpm/kg}.$$

Hence, the tangent line has formula $y = -0.25606(m - 68) + 69.647$.

63. The Clausius–Clapeyron Law relates the *vapor pressure* of water P (in atmospheres) to the temperature T (in kelvins):

$$\frac{dP}{dT} = k\frac{P}{T^2}$$

where k is a constant. Estimate dP/dT for $T = 303, 313, 323, 333, 343$ using the data and the approximation

$$\frac{dP}{dT} \approx \frac{P(T+10) - P(T-10)}{20}$$

T (K)	293	303	313	323	333	343	353
P (atm)	0.0278	0.0482	0.0808	0.1311	0.2067	0.3173	0.4754

Do your estimates seem to confirm the Clausius–Clapeyron Law? What is the approximate value of k?

SOLUTION Using the indicated approximation to the first derivative, we calculate

$$P'(303) \approx \frac{P(313) - P(293)}{20} = \frac{0.0808 - 0.0278}{20} = 0.00265 \text{ atm/K};$$

$$P'(313) \approx \frac{P(323) - P(303)}{20} = \frac{0.1311 - 0.0482}{20} = 0.004145 \text{ atm/K};$$

$$P'(323) \approx \frac{P(333) - P(313)}{20} = \frac{0.2067 - 0.0808}{20} = 0.006295 \text{ atm/K};$$

$$P'(333) \approx \frac{P(343) - P(323)}{20} = \frac{0.3173 - 0.1311}{20} = 0.00931 \text{ atm/K};$$

$$P'(343) \approx \frac{P(353) - P(333)}{20} = \frac{0.4754 - 0.2067}{20} = 0.013435 \text{ atm/K}$$

If the Clausius–Clapeyron law is valid, then $\frac{T^2}{P}\frac{dP}{dT}$ should remain constant as T varies. Using the data for vapor pressure and temperature and the approximate derivative values calculated above, we find

T (K)	303	313	323	333	343
$\frac{T^2}{P}\frac{dP}{dT}$	5047.59	5025.76	5009.54	4994.57	4981.45

These values are roughly constant, suggesting that the Clausius–Clapeyron law is valid, and that $k \approx 5000$.

65. In the setting of Exercise 64, show that the point of tangency is the midpoint of the segment of L lying in the first quadrant.

SOLUTION In the previous exercise, we saw that the tangent line to the hyperbola $xy = 1$ or $y = \frac{1}{x}$ at $x = a$ has y-intercept $P = (0, \frac{2}{a})$ and x-intercept $Q = (2a, 0)$. The midpoint of the line segment connecting P and Q is thus

$$\left(\frac{0 + 2a}{2}, \frac{\frac{2}{a} + 0}{2}\right) = \left(a, \frac{1}{a}\right),$$

which is the point of tangency.

67. Make a rough sketch of the graph of the derivative of the function in Figure 20(A).

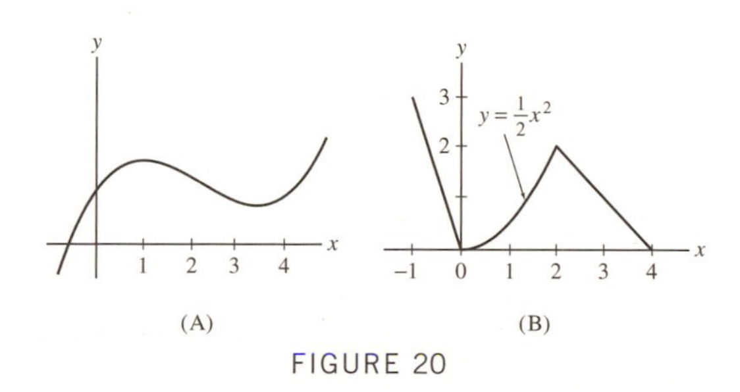

FIGURE 20

SOLUTION The graph has a tangent line with negative slope approximately on the interval $(1, 3.6)$, and has a tangent line with a positive slope elsewhere. This implies that the derivative must be negative on the interval $(1, 3.6)$ and positive elsewhere. The graph may therefore look like this:

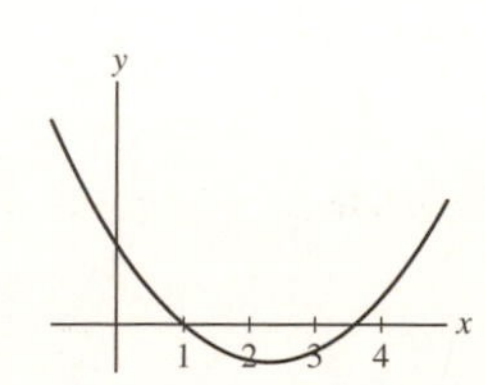

69. Sketch the graph of $f(x) = x\,|x|$. Then show that $f'(0)$ exists.

SOLUTION For $x < 0$, $f(x) = -x^2$, and $f'(x) = -2x$. For $x > 0$, $f(x) = x^2$, and $f'(x) = 2x$. At $x = 0$, we find

$$\lim_{h\to 0+} \frac{f(0+h) - f(0)}{h} = \lim_{h\to 0+} \frac{h^2}{h} = 0$$

and

$$\lim_{h\to 0-} \frac{f(0+h) - f(0)}{h} = \lim_{h\to 0-} \frac{-h^2}{h} = 0.$$

Because the two one-sided limits exist and are equal, it follows that $f'(0)$ exists and is equal to zero. Here is the graph of $f(x) = x|x|$.

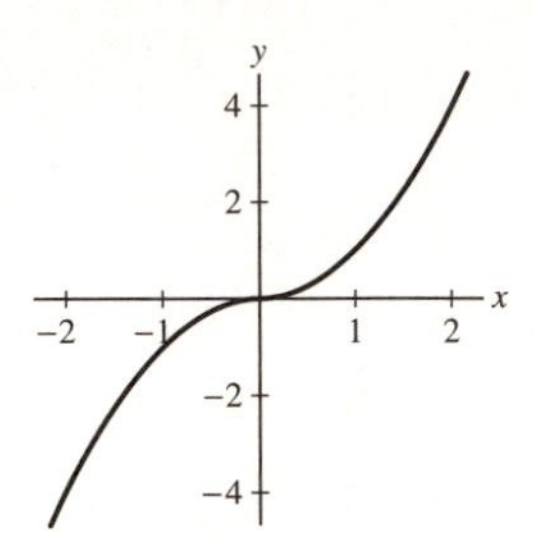

In Exercises 71–76, find the points c (if any) such that $f'(c)$ does not exist.

71. $f(x) = |x - 1|$

SOLUTION

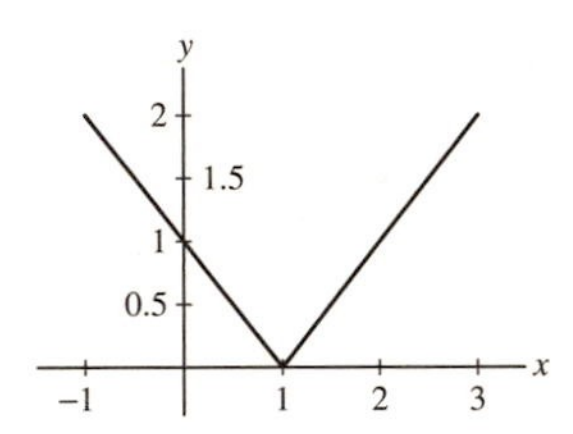

Here is the graph of $f(x) = |x - 1|$. Its derivative does not exist at $x = 1$. At that value of x there is a sharp corner.

73. $f(x) = x^{2/3}$

SOLUTION Here is the graph of $f(x) = x^{2/3}$. Its derivative does not exist at $x = 0$. At that value of x, there is a sharp corner or "cusp".

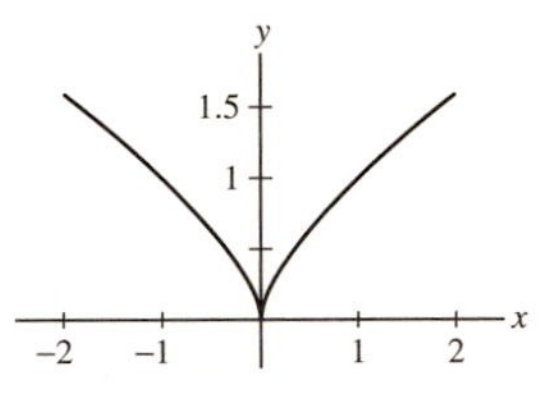

75. $f(x) = |x^2 - 1|$

SOLUTION Here is the graph of $f(x) = \left|x^2 - 1\right|$. Its derivative does not exist at $x = -1$ or at $x = 1$. At these values of x, the graph has sharp corners.

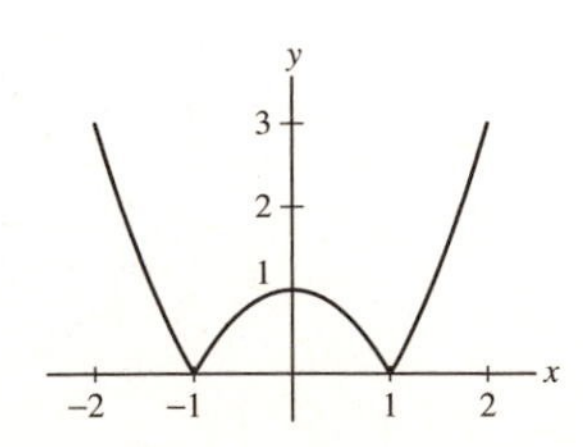

GU *In Exercises 77–82, zoom in on a plot of $f(x)$ at the point $(a, f(a))$ and state whether or not $f(x)$ appears to be differentiable at $x = a$. If it is nondifferentiable, state whether the tangent line appears to be vertical or does not exist.*

77. $f(x) = (x - 1)|x|, \quad a = 0$

SOLUTION The graph of $f(x) = (x - 1)|x|$ for x near 0 is shown below. Because the graph has a sharp corner at $x = 0$, it appears that f is not differentiable at $x = 0$. Moreover, the tangent line does not exist at this point.

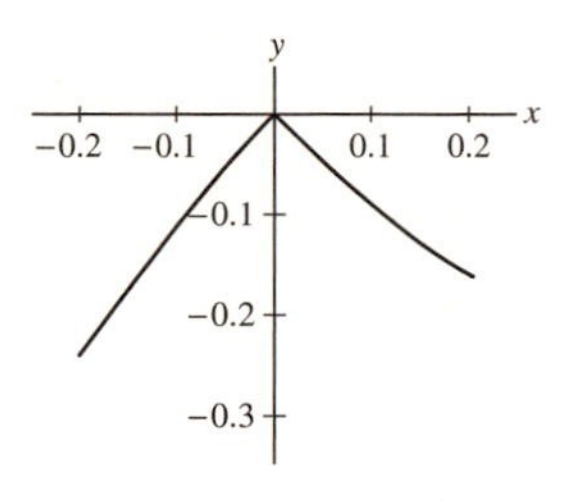

79. $f(x) = (x - 3)^{1/3}, \quad a = 3$

SOLUTION The graph of $f(x) = (x - 3)^{1/3}$ for x near 3 is shown below. From this graph, it appears that f is not differentiable at $x = 3$. Moreover, the tangent line appears to be vertical.

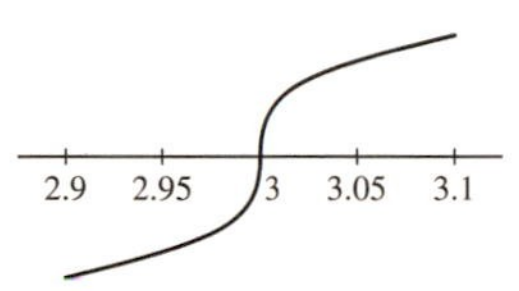

81. $f(x) = |\sin x|, \quad a = 0$

SOLUTION The graph of $f(x) = |\sin x|$ for x near 0 is shown below. Because the graph has a sharp corner at $x = 0$, it appears that f is not differentiable at $x = 0$. Moreover, the tangent line does not exist at this point.

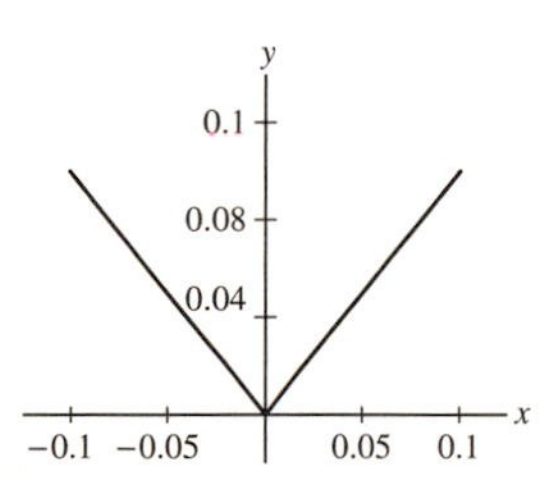

83. GU Plot the derivative $f'(x)$ of $f(x) = 2x^3 - 10x^{-1}$ for $x > 0$ (set the bounds of the viewing box appropriately) and observe that $f'(x) > 0$. What does the positivity of $f'(x)$ tell us about the graph of $f(x)$ itself? Plot $f(x)$ and confirm this conclusion.

SOLUTION Let $f(x) = 2x^3 - 10x^{-1}$. Then $f'(x) = 6x^2 + 10x^{-2}$. The graph of $f'(x)$ is shown in the figure below at the left and it is clear that $f'(x) > 0$ for all $x > 0$. The positivity of $f'(x)$ tells us that the graph of $f(x)$ is increasing for $x > 0$. This is confirmed in the figure below at the right, which shows the graph of $f(x)$.

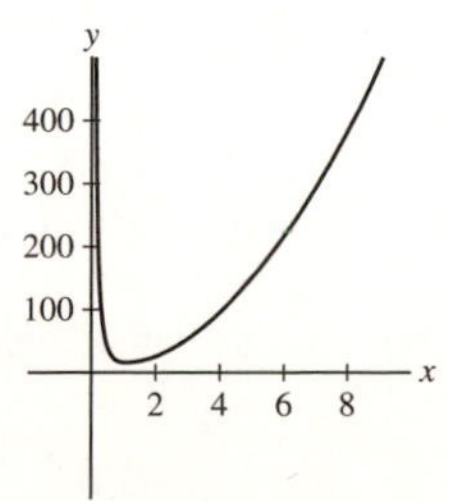

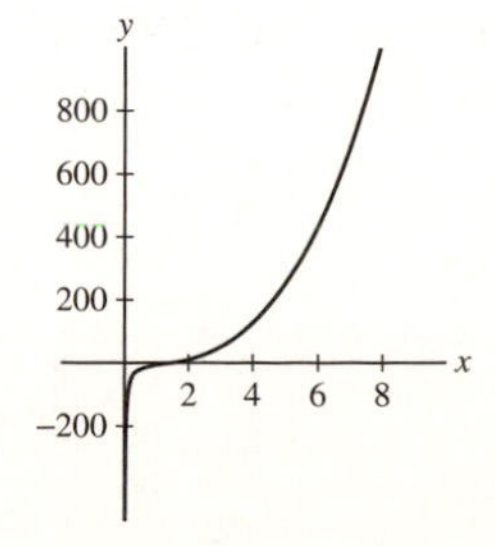

Exercises 85–88 refer to Figure 23. Length QR is called the subtangent *at P, and length RT is called the* subnormal.

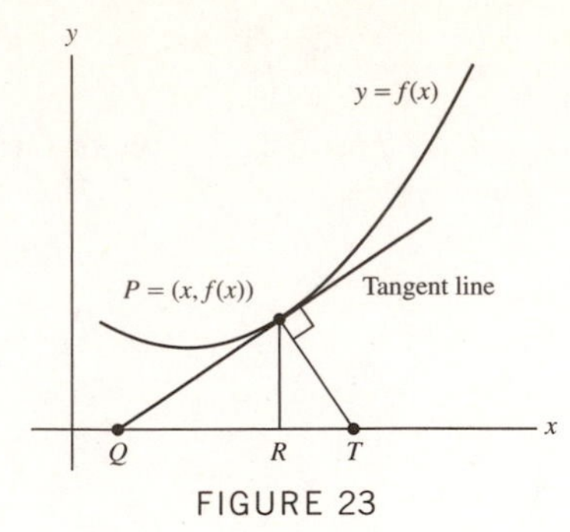

FIGURE 23

85. Calculate the subtangent of

$$f(x) = x^2 + 3x \quad \text{at } x = 2$$

SOLUTION Let $f(x) = x^2 + 3x$. Then $f'(x) = 2x + 3$, and the equation of the tangent line at $x = 2$ is

$$y = f'(2)(x-2) + f(2) = 7(x-2) + 10 = 7x - 4.$$

This line intersects the x-axis at $x = \frac{4}{7}$. Thus Q has coordinates $(\frac{4}{7}, 0)$, R has coordinates $(2, 0)$ and the subtangent is

$$2 - \frac{4}{7} = \frac{10}{7}.$$

87. Prove in general that the subnormal at P is $|f'(x)f(x)|$.

SOLUTION The slope of the tangent line at P is $f'(x)$. The slope of the line normal to the graph at P is then $-1/f'(x)$, and the normal line intersects the x-axis at the point T with coordinates $(x + f(x)f'(x), 0)$. The point R has coordinates $(x, 0)$, so the subnormal is

$$|x + f(x)f'(x) - x| = |f(x)f'(x)|.$$

89. Prove the following theorem of Apollonius of Perga (the Greek mathematician born in 262 BCE who gave the parabola, ellipse, and hyperbola their names): The subtangent of the parabola $y = x^2$ at $x = a$ is equal to $a/2$.

SOLUTION Let $f(x) = x^2$. The tangent line to f at $x = a$ is

$$y = f'(a)(x-a) + f(a) = 2a(x-a) + a^2 = 2ax - a^2.$$

The x-intercept of this line (where $y = 0$) is $\frac{a}{2}$ as claimed.

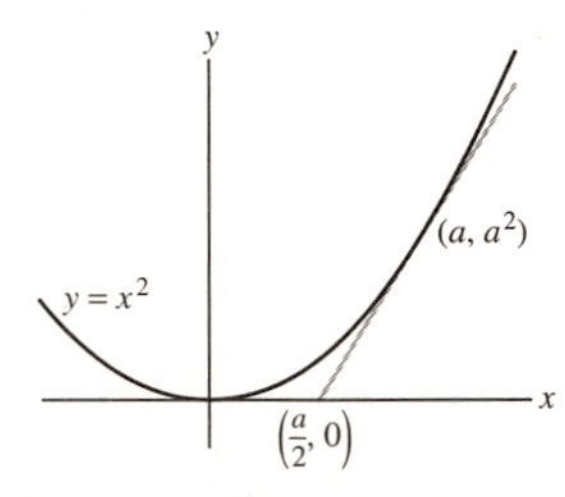

91. Formulate and prove a generalization of Exercise 90 for $y = x^n$.

SOLUTION Let $f(x) = x^n$. Then $f'(x) = nx^{n-1}$, and the equation of the tangent line t $x = a$ is

$$y = f'(a)(x-a) + f(a) = na^{n-1}(x-a) + a^n = na^{n-1}x - (n-1)a^n.$$

This line intersects the x-axis at $x = (n-1)a/n$. Thus, Q has coordinates $((n-1)a/n, 0)$, R has coordinates $(a, 0)$ and the subtangent is

$$a - \frac{n-1}{n}a = \frac{1}{n}a.$$

Further Insights and Challenges

93. A vase is formed by rotating $y = x^2$ around the y-axis. If we drop in a marble, it will either touch the bottom point of the vase or be suspended above the bottom by touching the sides (Figure 25). How small must the marble be to touch the bottom?

FIGURE 25

SOLUTION Suppose a circle is tangent to the parabola $y = x^2$ at the point (t, t^2). The slope of the parabola at this point is $2t$, so the slope of the radius of the circle at this point is $-\frac{1}{2t}$ (since it is perpendicular to the tangent line of the circle). Thus the center of the circle must be where the line given by $y = -\frac{1}{2t}(x - t) + t^2$ crosses the y-axis. We can find the y-coordinate by setting $x = 0$: we get $y = \frac{1}{2} + t^2$. Thus, the radius extends from $(0, \frac{1}{2} + t^2)$ to (t, t^2) and

$$r = \sqrt{\left(\frac{1}{2} + t^2 - t^2\right)^2 + t^2} = \sqrt{\frac{1}{4} + t^2}.$$

This radius is greater than $\frac{1}{2}$ whenever $t > 0$; so, if a marble has radius $> 1/2$ it sits on the edge of the vase, but if it has radius $\leq 1/2$ it rolls all the way to the bottom.

95. Negative Exponents Let n be a whole number. Use the Power Rule for x^n to calculate the derivative of $f(x) = x^{-n}$ by showing that

$$\frac{f(x+h) - f(x)}{h} = \frac{-1}{x^n(x+h)^n} \frac{(x+h)^n - x^n}{h}$$

SOLUTION Let $f(x) = x^{-n}$ where n is a positive integer.

- The difference quotient for f is

$$\frac{f(x+h) - f(x)}{h} = \frac{(x+h)^{-n} - x^{-n}}{h} = \frac{\frac{1}{(x+h)^n} - \frac{1}{x^n}}{h} = \frac{\frac{x^n - (x+h)^n}{x^n(x+h)^n}}{h}$$
$$= \frac{-1}{x^n(x+h)^n} \frac{(x+h)^n - x^n}{h}.$$

- Therefore,

$$f'(x) = \lim_{h \to 0} \frac{f(x+h) - f(x)}{h} = \lim_{h \to 0} \frac{-1}{x^n(x+h)^n} \frac{(x+h)^n - x^n}{h}$$
$$= \lim_{h \to 0} \frac{-1}{x^n(x+h)^n} \lim_{h \to 0} \frac{(x+h)^n - x^n}{h} = -x^{-2n} \frac{d}{dx}\left(x^n\right).$$

- From above, we continue: $f'(x) = -x^{-2n} \frac{d}{dx}\left(x^n\right) = -x^{-2n} \cdot nx^{n-1} = -nx^{-n-1}$. Since n is a positive integer, $k = -n$ is a negative integer and we have $\frac{d}{dx}\left(x^k\right) = \frac{d}{dx}\left(x^{-n}\right) = -nx^{-n-1} = kx^{k-1}$; i.e. $\frac{d}{dx}\left(x^k\right) = kx^{k-1}$ for negative integers k.

97. Infinitely Rapid Oscillations Define

$$f(x) = \begin{cases} x \sin \frac{1}{x} & x \neq 0 \\ 0 & x = 0 \end{cases}$$

Show that $f(x)$ is continuous at $x = 0$ but $f'(0)$ does not exist (see Figure 24).

SOLUTION Let $f(x) = \begin{cases} x\sin\left(\frac{1}{x}\right) & \text{if } x \neq 0 \\ 0 & \text{if } x = 0 \end{cases}$. As $x \to 0$,

$$|f(x) - f(0)| = \left|x\sin\left(\frac{1}{x}\right) - 0\right| = |x|\left|\sin\left(\frac{1}{x}\right)\right| \to 0$$

since the values of the sine lie between -1 and 1. Hence, by the Squeeze Theorem, $\lim_{x\to 0} f(x) = f(0)$ and thus f is continuous at $x = 0$.

As $x \to 0$, the difference quotient at $x = 0$,

$$\frac{f(x) - f(0)}{x - 0} = \frac{x\sin\left(\frac{1}{x}\right) - 0}{x - 0} = \sin\left(\frac{1}{x}\right)$$

does *not* converge to a limit since it oscillates infinitely through every value between -1 and 1. Accordingly, $f'(0)$ does not exist.

3.3 Product and Quotient Rules

Preliminary Questions

1. Are the following statements true or false? If false, state the correct version.

(a) fg denotes the function whose value at x is $f(g(x))$.

(b) f/g denotes the function whose value at x is $f(x)/g(x)$.

(c) The derivative of the product is the product of the derivatives.

(d) $\left.\frac{d}{dx}(fg)\right|_{x=4} = f(4)g'(4) - g(4)f'(4)$

(e) $\left.\frac{d}{dx}(fg)\right|_{x=0} = f(0)g'(0) + g(0)f'(0)$

SOLUTION

(a) False. The notation fg denotes the function whose value at x is $f(x)g(x)$.

(b) True.

(c) False. The derivative of a product fg is $f'(x)g(x) + f(x)g'(x)$.

(d) False. $\left.\frac{d}{dx}(fg)\right|_{x=4} = f(4)g'(4) + g(4)f'(4)$.

(e) True.

2. Find $(f/g)'(1)$ if $f(1) = f'(1) = g(1) = 2$ and $g'(1) = 4$.

SOLUTION $\frac{d}{dx}(f/g)|_{x=1} = [g(1)f'(1) - f(1)g'(1)]/g(1)^2 = [2(2) - 2(4)]/2^2 = -1$.

3. Find $g(1)$ if $f(1) = 0$, $f'(1) = 2$, and $(fg)'(1) = 10$.

SOLUTION $(fg)'(1) = f(1)g'(1) + f'(1)g(1)$, so $10 = 0 \cdot g'(1) + 2g(1)$ and $g(1) = 5$.

Exercises

In Exercises 1–6, use the Product Rule to calculate the derivative.

1. $f(x) = x^3(2x^2 + 1)$

SOLUTION Let $f(x) = x^3(2x^2 + 1)$. Then

$$f'(x) = x^3\frac{d}{dx}(2x^2 + 1) + (2x^2 + 1)\frac{d}{dx}x^3 = x^3(4x) + (2x^2 + 1)(3x^2) = 10x^4 + 3x^2.$$

3. $f(x) = x^2e^x$

SOLUTION Let $f(x) = x^2e^x$. Then

$$f'(x) = x^2\frac{d}{dx}e^x + e^x\frac{d}{dx}x^2 = x^2e^x + e^x(2x) = e^x(x^2 + 2x).$$

5. $\left.\dfrac{dh}{ds}\right|_{s=4}$, $h(s) = (s^{-1/2} + 2s)(7 - s^{-1})$

SOLUTION Let $h(s) = (s^{-1/2} + 2s)(7 - s^{-1})$. Then

$$\begin{aligned}\frac{dh}{ds} &= (s^{-1/2} + 2s)\frac{d}{dx}(7 - s^{-1}) + (7 - s^{-1})\frac{d}{ds}\left(s^{-1/2} + 2s\right)\\ &= (s^{-1/2} + 2s)(s^{-2}) + (7 - s^{-1})\left(-\frac{1}{2}s^{-3/2} + 2\right) = -\frac{7}{2}s^{-3/2} + \frac{3}{2}s^{-5/2} + 14.\end{aligned}$$

Therefore,

$$\left.\frac{dh}{ds}\right|_{s=4} = -\frac{7}{2}(4)^{-3/2} + \frac{3}{2}(4)^{-5/2} + 14 = \frac{871}{64}.$$

In Exercises 7–12, use the Quotient Rule to calculate the derivative.

7. $f(x) = \dfrac{x}{x - 2}$

SOLUTION Let $f(x) = \frac{x}{x-2}$. Then

$$f'(x) = \frac{(x-2)\frac{d}{dx}x - x\frac{d}{dx}(x-2)}{(x-2)^2} = \frac{(x-2) - x}{(x-2)^2} = \frac{-2}{(x-2)^2}.$$

9. $\left.\dfrac{dg}{dt}\right|_{t=-2}$, $g(t) = \dfrac{t^2 + 1}{t^2 - 1}$

SOLUTION Let $g(t) = \dfrac{t^2 + 1}{t^2 - 1}$. Then

$$\frac{dg}{dt} = \frac{(t^2 - 1)\frac{d}{dt}(t^2 + 1) - (t^2 + 1)\frac{d}{dt}(t^2 - 1)}{(t^2 - 1)^2} = \frac{(t^2 - 1)(2t) - (t^2 + 1)(2t)}{(t^2 - 1)^2} = -\frac{4t}{(t^2 - 1)^2}.$$

Therefore,

$$\left.\frac{dg}{dt}\right|_{t=-2} = -\frac{4(-2)}{((-2)^2 - 1)^2} = \frac{8}{9}.$$

11. $g(x) = \dfrac{1}{1 + e^x}$

SOLUTION Let $g(x) = \dfrac{1}{1 + e^x}$. Then

$$\frac{dg}{dx} = \frac{(1 + e^x)\frac{d}{dx}1 - 1\frac{d}{dx}(1 + e^x)}{(1 + e^x)^2} = \frac{(1 + e^x)(0) - e^x}{(1 + e^x)^2} = -\frac{e^x}{(1 + e^x)^2}.$$

In Exercises 13–16, calculate the derivative in two ways. First use the Product or Quotient Rule; then rewrite the function algebraically and apply the Power Rule directly.

13. $f(t) = (2t + 1)(t^2 - 2)$

SOLUTION Let $f(t) = (2t + 1)(t^2 - 2)$. Then, using the Product Rule,

$$f'(t) = (2t + 1)(2t) + (t^2 - 2)(2) = 6t^2 + 2t - 4.$$

Multiplying out first, we find $f(t) = 2t^3 + t^2 - 4t - 2$. Therefore, $f'(t) = 6t^2 + 2t - 4$.

15. $h(t) = \dfrac{t^2 - 1}{t - 1}$

SOLUTION Let $h(t) = \frac{t^2-1}{t-1}$. Using the quotient rule,

$$f'(t) = \frac{(t - 1)(2t) - (t^2 - 1)(1)}{(t - 1)^2} = \frac{t^2 - 2t + 1}{(t - 1)^2} = 1$$

for $t \neq 1$. Simplifying first, we find for $t \neq 1$,

$$h(t) = \frac{(t - 1)(t + 1)}{(t - 1)} = t + 1.$$

Hence $h'(t) = 1$ for $t \neq 1$.

In Exercises 17–38, calculate the derivative.

17. $f(x) = (x^3 + 5)(x^3 + x + 1)$

SOLUTION Let $f(x) = (x^3 + 5)(x^3 + x + 1)$. Then

$$f'(x) = (x^3 + 5)(3x^2 + 1) + (x^3 + x + 1)(3x^2) = 6x^5 + 4x^3 + 18x^2 + 5.$$

19. $\left.\frac{dy}{dx}\right|_{x=3}, \quad y = \frac{1}{x+10}$

SOLUTION Let $y = \frac{1}{x+10}$. Using the quotient rule:

$$\frac{dy}{dx} = \frac{(x+10)(0) - 1(1)}{(x+10)^2} = -\frac{1}{(x+10)^2}.$$

Therefore,

$$\left.\frac{dy}{dx}\right|_{x=3} = -\frac{1}{(3+10)^2} = -\frac{1}{169}.$$

21. $f(x) = (\sqrt{x} + 1)(\sqrt{x} - 1)$

SOLUTION Let $f(x) = (\sqrt{x} + 1)(\sqrt{x} - 1)$. Multiplying through first yields $f(x) = x - 1$ for $x \geq 0$. Therefore, $f'(x) = 1$ for $x \geq 0$. If we carry out the product rule on $f(x) = (x^{1/2} + 1)(x^{1/2} - 1)$, we get

$$f'(x) = (x^{1/2} + 1)\left(\frac{1}{2}(x^{-1/2})\right) + (x^{1/2} - 1)\left(\frac{1}{2}x^{-1/2}\right) = \frac{1}{2} + \frac{1}{2}x^{-1/2} + \frac{1}{2} - \frac{1}{2}x^{-1/2} = 1.$$

23. $\left.\frac{dy}{dx}\right|_{x=2}, \quad y = \frac{x^4 - 4}{x^2 - 5}$

SOLUTION Let $y = \frac{x^4 - 4}{x^2 - 5}$. Then

$$\frac{dy}{dx} = \frac{\left(x^2 - 5\right)\left(4x^3\right) - \left(x^4 - 4\right)(2x)}{\left(x^2 - 5\right)^2} = \frac{2x^5 - 20x^3 + 8x}{\left(x^2 - 5\right)^2}.$$

Therefore,

$$\left.\frac{dy}{dx}\right|_{x=2} = \frac{2(2)^5 - 20(2)^3 + 8(2)}{(2^2 - 5)^2} = -80.$$

25. $\left.\frac{dz}{dx}\right|_{x=1}, \quad z = \frac{1}{x^3 + 1}$

SOLUTION Let $z = \frac{1}{x^3+1}$. Using the quotient rule:

$$\frac{dz}{dx} = \frac{(x^3 + 1)(0) - 1(3x^2)}{(x^3 + 1)^2} = -\frac{3x^2}{(x^3 + 1)^2}.$$

Therefore,

$$\left.\frac{dz}{dx}\right|_{x=1} = -\frac{3(1)^2}{(1^3 + 1)^2} = -\frac{3}{4}.$$

27. $h(t) = \frac{t}{(t+1)(t^2+1)}$

SOLUTION Let $h(t) = \frac{t}{(t+1)(t^2+1)} = \frac{t}{t^3 + t^2 + t + 1}$. Then

$$h'(t) = \frac{\left(t^3 + t^2 + t + 1\right)(1) - t\left(3t^2 + 2t + 1\right)}{\left(t^3 + t^2 + t + 1\right)^2} = \frac{-2t^3 - t^2 + 1}{\left(t^3 + t^2 + t + 1\right)^2}.$$

29. $f(t) = 3^{1/2} \cdot 5^{1/2}$

SOLUTION Let $f(t) = \sqrt{3}\sqrt{5}$. Then $f'(t) = 0$, since $f(t)$ is a *constant* function!

31. $f(x) = (x+3)(x-1)(x-5)$

SOLUTION Let $f(x) = (x+3)(x-1)(x-5)$. Using the Product Rule inside the Product Rule with a first factor of $(x+3)$ and a second factor of $(x-1)(x-5)$, we find

$$f'(x) = (x+3)\left((x-1)(1) + (x-5)(1)\right) + (x-1)(x-5)(1) = 3x^2 - 6x - 13.$$

Alternatively,

$$f(x) = (x+3)\left(x^2 - 6x + 5\right) = x^3 - 3x^2 - 13x + 15.$$

Therefore, $f'(x) = 3x^2 - 6x - 13$.

33. $f(x) = \dfrac{e^x}{x+1}$

SOLUTION Let $f(x) = \dfrac{e^x}{(e^x+1)(x+1)}$. Then

$$f'(x) = \frac{(e^x+1)(x+1)e^x - e^x\left((e^x+1)(1) + (x+1)e^x\right)}{(e^x+1)^2(x+1)^2} = \frac{e^x(x-e^x)}{(e^x+1)^2(x+1)^2}.$$

35. $g(z) = \left(\dfrac{z^2-4}{z-1}\right)\left(\dfrac{z^2-1}{z+2}\right)$ *Hint:* Simplify first.

SOLUTION Let

$$g(z) = \left(\frac{z^2-4}{z-1}\right)\left(\frac{z^2-1}{z+2}\right) = \left(\frac{(z+2)(z-2)}{z-1}\right)\left(\frac{(z+1)(z-1)}{z+2}\right) = (z-2)(z+1)$$

for $z \neq -2$ and $z \neq 1$. Then,

$$g'(z) = (z+1)(1) + (z-2)(1) = 2z - 1.$$

37. $\dfrac{d}{dt}\left(\dfrac{xt-4}{t^2-x}\right)$ (x constant)

SOLUTION Let $f(t) = \frac{xt-4}{t^2-x}$. Using the quotient rule:

$$f'(t) = \frac{(t^2-x)(x) - (xt-4)(2t)}{(t^2-x)^2} = \frac{xt^2 - x^2 - 2xt^2 + 8t}{(t^2-x)^2} = \frac{-xt^2 + 8t - x^2}{(t^2-x)^2}.$$

In Exercises 39–42, calculate the derivative using the values:

$f(4)$	$f'(4)$	$g(4)$	$g'(4)$
10	−2	5	−1

39. $(fg)'(4)$ and $(f/g)'(4)$.

SOLUTION Let $h = fg$ and $H = f/g$. Then $h' = fg' + gf'$ and $H' = \frac{gf'-fg'}{g^2}$. Finally,

$$h'(4) = f(4)g'(4) + g(4)f'(4) = (10)(-1) + (5)(-2) = -20,$$

and

$$H'(4) = \frac{g(4)f'(4) - f(4)g'(4)}{(g(4))^2} = \frac{(5)(-2) - (10)(-1)}{(5)^2} = 0.$$

41. $G'(4)$, where $G(x) = g(x)^2$.

SOLUTION Let $G(x) = g(x)^2 = g(x)g(x)$. Then $G'(x) = g(x)g'(x) + g(x)g'(x) = 2g(x)g'(x)$, and

$$G'(4) = 2g(4)g'(4) = 2(5)(-1) = -10.$$

43. Calculate $F'(0)$, where

$$F(x) = \frac{x^9 + x^8 + 4x^5 - 7x}{x^4 - 3x^2 + 2x + 1}$$

Hint: Do not calculate $F'(x)$. Instead, write $F(x) = f(x)/g(x)$ and express $F'(0)$ directly in terms of $f(0)$, $f'(0)$, $g(0)$, $g'(0)$.

SOLUTION Taking the hint, let

$$f(x) = x^9 + x^8 + 4x^5 - 7x$$

and let

$$g(x) = x^4 - 3x^2 + 2x + 1.$$

Then $F(x) = \frac{f(x)}{g(x)}$. Now,

$$f'(x) = 9x^8 + 8x^7 + 20x^4 - 7 \quad \text{and} \quad g'(x) = 4x^3 - 6x + 2.$$

Moreover, $f(0) = 0$, $f'(0) = -7$, $g(0) = 1$, and $g'(0) = 2$.
Using the quotient rule:

$$F'(0) = \frac{g(0)f'(0) - f(0)g'(0)}{(g(0))^2} = \frac{-7 - 0}{1} = -7.$$

45. Use the Product Rule to calculate $\frac{d}{dx}e^{2x}$.

SOLUTION Note that $e^{2x} = e^x \cdot e^x$. Therefore

$$\frac{d}{dx}e^{2x} = \frac{d}{dx}(e^x \cdot e^x) = e^x \cdot e^x + e^x \cdot e^x = 2e^{2x}.$$

47. GU Plot $f(x) = x/(x^2 - 1)$ (in a suitably bounded viewing box). Use the plot to determine whether $f'(x)$ is positive or negative on its domain $\{x : x \neq \pm 1\}$. Then compute $f'(x)$ and confirm your conclusion algebraically.

SOLUTION Let $f(x) = \frac{x}{x^2 - 1}$. The graph of $f(x)$ is shown below. From this plot, we see that $f(x)$ is decreasing on its domain $\{x : x \neq \pm 1\}$. Consequently, $f'(x)$ must be negative. Using the quotient rule, we find

$$f'(x) = \frac{(x^2 - 1)(1) - x(2x)}{(x^2 - 1)^2} = -\frac{x^2 + 1}{(x^2 - 1)^2},$$

which is negative for all $x \neq \pm 1$.

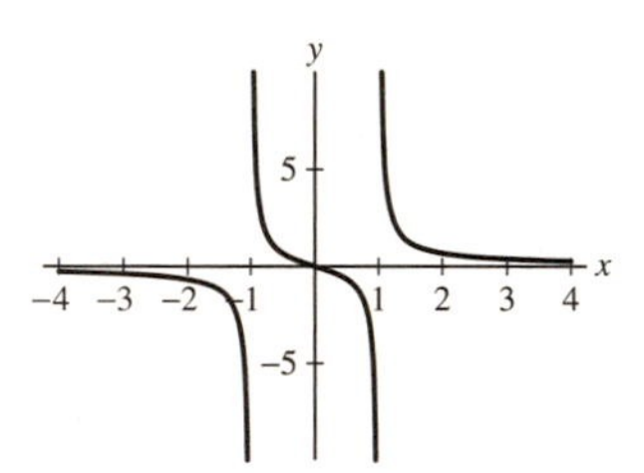

49. Find $a > 0$ such that the tangent line to the graph of

$$f(x) = x^2e^{-x} \quad \text{at } x = a$$

passes through the origin (Figure 4).

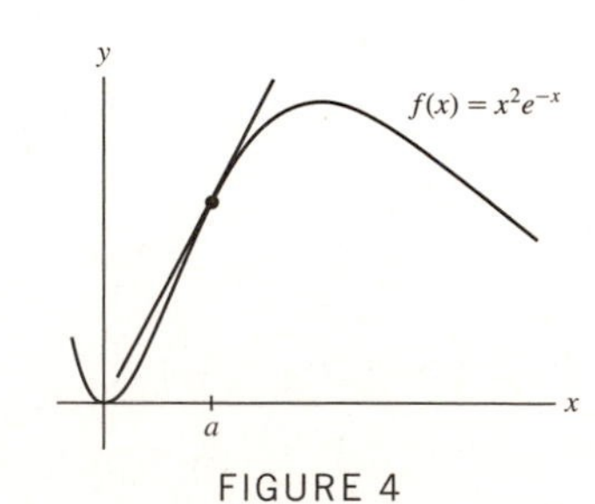

FIGURE 4

SOLUTION Let $f(x) = x^2e^{-x}$. Then $f(a) = a^2e^{-a}$,

$$f'(x) = -x^2e^{-x} + 2xe^{-x} = e^{-x}(2x - x^2),$$

$f'(a) = (2a - a^2)e^{-a}$, and the equation of the tangent line to f at $x = a$ is

$$y = f'(a)(x - a) + f(a) = (2a - a^2)e^{-a}(x - a) + a^2e^{-a}.$$

For this line to pass through the origin, we must have

$$0 = (2a - a^2)e^{-a}(-a) + a^2e^{-a} = e^{-a}\left(a^2 - 2a^2 + a^3\right) = a^2e^{-a}(a - 1).$$

Thus, $a = 0$ or $a = 1$. The only value $a > 0$ such that the tangent line to $f(x) = x^2e^{-x}$ passes through the origin is therefore $a = 1$.

51. The revenue per month earned by the Couture clothing chain at time t is $R(t) = N(t)S(t)$, where $N(t)$ is the number of stores and $S(t)$ is average revenue per store per month. Couture embarks on a two-part campaign: (A) to build new stores at a rate of 5 stores per month, and (B) to use advertising to increase average revenue per store at a rate of \$10,000 per month. Assume that $N(0) = 50$ and $S(0) = \$150{,}000$.

(a) Show that total revenue will increase at the rate

$$\frac{dR}{dt} = 5S(t) + 10{,}000N(t)$$

Note that the two terms in the Product Rule correspond to the separate effects of increasing the number of stores on the one hand, and the average revenue per store on the other.

(b) Calculate $\left.\dfrac{dR}{dt}\right|_{t=0}$.

(c) If Couture can implement only one leg (A or B) of its expansion at $t = 0$, which choice will grow revenue most rapidly?

SOLUTION

(a) Given $R(t) = N(t)S(t)$, it follows that

$$\frac{dR}{dt} = N(t)S'(t) + S(t)N'(t).$$

We are told that $N'(t) = 5$ stores per month and $S'(t) = 10{,}000$ dollars per month. Therefore,

$$\frac{dR}{dt} = 5S(t) + 10{,}000N(t).$$

(b) Using part (a) and the given values of $N(0)$ and $S(0)$, we find

$$\left.\frac{dR}{dt}\right|_{t=0} = 5(150{,}000) + 10{,}000(50) = 1{,}250{,}000.$$

(c) From part (b), we see that of the two terms contributing to total revenue growth, the term $5S(0)$ is larger than the term $10{,}000N(0)$. Thus, if only one leg of the campaign can be implemented, it should be part A: increase the number of stores by 5 per month.

53. The curve $y = 1/(x^2 + 1)$ is called the *witch of Agnesi* (Figure 6) after the Italian mathematician Maria Agnesi (1718–1799), who wrote one of the first books on calculus. This strange name is the result of a mistranslation of the Italian word *la versiera*, meaning "that which turns." Find equations of the tangent lines at $x = \pm 1$.

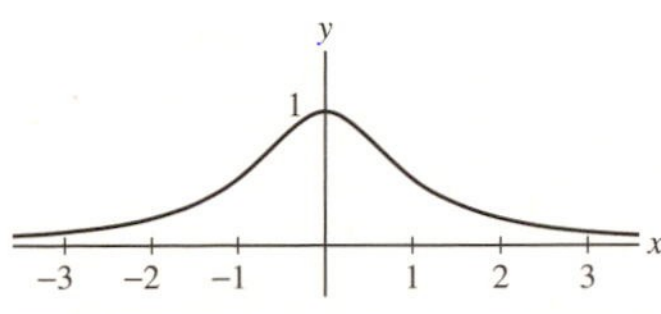

FIGURE 6 The witch of Agnesi.

SOLUTION Let $f(x) = \dfrac{1}{x^2 + 1}$. Then $f'(x) = \dfrac{(x^2 + 1)(0) - 1(2x)}{(x^2 + 1)^2} = -\dfrac{2x}{(x^2 + 1)^2}$.

- At $x = -1$, the tangent line is

$$y = f'(-1)(x + 1) + f(-1) = \frac{1}{2}(x + 1) + \frac{1}{2} = \frac{1}{2}x + 1.$$

- At $x = 1$, the tangent line is

$$y = f'(1)(x-1) + f(1) = -\frac{1}{2}(x-1) + \frac{1}{2} = -\frac{1}{2}x + 1.$$

55. Use the Product Rule to show that $(f^2)' = 2ff'$.

SOLUTION Let $g = f^2 = ff$. Then $g' = \left(f^2\right)' = (ff)' = ff' + ff' = 2ff'$.

Further Insights and Challenges

57. Let f, g, h be differentiable functions. Show that $(fgh)'(x)$ is equal to

$$f(x)g(x)h'(x) + f(x)g'(x)h(x) + f'(x)g(x)h(x)$$

Hint: Write fgh as $f(gh)$.

SOLUTION Let $p = fgh$. Then

$$p' = (fgh)' = f\left(gh' + hg'\right) + ghf' = f'gh + fg'h + fgh'.$$

59. Derivative of the Reciprocal Use the limit definition to prove

$$\frac{d}{dx}\left(\frac{1}{f(x)}\right) = -\frac{f'(x)}{f^2(x)} \qquad \boxed{7}$$

Hint: Show that the difference quotient for $1/f(x)$ is equal to

$$\frac{f(x) - f(x+h)}{hf(x)f(x+h)}$$

SOLUTION Let $g(x) = \frac{1}{f(x)}$. We then compute the derivative of $g(x)$ using the difference quotient:

$$g'(x) = \lim_{h\to 0}\frac{g(x+h) - g(x)}{h} = \lim_{h\to 0}\frac{1}{h}\left(\frac{1}{f(x+h)} - \frac{1}{f(x)}\right) = \lim_{h\to 0}\frac{1}{h}\left(\frac{f(x) - f(x+h)}{f(x)f(x+h)}\right)$$

$$= -\lim_{h\to 0}\left(\frac{f(x+h) - f(x)}{h}\right)\left(\frac{1}{f(x)f(x+h)}\right).$$

We can apply the rule of products for limits. The first parenthetical expression is the difference quotient definition of $f'(x)$. The second can be evaluated at $h = 0$ to give $\frac{1}{(f(x))^2}$. Hence

$$g'(x) = \frac{d}{dx}\left(\frac{1}{f(x)}\right) = -\frac{f'(x)}{f^2(x)}.$$

61. Use the limit definition of the derivative to prove the following special case of the Product Rule:

$$\frac{d}{dx}(xf(x)) = xf'(x) + f(x)$$

SOLUTION First note that because $f(x)$ is differentiable, it is also continuous. It follows that

$$\lim_{h\to 0} f(x+h) = f(x).$$

Now we tackle the derivative:

$$\frac{d}{dx}(xf(x)) = \lim_{h\to 0}\frac{(x+h)f(x+h) - f(x)}{h} = \lim_{h\to 0}\left(x\frac{f(x+h) - f(x)}{h} + f(x+h)\right)$$

$$= x\lim_{h\to 0}\frac{f(x+h) - f(x)}{h} + \lim_{h\to 0} f(x+h)$$

$$= xf'(x) + f(x).$$

63. The Power Rule Revisited If you are familiar with *proof by induction*, use induction to prove the Power Rule for all whole numbers n. Show that the Power Rule holds for $n = 1$; then write x^n as $x \cdot x^{n-1}$ and use the Product Rule.

SOLUTION Let k be a positive integer. If $k = 1$, then $x^k = x$. Note that

$$\frac{d}{dx}\left(x^1\right) = \frac{d}{dx}(x) = 1 = 1x^0.$$

Hence the Power Rule holds for $k = 1$. Assume it holds for $k = n$ where $n \geq 2$. Then for $k = n + 1$, we have

$$\frac{d}{dx}\left(x^k\right) = \frac{d}{dx}\left(x^{n+1}\right) = \frac{d}{dx}\left(x \cdot x^n\right) = x\frac{d}{dx}\left(x^n\right) + x^n\frac{d}{dx}(x)$$
$$= x \cdot nx^{n-1} + x^n \cdot 1 = (n+1)x^n = kx^{k-1}$$

Accordingly, the Power Rule holds for all positive integers by induction.

Exercises 64 and 65: A basic fact of algebra states that c is a root of a polynomial $f(x)$ if and only if $f(x) = (x - c)g(x)$ for some polynomial $g(x)$. We say that c is a ***multiple root*** *if $f(x) = (x - c)^2 h(x)$, where $h(x)$ is a polynomial.*

65. Use Exercise 64 to determine whether $c = -1$ is a multiple root:
(a) $x^5 + 2x^4 - 4x^3 - 8x^2 - x + 2$
(b) $x^4 + x^3 - 5x^2 - 3x + 2$

SOLUTION
(a) To show that -1 is a multiple root of

$$f(x) = x^5 + 2x^4 - 4x^3 - 8x^2 - x + 2,$$

it suffices to check that $f(-1) = f'(-1) = 0$. We have $f(-1) = -1 + 2 + 4 - 8 + 1 + 2 = 0$ and

$$f'(x) = 5x^4 + 8x^3 - 12x^2 - 16x - 1$$
$$f'(-1) = 5 - 8 - 12 + 16 - 1 = 0$$

(b) Let $f(x) = x^4 + x^3 - 5x^2 - 3x + 2$. Then $f'(x) = 4x^3 + 3x^2 - 10x - 3$. Because

$$f(-1) = 1 - 1 - 5 + 3 + 2 = 0$$

but

$$f'(-1) = -4 + 3 + 10 - 3 = 6 \neq 0,$$

it follows that $x = -1$ is a root of f, but not a multiple root.

67. According to Eq. (6) in Section 3.2, $\frac{d}{dx}b^x = m(b)\,b^x$. Use the Product Rule to show that $m(ab) = m(a) + m(b)$.

SOLUTION

$$m(ab)(ab)^x = \frac{d}{dx}(ab)^x = \frac{d}{dx}\left(a^x b^x\right) = a^x\frac{d}{dx}b^x + b^x\frac{d}{dx}a^x = m(b)a^x b^x + m(a)a^x b^x = (m(a) + m(b))(ab)^x.$$

Thus, $m(ab) = m(a) + m(b)$.

3.4 Rates of Change

Preliminary Questions

1. Which units might be used for each rate of change?
(a) Pressure (in atmospheres) in a water tank with respect to depth
(b) The rate of a chemical reaction (change in concentration with respect to time with concentration in moles per liter)

SOLUTION
(a) The rate of change of pressure with respect to depth might be measured in atmospheres/meter.
(b) The reaction rate of a chemical reaction might be measured in moles/(liter·hour).

2. Two trains travel from New Orleans to Memphis in 4 hours. The first train travels at a constant velocity of 90 mph, but the velocity of the second train varies. What was the second train's average velocity during the trip?

SOLUTION Since both trains travel the same distance in the same amount of time, they have the same average velocity: 90 mph.

3. Estimate $f(26)$, assuming that $f(25) = 43$, $f'(25) = 0.75$.

SOLUTION $f(x) \approx f(25) + f'(25)(x - 25)$, so $f(26) \approx 43 + 0.75(26 - 25) = 43.75$.

4. The population $P(t)$ of Freedonia in 2009 was $P(2009) = 5$ million.
(a) What is the meaning of $P'(2009)$?
(b) Estimate $P(2010)$ if $P'(2009) = 0.2$.

SOLUTION
(a) Because $P(t)$ measures the population of Freedonia as a function of time, the derivative $P'(2009)$ measures the rate of change of the population of Freedonia in the year 2009.
(b) $P(2010) \approx P(2009) + P'(2010)$. Thus, if $P'(2009) = 0.2$, then $P(2009) \approx 5.2$ million.

Exercises

In Exercises 1–8, find the rate of change.

1. Area of a square with respect to its side s when $s = 5$.

SOLUTION Let the area be $A = f(s) = s^2$. Then the rate of change of A with respect to s is $d/ds(s^2) = 2s$. When $s = 5$, the area changes at a rate of 10 square units per unit increase. (Draw a 5×5 square on graph paper and trace the area added by increasing each side length by 1, excluding the corner, to see what this means.)

3. Cube root $\sqrt[3]{x}$ with respect to x when $x = 1, 8, 27$.

SOLUTION Let $f(x) = \sqrt[3]{x}$. Writing $f(x) = x^{1/3}$, we see the rate of change of $f(x)$ with respect to x is given by $f'(x) = \frac{1}{3}x^{-2/3}$. The requested rates of change are given in the table that follows:

c	ROC of $f(x)$ with respect to x at $x = c$.
1	$f'(1) = \frac{1}{3}(1) = \frac{1}{3}$
8	$f'(8) = \frac{1}{3}(8^{-2/3}) = \frac{1}{3}(\frac{1}{4}) = \frac{1}{12}$
27	$f'(27) = \frac{1}{3}(27^{-2/3}) = \frac{1}{3}(\frac{1}{9}) = \frac{1}{27}$

5. The diameter of a circle with respect to radius.

SOLUTION The relationship between the diameter d of a circle and its radius r is $d = 2r$. The rate of change of the diameter with respect to the radius is then $d' = 2$.

7. Volume V of a cylinder with respect to radius if the height is equal to the radius.

SOLUTION The volume of the cylinder is $V = \pi r^2 h = \pi r^3$. Thus $dV/dr = 3\pi r^2$.

In Exercises 9–11, refer to Figure 10, the graph of distance $s(t)$ from the origin as a function of time for a car trip.

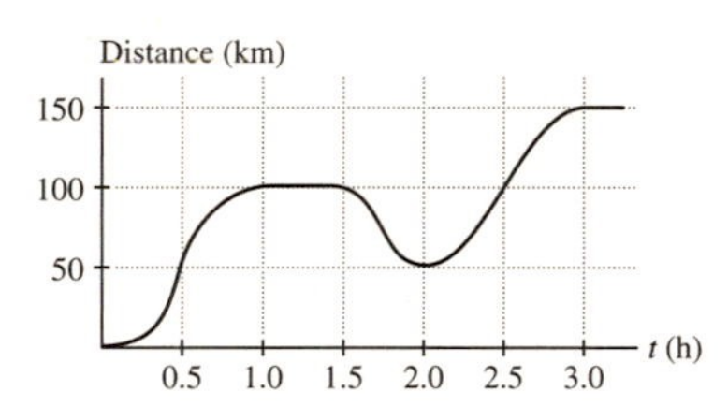

FIGURE 10 Distance from the origin versus time for a car trip.

9. Find the average velocity over each interval.

(a) $[0, 0.5]$ **(b)** $[0.5, 1]$ **(c)** $[1, 1.5]$ **(d)** $[1, 2]$

SOLUTION

(a) The average velocity over the interval $[0, 0.5]$ is

$$\frac{50 - 0}{0.5 - 0} = 100 \text{ km/hour.}$$

(b) The average velocity over the interval $[0.5, 1]$ is

$$\frac{100 - 50}{1 - 0.5} = 100 \text{ km/hour.}$$

(c) The average velocity over the interval $[1, 1.5]$ is

$$\frac{100 - 100}{1.5 - 1} = 0 \text{ km/hour.}$$

(d) The average velocity over the interval $[1, 2]$ is

$$\frac{50 - 100}{2 - 1} = -50 \text{ km/hour.}$$

11. Match the descriptions (i)–(iii) with the intervals (a)–(c).

(i) Velocity increasing

(ii) Velocity decreasing

(iii) Velocity negative

(a) [0, 0.5]
(b) [2.5, 3]
(c) [1.5, 2]

SOLUTION

(a) (i) : The distance curve is increasing, and is also *bending* upward, so that distance is increasing at an increasing rate.

(b) (ii) : Over the interval [2.5, 3], the distance curve is flattening, showing that the car is slowing down; that is, the velocity is decreasing.

(c) (iii) : The distance curve is decreasing, so the tangent line has negative slope; this means the velocity is negative.

13. Use Figure 3 from Example 1 to estimate the instantaneous rate of change of Martian temperature with respect to time (in degrees Celsius per hour) at $t = 4$ AM.

SOLUTION The segment of the temperature graph around $t = 4$ AM appears to be a straight line passing through roughly (1:36, −70) and (4:48, −75). The instantaneous rate of change of Martian temperature with respect to time at $t = 4$ AM is therefore approximately

$$\frac{dT}{dt} = \frac{-75 - (-70)}{3.2} = -1.5625°\text{C/hour}.$$

15. The velocity (in cm/s) of blood molecules flowing through a capillary of radius 0.008 cm is $v = 6.4 \times 10^{-8} - 0.001r^2$, where r is the distance from the molecule to the center of the capillary. Find the rate of change of velocity with respect to r when $r = 0.004$ cm.

SOLUTION The rate of change of the velocity of the blood molecules is $v'(r) = -0.002r$. When $r = 0.004$ cm, this rate is -8×10^{-6} 1/s.

17. Use Figure 12 to estimate dT/dh at $h = 30$ and 70, where T is atmospheric temperature (in degrees Celsius) and h is altitude (in kilometers). Where is dT/dh equal to zero?

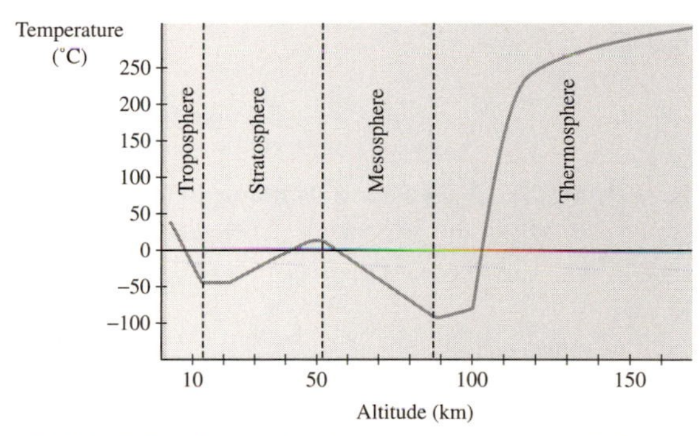

FIGURE 12 Atmospheric temperature versus altitude.

SOLUTION At $h = 30$ km, the graph of atmospheric temperature appears to be linear passing through the points (23, −50) and (40, 0). The slope of this segment of the graph is then

$$\frac{0 - (-50)}{40 - 23} = \frac{50}{17} = 2.94;$$

so

$$\left.\frac{dT}{dh}\right|_{h=30} \approx 2.94°\text{C/km}.$$

At $h = 70$ km, the graph of atmospheric temperature appears to be linear passing through the points (58, 0) and (88, −100). The slope of this segment of the graph is then

$$\frac{-100 - 0}{88 - 58} = \frac{-100}{30} = -3.33;$$

so

$$\left.\frac{dT}{dh}\right|_{h=70} \approx -3.33°\text{C/km}.$$

$\frac{dT}{dh} = 0$ at those points where the tangent line on the graph is horizontal. This appears to happen over the interval [13, 23], and near the points $h = 50$ and $h = 90$.

19. Calculate the rate of change of escape velocity $v_{\text{esc}} = (2.82 \times 10^7)r^{-1/2}$ m/s with respect to distance r from the center of the earth.

SOLUTION The rate that escape velocity changes is $v'_{\text{esc}}(r) = -1.41 \times 10^7 r^{-3/2}$.

21. The position of a particle moving in a straight line during a 5-s trip is $s(t) = t^2 - t + 10$ cm. Find a time t at which the instantaneous velocity is equal to the average velocity for the entire trip.

SOLUTION Let $s(t) = t^2 - t + 10, 0 \le t \le 5$, with s in centimeters (cm) and t in seconds (s). The average velocity over the t-interval $[0, 5]$ is

$$\frac{s(5) - s(0)}{5 - 0} = \frac{30 - 10}{5} = 4 \text{ cm/s}.$$

The (instantaneous) velocity is $v(t) = s'(t) = 2t - 1$. Solving $2t - 1 = 4$ yields $t = \frac{5}{2}$ s, the time at which the instantaneous velocity equals the calculated average velocity.

23. A particle moving along a line has position $s(t) = t^4 - 18t^2$ m at time t seconds. At which times does the particle pass through the origin? At which times is the particle instantaneously motionless (that is, it has zero velocity)?

SOLUTION The particle passes through the origin when $s(t) = t^4 - 18t^2 = t^2(t^2 - 18) = 0$. This happens when $t = 0$ seconds and when $t = 3\sqrt{2} \approx 4.24$ seconds. With $s(t) = t^4 - 18t^2$, it follows that $v(t) = s'(t) = 4t^3 - 36t = 4t(t^2 - 9)$. The particle is therefore instantaneously motionless when $t = 0$ seconds and when $t = 3$ seconds.

25. A bullet is fired in the air vertically from ground level with an initial velocity 200 m/s. Find the bullet's maximum velocity and maximum height.

SOLUTION We employ Galileo's formula, $s(t) = s_0 + v_0 t - \frac{1}{2}gt^2 = 200t - 4.9t^2$, where the time t is in seconds (s) and the height s is in meters (m). The velocity is $v(t) = 200 - 9.8t$. The maximum velocity of 200 m/s occurs at $t = 0$. This is the initial velocity. The bullet reaches its maximum height when $v(t) = 200 - 9.8t = 0$; i.e., when $t \approx 20.41$ s. At this point, the height is 2040.82 m.

27. A ball tossed in the air vertically from ground level returns to earth 4 s later. Find the initial velocity and maximum height of the ball.

SOLUTION Galileo's formula gives $s(t) = s_0 + v_0 t - \frac{1}{2}gt^2 = v_0 t - 4.9t^2$, where the time t is in seconds (s) and the height s is in meters (m). When the ball hits the ground after 4 seconds its height is 0. Solve $0 = s(4) = 4v_0 - 4.9(4)^2$ to obtain $v_0 = 19.6$ m/s. The ball reaches its maximum height when $s'(t) = 0$, that is, when $19.6 - 9.8t = 0$, or $t = 2$ s. At this time, $t = 2$ s,

$$s(2) = 0 + 19.6(2) - \frac{1}{2}(9.8)(4) = 19.6 \text{ m}.$$

29. Show that for an object falling according to Galileo's formula, the average velocity over any time interval $[t_1, t_2]$ is equal to the average of the instantaneous velocities at t_1 and t_2.

SOLUTION The simplest way to proceed is to compute both values and show that they are equal. The average velocity over $[t_1, t_2]$ is

$$\frac{s(t_2) - s(t_1)}{t_2 - t_1} = \frac{(s_0 + v_0 t_2 - \frac{1}{2}gt_2^2) - (s_0 + v_0 t_1 - \frac{1}{2}gt_1^2)}{t_2 - t_1} = \frac{v_0(t_2 - t_1) + \frac{g}{2}(t_2{}^2 - t_1{}^2)}{t_2 - t_1}$$

$$= \frac{v_0(t_2 - t_1)}{t_2 - t_1} - \frac{g}{2}(t_2 + t_1) = v_0 - \frac{g}{2}(t_2 + t_1)$$

Whereas the average of the instantaneous velocities at the beginning and end of $[t_1, t_2]$ is

$$\frac{s'(t_1) + s'(t_2)}{2} = \frac{1}{2}\Big((v_0 - gt_1) + (v_0 - gt_2)\Big) = \frac{1}{2}(2v_0) - \frac{g}{2}(t_2 + t_1) = v_0 - \frac{g}{2}(t_2 + t_1).$$

The two quantities are the same.

31. By Faraday's Law, if a conducting wire of length ℓ meters moves at velocity v m/s perpendicular to a magnetic field of strength B (in teslas), a voltage of size $V = -B\ell v$ is induced in the wire. Assume that $B = 2$ and $\ell = 0.5$.

(a) Calculate dV/dv.

(b) Find the rate of change of V with respect to time t if $v = 4t + 9$.

SOLUTION

(a) Assuming that $B = 2$ and $l = 0.5$, $V = -2(.5)v = -v$. Therefore,

$$\frac{dV}{dv} = -1.$$

(b) If $v = 4t + 9$, then $V = -2(.5)(4t + 9) = -(4t + 9)$. Therefore, $\frac{dV}{dt} = -4$.

33. Ethan finds that with h hours of tutoring, he is able to answer correctly $S(h)$ percent of the problems on a math exam. Which would you expect to be larger: $S'(3)$ or $S'(30)$? Explain.

SOLUTION One possible graph of $S(h)$ is shown in the figure below on the left. This graph indicates that in the early hours of working with the tutor, Ethan makes rapid progress in learning the material but eventually approaches either the limit of his ability to learn the material or the maximum possible score on the exam. In this scenario, $S'(3)$ would be larger than $S'(30)$.

An alternative graph of $S(h)$ is shown below on the right. Here, in the early hours of working with the tutor little progress is made (perhaps the tutor is assessing how much Ethan already knows, his learning style, his personality, etc.). This is followed by a period of rapid improvement and finally a leveling off as Ethan reaches his maximum score. In this scenario, $S'(3)$ and $S'(30)$ might be roughly equal.

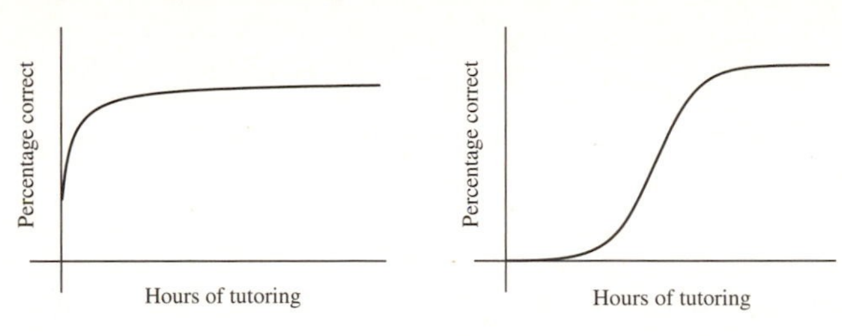

35. To determine drug dosages, doctors estimate a person's body surface area (BSA) (in meters squared) using the formula $\text{BSA} = \sqrt{hm}/60$, where h is the height in centimeters and m the mass in kilograms. Calculate the rate of change of BSA with respect to mass for a person of constant height $h = 180$. What is this rate at $m = 70$ and $m = 80$? Express your result in the correct units. Does BSA increase more rapidly with respect to mass at lower or higher body mass?

SOLUTION Assuming constant height $h = 180$ cm, let $f(m) = \sqrt{hm}/60 = \frac{\sqrt{5}}{10}\sqrt{m}$ be the formula for body surface area in terms of weight. The rate of change of BSA with respect to mass is

$$f'(m) = \frac{\sqrt{5}}{10}\left(\frac{1}{2}m^{-1/2}\right) = \frac{\sqrt{5}}{20\sqrt{m}}.$$

If $m = 70$ kg, this is

$$f'(70) = \frac{\sqrt{5}}{20\sqrt{70}} = \frac{\sqrt{14}}{280} \approx 0.0133631\frac{\text{m}^2}{\text{kg}}.$$

If $m = 80$ kg,

$$f'(80) = \frac{\sqrt{5}}{20\sqrt{80}} = \frac{1}{20\sqrt{16}} = \frac{1}{80}\frac{\text{m}^2}{\text{kg}}.$$

Because the rate of change of BSA depends on $1/\sqrt{m}$, it is clear that BSA increases more rapidly at lower body mass.

37. The tangent lines to the graph of $f(x) = x^2$ grow steeper as x increases. At what rate do the slopes of the tangent lines increase?

SOLUTION Let $f(x) = x^2$. The slopes s of the tangent lines are given by $s = f'(x) = 2x$. The rate at which these slopes are increasing is $ds/dx = 2$.

In Exercises 39–46, use Eq. (3) to estimate the unit change.

39. Estimate $\sqrt{2} - \sqrt{1}$ and $\sqrt{101} - \sqrt{100}$. Compare your estimates with the actual values.

SOLUTION Let $f(x) = \sqrt{x} = x^{1/2}$. Then $f'(x) = \frac{1}{2}(x^{-1/2})$. We are using the derivative to estimate the average rate of change. That is,

$$\frac{\sqrt{x+h} - \sqrt{x}}{h} \approx f'(x),$$

so that

$$\sqrt{x+h} - \sqrt{x} \approx hf'(x).$$

Thus, $\sqrt{2} - \sqrt{1} \approx 1f'(1) = \frac{1}{2}(1) = \frac{1}{2}$. The actual value, to six decimal places, is 0.414214. Also, $\sqrt{101} - \sqrt{100} \approx 1f'(100) = \frac{1}{2}\left(\frac{1}{10}\right) = 0.05$. The actual value, to six decimal places, is 0.0498756.

41. Let $F(s) = 1.1s + 0.05s^2$ be the stopping distance as in Example 3. Calculate $F(65)$ and estimate the increase in stopping distance if speed is increased from 65 to 66 mph. Compare your estimate with the actual increase.

SOLUTION Let $F(s) = 1.1s + .05s^2$ be as in Example 3. $F'(s) = 1.1 + 0.1s$.

- Then $F(65) = 282.75$ ft and $F'(65) = 7.6$ ft/mph.
- $F'(65) \approx F(66) - F(65)$ is approximately equal to the change in stopping distance per 1 mph increase in speed when traveling at 65 mph. Increasing speed from 65 to 66 therefore increases stopping distance by approximately 7.6 ft.
- The actual increase in stopping distance when speed increases from 65 mph to 66 mph is $F(66) - F(65) = 290.4 - 282.75 = 7.65$ feet, which differs by less than one percent from the estimate found using the derivative.

43. The dollar cost of producing x bagels is $C(x) = 300 + 0.25x - 0.5(x/1000)^3$. Determine the cost of producing 2000 bagels and estimate the cost of the 2001st bagel. Compare your estimate with the actual cost of the 2001st bagel.

SOLUTION Expanding the power of 3 yields

$$C(x) = 300 + 0.25x - 5 \times 10^{-10}x^3.$$

This allows us to get the derivative $C'(x) = 0.25 - 1.5 \times 10^{-9}x^2$. The cost of producing 2000 bagels is

$$C(2000) = 300 + 0.25(2000) - 0.5(2000/1000)^3 = 796$$

dollars. The cost of the 2001st bagel is, by definition, $C(2001) - C(2000)$. By the derivative estimate, $C(2001) - C(2000) \approx C'(2000)(1)$, so the cost of the 2001st bagel is approximately

$$C'(2000) = 0.25 - 1.5 \times 10^{-9}(2000^2) = \$0.244.$$

$C(2001) = 796.244$, so the *exact* cost of the 2001st bagel is indistinguishable from the estimated cost. The function is very nearly linear at this point.

45. Demand for a commodity generally decreases as the price is raised. Suppose that the demand for oil (per capita per year) is $D(p) = 900/p$ barrels, where p is the dollar price per barrel. Find the demand when $p = \$40$. Estimate the decrease in demand if p rises to \$41 and the increase if p declines to \$39.

SOLUTION $D(p) = 900p^{-1}$, so $D'(p) = -900p^{-2}$. When the price is \$40 a barrel, the per capita demand is $D(40) = 22.5$ barrels per year. With an increase in price from \$40 to \$41 a barrel, the change in demand $D(41) - D(40)$ is approximately $D'(40) = -900(40^{-2}) = -0.5625$ barrels a year. With a decrease in price from \$40 to \$39 a barrel, the change in demand $D(39) - D(40)$ is approximately $-D'(40) = +0.5625$. An increase in oil prices of a dollar leads to a decrease in demand of 0.5625 barrels a year, and a decrease of a dollar leads to an *increase* in demand of 0.5625 barrels a year.

47. According to Stevens' Law in psychology, the perceived magnitude of a stimulus is proportional (approximately) to a power of the actual intensity I of the stimulus. Experiments show that the *perceived brightness* B of a light satisfies $B = kI^{2/3}$, where I is the light intensity, whereas the *perceived heaviness* H of a weight W satisfies $H = kW^{3/2}$ (k is a constant that is different in the two cases). Compute dB/dI and dH/dW and state whether they are increasing or decreasing functions. Then explain the following statements:

(a) A one-unit increase in light intensity is felt more strongly when I is small than when I is large.

(b) Adding another pound to a load W is felt more strongly when W is large than when W is small.

SOLUTION

(a) $dB/dI = \dfrac{2k}{3}I^{-1/3} = \dfrac{2k}{3I^{1/3}}$.

As I increases, dB/dI shrinks, so that the rate of change of perceived intensity decreases as the actual intensity increases. Increased light intensity has a *diminished return* in perceived intensity. A sketch of B against I is shown: See that the height of the graph increases more slowly as you move to the right.

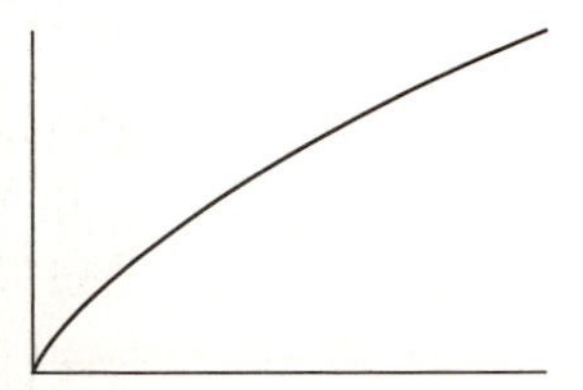

(b) $dH/dW = \frac{3k}{2}W^{1/2}$. As W increases, dH/dW increases as well, so that the rate of change of perceived weight increases as weight increases. A sketch of H against W is shown: See that the graph becomes steeper as you move to the right.

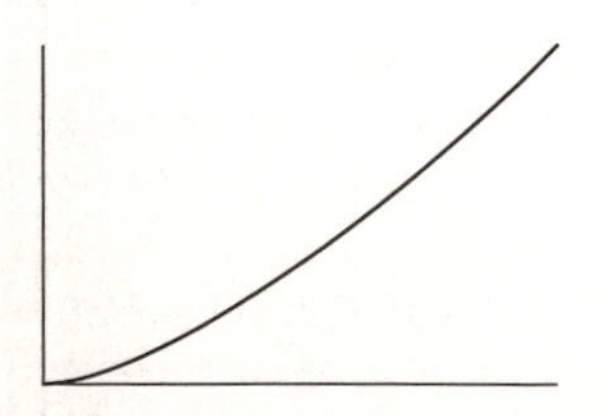

Further Insights and Challenges

*Exercises 49–51: The **Lorenz curve** $y = F(r)$ is used by economists to study income distribution in a given country (see Figure 14). By definition, $F(r)$ is the fraction of the total income that goes to the bottom rth part of the population, where $0 \le r \le 1$. For example, if $F(0.4) = 0.245$, then the bottom 40% of households receive 24.5% of the total income. Note that $F(0) = 0$ and $F(1) = 1$.*

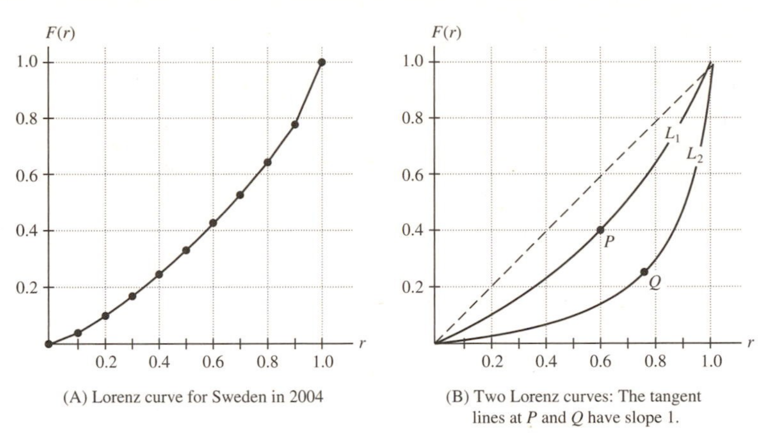

(A) Lorenz curve for Sweden in 2004

(B) Two Lorenz curves: The tangent lines at P and Q have slope 1.

FIGURE 14

49. Our goal is to find an interpretation for $F'(r)$. The average income for a group of households is the total income going to the group divided by the number of households in the group. The national average income is $A = T/N$, where N is the total number of households and T is the total income earned by the entire population.

(a) Show that the average income among households in the bottom rth part is equal to $(F(r)/r)A$.

(b) Show more generally that the average income of households belonging to an interval $[r, r + \Delta r]$ is equal to

$$\left(\frac{F(r + \Delta r) - F(r)}{\Delta r}\right) A$$

(c) Let $0 \le r \le 1$. A household belongs to the $100r$th percentile if its income is greater than or equal to the income of $100r$ % of all households. Pass to the limit as $\Delta r \to 0$ in (b) to derive the following interpretation: A household in the $100r$th percentile has income $F'(r)A$. In particular, a household in the $100r$th percentile receives more than the national average if $F'(r) > 1$ and less if $F'(r) < 1$.

(d) For the Lorenz curves L_1 and L_2 in Figure 14(B), what percentage of households have above-average income?

SOLUTION

(a) The total income among households in the bottom rth part is $F(r)T$ and there are rN households in this part of the population. Thus, the average income among households in the bottom rth part is equal to

$$\frac{F(r)T}{rN} = \frac{F(r)}{r} \cdot \frac{T}{N} = \frac{F(r)}{r} A.$$

(b) Consider the interval $[r, r + \Delta r]$. The total income among households between the bottom rth part and the bottom $r + \Delta r$-th part is $F(r + \Delta r)T - F(r)T$. Moreover, the number of households covered by this interval is $(r + \Delta r)N - rN = \Delta r N$. Thus, the average income of households belonging to an interval $[r, r + \Delta r]$ is equal to

$$\frac{F(r + \Delta r)T - F(r)T}{\Delta r N} = \frac{F(r + \Delta r) - F(r)}{\Delta r} \cdot \frac{T}{N} = \frac{F(r + \Delta r) - F(r)}{\Delta r} A.$$

(c) Take the result from part (b) and let $\Delta r \to 0$. Because

$$\lim_{\Delta r \to 0} \frac{F(r + \Delta r) - F(r)}{\Delta r} = F'(r),$$

we find that a household in the $100r$th percentile has income $F'(r)A$.

(d) The point P in Figure 14(B) has an r-coordinate of 0.6, while the point Q has an r-coordinate of roughly 0.75. Thus, on curve L_1, 40% of households have $F'(r) > 1$ and therefore have above-average income. On curve L_2, roughly 25% of households have above-average income.

51. Use Exercise 49 (c) to prove:

(a) $F'(r)$ is an increasing function of r.

(b) Income is distributed equally (all households have the same income) if and only if $F(r) = r$ for $0 \le r \le 1$.

SOLUTION

(a) Recall from Exercise 49 (c) that $F'(r)A$ is the income of a household in the $100r$-th percentile. Suppose $0 \le r_1 < r_2 \le 1$. Because $r_2 > r_1$, a household in the $100r_2$-th percentile must have income at least as large as a household in the $100r_1$-th percentile. Thus, $F'(r_2)A \ge F'(r_1)A$, or $F'(r_2) \ge F'(r_1)$. This implies $F'(r)$ is an increasing function of r.

(b) If $F(r) = r$ for $0 \le r \le 1$, then $F'(r) = 1$ and households in all percentiles have income equal to the national average; that is, income is distributed equally. Alternately, if income is distributed equally (all households have the same income), then $F'(r) = 1$ for $0 \le r \le 1$. Thus, F must be a linear function in r with slope 1. Moreover, the condition $F(0) = 0$ requires the F intercept of the line to be 0. Hence, $F(r) = 1 \cdot r + 0 = r$.

In Exercises 53 and 54, the average cost per unit at production level x is defined as $C_{avg}(x) = C(x)/x$, where $C(x)$ is the cost function. Average cost is a measure of the efficiency of the production process.

53. Show that $C_{avg}(x)$ is equal to the slope of the line through the origin and the point $(x, C(x))$ on the graph of $C(x)$. Using this interpretation, determine whether average cost or marginal cost is greater at points A, B, C, D in Figure 15.

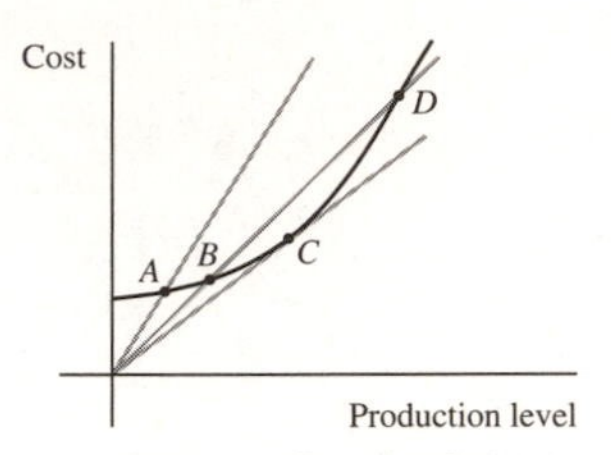

FIGURE 15 Graph of $C(x)$.

SOLUTION By definition, the slope of the line through the origin and $(x, C(x))$, that is, between $(0, 0)$ and $(x, C(x))$ is

$$\frac{C(x) - 0}{x - 0} = \frac{C(x)}{x} = C_{av}.$$

At point A, average cost is greater than marginal cost, as the line from the origin to A is steeper than the curve at this point (we see this because the line, tracing from the origin, crosses the curve from below). At point B, the average cost is still greater than the marginal cost. At the point C, the average cost and the marginal cost are nearly the same, since the tangent line and the line from the origin are nearly the same. The line from the origin to D crosses the cost curve from above, and so is less steep than the tangent line to the curve at D; the average cost at this point is less than the marginal cost.

3.5 Higher Derivatives

Preliminary Questions

1. On September 4, 2003, the *Wall Street Journal* printed the headline "Stocks Go Higher, Though the Pace of Their Gains Slows." Rephrase this headline as a statement about the first and second time derivatives of stock prices and sketch a possible graph.

SOLUTION Because stocks are going higher, stock prices are increasing and the first derivative of stock prices must therefore be positive. On the other hand, because the pace of gains is slowing, the second derivative of stock prices must be negative.

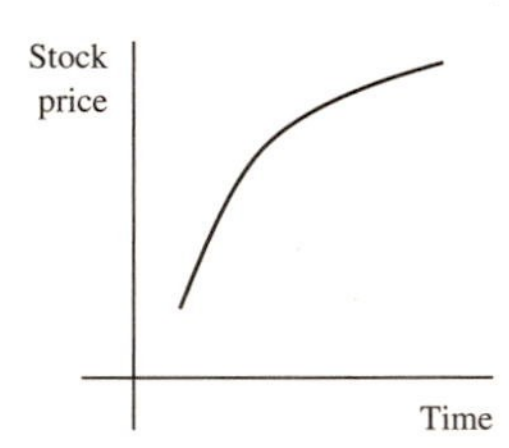

2. True or false? The third derivative of position with respect to time is zero for an object falling to earth under the influence of gravity. Explain.

SOLUTION This statement is true. The acceleration of an object falling to earth under the influence of gravity is constant; hence, the second derivative of position with respect to time is constant. Because the third derivative is just the derivative of the second derivative and the derivative of a constant is zero, it follows that the third derivative is zero.

3. Which type of polynomial satisfies $f'''(x) = 0$ for all x?

SOLUTION The third derivative of all quadratic polynomials (polynomials of the form $ax^2 + bx + c$ for some constants a, b and c) is equal to 0 for all x.

4. What is the millionth derivative of $f(x) = e^x$?

SOLUTION Every derivative of $f(x) = e^x$ is e^x.

Exercises

In Exercises 1–16, calculate y'' and y'''.

1. $y = 14x^2$

SOLUTION Let $y = 14x^2$. Then $y' = 28x$, $y'' = 28$, and $y''' = 0$.

3. $y = x^4 - 25x^2 + 2x$

SOLUTION Let $y = x^4 - 25x^2 + 2x$. Then $y' = 4x^3 - 50x + 2$, $y'' = 12x^2 - 50$, and $y''' = 24x$.

5. $y = \frac{4}{3}\pi r^3$

SOLUTION Let $y = \frac{4}{3}\pi r^3$. Then $y' = 4\pi r^2$, $y'' = 8\pi r$, and $y''' = 8\pi$.

7. $y = 20t^{4/5} - 6t^{2/3}$

SOLUTION Let $y = 20t^{4/5} - 6t^{2/3}$. Then $y' = 16t^{-1/5} - 4t^{-1/3}$, $y'' = -\frac{16}{5}t^{-6/5} + \frac{4}{3}t^{-4/3}$, and $y''' = \frac{96}{25}t^{-11/15} - \frac{16}{9}t^{-7/3}$.

9. $y = z - \frac{4}{z}$

SOLUTION Let $y = z - 4z^{-1}$. Then $y' = 1 + 4z^{-2}$, $y'' = -8z^{-3}$, and $y''' = 24z^{-4}$.

11. $y = \theta^2(2\theta + 7)$

SOLUTION Let $y = \theta^2(2\theta + 7) = 2\theta^3 + 7\theta^2$. Then $y' = 6\theta^2 + 14\theta$, $y'' = 12\theta + 14$, and $y''' = 12$.

13. $y = \frac{x - 4}{x}$

SOLUTION Let $y = \frac{x-4}{x} = 1 - 4x^{-1}$. Then $y' = 4x^{-2}$, $y'' = -8x^{-3}$, and $y''' = 24x^{-4}$.

15. $y = x^5 e^x$

SOLUTION Let $y = x^5 e^x$. Then

$$
\begin{aligned}
y' &= x^5 e^x + 5x^4 e^x = (x^5 + 5x^4)e^x \\
y'' &= (x^5 + 5x^4)e^x + (5x^4 + 20x^3)e^x = (x^5 + 10x^4 + 20x^3)e^x \\
y''' &= (x^5 + 10x^4 + 20x^3)e^x + (5x^4 + 40x^3 + 60x^2)e^x = (x^5 + 15x^4 + 60x^3 + 60x^2)e^x.
\end{aligned}
$$

In Exercises 17–26, calculate the derivative indicated.

17. $f^{(4)}(1)$, $\quad f(x) = x^4$

SOLUTION Let $f(x) = x^4$. Then $f'(x) = 4x^3$, $f''(x) = 12x^2$, $f'''(x) = 24x$, and $f^{(4)}(x) = 24$. Thus $f^{(4)}(1) = 24$.

19. $\left.\frac{d^2y}{dt^2}\right|_{t=1}$, $\quad y = 4t^{-3} + 3t^2$

SOLUTION Let $y = 4t^{-3} + 3t^2$. Then $\frac{dy}{dt} = -12t^{-4} + 6t$ and $\frac{d^2y}{dt^2} = 48t^{-5} + 6$. Hence

$$\left.\frac{d^2y}{dt^2}\right|_{t=1} = 48(1)^{-5} + 6 = 54.$$

21. $\left.\frac{d^4x}{dt^4}\right|_{t=16}$, $\quad x = t^{-3/4}$

SOLUTION Let $x(t) = t^{-3/4}$. Then $\frac{dx}{dt} = -\frac{3}{4}t^{-7/4}$, $\frac{d^2x}{dt^2} = \frac{21}{16}t^{-11/4}$, $\frac{d^3x}{dt^3} = -\frac{231}{64}t^{-15/4}$, and $\frac{d^4x}{dt^4} = \frac{3465}{256}t^{-19/4}$. Thus

$$\left.\frac{d^4x}{dt^4}\right|_{t=16} = \frac{3465}{256}16^{-19/4} = \frac{3465}{134217728}.$$

23. $f'''(-3)$, $\quad f(x) = 4e^x - x^3$

SOLUTION Let $f(x) = 4e^x - x^3$. Then $f'(x) = 4e^x - 3x^2$, $f''(x) = 4e^x - 6x$, $f'''(x) = 4e^x - 6$, and $f'''(-3) = 4e^{-3} - 6$.

25. $h''(1), \quad h(w) = \sqrt{w}e^w$

SOLUTION Let $h(w) = \sqrt{w}e^w = w^{1/2}e^w$. Then

$$h'(w) = w^{1/2}e^w + e^w\left(\frac{1}{2}w^{-1/2}\right) = \left(w^{1/2} + \frac{1}{2}w^{-1/2}\right)e^w$$

and

$$h''(w) = \left(w^{1/2} + \frac{1}{2}w^{-1/2}\right)e^w + e^w\left(\frac{1}{2}w^{-1/2} - \frac{1}{4}w^{-3/2}\right) = \left(w^{1/2} + w^{-1/2} - \frac{1}{4}w^{-3/2}\right)e^w.$$

Thus, $h''(1) = \frac{7}{4}e$.

27. Calculate $y^{(k)}(0)$ for $0 \le k \le 5$, where $y = x^4 + ax^3 + bx^2 + cx + d$ (with a, b, c, d the constants).

SOLUTION Applying the power, constant multiple, and sum rules at each stage, we get (note $y^{(0)}$ is y by convention):

k	$y^{(k)}$
0	$x^4 + ax^3 + bx^2 + cx + d$
1	$4x^3 + 3ax^2 + 2bx + c$
2	$12x^2 + 6ax + 2b$
3	$24x + 6a$
4	24
5	0

from which we get $y^{(0)}(0) = d$, $y^{(1)}(0) = c$, $y^{(2)}(0) = 2b$, $y^{(3)}(0) = 6a$, $y^{(4)}(0) = 24$, and $y^{(5)}(0) = 0$.

29. Use the result in Example 3 to find $\frac{d^6}{dx^6}x^{-1}$.

SOLUTION The equation in Example 3 indicates that

$$\frac{d^6}{dx^6}x^{-1} = (-1)^6 6! x^{-6-1}.$$

$(-1)^6 = 1$ and $6! = 6 \times 5 \times 4 \times 3 \times 2 \times 1 = 720$, so

$$\frac{d^6}{dx^6}x^{-1} = 720x^{-7}.$$

In Exercises 31–36, find a general formula for $f^{(n)}(x)$.

31. $f(x) = x^{-2}$

SOLUTION $f'(x) = -2x^{-3}$, $f''(x) = 6x^{-4}$, $f'''(x) = -24x^{-5}$, $f^{(4)}(x) = 5 \cdot 24x^{-6}$, ... From this we can conclude that the nth derivative can be written as $f^{(n)}(x) = (-1)^n(n+1)!x^{-(n+2)}$.

33. $f(x) = x^{-1/2}$

SOLUTION $f'(x) = \frac{-1}{2}x^{-3/2}$. We will avoid simplifying numerators and denominators to find the pattern:

$$f''(x) = \frac{-3}{2}\frac{-1}{2}x^{-5/2} = (-1)^2\frac{3 \times 1}{2^2}x^{-5/2}$$

$$f'''(x) = -\frac{5}{2}\frac{3 \times 1}{2^2}x^{-7/2} = (-1)^3\frac{5 \times 3 \times 1}{2^3}x^{-7/2}$$

$$\vdots$$

$$f^{(n)}(x) = (-1)^n\frac{(2n-1) \times (2n-3) \times \ldots \times 1}{2^n}x^{-(2n+1)/2}.$$

35. $f(x) = xe^{-x}$

SOLUTION Let $f(x) = xe^{-x}$. Then

$$f'(x) = x(-e^{-x}) + e^{-x} = (1-x)e^{-x} = -(x-1)e^{-x}$$
$$f''(x) = (1-x)(-e^{-x}) - e^{-x} = (x-2)e^{-x}$$
$$f'''(x) = (x-2)(-e^{-x}) + e^{-x} = (3-x)e^{-x} = -(x-3)e^{-x}$$

From this we conclude that the nth derivative can be written as $f^{(n)}(x) = (-1)^n(x-n)e^{-x}$.

37. (a) Find the acceleration at time $t = 5$ min of a helicopter whose height is $s(t) = 300t - 4t^3$ m.
(b) Plot the acceleration $h''(t)$ for $0 \le t \le 6$. How does this graph show that the helicopter is slowing down during this time interval?

SOLUTION

(a) Let $s(t) = 300t - 4t^3$, with t in minutes and s in meters. The velocity is $v(t) = s'(t) = 300 - 12t^2$ and acceleration is $a(t) = s''(t) = -24t$. Thus $a(5) = -120$ m/min^2.
(b) The acceleration of the helicopter for $0 \le t \le 6$ is shown in the figure below. As the acceleration of the helicopter is negative, the velocity of the helicopter must be decreasing. Because the velocity is positive for $0 \le t \le 6$, the helicopter is slowing down.

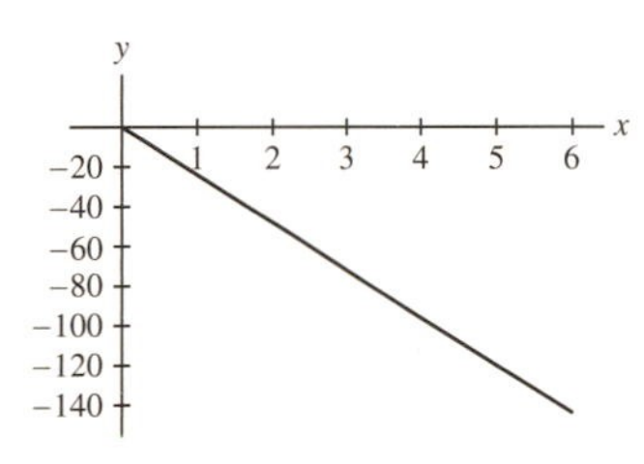

39. Figure 5 shows f, f', and f''. Determine which is which.

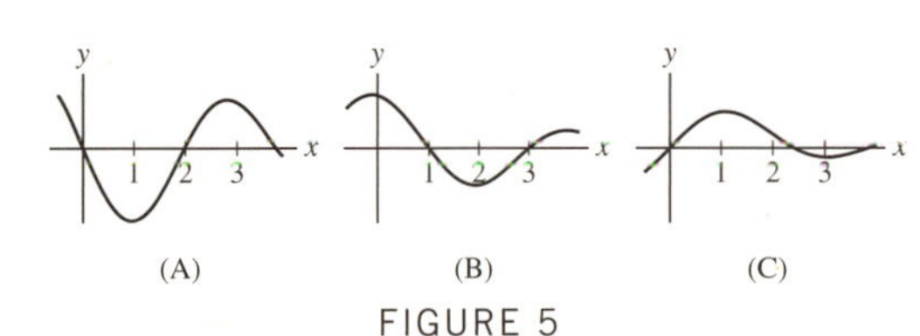

FIGURE 5

SOLUTION (a) f'' (b) f' (c) f.
The tangent line to (c) is horizontal at $x = 1$ and $x = 3$, where (b) has roots. The tangent line to (b) is horizontal at $x = 2$ and $x = 0$, where (a) has roots.

41. Figure 7 shows the graph of the position s of an object as a function of time t. Determine the intervals on which the acceleration is positive.

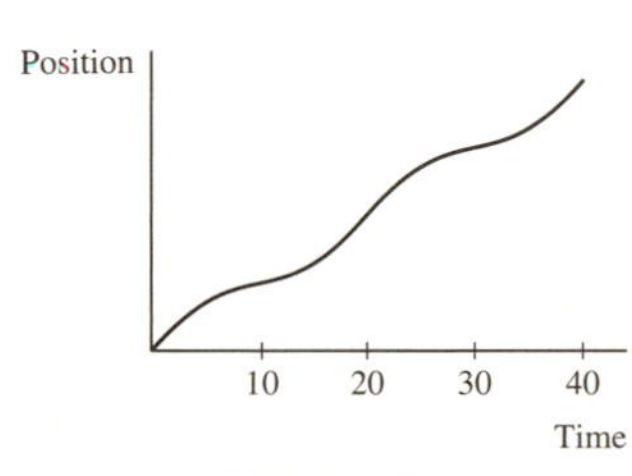

FIGURE 7

SOLUTION Roughly from time 10 to time 20 and from time 30 to time 40. The acceleration is positive over the same intervals over which the graph is bending upward.

43. Find all values of n such that $y = x^n$ satisfies

$$x^2y'' - 2xy' = 4y$$

SOLUTION We have $y' = nx^{n-1}$, $y'' = n(n-1)x^{n-2}$, so that

$$x^2y'' - 2xy' = x^2(n(n-1)x^{n-2}) - 2xnx^{n-1} = (n^2 - 3n)x^n = (n^2 - 3n)y$$

Thus the equation is satisfied if and only if $n^2 - 3n = 4$, so that $n^2 - 3n - 4 = 0$. This happens for $n = -1, 4$.

45. According to one model that takes into account air resistance, the acceleration $a(t)$ (in m/s^2) of a skydiver of mass m in free fall satisfies

$$a(t) = -9.8 + \frac{k}{m}v(t)^2$$

where $v(t)$ is velocity (negative since the object is falling) and k is a constant. Suppose that $m = 75$ kg and $k = 14$ kg/m.

(a) What is the object's velocity when $a(t) = -4.9$?

(b) What is the object's velocity when $a(t) = 0$? This velocity is the object's terminal velocity.

SOLUTION Solving $a(t) = -9.8 + \frac{k}{m}v(t)^2$ for the velocity and taking into account that the velocity is negative since the object is falling, we find

$$v(t) = -\sqrt{\frac{m}{k}(a(t) + 9.8)} = -\sqrt{\frac{75}{14}(a(t) + 9.8)}.$$

(a) Substituting $a(t) = -4.9$ into the above formula for the velocity, we find

$$v(t) = -\sqrt{\frac{75}{14}(4.9)} = -\sqrt{26.25} = -5.12 \text{ m/s}.$$

(b) When $a(t) = 0$,

$$v(t) = -\sqrt{\frac{75}{14}(9.8)} = -\sqrt{52.5} = -7.25 \text{ m/s}.$$

47. A servomotor controls the vertical movement of a drill bit that will drill a pattern of holes in sheet metal. The maximum vertical speed of the drill bit is 4 in./s, and while drilling the hole, it must move no more than 2.6 in./s to avoid warping the metal. During a cycle, the bit begins and ends at rest, quickly approaches the sheet metal, and quickly returns to its initial position after the hole is drilled. Sketch possible graphs of the drill bit's vertical velocity and acceleration. Label the point where the bit enters the sheet metal.

SOLUTION There will be multiple cycles, each of which will be more or less identical. Let $v(t)$ be the *downward* vertical velocity of the drill bit, and let $a(t)$ be the vertical acceleration. From the narrative, we see that $v(t)$ can be no greater than 4 and no greater than 2.6 while drilling is taking place. During each cycle, $v(t) = 0$ initially, $v(t)$ goes to 4 quickly. When the bit hits the sheet metal, $v(t)$ goes down to 2.6 quickly, at which it stays until the sheet metal is drilled through. As the drill pulls out, it reaches maximum non-drilling upward speed ($v(t) = -4$) quickly, and maintains this speed until it returns to rest. A possible plot follows:

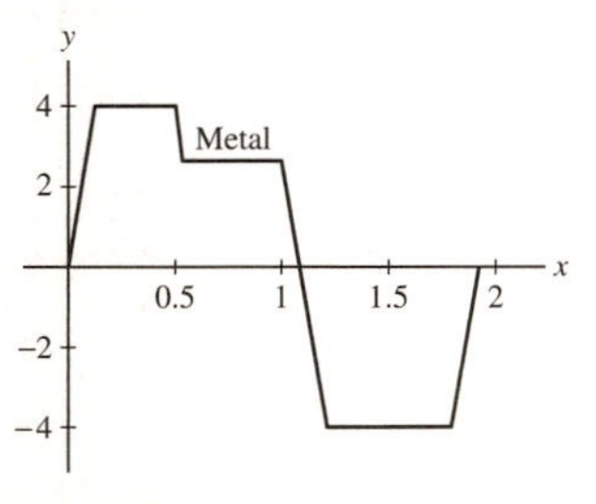

A graph of the acceleration is extracted from this graph:

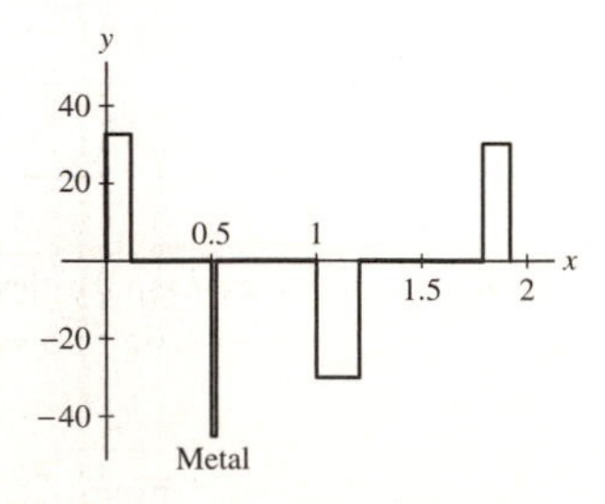

In Exercises 48 and 49, refer to the following. In a 1997 study, Boardman and Lave related the traffic speed S on a two-lane road to traffic density Q (number of cars per mile of road) by the formula

$$S = 2882Q^{-1} - 0.052Q + 31.73$$

for $60 \le Q \le 400$ (Figure 9).

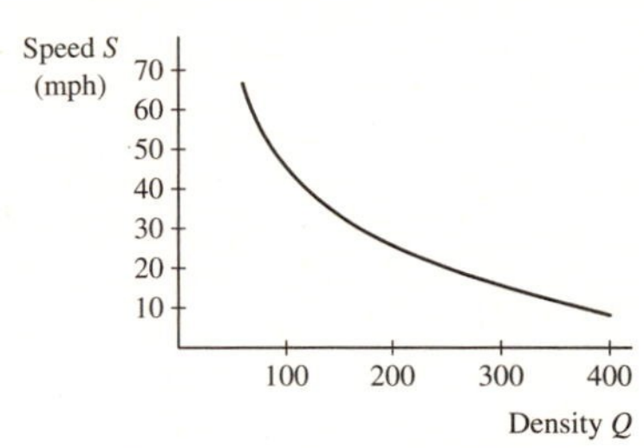

FIGURE 9 Speed as a function of traffic density.

49. (a) Explain intuitively why we should expect that $dS/dQ < 0$.

(b) Show that $d^2S/dQ^2 > 0$. Then use the fact that $dS/dQ < 0$ and $d^2S/dQ^2 > 0$ to justify the following statement: *A one-unit increase in traffic density slows down traffic more when Q is small than when Q is large.*

(c) GU Plot dS/dQ. Which property of this graph shows that $d^2S/dQ^2 > 0$?

SOLUTION

(a) Traffic speed must be reduced when the road gets more crowded so we expect dS/dQ to be negative. This is indeed the case since $dS/dQ = -0.052 - 2882/Q^2 < 0$.

(b) The decrease in speed due to a one-unit increase in density is approximately dS/dQ (a negative number). Since $d^2S/dQ^2 = 5764Q^{-3} > 0$ is positive, this tells us that dS/dQ gets larger as Q increases—and a negative number which gets larger is getting closer to zero. So the decrease in speed is smaller when Q is larger, that is, a one-unit increase in traffic density has a smaller effect when Q is large.

(c) dS/dQ is plotted below. The fact that this graph is increasing shows that $d^2S/dQ^2 > 0$.

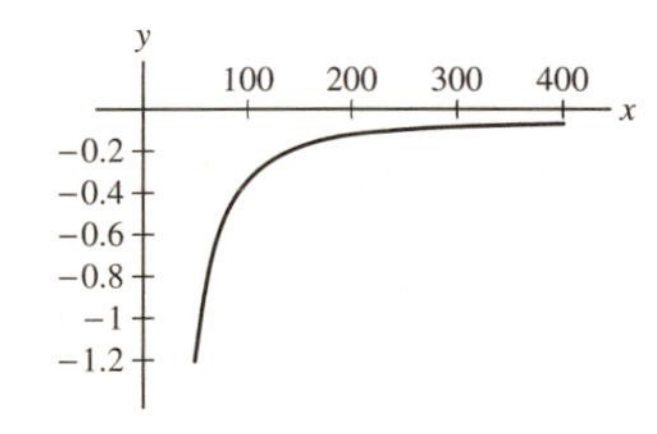

51. CAS Let $f(x) = \dfrac{x+2}{x-1}$. Use a computer algebra system to compute the $f^{(k)}(x)$ for $1 \le k \le 4$. Can you find a general formula for $f^{(k)}(x)$?

SOLUTION Let $f(x) = \dfrac{x+2}{x-1}$. Using a computer algebra system,

$$f'(x) = -\frac{3}{(x-1)^2} = (-1)^1 \frac{3 \cdot 1}{(x-1)^{1+1}};$$

$$f''(x) = \frac{6}{(x-1)^3} = (-1)^2 \frac{3 \cdot 2 \cdot 1}{(x-1)^{2+1}};$$

$$f'''(x) = -\frac{18}{(x-1)^4} = (-1)^3 \frac{3 \cdot 3!}{(x-1)^{3+1}}; \text{ and}$$

$$f^{(4)}(x) = \frac{72}{(x-1)^5} = (-1)^4 \frac{3 \cdot 4!}{(x-1)^{4+1}}.$$

From the pattern observed above, we conjecture

$$f^{(k)}(x) = (-1)^k \frac{3 \cdot k!}{(x-1)^{k+1}}.$$

Further Insights and Challenges

53. What is $p^{(99)}(x)$ for $p(x)$ as in Exercise 52?

SOLUTION First note that for any integer $n \leq 98$,

$$\frac{d^{99}}{dx^{99}}x^n = 0.$$

Now, if we expand $p(x)$, we find

$$p(x) = x^{99} + \text{ terms of degree at most 98;}$$

therefore,

$$\frac{d^{99}}{dx^{99}}p(x) = \frac{d^{99}}{dx^{99}}(x^{99} + \text{ terms of degree at most 98}) = \frac{d^{99}}{dx^{99}}x^{99}$$

Using logic similar to that used to compute the derivative in Example (3), we compute:

$$\frac{d^{99}}{dx^{99}}(x^{99}) = 99 \times 98 \times \ldots 1,$$

so that $\frac{d^{99}}{dx^{99}}p(x) = 99!$.

55. Use the Product Rule to find a formula for $(fg)'''$ and compare your result with the expansion of $(a+b)^3$. Then try to guess the general formula for $(fg)^{(n)}$.

SOLUTION Continuing from Exercise 54, we have

$$h''' = f''g' + gf''' + 2(f'g'' + g'f'') + fg''' + g''f' = f'''g + 3f''g' + 3f'g'' + fg'''$$

The binomial theorem gives

$$(a+b)^3 = a^3 + 3a^2b + 3ab^2 + b^3 = a^3b^0 + 3a^2b^1 + 3a^1b^2 + a^0b^3$$

and more generally

$$(a+b)^n = \sum_{k=0}^{n} \binom{n}{k} a^{n-k}b^k,$$

where the binomial coefficients are given by

$$\binom{n}{k} = \frac{k(k-1)\cdots(k-n+1)}{n!}.$$

Accordingly, the general formula for $(fg)^{(n)}$ is given by

$$(fg)^{(n)} = \sum_{k=0}^{n} \binom{n}{k} f^{(n-k)}g^{(k)},$$

where $p^{(k)}$ is the kth derivative of p (or p itself when $k = 0$).

3.6 Trigonometric Functions

Preliminary Questions

1. Determine the sign (+ or −) that yields the correct formula for the following:

(a) $\frac{d}{dx}(\sin x + \cos x) = \pm \sin x \pm \cos x$

(b) $\frac{d}{dx}\sec x = \pm \sec x \tan x$

(c) $\frac{d}{dx}\cot x = \pm \csc^2 x$

SOLUTION The correct formulas are

(a) $\frac{d}{dx}(\sin x + \cos x) = -\sin x + \cos x$

(b) $\frac{d}{dx}\sec x = \sec x \tan x$

(c) $\frac{d}{dx}\cot x = -\csc^2 x$

2. Which of the following functions can be differentiated using the rules we have covered so far?

(a) $y = 3\cos x \cot x$ **(b)** $y = \cos(x^2)$ **(c)** $y = e^x \sin x$

SOLUTION

(a) $3\cos x \cot x$ is a product of functions whose derivatives are known. This function can therefore be differentiated using the Product Rule.

(b) $\cos(x^2)$ is a composition of the functions $\cos x$ and x^2. We have not yet discussed how to differentiate composite functions.

(c) $x^2 \cos x$ is a product of functions whose derivatives are known. This function can therefore be differentiated using the Product Rule.

3. Compute $\frac{d}{dx}(\sin^2 x + \cos^2 x)$ without using the derivative formulas for $\sin x$ and $\cos x$.

SOLUTION Recall that $\sin^2 x + \cos^2 x = 1$ for all x. Thus,

$$\frac{d}{dx}(\sin^2 x + \cos^2 x) = \frac{d}{dx}1 = 0.$$

4. How is the addition formula used in deriving the formula $(\sin x)' = \cos x$?

SOLUTION The difference quotient for the function $\sin x$ involves the expression $\sin(x+h)$. The addition formula for the sine function is used to expand this expression as $\sin(x+h) = \sin x \cos h + \sin h \cos x$.

Exercises

In Exercises 1–4, find an equation of the tangent line at the point indicated.

1. $y = \sin x, \quad x = \frac{\pi}{4}$

SOLUTION Let $f(x) = \sin x$. Then $f'(x) = \cos x$ and the equation of the tangent line is

$$y = f'\left(\frac{\pi}{4}\right)\left(x - \frac{\pi}{4}\right) + f\left(\frac{\pi}{4}\right) = \frac{\sqrt{2}}{2}\left(x - \frac{\pi}{4}\right) + \frac{\sqrt{2}}{2} = \frac{\sqrt{2}}{2}x + \frac{\sqrt{2}}{2}\left(1 - \frac{\pi}{4}\right).$$

3. $y = \tan x, \quad x = \frac{\pi}{4}$

SOLUTION Let $f(x) = \tan x$. Then $f'(x) = \sec^2 x$ and the equation of the tangent line is

$$y = f'\left(\frac{\pi}{4}\right)\left(x - \frac{\pi}{4}\right) + f\left(\frac{\pi}{4}\right) = 2\left(x - \frac{\pi}{4}\right) + 1 = 2x + 1 - \frac{\pi}{2}.$$

In Exercises 5–24, compute the derivative.

5. $f(x) = \sin x \cos x$

SOLUTION Let $f(x) = \sin x \cos x$. Then

$$f'(x) = \sin x(-\sin x) + \cos x(\cos x) = -\sin^2 x + \cos^2 x.$$

7. $f(x) = \sin^2 x$

SOLUTION Let $f(x) = \sin^2 x = \sin x \sin x$. Then

$$f'(x) = \sin x(\cos x) + \sin x(\cos x) = 2\sin x \cos x.$$

9. $H(t) = \sin t \sec^2 t$

SOLUTION Let $H(t) = \sin t \sec^2 t$. Then

$$\begin{aligned}
H'(t) &= \sin t \frac{d}{dt}(\sec t \cdot \sec t) + \sec^2 t(\cos t) \\
&= \sin t(\sec t \sec t \tan t + \sec t \sec t \tan t) + \sec t \\
&= 2\sin t \sec^2 t \tan t + \sec t.
\end{aligned}$$

11. $f(\theta) = \tan\theta \sec\theta$

SOLUTION Let $f(\theta) = \tan\theta \sec\theta$. Then

$$f'(\theta) = \tan\theta \sec\theta \tan\theta + \sec\theta \sec^2\theta = \sec\theta \tan^2\theta + \sec^3\theta = \left(\tan^2\theta + \sec^2\theta\right)\sec\theta.$$

13. $f(x) = (2x^4 - 4x^{-1})\sec x$

SOLUTION Let $f(x) = (2x^4 - 4x^{-1})\sec x$. Then

$$f'(x) = (2x^4 - 4x^{-1})\sec x \tan x + \sec x(8x^3 + 4x^{-2}).$$

15. $y = \dfrac{\sec\theta}{\theta}$

SOLUTION Let $y = \dfrac{\sec\theta}{\theta}$. Then

$$y' = \frac{\theta \sec\theta \tan\theta - \sec\theta}{\theta^2}.$$

17. $R(y) = \dfrac{3\cos y - 4}{\sin y}$

SOLUTION Let $R(y) = \dfrac{3\cos y - 4}{\sin y}$. Then

$$R'(y) = \frac{\sin y(-3\sin y) - (3\cos y - 4)(\cos y)}{\sin^2 y} = \frac{4\cos y - 3(\sin^2 y + \cos^2 y)}{\sin^2 y} = \frac{4\cos y - 3}{\sin^2 y}.$$

19. $f(x) = \dfrac{1 + \tan x}{1 - \tan x}$

SOLUTION Let $f(x) = \dfrac{1 + \tan x}{1 - \tan x}$. Then

$$f'(x) = \frac{(1 - \tan x)\sec^2 x - (1 + \tan x)\left(-\sec^2 x\right)}{(1 - \tan x)^2} = \frac{2\sec^2 x}{(1 - \tan x)^2}.$$

21. $f(x) = e^x \sin x$

SOLUTION Let $f(x) = e^x \sin x$. Then $f'(x) = e^x \cos x + \sin x e^x = e^x(\cos x + \sin x)$.

23. $f(\theta) = e^\theta(5\sin\theta - 4\tan\theta)$

SOLUTION Let $f(\theta) = e^\theta(5\sin\theta - 4\tan\theta)$. Then

$$\begin{aligned} f'(\theta) &= e^\theta(5\cos\theta - 4\sec^2\theta) + e^\theta(5\sin\theta - 4\tan\theta) \\ &= e^\theta(5\sin\theta + 5\cos\theta - 4\tan\theta - 4\sec^2\theta). \end{aligned}$$

In Exercises 25–34, find an equation of the tangent line at the point specified.

25. $y = x^3 + \cos x, \quad x = 0$

SOLUTION Let $f(x) = x^3 + \cos x$. Then $f'(x) = 3x^2 - \sin x$ and $f'(0) = 0$. The tangent line at $x = 0$ is

$$y = f'(0)(x - 0) + f(0) = 0(x) + 1 = 1.$$

27. $y = \sin x + 3\cos x, \quad x = 0$

SOLUTION Let $f(x) = \sin x + 3\cos x$. Then $f'(x) = \cos x - 3\sin x$ and $f'(0) = 1$. The tangent line at $x = 0$ is

$$y = f'(0)(x - 0) + f(0) = x + 3.$$

29. $y = 2(\sin\theta + \cos\theta), \quad \theta = \frac{\pi}{3}$

SOLUTION Let $f(\theta) = 2(\sin\theta + \cos\theta)$. Then $f'(\theta) = 2(\cos\theta - \sin\theta)$ and $f'(\frac{\pi}{3}) = 1 - \sqrt{3}$. The tangent line at $x = \frac{\pi}{3}$ is

$$y = f'\left(\frac{\pi}{3}\right)\left(x - \frac{\pi}{3}\right) + f\left(\frac{\pi}{3}\right) = (1 - \sqrt{3})\left(x - \frac{\pi}{3}\right) + 1 + \sqrt{3}.$$

31. $y = e^x \cos x, \quad x = 0$

SOLUTION Let $f(x) = e^x \cos x$. Then

$$f'(x) = e^x(-\sin x) + e^x \cos x = e^x(\cos x - \sin x),$$

and $f'(0) = e^0(\cos 0 - \sin 0) = 1$. Thus, the equation of the tangent line is

$$y = f'(0)(x - 0) + f(0) = x + 1.$$

33. $y = e^t(1 - \cos t), \quad t = \frac{\pi}{2}$

SOLUTION Let $f(t) = e^t(1 - \cos t)$. Then

$$f'(t) = e^t \sin t + e^t(1 - \cos t) = e^t(1 + \sin t - \cos t),$$

and $f'(\frac{\pi}{2}) = 2e^{\pi/2}$. The tangent line at $x = \frac{\pi}{2}$ is

$$y = f'\left(\frac{\pi}{2}\right)\left(t - \frac{\pi}{2}\right) + f\left(\frac{\pi}{2}\right) = 2e^{\pi/2}\left(t - \frac{\pi}{2}\right) + e^{\pi/2}.$$

In Exercises 35–37, use Theorem 1 to verify the formula.

35. $\frac{d}{dx} \cot x = -\csc^2 x$

SOLUTION $\cot x = \frac{\cos x}{\sin x}$. Using the quotient rule and the derivative formulas, we compute:

$$\frac{d}{dx} \cot x = \frac{d}{dx} \frac{\cos x}{\sin x} = \frac{\sin x(-\sin x) - \cos x(\cos x)}{\sin^2 x} = \frac{-(\sin^2 x + \cos^2 x)}{\sin^2 x} = \frac{-1}{\sin^2 x} = -\csc^2 x.$$

37. $\frac{d}{dx} \csc x = -\csc x \cot x$

SOLUTION Since $\csc x = \frac{1}{\sin x}$, we can apply the quotient rule and the two known derivatives to get:

$$\frac{d}{dx} \csc x = \frac{d}{dx} \frac{1}{\sin x} = \frac{\sin x(0) - 1(\cos x)}{\sin^2 x} = \frac{-\cos x}{\sin^2 x} = -\frac{\cos x}{\sin x} \frac{1}{\sin x} = -\cot x \csc x.$$

In Exercises 39–42, calculate the higher derivative.

39. $f''(\theta), \quad f(\theta) = \theta \sin \theta$

SOLUTION Let $f(\theta) = \theta \sin \theta$. Then

$$f'(\theta) = \theta \cos \theta + \sin \theta$$
$$f''(\theta) = \theta(-\sin \theta) + \cos \theta + \cos \theta = -\theta \sin \theta + 2 \cos \theta.$$

41. $y'', \quad y''', \quad y = \tan x$

SOLUTION Let $y = \tan x$. Then $y' = \sec^2 x$ and by the Chain Rule,

$$y'' = = \frac{d}{dx} \sec^2 x = 2(\sec x)(\sec x \tan x) = 2 \sec^2 x \tan x$$
$$y''' = 2 \sec^2 x(\sec^2 x) + (2 \sec^2 x \tan x) \tan x = 2 \sec^4 + 4 \sec^4 x \tan^2 x$$

43. Calculate the first five derivatives of $f(x) = \cos x$. Then determine $f^{(8)}$ and $f^{(37)}$.

SOLUTION Let $f(x) = \cos x$.

- Then $f'(x) = -\sin x$, $f''(x) = -\cos x$, $f'''(x) = \sin x$, $f^{(4)}(x) = \cos x$, and $f^{(5)}(x) = -\sin x$.
- Accordingly, the successive derivatives of f cycle among

$$\{-\sin x, -\cos x, \sin x, \cos x\}$$

  in that order. Since 8 is a multiple of 4, we have $f^{(8)}(x) = \cos x$.
- Since 36 is a multiple of 4, we have $f^{(36)}(x) = \cos x$. Therefore, $f^{(37)}(x) = -\sin x$.

45. Find the values of x between 0 and 2π where the tangent line to the graph of $y = \sin x \cos x$ is horizontal.

SOLUTION Let $y = \sin x \cos x$. Then

$$y' = (\sin x)(-\sin x) + (\cos x)(\cos x) = \cos^2 x - \sin^2 x.$$

When $y' = 0$, we have $\sin x = \pm \cos x$. In the interval $[0, 2\pi]$, this occurs when $x = \frac{\pi}{4}, \frac{3\pi}{4}, \frac{5\pi}{4}, \frac{7\pi}{4}$.

47. GU Let $g(t) = t - \sin t$.

(a) Plot the graph of g with a graphing utility for $0 \le t \le 4\pi$.

(b) Show that the slope of the tangent line is nonnegative. Verify this on your graph.

(c) For which values of t in the given range is the tangent line horizontal?

SOLUTION Let $g(t) = t - \sin t$.

(a) Here is a graph of g over the interval $[0, 4\pi]$.

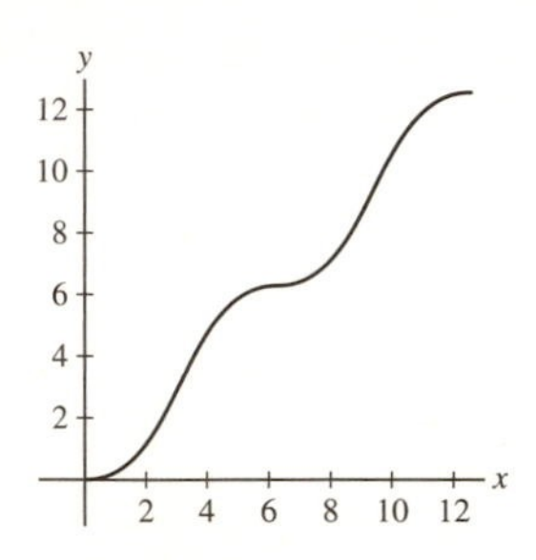

(b) Since $g'(t) = 1 - \cos t \ge 0$ for all t, the slope of the tangent line to g is always nonnegative.

(c) In the interval $[0, 4\pi]$, the tangent line is horizontal when $t = 0, 2\pi, 4\pi$.

49. Show that no tangent line to the graph of $f(x) = \tan x$ has zero slope. What is the least slope of a tangent line? Justify by sketching the graph of $(\tan x)'$.

SOLUTION Let $f(x) = \tan x$. Then $f'(x) = \sec^2 x = \frac{1}{\cos^2 x}$. Note that $f'(x) = \frac{1}{\cos^2 x}$ has numerator 1; the equation $f'(x) = 0$ therefore has no solution. Because the maximum value of $\cos^2 x$ is 1, the minimum value of $f'(x) = \frac{1}{\cos^2 x}$ is 1. Hence, the least slope for a tangent line to $\tan x$ is 1. Here is a graph of f'.

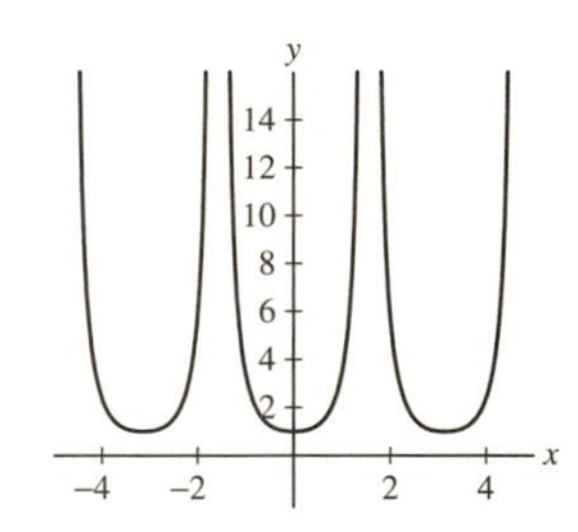

51. The horizontal range R of a projectile launched from ground level at an angle θ and initial velocity v_0 m/s is $R = (v_0^2/9.8)\sin\theta\cos\theta$. Calculate $dR/d\theta$. If $\theta = 7\pi/24$, will the range increase or decrease if the angle is increased slightly? Base your answer on the sign of the derivative.

SOLUTION Let $R(\theta) = (v_0^2/9.8)\sin\theta\cos\theta$.

$$\frac{dR}{d\theta} = R'(\theta) = (v_0^2/9.8)(-\sin^2\theta + \cos^2\theta).$$

If $\theta = 7\pi/24$, $\frac{\pi}{4} < \theta < \frac{\pi}{2}$, so $|\sin\theta| > |\cos\theta|$, and $dR/d\theta < 0$ (numerically, $dR/d\theta = -0.0264101v_0^2$). At this point, increasing the angle will *decrease* the range.

Further Insights and Challenges

53. Use the limit definition of the derivative and the addition law for the cosine function to prove that $(\cos x)' = -\sin x$.

SOLUTION Let $f(x) = \cos x$. Then

$$f'(x) = \lim_{h\to 0}\frac{\cos(x+h) - \cos x}{h} = \lim_{h\to 0}\frac{\cos x\cos h - \sin x\sin h - \cos x}{h}$$

$$= \lim_{h\to 0}\left((-\sin x)\frac{\sin h}{h} + (\cos x)\frac{\cos h - 1}{h}\right) = (-\sin x)\cdot 1 + (\cos x)\cdot 0 = -\sin x.$$

55. Verify the following identity and use it to give another proof of the formula $(\sin x)' = \cos x$.

$$\sin(x+h) - \sin x = 2\cos\left(x + \tfrac{1}{2}h\right)\sin\left(\tfrac{1}{2}h\right)$$

Hint: Use the addition formula to prove that $\sin(a+b) - \sin(a-b) = 2\cos a \sin b$.

SOLUTION Recall that

$$\sin(a+b) = \sin a \cos b + \cos a \sin b$$

and

$$\sin(a-b) = \sin a \cos b - \cos a \sin b.$$

Subtracting the second identity from the first yields

$$\sin(a+b) - \sin(a-b) = 2\cos a \sin b.$$

If we now set $a = x + \frac{h}{2}$ and $b = \frac{h}{2}$, then the previous equation becomes

$$\sin(x+h) - \sin x = 2\cos\left(x + \frac{h}{2}\right)\sin\left(\frac{h}{2}\right).$$

Finally, we use the limit definition of the derivative of $\sin x$ to obtain

$$\frac{d}{dx}\sin x = \lim_{h\to 0}\frac{\sin(x+h) - \sin x}{h} = \lim_{h\to 0}\frac{2\cos\left(x + \frac{h}{2}\right)\sin\left(\frac{h}{2}\right)}{h}$$

$$= \lim_{h\to 0}\cos\left(x + \frac{h}{2}\right) \cdot \lim_{h\to 0}\frac{\sin\left(\frac{h}{2}\right)}{\left(\frac{h}{2}\right)} = \cos x \cdot 1 = \cos x.$$

In other words, $\frac{d}{dx}(\sin x) = \cos x$.

57. Let $f(x) = x\sin x$ and $g(x) = x\cos x$.
(a) Show that $f'(x) = g(x) + \sin x$ and $g'(x) = -f(x) + \cos x$.
(b) Verify that $f''(x) = -f(x) + 2\cos x$ and $g''(x) = -g(x) - 2\sin x$.
(c) By further experimentation, try to find formulas for all higher derivatives of f and g. *Hint:* The kth derivative depends on whether $k = 4n, 4n+1, 4n+2$, or $4n+3$.

SOLUTION Let $f(x) = x\sin x$ and $g(x) = x\cos x$.
(a) We examine first derivatives: $f'(x) = x\cos x + (\sin x)\cdot 1 = g(x) + \sin x$ and $g'(x) = (x)(-\sin x) + (\cos x)\cdot 1 = -f(x) + \cos x$; i.e., $f'(x) = g(x) + \sin x$ and $g'(x) = -f(x) + \cos x$.
(b) Now look at second derivatives: $f''(x) = g'(x) + \cos x = -f(x) + 2\cos x$ and $g''(x) = -f'(x) - \sin x = -g(x) - 2\sin x$; i.e., $f''(x) = -f(x) + 2\cos x$ and $g''(x) = -g(x) - 2\sin x$.
(c)
- The third derivatives are $f'''(x) = -f'(x) - 2\sin x = -g(x) - 3\sin x$ and $g'''(x) = -g'(x) - 2\cos x = f(x) - 3\cos x$; i.e., $f'''(x) = -g(x) - 3\sin x$ and $g'''(x) = f(x) - 3\cos x$.
- The fourth derivatives are $f^{(4)}(x) = -g'(x) - 3\cos x = f(x) - 4\cos x$ and $g^{(4)}(x) = f'(x) + 3\sin x = g(x) + 4\sin x$; i.e., $f^{(4)} = f(x) - 4\cos x$ and $g^{(4)}(x) = g(x) + 4\sin x$.
- We can now see the pattern for the derivatives, which are summarized in the following table. Here $n = 0, 1, 2, \ldots$

k	$4n$	$4n+1$	$4n+2$	$4n+3$
$f^{(k)}(x)$	$f(x) - k\cos x$	$g(x) + k\sin x$	$-f(x) + k\cos x$	$-g(x) - k\sin x$
$g^{(k)}(x)$	$g(x) + k\sin x$	$-f(x) + k\cos x$	$-g(x) - k\sin x$	$f(x) - k\cos x$

3.7 The Chain Rule

Preliminary Questions

1. Identify the outside and inside functions for each of these composite functions.

(a) $y = \sqrt{4x + 9x^2}$ **(b)** $y = \tan(x^2 + 1)$

(c) $y = \sec^5 x$ **(d)** $y = (1 + e^x)^4$

SOLUTION

(a) The outer function is $\sqrt{x}$, and the inner function is $4x + 9x^2$.

(b) The outer function is $\tan x$, and the inner function is $x^2 + 1$.

(c) The outer function is x^5, and the inner function is $\sec x$.

(d) The outer function is x^4, and the inner function is $1 + e^x$.

2. Which of the following can be differentiated easily *without* using the Chain Rule?

(a) $y = \tan(7x^2 + 2)$

(b) $y = \dfrac{x}{x+1}$

(c) $y = \sqrt{x} \cdot \sec x$

(d) $y = \sqrt{x \cos x}$

(e) $y = xe^x$

(f) $y = e^{\sin x}$

SOLUTION The function $\frac{x}{x+1}$ can be differentiated using the Quotient Rule, and the functions $\sqrt{x} \cdot \sec x$ and xe^x can be differentiated using the Product Rule. The functions $\tan(7x^2 + 2)$, $\sqrt{x \cos x}$ and $e^{\sin x}$ require the Chain Rule.

3. Which is the derivative of $f(5x)$?

(a) $5f'(x)$

(b) $5f'(5x)$

(c) $f'(5x)$

SOLUTION The correct answer is **(b)**: $5f'(5x)$.

4. Suppose that $f'(4) = g(4) = g'(4) = 1$. Do we have enough information to compute $F'(4)$, where $F(x) = f(g(x))$? If not, what is missing?

SOLUTION If $F(x) = f(g(x))$, then $F'(x) = f'(g(x))g'(x)$ and $F'(4) = f'(g(4))g'(4)$. Thus, we do not have enough information to compute $F'(4)$. We are missing the value of $f'(1)$.

Exercises

In Exercises 1–4, fill in a table of the following type:

$f(g(x))$	$f'(u)$	$f'(g(x))$	$g'(x)$	$(f \circ g)'$

1. $f(u) = u^{3/2}$, $g(x) = x^4 + 1$

SOLUTION

$f(g(x))$	$f'(u)$	$f'(g(x))$	$g'(x)$	$(f \circ g)'$
$(x^4 + 1)^{3/2}$	$\frac{3}{2}u^{1/2}$	$\frac{3}{2}(x^4 + 1)^{1/2}$	$4x^3$	$6x^3(x^4 + 1)^{1/2}$

3. $f(u) = \tan u$, $g(x) = x^4$

SOLUTION

$f(g(x))$	$f'(u)$	$f'(g(x))$	$g'(x)$	$(f \circ g)'$
$\tan(x^4)$	$\sec^2 u$	$\sec^2(x^4)$	$4x^3$	$4x^3 \sec^2(x^4)$

In Exercises 5 and 6, write the function as a composite $f(g(x))$ and compute the derivative using the Chain Rule.

5. $y = (x + \sin x)^4$

SOLUTION Let $f(x) = x^4$, $g(x) = x + \sin x$, and $y = f(g(x)) = (x + \sin x)^4$. Then

$$\frac{dy}{dx} = f'(g(x))g'(x) = 4(x + \sin x)^3(1 + \cos x).$$

7. Calculate $\dfrac{d}{dx} \cos u$ for the following choices of $u(x)$:

(a) $u = 9 - x^2$

(b) $u = x^{-1}$

(c) $u = \tan x$

SOLUTION

(a) $\cos(u(x)) = \cos(9 - x^2)$.

$$\frac{d}{dx}\cos(u(x)) = -\sin(u(x))u'(x) = -\sin(9 - x^2)(-2x) = 2x\sin(9 - x^2).$$

(b) $\cos(u(x)) = \cos(x^{-1})$.

$$\frac{d}{dx}\cos(u(x)) = -\sin(u(x))u'(x) = -\sin(x^{-1})\left(-\frac{1}{x^2}\right) = \frac{\sin(x^{-1})}{x^2}.$$

(c) $\cos(u(x)) = \cos(\tan x)$.

$$\frac{d}{dx}\cos(u(x)) = -\sin(u(x))u'(x) = -\sin(\tan x)(\sec^2 x) = -\sec^2 x \sin(\tan x).$$

9. Compute $\frac{df}{dx}$ if $\frac{df}{du} = 2$ and $\frac{du}{dx} = 6$.

SOLUTION Assuming f is a function of u, which is in turn a function of x,

$$\frac{df}{dx} = \frac{df}{du} \cdot \frac{du}{dx} = 2(6) = 12.$$

In Exercises 11–22, use the General Power Rule or the Shifting and Scaling Rule to compute the derivative.

11. $y = (x^4 + 5)^3$

SOLUTION Using the General Power Rule,

$$\frac{d}{dx}(x^4+5)^3 = 3(x^4+5)^2\frac{d}{dx}(x^4+5) = 3(x^4+5)^2(4x^3) = 12x^3(x^4+5)^2.$$

13. $y = \sqrt{7x - 3}$

SOLUTION Using the Shifting and Scaling Rule

$$\frac{d}{dx}\sqrt{7x-3} = \frac{d}{dx}(7x-3)^{1/2} = \frac{1}{2}(7x-3)^{-1/2}(7) = \frac{7}{2\sqrt{7x-3}}.$$

15. $y = (x^2 + 9x)^{-2}$

SOLUTION Using the General Power Rule,

$$\frac{d}{dx}(x^2+9x)^{-2} = -2(x^2+9x)^{-3}\frac{d}{dx}(x^2+9x) = -2(x^2+9x)^{-3}(2x+9).$$

17. $y = \cos^4 \theta$

SOLUTION Using the General Power Rule,

$$\frac{d}{d\theta}\cos^4\theta = 4\cos^3\theta\frac{d}{d\theta}\cos\theta = -4\cos^3\theta\sin\theta.$$

19. $y = (2\cos\theta + 5\sin\theta)^9$

SOLUTION Using the General Power Rule,

$$\frac{d}{d\theta}(2\cos\theta+5\sin\theta)^9 = 9(2\cos\theta+5\sin\theta)^8\frac{d}{d\theta}(2\cos\theta+5\sin\theta) = 9(2\cos\theta+5\sin\theta)^8(5\cos\theta-2\sin\theta).$$

21. $y = e^{x-12}$

SOLUTION Using the Shifting and Scaling Rule,

$$\frac{d}{dx}e^{x-12} = (1)e^{x-12} = e^{x-12}.$$

In Exercises 23–26, compute the derivative of $f \circ g$.

23. $f(u) = \sin u, \quad g(x) = 2x + 1$

SOLUTION Let $h(x) = f(g(x)) = \sin(2x+1)$. Then, applying the shifting and scaling rule, $h'(x) = 2\cos(2x+1)$. Alternately,

$$\frac{d}{dx}f(g(x)) = f'(g(x))g'(x) = \cos(2x+1) \cdot 2 = 2\cos(2x+1).$$

25. $f(u) = e^u, \quad g(x) = x + x^{-1}$

SOLUTION Let $h(x) = f(g(x)) = e^{x+x^{-1}}$. Then

$$\frac{d}{dx} f(g(x)) = f'(g(x))g'(x) = e^{x+x^{-1}} \left(1 - x^{-2}\right).$$

In Exercises 27 and 28, find the derivatives of $f(g(x))$ and $g(f(x))$.

27. $f(u) = \cos u, \quad u = g(x) = x^2 + 1$

SOLUTION

$$\frac{d}{dx} f(g(x)) = f'(g(x))g'(x) = -\sin(x^2 + 1)(2x) = -2x \sin(x^2 + 1).$$

$$\frac{d}{dx} g(f(x)) = g'(f(x))f'(x) = 2(\cos x)(-\sin x) = -2 \sin x \cos x.$$

In Exercises 29–42, use the Chain Rule to find the derivative.

29. $y = \sin(x^2)$

SOLUTION Let $y = \sin\left(x^2\right)$. Then $y' = \cos\left(x^2\right) \cdot 2x = 2x \cos\left(x^2\right)$.

31. $y = \sqrt{t^2 + 9}$

SOLUTION Let $y = \sqrt{t^2 + 9} = (t^2 + 9)^{1/2}$. Then

$$y' = \frac{1}{2}(t^2 + 9)^{-1/2}(2t) = \frac{t}{\sqrt{t^2 + 9}}.$$

33. $y = (x^4 - x^3 - 1)^{2/3}$

SOLUTION Let $y = \left(x^4 - x^3 - 1\right)^{2/3}$. Then

$$y' = \frac{2}{3}\left(x^4 - x^3 - 1\right)^{-1/3}\left(4x^3 - 3x^2\right).$$

35. $y = \left(\frac{x+1}{x-1}\right)^4$

SOLUTION Let $y = \left(\frac{x+1}{x-1}\right)^4$. Then

$$y' = 4\left(\frac{x+1}{x-1}\right)^3 \cdot \frac{(x-1) \cdot 1 - (x+1) \cdot 1}{(x-1)^2} = -\frac{8\,(x+1)^3}{(x-1)^5} = \frac{8(1+x)^3}{(1-x)^5}.$$

37. $y = \sec \frac{1}{x}$

SOLUTION Let $f(x) = \sec\left(x^{-1}\right)$. Then

$$f'(x) = \sec\left(x^{-1}\right) \tan\left(x^{-1}\right) \cdot \left(-x^{-2}\right) = -\frac{\sec(1/x) \tan(1/x)}{x^2}.$$

39. $y = \tan(\theta + \cos\theta)$

SOLUTION Let $y = \tan(\theta + \cos\theta)$. Then

$$y' = \sec^2(\theta + \cos\theta) \cdot (1 - \sin\theta) = (1 - \sin\theta)\sec^2(\theta + \cos\theta).$$

41. $y = e^{2-9t^2}$

SOLUTION Let $y = e^{2-9t^2}$. Then

$$y' = e^{2-9t^2}(-18t) = -18te^{2-9t^2}.$$

In Exercises 43–72, find the derivative using the appropriate rule or combination of rules.

43. $y = \tan(x^2 + 4x)$

SOLUTION Let $y = \tan(x^2 + 4x)$. By the chain rule,

$$y' = \sec^2(x^2 + 4x) \cdot (2x + 4) = (2x + 4)\sec^2(x^2 + 4x).$$

45. $y = x\cos(1 - 3x)$

SOLUTION Let $y = x\cos(1 - 3x)$. Applying the product rule and then the scaling and shifting rule,

$$y' = x(-\sin(1 - 3x)) \cdot (-3) + \cos(1 - 3x) \cdot 1 = 3x\sin(1 - 3x) + \cos(1 - 3x).$$

47. $y = (4t + 9)^{1/2}$

SOLUTION Let $y = (4t + 9)^{1/2}$. By the shifting and scaling rule,

$$\frac{dy}{dt} = 4\left(\frac{1}{2}\right)(4t + 9)^{-1/2} = 2(4t + 9)^{-1/2}.$$

49. $y = (x^3 + \cos x)^{-4}$

SOLUTION Let $y = (x^3 + \cos x)^{-4}$. By the general power rule,

$$y' = -4(x^3 + \cos x)^{-5}(3x^2 - \sin x) = 4(\sin x - 3x^2)(x^3 + \cos x)^{-5}.$$

51. $y = \sqrt{\sin x \cos x}$

SOLUTION We start by using a trig identity to rewrite

$$y = \sqrt{\sin x \cos x} = \sqrt{\frac{1}{2}\sin 2x} = \frac{1}{\sqrt{2}}(\sin 2x)^{1/2}.$$

Then, after two applications of the chain rule,

$$y' = \frac{1}{\sqrt{2}} \cdot \frac{1}{2}(\sin 2x)^{-1/2} \cdot \cos 2x \cdot 2 = \frac{\cos 2x}{\sqrt{2\sin 2x}}.$$

53. $y = (\cos 6x + \sin x^2)^{1/2}$

SOLUTION Let $y = (\cos 6x + \sin(x^2))^{1/2}$. Applying the general power rule followed by both the scaling and shifting rule and the chain rule,

$$y' = \frac{1}{2}\left(\cos 6x + \sin(x^2)\right)^{-1/2}\left(-\sin 6x \cdot 6 + \cos(x^2) \cdot 2x\right) = \frac{x\cos(x^2) - 3\sin 6x}{\sqrt{\cos 6x + \sin(x^2)}}.$$

55. $y = \tan^3 x + \tan(x^3)$

SOLUTION Let $y = \tan^3 x + \tan(x^3) = (\tan x)^3 + \tan(x^3)$. Applying the general power rule to the first term and the chain rule to the second term,

$$y' = 3(\tan x)^2\sec^2 x + \sec^2(x^3) \cdot 3x^2 = 3\left(x^2\sec^2(x^3) + \sec^2 x\tan^2 x\right).$$

57. $y = \sqrt{\frac{z+1}{z-1}}$

SOLUTION Let $y = \left(\frac{z+1}{z-1}\right)^{1/2}$. Applying the general power rule followed by the quotient rule,

$$\frac{dy}{dz} = \frac{1}{2}\left(\frac{z+1}{z-1}\right)^{-1/2} \cdot \frac{(z-1) \cdot 1 - (z+1) \cdot 1}{(z-1)^2} = \frac{-1}{\sqrt{z+1}\,(z-1)^{3/2}}.$$

59. $y = \dfrac{\cos(1+x)}{1+\cos x}$

SOLUTION Let

$$y = \frac{\cos(1+x)}{1+\cos x}.$$

Then, applying the quotient rule and the shifting and scaling rule,

$$\frac{dy}{dx} = \frac{-(1+\cos x)\sin(1+x) + \cos(1+x)\sin x}{(1+\cos x)^2} = \frac{\cos(1+x)\sin x - \cos x \sin(1+x) - \sin(1+x)}{(1+\cos x)^2}$$
$$= \frac{\sin(-1) - \sin(1+x)}{(1+\cos x)^2}.$$

The last line follows from the identity

$$\sin(A - B) = \sin A \cos B - \cos A \sin B$$

with $A = x$ and $B = 1 + x$.

61. $y = \cot^7(x^5)$

SOLUTION Let $y = \cot^7\left(x^5\right)$. Applying the general power rule followed by the chain rule,

$$\frac{dy}{dx} = 7\cot^6\left(x^5\right) \cdot \left(-\csc^2\left(x^5\right)\right) \cdot 5x^4 = -35x^4 \cot^6\left(x^5\right)\csc^2\left(x^5\right).$$

63. $y = \left(1 + \cot^5(x^4+1)\right)^9$

SOLUTION Let $y = \left(1 + \cot^5\left(x^4+1\right)\right)^9$. Applying the general power rule, the chain rule, and the general power rule in succession,

$$\frac{dy}{dx} = 9\left(1 + \cot^5\left(x^4+1\right)\right)^8 \cdot 5\cot^4\left(x^4+1\right) \cdot \left(-\csc^2\left(x^4+1\right)\right) \cdot 4x^3$$
$$= -180x^3 \cot^4\left(x^4+1\right)\csc^2\left(x^4+1\right)\left(1+\cot^5\left(x^4+1\right)\right)^8.$$

65. $y = (2e^{3x} + 3e^{-2x})^4$

SOLUTION Let $y = (2e^{3x} + 3e^{-2x})^4$. Applying the general power rule followed by two applications of the chain rule, one for each exponential function, we find

$$\frac{dy}{dx} = 4(2e^{3x} + 3e^{-2x})^3(6e^{3x} - 6e^{-2x}) = 24(2e^{3x} + 3e^{-2x})^3(e^{3x} - e^{-2x}).$$

67. $y = e^{(x^2+2x+3)^2}$

SOLUTION Let $y = e^{(x^2+2x+3)^2}$. By the chain rule and the general power rule, we obtain

$$\frac{dy}{dx} = e^{(x^2+2x+3)^2} \cdot 2(x^2+2x+3)(2x+2) = 4(x+1)(x^2+2x+3)e^{(x^2+2x+3)^2}.$$

69. $y = \sqrt{1 + \sqrt{1 + \sqrt{x}}}$

SOLUTION Let $y = \left(1 + \left(1 + x^{1/2}\right)^{1/2}\right)^{1/2}$. Applying the general power rule twice,

$$\frac{dy}{dx} = \frac{1}{2}\left(1 + \left(1 + x^{1/2}\right)^{1/2}\right)^{-1/2} \cdot \frac{1}{2}\left(1 + x^{1/2}\right)^{-1/2} \cdot \frac{1}{2}x^{-1/2} = \frac{1}{8\sqrt{x}\sqrt{1+\sqrt{x}}\sqrt{1+\sqrt{1+\sqrt{x}}}}.$$

71. $y = (kx+b)^{-1/3}$; k and b any constants

SOLUTION Let $y = (kx+b)^{-1/3}$, where b and k are constants. By the scaling and shifting rule,

$$y' = -\frac{1}{3}(kx+b)^{-4/3} \cdot k = -\frac{k}{3}(kx+b)^{-4/3}.$$

In Exercises 73–76, compute the higher derivative.

73. $\dfrac{d^2}{dx^2}\sin(x^2)$

SOLUTION Let $f(x) = \sin\left(x^2\right)$. Then, by the chain rule, $f'(x) = 2x\cos\left(x^2\right)$ and, by the product rule and the chain rule,

$$f''(x) = 2x\left(-\sin\left(x^2\right)\cdot 2x\right) + 2\cos\left(x^2\right) = 2\cos\left(x^2\right) - 4x^2\sin\left(x^2\right).$$

75. $\dfrac{d^3}{dx^3}(9-x)^8$

SOLUTION Let $f(x) = (9-x)^8$. Then, by repeated use of the scaling and shifting rule,

$$f'(x) = 8(9-x)^7\cdot(-1) = -8(9-x)^7$$
$$f''(x) = -56(9-x)^6\cdot(-1) = 56(9-x)^6,$$
$$f'''(x) = 336(9-x)^5\cdot(-1) = -336(9-x)^5.$$

77. The average molecular velocity v of a gas in a certain container is given by $v = 29\sqrt{T}$ m/s, where T is the temperature in kelvins. The temperature is related to the pressure (in atmospheres) by $T = 200P$. Find $\left.\dfrac{dv}{dP}\right|_{P=1.5}$.

SOLUTION First note that when $P = 1.5$ atmospheres, $T = 200(1.5) = 300$K. Thus,

$$\left.\frac{dv}{dP}\right|_{P=1.5} = \left.\frac{dv}{dT}\right|_{T=300}\cdot\left.\frac{dT}{dP}\right|_{P=1.5} = \frac{29}{2\sqrt{300}}\cdot 200 = \frac{290\sqrt{3}}{3}\ \frac{\text{m}}{\text{s}\cdot\text{atmospheres}}.$$

Alternately, substituting $T = 200P$ into the equation for v gives $v = 290\sqrt{2P}$. Therefore,

$$\frac{dv}{dP} = \frac{290\sqrt{2}}{2\sqrt{P}} = \frac{290}{\sqrt{2P}},$$

so

$$\left.\frac{dv}{dP}\right|_{P=1.5} = \frac{290}{\sqrt{3}} = \frac{290\sqrt{3}}{3}\ \frac{\text{m}}{\text{s}\cdot\text{atmospheres}}.$$

79. An expanding sphere has radius $r = 0.4t$ cm at time t (in seconds). Let V be the sphere's volume. Find dV/dt when (a) $r = 3$ and (b) $t = 3$.

SOLUTION Let $r = 0.4t$, where t is in seconds (s) and r is in centimeters (cm). With $V = \frac{4}{3}\pi r^3$, we have

$$\frac{dV}{dr} = 4\pi r^2.$$

Thus

$$\frac{dV}{dt} = \frac{dV}{dr}\frac{dr}{dt} = 4\pi r^2\cdot(0.4) = 1.6\pi r^2.$$

(a) When $r = 3$, $\dfrac{dV}{dt} = 1.6\pi(3)^2 \approx 45.24$ cm/s.

(b) When $t = 3$, we have $r = 1.2$. Hence $\dfrac{dV}{dt} = 1.6\pi(1.2)^2 \approx 7.24$ cm/s.

81. A 1999 study by Starkey and Scarnecchia developed the following model for the average weight (in kilograms) at age t (in years) of channel catfish in the Lower Yellowstone River (Figure 3):

$$W(t) = (3.46293 - 3.32173e^{-0.03456t})^{3.4026}$$

Find the rate at which average weight is changing at age $t = 10$.

FIGURE 3 Average weight of channel catfish at age t

SOLUTION Let $W(t) = (3.46293 - 3.32173e^{-0.03456t})^{3.4026}$. Then

$$W'(t) = 3.4026(3.46293 - 3.32173e^{-0.03456t})^{2.4026}(3.32173)(0.03456)e^{-0.03456t}$$
$$= 0.3906(3.46293 - 3.32173e^{-0.03456t})^{2.4026}e^{-0.03456t}.$$

At age $t = 10$,

$$W'(10) = 0.3906(1.1118)^{2.4026}(0.7078) \approx 0.3566 \text{ kg/yr}.$$

83. With notation as in Example 7, calculate

(a) $\left.\frac{d}{d\theta}\underline{\sin}\,\theta\right|_{\theta=60^\circ}$ **(b)** $\left.\frac{d}{d\theta}\left(\theta + \underline{\tan}\,\theta\right)\right|_{\theta=45^\circ}$

SOLUTION

(a) $\left.\frac{d}{d\theta}\underline{\sin}\,\theta\right|_{\theta=60^\circ} = \left.\frac{d}{d\theta}\sin\left(\frac{\pi}{180}\theta\right)\right|_{\theta=60^\circ} = \left(\frac{\pi}{180}\right)\cos\left(\frac{\pi}{180}(60)\right) = \frac{\pi}{180}\frac{1}{2} = \frac{\pi}{360}.$

(b) $\left.\frac{d}{d\theta}\left(\theta + \underline{\tan}\,\theta\right)\right|_{\theta=45^\circ} = \left.\frac{d}{d\theta}\left(\theta + \tan\left(\frac{\pi}{180}\theta\right)\right)\right|_{\theta=45^\circ} = 1 + \frac{\pi}{180}\sec^2\left(\frac{\pi}{4}\right) = 1 + \frac{\pi}{90}.$

85. Compute the derivative of $h(\sin x)$ at $x = \frac{\pi}{6}$, assuming that $h'(0.5) = 10$.

SOLUTION Let $u = \sin x$ and suppose that $h'(0.5) = 10$. Then

$$\frac{d}{dx}(h(u)) = \frac{dh}{du}\frac{du}{dx} = \frac{dh}{du}\cos x.$$

When $x = \frac{\pi}{6}$, we have $u = .5$. Accordingly, the derivative of $h(\sin x)$ at $x = \frac{\pi}{6}$ is $10\cos\left(\frac{\pi}{6}\right) = 5\sqrt{3}$.

In Exercises 87–90, use the table of values to calculate the derivative of the function at the given point.

x	1	4	6
$f(x)$	4	0	6
$f'(x)$	5	7	4
$g(x)$	4	1	6
$g'(x)$	5	$\frac{1}{2}$	3

87. $f(g(x)),\quad x = 6$

SOLUTION $\left.\frac{d}{dx}f(g(x))\right|_{x=6} = f'(g(6))g'(6) = f'(6)g'(6) = 4 \times 3 = 12.$

89. $g(\sqrt{x}),\quad x = 16$

SOLUTION $\left.\frac{d}{dx}g(\sqrt{x})\right|_{x=16} = g'(4)\left(\frac{1}{2}\right)(1/\sqrt{16}) = \left(\frac{1}{2}\right)\left(\frac{1}{2}\right)\left(\frac{1}{4}\right) = \frac{1}{16}.$

91. The price (in dollars) of a computer component is $P = 2C - 18C^{-1}$, where C is the manufacturer's cost to produce it. Assume that cost at time t (in years) is $C = 9 + 3t^{-1}$. Determine the rate of change of price with respect to time at $t = 3$.

SOLUTION $\dfrac{dC}{dt} = -3t^{-2}$. $C(3) = 10$ and $C'(3) = -\frac{1}{3}$, so we compute:

$$\left.\frac{dP}{dt}\right|_{t=3} = 2C'(3) + \frac{18}{(C(3))^2}C'(3) = -\frac{2}{3} + \frac{18}{100}\left(-\frac{1}{3}\right) = -0.727\ \frac{\text{dollars}}{\text{year}}.$$

93. According to the U.S. standard atmospheric model, developed by the National Oceanic and Atmospheric Administration for use in aircraft and rocket design, atmospheric temperature T (in degrees Celsius), pressure P (kPa = 1,000 pascals), and altitude h (in meters) are related by these formulas (valid in the troposphere $h \le 11{,}000$):

$$T = 15.04 - 0.000649h, \qquad P = 101.29 + \left(\frac{T + 273.1}{288.08}\right)^{5.256}$$

Use the Chain Rule to calculate dP/dh. Then estimate the change in P (in pascals, Pa) per additional meter of altitude when $h = 3{,}000$.

SOLUTION

$$\frac{dP}{dT} = 5.256\left(\frac{T + 273.1}{288.08}\right)^{4.256}\left(\frac{1}{288.08}\right) = 6.21519 \times 10^{-13}\,(273.1 + T)^{4.256}$$

and $\frac{dT}{dh} = -0.000649°\text{C/m}$. $\frac{dP}{dh} = \frac{dP}{dT}\frac{dT}{dh}$, so

$$\frac{dP}{dh} = \left(6.21519 \times 10^{-13}\,(273.1 + T)^{4.256}\right)(-0.000649) = -4.03366 \times 10^{-16}\,(288.14 - 0.000649\,h)^{4.256}.$$

When $h = 3000$,

$$\frac{dP}{dh} = -4.03366 \times 10^{-16}(286.193)^{4.256} = -1.15 \times 10^{-5}\ \text{kPa/m};$$

therefore, for each additional meter of altitude,

$$\Delta P \approx -1.15 \times 10^{-5}\ \text{kPa} = -1.15 \times 10^{-2}\ \text{Pa}.$$

95. In the setting of Exercise 94, calculate the yearly rate of change of T if $T = 283$ K and R increases at a rate of 0.5 $\text{Js}^{-1}\text{m}^{-2}$ per year.

SOLUTION By the Chain Rule,

$$\frac{dR}{dt} = \frac{dR}{dT}\cdot\frac{dT}{dt} = 4\sigma T^3\frac{dT}{dt}.$$

Assuming $T = 283$ K and $\frac{dR}{dt} = 0.5\ \text{Js}^{-1}\text{m}^{-2}$ per year, it follows that author:

$$0.5 = 4\sigma(283)^3\frac{dT}{dt} \Rightarrow \frac{dT}{dt} = \frac{0.5}{4\sigma(283)^3} \approx 0.0973\ \text{kelvins/yr}$$

97. Use the Chain Rule to express the second derivative of $f \circ g$ in terms of the first and second derivatives of f and g.

SOLUTION Let $h(x) = f(g(x))$. Then

$$h'(x) = f'(g(x))g'(x)$$

and

$$h''(x) = f'(g(x))g''(x) + g'(x)f''(g(x))g'(x) = f'(g(x))g''(x) + f''(g(x))\left(g'(x)\right)^2.$$

Further Insights and Challenges

99. Show that if f, g, and h are differentiable, then

$$[f(g(h(x)))]' = f'(g(h(x)))g'(h(x))h'(x)$$

SOLUTION Let f, g, and h be differentiable. Let $u = h(x)$, $v = g(u)$, and $w = f(v)$. Then

$$\frac{dw}{dx} = \frac{df}{dv}\frac{dv}{dx} = \frac{df}{dv}\frac{dg}{du}\frac{du}{dx} = f'(g(h(x))g'(h(x))h'(x)$$

101. (a) Sketch a graph of any even function $f(x)$ and explain graphically why $f'(x)$ is odd.
(b) Suppose that $f'(x)$ is even. Is $f(x)$ necessarily odd? *Hint:* Check whether this is true for linear functions.

SOLUTION

(a) The graph of an even function is symmetric with respect to the y-axis. Accordingly, its image in the left half-plane is a mirror reflection of that in the right half-plane through the y-axis. If at $x = a \geq 0$, the slope of f exists and is equal to m, then by reflection its slope at $x = -a \leq 0$ is $-m$. That is, $f'(-a) = -f'(a)$. *Note:* This means that if $f'(0)$ exists, then it equals 0.

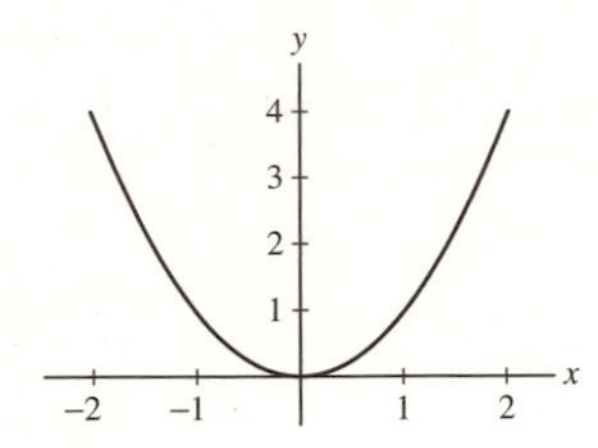

(b) Suppose that f' is even. Then f is not necessarily odd. Let $f(x) = 4x + 7$. Then $f'(x) = 4$, an even function. But f is not odd. For example, $f(2) = 15$, $f(-2) = -1$, but $f(-2) \neq -f(2)$.

103. Prove that for all whole numbers $n \geq 1$,

$$\frac{d^n}{dx^n}\sin x = \sin\left(x + \frac{n\pi}{2}\right)$$

Hint: Use the identity $\cos x = \sin\left(x + \frac{\pi}{2}\right)$.

SOLUTION We will proceed by induction on n. For $n = 1$, we find

$$\frac{d}{dx}\sin x = \cos x = \sin\left(x + \frac{\pi}{2}\right),$$

as required. Now, suppose that for some positive integer k,

$$\frac{d^k}{dx^k}\sin x = \sin\left(x + \frac{k\pi}{2}\right).$$

Then

$$\frac{d^{k+1}}{dx^{k+1}}\sin x = \frac{d}{dx}\sin\left(x + \frac{k\pi}{2}\right)$$
$$= \cos\left(x + \frac{k\pi}{2}\right) = \sin\left(x + \frac{(k+1)\pi}{2}\right).$$

105. Chain Rule This exercise proves the Chain Rule without the special assumption made in the text. For any number b, define a new function

$$F(u) = \frac{f(u) - f(b)}{u - b} \quad \text{for all } u \neq b$$

(a) Show that if we define $F(b) = f'(b)$, then $F(u)$ is continuous at $u = b$.
(b) Take $b = g(a)$. Show that if $x \neq a$, then for all u,

$$\frac{f(u) - f(g(a))}{x - a} = F(u)\frac{u - g(a)}{x - a} \qquad \boxed{2}$$

Note that both sides are zero if $u = g(a)$.
(c) Substitute $u = g(x)$ in Eq. (2) to obtain

$$\frac{f(g(x)) - f(g(a))}{x - a} = F(g(x))\,\frac{g(x) - g(a)}{x - a}$$

Derive the Chain Rule by computing the limit of both sides as $x \to a$.

SOLUTION For any differentiable function f and any number b, define

$$F(u) = \frac{f(u) - f(b)}{u - b}$$

for all $u \neq b$.

(a) Define $F(b) = f'(b)$. Then

$$\lim_{u\to b} F(u) = \lim_{u\to b} \frac{f(u) - f(b)}{u - b} = f'(b) = F(b),$$

i.e., $\lim_{u\to b} F(u) = F(b)$. Therefore, F is continuous at $u = b$.

(b) Let g be a differentiable function and take $b = g(a)$. Let x be a number distinct from a. If we substitute $u = g(a)$ into Eq. (2), both sides evaluate to 0, so equality is satisfied. On the other hand, if $u \neq g(a)$, then

$$\frac{f(u) - f(g(a))}{x - a} = \frac{f(u) - f(g(a))}{u - g(a)} \frac{u - g(a)}{x - a} = \frac{f(u) - f(b)}{u - b} \frac{u - g(a)}{x - a} = F(u) \frac{u - g(a)}{x - a}.$$

(c) Hence for all u, we have

$$\frac{f(u) - f(g(a))}{x - a} = F(u) \frac{u - g(a)}{x - a}.$$

(d) Substituting $u = g(x)$ in Eq. (2), we have

$$\frac{f(g(x)) - f(g(a))}{x - a} = F(g(x)) \frac{g(x) - g(a)}{x - a}.$$

Letting $x \to a$ gives

$$\lim_{x\to a} \frac{f(g(x)) - f(g(a))}{x - a} = \lim_{x\to a} \left(F(g(x)) \frac{g(x) - g(a)}{x - a} \right) = F(g(a))g'(a) = F(b)g'(a) = f'(b)g'(a)$$

$$= f'(g(a))g'(a)$$

Therefore $(f \circ g)'(a) = f'(g(a))g'(a)$, which is the Chain Rule.

3.8 Derivatives of Inverse Functions

Preliminary Questions

1. What is the slope of the line obtained by reflecting the line $y = \frac{x}{2}$ through the line $y = x$?

SOLUTION The line obtained by reflecting the line $y = x/2$ through the line $y = x$ has slope 2.

2. Suppose that $P = (2, 4)$ lies on the graph of $f(x)$ and that the slope of the tangent line through P is $m = 3$. Assuming that $f^{-1}(x)$ exists, what is the slope of the tangent line to the graph of $f^{-1}(x)$ at the point $Q = (4, 2)$?

SOLUTION The tangent line to the graph of $f^{-1}(x)$ at the point $Q = (4, 2)$ has slope $\frac{1}{3}$.

3. Which inverse trigonometric function $g(x)$ has the derivative $g'(x) = \dfrac{1}{x^2 + 1}$?

SOLUTION $g(x) = \tan^{-1} x$ has the derivative $g'(x) = \dfrac{1}{x^2 + 1}$.

4. What does the following identity tell us about the derivatives of $\sin^{-1} x$ and $\cos^{-1} x$?

$$\sin^{-1} x + \cos^{-1} x = \frac{\pi}{2}$$

SOLUTION Angles whose sine and cosine are x are complementary.

Exercises

1. Find the inverse $g(x)$ of $f(x) = \sqrt{x^2 + 9}$ with domain $x \geq 0$ and calculate $g'(x)$ in two ways: using Theorem 1 and by direct calculation.

SOLUTION To find a formula for $g(x) = f^{-1}(x)$, solve $y = \sqrt{x^2 + 9}$ for x. This yields $x = \pm\sqrt{y^2 - 9}$. Because the domain of f was restricted to $x \geq 0$, we must choose the positive sign in front of the radical. Thus

$$g(x) = f^{-1}(x) = \sqrt{x^2 - 9}.$$

Because $x^2 + 9 \geq 9$ for all x, it follows that $f(x) \geq 3$ for all x. Thus, the domain of $g(x) = f^{-1}(x)$ is $x \geq 3$. The range of g is the restricted domain of f: $y \geq 0$.

By Theorem 1,

$$g'(x) = \frac{1}{f'(g(x))}.$$

With

$$f'(x) = \frac{x}{\sqrt{x^2+9}},$$

it follows that

$$f'(g(x)) = \frac{\sqrt{x^2-9}}{\sqrt{(\sqrt{x^2-9})^2+9}} = \frac{\sqrt{x^2-9}}{\sqrt{x^2}} = \frac{\sqrt{x^2-9}}{x}$$

since the domain of g is $x \geq 3$. Thus,

$$g'(x) = \frac{1}{f'(g(x))} = \frac{x}{\sqrt{x^2-9}}.$$

This agrees with the answer we obtain by differentiating directly:

$$g'(x) = \frac{2x}{2\sqrt{x^2-9}} = \frac{x}{\sqrt{x^2-9}}.$$

In Exercises 3–8, use Theorem 1 to calculate $g'(x)$, where $g(x)$ is the inverse of $f(x)$.

3. $f(x) = 7x + 6$

SOLUTION Let $f(x) = 7x + 6$ then $f'(x) = 7$. Solving $y = 7x + 6$ for x and switching variables, we obtain the inverse $g(x) = (x - 6)/7$. Thus,

$$g'(x) = \frac{1}{f'(g(x))} = \frac{1}{7}.$$

5. $f(x) = x^{-5}$

SOLUTION Let $f(x) = x^{-5}$, then $f'(x) = -5x^{-6}$. Solving $y = x^{-5}$ for x and switching variables, we obtain the inverse $g(x) = x^{-1/5}$. Thus,

$$g'(x) = \frac{1}{-5(x^{-1/5})^{-6}} = -\frac{1}{5}x^{-6/5}.$$

7. $f(x) = \dfrac{x}{x+1}$

SOLUTION Let $f(x) = \frac{x}{x+1}$, then

$$f'(x) = \frac{(x+1) - x}{(x+1)^2} = \frac{1}{(x+1)^2}.$$

Solving $y = \frac{x}{x+1}$ for x and switching variables, we obtain the inverse $g(x) = \frac{x}{1-x}$. Thus

$$g'(x) = 1 \Big/ \frac{1}{(x/(1-x)+1)^2} = \frac{1}{(1-x)^2}.$$

9. Let $g(x)$ be the inverse of $f(x) = x^3 + 2x + 4$. Calculate $g(7)$ [without finding a formula for $g(x)$], and then calculate $g'(7)$.

SOLUTION Let $g(x)$ be the inverse of $f(x) = x^3 + 2x + 4$. Because

$$f(1) = 1^3 + 2(1) + 4 = 7,$$

it follows that $g(7) = 1$. Moreover, $f'(x) = 3x^2 + 2$, and

$$g'(7) = \frac{1}{f'(g(7))} = \frac{1}{f'(1)} = \frac{1}{5}.$$

In Exercises 11–16, calculate $g(b)$ and $g'(b)$, where g is the inverse of f (in the given domain, if indicated).

11. $f(x) = x + \cos x, \quad b = 1$

SOLUTION $f(0) = 1$, so $g(1) = 0$. $f'(x) = 1 - \sin x$ so $f'(g(1)) = f'(0) = 1 - \sin 0 = 1$. Thus, $g'(1) = 1/1 = 1$.

13. $f(x) = \sqrt{x^2 + 6x}$ for $x \geq 0$, $\quad b = 4$

SOLUTION To determine $g(4)$, we solve $f(x) = \sqrt{x^2 + 6x} = 4$ for x. This yields:

$$x^2 + 6x = 16$$
$$x^2 + 6x - 16 = 0$$
$$(x + 8)(x - 2) = 0$$

or $x = -8, 2$. Because the domain of f has been restricted to $x \geq 0$, we have $g(4) = 2$. With

$$f'(x) = \frac{x + 3}{\sqrt{x^2 + 6x}},$$

it then follows that

$$g'(4) = \frac{1}{f'(g(4))} = \frac{1}{f'(2)} = \frac{4}{5}.$$

15. $f(x) = \dfrac{1}{x + 1}$, $\quad b = \dfrac{1}{4}$

SOLUTION $f(3) = 1/4$, so $g(1/4) = 3$. $f'(x) = \frac{-1}{(x+1)^2}$ so $f'(g(1/4)) = f'(3) = \frac{-1}{(3+1)^2} = -1/16$. Thus, $g'(1/4) = -16$.

17. Let $f(x) = x^n$ and $g(x) = x^{1/n}$. Compute $g'(x)$ using Theorem 1 and check your answer using the Power Rule.

SOLUTION Note that $g(x) = f^{-1}(x)$. Therefore,

$$g'(x) = \frac{1}{f'(g(x))} = \frac{1}{n(g(x))^{n-1}} = \frac{1}{n(x^{1/n})^{n-1}} = \frac{1}{n(x^{1-1/n})} = \frac{x^{1/n-1}}{n} = \frac{1}{n}(x^{1/n-1})$$

which agrees with the Power Rule.

In Exercises 19–22, compute the derivative at the point indicated without using a calculator.

19. $y = \sin^{-1} x$, $\quad x = \frac{3}{5}$

SOLUTION Let $y = \sin^{-1} x$. Then $y' = \frac{1}{\sqrt{1-x^2}}$ and

$$y'\left(\frac{3}{5}\right) = \frac{1}{\sqrt{1 - 9/25}} = \frac{1}{4/5} = \frac{5}{4}.$$

21. $y = \sec^{-1} x$, $\quad x = 4$

SOLUTION Let $y = \sec^{-1} x$. Then $y' = \frac{1}{|x|\sqrt{x^2-1}}$ and

$$y'(4) = \frac{1}{4\sqrt{15}}.$$

In Exercises 23–36, find the derivative.

23. $y = \sin^{-1}(7x)$

SOLUTION $\dfrac{d}{dx}\sin^{-1}(7x) = \dfrac{1}{\sqrt{1 - (7x)^2}} \cdot \dfrac{d}{dx}7x = \dfrac{7}{\sqrt{1 - (7x)^2}}.$

25. $y = \cos^{-1}(x^2)$

SOLUTION $\dfrac{d}{dx}\cos^{-1}(x^2) = \dfrac{-1}{\sqrt{1 - x^4}} \cdot \dfrac{d}{dx}x^2 = \dfrac{-2x}{\sqrt{1 - x^4}}.$

27. $y = x \tan^{-1} x$

SOLUTION $\dfrac{d}{dx}x \tan^{-1} x = x\left(\dfrac{1}{1 + x^2}\right) + \tan^{-1} x.$

29. $y = \arcsin(e^x)$

SOLUTION $\dfrac{d}{dx}\sin^{-1}(e^x) = \dfrac{1}{\sqrt{1 - e^{2x}}} \cdot \dfrac{d}{dx}e^x = \dfrac{e^x}{\sqrt{1 - e^{2x}}}.$

31. $y = \sqrt{1 - t^2} + \sin^{-1} t$

SOLUTION $\dfrac{d}{dt}\left(\sqrt{1 - t^2} + \sin^{-1} t\right) = \dfrac{1}{2}(1 - t^2)^{-1/2}(-2t) + \dfrac{1}{\sqrt{1 - t^2}} = \dfrac{-t}{\sqrt{1 - t^2}} + \dfrac{1}{\sqrt{1 - t^2}} = \dfrac{1 - t}{\sqrt{1 - t^2}}.$

33. $y = (\tan^{-1} x)^3$

SOLUTION $\frac{d}{dx}\left((\tan^{-1} x)^3\right) = 3(\tan^{-1} x)^2 \frac{d}{dx} \tan^{-1} x = \frac{3(\tan^{-1} x)^2}{x^2+1}.$

35. $y = \cos^{-1} t^{-1} - \sec^{-1} t$

SOLUTION
$$\frac{d}{dx}(\cos^{-1} t^{-1} - \sec^{-1} t) = \frac{-1}{\sqrt{1-(1/t)^2}}\left(\frac{-1}{t^2}\right) - \frac{1}{|t|\sqrt{t^2-1}}$$
$$= \frac{1}{\sqrt{t^4-t^2}} - \frac{1}{|t|\sqrt{t^2-1}} = \frac{1}{|t|\sqrt{t^2-1}} - \frac{1}{|t|\sqrt{t^2-1}} = 0.$$

Alternately, let $t = \sec\theta$. Then $t^{-1} = \cos\theta$ and $\cos^{-1} t^{-1} - \sec^{-1} t = \theta - \theta = 0$. Consequently,

$$\frac{d}{dx}(\cos^{-1} t^{-1} - \sec^{-1} t) = 0.$$

37. Use Figure 5 to prove that $(\cos^{-1} x)' = -\frac{1}{\sqrt{1-x^2}}$.

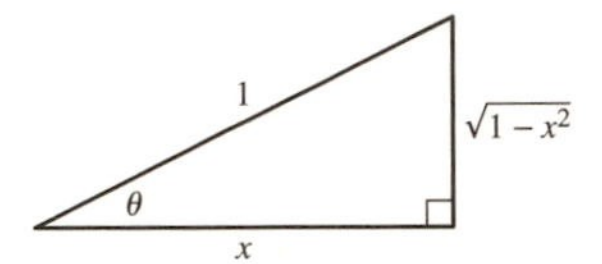

FIGURE 5 Right triangle with $\theta = \cos^{-1} x$.

SOLUTION Let $\theta = \cos^{-1} x$. Then $\cos\theta = x$ and

$$-\sin\theta \frac{d\theta}{dx} = 1 \quad \text{or} \quad \frac{d\theta}{dx} = -\frac{1}{\sin\theta} = -\frac{1}{\sin(\cos^{-1} x)}.$$

From Figure 5, we see that $\sin(\cos^{-1} x) = \sin\theta = \sqrt{1-x^2}$; hence,

$$\frac{d}{dx}\cos^{-1} x = \frac{1}{-\sin(\cos^{-1} x)} = -\frac{1}{\sqrt{1-x^2}}.$$

39. Let $\theta = \sec^{-1} x$. Show that $\tan\theta = \sqrt{x^2-1}$ if $x \ge 1$ and that $\tan\theta = -\sqrt{x^2-1}$ if $x \le -1$. *Hint:* $\tan\theta \ge 0$ on $\left(0, \frac{\pi}{2}\right)$ and $\tan\theta \le 0$ on $\left(\frac{\pi}{2}, \pi\right)$.

SOLUTION In general, $1 + \tan^2\theta = \sec^2\theta$, so $\tan\theta = \pm\sqrt{\sec^2\theta - 1}$. With $\theta = \sec^{-1} x$, it follows that $\sec\theta = x$, so $\tan\theta = \pm\sqrt{x^2-1}$. Finally, if $x \ge 1$ then $\theta = \sec^{-1} x \in [0, \pi/2)$ so $\tan\theta$ is positive; on the other hand, if $x \le 1$ then $\theta = \sec^{-1} x \in (-\pi/2, 0]$ so $\tan\theta$ is negative.

Further Insights and Challenges

41. Let $g(x)$ be the inverse of $f(x)$. Show that if $f'(x) = f(x)$, then $g'(x) = x^{-1}$. We will apply this in the next section to show that the inverse of $f(x) = e^x$ (the natural logarithm) has the derivative $f'(x) = x^{-1}$.

SOLUTION

$$g'(x) = \frac{1}{f'(g(x))} = \frac{1}{f'(f^{-1}(x))} = \frac{1}{f(f^{-1}(x))} = \frac{1}{x}.$$

3.9 Derivatives of General Exponential and Logarithmic Functions

Preliminary Questions

1. What is the slope of the tangent line to $y = 4^x$ at $x = 0$?

SOLUTION The slope of the tangent line to $y = 4^x$ at $x = 0$ is

$$\frac{d}{dx}4^x\bigg|_{x=0} = 4^x \ln 4\bigg|_{x=0} = \ln 4.$$

2. What is the rate of change of $y = \ln x$ at $x = 10$?

SOLUTION The rate of change of $y = \ln x$ at $x = 10$ is

$$\frac{d}{dx}\ln x\bigg|_{x=10} = \frac{1}{x}\bigg|_{x=10} = \frac{1}{10}.$$

3. What is $b > 0$ if the tangent line to $y = b^x$ at $x = 0$ has slope 2?

SOLUTION The tangent line to $y = b^x$ at $x = 0$ has slope

$$\frac{d}{dx}b^x\bigg|_{x=0} = b^x \ln b\bigg|_{x=0} = \ln b.$$

This slope will be equal to 2 when

$$\ln b = 2 \quad \text{or} \quad b = e^2.$$

4. What is b if $(\log_b x)' = \dfrac{1}{3x}$?

SOLUTION $(\log_b x)' = \left(\dfrac{\ln x}{\ln b}\right)' = \dfrac{1}{x \ln b}$. This derivative will equal $\frac{1}{3x}$ when

$$\ln b = 3 \quad \text{or} \quad b = e^3.$$

5. What are $y^{(100)}$ and $y^{(101)}$ for $y = \cosh x$?

SOLUTION Let $y = \cosh x$. Then $y' = \sinh x$, $y'' = \cosh x$, and this pattern repeats indefinitely. Thus, $y^{(100)} = \cosh x$ and $y^{(101)} = \sinh x$.

Exercises

In Exercises 1–20, find the derivative.

1. $y = x \ln x$

SOLUTION $\dfrac{d}{dx}x \ln x = \ln x + \dfrac{x}{x} = \ln x + 1.$

3. $y = (\ln x)^2$

SOLUTION $\dfrac{d}{dx}(\ln x)^2 = (2 \ln x)\dfrac{1}{x} = \dfrac{2}{x}\ln x.$

5. $y = \ln(9x^2 - 8)$

SOLUTION $\dfrac{d}{dx}\ln(9x^2 - 8) = \dfrac{18x}{9x^2 - 8}.$

7. $y = \ln(\sin t + 1)$

SOLUTION $\dfrac{d}{dt}\ln(\sin t + 1) = \dfrac{\cos t}{\sin t + 1}.$

9. $y = \dfrac{\ln x}{x}$

SOLUTION $\dfrac{d}{dx}\dfrac{\ln x}{x} = \dfrac{\frac{1}{x}(x) - \ln x}{x^2} = \dfrac{1 - \ln x}{x^2}.$

11. $y = \ln(\ln x)$

SOLUTION $\dfrac{d}{dx}\ln(\ln x) = \dfrac{1}{x \ln x}.$

13. $y = \big(\ln(\ln x)\big)^3$

SOLUTION $\dfrac{d}{dx}(\ln(\ln x))^3 = 3(\ln(\ln x))^2\left(\dfrac{1}{\ln x}\right)\left(\dfrac{1}{x}\right) = \dfrac{3(\ln(\ln x))^2}{x \ln x}.$

15. $y = \ln\big((x + 1)(2x + 9)\big)$

SOLUTION

$$\frac{d}{dx}\ln((x + 1)(2x + 9)) = \frac{1}{(x + 1)(2x + 9)} \cdot ((x + 1)2 + (2x + 9)) = \frac{4x + 11}{(x + 1)(2x + 9)}.$$

Alternately, because $\ln((x + 1)(2x + 9)) = \ln(x + 1) + \ln(2x + 9)$,

$$\frac{d}{dx}\ln((x + 1)(2x + 9)) = \frac{1}{x + 1} + \frac{2}{2x + 9} = \frac{4x + 11}{(x + 1)(2x + 9)}.$$

17. $y = 11^x$

SOLUTION $\dfrac{d}{dx}11^x = \ln 11 \cdot 11^x.$

19. $y = \dfrac{2^x - 3^{-x}}{x}$

SOLUTION $\dfrac{d}{dx}\dfrac{2^x - 3^{-x}}{x} = \dfrac{x(2^x \ln 2 + 3^{-x}\ln 3) - (2^x - 3^{-x})}{x^2}.$

In Exercises 21–24, compute the derivative.

21. $f'(x), \quad f(x) = \log_2 x$

SOLUTION $f(x) = \log_2 x = \dfrac{\ln x}{\ln 2}$. Thus, $f'(x) = \dfrac{1}{x}\cdot\dfrac{1}{\ln 2}$.

23. $\dfrac{d}{dt}\log_3(\sin t)$

SOLUTION $\dfrac{d}{dt}\log_3(\sin t) = \dfrac{d}{dt}\left(\dfrac{\ln(\sin t)}{\ln 3}\right) = \dfrac{1}{\ln 3}\cdot\dfrac{1}{\sin t}\cdot\cos t = \dfrac{\cot t}{\ln 3}.$

In Exercises 25–36, find an equation of the tangent line at the point indicated.

25. $f(x) = 6^x, \quad x = 2$

SOLUTION Let $f(x) = 6^x$. Then $f(2) = 36$, $f'(x) = 6^x \ln 6$ and $f'(2) = 36\ln 6$. The equation of the tangent line is therefore $y = 36\ln 6(x-2) + 36$.

27. $s(t) = 3^{9t}, \quad t = 2$

SOLUTION Let $s(t) = 3^{9t}$. Then $s(2) = 3^{18}$, $s'(t) = 3^{9t}9\ln 3$, and $s'(2) = 3^{18}\cdot 9\ln 3 = 3^{20}\ln 3$. The equation of the tangent line is therefore $y = 3^{20}\ln 3(t-2) + 3^{18}$.

29. $f(x) = 5^{x^2-2x}, \quad x = 1$

SOLUTION Let $f(x) = 5^{x^2-2x}$. Then $f(1) = 5^{-1}$, $f'(x) = \ln 5\cdot 5^{x^2-2x}(2x-2)$, and $f'(1) = \ln 5(0) = 0$. Therefore, the equation of the tangent line is $y = 5^{-1}$.

31. $s(t) = \ln(8-4t), \quad t = 1$

SOLUTION Let $s(t) = \ln(8-4t)$. Then $s(1) = \ln(8-4) = \ln 4$. $s'(t) = \frac{-4}{8-4t}$, so $s'(1) = -4/4 = -1$. Therefore the equation of the tangent line is $y = -1(t-1) + \ln 4$.

33. $R(z) = \log_5(2z^2+7), \quad z = 3$

SOLUTION Let $R(z) = \log_5(2z^2+7)$. Then $R(3) = \log_5(25) = 2$,

$$R'(z) = \frac{4z}{(2z^2+7)\ln 5}, \quad \text{and} \quad R'(3) = \frac{12}{25\ln 5}.$$

The equation of the tangent line is therefore

$$y = \frac{12}{25\ln 5}(z-3) + 2.$$

35. $f(w) = \log_2 w, \quad w = \frac{1}{8}$

SOLUTION Let $f(w) = \log_2 w$. Then

$$f\left(\frac{1}{8}\right) = \log_2\frac{1}{8} = \log_2 2^{-3} = -3,$$

$f'(w) = \frac{1}{w\ln 2}$, and

$$f'\left(\frac{1}{8}\right) = \frac{8}{\ln 2}.$$

The equation of the tangent line is therefore

$$y = \frac{8}{\ln 2}\left(w - \frac{1}{8}\right) - 3.$$

In Exercises 37–44, find the derivative using logarithmic differentiation as in Example 5.

37. $y = (x+5)(x+9)$

SOLUTION Let $y = (x+5)(x+9)$. Then $\ln y = \ln((x+5)(x+9)) = \ln(x+5) + \ln(x+9)$. By logarithmic differentiation

$$\frac{y'}{y} = \frac{1}{x+5} + \frac{1}{x+9}$$

or

$$y' = (x+5)(x+9)\left(\frac{1}{x+5} + \frac{1}{x+9}\right) = (x+9) + (x+5) = 2x + 14.$$

39. $y = (x-1)(x-12)(x+7)$

SOLUTION Let $y = (x-1)(x-12)(x+7)$. Then $\ln y = \ln(x-1) + \ln(x-12) + \ln(x+7)$. By logarithmic differentiation,

$$\frac{y'}{y} = \frac{1}{x-1} + \frac{1}{x-12} + \frac{1}{x+7}$$

or

$$y' = (x-12)(x+7) + (x-1)(x+7) + (x-1)(x-12) = 3x^2 - 12x + 79.$$

41. $y = \dfrac{x(x^2+1)}{\sqrt{x+1}}$

SOLUTION Let $y = \frac{x(x^2+1)}{\sqrt{x+1}}$. Then $\ln y = \ln x + \ln(x^2+1) - \frac{1}{2}\ln(x+1)$. By logarithmic differentiation

$$\frac{y'}{y} = \frac{1}{x} + \frac{2x}{x^2+1} - \frac{1}{2(x+1)},$$

so

$$y' = \frac{x(x^2+1)}{\sqrt{x+1}}\left(\frac{1}{x} + \frac{2x}{x^2+1} - \frac{1}{2(x+1)}\right).$$

43. $y = \sqrt{\dfrac{x(x+2)}{(2x+1)(3x+2)}}$

SOLUTION Let $y = \sqrt{\frac{x(x+2)}{(2x+1)(3x+2)}}$. Then $\ln y = \frac{1}{2}[\ln(x) + \ln(x+2) - \ln(2x+1) - \ln(3x+2)]$. By logarithmic differentiation

$$\frac{y'}{y} = \frac{1}{2}\left(\frac{1}{x} + \frac{1}{x+2} - \frac{2}{2x+1} - \frac{3}{3x+2}\right),$$

so

$$y' = \frac{1}{2}\sqrt{\frac{x(x+2)}{(2x+1)(3x+2)}} \cdot \left(\frac{1}{x} + \frac{1}{x+2} - \frac{2}{2x+1} - \frac{3}{3x+2}\right).$$

In Exercises 45–50, find the derivative using either method of Example 6.

45. $f(x) = x^{3x}$

SOLUTION Method 1: $x^{3x} = e^{3x\ln x}$, so

$$\frac{d}{dx}x^{3x} = e^{3x\ln x}(3 + 3\ln x) = x^{3x}(3 + 3\ln x).$$

Method 2: Let $y = x^{3x}$. Then, $\ln y = 3x\ln x$. By logarithmic differentiation

$$\frac{y'}{y} = 3x \cdot \frac{1}{x} + 3\ln x,$$

so

$$y' = y(3 + 3\ln x) = x^{3x}(3 + 3\ln x).$$

47. $f(x) = x^{e^x}$

SOLUTION Method 1: $x^{e^x} = e^{e^x\ln x}$, so

$$\frac{d}{dx}x^{e^x} = e^{e^x\ln x}\left(\frac{e^x}{x} + e^x\ln x\right) = x^{e^x}\left(\frac{e^x}{x} + e^x\ln x\right).$$

Method 2: Let $y = x^{e^x}$. Then $\ln y = e^x \ln x$. By logarithmic differentiation

$$\frac{y'}{y} = e^x \cdot \frac{1}{x} + e^x \ln x,$$

so

$$y' = y\left(\frac{e^x}{x} + e^x \ln x\right) = x^{e^x}\left(\frac{e^x}{x} + e^x \ln x\right).$$

49. $f(x) = x^{3^x}$

SOLUTION Method 1: $x^{3^x} = e^{3^x \ln x}$, so

$$\frac{d}{dx}x^{3^x} = e^{3^x \ln x}\left(\frac{3^x}{x} + (\ln x)(\ln 3)3^x\right) = x^{3^x}\left(\frac{3^x}{x} + (\ln x)(\ln 3)3^x\right).$$

Method 2: Let $y = x^{3^x}$. Then $\ln y = 3^x \ln x$. By logarithmic differentiation

$$\frac{y'}{y} = 3^x \frac{1}{x} + (\ln x)(\ln 3)3^x,$$

so

$$y' = x^{3^x}\left(\frac{3^x}{x} + (\ln x)(\ln 3)3^x\right).$$

In Exercises 51–74, calculate the derivative.

51. $y = \sinh(9x)$

SOLUTION $\frac{d}{dx}\sinh(9x) = 9\cosh(9x).$

53. $y = \cosh^2(9 - 3t)$

SOLUTION $\frac{d}{dt}\cosh^2(9 - 3t) = 2\cosh(9 - 3t) \cdot (-3\sinh(9 - 3t)) = -6\cosh(9 - 3t)\sinh(9 - 3t).$

55. $y = \sqrt{\cosh x + 1}$

SOLUTION $\frac{d}{dx}\sqrt{\cosh x + 1} = \frac{1}{2}(\cosh x + 1)^{-1/2}\sinh x.$

57. $y = \dfrac{\coth t}{1 + \tanh t}$

SOLUTION $\frac{d}{dt}\frac{\coth t}{1 + \tanh t} = \frac{-\operatorname{csch}^2 t(1 + \tanh t) - \coth t(\operatorname{sech}^2 t)}{(1 + \tanh t)^2} = -\frac{\operatorname{csch}^2 t + 2\operatorname{csch} t \operatorname{sech} t}{(1 + \tanh t^2)}$

59. $y = \sinh(\ln x)$

SOLUTION $\frac{d}{dx}\sinh(\ln x) = \frac{\cosh(\ln x)}{x}.$

61. $y = \tanh(e^x)$

SOLUTION $\frac{d}{dx}\tanh(e^x) = e^x \operatorname{sech}^2(e^x).$

63. $y = \operatorname{sech}(\sqrt{x})$

SOLUTION $\frac{d}{dx}\operatorname{sech}(\sqrt{x}) = -\frac{1}{2}x^{-1/2}\operatorname{sech}\sqrt{x}\tanh\sqrt{x}.$

65. $y = \operatorname{sech} x \coth x$

SOLUTION $\frac{d}{dx}\operatorname{sech} x \coth x = \frac{d}{dx}\operatorname{csch} x = -\operatorname{csch} x \coth x.$

67. $y = \cosh^{-1}(3x)$

SOLUTION $\frac{d}{dx}\cosh^{-1}(3x) = \frac{3}{\sqrt{9x^2 - 1}}.$

69. $y = (\sinh^{-1}(x^2))^3$

SOLUTION $\frac{d}{dx}(\sinh^{-1}(x^2))^3 = 3(\sinh^{-1}(x^2))^2\frac{2x}{\sqrt{x^4 + 1}}.$

71. $y = e^{\cosh^{-1} x}$

SOLUTION $\frac{d}{dx} e^{\cosh^{-1} x} = e^{\cosh^{-1} x}\left(\frac{1}{\sqrt{x^2-1}}\right).$

73. $y = \tanh^{-1}(\ln t)$

SOLUTION $\frac{d}{dt} \tanh^{-1}(\ln t) = \frac{1}{t(1-(\ln t)^2)}.$

In Exercises 75–77, prove the formula.

75. $\frac{d}{dx}(\coth x) = -\operatorname{csch}^2 x$

SOLUTION $\frac{d}{dx} \coth x = \frac{d}{dx}\frac{\cosh x}{\sinh x} = \frac{\sinh^2 x - \cosh^2 x}{\sinh^2 x} = \frac{-1}{\sinh^2 x} = -\operatorname{csch}^2 x.$

77. $\frac{d}{dt} \cosh^{-1} t = \frac{1}{\sqrt{t^2-1}}$ for $t > 1$

SOLUTION Let $x = \cosh^{-1} t$. Then $x \geq 0$, $t = \cosh x$ and

$$1 = \sinh x \frac{dx}{dt} \quad \text{or} \quad \frac{dx}{dt} = \frac{1}{\sinh x}.$$

Thus, for $t > 1$,

$$\frac{d}{dt} \cosh^{-1} t = \frac{1}{\sinh x},$$

where $\cosh x = t$. Working from the identity $\cosh^2 x - \sinh^2 x = 1$, we find $\sinh x = \pm\sqrt{\cosh^2 x - 1}$. Because $\sinh w \geq 0$ for $w \geq 0$, we know to choose the positive square root. Hence, $\sinh x = \sqrt{\cosh^2 x - 1} = \sqrt{t^2 - 1}$, and

$$\frac{d}{dt} \cosh^{-1} t = \frac{1}{\sinh x} = \frac{1}{\sqrt{t^2-1}}.$$

79. According to one simplified model, the purchasing power of a dollar in the year $2000 + t$ is equal to $P(t) = 0.68(1.04)^{-t}$ (in 1983 dollars). Calculate the predicted rate of decline in purchasing power (in cents per year) in the year 2020.

SOLUTION First, note that

$$P'(t) = -0.68(1.04)^{-t} \ln 1.04;$$

thus, the rate of change in the year 2020 is

$$P'(20) = -0.68(1.04)^{-20} \ln 1.04 = -0.0122.$$

That is, the rate of decline is 1.22 cents per year.

81. Show that for any constants M, k, and a, the function

$$y(t) = \frac{1}{2}M\left(1 + \tanh\left(\frac{k(t-a)}{2}\right)\right)$$

satisfies the **logistic equation**: $\frac{y'}{y} = k\left(1 - \frac{y}{M}\right).$

SOLUTION Let

$$y(t) = \frac{1}{2}M\left(1 + \tanh\left(\frac{k(t-a)}{2}\right)\right).$$

Then

$$1 - \frac{y(t)}{M} = \frac{1}{2}\left(1 - \tanh\left(\frac{k(t-a)}{2}\right)\right),$$

and

$$ky(t)\left(1 - \frac{y(t)}{M}\right) = \frac{1}{4}Mk\left(1 - \tanh^2\left(\frac{k(t-a)}{2}\right)\right)$$
$$= \frac{1}{4}Mk \operatorname{sech}^2\left(\frac{k(t-a)}{2}\right).$$

Finally,

$$y'(t) = \frac{1}{4}Mk\,\text{sech}^2\left(\frac{k(t-a)}{2}\right) = ky(t)\left(1 - \frac{y(t)}{M}\right).$$

83. The Palermo Technical Impact Hazard Scale P is used to quantify the risk associated with the impact of an asteroid colliding with the earth:

$$P = \log_{10}\left(\frac{p_i E^{0.8}}{0.03T}\right)$$

where p_i is the probability of impact, T is the number of years until impact, and E is the energy of impact (in megatons of TNT). The risk is greater than a random event of similar magnitude if $P > 0$.

(a) Calculate dP/dT, assuming that $p_i = 2 \times 10^{-5}$ and $E = 2$ megatons.

(b) Use the derivative to estimate the change in P if T increases from 8 to 9 years.

SOLUTION

(a) Observe that

$$P = \log_{10}\left(\frac{p_i E^{0.8}}{0.03T}\right) = \log_{10}\left(\frac{p_i E^{0.8}}{0.03}\right) - \log_{10} T,$$

so

$$\frac{dP}{dT} = -\frac{1}{T \ln 10}.$$

(b) If T increases to 9 years from 8 years, then

$$\Delta P \approx \left.\frac{dP}{dT}\right|_{T=8} \cdot \Delta T = -\frac{1}{(8\text{ yr})\ln 10} \cdot (1\text{ yr}) = -0.054$$

Further Insights and Challenges

85. Use the formula $\log_b x = \dfrac{\log_a x}{\log_a b}$ for $a, b > 0$ to verify the formula

$$\frac{d}{dx}\log_b x = \frac{1}{(\ln b)x}$$

SOLUTION $\dfrac{d}{dx}\log_b x = \dfrac{d}{dx}\dfrac{\ln x}{\ln b} = \dfrac{1}{(\ln b)x}.$

3.10 Implicit Differentiation

Preliminary Questions

1. Which differentiation rule is used to show $\dfrac{d}{dx}\sin y = \cos y\dfrac{dy}{dx}$?

SOLUTION The chain rule is used to show that $\frac{d}{dx}\sin y = \cos y\frac{dy}{dx}$.

2. One of (a)–(c) is incorrect. Find and correct the mistake.

(a) $\dfrac{d}{dy}\sin(y^2) = 2y\cos(y^2)$ **(b)** $\dfrac{d}{dx}\sin(x^2) = 2x\cos(x^2)$ **(c)** $\dfrac{d}{dx}\sin(y^2) = 2y\cos(y^2)$

SOLUTION

(a) This is correct. Note that the differentiation is with respect to the variable y.

(b) This is correct. Note that the differentiation is with respect to the variable x.

(c) This is incorrect. Because the differentiation is with respect to the variable x, the chain rule is needed to obtain

$$\frac{d}{dx}\sin(y^2) = 2y\cos(y^2)\frac{dy}{dx}.$$

3. On an exam, Jason was asked to differentiate the equation

$$x^2 + 2xy + y^3 = 7$$

Find the errors in Jason's answer: $2x + 2xy' + 3y^2 = 0$

SOLUTION There are two mistakes in Jason's answer. First, Jason should have applied the product rule to the second term to obtain

$$\frac{d}{dx}(2xy) = 2x\frac{dy}{dx} + 2y.$$

Second, he should have applied the general power rule to the third term to obtain

$$\frac{d}{dx}y^3 = 3y^2\frac{dy}{dx}.$$

4. Which of (a) or (b) is equal to $\frac{d}{dx}(x \sin t)$?

(a) $(x \cos t)\frac{dt}{dx}$ **(b)** $(x \cos t)\frac{dt}{dx} + \sin t$

SOLUTION Using the product rule and the chain rule we see that

$$\frac{d}{dx}(x \sin t) = x \cos t\frac{dt}{dx} + \sin t,$$

so the correct answer is **(b)**.

Exercises

1. Show that if you differentiate both sides of $x^2 + 2y^3 = 6$, the result is $2x + 6y^2\frac{dy}{dx} = 0$. Then solve for dy/dx and evaluate it at the point (2, 1).

SOLUTION

$$\begin{aligned} \frac{d}{dx}(x^2 + 2y^3) &= \frac{d}{dx}6 \\ 2x + 6y^2\frac{dy}{dx} &= 0 \\ 2x + 6y^2\frac{dy}{dx} &= 0 \\ 6y^2\frac{dy}{dx} &= -2x \\ \frac{dy}{dx} &= \frac{-2x}{6y^2}. \end{aligned}$$

At (2, 1), $\frac{dy}{dx} = \frac{-4}{6} = -\frac{2}{3}$.

In Exercises 3–8, differentiate the expression with respect to x, assuming that $y = f(x)$.

3. x^2y^3

SOLUTION Assuming that y depends on x, then

$$\frac{d}{dx}\left(x^2y^3\right) = x^2 \cdot 3y^2y' + y^3 \cdot 2x = 3x^2y^2y' + 2xy^3.$$

5. $(x^2 + y^2)^{3/2}$

SOLUTION Assuming that y depends on x, then

$$\frac{d}{dx}\left(\left(x^2 + y^2\right)^{3/2}\right) = \frac{3}{2}\left(x^2 + y^2\right)^{1/2}(2x + 2yy') = 3\left(x + yy'\right)\sqrt{x^2 + y^2}.$$

7. $\frac{y}{y+1}$

SOLUTION Assuming that y depends on x, then $\frac{d}{dx}\frac{y}{y+1} = \frac{(y+1)y' - yy'}{(y+1)^2} = \frac{y'}{(y+1)^2}$.

In Exercises 9–26, calculate the derivative with respect to x.

9. $3y^3 + x^2 = 5$

SOLUTION Let $3y^3 + x^2 = 5$. Then $9y^2y' + 2x = 0$, and $y' = -\frac{2x}{9y^2}$.

11. $x^2y + 2x^3y = x + y$

SOLUTION Let $x^2y + 2x^3y = x + y$. Then

$$\begin{aligned} x^2y' + 2xy + 2x^3y' + 6x^2y &= 1 + y' \\ x^2y' + 2x^3y' - y' &= 1 - 2xy - 6x^2y \\ y' &= \frac{1 - 2xy - 6x^2y}{x^2 + 2x^3 - 1}. \end{aligned}$$

13. $x^3R^5 = 1$

SOLUTION Let $x^3R^5 = 1$. Then $x^3 \cdot 5R^4R' + R^5 \cdot 3x^2 = 0$, and $R' = -\dfrac{3x^2R^5}{5x^3R^4} = -\dfrac{3R}{5x}$.

15. $\dfrac{y}{x} + \dfrac{x}{y} = 2y$

SOLUTION Let

$$\frac{y}{x} + \frac{x}{y} = 2y.$$

Then

$$\begin{aligned} \frac{xy' - y}{x^2} + \frac{y - xy'}{y^2} &= 2y' \\ \left(\frac{1}{x} - \frac{x}{y^2} - 2\right)y' &= \frac{y}{x^2} - \frac{1}{y} \\ \frac{y^2 - x^2 - 2xy^2}{xy^2}y' &= \frac{y^2 - x^2}{x^2y} \\ y' &= \frac{y(y^2 - x^2)}{x(y^2 - x^2 - 2xy^2)}. \end{aligned}$$

17. $y^{-2/3} + x^{3/2} = 1$

SOLUTION Let $y^{-2/3} + x^{3/2} = 1$. Then

$$-\frac{2}{3}y^{-5/3}y' + \frac{3}{2}x^{1/2} = 0 \quad \text{or} \quad y' = \frac{9}{4}x^{1/2}y^{5/3}.$$

19. $y + \dfrac{1}{y} = x^2 + x$

SOLUTION Let $y + \frac{1}{y} = x^2 + x$. Then

$$y' - \frac{1}{y^2}y' = 2x + 1 \quad \text{or} \quad y' = \frac{2x + 1}{1 - y^{-2}} = \frac{(2x + 1)y^2}{y^2 - 1}.$$

21. $\sin(x + y) = x + \cos y$

SOLUTION Let $\sin(x + y) = x + \cos y$. Then

$$\begin{aligned} (1 + y')\cos(x + y) &= 1 - y'\sin y \\ \cos(x + y) + y'\cos(x + y) &= 1 - y'\sin y \\ (\cos(x + y) + \sin y)\, y' &= 1 - \cos(x + y) \\ y' &= \frac{1 - \cos(x + y)}{\cos(x + y) + \sin y}. \end{aligned}$$

23. $xe^y = 2xy + y^3$

SOLUTION Let $xe^y = 2xy + y^3$. Then $xy'e^y + e^y = 2xy' + 2y + 3y^2y'$, whence

$$y' = \frac{e^y - 2y}{2x + 3y^2 - xe^y}.$$

25. $\ln x + \ln y = x - y$

SOLUTION Let $\ln x + \ln y = x - y$. Then

$$\frac{1}{x} + \frac{y'}{y} = 1 - y' \quad \text{or} \quad y' = \frac{1 - \frac{1}{x}}{1 + \frac{1}{y}} = \frac{xy - y}{xy + x}.$$

27. Show that $x + yx^{-1} = 1$ and $y = x - x^2$ define the same curve (except that $(0, 0)$ is not a solution of the first equation) and that implicit differentiation yields $y' = yx^{-1} - x$ and $y' = 1 - 2x$. Explain why these formulas produce the same values for the derivative.

SOLUTION Multiply the first equation by x and then isolate the y term to obtain

$$x^2 + y = x \quad \Rightarrow \quad y = x - x^2.$$

Implicit differentiation applied to the first equation yields

$$1 - yx^{-2} + x^{-1}y' = 0 \quad \text{or} \quad y' = yx^{-1} - x.$$

From the first equation, we find $yx^{-1} = 1 - x$; upon substituting this expression into the previous derivative, we find

$$y' = 1 - x - x = 1 - 2x,$$

which is the derivative of the second equation.

In Exercises 29 and 30, find dy/dx at the given point.

29. $(x + 2)^2 - 6(2y + 3)^2 = 3, \quad (1, -1)$

SOLUTION By the scaling and shifting rule,

$$2(x + 2) - 24(2y + 3)y' = 0.$$

If $x = 1$ and $y = -1$, then

$$2(3) - 24(1)y' = 0.$$

so that $24y' = 6$, or $y' = \frac{1}{4}$.

In Exercises 31–38, find an equation of the tangent line at the given point.

31. $xy + x^2y^2 = 5, \quad (2, 1)$

SOLUTION Taking the derivative of both sides of $xy + x^2y^2 = 5$ yields

$$xy' + y + 2xy^2 + 2x^2yy' = 0.$$

Substituting $x = 2$, $y = 1$, we find

$$2y' + 1 + 4 + 8y' = 0 \quad \text{or} \quad y' = -\frac{1}{2}.$$

Hence, the equation of the tangent line at $(2, 1)$ is $y - 1 = -\frac{1}{2}(x - 2)$ or $y = -\frac{1}{2}x + 2$.

33. $x^2 + \sin y = xy^2 + 1, \quad (1, 0)$

SOLUTION Taking the derivative of both sides of $x^2 + \sin y = xy^2 + 1$ yields

$$2x + \cos y\, y' = y^2 + 2xyy'.$$

Substituting $x = 1$, $y = 0$, we find

$$2 + y' = 0 \quad \text{or} \quad y' = -2.$$

Hence, the equation of the tangent line is $y - 0 = -2(x - 1)$ or $y = -2x + 2$.

35. $2x^{1/2} + 4y^{-1/2} = xy$, $(1, 4)$

SOLUTION Taking the derivative of both sides of $2x^{1/2} + 4y^{-1/2} = xy$ yields

$$x^{-1/2} - 2y^{-3/2}y' = xy' + y.$$

Substituting $x = 1$, $y = 4$, we find

$$1 - 2\left(\frac{1}{8}\right)y' = y' + 4 \quad \text{or} \quad y' = -\frac{12}{5}.$$

Hence, the equation of the tangent line is $y - 4 = -\frac{12}{5}(x - 1)$ or $y = -\frac{12}{5}x + \frac{32}{5}$.

37. $e^{2x-y} = \dfrac{x^2}{y}$, $(2, 4)$

SOLUTION taking the derivative of both sides of $e^{2x-y} = \dfrac{x^2}{y}$ yields

$$e^{2x-y}(2 - y') = \frac{2xy - x^2y'}{y^2}.$$

Substituting $x = 2$, $y = 4$, we find

$$e^0(2 - y') = \frac{16 - 4y'}{16} \quad \text{or} \quad y' = \frac{4}{3}.$$

Hence, the equation of the tangent line is $y - 4 = \frac{4}{3}(x - 2)$ or $y = \frac{4}{3}x + \frac{4}{3}$.

39. Find the points on the graph of $y^2 = x^3 - 3x + 1$ (Figure 6) where the tangent line is horizontal.

(a) First show that $2yy' = 3x^2 - 3$, where $y' = dy/dx$.

(b) Do not solve for y'. Rather, set $y' = 0$ and solve for x. This yields two values of x where the slope may be zero.

(c) Show that the positive value of x does not correspond to a point on the graph.

(d) The negative value corresponds to the two points on the graph where the tangent line is horizontal. Find their coordinates.

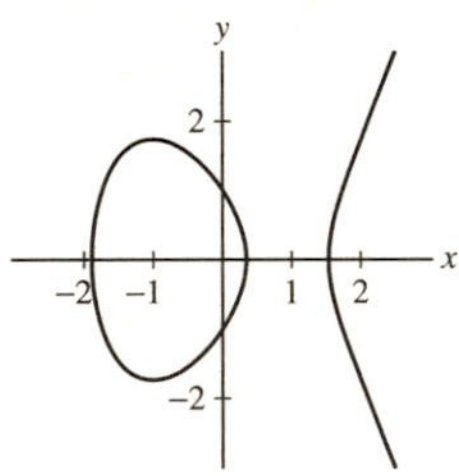

FIGURE 6 Graph of $y^2 = x^3 - 3x + 1$.

SOLUTION

(a) Applying implicit differentiation to $y^2 = x^3 - 3x + 1$, we have

$$2y\frac{dy}{dx} = 3x^2 - 3.$$

(b) Setting $y' = 0$ we have $0 = 3x^2 - 3$, so $x = 1$ or $x = -1$.

(c) If we return to the equation $y^2 = x^3 - 3x + 1$ and substitute $x = 1$, we obtain the equation $y^2 = -1$, which has no real solutions.

(d) Substituting $x = -1$ into $y^2 = x^3 - 3x + 1$ yields

$$y^2 = (-1)^3 - 3(-1) + 1 = -1 + 3 + 1 = 3,$$

so $y = \sqrt{3}$ or $-\sqrt{3}$. The tangent is horizontal at the points $(-1, \sqrt{3})$ and $(-1, -\sqrt{3})$.

41. Find all points on the graph of $3x^2 + 4y^2 + 3xy = 24$ where the tangent line is horizontal (Figure 7).

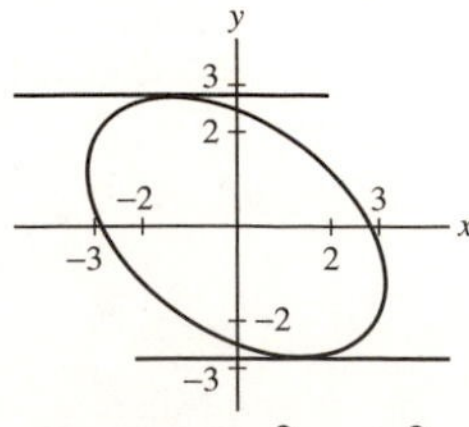

FIGURE 7 Graph of $3x^2 + 4y^2 + 3xy = 24$.

SOLUTION Differentiating the equation $3x^2 + 4y^2 + 3xy = 24$ implicitly yields

$$6x + 8yy' + 3xy' + 3y = 0,$$

so

$$y' = -\frac{6x + 3y}{8y + 3x}.$$

Setting $y' = 0$ leads to $6x + 3y = 0$, or $y = -2x$. Substituting $y = -2x$ into the equation $3x^2 + 4y^2 + 3xy = 24$ yields

$$3x^2 + 4(-2x)^2 + 3x(-2x) = 24,$$

or $13x^2 = 24$. Thus, $x = \pm 2\sqrt{78}/13$, and the coordinates of the two points on the graph of $3x^2 + 4y^2 + 3xy = 24$ where the tangent line is horizontal are

$$\left(\frac{2\sqrt{78}}{13}, -\frac{4\sqrt{78}}{13}\right) \quad \text{and} \quad \left(-\frac{2\sqrt{78}}{13}, \frac{4\sqrt{78}}{13}\right).$$

43. Figure 1 shows the graph of $y^4 + xy = x^3 - x + 2$. Find dy/dx at the two points on the graph with x-coordinate 0 and find an equation of the tangent line at $(1, 1)$.

SOLUTION Consider the equation $y^4 + xy = x^3 - x + 2$. Then $4y^3y' + xy' + y = 3x^2 - 1$, and

$$y' = \frac{3x^2 - y - 1}{x + 4y^3}.$$

- Substituting $x = 0$ into $y^4 + xy = x^3 - x + 2$ gives $y^4 = 2$, which has two real solutions, $y = \pm 2^{1/4}$. When $y = 2^{1/4}$, we have

$$y' = \frac{-2^{1/4} - 1}{4\left(2^{3/4}\right)} = -\frac{\sqrt{2} + \sqrt[4]{2}}{8} \approx -.3254.$$

When $y = -2^{1/4}$, we have

$$y' = \frac{2^{1/4} - 1}{-4\left(2^{3/4}\right)} = -\frac{\sqrt{2} - \sqrt[4]{2}}{8} \approx -.02813.$$

- At the point $(1, 1)$, we have $y' = \frac{1}{5}$. At this point the tangent line is $y - 1 = \frac{1}{5}(x - 1)$ or $y = \frac{1}{5}x + \frac{4}{5}$.

45. Find a point on the folium $x^3 + y^3 = 3xy$ other than the origin at which the tangent line is horizontal.

SOLUTION Using implicit differentiation, we find

$$\frac{d}{dx}\left(x^3 + y^3\right) = \frac{d}{dx}(3xy)$$
$$3x^2 + 3y^2y' = 3(xy' + y)$$

Setting $y' = 0$ in this equation yields $3x^2 = 3y$ or $y = x^2$. If we substitute this expression into the original equation $x^3 + y^3 = 3xy$, we obtain:

$$x^3 + x^6 = 3x(x^2) = 3x^3 \quad \text{or} \quad x^3(x^3 - 2) = 0.$$

One solution of this equation is $x = 0$ and the other is $x = 2^{1/3}$. Thus, the two points on the folium $x^3 + y^3 = 3xy$ at which the tangent line is horizontal are $(0, 0)$ and $(2^{1/3}, 2^{2/3})$.

47. Find the x-coordinates of the points where the tangent line is horizontal on the *trident curve* $xy = x^3 - 5x^2 + 2x - 1$, so named by Isaac Newton in his treatise on curves published in 1710 (Figure 9).
Hint: $2x^3 - 5x^2 + 1 = (2x - 1)(x^2 - 2x - 1)$.

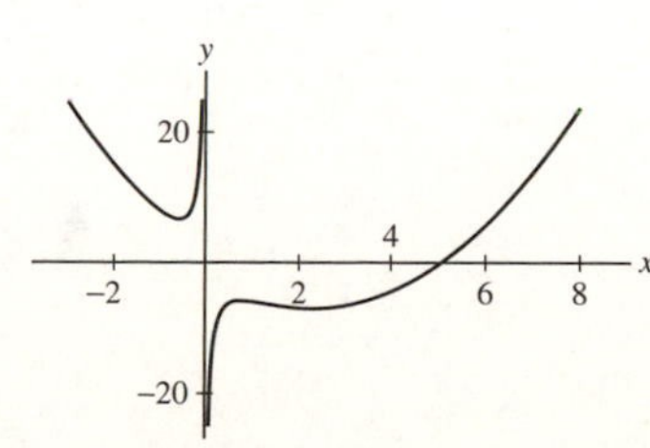

FIGURE 9 Trident curve: $xy = x^3 - 5x^2 + 2x - 1$.

SOLUTION Take the derivative of the equation of a trident curve:

$$xy = x^3 - 5x^2 + 2x - 1$$

to obtain

$$xy' + y = 3x^2 - 10x + 2.$$

Setting $y' = 0$ gives $y = 3x^2 - 10x + 2$. Substituting this into the equation of the trident, we have

$$xy = x(3x^2 - 10x + 2) = x^3 - 5x^2 + 2x - 1$$

or

$$3x^3 - 10x^2 + 2x = x^3 - 5x^2 + 2x - 1$$

Collecting like terms and setting to zero, we have

$$0 = 2x^3 - 5x^2 + 1 = (2x - 1)(x^2 - 2x - 1).$$

Hence, $x = \frac{1}{2}, 1 \pm \sqrt{2}$.

49. Find the derivative at the points where $x = 1$ on the folium $(x^2 + y^2)^2 = \frac{25}{4}xy^2$. See Figure 11.

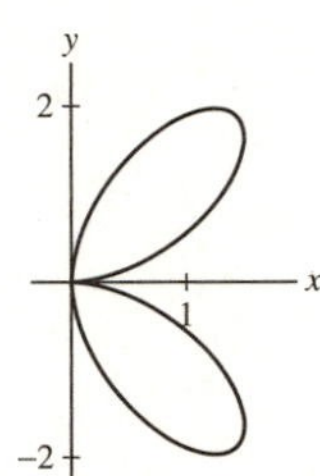

FIGURE 11 Folium curve: $(x^2 + y^2)^2 = \dfrac{25}{4}xy^2$

SOLUTION First, find the points $(1, y)$ on the curve. Setting $x = 1$ in the equation $(x^2 + y^2)^2 = \frac{25}{4}xy^2$ yields

$$\begin{aligned}
(1 + y^2)^2 &= \frac{25}{4}y^2 \\
y^4 + 2y^2 + 1 &= \frac{25}{4}y^2 \\
4y^4 + 8y^2 + 4 &= 25y^2 \\
4y^4 - 17y^2 + 4 &= 0 \\
(4y^2 - 1)(y^2 - 4) &= 0 \\
y^2 &= \frac{1}{4} \text{ or } y^2 = 4
\end{aligned}$$

Hence $y = \pm\frac{1}{2}$ or $y = \pm 2$. Taking $\frac{d}{dx}$ of both sides of the original equation yields

$$\begin{aligned}
2(x^2 + y^2)(2x + 2yy') &= \frac{25}{4}y^2 + \frac{25}{2}xyy' \\
4(x^2 + y^2)x + 4(x^2 + y^2)yy' &= \frac{25}{4}y^2 + \frac{25}{2}xyy' \\
(4(x^2 + y^2) - \frac{25}{2}x)yy' &= \frac{25}{4}y^2 - 4(x^2 + y^2)x \\
y' &= \frac{\frac{25}{4}y^2 - 4(x^2 + y^2)x}{y(4(x^2 + y^2) - \frac{25}{2}x)}
\end{aligned}$$

- At $(1, 2)$, $x^2 + y^2 = 5$, and

$$y' = \frac{\frac{25}{4}2^2 - 4(5)(1)}{2(4(5) - \frac{25}{2}(1))} = \frac{1}{3}.$$

- At $(1, -2)$, $x^2 + y^2 = 5$ as well, and

$$y' = \frac{\frac{25}{4}(-2)^2 - 4(5)(1)}{-2(4(5) - \frac{25}{2}(1))} = -\frac{1}{3}.$$

- At $(1, \frac{1}{2})$, $x^2 + y^2 = \frac{5}{4}$, and

$$y' = \frac{\frac{25}{4}\left(\frac{1}{2}\right)^2 - 4\left(\frac{5}{4}\right)(1)}{\frac{1}{2}\left(4\left(\frac{5}{4}\right) - \frac{25}{2}(1)\right)} = \frac{11}{12}.$$

- At $(1, -\frac{1}{2})$, $x^2 + y^2 = \frac{5}{4}$, and

$$y' = \frac{\frac{25}{4}\left(-\frac{1}{2}\right)^2 - 4\left(\frac{5}{4}\right)(1)}{-\frac{1}{2}\left(4\left(\frac{5}{4}\right) - \frac{25}{2}(1)\right)} = -\frac{11}{12}.$$

The folium and its tangent lines are plotted below:

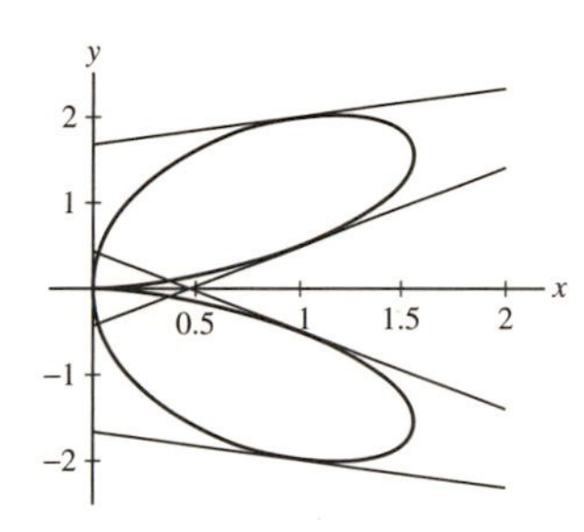

Exercises 51–53: If the derivative dx/dy (instead of $dy/dx = 0$) exists at a point and $dx/dy = 0$, then the tangent line at that point is vertical.

51. Calculate dx/dy for the equation $y^4 + 1 = y^2 + x^2$ and find the points on the graph where the tangent line is vertical.

SOLUTION Let $y^4 + 1 = y^2 + x^2$. Differentiating this equation with respect to y yields

$$4y^3 = 2y + 2x\frac{dx}{dy},$$

so

$$\frac{dx}{dy} = \frac{4y^3 - 2y}{2x} = \frac{y(2y^2 - 1)}{x}.$$

Thus, $\dfrac{dx}{dy} = 0$ when $y = 0$ and when $y = \pm\dfrac{\sqrt{2}}{2}$. Substituting $y = 0$ into the equation $y^4 + 1 = y^2 + x^2$ gives $1 = x^2$, so $x = \pm 1$. Substituting $y = \pm\dfrac{\sqrt{2}}{2}$, gives $x^2 = 3/4$, so $x = \pm\dfrac{\sqrt{3}}{2}$. Thus, there are six points on the graph of $y^4 + 1 = y^2 + x^2$ where the tangent line is vertical:

$$(1, 0), (-1, 0), \left(\frac{\sqrt{3}}{2}, \frac{\sqrt{2}}{2}\right), \left(-\frac{\sqrt{3}}{2}, \frac{\sqrt{2}}{2}\right), \left(\frac{\sqrt{3}}{2}, -\frac{\sqrt{2}}{2}\right), \left(-\frac{\sqrt{3}}{2}, -\frac{\sqrt{2}}{2}\right).$$

53. CAS Use a computer algebra system to plot $y^2 = x^3 - 4x$ for $-4 \le x \le 4, 4 \le y \le 4$. Show that if $dx/dy = 0$, then $y = 0$. Conclude that the tangent line is vertical at the points where the curve intersects the x-axis. Does your plot confirm this conclusion?

SOLUTION A plot of the curve $y^2 = x^3 - 4x$ is shown below.

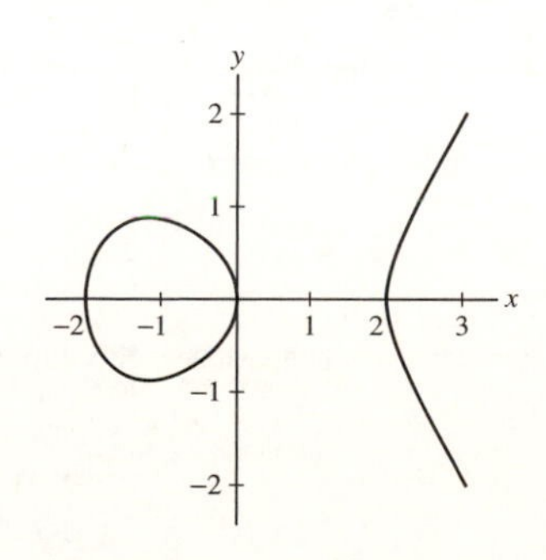

Differentiating the equation $y^2 = x^3 - 4x$ with respect to y yields

$$2y = 3x^2\frac{dx}{dy} - 4\frac{dx}{dy},$$

or

$$\frac{dx}{dy} = \frac{2y}{3x^2 - 4}.$$

From here, it follows that $\frac{dx}{dy} = 0$ when $y = 0$, so the tangent line to this curve is vertical at the points where the curve intersects the x-axis. This conclusion is confirmed by the plot of the curve shown above.

In Exercises 55–58, use implicit differentiation to calculate higher derivatives.

55. Consider the equation $y^3 - \frac{3}{2}x^2 = 1$.

(a) Show that $y' = x/y^2$ and differentiate again to show that

$$y'' = \frac{y^2 - 2xyy'}{y^4}$$

(b) Express y'' in terms of x and y using part (a).

SOLUTION

(a) Let $y^3 - \frac{3}{2}x^2 = 1$. Then $3y^2y' - 3x = 0$, and $y' = x/y^2$. Therefore,

$$y'' = \frac{y^2 \cdot 1 - x \cdot 2yy'}{y^4} = \frac{y^2 - 2xyy'}{y^4}.$$

(b) Substituting the expression for y' into the result for y'' gives

$$y'' = \frac{y^2 - 2xy\left(x/y^2\right)}{y^4} = \frac{y^3 - 2x^2}{y^5}.$$

57. Calculate y'' at the point $(1, 1)$ on the curve $xy^2 + y - 2 = 0$ by the following steps:

(a) Find y' by implicit differentiation and calculate y' at the point $(1, 1)$.

(b) Differentiate the expression for y' found in (a). Then compute y'' at $(1, 1)$ by substituting $x = 1$, $y = 1$, and the value of y' found in (a).

SOLUTION Let $xy^2 + y - 2 = 0$.

(a) Then $x \cdot 2yy' + y^2 \cdot 1 + y' = 0$, and $y' = -\dfrac{y^2}{2xy + 1}$. At $(x, y) = (1, 1)$, we have $y' = -\dfrac{1}{3}$.

(b) Therefore,

$$y'' = -\frac{(2xy+1)\left(2yy'\right) - y^2\left(2xy' + 2y\right)}{(2xy+1)^2} = -\frac{(3)\left(-\frac{2}{3}\right) - (1)\left(-\frac{2}{3} + 2\right)}{3^2} = -\frac{-6 + 2 - 6}{27} = \frac{10}{27}$$

given that $(x, y) = (1, 1)$ and $y' = -\frac{1}{3}$.

In Exercises 59–61, x and y are functions of a variable t and use implicit differentiation to relate dy/dt and dx/dt.

59. Differentiate $xy = 1$ with respect to t and derive the relation $\dfrac{dy}{dt} = -\dfrac{y}{x}\dfrac{dx}{dt}$.

SOLUTION Let $xy = 1$. Then $x\dfrac{dy}{dt} + y\dfrac{dx}{dt} = 0$, and $\dfrac{dy}{dt} = -\dfrac{y}{x}\dfrac{dx}{dt}$.

61. Calculate dy/dt in terms of dx/dt.

(a) $x^3 - y^3 = 1$

(b) $y^4 + 2xy + x^2 = 0$

SOLUTION

(a) Taking the derivative of both sides of the equation $x^3 - y^3 = 1$ with respect to t yields

$$3x^2\frac{dx}{dt} - 3y^2\frac{dy}{dt} = 0 \quad \text{or} \quad \frac{dy}{dt} = \frac{x^2}{y^2}\frac{dx}{dt}.$$

(b) Taking the derivative of both sides of the equation $y^4 + 2xy + x^2 = 0$ with respect to t yields

$$4y^3\frac{dy}{dt} + 2x\frac{dy}{dt} + 2y\frac{dx}{dt} + 2x\frac{dx}{dt} = 0,$$

or

$$\frac{dy}{dt} = -\frac{x+y}{2y^3+x}\frac{dx}{dt}.$$

Further Insights and Challenges

63. Show that if P lies on the intersection of the two curves $x^2 - y^2 = c$ and $xy = d$ (c, d constants), then the tangents to the curves at P are perpendicular.

SOLUTION Let $C1$ be the curve described by $x^2 - y^2 = c$, and let $C2$ be the curve described by $xy = d$. Suppose that $P = (x_0, y_0)$ lies on the intersection of the two curves $x^2 - y^2 = c$ and $xy = d$. Since $x^2 - y^2 = c$, the chain rule gives us $2x - 2yy' = 0$, so that $y' = \frac{2x}{2y} = \frac{x}{y}$. The slope to the tangent line to $C1$ is $\frac{x_0}{y_0}$. On the curve $C2$, since $xy = d$, the product rule yields that $xy' + y = 0$, so that $y' = -\frac{y}{x}$. Therefore the slope to the tangent line to $C2$ is $-\frac{y_0}{x_0}$. The two slopes are negative reciprocals of one another, hence the tangents to the two curves are perpendicular.

65. Divide the curve in Figure 15 into five branches, each of which is the graph of a function. Sketch the branches.

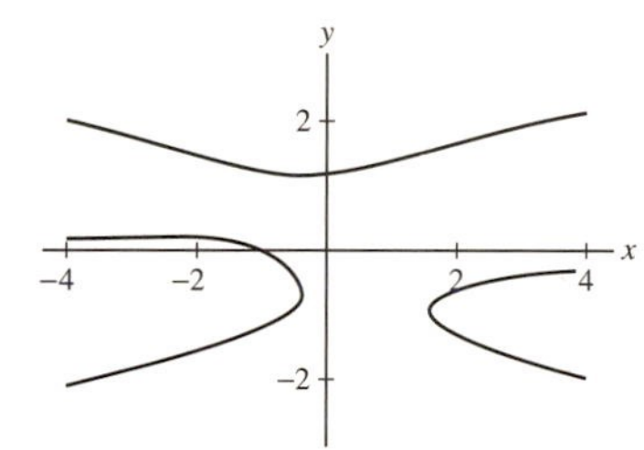

FIGURE 15 Graph of $y^5 - y = x^2y + x + 1$.

SOLUTION The branches are:

- Upper branch:

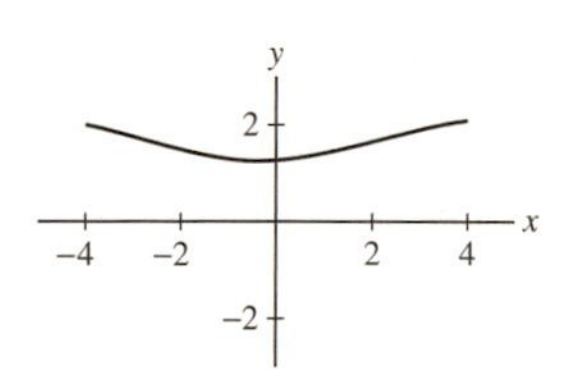

- Lower part of lower left curve:

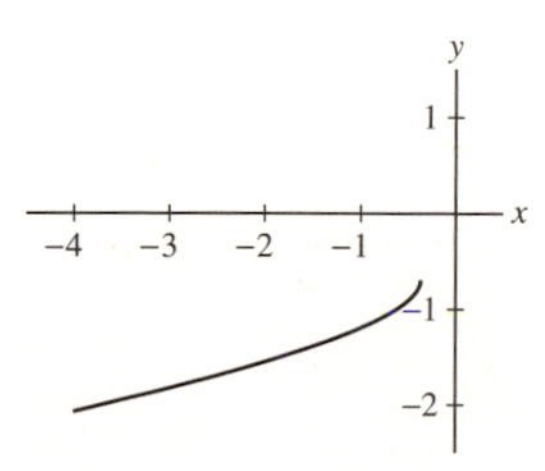

- Upper part of lower left curve:

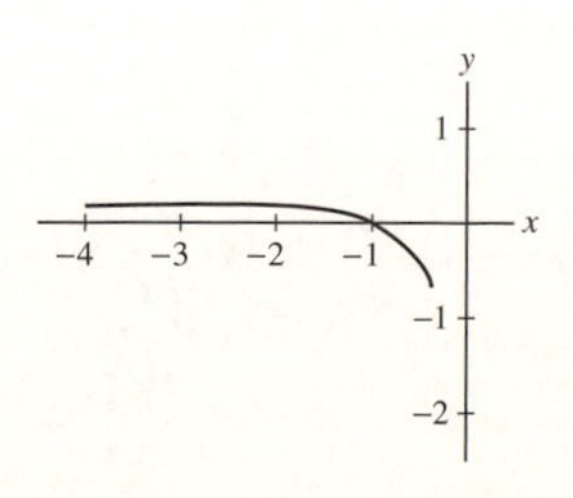

- Upper part of lower right curve:

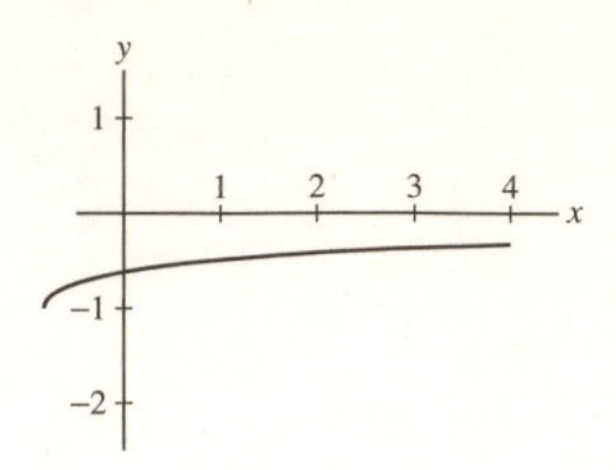

- Lower part of lower right curve:

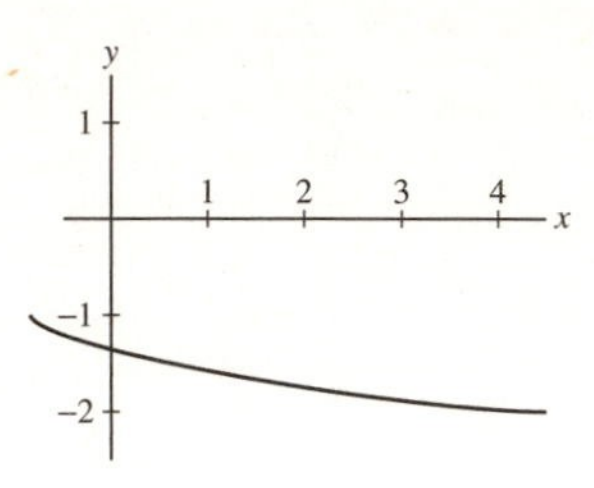

3.11 Related Rates

Preliminary Questions

1. Assign variables and restate the following problem in terms of known and unknown derivatives (but do not solve it): How fast is the volume of a cube increasing if its side increases at a rate of 0.5 cm/s?

SOLUTION Let s and V denote the length of the side and the corresponding volume of a cube, respectively. Determine $\frac{dV}{dt}$ if $\frac{ds}{dt} = 0.5$ cm/s.

2. What is the relation between dV/dt and dr/dt if $V = \left(\frac{4}{3}\right)\pi r^3$?

SOLUTION Applying the general power rule, we find $\frac{dV}{dt} = 4\pi r^2 \frac{dr}{dt}$. Therefore, the ratio is $4\pi r^2$.

In Questions 3 and 4, water pours into a cylindrical glass of radius 4 cm. Let V and h denote the volume and water level respectively, at time t.

3. Restate this question in terms of dV/dt and dh/dt: How fast is the water level rising if water pours in at a rate of 2 cm^3/min?

SOLUTION Determine $\frac{dh}{dt}$ if $\frac{dV}{dt} = 2$ cm^3/min.

4. Restate this question in terms of dV/dt and dh/dt: At what rate is water pouring in if the water level rises at a rate of 1 cm/min?

SOLUTION Determine $\frac{dV}{dt}$ if $\frac{dh}{dt} = 1$ cm/min.

Exercises

In Exercises 1 and 2, consider a rectangular bathtub whose base is 18 ft^2.

1. How fast is the water level rising if water is filling the tub at a rate of 0.7 ft^3/min?

SOLUTION Let h be the height of the water in the tub and V be the volume of the water. Then $V = 18h$ and $\frac{dV}{dt} = 18\frac{dh}{dt}$. Thus

$$\frac{dh}{dt} = \frac{1}{18}\frac{dV}{dt} = \frac{1}{18}(0.7) \approx 0.039 \text{ ft/min}.$$

3. The radius of a circular oil slick expands at a rate of 2 m/min.

(a) How fast is the area of the oil slick increasing when the radius is 25 m?

(b) If the radius is 0 at time $t = 0$, how fast is the area increasing after 3 min?

SOLUTION Let r be the radius of the oil slick and A its area.

(a) Then $A = \pi r^2$ and $\frac{dA}{dt} = 2\pi r\frac{dr}{dt}$. Substituting $r = 25$ and $\frac{dr}{dt} = 2$, we find

$$\frac{dA}{dt} = 2\pi\,(25)\,(2) = 100\pi \approx 314.16 \text{ m}^2/\text{min}.$$

(b) Since $\frac{dr}{dt} = 2$ and $r(0) = 0$, it follows that $r(t) = 2t$. Thus, $r(3) = 6$ and

$$\frac{dA}{dt} = 2\pi\,(6)\,(2) = 24\pi \approx 75.40 \text{ m}^2/\text{min}.$$

In Exercises 5–8, assume that the radius r of a sphere is expanding at a rate of 30 cm/min. *The volume of a sphere is* $V = \frac{4}{3}\pi r^3$ *and its surface area is* $4\pi r^2$. *Determine the given rate.*

5. Volume with respect to time when $r = 15$ cm.

SOLUTION As the radius is expanding at 30 centimeters per minute, we know that $\frac{dr}{dt} = 30$ cm/min. Taking $\frac{d}{dt}$ of the equation $V = \frac{4}{3}\pi r^3$ yields

$$\frac{dV}{dt} = \frac{4}{3}\pi\left(3r^2\frac{dr}{dt}\right) = 4\pi r^2\frac{dr}{dt}.$$

Substituting $r = 15$ and $\frac{dr}{dt} = 30$ yields

$$\frac{dV}{dt} = 4\pi(15)^2(30) = 27000\pi \text{ cm}^3/\text{min}.$$

7. Surface area with respect to time when $r = 40$ cm.

SOLUTION Taking the derivative of both sides of $A = 4\pi r^2$ with respect to t yields $\frac{dA}{dt} = 8\pi r\frac{dr}{dt}$. $\frac{dr}{dt} = 30$, so

$$\frac{dA}{dt} = 8\pi(40)(30) = 9600\pi \text{ cm}^2/\text{min}.$$

In Exercises 9–12, refer to a 5-meter ladder sliding down a wall, as in Figures 1 and 2. The variable h is the height of the ladder's top at time t, and x is the distance from the wall to the ladder's bottom.

9. Assume the bottom slides away from the wall at a rate of 0.8 m/s. Find the velocity of the top of the ladder at $t = 2$ s if the bottom is 1.5 m from the wall at $t = 0$ s.

SOLUTION Let x denote the distance from the base of the ladder to the wall, and h denote the height of the top of the ladder from the floor. The ladder is 5 m long, so $h^2 + x^2 = 5^2$. At any time t, $x = 1.5 + 0.8t$. Therefore, at time $t = 2$, the base is $x = 1.5 + 0.8(2) = 3.1$ m from the wall. Furthermore, we have

$$2h\frac{dh}{dt} + 2x\frac{dx}{dt} = 0 \quad \text{so} \quad \frac{dh}{dt} = -\frac{x}{h}\frac{dx}{dt}.$$

Substituting $x = 3.1$, $h = \sqrt{5^2 - 3.1^2}$ and $\frac{dx}{dt} = 0.8$, we obtain

$$\frac{dh}{dt} = -\frac{3.1}{\sqrt{5^2 - 3.1^2}}(0.8) \approx -0.632 \text{ m/s}.$$

11. Suppose that $h(0) = 4$ and the top slides down the wall at a rate of 1.2 m/s. Calculate x and dx/dt at $t = 2$ s.

SOLUTION Let h and x be the height of the ladder's top and the distance from the wall of the ladder's bottom, respectively. After 2 seconds, $h = 4 + 2\,(-1.2) = 1.6$ m. Since $h^2 + x^2 = 5^2$,

$$x = \sqrt{5^2 - 1.6^2} = 4.737 \text{ m}.$$

Furthermore, we have $2h\frac{dh}{dt} + 2x\frac{dx}{dt} = 0$, so that $\frac{dx}{dt} = -\frac{h}{x}\frac{dh}{dt}$. Substituting $h = 1.6$, $x = 4.737$, and $\frac{dh}{dt} = -1.2$, we find

$$\frac{dx}{dt} = -\frac{1.6}{4.737}(-1.2) \approx 0.405 \text{ m/s}.$$

13. A conical tank has height 3 m and radius 2 m at the top. Water flows in at a rate of 2 m³/min. How fast is the water level rising when it is 2 m?

SOLUTION Consider the cone of water in the tank at a certain instant. Let r be the radius of its (inverted) base, h its height, and V its volume. By similar triangles, $\frac{r}{h} = \frac{2}{3}$ or $r = \frac{2}{3}h$ and thus $V = \frac{1}{3}\pi r^2 h = \frac{4}{27}\pi h^3$. Therefore,

$$\frac{dV}{dt} = \frac{4}{9}\pi h^2\frac{dh}{dt},$$

and

$$\frac{dh}{dt} = \frac{9}{4\pi h^2}\frac{dV}{dt}.$$

Substituting $h = 2$ and $\frac{dV}{dt} = 2$ yields

$$\frac{dh}{dt} = \frac{9}{4\pi\,(2)^2} \times 2 = \frac{9}{8\pi} \approx -0.36 \text{ m/min}.$$

15. The radius r and height h of a circular cone change at a rate of 2 cm/s. How fast is the volume of the cone increasing when $r = 10$ and $h = 20$?

SOLUTION Let r be the radius, h be the height, and V be the volume of a right circular cone. Then $V = \frac{1}{3}\pi r^2 h$, and

$$\frac{dV}{dt} = \frac{1}{3}\pi\left(r^2\frac{dh}{dt} + 2hr\frac{dr}{dt}\right).$$

When $r = 10$, $h = 20$, and $\frac{dr}{dt} = \frac{dh}{dt} = 2$, we find

$$\frac{dV}{dt} = \frac{\pi}{3}\left(10^2 \cdot 2 + 2 \cdot 20 \cdot 10 \cdot 2\right) = \frac{1000\pi}{3} \approx 1047.20 \text{ cm}^3/\text{s}.$$

17. A man of height 1.8 meters walks away from a 5-meter lamppost at a speed of 1.2 m/s (Figure 9). Find the rate at which his shadow is increasing in length.

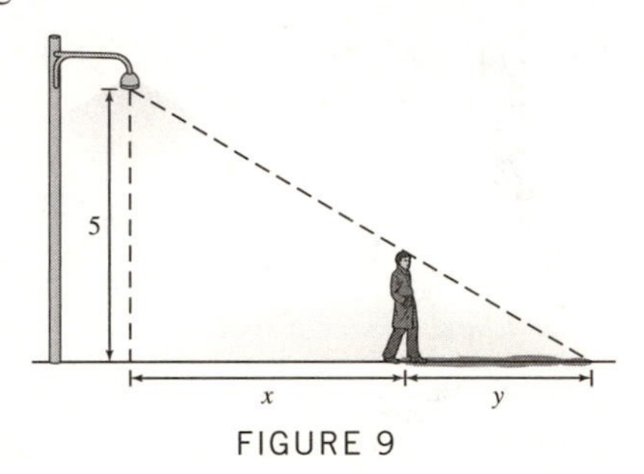

FIGURE 9

SOLUTION Since the man is moving at a rate of 1.2 m/s, his distance from the light post at any given time is $x = 1.2t$. Knowing the man is 1.8 meters tall and that the length of his shadow is denoted by y, we set up a proportion of similar triangles from the diagram:

$$\frac{y}{1.8} = \frac{1.2t + y}{5}.$$

Clearing fractions and solving for y yields

$$y = 0.675t.$$

Thus, $dy/dt = 0.675$ meters per second is the rate at which the length of the shadow is increasing.

19. At a given moment, a plane passes directly above a radar station at an altitude of 6 km.

(a) The plane's speed is 800 km/h. How fast is the distance between the plane and the station changing half a minute later?

(b) How fast is the distance between the plane and the station changing when the plane passes directly above the station?

SOLUTION Let x be the distance of the plane from the station along the ground and h the distance through the air.

(a) By the Pythagorean Theorem, we have

$$h^2 = x^2 + 6^2 = x^2 + 36.$$

Thus $2h\frac{dh}{dt} = 2x\frac{dx}{dt}$, and $\frac{dh}{dt} = \frac{x}{h}\frac{dx}{dt}$. After half a minute, $x = \frac{1}{2} \times \frac{1}{60} \times 800 = \frac{20}{3}$ kilometers. With $x = \frac{20}{3}$,

$$h = \sqrt{\left(\frac{20}{3}\right)^2 + 36} = \frac{1}{3}\sqrt{724} = \frac{2}{3}\sqrt{181} \approx 8.969 \text{ km},$$

and $\frac{dx}{dt} = 800$,

$$\frac{dh}{dt} = \frac{20}{3}\frac{3}{2\sqrt{181}} \times 800 = \frac{8000}{\sqrt{181}} \approx 594.64 \text{ km/h}.$$

(b) When the plane is directly above the station, $x = 0$, so the distance between the plane and the station is not changing, for at this instant we have

$$\frac{dh}{dt} = \frac{0}{6} \times 800 = 0 \text{ km/h}.$$

21. A hot air balloon rising vertically is tracked by an observer located 4 km from the lift-off point. At a certain moment, the angle between the observer's line of sight and the horizontal is $\frac{\pi}{5}$, and it is changing at a rate of 0.2 rad/min. How fast is the balloon rising at this moment?

SOLUTION Let y be the height of the balloon (in miles) and θ the angle between the line-of-sight and the horizontal. Via trigonometry, we have $\tan\theta = \dfrac{y}{4}$. Therefore,

$$\sec^2\theta \cdot \frac{d\theta}{dt} = \frac{1}{4}\frac{dy}{dt},$$

and

$$\frac{dy}{dt} = 4\frac{d\theta}{dt}\sec^2\theta.$$

Using $\frac{d\theta}{dt} = 0.2$ and $\theta = \frac{\pi}{5}$ yields

$$\frac{dy}{dt} = 4\,(0.2)\,\frac{1}{\cos^2(\pi/5)} \approx 1.22 \text{ km/min}.$$

23. A rocket travels vertically at a speed of 1200 km/h. The rocket is tracked through a telescope by an observer located 16 km from the launching pad. Find the rate at which the angle between the telescope and the ground is increasing 3 min after lift-off.

SOLUTION Let y be the height of the rocket and θ the angle between the telescope and the ground. Using trigonometry, we have $\tan\theta = \frac{y}{16}$. Therefore,

$$\sec^2\theta \cdot \frac{d\theta}{dt} = \frac{1}{16}\frac{dy}{dt},$$

and

$$\frac{d\theta}{dt} = \frac{\cos^2\theta}{16}\frac{dy}{dt}.$$

After the rocket has traveled for 3 minutes (or $\frac{1}{20}$ hour), its height is $\frac{1}{20} \times 1200 = 60$ km. At this instant, $\tan\theta = 60/16 = 15/4$ and thus

$$\cos\theta = \frac{4}{\sqrt{15^2 + 4^2}} = \frac{4}{\sqrt{241}}.$$

Finally,

$$\frac{d\theta}{dt} = \frac{16/241}{16}(1200) = \frac{1200}{241} \approx 4.98 \text{ rad/hr}.$$

25. A police car traveling south toward Sioux Falls at 160 km/h pursues a truck traveling east away from Sioux Falls, Iowa, at 140 km/h (Figure 11). At time $t = 0$, the police car is 20 km north and the truck is 30 km east of Sioux Falls. Calculate the rate at which the distance between the vehicles is changing:

(a) At time $t = 0$

(b) 5 minutes later

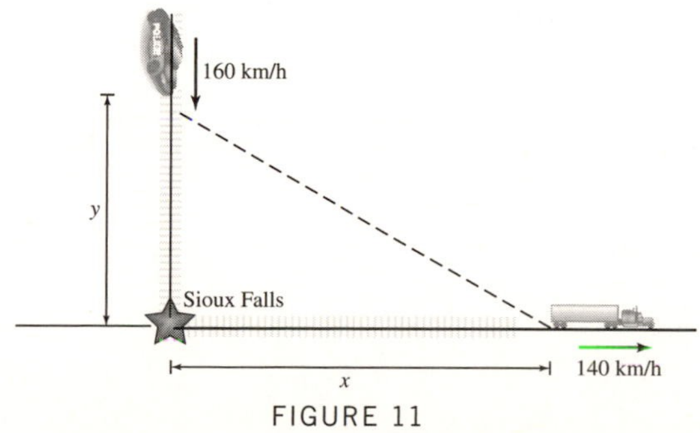

FIGURE 11

SOLUTION Let y denote the distance the police car is north of Sioux Falls and x the distance the truck is east of Sioux Falls. Then $y = 20 - 160t$ and $x = 30 + 140t$. If ℓ denotes the distance between the police car and the truck, then

$$\ell^2 = x^2 + y^2 = (30 + 140t)^2 + (20 - 160t)^2$$

and

$$\ell\frac{d\ell}{dt} = 140(30 + 140t) - 160(20 - 160t) = 1000 + 45200t.$$

(a) At $t = 0$, $\ell = \sqrt{30^2 + 20^2} = 10\sqrt{13}$, so

$$\frac{d\ell}{dt} = \frac{1000}{10\sqrt{13}} = \frac{100\sqrt{13}}{13} \approx 27.735 \text{ km/h}.$$

(b) At $t = 5$ minutes $= \frac{1}{12}$ hour,

$$\ell = \sqrt{\left(30 + 140 \cdot \frac{1}{12}\right)^2 + \left(20 - 160 \cdot \frac{1}{12}\right)^2} \approx 42.197 \text{ km},$$

and

$$\frac{d\ell}{dt} = \frac{1000 + 45200 \cdot \frac{1}{12}}{42.197} \approx 112.962 \text{ km/h}.$$

27. In the setting of Example 5, at a certain moment, the tractor's speed is 3 m/s and the bale is rising at 2 m/s. How far is the tractor from the bale at this moment?

SOLUTION From Example 5, we have the equation

$$\frac{x\frac{dx}{dt}}{\sqrt{x^2 + 4.5^2}} = \frac{dh}{dt},$$

where x denote the distance from the tractor to the bale and h denotes the height of the bale. Given

$$\frac{dx}{dt} = 3 \quad \text{and} \quad \frac{dh}{dt} = 2,$$

it follows that

$$\frac{3x}{\sqrt{4.5^2 + x^2}} = 2,$$

which yields $x = \sqrt{16.2} \approx 4.025$ m.

29. Julian is jogging around a circular track of radius 50 m. In a coordinate system with origin at the center of the track, Julian's x-coordinate is changing at a rate of -1.25 m/s when his coordinates are (40, 30). Find dy/dt at this moment.

SOLUTION We have $x^2 + y^2 = 50^2$, so

$$2x\frac{dx}{dt} + 2y\frac{dy}{dt} = 0 \quad \text{or} \quad \frac{dy}{dt} = -\frac{x}{y}\frac{dx}{dt}.$$

Given $x = 40$, $y = 30$ and $dx/dt = -1.25$, we find

$$\frac{dy}{dt} = -\frac{40}{30}(-1.25) = \frac{5}{3} \text{ m/s}.$$

In Exercises 31 and 32, assume that the pressure P (in kilopascals) and volume V (in cubic centimeters) of an expanding gas are related by $PV^b = C$, where b and C are constants (this holds in an adiabatic expansion, without heat gain or loss).

31. Find dP/dt if $b = 1.2$, $P = 8$ kPa, $V = 100$ cm^2, and $dV/dt = 20$ cm^3/min.

SOLUTION Let $PV^b = C$. Then

$$PbV^{b-1}\frac{dV}{dt} + V^b\frac{dP}{dt} = 0,$$

and

$$\frac{dP}{dt} = -\frac{Pb}{V}\frac{dV}{dt}.$$

Substituting $b = 1.2$, $P = 8$, $V = 100$, and $\frac{dV}{dt} = 20$, we find

$$\frac{dP}{dt} = -\frac{(8)(1.2)}{100}(20) = -1.92 \text{ kPa/min}.$$

33. The base x of the right triangle in Figure 14 increases at a rate of 5 cm/s, while the height remains constant at $h = 20$. How fast is the angle θ changing when $x = 20$?

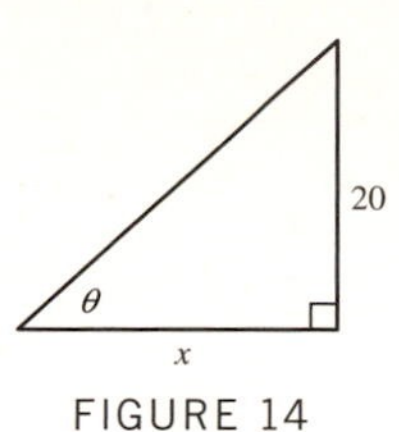

FIGURE 14

SOLUTION We have $\cot\theta = \frac{x}{20}$, from which

$$-\csc^2\theta \cdot \frac{d\theta}{dt} = \frac{1}{20}\frac{dx}{dt}$$

and thus

$$\frac{d\theta}{dt} = -\frac{\sin^2\theta}{20}\frac{dx}{dt}.$$

We are given $\frac{dx}{dt} = 5$ and when $x = h = 20$, $\theta = \frac{\pi}{4}$. Hence,

$$\frac{d\theta}{dt} = -\frac{\sin^2\left(\frac{\pi}{4}\right)}{20}(5) = -\frac{1}{8}\text{ rad/s}.$$

35. A particle travels along a curve $y = f(x)$ as in Figure 15. Let $L(t)$ be the particle's distance from the origin.

(a) Show that $\frac{dL}{dt} = \left(\frac{x + f(x)f'(x)}{\sqrt{x^2 + f(x)^2}}\right)\frac{dx}{dt}$ if the particle's location at time t is $P = (x, f(x))$.

(b) Calculate $L'(t)$ when $x = 1$ and $x = 2$ if $f(x) = \sqrt{3x^2 - 8x + 9}$ and $dx/dt = 4$.

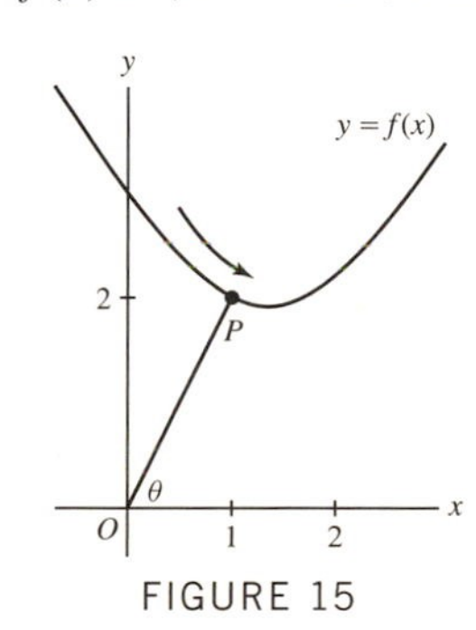

FIGURE 15

SOLUTION

(a) If the particle's location at time t is $P = (x, f(x))$, then

$$L(t) = \sqrt{x^2 + f(x)^2}.$$

Thus,

$$\frac{dL}{dt} = \frac{1}{2}(x^2 + f(x)^2)^{-1/2}\left(2x\frac{dx}{dt} + 2f(x)f'(x)\frac{dx}{dt}\right) = \left(\frac{x + f(x)f'(x)}{\sqrt{x^2 + f(x)^2}}\right)\frac{dx}{dt}.$$

(b) Given $f(x) = \sqrt{3x^2 - 8x + 9}$, it follows that

$$f'(x) = \frac{3x - 4}{\sqrt{3x^2 - 8x + 9}}.$$

Let's start with $x = 1$. Then $f(1) = 2$, $f'(1) = -\frac{1}{2}$ and

$$\frac{dL}{dt} = \left(\frac{1 - 1}{\sqrt{1^2 + 2^2}}\right)(4) = 0.$$

With $x = 2$, $f(2) = \sqrt{5}$, $f'(2) = 2/\sqrt{5}$ and

$$\frac{dL}{dt} = \frac{2 + 2}{\sqrt{2^2 + \sqrt{5}^2}}(4) = \frac{16}{3}.$$

Exercises 37 and 38 refer to the baseball diamond (a square of side 90 ft*) in Figure 16.*

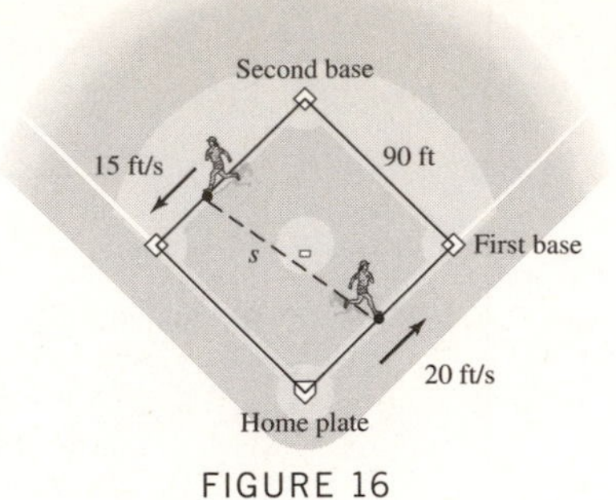

FIGURE 16

37. A baseball player runs from home plate toward first base at 20 ft/s. How fast is the player's distance from second base changing when the player is halfway to first base?

SOLUTION Let x be the distance of the player from home plate and h the player's distance from second base. Using the Pythagorean theorem, we have $h^2 = 90^2 + (90 - x)^2$. Therefore,

$$2h\frac{dh}{dt} = 2\,(90 - x)\left(-\frac{dx}{dt}\right),$$

and

$$\frac{dh}{dt} = -\frac{90 - x}{h}\frac{dx}{dt}.$$

We are given $\frac{dx}{dt} = 20$. When the player is halfway to first base, $x = 45$ and $h = \sqrt{90^2 + 45^2}$, so

$$\frac{dh}{dt} = -\frac{45}{\sqrt{90^2 + 45^2}}\,(20) = -4\sqrt{5} \approx -8.94 \text{ ft/s}.$$

39. The conical watering pail in Figure 17 has a grid of holes. Water flows out through the holes at a rate of kA m^3/min, where k is a constant and A is the surface area of the part of the cone in contact with the water. This surface area is $A = \pi r\sqrt{h^2 + r^2}$ and the volume is $V = \frac{1}{3}\pi r^2 h$. Calculate the rate dh/dt at which the water level changes at $h = 0.3$ m, assuming that $k = 0.25$ m.

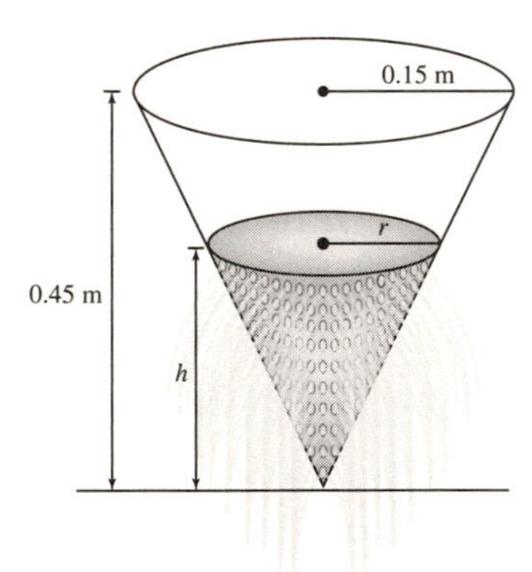

FIGURE 17

SOLUTION By similar triangles, we have

$$\frac{r}{h} = \frac{0.15}{0.45} = \frac{1}{3} \quad \text{so} \quad r = \frac{1}{3}h.$$

Substituting this expression for r into the formula for V yields

$$V = \frac{1}{3}\pi\left(\frac{1}{3}h\right)^2 h = \frac{1}{27}\pi h^3.$$

From here and the problem statement, it follows that

$$\frac{dV}{dt} = \frac{1}{9}\pi h^2\frac{dh}{dt} = -kA = -0.25\pi r\sqrt{h^2 + r^2}.$$

Solving for dh/dt gives

$$\frac{dh}{dt} = -\frac{9}{4}\frac{r}{h^2}\sqrt{h^2 + r^2}.$$

When $h = 0.3$, $r = 0.1$ and

$$\frac{dh}{dt} = -\frac{9}{4}\frac{0.1}{0.3^2}\sqrt{0.3^2 + 0.1^2} = -0.79 \text{ m/min}.$$

Further Insights and Challenges

41. A roller coaster has the shape of the graph in Figure 19. Show that when the roller coaster passes the point $(x, f(x))$, the vertical velocity of the roller coaster is equal to $f'(x)$ times its horizontal velocity.

FIGURE 19 Graph of $f(x)$ as a roller coaster track.

SOLUTION Let the equation $y = f(x)$ describe the shape of the roller coaster track. Taking $\frac{d}{dt}$ of both sides of this equation yields $\frac{dy}{dt} = f'(x)\frac{dx}{dt}$. In other words, the vertical velocity of a car moving along the track, $\frac{dy}{dt}$, is equal to $f'(x)$ times the horizontal velocity, $\frac{dx}{dt}$.

43. As the wheel of radius r cm in Figure 20 rotates, the rod of length L attached at point P drives a piston back and forth in a straight line. Let x be the distance from the origin to point Q at the end of the rod, as shown in the figure.

(a) Use the Pythagorean Theorem to show that

$$L^2 = (x - r\cos\theta)^2 + r^2\sin^2\theta \qquad \boxed{6}$$

(b) Differentiate Eq. (6) with respect to t to prove that

$$2(x - r\cos\theta)\left(\frac{dx}{dt} + r\sin\theta\frac{d\theta}{dt}\right) + 2r^2\sin\theta\cos\theta\frac{d\theta}{dt} = 0$$

(c) Calculate the speed of the piston when $\theta = \frac{\pi}{2}$, assuming that $r = 10$ cm, $L = 30$ cm, and the wheel rotates at 4 revolutions per minute.

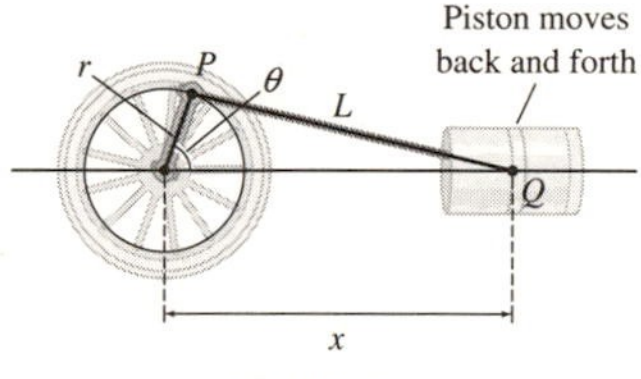

FIGURE 20

SOLUTION From the diagram, the coordinates of P are $(r\cos\theta, r\sin\theta)$ and those of Q are $(x, 0)$.

(a) The distance formula gives

$$L = \sqrt{(x - r\cos\theta)^2 + (-r\sin\theta)^2}.$$

Thus,

$$L^2 = (x - r\cos\theta)^2 + r^2\sin^2\theta.$$

Note that L (the length of the fixed rod) and r (the radius of the wheel) are constants.

(b) From (a) we have

$$0 = 2(x - r\cos\theta)\left(\frac{dx}{dt} + r\sin\theta\frac{d\theta}{dt}\right) + 2r^2\sin\theta\cos\theta\frac{d\theta}{dt}.$$

(c) Solving for dx/dt in (b) gives

$$\frac{dx}{dt} = \frac{r^2\sin\theta\cos\theta\frac{d\theta}{dt}}{r\cos\theta - x} - r\sin\theta\frac{d\theta}{dt} = \frac{rx\sin\theta\frac{d\theta}{dt}}{r\cos\theta - x}.$$

With $\theta = \frac{\pi}{2}$, $r = 10$, $L = 30$, and $\frac{d\theta}{dt} = 8\pi$,

$$\frac{dx}{dt} = \frac{(10)(x)\left(\sin\frac{\pi}{2}\right)(8\pi)}{(10)(0) - x} = -80\pi \approx -251.33 \text{ cm/min}$$

45. A cylindrical tank of radius R and length L lying horizontally as in Figure 21 is filled with oil to height h.

(a) Show that the volume $V(h)$ of oil in the tank is

$$V(h) = L\left(R^2\cos^{-1}\left(1 - \frac{h}{R}\right) - (R-h)\sqrt{2hR - h^2}\right)$$

(b) Show that $\frac{dV}{dh} = 2L\sqrt{h(2R-h)}$.

(c) Suppose that $R = 1.5$ m and $L = 10$ m and that the tank is filled at a constant rate of $0.6\ \text{m}^3/\text{min}$. How fast is the height h increasing when $h = 0.5$?

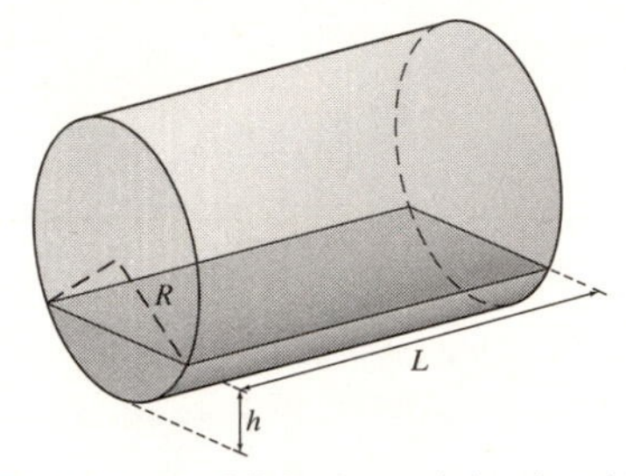

FIGURE 21 Oil in the tank has level h.

SOLUTION

(a) From Figure 21, we see that the volume of oil in the tank, $V(h)$, is equal to L times $A(h)$, the area of that portion of the circular cross section occupied by the oil. Now,

$$A(h) = \text{area of sector} - \text{area of triangle} = \frac{R^2\theta}{2} - \frac{R^2\sin\theta}{2},$$

where θ is the central angle of the sector. Referring to the diagram below,

$$\cos\frac{\theta}{2} = \frac{R-h}{R} \quad \text{and} \quad \sin\frac{\theta}{2} = \frac{\sqrt{2hR - h^2}}{R}.$$

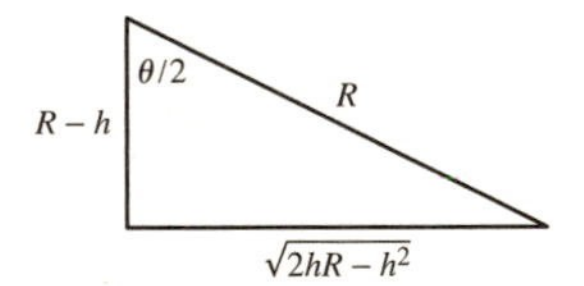

Thus,

$$\theta = 2\cos^{-1}\left(1 - \frac{h}{R}\right),$$

$$\sin\theta = 2\sin\frac{\theta}{2}\cos\frac{\theta}{2} = 2\frac{(R-h)\sqrt{2hR - h^2}}{R^2},$$

and

$$V(h) = L\left(R^2\cos^{-1}\left(1 - \frac{h}{R}\right) - (R-h)\sqrt{2hR - h^2}\right).$$

(b) Recalling that $\frac{d}{dx}\cos^{-1}u = -\frac{1}{\sqrt{1-x^2}}\frac{du}{dx}$,

$$\frac{dV}{dh} = L\left(\frac{d}{dh}\left(R^2\cos^{-1}\left(1 - \frac{h}{R}\right)\right) - \frac{d}{dh}\left((R-h)\sqrt{2hR - h^2}\right)\right)$$

$$= L\left(-R\frac{-1}{\sqrt{1 - (1 - (h/R))^2}} + \sqrt{2hR - h^2} - \frac{(R-h)^2}{\sqrt{2hR - h^2}}\right)$$

$$= L\left(\frac{R^2}{\sqrt{2hR - h^2}} + \sqrt{2hR - h^2} - \frac{R^2 - 2Rh + h^2}{\sqrt{2hR - h^2}}\right)$$

$$= L\left(\frac{R^2 + (2hR - h^2) - (R^2 - 2Rh + h^2)}{\sqrt{2hR - h^2}}\right)$$

$$= L\left(\frac{4hR - 2h^2}{\sqrt{2hR - h^2}}\right) = L\left(\frac{2(2hR - h^2)}{\sqrt{2hR - h^2}}\right) = 2L\sqrt{2hR - h^2}.$$

(c) $\frac{dV}{dt} = \frac{dV}{dh}\frac{dh}{dt}$, so $\frac{dh}{dt} = \frac{1}{dV/dh}\frac{dV}{dt}$. From part (b) with $R = 1.5$, $L = 10$ and $h = 0.5$,

$$\frac{dV}{dh} = 2(10)\sqrt{2(0.5)(1.5) - 0.5^2} = 10\sqrt{5}\text{ m}^2.$$

Thus,

$$\frac{dh}{dt} = \frac{1}{10\sqrt{5}}(0.6) = \frac{3\sqrt{5}}{2500} \approx 0.0027\text{ m/min}.$$

CHAPTER REVIEW EXERCISES

In Exercises 1–4, refer to the function $f(x)$ whose graph is shown in Figure 1.

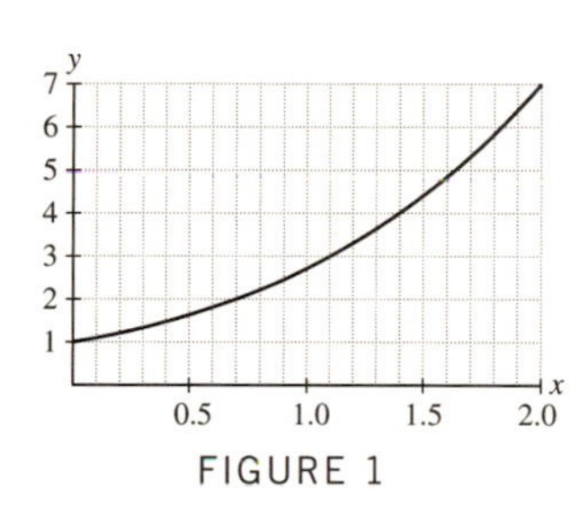

FIGURE 1

1. Compute the average rate of change of $f(x)$ over $[0, 2]$. What is the graphical interpretation of this average rate?

SOLUTION The average rate of change of $f(x)$ over $[0, 2]$ is

$$\frac{f(2) - f(0)}{2 - 0} = \frac{7 - 1}{2 - 0} = 3.$$

Graphically, this average rate of change represents the slope of the secant line through the points $(2, 7)$ and $(0, 1)$ on the graph of $f(x)$.

3. Estimate $\frac{f(0.7 + h) - f(0.7)}{h}$ for $h = 0.3$. Is this number larger or smaller than $f'(0.7)$?

SOLUTION For $h = 0.3$,

$$\frac{f(0.7 + h) - f(0.7)}{h} = \frac{f(1) - f(0.7)}{0.3} \approx \frac{2.8 - 2}{0.3} = \frac{8}{3}.$$

Because the curve is concave up, the slope of the secant line is larger than the slope of the tangent line, so the value of the difference quotient should be larger than the value of the derivative.

In Exercises 5–8, compute $f'(a)$ using the limit definition and find an equation of the tangent line to the graph of $f(x)$ at $x = a$.

5. $f(x) = x^2 - x$, $a = 1$

SOLUTION Let $f(x) = x^2 - x$ and $a = 1$. Then

$$f'(a) = \lim_{h\to 0}\frac{f(a + h) - f(a)}{h} = \lim_{h\to 0}\frac{(1 + h)^2 - (1 + h) - (1^2 - 1)}{h}$$

$$= \lim_{h\to 0}\frac{1 + 2h + h^2 - 1 - h}{h} = \lim_{h\to 0}(1 + h) = 1$$

and the equation of the tangent line to the graph of $f(x)$ at $x = a$ is

$$y = f'(a)(x - a) + f(a) = 1(x - 1) + 0 = x - 1.$$

7. $f(x) = x^{-1}, \quad a = 4$

SOLUTION Let $f(x) = x^{-1}$ and $a = 4$. Then

$$f'(a) = \lim_{h\to 0} \frac{f(a+h) - f(a)}{h} = \lim_{h\to 0} \frac{\frac{1}{4+h} - \frac{1}{4}}{h} = \lim_{h\to 0} \frac{4-(4+h)}{4h(4+h)}$$
$$= \lim_{h\to 0} \frac{-1}{4(4+h)} = -\frac{1}{4(4+0)} = -\frac{1}{16}$$

and the equation of the tangent line to the graph of $f(x)$ at $x = a$ is

$$y = f'(a)(x-a) + f(a) = -\frac{1}{16}(x-4) + \frac{1}{4} = -\frac{1}{16}x + \frac{1}{2}.$$

In Exercises 9–12, compute dy/dx using the limit definition.

9. $y = 4 - x^2$

SOLUTION Let $y = 4 - x^2$. Then

$$\frac{dy}{dx} = \lim_{h\to 0} \frac{4-(x+h)^2-(4-x^2)}{h} = \lim_{h\to 0} \frac{4-x^2-2xh-h^2-4+x^2}{h} = \lim_{h\to 0}(-2x-h) = -2x-0 = -2x.$$

11. $y = \dfrac{1}{2-x}$

SOLUTION Let $y = \dfrac{1}{2-x}$. Then

$$\frac{dy}{dx} = \lim_{h\to 0} \frac{\frac{1}{2-(x+h)} - \frac{1}{2-x}}{h} = \lim_{h\to 0} \frac{(2-x)-(2-x-h)}{h(2-x-h)(2-x)} = \lim_{h\to 0} \frac{1}{(2-x-h)(2-x)} = \frac{1}{(2-x)^2}.$$

In Exercises 13–16, express the limit as a derivative.

13. $\displaystyle\lim_{h\to 0} \frac{\sqrt{1+h}-1}{h}$

SOLUTION Let $f(x) = \sqrt{x}$. Then

$$\lim_{h\to 0} \frac{\sqrt{1+h}-1}{h} = \lim_{h\to 0} \frac{f(1+h)-f(1)}{h} = f'(1).$$

15. $\displaystyle\lim_{t\to\pi} \frac{\sin t \cos t}{t-\pi}$

SOLUTION Let $f(t) = \sin t \cos t$ and note that $f(\pi) = \sin\pi\cos\pi = 0$. Then

$$\lim_{t\to\pi} \frac{\sin t\cos t}{t-\pi} = \lim_{t\to\pi} \frac{f(t)-f(\pi)}{t-\pi} = f'(\pi).$$

17. Find $f(4)$ and $f'(4)$ if the tangent line to the graph of $f(x)$ at $x = 4$ has equation $y = 3x - 14$.

SOLUTION The equation of the tangent line to the graph of $f(x)$ at $x = 4$ is $y = f'(4)(x-4) + f(4) = f'(4)x + (f(4) - 4f'(4))$. Matching this to $y = 3x - 14$, we see that $f'(4) = 3$ and $f(4) - 4(3) = -14$, so $f(4) = -2$.

19. Is (A), (B), or (C) the graph of the derivative of the function $f(x)$ shown in Figure 3?

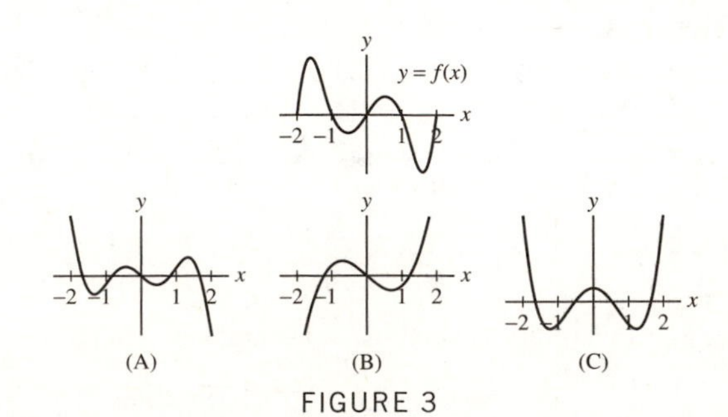

FIGURE 3

SOLUTION The graph of $f(x)$ has four horizontal tangent lines on $[-2, 2]$, so the graph of its derivative must have four x-intercepts on $[-2, 2]$. This eliminates (B). Moreover, $f(x)$ is increasing at both ends of the interval, so its derivative must be positive at both ends. This eliminates (A) and identifies (C) as the graph of $f'(x)$.

21. A girl's height $h(t)$ (in centimeters) is measured at time t (in years) for $0 \le t \le 14$:

52, 75.1, 87.5, 96.7, 104.5, 111.8, 118.7, 125.2,
131.5, 137.5, 143.3, 149.2, 155.3, 160.8, 164.7

(a) What is the average growth rate over the 14-year period?
(b) Is the average growth rate larger over the first half or the second half of this period?
(c) Estimate $h'(t)$ (in centimeters per year) for $t = 3, 8$.

SOLUTION

(a) The average growth rate over the 14-year period is

$$\frac{164.7 - 52}{14} = 8.05 \text{ cm/year.}$$

(b) Over the first half of the 14-year period, the average growth rate is

$$\frac{125.2 - 52}{7} \approx 10.46 \text{ cm/year,}$$

which is larger than the average growth rate over the second half of the 14-year period:

$$\frac{164.7 - 125.2}{7} \approx 5.64 \text{ cm/year.}$$

(c) For $t = 3$,

$$h'(3) \approx \frac{h(4) - h(3)}{4 - 3} = \frac{104.5 - 96.7}{1} = 7.8 \text{ cm/year;}$$

for $t = 8$,

$$h'(8) \approx \frac{h(9) - h(8)}{9 - 8} = \frac{137.5 - 131.5}{1} = 6.0 \text{ cm/year.}$$

In Exercises 23 and 24, use the following table of values for the number $A(t)$ of automobiles (in millions) manufactured in the United States in year t.

t	1970	1971	1972	1973	1974	1975	1976
$A(t)$	6.55	8.58	8.83	9.67	7.32	6.72	8.50

23. What is the interpretation of $A'(t)$? Estimate $A'(1971)$. Does $A'(1974)$ appear to be positive or negative?

SOLUTION Because $A(t)$ measures the number of automobiles manufactured in the United States in year t, $A'(t)$ measures the rate of change in automobile production in the United States. For $t = 1971$,

$$A'(1971) \approx \frac{A(1972) - A(1971)}{1972 - 1971} = \frac{8.83 - 8.58}{1} = 0.25 \text{ million automobiles/year.}$$

Because $A(t)$ decreases from 1973 to 1974 and from 1974 to 1975, it appears that $A'(1974)$ would be negative.

25. Which of the following is equal to $\frac{d}{dx}2^x$?

(a) 2^x **(b)** $(\ln 2)2^x$ **(c)** $x2^{x-1}$ **(d)** $\frac{1}{\ln 2}2^x$

SOLUTION The derivative of $f(x) = 2^x$ is

$$\frac{d}{dx}2^x = 2^x \ln 2.$$

Hence, the correct answer is **(b)**.

27. Show that if $f(x)$ is a function satisfying $f'(x) = f(x)^2$, then its inverse $g(x)$ satisfies $g'(x) = x^{-2}$.

SOLUTION

$$g'(x) = \frac{1}{f'(g(x))} = \frac{1}{f(g(x))^2} = \frac{1}{x^2} = x^{-2}.$$

In Exercises 29–80, compute the derivative.

29. $y = 3x^5 - 7x^2 + 4$

SOLUTION Let $y = 3x^5 - 7x^2 + 4$. Then

$$\frac{dy}{dx} = 15x^4 - 14x.$$

31. $y = t^{-7.3}$

SOLUTION Let $y = t^{-7.3}$. Then

$$\frac{dy}{dt} = -7.3t^{-8.3}.$$

33. $y = \dfrac{x+1}{x^2+1}$

SOLUTION Let $y = \dfrac{x+1}{x^2+1}$. Then

$$\frac{dy}{dx} = \frac{(x^2+1)(1) - (x+1)(2x)}{(x^2+1)^2} = \frac{1 - 2x - x^2}{(x^2+1)^2}.$$

35. $y = (x^4 - 9x)^6$

SOLUTION Let $y = (x^4 - 9x)^6$. Then

$$\frac{dy}{dx} = 6(x^4 - 9x)^5 \frac{d}{dx}(x^4 - 9x) = 6(4x^3 - 9)(x^4 - 9x)^5.$$

37. $y = (2 + 9x^2)^{3/2}$

SOLUTION Let $y = (2 + 9x^2)^{3/2}$. Then

$$\frac{dy}{dx} = \frac{3}{2}(2 + 9x^2)^{1/2} \frac{d}{dx}(2 + 9x^2) = 27x(2 + 9x^2)^{1/2}.$$

39. $y = \dfrac{z}{\sqrt{1-z}}$

SOLUTION Let $y = \dfrac{z}{\sqrt{1-z}}$. Then

$$\frac{dy}{dz} = \frac{\sqrt{1-z} - (-\frac{z}{2})\frac{1}{\sqrt{1-z}}}{1-z} = \frac{1 - z + \frac{z}{2}}{(1-z)^{3/2}} = \frac{2-z}{2(1-z)^{3/2}}.$$

41. $y = \dfrac{x^4 + \sqrt{x}}{x^2}$

SOLUTION Let

$$y = \frac{x^4 + \sqrt{x}}{x^2} = x^2 + x^{-3/2}.$$

Then

$$\frac{dy}{dx} = 2x - \frac{3}{2}x^{-5/2}.$$

43. $y = \sqrt{x + \sqrt{x + \sqrt{x}}}$

SOLUTION Let $y = \sqrt{x + \sqrt{x + \sqrt{x}}}$. Then

$$\begin{aligned}
\frac{dy}{dx} &= \frac{1}{2}\left(x + \sqrt{x + \sqrt{x}}\right)^{-1/2} \frac{d}{dx}\left(x + \sqrt{x + \sqrt{x}}\right) \\
&= \frac{1}{2}\left(x + \sqrt{x + \sqrt{x}}\right)^{-1/2} \left(1 + \frac{1}{2}(x + \sqrt{x})^{-1/2} \frac{d}{dx}(x + \sqrt{x})\right) \\
&= \frac{1}{2}\left(x + \sqrt{x + \sqrt{x}}\right)^{-1/2} \left(1 + \frac{1}{2}(x + \sqrt{x})^{-1/2}\left(1 + \frac{1}{2}x^{-1/2}\right)\right).
\end{aligned}$$

45. $y = \tan(t^{-3})$

SOLUTION Let $y = \tan(t^{-3})$. Then

$$\frac{dy}{dt} = \sec^2(t^{-3})\frac{d}{dt}t^{-3} = -3t^{-4}\sec^2(t^{-3}).$$

47. $y = \sin(2x)\cos^2 x$

SOLUTION Let $y = \sin(2x)\cos^2 x = 2\sin x\cos^3 x$. Then

$$\frac{dy}{dx} = -6\sin^2 x\cos^2 x + 2\cos^4 x.$$

49. $y = \dfrac{t}{1+\sec t}$

SOLUTION Let $y = \dfrac{t}{1+\sec t}$. Then

$$\frac{dy}{dt} = \frac{1+\sec t - t\sec t\tan t}{(1+\sec t)^2}.$$

51. $y = \dfrac{8}{1+\cot\theta}$

SOLUTION Let $y = \dfrac{8}{1+\cot\theta} = 8(1+\cot\theta)^{-1}$. Then

$$\frac{dy}{d\theta} = -8(1+\cot\theta)^{-2}\frac{d}{d\theta}(1+\cot\theta) = \frac{8\csc^2\theta}{(1+\cot\theta)^2}.$$

53. $y = \tan(\sqrt{1+\csc\theta})$

SOLUTION

$$\begin{aligned}\frac{dy}{dx} &= \sec^2(\sqrt{1+\csc\theta})\frac{d}{dx}\sqrt{1+\csc\theta} \\ &= \sec^2(\sqrt{1+\csc\theta})\cdot\frac{1}{2}(1+\csc\theta)^{-1/2}\frac{d}{dx}(1+\csc\theta) \\ &= -\frac{\sec^2(\sqrt{1+\csc\theta})\csc\theta\cot\theta}{2(\sqrt{1+\csc\theta})}.\end{aligned}$$

55. $f(x) = 9e^{-4x}$

SOLUTION $\dfrac{d}{dx}9e^{-4x} = -36e^{-4x}$.

57. $g(t) = e^{4t-t^2}$

SOLUTION $\dfrac{d}{dt}e^{4t-t^2} = (4-2t)e^{4t-t^2}$.

59. $f(x) = \ln(4x^2+1)$

SOLUTION $\dfrac{d}{dx}\ln(4x^2+1) = \dfrac{8x}{4x^2+1}$.

61. $G(s) = (\ln(s))^2$

SOLUTION $\dfrac{d}{ds}(\ln s)^2 = \dfrac{2\ln s}{s}$.

63. $f(\theta) = \ln(\sin\theta)$

SOLUTION $\dfrac{d}{d\theta}\ln(\sin\theta) = \dfrac{\cos\theta}{\sin\theta} = \cot\theta$.

65. $h(z) = \sec(z+\ln z)$

SOLUTION $\dfrac{d}{dz}\sec(z+\ln z) = \sec(z+\ln z)\tan(z+\ln z)\left(1+\dfrac{1}{z}\right)$.

67. $f(x) = 7^{-2x}$

SOLUTION $\dfrac{d}{dx}7^{-2x} = (-2\ln 7)(7^{-2x})$.

69. $g(x) = \tan^{-1}(\ln x)$

SOLUTION $\dfrac{d}{dx}\tan^{-1}(\ln x) = \dfrac{1}{1+(\ln x)^2}\cdot\dfrac{1}{x}.$

71. $f(x) = \ln(\csc^{-1} x)$

SOLUTION $\dfrac{d}{dx}\ln(\csc^{-1} x) = -\dfrac{1}{|x|\sqrt{x^2-1}\csc^{-1} x}.$

73. $R(s) = s^{\ln s}$

SOLUTION Rewrite

$$R(s) = \left(e^{\ln s}\right)^{\ln s} = e^{(\ln s)^2}.$$

Then

$$\frac{dR}{ds} = e^{(\ln s)^2}\cdot 2\ln s\cdot\frac{1}{s} = \frac{2\ln s}{s}s^{\ln s}.$$

Alternately, $R(s) = s^{\ln s}$ implies that $\ln R = \ln\left(s^{\ln s}\right) = (\ln s)^2$. Thus,

$$\frac{1}{R}\frac{dR}{ds} = 2\ln s\cdot\frac{1}{s} \quad\text{or}\quad \frac{dR}{ds} = \frac{2\ln s}{s}s^{\ln s}.$$

75. $G(t) = (\sin^2 t)^t$

SOLUTION Rewrite

$$G(t) = \left(e^{\ln\sin^2 t}\right)^t = e^{2t\ln\sin t}.$$

Then

$$\frac{dG}{dt} = e^{2t\ln\sin t}\left(2t\cdot\frac{\cos t}{\sin t} + 2\ln\sin t\right) = 2(\sin^2 t)^t(t\cot t + \ln\sin t).$$

Alternately, $G(t) = (\sin^2 t)^t$ implies that $\ln G = t\ln\sin^2 t = 2t\ln\sin t$. Thus,

$$\frac{1}{G}\frac{dG}{dt} = 2t\cdot\frac{\cos t}{\sin t} + 2\ln\sin t,$$

and

$$\frac{dG}{dt} = 2(\sin^2 t)^t(t\cot t + \ln\sin t).$$

77. $g(t) = \sinh(t^2)$

SOLUTION $\dfrac{d}{dt}\sinh(t^2) = 2t\cosh(t^2).$

79. $g(x) = \tanh^{-1}(e^x)$

SOLUTION $\dfrac{d}{dx}\tanh^{-1}(e^x) = \dfrac{1}{1-(e^x)^2}e^x = \dfrac{e^x}{1-e^{2x}}.$

81. For which values of α is $f(x) = |x|^\alpha$ differentiable at $x = 0$?

SOLUTION Let $f(x) = |x|^\alpha$. If $\alpha < 0$, then $f(x)$ is not continuous at $x = 0$ and therefore cannot be differentiable at $x = 0$. If $\alpha = 0$, then the function reduces to $f(x) = 1$, which is differentiable at $x = 0$. Now, suppose $\alpha > 0$ and consider the limit

$$\lim_{x\to 0}\frac{f(x)-f(0)}{x-0} = \lim_{x\to 0}\frac{|x|^\alpha}{x}.$$

If $0 < \alpha < 1$, then

$$\lim_{x\to 0-}\frac{|x|^\alpha}{x} = -\infty \quad\text{while}\quad \lim_{x\to 0+}\frac{|x|^\alpha}{x} = \infty$$

and $f'(0)$ does not exist. If $\alpha = 1$, then

$$\lim_{x\to 0-} \frac{|x|}{x} = -1 \qquad \text{while} \qquad \lim_{x\to 0+} \frac{|x|}{x} = 1$$

and $f'(0)$ again does not exist. Finally, if $\alpha > 1$, then

$$\lim_{x\to 0} \frac{|x|^\alpha}{x} = 0,$$

so $f'(0)$ does exist.

In summary, $f(x) = |x|^\alpha$ is differentiable at $x = 0$ when $\alpha = 0$ and when $\alpha > 1$.

In Exercises 83 and 84, let $f(x) = xe^{-x}$.

83. Show that $f(x)$ has an inverse on $[1, \infty)$. Let $g(x)$ be this inverse. Find the domain and range of $g(x)$ and compute $g'(2e^{-2})$.

SOLUTION Let $f(x) = xe^{-x}$. Then $f'(x) = e^{-x}(1 - x)$. On $[1, \infty)$, $f'(x) < 0$, so $f(x)$ is decreasing and therefore one-to-one. It follows that $f(x)$ has an inverse on $[1, \infty)$. Let $g(x)$ denote this inverse. Because $f(1) = e^{-1}$ and $f(x) \to 0$ as $x \to \infty$, the domain of $g(x)$ is $(0, e^{-1}]$, and the range is $[1, \infty)$.

To determine $g'(2e^{-2})$, we use the formula $g'(x) = 1/f'(g(x))$. Because $f(2) = 2e^{-2}$, it follows that $g(2e^{-2}) = 2$. Then,

$$g'(2e^{-2}) = \frac{1}{f'(g(2e^{-2}))} = \frac{1}{f'(2)} = \frac{1}{-e^{-2}} = -e^2.$$

In Exercises 85–90, use the following table of values to calculate the derivative of the given function at $x = 2$.

x	$f(x)$	$g(x)$	$f'(x)$	$g'(x)$
2	5	4	−3	9
4	3	2	−2	3

85. $S(x) = 3f(x) - 2g(x)$

SOLUTION Let $S(x) = 3f(x) - 2g(x)$. Then $S'(x) = 3f'(x) - 2g'(x)$ and

$$S'(2) = 3f'(2) - 2g'(2) = 3(-3) - 2(9) = -27.$$

87. $R(x) = \dfrac{f(x)}{g(x)}$

SOLUTION Let $R(x) = f(x)/g(x)$. Then

$$R'(x) = \frac{g(x)f'(x) - f(x)g'(x)}{g(x)^2}$$

and

$$R'(2) = \frac{g(2)f'(2) - f(2)g'(2)}{g(2)^2} = \frac{4(-3) - 5(9)}{4^2} = -\frac{57}{16}.$$

89. $F(x) = f(g(2x))$

SOLUTION Let $F(x) = f(g(2x))$. Then $F'(x) = 2f'(g(2x))g'(2x)$ and

$$F'(2) = 2f'(g(4))g'(4) = 2f'(2)g'(4) = 2(-3)(3) = -18.$$

91. Find the points on the graph of $f(x) = x^3 - 3x^2 + x + 4$ where the tangent line has slope 10.

SOLUTION Let $f(x) = x^3 - 3x^2 + x + 4$. Then $f'(x) = 3x^2 - 6x + 1$. The tangent line to the graph of $f(x)$ will have slope 10 when $f'(x) = 10$. Solving the quadratic equation $3x^2 - 6x + 1 = 10$ yields $x = -1$ and $x = 3$. Thus, the points on the graph of $f(x)$ where the tangent line has slope 10 are $(-1, -1)$ and $(3, 7)$.

93. Find a such that the tangent lines $y = x^3 - 2x^2 + x + 1$ at $x = a$ and $x = a + 1$ are parallel.

SOLUTION Let $f(x) = x^3 - 2x^2 + x + 1$. Then $f'(x) = 3x^2 - 4x + 1$ and the slope of the tangent line at $x = a$ is $f'(a) = 3a^2 - 4a + 1$, while the slope of the tangent line at $x = a + 1$ is

$$f'(a + 1) = 3(a + 1)^2 - 4(a + 1) + 1 = 3(a^2 + 2a + 1) - 4a - 4 + 1 = 3a^2 + 2a.$$

In order for the tangent lines at $x = a$ and $x = a + 1$ to have the same slope, we must have $f'(a) = f'(a+1)$, or

$$3a^2 - 4a + 1 = 3a^2 + 2a.$$

The only solution to this equation is $a = \frac{1}{6}$. The equation of the tangent line at $x = \frac{1}{6}$ is

$$y = f'\left(\frac{1}{6}\right)\left(x - \frac{1}{6}\right) + f\left(\frac{1}{6}\right) = \frac{5}{12}\left(x - \frac{1}{6}\right) + \frac{241}{216} = \frac{5}{12}x + \frac{113}{108},$$

and the equation of the tangent line at $x = \frac{7}{6}$ is

$$y = f'\left(\frac{7}{6}\right)\left(x - \frac{7}{6}\right) + f\left(\frac{7}{6}\right) = \frac{5}{12}\left(x - \frac{7}{6}\right) + \frac{223}{216} = \frac{5}{12}x + \frac{59}{108}.$$

The graphs of $f(x)$ and the two tangent lines appear below.

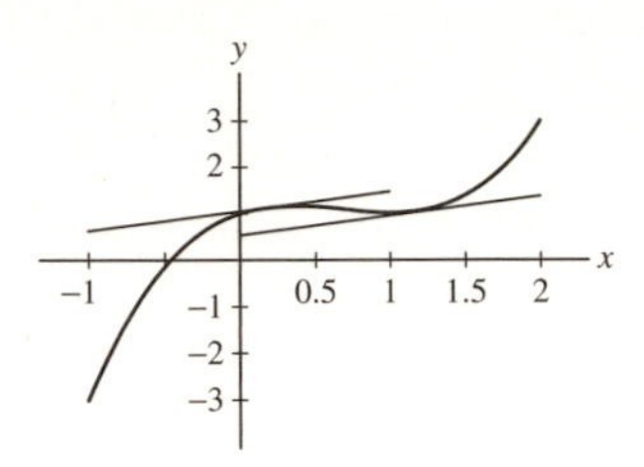

In Exercises 95–100, calculate y''.

95. $y = 12x^3 - 5x^2 + 3x$

SOLUTION Let $y = 12x^3 - 5x^2 + 3x$. Then

$$y' = 36x^2 - 10x + 3 \quad \text{and} \quad y'' = 72x - 10.$$

97. $y = \sqrt{2x+3}$

SOLUTION Let $y = \sqrt{2x+3} = (2x+3)^{1/2}$. Then

$$y' = \frac{1}{2}(2x+3)^{-1/2}\frac{d}{dx}(2x+3) = (2x+3)^{-1/2} \quad \text{and} \quad y'' = -\frac{1}{2}(2x+3)^{-3/2}\frac{d}{dx}(2x+3) = -(2x+3)^{-3/2}.$$

99. $y = \tan(x^2)$

SOLUTION Let $y = \tan(x^2)$. Then

$$y' = 2x\sec^2(x^2) \quad \text{and}$$

$$y'' = 2x\left(2\sec(x^2)\frac{d}{dx}\sec(x^2)\right) + 2\sec^2(x^2) = 8x^2\sec^2(x^2)\tan(x^2) + 2\sec^2(x^2).$$

In Exercises 101–106, compute $\frac{dy}{dx}$.

101. $x^3 - y^3 = 4$

SOLUTION Consider the equation $x^3 - y^3 = 4$. Differentiating with respect to x yields

$$3x^2 - 3y^2\frac{dy}{dx} = 0.$$

Therefore,

$$\frac{dy}{dx} = \frac{x^2}{y^2}.$$

103. $y = xy^2 + 2x^2$

SOLUTION Consider the equation $y = xy^2 + 2x^2$. Differentiating with respect to x yields

$$\frac{dy}{dx} = 2xy\frac{dy}{dx} + y^2 + 4x.$$

Therefore,

$$\frac{dy}{dx} = \frac{y^2 + 4x}{1 - 2xy}.$$

105. $y = \sin(x + y)$

SOLUTION Consider the equation $y = \sin(x + y)$. Differentiating with respect to x yields

$$\frac{dy}{dx} = \cos(x + y)\left(1 + \frac{dy}{dx}\right).$$

Therefore,

$$\frac{dy}{dx} = \frac{\cos(x + y)}{1 - \cos(x + y)}.$$

107. In Figure 5, label the graphs f, f', and f''.

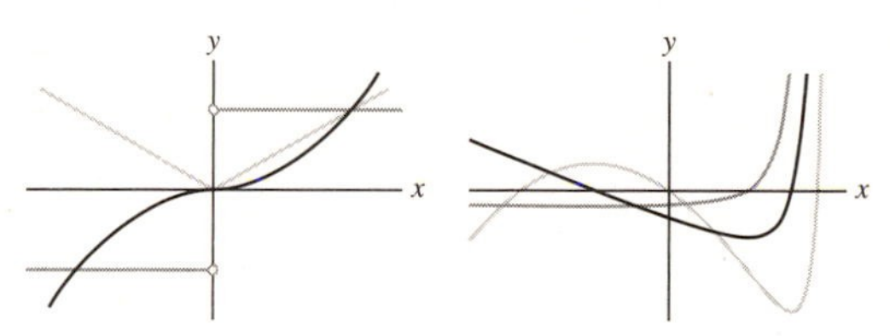

FIGURE 5

SOLUTION First consider the plot on the left. Observe that the green curve is nonnegative whereas the red curve is increasing, suggesting that the green curve is the derivative of the red curve. Moreover, the green curve is linear with negative slope for $x < 0$ and linear with positive slope for $x > 0$ while the blue curve is a negative constant for $x < 0$ and a positive constant for $x > 0$, suggesting the blue curve is the derivative of the green curve. Thus, the red, green and blue curves, respectively, are the graphs of f, f' and f''.

Now consider the plot on the right. Because the red curve is decreasing when the blue curve is negative and increasing when the blue curve is positive and the green curve is decreasing when the red curve is negative and increasing when the red curve is positive, it follows that the green, red and blue curves, respectively, are the graphs of f, f' and f''.

In Exercises 109–114, use logarithmic differentiation to find the derivative.

109. $y = \dfrac{(x + 1)^3}{(4x - 2)^2}$

SOLUTION Let $y = \dfrac{(x + 1)^3}{(4x - 2)^2}$. Then

$$\ln y = \ln\left(\frac{(x + 1)^3}{(4x - 2)^2}\right) = \ln\,(x + 1)^3 - \ln\,(4x - 2)^2 = 3\ln(x + 1) - 2\ln(4x - 2).$$

By logarithmic differentiation,

$$\frac{y'}{y} = \frac{3}{x + 1} - \frac{2}{4x - 2} \cdot 4 = \frac{3}{x + 1} - \frac{4}{2x - 1},$$

so

$$y' = \frac{(x + 1)^3}{(4x - 2)^2}\left(\frac{3}{x + 1} - \frac{4}{2x - 1}\right).$$

111. $y = e^{(x-1)^2}e^{(x-3)^2}$

SOLUTION Let $y = e^{(x-1)^2}e^{(x-3)^2}$. Then

$$\ln y = \ln\left(e^{(x-1)^2}e^{(x-3)^2}\right) = \ln\left(e^{(x-1)^2+(x-3)^2}\right) = (x - 1)^2 + (x - 3)^2.$$

By logarithmic differentiation,

$$\frac{y'}{y} = 2(x - 1) + 2(x - 3) = 4x - 8,$$

so

$$y' = 4e^{(x-1)^2}e^{(x-3)^2}(x - 2).$$

113. $y = \dfrac{e^{3x}(x-2)^2}{(x+1)^2}$

SOLUTION Let $y = \dfrac{e^{3x}(x-2)^2}{(x+1)^2}$. Then

$$\ln y = \ln\left(\frac{e^{3x}(x-2)^2}{(x+1)^2}\right) = \ln e^{3x} + \ln(x-2)^2 - \ln(x+1)^2$$
$$= 3x + 2\ln(x-2) - 2\ln(x+1).$$

By logarithmic differentiation,

$$\frac{y'}{y} = 3 + \frac{2}{x-2} - \frac{2}{x+1},$$

so

$$y = \frac{e^{3x}(x-2)^2}{(x+1)^2}\left(3 + \frac{2}{x-2} - \frac{2}{x+1}\right).$$

Exercises 115–117: Let q be the number of units of a product (cell phones, barrels of oil, etc.) that can be sold at the price p. The ***price elasticity of demand*** *E is defined as the percentage rate of change of q with respect to p. In terms of derivatives,*

$$E = \frac{p}{q}\frac{dq}{dp} = \lim_{\Delta p \to 0} \frac{(100\Delta q)/q}{(100\Delta p)/p}$$

115. Show that the total revenue $R = pq$ satisfies $\dfrac{dR}{dp} = q(1+E)$.

SOLUTION Let $R = pq$. Then

$$\frac{dR}{dp} = p\frac{dq}{dp} + q = q\frac{p}{q}\frac{dq}{dp} + q = q(E+1).$$

117. The monthly demand (in thousands) for flights between Chicago and St. Louis at the price p is $q = 40 - 0.2p$. Calculate the price elasticity of demand when $p = \$150$ and estimate the percentage increase in number of additional passengers if the ticket price is lowered by 1%.

SOLUTION Let $q = 40 - 0.2p$. Then $q'(p) = -0.2$ and

$$E(p) = \left(\frac{p}{q}\right)\frac{dq}{dp} = \frac{0.2p}{0.2p - 40}.$$

For $p = 150$,

$$E(150) = \frac{0.2(150)}{0.2(150) - 40} = -3,$$

so a 1% decrease in price increases demand by 3%. The demand when $p = 150$ is $q = 40 - 0.2(150) = 10$, or 10000 passengers. Therefore, a 1% increase in demand translates to 300 additional passengers.

119. The minute hand of a clock is 8 cm long, and the hour hand is 5 cm long. How fast is the distance between the tips of the hands changing at 3 o'clock?

SOLUTION Let S be the distance between the tips of the two hands. By the law of cosines

$$S^2 = 8^2 + 5^2 - 2 \cdot 8 \cdot 5\cos(\theta),$$

where θ is the angle between the hands. Thus

$$2S\frac{dS}{dt} = 80\sin(\theta)\frac{d\theta}{dt}.$$

At three o'clock $\theta = \pi/2$, $S = \sqrt{89}$, and

$$\frac{d\theta}{dt} = \left(\frac{\pi}{360} - \frac{\pi}{30}\right)\text{ rad/min} = -\frac{11\pi}{360}\text{ rad/min},$$

so

$$\frac{dS}{dt} = \frac{1}{2\sqrt{89}}(80)(1)\frac{-11\pi}{360} \approx -0.407\text{ cm/min}.$$

121. A bead slides down the curve $xy = 10$. Find the bead's horizontal velocity at time $t = 2$ s if its height at time t seconds is $y = 400 - 16t^2$ cm.

SOLUTION Let $xy = 10$. Then $x = 10/y$ and

$$\frac{dx}{dt} = -\frac{10}{y^2}\frac{dy}{dt}.$$

If $y = 400 - 16t^2$, then $\frac{dy}{dt} = -32t$ and

$$\frac{dx}{dt} = -\frac{10}{(400 - 16t^2)^2}(-32t) = \frac{320t}{(400 - 16t^2)^2}.$$

Thus, at $t = 2$,

$$\frac{dx}{dt} = \frac{640}{(336)^2} \approx 0.00567 \text{ cm/s}.$$

123. A light moving at 0.8 m/s approaches a man standing 4 m from a wall (Figure 9). The light is 1 m above the ground. How fast is the tip P of the man's shadow moving when the light is 7 m from the wall?

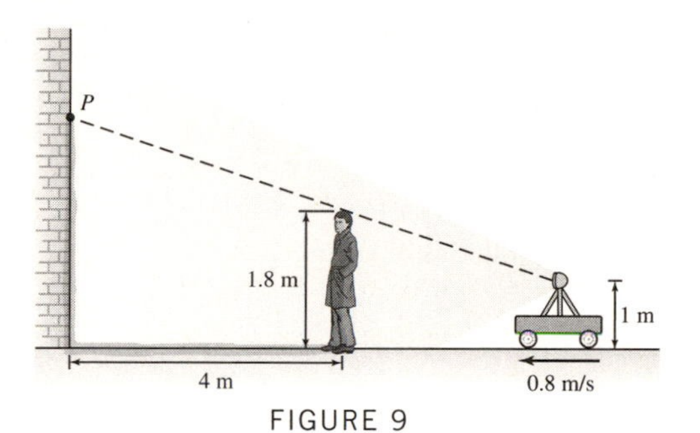

FIGURE 9

SOLUTION Let x denote the distance between the man and the light. Using similar triangles, we find

$$\frac{0.8}{x} = \frac{P - 1}{4 + x} \quad \text{or} \quad P = \frac{3.2}{x} + 1.8.$$

Therefore,

$$\frac{dP}{dt} = -\frac{3.2}{x^2}\frac{dx}{dt}.$$

When the light is 7 feet from the wall, $x = 3$. With $\frac{dx}{dt} = -0.8$, we have

$$\frac{dP}{dt} = -\frac{3.2}{3^2}(-0.8) = 0.284 \text{ m/s}.$$

4 APPLICATIONS OF THE DERIVATIVE

4.1 Linear Approximation and Applications

Preliminary Questions

1. True or False? The Linear Approximation says that the vertical change in the graph is approximately equal to the vertical change in the tangent line.

SOLUTION This statement is true. The linear approximation does say that the vertical change in the graph is approximately equal to the vertical change in the tangent line.

2. Estimate $g(1.2) - g(1)$ if $g'(1) = 4$.

SOLUTION Using the Linear Approximation,

$$g(1.2) - g(1) \approx g'(1)(1.2 - 1) = 4(0.2) = 0.8.$$

3. Estimate $f(2.1)$ if $f(2) = 1$ and $f'(2) = 3$.

SOLUTION Using the Linearization,

$$f(2.1) \approx f(2) + f'(2)(2.1 - 2) = 1 + 3(0.1) = 1.3$$

4. Complete the sentence: The Linear Approximation shows that up to a small error, the change in output Δf is directly proportional to ....

SOLUTION The Linear Approximation tells us that up to a small error, the change in output Δf is directly proportional to the change in input Δx when Δx is small.

Exercises

In Exercises 1–6, use Eq. (1) to estimate $\Delta f = f(3.02) - f(3)$.

1. $f(x) = x^2$

SOLUTION Let $f(x) = x^2$. Then $f'(x) = 2x$ and $\Delta f \approx f'(3)\Delta x = 6(0.02) = 0.12$.

3. $f(x) = x^{-1}$

SOLUTION Let $f(x) = x^{-1}$. Then $f'(x) = -x^{-2}$ and $\Delta f \approx f'(3)\Delta x = -\frac{1}{9}(0.02) = -0.00222$.

5. $f(x) = \sqrt{x+6}$

SOLUTION Let $f(x) = \sqrt{x+6}$. Then $f'(x) = \frac{1}{2}(x+6)^{-1/2}$ and

$$\Delta f \approx f'(3)\Delta x = \frac{1}{2}9^{-1/2}(0.02) = 0.003333.$$

7. The cube root of 27 is 3. How much larger is the cube root of 27.2? Estimate using the Linear Approximation.

SOLUTION Let $f(x) = x^{1/3}$, $a = 27$, and $\Delta x = 0.2$. Then $f'(x) = \frac{1}{3}x^{-2/3}$ and $f'(a) = f'(27) = \frac{1}{27}$. The Linear Approximation is

$$\Delta f \approx f'(a)\Delta x = \frac{1}{27}(0.2) = 0.0074074$$

In Exercises 9–12, use Eq. (1) to estimate Δf. Use a calculator to compute both the error and the percentage error.

9. $f(x) = \sqrt{1+x}, \quad a = 3, \quad \Delta x = 0.2$

SOLUTION Let $f(x) = (1+x)^{1/2}$, $a = 3$, and $\Delta x = 0.2$. Then $f'(x) = \frac{1}{2}(1+x)^{-1/2}$, $f'(a) = f'(3) = \frac{1}{4}$ and $\Delta f \approx f'(a)\Delta x = \frac{1}{4}(0.2) = 0.05$. The actual change is

$$\Delta f = f(a + \Delta x) - f(a) = f(3.2) - f(3) = \sqrt{4.2} - 2 \approx 0.049390.$$

The error in the Linear Approximation is therefore $|0.049390 - 0.05| = 0.000610$; in percentage terms, the error is

$$\frac{0.000610}{0.049390} \times 100\% \approx 1.24\%.$$

11. $f(x) = \dfrac{1}{1+x^2}, \quad a = 3, \quad \Delta x = 0.5$

SOLUTION Let $f(x) = \frac{1}{1+x^2}$, $a = 3$, and $\Delta x = .5$. Then $f'(x) = -\frac{2x}{(1+x^2)^2}$, $f'(a) = f'(3) = -0.06$ and $\Delta f \approx f'(a)\Delta x = -0.06(0.5) = -0.03$. The actual change is

$$\Delta f = f(a + \Delta x) - f(a) = f(3.5) - f(3) \approx -0.0245283.$$

The error in the Linear Approximation is therefore $|-0.0245283 - (-0.03)| = 0.0054717$; in percentage terms, the error is

$$\left|\frac{0.0054717}{-0.0245283}\right| \times 100\% \approx 22.31\%$$

In Exercises 13–16, estimate Δy using differentials [Eq. (3)].

13. $y = \cos x, \quad a = \frac{\pi}{6}, \quad dx = 0.014$

SOLUTION Let $f(x) = \cos x$. Then $f'(x) = -\sin x$ and

$$\Delta y \approx dy = f'(a)dx = -\sin\left(\frac{\pi}{6}\right)(0.014) = -0.007.$$

15. $y = \dfrac{10 - x^2}{2 + x^2}, \quad a = 1, \quad dx = 0.01$

SOLUTION Let $f(x) = \dfrac{10 - x^2}{2 + x^2}$. Then

$$f'(x) = \frac{(2 + x^2)(-2x) - (10 - x^2)(2x)}{(2 + x^2)^2} = -\frac{24x}{(2 + x^2)^2}$$

and

$$\Delta y \approx dy = f'(a)dx = -\frac{24}{9}(0.01) = -0.026667.$$

In Exercises 17–24, estimate using the Linear Approximation and find the error using a calculator.

17. $\sqrt{26} - \sqrt{25}$

SOLUTION Let $f(x) = \sqrt{x}$, $a = 25$, and $\Delta x = 1$. Then $f'(x) = \frac{1}{2}x^{-1/2}$ and $f'(a) = f'(25) = \frac{1}{10}$.

- The Linear Approximation is $\Delta f \approx f'(a)\Delta x = \frac{1}{10}(1) = 0.1$.
- The actual change is $\Delta f = f(a + \Delta x) - f(a) = f(26) - f(25) \approx 0.0990195$.
- The error in this estimate is $|0.0990195 - 0.1| = 0.000980486$.

19. $\dfrac{1}{\sqrt{101}} - \dfrac{1}{10}$

SOLUTION Let $f(x) = \frac{1}{\sqrt{x}}$, $a = 100$, and $\Delta x = 1$. Then $f'(x) = \frac{d}{dx}(x^{-1/2}) = -\frac{1}{2}x^{-3/2}$ and $f'(a) = -\frac{1}{2}(\frac{1}{1000}) = -0.0005$.

- The Linear Approximation is $\Delta f \approx f'(a)\Delta x = -0.0005(1) = -0.0005$.
- The actual change is

$$\Delta f = f(a + \Delta x) - f(a) = \frac{1}{\sqrt{101}} - \frac{1}{10} = -0.000496281.$$

- The error in this estimate is $|-0.0005 - (-0.000496281)| = 3.71902 \times 10^{-6}$.

21. $9^{1/3} - 2$

SOLUTION Let $f(x) = x^{1/3}$, $a = 8$, and $\Delta x = 1$. Then $f'(x) = \frac{1}{3}x^{-2/3}$ and $f'(a) = f'(8) = \frac{1}{12}$.

- The Linear Approximation is $\Delta f \approx f'(a)\Delta x = \frac{1}{12}(1) = 0.083333$.
- The actual change is $\Delta f = f(a + \Delta x) - f(a) = f(9) - f(8) = 0.080084$.
- The error in this estimate is $|0.080084 - 0.083333| \approx 3.25 \times 10^{-3}$.

23. $e^{-0.1} - 1$

SOLUTION Let $f(x) = e^x$, $a = 0$, and $\Delta x = -0.1$. Then $f'(x) = e^x$ and $f'(a) = f'(0) = 1$.

- The Linear Approximation is $\Delta f \approx f'(a)\Delta x = 1(-0.1) = -0.1$.
- The actual change is $\Delta f = f(a + \Delta x) - f(a) = f(-0.1) - f(0) = -0.095163$.
- The error in this estimate is $|-0.095163 - (-0.1)| \approx 4.84 \times 10^{-3}$.

25. Estimate $f(4.03)$ for $f(x)$ as in Figure 8.

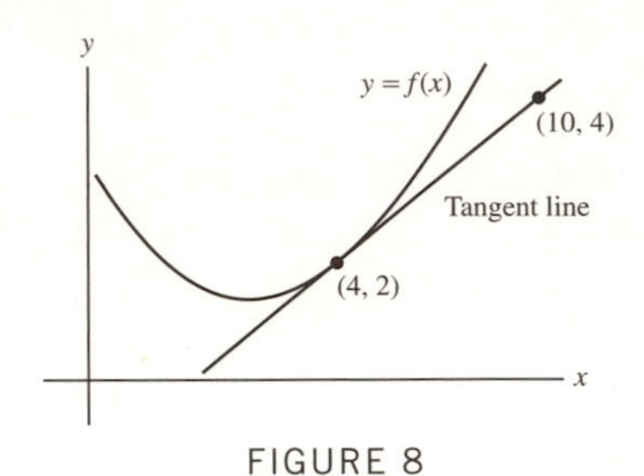

FIGURE 8

SOLUTION Using the Linear Approximation, $f(4.03) \approx f(4) + f'(4)(0.03)$. From the figure, we find that $f(4) = 2$ and

$$f'(4) = \frac{4-2}{10-4} = \frac{1}{3}.$$

Thus,

$$f(4.03) \approx 2 + \frac{1}{3}(0.03) = 2.01.$$

27. Which is larger: $\sqrt{2.1} - \sqrt{2}$ or $\sqrt{9.1} - \sqrt{9}$? Explain using the Linear Approximation.

SOLUTION Let $f(x) = \sqrt{x}$, and $\Delta x = 0.1$. Then $f'(x) = \frac{1}{2}x^{-1/2}$ and the Linear Approximation at $x = a$ gives

$$\Delta f = \sqrt{a + 0.1} - \sqrt{a} \approx f'(a)(0.1) = \frac{1}{2}a^{-1/2}(0.1) = \frac{0.05}{\sqrt{a}}$$

We see that Δf decreases as a increases. In particular

$$\sqrt{2.1} - \sqrt{2} \approx \frac{0.05}{\sqrt{2}} \quad \text{is larger than} \quad \sqrt{9.1} - \sqrt{9} \approx \frac{0.05}{3}$$

29. Box office revenue at a multiplex cinema in Paris is $R(p) = 3600p - 10p^3$ euros per showing when the ticket price is p euros. Calculate $R(p)$ for $p = 9$ and use the Linear Approximation to estimate ΔR if p is raised or lowered by 0.5 euros.

SOLUTION Let $R(p) = 3600p - 10p^3$. Then $R(9) = 3600(9) - 10(9)^3 = 25110$ euros. Moreover, $R'(p) = 3600 - 30p^2$, so by the Linear Approximation,

$$\Delta R \approx R'(9)\Delta p = 1170\Delta p.$$

If p is raised by 0.5 euros, then $\Delta R \approx 585$ euros; on the other hand, if p is lowered by 0.5 euros, then $\Delta R \approx -585$ euros.

31. A thin silver wire has length $L = 18$ cm when the temperature is $T = 30°$C. Estimate ΔL when T decreases to 25°C if the coefficient of thermal expansion is $k = 1.9 \times 10^{-5}\,°\text{C}^{-1}$ (see Example 3).

SOLUTION We have

$$\frac{dL}{dT} = kL = (1.9 \times 10^{-5})(18) = 3.42 \times 10^{-4} \text{ cm/°C}$$

The change in temperature is $\Delta T = -5°$ C, so by the Linear Approximation, the change in length is approximately

$$\Delta L \approx 3.42 \times 10^{-4}\Delta T = (3.42 \times 10^{-4})(-5) = -0.00171 \text{ cm}$$

At $T = 25°$ C, the length of the wire is approximately 17.99829 cm.

33. The atmospheric pressure at altitude h (kilometers) for $11 \le h \le 25$ is approximately

$$P(h) = 128e^{-0.157h} \text{ kilopascals.}$$

(a) Estimate ΔP at $h = 20$ when $\Delta h = 0.5$.

(b) Compute the actual change, and compute the percentage error in the Linear Approximation.

SOLUTION

(a) Let $P(h) = 128e^{-0.157h}$. Then $P'(h) = -20.096e^{-0.157h}$. Using the Linear Approximation,

$$\Delta P \approx P'(h)\Delta h = P'(20)(0.5) = -0.434906 \text{ kilopascals.}$$

(b) The actual change in pressure is

$$P(20.5) - P(20) = -0.418274 \text{ kilopascals.}$$

The percentage error in the Linear Approximation is

$$\left|\frac{-0.434906 - (-0.418274)}{-0.418274}\right| \times 100\% \approx 3.98\%.$$

35. Newton's Law of Gravitation shows that if a person weighs w pounds on the surface of the earth, then his or her weight at distance x from the center of the earth is

$$W(x) = \frac{wR^2}{x^2} \qquad (\text{for } x \geq R)$$

where $R = 3960$ miles is the radius of the earth (Figure 9).

(a) Show that the weight lost at altitude h miles above the earth's surface is approximately $\Delta W \approx -(0.0005w)h$. *Hint:* Use the Linear Approximation with $dx = h$.

(b) Estimate the weight lost by a 200-lb football player flying in a jet at an altitude of 7 miles.

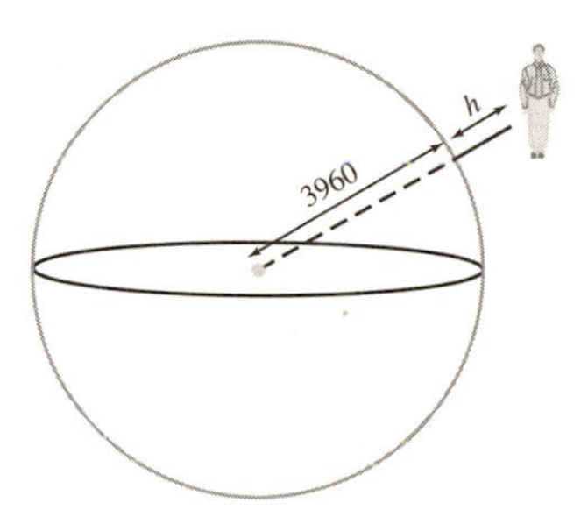

FIGURE 9 The distance to the center of the earth is $3960 + h$ miles.

SOLUTION

(a) Using the Linear Approximation

$$\Delta W \approx W'(R)\Delta x = -\frac{2wR^2}{R^3}h = -\frac{2wh}{R} \approx -0.0005wh.$$

(b) Substitute $w = 200$ and $h = 7$ into the result from part (a) to obtain

$$\Delta W \approx -0.0005(200)(7) = -0.7 \text{ pounds.}$$

37. A stone tossed vertically into the air with initial velocity v cm/s reaches a maximum height of $h = v^2/1960$ cm.

(a) Estimate Δh if $v = 700$ cm/s and $\Delta v = 1$ cm/s.

(b) Estimate Δh if $v = 1{,}000$ cm/s and $\Delta v = 1$ cm/s.

(c) In general, does a 1 cm/s increase in v lead to a greater change in h at low or high initial velocities? Explain.

SOLUTION A stone tossed vertically with initial velocity v cm/s attains a maximum height of $h(v) = v^2/1960$ cm. Thus, $h'(v) = v/980$.

(a) If $v = 700$ and $\Delta v = 1$, then $\Delta h \approx h'(v)\Delta v = \frac{1}{980}(700)(1) \approx 0.71$ cm.

(b) If $v = 1000$ and $\Delta v = 1$, then $\Delta h \approx h'(v)\Delta v = \frac{1}{980}(1000)(1) = 1.02$ cm.

(c) A one centimeter per second increase in initial velocity v increases the maximum height by approximately $v/980$ cm. Accordingly, there is a bigger effect at higher velocities.

In Exercises 39 and 40, use the following fact derived from Newton's Laws: An object released at an angle θ with initial velocity v ft/s travels a horizontal distance

$$s = \frac{1}{32}v^2 \sin 2\theta \text{ ft} \quad \text{(Figure 10)}$$

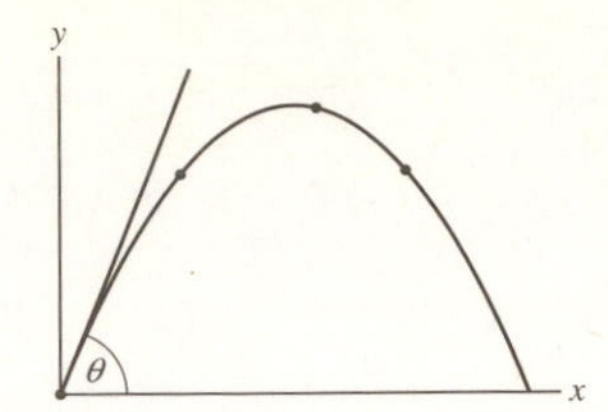

FIGURE 10 Trajectory of an object released at an angle θ.

39. A player located 18.1 ft from the basket launches a successful jump shot from a height of 10 ft (level with the rim of the basket), at an angle $\theta = 34°$ and initial velocity $v = 25$ ft/s.)

(a) Show that $\Delta s \approx 0.255\Delta\theta$ ft for a small change of $\Delta\theta$.

(b) Is it likely that the shot would have been successful if the angle had been off by 2°?

SOLUTION Using Newton's laws and the given initial velocity of $v = 25$ ft/s, the shot travels $s = \frac{1}{32}v^2 \sin 2t = \frac{625}{32} \sin 2t$ ft, where t is in radians.

(a) If $\theta = 34°$ (i.e., $t = \frac{17}{90}\pi$), then

$$\Delta s \approx s'(t)\Delta t = \frac{625}{16} \cos\left(\frac{17}{45}\pi\right) \Delta t = \frac{625}{16} \cos\left(\frac{17}{45}\pi\right) \Delta\theta \cdot \frac{\pi}{180} \approx 0.255\Delta\theta.$$

(b) If $\Delta\theta = 2°$, this gives $\Delta s \approx 0.51$ ft, in which case the shot would not have been successful, having been off half a foot.

41. The radius of a spherical ball is measured at $r = 25$ cm. Estimate the maximum error in the volume and surface area if r is accurate to within 0.5 cm.

SOLUTION The volume and surface area of the sphere are given by $V = \frac{4}{3}\pi r^3$ and $S = 4\pi r^2$, respectively. If $r = 25$ and $\Delta r = \pm 0.5$, then

$$\Delta V \approx V'(25)\Delta r = 4\pi(25)^2(0.5) \approx 3927 \text{ cm}^3,$$

and

$$\Delta S \approx S'(25)\Delta r = 8\pi(25)(0.5) \approx 314.2 \text{ cm}^2.$$

43. The volume (in liters) and pressure P (in atmospheres) of a certain gas satisfy $PV = 24$. A measurement yields $V = 4$ with a possible error of ± 0.3 L. Compute P and estimate the maximum error in this computation.

SOLUTION Given $PV = 24$ and $V = 4$, it follows that $P = 6$ atmospheres. Solving $PV = 24$ for P yields $P = 24V^{-1}$. Thus, $P' = -24V^{-2}$ and

$$\Delta P \approx P'(4)\Delta V = -24(4)^{-2}(\pm 0.3) = \pm 0.45 \text{ atmospheres}.$$

In Exercises 45–54, find the linearization at $x = a$.

45. $f(x) = x^4, \quad a = 1$

SOLUTION Let $f(x) = x^4$. Then $f'(x) = 4x^3$. The linearization at $a = 1$ is

$$L(x) = f'(a)(x - a) + f(a) = 4(x - 1) + 1 = 4x - 3.$$

47. $f(\theta) = \sin^2\theta, \quad a = \frac{\pi}{4}$

SOLUTION Let $f(\theta) = \sin^2\theta$. Then $f'(\theta) = 2\sin\theta\cos\theta = \sin 2\theta$. The linearization at $a = \frac{\pi}{4}$ is

$$L(\theta) = f'(a)(\theta - a) + f(a) = 1\left(\theta - \frac{\pi}{4}\right) + \frac{1}{2} = \theta - \frac{\pi}{4} + \frac{1}{2}.$$

49. $y = (1 + x)^{-1/2}, \quad a = 0$

SOLUTION Let $f(x) = (1 + x)^{-1/2}$. Then $f'(x) = -\frac{1}{2}(1 + x)^{-3/2}$. The linearization at $a = 0$ is

$$L(x) = f'(a)(x - a) + f(a) = -\frac{1}{2}x + 1.$$

51. $y = (1 + x^2)^{-1/2}, \quad a = 0$

SOLUTION Let $f(x) = (1 + x^2)^{-1/2}$. Then $f'(x) = -x(1 + x^2)^{-3/2}$, $f(a) = 1$ and $f'(a) = 0$, so the linearization at a is

$$L(x) = f'(a)(x - a) + f(a) = 1.$$

53. $y = e^{\sqrt{x}}, \quad a = 1$

SOLUTION Let $f(x) = e^{\sqrt{x}}$. Then

$$f'(x) = \frac{1}{2\sqrt{x}}e^{\sqrt{x}}, \quad f(a) = e, \text{ and } f'(a) = \frac{1}{2}e,$$

so the linearization of $f(x)$ at a is

$$L(x) = f'(a)(x-a) + f(a) = \frac{1}{2}e(x-1) + e = \frac{1}{2}e(x+1).$$

55. What is $f(2)$ if the linearization of $f(x)$ at $a = 2$ is $L(x) = 2x + 4$?

SOLUTION $f(2) = L(2) = 2(2) + 4 = 8$.

57. Estimate $\sqrt{16.2}$ using the linearization $L(x)$ of $f(x) = \sqrt{x}$ at $a = 16$. Plot $f(x)$ and $L(x)$ on the same set of axes and determine whether the estimate is too large or too small.

SOLUTION Let $f(x) = x^{1/2}, a = 16$, and $\Delta x = 0.2$. Then $f'(x) = \frac{1}{2}x^{-1/2}$ and $f'(a) = f'(16) = \frac{1}{8}$. The linearization to $f(x)$ is

$$L(x) = f'(a)(x-a) + f(a) = \frac{1}{8}(x-16) + 4 = \frac{1}{8}x + 2.$$

Thus, we have $\sqrt{16.2} \approx L(16.2) = 4.025$. Graphs of $f(x)$ and $L(x)$ are shown below. Because the graph of $L(x)$ lies above the graph of $f(x)$, we expect that the estimate from the Linear Approximation is too large.

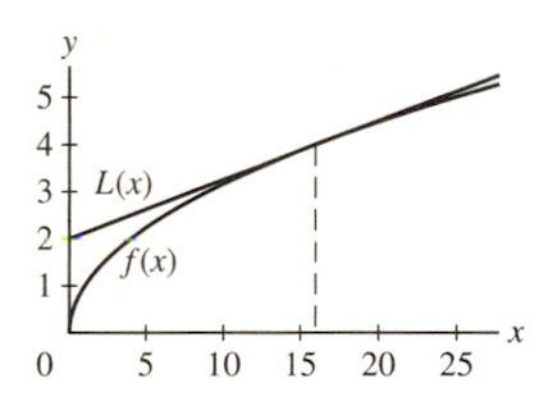

In Exercises 59–67, approximate using linearization and use a calculator to compute the percentage error.

59. $\dfrac{1}{\sqrt{17}}$

SOLUTION Let $f(x) = x^{-1/2}$, $a = 16$, and $\Delta x = 1$. Then $f'(x) = -\frac{1}{2}x^{-3/2}$, $f'(a) = f'(16) = -\frac{1}{128}$ and the linearization to $f(x)$ is

$$L(x) = f'(a)(x-a) + f(a) = -\frac{1}{128}(x-16) + \frac{1}{4} = -\frac{1}{128}x + \frac{3}{8}.$$

Thus, we have $\frac{1}{\sqrt{17}} \approx L(17) \approx 0.24219$. The percentage error in this estimate is

$$\left|\frac{\frac{1}{\sqrt{17}} - 0.24219}{\frac{1}{\sqrt{17}}}\right| \times 100\% \approx 0.14\%$$

61. $\dfrac{1}{(10.03)^2}$

SOLUTION Let $f(x) = x^{-2}$, $a = 10$ and $\Delta x = 0.03$. Then $f'(x) = -2x^{-3}$, $f'(a) = f'(10) = -0.002$ and the linearization to $f(x)$ is

$$L(x) = f'(a)(x-a) + f(a) = -0.002(x-10) + 0.01 = -0.002x + 0.03.$$

Thus, we have

$$\frac{1}{(10.03)^2} \approx L(10.03) = -0.002(10.03) + 0.03 = 0.00994.$$

The percentage error in this estimate is

$$\left|\frac{\frac{1}{(10.03)^2} - 0.00994}{\frac{1}{(10.03)^2}}\right| \times 100\% \approx 0.0027\%$$

63. $(64.1)^{1/3}$

SOLUTION Let $f(x) = x^{1/3}$, $a = 64$, and $\Delta x = 0.1$. Then $f'(x) = \frac{1}{3}x^{-2/3}$, $f'(a) = f'(64) = \frac{1}{48}$ and the linearization to $f(x)$ is

$$L(x) = f'(a)(x-a) + f(a) = \frac{1}{48}(x-64) + 4 = \frac{1}{48}x + \frac{8}{3}.$$

Thus, we have $(64.1)^{1/3} \approx L(64.1) \approx 4.002083$. The percentage error in this estimate is

$$\left|\frac{(64.1)^{1/3} - 4.002083}{(64.1)^{1/3}}\right| \times 100\% \approx 0.000019\%$$

65. $\cos^{-1}(0.52)$

SOLUTION Let $f(x) = \cos^{-1} x$ and $a = 0.5$. Then

$$f'(x) = -\frac{1}{\sqrt{1-x^2}}, \quad f'(a) = f'(0) = -\frac{2\sqrt{3}}{3},$$

and the linearization to $f(x)$ is

$$L(x) = f'(a)(x-a) + f(a) = -\frac{2\sqrt{3}}{3}(x-0.5) + \frac{\pi}{3}.$$

Thus, we have $\cos^{-1}(0.52) \approx L(0.02) = 1.024104$. The percentage error in this estimate is

$$\left|\frac{\cos^{-1}(0.52) - 1.024104}{\cos^{-1}(0.52)}\right| \times 100\% \approx 0.015\%.$$

67. $e^{-0.012}$

SOLUTION Let $f(x) = e^x$ and $a = 0$. Then $f'(x) = e^x$, $f'(a) = f'(0) = 1$ and the linearization to $f(x)$ is

$$L(x) = f'(a)(x-a) + f(a) = 1(x-0) + 1 = x + 1.$$

Thus, we have $e^{-0.012} \approx L(-0.012) = 1 - 0.012 = 0.988$. The percentage error in this estimate is

$$\left|\frac{e^{-0.012} - 0.988}{e^{-0.012}}\right| \times 100\% \approx 0.0073\%.$$

69. Show that the Linear Approximation to $f(x) = \sqrt{x}$ at $x = 9$ yields the estimate $\sqrt{9+h} - 3 \approx \frac{1}{6}h$. Set $K = 0.01$ and show that $|f''(x)| \le K$ for $x \ge 9$. Then verify numerically that the error E satisfies Eq. (5) for $h = 10^{-n}$, for $1 \le n \le 4$.

SOLUTION Let $f(x) = \sqrt{x}$. Then $f(9) = 3$, $f'(x) = \frac{1}{2}x^{-1/2}$ and $f'(9) = \frac{1}{6}$. Therefore, by the Linear Approximation,

$$f(9+h) - f(9) = \sqrt{9+h} - 3 \approx \frac{1}{6}h.$$

Moreover, $f''(x) = -\frac{1}{4}x^{-3/2}$, so $|f''(x)| = \frac{1}{4}x^{-3/2}$. Because this is a decreasing function, it follows that for $x \ge 9$,

$$K = \max |f''(x)| \le |f''(9)| = \frac{1}{108} < 0.01.$$

From the following table, we see that for $h = 10^{-n}$, $1 \le n \le 4$, $E \le \frac{1}{2}Kh^2$.

h	$E = \lvert\sqrt{9+h} - 3 - \frac{1}{6}h\rvert$	$\frac{1}{2}Kh^2$
10^{-1}	4.604×10^{-5}	5.00×10^{-5}
10^{-2}	4.627×10^{-7}	5.00×10^{-7}
10^{-3}	4.629×10^{-9}	5.00×10^{-9}
10^{-4}	4.627×10^{-11}	5.00×10^{-11}

Further Insights and Challenges

71. Compute dy/dx at the point $P = (2, 1)$ on the curve $y^3 + 3xy = 7$ and show that the linearization at P is $L(x) = -\frac{1}{3}x + \frac{5}{3}$. Use $L(x)$ to estimate the y-coordinate of the point on the curve where $x = 2.1$.

SOLUTION Differentiating both sides of the equation $y^3 + 3xy = 7$ with respect to x yields

$$3y^2\frac{dy}{dx} + 3x\frac{dy}{dx} + 3y = 0,$$

so

$$\frac{dy}{dx} = -\frac{y}{y^2 + x}.$$

Thus,

$$\left.\frac{dy}{dx}\right|_{(2,1)} = -\frac{1}{1^2 + 2} = -\frac{1}{3},$$

and the linearization at $P = (2, 1)$ is

$$L(x) = 1 - \frac{1}{3}(x - 2) = -\frac{1}{3}x + \frac{5}{3}.$$

Finally, when $x = 2.1$, we estimate that the y-coordinate of the point on the curve is

$$y \approx L(2.1) = -\frac{1}{3}(2.1) + \frac{5}{3} = 0.967.$$

73. Apply the method of Exercise 71 to $P = (-1, 2)$ on $y^4 + 7xy = 2$ to estimate the solution of $y^4 - 7.7y = 2$ near $y = 2$.

SOLUTION Differentiating both sides of the equation $y^4 + 7xy = 2$ with respect to x yields

$$4y^3\frac{dy}{dx} + 7x\frac{dy}{dx} + 7y = 0,$$

so

$$\frac{dy}{dx} = -\frac{7y}{4y^3 + 7x}.$$

Thus,

$$\left.\frac{dy}{dx}\right|_{(-1,2)} = -\frac{7(2)}{4(2)^3 + 7(-1)} = -\frac{14}{25},$$

and the linearization at $P = (-1, 2)$ is

$$L(x) = 2 - \frac{14}{25}(x + 1) = -\frac{14}{25}x + \frac{36}{25}.$$

Finally, the equation $y^4 - 7.7y = 2$ corresponds to $x = -1.1$, so we estimate the solution of this equation near $y = 2$ is

$$y \approx L(-1.1) = -\frac{14}{25}(-1.1) + \frac{36}{25} = 2.056.$$

75. Let $\Delta f = f(5 + h) - f(5)$, where $f(x) = x^2$. Verify directly that $E = |\Delta f - f'(5)h|$ satisfies (5) with $K = 2$.

SOLUTION Let $f(x) = x^2$. Then

$$\Delta f = f(5 + h) - f(5) = (5 + h)^2 - 5^2 = h^2 + 10h$$

and

$$E = |\Delta f - f'(5)h| = |h^2 + 10h - 10h| = h^2 = \frac{1}{2}(2)h^2 = \frac{1}{2}Kh^2.$$

4.2 Extreme Values

Preliminary Questions

1. What is the definition of a critical point?

SOLUTION A critical point is a value of the independent variable x in the domain of a function f at which either $f'(x) = 0$ or $f'(x)$ does not exist.

In Questions 2 and 3, choose the correct conclusion.

2. If $f(x)$ is not continuous on $[0, 1]$, then

(a) $f(x)$ has no extreme values on $[0, 1]$.

(b) $f(x)$ might not have any extreme values on $[0, 1]$.

SOLUTION The correct response is **(b)**: $f(x)$ might not have any extreme values on $[0, 1]$. Although $[0, 1]$ is closed, because f is not continuous, the function is not guaranteed to have any extreme values on $[0, 1]$.

3. If $f(x)$ is continuous but has no critical points in $[0, 1]$, then

(a) $f(x)$ has no min or max on $[0, 1]$.

(b) Either $f(0)$ or $f(1)$ is the minimum value on $[0, 1]$.

SOLUTION The correct response is **(b)**: either $f(0)$ or $f(1)$ is the minimum value on $[0, 1]$. Remember that extreme values occur either at critical points or endpoints. If a continuous function on a closed interval has no critical points, the extreme values must occur at the endpoints.

4. Fermat's Theorem *does not* claim that if $f'(c) = 0$, then $f(c)$ is a local extreme value (this is false). What *does* Fermat's Theorem assert?

SOLUTION Fermat's Theorem claims: If $f(c)$ is a local extreme value, then either $f'(c) = 0$ or $f'(c)$ does not exist.

Exercises

1. The following questions refer to Figure 15.

(a) How many critical points does $f(x)$ have on $[0, 8]$?

(b) What is the maximum value of $f(x)$ on $[0, 8]$?

(c) What are the local maximum values of $f(x)$?

(d) Find a closed interval on which both the minimum and maximum values of $f(x)$ occur at critical points.

(e) Find an interval on which the minimum value occurs at an endpoint.

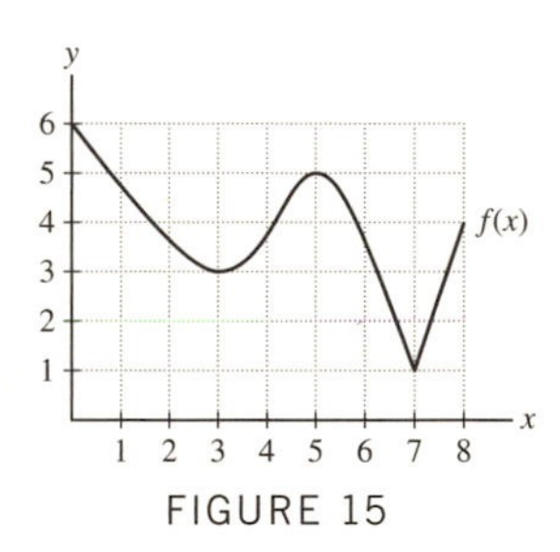

FIGURE 15

SOLUTION

(a) $f(x)$ has three critical points on the interval $[0, 8]$: at $x = 3$, $x = 5$ and $x = 7$. Two of these, $x = 3$ and $x = 5$, are where the derivative is zero and one, $x = 7$, is where the derivative does not exist.

(b) The maximum value of $f(x)$ on $[0, 8]$ is 6; the function takes this value at $x = 0$.

(c) $f(x)$ achieves a local maximum of 5 at $x = 5$.

(d) Answers may vary. One example is the interval $[4, 8]$. Another is $[2, 6]$.

(e) Answers may vary. The easiest way to ensure this is to choose an interval on which the graph takes no local minimum. One example is $[0, 2]$.

In Exercises 3–20, find all critical points of the function.

3. $f(x) = x^2 - 2x + 4$

SOLUTION Let $f(x) = x^2 - 2x + 4$. Then $f'(x) = 2x - 2 = 0$ implies that $x = 1$ is the lone critical point of f.

5. $f(x) = x^3 - \frac{9}{2}x^2 - 54x + 2$

SOLUTION Let $f(x) = x^3 - \frac{9}{2}x^2 - 54x + 2$. Then $f'(x) = 3x^2 - 9x - 54 = 3(x + 3)(x - 6) = 0$ implies that $x = -3$ and $x = 6$ are the critical points of f.

7. $f(x) = x^{-1} - x^{-2}$

SOLUTION Let $f(x) = x^{-1} - x^{-2}$. Then

$$f'(x) = -x^{-2} + 2x^{-3} = \frac{2-x}{x^3} = 0$$

implies that $x = 2$ is the only critical point of f. Though $f'(x)$ does not exist at $x = 0$, this is not a critical point of f because $x = 0$ is not in the domain of f.

9. $f(x) = \dfrac{x}{x^2+1}$

SOLUTION Let $f(x) = \dfrac{x}{x^2+1}$. Then $f'(x) = \dfrac{1-x^2}{(x^2+1)^2} = 0$ implies that $x = \pm 1$ are the critical points of f.

11. $f(t) = t - 4\sqrt{t+1}$

SOLUTION Let $f(t) = t - 4\sqrt{t+1}$. Then

$$f'(t) = 1 - \frac{2}{\sqrt{t+1}} = 0$$

implies that $t = 3$ is a critical point of f. Because $f'(t)$ does not exist at $t = -1$, this is another critical point of f.

13. $f(x) = x^2\sqrt{1-x^2}$

SOLUTION Let $f(x) = x^2\sqrt{1-x^2}$. Then

$$f'(x) = -\frac{x^3}{\sqrt{1-x^2}} + 2x\sqrt{1-x^2} = \frac{2x-3x^3}{\sqrt{1-x^2}}.$$

This derivative is 0 when $x = 0$ and when $x = \pm\sqrt{2/3}$; the derivative does not exist when $x = \pm 1$. All five of these values are critical points of f.

15. $g(\theta) = \sin^2\theta$

SOLUTION Let $g(\theta) = \sin^2\theta$. Then $g'(\theta) = 2\sin\theta\cos\theta = \sin 2\theta = 0$ implies that

$$\theta = \frac{n\pi}{2}$$

is a critical value of g for all integer values of n.

17. $f(x) = x\ln x$

SOLUTION Let $f(x) = x\ln x$. Then $f'(x) = 1 + \ln x = 0$ implies that $x = e^{-1} = \frac{1}{e}$ is the only critical point of f.

19. $f(x) = \sin^{-1}x - 2x$

SOLUTION Let $f(x) = \sin^{-1}x - 2x$. Then

$$f'(x) = \frac{1}{\sqrt{1-x^2}} - 2 = 0$$

implies that $x = \pm\frac{\sqrt{3}}{2}$ are the critical points of f.

21. Let $f(x) = x^2 - 4x + 1$.

(a) Find the critical point c of $f(x)$ and compute $f(c)$.
(b) Compute the value of $f(x)$ at the endpoints of the interval $[0, 4]$.
(c) Determine the min and max of $f(x)$ on $[0, 4]$.
(d) Find the extreme values of $f(x)$ on $[0, 1]$.

SOLUTION Let $f(x) = x^2 - 4x + 1$.

(a) Then $f'(c) = 2c - 4 = 0$ implies that $c = 2$ is the sole critical point of f. We have $f(2) = -3$.
(b) $f(0) = f(4) = 1$.
(c) Using the results from (a) and (b), we find the maximum value of f on $[0, 4]$ is 1 and the minimum value is -3.
(d) We have $f(1) = -2$. Hence the maximum value of f on $[0, 1]$ is 1 and the minimum value is -2.

23. Find the critical points of $f(x) = \sin x + \cos x$ and determine the extreme values on $\left[0, \frac{\pi}{2}\right]$.

SOLUTION

- Let $f(x) = \sin x + \cos x$. Then on the interval $\left[0, \frac{\pi}{2}\right]$, we have $f'(x) = \cos x - \sin x = 0$ at $x = \frac{\pi}{4}$, the only critical point of f in this interval.
- Since $f(\frac{\pi}{4}) = \sqrt{2}$ and $f(0) = f(\frac{\pi}{2}) = 1$, the maximum value of f on $\left[0, \frac{\pi}{2}\right]$ is $\sqrt{2}$, while the minimum value is 1.

25. GU Plot $f(x) = 2\sqrt{x} - x$ on $[0, 4]$ and determine the maximum value graphically. Then verify your answer using calculus.

SOLUTION The graph of $y = 2\sqrt{x} - x$ over the interval $[0, 4]$ is shown below. From the graph, we see that at $x = 1$, the function achieves its maximum value of 1.

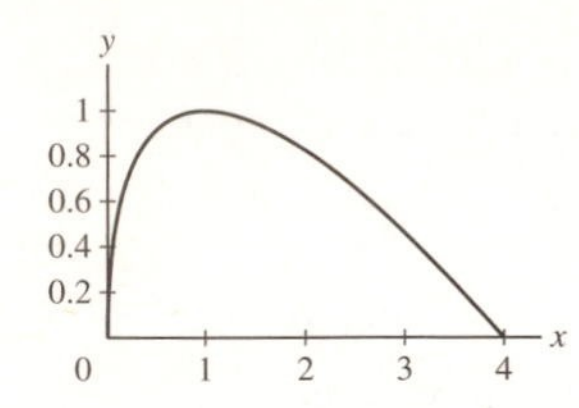

To verify the information obtained from the plot, let $f(x) = 2\sqrt{x} - x$. Then $f'(x) = x^{-1/2} - 1$. Solving $f'(x) = 0$ yields the critical points $x = 0$ and $x = 1$. Because $f(0) = f(4) = 0$ and $f(1) = 1$, we see that the maximum value of f on $[0, 4]$ is 1.

27. CAS Approximate the critical points of $g(x) = x\cos^{-1} x$ and estimate the maximum value of $g(x)$.

SOLUTION $g'(x) = \frac{-x}{\sqrt{1-x^2}} + \cos^{-1} x$, so $g'(x) = 0$ when $x \approx 0.652185$. Evaluating g at the endpoints of its domain, $x = \pm 1$, and at the critical point $x \approx 0.652185$, we find $g(-1) = -\pi$, $g(0.652185) \approx 0.561096$, and $g(1) = 0$. Hence, the maximum value of $g(x)$ is approximately 0.561096.

In Exercises 29–58, find the min and max of the function on the given interval by comparing values at the critical points and endpoints.

29. $y = 2x^2 + 4x + 5$, $[-2, 2]$

SOLUTION Let $f(x) = 2x^2 + 4x + 5$. Then $f'(x) = 4x + 4 = 0$ implies that $x = -1$ is the only critical point of f. The minimum of f on the interval $[-2, 2]$ is $f(-1) = 3$, whereas its maximum is $f(2) = 21$. (*Note:* $f(-2) = 5$.)

31. $y = 6t - t^2$, $[0, 5]$

SOLUTION Let $f(t) = 6t - t^2$. Then $f'(t) = 6 - 2t = 0$ implies that $t = 3$ is the only critical point of f. The minimum of f on the interval $[0, 5]$ is $f(0) = 0$, whereas the maximum is $f(3) = 9$. (*Note:* $f(5) = 5$.)

33. $y = x^3 - 6x^2 + 8$, $[1, 6]$

SOLUTION Let $f(x) = x^3 - 6x^2 + 8$. Then $f'(x) = 3x^2 - 12x = 3x(x - 4) = 0$ implies that $x = 0$ and $x = 4$ are the critical points of f. The minimum of f on the interval $[1, 6]$ is $f(4) = -24$, whereas the maximum is $f(6) = 8$. (*Note:* $f(1) = 3$ and the critical point $x = 0$ is not in the interval $[1, 6]$.)

35. $y = 2t^3 + 3t^2$, $[1, 2]$

SOLUTION Let $f(t) = 2t^3 + 3t^2$. Then $f'(t) = 6t^2 + 6t = 6t(t + 1) = 0$ implies that $t = 0$ and $t = -1$ are the critical points of f. The minimum of f on the interval $[1, 2]$ is $f(1) = 5$, whereas the maximum is $f(2) = 28$. (*Note:* Neither critical points are in the interval $[1, 2]$.)

37. $y = z^5 - 80z$, $[-3, 3]$

SOLUTION Let $f(z) = z^5 - 80z$. Then $f'(z) = 5z^4 - 80 = 5(z^4 - 16) = 5(z^2 + 4)(z + 2)(z - 2) = 0$ implies that $z = \pm 2$ are the critical points of f. The minimum value of f on the interval $[-3, 3]$ is $f(2) = -128$, whereas the maximum is $f(-2) = 128$. (*Note:* $f(-3) = 3$ and $f(3) = -3$.)

39. $y = \dfrac{x^2 + 1}{x - 4}$, $[5, 6]$

SOLUTION Let $f(x) = \dfrac{x^2 + 1}{x - 4}$. Then

$$f'(x) = \frac{(x - 4) \cdot 2x - (x^2 + 1) \cdot 1}{(x - 4)^2} = \frac{x^2 - 8x - 1}{(x - 4)^2} = 0$$

implies $x = 4 \pm \sqrt{17}$ are critical points of f. $x = 4$ is not a critical point because $x = 4$ is not in the domain of f. On the interval $[5, 6]$, the minimum of f is $f(6) = \frac{37}{2} = 18.5$, whereas the maximum of f is $f(5) = 26$. (*Note:* The critical points $x = 4 \pm \sqrt{17}$ are not in the interval $[5, 6]$.)

41. $y = x - \dfrac{4x}{x+1}, \quad [0, 3]$

SOLUTION Let $f(x) = x - \dfrac{4x}{x+1}$. Then

$$f'(x) = 1 - \frac{4}{(x+1)^2} = \frac{(x-1)(x+3)}{(x+1)^2} = 0$$

implies that $x = 1$ and $x = -3$ are critical points of f. $x = -1$ is not a critical point because $x = -1$ is not in the domain of f. The minimum of f on the interval $[0, 3]$ is $f(1) = -1$, whereas the maximum is $f(0) = f(3) = 0$. (*Note:* The critical point $x = -3$ is not in the interval $[0, 3]$.)

43. $y = (2+x)\sqrt{2+(2-x)^2}, \quad [0, 2]$

SOLUTION Let $f(x) = (2+x)\sqrt{2+(2-x)^2}$. Then

$$f'(x) = \sqrt{2+(2-x)^2} - (2+x)(2+(2-x)^2)^{-1/2}(2-x) = \frac{2(x-1)^2}{\sqrt{2+(2-x)^2}} = 0$$

implies that $x = 1$ is the critical point of f. On the interval $[0, 2]$, the minimum is $f(0) = 2\sqrt{6} \approx 4.9$ and the maximum is $f(2) = 4\sqrt{2} \approx 5.66$. (*Note:* $f(1) = 3\sqrt{3} \approx 5.2$.)

45. $y = \sqrt{x+x^2} - 2\sqrt{x}, \quad [0, 4]$

SOLUTION Let $f(x) = \sqrt{x+x^2} - 2\sqrt{x}$. Then

$$f'(x) = \frac{1}{2}(x+x^2)^{-1/2}(1+2x) - x^{-1/2} = \frac{1+2x-2\sqrt{1+x}}{2\sqrt{x}\sqrt{1+x}} = 0$$

implies that $x = 0$ and $x = \frac{\sqrt{3}}{2}$ are the critical points of f. Neither $x = -1$ nor $x = -\frac{\sqrt{3}}{2}$ is a critical point because neither is in the domain of f. On the interval $[0, 4]$, the minimum of f is $f\left(\frac{\sqrt{3}}{2}\right) \approx -0.589980$ and the maximum is $f(4) \approx 0.472136$. (*Note:* $f(0) = 0$.)

47. $y = \sin x \cos x, \quad \left[0, \frac{\pi}{2}\right]$

SOLUTION Let $f(x) = \sin x \cos x = \frac{1}{2}\sin 2x$. On the interval $\left[0, \frac{\pi}{2}\right]$, $f'(x) = \cos 2x = 0$ when $x = \frac{\pi}{4}$. The minimum of f on this interval is $f(0) = f(\frac{\pi}{2}) = 0$, whereas the maximum is $f(\frac{\pi}{4}) = \frac{1}{2}$.

49. $y = \sqrt{2}\theta - \sec\theta, \quad \left[0, \frac{\pi}{3}\right]$

SOLUTION Let $f(\theta) = \sqrt{2}\theta - \sec\theta$. On the interval $[0, \frac{\pi}{3}]$, $f'(\theta) = \sqrt{2} - \sec\theta\tan\theta = 0$ at $\theta = \frac{\pi}{4}$. The minimum value of f on this interval is $f(0) = -1$, whereas the maximum value over this interval is $f(\frac{\pi}{4}) = \sqrt{2}(\frac{\pi}{4} - 1) \approx -0.303493$. (*Note:* $f(\frac{\pi}{3}) = \sqrt{2}\frac{\pi}{3} - 2 \approx -.519039$.)

51. $y = \theta - 2\sin\theta, \quad [0, 2\pi]$

SOLUTION Let $g(\theta) = \theta - 2\sin\theta$. On the interval $[0, 2\pi]$, $g'(\theta) = 1 - 2\cos\theta = 0$ at $\theta = \frac{\pi}{3}$ and $\theta = \frac{5}{3}\pi$. The minimum of g on this interval is $g(\frac{\pi}{3}) = \frac{\pi}{3} - \sqrt{3} \approx -.685$ and the maximum is $g(\frac{5}{3}\pi) = \frac{5}{3}\pi + \sqrt{3} \approx 6.968$. (*Note:* $g(0) = 0$ and $g(2\pi) = 2\pi \approx 6.283$.)

53. $y = \tan x - 2x, \quad [0, 1]$

SOLUTION Let $f(x) = \tan x - 2x$. Then on the interval $[0, 1]$, $f'(x) = \sec^2 x - 2 = 0$ at $x = \frac{\pi}{4}$. The minimum of f is $f(\frac{\pi}{4}) = 1 - \frac{\pi}{2} \approx -0.570796$ and the maximum is $f(0) = 0$. (*Note:* $f(1) = \tan 1 - 2 \approx -0.442592$.)

55. $y = \dfrac{\ln x}{x}, \quad [1, 3]$

SOLUTION Let $f(x) = \frac{\ln x}{x}$. Then, on the interval $[1, 3]$,

$$f'(x) = \frac{1 - \ln x}{x^2} = 0$$

at $x = e$. The minimum of f on this interval is $f(1) = 0$ and the maximum is $f(e) = e^{-1} \approx 0.367879$. (*Note:* $f(3) = \frac{1}{3}\ln 3 \approx 0.366204$.)

57. $y = 5\tan^{-1} x - x, \quad [1, 5]$

SOLUTION Let $f(x) = 5\tan^{-1} x - x$. Then, on the interval $[1, 5]$,

$$f'(x) = 5\frac{1}{1+x^2} - 1 = 0$$

at $x = 2$. The minimum of f on this interval is $f(5) = 5\tan^{-1} 5 - 5 \approx 1.867004$ and the maximum is $f(2) = 5\tan^{-1} 2 - 2 \approx 3.535744$. (*Note:* $f(1) = \frac{5\pi}{4} - 1 \approx 2.926991$.)

59. Let $f(\theta) = 2\sin 2\theta + \sin 4\theta$.

(a) Show that θ is a critical point if $\cos 4\theta = -\cos 2\theta$.

(b) Show, using a unit circle, that $\cos\theta_1 = -\cos\theta_2$ if and only if $\theta_1 = \pi \pm \theta_2 + 2\pi k$ for an integer k.

(c) Show that $\cos 4\theta = -\cos 2\theta$ if and only if $\theta = \frac{\pi}{2} + \pi k$ or $\theta = \frac{\pi}{6} + \left(\frac{\pi}{3}\right)k$.

(d) Find the six critical points of $f(\theta)$ on $[0, 2\pi]$ and find the extreme values of $f(\theta)$ on this interval.

(e) GU Check your results against a graph of $f(\theta)$.

SOLUTION $f(\theta) = 2\sin 2\theta + \sin 4\theta$ is differentiable at all θ, so the way to find the critical points is to find all points such that $f'(\theta) = 0$.

(a) $f'(\theta) = 4\cos 2\theta + 4\cos 4\theta$. If $f'(\theta) = 0$, then $4\cos 4\theta = -4\cos 2\theta$, so $\cos 4\theta = -\cos 2\theta$.

(b) Given the point $(\cos\theta, \sin\theta)$ at angle θ on the unit circle, there are two points with x coordinate $-\cos\theta$. The graphic shows these two points, which are:

- The point $(\cos(\theta + \pi), \sin(\theta + \pi))$ on the opposite end of the unit circle.
- The point $(\cos(\pi - \theta), \sin(\theta - \pi))$ obtained by reflecting through the y axis.

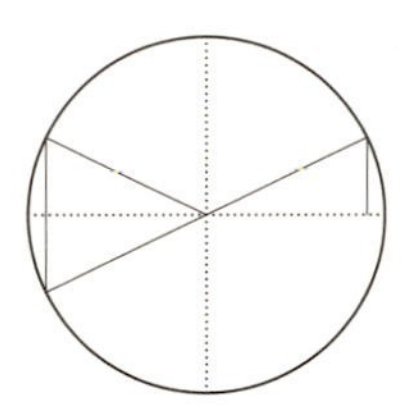

If we include all angles representing these points on the circle, we find that $\cos\theta_1 = -\cos\theta_2$ if and only if $\theta_1 = (\pi + \theta_2) + 2\pi k$ or $\theta_1 = (\pi - \theta_2) + 2\pi k$ for integers k.

(c) Using (b), we recognize that $\cos 4\theta = -\cos 2\theta$ if $4\theta = 2\theta + \pi + 2\pi k$ or $4\theta = \pi - 2\theta + 2\pi k$. Solving for θ, we obtain $\theta = \frac{\pi}{2} + k\pi$ or $\theta = \frac{\pi}{6} + \frac{\pi}{3}k$.

(d) To find all θ, $0 \le \theta < 2\pi$ indicated by (c), we use the following table:

k	0	1	2	3	4	5
$\frac{\pi}{2} + k\pi$	$\frac{\pi}{2}$	$\frac{3\pi}{2}$				
$\frac{\pi}{6} + \frac{\pi}{3}k$	$\frac{\pi}{6}$	$\frac{\pi}{2}$	$\frac{5\pi}{6}$	$\frac{7\pi}{6}$	$\frac{3\pi}{2}$	$\frac{11\pi}{6}$

The critical points in the range $[0, 2\pi]$ are $\frac{\pi}{6}, \frac{\pi}{2}, \frac{5\pi}{6}, \frac{7\pi}{6}, \frac{3\pi}{2}$, and $\frac{11\pi}{6}$. On this interval, the maximum value is $f(\frac{\pi}{6}) = f(\frac{7\pi}{6}) = \frac{3\sqrt{3}}{2}$ and the minimum value is $f(\frac{5\pi}{6}) = f(\frac{11\pi}{6}) = -\frac{3\sqrt{3}}{2}$.

(e) The graph of $f(\theta) = 2\sin 2\theta + \sin 4\theta$ is shown here:

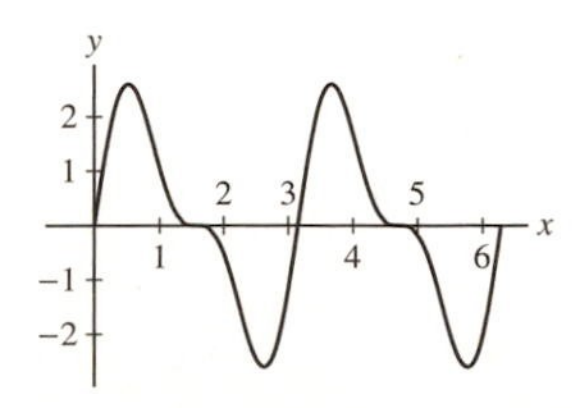

We can see that there are six flat points on the graph between 0 and 2π, as predicted. There are 4 local extrema, and two points at $(\frac{\pi}{2}, 0)$ and $(\frac{3\pi}{2}, 0)$ where the graph has neither a local maximum nor a local minimum.

In Exercises 61–64, find the critical points and the extreme values on [0, 4]. *In Exercises 63 and 64, refer to Figure 18.*

$y = |x^2 + 4x - 12|$ $\qquad$ $y = |\cos x|$

FIGURE 18

61. $y = |x - 2|$

SOLUTION Let $f(x) = |x - 2|$. For $x < 2$, we have $f'(x) = -1$. For $x > 2$, we have $f'(x) = 1$. Now as $x \to 2-$, we have $\dfrac{f(x) - f(2)}{x - 2} = \dfrac{(2 - x) - 0}{x - 2} \to -1$; whereas as $x \to 2+$, we have $\dfrac{f(x) - f(2)}{x - 2} = \dfrac{(x - 2) - 0}{x - 2} \to 1$. Therefore, $f'(2) = \lim\limits_{x \to 2} \dfrac{f(x) - f(2)}{x - 2}$ does not exist and the lone critical point of f is $x = 2$. Alternately, we examine the graph of $f(x) = |x - 2|$ shown below.

To find the extremum, we check the values of $f(x)$ at the critical point and the endpoints. $f(0) = 2$, $f(4) = 2$, and $f(2) = 0$. $f(x)$ takes its minimum value of 0 at $x = 2$, and its maximum of 2 at $x = 0$ and at $x = 4$.

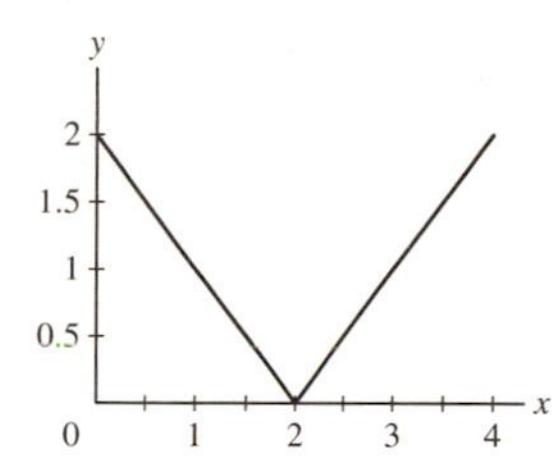

63. $y = |x^2 + 4x - 12|$

SOLUTION Let $f(x) = |x^2 + 4x - 12| = |(x + 6)(x - 2)|$. From the graph of f in Figure 18, we see that $f'(x)$ does not exist at $x = -6$ and at $x = 2$, so these are critical points of f. There is also a critical point between $x = -6$ and $x = 2$ at which $f'(x) = 0$. For $-6 < x < 2$, $f(x) = -x^2 - 4x + 12$, so $f'(x) = -2x - 4 = 0$ when $x = -2$. On the interval [0, 4] the minimum value of f is $f(2) = 0$ and the maximum value is $f(4) = 20$. (*Note:* $f(0) = 12$ and the critical points $x = -6$ and $x = -2$ are not in the interval.)

In Exercises 65–68, verify Rolle's Theorem for the given interval.

65. $f(x) = x + x^{-1}$, $\left[\frac{1}{2}, 2\right]$

SOLUTION Because f is continuous on $[\frac{1}{2}, 2]$, differentiable on $(\frac{1}{2}, 2)$ and

$$f\left(\frac{1}{2}\right) = \frac{1}{2} + \frac{1}{\frac{1}{2}} = \frac{5}{2} = 2 + \frac{1}{2} = f(2),$$

we may conclude from Rolle's Theorem that there exists a $c \in (\frac{1}{2}, 2)$ at which $f'(c) = 0$. Here, $f'(x) = 1 - x^{-2} = \frac{x^2 - 1}{x^2}$, so we may take $c = 1$.

67. $f(x) = \dfrac{x^2}{8x - 15}$, $\quad [3, 5]$

SOLUTION Because f is continuous on [3, 5], differentiable on (3, 5) and $f(3) = f(5) = 1$, we may conclude from Rolle's Theorem that there exists a $c \in (3, 5)$ at which $f'(c) = 0$. Here,

$$f'(x) = \frac{(8x - 15)(2x) - 8x^2}{(8x - 15)^2} = \frac{2x(4x - 15)}{(8x - 15)^2},$$

so we may take $c = \frac{15}{4}$.

69. Prove that $f(x) = x^5 + 2x^3 + 4x - 12$ has precisely one real root.

SOLUTION Let's first establish the $f(x) = x^5 + 2x^3 + 4x - 12$ has at least one root. Because f is a polynomial, it is continuous for all x. Moreover, $f(0) = -12 < 0$ and $f(2) = 44 > 0$. Therefore, by the Intermediate Value Theorem, there exists a $c \in (0, 2)$ such that $f(c) = 0$.

Next, we prove that this is the only root. We will use proof by contradiction. Suppose $f(x) = x^5 + 2x^3 + 4x - 12$ has two real roots, $x = a$ and $x = b$. Then $f(a) = f(b) = 0$ and Rolle's Theorem guarantees that there exists a $c \in (a, b)$ at which $f'(c) = 0$. However, $f'(x) = 5x^4 + 6x^2 + 4 \geq 4$ for all x, so there is no $c \in (a, b)$ at which $f'(c) = 0$. Based on this contradiction, we conclude that $f(x) = x^5 + 2x^3 + 4x - 12$ cannot have more than one real root. Finally, f must have precisely one real root.

71. Prove that $f(x) = x^4 + 5x^3 + 4x$ has no root c satisfying $c > 0$. *Hint:* Note that $x = 0$ is a root and apply Rolle's Theorem.

SOLUTION We will proceed by contradiction. Note that $f(0) = 0$ and suppose that there exists a $c > 0$ such that $f(c) = 0$. Then $f(0) = f(c) = 0$ and Rolle's Theorem guarantees that there exists a $d \in (0, c)$ such that $f'(d) = 0$. However, $f'(x) = 4x^3 + 15x^2 + 4 > 4$ for all $x > 0$, so there is no $d \in (0, c)$ such that $f'(d) = 0$. Based on this contradiction, we conclude that $f(x) = x^4 + 5x^3 + 4x$ has no root c satisfying $c > 0$.

73. The position of a mass oscillating at the end of a spring is $s(t) = A \sin \omega t$, where A is the amplitude and ω is the angular frequency. Show that the speed $|v(t)|$ is at a maximum when the acceleration $a(t)$ is zero and that $|a(t)|$ is at a maximum when $v(t)$ is zero.

SOLUTION Let $s(t) = A \sin \omega t$. Then

$$v(t) = \frac{ds}{dt} = A\omega \cos \omega t$$

and

$$a(t) = \frac{dv}{dt} = -A\omega^2 \sin \omega t.$$

Thus, the speed

$$|v(t)| = |A\omega \cos \omega t|$$

is a maximum when $|\cos \omega t| = 1$, which is precisely when $\sin \omega t = 0$; that is, the speed $|v(t)|$ is at a maximum when the acceleration $a(t)$ is zero. Similarly,

$$|a(t)| = |A\omega^2 \sin \omega t|$$

is a maximum when $|\sin \omega t| = 1$, which is precisely when $\cos \omega t = 0$; that is, $|a(t)|$ is at a maximum when $v(t)$ is zero.

75. CAS **Antibiotic Levels** A study shows that the concentration $C(t)$ (in micrograms per milliliter) of antibiotic in a patient's blood serum after t hours is $C(t) = 120(e^{-0.2t} - e^{-bt})$, where $b \geq 1$ is a constant that depends on the particular combination of antibiotic agents used. Solve numerically for the value of b (to two decimal places) for which maximum concentration occurs at $t = 1$ h. You may assume that the maximum occurs at a critical point as suggested by Figure 19.

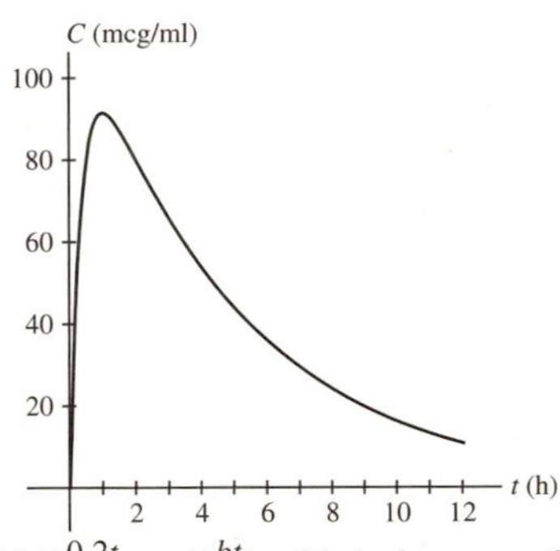

FIGURE 19 Graph of $C(t) = 120(e^{-0.2t} - e^{-bt})$ with b chosen so that the maximum occurs at $t = 1$ h.

SOLUTION Answer is $b = 2.86$. The max of $C(t)$ occurs at $t = \ln(5b)/(b - 0.2)$ so we solve $\ln(5b)/(b - 0.1) = 1$ numerically.

Let $C(t) = 120(e^{-0.2t} - e^{-bt})$. Then $C'(t) = 120(-0.2e^{-0.2t} + be^{-bt}) = 0$ when

$$t = \frac{\ln 5b}{b - 0.2}.$$

Substituting $t = 1$ and solving for b numerically yields $b \approx 2.86$.

77. In 1919, physicist Alfred Betz argued that the maximum efficiency of a wind turbine is around 59%. If wind enters a turbine with speed v_1 and exits with speed v_2, then the power extracted is the difference in kinetic energy per unit time:

$$P = \frac{1}{2}mv_1^2 - \frac{1}{2}mv_2^2 \quad \text{watts}$$

where m is the mass of wind flowing through the rotor per unit time (Figure 20). Betz assumed that $m = \rho A(v_1 + v_2)/2$, where ρ is the density of air and A is the area swept out by the rotor. Wind flowing undisturbed through the same area A would have mass per unit time $\rho A v_1$ and power $P_0 = \frac{1}{2}\rho A v_1^3$. The fraction of power extracted by the turbine is $F = P/P_0$.

(a) Show that F depends only on the ratio $r = v_2/v_1$ and is equal to $F(r) = \frac{1}{2}(1 - r^2)(1 + r)$, where $0 \le r \le 1$.

(b) Show that the maximum value of $F(r)$, called the **Betz Limit**, is $16/27 \approx 0.59$.

(c) Explain why Betz's formula for $F(r)$ is not meaningful for r close to zero. *Hint:* How much wind would pass through the turbine if v_2 were zero? Is this realistic?

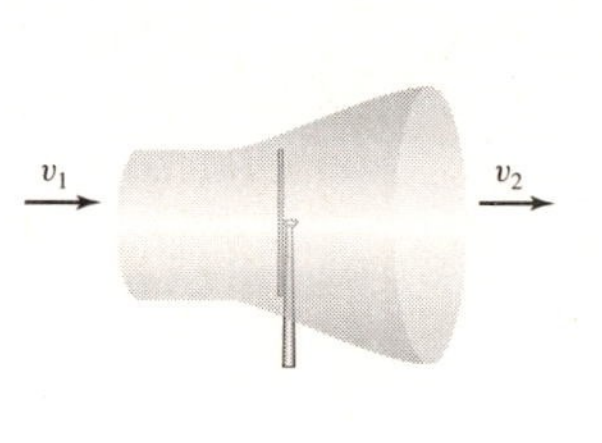

(A) Wind flowing through a turbine.

(B) F is the fraction of energy extracted by the turbine as a function of $r = v_2/v_1$.

FIGURE 20

SOLUTION

(a) We note that

$$F = \frac{P}{P_0} = \frac{\frac{1}{2}\frac{\rho A(v_1+v_2)}{2}(v_1^2 - v_2^2)}{\frac{1}{2}\rho A v_1^3}$$

$$= \frac{1}{2}\frac{v_1^2 - v_2^2}{v_1^2} \cdot \frac{v_1 + v_2}{v_1}$$

$$= \frac{1}{2}\left(1 - \frac{v_2^2}{v_1^2}\right)\left(1 + \frac{v_2}{v_1}\right)$$

$$= \frac{1}{2}(1 - r^2)(1 + r).$$

(b) Based on part (a),

$$F'(r) = \frac{1}{2}(1 - r^2) - r(1 + r) = -\frac{3}{2}r^2 - r + \frac{1}{2}.$$

The roots of this quadratic are $r = -1$ and $r = \frac{1}{3}$. Now, $F(0) = \frac{1}{2}$, $F(1) = 0$ and

$$F\left(\frac{1}{3}\right) = \frac{1}{2} \cdot \frac{8}{9} \cdot \frac{4}{3} = \frac{16}{27} \approx 0.59.$$

Thus, the Betz Limit is $16/27 \approx 0.59$.

(c) If v_2 were 0, then no air would be passing through the turbine, which is not realistic.

79. The response of a circuit or other oscillatory system to an input of frequency ω ("omega") is described by the function

$$\phi(\omega) = \frac{1}{\sqrt{(\omega_0^2 - \omega^2)^2 + 4D^2\omega^2}}$$

Both ω_0 (the natural frequency of the system) and D (the damping factor) are positive constants. The graph of ϕ is called a **resonance curve**, and the positive frequency $\omega_r > 0$, where ϕ takes its maximum value, if it exists, is called the **resonant frequency**. Show that $\omega_r = \sqrt{\omega_0^2 - 2D^2}$ if $0 < D < \omega_0/\sqrt{2}$ and that no resonant frequency exists otherwise (Figure 22).

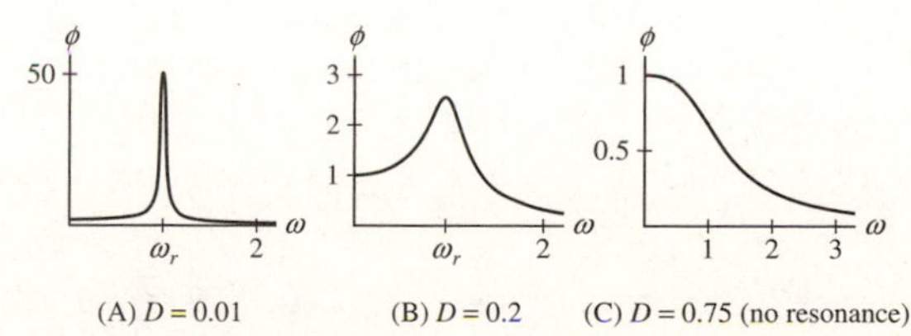

(A) $D = 0.01$ (B) $D = 0.2$ (C) $D = 0.75$ (no resonance)

FIGURE 22 Resonance curves with $\omega_0 = 1$.

SOLUTION Let $\phi(\omega) = ((\omega_0^2 - \omega^2)^2 + 4D^2\omega^2)^{-1/2}$. Then

$$\phi'(\omega) = \frac{2\omega((\omega_0^2 - \omega^2) - 2D^2)}{((\omega_0^2 - \omega^2)^2 + 4D^2\omega^2)^{3/2}}$$

and the non-negative critical points are $\omega = 0$ and $\omega = \sqrt{\omega_0^2 - 2D^2}$. The latter critical point is positive if and only if $\omega_0^2 - 2D^2 > 0$, and since we are given $D > 0$, this is equivalent to $0 < D < \omega_0/\sqrt{2}$.

Define $\omega_r = \sqrt{\omega_0^2 - 2D^2}$. Now, $\phi(0) = 1/\omega_0^2$ and $\phi(\omega) \to 0$ as $\omega \to \infty$. Finally,

$$\phi(\omega_r) = \frac{1}{2D\sqrt{\omega_0^2 - D^2}},$$

which, for $0 < D < \omega_0/\sqrt{2}$, is larger than $1/\omega_0^2$. Hence, the point $\omega = \sqrt{\omega_0^2 - 2D^2}$, if defined, is a local maximum.

81. Find the maximum of $y = x^a - x^b$ on $[0, 1]$ where $0 < a < b$. In particular, find the maximum of $y = x^5 - x^{10}$ on $[0, 1]$.

SOLUTION

- Let $f(x) = x^a - x^b$. Then $f'(x) = ax^{a-1} - bx^{b-1}$. Since $a < b$, $f'(x) = x^{a-1}(a - bx^{b-a}) = 0$ implies critical points $x = 0$ and $x = (\frac{a}{b})^{1/(b-a)}$, which is in the interval $[0, 1]$ as $a < b$ implies $\frac{a}{b} < 1$ and consequently $x = (\frac{a}{b})^{1/(b-a)} < 1$. Also, $f(0) = f(1) = 0$ and $a < b$ implies $x^a > x^b$ on the interval $[0, 1]$, which gives $f(x) > 0$ and thus the maximum value of f on $[0, 1]$ is

$$f\left(\left(\frac{a}{b}\right)^{1/(b-a)}\right) = \left(\frac{a}{b}\right)^{a/(b-a)} - \left(\frac{a}{b}\right)^{b/(b-a)}.$$

- Let $f(x) = x^5 - x^{10}$. Then by part (a), the maximum value of f on $[0, 1]$ is

$$f\left(\left(\frac{1}{2}\right)^{1/5}\right) = \left(\frac{1}{2}\right) - \left(\frac{1}{2}\right)^2 = \frac{1}{2} - \frac{1}{4} = \frac{1}{4}.$$

In Exercises 82–84, plot the function using a graphing utility and find its critical points and extreme values on $[-5, 5]$.

83. GU $y = \dfrac{1}{1 + |x - 1|} + \dfrac{1}{1 + |x - 4|}$

SOLUTION Let

$$f(x) = \frac{1}{1 + |x - 1|} + \frac{1}{1 + |x - 4|}.$$

The plot follows:

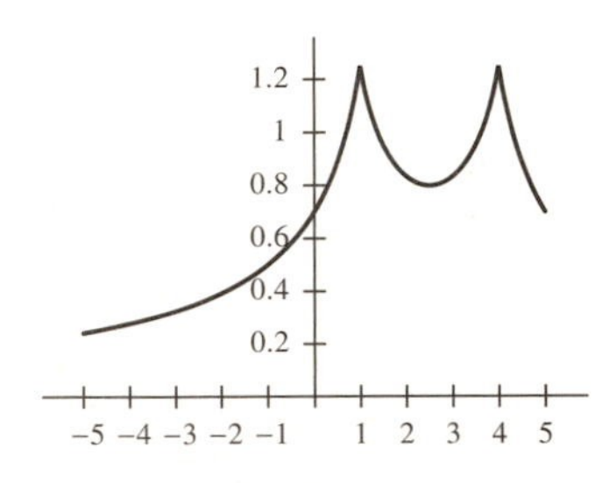

We can see on the plot that the critical points of $f(x)$ lie at the cusps at $x = 1$ and $x = 4$ and at the location of the horizontal tangent line at $x = \frac{5}{2}$. With $f(-5) = \frac{17}{70}$, $f(1) = f(4) = \frac{5}{4}$, $f(\frac{5}{2}) = \frac{4}{5}$ and $f(5) = \frac{7}{10}$, it follows that the maximum value of $f(x)$ on $[-5, 5]$ is $f(1) = f(4) = \frac{5}{4}$ and the minimum value is $f(-5) = \frac{17}{70}$.

85. **(a)** Use implicit differentiation to find the critical points on the curve $27x^2 = (x^2 + y^2)^3$.
(b) GU Plot the curve and the horizontal tangent lines on the same set of axes.

SOLUTION

(a) Differentiating both sides of the equation $27x^2 = (x^2 + y^2)^3$ with respect to x yields

$$54x = 3(x^2 + y^2)^2\left(2x + 2y\frac{dy}{dx}\right).$$

Solving for dy/dx we obtain

$$\frac{dy}{dx} = \frac{27x - 3x(x^2+y^2)^2}{3y(x^2+y^2)^2} = \frac{x(9-(x^2+y^2)^2)}{y(x^2+y^2)^2}.$$

Thus, the derivative is zero when $x^2 + y^2 = 3$. Substituting into the equation for the curve, this yields $x^2 = 1$, or $x = \pm 1$. There are therefore four points at which the derivative is zero:

$$(-1, -\sqrt{2}), (-1, \sqrt{2}), (1, -\sqrt{2}), (1, \sqrt{2}).$$

There are also critical points where the derivative does not exist. This occurs when $y = 0$ and gives the following points with vertical tangents:

$$(0, 0), (\pm\sqrt[4]{27}, 0).$$

(b) The curve $27x^2 = (x^2 + y^2)^3$ and its horizontal tangents are plotted below.

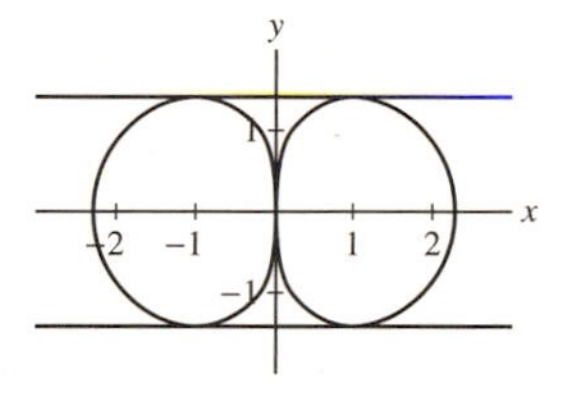

87. Sketch the graph of a continuous function on $(0, 4)$ having a local minimum but no absolute minimum.

SOLUTION Here is the graph of a function f on $(0, 4)$ with a local minimum value [between $x = 2$ and $x = 4$] but no absolute minimum [since $f(x) \to -\infty$ as $x \to 0+$].

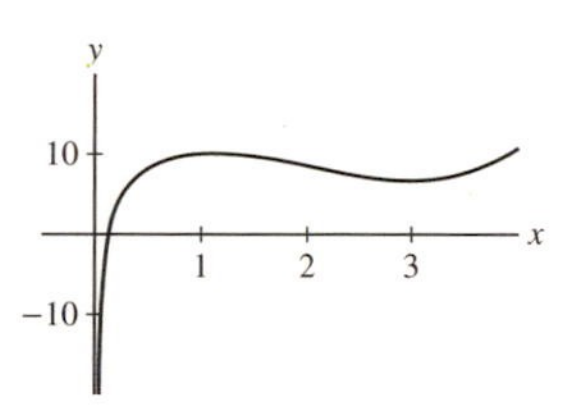

89. Sketch the graph of a function $f(x)$ on $[0, 4]$ with a discontinuity such that $f(x)$ has an absolute minimum but no absolute maximum.

SOLUTION Here is the graph of a function f on $[0, 4]$ that (a) has a discontinuity [at $x = 4$] and (b) has an absolute minimum [at $x = 0$] but no absolute maximum [since $f(x) \to \infty$ as $x \to 4-$].

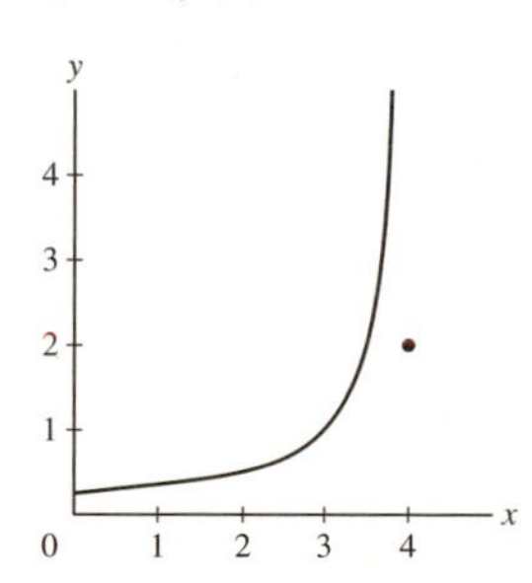

Further Insights and Challenges

91. Show that the extreme values of $f(x) = a \sin x + b \cos x$ are $\pm\sqrt{a^2 + b^2}$.

SOLUTION If $f(x) = a \sin x + b \cos x$, then $f'(x) = a \cos x - b \sin x$, so that $f'(x) = 0$ implies $a \cos x - b \sin x = 0$. This implies $\tan x = \frac{a}{b}$. Then,

$$\sin x = \frac{\pm a}{\sqrt{a^2+b^2}} \quad \text{and} \quad \cos x = \frac{\pm b}{\sqrt{a^2+b^2}}.$$

Therefore

$$f(x) = a \sin x + b \cos x = a\frac{\pm a}{\sqrt{a^2+b^2}} + b\frac{\pm b}{\sqrt{a^2+b^2}} = \pm\frac{a^2+b^2}{\sqrt{a^2+b^2}} = \pm\sqrt{a^2+b^2}.$$

93. Show that if the quadratic polynomial $f(x) = x^2 + rx + s$ takes on both positive and negative values, then its minimum value occurs at the midpoint between the two roots.

SOLUTION Let $f(x) = x^2 + rx + s$ and suppose that $f(x)$ takes on both positive and negative values. This will guarantee that f has two real roots. By the quadratic formula, the roots of f are

$$x = \frac{-r \pm \sqrt{r^2 - 4s}}{2}.$$

Observe that the midpoint between these roots is

$$\frac{1}{2}\left(\frac{-r + \sqrt{r^2 - 4s}}{2} + \frac{-r - \sqrt{r^2 - 4s}}{2}\right) = -\frac{r}{2}.$$

Next, $f'(x) = 2x + r = 0$ when $x = -\frac{r}{2}$ and, because the graph of $f(x)$ is an upward opening parabola, it follows that $f(-\frac{r}{2})$ is a minimum. Thus, f takes on its minimum value at the midpoint between the two roots.

95. A cubic polynomial may have a local min and max, or it may have neither (Figure 26). Find conditions on the coefficients a and b of

$$f(x) = \frac{1}{3}x^3 + \frac{1}{2}ax^2 + bx + c$$

that ensure that f has neither a local min nor a local max. *Hint:* Apply Exercise 92 to $f'(x)$.

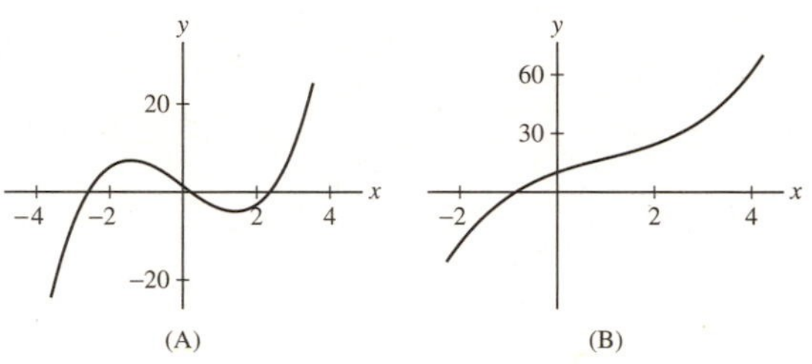

FIGURE 26 Cubic polynomials

SOLUTION Let $f(x) = \frac{1}{3}x^3 + \frac{1}{2}ax^2 + bx + c$. Using Exercise 92, we have $g(x) = f'(x) = x^2 + ax + b > 0$ for all x provided $b > \frac{1}{4}a^2$, in which case f has no critical points and hence no local extrema. (Actually $b \geq \frac{1}{4}a^2$ will suffice, since in this case [as we'll see in a later section] f has an inflection point but no local extrema.)

97. Prove that if f is continuous and $f(a)$ and $f(b)$ are local minima where $a < b$, then there exists a value c between a and b such that $f(c)$ is a local maximum. (*Hint:* Apply Theorem 1 to the interval $[a, b]$.) Show that continuity is a necessary hypothesis by sketching the graph of a function (necessarily discontinuous) with two local minima but no local maximum.

SOLUTION

- Let $f(x)$ be a continuous function with $f(a)$ and $f(b)$ local minima on the interval $[a, b]$. By Theorem 1, $f(x)$ must take on both a minimum and a maximum on $[a, b]$. Since local minima occur at $f(a)$ and $f(b)$, the maximum must occur at some other point in the interval, call it c, where $f(c)$ is a local maximum.
- The function graphed here is discontinuous at $x = 0$.

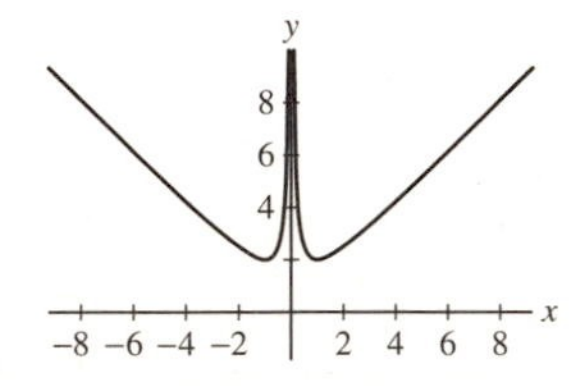

4.3 The Mean Value Theorem and Monotonicity

Preliminary Questions

1. For which value of m is the following statement correct? If $f(2) = 3$ and $f(4) = 9$, and $f(x)$ is differentiable, then f has a tangent line of slope m.

SOLUTION The Mean Value Theorem guarantees that the function has a tangent line with slope equal to

$$\frac{f(4) - f(2)}{4 - 2} = \frac{9 - 3}{4 - 2} = 3.$$

Hence, $m = 3$ makes the statement correct.

2. Assume f is differentiable. Which of the following statements does *not* follow from the MVT?
(a) If f has a secant line of slope 0, then f has a tangent line of slope 0.
(b) If $f(5) < f(9)$, then $f'(c) > 0$ for some $c \in (5, 9)$.
(c) If f has a tangent line of slope 0, then f has a secant line of slope 0.
(d) If $f'(x) > 0$ for all x, then every secant line has positive slope.

SOLUTION Conclusion **(c)** does not follow from the Mean Value Theorem. As a counterexample, consider the function $f(x) = x^3$. Note that $f'(0) = 0$, but no secant line has zero slope.

3. Can a function that takes on only negative values have a positive derivative? If so, sketch an example.

SOLUTION Yes. The figure below displays a function that takes on only negative values but has a positive derivative.

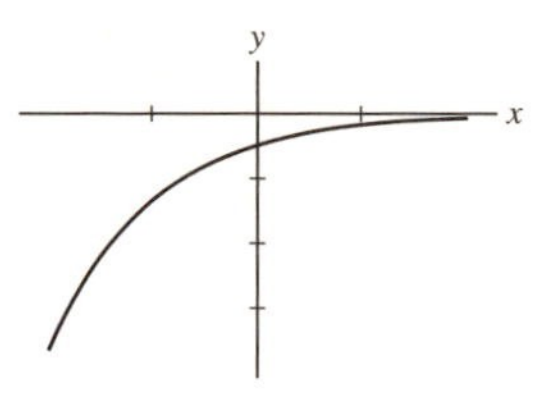

4. For $f(x)$ with derivative as in Figure 12:
(a) Is $f(c)$ a local minimum or maximum?
(b) Is $f(x)$ a decreasing function?

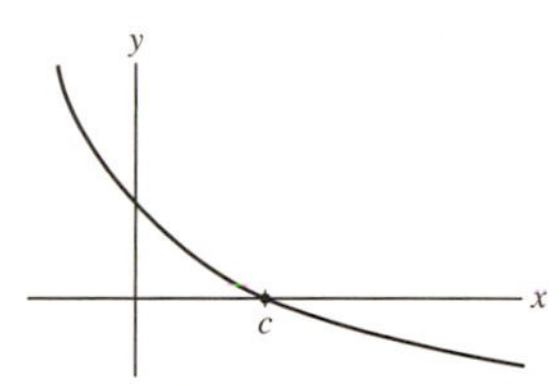

FIGURE 12 Graph of derivative $f'(x)$.

SOLUTION
(a) To the left of $x = c$, the derivative is positive, so f is increasing; to the right of $x = c$, the derivative is negative, so f is decreasing. Consequently, $f(c)$ must be a local maximum.
(b) No. The derivative is a decreasing function, but as noted in part (a), $f(x)$ is increasing for $x < c$ and decreasing for $x > c$.

Exercises

In Exercises 1–8, find a point c satisfying the conclusion of the MVT for the given function and interval.

1. $y = x^{-1}$, $[2, 8]$

SOLUTION Let $f(x) = x^{-1}$, $a = 2$, $b = 8$. Then $f'(x) = -x^{-2}$, and by the MVT, there exists a $c \in (2, 8)$ such that

$$-\frac{1}{c^2} = f'(c) = \frac{f(b) - f(a)}{b - a} = \frac{\frac{1}{8} - \frac{1}{2}}{8 - 2} = -\frac{1}{16}.$$

Thus $c^2 = 16$ and $c = \pm 4$. Choose $c = 4 \in (2, 8)$.

3. $y = \cos x - \sin x$, $[0, 2\pi]$

SOLUTION Let $f(x) = \cos x - \sin x$, $a = 0$, $b = 2\pi$. Then $f'(x) = -\sin x - \cos x$, and by the MVT, there exists a $c \in (0, 2\pi)$ such that

$$-\sin c - \cos c = f'(c) = \frac{f(b) - f(a)}{b - a} = \frac{1 - 1}{2\pi - 0} = 0.$$

Thus $-\sin c = \cos c$. Choose either $c = \frac{3\pi}{4}$ or $c = \frac{7\pi}{4} \in (0, 2\pi)$.

5. $y = x^3$, $[-4, 5]$

SOLUTION Let $f(x) = x^3$, $a = -4$, $b = 5$. Then $f'(x) = 3x^2$, and by the MVT, there exists a $c \in (-4, 5)$ such that

$$3c^2 = f'(c) = \frac{f(b) - f(a)}{b - a} = \frac{189}{9} = 21.$$

Solving for c yields $c^2 = 7$, so $c = \pm\sqrt{7}$. Both of these values are in the interval $[-4, 5]$, so either value can be chosen.

7. $y = e^{-2x}$, $[0, 3]$

SOLUTION Let $f(x) = e^{-2x}$, $a = 0$, $b = 3$. Then $f'(x) = -2e^{-2x}$, and by the MVT, there exists a $c \in (0, 3)$ such that

$$-2e^{-2c} = f'(c) = \frac{f(b) - f(a)}{b - a} = \frac{e^{-6} - 1}{3 - 0} = \frac{e^{-6} - 1}{3}.$$

Solving for c yields

$$c = -\frac{1}{2} \ln\left(\frac{1 - e^{-6}}{6}\right) \approx 0.8971 \in (0, 3).$$

9. GU Let $f(x) = x^5 + x^2$. The secant line between $x = 0$ and $x = 1$ has slope 2 (check this), so by the MVT, $f'(c) = 2$ for some $c \in (0, 1)$. Plot $f(x)$ and the secant line on the same axes. Then plot $y = 2x + b$ for different values of b until the line becomes tangent to the graph of f. Zoom in on the point of tangency to estimate x-coordinate c of the point of tangency.

SOLUTION Let $f(x) = x^5 + x^2$. The slope of the secant line between $x = 0$ and $x = 1$ is

$$\frac{f(1) - f(0)}{1 - 0} = \frac{2 - 0}{1} = 2.$$

A plot of $f(x)$, the secant line between $x = 0$ and $x = 1$, and the line $y = 2x - 0.764$ is shown below at the left. The line $y = 2x - 0.764$ appears to be tangent to the graph of $y = f(x)$. Zooming in on the point of tangency (see below at the right), it appears that the x-coordinate of the point of tangency is approximately 0.62.

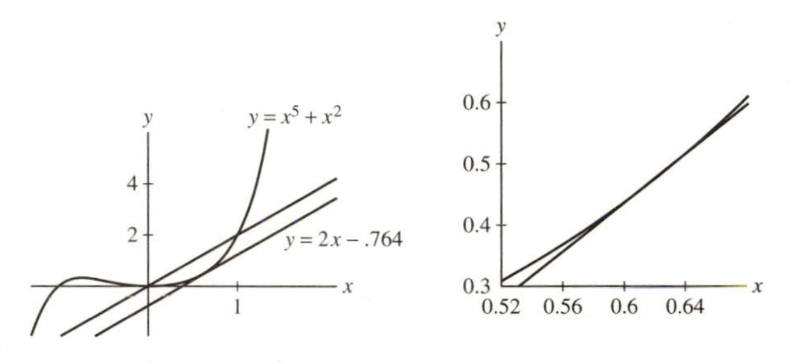

11. Determine the intervals on which $f'(x)$ is positive and negative, assuming that Figure 13 is the graph of $f(x)$.

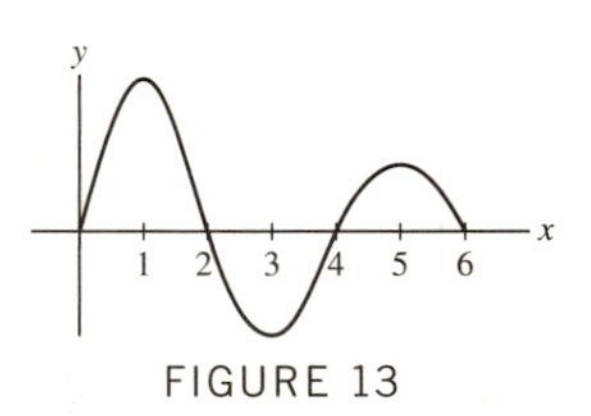

FIGURE 13

SOLUTION The derivative of f is positive on the intervals $(0, 1)$ and $(3, 5)$ where f is increasing; it is negative on the intervals $(1, 3)$ and $(5, 6)$ where f is decreasing.

13. State whether $f(2)$ and $f(4)$ are local minima or local maxima, assuming that Figure 13 is the graph of $f'(x)$.

SOLUTION

- $f'(x)$ makes a transition from positive to negative at $x = 2$, so $f(2)$ is a local maximum.
- $f'(x)$ makes a transition from negative to positive at $x = 4$, so $f(4)$ is a local minimum.

In Exercises 15–18, sketch the graph of a function $f(x)$ whose derivative $f'(x)$ has the given description.

15. $f'(x) > 0$ for $x > 3$ and $f'(x) < 0$ for $x < 3$

SOLUTION Here is the graph of a function f for which $f'(x) > 0$ for $x > 3$ and $f'(x) < 0$ for $x < 3$.

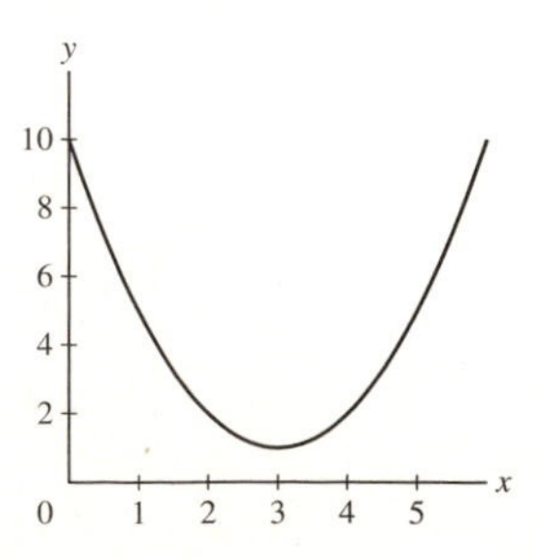

17. $f'(x)$ is negative on $(1, 3)$ and positive everywhere else.

SOLUTION Here is the graph of a function f for which $f'(x)$ is negative on $(1, 3)$ and positive elsewhere.

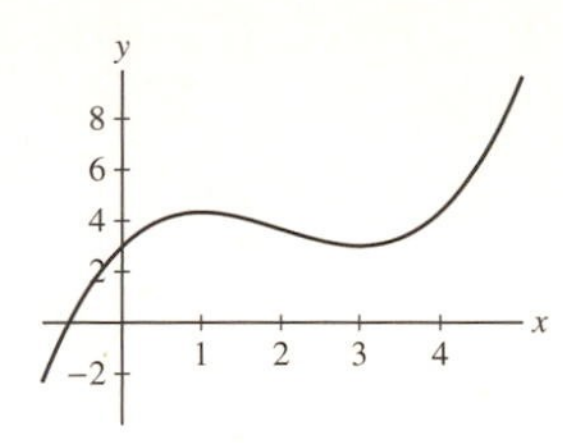

In Exercises 19–22, find all critical points of f and use the First Derivative Test to determine whether they are local minima or maxima.

19. $f(x) = 4 + 6x - x^2$

SOLUTION Let $f(x) = 4 + 6x - x^2$. Then $f'(x) = 6 - 2x = 0$ implies that $x = 3$ is the only critical point of f. As x increases through 3, $f'(x)$ makes the sign transition $+, -$. Therefore, $f(3) = 13$ is a local maximum.

21. $f(x) = \dfrac{x^2}{x+1}$

SOLUTION Let $f(x) = \dfrac{x^2}{x+1}$. Then

$$f'(x) = \frac{x(x+2)}{(x+1)^2} = 0$$

implies that $x = 0$ and $x = -2$ are critical points. Note that $x = -1$ is not a critical point because it is not in the domain of f. As x increases through -2, $f'(x)$ makes the sign transition $+, -$ so $f(-2) = -4$ is a local maximum. As x increases through 0, $f'(x)$ makes the sign transition $-, +$ so $f(0) = 0$ is a local minimum.

In Exercises 23–52, find the critical points and the intervals on which the function is increasing or decreasing. Use the First Derivative Test to determine whether the critical point is a local min or max (or neither).

SOLUTION *Here is a table legend for Exercises 23–44.*

SYMBOL	*MEANING*
$-$	*The entity is negative on the given interval.*
0	*The entity is zero at the specified point.*
$+$	*The entity is positive on the given interval.*
U	*The entity is undefined at the specified point.*
↗	*f is increasing on the given interval.*
↘	*f is decreasing on the given interval.*
M	*f has a local maximum at the specified point.*
m	*f has a local minimum at the specified point.*
¬	*There is no local extremum here.*

23. $y = -x^2 + 7x - 17$

SOLUTION Let $f(x) = -x^2 + 7x - 17$. Then $f'(x) = 7 - 2x = 0$ yields the critical point $c = \frac{7}{2}$.

x	$\left(-\infty, \frac{7}{2}\right)$	7/2	$\left(\frac{7}{2}, \infty\right)$
f'	$+$	0	$-$
f	↗	M	↘

25. $y = x^3 - 12x^2$

SOLUTION Let $f(x) = x^3 - 12x^2$. Then $f'(x) = 3x^2 - 24x = 3x(x - 8) = 0$ yields critical points $c = 0, 8$.

x	$(-\infty, 0)$	0	$(0, 8)$	8	$(8, \infty)$
f'	$+$	0	$-$	0	$+$
f	$\nearrow$	M	$\searrow$	m	$\nearrow$

27. $y = 3x^4 + 8x^3 - 6x^2 - 24x$

SOLUTION Let $f(x) = 3x^4 + 8x^3 - 6x^2 - 24x$. Then

$$\begin{aligned} f'(x) &= 12x^3 + 24x^2 - 12x - 24 \\ &= 12x^2(x + 2) - 12(x + 2) = 12(x + 2)(x^2 - 1) \\ &= 12\,(x - 1)\,(x + 1)\,(x + 2) = 0 \end{aligned}$$

yields critical points $c = -2, -1, 1$.

x	$(-\infty, -2)$	-2	$(-2, -1)$	-1	$(-1, 1)$	1	$(1, \infty)$
f'	$-$	0	$+$	0	$-$	0	$+$
f	$\searrow$	m	$\nearrow$	M	$\searrow$	m	$\nearrow$

29. $y = \frac{1}{3}x^3 + \frac{3}{2}x^2 + 2x + 4$

SOLUTION Let $f(x) = \frac{1}{3}x^3 + \frac{3}{2}x^2 + 2x + 4$. Then $f'(x) = x^2 + 3x + 2 = (x + 1)\,(x + 2) = 0$ yields critical points $c = -2, -1$.

x	$(-\infty, -2)$	-2	$(-2, -1)$	-1	$(-1, \infty)$
f'	$+$	0	$-$	0	$+$
f	$\nearrow$	M	$\searrow$	m	$\nearrow$

31. $y = x^5 + x^3 + 1$

SOLUTION Let $f(x) = x^5 + x^3 + 1$. Then $f'(x) = 5x^4 + 3x^2 = x^2(5x^2 + 3)$ yields a single critical point: $c = 0$.

x	$(-\infty, 0)$	0	$(0, \infty)$
f'	$+$	0	$+$
f	$\nearrow$	$\neg$	$\nearrow$

33. $y = x^4 - 4x^{3/2}$ $(x > 0)$

SOLUTION Let $f(x) = x^4 - 4x^{3/2}$ for $x > 0$. Then $f'(x) = 4x^3 - 6x^{1/2} = 2x^{1/2}(2x^{5/2} - 3) = 0$, which gives us the critical point $c = (\frac{3}{2})^{2/5}$. (*Note:* $c = 0$ is not in the interval under consideration.)

x	$\left(0, (\frac{3}{2})^{2/5}\right)$	$\frac{3}{2}^{2/5}$	$\left((\frac{3}{2})^{2/5}, \infty\right)$
f'	$-$	0	$+$
f	$\searrow$	m	$\nearrow$

35. $y = x + x^{-1}$ $(x > 0)$

SOLUTION Let $f(x) = x + x^{-1}$ for $x > 0$. Then $f'(x) = 1 - x^{-2} = 0$ yields the critical point $c = 1$. (*Note:* $c = -1$ is not in the interval under consideration.)

x	$(0, 1)$	1	$(1, \infty)$
f'	$-$	0	$+$
f	$\searrow$	m	$\nearrow$

37. $y = \dfrac{1}{x^2 + 1}$

SOLUTION Let $f(x) = \left(x^2 + 1\right)^{-1}$. Then $f'(x) = -2x\left(x^2 + 1\right)^{-2} = 0$ yields critical point $c = 0$.

x	$(-\infty, 0)$	0	$(0, \infty)$
f'	$+$	0	$-$
f	↗	M	↘

39. $y = \dfrac{x^3}{x^2 + 1}$

SOLUTION Let $f(x) = \dfrac{x^3}{x^2 + 1}$. Then

$$f'(x) = \frac{(x^2 + 1)(3x^2) - x^3(2x)}{(x^2 + 1)^2} = \frac{x^2(x^2 + 3)}{(x^2 + 1)^2} = 0$$

yields the single critical point $c = 0$.

x	$(-\infty, 0)$	0	$(0, \infty)$
f'	$+$	0	$+$
f	↗	¬	↗

41. $y = \theta + \sin\theta + \cos\theta$

SOLUTION Let $f(\theta) = \theta + \sin\theta + \cos\theta$. Then $f'(\theta) = 1 + \cos\theta - \sin\theta = 0$ yields the critical points $c = \frac{\pi}{2}$ and $c = \pi$.

θ	$\left(0, \frac{\pi}{2}\right)$	$\frac{\pi}{2}$	$\left(\frac{\pi}{2}, \pi\right)$	π	$(\pi, 2\pi)$
f'	$+$	0	$-$	0	$+$
f	↗	M	↘	m	↗

43. $y = \sin^2\theta + \sin\theta$

SOLUTION Let $f(\theta) = \sin^2\theta + \sin\theta$. Then $f'(\theta) = 2\sin\theta\cos\theta + \cos\theta = \cos\theta(2\sin\theta + 1) = 0$ yields the critical points $c = \frac{\pi}{2}, \frac{7\pi}{6}, \frac{3\pi}{2}$, and $\frac{11\pi}{6}$.

θ	$\left(0, \frac{\pi}{2}\right)$	$\frac{\pi}{2}$	$\left(\frac{\pi}{2}, \frac{7\pi}{6}\right)$	$\frac{7\pi}{6}$	$\left(\frac{7\pi}{6}, \frac{3\pi}{2}\right)$	$\frac{3\pi}{2}$	$\left(\frac{3\pi}{2}, \frac{11\pi}{6}\right)$	$\frac{11\pi}{6}$	$\left(\frac{11\pi}{6}, 2\pi\right)$
f'	$+$	0	$-$	0	$+$	0	$-$	0	$+$
f	↗	M	↘	m	↗	M	↘	m	↗

45. $y = x + e^{-x}$

SOLUTION Let $f(x) = x + e^{-x}$. Then $f'(x) = 1 - e^{-x}$, which yields $c = 0$ as the only critical point.

x	$(-\infty, 0)$	0	$(0, \infty)$
f'	$-$	0	$+$
f	↘	m	↗

47. $y = e^{-x}\cos x, \quad \left[-\frac{\pi}{2}, \frac{\pi}{2}\right]$

SOLUTION Let $f(x) = e^{-x}\cos x$. Then

$$f'(x) = -e^{-x}\sin x - e^{-x}\cos x = -e^{-x}(\sin x + \cos x),$$

which yields $c = -\frac{\pi}{4}$ as the only critical point on the interval $[-\frac{\pi}{2}, \frac{\pi}{2}]$.

x	$\left[-\frac{\pi}{2}, -\frac{\pi}{4}\right)$	$-\frac{\pi}{4}$	$\left(-\frac{\pi}{4}, \frac{\pi}{2}\right]$
f'	$+$	0	$-$
f	↗	M	↘

49. $y = \tan^{-1} x - \frac{1}{2}x$

SOLUTION Let $f(x) = \tan^{-1} x - \frac{1}{2}x$. Then

$$f'(x) = \frac{1}{1+x^2} - \frac{1}{2},$$

which yields $c = \pm 1$ as critical points.

x	$(-\infty, -1)$	-1	$(-1, 1)$	1	$(1, \infty)$
f'	$-$	0	$+$	0	$-$
f	↘	m	↗	M	↘

51. $y = x - \ln x \quad (x > 0)$

SOLUTION Let $f(x) = x - \ln x$. Then $f'(x) = 1 - x^{-1}$, which yields $c = 1$ as the only critical point.

x	$(0, 1)$	1	$(1, \infty)$
f'	$-$	0	$+$
f	↘	m	↗

53. Find the minimum value of $f(x) = x^x$ for $x > 0$.

SOLUTION Let $f(x) = x^x$. By logarithmic differentiation, we know that $f'(x) = x^x(1 + \ln x)$. Thus, $x = \frac{1}{e}$ is the only critical point. Because $f'(x) < 0$ for $0 < x < \frac{1}{e}$ and $f'(x) > 0$ for $x > \frac{1}{e}$,

$$f\left(\frac{1}{e}\right) = \left(\frac{1}{e}\right)^{1/e} \approx 0.692201$$

is the minimum value.

55. Show that $f(x) = x^3 - 2x^2 + 2x$ is an increasing function. *Hint:* Find the minimum value of $f'(x)$.

SOLUTION Let $f(x) = x^3 - 2x^2 + 2x$. For all x, we have

$$f'(x) = 3x^2 - 4x + 2 = 3\left(x - \frac{2}{3}\right)^2 + \frac{2}{3} \geq \frac{2}{3} > 0.$$

Since $f'(x) > 0$ for all x, the function f is everywhere increasing.

57. GU Let $h(x) = \dfrac{x(x^2-1)}{x^2+1}$ and suppose that $f'(x) = h(x)$. Plot $h(x)$ and use the plot to describe the local extrema and the increasing/decreasing behavior of $f(x)$. Sketch a plausible graph for $f(x)$ itself.

SOLUTION The graph of $h(x)$ is shown below at the left. Because $h(x)$ is negative for $x < -1$ and for $0 < x < 1$, it follows that $f(x)$ is decreasing for $x < -1$ and for $0 < x < 1$. Similarly, $f(x)$ is increasing for $-1 < x < 0$ and for $x > 1$ because $h(x)$ is positive on these intervals. Moreover, $f(x)$ has local minima at $x = -1$ and $x = 1$ and a local maximum at $x = 0$. A plausible graph for $f(x)$ is shown below at the right.

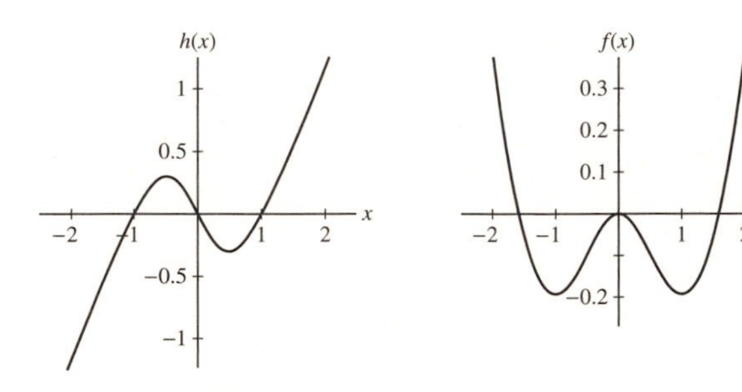

59. Determine where $f(x) = (1000 - x)^2 + x^2$ is decreasing. Use this to decide which is larger: $800^2 + 200^2$ or $600^2 + 400^2$.

SOLUTION If $f(x) = (1000 - x)^2 + x^2$, then $f'(x) = -2(1000 - x) + 2x = 4x - 2000$. $f'(x) < 0$ as long as $x < 500$. Therefore, $800^2 + 200^2 = f(200) > f(400) = 600^2 + 400^2$.

61. Which values of c satisfy the conclusion of the MVT on the interval $[a, b]$ if $f(x)$ is a linear function?

SOLUTION Let $f(x) = px + q$, where p and q are constants. Then the slope of *every* secant line and tangent line of f is p. Accordingly, considering the interval $[a, b]$, *every* point $c \in (a, b)$ satisfies $f'(c) = p = \dfrac{f(b) - f(a)}{b - a}$, the conclusion of the MVT.

63. Suppose that $f(0) = 2$ and $f'(x) \leq 3$ for $x > 0$. Apply the MVT to the interval $[0, 4]$ to prove that $f(4) \leq 14$. Prove more generally that $f(x) \leq 2 + 3x$ for all $x > 0$.

SOLUTION The MVT, applied to the interval $[0, 4]$, guarantees that there exists a $c \in (0, 4)$ such that

$$f'(c) = \frac{f(4) - f(0)}{4 - 0} \qquad \text{or} \qquad f(4) - f(0) = 4f'(c).$$

Because $c > 0$, $f'(c) \leq 3$, so $f(4) - f(0) \leq 12$. Finally, $f(4) \leq f(0) + 12 = 14$.

More generally, let $x > 0$. The MVT, applied to the interval $[0, x]$, guarantees there exists a $c \in (0, x)$ such that

$$f'(c) = \frac{f(x) - f(0)}{x - 0} \qquad \text{or} \qquad f(x) - f(0) = f'(c)x.$$

Because $c > 0$, $f'(c) \leq 3$, so $f(x) - f(0) \leq 3x$. Finally, $f(x) \leq f(0) + 3x = 3x + 2$.

65. Show that if $f(2) = 5$ and $f'(x) \geq 10$ for $x > 2$, then $f(x) \geq 10x - 15$ for all $x > 2$.

SOLUTION Let $x > 2$. The MVT, applied to the interval $[2, x]$, guarantees there exists a $c \in (2, x)$ such that

$$f'(c) = \frac{f(x) - f(2)}{x - 2} \qquad \text{or} \qquad f(x) - f(2) = (x - 2)f'(c).$$

Because $f'(x) \geq 10$, it follows that $f(x) - f(2) \geq 10(x - 2)$, or $f(x) \geq f(2) + 10(x - 2) = 10x - 15$.

Further Insights and Challenges

67. Prove that if $f(0) = g(0)$ and $f'(x) \leq g'(x)$ for $x \geq 0$, then $f(x) \leq g(x)$ for all $x \geq 0$. *Hint:* Show that $f(x) - g(x)$ is nonincreasing.

SOLUTION Let $h(x) = f(x) - g(x)$. By the sum rule, $h'(x) = f'(x) - g'(x)$. Since $f'(x) \leq g'(x)$ for all $x \geq 0$, $h'(x) \leq 0$ for all $x \geq 0$. This implies that h is nonincreasing. Since $h(0) = f(0) - g(0) = 0$, $h(x) \leq 0$ for all $x \geq 0$ (as h is nonincreasing, it cannot climb above zero). Hence $f(x) - g(x) \leq 0$ for all $x \geq 0$, and so $f(x) \leq g(x)$ for $x \geq 0$.

69. Use Exercise 67 and the inequality $\sin x \leq x$ for $x \geq 0$ (established in Theorem 3 of Section 2.6) to prove the following assertions for all $x \geq 0$ (each assertion follows from the previous one).

(a) $\cos x \geq 1 - \frac{1}{2}x^2$

(b) $\sin x \geq x - \frac{1}{6}x^3$

(c) $\cos x \leq 1 - \frac{1}{2}x^2 + \frac{1}{24}x^4$

(d) Can you guess the next inequality in the series?

SOLUTION

(a) We prove this using Exercise 67: Let $g(x) = \cos x$ and $f(x) = 1 - \frac{1}{2}x^2$. Then $f(0) = g(0) = 1$ and $g'(x) = -\sin x \geq -x = f'(x)$ for $x \geq 0$ by Exercise 68. Now apply Exercise 67 to conclude that $\cos x \geq 1 - \frac{1}{2}x^2$ for $x \geq 0$.

(b) Let $g(x) = \sin x$ and $f(x) = x - \frac{1}{6}x^3$. Then $f(0) = g(0) = 0$ and $g'(x) = \cos x \geq 1 - \frac{1}{2}x^2 = f'(x)$ for $x \geq 0$ by part (a). Now apply Exercise 67 to conclude that $\sin x \geq x - \frac{1}{6}x^3$ for $x \geq 0$.

(c) Let $g(x) = 1 - \frac{1}{2}x^2 + \frac{1}{24}x^4$ and $f(x) = \cos x$. Then $f(0) = g(0) = 1$ and $g'(x) = -x + \frac{1}{6}x^3 \geq -\sin x = f'(x)$ for $x \geq 0$ by part (b). Now apply Exercise 67 to conclude that $\cos x \leq 1 - \frac{1}{2}x^2 + \frac{1}{24}x^4$ for $x \geq 0$.

(d) The next inequality in the series is $\sin x \leq x - \frac{1}{6}x^3 + \frac{1}{120}x^5$, valid for $x \geq 0$. To construct (d) from (c), we note that the derivative of $\sin x$ is $\cos x$, and look for a polynomial (which we currently must do by educated guess) whose derivative is $1 - \frac{1}{2}x^2 + \frac{1}{24}x^4$. We know the derivative of x is 1, and that a term whose derivative is $-\frac{1}{2}x^2$ should be of the form Cx^3. $\frac{d}{dx}Cx^3 = 3Cx^2 = -\frac{1}{2}x^2$, so $C = -\frac{1}{6}$. A term whose derivative is $\frac{1}{24}x^4$ should be of the form Dx^5. From this, $\frac{d}{dx}Dx^5 = 5Dx^4 = \frac{1}{24}x^4$, so that $5D = \frac{1}{24}$, or $D = \frac{1}{120}$.

71. Assume that f'' exists and $f''(x) = 0$ for all x. Prove that $f(x) = mx + b$, where $m = f'(0)$ and $b = f(0)$.

SOLUTION

- Let $f''(x) = 0$ for all x. Then $f'(x) =$ constant for all x. Since $f'(0) = m$, we conclude that $f'(x) = m$ for all x.
- Let $g(x) = f(x) - mx$. Then $g'(x) = f'(x) - m = m - m = 0$ which implies that $g(x) =$ constant for all x and consequently $f(x) - mx =$ constant for all x. Rearranging the statement, $f(x) = mx +$ constant. Since $f(0) = b$, we conclude that $f(x) = mx + b$ for all x.

73. Suppose that $f(x)$ satisfies the following equation (an example of a **differential equation**):

$$f''(x) = -f(x) \tag{1}$$

(a) Show that $f(x)^2 + f'(x)^2 = f(0)^2 + f'(0)^2$ for all x. *Hint:* Show that the function on the left has zero derivative.
(b) Verify that $\sin x$ and $\cos x$ satisfy Eq. (1), and deduce that $\sin^2 x + \cos^2 x = 1$.

SOLUTION

(a) Let $g(x) = f(x)^2 + f'(x)^2$. Then

$$g'(x) = 2f(x)f'(x) + 2f'(x)f''(x) = 2f(x)f'(x) + 2f'(x)(-f(x)) = 0,$$

where we have used the fact that $f''(x) = -f(x)$. Because $g'(0) = 0$ for all x, $g(x) = f(x)^2 + f'(x)^2$ must be a constant function. In other words, $f(x)^2 + f'(x)^2 = C$ for some constant C. To determine the value of C, we can substitute any number for x. In particular, for this problem, we want to substitute $x = 0$ and find $C = f(0)^2 + f'(0)^2$. Hence,

$$f(x)^2 + f'(x)^2 = f(0)^2 + f'(0)^2.$$

(b) Let $f(x) = \sin x$. Then $f'(x) = \cos x$ and $f''(x) = -\sin x$, so $f''(x) = -f(x)$. Next, let $f(x) = \cos x$. Then $f'(x) = -\sin x$, $f''(x) = -\cos x$, and we again have $f''(x) = -f(x)$. Finally, if we take $f(x) = \sin x$, the result from part (a) guarantees that

$$\sin^2 x + \cos^2 x = \sin^2 0 + \cos^2 0 = 0 + 1 = 1.$$

75. Use Exercise 74 to prove: $f(x) = \sin x$ is the unique solution of Eq. (1) such that $f(0) = 0$ and $f'(0) = 1$; and $g(x) = \cos x$ is the unique solution such that $g(0) = 1$ and $g'(0) = 0$. This result can be used to develop all the properties of the trigonometric functions "analytically"—that is, without reference to triangles.

SOLUTION In part (b) of Exercise 73, it was shown that $f(x) = \sin x$ satisfies Eq. (1), and we can directly calculate that $f(0) = \sin 0 = 0$ and $f'(0) = \cos 0 = 1$. Suppose there is another function, call it $F(x)$, that satisfies Eq. (1) with the same initial conditions: $F(0) = 0$ and $F'(0) = 1$. By Exercise 74, it follows that $F(x) = \sin x$ for all x. Hence, $f(x) = \sin x$ is the unique solution of Eq. (1) satisfying $f(0) = 0$ and $f'(0) = 1$. The proof that $g(x) = \cos x$ is the unique solution of Eq. (1) satisfying $g(0) = 1$ and $g'(0) = 0$ is carried out in a similar manner.

4.4 The Shape of a Graph

Preliminary Questions

1. If f is concave up, then f' is (choose one):

(a) increasing **(b)** decreasing

SOLUTION The correct response is **(a)**: increasing. If the function is concave up, then f'' is positive. Since f'' is the derivative of f', it follows that the derivative of f' is positive and f' must therefore be increasing.

2. What conclusion can you draw if $f'(c) = 0$ and $f''(c) < 0$?

SOLUTION If $f'(c) = 0$ and $f''(c) < 0$, then $f(c)$ is a local maximum.

3. True or False? If $f(c)$ is a local min, then $f''(c)$ must be positive.

SOLUTION False. $f''(c)$ could be zero.

4. True or False? If $f''(x)$ changes from + to − at $x = c$, then f has a point of inflection at $x = c$.

SOLUTION False. f will have a point of inflection at $x = c$ only if $x = c$ is in the domain of f.

Exercises

1. Match the graphs in Figure 13 with the description:

(a) $f''(x) < 0$ for all x.
(b) $f''(x)$ goes from + to −.
(c) $f''(x) > 0$ for all x.
(d) $f''(x)$ goes from − to +.

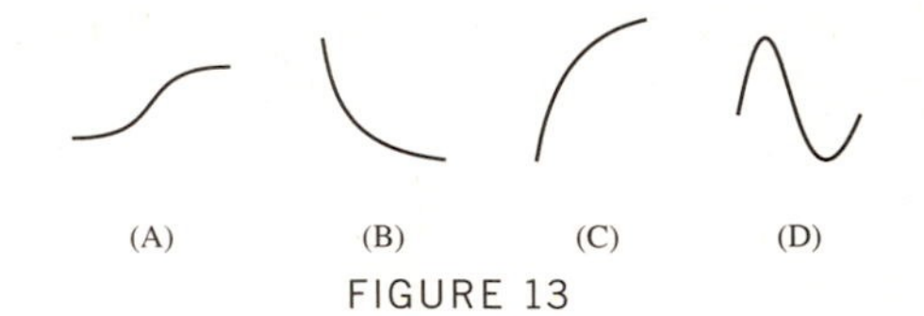

FIGURE 13

SOLUTION

(a) In C, we have $f''(x) < 0$ for all x.
(b) In A, $f''(x)$ goes from + to −.
(c) In B, we have $f''(x) > 0$ for all x.
(d) In D, $f''(x)$ goes from − to +.

In Exercises 3–18, determine the intervals on which the function is concave up or down and find the points of inflection.

3. $y = x^2 - 4x + 3$

SOLUTION Let $f(x) = x^2 - 4x + 3$. Then $f'(x) = 2x - 4$ and $f''(x) = 2 > 0$ for all x. Therefore, f is concave up everywhere, and there are no points of inflection.

5. $y = 10x^3 - x^5$

SOLUTION Let $f(x) = 10x^3 - x^5$. Then $f'(x) = 30x^2 - 5x^4$ and $f''(x) = 60x - 20x^3 = 20x(3 - x^2)$. Now, f is concave up for $x < -\sqrt{3}$ and for $0 < x < \sqrt{3}$ since $f''(x) > 0$ there. Moreover, f is concave down for $-\sqrt{3} < x < 0$ and for $x > \sqrt{3}$ since $f''(x) < 0$ there. Finally, because $f''(x)$ changes sign at $x = 0$ and at $x = \pm\sqrt{3}$, $f(x)$ has a point of inflection at $x = 0$ and at $x = \pm\sqrt{3}$.

7. $y = \theta - 2\sin\theta, \quad [0, 2\pi]$

SOLUTION Let $f(\theta) = \theta - 2\sin\theta$. Then $f'(\theta) = 1 - 2\cos\theta$ and $f''(\theta) = 2\sin\theta$. Now, f is concave up for $0 < \theta < \pi$ since $f''(\theta) > 0$ there. Moreover, f is concave down for $\pi < \theta < 2\pi$ since $f''(\theta) < 0$ there. Finally, because $f''(\theta)$ changes sign at $\theta = \pi$, $f(\theta)$ has a point of inflection at $\theta = \pi$.

9. $y = x(x - 8\sqrt{x}) \quad (x \geq 0)$

SOLUTION Let $f(x) = x(x - 8\sqrt{x}) = x^2 - 8x^{3/2}$. Then $f'(x) = 2x - 12x^{1/2}$ and $f''(x) = 2 - 6x^{-1/2}$. Now, f is concave down for $0 < x < 9$ since $f''(x) < 0$ there. Moreover, f is concave up for $x > 9$ since $f''(x) > 0$ there. Finally, because $f''(x)$ changes sign at $x = 9$, $f(x)$ has a point of inflection at $x = 9$.

11. $y = (x - 2)(1 - x^3)$

SOLUTION Let $f(x) = (x - 2)\left(1 - x^3\right) = x - x^4 - 2 + 2x^3$. Then $f'(x) = 1 - 4x^3 + 6x^2$ and $f''(x) = 12x - 12x^2 = 12x(1 - x) = 0$ at $x = 0$ and $x = 1$. Now, f is concave up on $(0, 1)$ since $f''(x) > 0$ there. Moreover, f is concave down on $(-\infty, 0) \cup (1, \infty)$ since $f''(x) < 0$ there. Finally, because $f''(x)$ changes sign at both $x = 0$ and $x = 1$, $f(x)$ has a point of inflection at both $x = 0$ and $x = 1$.

13. $y = \dfrac{1}{x^2 + 3}$

SOLUTION Let $f(x) = \dfrac{1}{x^2 + 3}$. Then $f'(x) = -\dfrac{2x}{(x^2 + 3)^2}$ and

$$f''(x) = -\frac{2(x^2 + 3)^2 - 8x^2(x^2 + 3)}{(x^2 + 3)^4} = \frac{6x^2 - 6}{(x^2 + 3)^3}.$$

Now, f is concave up for $|x| > 1$ since $f''(x) > 0$ there. Moreover, f is concave down for $|x| < 1$ since $f''(x) < 0$ there. Finally, because $f''(x)$ changes sign at both $x = -1$ and $x = 1$, $f(x)$ has a point of inflection at both $x = -1$ and $x = 1$.

15. $y = xe^{-3x}$

SOLUTION Let $f(x) = xe^{-3x}$. Then $f'(x) = -3xe^{-3x} + e^{-3x} = (1 - 3x)e^{-3x}$ and $f''(x) = -3(1 - 3x)e^{-3x} - 3e^{-3x} = (9x - 6)e^{-3x}$. Now, f is concave down for $x < \frac{2}{3}$ since $f''(x) < 0$ there. Moreover, f is concave up for $x > \frac{2}{3}$ since $f''(x) > 0$ there. Finally, because $f''(x)$ changes sign at $x = \frac{2}{3}$, $x = \frac{2}{3}$ is a point of inflection.

17. $y = 2x^2 + \ln x \quad (x > 0)$

SOLUTION Let $f(x) = 2x^2 + \ln x$. Then $f'(x) = 4x + x^{-1}$ and $f''(x) = 4 - x^{-2}$. Now, f is concave down for $x < \frac{1}{2}$ since $f''(x) < 0$ there. Moreover, f is concave up for $x > \frac{1}{2}$ since $f''(x) > 0$ there. Finally, because $f''(x)$ changes sign at $x = \frac{1}{2}$, f has a point of inflection at $x = \frac{1}{2}$.

19. The growth of a sunflower during the first 100 days after sprouting is modeled well by the *logistic curve* $y = h(t)$ shown in Figure 15. Estimate the growth rate at the point of inflection and explain its significance. Then make a rough sketch of the first and second derivatives of $h(t)$.

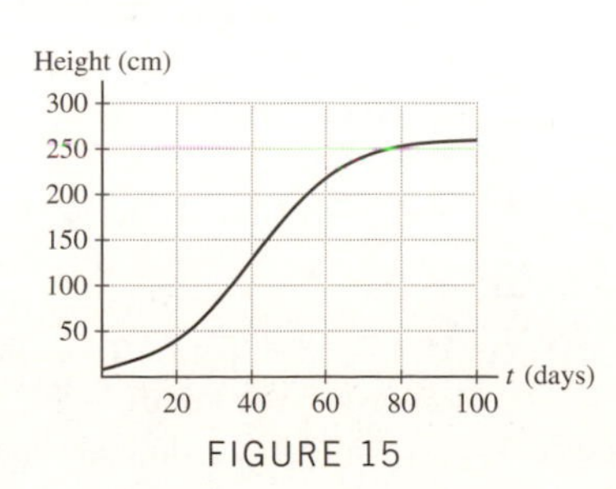

FIGURE 15

SOLUTION The point of inflection in Figure 15 appears to occur at $t = 40$ days. The graph below shows the logistic curve with an approximate tangent line drawn at $t = 40$. The approximate tangent line passes roughly through the points (20, 20) and (60, 240). The growth rate at the point of inflection is thus

$$\frac{240 - 20}{60 - 20} = \frac{220}{40} = 5.5 \text{ cm/day}.$$

Because the logistic curve changes from concave up to concave down at $t = 40$, the growth rate at this point is the maximum growth rate for the sunflower plant.

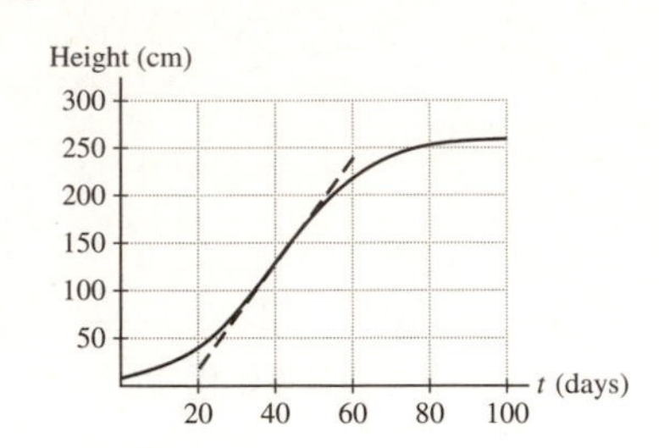

Sketches of the first and second derivative of $h(t)$ are shown below at the left and at the right, respectively.

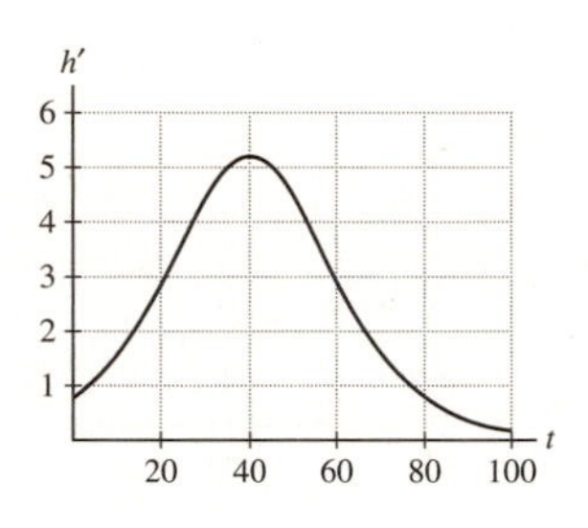

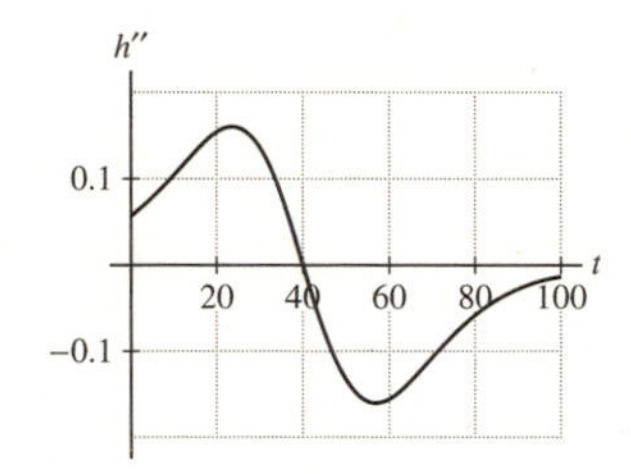

21. Repeat Exercise 20 but assume that Figure 16 is the graph of the *derivative* $f'(x)$.

SOLUTION Points of inflection occur when $f''(x)$ changes sign. Consequently, points of inflection occur when $f'(x)$ changes from increasing to decreasing or from decreasing to increasing. In Figure 16, this occurs at $x = b$ and at $x = e$; therefore, $f(x)$ has an inflection point at $x = b$ and another at $x = e$. The function $f(x)$ will be concave down when $f''(x) < 0$ or when $f'(x)$ is decreasing. Thus, $f(x)$ is concave down for $b < x < e$.

23. Figure 17 shows the *derivative* $f'(x)$ on [0, 1.2]. Locate the points of inflection of $f(x)$ and the points where the local minima and maxima occur. Determine the intervals on which $f(x)$ has the following properties:

(a) Increasing
(b) Decreasing
(c) Concave up
(d) Concave down

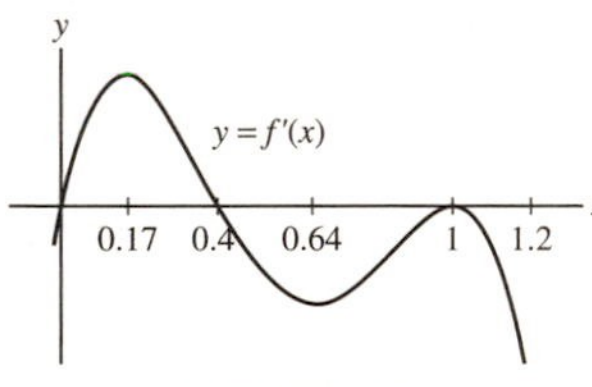

FIGURE 17

SOLUTION Recall that the graph is that of f', *not* f. The inflection points of f occur where f' changes from increasing to decreasing or vice versa because it is at these points that the sign of f'' changes. From the graph we conclude that f has points of inflection at $x = 0.17$, $x = 0.64$, and $x = 1$. The local extrema of f occur where f' changes sign. This occurs at $x = 0.4$. Because the sign of f' changes from + to −, $f(0.4)$ is a local maximum. There are no local minima.

(a) f is increasing when f' is positive. Hence, f is increasing on (0, 0.4).

(b) f is decreasing when f' is negative. Hence, f is decreasing on $(0.4, 1) \cup (1, 1.2)$.

(c) Now f is concave up where f' is increasing. This occurs on $(0, 0.17) \cup (0.64, 1)$.

(d) Moreover, f is concave down where f' is decreasing. This occurs on $(0.17, 0.64) \cup (1, 1.2)$.

In Exercises 25–38, find the critical points and apply the Second Derivative Test (or state that it fails).

25. $f(x) = x^3 - 12x^2 + 45x$

SOLUTION Let $f(x) = x^3 - 12x^2 + 45x$. Then $f'(x) = 3x^2 - 24x + 45 = 3(x-3)(x-5)$, and the critical points are $x = 3$ and $x = 5$. Moreover, $f''(x) = 6x - 24$, so $f''(3) = -6 < 0$ and $f''(5) = 6 > 0$. Therefore, by the Second Derivative Test, $f(3) = 54$ is a local maximum, and $f(5) = 50$ is a local minimum.

27. $f(x) = 3x^4 - 8x^3 + 6x^2$

SOLUTION Let $f(x) = 3x^4 - 8x^3 + 6x^2$. Then $f'(x) = 12x^3 - 24x^2 + 12x = 12x(x-1)^2 = 0$ at $x = 0, 1$ and $f''(x) = 36x^2 - 48x + 12$. Thus, $f''(0) > 0$, which implies $f(0)$ is a local minimum; however, $f''(1) = 0$, which is inconclusive.

29. $f(x) = \dfrac{x^2 - 8x}{x+1}$

SOLUTION Let $f(x) = \dfrac{x^2 - 8x}{x+1}$. Then

$$f'(x) = \frac{x^2 + 2x - 8}{(x+1)^2} \quad \text{and} \quad f''(x) = \frac{2(x+1)^2 - 2(x^2 + 2x - 8)}{(x+1)^3}.$$

Thus, the critical points are $x = -4$ and $x = 2$. Moreover, $f''(-4) < 0$ and $f''(2) > 0$. Therefore, by the second derivative test, $f(-4) = -16$ is a local maximum and $f(2) = -4$ is a local minimum.

31. $y = 6x^{3/2} - 4x^{1/2}$

SOLUTION Let $f(x) = 6x^{3/2} - 4x^{1/2}$. Then $f'(x) = 9x^{1/2} - 2x^{-1/2} = x^{-1/2}(9x - 2)$, so there are two critical points: $x = 0$ and $x = \frac{2}{9}$. Now,

$$f''(x) = \frac{9}{2}x^{-1/2} + x^{-3/2} = \frac{1}{2}x^{-3/2}(9x + 2).$$

Thus, $f''\left(\frac{2}{9}\right) > 0$, which implies $f\left(\frac{2}{9}\right)$ is a local minimum. $f''(x)$ is undefined at $x = 0$, so the Second Derivative Test cannot be applied there.

33. $f(x) = \sin^2 x + \cos x, \quad [0, \pi]$

SOLUTION Let $f(x) = \sin^2 x + \cos x$. Then $f'(x) = 2\sin x \cos x - \sin x = \sin x(2\cos x - 1)$. On the interval $[0, \pi]$, $f'(x) = 0$ at $x = 0$, $x = \frac{\pi}{3}$ and $x = \pi$. Now,

$$f''(x) = 2\cos^2 x - 2\sin^2 x - \cos x.$$

Thus, $f''(0) > 0$, so $f(0)$ is a local minimum. On the other hand, $f''(\frac{\pi}{3}) < 0$, so $f(\frac{\pi}{3})$ is a local maximum. Finally, $f''(\pi) > 0$, so $f(\pi)$ is a local minimum.

35. $f(x) = xe^{-x^2}$

SOLUTION Let $f(x) = xe^{-x^2}$. Then $f'(x) = -2x^2e^{-x^2} + e^{-x^2} = (1 - 2x^2)e^{-x^2}$, so there are two critical points: $x = \pm\frac{\sqrt{2}}{2}$. Now,

$$f''(x) = (4x^3 - 2x)e^{-x^2} - 4xe^{-x^2} = (4x^3 - 6x)e^{-x^2}.$$

Thus, $f''\left(\frac{\sqrt{2}}{2}\right) < 0$, so $f\left(\frac{\sqrt{2}}{2}\right)$ is a local maximum. On the other hand, $f''\left(-\frac{\sqrt{2}}{2}\right) > 0$, so $f\left(-\frac{\sqrt{2}}{2}\right)$ is a local minimum.

37. $f(x) = x^3 \ln x \quad (x > 0)$

SOLUTION Let $f(x) = x^3 \ln x$. Then $f'(x) = x^2 + 3x^2 \ln x = x^2(1 + 3\ln x)$, so there is only one critical point: $x = e^{-1/3}$. Now,

$$f''(x) = 3x + 2x(1 + 3\ln x) = x(5 + 6lnx).$$

Thus, $f''\left(e^{-1/3}\right) > 0$, so $f\left(e^{-1/3}\right)$ is a local minimum.

In Exercises 39–52, find the intervals on which f is concave up or down, the points of inflection, the critical points, and the local minima and maxima.

SOLUTION Here is a table legend for Exercises 39–49.

SYMBOL	MEANING
$-$	The entity is negative on the given interval.
0	The entity is zero at the specified point.
$+$	The entity is positive on the given interval.
U	The entity is undefined at the specified point.
↗	The function (f, g, etc.) is increasing on the given interval.
↘	The function (f, g, etc.) is decreasing on the given interval.
⌣	The function (f, g, etc.) is concave up on the given interval.
⌢	The function (f, g, etc.) is concave down on the given interval.
M	The function (f, g, etc.) has a local maximum at the specified point.
m	The function (f, g, etc.) has a local minimum at the specified point.
I	The function (f, g, etc.) has an inflection point here.
¬	There is no local extremum or inflection point here.

39. $f(x) = x^3 - 2x^2 + x$

SOLUTION Let $f(x) = x^3 - 2x^2 + x$.

- Then $f'(x) = 3x^2 - 4x + 1 = (x-1)(3x-1) = 0$ yields $x = 1$ and $x = \frac{1}{3}$ as candidates for extrema.
- Moreover, $f''(x) = 6x - 4 = 0$ gives a candidate for a point of inflection at $x = \frac{2}{3}$.

x	$(-\infty, \frac{1}{3})$	$\frac{1}{3}$	$(\frac{1}{3}, 1)$	1	$(1, \infty)$
f'	$+$	0	$-$	0	$+$
f	↗	M	↘	m	↗

x	$(-\infty, \frac{2}{3})$	$\frac{2}{3}$	$(\frac{2}{3}, \infty)$
f''	$-$	0	$+$
f	⌢	I	⌣

41. $f(t) = t^2 - t^3$

SOLUTION Let $f(t) = t^2 - t^3$.

- Then $f'(t) = 2t - 3t^2 = t(2-3t) = 0$ yields $t = 0$ and $t = \frac{2}{3}$ as candidates for extrema.
- Moreover, $f''(t) = 2 - 6t = 0$ gives a candidate for a point of inflection at $t = \frac{1}{3}$.

t	$(-\infty, 0)$	0	$(0, \frac{2}{3})$	$\frac{2}{3}$	$(\frac{2}{3}, \infty)$
f'	$-$	0	$+$	0	$-$
f	↘	m	↗	M	↘

t	$(-\infty, \frac{1}{3})$	$\frac{1}{3}$	$(\frac{1}{3}, \infty)$
f''	$+$	0	$-$
f	⌣	I	⌢

43. $f(x) = x^2 - 8x^{1/2}$ $(x \geq 0)$

SOLUTION Let $f(x) = x^2 - 8x^{1/2}$. Note that the domain of f is $x \geq 0$.

- Then $f'(x) = 2x - 4x^{-1/2} = x^{-1/2}\left(2x^{3/2} - 4\right) = 0$ yields $x = 0$ and $x = (2)^{2/3}$ as candidates for extrema.
- Moreover, $f''(x) = 2 + 2x^{-3/2} > 0$ for all $x \geq 0$, which means there are no inflection points.

x	0	$\left(0, (2)^{2/3}\right)$	$(2)^{2/3}$	$\left((2)^{2/3}, \infty\right)$
f'	U	$-$	0	$+$
f	M	↘	m	↗

45. $f(x) = \dfrac{x}{x^2+27}$

SOLUTION Let $f(x) = \dfrac{x}{x^2+27}$.

- Then $f'(x) = \dfrac{27-x^2}{\left(x^2+27\right)^2} = 0$ yields $x = \pm 3\sqrt{3}$ as candidates for extrema.
- Moreover, $f''(x) = \dfrac{-2x\left(x^2+27\right)^2 - (27-x^2)(2)\left(x^2+27\right)(2x)}{\left(x^2+27\right)^4} = \dfrac{2x\left(x^2-81\right)}{\left(x^2+27\right)^3} = 0$ gives candidates for a point of inflection at $x = 0$ and at $x = \pm 9$.

x	$\left(-\infty, -3\sqrt{3}\right)$	$-3\sqrt{3}$	$\left(-3\sqrt{3}, 3\sqrt{3}\right)$	$3\sqrt{3}$	$\left(3\sqrt{3}, \infty\right)$
f'	$-$	0	$+$	0	$-$
f	↘	m	↗	M	↘

x	$(-\infty, -9)$	-9	$(-9, 0)$	0	$(0, 9)$	9	$(9, \infty)$
f''	$-$	0	$+$	0	$-$	0	$+$
f	⌢	I	⌣	I	⌢	I	⌣

47. $f(\theta) = \theta + \sin\theta, \quad [0, 2\pi]$

SOLUTION Let $f(\theta) = \theta + \sin\theta$ on $[0, 2\pi]$.

- Then $f'(\theta) = 1 + \cos\theta = 0$ yields $\theta = \pi$ as a candidate for an extremum.
- Moreover, $f''(\theta) = -\sin\theta = 0$ gives candidates for a point of inflection at $\theta = 0$, at $\theta = \pi$, and at $\theta = 2\pi$.

θ	$(0, \pi)$	π	$(\pi, 2\pi)$
f'	$+$	0	$+$
f	↗	¬	↗

θ	0	$(0, \pi)$	π	$(\pi, 2\pi)$	2π
f''	0	$-$	0	$+$	0
f	¬	⌢	I	⌣	¬

49. $f(x) = \tan x, \quad \left[-\frac{\pi}{4}, \frac{\pi}{3}\right]$

SOLUTION Let $f(x) = \tan x$ on $\left[-\frac{\pi}{4}, \frac{\pi}{3}\right]$.

- Then $f'(x) = \sec^2 x \geq 1 > 0$ on $\left[-\frac{\pi}{4}, \frac{\pi}{3}\right]$.
- Moreover, $f''(x) = 2\sec x \cdot \sec x \tan x = 2\sec^2 x \tan x = 0$ gives a candidate for a point of inflection at $x = 0$.

x	$\left(-\frac{\pi}{4}, \frac{\pi}{3}\right)$
f'	$+$
f	↗

x	$\left(-\frac{\pi}{4}, 0\right)$	0	$\left(0, \frac{\pi}{3}\right)$
f''	$-$	0	$+$
f	⌢	I	⌣

51. $y = (x^2-2)e^{-x} \quad (x > 0)$

SOLUTION Let $f(x) = (x^2-2)e^{-x}$.

- Then $f'(x) = -(x^2-2x-2)e^{-x} = 0$ gives $x = 1+\sqrt{3}$ as a candidate for an extrema.
- Moreover, $f''(x) = (x^2-4x)e^{-x} = 0$ gives $x = 4$ as a candidate for a point of inflection.

x	$\left(0, 1+\sqrt{3}\right)$	$1+\sqrt{3}$	$\left(1+\sqrt{3}, \infty\right)$
f'	$+$	0	$-$
f	↗	M	↘

x	$(0, 4)$	4	$(4, \infty)$
f''	$-$	0	$+$
f	⌢	I	⌣

53. Sketch the graph of an increasing function such that $f''(x)$ changes from + to − at $x = 2$ and from − to + at $x = 4$. Do the same for a decreasing function.

SOLUTION The graph shown below at the left is an increasing function which changes from concave up to concave down at $x = 2$ and from concave down to concave up at $x = 4$. The graph shown below at the right is a decreasing function which changes from concave up to concave down at $x = 2$ and from concave down to concave up at $x = 4$.

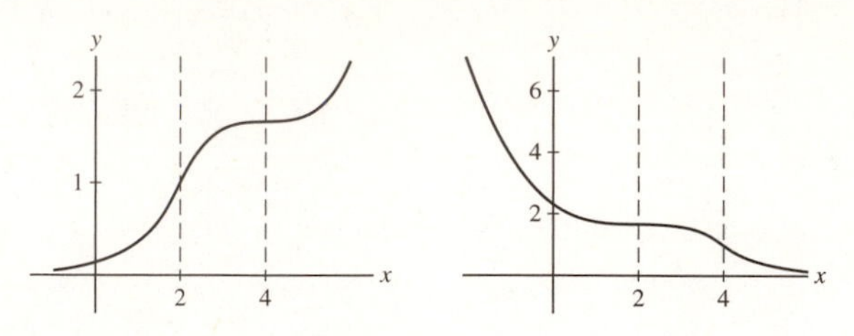

In Exercises 54–56, sketch the graph of a function $f(x)$ satisfying all of the given conditions.

55. (i) $f'(x) > 0$ for all x, and
(ii) $f''(x) < 0$ for $x < 0$ and $f''(x) > 0$ for $x > 0$.

SOLUTION Here is the graph of a function $f(x)$ satisfying **(i)** $f'(x) > 0$ for all x and **(ii)** $f''(x) < 0$ for $x < 0$ and $f''(x) > 0$ for $x > 0$.

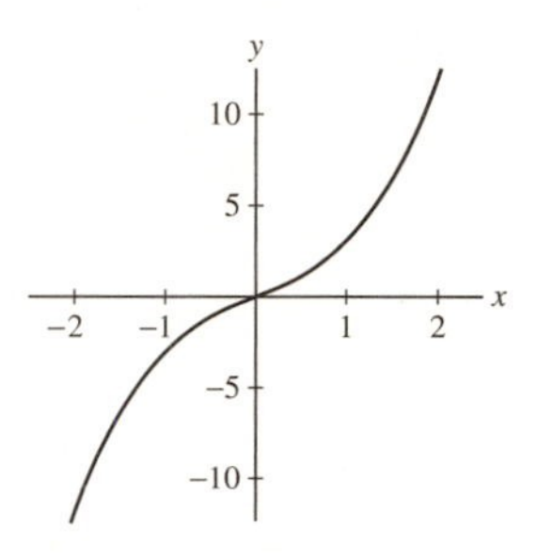

57. An infectious flu spreads slowly at the beginning of an epidemic. The infection process accelerates until a majority of the susceptible individuals are infected, at which point the process slows down.

(a) If $R(t)$ is the number of individuals infected at time t, describe the concavity of the graph of R near the beginning and end of the epidemic.

(b) Describe the status of the epidemic on the day that $R(t)$ has a point of inflection.

SOLUTION

(a) Near the beginning of the epidemic, the graph of R is concave up. Near the epidemic's end, R is concave down.

(b) "Epidemic subsiding: number of new cases declining."

59. Water is pumped into a sphere of radius R at a variable rate in such a way that the water level rises at a constant rate (Figure 18). Let $V(t)$ be the volume of water in the tank at time t. Sketch the graph $V(t)$ (approximately, but with the correct concavity). Where does the point of inflection occur?

SOLUTION Because water is entering the sphere in such a way that the water level rises at a constant rate, we expect the volume to increase more slowly near the bottom and top of the sphere where the sphere is not as "wide" and to increase more rapidly near the middle of the sphere. The graph of $V(t)$ should therefore start concave up and change to concave down when the sphere is half full; that is, the point of inflection should occur when the water level is equal to the radius of the sphere. A possible graph of $V(t)$ is shown below.

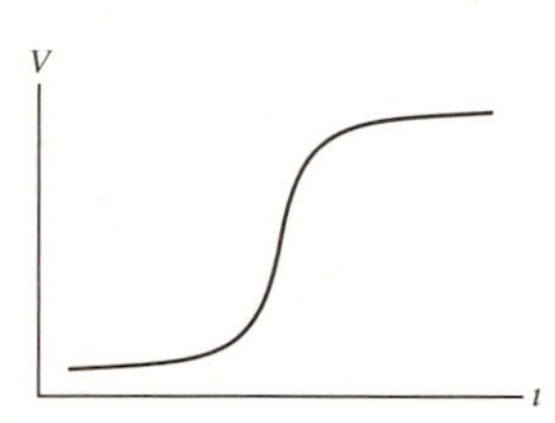

61. Image Processing The intensity of a pixel in a digital image is measured by a number u between 0 and 1. Often, images can be enhanced by rescaling intensities (Figure 19), where pixels of intensity u are displayed with intensity $g(u)$ for a suitable function $g(u)$. One common choice is the **sigmoidal correction**, defined for constants a, b by

$$g(u) = \frac{f(u) - f(0)}{f(1) - f(0)} \quad \text{where} \quad f(u) = \left(1 + e^{b(a-u)}\right)^{-1}$$

Figure 20 shows that $g(u)$ reduces the intensity of low-intensity pixels (where $g(u) < u$) and increases the intensity of high-intensity pixels.

(a) Verify that $f'(u) > 0$ and use this to show that $g(u)$ increases from 0 to 1 for $0 \le u \le 1$.

(b) Where does $g(u)$ have a point of inflection?

Original

Sigmoidal correction

FIGURE 19

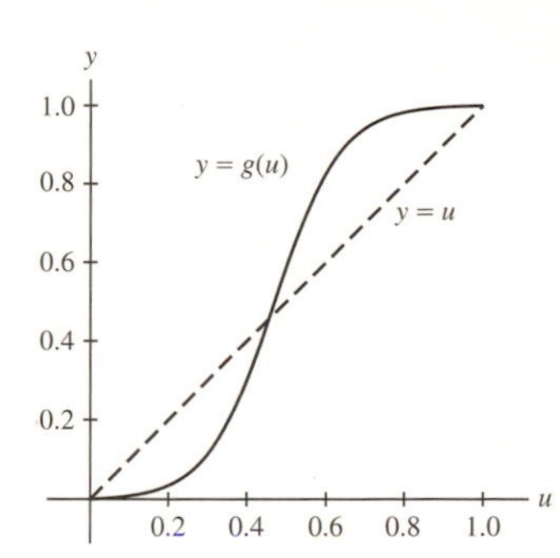

FIGURE 20 Sigmoidal correction with $a = 0.47$, $b = 12$.

SOLUTION

(a) With $f(u) = (1 + e^{b(a-u)})^{-1}$, it follows that

$$f'(u) = -(1 + e^{b(a-u)})^{-2} \cdot -be^{b(a-u)} = \frac{be^{b(a-u)}}{(1 + e^{b(a-u)})^2} > 0$$

for all u. Next, observe that

$$g(0) = \frac{f(0) - f(0)}{f(1) - f(0)} = 0, \quad g(1) = \frac{f(1) - f(0)}{f(1) - f(0)} = 1,$$

and

$$g'(u) = \frac{1}{f(1) - f(0)} f'(u) > 0$$

for all u. Thus, $g(u)$ increases from 0 to 1 for $0 \le u \le 1$.

(b) Working from part (a), we find

$$f''(u) = \frac{b^2 e^{b(a-u)} (2e^{b(a-u)} - 1)}{(1 + e^{b(a-u)})^3}.$$

Because

$$g''(u) = \frac{1}{f(1) - f(0)} f''(u),$$

it follows that $g(u)$ has a point of inflection when

$$2e^{b(a-u)} - 1 = 0 \quad \text{or} \quad u = a + \frac{1}{b} \ln 2.$$

Further Insights and Challenges

In Exercises 63–65, assume that $f(x)$ is differentiable.

63. Proof of the Second Derivative Test Let c be a critical point such that $f''(c) > 0$ (the case $f''(c) < 0$ is similar).

(a) Show that $f''(c) = \lim_{h \to 0} \frac{f'(c+h)}{h}$.

(b) Use (a) to show that there exists an open interval (a, b) containing c such that $f'(x) < 0$ if $a < x < c$ and $f'(x) > 0$ if $c < x < b$. Conclude that $f(c)$ is a local minimum.

SOLUTION

(a) Because c is a critical point, either $f'(c) = 0$ or $f'(c)$ does not exist; however, $f''(c)$ exists, so $f'(c)$ must also exist. Therefore, $f'(c) = 0$. Now, from the definition of the derivative, we have

$$f''(c) = \lim_{h \to 0} \frac{f'(c+h) - f'(c)}{h} = \lim_{h \to 0} \frac{f'(c+h)}{h}.$$

(b) We are given that $f''(c) > 0$. By part (a), it follows that

$$\lim_{h\to 0} \frac{f'(c+h)}{h} > 0;$$

in other words, for sufficiently small h,

$$\frac{f'(c+h)}{h} > 0.$$

Now, if h is sufficiently small but negative, then $f'(c+h)$ must also be negative (so that the ratio $f'(c+h)/h$ will be positive) and $c+h < c$. On the other hand, if h is sufficiently small but positive, then $f'(c+h)$ must also be positive and $c+h > c$. Thus, there exists an open interval (a, b) containing c such that $f'(x) < 0$ for $a < x < c$ and $f'(c) > 0$ for $c < x < b$. Finally, because $f'(x)$ changes from negative to positive at $x = c$, $f(c)$ must be a local minimum.

65. Assume that $f''(x)$ exists and let c be a point of inflection of $f(x)$.

(a) Use the method of Exercise 64 to prove that the tangent line at $x = c$ *crosses the graph* (Figure 21). *Hint:* Show that $G(x)$ changes sign at $x = c$.

(b) GU Verify this conclusion for $f(x) = \dfrac{x}{3x^2+1}$ by graphing $f(x)$ and the tangent line at each inflection point on the same set of axes.

FIGURE 21 Tangent line crosses graph at point of inflection.

SOLUTION

(a) Let $G(x) = f(x) - f'(c)(x-c) - f(c)$. Then, as in Exercise 63, $G(c) = G'(c) = 0$ and $G''(x) = f''(x)$. If $f''(x)$ changes from positive to negative at $x = c$, then so does $G''(x)$ and $G'(x)$ is increasing for $x < c$ and decreasing for $x > c$. This means that $G'(x) < 0$ for $x < c$ and $G'(x) < 0$ for $x > c$. This in turn implies that $G(x)$ is decreasing, so $G(x) > 0$ for $x < c$ but $G(x) < 0$ for $x > c$. On the other hand, if $f''(x)$ changes from negative to positive at $x = c$, then so does $G''(x)$ and $G'(x)$ is decreasing for $x < c$ and increasing for $x > c$. Thus, $G'(x) > 0$ for $x < c$ and $G'(x) > 0$ for $x > c$. This in turn implies that $G(x)$ is increasing, so $G(x) < 0$ for $x < c$ and $G(x) > 0$ for $x > c$. In either case, $G(x)$ changes sign at $x = c$, and the tangent line at $x = c$ crosses the graph of the function.

(b) Let $f(x) = \dfrac{x}{3x^2+1}$. Then

$$f'(x) = \frac{1-3x^2}{(3x^2+1)^2} \quad \text{and} \quad f''(x) = \frac{-18x(1-x^2)}{(3x^2+1)^3}.$$

Therefore $f(x)$ has a point of inflection at $x = 0$ and at $x = \pm 1$. The figure below shows the graph of $y = f(x)$ and its tangent lines at each of the points of inflection. It is clear that each tangent line crosses the graph of $f(x)$ at the inflection point.

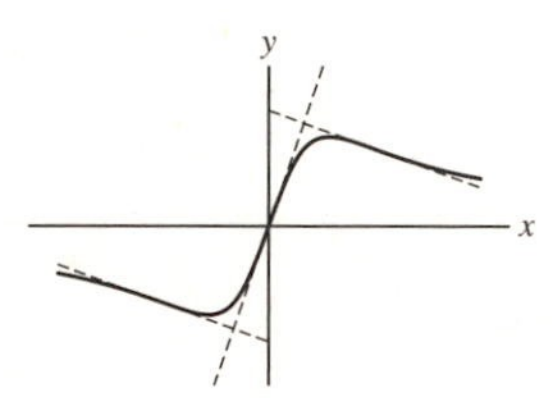

67. Let $f(x)$ be a polynomial of degree $n \geq 2$. Show that $f(x)$ has at least one point of inflection if n is odd. Then give an example to show that $f(x)$ need not have a point of inflection if n is even.

SOLUTION Let $f(x) = a_n x^n + a_{n-1}x^{n-1} + \cdots + a_1 x + a_0$ be a polynomial of degree n. Then $f'(x) = na_n x^{n-1} + (n-1)a_{n-1}x^{n-2} + \cdots + 2a_2 x + a_1$ and $f''(x) = n(n-1)a_n x^{n-2} + (n-1)(n-2)a_{n-1}x^{n-3} + \cdots + 6a_3 x + 2a_2$. If $n \geq 3$ and is odd, then $n-2$ is also odd and $f''(x)$ is a polynomial of odd degree. Therefore $f''(x)$ must take on both positive and negative values. It follows that $f''(x)$ has at least one root c such that $f''(x)$ changes sign at c. The function $f(x)$ will then have a point of inflection at $x = c$. On the other hand, the functions $f(x) = x^2, x^4$ and x^8 are polynomials of even degree that do not have any points of inflection.

4.5 L'Hôpital's Rule

Preliminary Questions

1. What is wrong with applying L'Hôpital's Rule to $\lim_{x\to 0} \frac{x^2-2x}{3x-2}$?

SOLUTION As $x \to 0$,

$$\frac{x^2-2x}{3x-2}$$

is not of the form $\frac{0}{0}$ or $\frac{\infty}{\infty}$, so L'Hôpital's Rule cannot be used.

2. Does L'Hôpital's Rule apply to $\lim_{x\to a} f(x)g(x)$ if $f(x)$ and $g(x)$ both approach ∞ as $x \to a$?

SOLUTION No. L'Hôpital's Rule only applies to limits of the form $\frac{0}{0}$ or $\frac{\infty}{\infty}$.

Exercises

In Exercises 1–10, use L'Hôpital's Rule to evaluate the limit, or state that L'Hôpital's Rule does not apply.

1. $\lim_{x\to 3} \frac{2x^2-5x-3}{x-4}$

SOLUTION Because the quotient is not indeterminate at $x=3$,

$$\left.\frac{2x^2-5x-3}{x-4}\right|_{x=3} = \frac{18-15-3}{3-4} = \frac{0}{-1},$$

L'Hôpital's Rule does not apply.

3. $\lim_{x\to 4} \frac{x^3-64}{x^2+16}$

SOLUTION Because the quotient is not indeterminate at $x=4$,

$$\left.\frac{x^3-64}{x^2+16}\right|_{x=4} = \frac{64-64}{16+16} = \frac{0}{32},$$

L'Hôpital's Rule does not apply.

5. $\lim_{x\to 9} \frac{x^{1/2}+x-6}{x^{3/2}-27}$

SOLUTION Because the quotient is not indeterminate at $x=9$,

$$\left.\frac{x^{1/2}+x-6}{x^{3/2}-27}\right|_{x=9} = \frac{3+9-6}{27-27} = \frac{6}{0},$$

L'Hôpital's Rule does not apply.

7. $\lim_{x\to 0} \frac{\sin 4x}{x^2+3x+1}$

SOLUTION Because the quotient is not indeterminate at $x=0$,

$$\left.\frac{\sin 4x}{x^2+3x+1}\right|_{x=0} = \frac{0}{0+0+1} = \frac{0}{1},$$

L'Hôpital's Rule does not apply.

9. $\lim_{x\to 0} \frac{\cos 2x-1}{\sin 5x}$

SOLUTION The functions $\cos 2x - 1$ and $\sin 5x$ are differentiable, but the quotient is indeterminate at $x=0$,

$$\left.\frac{\cos 2x-1}{\sin 5x}\right|_{x=0} = \frac{1-1}{0} = \frac{0}{0},$$

so L'Hôpital's Rule applies. We find

$$\lim_{x\to 0} \frac{\cos 2x-1}{\sin 5x} = \lim_{x\to 0} \frac{-2\sin 2x}{5\cos 5x} = \frac{0}{5} = 0.$$

In Exercises 11–16, show that L'Hôpital's Rule is applicable to the limit as $x \to \pm\infty$ and evaluate.

11. $\lim_{x\to\infty} \frac{9x+4}{3-2x}$

SOLUTION As $x \to \infty$, the quotient $\frac{9x+4}{3-2x}$ is of the form $\frac{\infty}{\infty}$, so L'Hôpital's Rule applies. We find

$$\lim_{x\to\infty} \frac{9x+4}{3-2x} = \lim_{x\to\infty} \frac{9}{-2} = -\frac{9}{2}.$$

13. $\lim_{x\to\infty} \frac{\ln x}{x^{1/2}}$

SOLUTION As $x \to \infty$, the quotient $\frac{\ln x}{x^{1/2}}$ is of the form $\frac{\infty}{\infty}$, so L'Hôpital's Rule applies. We find

$$\lim_{x\to\infty} \frac{\ln x}{x^{1/2}} = \lim_{x\to\infty} \frac{\frac{1}{x}}{\frac{1}{2}x^{-1/2}} = \lim_{x\to\infty} \frac{1}{2x^{1/2}} = 0.$$

15. $\lim_{x\to-\infty} \frac{\ln(x^4+1)}{x}$

SOLUTION As $x \to \infty$, the quotient $\frac{\ln(x^4+1)}{x}$ is of the form $\frac{\infty}{\infty}$, so L'Hôpital's Rule applies. Here, we use L'Hôpital's Rule twice to find

$$\lim_{x\to\infty} \frac{\ln(x^4+1)}{x} = \lim_{x\to\infty} \frac{\frac{4x^3}{x^4+1}}{1} = \lim_{x\to\infty} \frac{12x^2}{4x^3} = \lim_{x\to\infty} \frac{3}{x} = 0.$$

In Exercises 17–54, evaluate the limit.

17. $\lim_{x\to 1} \frac{\sqrt{8+x} - 3x^{1/3}}{x^2-3x+2}$

SOLUTION $\lim_{x\to 1} \frac{\sqrt{8+x} - 3x^{1/3}}{x^2-3x+2} = \lim_{x\to 1} \frac{\frac{1}{2}(8+x)^{-1/2} - x^{-2/3}}{2x-3} = \frac{\frac{1}{6}-1}{-1} = \frac{5}{6}.$

19. $\lim_{x\to-\infty} \frac{3x-2}{1-5x}$

SOLUTION $\lim_{x\to-\infty} \frac{3x-2}{1-5x} = \lim_{x\to-\infty} \frac{3}{-5} = -\frac{3}{5}.$

21. $\lim_{x\to-\infty} \frac{7x^2+4x}{9-3x^2}$

SOLUTION $\lim_{x\to-\infty} \frac{7x^2+4x}{9-3x^2} = \lim_{x\to-\infty} \frac{14x+4}{-6x} = \lim_{x\to-\infty} \frac{14}{-6} = -\frac{7}{3}.$

23. $\lim_{x\to 1} \frac{(1+3x)^{1/2}-2}{(1+7x)^{1/3}-2}$

SOLUTION Apply L'Hôpital's Rule once:

$$\lim_{x\to 1} \frac{(1+3x)^{1/2}-2}{(1+7x)^{1/3}-2} = \lim_{x\to 1} \frac{\frac{3}{2}(1+3x)^{-1/2}}{\frac{7}{3}(1+7x)^{-2/3}}$$

$$= \frac{(\frac{3}{2})\frac{1}{2}}{(\frac{7}{3})(\frac{1}{4})} = \frac{9}{7}$$

25. $\lim_{x\to 0} \frac{\sin 2x}{\sin 7x}$

SOLUTION $\lim_{x\to 0} \frac{\sin 2x}{\sin 7x} = \lim_{x\to 0} \frac{2\cos 2x}{7\cos 7x} = \frac{2}{7}.$

27. $\lim_{x\to 0} \frac{\tan x}{x}$

SOLUTION $\lim_{x\to 0} \frac{\tan x}{x} = \lim_{x\to 0} \frac{\sec^2 x}{1} = 1.$

29. $\lim_{x\to 0} \frac{\sin x - x\cos x}{x - \sin x}$

SOLUTION

$$\lim_{x\to 0} \frac{\sin x - x\cos x}{x - \sin x} = \lim_{x\to 0} \frac{x\sin x}{1-\cos x} = \lim_{x\to 0} \frac{\sin x + x\cos x}{\sin x} = \lim_{x\to 0} \frac{\cos x + \cos x - x\sin x}{\cos x} = 2.$$

31. $\lim_{x\to 0} \frac{\cos(x+\frac{\pi}{2})}{\sin x}$

SOLUTION $\lim_{x\to 0} \frac{\cos(x+\frac{\pi}{2})}{\sin x} = \lim_{x\to 0} \frac{-\sin(x+\frac{\pi}{2})}{\cos x} = -1.$

33. $\lim_{x\to\pi/2} \frac{\cos x}{\sin(2x)}$

SOLUTION $\lim_{x\to\pi/2} \frac{\cos x}{\sin(2x)} = \lim_{x\to\pi/2} \frac{-\sin x}{2\cos(2x)} = \frac{1}{2}.$

35. $\lim_{x\to\pi/2} (\sec x - \tan x)$

SOLUTION

$$\lim_{x\to\frac{\pi}{2}} (\sec x - \tan x) = \lim_{x\to\frac{\pi}{2}} \left(\frac{1}{\cos x} - \frac{\sin x}{\cos x}\right) = \lim_{x\to\frac{\pi}{2}} \left(\frac{1-\sin x}{\cos x}\right) = \lim_{x\to\frac{\pi}{2}} \left(\frac{-\cos x}{-\sin x}\right) = 0.$$

37. $\lim_{x\to 1} \tan\left(\frac{\pi x}{2}\right)\ln x$

SOLUTION $\lim_{x\to 1} \tan\left(\frac{\pi x}{2}\right)\ln x = \lim_{x\to 1} \frac{\ln x}{\cot(\frac{\pi x}{2})} = \lim_{x\to 1} \frac{\frac{1}{x}}{-\frac{\pi}{2}\csc^2(\frac{\pi x}{2})} = \lim_{x\to 1} \frac{-2}{\pi x}\sin^2\left(\frac{\pi}{2}x\right) = -\frac{2}{\pi}.$

39. $\lim_{x\to 0} \frac{e^x - 1}{\sin x}$

SOLUTION $\lim_{x\to 0} \frac{e^x - 1}{\sin x} = \lim_{x\to 0} \frac{e^x}{\cos x} = 1.$

41. $\lim_{x\to 0} \frac{e^{2x} - 1 - x}{x^2}$

SOLUTION $\lim_{x\to 0} \frac{e^{2x} - 1 - x}{x^2} = \lim_{x\to 0} \frac{2e^{2x} - 1}{2x}$ which does not exist.

43. $\lim_{t\to 0+} (\sin t)(\ln t)$

SOLUTION

$$\lim_{t\to 0+} (\sin t)(\ln t) = \lim_{t\to 0+} \frac{\ln t}{\csc t} = \lim_{t\to 0+} \frac{\frac{1}{t}}{-\csc t\cot t} = \lim_{t\to 0+} \frac{-\sin^2 t}{t\cos t} = \lim_{t\to 0+} \frac{-2\sin t\cos t}{\cos t - t\sin t} = 0.$$

45. $\lim_{x\to 0} \frac{a^x - 1}{x} \quad (a > 0)$

SOLUTION $\lim_{x\to 0} \frac{a^x - 1}{x} = \lim_{x\to 0} \frac{\ln a \cdot a^x}{1} = \ln a.$

47. $\lim_{x\to 1} (1+\ln x)^{1/(x-1)}$

SOLUTION $\lim_{x\to 1} \ln(1+\ln x)^{1/(x-1)} = \lim_{x\to 1} \frac{\ln(1+\ln x)}{x-1} = \lim_{x\to 1} \frac{1}{x(1+\ln x)} = 1.$ Hence,

$$\lim_{x\to 1} (1+\ln x)^{1/(x-1)} = \lim_{x\to 1} e^{(1+\ln x)^{1/(x-1)}} = e.$$

49. $\lim_{x\to 0} (\cos x)^{3/x^2}$

SOLUTION

$$\lim_{x\to 0} \ln(\cos x)^{3/x^2} = \lim_{x\to 0} \frac{3 \ln \cos x}{x^2}$$
$$= \lim_{x\to 0} -\frac{3 \tan x}{2x}$$
$$= \lim_{x\to 0} -\frac{3 \sec^2 x}{2} = -\frac{3}{2}.$$

Hence, $\lim_{x\to 0} (\cos x)^{3/x^2} = e^{-3/2}$.

51. $\lim_{x\to 0} \frac{\sin^{-1} x}{x}$

SOLUTION $\lim_{x\to 0} \frac{\sin^{-1} x}{x} = \lim_{x\to 0} \frac{\frac{1}{\sqrt{1-x^2}}}{1} = 1.$

53. $\lim_{x\to 1} \frac{\tan^{-1} x - \frac{\pi}{4}}{\tan \frac{\pi}{4}x - 1}$

SOLUTION $\lim_{x\to 1} \frac{\tan^{-1} x - \frac{\pi}{4}}{\tan(\pi x/4) - 1} = \lim_{x\to 1} \frac{\frac{1}{1+x^2}}{\frac{\pi}{4}\sec^2(\pi x/4)} = \frac{\frac{1}{2}}{\frac{\pi}{2}} = \frac{1}{\pi}.$

55. Evaluate $\lim_{x\to\pi/2} \frac{\cos mx}{\cos nx}$, where $m, n \neq 0$ are integers.

SOLUTION Suppose m and n are even. Then there exist integers k and l such that $m = 2k$ and $n = 2l$ and

$$\lim_{x\to\pi/2} \frac{\cos mx}{\cos nx} = \frac{\cos k\pi}{\cos l\pi} = (-1)^{k-l}.$$

Now, suppose m is even and n is odd. Then

$$\lim_{x\to\pi/2} \frac{\cos mx}{\cos nx}$$

does not exist (from one side the limit tends toward $-\infty$, while from the other side the limit tends toward $+\infty$). Third, suppose m is odd and n is even. Then

$$\lim_{x\to\pi/2} \frac{\cos mx}{\cos nx} = 0.$$

Finally, suppose m and n are odd. This is the only case when the limit is indeterminate. Then there exist integers k and l such that $m = 2k + 1$, $n = 2l + 1$ and, by L'Hôpital's Rule,

$$\lim_{x\to\pi/2} \frac{\cos mx}{\cos nx} = \lim_{x\to\pi/2} \frac{-m \sin mx}{-n \sin nx} = (-1)^{k-l}\frac{m}{n}.$$

To summarize,

$$\lim_{x\to\pi/2} \frac{\cos mx}{\cos nx} = \begin{cases} (-1)^{(m-n)/2}, & m, n \text{ even} \\ \text{does not exist}, & m \text{ even}, n \text{ odd} \\ 0 & m \text{ odd}, n \text{ even} \\ (-1)^{(m-n)/2}\frac{m}{n}, & m, n \text{ odd} \end{cases}$$

57. Prove the following limit formula for e:

$$e = \lim_{x\to 0} (1 + x)^{1/x}$$

Then find a value of x such that $|(1 + x)^{1/x} - e| \leq 0.001$.

SOLUTION Using L'Hôpital's Rule,

$$\lim_{x\to 0} \frac{\ln(1 + x)}{x} = \lim_{x\to 0} \frac{\frac{1}{1+x}}{1} = 1.$$

Thus,

$$\lim_{x\to 0} \ln\left((1+x)^{1/x}\right) = \lim_{x\to 0} \frac{1}{x}\ln(1+x) = \lim_{x\to 0} \frac{\ln(1+x)}{x} = 1,$$

and $\lim_{x\to 0}(1+x)^{1/x} = e^1 = e$. For $x = 0.0005$,

$$\left|(1+x)^{1/x} - e\right| = |(1.0005)^{2000} - e| \approx 6.79 \times 10^{-4} < 0.001.$$

59. Let $f(x) = x^{1/x}$ for $x > 0$.

(a) Calculate $\lim_{x\to 0+} f(x)$ and $\lim_{x\to\infty} f(x)$.

(b) Find the maximum value of $f(x)$, and determine the intervals on which $f(x)$ is increasing or decreasing.

SOLUTION

(a) Let $f(x) = x^{1/x}$. Note that $\lim_{x\to 0+} x^{1/x}$ is not indeterminate. As $x \to 0+$, the base of the function tends toward 0 and the exponent tends toward $+\infty$. Both of these factors force $x^{1/x}$ toward 0. Thus, $\lim_{x\to 0+} f(x) = 0$. On the other hand, $\lim_{x\to\infty} f(x)$ is indeterminate. We calculate this limit as follows:

$$\lim_{x\to\infty} \ln f(x) = \lim_{x\to\infty} \frac{\ln x}{x} = \lim_{x\to\infty} \frac{1}{x} = 0,$$

so $\lim_{x\to\infty} f(x) = e^0 = 1$.

(b) Again, let $f(x) = x^{1/x}$, so that $\ln f(x) = \frac{1}{x}\ln x$. To find the derivative f', we apply the derivative to both sides:

$$\frac{d}{dx}\ln f(x) = \frac{d}{dx}\left(\frac{1}{x}\ln x\right)$$

$$\frac{1}{f(x)} f'(x) = -\frac{\ln x}{x^2} + \frac{1}{x^2}$$

$$f'(x) = f(x)\left(-\frac{\ln x}{x^2} + \frac{1}{x^2}\right) = \frac{x^{1/x}}{x^2}(1 - \ln x)$$

Thus, f is increasing for $0 < x < e$, is decreasing for $x > e$ and has a maximum at $x = e$. The maximum value is $f(e) = e^{1/e} \approx 1.444668$.

61. Determine whether $f << g$ or $g << f$ (or neither) for the functions $f(x) = \log_{10} x$ and $g(x) = \ln x$.

SOLUTION Because

$$\lim_{x\to\infty} \frac{f(x)}{g(x)} = \lim_{x\to\infty} \frac{\log_{10} x}{\ln x} = \lim_{x\to\infty} \frac{\frac{\ln x}{\ln 10}}{\ln x} = \frac{1}{\ln 10},$$

neither $f << g$ or $g << f$ is satisfied.

63. Just as exponential functions are distinguished by their rapid rate of increase, the logarithm functions grow particularly slowly. Show that $\ln x << x^a$ for all $a > 0$.

SOLUTION Using L'Hôpital's Rule:

$$\lim_{x\to\infty} \frac{\ln x}{x^a} = \lim_{x\to\infty} \frac{x^{-1}}{ax^{a-1}} = \lim_{x\to\infty} \frac{1}{a}x^{-a} = 0;$$

hence, $\ln x << (x^a)$.

65. Determine whether $\sqrt{x} << e^{\sqrt{\ln x}}$ or $e^{\sqrt{\ln x}} << \sqrt{x}$. *Hint:* Use the substitution $u = \ln x$ instead of L'Hôpital's Rule.

SOLUTION Let $u = \ln x$, then $x = e^u$, and as $x \to \infty$, $u \to \infty$. So

$$\lim_{x\to\infty} \frac{e^{\sqrt{\ln x}}}{\sqrt{x}} = \lim_{u\to\infty} \frac{e^{\sqrt{u}}}{e^{u/2}} = \lim_{u\to\infty} e^{\sqrt{u}-\frac{u}{2}}.$$

We need to examine $\lim_{u\to\infty}(\sqrt{u} - \frac{u}{2})$. Since

$$\lim_{u\to\infty} \frac{u/2}{\sqrt{u}} = \lim_{u\to\infty} \frac{\frac{1}{2}}{\frac{1}{2\sqrt{u}}} = \lim_{u\to\infty} \sqrt{u} = \infty,$$

$\sqrt{u} = o(u/2)$ and $\lim_{u\to\infty}\left(\sqrt{u} - \frac{u}{2}\right) = -\infty$. Thus

$$\lim_{u\to\infty} e^{\sqrt{u}-\frac{u}{2}} = e^{-\infty} = 0 \quad \text{so} \quad \lim_{x\to\infty} \frac{e^{\sqrt{\ln x}}}{\sqrt{x}} = 0$$

and $e^{\sqrt{\ln x}} << \sqrt{x}$.

67. Assumptions Matter Let $f(x) = x(2 + \sin x)$ and $g(x) = x^2 + 1$.

(a) Show directly that $\lim_{x\to\infty} f(x)/g(x) = 0$.

(b) Show that $\lim_{x\to\infty} f(x) = \lim_{x\to\infty} g(x) = \infty$, but $\lim_{x\to\infty} f'(x)/g'(x)$ does not exist.

Do (a) and (b) contradict L'Hôpital's Rule? Explain.

SOLUTION

(a) $1 \le 2 + \sin x \le 3$, so

$$\frac{x}{x^2+1} \le \frac{x(2+\sin x)}{x^2+1} \le \frac{3x}{x^2+1}.$$

Since,

$$\lim_{x\to\infty} \frac{x}{x^2+1} = \lim_{x\to\infty} \frac{3x}{x^2+1} = 0,$$

it follows by the Squeeze Theorem that

$$\lim_{x\to\infty} \frac{x(2+\sin x)}{x^2+1} = 0.$$

(b) $\lim_{x\to\infty} f(x) = \lim_{x\to\infty} x(2+\sin x) \ge \lim_{x\to\infty} x = \infty$ and $\lim_{x\to\infty} g(x) = \lim_{x\to\infty} (x^2+1) = \infty$, but

$$\lim_{x\to\infty} \frac{f'(x)}{g'(x)} = \lim_{x\to\infty} \frac{x(\cos x) + (2+\sin x)}{2x}$$

does not exist since $\cos x$ oscillates. This does not violate L'Hôpital's Rule since the theorem clearly states

$$\lim_{x\to\infty} \frac{f(x)}{g(x)} = \lim_{x\to\infty} \frac{f'(x)}{g'(x)}$$

"provided the limit on the right exists."

69. Let $G(b) = \lim_{x\to\infty} (1 + b^x)^{1/x}$.

(a) Use the result of Exercise 68 to evaluate $G(b)$ for all $b > 0$.

(b) GU Verify your result graphically by plotting $y = (1 + b^x)^{1/x}$ together with the horizontal line $y = G(b)$ for the values $b = 0.25, 0.5, 2, 3$.

SOLUTION

(a) Using Exercise 68, we see that $G(b) = e^{H(b)}$. Thus, $G(b) = 1$ if $0 \le b \le 1$ and $G(b) = b$ if $b > 1$.

(b)

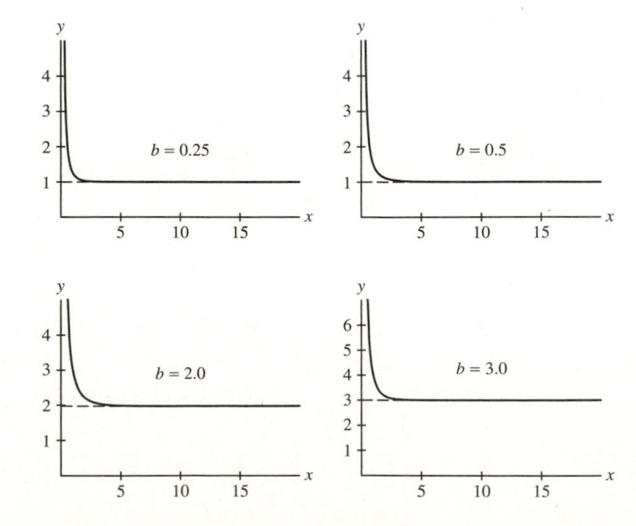

In Exercises 71–73, let

$$f(x) = \begin{cases} e^{-1/x^2} & \text{for } x \neq 0 \\ 0 & \text{for } x = 0 \end{cases}$$

These exercises show that $f(x)$ has an unusual property: All of its derivatives at $x = 0$ exist and are equal to zero.

71. Show that $\lim_{x\to 0} \frac{f(x)}{x^k} = 0$ for all k. *Hint:* Let $t = x^{-1}$ and apply the result of Exercise 70.

SOLUTION $\lim_{x\to 0} \frac{f(x)}{x^k} = \lim_{x\to 0} \frac{1}{x^k e^{1/x^2}}$. Let $t = 1/x$. As $x \to 0$, $t \to \infty$. Thus,

$$\lim_{x\to 0} \frac{1}{x^k e^{1/x^2}} = \lim_{t\to\infty} \frac{t^k}{e^{t^2}} = 0$$

by Exercise 70.

73. Show that for $k \geq 1$ and $x \neq 0$,

$$f^{(k)}(x) = \frac{P(x)e^{-1/x^2}}{x^r}$$

for some polynomial $P(x)$ and some exponent $r \geq 1$. Use the result of Exercise 71 to show that $f^{(k)}(0)$ exists and is equal to zero for all $k \geq 1$.

SOLUTION For $x \neq 0$, $f'(x) = e^{-1/x^2}\left(\frac{2}{x^3}\right)$. Here $P(x) = 2$ and $r = 3$. Assume $f^{(k)}(x) = \frac{P(x)e^{-1/x^2}}{x^r}$. Then

$$f^{(k+1)}(x) = e^{-1/x^2}\left(\frac{x^3P'(x) + (2 - rx^2)P(x)}{x^{r+3}}\right)$$

which is of the form desired.

Moreover, from Exercise 72, $f'(0) = 0$. Suppose $f^{(k)}(0) = 0$. Then

$$f^{(k+1)}(0) = \lim_{x\to 0} \frac{f^{(k)}(x) - f^{(k)}(0)}{x - 0} = \lim_{x\to 0} \frac{P(x)e^{-1/x^2}}{x^{r+1}} = P(0) \lim_{x\to 0} \frac{f(x)}{x^{r+1}} = 0.$$

Further Insights and Challenges

75. The Second Derivative Test for critical points fails if $f''(c) = 0$. This exercise develops a **Higher Derivative Test** based on the sign of the first nonzero derivative. Suppose that

$$f'(c) = f''(c) = \cdots = f^{(n-1)}(c) = 0, \quad \text{but} \quad f^{(n)}(c) \neq 0$$

(a) Show, by applying L'Hôpital's Rule n times, that

$$\lim_{x\to c} \frac{f(x) - f(c)}{(x - c)^n} = \frac{1}{n!} f^{(n)}(c)$$

where $n! = n(n-1)(n-2)\cdots(2)(1)$.

(b) Use (a) to show that if n is even, then $f(c)$ is a local minimum if $f^{(n)}(c) > 0$ and is a local maximum if $f^{(n)}(c) < 0$. *Hint:* If n is even, then $(x - c)^n > 0$ for $x \neq a$, so $f(x) - f(c)$ must be positive for x near c if $f^{(n)}(c) > 0$.

(c) Use (a) to show that if n is odd, then $f(c)$ is neither a local minimum nor a local maximum.

SOLUTION

(a) Repeated application of L'Hôpital's rule yields

$$\begin{aligned} \lim_{x\to c} \frac{f(x) - f(c)}{(x - c)^n} &= \lim_{x\to c} \frac{f'(x)}{n(x - c)^{n-1}} \\ &= \lim_{x\to c} \frac{f''(x)}{n(n-1)(x - c)^{n-2}} \\ &= \lim_{x\to c} \frac{f'''(x)}{n(n-1)(n-2)(x - c)^{n-3}} \\ &= \cdots \\ &= \frac{1}{n!} f^{(n)}(c) \end{aligned}$$

(b) Suppose n is even. Then $(x-c)^n > 0$ for all $x \neq c$. If $f^{(n)}(c) > 0$, it follows that $f(x) - f(c)$ must be positive for x near c. In other words, $f(x) > f(c)$ for x near c and $f(c)$ is a local minimum. On the other hand, if $f^{(n)}(c) < 0$, it follows that $f(x) - f(c)$ must be negative for x near c. In other words, $f(x) < f(c)$ for x near c and $f(c)$ is a local maximum.

(c) If n is odd, then $(x-c)^n > 0$ for $x > c$ but $(x-c)^n < 0$ for $x < c$. If $f^{(n)}(c) > 0$, it follows that $f(x) - f(c)$ must be positive for x near c and $x > c$ but is negative for x near c and $x < c$. In other words, $f(x) > f(c)$ for x near c and $x > c$ but $f(x) < f(c)$ for x near c and $x < c$. Thus, $f(c)$ is neither a local minimum nor a local maximum. We obtain a similar result if $f^{(n)}(c) < 0$.

77. We expended a lot of effort to evaluate $\lim\limits_{x\to 0} \frac{\sin x}{x}$ in Chapter 2. Show that we could have evaluated it easily using L'Hôpital's Rule. Then explain why this method would involve *circular reasoning*.

SOLUTION $\lim\limits_{x\to 0} \frac{\sin x}{x} = \lim\limits_{x\to 0} \frac{\cos x}{1} = 1$. To use L'Hôpital's Rule to evaluate $\lim\limits_{x\to 0} \frac{\sin x}{x}$, we must know that the derivative of $\sin x$ is $\cos x$, but to determine the derivative of $\sin x$, we must be able to evaluate $\lim\limits_{x\to 0} \frac{\sin x}{x}$.

79. Patience Required Use L'Hôpital's Rule to evaluate and check your answers numerically:

(a) $\lim\limits_{x\to 0+} \left(\frac{\sin x}{x}\right)^{1/x^2}$ **(b)** $\lim\limits_{x\to 0} \left(\frac{1}{\sin^2 x} - \frac{1}{x^2}\right)$

SOLUTION

(a) We start by evaluating

$$\lim_{x\to 0+} \ln\left(\frac{\sin x}{x}\right)^{1/x^2} = \lim_{x\to 0+} \frac{\ln(\sin x) - \ln x}{x^2}.$$

Repeatedly using L'Hôpital's Rule, we find

$$\begin{aligned}\lim_{x\to 0+} \ln\left(\frac{\sin x}{x}\right)^{1/x^2} &= \lim_{x\to 0+} \frac{\cot x - x^{-1}}{2x} = \lim_{x\to 0+} \frac{x\cos x - \sin x}{2x^2 \sin x} = \lim_{x\to 0+} \frac{-x\sin x}{2x^2\cos x + 4x\sin x} \\ &= \lim_{x\to 0+} \frac{-x\cos x - \sin x}{8x\cos x + 4\sin x - 2x^2\sin x} = \lim_{x\to 0+} \frac{-2\cos x + x\sin x}{12\cos x - 2x^2\cos x - 12x\sin x} \\ &= -\frac{2}{12} = -\frac{1}{6}.\end{aligned}$$

Therefore, $\lim\limits_{x\to 0+} \left(\frac{\sin x}{x}\right)^{1/x^2} = e^{-1/6}$. Numerically we find:

x	1	0.1	0.01
$\left(\frac{\sin x}{x}\right)^{1/x^2}$	0.841471	0.846435	0.846481

Note that $e^{-1/6} \approx 0.846481724$.

(b) Repeatedly using L'Hôpital's Rule and simplifying, we find

$$\begin{aligned}\lim_{x\to 0} \left(\frac{1}{\sin^2 x} - \frac{1}{x^2}\right) &= \lim_{x\to 0} \frac{x^2 - \sin^2 x}{x^2 \sin^2 x} = \lim_{x\to 0} \frac{2x - 2\sin x\cos x}{x^2(2\sin x\cos x) + 2x\sin^2 x} = \lim_{x\to 0} \frac{2x - 2\sin 2x}{x^2\sin 2x + 2x\sin^2 x} \\ &= \lim_{x\to 0} \frac{2 - 2\cos 2x}{2x^2\cos 2x + 2x\sin 2x + 4x\sin x\cos x + 2\sin^2 x} \\ &= \lim_{x\to 0} \frac{2 - 2\cos 2x}{2x^2\cos 2x + 4x\sin 2x + 2\sin^2 x} \\ &= \lim_{x\to 0} \frac{4\sin 2x}{-4x^2\sin 2x + 4x\cos 2x + 8x\cos 2x + 4\sin 2x + 4\sin x\cos x} \\ &= \lim_{x\to 0} \frac{4\sin 2x}{(6 - 4x^2)\sin 2x + 12x\cos 2x} \\ &= \lim_{x\to 0} \frac{8\cos 2x}{(12 - 8x^2)\cos 2x - 8x\sin 2x + 12\cos 2x - 24x\sin 2x} = \frac{1}{3}.\end{aligned}$$

Numerically we find:

x	1	0.1	0.01
$\frac{1}{\sin^2 x} - \frac{1}{x^2}$	0.412283	0.334001	0.333340

4.6 Graph Sketching and Asymptotes

Preliminary Questions

1. Sketch an arc where f' and f'' have the sign combination $++$. Do the same for $-+$.

SOLUTION An arc with the sign combination $++$ (increasing, concave up) is shown below at the left. An arc with the sign combination $-+$ (decreasing, concave up) is shown below at the right.

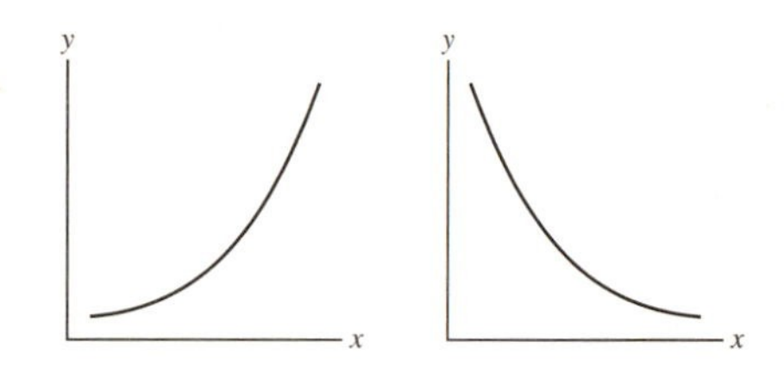

2. If the sign combination of f' and f'' changes from $++$ to $+-$ at $x = c$, then (choose the correct answer):

(a) $f(c)$ is a local min

(b) $f(c)$ is a local max

(c) c is a point of inflection

SOLUTION Because the sign of the second derivative changes at $x = c$, the correct response is **(c)**: c is a point of inflection.

3. The second derivative of the function $f(x) = (x-4)^{-1}$ is $f''(x) = 2(x-4)^{-3}$. Although $f''(x)$ changes sign at $x = 4$, $f(x)$ does not have a point of inflection at $x = 4$. Why not?

SOLUTION The function f does not have a point of inflection at $x = 4$ because $x = 4$ is not in the domain of f.

Exercises

1. Determine the sign combinations of f' and f'' for each interval A–G in Figure 16.

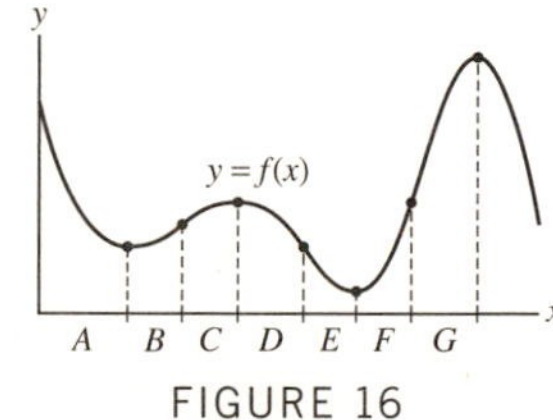

FIGURE 16

SOLUTION

- In A, f is decreasing and concave up, so $f' < 0$ and $f'' > 0$.
- In B, f is increasing and concave up, so $f' > 0$ and $f'' > 0$.
- In C, f is increasing and concave down, so $f' > 0$ and $f'' < 0$.
- In D, f is decreasing and concave down, so $f' < 0$ and $f'' < 0$.
- In E, f is decreasing and concave up, so $f' < 0$ and $f'' > 0$.
- In F, f is increasing and concave up, so $f' > 0$ and $f'' > 0$.
- In G, f is increasing and concave down, so $f' > 0$ and $f'' < 0$.

In Exercises 3–6, draw the graph of a function for which f' and f'' take on the given sign combinations.

3. $++$, $+-$, $--$

SOLUTION This function changes from concave up to concave down at $x = -1$ and from increasing to decreasing at $x = 0$.

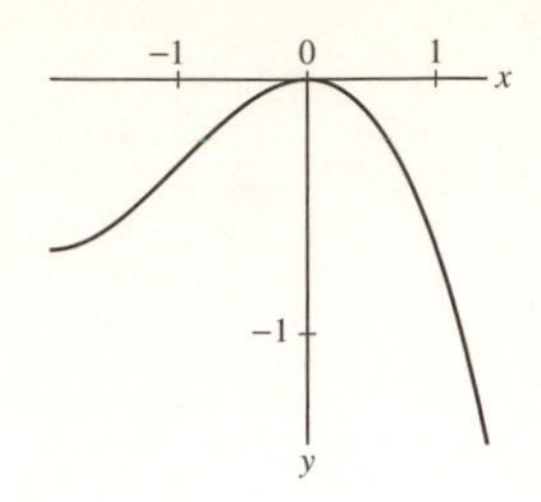

5. $-+,\quad --,\quad -+$

SOLUTION The function is decreasing everywhere and changes from concave up to concave down at $x = -1$ and from concave down to concave up at $x = -\frac{1}{2}$.

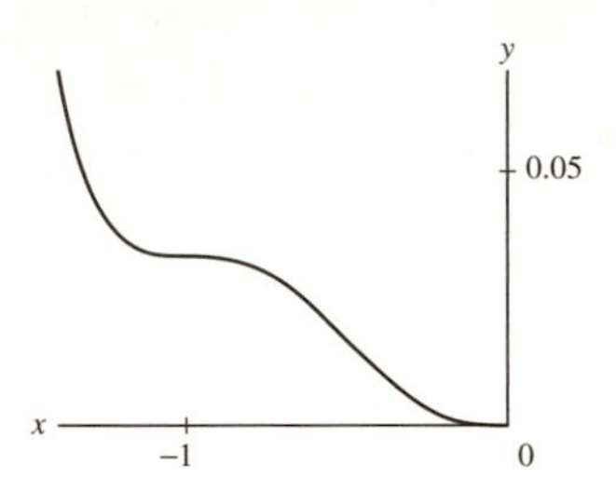

7. Sketch the graph of $y = x^2 - 5x + 4$.

SOLUTION Let $f(x) = x^2 - 5x + 4$. Then $f'(x) = 2x - 5$ and $f''(x) = 2$. Hence f is decreasing for $x < 5/2$, is increasing for $x > 5/2$, has a local minimum at $x = 5/2$ and is concave up everywhere.

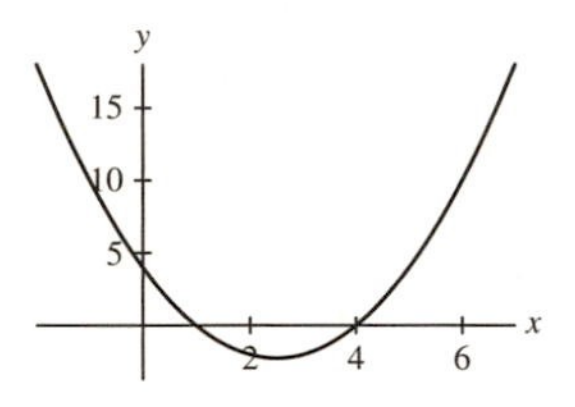

9. Sketch the graph of $f(x) = x^3 - 3x^2 + 2$. Include the zeros of $f(x)$, which are $x = 1$ and $1 \pm \sqrt{3}$ (approximately $-0.73, 2.73$).

SOLUTION Let $f(x) = x^3 - 3x^2 + 2$. Then $f'(x) = 3x^2 - 6x = 3x(x - 2) = 0$ yields $x = 0, 2$ and $f''(x) = 6x - 6$. Thus f is concave down for $x < 1$, is concave up for $x > 1$, has an inflection point at $x = 1$, is increasing for $x < 0$ and for $x > 2$, is decreasing for $0 < x < 2$, has a local maximum at $x = 0$, and has a local minimum at $x = 2$.

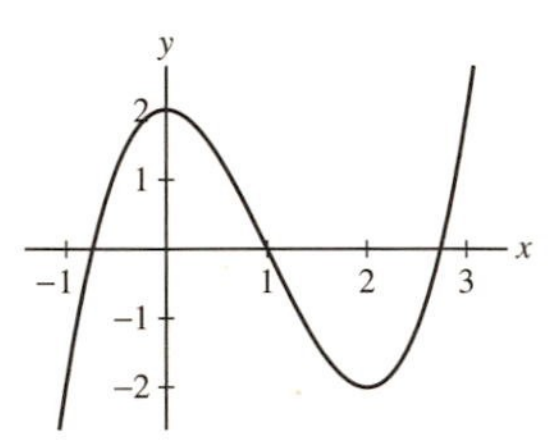

11. Extend the sketch of the graph of $f(x) = \cos x + \frac{1}{2}x$ in Example 4 to the interval $[0, 5\pi]$.

SOLUTION Let $f(x) = \cos x + \frac{1}{2}x$. Then $f'(x) = -\sin x + \frac{1}{2} = 0$ yields critical points at $x = \frac{\pi}{6}, \frac{5\pi}{6}, \frac{13\pi}{6}, \frac{17\pi}{6}, \frac{25\pi}{6}$, and $\frac{29\pi}{6}$. Moreover, $f''(x) = -\cos x$ so there are points of inflection at $x = \frac{\pi}{2}, \frac{3\pi}{2}, \frac{5\pi}{2}, \frac{7\pi}{2}$, and $\frac{9\pi}{2}$.

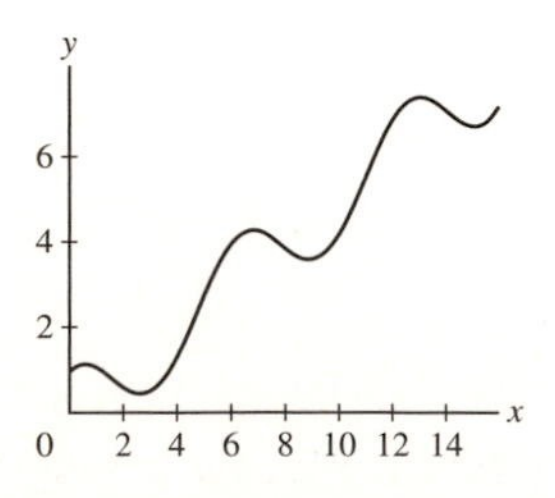

In Exercises 13–34, find the transition points, intervals of increase/decrease, concavity, and asymptotic behavior. Then sketch the graph, with this information indicated.

13. $y = x^3 + 24x^2$

SOLUTION Let $f(x) = x^3 + 24x^2$. Then $f'(x) = 3x^2 + 48x = 3x(x + 16)$ and $f''(x) = 6x + 48$. This shows that f has critical points at $x = 0$ and $x = -16$ and a candidate for an inflection point at $x = -8$.

Interval	$(-\infty, -16)$	$(-16, -8)$	$(-8, 0)$	$(0, \infty)$
Signs of f' and f''	$+-$	$--$	$-+$	$++$

Thus, there is a local maximum at $x = -16$, a local minimum at $x = 0$, and an inflection point at $x = -8$. Moreover,

$$\lim_{x\to-\infty} f(x) = -\infty \quad \text{and} \quad \lim_{x\to\infty} f(x) = \infty.$$

Here is a graph of f with these transition points highlighted as in the graphs in the textbook.

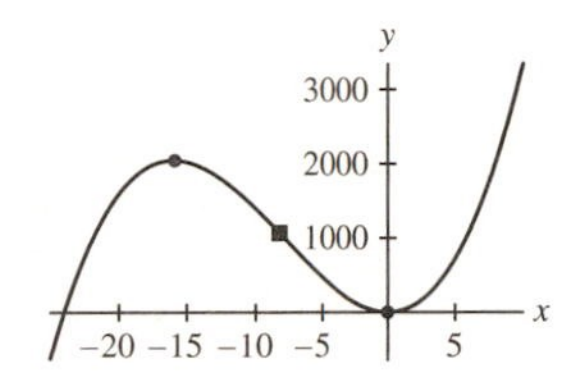

15. $y = x^2 - 4x^3$

SOLUTION Let $f(x) = x^2 - 4x^3$. Then $f'(x) = 2x - 12x^2 = 2x(1 - 6x)$ and $f''(x) = 2 - 24x$. Critical points are at $x = 0$ and $x = \frac{1}{6}$, and the sole candidate point of inflection is at $x = \frac{1}{12}$.

Interval	$(-\infty, 0)$	$(0, \frac{1}{12})$	$(\frac{1}{12}, \frac{1}{6})$	$(\frac{1}{6}, \infty)$
Signs of f' and f''	$-+$	$++$	$+-$	$--$

Thus, $f(0)$ is a local minimum, $f(\frac{1}{6})$ is a local maximum, and there is a point of inflection at $x = \frac{1}{12}$. Moreover,

$$\lim_{x\to\pm\infty} f(x) = \infty.$$

Here is the graph of f with transition points highlighted as in the textbook:

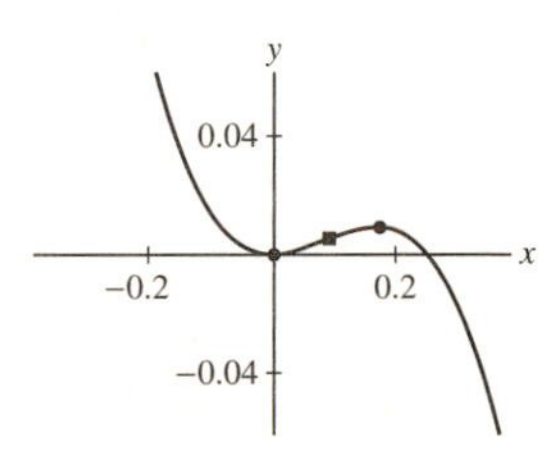

17. $y = 4 - 2x^2 + \frac{1}{6}x^4$

SOLUTION Let $f(x) = \frac{1}{6}x^4 - 2x^2 + 4$. Then $f'(x) = \frac{2}{3}x^3 - 4x = \frac{2}{3}x\left(x^2 - 6\right)$ and $f''(x) = 2x^2 - 4$. This shows that f has critical points at $x = 0$ and $x = \pm\sqrt{6}$ and has candidates for points of inflection at $x = \pm\sqrt{2}$.

Interval	$(-\infty, -\sqrt{6})$	$(-\sqrt{6}, -\sqrt{2})$	$(-\sqrt{2}, 0)$	$(0, \sqrt{2})$	$(\sqrt{2}, \sqrt{6})$	$(\sqrt{6}, \infty)$
Signs of f' and f''	$-+$	$++$	$+-$	$--$	$-+$	$++$

Thus, f has local minima at $x = \pm\sqrt{6}$, a local maximum at $x = 0$, and inflection points at $x = \pm\sqrt{2}$. Moreover,

$$\lim_{x\to\pm\infty} f(x) = \infty.$$

Here is a graph of f with transition points highlighted.

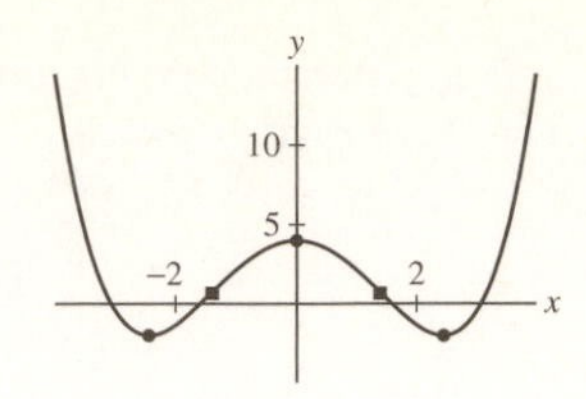

19. $y = x^5 + 5x$

SOLUTION Let $f(x) = x^5 + 5x$. Then $f'(x) = 5x^4 + 5 = 5(x^4 + 1)$ and $f''(x) = 20x^3$. $f'(x) > 0$ for all x, so the graph has no critical points and is always increasing. $f''(x) = 0$ at $x = 0$. Sign analyses reveal that $f''(x)$ changes from negative to positive at $x = 0$, so that the graph of $f(x)$ has an inflection point at $(0, 0)$. Moreover,

$$\lim_{x\to-\infty} f(x) = -\infty \quad \text{and} \quad \lim_{x\to\infty} f(x) = \infty.$$

Here is a graph of f with transition points highlighted.

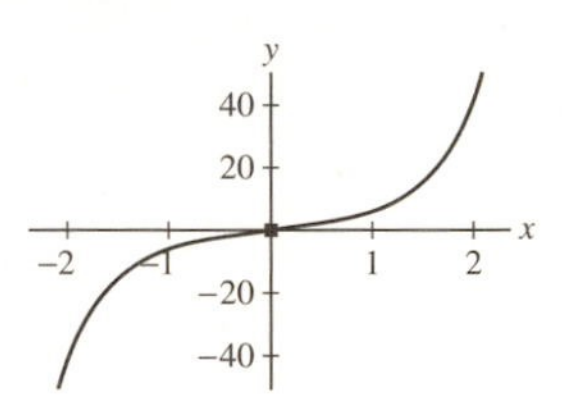

21. $y = x^4 - 3x^3 + 4x$

SOLUTION Let $f(x) = x^4 - 3x^3 + 4x$. Then $f'(x) = 4x^3 - 9x^2 + 4 = (4x^2 - x - 2)(x - 2)$ and $f''(x) = 12x^2 - 18x = 6x(2x - 3)$. This shows that f has critical points at $x = 2$ and $x = \dfrac{1 \pm \sqrt{33}}{8}$ and candidate points of inflection at $x = 0$ and $x = \frac{3}{2}$. Sign analyses reveal that $f'(x)$ changes from negative to positive at $x = \frac{1-\sqrt{33}}{8}$, from positive to negative at $x = \frac{1+\sqrt{33}}{8}$, and again from negative to positive at $x = 2$. Therefore, $f(\frac{1-\sqrt{33}}{8})$ and $f(2)$ are local minima of $f(x)$, and $f(\frac{1+\sqrt{33}}{8})$ is a local maximum. Further sign analyses reveal that $f''(x)$ changes from positive to negative at $x = 0$ and from negative to positive at $x = \frac{3}{2}$, so that there are points of inflection both at $x = 0$ and $x = \frac{3}{2}$. Moreover,

$$\lim_{x\to\pm\infty} f(x) = \infty.$$

Here is a graph of $f(x)$ with transition points highlighted.

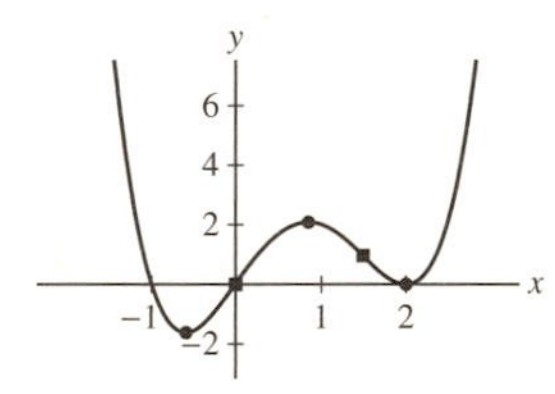

23. $y = x^7 - 14x^6$

SOLUTION Let $f(x) = x^7 - 14x^6$. Then $f'(x) = 7x^6 - 84x^5 = 7x^5(x - 12)$ and $f''(x) = 42x^5 - 420x^4 = 42x^4(x - 10)$. Critical points are at $x = 0$ and $x = 12$, and candidate inflection points are at $x = 0$ and $x = 10$. Sign analyses reveal that $f'(x)$ changes from positive to negative at $x = 0$ and from negative to positive at $x = 12$. Therefore $f(0)$ is a local maximum and $f(12)$ is a local minimum. Also, $f''(x)$ changes from negative to positive at $x = 10$. Therefore, there is a point of inflection at $x = 10$. Moreover,

$$\lim_{x\to-\infty} f(x) = -\infty \quad \text{and} \quad \lim_{x\to\infty} f(x) = \infty.$$

Here is a graph of f with transition points highlighted.

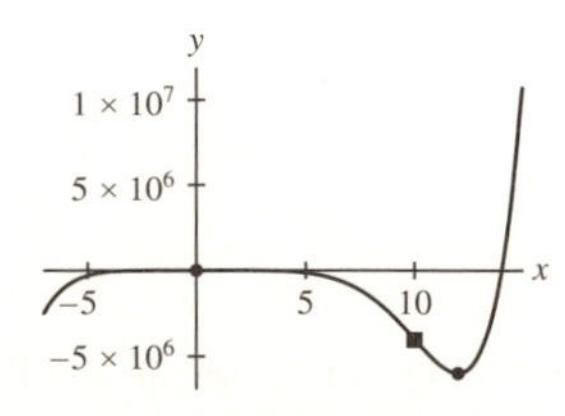

25. $y = x - 4\sqrt{x}$

SOLUTION Let $f(x) = x - 4\sqrt{x} = x - 4x^{1/2}$. Then $f'(x) = 1 - 2x^{-1/2}$. This shows that f has critical points at $x = 0$ (where the derivative does not exist) and at $x = 4$ (where the derivative is zero). Because $f'(x) < 0$ for $0 < x < 4$ and $f'(x) > 0$ for $x > 4$, $f(4)$ is a local minimum. Now $f''(x) = x^{-3/2} > 0$ for all $x > 0$, so the graph is always concave up. Moreover,

$$\lim_{x\to\infty} f(x) = \infty.$$

Here is a graph of f with transition points highlighted.

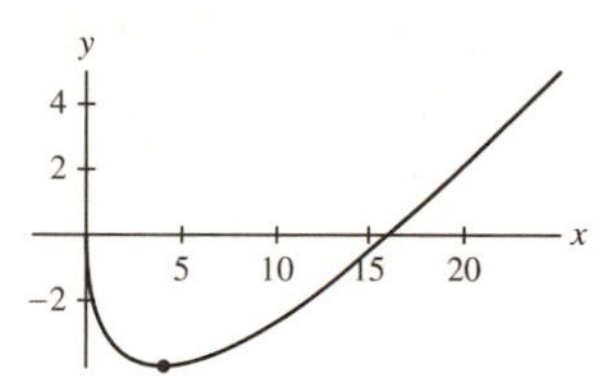

27. $y = x(8 - x)^{1/3}$

SOLUTION Let $f(x) = x\,(8 - x)^{1/3}$. Then

$$f'(x) = x \cdot \tfrac{1}{3}\,(8 - x)^{-2/3}\,(-1) + (8 - x)^{1/3} \cdot 1 = \frac{24 - 4x}{3\,(8 - x)^{2/3}}$$

and similarly

$$f''(x) = \frac{4x - 48}{9\,(8 - x)^{5/3}}.$$

Critical points are at $x = 8$ and $x = 6$, and candidate inflection points are $x = 8$ and $x = 12$. Sign analyses reveal that $f'(x)$ changes from positive to negative at $x = 6$ and $f'(x)$ remains negative on either side of $x = 8$. Moreover, $f''(x)$ changes from negative to positive at $x = 8$ and from positive to negative at $x = 12$. Therefore, f has a local maximum at $x = 6$ and inflection points at $x = 8$ and $x = 12$. Moreover,

$$\lim_{x\to\pm\infty} f(x) = -\infty.$$

Here is a graph of f with the transition points highlighted.

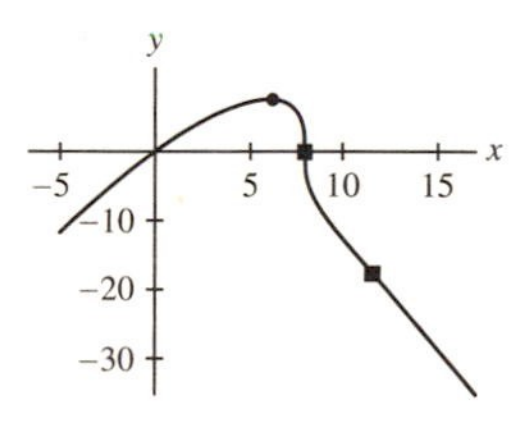

29. $y = xe^{-x^2}$

SOLUTION Let $f(x) = xe^{-x^2}$. Then

$$f'(x) = -2x^2e^{-x^2} + e^{-x^2} = (1 - 2x^2)e^{-x^2},$$

and

$$f''(x) = (4x^3 - 2x)e^{-x^2} - 4xe^{-x^2} = 2x(2x^2 - 3)e^{-x^2}.$$

There are critical points at $x = \pm\frac{\sqrt{2}}{2}$, and $x = 0$ and $x = \pm\frac{\sqrt{3}}{2}$ are candidates for inflection points. Sign analysis shows that $f'(x)$ changes from negative to positive at $x = -\frac{\sqrt{2}}{2}$ and from positive to negative at $x = \frac{\sqrt{2}}{2}$. Moreover, $f''(x)$ changes from negative to positive at both $x = \pm\frac{\sqrt{3}}{2}$ and from positive to negative at $x = 0$. Therefore, f has a local minimum at $x = -\frac{\sqrt{2}}{2}$, a local maximum at $x = \frac{\sqrt{2}}{2}$ and inflection points at $x = 0$ and at $x = \pm\frac{\sqrt{3}}{2}$. Moreover,

$$\lim_{x\to\pm\infty} f(x) = 0,$$

so the graph has a horizontal asymptote at $y = 0$. Here is a graph of f with the transition points highlighted.

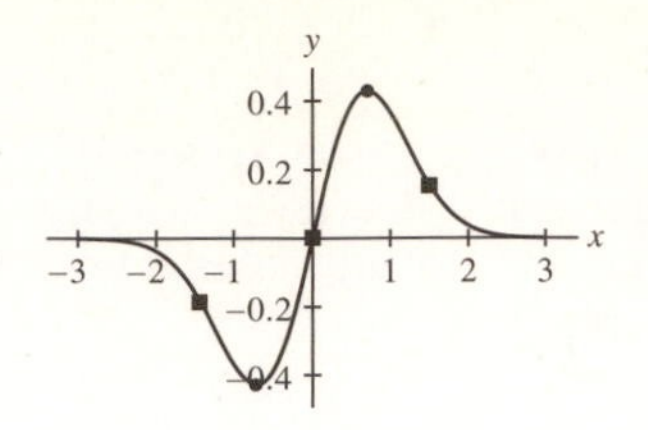

31. $y = x - 2\ln x$

SOLUTION Let $f(x) = x - 2\ln x$. Note that the domain of f is $x > 0$. Now,

$$f'(x) = 1 - \frac{2}{x} \quad \text{and} \quad f''(x) = \frac{2}{x^2}.$$

The only critical point is $x = 2$. Sign analysis shows that $f'(x)$ changes from negative to positive at $x = 2$, so $f(2)$ is a local minimum. Further, $f''(x) > 0$ for $x > 0$, so the graph is always concave up. Moreover,

$$\lim_{x\to\infty} f(x) = \infty.$$

Here is a graph of f with the transition points highlighted.

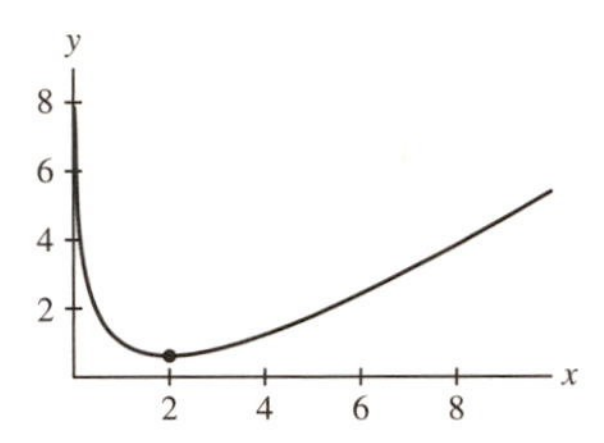

33. $y = x - x^2 \ln x$

SOLUTION Let $f(x) = x - x^2\ln x$. Then $f'(x) = 1 - x - 2x\ln x$ and $f''(x) = -3 - 2\ln x$. There is a critical point at $x = 1$, and $x = e^{-3/2} \approx 0.223$ is a candidate inflection point. Sign analysis shows that $f'(x)$ changes from positive to negative at $x = 1$ and that $f''(x)$ changes from positive to negative at $x = e^{-3/2}$. Therefore, f has a local maximum at $x = 1$ and a point of inflection at $x = e^{-3/2}$. Moreover,

$$\lim_{x\to\infty} f(x) = -\infty.$$

Here is a graph of f with the transition points highlighted.

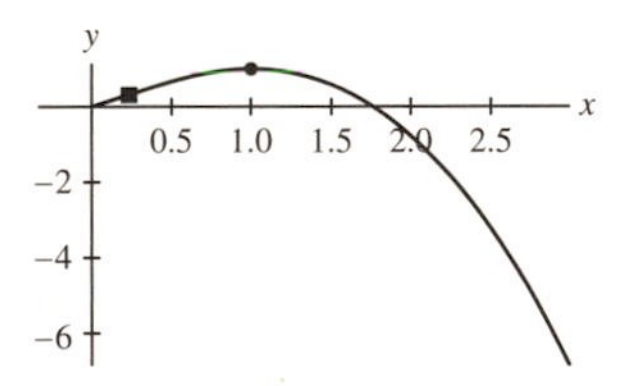

35. Sketch the graph of $f(x) = 18(x-3)(x-1)^{2/3}$ using the formulas

$$f'(x) = \frac{30\left(x - \frac{9}{5}\right)}{(x-1)^{1/3}}, \qquad f''(x) = \frac{20\left(x - \frac{3}{5}\right)}{(x-1)^{4/3}}$$

SOLUTION

$$f'(x) = \frac{30(x - \frac{9}{5})}{(x-1)^{1/3}}$$

yields critical points at $x = \frac{9}{5}$, $x = 1$.

$$f''(x) = \frac{20(x - \frac{3}{5})}{(x-1)^{4/3}}$$

yields potential inflection points at $x = \frac{3}{5}$, $x = 1$.

Interval	signs of f' and f''
$(-\infty, \frac{3}{5})$	$+-$
$(\frac{3}{5}, 1)$	$++$
$(1, \frac{9}{5})$	$-+$
$(\frac{9}{5}, \infty)$	$++$

The graph has an inflection point at $x = \frac{3}{5}$, a local maximum at $x = 1$ (at which the graph has a cusp), and a local minimum at $x = \frac{9}{5}$. The sketch looks something like this.

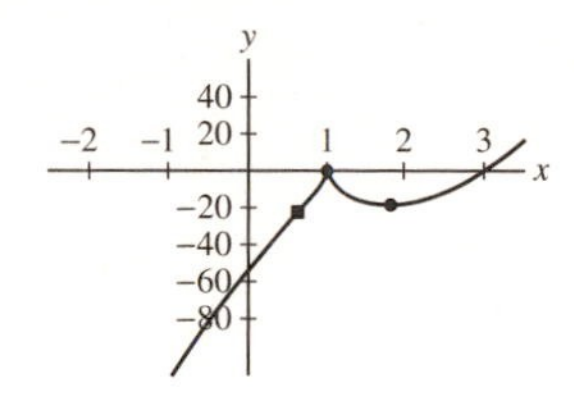

CAS *In Exercises 37–40, sketch the graph of the function, indicating all transition points. If necessary, use a graphing utility or computer algebra system to locate the transition points numerically.*

37. $y = x^2 - 10\ln(x^2 + 1)$

SOLUTION Let $f(x) = x^2 - 10\ln(x^2 + 1)$. Then $f'(x) = 2x - \dfrac{20x}{x^2 + 1}$, and

$$f''(x) = 2 - \frac{(x^2 + 1)(20) - (20x)(2x)}{(x^2 + 1)^2} = \frac{x^4 + 12x^2 - 9}{(x^2 + 1)^2}.$$

There are critical points at $x = 0$ and $x = \pm 3$, and $x = \pm\sqrt{-6 + 3\sqrt{5}}$ are candidates for inflection points. Sign analysis shows that $f'(x)$ changes from negative to positive at $x = \pm 3$ and from positive to negative at $x = 0$. Moreover, $f''(x)$ changes from positive to negative at $x = -\sqrt{-6 + 3\sqrt{5}}$ and from negative to positive at $x = \sqrt{-6 + 3\sqrt{5}}$. Therefore, f has a local maximum at $x = 0$, local minima at $x = \pm 3$ and points of inflection at $x = \pm\sqrt{-6 + 3\sqrt{5}}$. Here is a graph of f with the transition points highlighted.

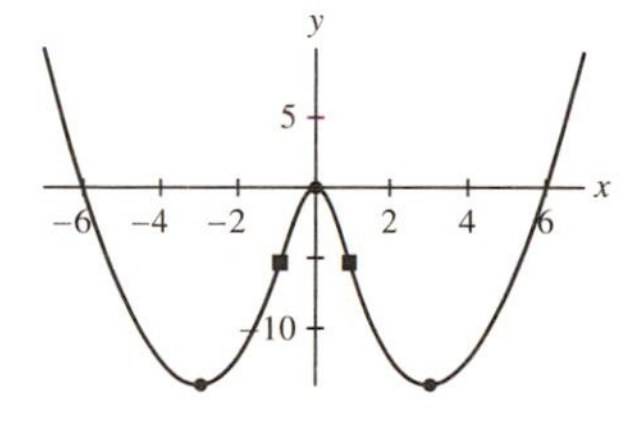

39. $y = x^4 - 4x^2 + x + 1$

SOLUTION Let $f(x) = x^4 - 4x^2 + x + 1$. Then $f'(x) = 4x^3 - 8x + 1$ and $f''(x) = 12x^2 - 8$. The critical points are $x = -1.473$, $x = 0.126$ and $x = 1.347$, while the candidates for points of inflection are $x = \pm\sqrt{\frac{2}{3}}$. Sign analysis reveals that $f'(x)$ changes from negative to positive at $x = -1.473$, from positive to negative at $x = 0.126$ and from negative to positive at $x = 1.347$. For the second derivative, $f''(x)$ changes from positive to negative at $x = -\sqrt{\frac{2}{3}}$ and from negative to positive at $x = \sqrt{\frac{2}{3}}$. Therefore, f has local minima at $x = -1.473$ and $x = 1.347$, a local maximum at $x = 0.126$ and points of inflection at $x = \pm\sqrt{\frac{2}{3}}$. Moreover,

$$\lim_{x\to\pm\infty} f(x) = \infty.$$

Here is a graph of f with the transition points highlighted.

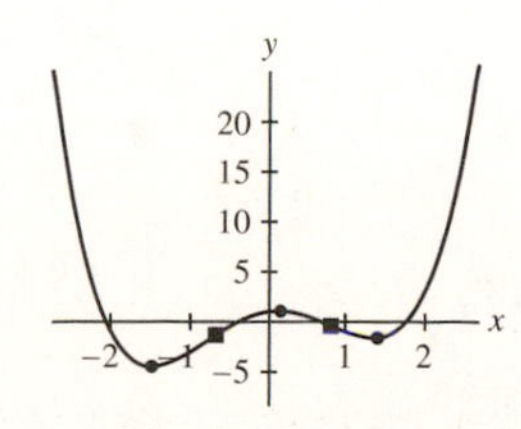

In Exercises 41–46, sketch the graph over the given interval, with all transition points indicated.

41. $y = x + \sin x$, $[0, 2\pi]$

SOLUTION Let $f(x) = x + \sin x$. Setting $f'(x) = 1 + \cos x = 0$ yields $\cos x = -1$, so that $x = \pi$ is the lone critical point on the interval $[0, 2\pi]$. Setting $f''(x) = -\sin x = 0$ yields potential points of inflection at $x = 0, \pi, 2\pi$ on the interval $[0, 2\pi]$.

Interval	signs of f' and f''
$(0, \pi)$	$+-$
$(\pi, 2\pi)$	$++$

The graph has an inflection point at $x = \pi$, and no local maxima or minima. Here is a sketch of the graph of $f(x)$:

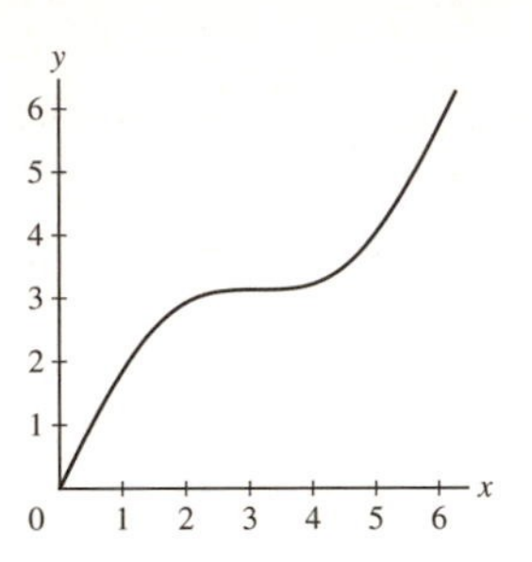

43. $y = 2\sin x - \cos^2 x$, $[0, 2\pi]$

SOLUTION Let $f(x) = 2\sin x - \cos^2 x$. Then $f'(x) = 2\cos x - 2\cos x\,(-\sin x) = \sin 2x + 2\cos x$ and $f''(x) = 2\cos 2x - 2\sin x$. Setting $f'(x) = 0$ yields $\sin 2x = -2\cos x$, so that $2\sin x\cos x = -2\cos x$. This implies $\cos x = 0$ or $\sin x = -1$, so that $x = \frac{\pi}{2}$ or $\frac{3\pi}{2}$. Setting $f''(x) = 0$ yields $2\cos 2x = 2\sin x$, so that $2\sin(\frac{\pi}{2} - 2x) = 2\sin x$, or $\frac{\pi}{2} - 2x = x \pm 2n\pi$. This yields $3x = \frac{\pi}{2} + 2n\pi$, or $x = \frac{\pi}{6}, \frac{5\pi}{6}, \frac{9\pi}{6} = \frac{3\pi}{2}$.

Interval	signs of f' and f''
$(0, \frac{\pi}{6})$	$++$
$(\frac{\pi}{6}, \frac{\pi}{2})$	$+-$
$(\frac{\pi}{2}, \frac{5\pi}{6})$	$--$
$(\frac{5\pi}{6}, \frac{3\pi}{2})$	$-+$
$(\frac{3\pi}{2}, 2\pi)$	$++$

The graph has a local maximum at $x = \frac{\pi}{2}$, a local minimum at $x = \frac{3\pi}{2}$, and inflection points at $x = \frac{\pi}{6}$ and $x = \frac{5\pi}{6}$. Here is a graph of f *without* transition points highlighted.

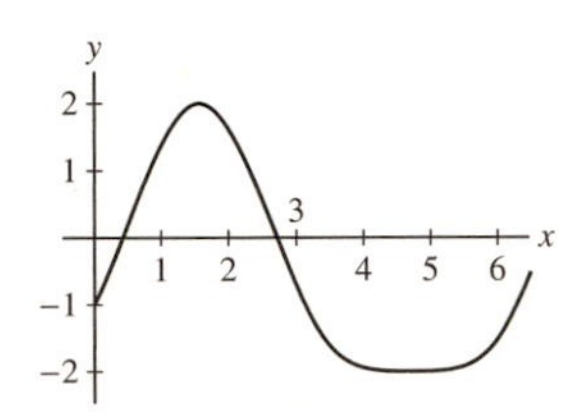

45. $y = \sin x + \sqrt{3}\cos x$, $[0, \pi]$

SOLUTION Let $f(x) = \sin x + \sqrt{3}\cos x$. Setting $f'(x) = \cos x - \sqrt{3}\sin x = 0$ yields $\tan x = \frac{1}{\sqrt{3}}$. In the interval $[0, \pi]$, the solution is $x = \frac{\pi}{6}$. Setting $f''(x) = -\sin x - \sqrt{3}\cos x = 0$ yields $\tan x = -\sqrt{3}$. In the interval $[0, \pi]$, the lone solution is $x = \frac{2\pi}{3}$.

Interval	signs of f' and f''
$(0, \pi/6)$	$+-$
$(\pi/6, 2\pi/3)$	$--$
$(2\pi/3, \pi)$	$-+$

The graph has a local maximum at $x = \frac{\pi}{6}$ and a point of inflection at $x = \frac{2\pi}{3}$. A plot without the transition points highlighted is given below:

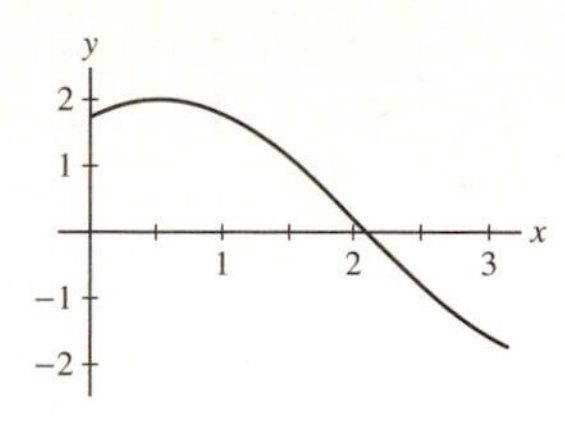

47. Are all sign transitions possible? Explain with a sketch why the transitions $++ \to -+$ and $-- \to +-$ do not occur if the function is differentiable. (See Exercise 76 for a proof.)

SOLUTION In both cases, there is a point where f is not differentiable at the transition from increasing to decreasing or decreasing to increasing.

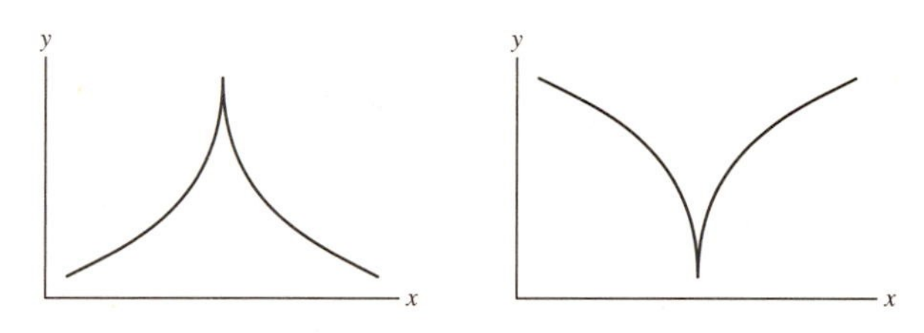

49. Which of the graphs in Figure 18 *cannot* be the graph of a polynomial? Explain.

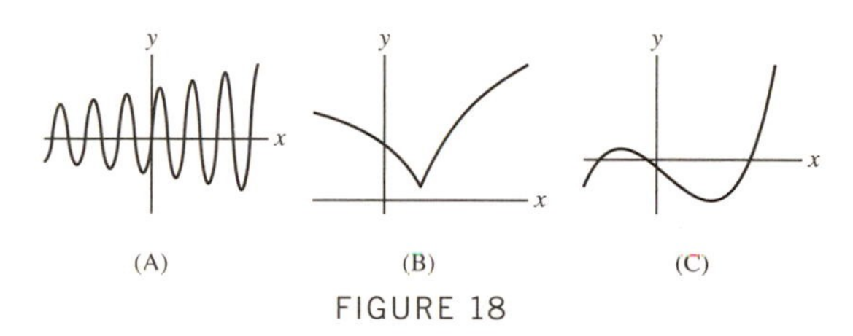

FIGURE 18

SOLUTION Polynomials are everywhere differentiable. Accordingly, graph (B) cannot be the graph of a polynomial, since the function in (B) has a cusp (sharp corner), signifying nondifferentiability at that point.

51. Match the graphs in Figure 20 with the two functions $y = \dfrac{3x}{x^2 - 1}$ and $y = \dfrac{3x^2}{x^2 - 1}$. Explain.

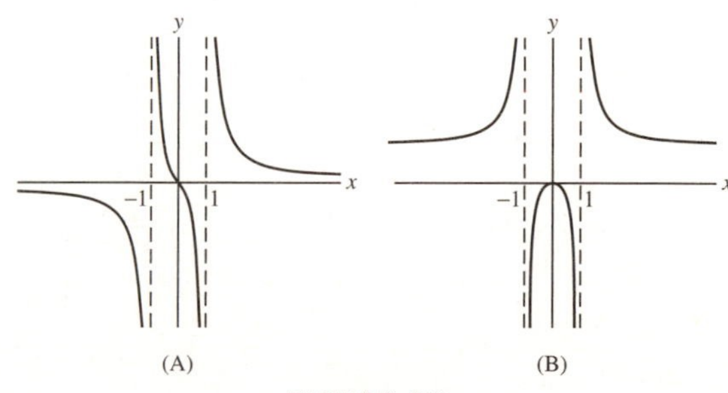

FIGURE 20

SOLUTION Since $\displaystyle\lim_{x\to\pm\infty} \frac{3x^2}{x^2 - 1} = \frac{3}{1} \cdot \lim_{x\to\pm\infty} 1 = 3$, the graph of $y = \dfrac{3x^2}{x^2 - 1}$ has a horizontal asymptote of $y = 3$; hence, the right curve is the graph of $f(x) = \dfrac{3x^2}{x^2 - 1}$. Since

$$\lim_{x\to\pm\infty} \frac{3x}{x^2 - 1} = \frac{3}{1} \cdot \lim_{x\to\pm\infty} x^{-1} = 0,$$

the graph of $y = \dfrac{3x}{x^2 - 1}$ has a horizontal asymptote of $y = 0$; hence, the left curve is the graph of $f(x) = \dfrac{3x}{x^2 - 1}$.

In Exercises 53–70, sketch the graph of the function. Indicate the transition points and asymptotes.

53. $y = \dfrac{1}{3x-1}$

SOLUTION Let $f(x) = \dfrac{1}{3x-1}$. Then $f'(x) = \dfrac{-3}{(3x-1)^2}$, so that f is decreasing for all $x \neq \frac{1}{3}$. Moreover, $f''(x) = \dfrac{18}{(3x-1)^3}$, so that f is concave up for $x > \frac{1}{3}$ and concave down for $x < \frac{1}{3}$. Because $\lim\limits_{x\to\pm\infty} \dfrac{1}{3x-1} = 0$, f has a horizontal asymptote at $y = 0$. Finally, f has a vertical asymptote at $x = \frac{1}{3}$ with

$$\lim_{x\to\frac{1}{3}-} \frac{1}{3x-1} = -\infty \quad \text{and} \quad \lim_{x\to\frac{1}{3}+} \frac{1}{3x-1} = \infty.$$

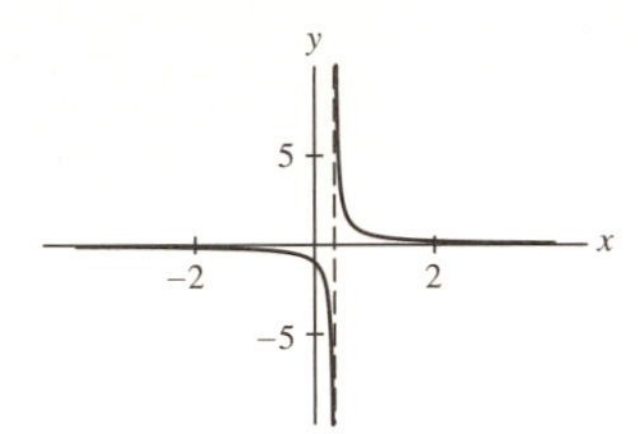

55. $y = \dfrac{x+3}{x-2}$

SOLUTION Let $f(x) = \dfrac{x+3}{x-2}$. Then $f'(x) = \dfrac{-5}{(x-2)^2}$, so that f is decreasing for all $x \neq 2$. Moreover, $f''(x) = \dfrac{10}{(x-2)^3}$, so that f is concave up for $x > 2$ and concave down for $x < 2$. Because $\lim\limits_{x\to\pm\infty} \dfrac{x+3}{x-2} = 1$, f has a horizontal asymptote at $y = 1$. Finally, f has a vertical asymptote at $x = 2$ with

$$\lim_{x\to 2-} \frac{x+3}{x-2} = -\infty \quad \text{and} \quad \lim_{x\to 2+} \frac{x+3}{x-2} = \infty.$$

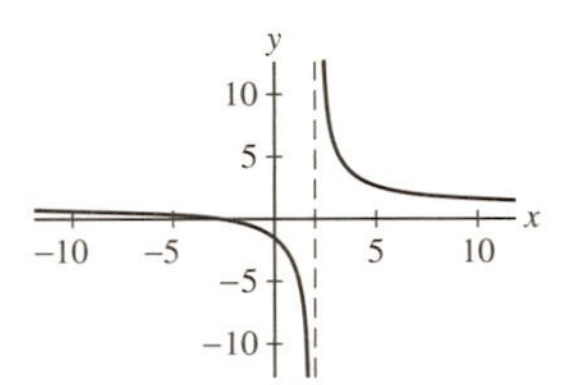

57. $y = \dfrac{1}{x} + \dfrac{1}{x-1}$

SOLUTION Let $f(x) = \dfrac{1}{x} + \dfrac{1}{x-1}$. Then $f'(x) = -\dfrac{2x^2-2x+1}{x^2(x-1)^2}$, so that f is decreasing for all $x \neq 0, 1$. Moreover,

$f''(x) = \dfrac{2\left(2x^3-3x^2+3x-1\right)}{x^3(x-1)^3}$, so that f is concave up for $0 < x < \frac{1}{2}$ and $x > 1$ and concave down for $x < 0$ and $\frac{1}{2} < x < 1$. Because $\lim\limits_{x\to\pm\infty} \left(\dfrac{1}{x} + \dfrac{1}{x-1}\right) = 0$, f has a horizontal asymptote at $y = 0$. Finally, f has vertical asymptotes at $x = 0$ and $x = 1$ with

$$\lim_{x\to 0-} \left(\frac{1}{x} + \frac{1}{x-1}\right) = -\infty \quad \text{and} \quad \lim_{x\to 0+} \left(\frac{1}{x} + \frac{1}{x-1}\right) = \infty$$

and

$$\lim_{x\to 1-} \left(\frac{1}{x} + \frac{1}{x-1}\right) = -\infty \quad \text{and} \quad \lim_{x\to 1+} \left(\frac{1}{x} + \frac{1}{x-1}\right) = \infty.$$

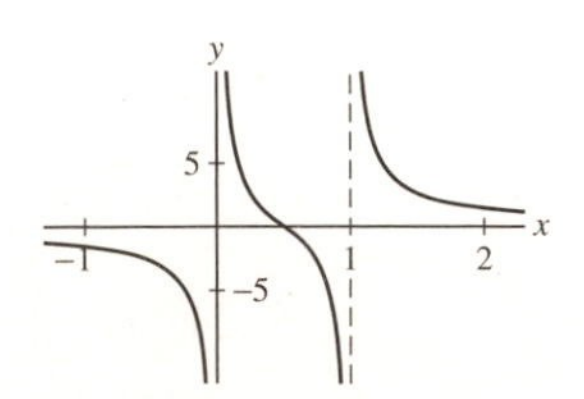

59. $y = \dfrac{1}{x(x-2)}$

SOLUTION Let $f(x) = \dfrac{1}{x(x-2)}$. Then $f'(x) = \dfrac{2(1-x)}{x^2(x-2)^2}$, so that f is increasing for $x < 0$ and $0 < x < 1$ and decreasing for $1 < x < 2$ and $x > 2$. Moreover, $f''(x) = \dfrac{2(3x^2 - 6x + 4)}{x^3(x-2)^3}$, so that f is concave up for $x < 0$ and $x > 2$ and concave down for $0 < x < 2$. Because $\lim_{x\to\pm\infty} \left(\dfrac{1}{x(x-2)}\right) = 0$, f has a horizontal asymptote at $y = 0$. Finally, f has vertical asymptotes at $x = 0$ and $x = 2$ with

$$\lim_{x\to 0-} \left(\frac{1}{x(x-2)}\right) = +\infty \quad \text{and} \quad \lim_{x\to 0+} \left(\frac{1}{x(x-2)}\right) = -\infty$$

and

$$\lim_{x\to 2-} \left(\frac{1}{x(x-2)}\right) = -\infty \quad \text{and} \quad \lim_{x\to 2+} \left(\frac{1}{x(x-2)}\right) = \infty.$$

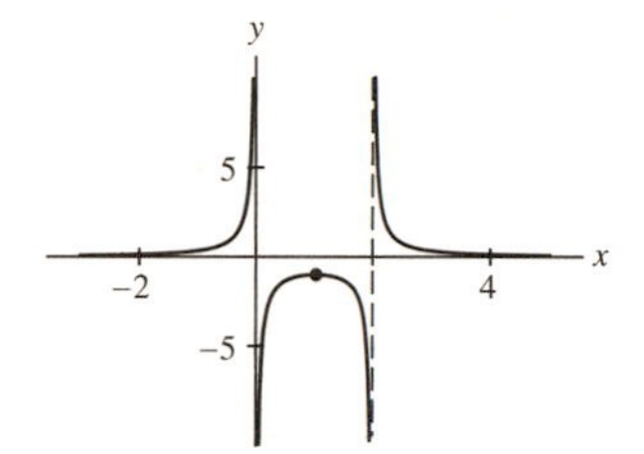

61. $y = \dfrac{1}{x^2 - 6x + 8}$

SOLUTION Let $f(x) = \dfrac{1}{x^2 - 6x + 8} = \dfrac{1}{(x-2)(x-4)}$. Then $f'(x) = \dfrac{6 - 2x}{\left(x^2 - 6x + 8\right)^2}$, so that f is increasing for $x < 2$ and for $2 < x < 3$, is decreasing for $3 < x < 4$ and for $x > 4$, and has a local maximum at $x = 3$. Moreover, $f''(x) = \dfrac{2\left(3x^2 - 18x + 28\right)}{\left(x^2 - 6x + 8\right)^3}$, so that f is concave up for $x < 2$ and for $x > 4$ and is concave down for $2 < x < 4$. Because $\lim_{x\to\pm\infty} \dfrac{1}{x^2 - 6x + 8} = 0$, f has a horizontal asymptote at $y = 0$. Finally, f has vertical asymptotes at $x = 2$ and $x = 4$, with

$$\lim_{x\to 2-} \left(\frac{1}{x^2 - 6x + 8}\right) = \infty \quad \text{and} \quad \lim_{x\to 2+} \left(\frac{1}{x^2 - 6x + 8}\right) = -\infty$$

and

$$\lim_{x\to 4-} \left(\frac{1}{x^2 - 6x + 8}\right) = -\infty \quad \text{and} \quad \lim_{x\to 4+} \left(\frac{1}{x^2 - 6x + 8}\right) = \infty.$$

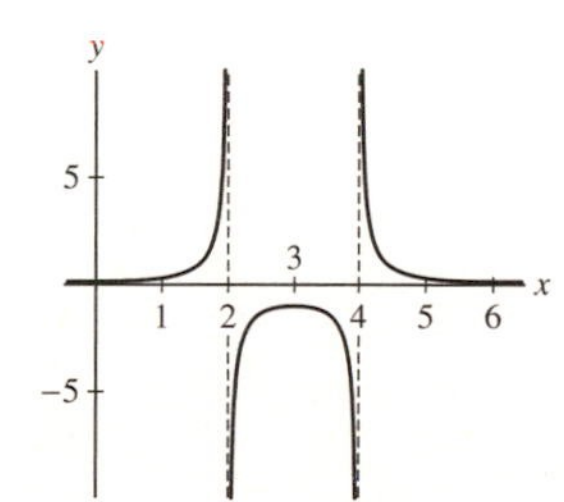

63. $y = 1 - \dfrac{3}{x} + \dfrac{4}{x^3}$

SOLUTION Let $f(x) = 1 - \dfrac{3}{x} + \dfrac{4}{x^3}$. Then

$$f'(x) = \frac{3}{x^2} - \frac{12}{x^4} = \frac{3(x-2)(x+2)}{x^4},$$

so that f is increasing for $|x| > 2$ and decreasing for $-2 < x < 0$ and for $0 < x < 2$. Moreover,

$$f''(x) = -\frac{6}{x^3} + \frac{48}{x^5} = \frac{6(8 - x^2)}{x^5},$$

so that f is concave down for $-2\sqrt{2} < x < 0$ and for $x > 2\sqrt{2}$, while f is concave up for $x < -2\sqrt{2}$ and for $0 < x < 2\sqrt{2}$. Because

$$\lim_{x\to\pm\infty}\left(1 - \frac{3}{x} + \frac{4}{x^3}\right) = 1,$$

f has a horizontal asymptote at $y = 1$. Finally, f has a vertical asymptote at $x = 0$ with

$$\lim_{x\to 0-}\left(1 - \frac{3}{x} + \frac{4}{x^3}\right) = -\infty \quad \text{and} \quad \lim_{x\to 0+}\left(1 - \frac{3}{x} + \frac{4}{x^3}\right) = \infty.$$

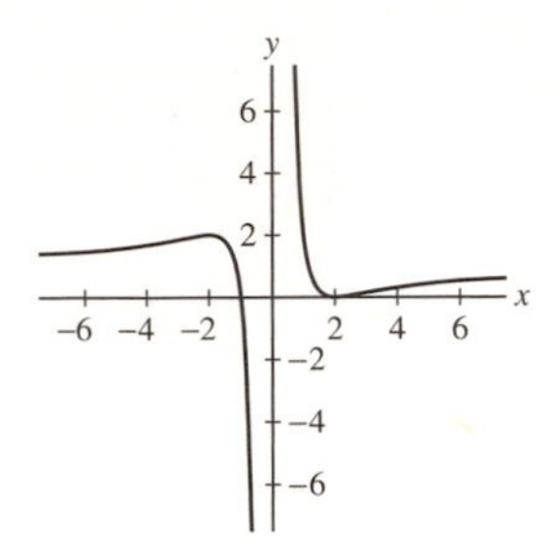

65. $y = \dfrac{1}{x^2} - \dfrac{1}{(x-2)^2}$

SOLUTION Let $f(x) = \dfrac{1}{x^2} - \dfrac{1}{(x-2)^2}$. Then $f'(x) = -2x^{-3} + 2(x-2)^{-3}$, so that f is increasing for $x < 0$ and for $x > 2$ and is decreasing for $0 < x < 2$. Moreover,

$$f''(x) = 6x^{-4} - 6(x-2)^{-4} = -\frac{48(x-1)(x^2-2x+2)}{x^4(x-2)^4},$$

so that f is concave up for $x < 0$ and for $0 < x < 1$, is concave down for $1 < x < 2$ and for $x > 2$, and has a point of inflection at $x = 1$. Because $\lim_{x\to\pm\infty}\left(\frac{1}{x^2} - \frac{1}{(x-2)^2}\right) = 0$, f has a horizontal asymptote at $y = 0$. Finally, f has vertical asymptotes at $x = 0$ and $x = 2$ with

$$\lim_{x\to 0-}\left(\frac{1}{x^2} - \frac{1}{(x-2)^2}\right) = \infty \quad \text{and} \quad \lim_{x\to 0+}\left(\frac{1}{x^2} - \frac{1}{(x-2)^2}\right) = \infty$$

and

$$\lim_{x\to 2-}\left(\frac{1}{x^2} - \frac{1}{(x-2)^2}\right) = -\infty \quad \text{and} \quad \lim_{x\to 2+}\left(\frac{1}{x^2} - \frac{1}{(x-2)^2}\right) = -\infty.$$

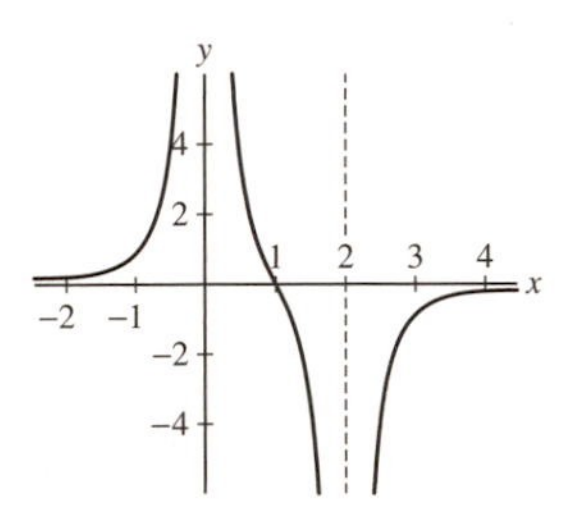

67. $y = \dfrac{1}{(x^2+1)^2}$

SOLUTION Let $f(x) = \dfrac{1}{(x^2+1)^2}$. Then $f'(x) = \dfrac{-4x}{(x^2+1)^3}$, so that f is increasing for $x < 0$, is decreasing for $x > 0$ and has a local maximum at $x = 0$. Moreover,

$$f''(x) = \frac{-4(x^2+1)^3 + 4x \cdot 3(x^2+1)^2 \cdot 2x}{(x^2+1)^6} = \frac{20x^2 - 4}{(x^2+1)^4},$$

so that f is concave up for $|x| > 1/\sqrt{5}$, is concave down for $|x| < 1/\sqrt{5}$, and has points of inflection at $x = \pm 1/\sqrt{5}$. Because $\lim_{x\to\pm\infty} \dfrac{1}{(x^2+1)^2} = 0$, f has a horizontal asymptote at $y = 0$. Finally, f has no vertical asymptotes.

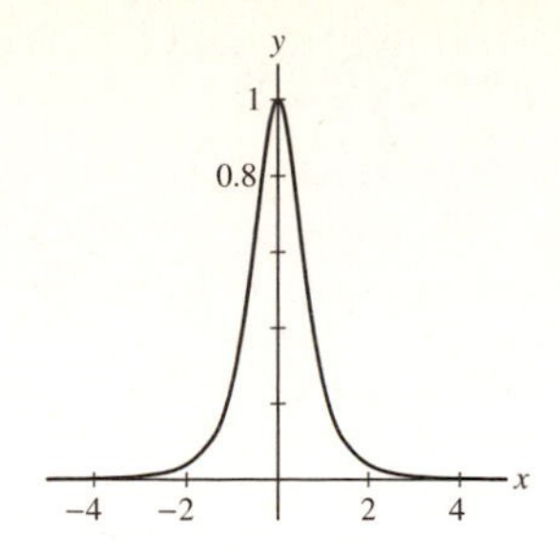

69. $y = \dfrac{1}{\sqrt{x^2+1}}$

SOLUTION Let $f(x) = \frac{1}{\sqrt{x^2+1}}$. Then

$$f'(x) = -\frac{x}{\sqrt{(x^2+1)^3}} = -x(x^2+1)^{-3/2},$$

so that f is increasing for $x < 0$ and decreasing for $x > 0$. Moreover,

$$f''(x) = -\frac{3}{2}x(x^2+1)^{-5/2}(-2x) - (x^2+1)^{-3/2} = (2x^2-1)(x^2+1)^{-5/2},$$

so that f is concave down for $|x| < \frac{\sqrt{2}}{2}$ and concave up for $|x| > \frac{\sqrt{2}}{2}$. Because

$$\lim_{x\to\pm\infty} \frac{1}{\sqrt{x^2+1}} = 0,$$

f has a horizontal asymptote at $y = 0$. Finally, f has no vertical asymptotes.

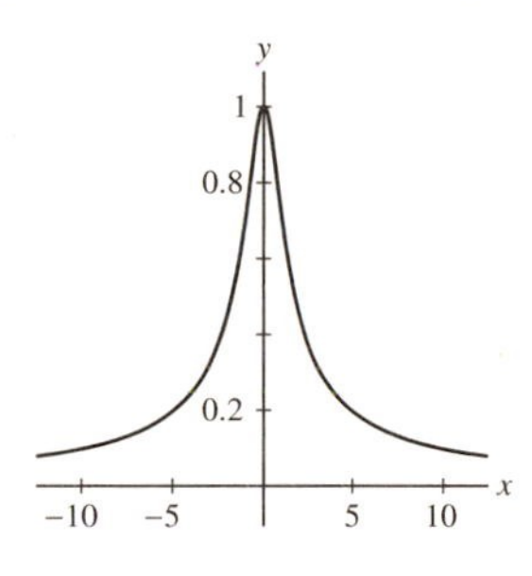

Further Insights and Challenges

*In Exercises 71–75, we explore functions whose graphs approach a nonhorizontal line as $x \to \infty$. A line $y = ax + b$ is called a **slant asymptote** if*

$$\lim_{x\to\infty} (f(x) - (ax+b)) = 0$$

or

$$\lim_{x\to-\infty} (f(x) - (ax+b)) = 0$$

71. Let $f(x) = \dfrac{x^2}{x-1}$ (Figure 22). Verify the following:

(a) $f(0)$ is a local max and $f(2)$ a local min.

(b) f is concave down on $(-\infty, 1)$ and concave up on $(1, \infty)$.

(c) $\lim_{x\to1-} f(x) = -\infty$ and $\lim_{x\to1+} f(x) = \infty$.

(d) $y = x + 1$ is a slant asymptote of $f(x)$ as $x \to \pm\infty$.

(e) The slant asymptote lies above the graph of $f(x)$ for $x < 1$ and below the graph for $x > 1$.

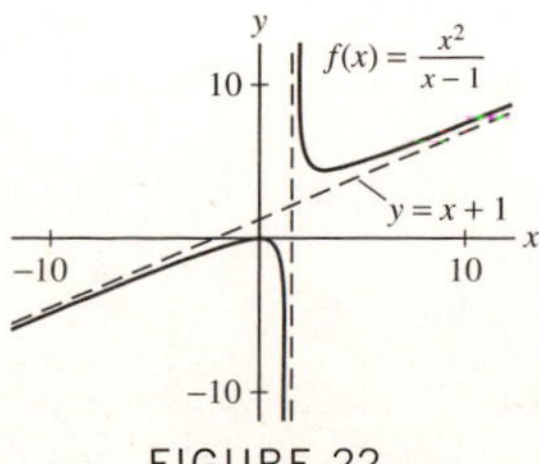

FIGURE 22

SOLUTION Let $f(x) = \frac{x^2}{x-1}$. Then $f'(x) = \frac{x(x-2)}{(x-1)^2}$ and $f''(x) = \frac{2}{(x-1)^3}$.

(a) Sign analysis of $f''(x)$ reveals that $f''(x) < 0$ on $(-\infty, 1)$ and $f''(x) > 0$ on $(1, \infty)$.

(b) Critical points of $f'(x)$ occur at $x = 0$ and $x = 2$. $x = 1$ is not a critical point because it is not in the domain of f. Sign analyses reveal that $x = 2$ is a local minimum of f and $x = 0$ is a local maximum.

(c)

$$\lim_{x\to 1-} f(x) = -1 \lim_{x\to 1-} \frac{1}{1-x} = -\infty \quad \text{and} \quad \lim_{x\to 1+} f(x) = 1 \lim_{x\to 1+} \frac{1}{x-1} = \infty.$$

(d) Note that using polynomial division, $f(x) = \frac{x^2}{x-1} = x + 1 + \frac{1}{x-1}$. Then $\lim_{x\to\pm\infty} (f(x) - (x+1)) = \lim_{x\to\pm\infty} x + 1 + \frac{1}{x-1} - (x+1) = \lim_{x\to\pm\infty} \frac{1}{x-1} = 0$.

(e) For $x > 1$, $f(x) - (x+1) = \frac{1}{x-1} > 0$, so $f(x)$ approaches $x + 1$ from above. Similarly, for $x < 1$, $f(x) - (x + 1) = \frac{1}{x-1} < 0$, so $f(x)$ approaches $x + 1$ from below.

73. Sketch the graph of

$$f(x) = \frac{x^2}{x+1}.$$

Proceed as in the previous exercise to find the slant asymptote.

SOLUTION Let $f(x) = \frac{x^2}{x+1}$. Then $f'(x) = \frac{x(x+2)}{(x+1)^2}$ and $f''(x) = \frac{2}{(x+1)^3}$. Thus, f is increasing for $x < -2$ and for $x > 0$, is decreasing for $-2 < x < -1$ and for $-1 < x < 0$, has a local minimum at $x = 0$, has a local maximum at $x = -2$, is concave down on $(-\infty, -1)$ and concave up on $(-1, \infty)$. Limit analyses give a vertical asymptote at $x = -1$, with

$$\lim_{x\to -1-} \frac{x^2}{x+1} = -\infty \quad \text{and} \quad \lim_{x\to -1+} \frac{x^2}{x+1} = \infty.$$

By polynomial division, $f(x) = x - 1 + \frac{1}{x+1}$ and

$$\lim_{x\to\pm\infty} \left(x - 1 + \frac{1}{x+1} - (x-1) \right) = 0,$$

which implies that the slant asymptote is $y = x - 1$. Notice that f approaches the slant asymptote as in exercise 71.

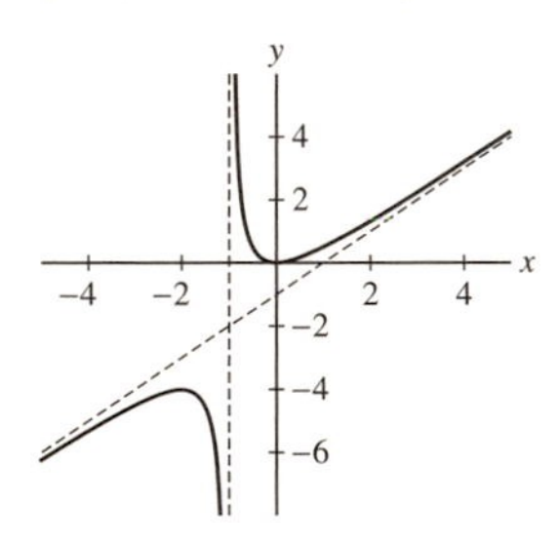

75. Sketch the graph of $f(x) = \frac{1-x^2}{2-x}$.

SOLUTION Let $f(x) = \frac{1-x^2}{2-x}$. Using polynomial division, $f(x) = x + 2 + \frac{3}{x-2}$. Then

$$\lim_{x\to\pm\infty} (f(x) - (x+2)) = \lim_{x\to\pm\infty} \left((x+2) + \frac{3}{x-2} - (x+2) \right) = \lim_{x\to\pm\infty} \frac{3}{x-2} = \frac{3}{1} \cdot \lim_{x\to\pm\infty} x^{-1} = 0$$

which implies that $y = x + 2$ is the slant asymptote of $f(x)$. Since $f(x) - (x+2) = \frac{3}{x-2} > 0$ for $x > 2$, $f(x)$ approaches the slant asymptote from above for $x > 2$; similarly, $\frac{3}{x-2} < 0$ for $x < 2$ so $f(x)$ approaches the slant asymptote from below for $x < 2$. Moreover, $f'(x) = \frac{x^2 - 4x + 1}{(2-x)^2}$ and $f''(x) = \frac{-6}{(2-x)^3}$. Sign analyses reveal a local minimum at $x = 2 + \sqrt{3}$, a local maximum at $x = 2 - \sqrt{3}$ and that f is concave down on $(-\infty, 2)$ and concave up on $(2, \infty)$. Limit analyses give a vertical asymptote at $x = 2$.

77. Assume that $f''(x)$ exists and $f''(x) > 0$ for all x. Show that $f(x)$ cannot be negative for all x. *Hint:* Show that $f'(b) \neq 0$ for some b and use the result of Exercise 64 in Section 4.4.

SOLUTION Let $f(x)$ be a function such that $f''(x)$ exists and $f''(x) > 0$ for all x. Since $f''(x) > 0$, there is at least one point $x = b$ such that $f'(b) \neq 0$. If not, $f'(x) = 0$ for all x, so $f''(x) = 0$. By the result of Exercise 64 in Section 4.4, $f(x) \geq f(b) + f'(b)(x - b)$. Now, if $f'(b) > 0$, we find that $f(b) + f'(b)(x - b) > 0$ whenever

$$x > \frac{bf'(b) - f(b)}{f'(b)},$$

a condition that must be met for some x sufficiently large. For such x, $f(x) > f(b) + f'(b)(x - b) > 0$. On the other hand, if $f'(b) < 0$, we find that $f(b) + f'(b)(x - b) > 0$ whenever

$$x < \frac{bf'(b) - f(b)}{f'(b)}.$$

For such an x, $f(x) > f(b) + f'(b)(x - b) > 0$.

4.7 Applied Optimization

Preliminary Questions

1. The problem is to find the right triangle of perimeter 10 whose area is as large as possible. What is the constraint equation relating the base b and height h of the triangle?

SOLUTION The perimeter of a right triangle is the sum of the lengths of the base, the height and the hypotenuse. If the base has length b and the height is h, then the length of the hypotenuse is $\sqrt{b^2 + h^2}$ and the perimeter of the triangle is $P = b + h + \sqrt{b^2 + h^2}$. The requirement that the perimeter be 10 translates to the constraint equation

$$b + h + \sqrt{b^2 + h^2} = 10.$$

2. Describe a way of showing that a continuous function on an open interval (a, b) has a minimum value.

SOLUTION If the function tends to infinity at the endpoints of the interval, then the function must take on a minimum value at a critical point.

3. Is there a rectangle of area 100 of largest perimeter? Explain.

SOLUTION No. Even by fixing the area at 100, we can take one of the dimensions as large as we like thereby allowing the perimeter to become as large as we like.

Exercises

1. Find the dimensions x and y of the rectangle of maximum area that can be formed using 3 meters of wire.

(a) What is the constraint equation relating x and y?

(b) Find a formula for the area in terms of x alone.

(c) What is the interval of optimization? Is it open or closed?

(d) Solve the optimization problem.

SOLUTION

(a) The perimeter of the rectangle is 3 meters, so $3 = 2x + 2y$, which is equivalent to $y = \frac{3}{2} - x$.

(b) Using part (a), $A = xy = x(\frac{3}{2} - x) = \frac{3}{2}x - x^2$.

(c) This problem requires optimization over the closed interval $[0, \frac{3}{2}]$, since both x and y must be non-negative.

(d) $A'(x) = \frac{3}{2} - 2x = 0$, which yields $x = \frac{3}{4}$ and consequently, $y = \frac{3}{4}$. Because $A(0) = A(3/2) = 0$ and $A(\frac{3}{4}) = 0.5625$, the maximum area $0.5625\ \text{m}^2$ is achieved with $x = y = \frac{3}{4}$ m.

3. Wire of length 12 m is divided into two pieces and the pieces are bent into a square and a circle. How should this be done in order to minimize the sum of their areas?

SOLUTION Suppose the wire is divided into one piece of length x m that is bent into a circle and a piece of length $12 - x$ m that is bent into a square. Because the circle has circumference x, it follows that the radius of the circle is $x/2\pi$; therefore, the area of the circle is

$$\pi\left(\frac{x}{2\pi}\right)^2 = \frac{x^2}{4\pi}.$$

As for the square, because the perimeter is $12 - x$, the length of each side is $3 - x/4$ and the area is $(3 - x/4)^2$. Then

$$A(x) = \frac{x^2}{4\pi} + \left(3 - \frac{1}{4}x\right)^2.$$

Now

$$A'(x) = \frac{x}{2\pi} - \frac{1}{2}\left(3 - \frac{1}{4}x\right) = 0$$

when

$$x = \frac{12\pi}{4+\pi}\text{ m} \approx 5.28\text{ m}.$$

Because $A(0) = 9\text{ m}^2$, $A(12) = 36/\pi \approx 11.46\text{ m}^2$, and

$$A\left(\frac{12\pi}{4+\pi}\right) \approx 5.04\text{ m}^2,$$

we see that the sum of the areas is minimized when approximately 5.28 m of the wire is allotted to the circle.

5. A flexible tube of length 4 m is bent into an L-shape. Where should the bend be made to minimize the distance between the two ends?

SOLUTION Let $x, y > 0$ be lengths of the side of the L. Since $x + y = 4$ or $y = 4 - x$, the distance between the ends of L is $h(x) = \sqrt{x^2 + y^2} = \sqrt{x^2 + (4-x)^2}$. We may equivalently minimize the square of the distance,

$$f(x) = x^2 + y^2 = x^2 + (4-x)^2$$

This is easier computationally (when working by hand). Solve $f'(x) = 4x - 8 = 0$ to obtain $x = 2$ m. Now $f(0) = f(4) = 16$, whereas $f(2) = 8$. Hence the distance between the two ends of the L is minimized when the bend is made at the middle of the wire.

7. A rancher will use 600 m of fencing to build a corral in the shape of a semicircle on top of a rectangle (Figure 9). Find the dimensions that maximize the area of the corral.

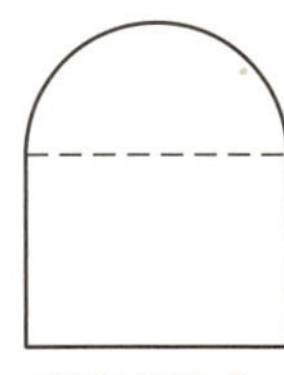

FIGURE 9

SOLUTION Let x be the width of the corral and therefore the diameter of the semicircle, and let y be the height of the rectangular section. Then the perimeter of the corral can be expressed by the equation $2y + x + \frac{\pi}{2}x = 2y + (1 + \frac{\pi}{2})x = 600$ m or equivalently, $y = \frac{1}{2}\left(600 - (1 + \frac{\pi}{2})x\right)$. Since x and y must both be nonnegative, it follows that x must be restricted to the interval $[0, \frac{600}{1+\pi/2}]$. The area of the corral is the sum of the area of the rectangle and semicircle, $A = xy + \frac{\pi}{8}x^2$. Making the substitution for y from the constraint equation,

$$A(x) = \frac{1}{2}x\left(600 - (1 + \frac{\pi}{2})x\right) + \frac{\pi}{8}x^2 = 300x - \frac{1}{2}\left(1 + \frac{\pi}{2}\right)x^2 + \frac{\pi}{8}x^2.$$

Now, $A'(x) = 300 - \left(1 + \frac{\pi}{2}\right)x + \frac{\pi}{4}x = 0$ implies $x = \frac{300}{(1+\frac{\pi}{4})} \approx 168.029746$ m. With $A(0) = 0\text{ m}^2$,

$$A\left(\frac{300}{1+\pi/4}\right) \approx 25204.5\text{ m}^2 \qquad \text{and} \qquad A\left(\frac{600}{1+\pi/2}\right) \approx 21390.8\text{ m}^2,$$

it follows that the corral of maximum area has dimensions

$$x = \frac{300}{1+\pi/4}\text{ m} \qquad \text{and} \qquad y = \frac{150}{1+\pi/4}\text{ m}.$$

9. Find the dimensions of the rectangle of maximum area that can be inscribed in a circle of radius $r = 4$ (Figure 11).

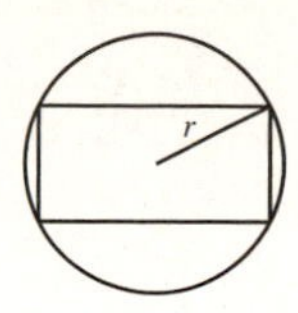

FIGURE 11

SOLUTION Place the center of the circle at the origin with the sides of the rectangle (of lengths $2x > 0$ and $2y > 0$) parallel to the coordinate axes. By the Pythagorean Theorem, $x^2 + y^2 = r^2 = 16$, so that $y = \sqrt{16 - x^2}$. Thus the area of the rectangle is $A(x) = 2x \cdot 2y = 4x\sqrt{16 - x^2}$. To guarantee both x and y are real and nonnegative, we must restrict x to the interval $[0, 4]$. Solve

$$A'(x) = 4\sqrt{16 - x^2} - \frac{4x^2}{\sqrt{16 - x^2}} = 0$$

for $x > 0$ to obtain $x = \frac{4}{\sqrt{2}} = 2\sqrt{2}$. Since $A(0) = A(4) = 0$ and $A(2\sqrt{2}) = 32$, the rectangle of maximum area has dimensions $2x = 2y = 4\sqrt{2}$.

11. Find the point on the line $y = x$ closest to the point $(1, 0)$. *Hint:* It is equivalent and easier to minimize the *square* of the distance.

SOLUTION With $y = x$, let's equivalently minimize the square of the distance, $f(x) = (x - 1)^2 + y^2 = 2x^2 - 2x + 1$, which is computationally easier (when working by hand). Solve $f'(x) = 4x - 2 = 0$ to obtain $x = \frac{1}{2}$. Since $f(x) \to \infty$ as $x \to \pm\infty$, $(\frac{1}{2}, \frac{1}{2})$ is the point on $y = x$ closest to $(1, 0)$.

13. CAS Find a good numerical approximation to the coordinates of the point on the graph of $y = \ln x - x$ closest to the origin (Figure 13).

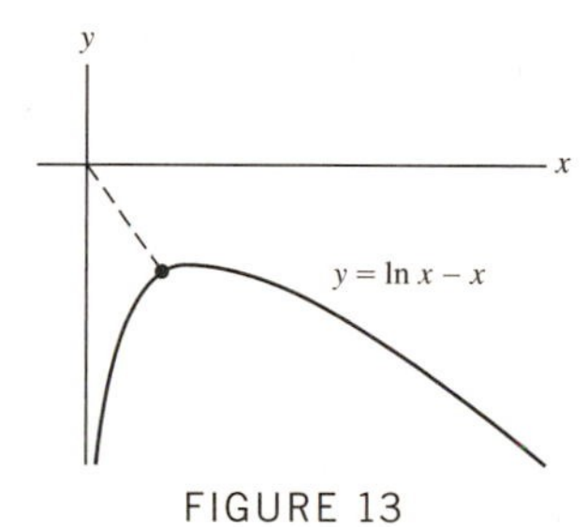

FIGURE 13

SOLUTION The distance from the origin to the point $(x, \ln x - x)$ on the graph of $y = \ln x - x$ is $d = \sqrt{x^2 + (\ln x - x)^2}$. As usual, we will minimize d^2. Let $d^2 = f(x) = x^2 + (\ln x - x)^2$. Then

$$f'(x) = 2x + 2(\ln x - x)\left(\frac{1}{x} - 1\right).$$

To determine x, we need to solve

$$4x + \frac{2\ln x}{x} - 2\ln x - 2 = 0.$$

This yields $x \approx .632784$. Thus, the point on the graph of $y = \ln x - x$ that is closest to the origin is approximately $(0.632784, -1.090410)$.

15. Find the angle θ that maximizes the area of the isosceles triangle whose legs have length ℓ (Figure 14).

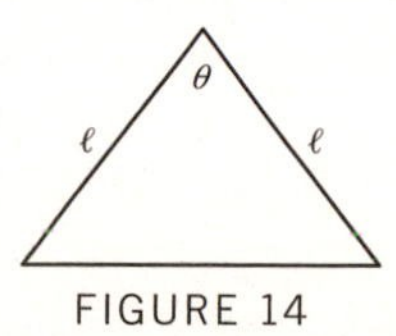

FIGURE 14

SOLUTION The area of the triangle is

$$A(\theta) = \frac{1}{2}\ell^2 \sin\theta,$$

where $0 \le \theta \le \pi$. Setting

$$A'(\theta) = \frac{1}{2}\ell^2 \cos\theta = 0$$

yields $\theta = \frac{\pi}{2}$. Since $A(0) = A(\pi) = 0$ and $A(\frac{\pi}{2}) = \frac{1}{2}\ell^2$, the angle that maximizes the area of the isosceles triangle is $\theta = \frac{\pi}{2}$.

17. Find the area of the largest isosceles triangle that can be inscribed in a circle of radius r.

SOLUTION Consider the following diagram:

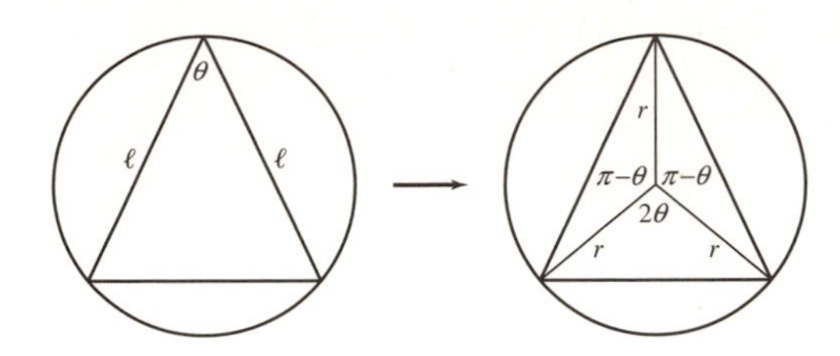

The area of the isosceles triangle is

$$A(\theta) = 2 \cdot \frac{1}{2}r^2 \sin(\pi - \theta) + \frac{1}{2}r^2 \sin(2\theta) = r^2 \sin\theta + \frac{1}{2}r^2 \sin(2\theta),$$

where $0 \le \theta \le \pi$. Solve

$$A'(\theta) = r^2 \cos\theta + r^2 \cos(2\theta) = 0$$

to obtain $\theta = \frac{\pi}{3}, \pi$. Since $A(0) = A(\pi) = 0$ and $A(\frac{\pi}{3}) = \frac{3\sqrt{3}}{4}r^2$, the area of the largest isosceles triangle that can be inscribed in a circle of radius r is $\frac{3\sqrt{3}}{4}r^2$.

19. A poster of area 6000 cm^2 has blank margins of width 10 cm on the top and bottom and 6 cm on the sides. Find the dimensions that maximize the printed area.

SOLUTION Let x be the width of the printed region, and let y be the height. The total printed area is $A = xy$. Because the total area of the poster is 6000 cm^2, we have the constraint $(x + 12)(y + 20) = 6000$, so that $xy + 12y + 20x + 240 = 6000$, or $y = \frac{5760-20x}{x+12}$. Therefore, $A(x) = 20\frac{288x-x^2}{x+12}$, where $0 \le x \le 288$.

$A(0) = A(288) = 0$, so we are looking for a critical point on the interval $[0, 288]$. Setting $A'(x) = 0$ yields

$$20\frac{(x + 12)(288 - 2x) - (288x - x^2)}{(x + 12)^2} = 0$$

$$\frac{-x^2 - 24x + 3456}{(x + 12)^2} = 0$$

$$x^2 + 24x - 3456 = 0$$

$$(x - 48)(x + 72) = 0$$

Therefore $x = 48$ or $x = -72$. $x = 48$ is the only critical point of $A(x)$ in the interval $[0, 288]$, so $A(48) = 3840$ is the maximum value of $A(x)$ in the interval $[0, 288]$. Now, $y = 20\frac{288-48}{48+12} = 80$ cm, so the poster with maximum printed area is $48 + 12 = 60$ cm. wide by $80 + 20 = 100$ cm. tall.

21. Kepler's Wine Barrel Problem In his work *Nova stereometria doliorum vinariorum* (New Solid Geometry of a Wine Barrel), published in 1615, astronomer Johannes Kepler stated and solved the following problem: Find the dimensions of the cylinder of largest volume that can be inscribed in a sphere of radius R. *Hint:* Show that an inscribed cylinder has volume $2\pi x(R^2 - x^2)$, where x is one-half the height of the cylinder.

SOLUTION Place the center of the sphere at the origin in three-dimensional space. Let the cylinder be of radius y and half-height x. The Pythagorean Theorem states, $x^2 + y^2 = R^2$, so that $y^2 = R^2 - x^2$. The volume of the cylinder is $V(x) = \pi y^2\,(2x) = 2\pi\left(R^2 - x^2\right)x = 2\pi R^2 x - 2\pi x^3$. Allowing for degenerate cylinders, we have $0 \le x \le R$. Solve $V'(x) = 2\pi R^2 - 6\pi x^2 = 0$ for $x \ge 0$ to obtain $x = \frac{R}{\sqrt{3}}$. Since $V(0) = V(R) = 0$, the largest volume is $V(\frac{R}{\sqrt{3}}) = \frac{4}{9}\pi\sqrt{3}R^3$ when $x = \frac{R}{\sqrt{3}}$ and $y = \sqrt{\frac{2}{3}}R$.

23. A landscape architect wishes to enclose a rectangular garden of area 1,000 m^2 on one side by a brick wall costing \$90/m and on the other three sides by a metal fence costing \$30/m. Which dimensions minimize the total cost?

SOLUTION Let x be the length of the brick wall and y the length of an adjacent side with $x, y > 0$. With $xy = 1000$ or $y = \frac{1000}{x}$, the total cost is

$$C(x) = 90x + 30(x + 2y) = 120x + 60000x^{-1}.$$

Solve $C'(x) = 120 - 60000x^{-2} = 0$ for $x > 0$ to obtain $x = 10\sqrt{5}$. Since $C(x) \to \infty$ as $x \to 0+$ and as $x \to \infty$, the minimum cost is $C(10\sqrt{5}) = 2400\sqrt{5} \approx \$5366.56$ when $x = 10\sqrt{5} \approx 22.36$ m and $y = 20\sqrt{5} \approx 44.72$ m.

25. Find the maximum area of a rectangle inscribed in the region bounded by the graph of $y = \dfrac{4 - x}{2 + x}$ and the axes (Figure 17).

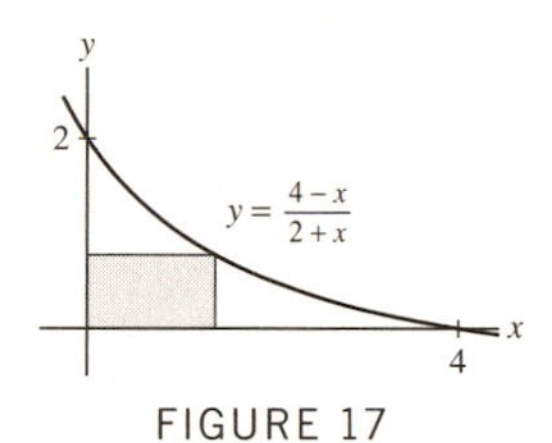

FIGURE 17

SOLUTION Let s be the width of the rectangle. The height of the rectangle is $h = \frac{4-s}{2+s}$, so that the area is

$$A(s) = s\frac{4 - s}{2 + s} = \frac{4s - s^2}{2 + s}.$$

We are maximizing on the closed interval $[0, 4]$. It is obvious from the pictures that $A(0) = A(4) = 0$, so we look for critical points of A.

$$A'(s) = \frac{(2 + s)(4 - 2s) - (4s - s^2)}{(2 + s)^2} = -\frac{s^2 + 4s - 8}{(s + 2)^2}.$$

The only point where $A'(s)$ doesn't exist is $s = -2$ which isn't under consideration.

Setting $A'(s) = 0$ gives, by the quadratic formula,

$$s = \frac{-4 \pm \sqrt{48}}{2} = -2 \pm 2\sqrt{3}.$$

Of these, only $-2 + 2\sqrt{3}$ is positive, so this is our lone critical point. $A(-2 + 2\sqrt{3}) \approx 1.0718 > 0$. Since we are finding the maximum over a closed interval and $-2 + 2\sqrt{3}$ is the only critical point, the maximum area is $A(-2 + 2\sqrt{3}) \approx 1.0718$.

27. Find the maximum area of a rectangle circumscribed around a rectangle of sides L and H. *Hint:* Express the area in terms of the angle θ (Figure 18).

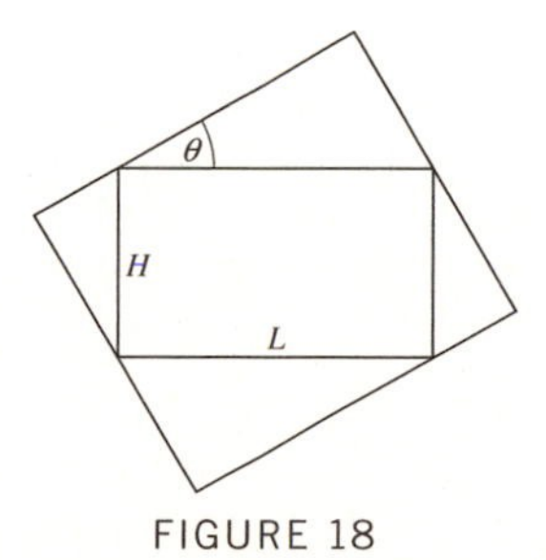

FIGURE 18

SOLUTION Position the $L \times H$ rectangle in the first quadrant of the xy-plane with its "northwest" corner at the origin. Let θ be the angle the base of the circumscribed rectangle makes with the positive x-axis, where $0 \le \theta \le \frac{\pi}{2}$. Then the area of the circumscribed rectangle is $A = LH + 2 \cdot \frac{1}{2}(H \sin\theta)(H \cos\theta) + 2 \cdot \frac{1}{2}(L \sin\theta)(L \cos\theta) = LH + \frac{1}{2}(L^2 + H^2)$ $\sin 2\theta$, which has a maximum value of $LH + \frac{1}{2}(L^2 + H^2)$ when $\theta = \frac{\pi}{4}$ because $\sin 2\theta$ achieves its maximum when $\theta = \frac{\pi}{4}$.

29. Find the equation of the line through $P = (4, 12)$ such that the triangle bounded by this line and the axes in the first quadrant has minimal area.

SOLUTION Let $P = (4, 12)$ be a point in the first quadrant and $y - 12 = m(x - 4), -\infty < m < 0$, be a line through P that cuts the positive x- and y-axes. Then $y = L(x) = m(x - 4) + 12$. The line $L(x)$ intersects the y-axis at $H\ (0, 12 - 4m)$ and the x-axis at $W\left(4 - \frac{12}{m}, 0\right)$. Hence the area of the triangle is

$$A(m) = \frac{1}{2}(12 - 4m)\left(4 - \frac{12}{m}\right) = 48 - 8m - 72m^{-1}.$$

Solve $A'(m) = 72m^{-2} - 8 = 0$ for $m < 0$ to obtain $m = -3$. Since $A \to \infty$ as $m \to -\infty$ or $m \to 0-$, we conclude that the minimal triangular area is obtained when $m = -3$. The equation of the line through $P = (4, 12)$ is $y = -3(x - 4) + 12 = -3x + 24$.

31. Archimedes' Problem A spherical cap (Figure 20) of radius r and height h has volume $V = \pi h^2\left(r - \frac{1}{3}h\right)$ and surface area $S = 2\pi rh$. Prove that the hemisphere encloses the largest volume among all spherical caps of fixed surface area S.

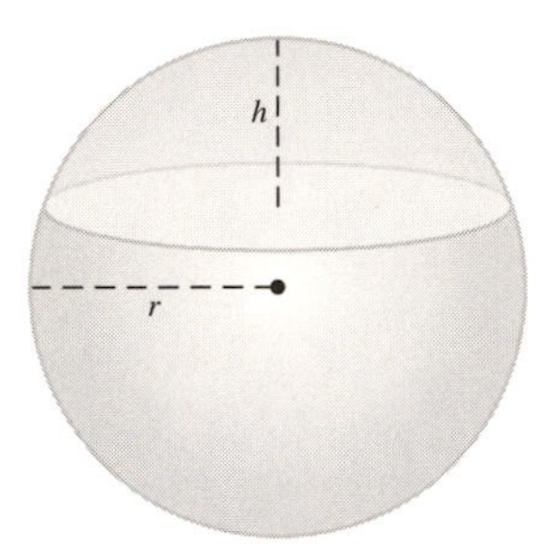

FIGURE 20

SOLUTION Consider all spherical caps of fixed surface area S. Because $S = 2\pi rh$, it follows that

$$r = \frac{S}{2\pi h}$$

and

$$V(h) = \pi h^2\left(\frac{S}{2\pi h} - \frac{1}{3}h\right) = \frac{S}{2}h - \frac{\pi}{3}h^3.$$

Now

$$V'(h) = \frac{S}{2} - \pi h^2 = 0$$

when

$$h^2 = \frac{S}{2\pi} \quad \text{or} \quad h = \frac{S}{2\pi h} = r.$$

Hence, the hemisphere encloses the largest volume among all spherical caps of fixed surface area S.

33. A box of volume 72 m^3 with square bottom and no top is constructed out of two different materials. The cost of the bottom is \$40/m^2 and the cost of the sides is \$30/m^2. Find the dimensions of the box that minimize total cost.

SOLUTION Let s denote the length of the side of the square bottom of the box and h denote the height of the box. Then

$$V = s^2h = 72 \quad \text{or} \quad h = \frac{72}{s^2}.$$

The cost of the box is

$$C = 40s^2 + 120sh = 40s^2 + \frac{8640}{s},$$

so

$$C'(s) = 80s - \frac{8640}{s^2} = 0$$

when $s = 3\sqrt[3]{4}$ m and $h = 2\sqrt[3]{4}$ m. Because $C \to \infty$ as $s \to 0-$ and as $s \to \infty$, we conclude that the critical point gives the minimum cost.

35. Your task is to design a rectangular industrial warehouse consisting of three separate spaces of equal size as in Figure 22. The wall materials cost \$500 per linear meter and your company allocates \$2,400,000 for the project.

(a) Which dimensions maximize the area of the warehouse?

(b) What is the area of each compartment in this case?

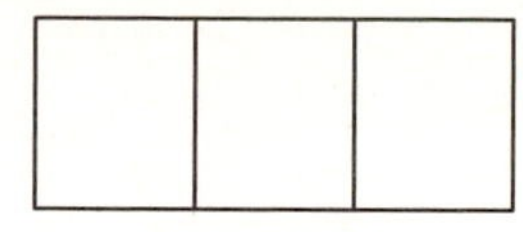

FIGURE 22

SOLUTION Let one compartment have length x and width y. Then total length of the wall of the warehouse is $P = 4x + 6y$ and the constraint equation is cost $= 2{,}400{,}000 = 500(4x + 6y)$, which gives $y = 800 - \frac{2}{3}x$.

(a) Area is given by $A = 3xy = 3x\left(800 - \frac{2}{3}x\right) = 2400x - 2x^2$, where $0 \le x \le 1200$. Then $A'(x) = 2400 - 4x = 0$ yields $x = 600$ and consequently $y = 400$. Since $A(0) = A(1200) = 0$ and $A(600) = 720,000$, the area of the warehouse is maximized when each compartment has length of 600 m and width of 400 m.

(b) The area of one compartment is $600 \cdot 400 = 240,000$ square meters.

37. According to a model developed by economists E. Heady and J. Pesek, if fertilizer made from N pounds of nitrogen and P pounds of phosphate is used on an acre of farmland, then the yield of corn (in bushels per acre) is

$$Y = 7.5 + 0.6N + 0.7P - 0.001N^2 - 0.002P^2 + 0.001NP$$

A farmer intends to spend \$30 per acre on fertilizer. If nitrogen costs 25 cents/lb and phosphate costs 20 cents/lb, which combination of N and L produces the highest yield of corn?

SOLUTION The farmer's budget for fertilizer is \$30 per acre, so we have the constraint equation

$$0.25N + 0.2P = 30 \qquad \text{or} \qquad P = 150 - 1.25N$$

Substituting for P in the equation for Y, we find

$$\begin{aligned} Y(N) &= 7.5 + 0.6N + 0.7(150 - 1.25N) - 0.001N^2 - 0.002(150 - 1.25N)^2 + 0.001N(150 - 1.25N) \\ &= 67.5 + 0.625N - 0.005375N^2 \end{aligned}$$

Both N and P must be nonnegative. Since $P = 150 - 1.25N \ge 0$, we require that $0 \le N \le 120$. Next,

$$\frac{dY}{dN} = 0.625 - 0.01075N = 0 \quad \Rightarrow \quad N = \frac{0.625}{0.01075} \approx 58.14 \text{ pounds.}$$

Now, $Y(0) = 67.5$, $Y(120) = 65.1$ and $Y(58.14) \approx 85.67$, so the maximum yield of corn occurs for $N \approx 58.14$ pounds and $P \approx 77.33$ pounds.

39. All units in a 100-unit apartment building are rented out when the monthly rent is set at $r = \$900$/month. Suppose that one unit becomes vacant with each \$10 increase in rent and that each occupied unit costs \$80/month in maintenance. Which rent r maximizes monthly profit?

SOLUTION Let n denote the number of \$10 increases in rent. Then the monthly profit is given by

$$P(n) = (100 - n)(900 + 10n - 80) = 82000 + 180n - 10n^2,$$

and

$$P'(n) = 180 - 20n = 0$$

when $n = 9$. We know this results in maximum profit because this gives the location of vertex of a downward opening parabola. Thus, monthly profit is maximized with a rent of \$990.

41. The monthly output of a Spanish light bulb factory is $P = 2LK^2$ (in millions), where L is the cost of labor and K is the cost of equipment (in millions of euros). The company needs to produce 1.7 million units per month. Which values of L and K would minimize the total cost $L + K$?

SOLUTION Since $P = 1.7$ and $P = 2LK^2$, we have $L = \dfrac{0.85}{K^2}$. Accordingly, the cost of production is

$$C(K) = L + K = K + \frac{0.85}{K^2}.$$

Solve $C'(K) = 1 - \dfrac{1.7}{K^3}$ for $K \ge 0$ to obtain $K = \sqrt[3]{1.7}$. Since $C(K) \to \infty$ as $K \to 0+$ and as $K \to \infty$, the minimum cost of production is achieved for $K = \sqrt[3]{1.7} \approx 1.2$ and $L = 0.6$. The company should invest 1.2 million euros in equipment and 600, 000 euros in labor.

43. Brandon is on one side of a river that is 50 m wide and wants to reach a point 200 m downstream on the opposite side as quickly as possible by swimming diagonally across the river and then running the rest of the way. Find the best route if Brandon can swim at 1.5 m/s and run at 4 m/s.

SOLUTION Let lengths be in meters, times in seconds, and speeds in m/s. Suppose that Brandon swims diagonally to a point located x meters downstream on the opposite side. Then Brandon then swims a distance $\sqrt{x^2+50^2}$ and runs a distance $200-x$. The total time of the trip is

$$f(x)=\frac{\sqrt{x^2+2500}}{1.5}+\frac{200-x}{4}, \qquad 0\le x\le 200.$$

Solve

$$f'(x)=\frac{2x}{3\sqrt{x^2+2500}}-\frac{1}{4}=0$$

to obtain $x=30\frac{5}{11}\approx 20.2$ and $f(20.2)\approx 80.9$. Since $f(0)\approx 83.3$ and $f(200)\approx 137.4$, we conclude that the minimal time is 80.9 s. This occurs when Brandon swims diagonally to a point located 20.2 m downstream and then runs the rest of the way.

In Exercises 45–47, a box (with no top) is to be constructed from a piece of cardboard of sides A and B by cutting out squares of length h from the corners and folding up the sides (Figure 26).

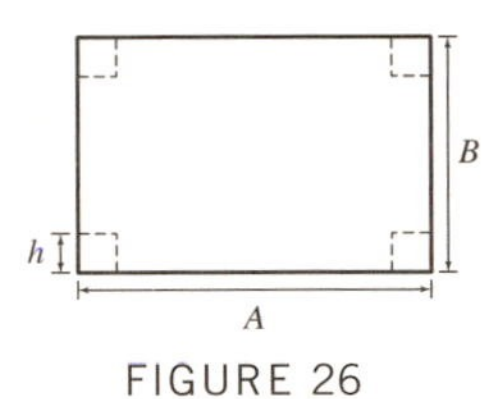

FIGURE 26

45. Find the value of h that maximizes the volume of the box if $A=15$ and $B=24$. What are the dimensions of this box?

SOLUTION Once the sides have been folded up, the base of the box will have dimensions $(A-2h)\times(B-2h)$ and the height of the box will be h. Thus

$$V(h)=h(A-2h)(B-2h)=4h^3-2(A+B)h^2+ABh.$$

When $A=15$ and $B=24$, this gives

$$V(h)=4h^3-78h^2+360h,$$

and we need to maximize over $0\le h\le \frac{15}{2}$. Now,

$$V'(h)=12h^2-156h+360=0$$

yields $h=3$ and $h=10$. Because $h=10$ is not in the domain of the problem and $V(0)=V(15/2)=0$ and $V(3)=486$, volume is maximized when $h=3$. The corresponding dimensions are $9\times 18\times 3$.

47. Which values of A and B maximize the volume of the box if $h=10$ cm and $AB=900$ cm.

SOLUTION With $h=10$ and $AB=900$ (which means that $B=900/A$), the volume of the box is

$$V(A)=10(A-20)\left(\frac{900}{A}-20\right)=13{,}000-200A-\frac{180{,}000}{A},$$

where $20\le A\le 45$. Now, solving

$$V'(A)=-200+\frac{180{,}000}{A^2}=0$$

yields $A=30$. Because $V(20)=V(45)=0$ and $V(30)=1000\text{ cm}^3$, maximum volume is achieved with $A=B=30$ cm.

49. A billboard of height b is mounted on the side of a building with its bottom edge at a distance h from the street as in Figure 27. At what distance x should an observer stand from the wall to maximize the angle of observation θ?

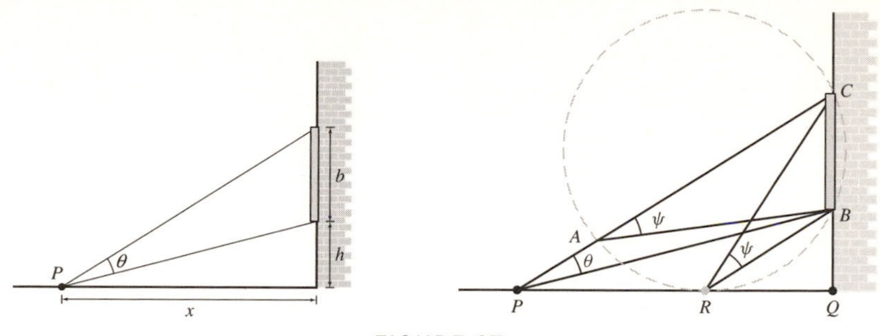

FIGURE 27

SOLUTION From the upper diagram in Figure 27 and the addition formula for the cotangent function, we see that

$$\cot\theta = \frac{1 + \frac{x}{b+h}\frac{x}{h}}{\frac{x}{h} - \frac{x}{b+h}} = \frac{x^2 + h(b+h)}{bx},$$

where b and h are constant. Now, differentiate with respect to x and solve

$$-\csc^2\theta\frac{d\theta}{dx} = \frac{x^2 - h(b+h)}{bx^2} = 0$$

to obtain $x = \sqrt{bh + h^2}$. Since this is the only critical point, and since $\theta \to 0$ as $x \to 0+$ and $\theta \to 0$ as $x \to \infty$, $\theta(x)$ reaches its maximum at $x = \sqrt{bh + h^2}$.

51. Optimal Delivery Schedule A gas station sells Q gallons of gasoline per year, which is delivered N times per year in equal shipments of Q/N gallons. The cost of each delivery is d dollars and the yearly storage costs are sQT, where T is the length of time (a fraction of a year) between shipments and s is a constant. Show that costs are minimized for $N = \sqrt{sQ/d}$. (*Hint:* $T = 1/N$.) Find the optimal number of deliveries if $Q = 2$ million gal, $d = \$8000$, and $s = 30$ cents/gal-yr. Your answer should be a whole number, so compare costs for the two integer values of N nearest the optimal value.

SOLUTION There are N shipments per year, so the time interval between shipments is $T = 1/N$ years. Hence, the total storage costs per year are sQ/N. The yearly delivery costs are dN and the total costs is $C(N) = dN + sQ/N$. Solving,

$$C'(N) = d - \frac{sQ}{N^2} = 0$$

for N yields $N = \sqrt{sQ/d}$. For the specific case $Q = 2{,}000{,}000$, $d = 8000$ and $s = 0.30$,

$$N = \sqrt{\frac{0.30(2{,}000{,}000)}{8000}} = 8.66.$$

With $C(8) = \$139{,}000$ and $C(9) = \$138{,}667$, the optimal number of deliveries per year is $N = 9$.

53. Let (a, b) be a fixed point in the first quadrant and let $S(d)$ be the sum of the distances from $(d, 0)$ to the points $(0, 0)$, (a, b), and $(a, -b)$.

(a) Find the value of d for which $S(d)$ is minimal. The answer depends on whether $b < \sqrt{3}a$ or $b \geq \sqrt{3}a$. *Hint:* Show that $d = 0$ when $b \geq \sqrt{3}a$.

(b) GU Let $a = 1$. Plot $S(d)$ for $b = 0.5, \sqrt{3}, 3$ and describe the position of the minimum.

SOLUTION

(a) If $d < 0$, then the distance from $(d, 0)$ to the other three points can all be reduced by increasing the value of d. Similarly, if $d > a$, then the distance from $(d, 0)$ to the other three points can all be reduced by decreasing the value of d. It follows that the minimum of $S(d)$ must occur for $0 \leq d \leq a$. Restricting attention to this interval, we find

$$S(d) = d + 2\sqrt{(d-a)^2 + b^2}.$$

Solving

$$S'(d) = 1 + \frac{2(d-a)}{\sqrt{(d-a)^2 + b^2}} = 0$$

yields the critical point $d = a - b/\sqrt{3}$. If $b < \sqrt{3}a$, then $d = a - b/\sqrt{3} > 0$ and the minimum occurs at this value of d. On the other hand, if $b \geq \sqrt{3}a$, then the minimum occurs at the endpoint $d = 0$.

(b) Let $a = 1$. Plots of $S(d)$ for $b = 0.5$, $b = \sqrt{3}$ and $b = 3$ are shown below. For $b = 0.5$, the results of (a) indicate the minimum should occur for $d = 1 - 0.5/\sqrt{3} \approx 0.711$, and this is confirmed in the plot. For both $b = \sqrt{3}$ and $b = 3$, the results of (a) indicate that the minimum should occur at $d = 0$, and both of these conclusions are confirmed in the plots.

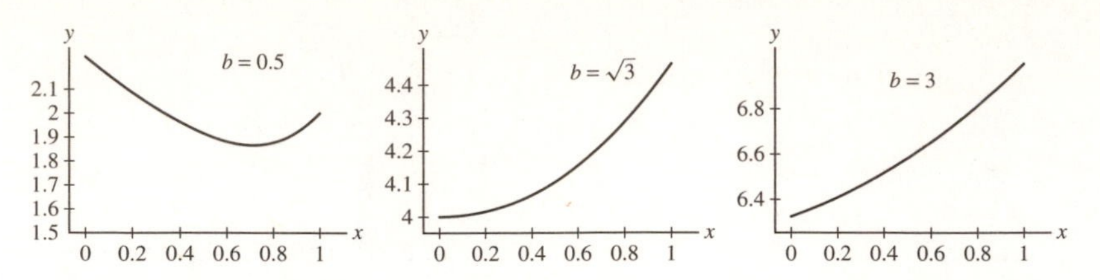

55. In the setting of Exercise 54, show that for any f the minimal force required is proportional to $1/\sqrt{1+f^2}$.

SOLUTION We minimize $F(\theta)$ by finding the maximum value $g(\theta) = \cos\theta + f\sin\theta$. The angle θ is restricted to the interval $[0, \frac{\pi}{2}]$. We solve for the critical points:

$$g'(\theta) = -\sin\theta + f\cos\theta = 0$$

We obtain

$$f\cos\theta = \sin\theta \Rightarrow \tan\theta = f$$

From the figure below we find that $\cos\theta = 1/\sqrt{1+f^2}$ and $\sin\theta = f/\sqrt{1+f^2}$. Hence

$$g(\theta) = \frac{1}{f} + \frac{f^2}{\sqrt{1+f^2}} = \frac{1+f^2}{\sqrt{1+f^2}} = \sqrt{1+f^2}$$

The values at the endpoints are

$$g(0) = 1, \qquad g\left(\frac{\pi}{2}\right) = f$$

Both of these values are less than $\sqrt{1+f^2}$. Therefore the maximum value of $g(\theta)$ is $\sqrt{1+f^2}$ and the minimum value of $F(\theta)$ is

$$F = \frac{fmg}{g(\theta)} = \frac{fmg}{\sqrt{1+f^2}}$$

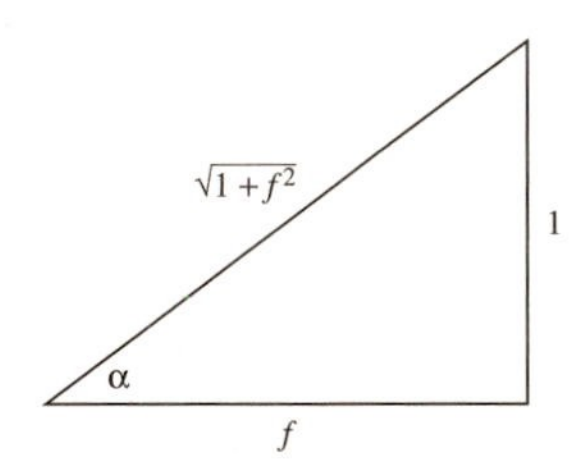

57. The problem is to put a "roof" of side s on an attic room of height h and width b. Find the smallest length s for which this is possible if $b = 27$ and $h = 8$ (Figure 31).

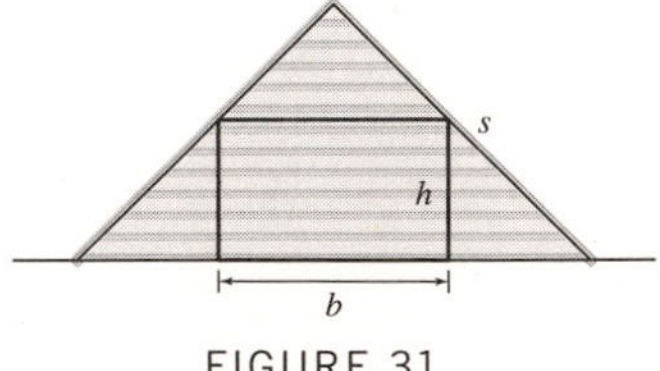

FIGURE 31

SOLUTION Consider the right triangle formed by the right half of the rectangle and its "roof". This triangle has hypotenuse s.

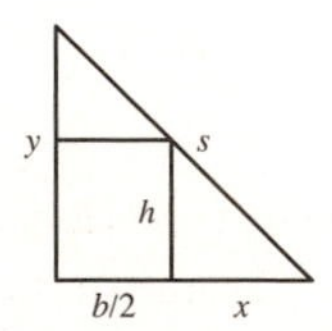

As shown, let y be the height of the roof, and let x be the distance from the right base of the rectangle to the base of the roof. By similar triangles applied to the smaller right triangles at the top and right of the larger triangle, we get:

$$\frac{y-8}{27/2} = \frac{8}{x} \quad \text{or} \quad y = \frac{108}{x} + 8.$$

s, y, and x are related by the Pythagorean Theorem:

$$s^2 = \left(\frac{27}{2} + x\right)^2 + y^2 = \left(\frac{27}{2} + x\right)^2 + \left(\frac{108}{x} + 8\right)^2.$$

Since $s > 0$, s^2 is least whenever s is least, so we can minimize s^2 instead of s. Setting the derivative equal to zero yields

$$2\left(\frac{27}{2} + x\right) + 2\left(\frac{108}{x} + 8\right)\left(-\frac{108}{x^2}\right) = 0$$

$$2\left(\frac{27}{2} + x\right) + 2\frac{8}{x}\left(\frac{27}{2} + x\right)\left(-\frac{108}{x^2}\right) = 0$$

$$2\left(\frac{27}{2} + x\right)\left(1 - \frac{864}{x^3}\right) = 0$$

The zeros are $x = -\frac{27}{2}$ (irrelevant) and $x = 6\sqrt[3]{4}$. Since this is the only critical point of s with $x > 0$, and since $s \to \infty$ as $x \to 0$ and $s \to \infty$ as $x \to \infty$, this is the point where s attains its minimum. For this value of x,

$$s^2 = \left(\frac{27}{2} + 6\sqrt[3]{4}\right)^2 + \left(9\sqrt[3]{2} + 8\right)^2 \approx 904.13,$$

so the smallest roof length is

$$s \approx 30.07.$$

59. Find the maximum length of a pole that can be carried horizontally around a corner joining corridors of widths $a = 24$ and $b = 3$ (Figure 32).

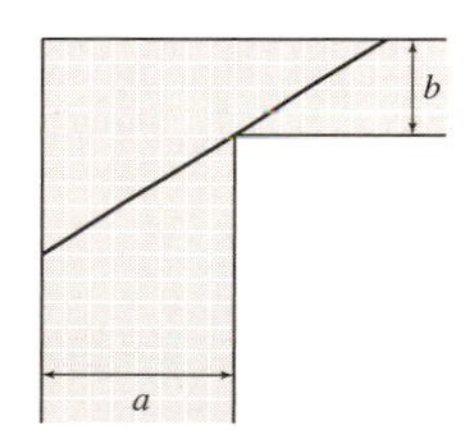

FIGURE 32

SOLUTION In order to find the length of the *longest* pole that can be carried around the corridor, we have to find the *shortest* length from the left wall to the top wall touching the corner of the inside wall. Any pole that does not fit in this shortest space cannot be carried around the corner, so an exact fit represents the longest possible pole.

Let θ be the angle between the pole and a horizontal line to the right. Let c_1 be the length of pole in the corridor of width 24 and let c_2 be the length of pole in the corridor of width 3. By the definitions of sine and cosine,

$$\frac{3}{c_2} = \sin\theta \quad \text{and} \quad \frac{24}{c_1} = \cos\theta,$$

so that $c_1 = \frac{24}{\cos\theta}$, $c_2 = \frac{3}{\sin\theta}$. What must be minimized is the total length, given by

$$f(\theta) = \frac{24}{\cos\theta} + \frac{3}{\sin\theta}.$$

Setting $f'(\theta) = 0$ yields

$$\frac{24\sin\theta}{\cos^2\theta} - \frac{3\cos\theta}{\sin^2\theta} = 0$$

$$\frac{24\sin\theta}{\cos^2\theta} = \frac{3\cos\theta}{\sin^2\theta}$$

$$24\sin^3\theta = 3\cos^3\theta$$

As $\theta < \frac{\pi}{2}$ (the pole is being turned around a corner, after all), we can divide both sides by $\cos^3\theta$, getting $\tan^3\theta = \frac{1}{8}$. This implies that $\tan\theta = \frac{1}{2}$ ($\tan\theta > 0$ as the angle is acute).

Since $f(\theta) \to \infty$ as $\theta \to 0+$ and as $\theta \to \frac{\pi}{2}-$, we can tell that the *minimum* is attained at θ_0 where $\tan\theta_0 = \frac{1}{2}$. Because

$$\tan\theta_0 = \frac{\text{opposite}}{\text{adjacent}} = \frac{1}{2},$$

we draw a triangle with opposite side 1 and adjacent side 2. By Pythagoras, $c = \sqrt{5}$, so

$$\sin\theta_0 = \frac{1}{\sqrt{5}} \quad \text{and} \quad \cos\theta_0 = \frac{2}{\sqrt{5}}.$$

From this, we get

$$f(\theta_0) = \frac{24}{\cos\theta_0} + \frac{3}{\sin\theta_0} = \frac{24}{2}\sqrt{5} + 3\sqrt{5} = 15\sqrt{5}.$$

61. Find the minimum length ℓ of a beam that can clear a fence of height h and touch a wall located b ft behind the fence (Figure 33).

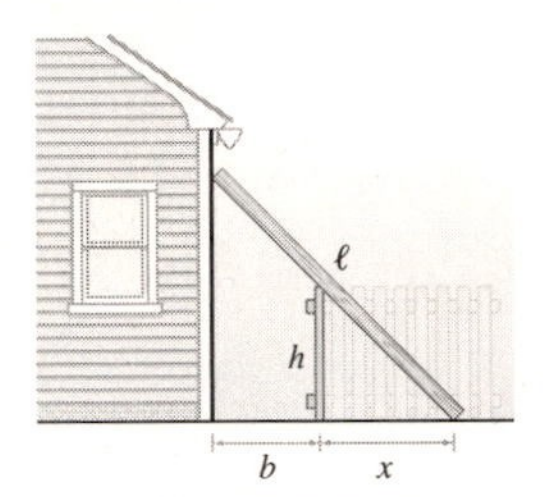

FIGURE 33

SOLUTION Let y be the height of the point where the beam touches the wall in feet. By similar triangles,

$$\frac{y-h}{b} = \frac{h}{x} \quad \text{or} \quad y = \frac{bh}{x} + h$$

and by Pythagoras:

$$\ell^2 = (b+x)^2 + \left(\frac{bh}{x} + h\right)^2.$$

We can minimize ℓ^2 rather than ℓ, so setting the derivative equal to zero gives:

$$2(b+x) + 2\left(\frac{bh}{x} + h\right)\left(-\frac{bh}{x^2}\right) = 2(b+x)\left(1 - \frac{h^2b}{x^3}\right) = 0.$$

The zeroes are $b = -x$ (irrelevant) and $x = \sqrt[3]{h^2b}$. Since $\ell^2 \to \infty$ as $x \to 0+$ and as $x \to \infty$, $x = \sqrt[3]{h^2b}$ corresponds to a minimum for ℓ^2. For this value of x, we have

$$\begin{aligned}\ell^2 &= (b + h^{2/3}b^{1/3})^2 + (h + h^{1/3}b^{2/3})^2 \\ &= b^{2/3}(b^{2/3} + h^{2/3})^2 + h^{2/3}(h^{2/3} + b^{2/3})^2 \\ &= (b^{2/3} + h^{2/3})^3\end{aligned}$$

and so

$$\ell = (b^{2/3} + h^{2/3})^{3/2}.$$

A beam that clears a fence of height h feet and touches a wall b feet behind the fence must have length at least $\ell = (b^{2/3} + h^{2/3})^{3/2}$ ft.

63. A basketball player stands d feet from the basket. Let h and α be as in Figure 34. Using physics, one can show that if the player releases the ball at an angle θ, then the initial velocity required to make the ball go through the basket satisfies

$$v^2 = \frac{16d}{\cos^2\theta(\tan\theta - \tan\alpha)}$$

(a) Explain why this formula is meaningful only for $\alpha < \theta < \frac{\pi}{2}$. Why does v approach infinity at the endpoints of this interval?

(b) GU Take $\alpha = \frac{\pi}{6}$ and plot v^2 as a function of θ for $\frac{\pi}{6} < \theta < \frac{\pi}{2}$. Verify that the minimum occurs at $\theta = \frac{\pi}{3}$.
(c) Set $F(\theta) = \cos^2\theta(\tan\theta - \tan\alpha)$. Explain why v is minimized for θ such that $F(\theta)$ is maximized.
(d) Verify that $F'(\theta) = \cos(\alpha - 2\theta)\sec\alpha$ (you will need to use the addition formula for cosine) and show that the maximum value of $F(\theta)$ on $\left[\alpha, \frac{\pi}{2}\right]$ occurs at $\theta_0 = \frac{\alpha}{2} + \frac{\pi}{4}$.
(e) For a given α, the optimal angle for shooting the basket is θ_0 because it minimizes v^2 and therefore minimizes the energy required to make the shot (energy is proportional to v^2). Show that the velocity $v_{\rm opt}$ at the optimal angle θ_0 satisfies

$$v_{\rm opt}^2 = \frac{32d\cos\alpha}{1-\sin\alpha} = \frac{32\,d^2}{-h+\sqrt{d^2+h^2}}$$

(f) GU Show with a graph that for fixed d (say, $d = 15$ ft, the distance of a free throw), $v_{\rm opt}^2$ is an increasing function of h. Use this to explain why taller players have an advantage and why it can help to jump while shooting.

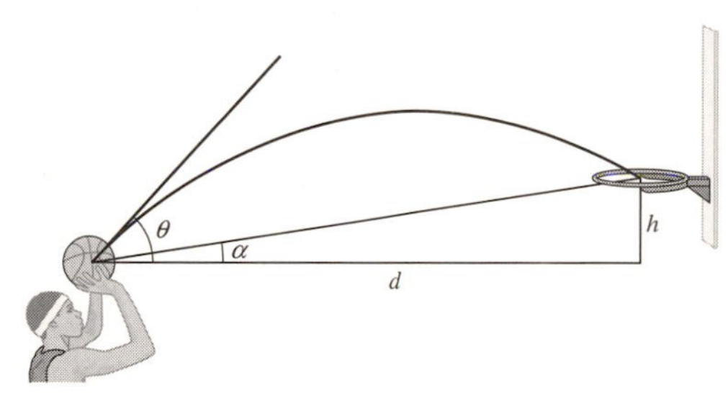

FIGURE 34

SOLUTION
(a) $\alpha = 0$ corresponds to shooting the ball directly at the basket while $\alpha = \pi/2$ corresponds to shooting the ball directly upward. In neither case is it possible for the ball to go into the basket.

If the angle α is extremely close to 0, the ball is shot almost directly at the basket, so that it must be launched with great speed, as it can only fall an extremely short distance on the way to the basket.

On the other hand, if the angle α is extremely close to $\pi/2$, the ball is launched almost vertically. This requires the ball to travel a great distance upward in order to travel the horizontal distance. In either one of these cases, the ball has to travel at an enormous speed.
(b)

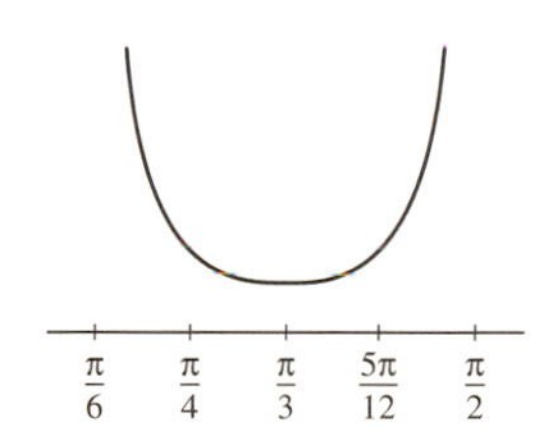

The minimum clearly occurs where $\theta = \pi/3$.
(c) If $F(\theta) = \cos^2\theta\,(\tan\theta - \tan\alpha)$,

$$v^2 = \frac{16d}{\cos^2\theta\,(\tan\theta - \tan\alpha)} = \frac{16d}{F(\theta)}.$$

Since $\alpha \le \theta$, $F(\theta) > 0$, hence v^2 is smallest whenever $F(\theta)$ is greatest.
(d) $F'(\theta) = -2\sin\theta\cos\theta\,(\tan\theta - \tan\alpha) + \cos^2\theta\left(\sec^2\theta\right) = -2\sin\theta\cos\theta\tan\theta + 2\sin\theta\cos\theta\tan\alpha + 1$. We will apply all the double angle formulas:

$$\cos(2\theta) = \cos^2\theta - \sin^2\theta = 1 - 2\sin^2\theta;\ \sin 2\theta = 2\sin\theta\cos\theta,$$

getting:

$$\begin{aligned}
F'(\theta) &= 2\sin\theta\cos\theta\tan\alpha - 2\sin\theta\cos\theta\tan\theta + 1\\
&= 2\sin\theta\cos\theta\frac{\sin\alpha}{\cos\alpha} - 2\sin\theta\cos\theta\frac{\sin\theta}{\cos\theta} + 1\\
&= \sec\alpha\left(-2\sin^2\theta\cos\alpha + 2\sin\theta\cos\theta\sin\alpha + \cos\alpha\right)\\
&= \sec\alpha\left(\cos\alpha\left(1 - 2\sin^2\theta\right) + \sin\alpha\,(2\sin\theta\cos\theta)\right)\\
&= \sec\alpha\,(\cos\alpha(\cos 2\theta) + \sin\alpha(\sin 2\theta))\\
&= \sec\alpha\cos(\alpha - 2\theta)
\end{aligned}$$

A critical point of $F(\theta)$ occurs where $\cos(\alpha - 2\theta) = 0$, so that $\alpha - 2\theta = -\frac{\pi}{2}$ (negative because $2\theta > \theta > \alpha$), and this gives us $\theta = \alpha/2 + \pi/4$. The minimum value $F(\theta_0)$ takes place at $\theta_0 = \alpha/2 + \pi/4$.

(e) Plug in $\theta_0 = \alpha/2 + \pi/4$. To find v_{opt}^2 we must simplify

$$\cos^2\theta_0(\tan\theta_0 - \tan\alpha) = \frac{\cos\theta_0(\sin\theta_0\cos\alpha - \cos\theta_0\sin\alpha)}{\cos\alpha}$$

By the addition law for sine:

$$\sin\theta_0\cos\alpha - \cos\theta_0\sin\alpha = \sin(\theta_0 - \alpha) = \sin(-\alpha/2 + \pi/4)$$

and so

$$\cos\theta_0(\sin\theta_0\cos\alpha - \cos\theta_0\sin\alpha) = \cos(\alpha/2 + \pi/4)\sin(-\alpha/2 + \pi/4)$$

Now use the identity (that follows from the addition law):

$$\sin x\cos y = \frac{1}{2}(\sin(x+y) + \sin(x-y))$$

to get

$$\cos(\alpha/2 + \pi/4)\sin(-\alpha/2 + \pi/4) = (1/2)(1 - \sin\alpha)$$

So we finally get

$$\cos^2\theta_0(\tan\theta_0 - \tan\alpha) = \frac{(1/2)(1 - \sin\alpha)}{\cos\alpha}$$

and therefore

$$v_{\text{opt}}^2 = \frac{32d\cos\alpha}{1 - \sin\alpha}$$

as claimed. From Figure 34 we see that

$$\cos\alpha = \frac{d}{\sqrt{d^2 + h^2}} \qquad \text{and} \qquad \sin\alpha = \frac{h}{\sqrt{d^2 + h^2}}.$$

Substituting these values into the expression for v_{opt}^2 yields

$$v_{\text{opt}}^2 = \frac{32d^2}{-h + \sqrt{d^2 + h^2}}.$$

(f) A sketch of the graph of v_{opt}^2 versus h for $d = 15$ feet is given below: v_{opt}^2 increases with respect to basket height relative to the shooter. This shows that the minimum velocity required to launch the ball to the basket drops as shooter height increases. This shows one of the ways height is an advantage in free throws; a taller shooter need not shoot the ball as hard to reach the basket.

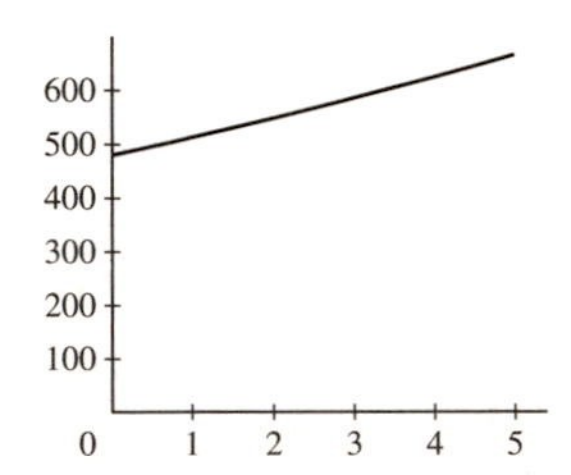

Further Insights and Challenges

65. Tom and Ali drive along a highway represented by the graph of $f(x)$ in Figure 36. During the trip, Ali views a billboard represented by the segment $\overline{BC}$ along the y-axis. Let Q be the y-intercept of the tangent line to $y = f(x)$. Show that θ is maximized at the value of x for which the angles $\angle QPB$ and $\angle QCP$ are equal. This generalizes Exercise 50 (c) (which corresponds to the case $f(x) = 0$). *Hints:*

(a) Show that $d\theta/dx$ is equal to

$$(b-c)\cdot\frac{(x^2 + (xf'(x))^2) - (b - (f(x) - xf'(x)))(c - (f(x) - xf'(x)))}{(x^2 + (b - f(x))^2)(x^2 + (c - f(x))^2)}$$

(b) Show that the y-coordinate of Q is $f(x) - xf'(x)$.

(c) Show that the condition $d\theta/dx = 0$ is equivalent to

$$PQ^2 = BQ \cdot CQ$$

(d) Conclude that ΔQPB and ΔQCP are similar triangles.

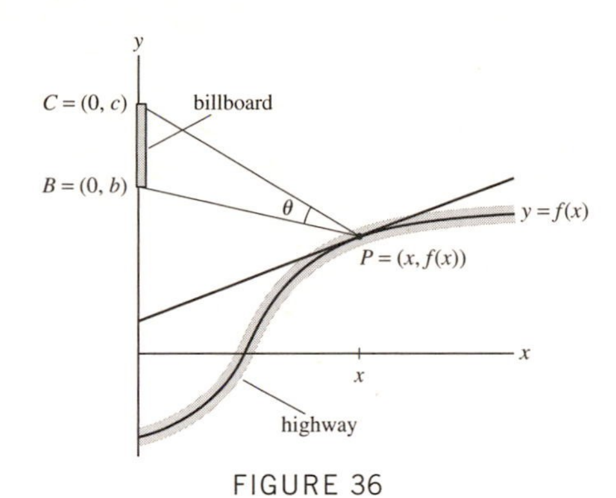

FIGURE 36

SOLUTION

(a) From the figure, we see that

$$\theta(x) = \tan^{-1}\frac{c - f(x)}{x} - \tan^{-1}\frac{b - f(x)}{x}.$$

Then

$$\begin{aligned}
\theta'(x) &= \frac{b - (f(x) - xf'(x))}{x^2 + (b - f(x))^2} - \frac{c - (f(x) - xf'(x))}{x^2 + (c - f(x))^2} \\
&= (b - c)\frac{x^2 - bc + (b + c)(f(x) - xf'(x)) - (f(x))^2 + 2xf(x)f'(x)}{(x^2 + (b - f(x))^2)(x^2 + (c - f(x))^2)} \\
&= (b - c)\frac{(x^2 + (xf'(x))^2 - (bc - (b + c)(f(x) - xf'(x)) + (f(x) - xf'(x))^2)}{(x^2 + (b - f(x))^2)(x^2 + (c - f(x))^2)} \\
&= (b - c)\frac{(x^2 + (xf'(x))^2 - (b - (f(x) - xf'(x)))(c - (f(x) - xf'(x)))}{(x^2 + (b - f(x))^2)(x^2 + (c - f(x))^2)}.
\end{aligned}$$

(b) The point Q is the y-intercept of the line tangent to the graph of $f(x)$ at point P. The equation of this tangent line is

$$Y - f(x) = f'(x)(X - x).$$

The y-coordinate of Q is then $f(x) - xf'(x)$.

(c) From the figure, we see that

$$\begin{aligned}
BQ &= b - (f(x) - xf'(x)), \\
CQ &= c - (f(x) - xf'(x))
\end{aligned}$$

and

$$PQ = \sqrt{x^2 + (f(x) - (f(x) - xf'(x)))^2} = \sqrt{x^2 + (xf'(x))^2}.$$

Comparing these expressions with the numerator of $d\theta/dx$, it follows that $\dfrac{d\theta}{dx} = 0$ is equivalent to

$$PQ^2 = BQ \cdot CQ.$$

(d) The equation $PQ^2 = BQ \cdot CQ$ is equivalent to

$$\frac{PQ}{BQ} = \frac{CQ}{PQ}.$$

In other words, the sides CQ and PQ from the triangle ΔQCP are proportional in length to the sides PQ and BQ from the triangle ΔQPB. As $\angle PQB = \angle CQP$, it follows that triangles ΔQCP and ΔQPB are similar.

Seismic Prospecting *Exercises 66–68 are concerned with determining the thickness d of a layer of soil that lies on top of a rock formation. Geologists send two sound pulses from point A to point D separated by a distance s. The first pulse travels directly from A to D along the surface of the earth. The second pulse travels down to the rock formation, then along its surface, and then back up to D (path ABCD), as in Figure 37. The pulse travels with velocity v_1 in the soil and v_2 in the rock.*

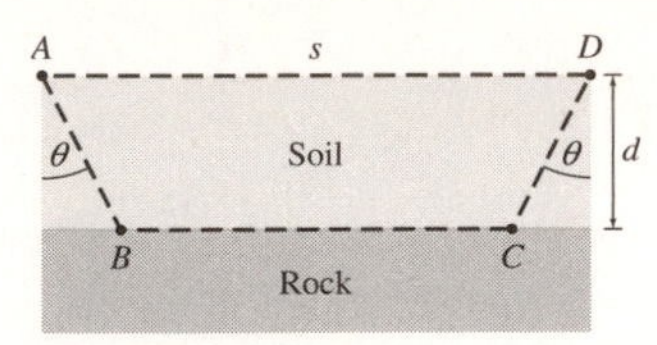

FIGURE 37

67. In this exercise, assume that $v_2/v_1 \geq \sqrt{1+4(d/s)^2}$.

(a) Show that inequality (2) holds if $\sin\theta = v_1/v_2$.

(b) Show that the minimal time for the second pulse is

$$t_2 = \frac{2d}{v_1}(1-k^2)^{1/2} + \frac{s}{v_2}$$

where $k = v_1/v_2$.

(c) Conclude that $\dfrac{t_2}{t_1} = \dfrac{2d(1-k^2)^{1/2}}{s} + k$.

SOLUTION

(a) If $\sin\theta = \frac{v_1}{v_2}$, then

$$\tan\theta = \frac{v_1}{\sqrt{v_2^2 - v_1^2}} = \frac{1}{\sqrt{\left(\frac{v_2}{v_1}\right)^2 - 1}}.$$

Because $\frac{v_2}{v_1} \geq \sqrt{1+4(\frac{d}{s})^2}$, it follows that

$$\sqrt{\left(\frac{v_2}{v_1}\right)^2 - 1} \geq \sqrt{1+4\left(\frac{d}{s}\right)^2 - 1} = \frac{2d}{s}.$$

Hence, $\tan\theta \leq \frac{s}{2d}$ as required.

(b) For the time-minimizing choice of θ, we have $\sin\theta = \dfrac{v_1}{v_2}$ from which $\sec\theta = \dfrac{v_2}{\sqrt{v_2^2 - v_1^2}}$ and $\tan\theta = \dfrac{v_1}{\sqrt{v_2^2 - v_1^2}}$.

Thus

$$t_2 = \frac{2d}{v_1}\sec\theta + \frac{s - 2d\tan\theta}{v_2} = \frac{2d}{v_1}\frac{v_2}{\sqrt{v_2^2 - v_1^2}} + \frac{s - 2d\frac{v_1}{\sqrt{v_2^2 - v_1^2}}}{v_2}$$

$$= \frac{2d}{v_1}\left(\frac{v_2}{\sqrt{v_2^2 - v_1^2}} - \frac{v_1^2}{v_2\sqrt{v_2^2 - v_1^2}}\right) + \frac{s}{v_2}$$

$$= \frac{2d}{v_1}\left(\frac{v_2^2 - v_1^2}{v_2\sqrt{v_2^2 - v_1^2}}\right) + \frac{s}{v_2} = \frac{2d}{v_1}\left(\frac{\sqrt{v_2^2 - v_1^2}}{\sqrt{v_2^2}}\right) + \frac{s}{v_2}$$

$$= \frac{2d}{v_1}\sqrt{1 - \left(\frac{v_1}{v_2}\right)^2} + \frac{s}{v_2} = \frac{2d\left(1-k^2\right)^{1/2}}{v_1} + \frac{s}{v_2}.$$

(c) Recall that $t_1 = \dfrac{s}{v_1}$. We therefore have

$$\frac{t_2}{t_1} = \frac{\frac{2d(1-k^2)^{1/2}}{v_1} + \frac{s}{v_2}}{\frac{s}{v_1}}$$

$$= \frac{2d\left(1-k^2\right)^{1/2}}{s} + \frac{v_1}{v_2} = \frac{2d\left(1-k^2\right)^{1/2}}{s} + k.$$

69. In this exercise we use Figure 38 to prove Heron's principle of Example 6 without calculus. By definition, C is the reflection of B across the line $\overline{MN}$ (so that $\overline{BC}$ is perpendicular to $\overline{MN}$ and $BN = CN$. Let P be the intersection of $\overline{AC}$ and $\overline{MN}$. Use geometry to justify:

(a) $\triangle PNB$ and $\triangle PNC$ are congruent and $\theta_1 = \theta_2$.

(b) The paths APB and APC have equal length.

(c) Similarly AQB and AQC have equal length.

(d) The path APC is shorter than AQC for all $Q \neq P$.

Conclude that the shortest path AQB occurs for $Q = P$.

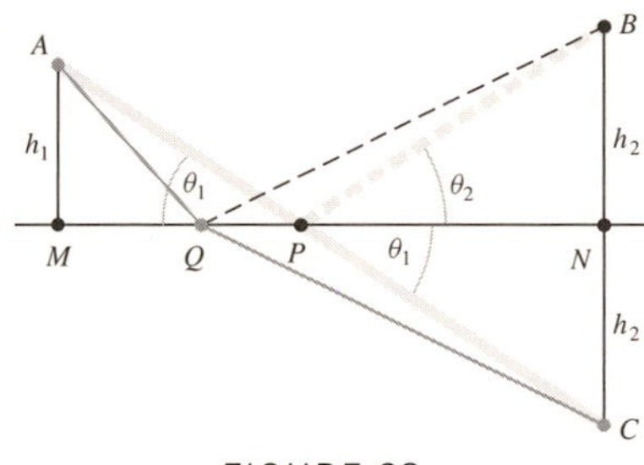

FIGURE 38

SOLUTION

(a) By definition, $\overline{BC}$ is orthogonal to $\overline{QM}$, so triangles $\triangle PNB$ and $\triangle PNC$ are congruent by side–angle–side. Therefore $\theta_1 = \theta_2$

(b) Because $\triangle PNB$ and $\triangle PNC$ are congruent, it follows that $\overline{PB}$ and $\overline{PC}$ are of equal length. Thus, paths APB and APC have equal length.

(c) The same reasoning used in parts (a) and (b) lead us to conclude that $\triangle QNB$ and $\triangle QNC$ are congruent and that $\overline{PB}$ and $\overline{PC}$ are of equal length. Thus, paths AQB and AQC are of equal length.

(d) Consider triangle $\triangle AQC$. By the triangle inequality, the length of side $\overline{AC}$ is less than or equal to the sum of the lengths of the sides $\overline{AQ}$ and $\overline{QC}$. Thus, the path APC is shorter than AQC for all $Q \neq P$.

Finally, the shortest path AQB occurs for $Q = P$.

4.8 Newton's Method

Preliminary Questions

1. How many iterations of Newton's Method are required to compute a root if $f(x)$ is a linear function?

SOLUTION Remember that Newton's Method uses the linear approximation of a function to estimate the location of a root. If the original function is linear, then only one iteration of Newton's Method will be required to compute the root.

2. What happens in Newton's Method if your initial guess happens to be a zero of f?

SOLUTION If x_0 happens to be a zero of f, then

$$x_1 = x_0 - \frac{f(x_0)}{f'(x_0)} = x_0 - 0 = x_0;$$

in other words, every term in the Newton's Method sequence will remain x_0.

3. What happens in Newton's Method if your initial guess happens to be a local min or max of f?

SOLUTION Assuming that the function is differentiable, then the derivative is zero at a local maximum or a local minimum. If Newton's Method is started with an initial guess such that $f'(x_0) = 0$, then Newton's Method will fail in the sense that x_1 will not be defined. That is, the tangent line will be parallel to the x-axis and will never intersect it.

4. Is the following a reasonable description of Newton's Method: "A root of the equation of the tangent line to $f(x)$ is used as an approximation to a root of $f(x)$ itself"? Explain.

SOLUTION Yes, that is a reasonable description. The iteration formula for Newton's Method was derived by solving the equation of the tangent line to $y = f(x)$ at x_0 for its x-intercept.

Exercises

In this exercise set, all approximations should be carried out using Newton's Method.

In Exercises 1–6, apply Newton's Method to $f(x)$ and initial guess x_0 to calculate x_1, x_2, x_3.

1. $f(x) = x^2 - 6, \quad x_0 = 2$

SOLUTION Let $f(x) = x^2 - 6$ and define

$$x_{n+1} = x_n - \frac{f(x_n)}{f'(x_n)} = x_n - \frac{x_n^2 - 6}{2x_n}.$$

With $x_0 = 2$, we compute

n	1	2	3
x_n	2.5	2.45	2.44948980

3. $f(x) = x^3 - 10, \quad x_0 = 2$

SOLUTION Let $f(x) = x^3 - 10$ and define

$$x_{n+1} = x_n - \frac{f(x_n)}{f'(x_n)} = x_n - \frac{x_n^3 - 10}{3x_n^2}.$$

With $x_0 = 2$ we compute

n	1	2	3
x_n	2.16666667	2.15450362	2.15443469

5. $f(x) = \cos x - 4x, \quad x_0 = 1$

SOLUTION Let $f(x) = \cos x - 4x$ and define

$$x_{n+1} = x_n - \frac{f(x_n)}{f'(x_n)} = x_n - \frac{\cos x_n - 4x_n}{-\sin x_n - 4}.$$

With $x_0 = 1$ we compute

n	1	2	3
x_n	0.28540361	0.24288009	0.24267469

7. Use Figure 6 to choose an initial guess x_0 to the unique real root of $x^3 + 2x + 5 = 0$ and compute the first three Newton iterates.

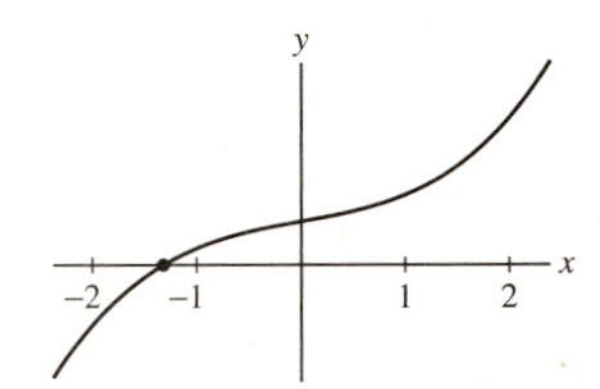

FIGURE 6 Graph of $y = x^3 + 2x + 5$.

SOLUTION Let $f(x) = x^3 + 2x + 5$ and define

$$x_{n+1} = x_n - \frac{f(x_n)}{f'(x_n)} = x_n - \frac{x_n^3 + 2x_n + 5}{3x_n^2 + 2}.$$

We take $x_0 = -1.4$, based on the figure, and then calculate

n	1	2	3
x_n	−1.330964467	−1.328272820	−1.328268856

9. Approximate both solutions of $e^x = 5x$ to three decimal places (Figure 7).

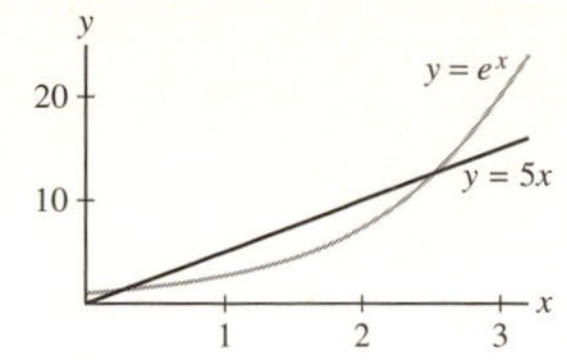

FIGURE 7 Graphs of e^x and $5x$.

SOLUTION We need to solve $e^x - 5x = 0$, so let $f(x) = e^x - 5x$. Then $f'(x) = e^x - 5$. With an initial guess of $x_0 = 0.2$, we calculate

Newton's Method (First root)	$x_0 = 0.2$ (guess)
$x_1 = 0.2 - \dfrac{f(0.2)}{f'(0.2)}$	$x_1 \approx 0.25859$
$x_2 = 0.25859 - \dfrac{f(0.25859)}{f'(0.25859)}$	$x_2 \approx 0.25917$
$x_3 = 0.25917 - \dfrac{f(0.25917)}{f'(0.25917)}$	$x_3 \approx 0.25917$

For the second root, we use an initial guess of $x_0 = 2.5$.

Newton's Method (Second root)	$x_0 = 2.5$ (guess)
$x_1 = 2.5 - \dfrac{f(2.5)}{f'(2.5)}$	$x_1 \approx 2.54421$
$x_2 = 2.54421 - \dfrac{f(2.54421)}{f'(2.54421)}$	$x_2 \approx 2.54264$
$x_3 = 2.54264 - \dfrac{f(2.54264)}{f'(2.54264)}$	$x_3 \approx 2.54264$

Thus the two solutions of $e^x = 5x$ are approximately $r_1 \approx 0.25917$ and $r_2 \approx 2.54264$.

In Exercises 11–14, approximate to three decimal places using Newton's Method and compare with the value from a calculator.

11. $\sqrt{11}$

SOLUTION Let $f(x) = x^2 - 11$, and let $x_0 = 3$. Newton's Method yields:

n	1	2	3
x_n	3.33333333	3.31666667	3.31662479

A calculator yields 3.31662479.

13. $2^{7/3}$

SOLUTION Note that $2^{7/3} = 4 \cdot 2^{1/3}$. Let $f(x) = x^3 - 2$, and let $x_0 = 1$. Newton's Method yields:

n	1	2	3
x_n	1.33333333	1.26388889	1.25993349

Thus, $2^{7/3} \approx 4 \cdot 1.25993349 = 5.03973397$. A calculator yields 5.0396842.

15. Approximate the largest positive root of $f(x) = x^4 - 6x^2 + x + 5$ to within an error of at most 10^{-4}. Refer to Figure 5.

SOLUTION Figure 5 from the text suggests the largest positive root of $f(x) = x^4 - 6x^2 + x + 5$ is near 2. So let $f(x) = x^4 - 6x^2 + x + 5$ and take $x_0 = 2$.

n	1	2	3	4
x_n	2.111111111	2.093568458	2.093064768	2.093064358

The largest positive root of $x^4 - 6x^2 + x + 5$ is approximately 2.093064358.

GU *In Exercises 16–19, approximate the root specified to three decimal places using Newton's Method. Use a plot to choose an initial guess.*

17. Negative root of $f(x) = x^5 - 20x + 10$.

SOLUTION Let $f(x) = x^5 - 20x + 10$. The graph of $f(x)$ shown below suggests taking $x_0 = -2.2$. Starting from $x_0 = -2.2$, the first three iterates of Newton's Method are:

n	1	2	3
x_n	−2.22536529	−2.22468998	−2.22468949

Thus, to three decimal places, the negative root of $f(x) = x^5 - 20x + 10$ is −2.225.

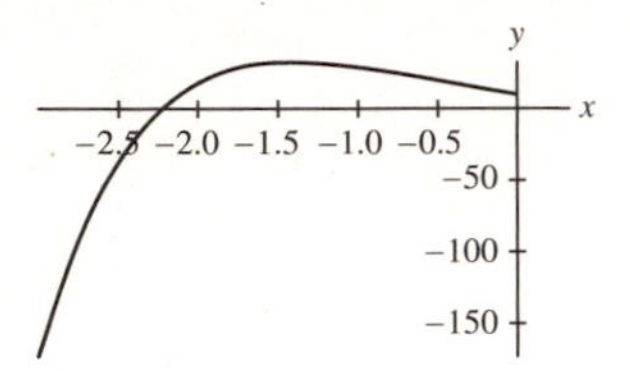

19. Solution of $\ln(x + 4) = x$.

SOLUTION From the graph below, we see that the positive solution to the equation $\ln(x + 4) = x$ is approximately $x = 2$. Now, let $f(x) = \ln(x + 4) - x$ and define

$$x_{n+1} = x_n - \frac{f(x_n)}{f'(x_n)} = x_n - \frac{\ln(x_n + 4) - x_n}{\frac{1}{x_n+4} - 1}.$$

With $x_0 = 2$ we find

n	1	2	3
x_n	1.750111363	1.749031407	1.749031386

Thus, to three decimal places, the positive solution to the equation $\ln(x + 4) = x$ is 1.749.

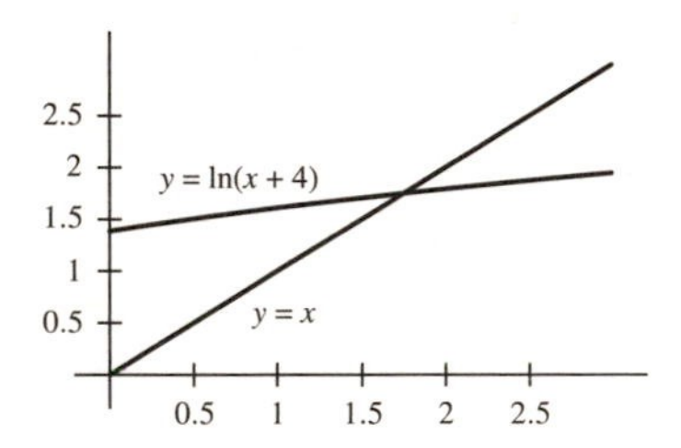

21. GU Find the smallest positive value of x at which $y = x$ and $y = \tan x$ intersect. *Hint:* Draw a plot.

SOLUTION Here is a plot of $\tan x$ and x on the same axes:

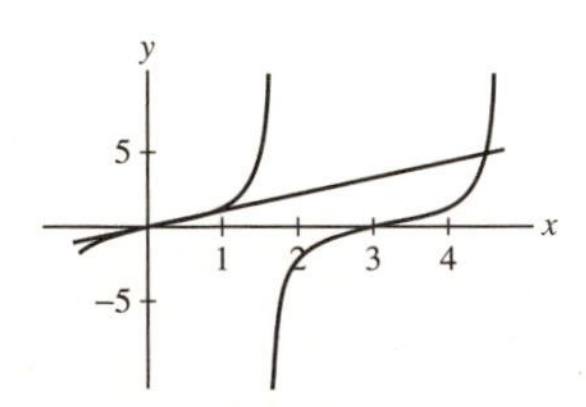

The first intersection with $x > 0$ lies on the second "branch" of $y = \tan x$, between $x = \frac{5\pi}{4}$ and $x = \frac{3\pi}{2}$. Let $f(x) = \tan x - x$. The graph suggests an initial guess $x_0 = \frac{5\pi}{4}$, from which we get the following table:

n	1	2	3	4
x_n	6.85398	21.921	4480.8	7456.27

This is clearly leading nowhere, so we need to try a better initial guess. *Note: This happens with Newton's Method—it is sometimes difficult to choose an initial guess.* We try the point directly between $\frac{5\pi}{4}$ and $\frac{3\pi}{2}$, $x_0 = \frac{11\pi}{8}$:

n	1	2	3	4	5	6	7
x_n	4.64662	4.60091	4.54662	4.50658	4.49422	4.49341	4.49341

The first point where $y = x$ and $y = \tan x$ cross is at approximately $x = 4.49341$, which is approximately 1.4303π.

23. Find (to two decimal places) the coordinates of the point P in Figure 9 where the tangent line to $y = \cos x$ passes through the origin.

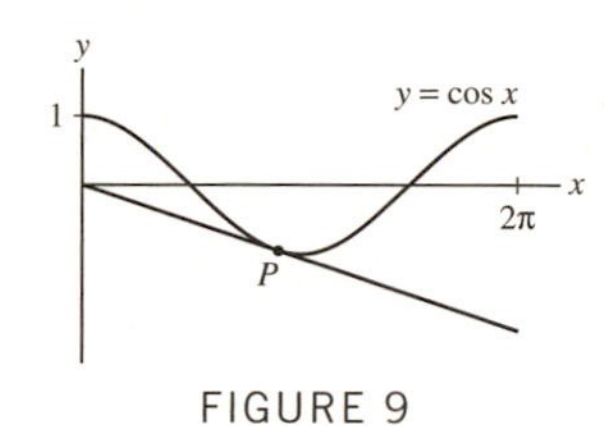

FIGURE 9

SOLUTION Let $(x_r, \cos(x_r))$ be the coordinates of the point P. The slope of the tangent line is $-\sin(x_r)$, so we are looking for a tangent line:

$$y = -\sin(x_r)(x - x_r) + \cos(x_r)$$

such that $y = 0$ when $x = 0$. This gives us the equation:

$$-\sin(x_r)(-x_r) + \cos(x_r) = 0.$$

Let $f(x) = \cos x + x \sin x$. We are looking for the first point $x = r$ where $f(r) = 0$. The sketch given indicates that $x_0 = 3\pi/4$ would be a good initial guess. The following table gives successive Newton's Method approximations:

n	1	2	3	4
x_n	2.931781309	2.803636974	2.798395826	2.798386046

The point P has approximate coordinates (2.7984, −0.941684).

Newton's Method is often used to determine interest rates in financial calculations. In Exercises 24–26, r denotes a yearly interest rate expressed as a decimal (rather than as a percent).

25. If you borrow L dollars for N years at a yearly interest rate r, your monthly payment of P dollars is calculated using the equation

$$L = P\left(\frac{1 - b^{-12N}}{b - 1}\right) \qquad \text{where } b = 1 + \frac{r}{12}$$

(a) Find P if $L = \$5000$, $N = 3$, and $r = 0.08$ (8%).
(b) You are offered a loan of $L = \$5000$ to be paid back over 3 years with monthly payments of $P = \$200$. Use Newton's Method to compute b and find the implied interest rate r of this loan. *Hint:* Show that $(L/P)b^{12N+1} - (1 + L/P)b^{12N} + 1 = 0$.

SOLUTION

(a) $b = (1 + 0.08/12) = 1.00667$

$$P = L\left(\frac{b - 1}{1 - b^{-12N}}\right) = 5000\left(\frac{1.00667 - 1}{1 - 1.00667^{-36}}\right) \approx \$156.69$$

(b) Starting from

$$L = P\left(\frac{1 - b^{-12N}}{b - 1}\right),$$

divide by P, multiply by $b - 1$, multiply by b^{12N} and collect like terms to arrive at

$$(L/P)b^{12N+1} - (1 + L/P)b^{12N} + 1 = 0.$$

Since $L/P = 5000/200 = 25$, we must solve

$$25b^{37} - 26b^{36} + 1 = 0.$$

Newton's Method gives $b \approx 1.02121$ and

$$r = 12(b - 1) = 12(0.02121) \approx 0.25452$$

So the interest rate is around 25.45%.

27. There is no simple formula for the position at time t of a planet P in its orbit (an ellipse) around the sun. Introduce the auxiliary circle and angle θ in Figure 10 (note that P determines θ because it is the central angle of point B on the circle). Let $a = OA$ and $e = OS/OA$ (the eccentricity of the orbit).

(a) Show that sector BSA has area $(a^2/2)(\theta - e\sin\theta)$.

(b) By Kepler's Second Law, the area of sector BSA is proportional to the time t elapsed since the planet passed point A, and because the circle has area πa^2, BSA has area $(\pi a^2)(t/T)$, where T is the period of the orbit. Deduce **Kepler's Equation**:

$$\frac{2\pi t}{T} = \theta - e\sin\theta$$

(c) The eccentricity of Mercury's orbit is approximately $e = 0.2$. Use Newton's Method to find θ after a quarter of Mercury's year has elapsed ($t = T/4$). Convert θ to degrees. Has Mercury covered more than a quarter of its orbit at $t = T/4$?

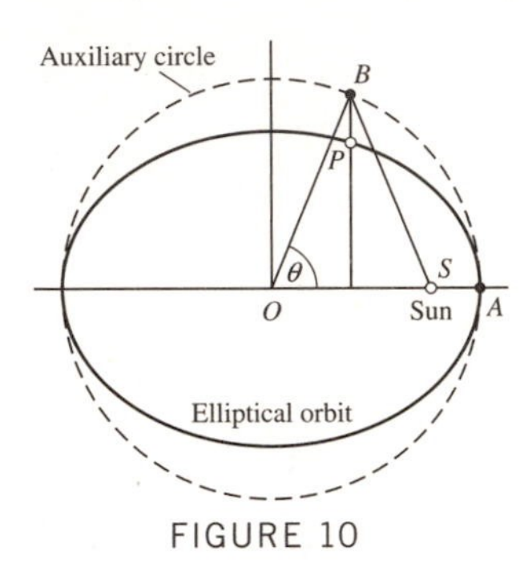

FIGURE 10

SOLUTION

(a) The sector SAB is the slice OAB with the triangle OPS removed. OAB is a central sector with arc θ and radius $\overline{OA} = a$, and therefore has area $\frac{a^2\theta}{2}$. OPS is a triangle with height $a\sin\theta$ and base length $\overline{OS} = ea$. Hence, the area of the sector is

$$\frac{a^2}{2}\theta - \frac{1}{2}ea^2\sin\theta = \frac{a^2}{2}(\theta - e\sin\theta).$$

(b) Since Kepler's second law indicates that the area of the sector is proportional to the time t since the planet passed point A, we get

$$\pi a^2\,(t/T) = a^2/2\,(\theta - e\sin\theta)$$

$$2\pi\frac{t}{T} = \theta - e\sin\theta.$$

(c) If $t = T/4$, the last equation in (b) gives:

$$\frac{\pi}{2} = \theta - e\sin\theta = \theta - .2\sin\theta.$$

Let $f(\theta) = \theta - .2\sin\theta - \frac{\pi}{2}$. We will use Newton's Method to find the point where $f(\theta) = 0$. Since a quarter of the year on Mercury has passed, a good first estimate θ_0 would be $\frac{\pi}{2}$.

n	1	2	3	4
x_n	1.7708	1.76696	1.76696	1.76696

From the point of view of the Sun, Mercury has traversed an angle of approximately 1.76696 radians $= 101.24^\circ$. Mercury has therefore traveled more than one fourth of the way around (from the point of view of central angle) during this time.

29. What happens when you apply Newton's Method to find a zero of $f(x) = x^{1/3}$? Note that $x = 0$ is the only zero.

SOLUTION Let $f(x) = x^{1/3}$. Define

$$x_{n+1} = x_n - \frac{f(x_n)}{f'(x_n)} = x_n - \frac{x_n^{1/3}}{\frac{1}{3}x_n^{-2/3}} = x_n - 3x_n = -2x_n.$$

Take $x_0 = 0.5$. Then the sequence of iterates is $-1, 2, -4, 8, -16, 32, -64, \ldots$ That is, for any nonzero starting value, the sequence of iterates diverges spectacularly, since $x_n = (-2)^n x_0$. Thus $\lim_{n\to\infty}|x_n| = \lim_{n\to\infty} 2^n\,|x_0| = \infty$.

Further Insights and Challenges

31. Newton's Method can be used to compute reciprocals without performing division. Let $c > 0$ and set $f(x) = x^{-1} - c$.

(a) Show that $x - (f(x)/f'(x)) = 2x - cx^2$.

(b) Calculate the first three iterates of Newton's Method with $c = 10.3$ and the two initial guesses $x_0 = 0.1$ and $x_0 = 0.5$.

(c) Explain graphically why $x_0 = 0.5$ does not yield a sequence converging to $1/10.3$.

SOLUTION

(a) Let $f(x) = \frac{1}{x} - c$. Then

$$x - \frac{f(x)}{f'(x)} = x - \frac{\frac{1}{x} - c}{-x^{-2}} = 2x - cx^2.$$

(b) For $c = 10.3$, we have $f(x) = \frac{1}{x} - 10.3$ and thus $x_{n+1} = 2x_n - 10.3x_n^2$.

- Take $x_0 = 0.1$.

n	1	2	3
x_n	0.097	0.0970873	0.09708738

- Take $x_0 = 0.5$.

n	1	2	3
x_n	−1.575	−28.7004375	−8541.66654

(c) The graph is disconnected. If $x_0 = .5$, $(x_1, f(x_1))$ is on the other portion of the graph, which will never converge to any point under Newton's Method.

In Exercises 32 and 33, consider a metal rod of length L fastened at both ends. If you cut the rod and weld on an additional segment of length m, leaving the ends fixed, the rod will bow up into a circular arc of radius R (unknown), as indicated in Figure 12.

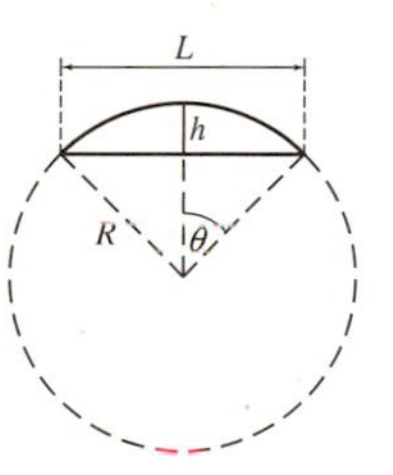

FIGURE 12 The bold circular arc has length $L + m$.

33. Let $L = 3$ and $m = 1$. Apply Newton's Method to Eq. (2) to estimate θ, and use this to estimate h.

SOLUTION We let $L = 3$ and $m = 1$. We want the solution of:

$$\frac{\sin\theta}{\theta} = \frac{L}{L+m}$$

$$\frac{\sin\theta}{\theta} - \frac{L}{L+m} = 0$$

$$\frac{\sin\theta}{\theta} - \frac{3}{4} = 0.$$

Let $f(\theta) = \frac{\sin\theta}{\theta} - \frac{3}{4}$.

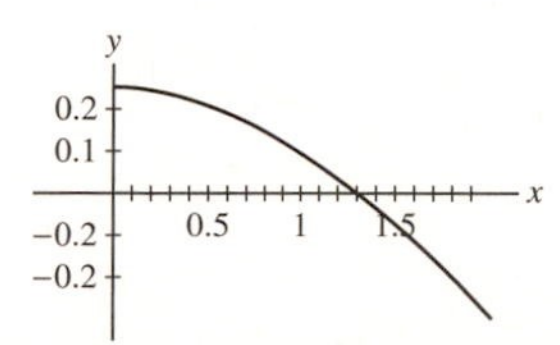

The figure above suggests that $\theta_0 = 1.5$ would be a good initial guess. The Newton's Method approximations for the solution follow:

n	1	2	3	4
θ_n	1.2854388	1.2757223	1.2756981	1.2756981

The angle where $\frac{\sin\theta}{\theta} = \frac{L}{L+m}$ is approximately 1.2757. Hence

$$h = L\frac{1-\cos\theta}{2\sin\theta} \approx 1.11181.$$

In Exercises 35–37, a flexible chain of length L is suspended between two poles of equal height separated by a distance $2M$ (Figure 13). By Newton's laws, the chain describes a ***catenary*** $y = a\cosh\left(\frac{x}{a}\right)$, *where a is the number such that $L = 2a\sinh\left(\frac{M}{a}\right)$. The sag s is the vertical distance from the highest to the lowest point on the chain.*

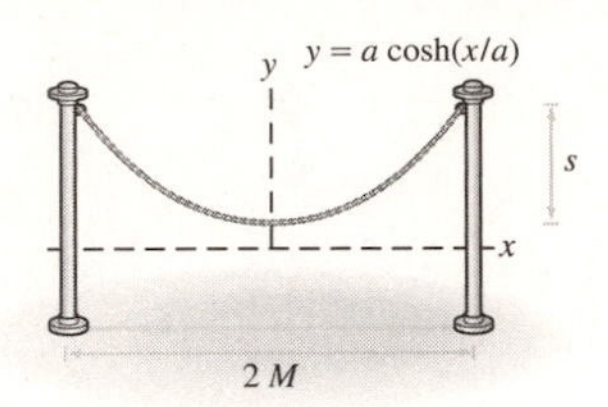

FIGURE 13 Chain hanging between two poles.

35. Suppose that $L = 120$ and $M = 50$.

(a) Use Newton's Method to find a value of a (to two decimal places) satisfying $L = 2a\sinh(M/a)$.

(b) Compute the sag s.

SOLUTION

(a) Let

$$f(a) = 2a\sinh\left(\frac{50}{a}\right) - 120.$$

The graph of f shown below suggests $a \approx 47$ is a root of f. Starting with $a_0 = 47$, we find the following approximations using Newton's method:

$$a_1 = 46.95408 \quad \text{and} \quad a_2 = 46.95415$$

Thus, to two decimal places, $a = 46.95$.

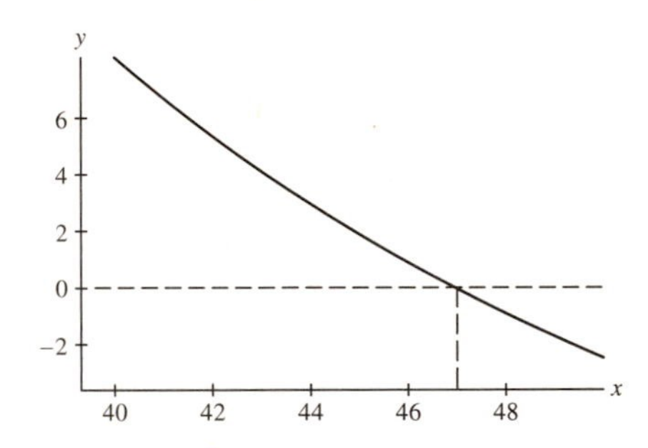

(b) The sag is given by

$$s = y(M) - y(0) = \left(a\cosh\frac{M}{a} + C\right) - \left(a\cosh\frac{0}{a} + C\right) = a\cosh\frac{M}{a} - a.$$

Using $M = 50$ and $a = 46.95$, we find $s = 29.24$.

37. Suppose that $L = 160$ and $M = 50$.

(a) Use Newton's Method to find a value of a (to two decimal places) satisfying $L = 2a\sinh(M/a)$.

(b) Use Eq. (3) and the Linear Approximation to estimate the increase in sag Δs for changes in length $\Delta L = 1$ and $\Delta L = 5$.

(c) CAS Compute $s(161) - s(160)$ and $s(165) - s(160)$ directly and compare with your estimates in (b).

SOLUTION

(a) Let $f(x) = 2x\sinh(50/x) - 160$. Using the graph below, we select an initial guess of $x_0 = 30$. Newton's Method then yields:

n	1	2	3
x_n	28.30622107	28.45653356	28.45797517

Thus, to two decimal places, $a \approx 28.46$.

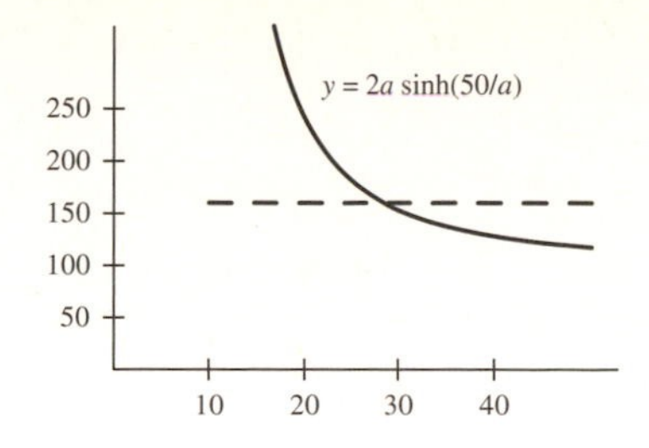

(b) With $M = 50$ and $a \approx 28.46$, we find using Eq. (3) that

$$\frac{ds}{dL} = 0.61.$$

By the Linear Approximation,

$$\Delta s \approx \frac{ds}{dL} \cdot \Delta L.$$

If L increases from 160 to 161, then $\Delta L = 1$ and $\Delta s \approx 0.61$; if L increases from 160 to 165, then $\Delta L = 5$ and $\Delta s \approx 3.05$.

(c) When $L = 160$, $a \approx 28.46$ and

$$s(160) = 28.46 \cosh\left(\frac{50}{28.46}\right) - 28.46 \approx 56.45;$$

whereas, when $L = 161$, $a \approx 28.25$ and

$$s(161) = 28.25 \cosh\left(\frac{50}{28.25}\right) - 28.25 \approx 57.07.$$

Therefore, $s(161) - s(160) = 0.62$, very close to the approximation obtained from the Linear Approximation. Moreover, when $L = 165$, $a \approx 27.49$ and

$$s(165) = 27.49 \cosh\left(\frac{50}{27.49}\right) - 27.49 \approx 59.47;$$

thus, $s(165) - s(160) = 3.02$, again very close to the approximation obtained from the Linear Approximation.

4.9 Antiderivatives

Preliminary Questions

1. Find an antiderivative of the function $f(x) = 0$.

SOLUTION Since the derivative of any constant is zero, any constant function is an antiderivative for the function $f(x) = 0$.

2. Is there a difference between finding the general antiderivative of a function $f(x)$ and evaluating $\int f(x)\,dx$?

SOLUTION No difference. The indefinite integral is the symbol for denoting the general antiderivative.

3. Jacques was told that $f(x)$ and $g(x)$ have the same derivative, and he wonders whether $f(x) = g(x)$. Does Jacques have sufficient information to answer his question?

SOLUTION No. Knowing that the two functions have the same derivative is only good enough to tell Jacques that the functions may differ by at most an additive constant. To determine whether the functions are equal for all x, Jacques needs to know the value of each function for a single value of x. If the two functions produce the same output value for a single input value, they must take the same value for all input values.

4. Suppose that $F'(x) = f(x)$ and $G'(x) = g(x)$. Which of the following statements are true? Explain.

(a) If $f = g$, then $F = G$.

(b) If F and G differ by a constant, then $f = g$.

(c) If f and g differ by a constant, then $F = G$.

SOLUTION

(a) False. Even if $f(x) = g(x)$, the antiderivatives F and G may differ by an additive constant.

(b) True. This follows from the fact that the derivative of any constant is 0.

(c) False. If the functions f and g are different, then the antiderivatives F and G differ by a linear function: $F(x) - G(x) = ax + b$ for some constants a and b.

5. Is $y = x$ a solution of the following Initial Value Problem?

$$\frac{dy}{dx} = 1, \qquad y(0) = 1$$

SOLUTION Although $\frac{d}{dx}x = 1$, the function $f(x) = x$ takes the value 0 when $x = 0$, so $y = x$ is *not* a solution of the indicated initial value problem.

Exercises

In Exercises 1–8, find the general antiderivative of $f(x)$ and check your answer by differentiating.

1. $f(x) = 18x^2$

SOLUTION

$$\int 18x^2\,dx = 18\int x^2\,dx = 18\cdot\frac{1}{3}x^3 + C = 6x^3 + C.$$

As a check, we have

$$\frac{d}{dx}(6x^3 + C) = 18x^2$$

as needed.

3. $f(x) = 2x^4 - 24x^2 + 12x^{-1}$

SOLUTION

$$\begin{aligned}\int (2x^4 - 24x^2 + 12x^{-1})\,dx &= 2\int x^4\,dx - 24\int x^2\,dx + 12\int \frac{1}{x}\,dx\\ &= 2\cdot\frac{1}{5}x^5 - 24\cdot\frac{1}{3}x^3 + 12\ln|x| + C\\ &= \frac{2}{5}x^5 - 8x^3 + 12\ln|x| + C.\end{aligned}$$

As a check, we have

$$\frac{d}{dx}\left(\frac{2}{5}x^5 - 8x^3 + 12\ln|x| + C\right) = 2x^4 - 24x^2 + 12x^{-1}$$

as needed.

5. $f(x) = 2\cos x - 9\sin x$

SOLUTION

$$\begin{aligned}\int (2\cos x - 9\sin x)\,dx &= 2\int \cos x\,dx - 9\int \sin x\,dx\\ &= 2\sin x - 9(-\cos x) + C = 2\sin x + 9\cos x + C\end{aligned}$$

As a check, we have

$$\frac{d}{dx}(2\sin x + 9\cos x + C) = 2\cos x + 9(-\sin x) = 2\cos x - 9\sin x$$

as needed.

7. $f(x) = 12e^x - 5x^{-2}$

SOLUTION

$$\int (12e^x - 5x^{-2})\,dx = 12\int e^x\,dx - 5\int x^{-2}\,dx = 12e^x - 5(-x^{-1}) + C = 12e^x + 5x^{-1} + C.$$

As a check, we have

$$\frac{d}{dx}\left(12e^x + 5x^{-1} + C\right) = 12e^x + 5(-x^{-2}) = 12e^x - 5x^{-2}$$

as needed.

9. Match functions (a)–(d) with their antiderivatives (i)–(iv).

(a) $f(x) = \sin x$	(i) $F(x) = \cos(1 - x)$
(b) $f(x) = x\sin(x^2)$	(ii) $F(x) = -\cos x$
(c) $f(x) = \sin(1 - x)$	(iii) $F(x) = -\frac{1}{2}\cos(x^2)$
(d) $f(x) = x\sin x$	(iv) $F(x) = \sin x - x\cos x$

SOLUTION

(a) An antiderivative of $\sin x$ is $-\cos x$, which is **(ii)**. As a check, we have $\frac{d}{dx}(-\cos x) = -(-\sin x) = \sin x$.

(b) An antiderivative of $x\sin(x^2)$ is $-\frac{1}{2}\cos(x^2)$, which is **(iii)**. This is because, by the Chain Rule, we have $\frac{d}{dx}\left(-\frac{1}{2}\cos(x^2)\right) = -\frac{1}{2}\left(-\sin(x^2)\right)\cdot 2x = x\sin(x^2)$.

(c) An antiderivative of $\sin(1 - x)$ is $\cos(1 - x)$ or **(i)**. As a check, we have $\frac{d}{dx}\cos(1 - x) = -\sin(1 - x)\cdot(-1) = \sin(1 - x)$.

(d) An antiderivative of $x\sin x$ is $\sin x - x\cos x$, which is **(iv)**. This is because

$$\frac{d}{dx}(\sin x - x\cos x) = \cos x - (x(-\sin x) + \cos x\cdot 1) = x\sin x$$

In Exercises 10–39, evaluate the indefinite integral.

11. $\int (4 - 18x)\,dx$

SOLUTION $\int (4 - 18x)\,dx = 4x - 9x^2 + C.$

13. $\int t^{-6/11}\,dt$

SOLUTION $\int t^{-6/11}\,dt = \frac{t^{5/11}}{5/11} + C = \frac{11}{5}t^{5/11} + C.$

15. $\int (18t^5 - 10t^4 - 28t)\,dt$

SOLUTION $\int (18t^5 - 10t^4 - 28t)\,dt = 3t^6 - 2t^5 - 14t^2 + C.$

17. $\int (z^{-4/5} - z^{2/3} + z^{5/4})\,dz$

SOLUTION $\int ((z^{-4/5} - z^{2/3} + z^{5/4})\,dz = \frac{z^{1/5}}{1/5} - \frac{z^{5/3}}{5/3} + \frac{z^{9/4}}{9/4} + C = 5z^{1/5} - \frac{3}{5}z^{5/3} + \frac{4}{9}z^{9/4} + C.$

19. $\int \frac{1}{\sqrt[3]{x}}\,dx$

SOLUTION $\int \frac{1}{\sqrt[3]{x}}\,dx = \int x^{-1/3}\,dx = \frac{x^{2/3}}{2/3} + C = \frac{3}{2}x^{2/3} + C.$

21. $\int \frac{36\,dt}{t^3}$

SOLUTION $\int \frac{36}{t^3}\,dt = \int 36t^{-3}\,dt = 36\frac{t^{-2}}{-2} + C = -\frac{18}{t^2} + C.$

23. $\int (t^{1/2} + 1)(t + 1)\,dt$

SOLUTION

$$\begin{aligned}\int (t^{1/2} + 1)(t + 1)\,dt &= \int (t^{3/2} + t + t^{1/2} + 1)\,dt \\ &= \frac{t^{5/2}}{5/2} + \frac{1}{2}t^2 + \frac{t^{3/2}}{3/2} + t + C \\ &= \frac{2}{5}t^{5/2} + \frac{1}{2}t^2 + \frac{2}{3}t^{3/2} + t + C\end{aligned}$$

25. $\displaystyle\int \frac{x^3 + 3x - 4}{x^2}\,dx$

SOLUTION

$$\int \frac{x^3 + 3x - 4}{x^2}\,dx = \int (x + 3x^{-1} - 4x^{-2})\,dx$$
$$= \frac{1}{2}x^2 + 3\ln|x| + 4x^{-1} + C$$

27. $\displaystyle\int 12\sec x \tan x\,dx$

SOLUTION $\displaystyle\int 12\sec x \tan x\,dx = 12\sec x + C.$

29. $\displaystyle\int (\csc t \cot t)\,dt$

SOLUTION $\displaystyle\int (\csc t \cot t)\,dt = -\csc t + C.$

31. $\displaystyle\int \sec^2(7 - 3\theta)\,d\theta$

SOLUTION $\displaystyle\int \sec^2(7 - 3\theta)\,d\theta = -\frac{1}{3}\tan(7 - 3\theta) + C.$

33. $\displaystyle\int 25\sec^2(3z + 1)\,dz$

SOLUTION $\displaystyle\int 25\sec^2(3z + 1)\,dz = \frac{25}{3}\tan(3z + 1) + C.$

35. $\displaystyle\int \left(\cos(3\theta) - \frac{1}{2}\sec^2\left(\frac{\theta}{4}\right)\right) d\theta$

SOLUTION $\displaystyle\int \left(\cos(3\theta) - \frac{1}{2}\sec^2\left(\frac{\theta}{4}\right)\right) d\theta = \frac{1}{3}\sin(3\theta) - 2\tan\left(\frac{\theta}{4}\right) + C.$

37. $\displaystyle\int (3e^{5x})\,dx$

SOLUTION $\displaystyle\int (3e^{5x})\,dx = \frac{3}{5}e^{5x} + C.$

39. $\displaystyle\int (8x - 4e^{5-2x})\,dx$

SOLUTION $\displaystyle\int (8x - 4e^{5-2x})\,dx = 4x^2 + 2e^{5-2x} + C.$

41. In Figure 4, which of graphs (A), (B), and (C) is *not* the graph of an antiderivative of $f(x)$? Explain.

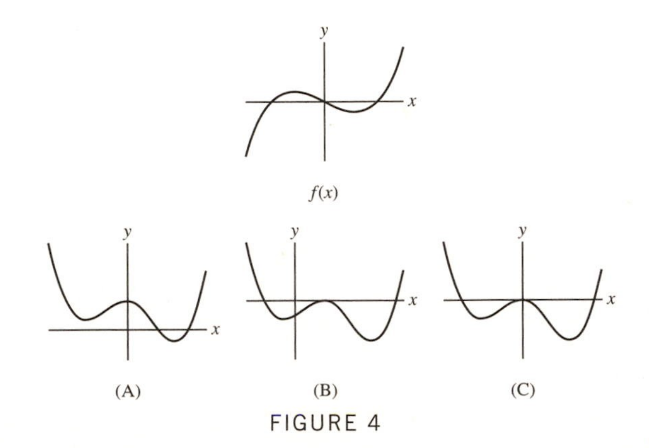

FIGURE 4

SOLUTION Let $F(x)$ be an antiderivative of $f(x)$. Notice that $f(x) = F'(x)$ changes sign from $-$ to $+$ to $-$ to $+$. Hence, $F(x)$ must transition from decreasing to increasing to decreasing to increasing.

- Both graph (A) and graph (C) meet the criteria discussed above and only differ by an additive constant. Thus either could be an antiderivative of $f(x)$.
- Graph (B) does not have the same local extrema as indicated by $f(x)$ and therefore is *not* an antiderivative of $f(x)$.

In Exercises 43–46, verify by differentiation.

43. $\int (x+13)^6\,dx = \frac{1}{7}(x+13)^7 + C$

SOLUTION $\frac{d}{dx}\left(\frac{1}{7}(x+13)^7 + C\right) = (x+13)^6$ as required.

45. $\int (4x+13)^2\,dx = \frac{1}{12}(4x+13)^3 + C$

SOLUTION $\frac{d}{dx}\left(\frac{1}{12}(4x+13)^3 + C\right) = \frac{1}{4}(4x+13)^2(4) = (4x+13)^2$ as required.

In Exercises 47–62, solve the initial value problem.

47. $\frac{dy}{dx} = x^3, \quad y(0) = 4$

SOLUTION Since $\frac{dy}{dx} = x^3$, we have

$$y = \int x^3\,dx = \frac{1}{4}x^4 + C.$$

Thus,

$$4 = y(0) = \frac{1}{4}0^4 + C = C,$$

so that $C = 4$. Therefore, $y = \frac{1}{4}x^4 + 4$.

49. $\frac{dy}{dt} = 2t + 9t^2, \quad y(1) = 2$

SOLUTION Since $\frac{dy}{dt} = 2t + 9t^2$, we have

$$y = \int (2t + 9t^2)\,dt = t^2 + 3t^3 + C.$$

Thus,

$$2 = y(1) = 1^2 + 3(1)^3 + C,$$

so that $C = -2$. Therefore $y = t^2 + 3t^3 - 2$.

51. $\frac{dy}{dt} = \sqrt{t}, \quad y(1) = 1$

SOLUTION Since $\frac{dy}{dt} = \sqrt{t} = t^{1/2}$, we have

$$y = \int t^{1/2}\,dt = \frac{2}{3}t^{3/2} + C.$$

Thus

$$1 = y(1) = \frac{2}{3} + C,$$

so that $C = \frac{1}{3}$. Therefore, $y = \frac{2}{3}t^{3/2} + \frac{1}{3}$.

53. $\frac{dy}{dx} = (3x+2)^3, \quad y(0) = 1$

SOLUTION Since $\frac{dy}{dx} = (3x+2)^3$, we have

$$y = \int (3x+2)^3\,dx = \frac{1}{4}\cdot\frac{1}{3}(3x+2)^4 + C = \frac{1}{12}(3x+2)^4 + C.$$

Thus,

$$1 = y(0) = \frac{1}{12}(2)^4 + C,$$

so that $C = 1 - \frac{4}{3} = -\frac{1}{3}$. Therefore, $y = \frac{1}{12}(3x+2)^4 - \frac{1}{3}$.

55. $\dfrac{dy}{dx} = \sin x, \; y\left(\dfrac{\pi}{2}\right) = 1$

SOLUTION Since $\frac{dy}{dx} = \sin x$, we have

$$y = \int \sin x\, dx = -\cos x + C.$$

Thus

$$1 = y\left(\frac{\pi}{2}\right) = 0 + C,$$

so that $C = 1$. Therefore, $y = 1 - \cos x$.

57. $\dfrac{dy}{dx} = \cos 5x, \; y(\pi) = 3$

SOLUTION Since $\frac{dy}{dx} = \cos 5x$, we have

$$y = \int \cos 5x\, dx = \frac{1}{5}\sin 5x + C.$$

Thus $3 = y(\pi) = 0 + C$, so that $C = 3$. Therefore, $y = 3 + \frac{1}{5}\sin 5x$.

59. $\dfrac{dy}{dx} = e^x, \; y(2) = 0$

SOLUTION Since $\frac{dy}{dx} = e^x$, we have

$$y = \int e^x\, dx = e^x + C.$$

Thus,

$$0 = y(2) = e^2 + C,$$

so that $C = -e^2$. Therefore, $y = e^x - e^2$.

61. $\dfrac{dy}{dt} = 9e^{12-3t}, \; y(4) = 7$

SOLUTION Since $\frac{dy}{dt} = 9e^{12-3t}$, we have

$$y = \int 9e^{12-3t}\, dt = -3e^{12-3t} + C.$$

Thus,

$$7 = y(4) = -3e^0 + C,$$

so that $C = 10$. Therefore, $y = -3e^{12-3t} + 10$.

In Exercises 63–69, first find f' and then find f.

63. $f''(x) = 12x, \quad f'(0) = 1, \quad f(0) = 2$

SOLUTION Let $f''(x) = 12x$. Then $f'(x) = 6x^2 + C$. Given $f'(0) = 1$, it follows that $1 = 6(0)^2 + C$ and $C = 1$. Thus, $f'(x) = 6x^2 + 1$. Next, $f(x) = 2x^3 + x + C$. Given $f(0) = 2$, it follows that $2 = 2(0)^3 + 0 + C$ and $C = 2$. Finally, $f(x) = 2x^3 + x + 2$.

65. $f''(x) = x^3 - 2x + 1, \quad f'(0) = 1, \quad f(0) = 0$

SOLUTION Let $g(x) = f'(x)$. The statement gives us $g'(x) = x^3 - 2x + 1$, $g(0) = 1$. From this, we get $g(x) = \frac{1}{4}x^4 - x^2 + x + C$. $g(0) = 1$ gives us $1 = C$, so $f'(x) = g(x) = \frac{1}{4}x^4 - x^2 + x + 1$. $f'(x) = \frac{1}{4}x^4 - x^2 + x + 1$, so $f(x) = \frac{1}{20}x^5 - \frac{1}{3}x^3 + \frac{1}{2}x^2 + x + C$. $f(0) = 0$ gives $C = 0$, so

$$f(x) = \frac{1}{20}x^5 - \frac{1}{3}x^3 + \frac{1}{2}x^2 + x.$$

67. $f''(t) = t^{-3/2}, \quad f'(4) = 1, \quad f(4) = 4$

SOLUTION Let $g(t) = f'(t)$. The problem statement is $g'(t) = t^{-3/2}$, $g(4) = 1$. From $g'(t)$ we get $g(t) = \frac{1}{-1/2}t^{-1/2} + C = -2t^{-1/2} + C$. From $g(4) = 1$ we get $-1 + C = 1$ so that $C = 2$. Hence $f'(t) = g(t) = -2t^{-1/2} + 2$. From $f'(t)$ we get $f(t) = -2\frac{1}{1/2}t^{1/2} + 2t + C = -4t^{1/2} + 2t + C$. From $f(4) = 4$ we get $-8 + 8 + C = 4$, so that $C = 4$. Hence, $f(t) = -4t^{1/2} + 2t + 4$.

69. $f''(t) = t - \cos t, \quad f'(0) = 2, \quad f(0) = -2$

SOLUTION Let $g(t) = f'(t)$. The problem statement gives

$$g'(t) = t - \cos t, \qquad g(0) = 2.$$

From $g'(t)$, we get $g(t) = \frac{1}{2}t^2 - \sin t + C$. From $g(0) = 2$, we get $C = 2$. Hence $f'(t) = g(t) = \frac{1}{2}t^2 - \sin t + 2$. From $f'(t)$, we get $f(t) = \frac{1}{2}(\frac{1}{3}t^3) + \cos t + 2t + C$. From $f(0) = -2$, we get $1 + C = -2$, hence $C = -3$, and

$$f(t) = \frac{1}{6}t^3 + \cos t + 2t - 3.$$

71. A particle located at the origin at $t = 1$ s moves along the x-axis with velocity $v(t) = (6t^2 - t)$ m/s. State the differential equation with initial condition satisfied by the position $s(t)$ of the particle, and find $s(t)$.

SOLUTION The differential equation satisfied by $s(t)$ is

$$\frac{ds}{dt} = v(t) = 6t^2 - t,$$

and the associated initial condition is $s(1) = 0$. From the differential equation, we find

$$s(t) = \int (6t^2 - t)\, dt = 2t^3 - \frac{1}{2}t^2 + C.$$

Using the initial condition, it follows that

$$0 = s(1) = 2 - \frac{1}{2} + C \quad \text{so} \quad C = -\frac{3}{2}.$$

Finally,

$$s(t) = 2t^3 - \frac{1}{2}t^2 - \frac{3}{2}.$$

73. A mass oscillates at the end of a spring. Let $s(t)$ be the displacement of the mass from the equilibrium position at time t. Assuming that the mass is located at the origin at $t = 0$ and has velocity $v(t) = \sin(\pi t/2)$ m/s, state the differential equation with initial condition satisfied by $s(t)$, and find $s(t)$.

SOLUTION The differential equation satisfied by $s(t)$ is

$$\frac{ds}{dt} = v(t) = \sin(\pi t/2),$$

and the associated initial condition is $s(0) = 0$. From the differential equation, we find

$$s(t) = \int \sin(\pi t/2)\, dt = -\frac{2}{\pi}\cos(\pi t/2) + C.$$

Using the initial condition, it follows that

$$0 = s(0) = -\frac{2}{\pi} + C \quad \text{so} \quad C = \frac{2}{\pi}.$$

Finally,

$$s(t) = \frac{2}{\pi}(1 - \cos(\pi t/2)).$$

75. A car traveling 25 m/s begins to decelerate at a constant rate of 4 m/s^2. After how many seconds does the car come to a stop and how far will the car have traveled before stopping?

SOLUTION Since the acceleration of the car is a constant -4m/s^2, v is given by the differential equation:

$$\frac{dv}{dt} = -4, \qquad v(0) = 25.$$

From $\frac{dv}{dt}$, we get $v(t) = \int -4\, dt = -4t + C$. Since $v(0)25$, $C = 25$. From this, $v(t) = -4t + 25\ \frac{\text{m}}{\text{s}}$. To find the time until the car stops, we must solve $v(t) = 0$:

$$\begin{aligned} -4t + 25 &= 0 \\ 4t &= 25 \\ t &= 25/4 = 6.25 \text{ s.} \end{aligned}$$

Now we have a differential equation for $s(t)$. Since we want to know how far the car has traveled from the beginning of its deceleration at time $t = 0$, we have $s(0) = 0$ by definition, so:

$$\frac{ds}{dt} = v(t) = -4t + 25, \qquad s(0) = 0.$$

From this, $s(t) = \int (-4t + 25)\, dt = -2t^2 + 25t + C$. Since $s(0) = 0$, we have $C = 0$, and

$$s(t) = -2t^2 + 25t.$$

At stopping time $t = 0.25$ s, the car has traveled

$$s(6.25) = -2(6.25)^2 + 25(6.25) = 78.125 \text{ m}.$$

77. A 900-kg rocket is released from a space station. As it burns fuel, the rocket's mass decreases and its velocity increases. Let $v(m)$ be the velocity (in meters per second) as a function of mass m. Find the velocity when $m = 729$ if $dv/dm = -50m^{-1/2}$. Assume that $v(900) = 0$.

SOLUTION Since $\frac{dv}{dm} = -50m^{-1/2}$, we have $v(m) = \int -50m^{-1/2}\, dm = -100m^{1/2} + C$. Thus $0 = v(900) = -100\sqrt{900} + C = -3000 + C$, and $C = 3000$. Therefore, $v(m) = 3000 - 100\sqrt{m}$. Accordingly,

$$v(729) = 3000 - 100\sqrt{729} = 3000 - 100(27) = 300 \text{ meters/sec}.$$

79. Verify the linearity properties of the indefinite integral stated in Theorem 4.

SOLUTION To verify the Sum Rule, let $F(x)$ and $G(x)$ be any antiderivatives of $f(x)$ and $g(x)$, respectively. Because

$$\frac{d}{dx}(F(x) + G(x)) = \frac{d}{dx}F(x) + \frac{d}{dx}G(x) = f(x) + g(x),$$

it follows that $F(x) + G(x)$ is an antiderivative of $f(x) + g(x)$; i.e.,

$$\int (f(x) + g(x))\, dx = \int f(x)\, dx + \int g(x)\, dx.$$

To verify the Multiples Rule, again let $F(x)$ be any antiderivative of $f(x)$ and let c be a constant. Because

$$\frac{d}{dx}(cF(x)) = c\frac{d}{dx}F(x) = cf(x),$$

it follows that $cF(x)$ is and antiderivative of $cf(x)$; i.e.,

$$\int (cf(x))\, dx = c\int f(x)\, dx.$$

Further Insights and Challenges

81. Find constants c_1 and c_2 such that $F(x) = c_1xe^x + c_2e^x$ is an antiderivative of $f(x) = xe^x$.

SOLUTION Let $F(x) = c_1xe^x + c_2e^x$. If $F(x)$ is to be an antiderivative of $f(x) = xe^x$, we must have $F'(x) = f(x)$ for all x. Hence,

$$c_1xe^x + (c_1 + c_2)e^x = xe^x = 1 \cdot xe^x + 0 \cdot e^x.$$

Equating coefficients of like terms we have $c_1 = 1$ and $c_1 + c_2 = 0$. Thus, $c_1 = 1$ and $c_2 = -1$.

83. Suppose that $F'(x) = f(x)$.

(a) Show that $\frac{1}{2}F(2x)$ is an antiderivative of $f(2x)$.

(b) Find the general antiderivative of $f(kx)$ for $k \neq 0$.

SOLUTION Let $F'(x) = f(x)$.

(a) By the Chain Rule, we have

$$\frac{d}{dx}\left(\frac{1}{2}F(2x)\right) = \frac{1}{2}F'(2x) \cdot 2 = F'(2x) = f(2x).$$

Thus $\frac{1}{2}F(2x)$ is an antiderivative of $f(2x)$.

(b) For nonzero constant k, the Chain Rules gives

$$\frac{d}{dx}\left(\frac{1}{k}F(kx)\right) = \frac{1}{k}F'(kx)\cdot k = F'(kx) = f(kx).$$

Thus $\frac{1}{k}F(kx)$ is an antiderivative of $f(kx)$. Hence the general antiderivative of $f(kx)$ is $\frac{1}{k}F(kx) + C$, where C is a constant.

85. Using Theorem 1, prove that $F'(x) = f(x)$ where $f(x)$ is a polynomial of degree $n-1$, then $F(x)$ is a polynomial of degree n. Then prove that if $g(x)$ is any function such that $g^{(n)}(x) = 0$, then $g(x)$ is a polynomial of degree at most n.

SOLUTION Suppose $F'(x) = f(x)$ where $f(x)$ is a polynomial of degree $n-1$. Now, we know that the derivative of a polynomial of degree n is a polynomial of degree $n-1$, and hence an antiderivative of a polynomial of degree $n-1$ is a polynomial of degree n. Thus, by Theorem 1, $F(x)$ can differ from a polynomial of degree n by at most a constant term, which is still a polynomial of degree n. Now, suppose that $g(x)$ is any function such that $g^{(n+1)}(x) = 0$. We know that the $n+1$-st derivative of any polynomial of degree at most n is zero, so by repeated application of Theorem 1, $g(x)$ can differ from a polynomial of degree at most n by at most a constant term. Hence, $g(x)$ is a polynomial of degree at most n.

CHAPTER REVIEW EXERCISES

In Exercises 1–6, estimate using the Linear Approximation or linearization, and use a calculator to estimate the error.

1. $8.1^{1/3} - 2$

SOLUTION Let $f(x) = x^{1/3}$, $a = 8$ and $\Delta x = 0.1$. Then $f'(x) = \frac{1}{3}x^{-2/3}$, $f'(a) = \frac{1}{12}$ and, by the Linear Approximation,

$$\Delta f = 8.1^{1/3} - 2 \approx f'(a)\Delta x = \frac{1}{12}(0.1) = 0.00833333.$$

Using a calculator, $8.1^{1/3} - 2 = 0.00829885$. The error in the Linear Approximation is therefore

$$|0.00829885 - 0.00833333| = 3.445 \times 10^{-5}.$$

3. $625^{1/4} - 624^{1/4}$

SOLUTION Let $f(x) = x^{1/4}$, $a = 625$ and $\Delta x = -1$. Then $f'(x) = \frac{1}{4}x^{-3/4}$, $f'(a) = \frac{1}{500}$ and, by the Linear Approximation,

$$\Delta f = 624^{1/4} - 625^{1/4} \approx f'(a)\Delta x = \frac{1}{500}(-1) = -0.002.$$

Thus $625^{1/4} - 624^{1/4} \approx 0.002$. Using a calculator,

$$625^{1/4} - 624^{1/4} = 0.00200120.$$

The error in the Linear Approximation is therefore

$$|0.00200120 - (0.002)| = 1.201 \times 10^{-6}.$$

5. $\dfrac{1}{1.02}$

SOLUTION Let $f(x) = x^{-1}$ and $a = 1$. Then $f(a) = 1$, $f'(x) = -x^{-2}$ and $f'(a) = -1$. The linearization of $f(x)$ at $a = 1$ is therefore

$$L(x) = f(a) + f'(a)(x - a) = 1 - (x - 1) = 2 - x,$$

and $\frac{1}{1.02} \approx L(1.02) = 0.98$. Using a calculator, $\frac{1}{1.02} = 0.980392$, so the error in the Linear Approximation is

$$|0.980392 - 0.98| = 3.922 \times 10^{-4}.$$

In Exercises 7–12, find the linearization at the point indicated.

7. $y = \sqrt{x}$, $a = 25$

SOLUTION Let $y = \sqrt{x}$ and $a = 25$. Then $y(a) = 5$, $y' = \frac{1}{2}x^{-1/2}$ and $y'(a) = \frac{1}{10}$. The linearization of y at $a = 25$ is therefore

$$L(x) = y(a) + y'(a)(x - 25) = 5 + \frac{1}{10}(x - 25).$$

9. $A(r) = \frac{4}{3}\pi r^3, \quad a = 3$

SOLUTION Let $A(r) = \frac{4}{3}\pi r^3$ and $a = 3$. Then $A(a) = 36\pi$, $A'(r) = 4\pi r^2$ and $A'(a) = 36\pi$. The linearization of $A(r)$ at $a = 3$ is therefore

$$L(r) = A(a) + A'(a)(r - a) = 36\pi + 36\pi(r - 3) = 36\pi(r - 2).$$

11. $P(x) = e^{-x^2/2}, \quad a = 1$

SOLUTION Let $P(x) = e^{-x^2/2}$ and $a = 1$. Then $P(a) = e^{-1/2}$, $P'(x) = -xe^{-x^2/2}$, and $P'(a) = -e^{-1/2}$. The linearization of $P(x)$ at $a = 1$ is therefore

$$L(x) = P(a) + P'(a)(x - a) = e^{-1/2} - e^{-1/2}(x - 1) = \frac{1}{\sqrt{e}}(2 - x).$$

In Exercises 13–18, use the Linear Approximation.

13. The position of an object in linear motion at time t is $s(t) = 0.4t^2 + (t+1)^{-1}$. Estimate the distance traveled over the time interval [4, 4.2].

SOLUTION Let $s(t) = 0.4t^2 + (t+1)^{-1}$, $a = 4$ and $\Delta t = 0.2$. Then $s'(t) = 0.8t - (t+1)^{-2}$ and $s'(a) = 3.16$. Using the Linear Approximation, the distance traveled over the time interval [4, 4.2] is approximately

$$\Delta s = s(4.2) - s(4) \approx s'(a)\Delta t = 3.16(0.2) = 0.632.$$

15. When a bus pass from Albuquerque to Los Alamos is priced at p dollars, a bus company takes in a monthly revenue of $R(p) = 1.5p - 0.01p^2$ (in thousands of dollars).

(a) Estimate ΔR if the price rises from \$50 to \$53.

(b) If $p = 80$, how will revenue be affected by a small increase in price? Explain using the Linear Approximation.

SOLUTION

(a) If the price is raised from \$50 to \$53, then $\Delta p = 3$ and

$$\Delta R \approx R'(50)\Delta p = (1.5 - 0.02(50))(3) = 1.5$$

We therefore estimate an increase of \$1500 in revenue.

(b) Because $R'(80) = 1.5 - 0.02(80) = -0.1$, the Linear Approximation gives $\Delta R \approx -0.1\Delta p$. A small increase in price would thus result in a decrease in revenue.

17. The circumference of a sphere is measured at $C = 100$ cm. Estimate the maximum percentage error in V if the error in C is at most 3 cm.

SOLUTION The volume of a sphere is $V = \frac{4}{3}\pi r^3$ and the circumference is $C = 2\pi r$, where r is the radius of the sphere. Thus, $r = \frac{1}{2\pi}C$ and

$$V = \frac{4}{3}\pi\left(\frac{C}{2\pi}\right)^3 = \frac{1}{6\pi^2}C^3.$$

Using the Linear Approximation,

$$\Delta V \approx \frac{dV}{dC}\Delta C = \frac{1}{2\pi^2}C^2\Delta C,$$

so

$$\frac{\Delta V}{V} \approx \frac{\frac{1}{2\pi^2}C^2\Delta C}{\frac{1}{6\pi^2}C^3} = 3\frac{\Delta C}{C}.$$

With $C = 100$ cm and ΔC at most 3 cm, we estimate that the maximum percentage error in V is $3\frac{3}{100} = 0.09$, or 9%.

19. Use the Intermediate Value Theorem to prove that $\sin x - \cos x = 3x$ has a solution, and use Rolle's Theorem to show that this solution is unique.

SOLUTION Let $f(x) = \sin x - \cos x - 3x$, and observe that each root of this function corresponds to a solution of the equation $\sin x - \cos x = 3x$. Now,

$$f\left(-\frac{\pi}{2}\right) = -1 + \frac{3\pi}{2} > 0 \qquad \text{and} \qquad f(0) = -1 < 0.$$

Because f is continuous on $(-\frac{\pi}{2}, 0)$ and $f(-\frac{\pi}{2})$ and $f(0)$ are of opposite sign, the Intermediate Value Theorem guarantees there exists a $c \in (-\frac{\pi}{2}, 0)$ such that $f(c) = 0$. Thus, the equation $\sin x - \cos x = 3x$ has at least one solution.

Next, suppose that the equation $\sin x - \cos x = 3x$ has two solutions, and therefore $f(x)$ has two roots, say a and b. Because f is continuous on $[a, b]$, differentiable on (a, b) and $f(a) = f(b) = 0$, Rolle's Theorem guarantees there exists $c \in (a, b)$ such that $f'(c) = 0$. However,

$$f'(x) = \cos x + \sin x - 3 \le -1$$

for all x. We have reached a contradiction. Consequently, $f(x)$ has a unique root and the equation $\sin x - \cos x = 3x$ has a unique solution.

21. Verify the MVT for $f(x) = \ln x$ on $[1, 4]$.

SOLUTION Let $f(x) = \ln x$. On the interval $[1, 4]$, this function is continuous and differentiable, so the MVT applies. Now, $f'(x) = \frac{1}{x}$, so

$$\frac{1}{c} = f'(c) = \frac{f(b) - f(a)}{b - a} = \frac{\ln 4 - \ln 1}{4 - 1} = \frac{1}{3}\ln 4,$$

or

$$c = \frac{3}{\ln 4} \approx 2.164 \in (1, 4).$$

23. Use the MVT to prove that if $f'(x) \le 2$ for $x > 0$ and $f(0) = 4$, then $f(x) \le 2x + 4$ for all $x \ge 0$.

SOLUTION Let $x > 0$. Because f is continuous on $[0, x]$ and differentiable on $(0, x)$, the Mean Value Theorem guarantees there exists a $c \in (0, x)$ such that

$$f'(c) = \frac{f(x) - f(0)}{x - 0} \quad \text{or} \quad f(x) = f(0) + xf'(c).$$

Now, we are given that $f(0) = 4$ and that $f'(x) \le 2$ for $x > 0$. Therefore, for all $x \ge 0$,

$$f(x) \le 4 + x(2) = 2x + 4.$$

In Exercises 25–30, find the critical points and determine whether they are minima, maxima, or neither.

25. $f(x) = x^3 - 4x^2 + 4x$

SOLUTION Let $f(x) = x^3 - 4x^2 + 4x$. Then $f'(x) = 3x^2 - 8x + 4 = (3x - 2)(x - 2)$, so that $x = \frac{2}{3}$ and $x = 2$ are critical points. Next, $f''(x) = 6x - 8$, so $f''(\frac{2}{3}) = -4 < 0$ and $f''(2) = 4 > 0$. Therefore, by the Second Derivative Test, $f(\frac{2}{3})$ is a local maximum while $f(2)$ is a local minimum.

27. $f(x) = x^2(x + 2)^3$

SOLUTION Let $f(x) = x^2(x + 2)^3$. Then

$$f'(x) = 3x^2(x + 2)^2 + 2x(x + 2)^3 = x(x + 2)^2(3x + 2x + 4) = x(x + 2)^2(5x + 4),$$

so that $x = 0$, $x = -2$ and $x = -\frac{4}{5}$ are critical points. The sign of the first derivative on the intervals surrounding the critical points is indicated in the table below. Based on this information, $f(-2)$ is neither a local maximum nor a local minimum, $f(-\frac{4}{5})$ is a local maximum and $f(0)$ is a local minimum.

Interval	$(-\infty, -2)$	$(-2, -\frac{4}{5})$	$(-\frac{4}{5}, 0)$	$(0, \infty)$
Sign of f'	+	+	−	+

29. $g(\theta) = \sin^2\theta + \theta$

SOLUTION Let $g(\theta) = \sin^2\theta + \theta$. Then

$$g'(\theta) = 2\sin\theta\cos\theta + 1 = 2\sin 2\theta + 1,$$

so the critical points are

$$\theta = \frac{3\pi}{4} + n\pi$$

for all integers n. Because $g'(\theta) \ge 0$ for all θ, it follows that $g\left(\frac{3\pi}{4} + n\pi\right)$ is neither a local maximum nor a local minimum for all integers n.

In Exercises 31–38, find the extreme values on the interval.

31. $f(x) = x(10 - x)$, $[-1, 3]$

SOLUTION Let $f(x) = x(10 - x) = 10x - x^2$. Then $f'(x) = 10 - 2x$, so that $x = 5$ is the only critical point. As this critical point is not in the interval $[-1, 3]$, we only need to check the value of f at the endpoints to determine the extreme values. Because $f(-1) = -11$ and $f(3) = 21$, the maximum value of $f(x) = x(10 - x)$ on the interval $[-1, 3]$ is 21 while the minimum value is -11.

33. $g(\theta) = \sin^2\theta - \cos\theta$, $[0, 2\pi]$

SOLUTION Let $g(\theta) = \sin^2\theta - \cos\theta$. Then

$$g'(\theta) = 2\sin\theta\cos\theta + \sin\theta = \sin\theta(2\cos\theta + 1) = 0$$

when $\theta = 0, \frac{2\pi}{3}, \pi, \frac{4\pi}{3}, 2\pi$. The table below lists the value of g at each of the critical points and the endpoints of the interval $[0, 2\pi]$. Based on this information, the minimum value of $g(\theta)$ on the interval $[0, 2\pi]$ is -1 and the maximum value is $\frac{5}{4}$.

θ	0	$2\pi/3$	π	$4\pi/3$	2π
$g(\theta)$	-1	5/4	1	5/4	-1

35. $f(x) = x^{2/3} - 2x^{1/3}$, $[-1, 3]$

SOLUTION Let $f(x) = x^{2/3} - 2x^{1/3}$. Then $f'(x) = \frac{2}{3}x^{-1/3} - \frac{2}{3}x^{-2/3} = \frac{2}{3}x^{-2/3}(x^{1/3} - 1)$, so that the critical points are $x = 0$ and $x = 1$. With $f(-1) = 3$, $f(0) = 0$, $f(1) = -1$ and $f(3) = \sqrt[3]{9} - 2\sqrt[3]{3} \approx -0.804$, it follows that the minimum value of $f(x)$ on the interval $[-1, 3]$ is -1 and the maximum value is 3.

37. $f(x) = x - 12\ln x$, $[5, 40]$

SOLUTION Let $f(x) = x - 12\ln x$. Then $f'(x) = 1 - \frac{12}{x}$, whence $x = 12$ is the only critical point. The minimum value of f is then $12 - 12\ln 12 \approx -17.818880$, and the maximum value is $40 - 12\ln 40 \approx -4.266553$. Note that $f(5) = 5 - 12\ln 5 \approx -14.313255$.

39. Find the critical points and extreme values of $f(x) = |x - 1| + |2x - 6|$ in $[0, 8]$.

SOLUTION Let

$$f(x) = |x - 1| + |2x - 6| = \begin{cases} 7 - 3x, & x < 1 \\ 5 - x, & 1 \le x < 3 \\ 3x - 7, & x \ge 3 \end{cases}.$$

The derivative of $f(x)$ is never zero but does not exist at the transition points $x = 1$ and $x = 3$. Thus, the critical points of f are $x = 1$ and $x = 3$. With $f(0) = 7$, $f(1) = 4$, $f(3) = 2$ and $f(8) = 17$, it follows that the minimum value of $f(x)$ on the interval $[0, 8]$ is 2 and the maximum value is 17.

In Exercises 41–46, find the points of inflection.

41. $y = x^3 - 4x^2 + 4x$

SOLUTION Let $y = x^3 - 4x^2 + 4x$. Then $y' = 3x^2 - 8x + 4$ and $y'' = 6x - 8$. Thus, $y'' > 0$ and y is concave up for $x > \frac{4}{3}$, while $y'' < 0$ and y is concave down for $x < \frac{4}{3}$. Hence, there is a point of inflection at $x = \frac{4}{3}$.

43. $y = \dfrac{x^2}{x^2 + 4}$

SOLUTION Let $y = \dfrac{x^2}{x^2+4} = 1 - \dfrac{4}{x^2+4}$. Then $y' = \dfrac{8x}{(x^2+4)^2}$ and

$$y'' = \frac{(x^2+4)^2(8) - 8x(2)(2x)(x^2+4)}{(x^2+4)^4} = \frac{8(4 - 3x^2)}{(x^2+4)^3}.$$

Thus, $y'' > 0$ and y is concave up for

$$-\frac{2}{\sqrt{3}} < x < \frac{2}{\sqrt{3}},$$

while $y'' < 0$ and y is concave down for

$$|x| \geq \frac{2}{\sqrt{3}}.$$

Hence, there are points of inflection at

$$x = \pm\frac{2}{\sqrt{3}}.$$

45. $f(x) = (x^2 - x)e^{-x}$

SOLUTION Let $f(x) = (x^2 - x)e^{-x}$. Then

$$y' = -(x^2 - x)e^{-x} + (2x - 1)e^{-x} = -(x^2 - 3x + 1)e^{-x},$$

and

$$y'' = (x^2 - 3x + 1)e^{-x} - (2x - 3)e^{-x} = e^{-x}(x^2 - 5x + 4) = e^{-x}(x - 1)(x - 4).$$

Thus, $y'' > 0$ and y is concave up for $x < 1$ and for $x > 4$, while $y'' < 0$ and y is concave down for $1 < x < 4$. Hence, there are points of inflection at $x = 1$ and $x = 4$.

In Exercises 47–56, sketch the graph, noting the transition points and asymptotic behavior.

47. $y = 12x - 3x^2$

SOLUTION Let $y = 12x - 3x^2$. Then $y' = 12 - 6x$ and $y'' = -6$. It follows that the graph of $y = 12x - 3x^2$ is increasing for $x < 2$, decreasing for $x > 2$, has a local maximum at $x = 2$ and is concave down for all x. Because

$$\lim_{x\to\pm\infty} (12x - 3x^2) = -\infty,$$

the graph has no horizontal asymptotes. There are also no vertical asymptotes. The graph is shown below.

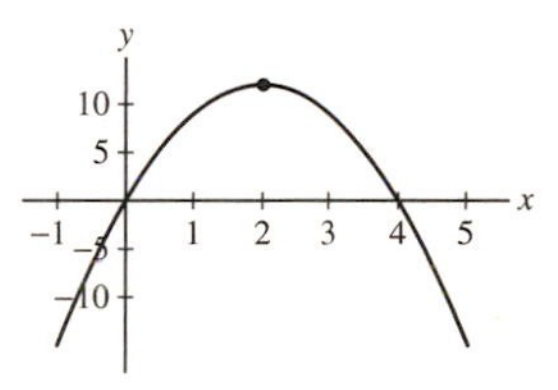

49. $y = x^3 - 2x^2 + 3$

SOLUTION Let $y = x^3 - 2x^2 + 3$. Then $y' = 3x^2 - 4x$ and $y'' = 6x - 4$. It follows that the graph of $y = x^3 - 2x^2 + 3$ is increasing for $x < 0$ and $x > \frac{4}{3}$, is decreasing for $0 < x < \frac{4}{3}$, has a local maximum at $x = 0$, has a local minimum at $x = \frac{4}{3}$, is concave up for $x > \frac{2}{3}$, is concave down for $x < \frac{2}{3}$ and has a point of inflection at $x = \frac{2}{3}$. Because

$$\lim_{x\to-\infty} (x^3 - 2x^2 + 3) = -\infty \quad \text{and} \quad \lim_{x\to\infty} (x^3 - 2x^2 + 3) = \infty,$$

the graph has no horizontal asymptotes. There are also no vertical asymptotes. The graph is shown below.

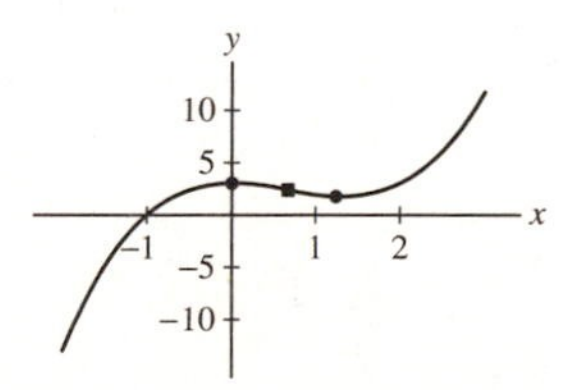

51. $y = \dfrac{x}{x^3 + 1}$

SOLUTION Let $y = \dfrac{x}{x^3 + 1}$. Then

$$y' = \frac{x^3 + 1 - x(3x^2)}{(x^3 + 1)^2} = \frac{1 - 2x^3}{(x^3 + 1)^2}$$

and

$$y'' = \frac{(x^3 + 1)^2(-6x^2) - (1 - 2x^3)(2)(x^3 + 1)(3x^2)}{(x^3 + 1)^4} = -\frac{6x^2(2 - x^3)}{(x^3 + 1)^3}.$$

It follows that the graph of $y = \dfrac{x}{x^3+1}$ is increasing for $x < -1$ and $-1 < x < \sqrt[3]{\frac{1}{2}}$, is decreasing for $x > \sqrt[3]{\frac{1}{2}}$, has a local maximum at $x = \sqrt[3]{\frac{1}{2}}$, is concave up for $x < -1$ and $x > \sqrt[3]{2}$, is concave down for $-1 < x < 0$ and $0 < x < \sqrt[3]{2}$ and has a point of inflection at $x = \sqrt[3]{2}$. Note that $x = -1$ is not an inflection point because $x = -1$ is not in the domain of the function. Now,

$$\lim_{x\to\pm\infty} \frac{x}{x^3+1} = 0,$$

so $y = 0$ is a horizontal asymptote. Moreover,

$$\lim_{x\to-1-} \frac{x}{x^3+1} = \infty \quad \text{and} \quad \lim_{x\to-1+} \frac{x}{x^3+1} = -\infty,$$

so $x = -1$ is a vertical asymptote. The graph is shown below.

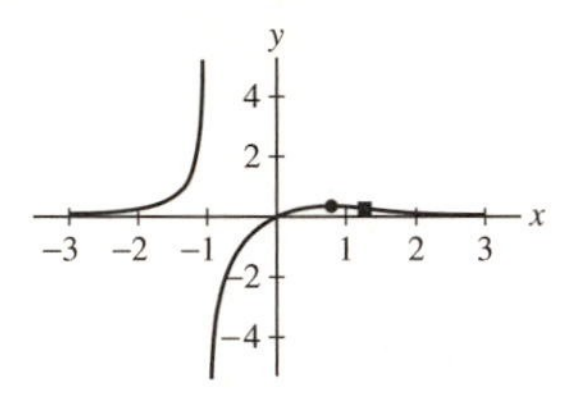

53. $y = \dfrac{1}{|x+2|+1}$

SOLUTION Let $y = \dfrac{1}{|x+2|+1}$. Because

$$\lim_{x\to\pm\infty} \frac{1}{|x+2|+1} = 0,$$

the graph of this function has a horizontal asymptote of $y = 0$. The graph has no vertical asymptotes as $|x+2|+1 \geq 1$ for all x. The graph is shown below. From this graph we see there is a local maximum at $x = -2$.

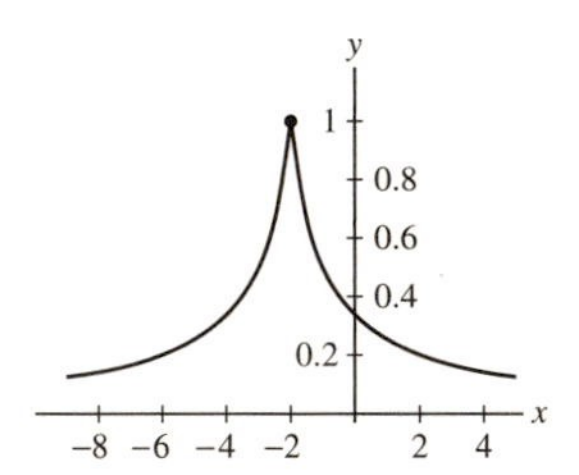

55. $y = \sqrt{3}\sin x - \cos x$ on $[0, 2\pi]$

SOLUTION Let $y = \sqrt{3}\sin x - \cos x$. Then $y' = \sqrt{3}\cos x + \sin x$ and $y'' = -\sqrt{3}\sin x + \cos x$. It follows that the graph of $y = \sqrt{3}\sin x - \cos x$ is increasing for $0 < x < 5\pi/6$ and $11\pi/6 < x < 2\pi$, is decreasing for $5\pi/6 < x < 11\pi/6$, has a local maximum at $x = 5\pi/6$, has a local minimum at $x = 11\pi/6$, is concave up for $0 < x < \pi/3$ and $4\pi/3 < x < 2\pi$, is concave down for $\pi/3 < x < 4\pi/3$ and has points of inflection at $x = \pi/3$ and $x = 4\pi/3$. The graph is shown below.

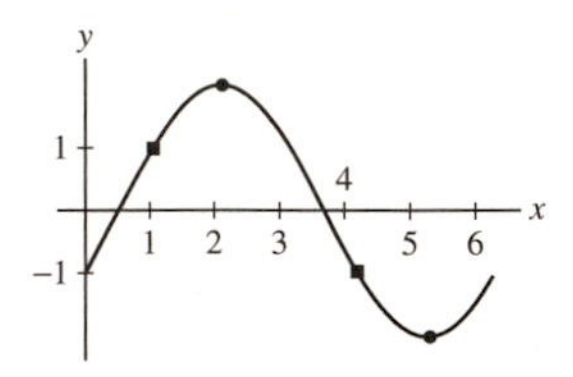

57. Draw a curve $y = f(x)$ for which f' and f'' have signs as indicated in Figure 2.

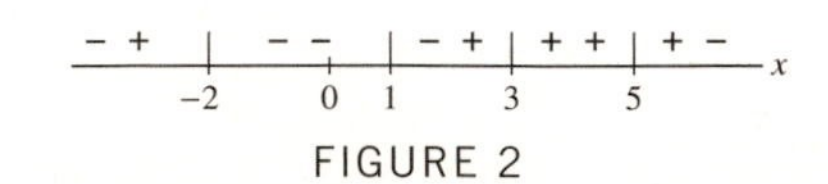

FIGURE 2

SOLUTION The figure below depicts a curve for which $f'(x)$ and $f''(x)$ have the required signs.

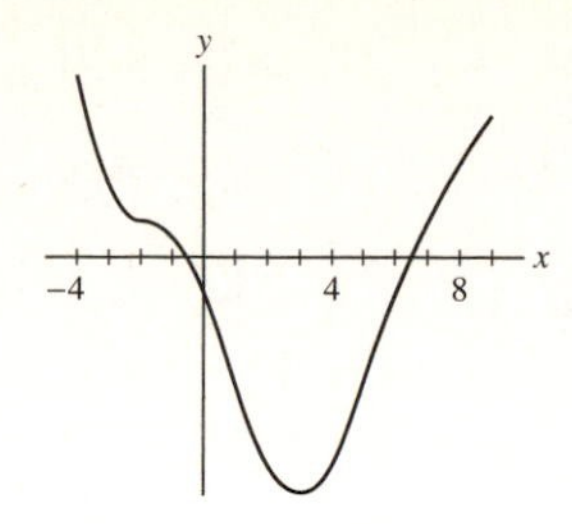

59. A rectangular box of height h with square base of side b has volume $V = 4 \text{ m}^3$. Two of the side faces are made of material costing \$40/m^2. The remaining sides cost \$20/m^2. Which values of b and h minimize the cost of the box?

SOLUTION Because the volume of the box is

$$V = b^2 h = 4 \quad \text{it follows that} \quad h = \frac{4}{b^2}.$$

Now, the cost of the box is

$$C = 40(2bh) + 20(2bh) + 20b^2 = 120bh + 20b^2 = \frac{480}{b} + 20b^2.$$

Thus,

$$C'(b) = -\frac{480}{b^2} + 40b = 0$$

when $b = \sqrt[3]{12}$ meters. Because $C(b) \to \infty$ as $b \to 0+$ and as $b \to \infty$, it follows that cost is minimized when $b = \sqrt[3]{12}$ meters and $h = \frac{1}{3}\sqrt[3]{12}$ meters.

61. Let $N(t)$ be the size of a tumor (in units of 10^6 cells) at time t (in days). According to the **Gompertz Model**, $dN/dt = N(a - b \ln N)$ where a, b are positive constants. Show that the maximum value of N is $e^{\frac{a}{b}}$ and that the tumor increases most rapidly when $N = e^{\frac{a}{b}-1}$.

SOLUTION Given $dN/dt = N(a - b \ln N)$, the critical points of N occur when $N = 0$ and when $N = e^{a/b}$. The sign of $N'(t)$ changes from positive to negative at $N = e^{a/b}$ so the maximum value of N is $e^{a/b}$. To determine when N changes most rapidly, we calculate

$$N''(t) = N\left(-\frac{b}{N}\right) + a - b \ln N = (a - b) - b \ln N.$$

Thus, $N'(t)$ is increasing for $N < e^{a/b-1}$, is decreasing for $N > e^{a/b-1}$ and is therefore maximum when $N = e^{a/b-1}$. Therefore, the tumor increases most rapidly when $N = e^{\frac{a}{b}-1}$.

63. Find the maximum volume of a right-circular cone placed upside-down in a right-circular cone of radius $R = 3$ and height $H = 4$ as in Figure 3. A cone of radius r and height h has volume $\frac{1}{3}\pi r^2 h$.

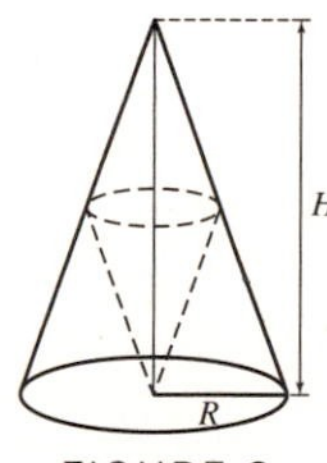

FIGURE 3

SOLUTION Let r denote the radius and h the height of the upside down cone. By similar triangles, we obtain the relation

$$\frac{4-h}{r} = \frac{4}{3} \quad \text{so} \quad h = 4\left(1 - \frac{r}{3}\right)$$

and the volume of the upside down cone is

$$V(r) = \frac{1}{3}\pi r^2 h = \frac{4}{3}\pi\left(r^2 - \frac{r^3}{3}\right)$$

for $0 \le r \le 3$. Thus,

$$\frac{dV}{dr} = \frac{4}{3}\pi\left(2r - r^2\right),$$

and the critical points are $r = 0$ and $r = 2$. Because $V(0) = V(3) = 0$ and

$$V(2) = \frac{4}{3}\pi\left(4 - \frac{8}{3}\right) = \frac{16}{9}\pi,$$

the maximum volume of a right-circular cone placed upside down in a right-circular cone of radius 3 and height 4 is

$$\frac{16}{9}\pi.$$

65. Show that the maximum area of a parallelogram $ADEF$ that is inscribed in a triangle ABC, as in Figure 4, is equal to one-half the area of $\triangle ABC$.

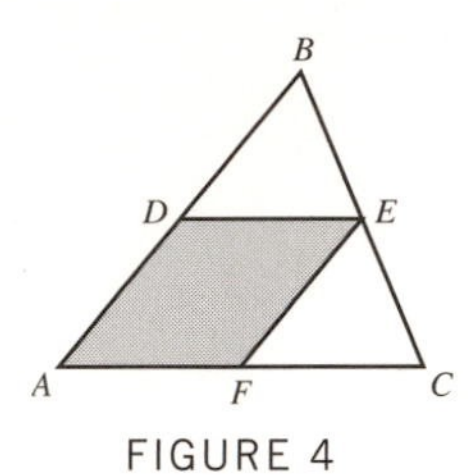

FIGURE 4

SOLUTION Let θ denote the measure of angle BAC. Then the area of the parallelogram is given by $\overline{AD} \cdot \overline{AF} \sin\theta$. Now, suppose that

$$\overline{BE}/\overline{BC} = x.$$

Then, by similar triangles, $\overline{AD} = (1 - x)\overline{AB}$, $\overline{AF} = \overline{DE} = x\overline{AC}$, and the area of the parallelogram becomes $\overline{AB} \cdot \overline{AC}x(1 - x)\sin\theta$. The function $x(1 - x)$ achieves its maximum value of $\frac{1}{4}$ when $x = \frac{1}{2}$. Thus, the maximum area of a parallelogram inscribed in a triangle ΔABC is

$$\frac{1}{4}\overline{AB} \cdot \overline{AC}\sin\theta = \frac{1}{2}\left(\frac{1}{2}\overline{AB} \cdot \overline{AC}\sin\theta\right) = \frac{1}{2}\,(\text{area of } \Delta ABC).$$

67. Let $f(x)$ be a function whose graph does not pass through the x-axis and let $Q = (a, 0)$. Let $P = (x_0, f(x_0))$ be the point on the graph closest to Q (Figure 5). Prove that $\overline{PQ}$ is perpendicular to the tangent line to the graph of x_0. *Hint:* Find the minimum value of the *square* of the distance from $(x, f(x))$ to $(a, 0)$.

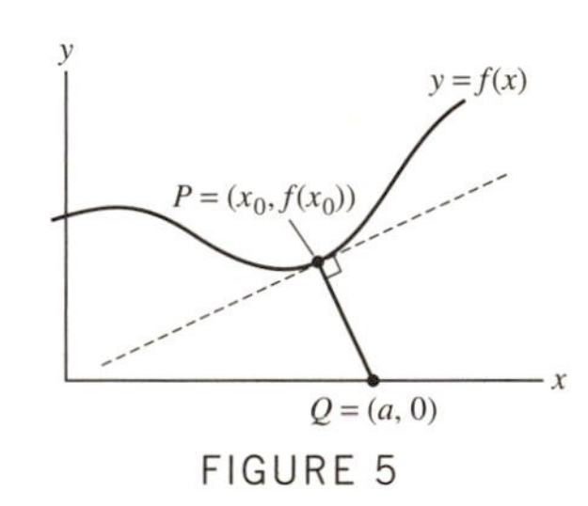

FIGURE 5

SOLUTION Let $P = (a, 0)$ and let $Q = (x_0, f(x_0))$ be the point on the graph of $y = f(x)$ closest to P. The slope of the segment joining P and Q is then

$$\frac{f(x_0)}{x_0 - a}.$$

Now, let

$$q(x) = \sqrt{(x - a)^2 + (f(x))^2},$$

the distance from the arbitrary point $(x, f(x))$ on the graph of $y = f(x)$ to the point P. As $(x_0, f(x_0))$ is the point closest to P, we must have

$$q'(x_0) = \frac{2(x_0 - a) + 2f(x_0)f'(x_0)}{\sqrt{(x_0 - a)^2 + (f(x_0))^2}} = 0.$$

Thus,

$$f'(x_0) = -\frac{x_0 - a}{f(x_0)} = -\left(\frac{f(x_0)}{x_0 - a}\right)^{-1}.$$

In other words, the slope of the segment joining P and Q is the negative reciprocal of the slope of the line tangent to the graph of $y = f(x)$ at $x = x_0$; hence; the two lines are perpendicular.

69. Use Newton's Method to estimate $\sqrt[3]{25}$ to four decimal places.

SOLUTION Let $f(x) = x^3 - 25$ and define

$$x_{n+1} = x_n - \frac{f(x_n)}{f'(x_n)} = x_n - \frac{x_n^3 - 25}{3x_n^2}.$$

With $x_0 = 3$, we find

n	1	2	3
x_n	2.925925926	2.924018982	2.924017738

Thus, to four decimal places $\sqrt[3]{25} = 2.9240$.

In Exercises 71–84, calculate the indefinite integral.

71. $\int (4x^3 - 2x^2)\, dx$

SOLUTION $\int (4x^3 - 2x^2)\, dx = x^4 - \frac{2}{3}x^3 + C.$

73. $\int \sin(\theta - 8)\, d\theta$

SOLUTION $\int \sin(\theta - 8)\, d\theta = -\cos(\theta - 8) + C.$

75. $\int (4t^{-3} - 12t^{-4})\, dt$

SOLUTION $\int (4t^{-3} - 12t^{-4})\, dt = -2t^{-2} + 4t^{-3} + C.$

77. $\int \sec^2 x\, dx$

SOLUTION $\int \sec^2 x\, dx = \tan x + C.$

79. $\int (y + 2)^4\, dy$

SOLUTION $\int (y + 2)^4\, dy = \frac{1}{5}(y + 2)^5 + C.$

81. $\int (e^x - x)\, dx$

SOLUTION $\int (e^x - x)\, dx = e^x - \frac{1}{2}x^2 + C.$

83. $\int 4x^{-1}\, dx$

SOLUTION $\int 4x^{-1}\, dx = 4 \ln |x| + C.$

In Exercises 85–90, solve the differential equation with the given initial condition.

85. $\frac{dy}{dx} = 4x^3, \quad y(1) = 4$

SOLUTION Let $\frac{dy}{dx} = 4x^3$. Then

$$y(x) = \int 4x^3\, dx = x^4 + C.$$

Using the initial condition $y(1) = 4$, we find $y(1) = 1^4 + C = 4$, so $C = 3$. Thus, $y(x) = x^4 + 3$.

87. $\frac{dy}{dx} = x^{-1/2}, \quad y(1) = 1$

SOLUTION Let $\frac{dy}{dx} = x^{-1/2}$. Then

$$y(x) = \int x^{-1/2}\, dx = 2x^{1/2} + C.$$

Using the initial condition $y(1) = 1$, we find $y(1) = 2\sqrt{1} + C = 1$, so $C = -1$. Thus, $y(x) = 2x^{1/2} - 1$.

89. $\dfrac{dy}{dx} = e^{-x}, \quad y(0) = 3$

SOLUTION Let $\frac{dy}{dx} = e^{-x}$. Then

$$y(x) = \int e^{-x}\,dx = -e^{-x} + C.$$

Using the initial condition $y(0) = 3$, we find $y(0) = -e^0 + C = 3$, so $C = 4$. Thus, $y(x) = 4 - e^{-x}$.

91. Find $f(t)$ if $f''(t) = 1 - 2t$, $f(0) = 2$, and $f'(0) = -1$.

SOLUTION Suppose $f''(t) = 1 - 2t$. Then

$$f'(t) = \int f''(t)\,dt = \int (1 - 2t)\,dt = t - t^2 + C.$$

Using the initial condition $f'(0) = -1$, we find $f'(0) = 0 - 0^2 + C = -1$, so $C = -1$. Thus, $f'(t) = t - t^2 - 1$. Now,

$$f(t) = \int f'(t)\,dt = \int (t - t^2 - 1)\,dt = \frac{1}{2}t^2 - \frac{1}{3}t^3 - t + C.$$

Using the initial condition $f(0) = 2$, we find $f(0) = \frac{1}{2}0^2 - \frac{1}{3}0^3 - 0 + C = 2$, so $C = 2$. Thus,

$$f(t) = \frac{1}{2}t^2 - \frac{1}{3}t^3 - t + 2.$$

93. Find the local extrema of $f(x) = \dfrac{e^{2x} + 1}{e^{x+1}}$.

SOLUTION To simplify the differentiation, we first rewrite $f(x) = \frac{e^{2x}+1}{e^{x+1}}$ using the Laws of Exponents:

$$f(x) = \frac{e^{2x}}{e^{x+1}} + \frac{1}{e^{x+1}} = e^{2x-(x+1)} + e^{-(x+1)} = e^{x-1} + e^{-x-1}.$$

Now,

$$f'(x) = e^{x-1} - e^{-x-1}.$$

Setting the derivative equal to zero yields

$$e^{x-1} - e^{-x-1} = 0 \quad \text{or} \quad e^{x-1} = e^{-x-1}.$$

Thus,

$$x - 1 = -x - 1 \quad \text{or} \quad x = 0.$$

Next, we use the Second Derivative Test. With $f''(x) = e^{x-1} + e^{-x-1}$, it follows that

$$f''(0) = e^{-1} + e^{-1} = \frac{2}{e} > 0.$$

Hence, $x = 0$ is a local minimum. Since $f(0) = e^{0-1} + e^{-0-1} = \frac{2}{e}$, we conclude that the point $(0, \frac{2}{e})$ is a local minimum.

In Exercises 95–98, find the local extrema and points of inflection, and sketch the graph. Use L'Hôpital's Rule to determine the limits as $x \to 0+$ or $x \to \pm\infty$ if necessary.

95. $y = x\ln x \quad (x > 0)$

SOLUTION Let $y = x\ln x$. Then

$$y' = \ln x + x\left(\frac{1}{x}\right) = 1 + \ln x,$$

and $y'' = \frac{1}{x}$. Solving $y' = 0$ yields the critical point $x = e^{-1}$. Since $y''(e^{-1}) = e > 0$, the function has a local minimum at $x = e^{-1}$. y'' is positive for $x > 0$, hence the function is concave up for $x > 0$ and there are no points of inflection. As $x \to 0+$ and as $x \to \infty$, we find

$$\lim_{x\to 0+} x\ln x = \lim_{x\to 0+} \frac{\ln x}{x^{-1}} = \lim_{x\to 0+} \frac{x^{-1}}{-x^{-2}} = \lim_{x\to 0+} (-x) = 0;$$

$$\lim_{x\to\infty} x\ln x = \infty.$$

The graph is shown below:

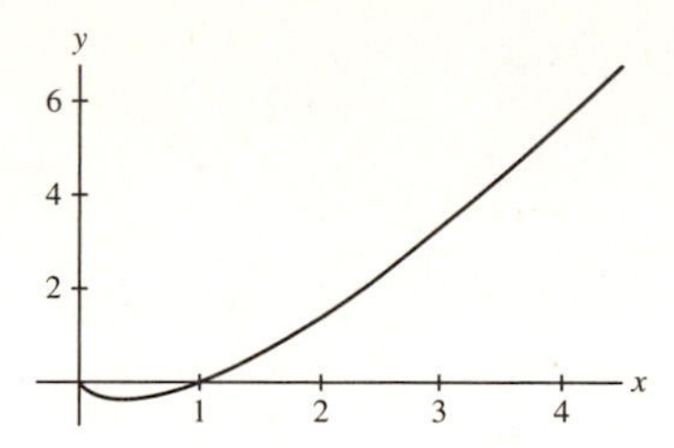

97. $y = x(\ln x)^2 \quad (x > 0)$

SOLUTION Let $y = x(\ln x)^2$. Then

$$y' = x\frac{2\ln x}{x} + (\ln x)^2 = 2\ln x + (\ln x)^2 = \ln x(2 + \ln x),$$

and

$$y'' = \frac{2}{x} + \frac{2\ln x}{x} = \frac{2}{x}(1 + \ln x).$$

Solving $y' = 0$ yields the critical points $x = e^{-2}$ and $x = 1$. Since $y''(e^{-2}) = -2e^2 < 0$ and $y''(1) = 2 > 0$, the function has a local maximum at $x = e^{-2}$ and a local minimum at $x = 1$. $y'' < 0$ and the function is concave down for $x < e^{-1}$, whereas $y'' > 0$ and the function is concave up for $x > e^{-1}$; hence, there is a point of inflection at $x = e^{-1}$. As $x \to 0+$ and as $x \to \infty$, we find

$$\lim_{x\to 0+} x(\ln x)^2 = \lim_{x\to 0+} \frac{(\ln x)^2}{x^{-1}} = \lim_{x\to 0+} \frac{2\ln x \cdot x^{-1}}{-x^{-2}} = \lim_{x\to 0+} \frac{2\ln x}{-x^{-1}} = \lim_{x\to 0+} \frac{2x^{-1}}{x^{-2}} = \lim_{x\to 0+} 2x = 0;$$

$$\lim_{x\to\infty} x(\ln x)^2 = \infty.$$

The graph is shown below:

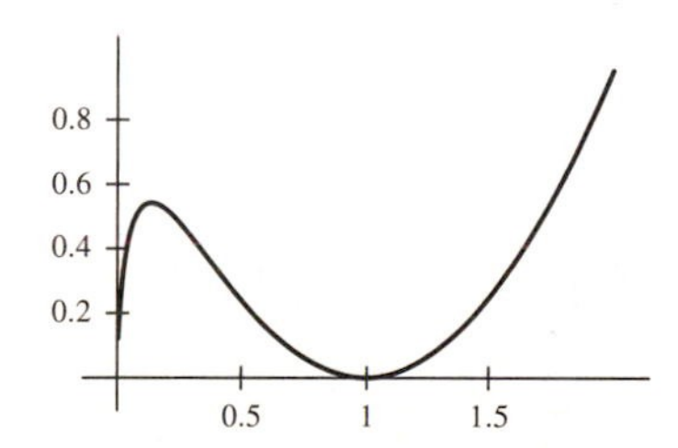

99. Explain why L'Hôpital's Rule gives no information about $\lim_{x\to\infty} \frac{2x - \sin x}{3x + \cos 2x}$. Evaluate the limit by another method.

SOLUTION As $x \to \infty$, both $2x - \sin x$ and $3x + \cos 2x$ tend toward infinity, so L'Hôpital's Rule applies to $\lim_{x\to\infty} \frac{2x - \sin x}{3x + \cos 2x}$; however, the resulting limit, $\lim_{x\to\infty} \frac{2 - \cos x}{3 - 2\sin 2x}$, does not exist due to the oscillation of $\sin x$ and $\cos x$ and further applications of L'Hôpital's rule will not change this situation.

To evaluate the limit, we note

$$\lim_{x\to\infty} \frac{2x - \sin x}{3x + \cos 2x} = \lim_{x\to\infty} \frac{2 - \frac{\sin x}{x}}{3 + \frac{\cos 2x}{x}} = \frac{2}{3}.$$

In Exercises 101–112, verify that L'Hôpital's Rule applies and evaluate the limit.

101. $\lim_{x\to 3} \frac{4x - 12}{x^2 - 5x + 6}$

SOLUTION The given expression is an indeterminate form of type $\frac{0}{0}$, therefore L'Hôpital's Rule applies. We find

$$\lim_{x\to 3} \frac{4x - 12}{x^2 - 5x + 6} = \lim_{x\to 3} \frac{4}{2x - 5} = 4.$$

103. $\lim_{x\to 0+} x^{1/2}\ln x$

SOLUTION First rewrite

$$x^{1/2}\ln x \quad \text{as} \quad \frac{\ln x}{x^{-1/2}}.$$

The rewritten expression is an indeterminate form of type $\frac{\infty}{\infty}$, therefore L'Hôpital's Rule applies. We find

$$\lim_{x\to 0+} x^{1/2}\ln x = \lim_{x\to 0+}\frac{\ln x}{x^{-1/2}} = \lim_{x\to 0+}\frac{1/x}{-\frac{1/2^{-3/2}}{x}} = \lim_{x\to 0+} -\frac{x^{1/2}}{2} = 0.$$

105. $\lim_{\theta\to 0}\frac{2\sin\theta - \sin 2\theta}{\sin\theta - \theta\cos\theta}$

SOLUTION The given expression is an indeterminate form of type $\frac{0}{0}$; hence, we may apply L'Hôpital's Rule. We find

$$\lim_{\theta\to 0}\frac{2\sin\theta - \sin 2\theta}{\sin\theta - \theta\cos\theta} = \lim_{\theta\to 0}\frac{2\cos\theta - 2\cos 2\theta}{\cos\theta - (\cos\theta - \theta\sin\theta)} = \lim_{\theta\to 0}\frac{2\cos\theta - 2\cos 2\theta}{\theta\sin\theta}$$
$$= \lim_{\theta\to 0}\frac{-2\sin\theta + 4\sin 2\theta}{\sin\theta + \theta\cos\theta} = \lim_{\theta\to 0}\frac{-2\cos\theta + 8\cos 2\theta}{\cos\theta + \cos\theta - \theta\sin\theta} = \frac{-2+8}{1+1-0} = 3.$$

107. $\lim_{t\to\infty}\frac{\ln(t+2)}{\log_2 t}$

SOLUTION The limit is an indeterminate form of type $\frac{\infty}{\infty}$; hence, we may apply L'Hôpital's Rule. We find

$$\lim_{t\to\infty}\frac{\ln(t+2)}{\log_2 t} = \lim_{t\to\infty}\frac{\frac{1}{t+2}}{\frac{1}{t\ln 2}} = \lim_{t\to\infty}\frac{t\ln 2}{t+2} = \lim_{t\to\infty}\frac{\ln 2}{1} = \ln 2.$$

109. $\lim_{y\to 0}\frac{\sin^{-1}y - y}{y^3}$

SOLUTION The limit is an indeterminate form of type $\frac{0}{0}$; hence, we may apply L'Hôpital's Rule. We find

$$\lim_{y\to 0}\frac{\sin^{-1}y - y}{y^3} = \lim_{y\to 0}\frac{\frac{1}{\sqrt{1-y^2}} - 1}{3y^2} = \lim_{y\to 0}\frac{y(1-y^2)^{-3/2}}{6y} = \lim_{y\to 0}\frac{(1-y^2)^{-3/2}}{6} = \frac{1}{6}.$$

111. $\lim_{x\to 0}\frac{\sinh(x^2)}{\cosh x - 1}$

SOLUTION The limit is an indeterminate form of type $\frac{0}{0}$; hence, we may apply L'Hôpital's Rule. We find

$$\lim_{x\to 0}\frac{\sinh(x^2)}{\cosh x - 1} = \lim_{x\to 0}\frac{2x\cosh(x^2)}{\sinh x} = \lim_{x\to 0}\frac{2\cosh(x^2) + 4x^2\sinh(x^2)}{\cosh x} = \frac{2+0}{1} = 2.$$

113. Let $f(x) = e^{-Ax^2/2}$, where $A > 0$. Given any n numbers $a_1, a_2, \dots, a_n$, set

$$\Phi(x) = f(x-a_1)f(x-a_2)\cdots f(x-a_n)$$

(a) Assume $n = 2$ and prove that $\Phi(x)$ attains its maximum value at the average $x = \frac{1}{2}(a_1 + a_2)$. *Hint:* Calculate $\Phi'(x)$ using logarithmic differentiation.

(b) Show that for any n, $\Phi(x)$ attains its maximum value at $x = \frac{1}{n}(a_1 + a_2 + \cdots + a_n)$. This fact is related to the role of $f(x)$ (whose graph is a bell-shaped curve) in statistics.

SOLUTION

(a) For $n = 2$ we have,

$$\Phi(x) = f(x-a_1)\,f(x-a_2) = e^{-\frac{A}{2}(x-a_1)^2}\cdot e^{-\frac{A}{2}(x-a_2)^2} = e^{-\frac{A}{2}\left((x-a_1)^2+(x-a_2)^2\right)}.$$

Since $e^{-\frac{A}{2}y}$ is a decreasing function of y, it attains its maximum value where y is minimum. Therefore, we must find the minimum value of

$$y = (x-a_1)^2 + (x-a_2)^2 = 2x^2 - 2(a_1+a_2)x + a_1^2 + a_2^2.$$

Now, $y' = 4x - 2(a_1 + a_2) = 0$ when

$$x = \frac{a_1 + a_2}{2}.$$

We conclude that $\Phi(x)$ attains a maximum value at this point.

(b) We have

$$\Phi(x) = e^{-\frac{A}{2}(x-a_1)^2} \cdot e^{-\frac{A}{2}(x-a_2)^2} \cdot \cdots \cdot e^{-\frac{A}{2}(x-a_n)^2} = e^{-\frac{A}{2}\left((x-a_1)^2+\cdots+(x-a_n)^2\right)}.$$

Since the function $e^{-\frac{A}{2}y}$ is a decreasing function of y, it attains a maximum value where y is minimum. Therefore we must minimize the function

$$y = (x - a_1)^2 + (x - a_2)^2 + \cdots + (x - a_n)^2.$$

We find the critical points by solving:

$$y' = 2(x - a_1) + 2(x - a_2) + \cdots + 2(x - a_n) = 0$$

$$2nx = 2(a_1 + a_2 + \cdots + a_n)$$

$$x = \frac{a_1 + \cdots + a_n}{n}.$$

We verify that this point corresponds the minimum value of y by examining the sign of y'' at this point: $y'' = 2n > 0$. We conclude that y attains a minimum value at the point $x = \frac{a_1+\cdots+a_n}{n}$, hence $\Phi(x)$ attains a maximum value at this point.

5 THE INTEGRAL

5.1 Approximating and Computing Area

Preliminary Questions

1. What are the right and left endpoints if [2, 5] is divided into six subintervals?

SOLUTION If the interval [2, 5] is divided into six subintervals, the length of each subinterval is $\frac{5-2}{6} = \frac{1}{2}$. The right endpoints of the subintervals are then $\frac{5}{2}, 3, \frac{7}{2}, 4, \frac{9}{2}, 5$, while the left endpoints are $2, \frac{5}{2}, 3, \frac{7}{2}, 4, \frac{9}{2}$.

2. The interval [1, 5] is divided into eight subintervals.

(a) What is the left endpoint of the last subinterval?

(b) What are the right endpoints of the first two subintervals?

SOLUTION Note that each of the 8 subintervals has length $\frac{5-1}{8} = \frac{1}{2}$.

(a) The left endpoint of the last subinterval is $5 - \frac{1}{2} = \frac{9}{2}$.

(b) The right endpoints of the first two subintervals are $1 + \frac{1}{2} = \frac{3}{2}$ and $1 + 2\left(\frac{1}{2}\right) = 2$.

3. Which of the following pairs of sums are *not* equal?

(a) $\sum_{i=1}^{4} i, \quad \sum_{\ell=1}^{4} \ell$

(b) $\sum_{j=1}^{4} j^2, \quad \sum_{k=2}^{5} k^2$

(c) $\sum_{j=1}^{4} j, \quad \sum_{i=2}^{5} (i-1)$

(d) $\sum_{i=1}^{4} i(i+1), \quad \sum_{j=2}^{5} (j-1)j$

SOLUTION

(a) Only the name of the index variable has been changed, so these two sums *are* the same.

(b) These two sums are *not* the same; the second squares the numbers two through five while the first squares the numbers one through four.

(c) These two sums *are* the same. Note that when i ranges from two through five, the expression $i - 1$ ranges from one through four.

(d) These two sums *are* the same. Both sums are $1 \cdot 2 + 2 \cdot 3 + 3 \cdot 4 + 4 \cdot 5$.

4. Explain: $\sum_{j=1}^{100} j = \sum_{j=0}^{100} j$ but $\sum_{j=1}^{100} 1$ is not equal to $\sum_{j=0}^{100} 1$.

SOLUTION The first term in the sum $\sum_{j=0}^{100} j$ is equal to zero, so it may be dropped. More specifically,

$$\sum_{j=0}^{100} j = 0 + \sum_{j=1}^{100} j = \sum_{j=1}^{100} j.$$

On the other hand, the first term in $\sum_{j=0}^{100} 1$ is not zero, so this term cannot be dropped. In particular,

$$\sum_{j=0}^{100} 1 = 1 + \sum_{j=1}^{100} 1 \neq \sum_{j=1}^{100} 1.$$

5. Explain why $L_{100} \geq R_{100}$ for $f(x) = x^{-2}$ on [3, 7].

SOLUTION On [3, 7], the function $f(x) = x^{-2}$ is a decreasing function; hence, for any subinterval of [3, 7], the function value at the left endpoint is larger than the function value at the right endpoint. Consequently, L_{100} must be larger than R_{100}.

Exercises

1. Figure 15 shows the velocity of an object over a 3-min interval. Determine the distance traveled over the intervals $[0, 3]$ and $[1, 2.5]$ (remember to convert from km/h to km/min).

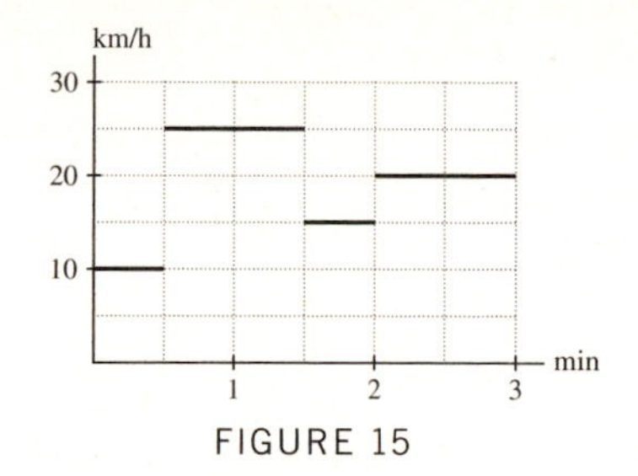

FIGURE 15

SOLUTION The distance traveled by the object can be determined by calculating the area underneath the velocity graph over the specified interval. During the interval $[0, 3]$, the object travels

$$\left(\frac{10}{60}\right)\left(\frac{1}{2}\right)+\left(\frac{25}{60}\right)(1)+\left(\frac{15}{60}\right)\left(\frac{1}{2}\right)+\left(\frac{20}{60}\right)(1)=\frac{23}{24}\approx 0.96\text{ km}.$$

During the interval $[1, 2.5]$, it travels

$$\left(\frac{25}{60}\right)\left(\frac{1}{2}\right)+\left(\frac{15}{60}\right)\left(\frac{1}{2}\right)+\left(\frac{20}{60}\right)\left(\frac{1}{2}\right)=\frac{1}{2}=0.5\text{ km}.$$

3. A rainstorm hit Portland, Maine, in October 1996, resulting in record rainfall. The rainfall rate $R(t)$ on October 21 is recorded, in centimeters per hour, in the following table, where t is the number of hours since midnight. Compute the total rainfall during this 24-hour period and indicate on a graph how this quantity can be interpreted as an area.

t (h)	0–2	2–4	4–9	9–12	12–20	20–24
$R(t)$ (cm)	0.5	0.3	1.0	2.5	1.5	0.6

SOLUTION Over each interval, the total rainfall is the time interval in hours times the rainfall in centimeters per hour. Thus

$$R = 2(0.5) + 2(0.3) + 5(1.0) + 3(2.5) + 8(1.5) + 4(0.6) = 28.5\text{ cm}.$$

The figure below is a graph of the rainfall as a function of time. The area of the shaded region represents the total rainfall.

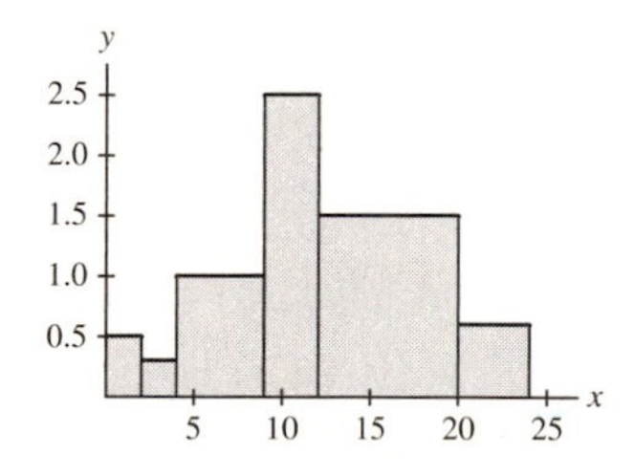

5. Compute R_5 and L_5 over $[0, 1]$ using the following values.

x	0	0.2	0.4	0.6	0.8	1
$f(x)$	50	48	46	44	42	40

SOLUTION $\Delta x = \frac{1-0}{5} = 0.2$. Thus,

$$L_5 = 0.2\,(50 + 48 + 46 + 44 + 42) = 0.2(230) = 46,$$

and

$$R_5 = 0.2\,(48 + 46 + 44 + 42 + 40) = 0.2(220) = 44.$$

The average is

$$\frac{46 + 44}{2} = 45.$$

This estimate is frequently referred to as the *Trapezoidal Approximation*.

7. Let $f(x) = 2x + 3$.

(a) Compute R_6 and L_6 over $[0, 3]$.

(b) Use geometry to find the exact area A and compute the errors $|A - R_6|$ and $|A - L_6|$ in the approximations.

SOLUTION Let $f(x) = 2x + 3$ on $[0, 3]$.

(a) We partition $[0, 3]$ into 6 equally-spaced subintervals. The left endpoints of the subintervals are $\left\{0, \frac{1}{2}, 1, \frac{3}{2}, 2, \frac{5}{2}\right\}$ whereas the right endpoints are $\left\{\frac{1}{2}, 1, \frac{3}{2}, 2, \frac{5}{2}, 3\right\}$.

- Let $a = 0$, $b = 3$, $n = 6$, $\Delta x = (b - a)/n = \frac{1}{2}$, and $x_k = a + k\Delta x$, $k = 0, 1, \ldots, 5$ (left endpoints). Then

$$L_6 = \sum_{k=0}^{5} f(x_k)\Delta x = \Delta x \sum_{k=0}^{5} f(x_k) = \frac{1}{2}(3 + 4 + 5 + 6 + 7 + 8) = 16.5.$$

- With $x_k = a + k\Delta x$, $k = 1, 2, \ldots, 6$ (right endpoints), we have

$$R_6 = \sum_{k=1}^{6} f(x_k)\Delta x = \Delta x \sum_{k=1}^{6} f(x_k) = \frac{1}{2}(4 + 5 + 6 + 7 + 8 + 9) = 19.5.$$

(b) Via geometry (see figure below), the exact area is $A = \frac{1}{2}(3)(6) + 3^2 = 18$. Thus, L_6 underestimates the true area $(L_6 - A = -1.5)$, while R_6 overestimates the true area $(R_6 - A = +1.5)$.

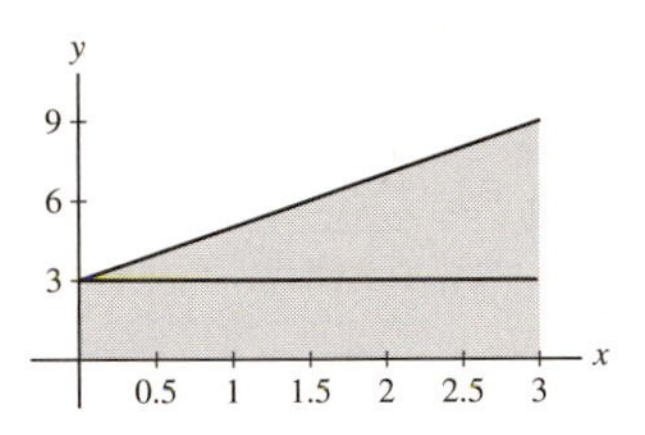

9. Calculate R_3 and L_3

$$\text{for } f(x) = x^2 - x + 4 \quad \text{over } [1, 4]$$

Then sketch the graph of f and the rectangles that make up each approximation. Is the area under the graph larger or smaller than R_3? Is it larger or smaller than L_3?

SOLUTION Let $f(x) = x^2 - x + 4$ and set $a = 1$, $b = 4$, $n = 3$, $\Delta x = (b - a)/n = (4 - 1)/3 = 1$.

(a) Let $x_k = a + k\Delta x$, $k = 0, 1, 2, 3$.

- Selecting the left endpoints of the subintervals, x_k, $k = 0, 1, 2$, or $\{1, 2, 3\}$, we have

$$L_3 = \sum_{k=0}^{2} f(x_k)\Delta x = \Delta x \sum_{k=0}^{2} f(x_k) = (1)(4 + 6 + 10) = 20.$$

- Selecting the right endpoints of the subintervals, x_k, $k = 1, 2, 3$, or $\{2, 3, 4\}$, we have

$$R_3 = \sum_{k=1}^{3} f(x_k)\Delta x = \Delta x \sum_{k=1}^{3} f(x_k) = (1)(6 + 10 + 16) = 32.$$

(b) Here are figures of the three rectangles that approximate the area under the curve $f(x)$ over the interval $[1, 4]$. Clearly, the area under the graph is larger than L_3 but smaller than R_3.

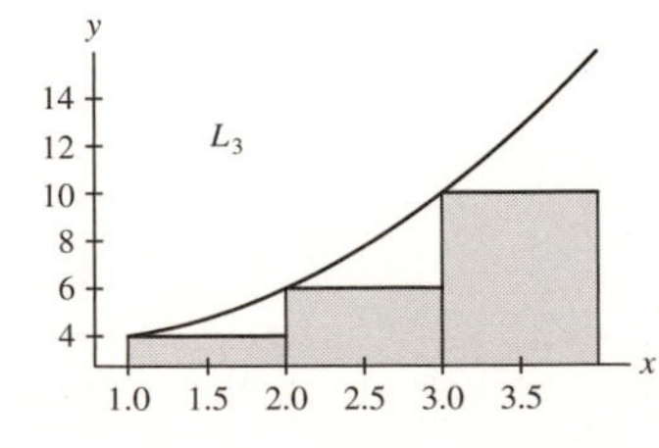

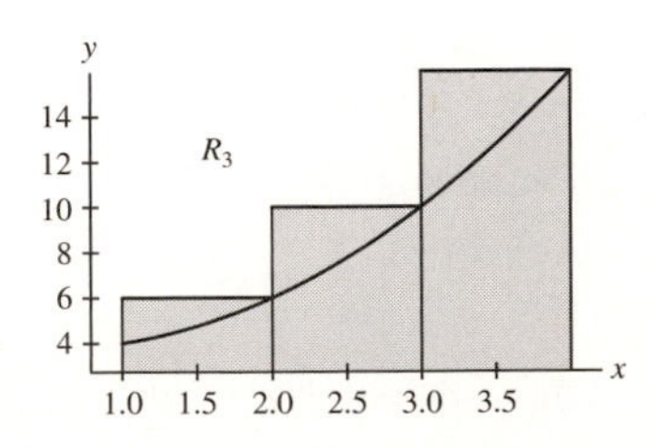

11. Estimate R_3, M_3, and L_6 over [0, 1.5] for the function in Figure 17.

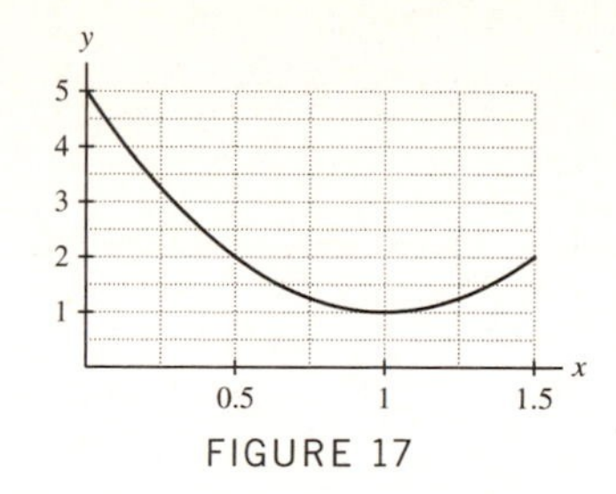

FIGURE 17

SOLUTION Let $f(x)$ on $[0, \frac{3}{2}]$ be given by Figure 17. For $n = 3$, $\Delta x = (\frac{3}{2} - 0)/3 = \frac{1}{2}$, $\{x_k\}_{k=0}^{3} = \left\{0, \frac{1}{2}, 1, \frac{3}{2}\right\}$. Therefore

$$R_3 = \frac{1}{2}\sum_{k=1}^{3} f(x_k) = \frac{1}{2}(2 + 1 + 2) = 2.5,$$

$$M_3 = \frac{1}{2}\sum_{k=1}^{6} f\left(x_k - \frac{1}{2}\Delta x\right) = \frac{1}{2}(3.25 + 1.25 + 1.25) = 2.875.$$

For $n = 6$, $\Delta x = (\frac{3}{2} - 0)/6 = \frac{1}{4}$, $\{x_k\}_{k=0}^{6} = \left\{0, \frac{1}{4}, \frac{1}{2}, \frac{3}{4}, 1, \frac{5}{4}, \frac{3}{2}\right\}$. Therefore

$$L_6 = \frac{1}{4}\sum_{k=0}^{5} f(x_k) = \frac{1}{4}(5 + 3.25 + 2 + 1.25 + 1 + 1.25) = 3.4375.$$

In Exercises 13–20, calculate the approximation for the given function and interval.

13. R_3, $f(x) = 7 - x$, $[3, 5]$

SOLUTION Let $f(x) = 7 - x$ on [3, 5]. For $n = 3$, $\Delta x = (5 - 3)/3 = \frac{2}{3}$, and $\{x_k\}_{k=0}^{3} = \left\{3, \frac{11}{3}, \frac{13}{3}, 5\right\}$. Therefore

$$R_3 = \frac{2}{3}\sum_{k=1}^{3}(7 - x_k)$$

$$= \frac{2}{3}\left(\frac{10}{3} + \frac{8}{3} + 2\right) = \frac{2}{3}(8) = \frac{16}{3}.$$

15. M_6, $f(x) = 4x + 3$, $[5, 8]$

SOLUTION Let $f(x) = 4x + 3$ on [5, 8]. For $n = 6$, $\Delta x = (8 - 5)/6 = \frac{1}{2}$, and $\{x_k^*\}_{k=0}^{5} = \{5.25, 5.75, 6.25, 6.75, 7.25, 7.75\}$. Therefore,

$$M_6 = \frac{1}{2}\sum_{k=0}^{5}(4x_k^* + 3)$$

$$= \frac{1}{2}(24 + 26 + 28 + 30 + 32 + 34)$$

$$= \frac{1}{2}(174) = 87.$$

17. L_6, $f(x) = x^2 + 3|x|$, $[-2, 1]$

SOLUTION Let $f(x) = x^2 + 3|x|$ on $[-2, 1]$. For $n = 6$, $\Delta x = (1 - (-2))/6 = \frac{1}{2}$, and $\{x_k\}_{k=0}^{6} = \{-2, -1.5, -1, -0.5, 0, 0.5, 1\}$. Therefore

$$L_6 = \frac{1}{2}\sum_{k=0}^{5}(x_k^2 + 3|x_k|) = \frac{1}{2}(10 + 6.75 + 4 + 1.75 + 0 + 1.75) = 12.125.$$

19. L_4, $f(x) = \cos^2 x$, $\left[\frac{\pi}{6}, \frac{\pi}{2}\right]$

SOLUTION Let $f(x) = \cos^2 x$ on $[\frac{\pi}{6}, \frac{\pi}{2}]$. For $n = 4$,

$$\Delta x = \frac{(\pi/2 - \pi/6)}{4} = \frac{\pi}{12} \quad \text{and} \quad \{x_k\}_{k=0}^{4} = \left\{\frac{\pi}{6}, \frac{\pi}{4}, \frac{\pi}{3}, \frac{5\pi}{12}, \frac{\pi}{2}\right\}.$$

Therefore

$$L_4 = \frac{\pi}{12}\sum_{k=0}^{3}\cos^2 x_k \approx 0.410236.$$

In Exercises 21–26, write the sum in summation notation.

21. $4^7 + 5^7 + 6^7 + 7^7 + 8^7$

SOLUTION The first term is 4^7, and the last term is 8^7, so it seems the kth term is k^7. Therefore, the sum is:

$$\sum_{k=4}^{8} k^7.$$

23. $(2^2 + 2) + (2^3 + 2) + (2^4 + 2) + (2^5 + 2)$

SOLUTION The first term is $2^2 + 2$, and the last term is $2^5 + 2$, so it seems the sum limits are 2 and 5, and the kth term is $2^k + 2$. Therefore, the sum is:

$$\sum_{k=2}^{5}(2^k + 2).$$

25. $\frac{1}{2\cdot 3} + \frac{2}{3\cdot 4} + \cdots + \frac{n}{(n+1)(n+2)}$

SOLUTION The first summand is $\frac{1}{(1+1)\cdot(1+2)}$. This shows us

$$\frac{1}{2\cdot 3} + \frac{2}{3\cdot 4} + \cdots + \frac{n}{(n+1)(n+2)} = \sum_{i=1}^{n}\frac{i}{(i+1)(i+2)}.$$

27. Calculate the sums:

(a) $\sum_{i=1}^{5} 9$ **(b)** $\sum_{i=0}^{5} 4$ **(c)** $\sum_{k=2}^{4} k^3$

SOLUTION

(a) $\sum_{i=1}^{5} 9 = 9 + 9 + 9 + 9 + 9 = 45$. Alternatively, $\sum_{i=1}^{5} 9 = 9\sum_{i=1}^{5} 1 = (9)(5) = 45$.

(b) $\sum_{i=0}^{5} 4 = 4 + 4 + 4 + 4 + 4 + 4 = 24$. Alternatively, $\sum_{i=0}^{5} 4 = 4\sum_{i=0}^{5} = (4)(6) = 24$.

(c) $\sum_{k=2}^{4} k^3 = 2^3 + 3^3 + 4^3 = 99$. Alternatively,

$$\sum_{k=2}^{4} k^3 = \left(\sum_{k=1}^{4} k^3\right) - \left(\sum_{k=1}^{1} k^3\right) = \left(\frac{4^4}{4} + \frac{4^3}{2} + \frac{4^2}{4}\right) - \left(\frac{1^4}{4} + \frac{1^3}{2} + \frac{1^2}{4}\right) = 99.$$

29. Let $b_1 = 4$, $b_2 = 1$, $b_3 = 2$, and $b_4 = -4$. Calculate:

(a) $\sum_{i=2}^{4} b_i$ **(b)** $\sum_{j=1}^{2}(2^{b_j} - b_j)$ **(c)** $\sum_{k=1}^{3} kb_k$

SOLUTION

(a) $\sum_{i=2}^{4} b_i = b_2 + b_3 + b_4 = 1 + 2 + (-4) = -1$.

(b) $\sum_{j=1}^{2}\left(2^{b_j} - b_j\right) = (2^4 - 4) + (2^1 - 1) = 13$.

(c) $\sum_{k=1}^{3} kb_k = 1(4) + 2(1) + 3(2) = 12$.

31. Calculate $\sum_{j=101}^{200} j$. *Hint:* Write as a difference of two sums and use formula (3).

SOLUTION

$$\sum_{j=101}^{200} j = \sum_{j=1}^{200} j - \sum_{j=1}^{100} j = \left(\frac{200^2}{2} + \frac{200}{2}\right) - \left(\frac{100^2}{2} + \frac{100}{2}\right) = 20100 - 5050 = 15050.$$

In Exercises 33–40, use linearity and formulas (3)–(5) to rewrite and evaluate the sums.

33. $\sum_{j=1}^{20} 8j^3$

SOLUTION $\sum_{j=1}^{20} 8j^3 = 8\sum_{j=1}^{20} j^3 = 8\left(\frac{20^4}{4} + \frac{20^3}{2} + \frac{20^2}{4}\right) = 8(44,100) = 352,800.$

35. $\sum_{n=51}^{150} n^2$

SOLUTION

$$\sum_{n=51}^{150} n^2 = \sum_{n=1}^{150} n^2 - \sum_{n=1}^{50} n^2$$

$$= \left(\frac{150^3}{3} + \frac{150^2}{2} + \frac{150}{6}\right) - \left(\frac{50^3}{3} + \frac{50^2}{2} + \frac{50}{6}\right)$$

$$= 1,136,275 - 42,925 = 1,093,350.$$

37. $\sum_{j=0}^{50} j(j-1)$

SOLUTION

$$\sum_{j=0}^{50} j(j-1) = \sum_{j=0}^{50} (j^2 - j) = \sum_{j=0}^{50} j^2 - \sum_{j=0}^{50} j$$

$$= \left(\frac{50^3}{3} + \frac{50^2}{2} + \frac{50}{6}\right) - \left(\frac{50^2}{2} + \frac{50}{2}\right) = \frac{50^3}{3} - \frac{50}{3} = \frac{124,950}{3} = 41,650.$$

The power sum formula is usable because $\sum_{j=0}^{50} j(j-1) = \sum_{j=1}^{50} j(j-1)$.

39. $\sum_{m=1}^{30} (4-m)^3$

SOLUTION

$$\sum_{m=1}^{30} (4-m)^3 = \sum_{m=1}^{30} (64 - 48m + 12m^2 - m^3)$$

$$= 64\sum_{m=1}^{30} 1 - 48\sum_{m=1}^{30} m + 12\sum_{m=1}^{30} m^2 - \sum_{m=1}^{30} m^3$$

$$= 64(30) - 48\frac{(30)(31)}{2} + 12\left(\frac{30^3}{3} + \frac{30^2}{2} + \frac{30}{6}\right) - \left(\frac{30^4}{4} + \frac{30^3}{2} + \frac{30^2}{4}\right)$$

$$= 1920 - 22,320 + 113,460 - 216,225 = -123,165.$$

In Exercises 41–44, use formulas (3)–(5) to evaluate the limit.

41. $\lim_{N\to\infty} \sum_{i=1}^{N} \frac{i}{N^2}$

SOLUTION Let $s_N = \sum_{i=1}^{N} \frac{i}{N^2}$. Then,

$$s_N = \sum_{i=1}^{N} \frac{i}{N^2} = \frac{1}{N^2}\sum_{i=1}^{N} i = \frac{1}{N^2}\left(\frac{N^2}{2} + \frac{N}{2}\right) = \frac{1}{2} + \frac{1}{2N}.$$

Therefore, $\lim_{N\to\infty} s_N = \frac{1}{2}$.

43. $\lim_{N\to\infty} \sum_{i=1}^{N} \frac{i^2 - i + 1}{N^3}$

SOLUTION Let $s_N = \sum_{i=1}^{N} \frac{i^2 - i + 1}{N^3}$. Then

$$s_N = \sum_{i=1}^{N} \frac{i^2 - i + 1}{N^3} = \frac{1}{N^3}\left[\left(\sum_{i=1}^{N} i^2\right) - \left(\sum_{i=1}^{N} i\right) + \left(\sum_{i=1}^{N} 1\right)\right]$$

$$= \frac{1}{N^3}\left[\left(\frac{N^3}{3} + \frac{N^2}{2} + \frac{N}{6}\right) - \left(\frac{N^2}{2} + \frac{N}{2}\right) + N\right] = \frac{1}{3} + \frac{2}{3N^2}.$$

Therefore, $\lim_{N\to\infty} s_N = \frac{1}{3}$.

In Exercises 45–50, calculate the limit for the given function and interval. Verify your answer by using geometry.

45. $\lim_{N\to\infty} R_N, \quad f(x) = 9x, \quad [0, 2]$

SOLUTION Let $f(x) = 9x$ on $[0, 2]$. Let N be a positive integer and set $a = 0$, $b = 2$, and $\Delta x = (b - a)/N = (2 - 0)/N = 2/N$. Also, let $x_k = a + k\Delta x = 2k/N$, $k = 1, 2, \dots, N$ be the right endpoints of the N subintervals of $[0, 2]$. Then

$$R_N = \Delta x \sum_{k=1}^{N} f(x_k) = \frac{2}{N}\sum_{k=1}^{N} 9\left(\frac{2k}{N}\right) = \frac{36}{N^2}\sum_{k=1}^{N} k = \frac{36}{N^2}\left(\frac{N^2}{2} + \frac{N}{2}\right) = 18 + \frac{18}{N}.$$

The area under the graph is

$$\lim_{N\to\infty} R_N = \lim_{N\to\infty}\left(18 + \frac{18}{N}\right) = 18.$$

The region under the graph is a triangle with base 2 and height 18. The area of the region is then $\frac{1}{2}(2)(18) = 18$, which agrees with the value obtained from the limit of the right-endpoint approximations.

47. $\lim_{N\to\infty} L_N, \quad f(x) = \frac{1}{2}x + 2, \quad [0, 4]$

SOLUTION Let $f(x) = \frac{1}{2}x + 2$ on $[0, 4]$. Let $N > 0$ be an integer, and set $a = 0$, $b = 4$, and $\Delta x = (4 - 0)/N = \frac{4}{N}$. Also, let $x_k = 0 + k\Delta x = \frac{4k}{N}$, $k = 0, 1, \dots, N - 1$ be the left endpoints of the N subintervals. Then

$$L_N = \Delta x \sum_{k=0}^{N-1} f(x_k) = \frac{4}{N}\sum_{k=0}^{N-1}\left(\frac{1}{2}\left(\frac{4k}{N}\right) + 2\right) = \frac{8}{N}\sum_{k=0}^{N-1} 1 + \frac{8}{N^2}\sum_{k=0}^{N-1} k$$

$$= 8 + \frac{8}{N^2}\left(\frac{(N-1)^2}{2} + \frac{N-1}{2}\right) = 12 - \frac{4}{N}.$$

The area under the graph is

$$\lim_{N\to\infty} L_N = 12.$$

The region under the curve over [0, 4] is a trapezoid with base width 4 and heights 2 and 4. From this, we get that the area is $\frac{1}{2}(4)(2+4) = 12$, which agrees with the answer obtained from the limit of the left-endpoint approximations.

49. $\lim_{N\to\infty} M_N$, $f(x) = x$, $[0, 2]$

SOLUTION Let $f(x) = x$ on $[0, 2]$. Let $N > 0$ be an integer and set $a = 0$, $b = 2$, and $\Delta x = (b-a)/N = \frac{2}{N}$. Also, let $x_k^* = 0 + (k - \frac{1}{2})\Delta x = \frac{2k-1}{N}$, $k = 1, 2, \dots N$, be the midpoints of the N subintervals of $[0, 2]$. Then

$$M_N = \Delta x \sum_{k=1}^{N} f(x_k^*) = \frac{2}{N} \sum_{k=1}^{N} \frac{2k-1}{N} = \frac{2}{N^2} \sum_{k=1}^{N} (2k-1)$$

$$= \frac{2}{N^2}\left(2\sum_{k=1}^{N} k - N\right) = \frac{4}{N^2}\left(\frac{N^2}{2} + \frac{N}{2}\right) - \frac{2}{N} = 2.$$

The area under the curve over $[0, 2]$ is

$$\lim_{N\to\infty} M_N = 2.$$

The region under the curve over [0, 2] is a triangle with base and height 2, and thus area 2, which agrees with the answer obtained from the limit of the midpoint approximations.

51. Show, for $f(x) = 3x^2 + 4x$ over $[0, 2]$, that

$$R_N = \frac{2}{N} \sum_{j=1}^{N} \left(\frac{24j^2}{N^2} + \frac{16j}{N}\right)$$

Then evaluate $\lim_{N\to\infty} R_N$.

SOLUTION Let $f(x) = 3x^2 + 4x$ on $[0, 2]$. Let N be a positive integer and set $a = 0$, $b = 2$, and $\Delta x = (b-a)/N = (2-0)/N = 2/N$. Also, let $x_j = a + j\Delta x = 2j/N$, $j = 1, 2, \dots, N$ be the right endpoints of the N subintervals of $[0, 3]$. Then

$$R_N = \Delta x \sum_{j=1}^{N} f(x_j) = \frac{2}{N} \sum_{j=1}^{N} \left(3\left(\frac{2j}{N}\right)^2 + 4\frac{2j}{N}\right)$$

$$= \frac{2}{N} \sum_{j=1}^{N} \left(\frac{12j^2}{N^2} + \frac{8j}{N}\right)$$

Continuing, we find

$$R_N = \frac{24}{N^3} \sum_{j=1}^{N} j^2 + \frac{16}{N^2} \sum_{j=1}^{N} j$$

$$= \frac{24}{N^3}\left(\frac{N^3}{3} + \frac{N^2}{2} + \frac{N}{6}\right) + \frac{16}{N^2}\left(\frac{N^2}{2} + \frac{N}{2}\right)$$

$$= 16 + \frac{20}{N} + \frac{4}{N^2}$$

Thus,

$$\lim_{N\to\infty} R_N = \lim_{N\to\infty} \left(16 + \frac{20}{N} + \frac{4}{N^2}\right) = 16.$$

In Exercises 53–60, find a formula for R_N and compute the area under the graph as a limit.

53. $f(x) = x^2$, $[0, 1]$

SOLUTION Let $f(x) = x^2$ on the interval $[0, 1]$. Then $\Delta x = \dfrac{1-0}{N} = \dfrac{1}{N}$ and $a = 0$. Hence,

$$R_N = \Delta x \sum_{j=1}^{N} f(0 + j\Delta x) = \frac{1}{N} \sum_{j=1}^{N} j^2 \frac{1}{N^2} = \frac{1}{N^3}\left(\frac{N^3}{3} + \frac{N^2}{2} + \frac{N}{6}\right) = \frac{1}{3} + \frac{1}{2N} + \frac{1}{6N^2}$$

and

$$\lim_{N\to\infty} R_N = \lim_{N\to\infty}\left(\frac{1}{3}+\frac{1}{2N}+\frac{1}{6N^2}\right)=\frac{1}{3}.$$

55. $f(x)=6x^2-4,\quad [2,5]$

SOLUTION Let $f(x)=6x^2-4$ on the interval $[2,5]$. Then $\Delta x=\dfrac{5-2}{N}=\dfrac{3}{N}$ and $a=2$. Hence,

$$R_N=\Delta x\sum_{j=1}^{N} f(2+j\Delta x)=\frac{3}{N}\sum_{j=1}^{N}\left(6\left(2+\frac{3j}{N}\right)^2-4\right)=\frac{3}{N}\sum_{j=1}^{N}\left(20+\frac{72j}{N}+\frac{54j^2}{N^2}\right)$$

$$=60+\frac{216}{N^2}\sum_{j=1}^{N} j+\frac{162}{N^3}\sum_{j=1}^{N} j^2$$

$$=60+\frac{216}{N^2}\left(\frac{N^2}{2}+\frac{N}{2}\right)+\frac{162}{N^3}\left(\frac{N^3}{3}+\frac{N^2}{2}+\frac{N}{6}\right)$$

$$=222+\frac{189}{N}+\frac{27}{N^2}$$

and

$$\lim_{N\to\infty} R_N=\lim_{N\to\infty}\left(222+\frac{189}{N}+\frac{27}{N^2}\right)=222.$$

57. $f(x)=x^3-x,\quad [0,2]$

SOLUTION Let $f(x)=x^3-x$ on the interval $[0,2]$. Then $\Delta x=\dfrac{2-0}{N}=\dfrac{2}{N}$ and $a=0$. Hence,

$$R_N=\Delta x\sum_{j=1}^{N} f(0+j\Delta x)=\frac{2}{N}\sum_{j=1}^{N}\left(\left(\frac{2j}{N}\right)^3-\frac{2j}{N}\right)=\frac{2}{N}\sum_{j=1}^{N}\left(\frac{8j^3}{N^3}-\frac{2j}{N}\right)$$

$$=\frac{16}{N^4}\sum_{j=1}^{N} j^3-\frac{4}{N^2}\sum_{j=1}^{N} j$$

$$=\frac{16}{N^4}\left(\frac{N^4}{4}+\frac{N^3}{2}+\frac{N^2}{2}\right)-\frac{4}{N^2}\left(\frac{N^2}{2}+\frac{N}{2}\right)$$

$$=2+\frac{6}{N}+\frac{8}{N^2}$$

and

$$\lim_{N\to\infty} R_N=\lim_{N\to\infty}\left(2+\frac{6}{N}+\frac{8}{N^2}\right)=2.$$

59. $f(x)=2x+1,\quad [a,b]\quad$ (a,b constants with $a<b$)

SOLUTION Let $f(x)=2x+1$ on the interval $[a,b]$. Then $\Delta x=\dfrac{b-a}{N}$. Hence,

$$R_N=\Delta x\sum_{j=1}^{N} f(a+j\Delta x)=\frac{(b-a)}{N}\sum_{j=1}^{N}\left(2\left(a+j\frac{(b-a)}{N}\right)+1\right)$$

$$=\frac{(b-a)}{N}(2a+1)\sum_{j=1}^{N} 1+\frac{2(b-a)^2}{N^2}\sum_{j=1}^{N} j$$

$$=\frac{(b-a)}{N}(2a+1)N+\frac{2(b-a)^2}{N^2}\left(\frac{N^2}{2}+\frac{N}{2}\right)$$

$$=(b-a)(2a+1)+(b-a)^2+\frac{(b-a)^2}{N}$$

and

$$\lim_{N\to\infty} R_N = \lim_{N\to\infty}\left((b-a)(2a+1)+(b-a)^2+\frac{(b-a)^2}{N}\right)$$
$$= (b-a)(2a+1)+(b-a)^2 = (b^2+b)-(a^2+a).$$

In Exercises 61–64, describe the area represented by the limits.

61. $\lim_{N\to\infty} \frac{1}{N}\sum_{j=1}^{N}\left(\frac{j}{N}\right)^4$

SOLUTION The limit

$$\lim_{N\to\infty} R_N = \lim_{N\to\infty} \frac{1}{N}\sum_{j=1}^{N}\left(\frac{j}{N}\right)^4$$

represents the area between the graph of $f(x) = x^4$ and the x-axis over the interval $[0, 1]$.

63. $\lim_{N\to\infty} \frac{5}{N}\sum_{j=0}^{N-1} e^{-2+5j/N}$

SOLUTION The limit

$$\lim_{N\to\infty} L_N = \lim_{N\to\infty} \frac{5}{N}\sum_{j=0}^{N-1} e^{-2+5j/N}$$

represents the area between the graph of $y = e^x$ and the x-axis over the interval $[-2, 3]$.

In Exercises 65–70, express the area under the graph as a limit using the approximation indicated (in summation notation), but do not evaluate.

65. R_N, $f(x) = \sin x$ over $[0, \pi]$

SOLUTION Let $f(x) = \sin x$ over $[0, \pi]$ and set $a = 0$, $b = \pi$, and $\Delta x = (b-a)/N = \pi/N$. Then

$$R_N = \Delta x \sum_{k=1}^{N} f(x_k) = \frac{\pi}{N}\sum_{k=1}^{N} \sin\left(\frac{k\pi}{N}\right).$$

Hence

$$\lim_{N\to\infty} R_N = \lim_{N\to\infty} \frac{\pi}{N}\sum_{k=1}^{N} \sin\left(\frac{k\pi}{N}\right)$$

is the area between the graph of $f(x) = \sin x$ and the x-axis over $[0, \pi]$.

67. L_N, $f(x) = \sqrt{2x+1}$ over $[7, 11]$

SOLUTION Let $f(x) = \sqrt{2x+1}$ over the interval $[7, 11]$. Then $\Delta x = \frac{11-7}{N} = \frac{4}{N}$ and $a = 7$. Hence,

$$L_N = \Delta x \sum_{j=0}^{N-1} f(7 + j\Delta x) = \frac{4}{N}\sum_{j=0}^{N-1}\sqrt{2(7 + j\frac{4}{N}) + 1}$$

and

$$\lim_{N\to\infty} L_N = \lim_{N\to\infty} \frac{4}{N}\sum_{j=0}^{N-1}\sqrt{15 + \frac{8j}{N}}$$

is the area between the graph of $f(x) = \sqrt{2x+1}$ and the x-axis over $[7, 11]$.

69. M_N, $f(x) = \tan x$ over $\left[\frac{1}{2}, 1\right]$

SOLUTION Let $f(x) = \tan x$ over the interval $[\frac{1}{2}, 1]$. Then $\Delta x = \frac{1-\frac{1}{2}}{N} = \frac{1}{2N}$ and $a = \frac{1}{2}$. Hence

$$M_N = \Delta x \sum_{j=1}^{N} f\left(\frac{1}{2} + \left(j - \frac{1}{2}\right)\Delta x\right) = \frac{1}{2N}\sum_{j=1}^{N} \tan\left(\frac{1}{2} + \frac{1}{2N}\left(j - \frac{1}{2}\right)\right)$$

and so

$$\lim_{N\to\infty} M_N = \lim_{N\to\infty} \frac{1}{2N}\sum_{j=1}^{N} \tan\left(\frac{1}{2} + \frac{1}{2N}\left(j - \frac{1}{2}\right)\right)$$

is the area between the graph of $f(x) = \tan x$ and the x-axis over $[\frac{1}{2}, 1]$.

71. Evaluate $\lim_{N\to\infty} \frac{1}{N}\sum_{j=1}^{N}\sqrt{1-\left(\frac{j}{N}\right)^2}$ by interpreting it as the area of part of a familiar geometric figure.

SOLUTION The limit

$$\lim_{N\to\infty} R_N = \lim_{N\to\infty} \frac{1}{N}\sum_{j=1}^{N}\sqrt{1-\left(\frac{j}{N}\right)^2}$$

represents the area between the graph of $y = f(x) = \sqrt{1-x^2}$ and the x-axis over the interval $[0, 1]$. This is the portion of the circular disk $x^2 + y^2 \le 1$ that lies in the first quadrant. Accordingly, its area is $\frac{1}{4}\pi\,(1)^2 = \frac{\pi}{4}$.

In Exercises 72–74, let $f(x) = x^2$ and let R_N, L_N, and M_N be the approximations for the interval $[0, 1]$.

73. Show that

$$L_N = \frac{1}{3} - \frac{1}{2N} + \frac{1}{6N^2}, \qquad M_N = \frac{1}{3} - \frac{1}{12N^2}$$

Then rank the three approximations R_N, L_N, and M_N in order of increasing accuracy (use Exercise 72).

SOLUTION Let $f(x) = x^2$ on $[0, 1]$. Let N be a positive integer and set $a = 0$, $b = 1$, and $\Delta x = (b-a)/N = 1/N$. Let $x_k = a + k\Delta x = k/N$, $k = 0, 1, \ldots, N$ and let $x_k^* = a + (k + \frac{1}{2})\Delta x = (k + \frac{1}{2})/N$, $k = 0, 1, \ldots, N-1$. Then

$$L_N = \Delta x\sum_{k=0}^{N-1} f(x_k) = \frac{1}{N}\sum_{k=0}^{N-1}\left(\frac{k}{N}\right)^2 = \frac{1}{N^3}\sum_{k=1}^{N-1} k^2$$

$$= \frac{1}{N^3}\left(\frac{(N-1)^3}{3} + \frac{(N-1)^2}{2} + \frac{N-1}{6}\right) = \frac{1}{3} - \frac{1}{2N} + \frac{1}{6N^2}$$

$$M_N = \Delta x\sum_{k=0}^{N-1} f(x_k^*) = \frac{1}{N}\sum_{k=0}^{N-1}\left(\frac{k+\frac{1}{2}}{N}\right)^2 = \frac{1}{N^3}\sum_{k=0}^{N-1}\left(k^2 + k + \frac{1}{4}\right)$$

$$= \frac{1}{N^3}\left(\left(\sum_{k=1}^{N-1} k^2\right) + \left(\sum_{k=1}^{N-1} k\right) + \frac{1}{4}\left(\sum_{k=0}^{N-1} 1\right)\right)$$

$$= \frac{1}{N^3}\left(\left(\frac{(N-1)^3}{3} + \frac{(N-1)^2}{2} + \frac{N-1}{6}\right) + \left(\frac{(N-1)^2}{2} + \frac{N-1}{2}\right) + \frac{1}{4}N\right) = \frac{1}{3} - \frac{1}{12N^2}$$

The error of R_N is given by $\frac{1}{2N} + \frac{1}{6N^2}$, the error of L_N is given by $-\frac{1}{2N} + \frac{1}{6N^2}$ and the error of M_N is given by $-\frac{1}{12N^2}$. Of the three approximations, R_N is the least accurate, then L_N and finally M_N is the most accurate.

In Exercises 75–80, use the Graphical Insight on page 291 to obtain bounds on the area.

75. Let A be the area under $f(x) = \sqrt{x}$ over $[0, 1]$. Prove that $0.51 \le A \le 0.77$ by computing R_4 and L_4. Explain your reasoning.

SOLUTION For $n = 4$, $\Delta x = \frac{1-0}{4} = \frac{1}{4}$ and $\{x_i\}_{i=0}^{4} = \{0 + i\Delta x\} = \{0, \frac{1}{4}, \frac{1}{2}, \frac{3}{4}, 1\}$. Therefore,

$$R_4 = \Delta x\sum_{i=1}^{4} f(x_i) = \frac{1}{4}\left(\frac{1}{2} + \frac{\sqrt{2}}{2} + \frac{\sqrt{3}}{2} + 1\right) \approx 0.768$$

$$L_4 = \Delta x\sum_{i=0}^{3} f(x_i) = \frac{1}{4}\left(0 + \frac{1}{2} + \frac{\sqrt{2}}{2} + \frac{\sqrt{3}}{2}\right) \approx 0.518.$$

In the plot below, you can see the rectangles whose area is represented by L_4 under the graph and the top of those whose area is represented by R_4 above the graph. The area A under the curve is somewhere between L_4 and R_4, so

$$0.518 \le A \le 0.768.$$

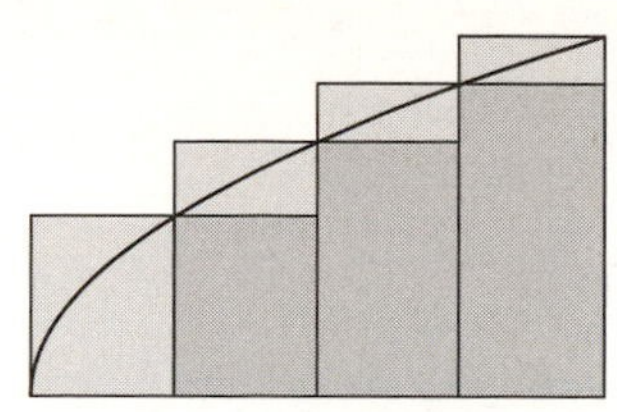

L_4, R_4 and the graph of $f(x)$.

77. Use R_4 and L_4 to show that the area A under the graph of $y = \sin x$ over $\left[0, \frac{\pi}{2}\right]$ satisfies $0.79 \le A \le 1.19$.

SOLUTION Let $f(x) = \sin x$. $f(x)$ is increasing over the interval $[0, \pi/2]$, so the Insight on page 291 applies, which indicates that $L_4 \le A \le R_4$. For $n = 4$, $\Delta x = \frac{\pi/2-0}{4} = \frac{\pi}{8}$ and $\{x_i\}_{i=0}^4 = \{0 + i\Delta x\}_{i=0}^4 = \{0, \frac{\pi}{8}, \frac{\pi}{4}, \frac{3\pi}{8}, \frac{\pi}{2}\}$. From this,

$$L_4 = \frac{\pi}{8}\sum_{i=0}^{3} f(x_i) \approx 0.79, \qquad R_4 = \frac{\pi}{8}\sum_{i=1}^{4} f(x_i) \approx 1.18.$$

Hence A is between 0.79 and 1.19.

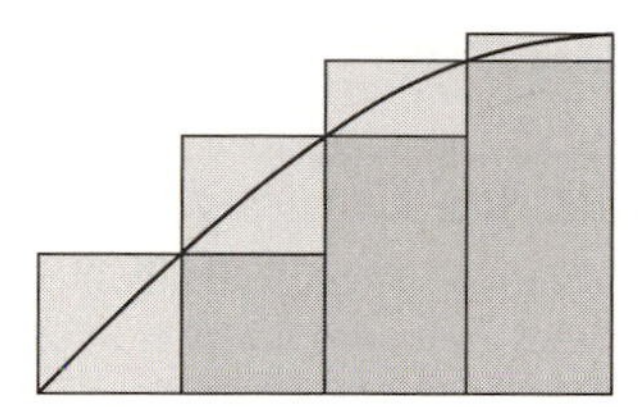

Left and Right endpoint approximations to A.

79. CAS Show that the area A under $y = x^{1/4}$ over $[0, 1]$ satisfies $L_N \le A \le R_N$ for all N. Use a computer algebra system to calculate L_N and R_N for $N = 100$ and 200, and determine A to two decimal places.

SOLUTION On $[0, 1]$, $f(x) = x^{1/4}$ is an increasing function; therefore, $L_N \le A \le R_N$ for all N. We find

$$L_{100} = 0.793988 \quad \text{and} \quad R_{100} = 0.80399,$$

while

$$L_{200} = 0.797074 \quad \text{and} \quad R_{200} = 0.802075.$$

Thus, $A = 0.80$ to two decimal places.

81. In this exercise, we evaluate the area A under the graph of $y = e^x$ over $[0, 1]$ [Figure 19(A)] using the formula for a geometric sum (valid for $r \ne 1$):

$$1 + r + r^2 + \cdots + r^{N-1} = \sum_{j=0}^{N-1} r^j = \frac{r^N - 1}{r - 1} \qquad \boxed{8}$$

(a) Show that $L_N = \frac{1}{N}\sum_{j=0}^{N-1} e^{j/N}$.

(b) Apply Eq. (8) with $r = e^{1/N}$ to prove $L_N = \frac{e-1}{N(e^{1/N}-1)}$.

(c) Compute $A = \lim_{N\to\infty} L_N$ using L'Hôpital's Rule.

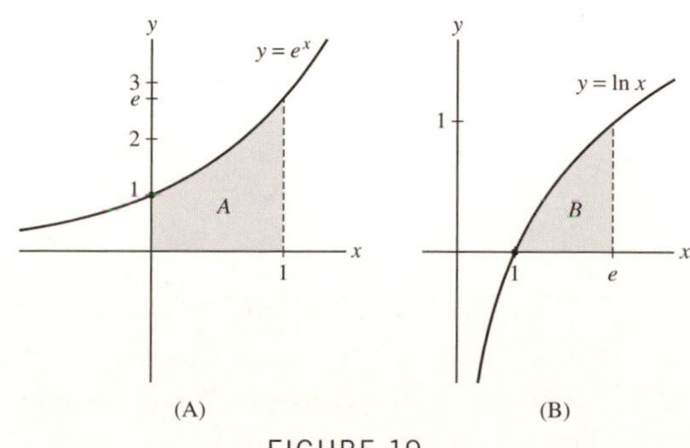

FIGURE 19

SOLUTION

(a) Let $f(x) = e^x$ on $[0, 1]$. With $n = N$, $\Delta x = (1-0)/N = 1/N$ and

$$x_j = a + j\Delta x = \frac{j}{N}$$

for $j = 0, 1, 2, \ldots, N$. Therefore,

$$L_N = \Delta x \sum_{j=0}^{N-1} f(x_j) = \frac{1}{N} \sum_{j=0}^{N-1} e^{j/N}.$$

(b) Applying Eq. (8) with $r = e^{1/N}$, we have

$$L_N = \frac{1}{N} \frac{(e^{1/N})^N - 1}{e^{1/N} - 1} = \frac{e-1}{N(e^{1/N} - 1)}.$$

Therefore,

$$A = \lim_{N\to\infty} L_N = (e-1) \lim_{N\to\infty} \frac{1}{N(e^{1/N} - 1)}.$$

(c) Using L'Hôpital's Rule,

$$A = (e-1) \lim_{N\to\infty} \frac{N^{-1}}{e^{1/N} - 1} = (e-1) \lim_{N\to\infty} \frac{-N^{-2}}{-N^{-2} e^{1/N}} = (e-1) \lim_{N\to\infty} e^{-1/N} = e - 1.$$

Further Insights and Challenges

83. Although the accuracy of R_N generally improves as N increases, this need not be true for small values of N. Draw the graph of a positive continuous function $f(x)$ on an interval such that R_1 is closer than R_2 to the exact area under the graph. Can such a function be monotonic?

SOLUTION Let δ be a small positive number less than $\frac{1}{4}$. (In the figures below, $\delta = \frac{1}{10}$. But imagine δ being *very* tiny.) Define $f(x)$ on $[0, 1]$ by

$$f(x) = \begin{cases} 1 & \text{if } 0 \le x < \frac{1}{2} - \delta \\ \frac{1}{2\delta} - \frac{x}{\delta} & \text{if } \frac{1}{2} - \delta \le x < \frac{1}{2} \\ \frac{x}{\delta} - \frac{1}{2\delta} & \text{if } \frac{1}{2} \le x < \frac{1}{2} + \delta \\ 1 & \text{if } \frac{1}{2} + \delta \le x \le 1 \end{cases}$$

Then f is continuous on $[0, 1]$. (Again, just look at the figures.)

- The exact area between f and the x-axis is $A = 1 - \frac{1}{2}bh = 1 - \frac{1}{2}(2\delta)(1) = 1 - \delta$. (For $\delta = \frac{1}{10}$, we have $A = \frac{9}{10}$.)
- With $R_1 = 1$, the absolute error is $|E_1| = |R_1 - A| = |1 - (1-\delta)| = \delta$. (For $\delta = \frac{1}{10}$, this absolute error is $|E_1| = \frac{1}{10}$.)
- With $R_2 = \frac{1}{2}$, the absolute error is $|E_2| = |R_2 - A| = \left|\frac{1}{2} - (1-\delta)\right| = \left|\delta - \frac{1}{2}\right| = \frac{1}{2} - \delta$. (For $\delta = \frac{1}{10}$, we have $|E_2| = \frac{2}{5}$.)
- Accordingly, R_1 is closer to the exact area A than is R_2. Indeed, the tinier δ is, the more dramatic the effect.
- For a monotonic function, this phenomenon cannot occur. Successive approximations from either side get progressively more accurate.

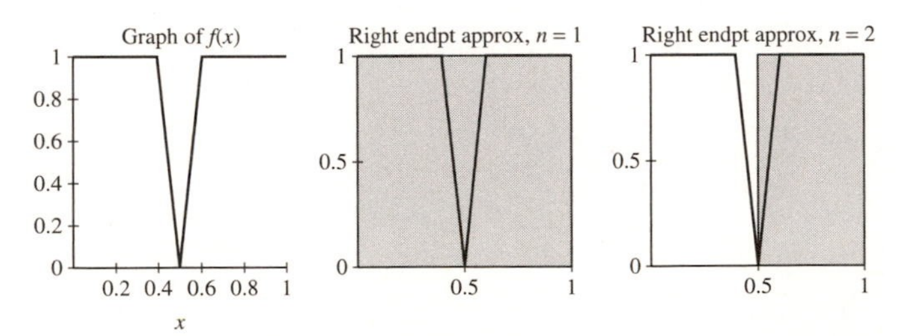

85. Explain graphically: *The endpoint approximations are less accurate when $f'(x)$ is large.*

SOLUTION When f' is large, the graph of f is steeper and hence there is more gap between f and L_N or R_N. Recall that the top line segments of the rectangles involved in an endpoint approximation constitute a piecewise constant function. If f' is large, then f is increasing more rapidly and hence is less like a constant function.

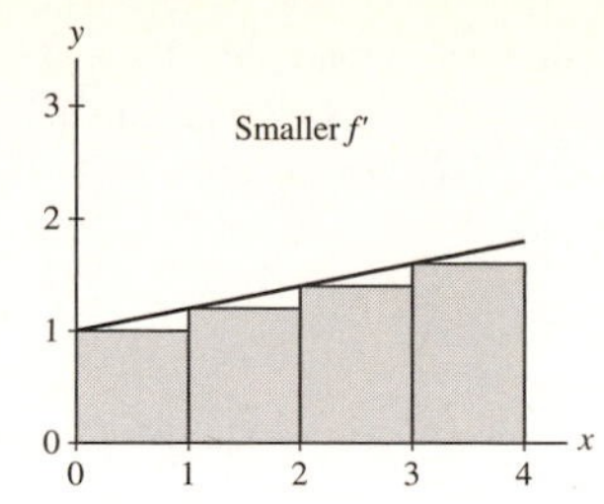

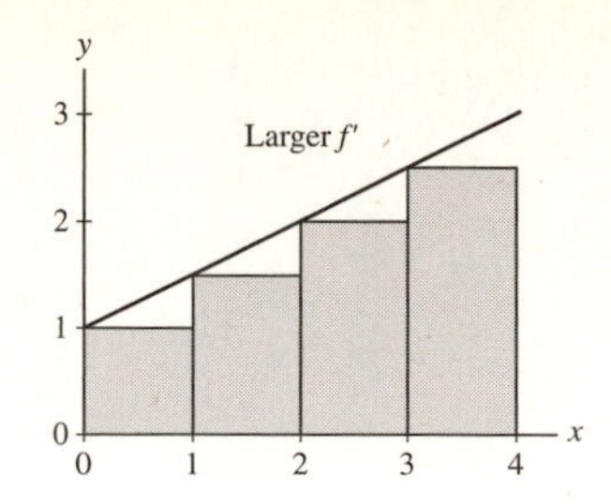

87. In this exercise, we prove that $\lim_{N\to\infty} R_N$ and $\lim_{N\to\infty} L_N$ exist and are equal if $f(x)$ is increasing [the case of $f(x)$ decreasing is similar]. We use the concept of a least upper bound discussed in Appendix B.

(a) Explain with a graph why $L_N \le R_M$ for all $N, M \ge 1$.

(b) By (a), the sequence $\{L_N\}$ is bounded, so it has a least upper bound L. By definition, L is the smallest number such that $L_N \le L$ for all N. Show that $L \le R_M$ for all M.

(c) According to (b), $L_N \le L \le R_N$ for all N. Use Eq. (9) to show that $\lim_{N\to\infty} L_N = L$ and $\lim_{N\to\infty} R_N = L$.

SOLUTION

(a) Let $f(x)$ be positive and increasing, and let N and M be positive integers. From the figure below at the left, we see that L_N underestimates the area under the graph of $y = f(x)$, while from the figure below at the right, we see that R_M overestimates the area under the graph. Thus, for all $N, M \ge 1$, $L_N \le R_M$.

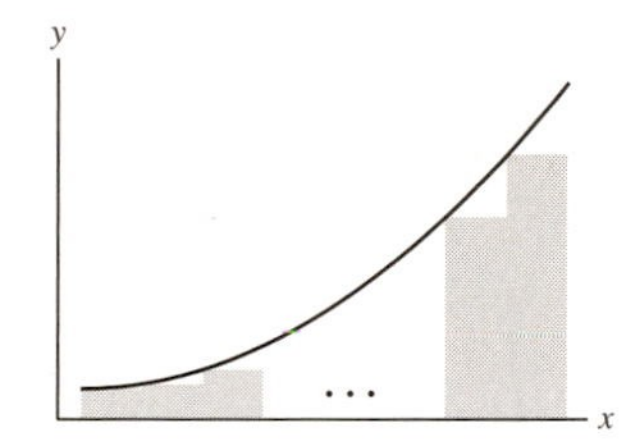

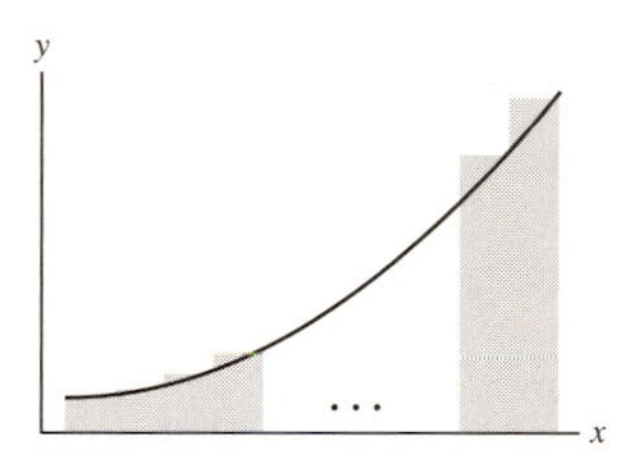

(b) Because the sequence $\{L_N\}$ is bounded above by R_M for any M, each R_M is an upper bound for the sequence. Furthermore, the sequence $\{L_N\}$ must have a least upper bound, call it L. By definition, the least upper bound must be no greater than any other upper bound; consequently, $L \le R_M$ for all M.

(c) Since $L_N \le L \le R_N$, $R_N - L \le R_N - L_N$, so $|R_N - L| \le |R_N - L_N|$. From this,

$$\lim_{N\to\infty} |R_N - L| \le \lim_{N\to\infty} |R_N - L_N|.$$

By Eq. (9),

$$\lim_{N\to\infty} |R_N - L_N| = \lim_{N\to\infty} \frac{1}{N}|(b-a)(f(b) - f(a))| = 0,$$

so $\lim_{N\to\infty} |R_N - L| \le |R_N - L_N| = 0$, hence $\lim_{N\to\infty} R_N = L$.

Similarly, $|L_N - L| = L - L_N \le R_N - L_N$, so

$$|L_N - L| \le |R_N - L_N| = \frac{(b-a)}{N}(f(b) - f(a)).$$

This gives us that

$$\lim_{N\to\infty} |L_N - L| \le \lim_{N\to\infty} \frac{1}{N}|(b-a)(f(b) - f(a))| = 0,$$

so $\lim_{N\to\infty} L_N = L$.

This proves $\lim_{N\to\infty} L_N = \lim_{N\to\infty} R_N = L$.

In Exercises 89 and 90, use Eq. (10) to find a value of N such that $|R_N - A| < 10^{-4}$ for the given function and interval.

89. $f(x) = \sqrt{x}$, $[1, 4]$

SOLUTION Let $f(x) = \sqrt{x}$ on $[1, 4]$. Then $b = 4$, $a = 1$, and

$$|R_N - A| \le \frac{4-1}{N}(f(4) - f(1)) = \frac{3}{N}(2-1) = \frac{3}{N}.$$

We need $\frac{3}{N} < 10^{-4}$, which gives $N > 30{,}000$. Thus $|R_{30{,}001} - A| < 10^{-4}$ for $f(x) = \sqrt{x}$ on $[1, 4]$.

91. Prove that if $f(x)$ is positive and monotonic, then M_N lies between R_N and L_N and is closer to the actual area under the graph than both R_N and L_N. *Hint:* In the case that $f(x)$ is increasing, Figure 20 shows that the part of the error in R_N due to the ith rectangle is the sum of the areas $A + B + D$, and for M_N it is $|B - E|$. On the other hand, $A \geq E$.

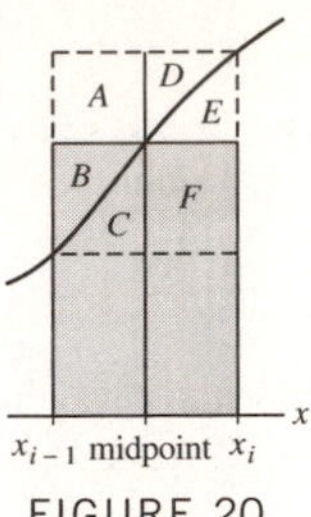

FIGURE 20

SOLUTION Suppose $f(x)$ is monotonic increasing on the interval $[a, b]$, $\Delta x = \dfrac{b-a}{N}$,

$$\{x_k\}_{k=0}^{N} = \{a, a + \Delta x, a + 2\Delta x, \dots, a + (N-1)\Delta x, b\}$$

and

$$\{x_k^*\}_{k=0}^{N-1} = \left\{\frac{a + (a + \Delta x)}{2}, \frac{(a + \Delta x) + (a + 2\Delta x)}{2}, \dots, \frac{(a + (N-1)\Delta x) + b}{2}\right\}.$$

Note that $x_i < x_i^* < x_{i+1}$ implies $f(x_i) < f(x_i^*) < f(x_{i+1})$ for all $0 \leq i < N$ because $f(x)$ is monotone increasing. Then

$$\left(L_N = \frac{b-a}{N}\sum_{k=0}^{N-1} f(x_k)\right) < \left(M_N = \frac{b-a}{N}\sum_{k=0}^{N-1} f(x_k^*)\right) < \left(R_N = \frac{b-a}{N}\sum_{k=1}^{N} f(x_k)\right)$$

Similarly, if $f(x)$ is monotone decreasing,

$$\left(L_N = \frac{b-a}{N}\sum_{k=0}^{N-1} f(x_k)\right) > \left(M_N = \frac{b-a}{N}\sum_{k=0}^{N-1} f(x_k^*)\right) > \left(R_N = \frac{b-a}{N}\sum_{k=1}^{N} f(x_k)\right)$$

Thus, if $f(x)$ is monotonic, then M_N always lies in between R_N and L_N.

Now, as in Figure 20, consider the typical subinterval $[x_{i-1}, x_i]$ and its midpoint x_i^*. We let A, B, C, D, E, and F be the areas as shown in Figure 20. Note that, by the fact that x_i^* is the midpoint of the interval, $A = D + E$ and $F = B + C$. Let E_R represent the right endpoint approximation error ($= A + B + D$), let E_L represent the left endpoint approximation error ($= C + F + E$) and let E_M represent the midpoint approximation error ($= |B - E|$).

- If $B > E$, then $E_M = B - E$. In this case,

$$E_R - E_M = A + B + D - (B - E) = A + D + E > 0,$$

  so $E_R > E_M$, while

$$E_L - E_M = C + F + E - (B - E) = C + (B + C) + E - (B - E) = 2C + 2E > 0,$$

  so $E_L > E_M$. Therefore, the midpoint approximation is more accurate than either the left or the right endpoint approximation.

- If $B < E$, then $E_M = E - B$. In this case,

$$E_R - E_M = A + B + D - (E - B) = D + E + D - (E - B) = 2D + B > 0,$$

  so that $E_R > E_M$ while

$$E_L - E_M = C + F + E - (E - B) = C + F + B > 0,$$

  so $E_L > E_M$. Therefore, the midpoint approximation is more accurate than either the right or the left endpoint approximation.

- If $B = E$, the midpoint approximation is exactly equal to the area.

Hence, for $B < E$, $B > E$, or $B = E$, the midpoint approximation is more accurate than either the left endpoint or the right endpoint approximation.

5.2 The Definite Integral

Preliminary Questions

1. What is $\int_3^5 dx$ [the function is $f(x) = 1$]?

SOLUTION $\int_3^5 dx = \int_3^5 1 \cdot dx = 1(5 - 3) = 2$.

2. Let $I = \int_2^7 f(x)\,dx$, where $f(x)$ is continuous. State whether true or false:

(a) I is the area between the graph and the x-axis over $[2, 7]$.

(b) If $f(x) \geq 0$, then I is the area between the graph and the x-axis over $[2, 7]$.

(c) If $f(x) \leq 0$, then $-I$ is the area between the graph of $f(x)$ and the x-axis over $[2, 7]$.

SOLUTION

(a) False. $\int_a^b f(x)\,dx$ is the *signed* area between the graph and the x-axis.

(b) True.

(c) True.

3. Explain graphically: $\int_0^\pi \cos x\,dx = 0$.

SOLUTION Because $\cos(\pi - x) = -\cos x$, the "negative" area between the graph of $y = \cos x$ and the x-axis over $[\frac{\pi}{2}, \pi]$ exactly cancels the "positive" area between the graph and the x-axis over $[0, \frac{\pi}{2}]$.

4. Which is negative, $\int_{-1}^{-5} 8\,dx$ or $\int_{-5}^{-1} 8\,dx$?

SOLUTION Because $-5 - (-1) = -4$, $\int_{-1}^{-5} 8\,dx$ is negative.

Exercises

In Exercises 1–10, draw a graph of the signed area represented by the integral and compute it using geometry.

1. $\int_{-3}^{3} 2x\,dx$

SOLUTION The region bounded by the graph of $y = 2x$ and the x-axis over the interval $[-3, 3]$ consists of two right triangles. One has area $\frac{1}{2}(3)(6) = 9$ below the axis, and the other has area $\frac{1}{2}(3)(6) = 9$ above the axis. Hence,

$$\int_{-3}^{3} 2x\,dx = 9 - 9 = 0.$$

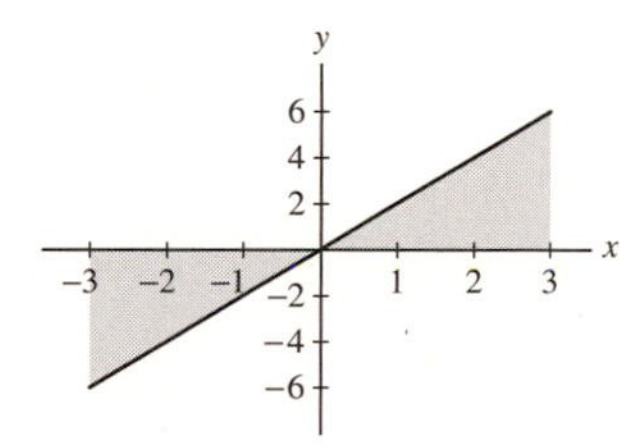

3. $\int_{-2}^{1} (3x + 4)\,dx$

SOLUTION The region bounded by the graph of $y = 3x + 4$ and the x-axis over the interval $[-2, 1]$ consists of two right triangles. One has area $\frac{1}{2}(\frac{2}{3})(2) = \frac{2}{3}$ below the axis, and the other has area $\frac{1}{2}(\frac{7}{3})(7) = \frac{49}{6}$ above the axis. Hence,

$$\int_{-2}^{1} (3x + 4)\,dx = \frac{49}{6} - \frac{2}{3} = \frac{15}{2}.$$

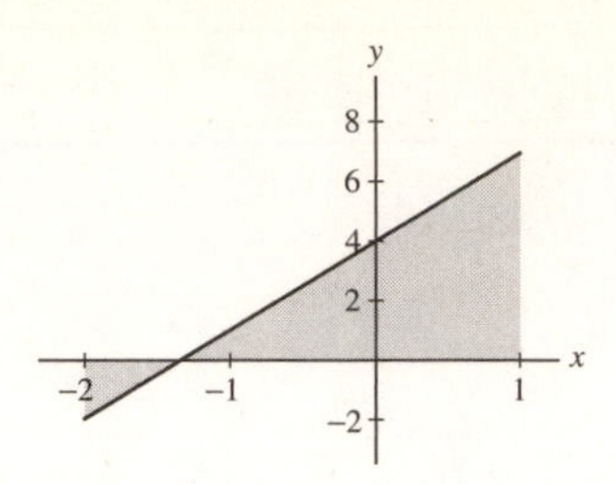

5. $\int_6^8 (7 - x)\,dx$

SOLUTION The region bounded by the graph of $y = 7 - x$ and the x-axis over the interval $[6, 8]$ consists of two right triangles. One triangle has area $\frac{1}{2}(1)(1) = \frac{1}{2}$ above the axis, and the other has area $\frac{1}{2}(1)(1) = \frac{1}{2}$ below the axis. Hence,

$$\int_6^8 (7 - x)\,dx = \frac{1}{2} - \frac{1}{2} = 0.$$

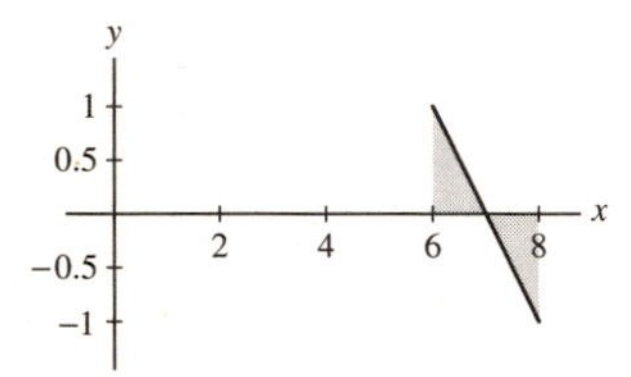

7. $\int_0^5 \sqrt{25 - x^2}\,dx$

SOLUTION The region bounded by the graph of $y = \sqrt{25 - x^2}$ and the x-axis over the interval $[0, 5]$ is one-quarter of a circle of radius 5. Hence,

$$\int_0^5 \sqrt{25 - x^2}\,dx = \frac{1}{4}\pi(5)^2 = \frac{25\pi}{4}.$$

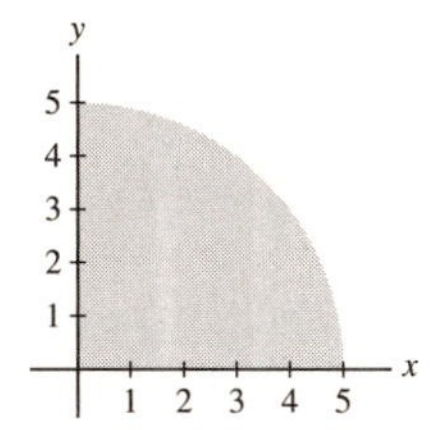

9. $\int_{-2}^2 (2 - |x|)\,dx$

SOLUTION The region bounded by the graph of $y = 2 - |x|$ and the x-axis over the interval $[-2, 2]$ is a triangle above the axis with base 4 and height 2. Consequently,

$$\int_{-2}^2 (2 - |x|)\,dx = \frac{1}{2}(2)(4) = 4.$$

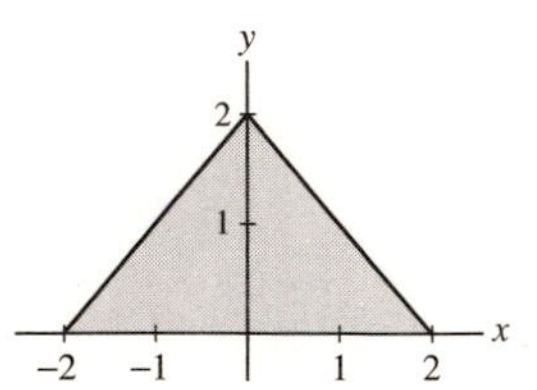

11. Calculate $\int_0^{10} (8 - x)\,dx$ in two ways:

(a) As the limit $\lim_{N\to\infty} R_N$

(b) By sketching the relevant signed area and using geometry

SOLUTION Let $f(x) = 8 - x$ over $[0, 10]$. Consider the integral $\int_0^{10} f(x)\,dx = \int_0^{10} (8 - x)\,dx$.

(a) Let N be a positive integer and set $a = 0$, $b = 10$, $\Delta x = (b - a)/N = 10/N$. Also, let $x_k = a + k\Delta x = 10k/N$, $k = 1, 2, \ldots, N$ be the right endpoints of the N subintervals of $[0, 10]$. Then

$$R_N = \Delta x \sum_{k=1}^{N} f(x_k) = \frac{10}{N} \sum_{k=1}^{N} \left(8 - \frac{10k}{N}\right) = \frac{10}{N}\left(8\left(\sum_{k=1}^{N} 1\right) - \frac{10}{N}\left(\sum_{k=1}^{N} k\right)\right)$$

$$= \frac{10}{N}\left(8N - \frac{10}{N}\left(\frac{N^2}{2} + \frac{N}{2}\right)\right) = 30 - \frac{50}{N}.$$

Hence $\lim_{N\to\infty} R_N = \lim_{N\to\infty} \left(30 - \frac{50}{N}\right) = 30.$

(b) The region bounded by the graph of $y = 8 - x$ and the x-axis over the interval $[0, 10]$ consists of two right triangles. One triangle has area $\frac{1}{2}(8)(8) = 32$ above the axis, and the other has area $\frac{1}{2}(2)(2) = 2$ below the axis. Hence,

$$\int_0^{10} (8 - x)\, dx = 32 - 2 = 30.$$

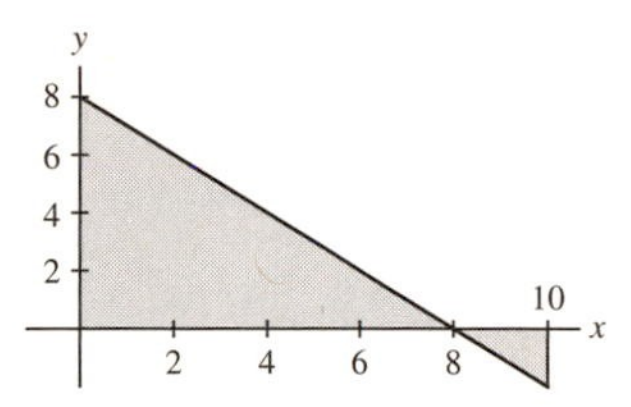

In Exercises 13 and 14, refer to Figure 14.

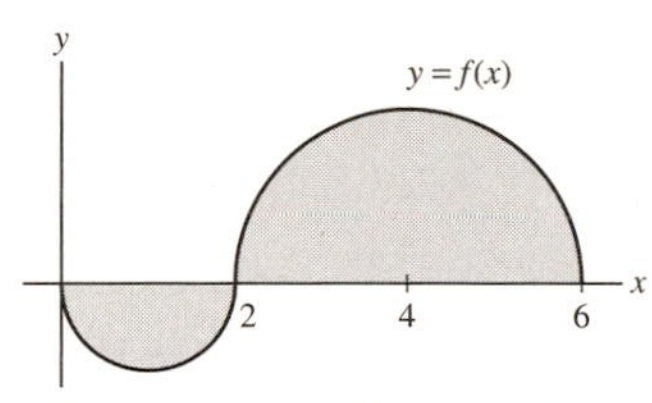

FIGURE 14 The two parts of the graph are semicircles.

13. Evaluate: (a) $\int_0^2 f(x)\, dx$ (b) $\int_0^6 f(x)\, dx$

SOLUTION Let $f(x)$ be given by Figure 14.

(a) The definite integral $\int_0^2 f(x)\, dx$ is the signed area of a semicircle of radius 1 which lies below the x-axis. Therefore,

$$\int_0^2 f(x)\, dx = -\frac{1}{2}\pi\,(1)^2 = -\frac{\pi}{2}.$$

(b) The definite integral $\int_0^6 f(x)\, dx$ is the signed area of a semicircle of radius 1 which lies below the x-axis and a semicircle of radius 2 which lies above the x-axis. Therefore,

$$\int_0^6 f(x)\, dx = \frac{1}{2}\pi\,(2)^2 - \frac{1}{2}\pi\,(1)^2 = \frac{3\pi}{2}.$$

In Exercises 15 and 16, refer to Figure 15.

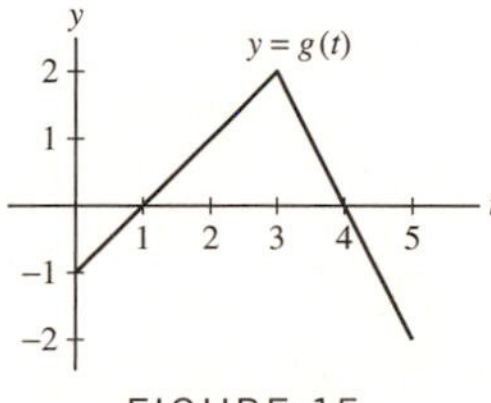

FIGURE 15

15. Evaluate $\int_0^3 g(t)\, dt$ and $\int_3^5 g(t)\, dt$.

SOLUTION

- The region bounded by the curve $y = g(x)$ and the x-axis over the interval $[0, 3]$ is comprised of two right triangles, one with area $\frac{1}{2}$ below the axis, and one with area 2 above the axis. The definite integral is therefore equal to $2 - \frac{1}{2} = \frac{3}{2}$.

- The region bounded by the curve $y = g(x)$ and the x-axis over the interval [3, 5] is comprised of another two right triangles, one with area 1 above the axis and one with area 1 below the axis. The definite integral is therefore equal to 0.

17. Describe the partition P and the set of sample points C for the Riemann sum shown in Figure 16. Compute the value of the Riemann sum.

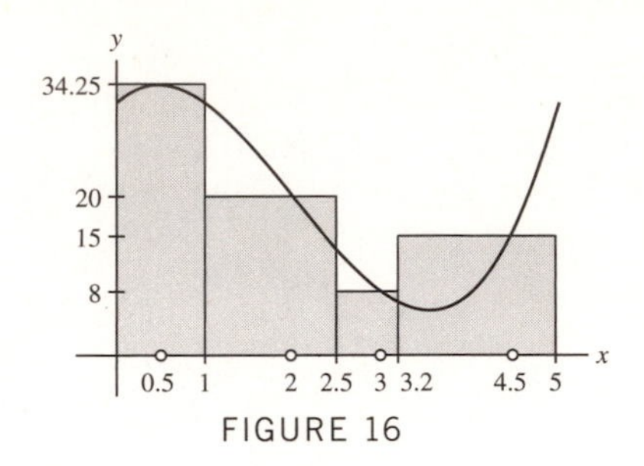

FIGURE 16

SOLUTION The partition P is defined by

$$x_0 = 0 \quad < \quad x_1 = 1 \quad < \quad x_2 = 2.5 \quad < \quad x_3 = 3.2 \quad < \quad x_4 = 5$$

The set of sample points is given by $C = \{c_1 = 0.5, c_2 = 2, c_3 = 3, c_4 = 4.5\}$. Finally, the value of the Riemann sum is

$$34.25(1 - 0) + 20(2.5 - 1) + 8(3.2 - 2.5) + 15(5 - 3.2) = 96.85.$$

In Exercises 19–22, calculate the Riemann sum $R(f, P, C)$ for the given function, partition, and choice of sample points. Also, sketch the graph of f and the rectangles corresponding to $R(f, P, C)$.

19. $f(x) = x, \quad P = \{1, 1.2, 1.5, 2\}, \quad C = \{1.1, 1.4, 1.9\}$

SOLUTION Let $f(x) = x$. With

$$P = \{x_0 = 1, x_1 = 1.2, x_2 = 1.5, x_3 = 2\} \qquad \text{and} \qquad C = \{c_1 = 1.1, c_2 = 1.4, c_3 = 1.9\},$$

we get

$$\begin{aligned} R(f, P, C) &= \Delta x_1 f(c_1) + \Delta x_2 f(c_2) + \Delta x_3 f(c_3) \\ &= (1.2 - 1)(1.1) + (1.5 - 1.2)(1.4) + (2 - 1.5)(1.9) = 1.59. \end{aligned}$$

Here is a sketch of the graph of f and the rectangles.

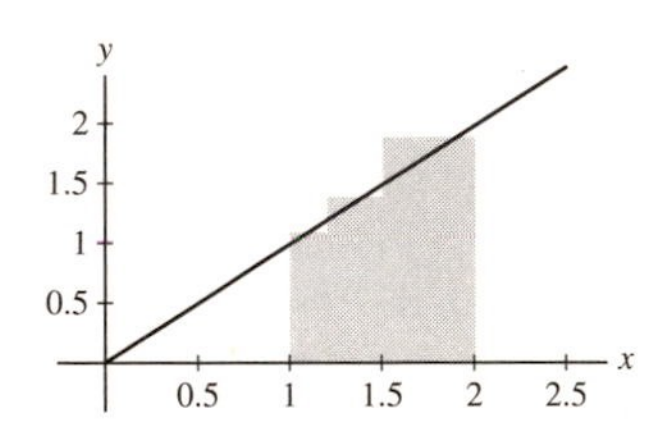

21. $f(x) = x^2 + x, \quad P = \{2, 3, 4.5, 5\}, \quad C = \{2, 3.5, 5\}$

SOLUTION Let $f(x) = x^2 + x$. With

$$P = \{x_0 = 2, x_1 = 3, x_3 = 4.5, x_4 = 5\} \qquad \text{and} \qquad C = \{c_1 = 2, c_2 = 3.5, c_3 = 5\},$$

we get

$$\begin{aligned} R(f, P, C) &= \Delta x_1 f(c_1) + \Delta x_2 f(c_2) + \Delta x_3 f(c_3) \\ &= (3 - 2)(6) + (4.5 - 3)(15.75) + (5 - 4.5)(30) = 44.625. \end{aligned}$$

Here is a sketch of the graph of f and the rectangles.

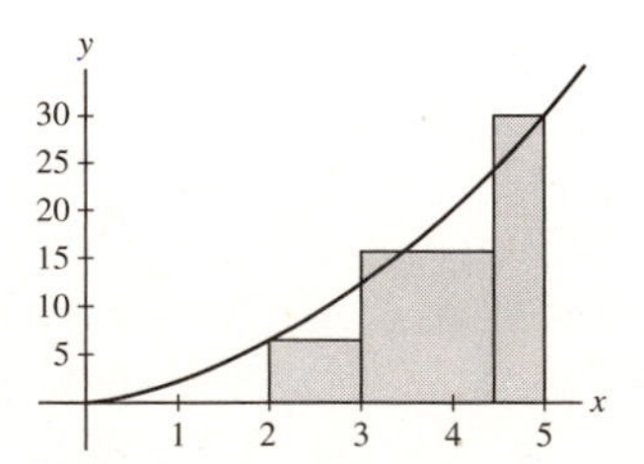

In Exercises 23–28, sketch the signed area represented by the integral. Indicate the regions of positive and negative area.

23. $\int_0^5 (4x - x^2)\,dx$

SOLUTION Here is a sketch of the signed area represented by the integral $\int_0^5 (4x - x^2)\,dx$.

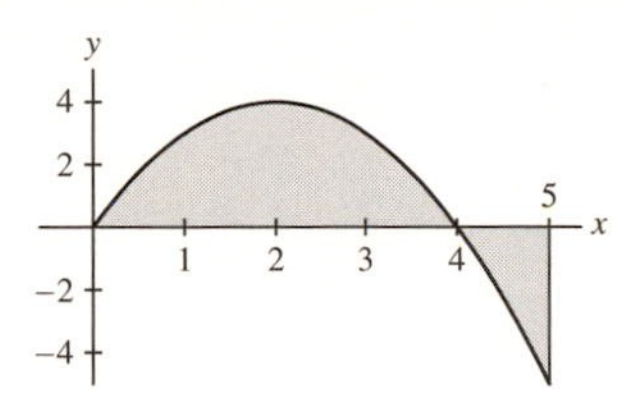

25. $\int_\pi^{2\pi} \sin x\,dx$

SOLUTION Here is a sketch of the signed area represented by the integral $\int_\pi^{2\pi} \sin x\,dx$.

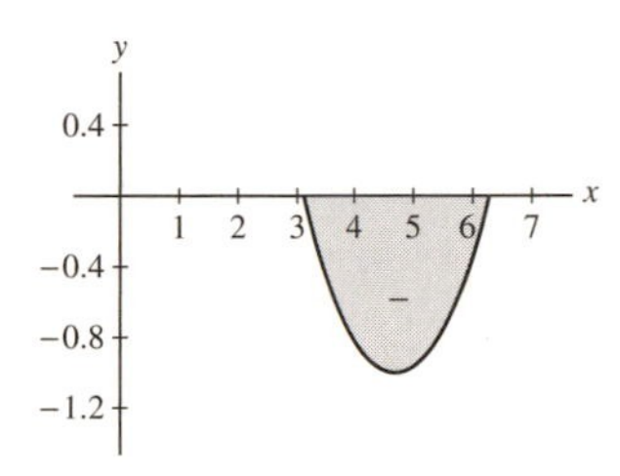

27. $\int_{1/2}^{2} \ln x\,dx$

SOLUTION Here is a sketch of the signed area represented by the integral $\int_{1/2}^{2} \ln x\,dx$.

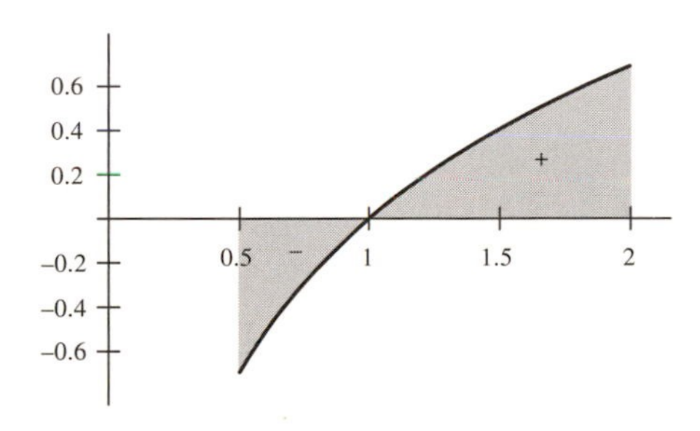

In Exercises 29–32, determine the sign of the integral without calculating it. Draw a graph if necessary.

29. $\int_{-2}^{1} x^4\,dx$

SOLUTION The integrand is always positive. The integral must therefore be positive, since the signed area has only positive part.

31. GU $\int_0^{2\pi} x \sin x\,dx$

SOLUTION As you can see from the graph below, the area below the axis is greater than the area above the axis. Thus, the definite integral is negative.

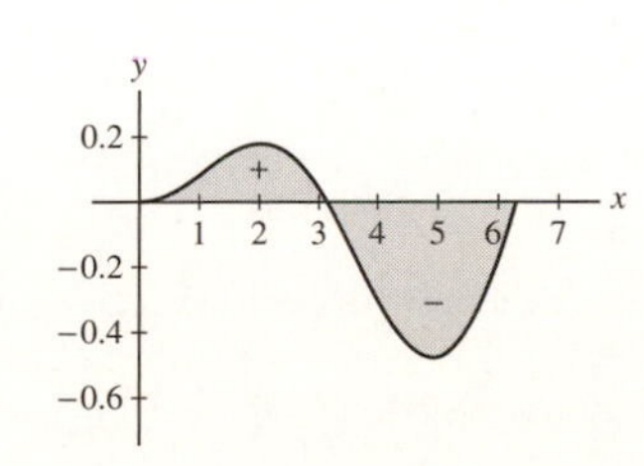

In Exercises 33–42, use properties of the integral and the formulas in the summary to calculate the integrals.

33. $\int_0^4 (6t-3)\,dt$

SOLUTION $\int_0^4 (6t-3)\,dt = 6\int_0^4 t\,dt - 3\int_0^4 1\,dt = 6\cdot\frac{1}{2}(4)^2 - 3(4-0) = 36.$

35. $\int_0^9 x^2\,dx$

SOLUTION By formula (5), $\int_0^9 x^2\,dx = \frac{1}{3}(9)^3 = 243.$

37. $\int_0^1 (u^2-2u)\,du$

SOLUTION

$$\int_0^1 (u^2-2u)\,du = \int_0^1 u^2\,du - 2\int_0^1 u\,du = \frac{1}{3}(1)^3 - 2\left(\frac{1}{2}\right)(1)^2 = \frac{1}{3} - 1 = -\frac{2}{3}.$$

39. $\int_{-3}^1 (7t^2+t+1)\,dt$

SOLUTION First, write

$$\int_{-3}^1 (7t^2+t+1)\,dt = \int_{-3}^0 (7t^2+t+1)\,dt + \int_0^1 (7t^2+t+1)\,dt$$
$$= -\int_0^{-3} (7t^2+t+1)\,dt + \int_0^1 (7t^2+t+1)\,dt$$

Then,

$$\int_{-3}^1 (7t^2+t+1)\,dt = -\left(7\cdot\frac{1}{3}(-3)^3 + \frac{1}{2}(-3)^2 - 3\right) + \left(7\cdot\frac{1}{3}1^3 + \frac{1}{2}1^2 + 1\right)$$
$$= -\left(-63 + \frac{9}{2} - 3\right) + \left(\frac{7}{3} + \frac{1}{2} + 1\right) = \frac{196}{3}.$$

41. $\int_{-a}^1 (x^2+x)\,dx$

SOLUTION First, $\int_0^b (x^2+x)\,dx = \int_0^b x^2\,dx + \int_0^b x\,dx = \frac{1}{3}b^3 + \frac{1}{2}b^2$. Therefore

$$\int_{-a}^1 (x^2+x)\,dx = \int_{-a}^0 (x^2+x)\,dx + \int_0^1 (x^2+x)\,dx = \int_0^1 (x^2+x)\,dx - \int_0^{-a} (x^2+x)\,dx$$
$$= \left(\frac{1}{3}\cdot 1^3 + \frac{1}{2}\cdot 1^2\right) - \left(\frac{1}{3}(-a)^3 + \frac{1}{2}(-a)^2\right) = \frac{1}{3}a^3 - \frac{1}{2}a^2 + \frac{5}{6}.$$

In Exercises 43–47, calculate the integral, assuming that

$$\int_0^5 f(x)\,dx = 5, \qquad \int_0^5 g(x)\,dx = 12$$

43. $\int_0^5 (f(x)+g(x))\,dx$

SOLUTION $\int_0^5 (f(x)+g(x))\,dx = \int_0^5 f(x)\,dx + \int_0^5 g(x)\,dx = 5 + 12 = 17.$

45. $\int_5^0 g(x)\,dx$

SOLUTION $\int_5^0 g(x)\,dx = -\int_0^5 g(x)\,dx = -12.$

47. Is it possible to calculate $\int_0^5 g(x)f(x)\,dx$ from the information given?

SOLUTION It is not possible to calculate $\int_0^5 g(x)f(x)\,dx$ from the information given.

In Exercises 49–54, evaluate the integral using the formulas in the summary and Eq. (9).

49. $\int_0^3 x^3\,dx$

SOLUTION By Eq. (9), $\int_0^3 x^3\,dx = \frac{3^4}{4} = \frac{81}{4}$.

51. $\int_0^3 (x - x^3)\,dx$

SOLUTION $\int_0^3 (x - x^3)\,dx = \int_0^3 x\,dx - \int_0^3 x^3\,dx = \frac{1}{2}3^2 - \frac{1}{4}3^4 = -\frac{63}{4}$.

53. $\int_0^1 (12x^3 + 24x^2 - 8x)\,dx$

SOLUTION

$$\begin{aligned}\int_0^1 (12x^3 + 24x^2 - 8x)\,dx &= 12\int_0^1 x^3\,dx + 24\int_0^1 x^2 - 8\int_0^1 x\,dx\\ &= 12\cdot\frac{1}{4}1^4 + 24\cdot\frac{1}{3}1^3 - 8\cdot\frac{1}{2}1^2\\ &= 3 + 8 - 4 = 7\end{aligned}$$

In Exercises 55–58, calculate the integral, assuming that

$$\int_0^1 f(x)\,dx = 1, \qquad \int_0^2 f(x)\,dx = 4, \qquad \int_1^4 f(x)\,dx = 7$$

55. $\int_0^4 f(x)\,dx$

SOLUTION $\int_0^4 f(x)\,dx = \int_0^1 f(x)\,dx + \int_1^4 f(x)\,dx = 1 + 7 = 8$.

57. $\int_4^1 f(x)\,dx$

SOLUTION $\int_4^1 f(x)\,dx = -\int_1^4 f(x)\,dx = -7$.

In Exercises 59–62, express each integral as a single integral.

59. $\int_0^3 f(x)\,dx + \int_3^7 f(x)\,dx$

SOLUTION $\int_0^3 f(x)\,dx + \int_3^7 f(x)\,dx = \int_0^7 f(x)\,dx$.

61. $\int_2^9 f(x)\,dx - \int_2^5 f(x)\,dx$

SOLUTION $\int_2^9 f(x)\,dx - \int_2^5 f(x)\,dx = \left(\int_2^5 f(x)\,dx + \int_5^9 f(x)\,dx\right) - \int_2^5 f(x)\,dx = \int_5^9 f(x)\,dx$.

In Exercises 63–66, calculate the integral, assuming that f is integrable and $\int_1^b f(x)\,dx = 1 - b^{-1}$ for all $b > 0$.

63. $\int_1^5 f(x)\,dx$

SOLUTION $\int_1^5 f(x)\,dx = 1 - 5^{-1} = \frac{4}{5}$.

65. $\int_1^6 (3f(x) - 4)\,dx$

SOLUTION $\int_1^6 (3f(x) - 4)\,dx = 3\int_1^6 f(x)\,dx - 4\int_1^6 1\,dx = 3(1 - 6^{-1}) - 4(6 - 1) = -\frac{35}{2}$.

67. Explain the difference in graphical interpretation between $\int_a^b f(x)\,dx$ and $\int_a^b |f(x)|\,dx$.

SOLUTION When $f(x)$ takes on both positive and negative values on $[a, b]$, $\int_a^b f(x)\,dx$ represents the signed area between $f(x)$ and the x-axis, whereas $\int_a^b |f(x)|\,dx$ represents the total (unsigned) area between $f(x)$ and the x-axis. Any negatively signed areas that were part of $\int_a^b f(x)\,dx$ are regarded as positive areas in $\int_a^b |f(x)|\,dx$. Here is a graphical example of this phenomenon.

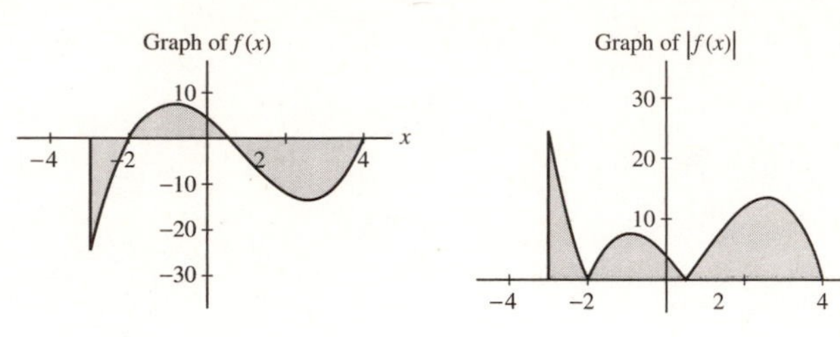

69. Let $f(x) = x$. Find an interval $[a, b]$ such that

$$\left|\int_a^b f(x)\,dx\right| = \frac{1}{2} \quad \text{and} \quad \int_a^b |f(x)|\,dx = \frac{3}{2}$$

SOLUTION If $a > 0$, then $f(x) \geq 0$ for all $x \in [a, b]$, so

$$\left|\int_a^b f(x)\,dx\right| = \int_a^b |f(x)|\,dx$$

by the previous exercise. We find a similar result if $b < 0$. Thus, we must have $a < 0$ and $b > 0$. Now,

$$\int_a^b |f(x)|\,dx = \frac{1}{2}a^2 + \frac{1}{2}b^2.$$

Because

$$\int_a^b f(x)\,dx = \frac{1}{2}b^2 - \frac{1}{2}a^2,$$

then

$$\left|\int_a^b f(x)\,dx\right| = \frac{1}{2}|b^2 - a^2|.$$

If $b^2 > a^2$, then

$$\frac{1}{2}a^2 + \frac{1}{2}b^2 = \frac{3}{2} \quad \text{and} \quad \frac{1}{2}(b^2 - a^2) = \frac{1}{2}$$

yield $a = -1$ and $b = \sqrt{2}$. On the other hand, if $b^2 < a^2$, then

$$\frac{1}{2}a^2 + \frac{1}{2}b^2 = \frac{3}{2} \quad \text{and} \quad \frac{1}{2}(a^2 - b^2) = \frac{1}{2}$$

yield $a = -\sqrt{2}$ and $b = 1$.

In Exercises 71–74, calculate the integral.

71. $\int_0^6 |3 - x|\,dx$

SOLUTION Over the interval, the region between the curve and the interval [0, 6] consists of two triangles above the x axis, each of which has height 3 and width 3, and so area $\frac{9}{2}$. The total area, hence the definite integral, is 9.

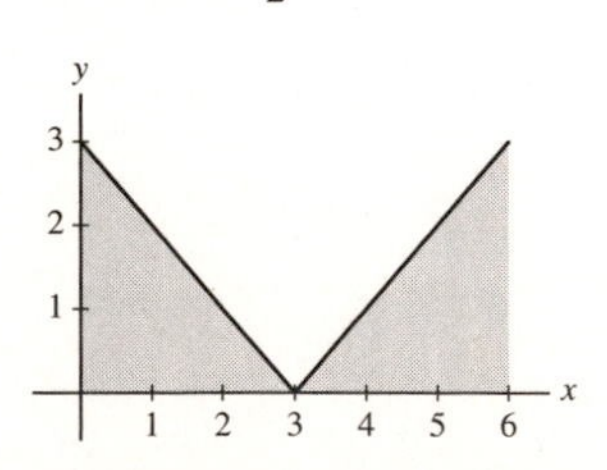

Alternately,

$$\int_0^6 |3-x|\,dx = \int_0^3 (3-x)\,dx + \int_3^6 (x-3)\,dx$$
$$= 3\int_0^3 dx - \int_0^3 x\,dx + \left(\int_0^6 x\,dx - \int_0^3 x\,dx\right) - 3\int_3^6 dx$$
$$= 9 - \frac{1}{2}3^2 + \frac{1}{2}6^2 - \frac{1}{2}3^2 - 9 = 9.$$

73. $\displaystyle\int_{-1}^{1} |x^3|\,dx$

SOLUTION

$$|x^3| = \begin{cases} x^3 & x \geq 0 \\ -x^3 & x < 0. \end{cases}$$

Therefore,

$$\int_{-1}^{1} |x^3|\,dx = \int_{-1}^{0} -x^3\,dx + \int_0^1 x^3\,dx = \int_0^{-1} x^3\,dx + \int_0^1 x^3\,dx = \frac{1}{4}(-1)^4 + \frac{1}{4}(1)^4 = \frac{1}{2}.$$

75. Use the Comparison Theorem to show that

$$\int_0^1 x^5\,dx \leq \int_0^1 x^4\,dx, \qquad \int_1^2 x^4\,dx \leq \int_1^2 x^5\,dx$$

SOLUTION On the interval $[0, 1]$, $x^5 \leq x^4$, so, by Theorem 5,

$$\int_0^1 x^5\,dx \leq \int_0^1 x^4\,dx.$$

On the other hand, $x^4 \leq x^5$ for $x \in [1, 2]$, so, by the same Theorem,

$$\int_1^2 x^4\,dx \leq \int_1^2 x^5\,dx.$$

77. Prove that $0.0198 \leq \int_{0.2}^{0.3} \sin x\,dx \leq 0.0296$. *Hint:* Show that $0.198 \leq \sin x \leq 0.296$ for x in $[0.2, 0.3]$.

SOLUTION For $0 \leq x \leq \frac{\pi}{6} \approx 0.52$, we have $\frac{d}{dx}(\sin x) = \cos x > 0$. Hence $\sin x$ is increasing on $[0.2, 0.3]$. Accordingly, for $0.2 \leq x \leq 0.3$, we have

$$m = 0.198 \leq 0.19867 \approx \sin 0.2 \leq \sin x \leq \sin 0.3 \approx 0.29552 \leq 0.296 = M$$

Therefore, by the Comparison Theorem, we have

$$0.0198 = m(0.3-0.2) = \int_{0.2}^{0.3} m\,dx \leq \int_{0.2}^{0.3} \sin x\,dx \leq \int_{0.2}^{0.3} M\,dx = M(0.3-0.2) = 0.0296.$$

79. Prove that $\displaystyle 0 \leq \int_{\pi/4}^{\pi/2} \frac{\sin x}{x}\,dx \leq \frac{\sqrt{2}}{2}$.

SOLUTION Let

$$f(x) = \frac{\sin x}{x}.$$

As we can see in the sketch below, $f(x)$ is decreasing on the interval $[\pi/4, \pi/2]$. Therefore $f(x) \leq f(\pi/4)$ for all x in $[\pi/4, \pi/2]$. $f(\pi/4) = \frac{2\sqrt{2}}{\pi}$, so:

$$\int_{\pi/4}^{\pi/2} \frac{\sin x}{x}\,dx \leq \int_{\pi/4}^{\pi/2} \frac{2\sqrt{2}}{\pi}\,dx = \frac{\pi}{4}\frac{2\sqrt{2}}{\pi} = \frac{\sqrt{2}}{2}.$$

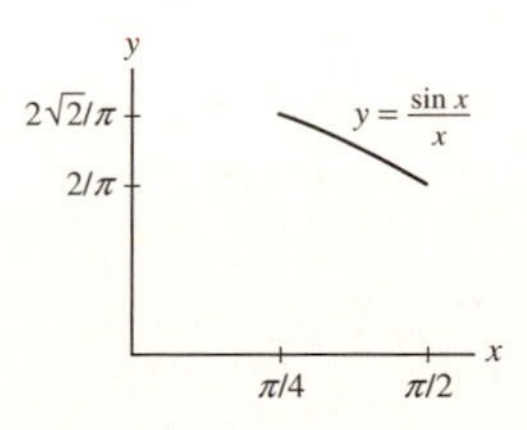

81. Suppose that $f(x) \le g(x)$ on $[a, b]$. By the Comparison Theorem, $\int_a^b f(x)\,dx \le \int_a^b g(x)\,dx$. Is it also true that $f'(x) \le g'(x)$ for $x \in [a, b]$? If not, give a counterexample.

SOLUTION The assertion $f'(x) \le g'(x)$ is false. Consider $a = 0$, $b = 1$, $f(x) = x$, $g(x) = 2$. $f(x) \le g(x)$ for all x in the interval $[0, 1]$, but $f'(x) = 1$ while $g'(x) = 0$ for all x.

Further Insights and Challenges

83. Explain graphically: If $f(x)$ is an odd function, then

$$\int_{-a}^{a} f(x)\,dx = 0.$$

SOLUTION If f is an odd function, then $f(-x) = -f(x)$ for all x. Accordingly, for every positively signed area in the right half-plane where f is above the x-axis, there is a corresponding negatively signed area in the left half-plane where f is below the x-axis. Similarly, for every negatively signed area in the right half-plane where f is below the x-axis, there is a corresponding positively signed area in the left half-plane where f is above the x-axis. We conclude that the net area between the graph of f and the x-axis over $[-a, a]$ is 0, since the positively signed areas and negatively signed areas cancel each other out exactly.

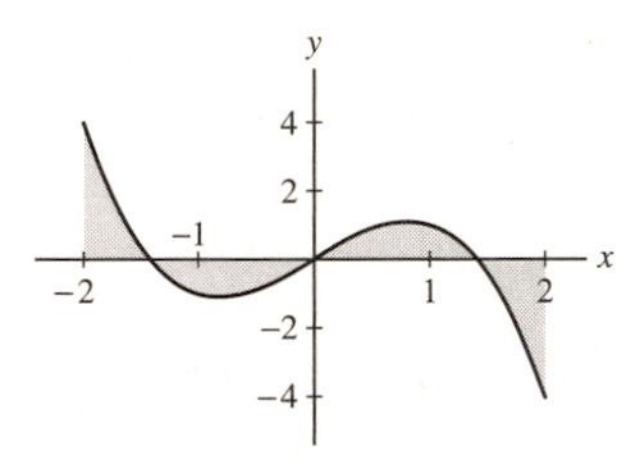

85. Let k and b be positive. Show, by comparing the right-endpoint approximations, that

$$\int_0^b x^k\,dx = b^{k+1}\int_0^1 x^k\,dx$$

SOLUTION Let k and b be any positive numbers. Let $f(x) = x^k$ on $[0, b]$. Since f is continuous, both $\int_0^b f(x)\,dx$ and $\int_0^1 f(x)\,dx$ exist. Let N be a positive integer and set $\Delta x = (b - 0)/N = b/N$. Let $x_j = a + j\Delta x = bj/N$, $j = 1, 2, \ldots, N$ be the right endpoints of the N subintervals of $[0, b]$. Then the right-endpoint approximation to $\int_0^b f(x)\,dx = \int_0^b x^k\,dx$ is

$$R_N = \Delta x \sum_{j=1}^{N} f(x_j) = \frac{b}{N}\sum_{j=1}^{N}\left(\frac{bj}{N}\right)^k = b^{k+1}\left(\frac{1}{N^{k+1}}\sum_{j=1}^{N} j^k\right).$$

In particular, if $b = 1$ above, then the right-endpoint approximation to $\int_0^1 f(x)\,dx = \int_0^1 x^k\,dx$ is

$$S_N = \Delta x \sum_{j=1}^{N} f(x_j) = \frac{1}{N}\sum_{j=1}^{N}\left(\frac{j}{N}\right)^k = \frac{1}{N^{k+1}}\sum_{j=1}^{N} j^k = \frac{1}{b^{k+1}}R_N$$

In other words, $R_N = b^{k+1}S_N$. Therefore,

$$\int_0^b x^k\,dx = \lim_{N\to\infty} R_N = \lim_{N\to\infty} b^{k+1}S_N = b^{k+1}\lim_{N\to\infty} S_N = b^{k+1}\int_0^1 x^k\,dx.$$

87. Suppose that f and g are continuous functions such that, *for all* a,

$$\int_{-a}^{a} f(x)\,dx = \int_{-a}^{a} g(x)\,dx$$

Give an *intuitive* argument showing that $f(0) = g(0)$. Explain your idea with a graph.

SOLUTION Let $c = -b$. Since $b < 0$, $c > 0$, so by Eq. (5),

$$\int_0^c x^2\,dx = \frac{1}{3}c^3.$$

Furthermore, x^2 is an even function, so symmetry of the areas gives

$$\int_{-c}^{0} x^2\,dx = \int_0^c x^2\,dx.$$

Finally,

$$\int_0^b x^2\,dx = \int_0^{-c} x^2\,dx = -\int_{-c}^0 x^2\,dx = -\int_0^c x^2\,dx = -\frac{1}{3}c^3 = \frac{1}{3}b^3.$$

5.3 The Fundamental Theorem of Calculus, Part I

Preliminary Questions

1. Suppose that $F'(x) = f(x)$ and $F(0) = 3$, $F(2) = 7$.

(a) What is the area under $y = f(x)$ over $[0, 2]$ if $f(x) \ge 0$?

(b) What is the graphical interpretation of $F(2) - F(0)$ if $f(x)$ takes on both positive and negative values?

SOLUTION

(a) If $f(x) \ge 0$ over $[0, 2]$, then the area under $y = f(x)$ is $F(2) - F(0) = 7 - 3 = 4$.

(b) If $f(x)$ takes on both positive and negative values, then $F(2) - F(0)$ gives the signed area between $y = f(x)$ and the x-axis.

2. Suppose that $f(x)$ is a *negative* function with antiderivative F such that $F(1) = 7$ and $F(3) = 4$. What is the area (a positive number) between the x-axis and the graph of $f(x)$ over $[1, 3]$?

SOLUTION $\int_1^3 f(x)\,dx$ represents the *signed* area bounded by the curve and the interval $[1, 3]$. Since $f(x)$ is negative on $[1, 3]$, $\int_1^3 f(x)\,dx$ is the negative of the area. Therefore, if A is the area between the x-axis and the graph of $f(x)$, we have:

$$A = -\int_1^3 f(x)\,dx = -\left(F(3) - F(1)\right) = -(4 - 7) = -(-3) = 3.$$

3. Are the following statements true or false? Explain.

(a) FTC I is valid only for positive functions.

(b) To use FTC I, you have to choose the right antiderivative.

(c) If you cannot find an antiderivative of $f(x)$, then the definite integral does not exist.

SOLUTION

(a) False. The FTC I is valid for continuous functions.

(b) False. The FTC I works for any antiderivative of the integrand.

(c) False. If you cannot find an antiderivative of the integrand, you cannot use the FTC I to evaluate the definite integral, but the definite integral may still exist.

4. Evaluate $\int_2^9 f'(x)\,dx$ where $f(x)$ is differentiable and $f(2) = f(9) = 4$.

SOLUTION Because f is differentiable, $\int_2^9 f'(x)\,dx = f(9) - f(2) = 4 - 4 = 0$.

Exercises

In Exercises 1–4, sketch the region under the graph of the function and find its area using FTC I.

1. $f(x) = x^2$, $[0, 1]$

SOLUTION

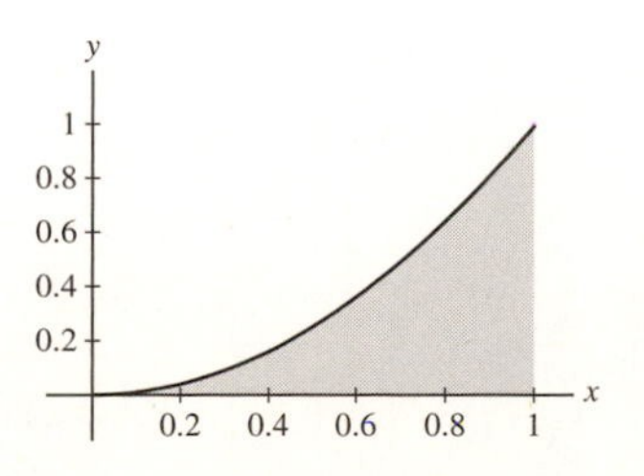

We have the area

$$A = \int_0^1 x^2\,dx = \frac{1}{3}x^3\Big|_0^1 = \frac{1}{3}.$$

3. $f(x) = x^{-2}$, $[1, 2]$

SOLUTION

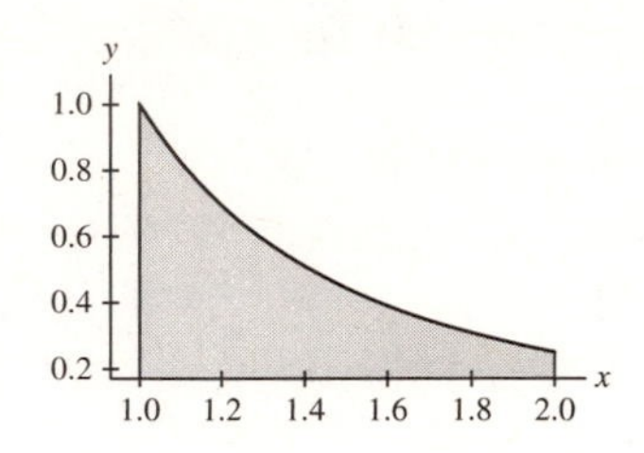

We have the area

$$A = \int_1^2 x^{-2}\,dx = \frac{x^{-1}}{-1}\Big|_1^2 = -\frac{1}{2} + 1 = \frac{1}{2}.$$

In Exercises 5–42, evaluate the integral using FTC I.

5. $\displaystyle\int_3^6 x\,dx$

SOLUTION $\displaystyle\int_3^6 x\,dx = \frac{1}{2}x^2\Big|_3^6 = \frac{1}{2}(6)^2 - \frac{1}{2}(3)^2 = \frac{27}{2}.$

7. $\displaystyle\int_0^1 (4x - 9x^2)\,dx$

SOLUTION $\displaystyle\int_0^1 (4x - 9x^2)\,dx = (2x^2 - 3x^3)\Big|_0^1 = (2-3) - (0-0) = -1.$

9. $\displaystyle\int_0^2 (12x^5 + 3x^2 - 4x)\,dx$

SOLUTION $\displaystyle\int_0^2 (12x^5 + 3x^2 - 4x)\,dx = (2x^6 + x^3 - 2x^2)\Big|_0^2 = (128 + 8 - 8) - (0 + 0 - 0) = 128.$

11. $\displaystyle\int_3^0 (2t^3 - 6t^2)\,dt$

SOLUTION $\displaystyle\int_3^0 (2t^3 - 6t^2)\,dt = \left(\frac{1}{2}t^4 - 2t^3\right)\Big|_3^0 = (0-0) - \left(\frac{81}{2} - 54\right) = \frac{27}{2}.$

13. $\displaystyle\int_0^4 \sqrt{y}\,dy$

SOLUTION $\displaystyle\int_0^4 \sqrt{y}\,dy = \int_0^4 y^{1/2}\,dy = \frac{2}{3}y^{3/2}\Big|_0^4 = \frac{2}{3}(4)^{3/2} - \frac{2}{3}(0)^{3/2} = \frac{16}{3}.$

15. $\displaystyle\int_{1/16}^1 t^{1/4}\,dt$

SOLUTION $\displaystyle\int_{1/16}^1 t^{1/4}\,dt = \frac{4}{5}t^{5/4}\Big|_{1/16}^1 = \frac{4}{5} - \frac{1}{40} = \frac{31}{40}.$

17. $\displaystyle\int_1^3 \frac{dt}{t^2}$

SOLUTION $\displaystyle\int_1^3 \frac{dt}{t^2} = \int_1^3 t^{-2}\,dt = -t^{-1}\Big|_1^3 = -\frac{1}{3} + 1 = \frac{2}{3}.$

19. $\displaystyle\int_{1/2}^1 \frac{8}{x^3}\,dx$

SOLUTION $\displaystyle\int_{1/2}^1 \frac{8}{x^3}\,dx = \int_{1/2}^1 8x^{-3}\,dx = -4x^{-2}\Big|_{1/2}^1 = -4 + 16 = 12.$

21. $\int_1^2 (x^2 - x^{-2})\,dx$

SOLUTION $\int_1^2 (x^2 - x^{-2})\,dx = \left(\frac{1}{3}x^3 + x^{-1}\right)\Big|_1^2 = \left(\frac{8}{3} + \frac{1}{2}\right) - \left(\frac{1}{3} + 1\right) = \frac{11}{6}.$

23. $\int_1^{27} \frac{t+1}{\sqrt{t}}\,dt$

SOLUTION

$$\int_1^{27} \frac{t+1}{\sqrt{t}}\,dt = \int_1^{27} (t^{1/2} + t^{-1/2})\,dt = \left(\frac{2}{3}t^{3/2} + 2t^{1/2}\right)\Big|_1^{27}$$
$$= \left(\frac{2}{3}(81\sqrt{3}) + 6\sqrt{3}\right) - \left(\frac{2}{3} + 2\right) = 60\sqrt{3} - \frac{8}{3}.$$

25. $\int_{\pi/4}^{3\pi/4} \sin\theta\,d\theta$

SOLUTION $\int_{\pi/4}^{3\pi/4} \sin\theta\,d\theta = -\cos\theta\Big|_{\pi/4}^{3\pi/4} = \frac{\sqrt{2}}{2} + \frac{\sqrt{2}}{2} = \sqrt{2}.$

27. $\int_0^{\pi/2} \cos\left(\frac{1}{3}\theta\right)d\theta$

SOLUTION $\int_0^{\pi/2} \cos\left(\frac{1}{3}\theta\right)d\theta = 3\sin\left(\frac{1}{3}\theta\right)\Big|_0^{\pi/2} = \frac{3}{2}.$

29. $\int_0^{\pi/6} \sec^2\left(3t - \frac{\pi}{6}\right)dt$

SOLUTION $\int_0^{\pi/6} \sec^2\left(3t - \frac{\pi}{6}\right)dt = \frac{1}{3}\tan\left(3t - \frac{\pi}{6}\right)\Big|_0^{\pi/6} = \frac{1}{3}\left(\sqrt{3} + \frac{1}{\sqrt{3}}\right) = \frac{4}{3\sqrt{3}}.$

31. $\int_{\pi/20}^{\pi/10} \csc 5x \cot 5x\,dx$

SOLUTION $\int_{\pi/20}^{\pi/10} \csc 5x \cot 5x\,dx = -\frac{1}{5}\csc 5x\Big|_{\pi/20}^{\pi/10} = -\frac{1}{5}\left(1 - \sqrt{2}\right) = \frac{1}{5}(\sqrt{2} - 1).$

33. $\int_0^1 e^x\,dx$

SOLUTION $\int_0^1 e^x\,dx = e^x\Big|_0^1 = e - 1.$

35. $\int_0^3 e^{1-6t}\,dt$

SOLUTION $\int_0^3 e^{1-6t}\,dt = -\frac{1}{6}e^{1-6t}\Big|_0^3 = -\frac{1}{6}e^{-17} + \frac{1}{6}e = \frac{1}{6}(e - e^{-17}).$

37. $\int_2^{10} \frac{dx}{x}$

SOLUTION $\int_2^{10} \frac{dx}{x} = \ln|x|\Big|_2^{10} = \ln 10 - \ln 2 = \ln 5.$

39. $\int_0^1 \frac{dt}{t+1}$

SOLUTION $\int_0^1 \frac{dt}{t+1} = \ln|t+1|\Big|_0^1 = \ln 2 - \ln 1 = \ln 2.$

41. $\int_{-2}^0 (3x - 9e^{3x})\,dx$

SOLUTION $\int_{-2}^0 (3x - 9e^{3x})\,dx = \left(\frac{3}{2}x^2 - 3e^{3x}\right)\Big|_{-2}^0 = (0 - 3) - (6 - 3e^{-6}) = 3e^{-6} - 9.$

In Exercises 43–48, write the integral as a sum of integrals without absolute values and evaluate.

43. $\int_{-2}^{1} |x|\, dx$

SOLUTION

$$\int_{-2}^{1} |x|\, dx = \int_{-2}^{0} (-x)\, dx + \int_{0}^{1} x\, dx = -\frac{1}{2}x^2\Big|_{-2}^{0} + \frac{1}{2}x^2\Big|_{0}^{1} = 0 - \left(-\frac{1}{2}(4)\right) + \frac{1}{2} = \frac{5}{2}.$$

45. $\int_{-2}^{3} |x^3|\, dx$

SOLUTION

$$\int_{-2}^{3} |x^3|\, dx = \int_{-2}^{0} (-x^3)\, dx + \int_{0}^{3} x^3\, dx = -\frac{1}{4}x^4\Big|_{-2}^{0} + \frac{1}{4}x^4\Big|_{0}^{3}$$
$$= 0 + \frac{1}{4}(-2)^4 + \frac{1}{4}3^4 - 0 = \frac{97}{4}.$$

47. $\int_{0}^{\pi} |\cos x|\, dx$

SOLUTION

$$\int_{0}^{\pi} |\cos x|\, dx = \int_{0}^{\pi/2} \cos x\, dx + \int_{\pi/2}^{\pi} (-\cos x)\, dx = \sin x\Big|_{0}^{\pi/2} - \sin x\Big|_{\pi/2}^{\pi} = 1 - 0 - (-1 - 0) = 2.$$

In Exercises 49–54, evaluate the integral in terms of the constants.

49. $\int_{1}^{b} x^3\, dx$

SOLUTION $\int_{1}^{b} x^3\, dx = \frac{1}{4}x^4\Big|_{1}^{b} = \frac{1}{4}b^4 - \frac{1}{4}(1)^4 = \frac{1}{4}\left(b^4 - 1\right)$ for any number b.

51. $\int_{1}^{b} x^5\, dx$

SOLUTION $\int_{1}^{b} x^5\, dx = \frac{1}{6}x^6\Big|_{1}^{b} = \frac{1}{6}b^6 - \frac{1}{6}(1)^6 = \frac{1}{6}(b^6 - 1)$ for any number b.

53. $\int_{a}^{5a} \frac{dx}{x}$

SOLUTION $\int_{a}^{5a} \frac{dx}{x} = \ln|x|\Big|_{a}^{5a} = \ln|5a| - \ln|a| = \ln 5.$

55. Calculate $\int_{-2}^{3} f(x)\, dx$, where

$$f(x) = \begin{cases} 12 - x^2 & \text{for } x \le 2 \\ x^3 & \text{for } x > 2 \end{cases}$$

SOLUTION

$$\int_{-2}^{3} f(x)\, dx = \int_{-2}^{2} f(x)\, dx + \int_{2}^{3} f(x)\, dx = \int_{-2}^{2} (12 - x^2)\, dx + \int_{2}^{3} x^3\, dx$$
$$= \left(12x - \frac{1}{3}x^3\right)\Big|_{-2}^{2} + \frac{1}{4}x^4\Big|_{2}^{3}$$
$$= \left(12(2) - \frac{1}{3}2^3\right) - \left(12(-2) - \frac{1}{3}(-2)^3\right) + \frac{1}{4}3^4 - \frac{1}{4}2^4$$
$$= \frac{128}{3} + \frac{65}{4} = \frac{707}{12}.$$

57. Use FTC I to show that $\int_{-1}^{1} x^n\,dx = 0$ if n is an odd whole number. Explain graphically.

SOLUTION We have

$$\int_{-1}^{1} x^n\,dx = \left.\frac{x^{n+1}}{n+1}\right|_{-1}^{1} = \frac{(1)^{n+1}}{n+1} - \frac{(-1)^{n+1}}{n+1}.$$

Because n is odd, $n+1$ is even, which means that $(-1)^{n+1} = (1)^{n+1} = 1$. Hence

$$\frac{(1)^{n+1}}{n+1} - \frac{(-1)^{n+1}}{n+1} = \frac{1}{n+1} - \frac{1}{n+1} = 0.$$

Graphically speaking, for an odd function such as x^3 shown here, the positively signed area from $x = 0$ to $x = 1$ cancels the negatively signed area from $x = -1$ to $x = 0$.

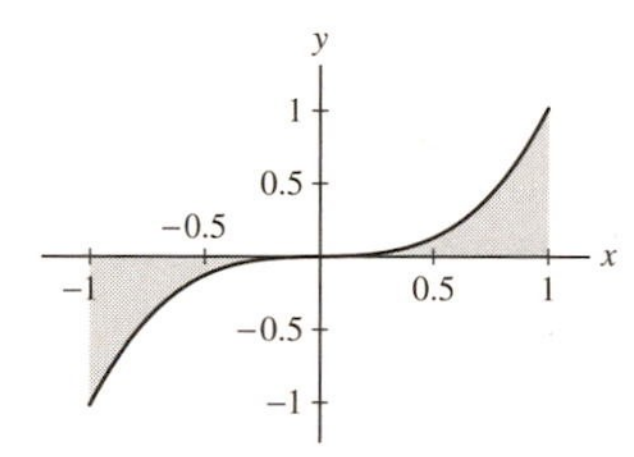

59. Calculate $F(4)$ given that $F(1) = 3$ and $F'(x) = x^2$. *Hint:* Express $F(4) - F(1)$ as a definite integral.

SOLUTION By FTC I,

$$F(4) - F(1) = \int_1^4 x^2\,dx = \frac{4^3 - 1^3}{3} = 21$$

Therefore $F(4) = F(1) + 21 = 3 + 21 = 24$.

61. Does $\int_0^1 x^n\,dx$ get larger or smaller as n increases? Explain graphically.

SOLUTION Let $n \ge 0$ and consider $\int_0^1 x^n\,dx$. (Note: for $n < 0$ the integrand $x^n \to \infty$ as $x \to 0+$, so we exclude this possibility.) Now

$$\int_0^1 x^n\,dx = \left.\left(\frac{1}{n+1}x^{n+1}\right)\right|_0^1 = \left(\frac{1}{n+1}(1)^{n+1}\right) - \left(\frac{1}{n+1}(0)^{n+1}\right) = \frac{1}{n+1},$$

which decreases as n increases. Recall that $\int_0^1 x^n\,dx$ represents the area between the positive curve $f(x) = x^n$ and the x-axis over the interval $[0, 1]$. Accordingly, this area gets smaller as n gets larger. This is readily evident in the following graph, which shows curves for several values of n.

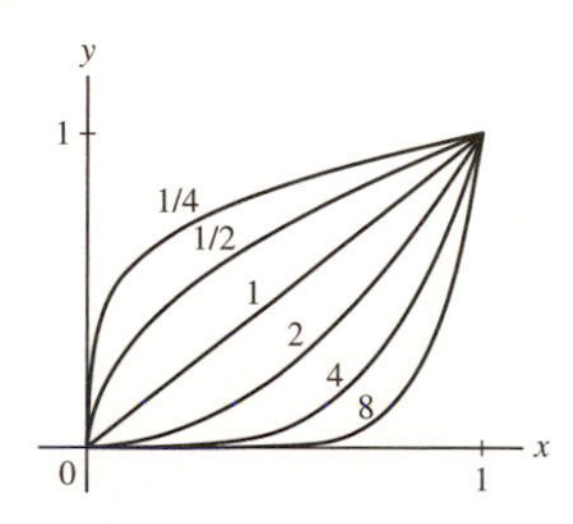

Further Insights and Challenges

63. Prove a famous result of Archimedes (generalizing Exercise 62): For $r < s$, the area of the shaded region in Figure 9 is equal to four-thirds the area of triangle $\triangle ACE$, where C is the point on the parabola at which the tangent line is parallel to secant line $\overline{AE}$.

(a) Show that C has x-coordinate $(r+s)/2$.

(b) Show that $ABDE$ has area $(s-r)^3/4$ by viewing it as a parallelogram of height $s - r$ and base of length $\overline{CF}$.

(c) Show that $\triangle ACE$ has area $(s-r)^3/8$ by observing that it has the same base and height as the parallelogram.

(d) Compute the shaded area as the area under the graph minus the area of a trapezoid, and prove Archimedes' result.

FIGURE 9 Graph of $f(x) = (x-a)(b-x)$.

SOLUTION

(a) The slope of the secant line $\overline{AE}$ is

$$\frac{f(s)-f(r)}{s-r} = \frac{(s-a)(b-s)-(r-a)(b-r)}{s-r} = a+b-(r+s)$$

and the slope of the tangent line along the parabola is

$$f'(x) = a+b-2x.$$

If C is the point on the parabola at which the tangent line is parallel to the secant line $\overline{AE}$, then its x-coordinate must satisfy

$$a+b-2x = a+b-(r+s) \qquad \text{or} \qquad x = \frac{r+s}{2}.$$

(b) Parallelogram $ABDE$ has height $s-r$ and base of length $\overline{CF}$. Since the equation of the secant line $\overline{AE}$ is

$$y = [a+b-(r+s)](x-r)+(r-a)(b-r),$$

the length of the segment $\overline{CF}$ is

$$\left(\frac{r+s}{2}-a\right)\left(b-\frac{r+s}{2}\right)-[a+b-(r+s)]\left(\frac{r+s}{2}-r\right)-(r-a)(b-r) = \frac{(s-r)^2}{4}.$$

Thus, the area of $ABDE$ is $\frac{(s-r)^3}{4}$.

(c) Triangle ACE is comprised of ΔACF and ΔCEF. Each of these smaller triangles has height $\frac{s-r}{2}$ and base of length $\frac{(s-r)^2}{4}$. Thus, the area of ΔACE is

$$\frac{1}{2}\frac{s-r}{2}\cdot\frac{(s-r)^2}{4}+\frac{1}{2}\frac{s-r}{2}\cdot\frac{(s-r)^2}{4} = \frac{(s-r)^3}{8}.$$

(d) The area under the graph of the parabola between $x=r$ and $x=s$ is

$$\begin{aligned}\int_r^s (x-a)(b-x)\,dx &= \left(-abx+\frac{1}{2}(a+b)x^2-\frac{1}{3}x^3\right)\Bigg|_r^s \\ &= -abs+\frac{1}{2}(a+b)s^2-\frac{1}{3}s^3+abr-\frac{1}{2}(a+b)r^2+\frac{1}{3}r^3 \\ &= ab(r-s)+\frac{1}{2}(a+b)(s-r)(s+r)+\frac{1}{3}(r-s)(r^2+rs+s^2),\end{aligned}$$

while the area of the trapezoid under the shaded region is

$$\begin{aligned}&\frac{1}{2}(s-r)\left[(s-a)(b-s)+(r-a)(b-r)\right] \\ &\quad = \frac{1}{2}(s-r)\left[-2ab+(a+b)(r+s)-r^2-s^2\right] \\ &\quad = ab(r-s)+\frac{1}{2}(a+b)(s-r)(r+s)+\frac{1}{2}(r-s)(r^2+s^2).\end{aligned}$$

Thus, the area of the shaded region is

$$(r-s)\left(\frac{1}{3}r^2+\frac{1}{3}rs+\frac{1}{3}s^2-\frac{1}{2}r^2-\frac{1}{2}s^2\right) = (s-r)\left(\frac{1}{6}r^2-\frac{1}{3}rs+\frac{1}{6}s^2\right) = \frac{1}{6}(s-r)^3,$$

which is four-thirds the area of the triangle ACE.

65. Use the method of Exercise 64 to prove that

$$1 - \frac{x^2}{2} \le \cos x \le 1 - \frac{x^2}{2} + \frac{x^4}{24}$$

$$x - \frac{x^3}{6} \le \sin x \le x - \frac{x^3}{6} + \frac{x^5}{120} \qquad (\text{for } x \ge 0)$$

Verify these inequalities for $x = 0.1$. Why have we specified $x \ge 0$ for $\sin x$ but not for $\cos x$?

SOLUTION By Exercise 64, $t - \frac{1}{6}t^3 \le \sin t \le t$ for $t > 0$. Integrating this inequality over the interval $[0, x]$, and then solving for $\cos x$, yields:

$$\frac{1}{2}x^2 - \frac{1}{24}x^4 \le 1 - \cos x \le \frac{1}{2}x^2$$

$$1 - \frac{1}{2}x^2 \le \cos x \le 1 - \frac{1}{2}x^2 + \frac{1}{24}x^4.$$

These inequalities apply for $x \ge 0$. Since $\cos x$, $1 - \frac{x^2}{2}$, and $1 - \frac{x^2}{2} + \frac{x^4}{24}$ are all even functions, they also apply for $x \le 0$.

Having established that

$$1 - \frac{t^2}{2} \le \cos t \le 1 - \frac{t^2}{2} + \frac{t^4}{24},$$

for all $t \ge 0$, we integrate over the interval $[0, x]$, to obtain:

$$x - \frac{x^3}{6} \le \sin x \le x - \frac{x^3}{6} + \frac{x^5}{120}.$$

The functions $\sin x$, $x - \frac{1}{6}x^3$ and $x - \frac{1}{6}x^3 + \frac{1}{120}x^5$ are all odd functions, so the inequalities are reversed for $x < 0$.

Evaluating these inequalities at $x = 0.1$ yields

$$0.995000000 \le 0.995004165 \le 0.995004167$$

$$0.0998333333 \le 0.0998334166 \le 0.0998334167,$$

both of which are true.

67. Use FTC I to prove that if $|f'(x)| \le K$ for $x \in [a, b]$, then $|f(x) - f(a)| \le K|x - a|$ for $x \in [a, b]$.

SOLUTION Let $a > b$ be real numbers, and let $f(x)$ be such that $|f'(x)| \le K$ for $x \in [a, b]$. By FTC,

$$\int_a^x f'(t)\,dt = f(x) - f(a).$$

Since $f'(x) \ge -K$ for all $x \in [a, b]$, we get:

$$f(x) - f(a) = \int_a^x f'(t)\,dt \ge -K(x - a).$$

Since $f'(x) \le K$ for all $x \in [a, b]$, we get:

$$f(x) - f(a) = \int_a^x f'(t)\,dt \le K(x - a).$$

Combining these two inequalities yields

$$-K(x - a) \le f(x) - f(a) \le K(x - a),$$

so that, by definition,

$$|f(x) - f(a)| \le K|x - a|.$$

5.4 The Fundamental Theorem of Calculus, Part II

Preliminary Questions

1. Let $G(x) = \int_4^x \sqrt{t^3 + 1}\,dt$.

(a) Is the FTC needed to calculate $G(4)$?

(b) Is the FTC needed to calculate $G'(4)$?

SOLUTION

(a) No. $G(4) = \int_4^4 \sqrt{t^3 + 1}\,dt = 0$.

(b) Yes. By the FTC II, $G'(x) = \sqrt{x^3 + 1}$, so $G'(4) = \sqrt{65}$.

2. Which of the following is an antiderivative $F(x)$ of $f(x) = x^2$ satisfying $F(2) = 0$?

(a) $\int_2^x 2t\,dt$ **(b)** $\int_0^2 t^2\,dt$ **(c)** $\int_2^x t^2\,dt$

SOLUTION The correct answer is **(c)**: $\int_2^x t^2\,dt$.

3. Does every continuous function have an antiderivative? Explain.

SOLUTION Yes. All continuous functions have an antiderivative, namely $\int_a^x f(t)\,dt$.

4. Let $G(x) = \int_4^{x^3} \sin t\,dt$. Which of the following statements are correct?

(a) $G(x)$ is the composite function $\sin(x^3)$.

(b) $G(x)$ is the composite function $A(x^3)$, where

$$A(x) = \int_4^x \sin(t)\,dt$$

(c) $G(x)$ is too complicated to differentiate.

(d) The Product Rule is used to differentiate $G(x)$.

(e) The Chain Rule is used to differentiate $G(x)$.

(f) $G'(x) = 3x^2 \sin(x^3)$.

SOLUTION Statements **(b)**, **(e)**, and **(f)** are correct.

Exercises

1. Write the area function of $f(x) = 2x + 4$ with lower limit $a = -2$ as an integral and find a formula for it.

SOLUTION Let $f(x) = 2x + 4$. The area function with lower limit $a = -2$ is

$$A(x) = \int_a^x f(t)\,dt = \int_{-2}^x (2t + 4)\,dt.$$

Carrying out the integration, we find

$$\int_{-2}^x (2t + 4)\,dt = (t^2 + 4t)\Big|_{-2}^x = (x^2 + 4x) - ((-2)^2 + 4(-2)) = x^2 + 4x + 4$$

or $(x + 2)^2$. Therefore, $A(x) = (x + 2)^2$.

3. Let $G(x) = \int_1^x (t^2 - 2)\,dt$. Calculate $G(1)$, $G'(1)$ and $G'(2)$. Then find a formula for $G(x)$.

SOLUTION Let $G(x) = \int_1^x (t^2 - 2)\,dt$. Then $G(1) = \int_1^1 (t^2 - 2)\,dt = 0$. Moreover, $G'(x) = x^2 - 2$, so that $G'(1) = -1$ and $G'(2) = 2$. Finally,

$$G(x) = \int_1^x (t^2 - 2)\,dt = \left(\frac{1}{3}t^3 - 2t\right)\Big|_1^x = \left(\frac{1}{3}x^3 - 2x\right) - \left(\frac{1}{3}(1)^3 - 2(1)\right) = \frac{1}{3}x^3 - 2x + \frac{5}{3}.$$

5. Find $G(1)$, $G'(0)$, and $G'(\pi/4)$, where $G(x) = \int_1^x \tan t\,dt$.

SOLUTION By definition, $G(1) = \int_1^1 \tan t\,dt = 0$. By FTC, $G'(x) = \tan x$, so that $G'(0) = \tan 0 = 0$ and $G'(\frac{\pi}{4}) = \tan\frac{\pi}{4} = 1$.

In Exercises 7–16, find formulas for the functions represented by the integrals.

7. $\int_2^x u^4\,du$

SOLUTION $F(x) = \int_2^x u^4\,du = \frac{1}{5}u^5\Big|_2^x = \frac{1}{5}x^5 - \frac{32}{5}.$

9. $\int_0^x \sin u\,du$

SOLUTION $F(x) = \int_0^x \sin u\,du = (-\cos u)\Big|_0^x = 1 - \cos x.$

11. $\int_4^x e^{3u}\,du$

SOLUTION $F(x) = \int_4^x e^{3u}\,du = \frac{1}{3}e^{3u}\Big|_4^x = \frac{1}{3}e^{3x} - \frac{1}{3}e^{12}.$

13. $\int_{1}^{x^2} t\,dt$

SOLUTION $F(x) = \int_{1}^{x^2} t\,dt = \frac{1}{2}t^2\Big|_1^{x^2} = \frac{1}{2}x^4 - \frac{1}{2}.$

15. $\int_{3x}^{9x+2} e^{-u}\,du$

SOLUTION $F(x) = \int_{3x}^{9x+2} e^{-u}\,du = -e^{-u}\Big|_{3x}^{9x+2} = -e^{-9x-2} + e^{-3x}.$

In Exercises 17–20, express the antiderivative $F(x)$ of $f(x)$ satisfying the given initial condition as an integral.

17. $f(x) = \sqrt{x^3+1}, \quad F(5) = 0$

SOLUTION The antiderivative $F(x)$ of $\sqrt{x^3+1}$ satisfying $F(5) = 0$ is $F(x) = \int_5^x \sqrt{t^3+1}\,dt.$

19. $f(x) = \sec x, \quad F(0) = 0$

SOLUTION The antiderivative $F(x)$ of $f(x) = \sec x$ satisfying $F(0) = 0$ is $F(x) = \int_0^x \sec t\,dt.$

In Exercises 21–24, calculate the derivative.

21. $\frac{d}{dx}\int_0^x (t^5 - 9t^3)\,dt$

SOLUTION By FTC II, $\frac{d}{dx}\int_0^x (t^5 - 9t^3)\,dt = x^5 - 9x^3.$

23. $\frac{d}{dt}\int_{100}^{t} \sec(5x-9)\,dx$

SOLUTION By FTC II, $\frac{d}{dt}\int_{100}^{t} \sec(5x-9)\,dx = \sec(5t-9).$

25. Let $A(x) = \int_0^x f(t)\,dt$ for $f(x)$ in Figure 8.

(a) Calculate $A(2)$, $A(3)$, $A'(2)$, and $A'(3)$.

(b) Find formulas for $A(x)$ on $[0, 2]$ and $[2, 4]$ and sketch the graph of $A(x)$.

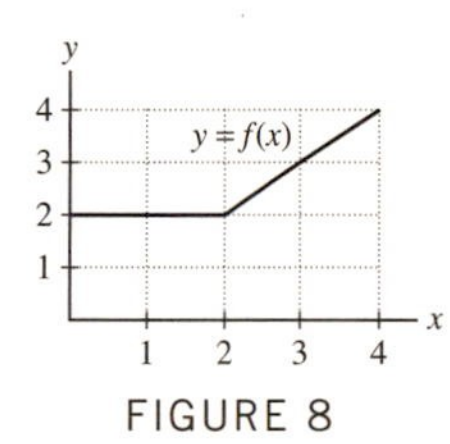

FIGURE 8

SOLUTION

(a) $A(2) = 2 \cdot 2 = 4$, the area under $f(x)$ from $x = 0$ to $x = 2$, while $A(3) = 2 \cdot 3 + \frac{1}{2} = 6.5$, the area under $f(x)$ from $x = 0$ to $x = 3$. By the FTC, $A'(x) = f(x)$ so $A'(2) = f(2) = 2$ and $A'(3) = f(3) = 3$.

(b) For each $x \in [0, 2]$, the region under the graph of $y = f(x)$ is a rectangle of length x and height 2; for each $x \in [2, 4]$, the region is comprised of a square of side length 2 and a trapezoid of height $x - 2$ and bases 2 and x. Hence,

$$A(x) = \begin{cases} 2x, & 0 \le x < 2 \\ \frac{1}{2}x^2 + 2, & 2 \le x \le 4 \end{cases}$$

A graph of the area function $A(x)$ is shown below.

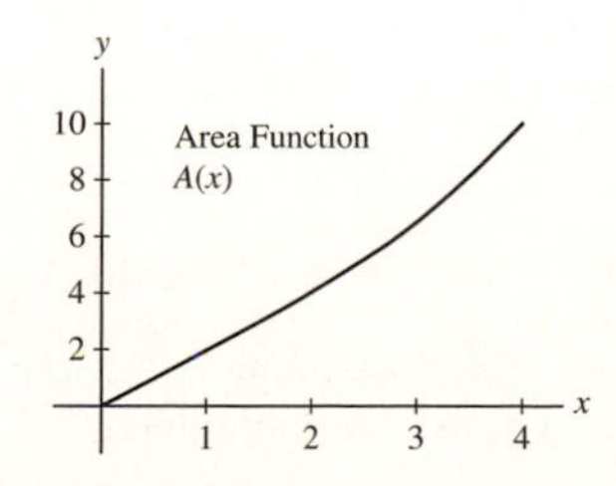

27. Verify: $\int_0^x |t|\,dt = \frac{1}{2}x|x|$. *Hint:* Consider $x \geq 0$ and $x \leq 0$ separately.

SOLUTION Let $f(t) = |t| = \begin{cases} t & \text{for } t \geq 0 \\ -t & \text{for } t < 0 \end{cases}$. Then

$$F(x) = \int_0^x f(t)\,dt = \begin{cases} \int_0^x t\,dt & \text{for } x \geq 0 \\ \int_0^x -t\,dt & \text{for } x < 0 \end{cases} = \begin{cases} \frac{1}{2}t^2\Big|_0^x = \frac{1}{2}x^2 & \text{for } x \geq 0 \\ \left(-\frac{1}{2}t^2\right)\Big|_0^x = -\frac{1}{2}x^2 & \text{for } x < 0 \end{cases}$$

For $x \geq 0$, we have $F(x) = \frac{1}{2}x^2 = \frac{1}{2}x\,|x|$ since $|x| = x$, while for $x < 0$, we have $F(x) = -\frac{1}{2}x^2 = \frac{1}{2}x\,|x|$ since $|x| = -x$. Therefore, for all real x we have $F(x) = \frac{1}{2}x\,|x|$.

In Exercises 29–34, calculate the derivative.

29. $\frac{d}{dx}\int_0^{x^2} \frac{t\,dt}{t+1}$

SOLUTION By the Chain Rule and the FTC, $\frac{d}{dx}\int_0^{x^2} \frac{t\,dt}{t+1} = \frac{x^2}{x^2+1}\cdot 2x = \frac{2x^3}{x^2+1}$.

31. $\frac{d}{ds}\int_{-6}^{\cos s} u^4\,du$

SOLUTION By the Chain Rule and the FTC, $\frac{d}{ds}\int_{-6}^{\cos s} u^4\,du = \cos^4 s(-\sin s) = -\cos^4 s \sin s$.

33. $\frac{d}{dx}\int_{\sqrt{x}}^{x^2} \tan t\,dt$

SOLUTION Let

$$G(x) = \int_{\sqrt{x}}^{x^2} \tan t\,dt = \int_0^{x^2} \tan t\,dt - \int_0^{\sqrt{x}} \tan t\,dt.$$

Applying the Chain Rule combined with FTC twice, we have

$$G'(x) = \tan(x^2)\cdot 2x - \tan(\sqrt{x})\cdot \frac{1}{2}x^{-1/2} = 2x\tan(x^2) - \frac{\tan(\sqrt{x})}{2\sqrt{x}}.$$

In Exercises 35–38, with $f(x)$ as in Figure 10 let

$$A(x) = \int_0^x f(t)\,dt \quad \textit{and} \quad B(x) = \int_2^x f(t)\,dt.$$

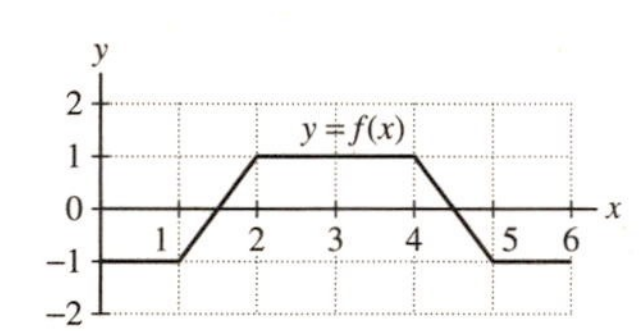

FIGURE 10

35. Find the min and max of $A(x)$ on $[0, 6]$.

SOLUTION The minimum values of $A(x)$ on $[0, 6]$ occur where $A'(x) = f(x)$ goes from negative to positive. This occurs at one place, where $x = 1.5$. The minimum value of $A(x)$ is therefore $A(1.5) = -1.25$. The maximum values of $A(x)$ on $[0, 6]$ occur where $A'(x) = f(x)$ goes from positive to negative. This occurs at one place, where $x = 4.5$. The maximum value of $A(x)$ is therefore $A(4.5) = 1.25$.

37. Find formulas for $A(x)$ and $B(x)$ valid on $[2, 4]$.

SOLUTION On the interval $[2, 4]$, $A'(x) = B'(x) = f(x) = 1$. $A(2) = \int_0^2 f(t)\,dt = -1$ and $B(2) = \int_2^2 f(t)\,dt = 0$. Hence $A(x) = (x-2) - 1$ and $B(x) = (x-2)$.

39. Let $A(x) = \int_0^x f(t)\,dt$, with $f(x)$ as in Figure 11.

(a) Does $A(x)$ have a local maximum at P?

(b) Where does $A(x)$ have a local minimum?

(c) Where does $A(x)$ have a local maximum?
(d) True or false? $A(x) < 0$ for all x in the interval shown.

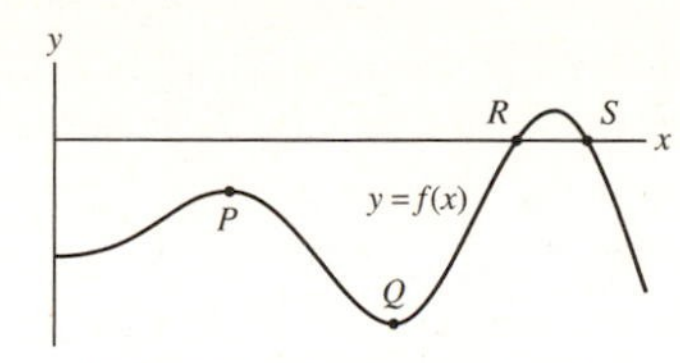

FIGURE 11 Graph of $f(x)$.

SOLUTION

(a) In order for $A(x)$ to have a local maximum, $A'(x) = f(x)$ must transition from positive to negative. As this does not happen at P, $A(x)$ does not have a local maximum at P.
(b) $A(x)$ will have a local minimum when $A'(x) = f(x)$ transitions from negative to positive. This happens at R, so $A(x)$ has a local minimum at R.
(c) $A(x)$ will have a local maximum when $A'(x) = f(x)$ transitions from positive to negative. This happens at S, so $A(x)$ has a local maximum at S.
(d) It is true that $A(x) < 0$ on I since the signed area from 0 to x is clearly always negative from the figure.

41. Determine the function $g(x)$ and all values of c such that

$$\int_c^x g(t)\,dt = x^2 + x - 6$$

SOLUTION By the FTC II we have

$$g(x) = \frac{d}{dx}(x^2 + x - 6) = 2x + 1$$

and therefore,

$$\int_c^x g(t)\,dt = x^2 + x - (c^2 + c)$$

We must choose c so that $c^2 + c = 6$. We can take $c = 2$ or $c = -3$.

In Exercises 43 and 44, let $A(x) = \int_a^x f(t)\,dt$.

43. **Area Functions and Concavity** Explain why the following statements are true. Assume $f(x)$ is differentiable.
(a) If c is an inflection point of $A(x)$, then $f'(c) = 0$.
(b) $A(x)$ is concave up if $f(x)$ is increasing.
(c) $A(x)$ is concave down if $f(x)$ is decreasing.

SOLUTION

(a) If $x = c$ is an inflection point of $A(x)$, then $A''(c) = f'(c) = 0$.
(b) If $A(x)$ is concave up, then $A''(x) > 0$. Since $A(x)$ is the area function associated with $f(x)$, $A'(x) = f(x)$ by FTC II, so $A''(x) = f'(x)$. Therefore $f'(x) > 0$, so $f(x)$ is increasing.
(c) If $A(x)$ is concave down, then $A''(x) < 0$. Since $A(x)$ is the area function associated with $f(x)$, $A'(x) = f(x)$ by FTC II, so $A''(x) = f'(x)$. Therefore, $f'(x) < 0$ and so $f(x)$ is decreasing.

45. Let $A(x) = \int_0^x f(t)\,dt$, with $f(x)$ as in Figure 12. Determine:
(a) The intervals on which $A(x)$ is increasing and decreasing
(b) The values x where $A(x)$ has a local min or max
(c) The inflection points of $A(x)$
(d) The intervals where $A(x)$ is concave up or concave down

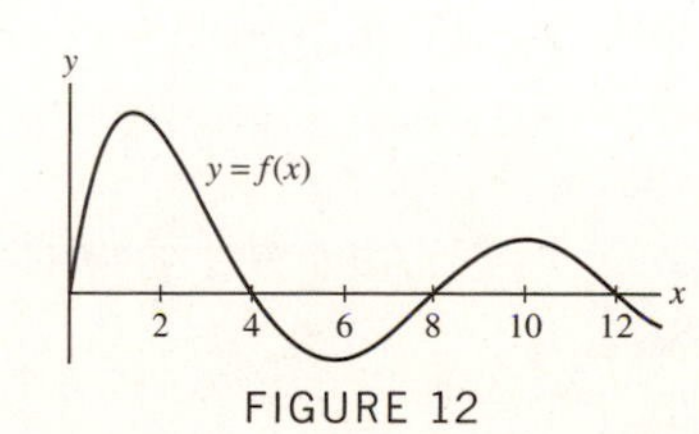

FIGURE 12

SOLUTION

(a) $A(x)$ is increasing when $A'(x) = f(x) > 0$, which corresponds to the intervals $(0, 4)$ and $(8, 12)$. $A(x)$ is decreasing when $A'(x) = f(x) < 0$, which corresponds to the intervals $(4, 8)$ and $(12, \infty)$.

(b) $A(x)$ has a local minimum when $A'(x) = f(x)$ changes from $-$ to $+$, corresponding to $x = 8$. $A(x)$ has a local maximum when $A'(x) = f(x)$ changes from $+$ to $-$, corresponding to $x = 4$ and $x = 12$.

(c) Inflection points of $A(x)$ occur where $A''(x) = f'(x)$ changes sign, or where f changes from increasing to decreasing or vice versa. Consequently, $A(x)$ has inflection points at $x = 2$, $x = 6$, and $x = 10$.

(d) $A(x)$ is concave up when $A''(x) = f'(x)$ is positive or $f(x)$ is increasing, which corresponds to the intervals $(0, 2)$ and $(6, 10)$. Similarly, $A(x)$ is concave down when $f(x)$ is decreasing, which corresponds to the intervals $(2, 6)$ and $(10, \infty)$.

47. Sketch the graph of an increasing function $f(x)$ such that both $f'(x)$ and $A(x) = \int_0^x f(t)\,dt$ are decreasing.

SOLUTION If $f'(x)$ is decreasing, then $f''(x)$ must be negative. Furthermore, if $A(x) = \displaystyle\int_0^x f(t)\,dt$ is decreasing, then $A'(x) = f(x)$ must also be negative. Thus, we need a function which is negative but increasing and concave down. The graph of one such function is shown below.

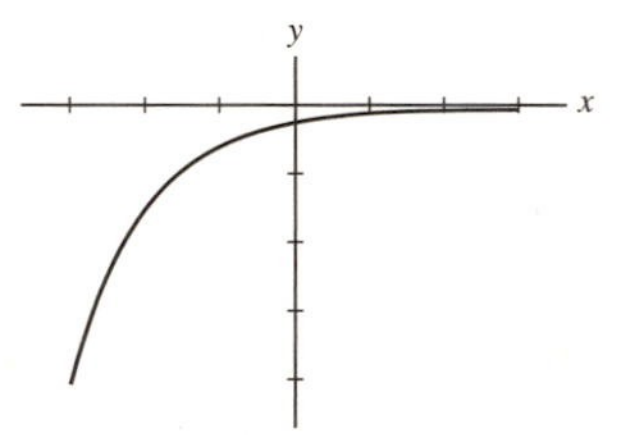

49. GU Find the smallest positive critical point of

$$F(x) = \int_0^x \cos(t^{3/2})\,dt$$

and determine whether it is a local min or max. Then find the smallest positive inflection point of $F(x)$ and use a graph of $y = \cos(x^{3/2})$ to determine whether the concavity changes from up to down or from down to up.

SOLUTION A critical point of $F(x)$ occurs where $F'(x) = \cos(x^{3/2}) = 0$. The smallest positive critical points occurs where $x^{3/2} = \pi/2$, so that $x = (\pi/2)^{2/3}$. $F'(x)$ goes from positive to negative at this point, so $x = (\pi/2)^{2/3}$ corresponds to a local maximum..

Candidate inflection points of $F(x)$ occur where $F''(x) = 0$. By FTC, $F'(x) = \cos(x^{3/2})$, so $F''(x) = -(3/2)x^{1/2}\sin(x^{3/2})$. Finding the smallest positive solution of $F''(x) = 0$, we get:

$$\begin{aligned} -(3/2)x^{1/2}\sin(x^{3/2}) &= 0 \\ \sin(x^{3/2}) &= 0 \qquad (\text{since } x > 0) \\ x^{3/2} &= \pi \\ x &= \pi^{2/3} \approx 2.14503. \end{aligned}$$

From the plot below, we see that $F'(x) = \cos(x^{3/2})$ changes from decreasing to increasing at $\pi^{2/3}$, so $F(x)$ changes from concave down to concave up at that point.

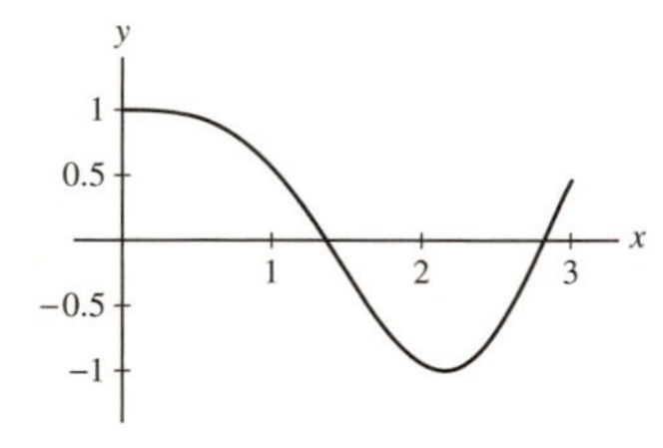

Further Insights and Challenges

51. Proof of FTC I FTC I asserts that $\int_a^b f(t)\,dt = F(b) - F(a)$ if $F'(x) = f(x)$. Use FTC II to give a new proof of FTC I as follows. Set $A(x) = \int_a^x f(t)\,dt$.

(a) Show that $F(x) = A(x) + C$ for some constant.

(b) Show that $F(b) - F(a) = A(b) - A(a) = \displaystyle\int_a^b f(t)\,dt$.

SOLUTION Let $F'(x) = f(x)$ and $A(x) = \int_a^x f(t)\,dt$.

(a) Then by the FTC, Part II, $A'(x) = f(x)$ and thus $A(x)$ and $F(x)$ are both antiderivatives of $f(x)$. Hence $F(x) = A(x) + C$ for some constant C.

(b)

$$\begin{aligned} F(b) - F(a) &= (A(b) + C) - (A(a) + C) = A(b) - A(a) \\ &= \int_a^b f(t)\,dt - \int_a^a f(t)\,dt = \int_a^b f(t)\,dt - 0 = \int_a^b f(t)\,dt \end{aligned}$$

which proves the FTC, Part I.

53. Prove the formula

$$\frac{d}{dx}\int_{u(x)}^{v(x)} f(t)\,dt = f(v(x))v'(x) - f(u(x))u'(x)$$

SOLUTION Write

$$\int_{u(x)}^{v(x)} f(x)\,dx = \int_{u(x)}^{0} f(x)\,dx + \int_0^{v(x)} f(x)\,dx = \int_0^{v(x)} f(x)\,dx - \int_0^{u(x)} f(x)\,dx.$$

Then, by the Chain Rule and the FTC,

$$\begin{aligned} \frac{d}{dx}\int_{u(x)}^{v(x)} f(x)\,dx &= \frac{d}{dx}\int_0^{v(x)} f(x)\,dx - \frac{d}{dx}\int_0^{u(x)} f(x)\,dx \\ &= f(v(x))v'(x) - f(u(x))u'(x). \end{aligned}$$

5.5 Net Change as the Integral of a Rate

Preliminary Questions

1. A hot metal object is submerged in cold water. The rate at which the object cools (in degrees per minute) is a function $f(t)$ of time. Which quantity is represented by the integral $\int_0^T f(t)\,dt$?

SOLUTION The definite integral $\int_0^T f(t)\,dt$ represents the total drop in temperature of the metal object in the first T minutes after being submerged in the cold water.

2. A plane travels 560 km from Los Angeles to San Francisco in 1 hour. If the plane's velocity at time t is $v(t)$ km/h, what is the value of $\int_0^1 v(t)\,dt$?

SOLUTION The definite integral $\int_0^1 v(t)\,dt$ represents the total distance traveled by the airplane during the one hour flight from Los Angeles to San Francisco. Therefore the value of $\int_0^1 v(t)\,dt$ is 560 km.

3. Which of the following quantities would be naturally represented as derivatives and which as integrals?

(a) Velocity of a train

(b) Rainfall during a 6-month period

(c) Mileage per gallon of an automobile

(d) Increase in the U.S. population from 1990 to 2010

SOLUTION Quantities **(a)** and **(c)** involve rates of change, so these would naturally be represented as derivatives. Quantities **(b)** and **(d)** involve an accumulation, so these would naturally be represented as integrals.

Exercises

1. Water flows into an empty reservoir at a rate of $3000 + 20t$ liters per hour. What is the quantity of water in the reservoir after 5 hours?

SOLUTION The quantity of water in the reservoir after five hours is

$$\int_0^5 (3000 + 20t)\,dt = \left(3000t + 10t^2\right)\Big|_0^5 = 15{,}250 \text{ gallons}.$$

3. A survey shows that a mayoral candidate is gaining votes at a rate of $2000t + 1000$ votes per day, where t is the number of days since she announced her candidacy. How many supporters will the candidate have after 60 days, assuming that she had no supporters at $t = 0$?

SOLUTION The number of supporters the candidate has after 60 days is

$$\int_0^{60} (2000t + 1000)\,dt = (1000t^2 + 1000t)\Big|_0^{60} = 3{,}660{,}000.$$

5. Find the displacement of a particle moving in a straight line with velocity $v(t) = 4t - 3$ m/s over the time interval $[2, 5]$.

SOLUTION The displacement is given by

$$\int_2^5 (4t - 3)\,dt = (2t^2 - 3t)\Big|_2^5 = (50 - 15) - (8 - 6) = 33\text{m}.$$

7. A cat falls from a tree (with zero initial velocity) at time $t = 0$. How far does the cat fall between $t = 0.5$ and $t = 1$ s? Use Galileo's formula $v(t) = -9.8t$ m/s.

SOLUTION Given $v(t) = -9.8t$ m/s, the total distance the cat falls during the interval $[\frac{1}{2}, 1]$ is

$$\int_{1/2}^1 |v(t)|\,dt = \int_{1/2}^1 9.8t\,dt = 4.9t^2\Big|_{1/2}^1 = 4.9 - 1.225 = 3.675\text{ m}.$$

In Exercises 9–12, a particle moves in a straight line with the given velocity (in m/s). Find the displacement and distance traveled over the time interval, and draw a motion diagram like Figure 3 (with distance and time labels).

9. $v(t) = 12 - 4t$, $[0, 5]$

SOLUTION Displacement is given by $\int_0^5 (12 - 4t)\,dt = (12t - 2t^2)\Big|_0^5 = 10$ ft, while total distance is given by

$$\int_0^5 |12 - 4t|\,dt = \int_0^3 (12 - 4t)\,dt + \int_3^5 (4t - 12)\,dt = (12t - 2t^2)\Big|_0^3 + (2t^2 - 12t)\Big|_3^5 = 26\text{ ft}.$$

The displacement diagram is given here.

$t = 5$
$t = 3$
$t = 0$
0 10 18 Distance

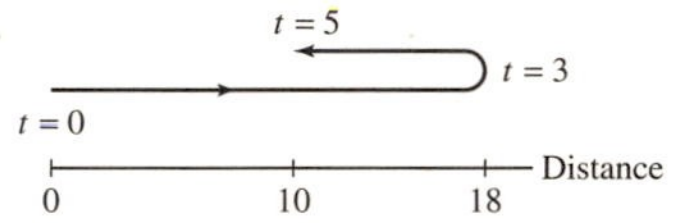

11. $v(t) = t^{-2} - 1$, $[0.5, 2]$

SOLUTION Displacement is given by $\int_{0.5}^2 (t^{-2} - 1)\,dt = (-t^{-1} - t)\Big|_{0.5}^2 = 0$ m, while total distance is given by

$$\int_{0.5}^2 \left|t^{-2} - 1\right|\,dt = \int_{0.5}^1 (t^{-2} - 1)\,dt + \int_1^2 (1 - t^{-2})\,dt = (-t^{-1} - t)\Big|_{0.5}^1 + (t + t^{-1})\Big|_1^2 = 1\text{ m}.$$

The displacement diagram is given here.

$t = 2$
$t = 1$
$t = 0$
0 0.5 Distance

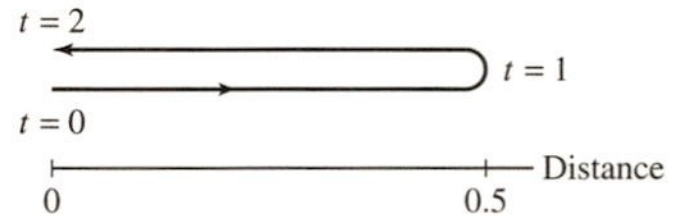

13. Find the net change in velocity over $[1, 4]$ of an object with $a(t) = 8t - t^2$ m/s^2.

SOLUTION The net change in velocity is

$$\int_1^4 a(t)\,dt = \int_1^4 (8t - t^2)\,dt = \left(4t^2 - \frac{1}{3}t^3\right)\Big|_1^4 = 39\text{ m/s}.$$

15. The traffic flow rate past a certain point on a highway is $q(t) = 3000 + 2000t - 300t^2$ (t in hours), where $t = 0$ is 8 AM. How many cars pass by in the time interval from 8 to 10 AM?

SOLUTION The number of cars is given by

$$\int_0^2 q(t)\,dt = \int_0^2 (3000 + 2000t - 300t^2)\,dt = \left(3000t + 1000t^2 - 100t^3\right)\Big|_0^2$$
$$= 3000(2) + 1000(4) - 100(8) = 9200\text{ cars}.$$

17. A small boutique produces wool sweaters at a marginal cost of $40 - 5[[x/5]]$ for $0 \le x \le 20$, where $[[x]]$ is the greatest integer function. Find the cost of producing 20 sweaters. Then compute the average cost of the first 10 sweaters and the last 10 sweaters.

SOLUTION The total cost of producing 20 sweaters is

$$\int_0^{20} (40 - 5[[x/5]])\,dx = \int_0^5 40\,dx + \int_5^{10} 35\,dx + \int_{10}^{15} 30\,dx + \int_{15}^{20} 25\,dx$$
$$= 40(5) + 35(5) + 30(5) + 25(5) = 650 \text{ dollars.}$$

From this calculation, we see that the cost of the first 10 sweaters is \$375 and the cost of the last ten sweaters is \$275; thus, the average cost of the first ten sweaters is \$37.50 and the average cost of the last ten sweaters is \$27.50.

19. The velocity of a car is recorded at half-second intervals (in feet per second). Use the average of the left- and right-endpoint approximations to estimate the total distance traveled during the first 4 seconds.

t	0	0.5	1	1.5	2	2.5	3	3.5	4
$v(t)$	0	12	20	29	38	44	32	35	30

SOLUTION Let $\Delta t = .5$. Then

$$R_N = 0.5 \cdot (12 + 20 + 29 + 38 + 44 + 32 + 35 + 30) = 120 \text{ ft.}$$
$$L_N = 0.5 \cdot (0 + 12 + 20 + 29 + 38 + 44 + 32 + 35) = 105 \text{ ft.}$$

The average of R_N and L_N is 112.5 ft.

21. A megawatt of power is 10^6 W, or 3.6×10^9 J/hour. Which quantity is represented by the area under the graph in Figure 5? Estimate the energy (in joules) consumed during the period 4 PM to 8 PM.

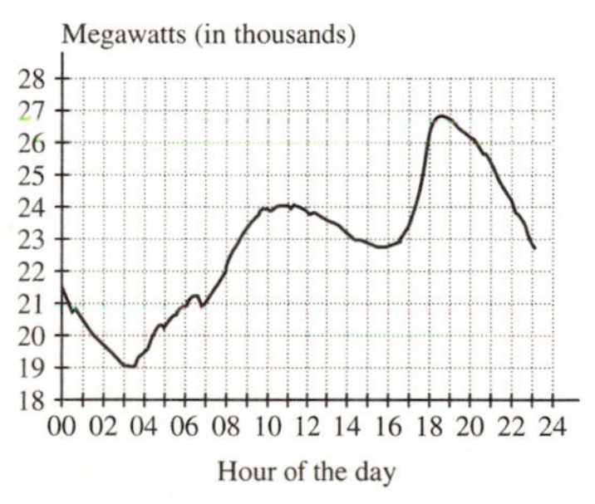

FIGURE 5 Power consumption over 1-day period in California (February 2010).

SOLUTION The area under the graph in Figure 5 represents the total power consumption over one day in California. Assuming $t = 0$ corresponds to midnight, the period 4 PM to 8 PM corresponds to $t = 16$ to $t = 20$. The left and right endpoint approximations are

$$L = 1(22.8 + 23.5 + 26.1 + 26.7) = 99.1\text{megawatt} \cdot \text{hours}$$
$$R = 1(23.5 + 26.1 + 26.7 + 26.1) = 102.4\text{megawatt} \cdot \text{hours}$$

The average of these values is

$$100.75\text{megawatt} \cdot \text{hours} = 3.627 \times 10^{11} \text{ joules.}$$

23. Let $N(d)$ be the number of asteroids of diameter $\le d$ kilometers. Data suggest that the diameters are distributed according to a piecewise power law:

$$N'(d) = \begin{cases} 1.9 \times 10^9 d^{-2.3} & \text{for } d < 70 \\ 2.6 \times 10^{12} d^{-4} & \text{for } d \ge 70 \end{cases}$$

(a) Compute the number of asteroids with diameter between 0.1 and 100 km.

(b) Using the approximation $N(d + 1) - N(d) \approx N'(d)$, estimate the number of asteroids of diameter 50 km.

SOLUTION

(a) The number of asteroids with diameter between 0.1 and 100 km is

$$\int_{0.1}^{100} N'(d)\,dd = \int_{0.1}^{70} 1.9 \times 10^9 d^{-2.3}\,dd + \int_{70}^{100} 2.6 \times 10^{12} d^{-4}\,dd$$

$$= -\frac{1.9 \times 10^9}{1.3} d^{-1.3}\Big|_{0.1}^{70} - \frac{2.6 \times 10^{12}}{3} d^{-3}\Big|_{70}^{100}$$

$$= 2.916 \times 10^{10} + 1.66 \times 10^6 \approx 2.916 \times 10^{10}.$$

(b) Taking $d = 49.5$,

$$N(50.5) - N(49.5) \approx N'(49.5) = 1.9 \times 10^9 49.5^{-2.3} = 240{,}525.79.$$

Thus, there are approximately 240,526 asteroids of diameter 50 km.

25. Figure 7 shows the rate $R(t)$ of natural gas consumption (in billions of cubic feet per day) in the mid-Atlantic states (New York, New Jersey, Pennsylvania). Express the total quantity of natural gas consumed in 2009 as an integral (with respect to time t in days). Then estimate this quantity, given the following monthly values of $R(t)$:

3.18, 2.86, 2.39, 1.49, 1.08, 0.80,
1.01, 0.89, 0.89, 1.20, 1.64, 2.52

Keep in mind that the number of days in a month varies with the month.

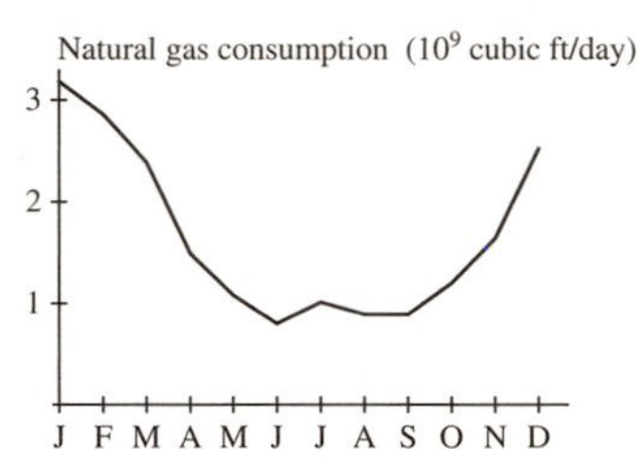

FIGURE 7 Natural gas consumption in 2009 in the mid-Atlantic states

SOLUTION The total quantity of natural gas consumed is given by

$$\int_0^{365} R(t)\,dt.$$

With the given data, we find

$$\int_0^{365} R(t)\,dt \approx 31(3.18) + 28(2.86) + 31(2.39) + 30(1.49) + 31(1.08) + 30(0.80)$$

$$+31(1.01) + 31(0.89) + 30(0.89) + 31(1.20) + 30(1.64) + 31(2.52)$$

$$= 605.05 \text{ billion cubic feet.}$$

Exercises 27 and 28: A study suggests that the extinction rate $r(t)$ of marine animal families during the Phanerozoic Eon can be modeled by the function $r(t) = 3130/(t + 262)$ for $0 \le t \le 544$, where t is time elapsed (in millions of years) since the beginning of the eon 544 million years ago. Thus, $t = 544$ refers to the present time, $t = 540$ is 4 million years ago, and so on.

27. Compute the average of R_N and L_N with $N = 5$ to estimate the total number of families that became extinct in the periods $100 \le t \le 150$ and $350 \le t \le 400$.

SOLUTION

- $(100 \le t \le 150)$ For $N = 5$,

$$\Delta t = \frac{150 - 100}{5} = 10.$$

The table of values $\{r(t_i)\}_{i=0\ldots5}$ is given below:

t_i	100	110	120	130	140	150
$r(t_i)$	8.64641	8.41398	8.19372	7.98469	7.78607	7.59709

The endpoint approximations are:

$$R_N = 10(8.41398 + 8.19372 + 7.98469 + 7.78607 + 7.59709) \approx 399.756 \text{ families}$$

$$L_N = 10(8.64641 + 8.41398 + 8.19372 + 7.98469 + 7.78607) \approx 410.249 \text{ families}$$

The right endpoint approximation estimates 399.756 families became extinct in the period $100 \le t \le 150$, the left endpoint approximation estimates 410.249 families became extinct during this time. The average of the two is 405.362 families.

- $(350 \le t \le 400)$ For $N = 10$,

$$\Delta t = \frac{400 - 350}{5} = 19.$$

The table of values $\{r(t_i)\}_{i=0...5}$ is given below:

t_i	350	360	370	380	390	400
$r(t_i)$	5.11438	5.03215	4.95253	4.87539	4.80061	4.72810

The endpoint approximations are:

$$R_N = 10(5.03215 + 4.95253 + 4.87539 + 4.80061 + 4.72810) \approx 243.888 \text{ families}$$

$$L_N = 10(5.11438 + 5.03215 + 4.95253 + 4.87539 + 4.80061) \approx 247.751 \text{ families}$$

The right endpoint approximation estimates 243.888 families became extinct in the period $350 \le t \le 400$, the left endpoint approximation estimates 247.751 families became extinct during this time. The average of the two is 245.820 families.

Further Insights and Challenges

29. Show that a particle, located at the origin at $t = 1$ and moving along the x-axis with velocity $v(t) = t^{-2}$, will never pass the point $x = 2$.

SOLUTION The particle's velocity is $v(t) = s'(t) = t^{-2}$, an antiderivative for which is $F(t) = -t^{-1}$. Hence, the particle's position at time t is

$$s(t) = \int_1^t s'(u)\,du = F(u)\Big|_1^t = F(t) - F(1) = 1 - \frac{1}{t} < 1$$

for all $t \ge 1$. Thus, the particle will never pass $x = 1$, which implies it will never pass $x = 2$ either.

5.6 Substitution Method

Preliminary Questions

1. Which of the following integrals is a candidate for the Substitution Method?

(a) $\int 5x^4 \sin(x^5)\,dx$ **(b)** $\int \sin^5 x \cos x\,dx$ **(c)** $\int x^5 \sin x\,dx$

SOLUTION The function in **(c)**: $x^5 \sin x$ is not of the form $g(u(x))u'(x)$. The function in **(a)** meets the prescribed pattern with $g(u) = \sin u$ and $u(x) = x^5$. Similarly, the function in **(b)** meets the prescribed pattern with $g(u) = u^5$ and $u(x) = \sin x$.

2. Find an appropriate choice of u for evaluating the following integrals by substitution:

(a) $\int x(x^2+9)^4\,dx$ **(b)** $\int x^2 \sin(x^3)\,dx$ **(c)** $\int \sin x \cos^2 x\,dx$

SOLUTION

(a) $x(x^2+9)^4 = \frac{1}{2}(2x)(x^2+9)^4$; hence, $c = \frac{1}{2}$, $f(u) = u^4$, and $u(x) = x^2 + 9$.
(b) $x^2 \sin(x^3) = \frac{1}{3}(3x^2)\sin(x^3)$; hence, $c = \frac{1}{3}$, $f(u) = \sin u$, and $u(x) = x^3$.
(c) $\sin x \cos^2 x = -(-\sin x)\cos^2 x$; hence, $c = -1$, $f(u) = u^2$, and $u(x) = \cos x$.

3. Which of the following is equal to $\int_0^2 x^2(x^3+1)\,dx$ for a suitable substitution?

(a) $\frac{1}{3}\int_0^2 u\,du$ **(b)** $\int_0^9 u\,du$ **(c)** $\frac{1}{3}\int_1^9 u\,du$

SOLUTION With the substitution $u = x^3 + 1$, the definite integral $\int_0^2 x^2(x^3+1)\,dx$ becomes $\frac{1}{3}\int_1^9 u\,du$. The correct answer is **(c)**.

Exercises

In Exercises 1–6, calculate du.

1. $u = x^3 - x^2$

SOLUTION Let $u = x^3 - x^2$. Then $du = (3x^2 - 2x)\,dx$.

3. $u = \cos(x^2)$

SOLUTION Let $u = \cos(x^2)$. Then $du = -\sin(x^2)\cdot 2x\,dx = -2x\sin(x^2)\,dx$.

5. $u = e^{4x+1}$

SOLUTION Let $u = e^{4x+1}$. Then $du = 4e^{4x+1}\,dx$.

In Exercises 7–22, write the integral in terms of u and du. Then evaluate.

7. $\int (x-7)^3\,dx, \quad u = x - 7$

SOLUTION Let $u = x - 7$. Then $du = dx$. Hence

$$\int (x-7)^3\,dx = \int u^3\,du = \frac{1}{4}u^4 + C = \frac{1}{4}(x-7)^4 + C.$$

9. $\int t\sqrt{t^2+1}\,dt, \quad u = t^2 + 1$

SOLUTION Let $u = t^2 + 1$. Then $du = 2t\,dt$. Hence,

$$\int t\sqrt{t^2+1}\,dt = \frac{1}{2}\int u^{1/2}\,du = \frac{1}{3}u^{3/2} + C = \frac{1}{3}(t^2+1)^{3/2} + C.$$

11. $\int \frac{t^3}{(4-2t^4)^{11}}\,dt, \quad u = 4 - 2t^4$

SOLUTION Let $u = 4 - 2t^4$. Then $du = -8t^3\,dt$. Hence,

$$\int \frac{t^3}{(4-2t^4)^{11}}\,dt = -\frac{1}{8}\int u^{-11}\,du = \frac{1}{80}u^{-10} + C = \frac{1}{80}(4-2t^4)^{-10} + C.$$

13. $\int x(x+1)^9\,dx, \quad u = x + 1$

SOLUTION Let $u = x + 1$. Then $x = u - 1$ and $du = dx$. Hence

$$\int x(x+1)^9\,dx = \int (u-1)u^9\,du = \int (u^{10} - u^9)\,du$$
$$= \frac{1}{11}u^{11} - \frac{1}{10}u^{10} + C = \frac{1}{11}(x+1)^{11} - \frac{1}{10}(x+1)^{10} + C.$$

15. $\int x^2\sqrt{x+1}\,dx, \quad u = x + 1$

SOLUTION Let $u = x + 1$. Then $x = u - 1$ and $du = dx$. Hence

$$\int x^2\sqrt{x+1}\,dx = \int (u-1)^2 u^{1/2}\,du = \int (u^{5/2} - 2u^{3/2} + u^{1/2})\,du$$
$$= \frac{2}{7}u^{7/2} - \frac{4}{5}u^{5/2} + \frac{2}{3}u^{3/2} + C$$
$$= \frac{2}{7}(x+1)^{7/2} - \frac{4}{5}(x+1)^{5/2} + \frac{2}{3}(x+1)^{3/2} + C.$$

17. $\int \sin^2\theta\cos\theta\,d\theta, \quad u = \sin\theta$

SOLUTION Let $u = \sin\theta$. Then $du = \cos\theta\,d\theta$. Hence,

$$\int \sin^2\theta\cos\theta\,d\theta = \int u^2\,du = \frac{1}{3}u^3 + C = \frac{1}{3}\sin^3\theta + C.$$

19. $\int xe^{-x^2}\,dx, \quad u = -x^2$

SOLUTION Let $u = -x^2$. Then $du = -2x\,dx$ or $-\frac{1}{2}\,du = x\,dx$. Hence,

$$\int xe^{-x^2}\,dx = -\frac{1}{2}\int e^u\,du = -\frac{1}{2}e^u + C = -\frac{1}{2}e^{-x^2} + C.$$

21. $\int \frac{(\ln x)^2\,dx}{x}, \quad u = \ln x$

SOLUTION Let $u = \ln x$. Then $du = \frac{1}{x}\,dx$, and

$$\int \frac{(\ln x)^2}{x}\,dx = \int u^2\,du = \frac{1}{3}u^3 + C = \frac{1}{3}(\ln x)^3 + C.$$

In Exercises 23–26, evaluate the integral in the form $a\sin(u(x)) + C$ for an appropriate choice of $u(x)$ and constant a.

23. $\int x^3\cos(x^4)\,dx$

SOLUTION Let $u = x^4$. Then $du = 4x^3\,dx$ or $\frac{1}{4}\,du = x^3 dx$. Hence

$$\int x^3\cos(x^4)\,dx = \frac{1}{4}\int \cos u\,du = \frac{1}{4}\sin u + C = \frac{1}{4}\sin(x^4) + C.$$

25. $\int x^{1/2}\cos(x^{3/2})\,dx$

SOLUTION Let $u = x^{3/2}$. Then $du = \frac{3}{2}x^{1/2}\,dx$ or $\frac{2}{3}\,du = x^{1/2}\,dx$. Hence

$$\int x^{1/2}\cos(x^{3/2})\,dx = \frac{2}{3}\int \cos u\,du = \frac{2}{3}\sin u + C = \frac{2}{3}\sin(x^{3/2}) + C.$$

In Exercises 27–72, evaluate the indefinite integral.

27. $\int (4x+5)^9\,dx$

SOLUTION Let $u = 4x + 5$. Then $du = 4\,dx$ and

$$\int (4x+5)^9\,dx = \frac{1}{4}\int u^9\,du = \frac{1}{40}u^{10} + C = \frac{1}{40}(4x+5)^{10} + C.$$

29. $\int \frac{dt}{\sqrt{t+12}}$

SOLUTION Let $u = t + 12$. Then $du = dt$ and

$$\int \frac{dt}{\sqrt{t+12}} = \int u^{-1/2}\,du = 2u^{1/2} + C = 2\sqrt{t+12} + C.$$

31. $\int \frac{x+1}{(x^2+2x)^3}\,dx$

SOLUTION Let $u = x^2 + 2x$. Then $du = (2x+2)\,dx$ or $\frac{1}{2}du = (x+1)\,dx$. Hence

$$\int \frac{x+1}{(x^2+2x)^3}\,dx = \frac{1}{2}\int \frac{1}{u^3}\,du = \frac{1}{2}\left(-\frac{1}{2}u^{-2}\right) + C = -\frac{1}{4}(x^2+2x)^{-2} + C = \frac{-1}{4(x^2+2x)^2} + C.$$

33. $\int \frac{x}{\sqrt{x^2+9}}\,dx$

SOLUTION Let $u = x^2 + 9$. Then $du = 2x\,dx$ or $\frac{1}{2}du = x\,dx$. Hence

$$\int \frac{x}{\sqrt{x^2+9}}\,dx = \frac{1}{2}\int \frac{1}{\sqrt{u}}\,du = \frac{1}{2}\frac{\sqrt{u}}{\frac{1}{2}} + C = \sqrt{x^2+9} + C.$$

35. $\int (3x^2+1)(x^3+x)^2\,dx$

SOLUTION Let $u = x^3 + x$. Then $du = (3x^2+1)\,dx$. Hence

$$\int (3x^2+1)(x^3+x)^2\,dx = \int u^2\,du = \frac{1}{3}u^3 + C = \frac{1}{3}(x^3+x)^3 + C.$$

37. $\int (3x+8)^{11}\,dx$

SOLUTION Let $u = 3x + 8$. Then $du = 3\,dx$ and

$$\int (3x+8)^{11}\,dx = \frac{1}{3}\int u^{11}\,du = \frac{1}{36}u^{12} + C = \frac{1}{36}(3x+8)^{12} + C.$$

39. $\int x^2\sqrt{x^3+1}\,dx$

SOLUTION Let $u = x^3 + 1$. Then $du = 3x^2\,dx$ and

$$\int x^2\sqrt{x^3+1}\,dx = \frac{1}{3}\int u^{1/2}\,du = \frac{2}{9}u^{3/2} + C = \frac{2}{9}(x^3+1)^{3/2} + C.$$

41. $\int \frac{dx}{(x+5)^3}$

SOLUTION Let $u = x + 5$. Then $du = dx$ and

$$\int \frac{dx}{(x+5)^3} = \int u^{-3}\,du = -\frac{1}{2}u^{-2} + C = -\frac{1}{2}(x+5)^{-2} + C.$$

43. $\int z^2(z^3+1)^{12}\,dz$

SOLUTION Let $u = z^3 + 1$. Then $du = 3z^2\,dz$ and

$$\int z^2(z^3+1)^{12}\,dz = \frac{1}{3}\int u^{12}\,du = \frac{1}{39}u^{13} + C = \frac{1}{39}(z^3+1)^{13} + C.$$

45. $\int (x+2)(x+1)^{1/4}\,dx$

SOLUTION Let $u = x + 1$. Then $x = u - 1$, $du = dx$ and

$$\begin{aligned}\int (x+2)(x+1)^{1/4}\,dx &= \int (u+1)u^{1/4}\,du = \int (u^{5/4} + u^{1/4})\,du \\ &= \frac{4}{9}u^{9/4} + \frac{4}{5}u^{5/4} + C \\ &= \frac{4}{9}(x+1)^{9/4} + \frac{4}{5}(x+1)^{5/4} + C.\end{aligned}$$

47. $\int \sin(8-3\theta)\,d\theta$

SOLUTION Let $u = 8 - 3\theta$. Then $du = -3\,d\theta$ and

$$\int \sin(8-3\theta)\,d\theta = -\frac{1}{3}\int \sin u\,du = \frac{1}{3}\cos u + C = \frac{1}{3}\cos(8-3\theta) + C.$$

49. $\int \frac{\cos\sqrt{t}}{\sqrt{t}}\,dt$

SOLUTION Let $u = \sqrt{t} = t^{1/2}$. Then $du = \frac{1}{2}t^{-1/2}\,dt$ and

$$\int \frac{\cos\sqrt{t}}{\sqrt{t}}\,dt = 2\int \cos u\,du = 2\sin u + C = 2\sin\sqrt{t} + C.$$

51. $\int \tan(4\theta + 9)\, d\theta$

SOLUTION Let $u = 4\theta + 9$. Then $du = 4\, d\theta$ and

$$\int \tan(4\theta + 9)\, d\theta = \frac{1}{4}\int \tan u\, du = \frac{1}{4}\ln|\sec u| + C = \frac{1}{4}\ln|\sec(4\theta + 9)| + C.$$

53. $\int \cot x\, dx$

SOLUTION Let $u = \sin x$. Then $du = \cos x\, dx$, and

$$\int \cot x\, dx = \int \frac{\cos x}{\sin x}\, dx = \int \frac{du}{u} = \ln|u| + C = \ln|\sin x| + C.$$

55. $\int \sec^2(4x + 9)\, dx$

SOLUTION Let $u = 4x + 9$. Then $du = 4\, dx$ or $\frac{1}{4}\, du = dx$. Hence

$$\int \sec^2(4x + 9)\, dx = \frac{1}{4}\int \sec^2 u\, du = \frac{1}{4}\tan u + C = \frac{1}{4}\tan(4x + 9) + C.$$

57. $\int \frac{\sec^2(\sqrt{x})\, dx}{\sqrt{x}}$

SOLUTION Let $u = \sqrt{x}$. Then $du = \frac{1}{2\sqrt{x}}\, dx$ or $2\, du = \frac{1}{\sqrt{x}}\, dx$. Hence,

$$\int \frac{\sec^2(\sqrt{x})\, dx}{\sqrt{x}} = 2\int \sec^2 u\, dx = 2\tan u + C = 2\tan(\sqrt{x}) + C.$$

59. $\int \sin 4x\sqrt{\cos 4x + 1}\, dx$

SOLUTION Let $u = \cos 4x + 1$. Then $du = -4\sin 4x$ or $-\frac{1}{4}du = \sin 4x$. Hence

$$\int \sin 4x\sqrt{\cos 4x + 1}\, dx = -\frac{1}{4}\int u^{1/2}\, du = -\frac{1}{4}\left(\frac{2}{3}u^{3/2}\right) + C = -\frac{1}{6}(\cos 4x + 1)^{3/2} + C.$$

61. $\int \sec\theta\tan\theta(\sec\theta - 1)\, d\theta$

SOLUTION Let $u = \sec\theta - 1$. Then $du = \sec\theta\tan\theta\, d\theta$ and

$$\int \sec\theta\tan\theta(\sec\theta - 1)\, d\theta = \int u\, du = \frac{1}{2}u^2 + C = \frac{1}{2}(\sec\theta - 1)^2 + C.$$

63. $\int e^{14x-7}\, dx$

SOLUTION Let $u = 14x - 7$. Then $du = 14\, dx$ or $\frac{1}{14}\, du = dx$. Hence,

$$\int e^{14x-7}\, dx = \frac{1}{14}\int e^u\, du = \frac{1}{14}e^u + C = \frac{1}{14}e^{14x-7} + C.$$

65. $\int \frac{e^x\, dx}{(e^x + 1)^4}$

SOLUTION Let $u = e^x + 1$. Then $du = e^x\, dx$, and

$$\int \frac{e^x}{(e^x + 1)^4}\, dx = \int u^{-4}\, du = -\frac{1}{3u^3} + C = -\frac{1}{3(e^x + 1)^3} + C.$$

67. $\int \frac{e^t\, dt}{e^{2t} + 2e^t + 1}$

SOLUTION Let $u = e^t$. Then $du = e^t\, dt$, and

$$\int \frac{e^t\, dt}{e^{2t} + 2e^t + 1} = \int \frac{du}{u^2 + 2u + 1} = \int \frac{du}{(u + 1)^2} = -\frac{1}{u + 1} + C = -\frac{1}{e^t + 1} + C.$$

69. $\displaystyle\int \frac{(\ln x)^4\,dx}{x}$

SOLUTION Let $u = \ln x$. Then $du = \frac{1}{x}\,dx$, and

$$\int \frac{(\ln x)^4}{x}\,dx = \int u^4\,du = \frac{1}{5}u^5 + C = \frac{1}{5}(\ln x)^5 + C.$$

71. $\displaystyle\int \frac{\tan(\ln x)}{x}\,dx$

SOLUTION Let $u = \cos(\ln x)$. Then $du = -\frac{1}{x}\sin(\ln x)\,dx$ or $-du = \frac{1}{x}\sin(\ln x)\,dx$. Hence,

$$\int \frac{\tan(\ln x)}{x}\,dx = \int \frac{\sin(\ln x)}{x\cos(\ln x)}\,dx = -\int \frac{du}{u} = -\ln|u| + C = -\ln|\cos(\ln x)| + C.$$

73. Evaluate $\displaystyle\int \frac{dx}{(1+\sqrt{x})^3}$ using $u = 1 + \sqrt{x}$. *Hint:* Show that $dx = 2(u-1)du$.

SOLUTION Let $u = 1 + \sqrt{x}$. Then

$$du = \frac{1}{2\sqrt{x}}\,dx \quad \text{or} \quad dx = 2\sqrt{x}\,du = 2(u-1)\,du.$$

Hence,

$$\begin{aligned}\int \frac{dx}{(1+\sqrt{x})^3} &= 2\int \frac{u-1}{u^3}\,du = 2\int (u^{-2} - u^{-3})\,du \\ &= -2u^{-1} + u^{-2} + C = -\frac{2}{1+\sqrt{x}} + \frac{1}{(1+\sqrt{x})^2} + C.\end{aligned}$$

75. Evaluate $\int \sin x \cos x\,dx$ using substitution in two different ways: first using $u = \sin x$ and then using $u = \cos x$. Reconcile the two different answers.

SOLUTION First, let $u = \sin x$. Then $du = \cos x\,dx$ and

$$\int \sin x \cos x\,dx = \int u\,du = \frac{1}{2}u^2 + C_1 = \frac{1}{2}\sin^2 x + C_1.$$

Next, let $u = \cos x$. Then $du = -\sin x\,dx$ or $-du = \sin x\,dx$. Hence,

$$\int \sin x \cos x\,dx = -\int u\,du = -\frac{1}{2}u^2 + C_2 = -\frac{1}{2}\cos^2 x + C_2.$$

To reconcile these two seemingly different answers, recall that any two antiderivatives of a specified function differ by a constant. To show that this is true here, note that $(\frac{1}{2}\sin^2 x + C_1) - (-\frac{1}{2}\cos^2 x + C_2) = \frac{1}{2} + C_1 - C_2$, a constant. Here we used the trigonometric identity $\sin^2 x + \cos^2 x = 1$.

77. What are the new limits of integration if we apply the substitution $u = 3x + \pi$ to the integral $\int_0^{\pi} \sin(3x+\pi)\,dx$?

SOLUTION The new limits of integration are $u(0) = 3\cdot 0 + \pi = \pi$ and $u(\pi) = 3\pi + \pi = 4\pi$.

In Exercises 79–90, use the Change-of-Variables Formula to evaluate the definite integral.

79. $\displaystyle\int_1^3 (x+2)^3\,dx$

SOLUTION Let $u = x + 2$. Then $du = dx$. Hence

$$\int_1^3 (x+2)^3\,dx = \int_3^5 u^3\,du = \left.\frac{1}{4}u^4\right|_3^5 = \frac{5^4}{4} - \frac{3^4}{4} = 136.$$

81. $\displaystyle\int_0^1 \frac{x}{(x^2+1)^3}\,dx$

SOLUTION Let $u = x^2 + 1$. Then $du = 2x\,dx$ or $\frac{1}{2}\,du = x\,dx$. Hence

$$\int_0^1 \frac{x}{(x^2+1)^3}\,dx = \frac{1}{2}\int_1^2 \frac{1}{u^3}\,du = \frac{1}{2}\left(-\frac{1}{2}u^{-2}\right)\Bigg|_1^2 = -\frac{1}{16} + \frac{1}{4} = \frac{3}{16} = 0.1875.$$

83. $\int_0^4 x\sqrt{x^2+9}\,dx$

SOLUTION Let $u = x^2 + 9$. Then $du = 2x\,dx$ or $\frac{1}{2}\,du = x\,dx$. Hence

$$\int_0^4 \sqrt{x^2+9}\,dx = \frac{1}{2}\int_9^{25} \sqrt{u}\,du = \frac{1}{2}\left(\frac{2}{3}u^{3/2}\right)\Big|_9^{25} = \frac{1}{3}(125-27) = \frac{98}{3}.$$

85. $\int_0^1 (x+1)(x^2+2x)^5\,dx$

SOLUTION Let $u = x^2 + 2x$. Then $du = (2x+2)\,dx = 2(x+1)\,dx$, and

$$\int_0^1 (x+1)(x^2+2x)^5\,dx = \frac{1}{2}\int_0^3 u^5\,du = \frac{1}{12}u^6\Big|_0^3 = \frac{729}{12} = \frac{243}{4}.$$

87. $\int_0^1 \theta\tan(\theta^2)\,d\theta$

SOLUTION Let $u = \cos\theta^2$. Then $du = -2\theta\sin\theta^2\,d\theta$ or $-\frac{1}{2}du = \theta\sin\theta^2\,d\theta$. Hence,

$$\int_0^1 \theta\tan(\theta^2)\,d\theta = \int_0^1 \frac{\theta\sin(\theta^2)}{\cos(\theta^2)}\,d\theta = -\frac{1}{2}\int_1^{\cos 1}\frac{du}{u} = -\frac{1}{2}\ln|u|\Big|_1^{\cos 1} = -\frac{1}{2}\left[\ln(\cos 1) + \ln 1\right] = \frac{1}{2}\ln(\sec 1).$$

89. $\int_0^{\pi/2} \cos^3 x\sin x\,dx$

SOLUTION Let $u = \cos x$. Then $du = -\sin x\,dx$. Hence

$$\int_0^{\pi/2} \cos^3 x\sin x\,dx = -\int_1^0 u^3\,du = \int_0^1 u^3\,du = \frac{1}{4}u^4\Big|_0^1 = \frac{1}{4} - 0 = \frac{1}{4}.$$

91. Evaluate $\int_0^2 r\sqrt{5-\sqrt{4-r^2}}\,dr$.

SOLUTION Let $u = 5 - \sqrt{4-r^2}$. Then

$$du = \frac{r\,dr}{\sqrt{4-r^2}} = \frac{r\,dr}{5-u}$$

so that

$$r\,dr = (5-u)\,du.$$

Hence, the integral becomes:

$$\int_0^2 r\sqrt{5-\sqrt{4-r^2}}\,dr = \int_3^5 \sqrt{u}(5-u)\,du = \int_3^5 \left(5u^{1/2} - u^{3/2}\right)du = \left(\frac{10}{3}u^{3/2} - \frac{2}{5}u^{5/2}\right)\Big|_3^5$$
$$= \left(\frac{50}{3}\sqrt{5} - 10\sqrt{5}\right) - \left(10\sqrt{3} - \frac{18}{5}\sqrt{3}\right) = \frac{20}{3}\sqrt{5} - \frac{32}{5}\sqrt{3}.$$

93. Wind engineers have found that wind speed v (in meters/second) at a given location follows a **Rayleigh distribution** of the type

$$W(v) = \frac{1}{32}ve^{-v^2/64}$$

This means that at a given moment in time, the probability that v lies between a and b is equal to the shaded area in Figure 4.

(a) Show that the probability that $v \in [0, b]$ is $1 - e^{-b^2/64}$.

(b) Calculate the probability that $v \in [2, 5]$.

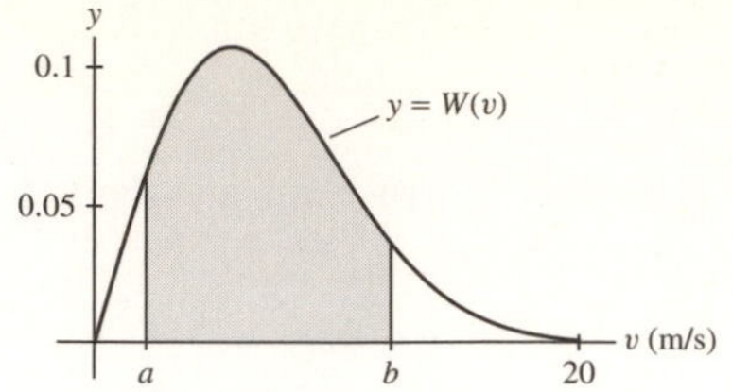

FIGURE 4 The shaded area is the probability that v lies betweena and b.

SOLUTION

(a) The probability that $v \in [0, b]$ is

$$\int_0^b \frac{1}{32} v e^{-v^2/64}\, dv.$$

Let $u = -v^2/64$. Then $du = -v/32\, dv$ and

$$\int_0^b \frac{1}{32} v e^{-v^2/64}\, dv = -\int_0^{-b^2/64} e^u\, du = -e^u \Big|_0^{-b^2/64} = -e^{-b^2/64} + 1.$$

(b) The probability that $v \in [2, 5]$ is the probability that $v \in [0, 5]$ minus the probability that $v \in [0, 2]$. By part (a), the probability that $v \in [2, 5]$ is

$$\left(1 - e^{-25/64}\right) - \left(1 - e^{-1/16}\right) = e^{-1/16} - e^{-25/64}.$$

In Exercises 95–96, use substitution to evaluate the integral in terms of $f(x)$.

95. $\int f(x)^3 f'(x)\, dx$

SOLUTION Let $u = f(x)$. Then $du = f'(x)\, dx$. Hence

$$\int f(x)^3 f'(x)\, dx = \int u^3\, du = \frac{1}{4} u^4 + C = \frac{1}{4} f(x)^4 + C.$$

97. Show that $\int_0^{\pi/6} f(\sin\theta)\, d\theta = \int_0^{1/2} f(u) \frac{1}{\sqrt{1-u^2}}\, du.$

SOLUTION Let $u = \sin\theta$. Then $u(\pi/6) = 1/2$ and $u(0) = 0$, as required. Furthermore, $du = \cos\theta\, d\theta$, so that

$$d\theta = \frac{du}{\cos\theta}.$$

If $\sin\theta = u$, then $u^2 + \cos^2\theta = 1$, so that $\cos\theta = \sqrt{1-u^2}$. Therefore $d\theta = du/\sqrt{1-u^2}$. This gives

$$\int_0^{\pi/6} f(\sin\theta)\, d\theta = \int_0^{1/2} f(u) \frac{1}{\sqrt{1-u^2}}\, du.$$

Further Insights and Challenges

99. Evaluate $I = \int_0^{\pi/2} \frac{d\theta}{1 + \tan^{6,000}\theta}$. *Hint:* Use substitution to show that I is equal to $J = \int_0^{\pi/2} \frac{d\theta}{1 + \cot^{6,000}\theta}$ and then check that $I + J = \int_0^{\pi/2} d\theta.$

SOLUTION To evaluate

$$I = \int_0^{\pi/2} \frac{dx}{1 + \tan^{6000} x},$$

we substitute $t = \pi/2 - x$. Then $dt = -dx$, $x = \pi/2 - t$, $t(0) = \pi/2$, and $t(\pi/2) = 0$. Hence,

$$I = \int_0^{\pi/2} \frac{dx}{1 + \tan^{6000} x} = -\int_{\pi/2}^{0} \frac{dt}{1 + \tan^{6000}(\pi/2 - t)} = \int_0^{\pi/2} \frac{dt}{1 + \cot^{6000} t}.$$

Let $J = \int_0^{\pi/2} \dfrac{dt}{1+\cot^{6000}(t)}$. We know $I = J$, so $I + J = 2I$. On the other hand, by the definition of I and J and the linearity of the integral,

$$
\begin{aligned}
I + J &= \int_0^{\pi/2} \frac{dx}{1+\tan^{6000} x} + \frac{dx}{1+\cot^{6000} x} = \int_0^{\pi/2} \left(\frac{1}{1+\tan^{6000} x} + \frac{1}{1+\cot^{6000} x} \right) dx \\
&= \int_0^{\pi/2} \left(\frac{1}{1+\tan^{6000} x} + \frac{1}{1+(1/\tan^{6000} x)} \right) dx \\
&= \int_0^{\pi/2} \left(\frac{1}{1+\tan^{6000} x} + \frac{1}{(\tan^{6000} x + 1)/\tan^{6000} x} \right) dx \\
&= \int_0^{\pi/2} \left(\frac{1}{1+\tan^{6000} x} + \frac{\tan^{6000} x}{1+\tan^{6000} x} \right) dx \\
&= \int_0^{\pi/2} \left(\frac{1+\tan^{6000} x}{1+\tan^{6000} x} \right) dx = \int_0^{\pi/2} 1\, dx = \pi/2.
\end{aligned}
$$

Hence, $I + J = 2I = \pi/2$, so $I = \pi/4$.

101. Prove that $\int_a^b \frac{1}{x}\,dx = \int_1^{b/a} \frac{1}{x}\,dx$ for $a, b > 0$. Then show that the regions under the hyperbola over the intervals $[1, 2]$, $[2, 4]$, $[4, 8]$, ... all have the same area (Figure 5).

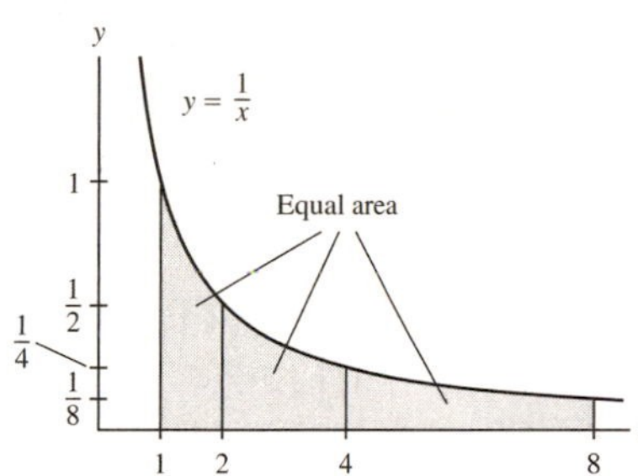

FIGURE 5 The area under $y = \frac{1}{x}$ over $[2^n, 2^{n+1}]$ is the same for all $n = 0, 1, 2, \dots$.

SOLUTION

(a) Let $u = \frac{x}{a}$. Then $au = x$ and $du = \frac{1}{a}\,dx$ or $a\,du = dx$. Hence

$$\int_a^b \frac{1}{x}\,dx = \int_1^{b/a} \frac{a}{au}\,du = \int_1^{b/a} \frac{1}{u}\,du.$$

Note that $\displaystyle\int_1^{b/a} \frac{1}{u}\,du = \int_1^{b/a} \frac{1}{x}\,dx$ after the substitution $x = u$.

(b) The area under the hyperbola over the interval $[1, 2]$ is given by the definite integral $\int_1^2 \frac{1}{x}\,dx$. Denote this definite integral by A. Using the result from part (a), we find the area under the hyperbola over the interval $[2, 4]$ is

$$\int_2^4 \frac{1}{x}\,dx = \int_1^{4/2} \frac{1}{x}\,dx = \int_1^2 \frac{1}{x}\,dx = A.$$

Similarly, the area under the hyperbola over the interval $[4, 8]$ is

$$\int_4^8 \frac{1}{x}\,dx = \int_1^{8/4} \frac{1}{x}\,dx = \int_1^2 \frac{1}{x}\,dx = A.$$

In general, the area under the hyperbola over the interval $[2^n, 2^{n+1}]$ is

$$\int_{2^n}^{2^{n+1}} \frac{1}{x}\,dx = \int_1^{2^{n+1}/2^n} \frac{1}{x}\,dx = \int_1^2 \frac{1}{x}\,dx = A.$$

103. Area of an Ellipse Prove the formula $A = \pi ab$ for the area of the ellipse with equation (Figure 7)

$$\frac{x^2}{a^2} + \frac{y^2}{b^2} = 1$$

Hint: Use a change of variables to show that A is equal to ab times the area of the unit circle.

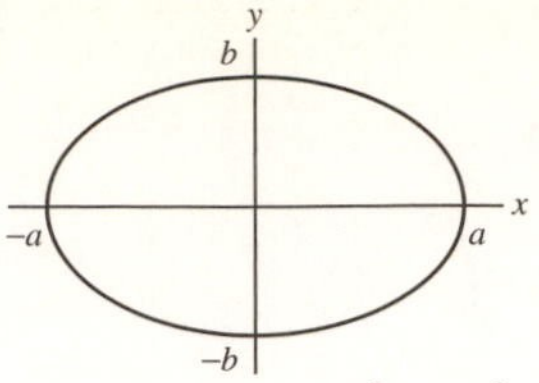

FIGURE 7 Graph of $\frac{x^2}{a^2} + \frac{y^2}{b^2} = 1$.

SOLUTION Consider the ellipse with equation $\frac{x^2}{a^2} + \frac{y^2}{b^2} = 1$; here $a, b > 0$. The area between the part of the ellipse in the upper half-plane, $y = f(x) = \sqrt{b^2\left(1 - \frac{x^2}{a^2}\right)}$, and the x-axis is $\int_{-a}^{a} f(x)\,dx$. By symmetry, the part of the elliptical region in the lower half-plane has the same area. Accordingly, the area enclosed by the ellipse is

$$2\int_{-a}^{a} f(x)\,dx = 2\int_{-a}^{a} \sqrt{b^2\left(1 - \frac{x^2}{a^2}\right)}\,dx = 2b\int_{-a}^{a} \sqrt{1 - (x/a)^2}\,dx$$

Now, let $u = x/a$. Then $x = au$ and $a\,du = dx$. Accordingly,

$$2b\int_{-a}^{a} \sqrt{1 - \left(\frac{x}{a}\right)^2}\,dx = 2ab\int_{-1}^{1} \sqrt{1 - u^2}\,du = 2ab\left(\frac{\pi}{2}\right) = \pi ab$$

Here we recognized that $\int_{-1}^{1} \sqrt{1 - u^2}\,du$ represents the area of the upper unit semicircular disk, which by Exercise 102 is $2(\frac{\pi}{4}) = \frac{\pi}{2}$.

5.7 Further Transcendental Functions

Preliminary Questions

1. Find b such that $\int_1^b \frac{dx}{x}$ is equal to

(a) $\ln 3$

(b) 3

SOLUTION For $b > 0$,

$$\int_1^b \frac{dx}{x} = \ln|x|\Big|_1^b = \ln b - \ln 1 = \ln b.$$

(a) For the value of the definite integral to equal $\ln 3$, we must have $b = 3$.

(b) For the value of the definite integral to equal 3, we must have $b = e^3$.

2. Find b such that $\int_0^b \frac{dx}{1+x^2} = \frac{\pi}{3}$.

SOLUTION In general,

$$\int_0^b \frac{dx}{1+x^2} = \tan^{-1} x\Big|_0^b = \tan^{-1} b - \tan^{-1} 0 = \tan^{-1} b.$$

For the value of the definite integral to equal $\frac{\pi}{3}$, we must have

$$\tan^{-1} b = \frac{\pi}{3} \quad \text{or} \quad b = \tan\frac{\pi}{3} = \sqrt{3}.$$

3. Which integral should be evaluated using substitution?

(a) $\int \frac{9\,dx}{1+x^2}$

(b) $\int \frac{dx}{1+9x^2}$

SOLUTION Use the substitution $u = 3x$ on the integral in **(b)**.

4. Which relation between x and u yields $\sqrt{16+x^2} = 4\sqrt{1+u^2}$?

SOLUTION To transform $\sqrt{16+x^2}$ into $4\sqrt{1+u^2}$, make the substitution $x = 4u$.

Exercises

In Exercises 1–10, evaluate the definite integral.

1. $\displaystyle\int_1^9 \frac{dx}{x}$

SOLUTION $\displaystyle\int_1^9 \frac{1}{x}\,dx = \ln|x|\Big|_1^9 = \ln 9 - \ln 1 = \ln 9.$

3. $\displaystyle\int_1^{e^3} \frac{1}{t}\,dt$

SOLUTION $\displaystyle\int_1^{e^3} \frac{1}{t}\,dt = \ln|t|\Big|_1^{e^3} = \ln e^3 - \ln 1 = 3.$

5. $\displaystyle\int_2^{12} \frac{dt}{3t+4}$

SOLUTION Let $u = 3t+4$. Then $du = 3\,dt$ and

$$\int_2^{12} \frac{dt}{3t+4} = \frac{1}{3}\int_{10}^{40} \frac{du}{u} = \frac{1}{3}\ln|u|\Big|_{10}^{40} = \frac{1}{3}(\ln 40 - \ln 10) = \frac{1}{3}\ln 4.$$

7. $\displaystyle\int_{\tan 1}^{\tan 8} \frac{dx}{x^2+1}$

SOLUTION $\displaystyle\int_{\tan 1}^{\tan 8} \frac{dx}{1+x^2} = \tan^{-1} x\Big|_{\tan 1}^{\tan 8} = \tan^{-1}(\tan 8) - \tan^{-1}(\tan 1) = 8 - 1 = 7.$

9. $\displaystyle\int_0^{1/2} \frac{dx}{\sqrt{1-x^2}}$

SOLUTION $\displaystyle\int_0^{1/2} \frac{dx}{\sqrt{1-x^2}} = \sin^{-1} x\Big|_0^{1/2} = \sin^{-1}\frac{1}{2} - \sin^{-1} 0 = \frac{\pi}{6}.$

11. Use the substitution $u = x/3$ to prove

$$\int \frac{dx}{9+x^2} = \frac{1}{3}\tan^{-1}\frac{x}{3} + C$$

SOLUTION Let $u = x/3$. Then, $x = 3u$, $dx = 3\,du$, $9+x^2 = 9(1+u^2)$, and

$$\int \frac{dx}{9+x^2} = \int \frac{3\,du}{9(1+u^2)} = \frac{1}{3}\int \frac{du}{1+u^2} = \frac{1}{3}\tan^{-1} u + C = \frac{1}{3}\tan^{-1}\frac{x}{3} + C.$$

In Exercises 13–32, calculate the integral.

13. $\displaystyle\int_0^3 \frac{dx}{x^2+3}$

SOLUTION Let $x = \sqrt{3}u$. Then $dx = \sqrt{3}\,du$ and

$$\int_0^3 \frac{dx}{x^2+3} = \frac{1}{\sqrt{3}}\int_0^{\sqrt{3}} \frac{du}{u^2+1} = \frac{1}{\sqrt{3}}\tan^{-1} u\Big|_0^{\sqrt{3}} = \frac{1}{\sqrt{3}}(\tan^{-1}\sqrt{3} - \tan^{-1} 0) = \frac{\pi}{3\sqrt{3}}.$$

15. $\displaystyle\int \frac{dt}{\sqrt{1-16t^2}}$

SOLUTION Let $u = 4t$. Then $du = 4\,dt$, and

$$\int \frac{dt}{\sqrt{1-16t^2}} = \int \frac{du}{4\sqrt{1-u^2}} = \frac{1}{4}\sin^{-1} u + C = \frac{1}{4}\sin^{-1}(4t) + C.$$

17. $\displaystyle\int \frac{dt}{\sqrt{5-3t^2}}$

SOLUTION Let $t = \sqrt{5/3}u$. Then $dt = \sqrt{5/3}\,du$ and

$$\int \frac{dt}{\sqrt{5-3t^2}} = \int \frac{\sqrt{5/3}\,du}{\sqrt{5}\sqrt{1-t^2}} = \frac{1}{\sqrt{3}}\int \frac{du}{\sqrt{1-u^2}} = \frac{1}{\sqrt{3}}\sin^{-1} u + C = \frac{1}{\sqrt{3}}\sin^{-1}\sqrt{\frac{3}{5}}t + C.$$

19. $\displaystyle\int \frac{dx}{x\sqrt{12x^2-3}}$

SOLUTION Let $u = 2x$. Then $du = 2\,dx$ and

$$\int \frac{dx}{x\sqrt{12x^2-3}} = \frac{1}{\sqrt{3}}\int \frac{du}{u\sqrt{u^2-1}} = \frac{1}{\sqrt{3}}\sec^{-1} u + C = \frac{1}{\sqrt{3}}\sec^{-1}(2x) + C.$$

21. $\displaystyle\int \frac{dx}{x\sqrt{x^4-1}}$

SOLUTION Let $u = x^2$. Then $du = 2x\,dx$, and

$$\int \frac{dx}{x\sqrt{x^4-1}} = \int \frac{du}{2u\sqrt{u^2-1}} = \frac{1}{2}\sec^{-1} u + C = \frac{1}{2}\sec^{-1} x^2 + C.$$

23. $\displaystyle\int_{-\ln 2}^{0} \frac{e^x\,dx}{1+e^{2x}}$

SOLUTION Let $u = e^x$. Then $du = e^x\,dx$, and

$$\int_{-\ln 2}^{0} \frac{e^x\,dx}{1+e^{2x}} = \int_{1/2}^{1} \frac{du}{1+u^2} = \tan^{-1} u\Big|_{1/2}^{1} = \frac{\pi}{4} - \tan^{-1}(1/2).$$

25. $\displaystyle\int \frac{\tan^{-1} x\,dx}{1+x^2}$

SOLUTION Let $u = \tan^{-1} x$. Then $du = \dfrac{dx}{1+x^2}$, and

$$\int \frac{\tan^{-1} x\,dx}{1+x^2} = \int u\,du = \frac{1}{2}u^2 + C = \frac{(\tan^{-1} x)^2}{2} + C.$$

27. $\displaystyle\int_0^1 3^x\,dx$

SOLUTION $\displaystyle\int_0^1 3^x\,dx = \frac{3^x}{\ln 3}\Big|_0^1 = \frac{1}{\ln 3}(3-1) = \frac{2}{\ln 3}.$

29. $\displaystyle\int_0^{\log_4(3)} 4^x\,dx$

SOLUTION $\displaystyle\int_0^{\log_4(3)} 4^x\,dx = \frac{4^x}{\ln 4}\Big|_0^{\log_4 3} = \frac{1}{\ln 4}(3-1) = \frac{2}{\ln 4} = \frac{1}{\ln 2}.$

31. $\displaystyle\int 9^x \sin(9^x)\,dx$

SOLUTION Let $u = 9^x$. Then $du = 9^x \ln 9\,dx$ and

$$\int 9^x \sin(9^x)\,dx = \frac{1}{\ln 9}\int \sin u\,du = -\frac{1}{\ln 9}\cos u + C = -\frac{1}{\ln 9}\cos(9^x) + C.$$

In Exercises 33–70, evaluate the integral using the methods covered in the text so far.

33. $\displaystyle\int y e^{y^2}\,dy$

SOLUTION Use the substitution $u = y^2$, $du = 2y\,dy$. Then

$$\int y e^{y^2}\,dy = \frac{1}{2}\int e^u\,du = \frac{1}{2}e^u + C = \frac{1}{2}e^{y^2} + C.$$

35. $\displaystyle\int \frac{x\,dx}{\sqrt{4x^2+9}}$

SOLUTION Let $u = 4x^2+9$. Then $du = 8x\,dx$ and

$$\int \frac{x}{\sqrt{4x^2+9}}\,dx = \frac{1}{8}\int u^{-1/2}\,du = \frac{1}{4}u^{1/2} + C = \frac{1}{4}\sqrt{4x^2+9} + C.$$

37. $\int 7^{-x}\,dx$

SOLUTION Let $u = -x$. Then $du = -dx$ and

$$\int 7^{-x}\,dx = -\int 7^u\,du = -\frac{7^u}{\ln 7} + C = -\frac{7^{-x}}{\ln 7} + C.$$

39. $\int \sec^2\theta \tan^7\theta\,d\theta$

SOLUTION Let $u = \tan\theta$. Then $du = \sec^2\theta\,d\theta$ and

$$\int \sec^2\theta \tan^7\theta\,d\theta = \int u^7\,du = \frac{1}{8}u^8 + C = \frac{1}{8}\tan^8\theta + C.$$

41. $\int \frac{t\,dt}{\sqrt{7-t^2}}$

SOLUTION Let $u = 7 - t^2$. Then $du = -2t\,dt$ and

$$\int \frac{t\,dt}{\sqrt{7-t^2}} = -\frac{1}{2}\int u^{-1/2}\,du = -u^{1/2} + C = -\sqrt{7-t^2} + C.$$

43. $\int \frac{(3x+2)\,dx}{x^2+4}$

SOLUTION Write

$$\int \frac{(3x+2)\,dx}{x^2+4} = \int \frac{3x\,dx}{x^2+4} + \int \frac{2\,dx}{x^2+4}.$$

In the first integral, let $u = x^2 + 4$. Then $du = 2x\,dx$ and

$$\int \frac{3x\,dx}{x^2+4} = \frac{3}{2}\int \frac{du}{u} = \frac{3}{2}\ln|u| + C_1 = \frac{3}{2}\ln(x^2+4) + C_1.$$

For the second integral, let $x = 2u$. Then $dx = 2\,du$ and

$$\int \frac{2\,dx}{x^2+4} = \int \frac{du}{u^2+1} = \tan^{-1}u + C_2 = \tan^{-1}(x/2) + C_2.$$

Combining these two results yields

$$\int \frac{(3x+2)\,dx}{x^2+4} = \frac{3}{2}\ln(x^2+4) + \tan^{-1}(x/2) + C.$$

45. $\int \frac{dx}{\sqrt{1-16x^2}}$

SOLUTION Let $u = 4x$. Then $du = 4\,dx$ and

$$\int \frac{dx}{\sqrt{1-16x^2}} = \frac{1}{4}\int \frac{du}{\sqrt{1-u^2}} = \frac{1}{4}\sin^{-1}u + C = \frac{1}{4}\sin^{-1}(4x) + C.$$

47. $\int (e^{-x} - 4x)\,dx$

SOLUTION First, observe that

$$\int (e^{-x} - 4x)\,dx = \int e^{-x}\,dx - \int 4x\,dx = \int e^{-x}\,dx - 2x^2.$$

In the remaining integral, use the substitution $u = -x$, $du = -dx$. Then

$$\int e^{-x}\,dx = -\int e^u\,du = -e^u + C = -e^{-x} + C.$$

Finally,

$$\int (e^{-x} - 4x)\,dx = -e^{-x} - 2x^2 + C.$$

49. $\displaystyle\int \frac{e^{2x} - e^{4x}}{e^x}\,dx$

SOLUTION

$$\int \left(\frac{e^{2x} - e^{4x}}{e^x}\right) dx = \int (e^x - e^{3x})\,dx = e^x - \frac{e^{3x}}{3} + C.$$

51. $\displaystyle\int \frac{(x+5)\,dx}{\sqrt{4-x^2}}$

SOLUTION Write

$$\int \frac{(x+5)\,dx}{\sqrt{4-x^2}} = \int \frac{x\,dx}{\sqrt{4-x^2}} + \int \frac{5\,dx}{\sqrt{4-x^2}}.$$

In the first integral, let $u = 4 - x^2$. Then $du = -2x\,dx$ and

$$\int \frac{x\,dx}{\sqrt{4-x^2}} = -\frac{1}{2}\int u^{-1/2}\,du = -u^{1/2} + C_1 = -\sqrt{4-x^2} + C_1.$$

In the second integral, let $x = 2u$. Then $dx = 2\,du$ and

$$\int \frac{5\,dx}{\sqrt{4-x^2}} = 5\int \frac{du}{\sqrt{1-u^2}} = 5\sin^{-1} u + C_2 = 5\sin^{-1}(x/2) + C_2.$$

Combining these two results yields

$$\int \frac{(x+5)\,dx}{\sqrt{4-x^2}} = -\sqrt{4-x^2} + 5\sin^{-1}(x/2) + C.$$

53. $\displaystyle\int e^x \cos(e^x)\,dx$

SOLUTION Use the substitution $u = e^x$, $du = e^x\,dx$. Then

$$\int e^x \cos(e^x)\,dx = \int \cos u\,du = \sin u + C = \sin(e^x) + C.$$

55. $\displaystyle\int \frac{dx}{\sqrt{9-16x^2}}$

SOLUTION First rewrite

$$\int \frac{dx}{\sqrt{9-16x^2}} = \frac{1}{3}\int \frac{dx}{\sqrt{1-\left(\frac{4}{3}x\right)^2}}.$$

Now, let $u = \frac{4}{3}x$. Then $du = \frac{4}{3}\,dx$ and

$$\int \frac{dx}{\sqrt{9-16x^2}} = \frac{1}{4}\int \frac{du}{\sqrt{1-u^2}} = \frac{1}{4}\sin^{-1} u + C = \frac{1}{4}\sin^{-1}\left(\frac{4x}{3}\right) + C.$$

57. $\displaystyle\int e^x (e^{2x} + 1)^3\,dx$

SOLUTION Use the substitution $u = e^x$, $du = e^x\,dx$. Then

$$\begin{aligned}\int e^x (e^{2x}+1)^3\,dx &= \int \left(u^2+1\right)^3 du = \int \left(u^6 + 3u^4 + 3u^2 + 1\right) du \\ &= \frac{1}{7}u^7 + \frac{3}{5}u^5 + u^3 + u + C = \frac{1}{7}(e^x)^7 + \frac{3}{5}(e^x)^5 + (e^x)^3 + e^x + C \\ &= \frac{e^{7x}}{7} + \frac{3e^{5x}}{5} + e^{3x} + e^x + C.\end{aligned}$$

59. $\displaystyle\int \frac{x^2\,dx}{x^3+2}$

SOLUTION Let $u = x^3 + 2$. Then $du = 3x^2\,dx$, and

$$\int \frac{x^2\,dx}{x^3+2} = \frac{1}{3}\int \frac{du}{u} = \frac{1}{3}\ln|x^3+2| + C.$$

61. $\displaystyle\int \cot x\,dx$

SOLUTION We rewrite $\int \cot x\,dx$ as $\int \frac{\cos x}{\sin x}\,dx$. Let $u = \sin x$. Then $du = \cos x\,dx$, and

$$\int \frac{\cos x}{\sin x}\,dx = \int \frac{du}{u} = \ln|\sin x| + C.$$

63. $\displaystyle\int \frac{4\ln x + 5}{x}\,dx$

SOLUTION Let $u = 4\ln x + 5$. Then $du = (4/x)dx$, and

$$\int \frac{4\ln x + 5}{x} dx = \frac{1}{4}\int u\,du = \frac{1}{8}u^2 + C = \frac{1}{8}(4\ln x + 5)^2 + C.$$

65. $\displaystyle\int x3^{x^2}\,dx$

SOLUTION Let $u = x^2$. Then $du = 2x\,dx$, and

$$\int x3^{x^2}\,dx = \frac{1}{2}\int 3^u du = \frac{1}{2}\frac{3^u}{\ln 3} + C = \frac{3^{x^2}}{2\ln 3} + C.$$

67. $\displaystyle\int \cot x \ln(\sin x)\,dx$

SOLUTION Let $u = \ln(\sin x)$. Then

$$du = \frac{1}{\sin x} \cdot \cos x\,dx = \cot x\,dx,$$

and

$$\int \cot x \ln(\sin x)\,dx = \int u\,du = \frac{u^2}{2} + C = \frac{(\ln(\sin x))^2}{2} + C.$$

69. $\displaystyle\int t^2\sqrt{t-3}\,dt$

SOLUTION Let $u = t - 3$. Then $t = u + 3$, $du = dt$ and

$$\begin{aligned}
\int t^2\sqrt{t-3}\,dt &= \int (u+3)^2\sqrt{u}\,du \\
&= \int (u^2 + 6u + 9)\sqrt{u}\,du = \int (u^{5/2} + 6u^{3/2} + 9u^{1/2})\,du \\
&= \frac{2}{7}u^{7/2} + \frac{12}{5}u^{5/2} + 6u^{3/2} + C \\
&= \frac{2}{7}(t-3)^{7/2} + \frac{12}{5}(t-3)^{5/2} + 6(t-3)^{3/2} + C.
\end{aligned}$$

71. Use Figure 4 to prove

$$\int_0^x \sqrt{1-t^2}\,dt = \frac{1}{2}x\sqrt{1-x^2} + \frac{1}{2}\sin^{-1} x$$

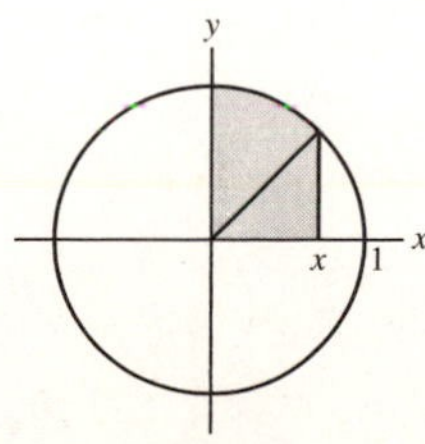

FIGURE 4

SOLUTION The definite integral $\int_0^x \sqrt{1-t^2}\,dt$ represents the area of the region under the upper half of the unit circle from 0 to x. The region consists of a sector of the circle and a right triangle. The sector has a central angle of $\frac{\pi}{2}-\theta$, where $\cos\theta = x$. Hence, the sector has an area of

$$\frac{1}{2}(1)^2\left(\frac{\pi}{2}-\cos^{-1}x\right) = \frac{1}{2}\sin^{-1}x.$$

The right triangle has a base of length x, a height of $\sqrt{1-x^2}$, and hence an area of $\frac{1}{2}x\sqrt{1-x^2}$. Thus,

$$\int_0^x \sqrt{1-t^2}\,dt = \frac{1}{2}x\sqrt{1-x^2} + \frac{1}{2}\sin^{-1}x.$$

73. Prove:

$$\int \sin^{-1}t\,dt = \sqrt{1-t^2} + t\sin^{-1}t.$$

SOLUTION Let $G(t) = \sqrt{1-t^2} + t\sin^{-1}t$. Then

$$G'(t) = \frac{d}{dt}\sqrt{1-t^2} + \frac{d}{dt}\left(t\sin^{-1}t\right) = \frac{-t}{\sqrt{1-t^2}} + \left(t\cdot\frac{d}{dt}\sin^{-1}t + \sin^{-1}t\right)$$

$$= \frac{-t}{\sqrt{1-t^2}} + \left(\frac{t}{\sqrt{1-t^2}} + \sin^{-1}t\right) = \sin^{-1}t.$$

This proves the formula $\int \sin^{-1}t\,dt = \sqrt{1-t^2} + t\sin^{-1}t$.

Further Insights and Challenges

75. Recall that if $f(t) \ge g(t)$ for $t \ge 0$, then for all $x \ge 0$,

$$\int_0^x f(t)\,dt \ge \int_0^x g(t)\,dt \qquad \boxed{7}$$

The inequality $e^t \ge 1$ holds for $t \ge 0$ because $e > 1$. Use Eq. (7) to prove that $e^x \ge 1 + x$ for $x \ge 0$. Then prove, by successive integration, the following inequalities (for $x \ge 0$):

$$e^x \ge 1 + x + \frac{1}{2}x^2, \qquad e^x \ge 1 + x + \frac{1}{2}x^2 + \frac{1}{6}x^3$$

SOLUTION Integrating both sides of the inequality $e^t \ge 1$ yields

$$\int_0^x e^t\,dt = e^x - 1 \ge x \quad \text{or} \quad e^x \ge 1 + x.$$

Integrating both sides of this new inequality then gives

$$\int_0^x e^t\,dt = e^x - 1 \ge x + x^2/2 \quad \text{or} \quad e^x \ge 1 + x + x^2/2.$$

Finally, integrating both sides again gives

$$\int_0^x e^t\,dt = e^x - 1 \ge x + x^2/2 + x^3/6 \quad \text{or} \quad e^x \ge 1 + x + x^2/2 + x^3/6$$

as requested.

77. Use Exercise 75 to show that $e^x/x^2 \ge x/6$ and conclude that $\lim_{x\to\infty} e^x/x^2 = \infty$. Then use Exercise 76 to prove more generally that $\lim_{x\to\infty} e^x/x^n = \infty$ for all n.

SOLUTION By Exercise 75, $e^x \ge 1 + x + \frac{x^2}{2} + \frac{x^3}{6}$. Thus

$$\frac{e^x}{x^2} \ge \frac{1}{x^2} + \frac{1}{x} + \frac{1}{2} + \frac{x}{6} \ge \frac{x}{6}.$$

Since $\lim_{x\to\infty} x/6 = \infty$, $\lim_{x\to\infty} e^x/x^2 = \infty$. More generally, by Exercise 76,

$$e^x \ge 1 + \frac{x^2}{2} + \cdots + \frac{x^{n+1}}{(n+1)!}.$$

Thus

$$\frac{e^x}{x^n} \geq \frac{1}{x^n} + \cdots + \frac{x}{(n+1)!} \geq \frac{x}{(n+1)!}.$$

Since $\lim_{x\to\infty} \frac{x}{(n+1)!} = \infty$, $\lim_{x\to\infty} \frac{e^x}{x^n} = \infty$.

Exercises 78–80 develop an elegant approach to the exponential and logarithm functions. Define a function $G(x)$ for $x > 0$:

$$G(x) = \int_1^x \frac{1}{t}\,dt$$

79. Defining e^x Use Exercise 78 to prove the following statements.

(a) $G(x)$ has an inverse with domain $\mathbf{R}$ and range $\{x : x > 0\}$. Denote the inverse by $F(x)$.

(b) $F(x+y) = F(x)F(y)$ for all x, y. *Hint:* It suffices to show that $G(F(x)F(y)) = G(F(x+y))$.

(c) $F(r) = E^r$ for all numbers. In particular, $F(0) = 1$.

(d) $F'(x) = F(x)$. *Hint:* Use the formula for the derivative of an inverse function.

This shows that $E = e$ and $F(x)$ is the function e^x as defined in the text.

SOLUTION

(a) The domain of $G(x)$ is $x > 0$ and, by part (i) of the previous exercise, the range of $G(x)$ is $\mathbf{R}$. Now,

$$G'(x) = \frac{1}{x} > 0$$

for all $x > 0$. Thus, $G(x)$ is increasing on its domain, which implies that $G(x)$ has an inverse. The domain of the inverse is $\mathbf{R}$ and the range is $\{x : x > 0\}$. Let $F(x)$ denote the inverse of $G(x)$.

(b) Let x and y be real numbers and suppose that $x = G(w)$ and $y = G(z)$ for some positive real numbers w and z. Then, using part (b) of the previous exercise

$$F(x+y) = F(G(w) + G(z)) = F(G(wz)) = wz = F(x) + F(y).$$

(c) Let r be any real number. By part (k) of the previous exercise, $G(E^r) = r$. By definition of an inverse function, it then follows that $F(r) = E^r$.

(d) By the formula for the derivative of an inverse function

$$F'(x) = \frac{1}{G'(F(x))} = \frac{1}{1/F(x)} = F(x).$$

81. The formula $\int x^n\,dx = \frac{x^{n+1}}{n+1} + C$ is valid for $n \neq -1$. Show that the exceptional case $n = -1$ is a limit of the general case by applying L'Hôpital's Rule to the limit on the left.

$$\lim_{n\to -1} \int_1^x t^n\,dt = \int_1^x t^{-1}\,dt \qquad \text{(for fixed } x > 0\text{)}$$

Note that the integral on the left is equal to $\frac{x^{n+1} - 1}{n+1}$.

SOLUTION

$$\lim_{n\to -1} \int_1^x t^n\,dt = \lim_{n\to -1} \left.\frac{t^{n+1}}{n+1}\right|_1^x = \lim_{n\to -1} \left(\frac{x^{n+1}}{n+1} - \frac{1^{n+1}}{n+1}\right)$$

$$= \lim_{n\to -1} \frac{x^{n+1} - 1}{n+1} = \lim_{n\to -1} (x^{n+1}) \ln x = \ln x = \int_1^x t^{-1}\,dt$$

Note that when using L'Hôpital's Rule in the second line, we need to differentiate with respect to n.

83. **(a)** Explain why the shaded region in Figure 5 has area $\int_0^{\ln a} e^y\,dy$.

(b) Prove the formula $\int_1^a \ln x\,dx = a\ln a - \int_0^{\ln a} e^y\,dy$.

(c) Conclude that $\int_1^a \ln x\,dx = a\ln a - a + 1$.

(d) Use the result of (a) to find an antiderivative of $\ln x$.

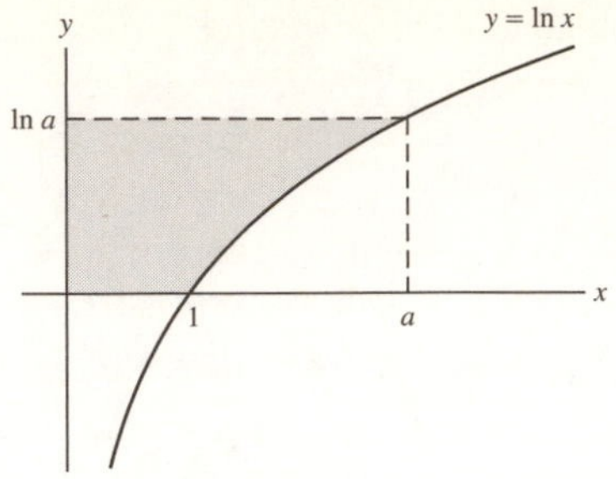

FIGURE 5

SOLUTION

(a) Interpreting the graph with y as the independent variable, we see that the function is $x = e^y$. Integrating in y then gives the area of the shaded region as $\int_0^{\ln a} e^y\,dy$

(b) We can obtain the area under the graph of $y = \ln x$ from $x = 1$ to $x = a$ by computing the area of the rectangle extending from $x = 0$ to $x = a$ horizontally and from $y = 0$ to $y = \ln a$ vertically and then subtracting the area of the shaded region. This yields

$$\int_1^a \ln x\,dx = a\ln a - \int_0^{\ln a} e^y\,dy.$$

(c) By direct calculation

$$\int_0^{\ln a} e^y\,dy = e^y\Big|_0^{\ln a} = a - 1.$$

Thus,

$$\int_1^a \ln x\,dx = a\ln a - (a-1) = a\ln a - a + 1.$$

(d) Based on these results it appears that

$$\int \ln x\,dx = x\ln x - x + C.$$

5.8 Exponential Growth and Decay

Preliminary Questions

1. Two quantities increase exponentially with growth constants $k = 1.2$ and $k = 3.4$, respectively. Which quantity doubles more rapidly?

SOLUTION Doubling time is inversely proportional to the growth constant. Consequently, the quantity with $k = 3.4$ doubles more rapidly.

2. A cell population grows exponentially beginning with one cell. Which takes longer: increasing from one to two cells or increasing from 15 million to 20 million cells?

SOLUTION It takes longer for the population to increase from one cell to two cells, because this requires doubling the population. Increasing from 15 million to 20 million is less than doubling the population.

3. Referring to his popular book *A Brief History of Time*, the renowned physicist Stephen Hawking said, "Someone told me that each equation I included in the book would halve its sales." Find a differential equation satisfied by the function $S(n)$, the number of copies sold if the book has n equations.

SOLUTION Let $S(0)$ denote the sales with no equations in the book. Translating Hawking's observation into an equation yields

$$S(n) = \frac{S(0)}{2^n}.$$

Differentiating with respect to n then yields

$$\frac{dS}{dn} = S(0)\frac{d}{dn}2^{-n} = -\ln 2S(0)2^{-n} = -\ln 2S(n).$$

4. The PV of N dollars received at time T is (choose the correct answer):

(a) The value at time T of N dollars invested today

(b) The amount you would have to invest today in order to receive N dollars at time T

SOLUTION The correct response is **(b)**: the PV of N dollars received at time T is the amount you would have to invest today in order to receive N dollars at time T.

5. In one year, you will be paid \$1. Will the PV increase or decrease if the interest rate goes up?

SOLUTION If the interest rate goes up, the present value of \$1 a year from now will decrease.

Exercises

1. A certain population P of bacteria obeys the exponential growth law $P(t) = 2000e^{1.3t}$ (t in hours).

(a) How many bacteria are present initially?

(b) At what time will there be 10,000 bacteria?

SOLUTION

(a) $P(0) = 2000e^0 = 2000$ bacteria initially.

(b) We solve $2000e^{1.3t} = 10,000$ for t. Thus, $e^{1.3t} = 5$ or

$$t = \frac{1}{1.3} \ln 5 \approx 1.24 \text{ hours.}$$

3. Write $f(t) = 5(7)^t$ in the form $f(t) = P_0 e^{kt}$ for some P_0 and k.

SOLUTION Because $7 = e^{\ln 7}$, it follows that

$$f(t) = 5(7)^t = 5(e^{\ln 7})^t = 5e^{t \ln 7}.$$

Thus, $P_0 = 5$ and $k = \ln 7$.

5. A certain RNA molecule replicates every 3 minutes. Find the differential equation for the number $N(t)$ of molecules present at time t (in minutes). How many molecules will be present after one hour if there is one molecule at $t = 0$?

SOLUTION The doubling time is $\frac{\ln 2}{k}$ so $k = \frac{\ln 2}{\text{doubling time}}$. Thus, the differential equation is $N'(t) = kN(t) = \frac{\ln 2}{3} N(t)$. With one molecule initially,

$$N(t) = e^{(\ln 2/3)t} = 2^{t/3}.$$

Thus, after one hour, there are

$$N(60) = 2^{60/3} = 1{,}048{,}576$$

molecules present.

7. Find all solutions to the differential equation $y' = -5y$. Which solution satisfies the initial condition $y(0) = 3.4$?

SOLUTION $y' = -5y$, so $y(t) = Ce^{-5t}$ for some constant C. The initial condition $y(0) = 3.4$ determines $C = 3.4$. Therefore, $y(t) = 3.4e^{-5t}$.

9. Find the solution to $y' = 3y$ satisfying $y(2) = 1000$.

SOLUTION $y' = 3y$, so $y(t) = Ce^{3t}$ for some constant C. The initial condition $y(2) = 1000$ determines $C = \frac{1000}{e^6}$. Therefore, $y(t) = \frac{1000}{e^6} e^{3t} = 1000e^{3(t-2)}$.

11. The decay constant of cobalt-60 is 0.13 year^{-1}. Find its half-life.

SOLUTION Half-life $= \frac{\ln 2}{0.13} \approx 5.33$ years.

13. One of the world's smallest flowering plants, *Wolffia globosa* (Figure 13), has a doubling time of approximately 30 hours. Find the growth constant k and determine the initial population if the population grew to 1000 after 48 hours.

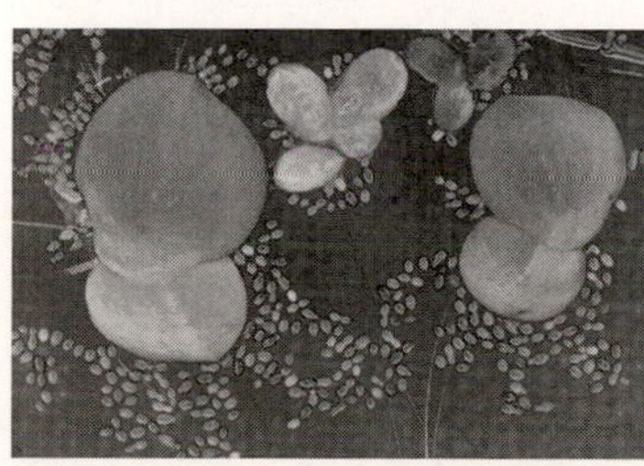

FIGURE 13 The tiny plants are *Wolffia*, with plant bodies smaller than the head of a pin.

SOLUTION By the formula for the doubling time, $30 = \frac{\ln 2}{k}$. Therefore,

$$k = \frac{\ln 2}{30} \approx 0.023 \text{ hours}^{-1}.$$

The plant population after t hours is $P(t) = P_0 e^{0.023t}$. If $P(48) = 1000$, then

$$P_0 e^{(0.023)48} = 1000 \quad \Rightarrow \quad P_0 = 1000e^{-(0.023)48} \approx 332$$

15. The population of a city is $P(t) = 2 \cdot e^{0.06t}$ (in millions), where t is measured in years. Calculate the time it takes for the population to double, to triple, and to increase seven-fold.

SOLUTION Since $k = 0.06$, the doubling time is

$$\frac{\ln 2}{k} \approx 11.55 \text{ years.}$$

The tripling time is calculated in the same way as the doubling time. Solve for Δ in the equation

$$\begin{aligned} P(t+\Delta) &= 3P(t) \\ 2 \cdot e^{0.06(t+\Delta)} &= 3(2e^{0.06t}) \\ 2 \cdot e^{0.06t} e^{0.06\Delta} &= 3(2e^{0.06t}) \\ e^{0.06\Delta} &= 3 \\ 0.06\Delta &= \ln 3, \end{aligned}$$

or $\Delta = \ln 3/0.06 \approx 18.31$ years. Working in a similar fashion, we find that the time required for the population to increase seven-fold is

$$\frac{\ln 7}{k} = \frac{\ln 7}{0.06} \approx 32.43 \text{ years.}$$

17. The decay constant for a certain drug is $k = 0.35$ day^{-1}. Calculate the time it takes for the quantity present in the bloodstream to decrease by half, by one-third, and by one-tenth.

SOLUTION The time required for the quantity present in the bloodstream to decrease by half is

$$\frac{\ln 2}{k} = \frac{\ln 2}{0.35} \approx 1.98 \text{ days.}$$

To decay by one-third, the time is

$$\frac{\ln 3}{k} = \frac{\ln 3}{0.35} \approx 3.14 \text{ days.}$$

Finally, to decay by one-tenth, the time is

$$\frac{\ln 10}{k} = \frac{\ln 10}{0.35} \approx 6.58 \text{ days.}$$

19. Assuming that population growth is approximately exponential, which of the following two sets of data is most likely to represent the population (in millions) of a city over a 5-year period?

Year	2000	2001	2002	2003	2004
Set I	3.14	3.36	3.60	3.85	4.11
Set II	3.14	3.24	3.54	4.04	4.74

SOLUTION If the population growth is approximately exponential, then the ratio between successive years' data needs to be approximately the same.

Year	2000		2001		2002		2003		2004
Data I	3.14		3.36		3.60		3.85		4.11
Ratios		1.07006		1.07143		1.06944		1.06753	
Data II	3.14		3.24		3.54		4.04		4.74
Ratios		1.03185		1.09259		1.14124		1.17327	

As you can see, the ratio of successive years in the data from "Data I" is very close to 1.07. Therefore, we would expect exponential growth of about $P(t) \approx (3.14)(1.07^t)$.

21. Degrees in Physics One study suggests that from 1955 to 1970, the number of bachelor's degrees in physics awarded per year by U.S. universities grew exponentially, with growth constant $k = 0.1$.

(a) If exponential growth continues, how long will it take for the number of degrees awarded per year to increase 14-fold?

(b) If 2500 degrees were awarded in 1955, in which year were 10,000 degrees awarded?

SOLUTION

(a) The time required for the number of degrees to increase 14-fold is

$$\frac{\ln 14}{k} = \frac{\ln 14}{0.1} \approx 26.39 \text{ years.}$$

(b) The doubling time is $(\ln 2)/0.1 \approx 0.693/0.1 = 6.93$ years. Since degrees are usually awarded once a year, we round off the doubling time to 7 years. The number quadruples after 14 years, so 10, 000 degrees would be awarded in 1969.

23. A sample of sheepskin parchment discovered by archaeologists had a C^{14}-to-C^{12} ratio equal to 40% of that found in the atmosphere. Approximately how old is the parchment?

SOLUTION The ratio of C^{14} to C^{12} is $Re^{-0.000121t} = 0.4R$ so $-0.000121t = \ln(0.4)$ or $t = 7572.65 \approx 7600$ years.

25. A paleontologist discovers remains of animals that appear to have died at the onset of the Holocene ice age, between 10,000 and 12,000 years ago. What range of C^{14}-to-C^{12} ratio would the scientist expect to find in the animal remains?

SOLUTION The scientist would expect to find C^{14}-C^{12} ratios ranging from

$$10^{-12}e^{-0.000121(12,000)} \approx 2.34 \times 10^{-13}$$

to

$$10^{-12}e^{-0.000121(10,000)} \approx 2.98 \times 10^{-13}.$$

27. Continuing with Exercise 26, suppose that 50 grams of sugar are dissolved in a container of water. After how many hours will 20 grams of invert sugar be present?

SOLUTION If there are 20 grams of invert sugar present, then there are 30 grams of unconverted sugar. This means that $f = 60$. Solving

$$100e^{-0.2t} = 60$$

for t yields

$$t = -\frac{1}{0.2}\ln 0.6 \approx 2.55 \text{ hours.}$$

29. Moore's Law In 1965, Gordon Moore predicted that the number N of transistors on a microchip would increase exponentially.

(a) Does the table of data below confirm Moore's prediction for the period from 1971 to 2000? If so, estimate the growth constant k.

(b) CAS Plot the data in the table.

(c) Let $N(t)$ be the number of transistors t years after 1971. Find an approximate formula $N(t) \approx Ce^{kt}$, where t is the number of years after 1971.

(d) Estimate the doubling time in Moore's Law for the period from 1971 to 2000.

(e) How many transistors will a chip contain in 2015 if Moore's Law continues to hold?

(f) Can Moore have expected his prediction to hold indefinitely?

Processor	Year	No. Transistors
4004	1971	2250
8008	1972	2500
8080	1974	5000
8086	1978	29,000
286	1982	120,000
386 processor	1985	275,000
486 DX processor	1989	1,180,000
Pentium processor	1993	3,100,000
Pentium II processor	1997	7,500,000
Pentium III processor	1999	24,000,000
Pentium 4 processor	2000	42,000,000
Xeon processor	2008	1,900,000,000

SOLUTION

(a) Yes, the graph looks like an exponential graph especially towards the latter years. We estimate the growth constant by setting 1971 as our starting point, so $P_0 = 2250$. Therefore, $P(t) = 2250e^{kt}$. In 2008, $t = 37$. Therefore, $P(37) = 2250e^{37k} = 1{,}900{,}000{,}000$, so $k = \frac{\ln 844{,}444.444}{37} \approx 0.369$. Note: A better estimate can be found by calculating k for each time period and then averaging the k values.

(b)

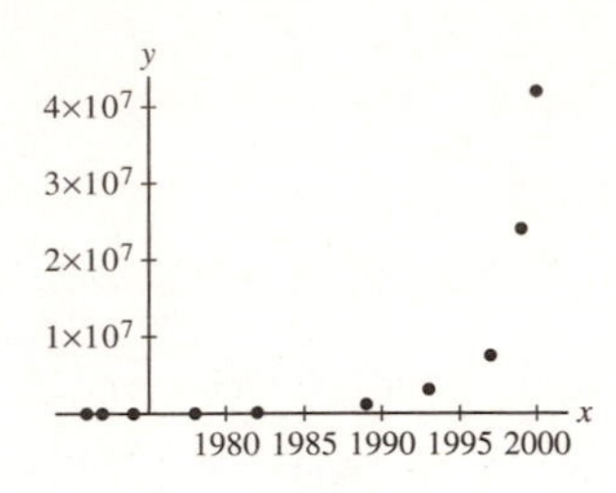

(c) $N(t) = 2250e^{0.369t}$

(d) The doubling time is $\ln 2/0.369 \approx 1.88$ years.

(e) In 2015, $t = 44$ years. Therefore, $N(44) = 2250e^{0.369(44)} \approx 2.53 \times 10^{10}$.

(f) No, you can't make a microchip smaller than an atom.

31. The only functions with a *constant* doubling time are the exponential functions P_0e^{kt} with $k > 0$. Show that the doubling time of linear function $f(t) = at + b$ at time t_0 is $t_0 + b/a$ (which increases with t_0). Compute the doubling times of $f(t) = 3t + 12$ at $t_0 = 10$ and $t_0 = 20$.

SOLUTION Let $f(t) = at + b$ and suppose $f(t_0) = P_0$. The time it takes for the value of f to double is the solution of the equation

$$2P_0 = 2(at_0 + b) = at + b \quad \text{or} \quad t = 2t_0 + b/a.$$

For the function $f(t) = 3t + 12$, $a = 3$, $b = 12$ and $b/a = 4$. With $t_0 = 10$, the doubling time is then 24; with $t_0 = 20$, the doubling time is 44.

33. Compute the balance after 10 years if \$2000 is deposited in an account paying 9% interest and interest is compounded (a) quarterly, (b) monthly, and (c) continuously.

SOLUTION

(a) $P(10) = 2000(1 + 0.09/4)^{4(10)} = \$4870.38$

(b) $P(10) = 2000(1 + 0.09/12)^{12(10)} = \$4902.71$

(c) $P(10) = 2000e^{0.09(10)} = \$4919.21$

35. A bank pays interest at a rate of 5%. What is the yearly multiplier if interest is compounded

(a) three times a year? **(b)** continuously?

SOLUTION

(a) $P(t) = P_0\left(1 + \frac{0.05}{3}\right)^{3t}$, so the yearly multiplier is $\left(1 + \frac{0.05}{3}\right)^{3} \approx 1.0508$.

(b) $P(t) = P_0e^{0.05t}$, so the yearly multiplier is $e^{0.05} \approx 1.0513$.

37. How much must one invest today in order to receive \$20,000 after 5 years if interest is compounded continuously at the rate $r = 9\%$?

SOLUTION Solving $20{,}000 = P_0e^{0.09(5)}$ for P_0 yields

$$P_0 = \frac{20{,}000}{e^{0.45}} \approx \$12{,}752.56.$$

39. Compute the PV of \$5000 received in 3 years if the interest rate is (a) 6% and (b) 11%. What is the PV in these two cases if the sum is instead received in 5 years?

SOLUTION In 3 years:

(a) $PV = 5000e^{-0.06(3)} = \$4176.35$

(b) $PV = 5000e^{-0.11(3)} = \$3594.62$

In 5 years:

(a) $PV = 5000e^{-0.06(5)} = \$3704.09$

(b) $PV = 5000e^{-0.11(5)} = \$2884.75$

41. Find the interest rate r if the PV of \$8000 to be received in 1 year is \$7300.

SOLUTION Solving $7300 = 8000e^{-r(1)}$ for r yields

$$r = -\ln\left(\frac{7300}{8000}\right) = 0.0916,$$

or 9.16%.

43. A new computer system costing \$25,000 will reduce labor costs by \$7000/year for 5 years.
(a) Is it a good investment if $r = 8\%$?
(b) How much money will the company actually save?

SOLUTION
(a) The present value of the reduced labor costs is

$$7000(e^{-0.08} + e^{-0.16} + e^{-0.24} + e^{-0.32} + e^{-0.4}) = \$27{,}708.50.$$

This is more than the \$25,000 cost of the computer system, so the computer system should be purchased.
(b) The present value of the savings is

$$\$27{,}708.50 - \$25{,}000 = \$2708.50.$$

45. Use Eq. (3) to compute the PV of an income stream paying out $R(t) = \$5000/\text{year}$ continuously for 10 years, assuming $r = 0.05$.

SOLUTION $PV = \int_0^{10} 5000e^{-0.05t}\,dt = -100{,}000e^{-0.05t}\Big|_0^{10} = \$39{,}346.93.$

47. Find the PV of an income stream that pays out continuously at a rate $R(t) = \$5000e^{0.1t}/\text{year}$ for 7 years, assuming $r = 0.05$.

SOLUTION $PV = \int_0^{7} 5000e^{0.1t}e^{-0.05t}\,dt = \int_0^7 5000e^{0.05t}\,dt = 100{,}000e^{0.05t}\Big|_0^{7} = \$41{,}906.75.$

49. Show that an investment that pays out R dollars per year continuously for T years has a PV of $R(1 - e^{-rT})/r$.

SOLUTION The present value of an investment that pays out R dollars/year continuously for T years is

$$PV = \int_0^T Re^{-rt}\,dt.$$

Let $u = -rt,\ du = -r\,dt$. Then

$$PV = -\frac{1}{r}\int_0^{-rT} Re^u\,du = -\frac{R}{r}e^u\Big|_0^{-rT} = -\frac{R}{r}(e^{-rT} - 1) = \frac{R}{r}(1 - e^{-rT}).$$

51. Suppose that $r = 0.06$. Use the result of Exercise 50 to estimate the payout rate R needed to produce an income stream whose PV is \$20,000, assuming that the stream continues for a large number of years.

SOLUTION From Exercise 50, $PV = \frac{R}{r}$ so $20{,}000 = \frac{R}{0.06}$ or $R = \$1200$.

53. Use Eq. (5) to compute the PV of an investment that pays out income continuously at a rate $R(t) = (5000 + 1000t)e^{0.02t}$ dollars per year for 10 years, assuming $r = 0.08$.

SOLUTION

$$\begin{aligned} PV &= \int_0^{10} (5000 + 1000t)(e^{0.02t})e^{-0.08t}\,dt = \int_0^{10} 5000e^{-0.06t}\,dt + \int_0^{10} 1000te^{-0.06t}\,dt \\ &= \frac{5000}{-0.06}(e^{-0.06(10)} - 1) - 1000\left(\frac{e^{-0.06(10)}(1 + 0.06(10))}{(0.06)^2}\right) + 1000\frac{1}{(0.06)^2} \\ &= 37{,}599.03 - 243{,}916.28 + 277{,}777.78 \approx \$71{,}460.53. \end{aligned}$$

55. **Drug Dosing Interval** Let $y(t)$ be the drug concentration (in mg/kg) in a patient's body at time t. The initial concentration is $y(0) = L$. Additional doses that increase the concentration by an amount d are administered at regular time intervals of length T. In between doses, $y(t)$ decays exponentially—that is, $y' = -ky$. Find the value of T (in terms of k and d) for which the the concentration varies between L and $L - d$ as in Figure 15.

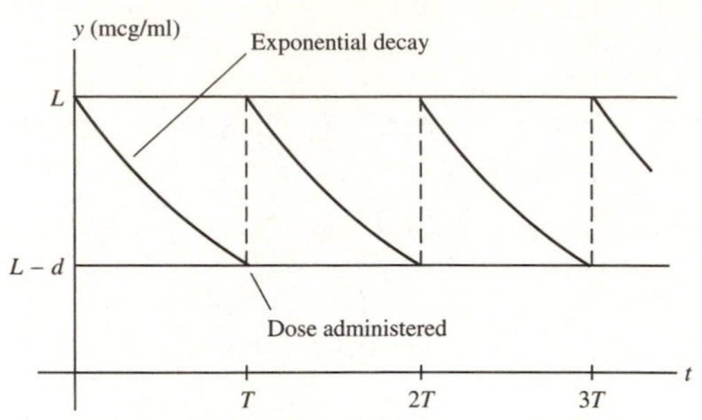

FIGURE 15 Drug concentration with periodic doses.

SOLUTION Because $y' = -ky$ and $y(0) = L$, it follows that $y(t) = Le^{-kt}$. We want $y(T) = L - d$, thus

$$Le^{-kT} = L - d \quad \text{or} \quad T = -\frac{1}{k}\ln\left(1 - \frac{d}{L}\right).$$

*Exercises 56 and 57: The **Gompertz differential equation***

$$\frac{dy}{dt} = ky\ln\left(\frac{y}{M}\right) \tag{6}$$

(where M and k are constants) was introduced in 1825 by the English mathematician Benjamin Gompertz and is still used today to model aging and mortality.

57. To model mortality in a population of 200 laboratory rats, a scientist assumes that the number $P(t)$ of rats alive at time t (in months) satisfies Eq. (6) with $M = 204$ and $k = 0.15$ month^{-1} (Figure 16). Find $P(t)$ [note that $P(0) = 200$] and determine the population after 20 months.

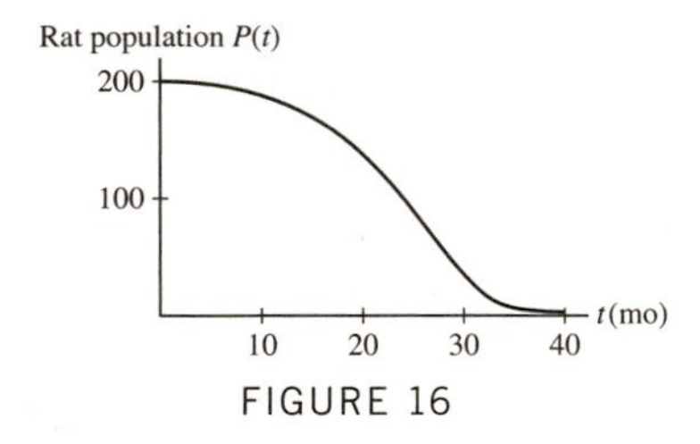

FIGURE 16

SOLUTION The solution to the Gompertz equation with $M = 204$ and $k = 0.15$ is of the form:

$$P(t) = 204e^{ae^{0.15t}}$$

Applying the initial condition allows us to solve for a:

$$200 = 204e^{a}$$
$$\frac{200}{204} = e^{a}$$
$$\ln\left(\frac{200}{204}\right) = a$$

so that $a \approx -0.02$. After $t = 20$ months,

$$P(20) = 204e^{-0.02e^{0.15(20)}} = 136.51,$$

so there are 136 rats.

59. Let $P = P(t)$ be a quantity that obeys an exponential growth law with growth constant k. Show that P increases m-fold after an interval of $(\ln m)/k$ years.

SOLUTION For m-fold growth, $P(t) = mP_0$ for some t. Solving $mP_0 = P_0e^{kt}$ for t, we find $t = \dfrac{\ln m}{k}$

Further Insights and Challenges

61. Modify the proof of the relation $e = \lim_{n\to\infty}\left(1+\frac{1}{n}\right)^n$ given in the text to prove $e^x = \lim_{n\to\infty}\left(1+\frac{x}{n}\right)^n$. *Hint:* Express $\ln(1+xn^{-1})$ as an integral and estimate above and below by rectangles.

SOLUTION Start by expressing

$$\ln\left(1+\frac{x}{n}\right) = \int_1^{1+x/n}\frac{dt}{t}.$$

Following the proof in the text, we note that

$$\frac{x}{n+x} \le \ln\left(1+\frac{x}{n}\right) \le \frac{x}{n}$$

provided $x > 0$, while

$$\frac{x}{n} \le \ln\left(1+\frac{x}{n}\right) \le \frac{x}{n+x}$$

when $x < 0$. Multiplying both sets of inequalities by n and passing to the limit as $n \to \infty$, the squeeze theorem guarantees that

$$\lim_{n\to\infty}\left(\ln\left(1+\frac{x}{n}\right)\right)^n = x.$$

Finally,

$$\lim_{n\to\infty}\left(1+\frac{x}{n}\right)^n = e^x.$$

63. A bank pays interest at the rate r, compounded M times yearly. The **effective interest rate** r_e is the rate at which interest, if compounded annually, would have to be paid to produce the same yearly return.

(a) Find r_e if $r = 9\%$ compounded monthly.

(b) Show that $r_e = (1+r/M)^M - 1$ and that $r_e = e^r - 1$ if interest is compounded continuously.

(c) Find r_e if $r = 11\%$ compounded continuously.

(d) Find the rate r that, compounded weekly, would yield an effective rate of 20%.

SOLUTION

(a) Compounded monthly, $P(t) = P_0(1+r/12)^{12t}$. By the definition of r_e,

$$P_0(1+0.09/12)^{12t} = P_0(1+r_e)^t$$

so

$$(1+0.09/12)^{12t} = (1+r_e)^t \quad \text{or} \quad r_e = (1+0.09/12)^{12} - 1 = 0.0938,$$

or 9.38%

(b) In general,

$$P_0(1+r/M)^{Mt} = P_0(1+r_e)^t,$$

so $(1+r/M)^{Mt} = (1+r_e)^t$ or $r_e = (1+r/M)^M - 1$. If interest is compounded continuously, then $P_0e^{rt} = P_0(1+r_e)^t$ so $e^{rt} = (1+r_e)^t$ or $r_e = e^r - 1$.

(c) Using part (b), $r_e = e^{0.11} - 1 \approx 0.1163$ or 11.63%.

(d) Solving

$$0.20 = \left(1+\frac{r}{52}\right)^{52} - 1$$

for r yields $r = 52(1.2^{1/52} - 1) = 0.1826$ or 18.26%.

CHAPTER REVIEW EXERCISES

In Exercises 1–4, refer to the function $f(x)$ whose graph is shown in Figure 1.

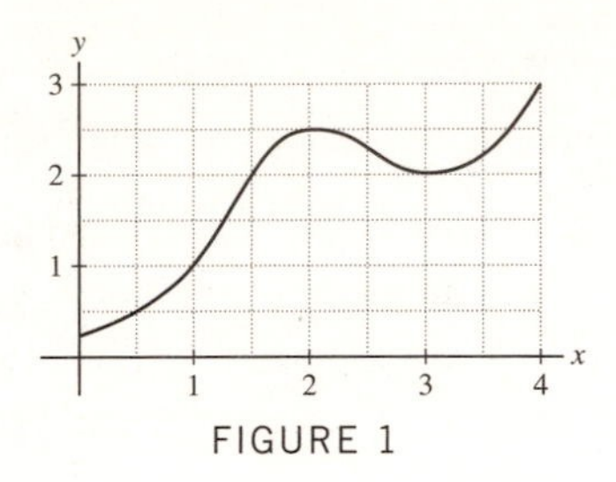

FIGURE 1

1. Estimate L_4 and M_4 on $[0, 4]$.

SOLUTION With $n = 4$ and an interval of $[0, 4]$, $\Delta x = \frac{4-0}{4} = 1$. Then,

$$L_4 = \Delta x(f(0) + f(1) + f(2) + f(3)) = 1\left(\frac{1}{4} + 1 + \frac{5}{2} + 2\right) = \frac{23}{4}$$

and

$$M_4 = \Delta x\left(f\left(\frac{1}{2}\right) + f\left(\frac{3}{2}\right) + f\left(\frac{5}{2}\right) + f\left(\frac{7}{2}\right)\right) = 1\left(\frac{1}{2} + 2 + \frac{9}{4} + \frac{9}{4}\right) = 7.$$

3. Find an interval $[a, b]$ on which R_4 is larger than $\int_a^b f(x)\,dx$. Do the same for L_4.

SOLUTION In general, R_N is larger than $\int_a^b f(x)\,dx$ on any interval $[a, b]$ over which $f(x)$ is increasing. Given the graph of $f(x)$, we may take $[a, b] = [0, 2]$. In order for L_4 to be larger than $\int_a^b f(x)\,dx$, $f(x)$ must be decreasing over the interval $[a, b]$. We may therefore take $[a, b] = [2, 3]$.

In Exercises 5–8, let $f(x) = x^2 + 3x$.

5. Calculate R_6, M_6, and L_6 for $f(x)$ on the interval $[2, 5]$. Sketch the graph of $f(x)$ and the corresponding rectangles for each approximation.

SOLUTION Let $f(x) = x^2 + 3x$. A uniform partition of $[2, 5]$ with $N = 6$ subintervals has

$$\Delta x = \frac{5-2}{6} = \frac{1}{2}, \qquad x_j = a + j\Delta x = 2 + \frac{j}{2},$$

and

$$x_j^* = a + \left(j - \frac{1}{2}\right)\Delta x = \frac{7}{4} + \frac{j}{2}.$$

Now,

$$R_6 = \Delta x \sum_{j=1}^{6} f(x_j) = \frac{1}{2}\left(f\left(\frac{5}{2}\right) + f(3) + f\left(\frac{7}{2}\right) + f(4) + f\left(\frac{9}{2}\right) + f(5)\right)$$

$$= \frac{1}{2}\left(\frac{55}{4} + 18 + \frac{91}{4} + 28 + \frac{135}{4} + 40\right) = \frac{625}{8}.$$

The rectangles corresponding to this approximation are shown below.

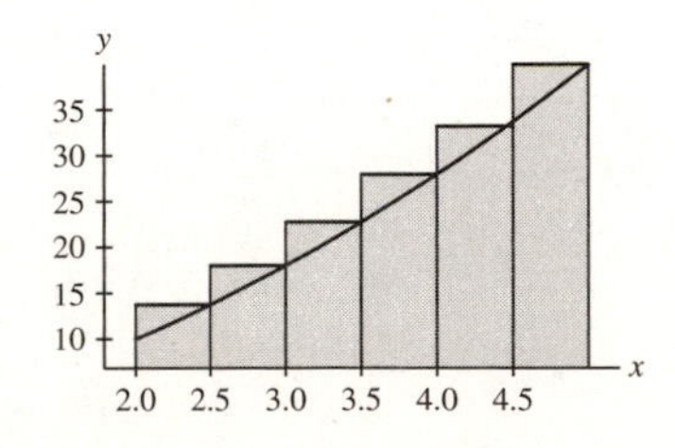

Next,

$$M_6 = \Delta x \sum_{j=1}^{6} f(x_j^*) = \frac{1}{2}\left(f\left(\frac{9}{4}\right) + f\left(\frac{11}{4}\right) + f\left(\frac{13}{4}\right) + f\left(\frac{15}{4}\right) + f\left(\frac{17}{4}\right) + f\left(\frac{19}{4}\right)\right)$$

$$= \frac{1}{2}\left(\frac{189}{16} + \frac{253}{16} + \frac{325}{16} + \frac{405}{16} + \frac{493}{16} + \frac{589}{16}\right) = \frac{2254}{32} = \frac{1127}{16}.$$

The rectangles corresponding to this approximation are shown below.

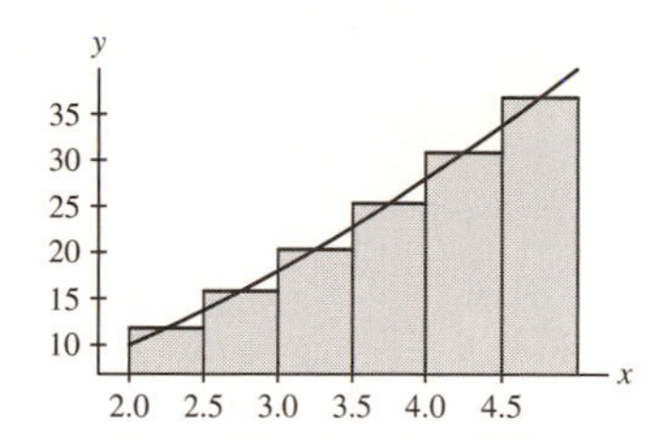

Finally,

$$L_6 = \Delta x \sum_{j=0}^{5} f(x_j) = \frac{1}{2}\left(f(2) + f\left(\frac{5}{2}\right) + f(3) + f\left(\frac{7}{2}\right) + f(4) + f\left(\frac{9}{2}\right)\right)$$

$$= \frac{1}{2}\left(10 + \frac{55}{4} + 18 + \frac{91}{4} + 28 + \frac{135}{4}\right) = \frac{505}{8}.$$

The rectangles corresponding to this approximation are shown below.

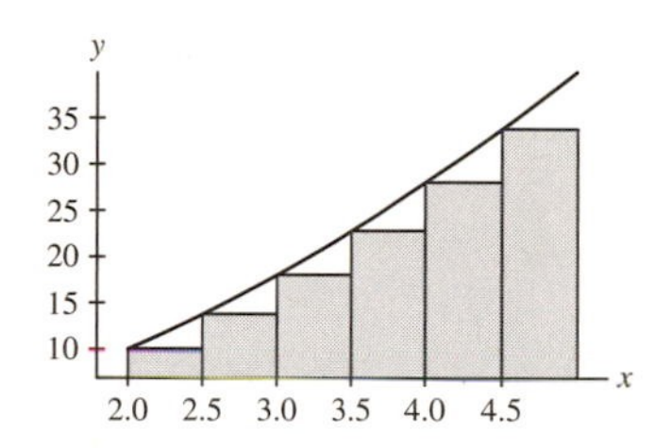

7. Find a formula for R_N for $f(x)$ on [2, 5] and compute $\int_2^5 f(x)\,dx$ by taking the limit.

SOLUTION Let $f(x) = x^2 + 3x$ on the interval [2, 5]. Then $\Delta x = \frac{5-2}{N} = \frac{3}{N}$ and $a = 2$. Hence,

$$R_N = \Delta x \sum_{j=1}^{N} f(2 + j\Delta x) = \frac{3}{N}\sum_{j=1}^{N}\left(\left(2 + \frac{3j}{N}\right)^2 + 3\left(2 + \frac{3j}{N}\right)\right) = \frac{3}{N}\sum_{j=1}^{N}\left(10 + \frac{21j}{N} + \frac{9j^2}{N^2}\right)$$

$$= 30 + \frac{63}{N^2}\sum_{j=1}^{N} j + \frac{27}{N^3}\sum_{j=1}^{N} j^2$$

$$= 30 + \frac{63}{N^2}\left(\frac{N^2}{2} + \frac{N}{2}\right) + \frac{27}{N^3}\left(\frac{N^3}{3} + \frac{N^2}{2} + \frac{N}{6}\right)$$

$$= \frac{141}{2} + \frac{45}{N} + \frac{9}{2N^2}$$

and

$$\lim_{N\to\infty} R_N = \lim_{N\to\infty}\left(\frac{141}{2} + \frac{45}{N} + \frac{9}{2N^2}\right) = \frac{141}{2}.$$

9. Calculate R_5, M_5, and L_5 for $f(x) = (x^2+1)^{-1}$ on the interval [0, 1].

SOLUTION Let $f(x) = (x^2+1)^{-1}$. A uniform partition of [0, 1] with $N = 5$ subintervals has

$$\Delta x = \frac{1-0}{5} = \frac{1}{5}, \qquad x_j = a + j\Delta x = \frac{j}{5},$$

and

$$x_j^* = a + \left(j - \frac{1}{2}\right)\Delta x = \frac{2j-1}{10}.$$

Now,

$$R_5 = \Delta x \sum_{j=1}^{5} f(x_j) = \frac{1}{5}\left(f\left(\frac{1}{5}\right) + f\left(\frac{2}{5}\right) + f\left(\frac{3}{5}\right) + f\left(\frac{4}{5}\right) + f(1)\right)$$

$$= \frac{1}{5}\left(\frac{25}{26} + \frac{25}{29} + \frac{25}{34} + \frac{25}{41} + \frac{1}{2}\right) \approx 0.733732.$$

Next,

$$M_5 = \Delta x \sum_{j=1}^{5} f(x_j^*) = \frac{1}{5}\left(f\left(\frac{1}{10}\right) + f\left(\frac{3}{10}\right) + f\left(\frac{1}{2}\right) + f\left(\frac{7}{10}\right) + f\left(\frac{9}{10}\right)\right)$$

$$= \frac{1}{5}\left(\frac{100}{101} + \frac{100}{109} + \frac{4}{5} + \frac{100}{149} + \frac{100}{181}\right) \approx 0.786231.$$

Finally,

$$L_5 = \Delta x \sum_{j=0}^{4} f(x_j) = \frac{1}{5}\left(f(0) + f\left(\frac{1}{5}\right) + f\left(\frac{2}{5}\right) + f\left(\frac{3}{5}\right) + f\left(\frac{4}{5}\right)\right)$$

$$= \frac{1}{5}\left(1 + \frac{25}{26} + \frac{25}{29} + \frac{25}{34} + \frac{25}{41}\right) \approx 0.833732.$$

11. Which approximation to the area is represented by the shaded rectangles in Figure 3? Compute R_5 and L_5.

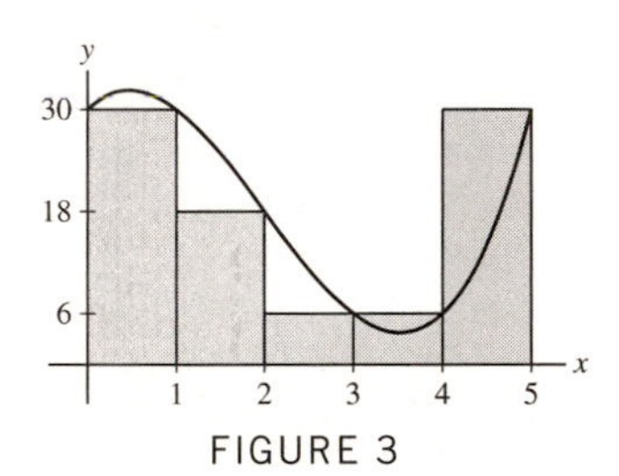

FIGURE 3

SOLUTION There are five rectangles and the height of each is given by the function value at the right endpoint of the subinterval. Thus, the area represented by the shaded rectangles is R_5.

From the figure, we see that $\Delta x = 1$. Then

$$R_5 = 1(30 + 18 + 6 + 6 + 30) = 90 \quad \text{and} \quad L_5 = 1(30 + 30 + 18 + 6 + 6) = 90.$$

In Exercises 13–16, express the limit as an integral (or multiple of an integral) and evaluate.

13. $\lim_{N\to\infty} \frac{\pi}{6N} \sum_{j=1}^{N} \sin\left(\frac{\pi}{3} + \frac{\pi j}{6N}\right)$

SOLUTION Let $f(x) = \sin x$ and N be a positive integer. A uniform partition of the interval $[\pi/3, \pi/2]$ with N subintervals has

$$\Delta x = \frac{\pi}{6N} \quad \text{and} \quad x_j = \frac{\pi}{3} + \frac{\pi j}{6N}$$

for $0 \le j \le N$. Then

$$\frac{\pi}{6N} \sum_{j=1}^{N} \sin\left(\frac{\pi}{3} + \frac{\pi j}{6N}\right) = \Delta x \sum_{j=1}^{N} f(x_j) = R_N;$$

consequently,

$$\lim_{N\to\infty} \frac{\pi}{6N} \sum_{j=1}^{N} \sin\left(\frac{\pi}{3} + \frac{\pi j}{6N}\right) = \int_{\pi/3}^{\pi/2} \sin x\, dx = -\cos x\Big|_{\pi/3}^{\pi/2} = 0 + \frac{1}{2} = \frac{1}{2}.$$

15. $\lim_{N\to\infty} \frac{5}{N}\sum_{j=1}^{N}\sqrt{4+5j/N}$

SOLUTION Let $f(x)=\sqrt{x}$ and N be a positive integer. A uniform partition of the interval [4, 9] with N subintervals has

$$\Delta x = \frac{5}{N} \quad \text{and} \quad x_j = 4 + \frac{5j}{N}$$

for $0 \le j \le N$. Then

$$\frac{5}{N}\sum_{j=1}^{N}\sqrt{4+5j/N} = \Delta x \sum_{j=1}^{N} f(x_j) = R_N;$$

consequently,

$$\lim_{N\to\infty} \frac{5}{N}\sum_{j=1}^{N}\sqrt{4+5j/N} = \int_4^9 \sqrt{x}\,dx = \frac{2}{3}x^{3/2}\Big|_4^9 = \frac{54}{3} - \frac{16}{3} = \frac{38}{3}.$$

In Exercises 17–20, use the given substitution to evaluate the integral.

17. $\int_0^2 \frac{dt}{4t+12}$, $\quad u = 4t+12$

SOLUTION Let $u = 4t+12$. Then $du = 4dt$, and the new limits of integration are $u = 12$ and $u = 20$. Thus,

$$\int_0^2 \frac{dt}{4t+12} = \frac{1}{4}\int_{12}^{20}\frac{du}{u} = \frac{1}{4}\ln u\Big|_{12}^{20} = \frac{1}{4}(\ln 20 - \ln 12) = \frac{1}{4}\ln\frac{20}{12} = \frac{1}{4}\ln\frac{5}{3}.$$

19. $\int_0^{\pi/6} \sin x \cos^4 x\,dx$, $\quad u = \cos x$

SOLUTION Let $u = \cos x$. Then $du = -\sin x\,dx$ and the new limits of integration are $u = 1$ and $u = \sqrt{3}/2$. Thus,

$$\begin{aligned}\int_0^{\pi/6} \sin x\cos^4 x\,dx &= -\int_1^{\sqrt{3}/2} u^4\,du \\ &= -\frac{1}{5}u^5\Big|_1^{\sqrt{3}/2} \\ &= \frac{1}{5}\left(1 - \frac{9\sqrt{3}}{32}\right).\end{aligned}$$

In Exercises 21–70, evaluate the integral.

21. $\int (20x^4 - 9x^3 - 2x)\,dx$

SOLUTION $\int (20x^4 - 9x^3 - 2x)\,dx = 4x^5 - \frac{9}{4}x^4 - x^2 + C.$

23. $\int (2x^2 - 3x)^2\,dx$

SOLUTION $\int (2x^2 - 3x)^2\,dx = \int (4x^4 - 12x^3 + 9x^2)\,dx = \frac{4}{5}x^5 - 3x^4 + 3x^3 + C.$

25. $\int \frac{x^5 + 3x^4}{x^2}\,dx$

SOLUTION $\int \frac{x^5 + 3x^4}{x^2}\,dx = \int (x^3 + 3x^2)\,dx = \frac{1}{4}x^4 + x^3 + C.$

27. $\displaystyle\int_{-3}^{3} |x^2 - 4|\,dx$

SOLUTION

$$\begin{aligned}\int_{-3}^{3} |x^2 - 4|\,dx &= \int_{-3}^{-2} (x^2 - 4)\,dx + \int_{-2}^{2} (4 - x^2)\,dx + \int_{2}^{3} (x^2 - 4)\,dx \\ &= \left(\frac{1}{3}x^3 - 4x\right)\Big|_{-3}^{-2} + \left(4x - \frac{1}{3}x^3\right)\Big|_{-2}^{2} + \left(\frac{1}{3}x^3 - 4x\right)\Big|_{2}^{3} \\ &= \left(\frac{16}{3} - 3\right) + \left(\frac{16}{3} + \frac{16}{3}\right) + \left(-3 + \frac{16}{3}\right) \\ &= \frac{46}{3}.\end{aligned}$$

29. $\displaystyle\int_{1}^{3} [t]\,dt$

SOLUTION

$$\int_1^3 [t]\,dt = \int_1^2 [t]\,dt + \int_2^3 [t]\,dt = \int_1^2 dt + \int_2^3 2\,dt = t\Big|_1^2 + 2t\Big|_2^3 = (2 - 1) + (6 - 4) = 3.$$

31. $\displaystyle\int (10t - 7)^{14}\,dt$

SOLUTION Let $u = 10t - 7$. Then $du = 10dt$ and

$$\int (10t - 7)^{14}\,dt = \frac{1}{10}\int u^{14}\,du = \frac{1}{150}u^{15} + C = \frac{1}{150}(10t - 7)^{15} + C.$$

33. $\displaystyle\int \frac{(2x^3 + 3x)\,dx}{(3x^4 + 9x^2)^5}$

SOLUTION Let $u = 3x^4 + 9x^2$. Then $du = (12x^3 + 18x)\,dx = 6(2x^3 + 3x)\,dx$ and

$$\int \frac{(2x^3 + 3x)\,dx}{(3x^4 + 9x^2)^5} = \frac{1}{6}\int u^{-5}\,du = -\frac{1}{24}u^{-4} + C = -\frac{1}{24}(3x^4 + 9x^2)^{-4} + C.$$

35. $\displaystyle\int_0^5 15x\sqrt{x + 4}\,dx$

SOLUTION Let $u = x + 4$. Then $x = u - 4$, $du = dx$ and the new limits of integration are $u = 4$ and $u = 9$. Thus,

$$\begin{aligned}\int_0^5 15x\sqrt{x + 4}\,dx &= \int_4^9 15(u - 4)\sqrt{u}\,du \\ &= 15\int_4^9 (u^{3/2} - 4u^{1/2})\,du \\ &= 15\left(\frac{2}{5}u^{5/2} - \frac{8}{3}u^{3/2}\right)\Big|_4^9 \\ &= 15\left(\left(\frac{486}{5} - 72\right) - \left(\frac{64}{5} - \frac{64}{3}\right)\right) \\ &= 506.\end{aligned}$$

37. $\displaystyle\int_0^1 \cos\left(\frac{\pi}{3}(t + 2)\right)\,dt$

SOLUTION $\displaystyle\int_0^1 \cos\left(\frac{\pi}{3}(t + 2)\right)\,dt = \frac{3}{\pi}\sin\left(\frac{\pi}{3}(t + 2)\right)\Big|_0^1 = -\frac{3\sqrt{3}}{2\pi}.$

39. $\displaystyle\int t^2\sec^2(9t^3 + 1)\,dt$

SOLUTION Let $u = 9t^3 + 1$. Then $du = 27t^2\,dt$ and

$$\int t^2\sec^2(9t^3 + 1)\,dt = \frac{1}{27}\int \sec^2 u\,du = \frac{1}{27}\tan u + C = \frac{1}{27}\tan(9t^3 + 1) + C.$$

41. $\int \csc^2(9-2\theta)\,d\theta$

SOLUTION Let $u = 9 - 2\theta$. Then $du = -2\,d\theta$ and

$$\int \csc^2(9-2\theta)\,d\theta = -\frac{1}{2}\int \csc^2 u\,du = \frac{1}{2}\cot u + C = \frac{1}{2}\cot(9-2\theta) + C.$$

43. $\int_0^{\pi/3} \frac{\sin\theta}{\cos^{2/3}\theta}\,d\theta$

SOLUTION Let $u = \cos\theta$. Then $du = -\sin\theta\,d\theta$ and when $\theta = 0$, $u = 1$ and when $\theta = \frac{\pi}{3}$, $u = \frac{1}{2}$. Finally,

$$\int_0^{\pi/3} \frac{\sin\theta}{\cos^{2/3}\theta}\,d\theta = -\int_1^{1/2} u^{-2/3}\,du = -3u^{1/3}\Big|_1^{1/2} = -3(2^{-1/3}-1) = 3 - \frac{3\sqrt[3]{4}}{2}.$$

45. $\int e^{9-2x}\,dx$

SOLUTION Let $u = 9 - 2x$. Then $du = -2\,dx$, and

$$\int e^{9-2x}\,dx = -\frac{1}{2}\int e^u\,du = -\frac{1}{2}e^u + C = -\frac{1}{2}e^{9-2x} + C.$$

47. $\int x^2 e^{x^3}\,dx$

SOLUTION Let $u = x^3$. Then $du = 3x^2\,dx$, and

$$\int x^2 e^{x^3}dx = \frac{1}{3}\int e^u du = \frac{1}{3}e^u + C = \frac{1}{3}e^{x^3} + C.$$

49. $\int e^x 10^x\,dx$

SOLUTION $\int e^x 10^x\,dx = \int (10e)^x\,dx = \frac{(10e)^x}{\ln(10e)} + C = \frac{(10e)^x}{\ln 10 + \ln e} + C = \frac{10^x e^x}{\ln 10 + 1} + C.$

51. $\int \frac{e^{-x}\,dx}{(e^{-x}+2)^3}$

SOLUTION Let $u = e^{-x} + 2$. Then $du = -e^{-x}\,dx$ and

$$\int \frac{e^{-x}\,dx}{(e^{-x}+2)^3} = -\int u^{-3}\,du = \frac{1}{2u^2} + C = \frac{1}{2(e^{-x}+2)^2} + C.$$

53. $\int_0^{\pi/6} \tan 2\theta\,d\theta$

SOLUTION $\int_0^{\pi/6} \tan 2\theta\,d\theta = \frac{1}{2}\ln|\sec 2\theta|\Big|_0^{\pi/6} = \frac{1}{2}\ln 2.$

55. $\int \frac{dt}{t(1+(\ln t)^2)}$

SOLUTION Let $u = \ln t$. Then, $du = \frac{1}{t}\,dt$ and

$$\int \frac{dt}{t(1+(\ln t)^2)} = \int \frac{du}{1+u^2} = \tan^{-1}u + C = \tan^{-1}(\ln t) + C.$$

57. $\int_1^e \frac{\ln x\,dx}{x}$

SOLUTION Let $u = \ln x$. Then $du = \frac{dx}{x}$ and the new limits of integration are $u = \ln 1 = 0$ and $u = \ln e = 1$. Thus,

$$\int_1^e \frac{\ln x\,dx}{x} = \int_0^1 u\,du = \frac{1}{2}u^2\Big|_0^1 = \frac{1}{2}.$$

59. $\displaystyle\int \frac{dx}{4x^2+9}$

SOLUTION Let $u = \frac{2x}{3}$. Then $x = \frac{3}{2}u$, $dx = \frac{3}{2}\,du$, and

$$\int \frac{dx}{4x^2+9} = \int \frac{\frac{3}{2}\,du}{4\cdot\frac{9}{4}u^2+9} = \frac{1}{6}\int \frac{du}{u^2+1} = \frac{1}{6}\tan^{-1}u + C = \frac{1}{6}\tan^{-1}\left(\frac{2x}{3}\right) + C.$$

61. $\displaystyle\int_4^{12} \frac{dx}{x\sqrt{x^2-1}}$

SOLUTION $\displaystyle\int_4^{12} \frac{dx}{x\sqrt{x^2-1}} = \sec^{-1}x\Big|_4^{12} = \sec^{-1}12 - \sec^{-1}4.$

63. $\displaystyle\int_0^3 \frac{dx}{x^2+9}$

SOLUTION Let $u = \frac{x}{3}$. Then $du = \frac{dx}{3}$, and the new limits of integration are $u = 0$ and $u = 1$. Thus,

$$\int_0^3 \frac{dx}{x^2+9} = \frac{1}{3}\int_0^1 \frac{dt}{t^2+1} = \frac{1}{3}\tan^{-1}t\Big|_0^1 = \frac{1}{3}(\tan^{-1}1 - \tan^{-1}0) = \frac{1}{3}\left(\frac{\pi}{4} - 0\right) = \frac{\pi}{12}.$$

65. $\displaystyle\int \frac{x\,dx}{\sqrt{1-x^4}}$

SOLUTION Let $u = x^2$. Then $du = 2x\,dx$, and $\sqrt{1-x^4} = \sqrt{1-u^2}$. Thus,

$$\int \frac{x\,dx}{\sqrt{1-x^4}} = \frac{1}{2}\int \frac{du}{\sqrt{1-u^2}} = \frac{1}{2}\sin^{-1}u + C = \frac{1}{2}\sin^{-1}(x^2) + C.$$

67. $\displaystyle\int_0^4 \frac{dx}{2x^2+1}$

SOLUTION Let $u = \sqrt{2}x$. Then $du = \sqrt{2}\,dx$, and the new limits of integration are $u = 0$ and $u = 4\sqrt{2}$. Thus,

$$\int_0^4 \frac{dx}{2x^2+1} = \int_0^{4\sqrt{2}} \frac{\frac{1}{\sqrt{2}}\,du}{u^2+1} = \frac{1}{\sqrt{2}}\int_0^{4\sqrt{2}} \frac{du}{u^2+1}$$

$$= \frac{1}{\sqrt{2}}\tan^{-1}u\Big|_0^{4\sqrt{2}} = \frac{1}{\sqrt{2}}\left(\tan^{-1}(4\sqrt{2}) - \tan^{-1}0\right) = \frac{1}{\sqrt{2}}\tan^{-1}(4\sqrt{2}).$$

69. $\displaystyle\int_0^1 \frac{(\tan^{-1}x)^3\,dx}{1+x^2}$

SOLUTION Let $u = \tan^{-1}x$. Then

$$du = \frac{1}{1+x^2}\,dx$$

and

$$\int_0^1 \frac{(\tan^{-1}x)^3\,dx}{1+x^2} = \int_0^{\pi/4} u^3\,du = \frac{1}{4}u^4\Big|_0^{\pi/4} = \frac{1}{4}\left(\frac{\pi}{4}\right)^4 = \frac{\pi^4}{1024}.$$

71. Combine to write as a single integral:

$$\int_0^8 f(x)\,dx + \int_{-2}^0 f(x)\,dx + \int_8^6 f(x)\,dx$$

SOLUTION First, rewrite

$$\int_0^8 f(x)\,dx = \int_0^6 f(x)\,dx + \int_6^8 f(x)\,dx$$

and observe that

$$\int_8^6 f(x)\,dx = -\int_6^8 f(x)\,dx.$$

Thus,

$$\int_0^8 f(x)\,dx + \int_8^6 f(x)\,dx = \int_0^6 f(x)\,dx.$$

Finally,

$$\int_0^8 f(x)\,dx + \int_{-2}^0 f(x)\,dx + \int_8^6 f(x)\,dx = \int_0^6 f(x)\,dx + \int_{-2}^0 f(x)\,dx = \int_{-2}^6 f(x)\,dx.$$

73. Find the local minima, the local maxima, and the inflection points of $A(x) = \int_3^x \frac{t\,dt}{t^2+1}$.

SOLUTION Let

$$A(x) = \int_3^x \frac{t\,dt}{t^2+1}.$$

Then

$$A'(x) = \frac{x}{x^2+1}$$

and

$$A''(x) = \frac{(x^2+1)(1) - x(2x)}{(x^2+1)^2} = \frac{1-x^2}{(x^2+1)^2}.$$

Now, $x = 0$ is the only critical point of A; because $A''(0) > 0$, it follows that A has a local minimum at $x = 0$. There are no local maxima. Moreover, $A(x)$ is concave down for $|x| > 1$ and concave up for $|x| < 1$. $A(x)$ therefore has inflection points at $x = \pm 1$.

75. On a typical day, a city consumes water at the rate of $r(t) = 100 + 72t - 3t^2$ (in thousands of gallons per hour), where t is the number of hours past midnight. What is the daily water consumption? How much water is consumed between 6 PM and midnight?

SOLUTION With a consumption rate of $r(t) = 100 + 72t - 3t^2$ thousand gallons per hour, the daily consumption of water is

$$\int_0^{24} (100 + 72t - 3t^2)\,dt = \left(100t + 36t^2 - t^3\right)\Big|_0^{24} = 100(24) + 36(24)^2 - (24)^3 = 9312,$$

or 9.312 million gallons. From 6 PM to midnight, the water consumption is

$$\begin{aligned}\int_{18}^{24} (100 + 72t - 3t^2)\,dt &= \left(100t + 36t^2 - t^3\right)\Big|_{18}^{24}\\ &= 100(24) + 36(24)^2 - (24)^3 - \left(100(18) + 36(18)^2 - (18)^3\right)\\ &= 9312 - 7632 = 1680,\end{aligned}$$

or 1.68 million gallons.

77. Cost engineers at NASA have the task of projecting the cost P of major space projects. It has been found that the cost C of developing a projection increases with P at the rate $dC/dP \approx 21P^{-0.65}$, where C is in thousands of dollars and P in millions of dollars. What is the cost of developing a projection for a project whose cost turns out to be $P = \$35$ million?

SOLUTION Assuming it costs nothing to develop a projection for a project with a cost of \$0, the cost of developing a projection for a project whose cost turns out to be \$35 million is

$$\int_0^{35} 21P^{-0.65}\,dP = 60P^{0.35}\Big|_0^{35} = 60(35)^{0.35} \approx 208.245,$$

or \$208,245.

79. Evaluate $\int_{-8}^8 \frac{x^{15}\,dx}{3+\cos^2 x}$, using the properties of odd functions.

SOLUTION Let $f(x) = \frac{x^{15}}{3+\cos^2 x}$ and note that

$$f(-x) = \frac{(-x)^{15}}{3+\cos^2(-x)} = -\frac{x^{15}}{\cos^2 x} = -f(x).$$

Because $f(x)$ is an odd function and the interval $-8 \le x \le 8$ is symmetric about $x = 0$, it follows that

$$\int_{-8}^{8} \frac{x^{15}\,dx}{3+\cos^2 x} = 0.$$

81. GU Plot the graph of $f(x) = \sin mx \sin nx$ on $[0, \pi]$ for the pairs $(m, n) = (2, 4), (3, 5)$ and in each case guess the value of $I = \int_0^\pi f(x)\,dx$. Experiment with a few more values (including two cases with $m = n$) and formulate a conjecture for when I is zero.

SOLUTION The graphs of $f(x) = \sin mx \sin nx$ with $(m, n) = (2, 4)$ and $(m, n) = (3, 5)$ are shown below. It appears as if the positive areas balance the negative areas, so we expect that

$$I = \int_0^\pi f(x)\,dx = 0$$

in these cases.

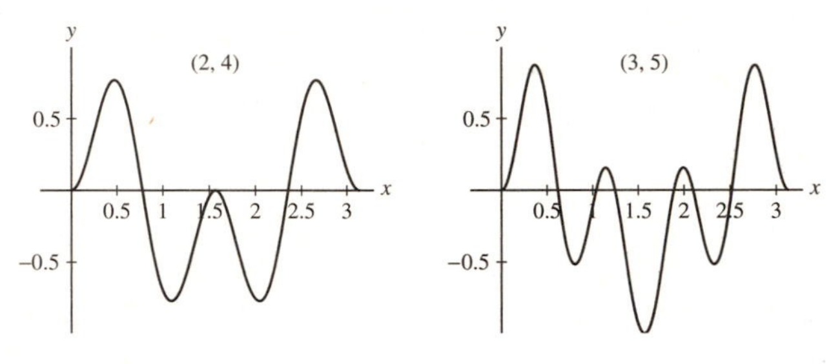

We arrive at the same conclusion for the cases $(m, n) = (4, 1)$ and $(m, n) = (5, 2)$.

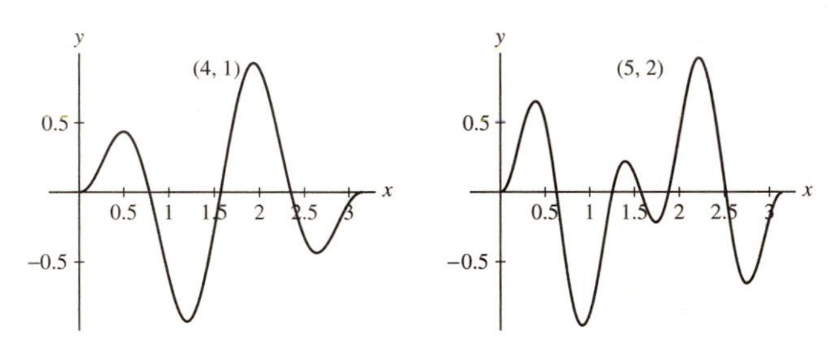

However, when $(m, n) = (3, 3)$ and when $(m, n) = (5, 5)$, the value of

$$I = \int_0^\pi f(x)\,dx$$

is clearly not zero as there is no negative area.

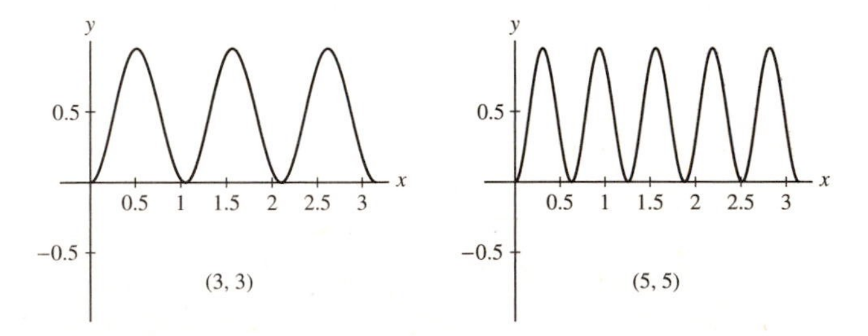

We therefore conjecture that I is zero whenever $m \ne n$.

83. Prove

$$2 \le \int_1^2 2^x\,dx \le 4 \qquad \text{and} \qquad \frac{1}{9} \le \int_1^2 3^{-x}\,dx \le \frac{1}{3}$$

SOLUTION The function $f(x) = 2^x$ is increasing, so $1 \le x \le 2$ implies that $2 = 2^1 \le 2^x \le 2^2 = 4$. Consequently,

$$2 = \int_1^2 2\,dx \le \int_1^2 2^x\,dx \le \int_1^2 4\,dx = 4.$$

On the other hand, the function $f(x) = 3^{-x}$ is decreasing, so $1 \le x \le 2$ implies that

$$\frac{1}{9} = 3^{-2} \le 3^{-x} \le 3^{-1} = \frac{1}{3}.$$

It then follows that

$$\frac{1}{9} = \int_1^2 \frac{1}{9}\,dx \le \int_1^2 3^{-x}\,dx \le \int_1^2 \frac{1}{3}\,dx = \frac{1}{3}.$$

85. Find upper and lower bounds for $\int_0^1 f(x)\,dx$, for $f(x)$ in Figure 6.

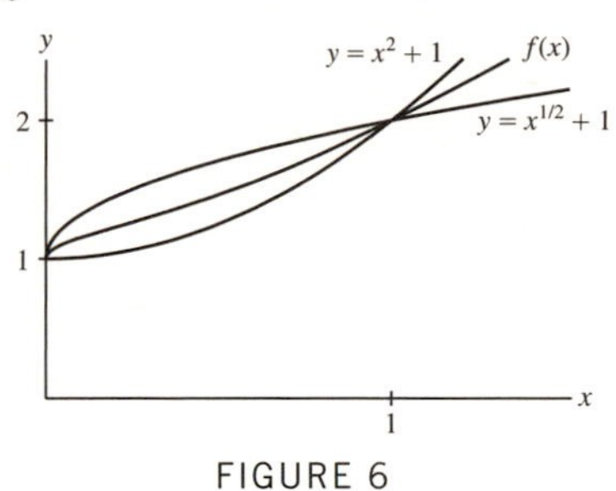

FIGURE 6

SOLUTION From the figure, we see that the inequalities $x^2 + 1 \le f(x) \le \sqrt{x} + 1$ hold for $0 \le x \le 1$. Because

$$\int_0^1 (x^2 + 1)\,dx = \left(\frac{1}{3}x^3 + x\right)\Big|_0^1 = \frac{4}{3}$$

and

$$\int_0^1 (\sqrt{x} + 1)\,dx = \left(\frac{2}{3}x^{3/2} + x\right)\Big|_0^1 = \frac{5}{3},$$

it follows that

$$\frac{4}{3} \le \int_0^1 f(x)\,dx \le \frac{5}{3}.$$

In Exercises 86–91, find the derivative.

87. $A'(\pi)$, where $A(x) = \int_2^x \frac{\cos t}{1+t}\,dt$

SOLUTION Let $A(x) = \int_2^x \frac{\cos t}{1+t}\,dt$. Then $A'(x) = \frac{\cos x}{1+x}$ and

$$A'(\pi) = \frac{\cos \pi}{1+\pi} = -\frac{1}{1+\pi}.$$

89. $G'(x)$, where $G(x) = \int_{-2}^{\sin x} t^3\,dt$

SOLUTION Let $G(x) = \int_{-2}^{\sin x} t^3\,dt$. Then

$$G'(x) = \sin^3 x \frac{d}{dx} \sin x = \sin^3 x \cos x.$$

91. $H'(1)$, where $H(x) = \int_{4x^2}^{9} \frac{1}{t}\,dt$

SOLUTION Let $H(x) = \int_{4x^2}^{9} \frac{1}{t}\,dt = -\int_9^{4x^2} \frac{1}{t}\,dt$. Then

$$H'(x) = -\frac{1}{4x^2}\frac{d}{dx}4x^2 = -\frac{8x}{4x^2} = -\frac{2}{x}$$

and $H'(1) = -2$.

93. Explain with a graph: If $f(x)$ is linear on $[a, b]$, then the $\int_a^b f(x)\,dx = \frac{1}{2}(R_N + L_N)$ for all N.

SOLUTION Consider the figure below, which displays a portion of the graph of a linear function.

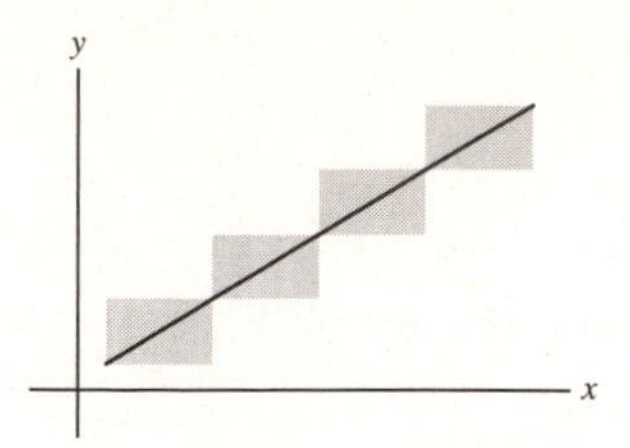

The shaded rectangles represent the differences between the right-endpoint approximation R_N and the left-endpoint approximation L_N. In particular, the portion of each rectangle that lies below the graph of $y = f(x)$ is the amount by which L_N underestimates the area under the graph, whereas the portion of each rectangle that lies above the graph of $y = f(x)$ is the amount by which R_N overestimates the area. Because the graph of $y = f(x)$ is a line, the lower portion of each shaded rectangle is exactly the same size as the upper portion. Therefore, if we average L_N and R_N, the error in the two approximations will exactly cancel, leaving

$$\frac{1}{2}(R_N + L_N) = \int_a^b f(x)\,dx.$$

95. Let

$$F(x) = x\sqrt{x^2 - 1} - 2\int_1^x \sqrt{t^2 - 1}\,dt$$

Prove that $F(x)$ and $\cosh^{-1} x$ differ by a constant by showing that they have the same derivative. Then prove they are equal by evaluating both at $x = 1$.

SOLUTION Let

$$F(x) = x\sqrt{x^2 - 1} - 2\int_1^x \sqrt{t^2 - 1}\,dt.$$

Then

$$\frac{dF}{dx} = \sqrt{x^2 - 1} + \frac{x^2}{\sqrt{x^2 - 1}} - 2\sqrt{x^2 - 1} = \frac{x^2}{\sqrt{x^2 - 1}} - \sqrt{x^2 - 1} = \frac{1}{\sqrt{x^2 - 1}}.$$

Also, $\frac{d}{dx}(\cosh^{-1}x) = \frac{1}{\sqrt{x^2-1}}$; therefore, $F(x)$ and $\cosh^{-1}x$ have the same derivative. We conclude that $F(x)$ and $\cosh^{-1}x$ differ by a constant:

$$F(x) = \cosh^{-1}x + C.$$

Now, let $x = 1$. Because $F(1) = 0$ and $\cosh^{-1} 1 = 0$, it follows that $C = 0$. Therefore,

$$F(x) = \cosh^{-1}x.$$

97. How can we interpret the quantity I in Eq. (2) if $a < b \le 0$? Explain with a graph.

SOLUTION We will consider each term on the right-hand side of (2) separately. For convenience, let **I**, **II**, **III** and **IV** denote the area of the similarly labeled region in the diagram below.

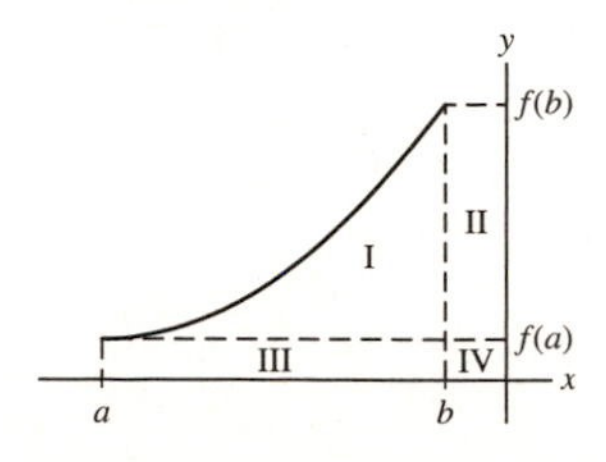

Because $b < 0$, the expression $bf(b)$ is the opposite of the area of the rectangle along the right; that is,

$$bf(b) = -\mathbf{II} - \mathbf{IV}.$$

Similarly,

$$-af(a) = \mathbf{III} + \mathbf{IV} \qquad \text{and} \qquad -\int_a^b f(x)\,dx = -\mathbf{I} - \mathbf{III}.$$

Therefore,

$$bf(b) - af(a) - \int_a^b f(x)\,dx = -\mathbf{I} - \mathbf{II};$$

that is, the opposite of the area of the shaded region shown below.

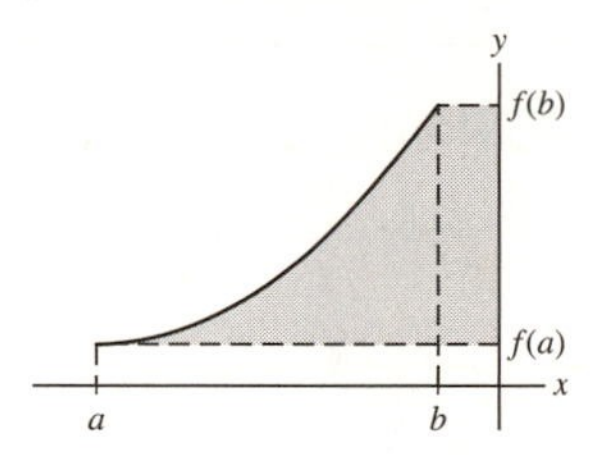

99. The Oldest Snack Food? In Bat Cave, New Mexico, archaeologists found ancient human remains, including cobs of popping corn whose C^{14}-to-C^{12} ratio was approximately 48% of that found in living matter. Estimate the age of the corn cobs.

SOLUTION Let t be the age of the corn cobs. The C^{14} to C^{12} ratio decreased by a factor of $e^{-0.000121t}$ which is equal to 0.48. That is:

$$e^{-0.000121t} = 0.48,$$

so

$$-0.000121t = \ln 0.48,$$

and

$$t = -\frac{1}{0.000121}\ln 0.48 \approx 6065.9.$$

We conclude that the age of the corn cobs is approximately 6065.9 years.

101. What is the interest rate if the PV of \$50,000 to be delivered in 3 years is \$43,000?

SOLUTION Let r denote the interest rate. The present value of \$50,000 received in 3 years with an interest rate of r is $50{,}000e^{-3r}$. Thus, we need to solve

$$43{,}000 = 50{,}000e^{-3r}$$

for r. This yields

$$r = -\frac{1}{3}\ln\frac{43}{50} = 0.0503.$$

Thus, the interest rate is 5.03%.

103. Find the PV of an income stream paying out continuously at a rate of $5000e^{-0.1t}$ dollars per year for 5 years, assuming an interest rate of $r = 4\%$.

SOLUTION $PV = \int_0^5 5000e^{-0.1t}e^{-0.04t}\,dt = \int_0^5 5000e^{-0.14t}\,dt = \frac{5000}{-0.14}e^{-0.14t}\Big|_0^5 = \$17,979.10.$

6 APPLICATIONS OF THE INTEGRAL

6.1 Area Between Two Curves

Preliminary Questions

1. What is the area interpretation of $\int_a^b \big(f(x) - g(x)\big)\,dx$ if $f(x) \geq g(x)$?

SOLUTION Because $f(x) \geq g(x)$, $\int_a^b (f(x) - g(x))\,dx$ represents the area of the region bounded between the graphs of $y = f(x)$ and $y = g(x)$, bounded on the left by the vertical line $x = a$ and on the right by the vertical line $x = b$.

2. Is $\int_a^b \big(f(x) - g(x)\big)\,dx$ still equal to the area between the graphs of f and g if $f(x) \geq 0$ but $g(x) \leq 0$?

SOLUTION Yes. Since $f(x) \geq 0$ and $g(x) \leq 0$, it follows that $f(x) - g(x) \geq 0$.

3. Suppose that $f(x) \geq g(x)$ on $[0, 3]$ and $g(x) \geq f(x)$ on $[3, 5]$. Express the area between the graphs over $[0, 5]$ as a sum of integrals.

SOLUTION Remember that to calculate an area between two curves, one must subtract the equation for the lower curve from the equation for the upper curve. Over the interval $[0, 3]$, $y = f(x)$ is the upper curve. On the other hand, over the interval $[3, 5]$, $y = g(x)$ is the upper curve. The area between the graphs over the interval $[0, 5]$ is therefore given by

$$\int_0^3 (f(x) - g(x))\,dx + \int_3^5 (g(x) - f(x))\,dx.$$

4. Suppose that the graph of $x = f(y)$ lies to the left of the y-axis. Is $\int_a^b f(y)\,dy$ positive or negative?

SOLUTION If the graph of $x = f(y)$ lies to the left of the y-axis, then for each value of y, the corresponding value of x is less than zero. Hence, the value of $\int_a^b f(y)\,dy$ is negative.

Exercises

1. Find the area of the region between $y = 3x^2 + 12$ and $y = 4x + 4$ over $[-3, 3]$ (Figure 9).

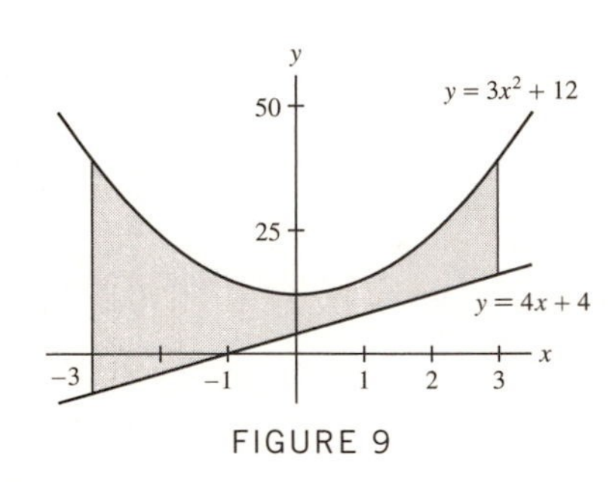

FIGURE 9

SOLUTION As the graph of $y = 3x^2 + 12$ lies above the graph of $y = 4x + 4$ over the interval $[-3, 3]$, the area between the graphs is

$$\int_{-3}^3 \Big((3x^2 + 12) - (4x + 4)\Big)\,dx = \int_{-3}^3 (3x^2 - 4x + 8)\,dx = \Big(x^3 - 2x^2 + 8x\Big)\Big|_{-3}^3 = 102.$$

3. Find the area of the region enclosed by the graphs of $f(x) = x^2 + 2$ and $g(x) = 2x + 5$ (Figure 10).

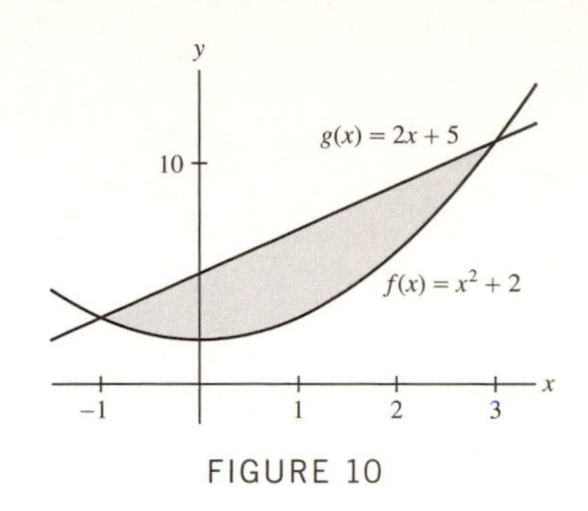

FIGURE 10

SOLUTION From the figure, we see that the graph of $g(x) = 2x + 5$ lies above the graph of $f(x) = x^2 + 2$ over the interval $[-1, 3]$. Thus, the area between the graphs is

$$\int_{-1}^{3} \left[(2x+5) - \left(x^2+2\right)\right] dx = \int_{-1}^{3} \left(-x^2 + 2x + 3\right) dx$$
$$= \left(-\frac{1}{3}x^3 + x^2 + 3x\right)\Big|_{-1}^{3}$$
$$= 9 - \left(-\frac{5}{3}\right) = \frac{32}{3}.$$

In Exercises 5 and 6, sketch the region between $y = \sin x$ and $y = \cos x$ over the interval and find its area.

5. $\left[\frac{\pi}{4}, \frac{\pi}{2}\right]$

SOLUTION Over the interval $[\frac{\pi}{4}, \frac{\pi}{2}]$, the graph of $y = \cos x$ lies below that of $y = \sin x$ (see the sketch below). Hence, the area between the two curves is

$$\int_{\pi/4}^{\pi/2} (\sin x - \cos x)\, dx = (-\cos x - \sin x)\Big|_{\pi/4}^{\pi/2} = (0 - 1) - \left(-\frac{\sqrt{2}}{2} - \frac{\sqrt{2}}{2}\right) = \sqrt{2} - 1.$$

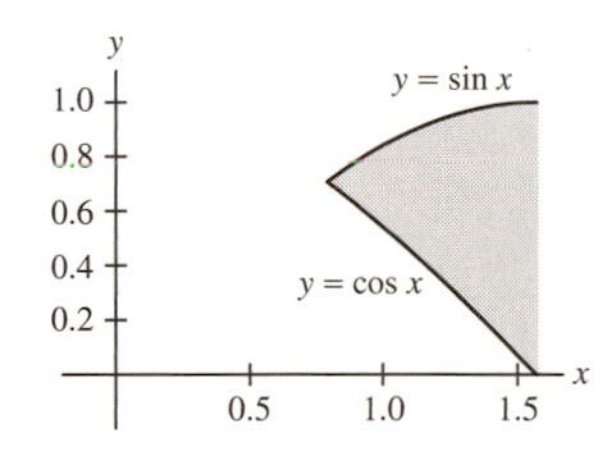

In Exercises 7 and 8, let $f(x) = 20 + x - x^2$ and $g(x) = x^2 - 5x$.

7. Sketch the region enclosed by the graphs of $f(x)$ and $g(x)$ and compute its area.

SOLUTION Setting $f(x) = g(x)$ gives $20 + x - x^2 = x^2 - 5x$, which simplifies to

$$0 = 2x^2 - 6x - 20 = 2(x-5)(x+2).$$

Thus, the curves intersect at $x = -2$ and $x = 5$. With $y = 20 + x - x^2$ being the upper curve (see the sketch below), the area between the two curves is

$$\int_{-2}^{5} \left((20 + x - x^2) - (x^2 - 5x)\right) dx = \int_{-2}^{5} \left(20 + 6x - 2x^2\right) dx = \left(20x + 3x^2 - \frac{2}{3}x^3\right)\Big|_{-2}^{5} = \frac{343}{3}.$$

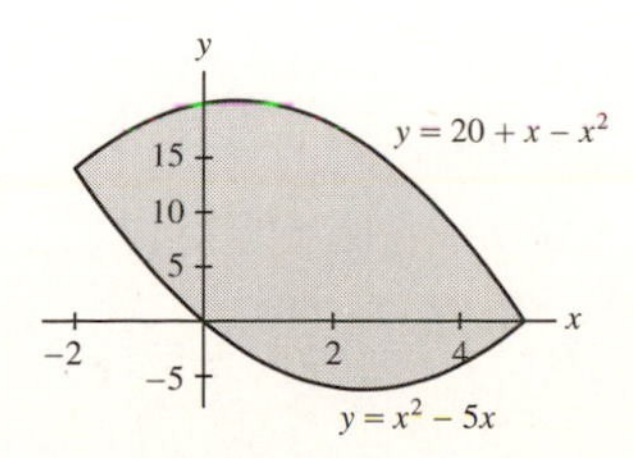

9. Find the area between $y = e^x$ and $y = e^{2x}$ over [0, 1].

SOLUTION As the graph of $y = e^{2x}$ lies above the graph of $y = e^x$ over the interval [0, 1], the area between the graphs is

$$\int_0^1 (e^{2x} - e^x)\,dx = \left(\frac{1}{2}e^{2x} - e^x\right)\Big|_0^1 = \frac{1}{2}e^2 - e - \left(\frac{1}{2} - 1\right) = \frac{1}{2}e^2 - e + \frac{1}{2}.$$

11. Sketch the region bounded by the line $y = 2$ and the graph of $y = \sec^2 x$ for $-\frac{\pi}{2} < x < \frac{\pi}{2}$ and find its area.

SOLUTION A sketch of the region bounded by $y = \sec^2 x$ and $y = 2$ is shown below. Note the region extends from $x = -\frac{\pi}{4}$ on the left to $x = \frac{\pi}{4}$ on the right. As the graph of $y = 2$ lies above the graph of $y = \sec^2 x$, the area between the graphs is

$$\int_{-\pi/4}^{\pi/4} (2 - \sec^2 x)\,dx = (2x - \tan x)\Big|_{-\pi/4}^{\pi/4} = \left(\frac{\pi}{2} - 1\right) - \left(-\frac{\pi}{2} + 1\right) = \pi - 2.$$

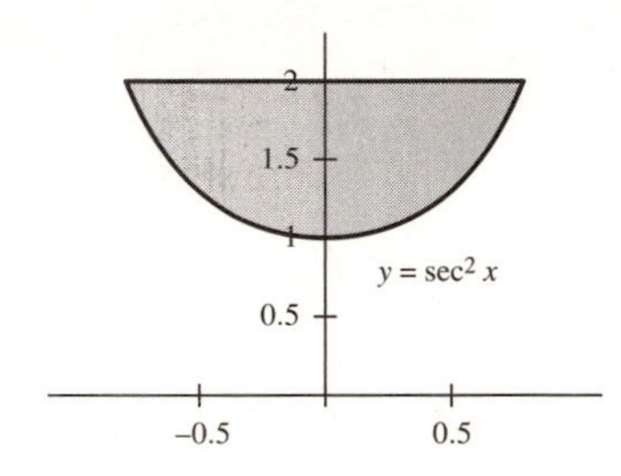

In Exercises 13–16, find the area of the shaded region in Figures 12–15.

13.

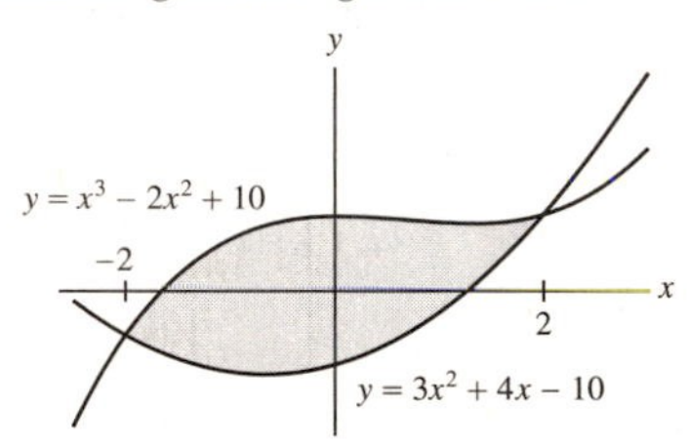

FIGURE 12

SOLUTION As the graph of $y = x^3 - 2x^2 + 10$ lies above the graph of $y = 3x^2 + 4x - 10$, the area of the shaded region is

$$\int_{-2}^{2} \left((x^3 - 2x^2 + 10) - (3x^2 + 4x - 10)\right) dx = \int_{-2}^{2} \left(x^3 - 5x^2 - 4x + 20\right) dx$$

$$= \left(\frac{1}{4}x^4 - \frac{5}{3}x^3 - 2x^2 + 20x\right)\Big|_{-2}^{2} = \frac{160}{3}.$$

15.

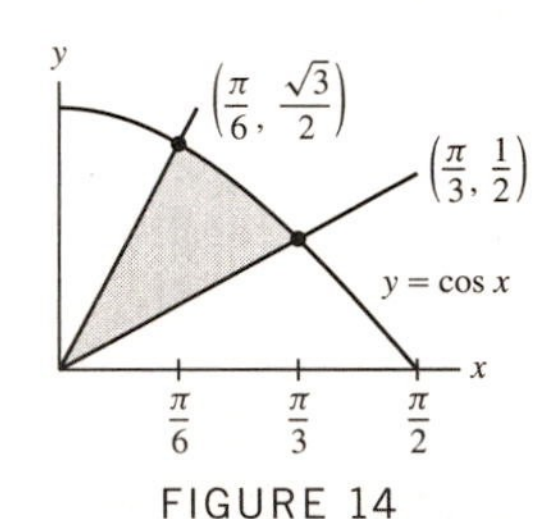

FIGURE 14

SOLUTION The line on the top-left has equation $y = \frac{3\sqrt{3}}{\pi}x$, and the line on the bottom-right has equation $y = \frac{3}{2\pi}x$. Thus, the area to the left of $x = \frac{\pi}{6}$ is

$$\int_0^{\pi/6} \left(\frac{3\sqrt{3}}{\pi}x - \frac{3}{2\pi}x\right) dx = \left(\frac{3\sqrt{3}}{2\pi}x^2 - \frac{3}{4\pi}x^2\right)\Big|_0^{\pi/6} = \frac{3\sqrt{3}}{2\pi}\frac{\pi^2}{36} - \frac{3}{4\pi}\frac{\pi^2}{36} = \frac{(2\sqrt{3} - 1)\pi}{48}.$$

The area to the right of $x = \frac{\pi}{6}$ is

$$\int_{\pi/6}^{\pi/3} \left(\cos x - \frac{3}{2\pi}x\right) dx = \left(\sin x - \frac{3}{4\pi}x^2\right)\Big|_{\pi/6}^{\pi/3} = \frac{8\sqrt{3} - 8 - \pi}{16}.$$

The entire area is then

$$\frac{(2\sqrt{3}-1)\pi}{48} + \frac{8\sqrt{3}-8-\pi}{16} = \frac{12\sqrt{3}-12+(\sqrt{3}-2)\pi}{24}.$$

In Exercises 17 and 18, find the area between the graphs of $x = \sin y$ and $x = 1 - \cos y$ over the given interval (Figure 16).

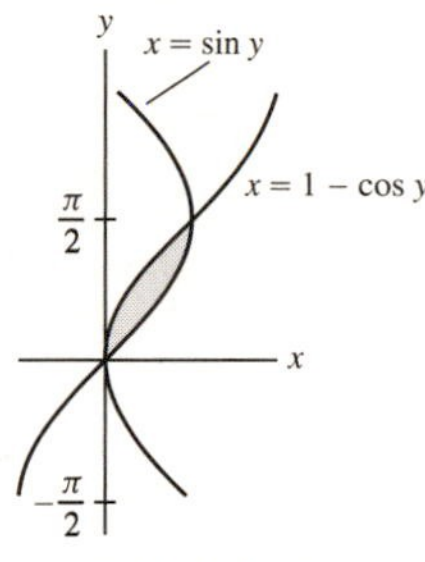

FIGURE 16

17. $0 \le y \le \frac{\pi}{2}$

SOLUTION As shown in the figure, the graph on the right is $x = \sin y$ and the graph on the left is $x = 1 - \cos y$. Therefore, the area between the two curves is given by

$$\int_0^{\pi/2} (\sin y - (1 - \cos y))\,dy = \left.(-\cos y - y + \sin y)\right|_0^{\pi/2} = \left(-\frac{\pi}{2} + 1\right) - (-1) = 2 - \frac{\pi}{2}.$$

19. Find the area of the region lying to the right of $x = y^2 + 4y - 22$ and to the left of $x = 3y + 8$.

SOLUTION Setting $y^2 + 4y - 22 = 3y + 8$ yields

$$0 = y^2 + y - 30 = (y+6)(y-5),$$

so the two curves intersect at $y = -6$ and $y = 5$. The area in question is then given by

$$\int_{-6}^{5} \left((3y+8) - (y^2+4y-22)\right) dy = \int_{-6}^{5} \left(-y^2 - y + 30\right) dy = \left.\left(-\frac{y^3}{3} - \frac{y^2}{2} + 30y\right)\right|_{-6}^{5} = \frac{1331}{6}.$$

21. Figure 17 shows the region enclosed by $x = y^3 - 26y + 10$ and $x = 40 - 6y^2 - y^3$. Match the equations with the curves and compute the area of the region.

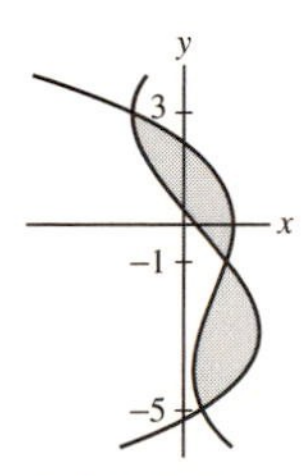

FIGURE 17

SOLUTION Substituting $y = 0$ into the equations for both curves indicates that the graph of $x = y^3 - 26y + 10$ passes through the point $(10, 0)$ while the graph of $x = 40 - 6y^2 - y^3$ passes through the point $(40, 0)$. Therefore, over the y-interval $[-1, 3]$, the graph of $x = 40 - 6y^2 - y^3$ lies to the right of the graph of $x = y^3 - 26y + 10$. The orientation of the two graphs is reversed over the y-interval $[-5, -1]$. Hence, the area of the shaded region is

$$\int_{-5}^{-1} \left((y^3 - 26y + 10) - (40 - 6y^2 - y^3)\right) dy + \int_{-1}^{3} \left((40 - 6y^2 - y^3) - (y^3 - 26y + 10)\right) dy$$

$$= \int_{-5}^{-1} \left(2y^3 + 6y^2 - 26y - 30\right) dy + \int_{-1}^{3} \left(-2y^3 - 6y^2 + 26y + 30\right) dy$$

$$= \left.\left(\frac{1}{2}y^4 + 2y^3 - 13y^2 - 30y\right)\right|_{-5}^{-1} + \left.\left(-\frac{1}{2}y^4 - 2y^3 + 13y^2 + 30y\right)\right|_{-1}^{3} = 256.$$

In Exercises 23 and 24, find the area enclosed by the graphs in two ways: by integrating along the x-axis and by integrating along the y-axis.

23. $x = 9 - y^2, \quad x = 5$

SOLUTION Along the y-axis, we have points of intersection at $y = \pm 2$. Therefore, the area enclosed by the two curves is

$$\int_{-2}^{2} \left(9 - y^2 - 5\right) dy = \int_{-2}^{2} \left(4 - y^2\right) dy = \left(4y - \frac{1}{3}y^3\right)\Big|_{-2}^{2} = \frac{32}{3}.$$

Along the x-axis, we have integration limits of $x = 5$ and $x = 9$. Therefore, the area enclosed by the two curves is

$$\int_{5}^{9} 2\sqrt{9 - x}\, dx = -\frac{4}{3}(9 - x)^{3/2}\Big|_{5}^{9} = 0 - \left(-\frac{32}{3}\right) = \frac{32}{3}.$$

In Exercises 25 and 26, find the area of the region using the method (integration along either the x- or the y-axis) that requires you to evaluate just one integral.

25. Region between $y^2 = x + 5$ and $y^2 = 3 - x$

SOLUTION From the figure below, we see that integration along the x-axis would require two integrals, but integration along the y-axis requires only one integral. Setting $y^2 - 5 = 3 - y^2$ yields points of intersection at $y = \pm 2$. Thus, the area is given by

$$\int_{-2}^{2} \left((3 - y^2) - (y^2 + 5)\right) dy = \int_{-2}^{2} \left(8 - 2y^2\right) dy = \left(8y - \frac{2}{3}y^3\right)\Big|_{-2}^{2} = \frac{64}{3}.$$

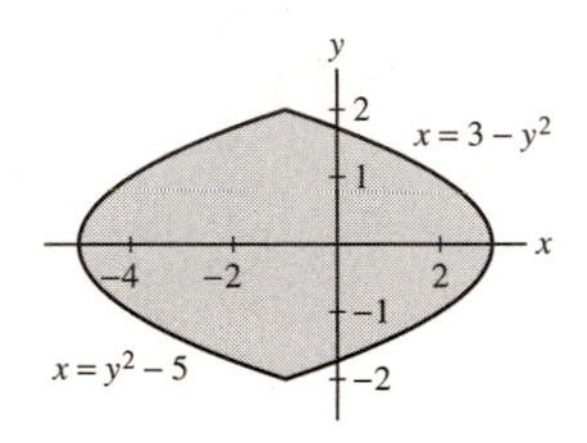

In Exercises 27–44, sketch the region enclosed by the curves and compute its area as an integral along the x- or y-axis.

27. $y = 4 - x^2, \quad y = x^2 - 4$

SOLUTION Setting $4 - x^2 = x^2 - 4$ yields $2x^2 = 8$ or $x^2 = 4$. Thus, the curves $y = 4 - x^2$ and $y = x^2 - 4$ intersect at $x = \pm 2$. From the figure below, we see that $y = 4 - x^2$ lies above $y = x^2 - 4$ over the interval $[-2, 2]$; hence, the area of the region enclosed by the curves is

$$\int_{-2}^{2} \left((4 - x^2) - (x^2 - 4)\right) dx = \int_{-2}^{2} (8 - 2x^2)\, dx = \left(8x - \frac{2}{3}x^3\right)\Big|_{-2}^{2} = \frac{64}{3}.$$

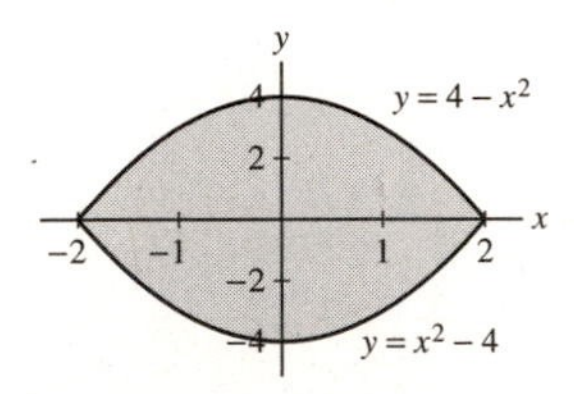

29. $x + y = 4, \quad x - y = 0, \quad y + 3x = 4$

SOLUTION From the graph below, we see that the top of the region enclosed by the three lines is always bounded by $x + y = 4$. On the other hand, the bottom of the region is bounded by $y + 3x = 4$ for $0 \le x \le 1$ and by $x - y = 0$ for $1 \le x \le 2$. The total area of the region is then

$$\int_{0}^{1} ((4 - x) - (4 - 3x))\, dx + \int_{1}^{2} ((4 - x) - x)\, dx = \int_{0}^{1} 2x\, dx + \int_{1}^{2} (4 - 2x)\, dx$$

$$= x^2\Big|_{0}^{1} + (4x - x^2)\Big|_{1}^{2} = 1 + (8 - 4) - (4 - 1) = 2.$$

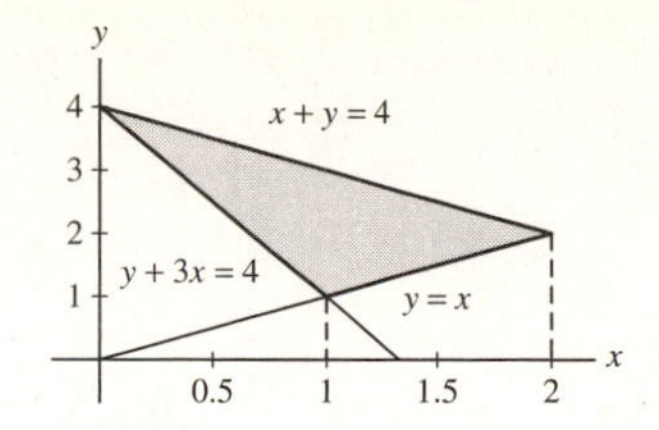

31. $y = 8 - \sqrt{x}, \quad y = \sqrt{x}, \quad x = 0$

SOLUTION Setting $8 - \sqrt{x} = \sqrt{x}$ yields $\sqrt{x} = 4$ or $x = 16$. Using the graph shown below, we see that $y = 8 - \sqrt{x}$ lies above $y = \sqrt{x}$ over the interval $[0, 16]$. The area of the region enclosed by these two curves and the y-axis is then

$$\int_0^{16} (8 - \sqrt{x} - \sqrt{x})\, dx = \int_0^{16} (8 - 2\sqrt{x})\, dx = \left(8x - \frac{4}{3}x^{3/2}\right)\Big|_0^{16} = \frac{128}{3}.$$

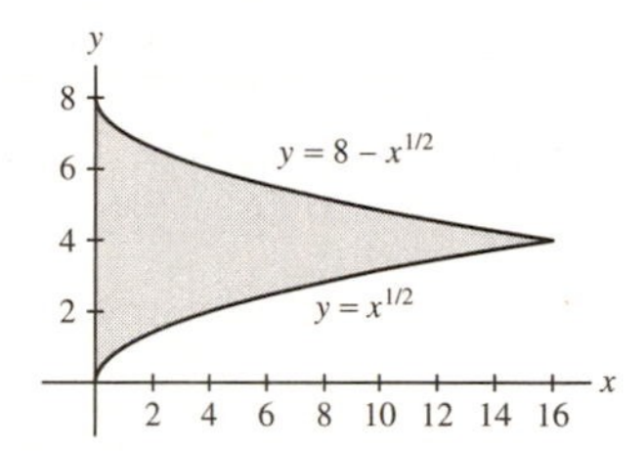

33. $x = |y|, \quad x = 1 - |y|$

SOLUTION From the graph below, we see that the region enclosed by the curves $x = |y|$ and $x = 1 - |y|$ is symmetric with respect to the x-axis. We can therefore determine the total area by doubling the area in the first quadrant. For $y > 0$, setting $y = 1 - y$ yields $y = \frac{1}{2}$ as the point of intersection. Moreover, $x = 1 - |y| = 1 - y$ lies to the right of $x = |y| = y$, so the total area of the region is

$$2\int_0^{1/2} ((1 - y) - y)\, dy = 2(y - y^2)\Big|_0^{1/2} = 2\left(\frac{1}{2} - \frac{1}{4}\right) = \frac{1}{2}.$$

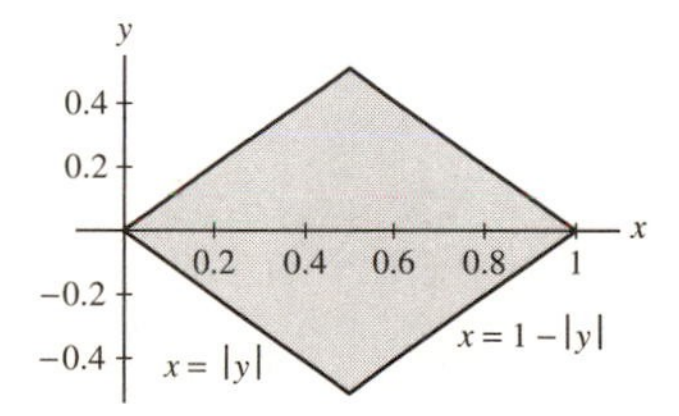

35. $x = y^3 - 18y, \quad y + 2x = 0$

SOLUTION Setting $y^3 - 18y = -\frac{y}{2}$ yields

$$0 = y^3 - \frac{35}{2}y = y\left(y^2 - \frac{35}{2}\right),$$

so the points of intersection occur at $y = 0$ and $y = \pm\frac{\sqrt{70}}{2}$. From the graph below, we see that both curves are symmetric with respect to the origin. It follows that the portion of the region enclosed by the curves in the second quadrant is identical to the region enclosed in the fourth quadrant. We can therefore determine the total area enclosed by the two curves by doubling the area enclosed in the second quadrant. In the second quadrant, $y + 2x = 0$ lies to the right of $x = y^3 - 18y$, so the total area enclosed by the two curves is

$$2\int_0^{\sqrt{70}/2} \left(-\frac{y}{2} - (y^3 - 18y)\right) dy = 2\left(\frac{35}{4}y^2 - \frac{1}{4}y^4\right)\Big|_0^{\sqrt{70}/2} = 2\left(\frac{1225}{8} - \frac{1225}{16}\right) = \frac{1225}{8}.$$

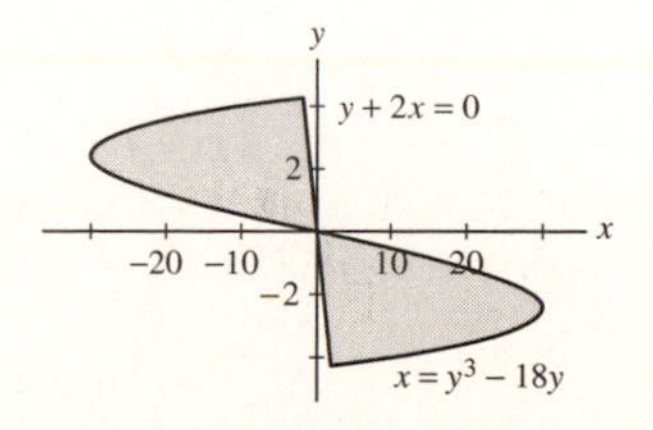

37. $x = 2y, \quad x + 1 = (y - 1)^2$

SOLUTION Setting $2y = (y - 1)^2 - 1$ yields

$$0 = y^2 - 4y = y(y - 4),$$

so the two curves intersect at $y = 0$ and at $y = 4$. From the graph below, we see that $x = 2y$ lies to the right of $x + 1 = (y - 1)^2$ over the interval $[0, 4]$ along the y-axis. Thus, the area of the region enclosed by the two curves is

$$\int_0^4 \left(2y - ((y - 1)^2 - 1)\right) dy = \int_0^4 \left(4y - y^2\right) dy = \left(2y^2 - \frac{1}{3}y^3\right)\Big|_0^4 = \frac{32}{3}.$$

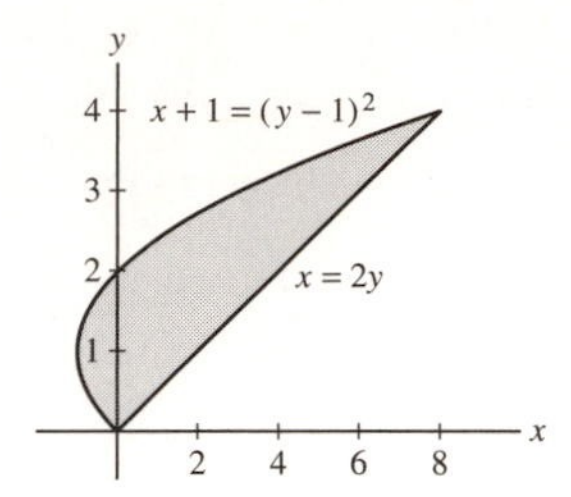

39. $y = \cos x, \quad y = \cos 2x, \quad x = 0, \quad x = \dfrac{2\pi}{3}$

SOLUTION From the graph below, we see that $y = \cos x$ lies above $y = \cos 2x$ over the interval $[0, \frac{2\pi}{3}]$. The area of the region enclosed by the two curves is therefore

$$\int_0^{2\pi/3} (\cos x - \cos 2x)\, dx = \left(\sin x - \frac{1}{2}\sin 2x\right)\Big|_0^{2\pi/3} = \frac{3\sqrt{3}}{4}.$$

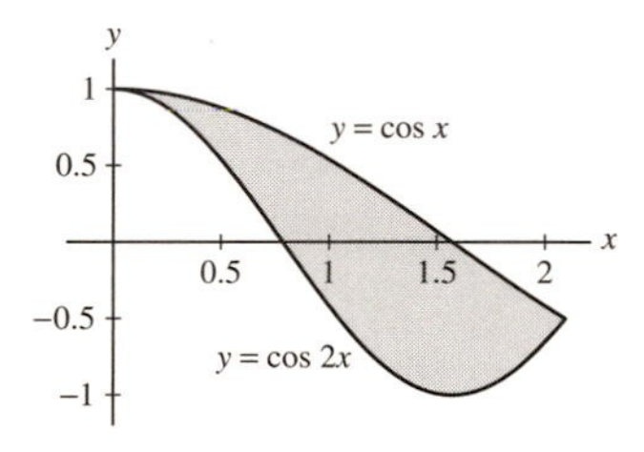

41. $y = \sin x, \quad y = \csc^2 x, \quad x = \dfrac{\pi}{4}$

SOLUTION Over the interval $[\frac{\pi}{4}, \frac{\pi}{2}]$, $y = \csc^2 x$ lies above $y = \sin x$. The area of the region enclosed by the two curves is then

$$\int_{\pi/4}^{\pi/2} \left(\csc^2 x - \sin x\right) dx = \left(-\cot x + \cos x\right)\Big|_{\pi/4}^{\pi/2} = (0 - 0) - \left(-1 + \frac{\sqrt{2}}{2}\right) = 1 - \frac{\sqrt{2}}{2}.$$

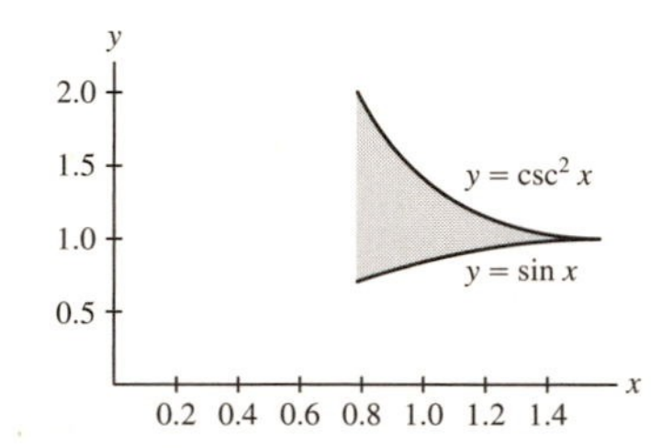

43. $y = e^x, \quad y = e^{-x}, \quad y = 2$

SOLUTION From the figure below, we see that integration in y would be most appropriate - unfortunately, we have not yet learned how to integrate $\ln y$. Consequently, we will calculate the area using two integrals in x:

$$\begin{aligned} A &= \int_{-\ln 2}^{0} (2 - e^{-x})\, dx + \int_0^{\ln 2} (2 - e^x)\, dx \\ &= \left(2x + e^{-x}\right)\Big|_{-\ln 2}^{0} + \left(2x - e^x\right)\Big|_0^{\ln 2} \\ &= 1 - (-2\ln 2 + 2) + (2\ln 2 - 2) - (-1) = 4\ln 2 - 2. \end{aligned}$$

45. CAS Plot

$$y = \frac{x}{\sqrt{x^2+1}} \quad \text{and} \quad y = (x-1)^2$$

on the same set of axes. Use a computer algebra system to find the points of intersection numerically and compute the area between the curves.

SOLUTION Using a computer algebra system, we find that the curves

$$y = \frac{x}{\sqrt{x^2+1}} \qquad \text{and} \qquad y = (x-1)^2$$

intersect at $x = 0.3943285581$ and at $x = 1.942944418$. From the graph below, we see that $y = \frac{x}{\sqrt{x^2+1}}$ lies above $y = (x-1)^2$, so the area of the region enclosed by the two curves is

$$\int_{0.3943285581}^{1.942944418} \left(\frac{x}{\sqrt{x^2+1}} - (x-1)^2 \right) dx = 0.7567130951$$

The value of the definite integral was also obtained using a computer algebra system.

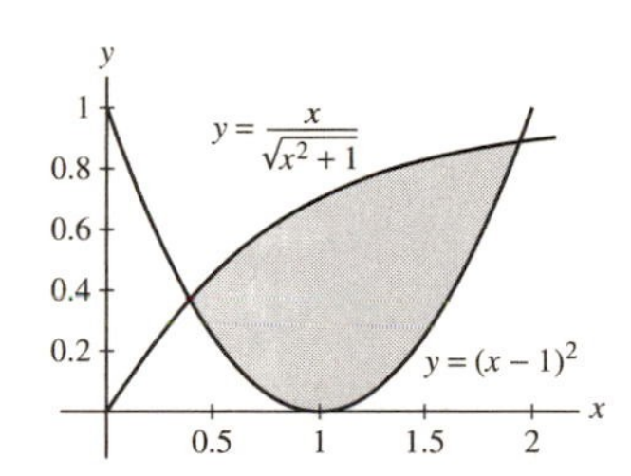

47. Athletes 1 and 2 run along a straight track with velocities $v_1(t)$ and $v_2(t)$ (in m/s) as shown in Figure 19.

(a) Which of the following is represented by the area of the shaded region over [0, 10]?

i. The distance between athletes 1 and 2 at time $t = 10$ s.

ii. The difference in the distance traveled by the athletes over the time interval [0, 10].

(b) Does Figure 19 give us enough information to determine who is ahead at time $t = 10$ s?

(c) If the athletes begin at the same time and place, who is ahead at $t = 10$ s? At $t = 25$ s?

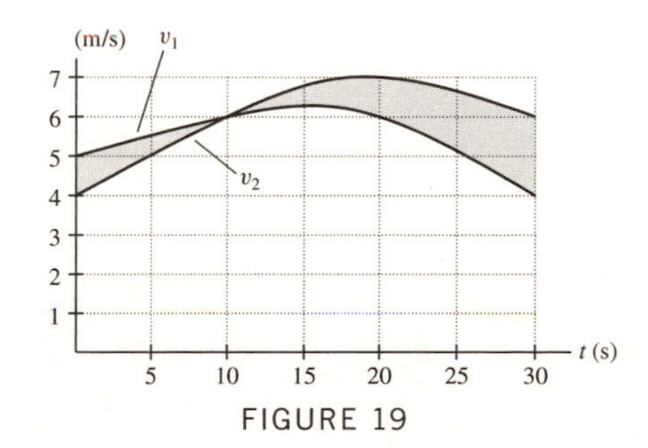

FIGURE 19

SOLUTION

(a) The area of the shaded region over [0, 10] represents **(ii)**: the difference in the distance traveled by the athletes over the time interval [0, 10].

(b) No, Figure 19 does not give us enough information to determine who is ahead at time $t = 10$ s. We would additionally need to know the relative position of the runners at $t = 0$ s.

(c) If the athletes begin at the same time and place, then athlete 1 is ahead at $t = 10$ s because the velocity graph for athlete 1 lies above the velocity graph for athlete 2 over the interval [0, 10]. Over the interval [10, 25], the velocity graph for athlete 2 lies above the velocity graph for athlete 1 and appears to have a larger area than the area between the graphs over [0, 10]. Thus, it appears that athlete 2 is ahead at $t = 25$ s.

49. Find the area enclosed by the curves $y = c - x^2$ and $y = x^2 - c$ as a function of c. Find the value of c for which this area is equal to 1.

SOLUTION The curves intersect at $x = \pm\sqrt{c}$, with $y = c - x^2$ above $y = x^2 - c$ over the interval $[-\sqrt{c}, \sqrt{c}]$. The area of the region enclosed by the two curves is then

$$\int_{-\sqrt{c}}^{\sqrt{c}} \left((c - x^2) - (x^2 - c) \right) dx = \int_{-\sqrt{c}}^{\sqrt{c}} \left(2c - 2x^2 \right) dx = \left(2cx - \frac{2}{3}x^3 \right)\Big|_{-\sqrt{c}}^{\sqrt{c}} = \frac{8}{3}c^{3/2}.$$

In order for the area to equal 1, we must have $\frac{8}{3}c^{3/2} = 1$, which gives

$$c = \frac{9^{1/3}}{4} \approx 0.520021.$$

51. Set up (but do not evaluate) an integral that expresses the area between the graphs of $y = (1 + x^2)^{-1}$ and $y = x^2$.

SOLUTION Setting $(1 + x^2)^{-1} = x^2$ yields $x^4 + x^2 - 1 = 0$. This is a quadratic equation in the variable x^2. By the quadratic formula,

$$x^2 = \frac{-1 \pm \sqrt{1 - 4(-1)}}{2} = \frac{-1 \pm \sqrt{5}}{2}.$$

As x^2 must be nonnegative, we discard $\frac{-1-\sqrt{5}}{2}$. Finally, we find the two curves intersect at $x = \pm\sqrt{\frac{-1+\sqrt{5}}{2}}$. From the graph below, we see that $y = (1 + x^2)^{-1}$ lies above $y = x^2$. The area enclosed by the two curves is then

$$\int_{-\sqrt{\frac{-1+\sqrt{5}}{2}}}^{\sqrt{\frac{-1+\sqrt{5}}{2}}} \left((1 + x^2)^{-1} - x^2 \right) dx.$$

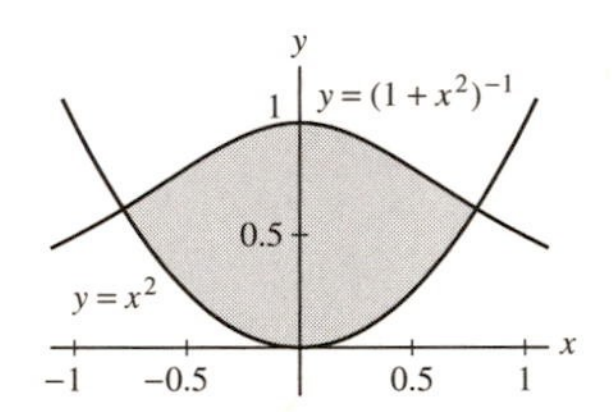

53. CAS Find a numerical approximation to the area above $y = |x|$ and below $y = \cos x$.

SOLUTION The region in question is shown in the figure below. We see that the region is symmetric with respect to the y-axis, so we can determine the total area of the region by doubling the area of the portion in the first quadrant. Using a computer algebra system, we find that $y = \cos x$ and $y = |x|$ intersect at $x = 0.7390851332$. The area of the region between the two curves is then

$$2\int_0^{0.7390851332} (\cos x - x)\, dx = 0.8009772242,$$

where the definite integral was evaluated using a computer algebra system.

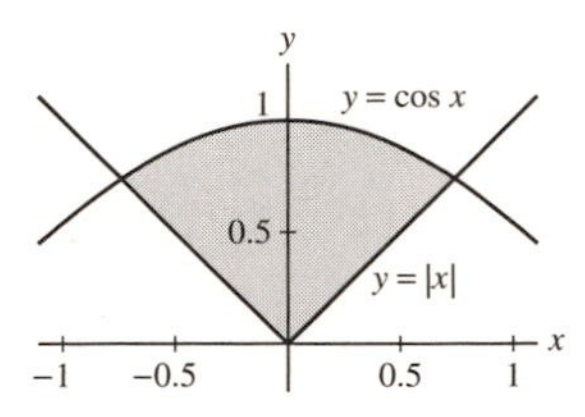

55. The back of Jon's guitar (Figure 21) is 19 inches long. Jon measured the width at 1-in. intervals, beginning and ending $\frac{1}{2}$ in. from the ends, obtaining the results

6, 9, 10.25, 10.75, 10.75, 10.25, 9.75, 9.5, 10, 11.25,

12.75, 13.75, 14.25, 14.5, 14.5, 14, 13.25, 11.25, 9

Use the midpoint rule to estimate the area of the back.

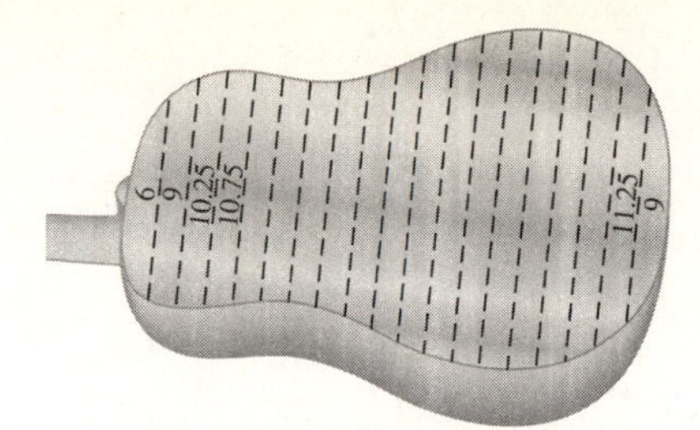

FIGURE 21 Back of guitar.

SOLUTION Note that the measurements were taken at the midpoint of each one-inch section of the guitar. For example, in the 0 to 1 inch section, the midpoint would be at $\frac{1}{2}$ inch, and thus the approximate area of the first rectangle would be $1 \cdot 6$ inches^2. An approximation for the entire area is then

$$\begin{aligned} A &= 1(6 + 9 + 10.25 + 10.75 + 10.75 + 10.25 + 9.75 + 9.5 + 10 + 11.25 \\ &\quad + 12.75 + 13.75 + 14.25 + 14.5 + 14.5 + 14 + 13.25 + 11.25 + 9) \\ &= 214.75 \text{ in}^2. \end{aligned}$$

Exercises 57 and 58 use the notation and results of Exercises 49–51 of Section 3.4. For a given country, $F(r)$ is the fraction of total income that goes to the bottom rth fraction of households. The graph of $y = F(r)$ is called the Lorenz curve.

57. Let A be the area between $y = r$ and $y = F(r)$ over the interval $[0, 1]$ (Figure 22). The **Gini index** is the ratio $G = A/B$, where B is the area under $y = r$ over $[0, 1]$.

(a) Show that $G = 2\int_0^1 (r - F(r))\,dr$.

(b) Calculate G if

$$F(r) = \begin{cases} \frac{1}{3}r & \text{for } 0 \le r \le \frac{1}{2} \\ \frac{5}{3}r - \frac{2}{3} & \text{for } \frac{1}{2} \le r \le 1 \end{cases}$$

(c) The Gini index is a measure of income distribution, with a lower value indicating a more equal distribution. Calculate G if $F(r) = r$ (in this case, all households have the same income by Exercise 51(b) of Section 3.4).

(d) What is G if all of the income goes to one household? *Hint:* In this extreme case, $F(r) = 0$ for $0 \le r < 1$.

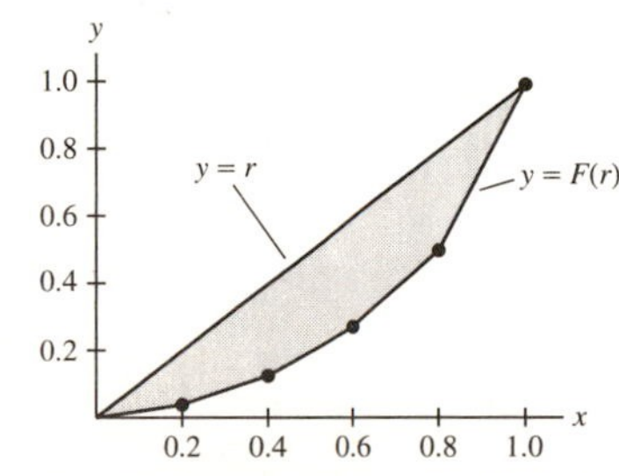

FIGURE 22 Lorenz Curve for U.S. in 2001.

SOLUTION

(a) Because the graph of $y = r$ lies above the graph of $y = F$'' in Figure 22,

$$A = \int_0^1 (r - F(r))\,dr.$$

Moreover,

$$B = \int_0^1 r\,dr = \frac{1}{2}r^2\bigg|_0^1 = \frac{1}{2}.$$

Thus,

$$G = \frac{A}{B} = 2\int_0^1 (r - F(r))\,dr.$$

(b) With the given $F(r)$,

$$\begin{aligned} G &= 2\int_0^{1/2}\left(r - \frac{1}{3}r\right)dr + 2\int_{1/2}^{1}\left(r - \left(\frac{5}{3}r - \frac{2}{3}\right)\right)dr \\ &= \frac{4}{3}\int_0^{1/2} r\,dr - \frac{4}{3}\int_{1/2}^{1}(r-1)\,dr \\ &= \frac{2}{3}r^2\Big|_0^{1/2} - \frac{4}{3}\left(\frac{1}{2}r^2 - r\right)\Big|_{1/2}^{1} \\ &= \frac{1}{6} - \frac{4}{3}\left(-\frac{1}{2}\right) + \frac{4}{3}\left(-\frac{3}{8}\right) = \frac{1}{3}. \end{aligned}$$

(c) If $F(r) = r$, then

$$G = 2\int_0^1 (r - r)\,dr = 0.$$

(d) If $F(r) = 0$ for $0 \le r < 1$, then

$$G = 2\int_0^1 (r - 0)\,dr = 2\left(\frac{1}{2}r^2\right)\Big|_0^1 = 2\left(\frac{1}{2}\right) = 1.$$

Further Insights and Challenges

59. Find the line $y = mx$ that divides the area under the curve $y = x(1-x)$ over $[0, 1]$ into two regions of equal area.

SOLUTION First note that

$$\int_0^1 x(1-x)\,dx = \int_0^1 \left(x - x^2\right)dx = \left(\frac{1}{2}x^2 - \frac{1}{3}x^3\right)\Big|_0^1 = \frac{1}{6}.$$

Now, the line $y = mx$ and the curve $y = x(1-x)$ intersect when $mx = x(1-x)$, or at $x = 0$ and at $x = 1 - m$. The area of the region enclosed by the two curves is then

$$\int_0^{1-m} (x(1-x) - mx)\,dx = \int_0^{1-m}\left((1-m)x - x^2\right)dx = \left((1-m)\frac{x^2}{2} - \frac{1}{3}x^3\right)\Bigg|_0^{1-m} = \frac{1}{6}(1-m)^3.$$

To have $\frac{1}{6}(1-m)^3 = \frac{1}{2}\cdot\frac{1}{6}$ requires

$$m = 1 - \left(\frac{1}{2}\right)^{1/3} \approx 0.206299.$$

61. Explain geometrically (without calculation):

$$\int_0^1 x^n\,dx + \int_0^1 x^{1/n}\,dx = 1 \qquad (\text{for } n > 0)$$

SOLUTION Let A_1 denote the area of region 1 in the figure below. Define A_2 and A_3 similarly. It is clear from the figure that

$$A_1 + A_2 + A_3 = 1.$$

Now, note that x^n and $x^{1/n}$ are inverses of each other. Therefore, the graphs of $y = x^n$ and $y = x^{1/n}$ are symmetric about the line $y = x$, so regions 1 and 3 are also symmetric about $y = x$. This guarantees that $A_1 = A_3$. Finally,

$$\int_0^1 x^n\,dx + \int_0^1 x^{1/n}\,dx = A_3 + (A_2 + A_3) = A_1 + A_2 + A_3 = 1.$$

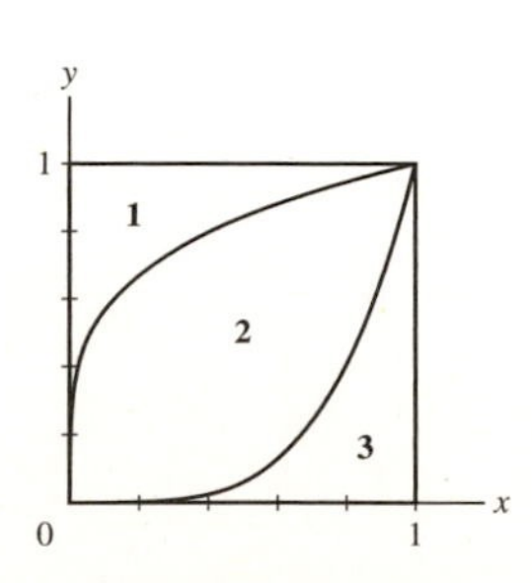

6.2 Setting Up Integrals: Volume, Density, Average Value

Preliminary Questions

1. What is the average value of $f(x)$ on $[0, 4]$ if the area between the graph of $f(x)$ and the x-axis is equal to 12?

SOLUTION Assuming that $f(x) \geq 0$ over the interval $[1, 4]$, the fact that the area between the graph of f and the x-axis is equal to 9 indicates that $\int_1^4 f(x)\,dx = 9$. The average value of f over the interval $[1, 4]$ is then

$$\frac{\int_1^4 f(x)\,dx}{4-1} = \frac{9}{3} = 3.$$

2. Find the volume of a solid extending from $y = 2$ to $y = 5$ if every cross section has area $A(y) = 5$.

SOLUTION Because the cross-sectional area of the solid is constant, the volume is simply the cross-sectional area times the length, or $5 \times 3 = 15$.

3. What is the definition of flow rate?

SOLUTION The flow rate of a fluid is the volume of fluid that passes through a cross-sectional area at a given point per unit time.

4. Which assumption about fluid velocity did we use to compute the flow rate as an integral?

SOLUTION To express flow rate as an integral, we assumed that the fluid velocity depended only on the radial distance from the center of the tube.

5. The average value of $f(x)$ on $[1, 4]$ is 5. Find $\displaystyle\int_1^4 f(x)\,dx$.

SOLUTION

$$\begin{aligned}\int_1^4 f(x)\,dx &= \text{average value on } [1,4] \times \text{ length of } [1,4] \\ &= 5 \times 3 = 15.\end{aligned}$$

Exercises

1. Let V be the volume of a pyramid of height 20 whose base is a square of side 8.

(a) Use similar triangles as in Example 1 to find the area of the horizontal cross section at a height y.

(b) Calculate V by integrating the cross-sectional area.

SOLUTION

(a) We can use similar triangles to determine the side length, s, of the square cross section at height y. Using the diagram below, we find

$$\frac{8}{20} = \frac{s}{20-y} \quad \text{or} \quad s = \frac{2}{5}(20-y).$$

The area of the cross section at height y is then given by $\frac{4}{25}(20-y)^2$.

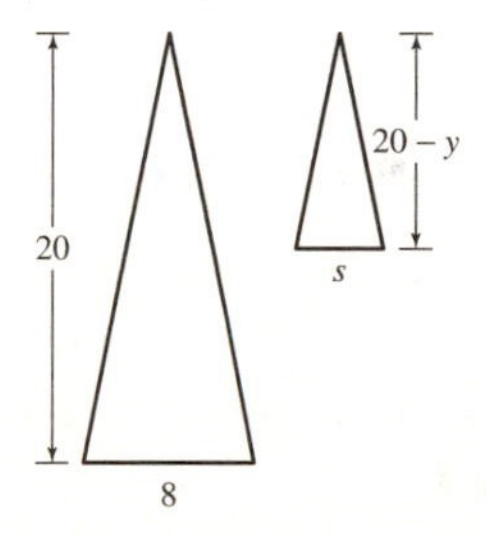

(b) The volume of the pyramid is

$$\int_0^{20} \frac{4}{25}(20-y)^2\,dy = -\frac{4}{75}(20-y)^3\Big|_0^{20} = \frac{1280}{3}.$$

3. Use the method of Exercise 2 to find the formula for the volume of a right circular cone of height h whose base is a circle of radius R [Figure 17(B)].

SOLUTION

(a) From similar triangles (see Figure 17),

$$\frac{h}{h-y} = \frac{R}{r_0},$$

where r_0 is the radius of the cone at a height of y. Thus, $r_0 = R - \frac{Ry}{h}$.

(b) The volume of the cone is

$$\pi \int_0^h \left(R - \frac{Ry}{h}\right)^2 dy = \frac{-h\pi}{R} \frac{\left(R - \frac{Ry}{h}\right)^3}{3}\Bigg|_0^h = \frac{h\pi}{R}\frac{R^3}{3} = \frac{\pi R^2 h}{3}.$$

5. Find the volume of liquid needed to fill a sphere of radius R to height h (Figure 19).

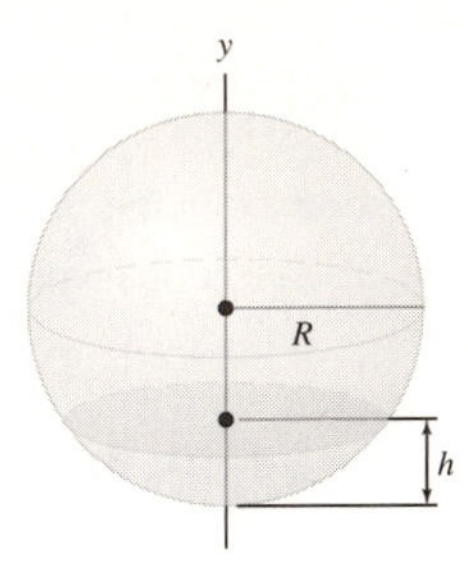

FIGURE 19 Sphere filled with liquid to height h.

SOLUTION The radius r at any height y is given by $r = \sqrt{R^2 - (R - y)^2}$. Thus, the volume of the filled portion of the sphere is

$$\pi \int_0^h r^2\,dy = \pi \int_0^h \left(R^2 - (R - y)^2\right) dy = \pi \int_0^h (2Ry - y^2)\,dy = \pi \left(Ry^2 - \frac{y^3}{3}\right)\Bigg|_0^h = \pi \left(Rh^2 - \frac{h^3}{3}\right).$$

7. Derive a formula for the volume of the wedge in Figure 20(B) in terms of the constants a, b, and c.

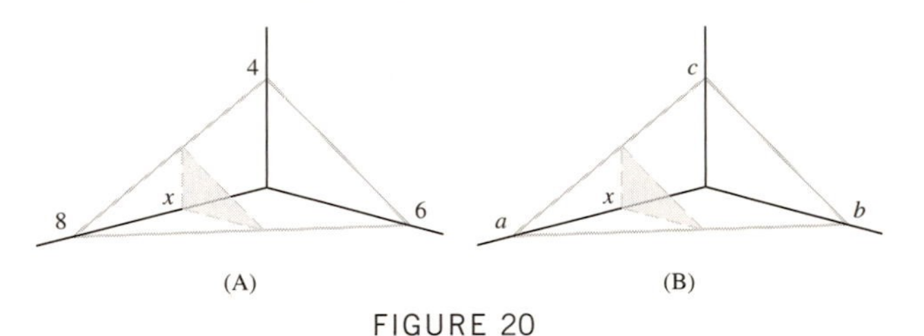

FIGURE 20

SOLUTION The line from c to a is given by the equation $(z/c) + (x/a) = 1$ and the line from b to a is given by $(y/b) + (x/a) = 1$. The cross sections perpendicular to the x-axis are right triangles with height $c(1 - x/a)$ and base $b(1 - x/a)$. Thus we have

$$\int_0^a \frac{1}{2} bc\,(1 - x/a)^2\,dx = -\frac{1}{6}abc\left(1 - \frac{x}{a}\right)^3\Bigg|_0^a = \frac{1}{6}abc.$$

In Exercises 9–14, find the volume of the solid with the given base and cross sections.

9. The base is the unit circle $x^2 + y^2 = 1$, and the cross sections perpendicular to the x-axis are triangles whose height and base are equal.

SOLUTION At each location x, the side of the triangular cross section that lies in the base of the solid extends from the top half of the unit circle (with $y = \sqrt{1 - x^2}$) to the bottom half (with $y = -\sqrt{1 - x^2}$). The triangle therefore has base and height equal to $2\sqrt{1 - x^2}$ and area $2(1 - x^2)$. The volume of the solid is then

$$\int_{-1}^1 2(1 - x^2)\,dx = 2\left(x - \frac{1}{3}x^3\right)\Bigg|_{-1}^1 = \frac{8}{3}.$$

11. The base is the semicircle $y = \sqrt{9 - x^2}$, where $-3 \le x \le 3$. The cross sections perpendicular to the x-axis are squares.

SOLUTION For each x, the base of the square cross section extends from the semicircle $y = \sqrt{9 - x^2}$ to the x-axis. The square therefore has a base with length $\sqrt{9 - x^2}$ and an area of $\left(\sqrt{9 - x^2}\right)^2 = 9 - x^2$. The volume of the solid is then

$$\int_{-3}^3 \left(9 - x^2\right) dx = \left(9x - \frac{1}{3}x^3\right)\Bigg|_{-3}^3 = 36.$$

13. The base is the region enclosed by $y = x^2$ and $y = 3$. The cross sections perpendicular to the y-axis are squares.

SOLUTION At any location y, the distance to the parabola from the y-axis is $\sqrt{y}$. Thus the base of the square will have length $2\sqrt{y}$. Therefore the volume is

$$\int_0^3 (2\sqrt{y})\,(2\sqrt{y})\,dy = \int_0^3 4y\,dy = 2y^2\Big|_0^3 = 18.$$

15. Find the volume of the solid whose base is the region $|x| + |y| \le 1$ and whose vertical cross sections perpendicular to the y-axis are semicircles (with diameter along the base).

SOLUTION The region R in question is a diamond shape connecting the points $(1, 0)$, $(0, -1)$, $(-1, 0)$, and $(0, 1)$. Thus, in the lower half of the xy-plane, the radius of the circles is $y + 1$ and in the upper half, the radius is $1 - y$. Therefore, the volume is

$$\frac{\pi}{2}\int_{-1}^{0} (y+1)^2\,dy + \frac{\pi}{2}\int_0^1 (1-y)^2\,dy = \frac{\pi}{2}\left(\frac{1}{3} + \frac{1}{3}\right) = \frac{\pi}{3}.$$

17. The area of an ellipse is πab, where a and b are the lengths of the semimajor and semiminor axes (Figure 21). Compute the volume of a cone of height 12 whose base is an ellipse with semimajor axis $a = 6$ and semiminor axis $b = 4$.

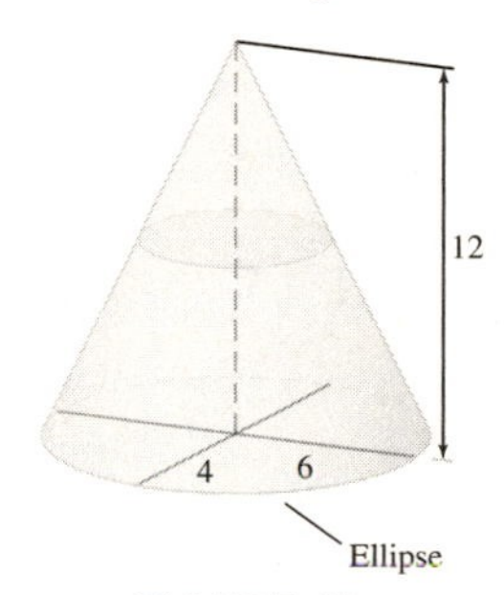

FIGURE 21

SOLUTION At each height y, the elliptical cross section has major axis $\frac{1}{2}(12 - y)$ and minor axis $\frac{1}{3}(12 - y)$. The cross-sectional area is then $\frac{\pi}{6}(12 - y)^2$, and the volume is

$$\int_0^{12} \frac{\pi}{6}(12 - y)^2\,dy = -\frac{\pi}{18}(12 - y)^3\Big|_0^{12} = 96\pi.$$

19. A frustum of a pyramid is a pyramid with its top cut off [Figure 23(A)]. Let V be the volume of a frustum of height h whose base is a square of side a and whose top is a square of side b with $a > b \ge 0$.

(a) Show that if the frustum were continued to a full pyramid, it would have height $ha/(a - b)$ [Figure 23(B)].

(b) Show that the cross section at height x is a square of side $(1/h)(a(h - x) + bx)$.

FIGURE 22

(c) Show that $V = \frac{1}{3}h(a^2 + ab + b^2)$. A papyrus dating to the year 1850 BCE indicates that Egyptian mathematicians had discovered this formula almost 4000 years ago.

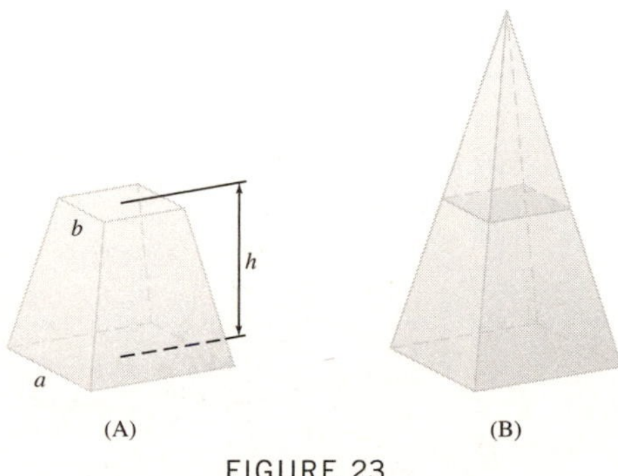

FIGURE 23

SOLUTION

(a) Let H be the height of the full pyramid. Using similar triangles, we have the proportion

$$\frac{H}{a} = \frac{H - h}{b}$$

which gives

$$H = \frac{ha}{a - b}.$$

(b) Let w denote the side length of the square cross section at height x. By similar triangles, we have

$$\frac{a}{H} = \frac{w}{H-x}.$$

Substituting the value for H from part (a) gives

$$w = \frac{a(h-x)+bx}{h}.$$

(c) The volume of the frustrum is

$$\int_0^h \left(\frac{1}{h}(a(h-x)+bx)\right)^2 dx = \frac{1}{h^2}\int_0^h \left(a^2(h-x)^2 + 2ab(h-x)x + b^2x^2\right) dx$$

$$= \frac{1}{h^2}\left(-\frac{a^2}{3}(h-x)^3 + abhx^2 - \frac{2}{3}abx^3 + \frac{1}{3}b^2x^3\right)\Bigg|_0^h = \frac{h}{3}\left(a^2+ab+b^2\right).$$

21. The solid S in Figure 25 is the intersection of two cylinders of radius r whose axes are perpendicular.

(a) The horizontal cross section of each cylinder at distance y from the central axis is a rectangular strip. Find the strip's width.

(b) Find the area of the horizontal cross section of S at distance y.

(c) Find the volume of S as a function of r.

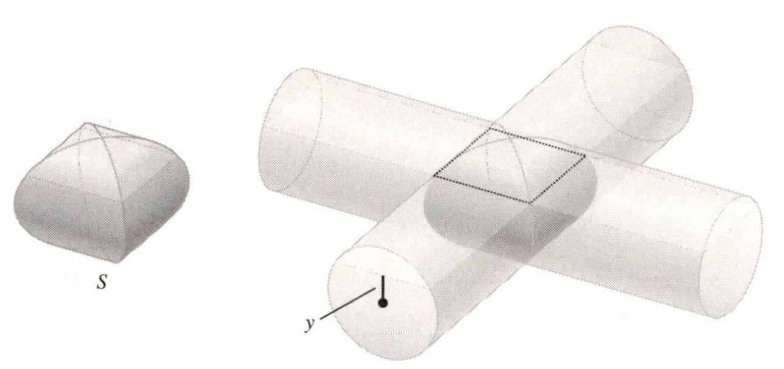

FIGURE 25 Two cylinders intersecting at right angles.

SOLUTION

(a) The horizontal cross section at distance y from the central axis (for $-r \le y \le r$) is a square of width $w = 2\sqrt{r^2-y^2}$.

(b) The area of the horizontal cross section of S at distance y from the central axis is $w^2 = 4(r^2-y^2)$.

(c) The volume of the solid S is then

$$4\int_{-r}^{r}\left(r^2-y^2\right) dy = 4\left(r^2y - \frac{1}{3}y^3\right)\Bigg|_{-r}^{r} = \frac{16}{3}r^3.$$

23. Calculate the volume of a cylinder inclined at an angle $\theta = 30°$ with height 10 and base of radius 4 (Figure 26).

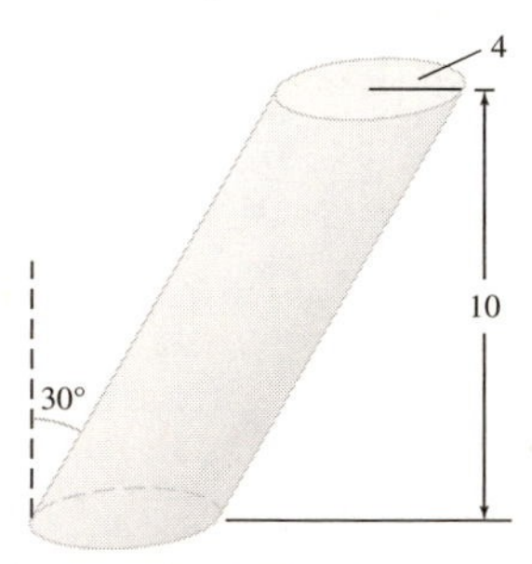

FIGURE 26 Cylinder inclined at an angle $\theta = 30°$.

SOLUTION The area of each circular cross section is $\pi(4)^2 = 16\pi$, hence the volume of the cylinder is

$$\int_0^{10} 16\pi\, dx = (16\pi x)\Big|_0^{10} = 160\pi$$

25. Find the total mass of a 1-m rod whose linear density function is $\rho(x) = 10(x+1)^{-2}$ kg/m for $0 \le x \le 1$.

SOLUTION The total mass of the rod is

$$\int_0^1 \rho(x)\, dx = \int_0^1 \left(10(x+1)^{-2}\right) dx = \left(-10(x+1)^{-1}\right)\Big|_0^1 = 5 \text{ kg}.$$

27. A mineral deposit along a strip of length 6 cm has density $s(x) = 0.01x(6 - x)$ g/cm for $0 \le x \le 6$. Calculate the total mass of the deposit.

SOLUTION The total mass of the deposit is

$$\int_0^6 s(x)\,dx = \int_0^6 0.01x(6-x)\,dx = \left(0.03x^2 - \frac{0.01}{3}x^3\right)\Big|_0^6 = 0.36 \text{ g}.$$

29. Calculate the population within a 10-mile radius of the city center if the radial population density is $\rho(r) = 4(1 + r^2)^{1/3}$ (in thousands per square mile).

SOLUTION The total population is

$$2\pi \int_0^{10} r \cdot \rho(r)\,dr = 2\pi \int_0^{10} 4r(1+r^2)^{1/3}\,dr = 3\pi(1+r^2)^{4/3}\Big|_0^{10}$$

$$\approx 4423.59 \text{ thousand} \approx 4.4 \text{ million}.$$

31. Table 1 lists the population density (in people per square kilometer) as a function of distance r (in kilometers) from the center of a rural town. Estimate the total population within a 1.2-km radius of the center by taking the average of the left- and right-endpoint approximations.

TABLE 1 **Population Density**

r	$\rho(r)$	r	$\rho(r)$
0.0	125.0	0.8	56.2
0.2	102.3	1.0	46.0
0.4	83.8	1.2	37.6
0.6	68.6		

SOLUTION The total population is given by

$$2\pi \int_0^{1.2} r \cdot \rho(r)\,dr.$$

With $\Delta r = 0.2$, the left- and right-endpoint approximations to the required definite integral are

$$L_6 = 0.2(2\pi)[0(125) + (0.2)(102.3) + (0.4)(83.8) + (0.6)(68.6) + (0.8)(56.2) + (1)(46)]$$
$$= 233.86;$$
$$R_{10} = 0.2(2\pi)[(0.2)(102.3) + (0.4)(83.8) + (0.6)(68.6) + (0.8)(56.2) + (1)(46) + (1.2)(37.6)]$$
$$= 290.56.$$

This gives an average of 262.21. Thus, there are roughly 262 people within a 1.2-km radius of the town center.

33. The density of deer in a forest is the radial function $\rho(r) = 150(r^2 + 2)^{-2}$ deer per square kilometer, where r is the distance (in kilometers) to a small meadow. Calculate the number of deer in the region $2 \le r \le 5$ km.

SOLUTION The number of deer in the region $2 \le r \le 5$ km is

$$2\pi \int_2^5 r\,(150)\left(r^2+2\right)^{-2}\,dr = -150\pi\left(\frac{1}{r^2+2}\right)\Big|_2^5 = -150\pi\left(\frac{1}{27} - \frac{1}{6}\right) \approx 61 \text{ deer}.$$

35. Find the flow rate through a tube of radius 4 cm, assuming that the velocity of fluid particles at a distance r cm from the center is $v(r) = (16 - r^2)$ cm/s.

SOLUTION The flow rate is

$$2\pi \int_0^R r v(r)\,dr = 2\pi \int_0^4 r\left(16 - r^2\right)\,dr = 2\pi\left(8r^2 - \frac{1}{4}r^4\right)\Big|_0^4 = 128\pi\,\frac{\text{cm}^3}{\text{s}}.$$

37. A solid rod of radius 1 cm is placed in a pipe of radius 3 cm so that their axes are aligned. Water flows through the pipe and around the rod. Find the flow rate if the velocity of the water is given by the radial function $v(r) = 0.5(r - 1)(3 - r)$ cm/s.

SOLUTION The flow rate is

$$2\pi \int_1^3 r(0.5)(r-1)(3-r)\,dr = \pi \int_1^3 \left(-r^3 + 4r^2 - 3r\right)\,dr = \pi\left(-\frac{1}{4}r^4 + \frac{4}{3}r^3 - \frac{3}{2}r^2\right)\Big|_1^3 = \frac{8\pi}{3}\,\frac{\text{cm}^3}{\text{s}}.$$

In Exercises 39–48, calculate the average over the given interval.

39. $f(x) = x^3$, $[0, 4]$

SOLUTION The average is

$$\frac{1}{4-0}\int_0^4 x^3\,dx = \frac{1}{4}\int_0^4 x^3\,dx = \frac{1}{16}x^4\bigg|_0^4 = 16.$$

41. $f(x) = \cos x$, $\left[0, \frac{\pi}{6}\right]$

SOLUTION The average is

$$\frac{1}{\pi/6 - 0}\int_0^{\pi/6} \cos x\,dx = \frac{6}{\pi}\int_0^{\pi/6} \cos x\,dx = \frac{6}{\pi}\sin x\bigg|_0^{\pi/6} = \frac{3}{\pi}.$$

43. $f(s) = s^{-2}$, $[2, 5]$

SOLUTION The average is

$$\frac{1}{5-2}\int_2^5 s^{-2}\,ds = -\frac{1}{3}s^{-1}\bigg|_2^5 = \frac{1}{10}.$$

45. $f(x) = 2x^3 - 6x^2$, $[-1, 3]$

SOLUTION The average is

$$\frac{1}{3-(-1)}\int_{-1}^3 (2x^3 - 6x^2)\,dx = \frac{1}{4}\int_{-1}^3 (2x^3 - 6x^2)\,dx = \frac{1}{4}\left(\frac{1}{2}x^4 - 2x^3\right)\bigg|_{-1}^3 = \frac{1}{4}\left(-\frac{27}{2} - \frac{5}{2}\right) = -4.$$

47. $f(x) = x^n$ for $n \geq 0$, $[0, 1]$

SOLUTION For $n > -1$, the average is

$$\frac{1}{1-0}\int_0^1 x^n\,dx = \int_0^1 x^n\,dx = \frac{1}{n+1}x^{n+1}\bigg|_0^1 = \frac{1}{n+1}.$$

49. The temperature (in °C) at time t (in hours) in an art museum varies according to $T(t) = 20 + 5\cos\left(\frac{\pi}{12}t\right)$. Find the average over the time periods $[0, 24]$ and $[2, 6]$.

SOLUTION

- The average temperature over the 24-hour period is

$$\frac{1}{24-0}\int_0^{24}\left(20 + 5\cos\left(\frac{\pi}{12}t\right)\right)dt = \frac{1}{24}\left(20t + \frac{60}{\pi}\sin\left(\frac{\pi}{12}t\right)\right)\bigg|_0^{24} = 20°\text{C}.$$

- The average temperature over the 4-hour period is

$$\frac{1}{6-2}\int_2^6\left(20 + 5\cos\left(\frac{\pi}{12}t\right)\right)dt = \frac{1}{4}\left(20t + \frac{60}{\pi}\sin\left(\frac{\pi}{12}t\right)\right)\bigg|_2^6 = 22.4°\text{C}.$$

51. Find the average speed over the time interval $[1, 5]$ of a particle whose position at time t is $s(t) = t^3 - 6t^2$ m/s.

SOLUTION The average speed over the time interval $[1, 5]$ is

$$\frac{1}{5-1}\int_1^5 |s'(t)|\,dt.$$

Because $s'(t) = 3t^2 - 12t = 3t(t-4)$, it follows that

$$\begin{aligned}\int_1^5 |s'(t)|\,dt &= \int_1^4 (12t - 3t^2)\,dt + \int_4^5 (3t^2 - 12t)\,dt \\ &= (6t^2 - t^3)\bigg|_1^4 + (t^3 - 6t^2)\bigg|_4^5 \\ &= (96 - 64) - (6 - 1) + (125 - 150) - (64 - 96) \\ &= 34.\end{aligned}$$

Thus, the average speed is

$$\frac{34}{4} = \frac{17}{2} \text{ m/s}.$$

53. The acceleration of a particle is $a(t) = 60t - 4t^3$ m/s^2. Compute the average acceleration and the average speed over the time interval $[2, 6]$, assuming that the particle's initial velocity is zero.

SOLUTION The average acceleration over the time interval $[2, 6]$ is

$$\begin{aligned}\frac{1}{6-2}\int_2^6 (60t - 4t^3)\,dt &= \frac{1}{4}(30t^2 - t^4)\Big|_2^6 \\ &= \frac{1}{4}[(1080 - 1296) - (120 - 16)] \\ &= -\frac{320}{4} = -80 \text{ m/s}^2.\end{aligned}$$

Given $a(t) = 60t - 4t^3$ and $v(0) = 0$, it follows that $v(t) = 30t^2 - t^4$. Now, average speed is given by

$$\frac{1}{6-2}\int_2^6 |v(t)|\,dt.$$

Based on the formula for $v(t)$,

$$\begin{aligned}\int_2^6 |v(t)|\,dt &= \int_2^{\sqrt{30}} (30t^2 - t^4)\,dt + \int_{\sqrt{30}}^6 (t^4 - 30t^2)\,dt \\ &= \left(10t^3 - \frac{1}{5}t^5\right)\Big|_2^{\sqrt{30}} + \left(\frac{1}{5}t^5 - 10t^3\right)\Big|_{\sqrt{30}}^6 \\ &= 120\sqrt{30} - \frac{368}{5} - \frac{3024}{5} + 120\sqrt{30} \\ &= 240\sqrt{30} - \frac{3392}{5}.\end{aligned}$$

Finally, the average speed is

$$\frac{1}{4}\left(240\sqrt{30} - \frac{3392}{5}\right) = 60\sqrt{30} - \frac{848}{5} \approx 159.03 \text{ m/s}.$$

55. Let M be the average value of $f(x) = x^4$ on $[0, 3]$. Find a value of c in $[0, 3]$ such that $f(c) = M$.

SOLUTION We have

$$M = \frac{1}{3-0}\int_0^3 x^4\,dx = \frac{1}{3}\int_0^3 x^4\,dx = \frac{1}{15}x^5\Big|_0^3 = \frac{81}{5}.$$

Then $M = f(c) = c^4 = \frac{81}{5}$ implies $c = \frac{3}{5^{1/4}} = 2.006221$.

57. Let M be the average value of $f(x) = x^3$ on $[0, A]$, where $A > 0$. Which theorem guarantees that $f(c) = M$ has a solution c in $[0, A]$? Find c.

SOLUTION The Mean Value Theorem for Integrals guarantees that $f(c) = M$ has a solution c in $[0, A]$. With $f(x) = x^3$ on $[0, A]$,

$$M = \frac{1}{A-0}\int_0^A x^3\,dx = \frac{1}{A}\,\frac{1}{4}x^4\Big|_0^A = \frac{A^3}{4}.$$

Solving $f(c) = c^3 = \frac{A^3}{4}$ for c yields

$$c = \frac{A}{\sqrt[3]{4}}.$$

59. Which of $f(x) = x\sin^2 x$ and $g(x) = x^2\sin^2 x$ has a larger average value over $[0, 1]$? Over $[1, 2]$?

SOLUTION The functions f and g differ only in the power of x multiplying $\sin^2 x$. It is also important to note that $\sin^2 x \ge 0$ for all x. Now, for each $x \in (0, 1)$, $x > x^2$ so

$$f(x) = x\sin^2 x > x^2\sin^2 x = g(x).$$

Thus, over $[0, 1]$, $f(x)$ will have a larger average value than $g(x)$. On the other hand, for each $x \in (1, 2)$, $x^2 > x$, so

$$g(x) = x^2 \sin^2 x > x \sin^2 x = f(x).$$

Thus, over $[1, 2]$, $g(x)$ will have the larger average value.

61. Sketch the graph of a function $f(x)$ such that $f(x) \geq 0$ on $[0, 1]$ and $f(x) \leq 0$ on $[1, 2]$, whose average on $[0, 2]$ is negative.

SOLUTION Many solutions will exist. One could be

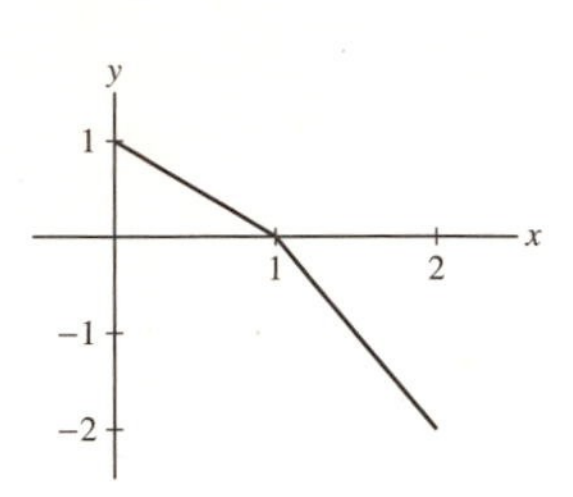

Further Insights and Challenges

63. An object is tossed into the air vertically from ground level with initial velocity v_0 ft/s at time $t = 0$. Find the average speed of the object over the time interval $[0, T]$, where T is the time the object returns to earth.

SOLUTION The height is given by $h(t) = v_0 t - 16t^2$. The ball is at ground level at time $t = 0$ and $T = v_0/16$. The velocity is given by $v(t) = v_0 - 32t$ and thus the speed is given by $s(t) = |v_0 - 32t|$. The average speed is

$$\frac{1}{v_0/16 - 0} \int_0^{v_0/16} |v_0 - 32t| \, dt = \frac{16}{v_0} \int_0^{v_0/32} (v_0 - 32t) \, dt + \frac{16}{v_0} \int_{v_0/32}^{v_0/16} (32t - v_0) \, dt$$
$$= \frac{16}{v_0} \left(v_0 t - 16t^2 \right) \Big|_0^{v_0/32} + \frac{16}{v_0} \left(16t^2 - v_0 t \right) \Big|_{v_0/32}^{v_0/16} = v_0/2.$$

6.3 Volumes of Revolution

Preliminary Questions

1. Which of the following is a solid of revolution?

(a) Sphere **(b)** Pyramid **(c)** Cylinder **(d)** Cube

SOLUTION The sphere and the cylinder have circular cross sections; hence, these are solids of revolution. The pyramid and cube do not have circular cross sections, so these are not solids of revolution.

2. True or false? When the region under a single graph is rotated about the x-axis, the cross sections of the solid perpendicular to the x-axis are circular disks.

SOLUTION True. The cross sections will be disks with radius equal to the value of the function.

3. True or false? When the region between two graphs is rotated about the x-axis, the cross sections to the solid perpendicular to the x-axis are circular disks.

SOLUTION False. The cross sections may be washers.

4. Which of the following integrals expresses the volume obtained by rotating the area between $y = f(x)$ and $y = g(x)$ over $[a, b]$ around the x-axis? [Assume $f(x) \geq g(x) \geq 0$.]

(a) $\pi \int_a^b (f(x) - g(x))^2 \, dx$

(b) $\pi \int_a^b \left(f(x)^2 - g(x)^2\right) dx$

SOLUTION The correct answer is **(b)**. Cross sections of the solid will be washers with outer radius $f(x)$ and inner radius $g(x)$. The area of the washer is then $\pi f(x)^2 - \pi g(x)^2 = \pi(f(x)^2 - g(x)^2)$.

Exercises

In Exercises 1–4, (a) sketch the solid obtained by revolving the region under the graph of $f(x)$ about the x-axis over the given interval, (b) describe the cross section perpendicular to the x-axis located at x, and (c) calculate the volume of the solid.

1. $f(x) = x + 1$, $[0, 3]$

SOLUTION

(a) A sketch of the solid of revolution is shown below:

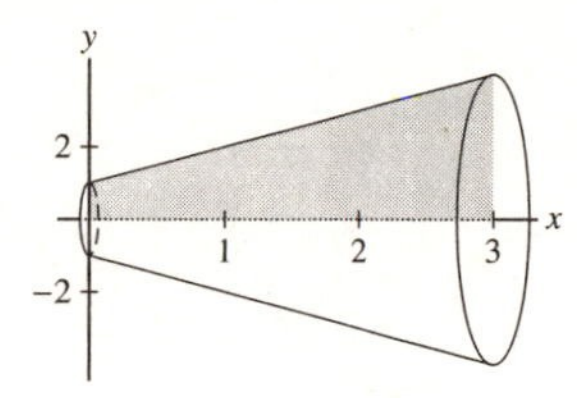

(b) Each cross section is a disk with radius $x + 1$.

(c) The volume of the solid of revolution is

$$\pi \int_0^3 (x+1)^2\,dx = \pi \int_0^3 (x^2 + 2x + 1)\,dx = \pi \left(\frac{1}{3}x^3 + x^2 + x\right)\Bigg|_0^3 = 21\pi.$$

3. $f(x) = \sqrt{x+1}$, $[1, 4]$

SOLUTION

(a) A sketch of the solid of revolution is shown below:

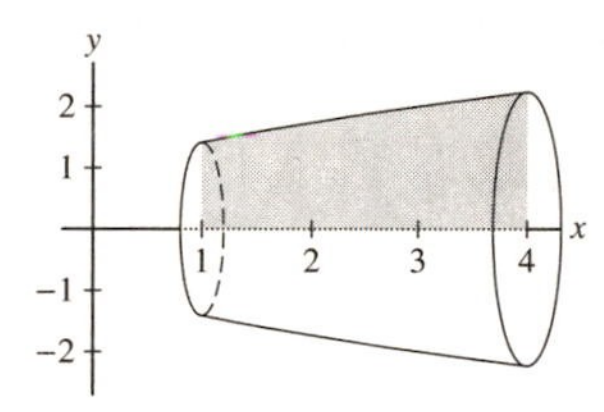

(b) Each cross section is a disk with radius $\sqrt{x+1}$.

(c) The volume of the solid of revolution is

$$\pi \int_1^4 (\sqrt{x+1})^2\,dx = \pi \int_1^4 (x+1)\,dx = \pi \left(\frac{1}{2}x^2 + x\right)\Bigg|_1^4 = \frac{21\pi}{2}.$$

In Exercises 5–12, find the volume of revolution about the x-axis for the given function and interval.

5. $f(x) = 3x - x^2$, $[0, 3]$

SOLUTION The volume of the solid of revolution is

$$\pi \int_0^3 (3x - x^2)^2\,dx = \pi \int_0^3 (9x^2 - 6x^3 + x^4)\,dx = \pi \left(3x^3 - \frac{3}{2}x^4 + \frac{1}{5}x^5\right)\Bigg|_0^3 = \frac{81\pi}{10}.$$

7. $f(x) = x^{5/3}$, $[1, 8]$

SOLUTION The volume of the solid of revolution is

$$\pi \int_1^8 (x^{5/3})^2\,dx = \pi \int_1^8 x^{10/3}\,dx = \frac{3\pi}{13}x^{13/3}\Bigg|_1^8 = \frac{3\pi}{13}(2^{13} - 1) = \frac{24573\pi}{13}.$$

9. $f(x) = \dfrac{2}{x+1}$, $[1, 3]$

SOLUTION The volume of the solid of revolution is

$$\pi \int_1^3 \left(\frac{2}{x+1}\right)^2 dx = 4\pi \int_1^3 (x+1)^{-2}\,dx = -4\pi\,(x+1)^{-1}\Bigg|_1^3 = \pi.$$

11. $f(x) = e^x$, $[0, 1]$

SOLUTION The volume of the solid of revolution is

$$\pi \int_0^1 (e^x)^2\,dx = \frac{1}{2}\pi e^{2x}\Bigg|_0^1 = \frac{1}{2}\pi(e^2 - 1).$$

In Exercises 13 and 14, R is the shaded region in Figure 11.

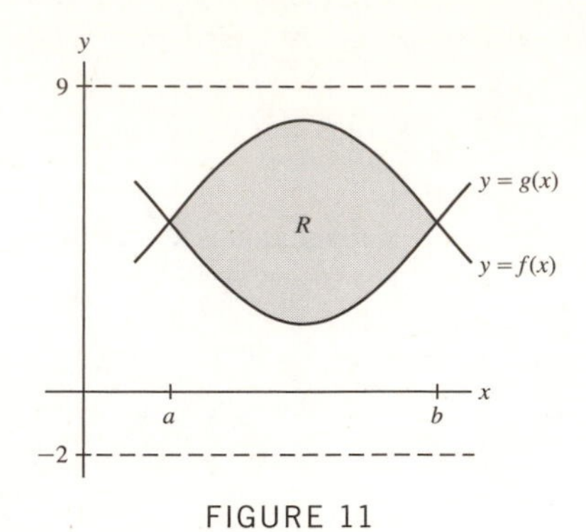

FIGURE 11

13. Which of the integrands (i)–(iv) is used to compute the volume obtained by rotating region R about $y = -2$?

(i) $(f(x)^2 + 2^2) - (g(x)^2 + 2^2)$
(ii) $(f(x) + 2)^2 - (g(x) + 2)^2$
(iii) $(f(x)^2 - 2^2) - (g(x)^2 - 2^2)$
(iv) $(f(x) - 2)^2 - (g(x) - 2)^2$

SOLUTION when the region R is rotated about $y = -2$, the outer radius is $f(x) - (-2) = f(x) + 2$ and the inner radius is $g(x) - (-2) = g(x) + 2$. Thus, the appropriate integrand is **(ii)**: $(f(x) + 2)^2 - (g(x) + 2)^2$.

In Exercises 15–20, (a) sketch the region enclosed by the curves, (b) describe the cross section perpendicular to the x-axis located at x, and (c) find the volume of the solid obtained by rotating the region about the x-axis.

15. $y = x^2 + 2, \quad y = 10 - x^2$

SOLUTION

(a) Setting $x^2 + 2 = 10 - x^2$ yields $2x^2 = 8$, or $x^2 = 4$. The two curves therefore intersect at $x = \pm 2$. The region enclosed by the two curves is shown in the figure below.

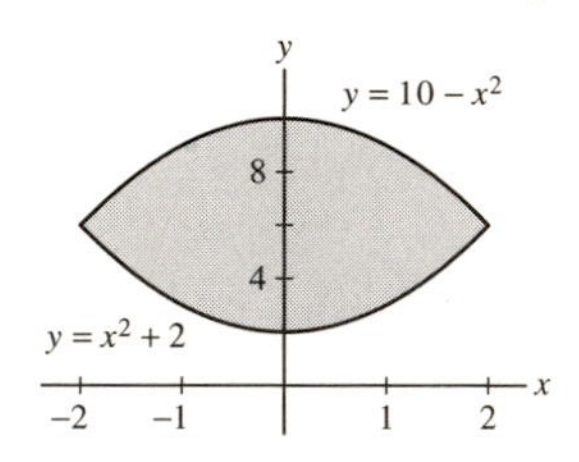

(b) When the region is rotated about the x-axis, each cross section is a washer with outer radius $R = 10 - x^2$ and inner radius $r = x^2 + 2$.

(c) The volume of the solid of revolution is

$$\pi \int_{-2}^{2} \left((10 - x^2)^2 - (x^2 + 2)^2\right) dx = \pi \int_{-2}^{2} (96 - 24x^2)\, dx = \pi \left(96x - 8x^3\right)\Big|_{-2}^{2} = 256\pi.$$

17. $y = 16 - x, \quad y = 3x + 12, \quad x = -1$

SOLUTION

(a) Setting $16 - x = 3x + 12$, we find that the two lines intersect at $x = 1$. The region enclosed by the two curves is shown in the figure below.

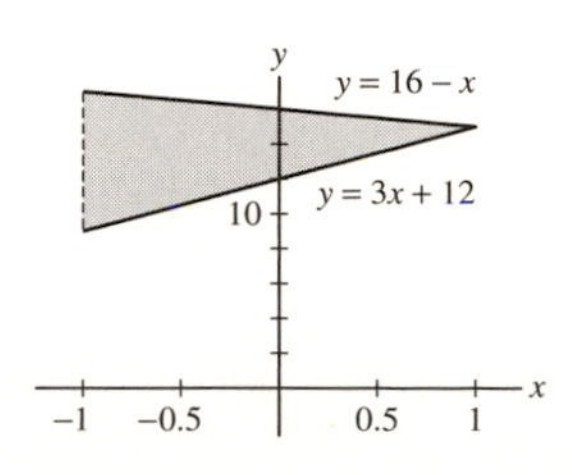

(b) When the region is rotated about the x-axis, each cross section is a washer with outer radius $R = 16 - x$ and inner radius $r = 3x + 12$.

(c) The volume of the solid of revolution is

$$\pi \int_{-1}^{1} \left((16-x)^2 - (3x+12)^2\right) dx = \pi \int_{-1}^{1} (112 - 104x - 8x^2)\, dx = \pi \left(112x - 52x^2 - \frac{8}{3}x^3\right)\Bigg|_{-1}^{1} = \frac{656\pi}{3}.$$

19. $y = \sec x, \quad y = 0, \quad x = -\frac{\pi}{4}, \quad x = \frac{\pi}{4}$

SOLUTION

(a) The region in question is shown in the figure below.

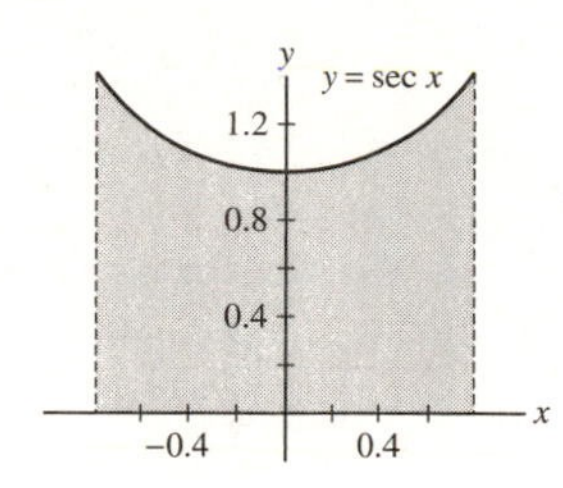

(b) When the region is rotated about the x-axis, each cross section is a circular disk with radius $R = \sec x$.

(c) The volume of the solid of revolution is

$$\pi \int_{-\pi/4}^{\pi/4} (\sec x)^2\, dx = \pi\, (\tan x)\Big|_{-\pi/4}^{\pi/4} = 2\pi.$$

In Exercises 21–24, find the volume of the solid obtained by rotating the region enclosed by the graphs about the y-axis over the given interval.

21. $x = \sqrt{y}, \quad x = 0; \quad 1 \le y \le 4$

SOLUTION When the region in question (shown in the figure below) is rotated about the y-axis, each cross section is a disk with radius $\sqrt{y}$. The volume of the solid of revolution is

$$\pi \int_{1}^{4} \left(\sqrt{y}\right)^2 dy = \frac{\pi y^2}{2}\Bigg|_{1}^{4} = \frac{15\pi}{2}.$$

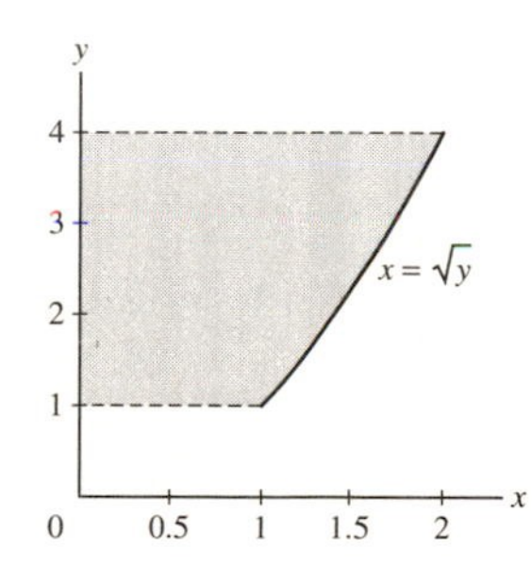

23. $x = y^2, \quad x = \sqrt{y}$

SOLUTION Setting $y^2 = \sqrt{y}$ and then squaring both sides yields

$$y^4 = y \quad \text{or} \quad y^4 - y = y(y^3 - 1) = 0,$$

so the two curves intersect at $y = 0$ and $y = 1$. When the region in question (shown in the figure below) is rotated about the y-axis, each cross section is a washer with outer radius $R = \sqrt{y}$ and inner radius $r = y^2$. The volume of the solid of revolution is

$$\pi \int_{0}^{1} \left((\sqrt{y})^2 - (y^2)^2\right) dy = \pi \left(\frac{y^2}{2} - \frac{y^5}{5}\right)\Bigg|_{0}^{1} = \frac{3\pi}{10}.$$

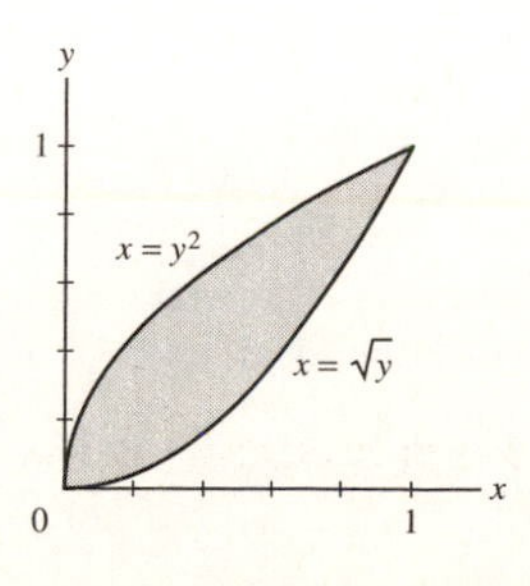

25. Rotation of the region in Figure 12 about the y-axis produces a solid with two types of different cross sections. Compute the volume as a sum of two integrals, one for $-12 \le y \le 4$ and one for $4 \le y \le 12$.

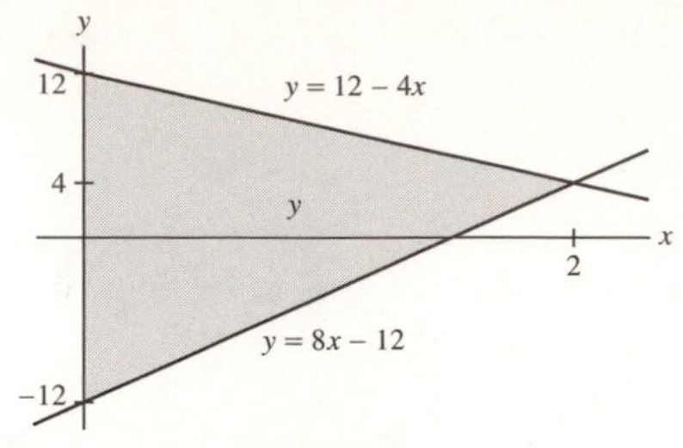

FIGURE 12

SOLUTION For $-12 \le y \le 4$, the cross section is a disk with radius $\frac{1}{8}(y+12)$; for $4 \le y \le 12$, the cross section is a disk with radius $\frac{1}{4}(12-y)$. Therefore, the volume of the solid of revolution is

$$\begin{aligned} V &= \frac{\pi}{8}\int_{-12}^{4}(y+12)^2\,dy + \frac{\pi}{4}\int_{4}^{12}(12-y)^2\,dy \\ &= \frac{\pi}{24}(y+12)^3\Big|_{-12}^{4} - \frac{\pi}{12}(12-y)^3\Big|_{4}^{12} \\ &= \frac{512\pi}{3} + \frac{128\pi}{3} = \frac{640\pi}{3}. \end{aligned}$$

In Exercises 27–32, find the volume of the solid obtained by rotating region A in Figure 13 about the given axis.

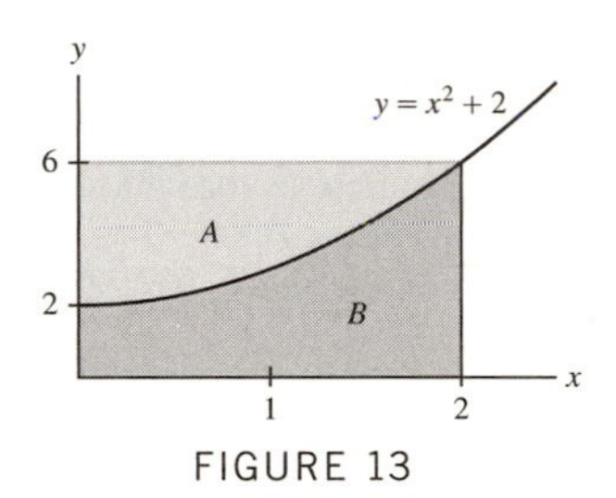

FIGURE 13

27. x-axis

SOLUTION Rotating region A about the x-axis produces a solid whose cross sections are washers with outer radius $R = 6$ and inner radius $r = x^2 + 2$. The volume of the solid of revolution is

$$\pi\int_0^2\left((6)^2 - (x^2+2)^2\right)dx = \pi\int_0^2(32 - 4x^2 - x^4)\,dx = \pi\left(32x - \frac{4}{3}x^3 - \frac{1}{5}x^5\right)\Big|_0^2 = \frac{704\pi}{15}.$$

29. $y = 2$

SOLUTION Rotating the region A about $y = 2$ produces a solid whose cross sections are washers with outer radius $R = 6 - 2 = 4$ and inner radius $r = x^2 + 2 - 2 = x^2$. The volume of the solid of revolution is

$$\pi\int_0^2\left(4^2 - (x^2)^2\right)dx = \pi\left(16x - \frac{1}{5}x^5\right)\Big|_0^2 = \frac{128\pi}{5}.$$

31. $x = -3$

SOLUTION Rotating region A about $x = -3$ produces a solid whose cross sections are washers with outer radius $R = \sqrt{y-2} - (-3) = \sqrt{y-2} + 3$ and inner radius $r = 0 - (-3) = 3$. The volume of the solid of revolution is

$$\pi\int_2^6\left((3+\sqrt{y-2})^2 - (3)^2\right)dy = \pi\int_2^6(6\sqrt{y-2} + y - 2)\,dy = \pi\left(4(y-2)^{3/2} + \frac{1}{2}y^2 - 2y\right)\Big|_2^6 = 40\pi.$$

In Exercises 33–38, find the volume of the solid obtained by rotating region B in Figure 13 about the given axis.

33. x-axis

SOLUTION Rotating region B about the x-axis produces a solid whose cross sections are disks with radius $R = x^2 + 2$. The volume of the solid of revolution is

$$\pi\int_0^2(x^2+2)^2\,dx = \pi\int_0^2(x^4 + 4x^2 + 4)\,dx = \pi\left(\frac{1}{5}x^5 + \frac{4}{3}x^3 + 4x\right)\Big|_0^2 = \frac{376\pi}{15}.$$

35. $y = 6$

SOLUTION Rotating region B about $y = 6$ produces a solid whose cross sections are washers with outer radius $R = 6 - 0 = 6$ and inner radius $r = 6 - (x^2 + 2) = 4 - x^2$. The volume of the solid of revolution is

$$\pi \int_0^2 \left(6^2 - (4 - x^2)^2\right) dy = \pi \int_0^2 \left(20 + 8x^2 - x^4\right) dy = \pi \left(20x + \frac{8}{3}x^3 - \frac{1}{5}x^5\right)\Big|_0^2 = \frac{824\pi}{15}.$$

37. $x = 2$

SOLUTION Rotating region B about $x = 2$ produces a solid with two different cross sections. For each $y \in [0, 2]$, the cross section is a disk with radius $R = 2$; for each $y \in [2, 6]$, the cross section is a disk with radius $R = 2 - \sqrt{y - 2}$. The volume of the solid of revolution is

$$\pi \int_0^2 (2)^2\, dy + \pi \int_2^6 (2 - \sqrt{y-2})^2\, dy = \pi \int_0^2 4\, dy + \pi \int_2^6 (2 + y - 4\sqrt{y-2})\, dy$$

$$= \pi\,(4y)\Big|_0^2 + \pi \left(2y + \frac{1}{2}y^2 - \frac{8}{3}(y-2)^{3/2}\right)\Big|_2^6 = \frac{32\pi}{3}.$$

In Exercises 39–52, find the volume of the solid obtained by rotating the region enclosed by the graphs about the given axis.

39. $y = x^2, \quad y = 12 - x, \quad x = 0, \quad$ about $y = -2$

SOLUTION Rotating the region enclosed by $y = x^2$, $y = 12 - x$ and the y-axis (shown in the figure below) about $y = -2$ produces a solid whose cross sections are washers with outer radius $R = 12 - x - (-2) = 14 - x$ and inner radius $r = x^2 - (-2) = x^2 + 2$. The volume of the solid of revolution is

$$\pi \int_0^3 \left((14 - x)^2 - (x^2 + 2)^2\right) dx = \pi \int_0^3 (192 - 28x - 3x^2 - x^4)\, dx$$

$$= \pi \left(192x - 14x^2 - x^3 - \frac{1}{5}x^5\right)\Big|_0^3 = \frac{1872\pi}{5}.$$

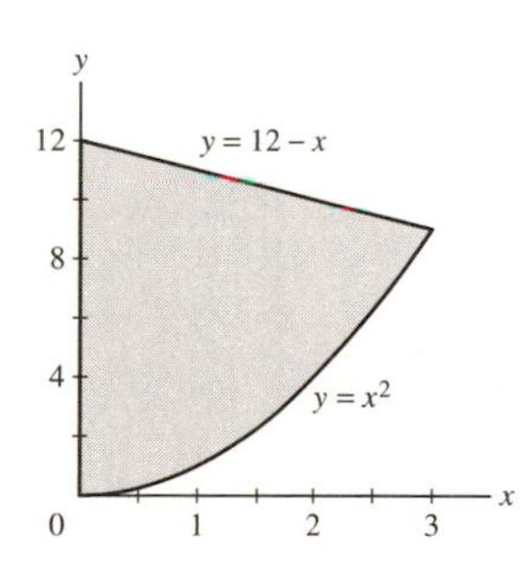

41. $y = 16 - 2x, \quad y = 6, \quad x = 0, \quad$ about x-axis

SOLUTION Rotating the region enclosed by $y = 16 - 2x$, $y = 6$ and the y-axis (shown in the figure below) about the x-axis produces a solid whose cross sections are washers with outer radius $R = 16 - 2x$ and inner radius $r = 6$. The volume of the solid of revolution is

$$\pi \int_0^5 \left((16 - 2x)^2 - 6^2\right) dx = \pi \int_0^5 (220 - 64x + 4x^2)\, dx$$

$$= \pi \left(220x - 32x^2 + \frac{4}{3}x^3\right)\Big|_0^5 = \frac{1400\pi}{3}.$$

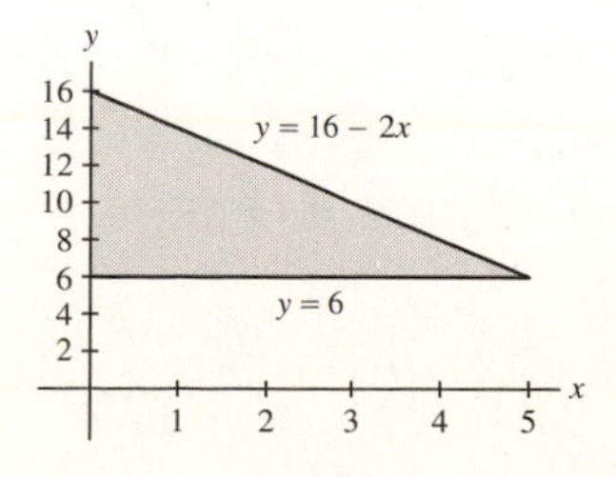

43. $y = \sec x, \quad y = 1 + \dfrac{3}{\pi}x, \quad$ about x-axis

SOLUTION We first note that $y = \sec x$ and $y = 1 + (3/\pi)x$ intersect at $x = 0$ and $x = \pi/3$. Rotating the region enclosed by $y = \sec x$ and $y = 1 + (3/\pi)x$ (shown in the figure below) about the x-axis produces a cross section that is a washer with outer radius $R = 1 + (3/\pi)x$ and inner radius $r = \sec x$. The volume of the solid of revolution is

$$V = \pi \int_0^{\pi/3} \left(\left(1 + \frac{3}{\pi}x\right)^2 - \sec^2 x \right) dx$$

$$= \pi \int_0^{\pi/3} \left(1 + \frac{6}{\pi}x + \frac{9}{\pi^2}x^2 - \sec^2 x \right) dx$$

$$= \pi \left(x + \frac{3}{\pi}x^2 + \frac{3}{\pi^2}x^3 - \tan x \right)\Bigg|_0^{\pi/3}$$

$$= \pi \left(\frac{\pi}{3} + \frac{\pi}{3} + \frac{\pi}{9} - \sqrt{3} \right) = \frac{7\pi^2}{9} - \sqrt{3}\pi.$$

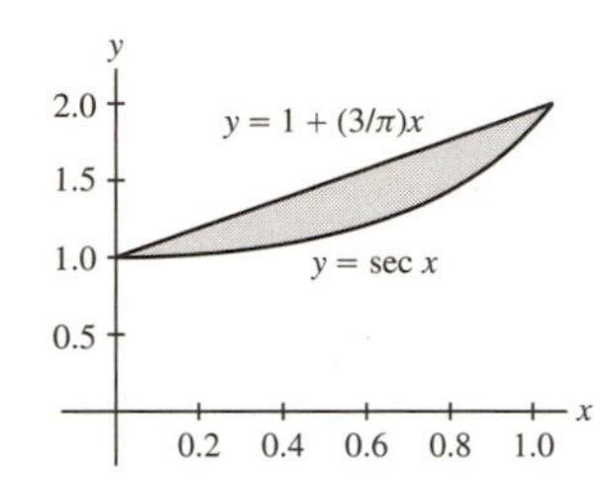

45. $y = 2\sqrt{x}, \quad y = x, \quad$ about $x = -2$

SOLUTION Setting $2\sqrt{x} = x$ and squaring both sides yields

$$4x = x^2 \quad \text{or} \quad x(x-4) = 0,$$

so the two curves intersect at $x = 0$ and $x = 4$. Rotating the region enclosed by $y = 2\sqrt{x}$ and $y = x$ (see the figure below) about $x = -2$ produces a solid whose cross sections are washers with outer radius $R = y - (-2) = y + 2$ and inner radius $r = \frac{1}{4}y^2 - (-2) = \frac{1}{4}y^2 + 2$. The volume of the solid of revolution is

$$V = \pi \int_0^4 \left((y+2)^2 - \left(\frac{1}{4}y^2 + 2\right)^2 \right) dy$$

$$= \pi \int_0^4 \left(4y - \frac{1}{16}y^4 \right) dy$$

$$= \pi \left(2y^2 - \frac{1}{80}y^5 \right)\Bigg|_0^4$$

$$= \pi \left(32 - \frac{64}{5} \right) = \frac{96\pi}{5}.$$

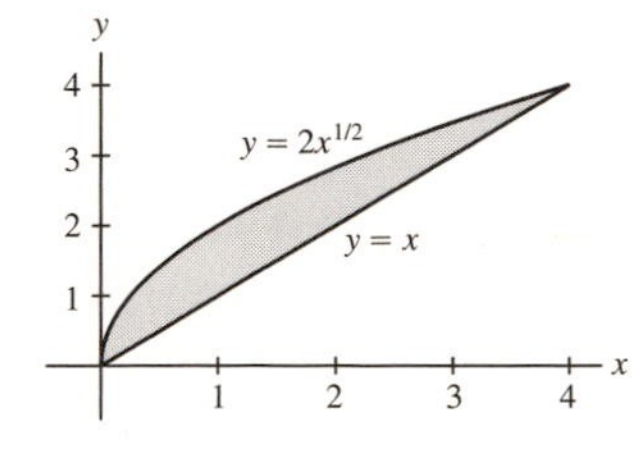

47. $y = x^3, \quad y = x^{1/3}, \quad$ for $x \ge 0, \quad$ about y-axis

SOLUTION Rotating the region enclosed by $y = x^3$ and $y = x^{1/3}$ (shown in the figure below) about the y-axis produces a solid whose cross sections are washers with outer radius $R = y^{1/3}$ and inner radius $r = y^3$. The volume of the solid of revolution is

$$\pi \int_0^1 \left((y^{1/3})^2 - (y^3)^2 \right) dy = \pi \int_0^1 (y^{2/3} - y^6)\, dy = \pi \left(\frac{3}{5}y^{5/3} - \frac{1}{7}y^7 \right)\Bigg|_0^1 = \frac{16\pi}{35}.$$

49. $y = \dfrac{9}{x^2}, \quad y = 10 - x^2, \quad x \ge 0, \quad$ about $y = 12$

SOLUTION The region enclosed by the two curves is shown in the figure below. Rotating this region about $y = 12$ produces a solid whose cross sections are washers with outer radius $R = 12 - 9x^{-2}$ and inner radius $r = 12 - (10 - x^2) = 2 + x^2$. The volume of the solid of revolution is

$$\pi \int_1^3 \left((12 - 9x^{-2})^2 - (x^2 + 2)^2 \right) dx = \pi \int_1^3 \left(140 - 4x^2 - x^4 - 216x^{-2} + 81x^{-4} \right) dx$$

$$= \pi \left(140x - \frac{4}{3}x^3 - \frac{1}{5}x^5 + 216x^{-1} - 27x^{-3} \right) \Bigg|_1^3 = \frac{1184\pi}{15}.$$

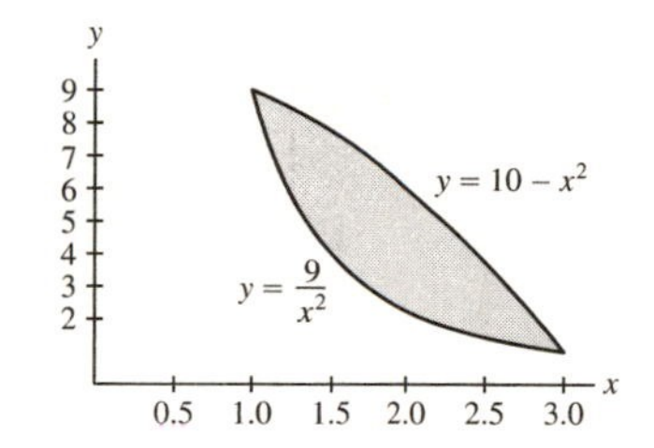

51. $y = e^{-x}, \quad y = 1 - e^{-x}, \quad x = 0, \quad$ about $y = 4$

SOLUTION Rotating the region enclosed by $y = 1 - e^{-x}$, $y = e^{-x}$ and the y-axis (shown in the figure below) about the line $y = 4$ produces a solid whose cross sections are washers with outer radius $R = 4 - (1 - e^{-x}) = 3 + e^{-x}$ and inner radius $r = 4 - e^{-x}$. The volumc of the solid of revolution is

$$\pi \int_0^{\ln 2} \left((3 + e^{-x})^2 - (4 - e^{-x})^2 \right) dx = \pi \int_0^{\ln 2} (14e^{-x} - 7)\, dx = \pi(-14e^{-x} - 7x) \Big|_0^{\ln 2}$$

$$= \pi(-7 - 7\ln 2 + 14) = 7\pi(1 - \ln 2).$$

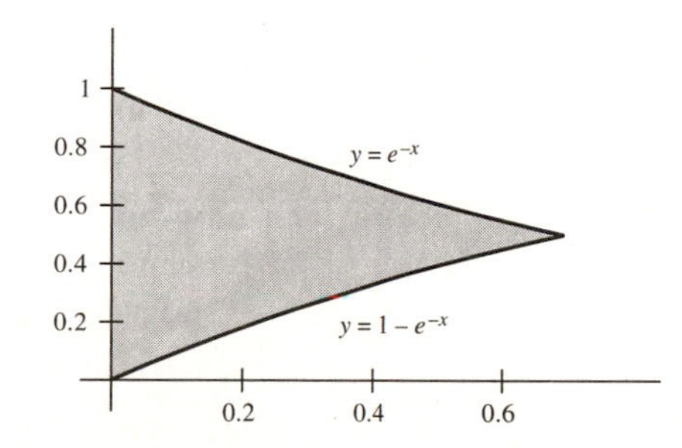

53. The bowl in Figure 14(A) is 21 cm high, obtained by rotating the curve in Figure 14(B) as indicated. Estimate the volume capacity of the bowl shown by taking the average of right- and left-endpoint approximations to the integral with $N = 7$. The inner radii (in cm) starting from the top are 0, 4, 7, 8, 10, 13, 14, 20.

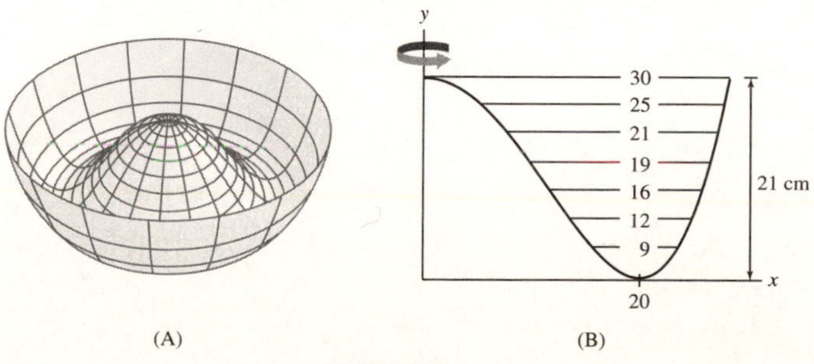

FIGURE 14

SOLUTION Using the given values for the inner radii and the values in Figure 14(B), which indicate the difference between the inner and outer radii, we find

$$R_7 = 3\pi\left((23^2 - 14^2) + (25^2 - 13^2) + (26^2 - 10^2) + (27^2 - 8^2) + (28^2 - 7^2) + (29^2 - 4^2) + (30^2 - 0^2)\right)$$
$$= 3\pi(4490) = 13470\pi$$

and

$$L_7 = 3\pi\left((20^2 - 20^2) + (23^2 - 14^2) + (25^2 - 13^2) + (26^2 - 10^2) + (27^2 - 8^2) + (28^2 - 7^2) + (29^2 - 4^2)\right)$$
$$= 3\pi(3590) = 10770\pi$$

Averaging these two values, we estimate that the volume capacity of the bowl is

$$V = 12120\pi \approx 38076.1 \text{ cm}^3.$$

55. Find the volume of the cone obtained by rotating the region under the segment joining $(0, h)$ and $(r, 0)$ about the y-axis.

SOLUTION The segment joining $(0, h)$ and $(r, 0)$ has the equation

$$y = -\frac{h}{r}x + h \quad \text{or} \quad x = \frac{r}{h}(h - y).$$

Rotating the region under this segment about the y-axis produces a cone with volume

$$\frac{\pi r^2}{h^2}\int_0^h (h - y)^2\, dx = -\frac{\pi r^2}{3h^2}(h - y)^3\Big|_0^h$$
$$= \frac{1}{3}\pi r^2 h.$$

57. GU Sketch the hypocycloid $x^{2/3} + y^{2/3} = 1$ and find the volume of the solid obtained by revolving it about the x-axis.

SOLUTION A sketch of the hypocycloid is shown below.

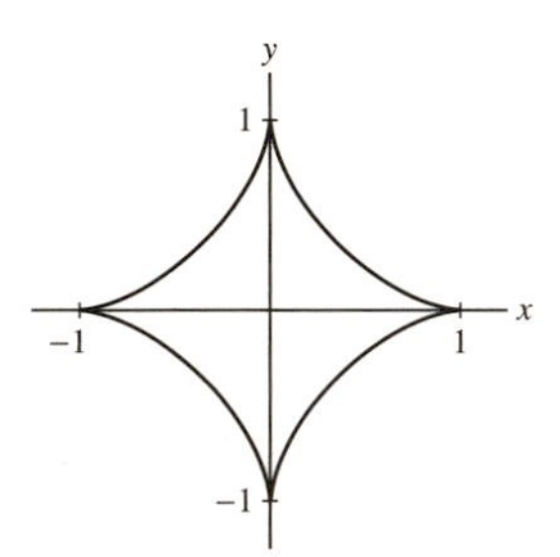

For the hypocycloid, $y = \pm\left(1 - x^{2/3}\right)^{3/2}$. Rotating this region about the x-axis will produce a solid whose cross sections are disks with radius $R = \left(1 - x^{2/3}\right)^{3/2}$. Thus the volume of the solid of revolution will be

$$\pi\int_{-1}^{1}\left((1 - x^{2/3})^{3/2}\right)^2 dx = \pi\left(\frac{-x^3}{3} + \frac{9}{7}x^{7/3} - \frac{9}{5}x^{5/3} + x\right)\Bigg|_{-1}^{1} = \frac{32\pi}{105}.$$

59. A "bead" is formed by removing a cylinder of radius r from the center of a sphere of radius R (Figure 17). Find the volume of the bead with $r = 1$ and $R = 2$.

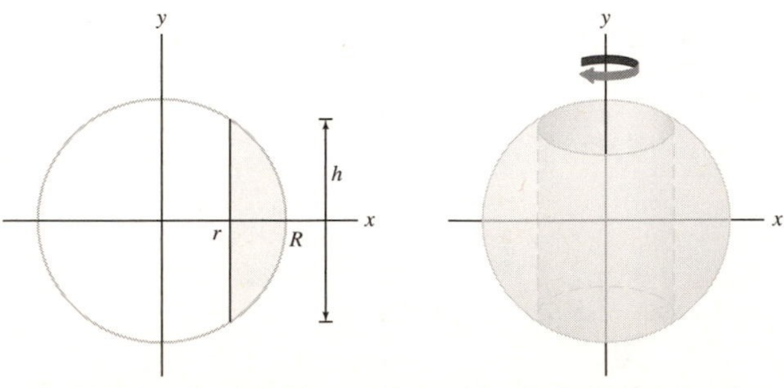

FIGURE 17 A bead is a sphere with a cylinder removed.

SOLUTION The equation of the outer circle is $x^2 + y^2 = 2^2$, and the inner cylinder intersects the sphere when $y = \pm\sqrt{3}$. Each cross section of the bead is a washer with outer radius $\sqrt{4 - y^2}$ and inner radius 1, so the volume is given by

$$\pi \int_{-\sqrt{3}}^{\sqrt{3}} \left(\left(\sqrt{4 - y^2} \right)^2 - 1^2 \right) dy = \pi \int_{-\sqrt{3}}^{\sqrt{3}} \left(3 - y^2 \right) dy = 4\pi\sqrt{3}.$$

Further Insights and Challenges

61. The solid generated by rotating the region inside the ellipse with equation $\left(\frac{x}{a}\right)^2 + \left(\frac{y}{b}\right)^2 = 1$ around the x-axis is called an **ellipsoid**. Show that the ellipsoid has volume $\frac{4}{3}\pi ab^2$. What is the volume if the ellipse is rotated around the y-axis?

SOLUTION

- Rotating the ellipse about the x-axis produces an ellipsoid whose cross sections are disks with radius $R = b\sqrt{1 - (x/a)^2}$. The volume of the ellipsoid is then

$$\pi \int_{-a}^{a} \left(b\sqrt{1 - (x/a)^2} \right)^2 dx = b^2\pi \int_{-a}^{a} \left(1 - \frac{1}{a^2}x^2 \right) dx = b^2\pi \left(x - \frac{1}{3a^2}x^3 \right) \Big|_{-a}^{a} = \frac{4}{3}\pi ab^2.$$

- Rotating the ellipse about the y-axis produces an ellipsoid whose cross sections are disks with radius $R = a\sqrt{1 - (y/b)^2}$. The volume of the ellipsoid is then

$$\int_{-b}^{b} \left(a\sqrt{1 - (y/b)^2} \right)^2 dy = a^2\pi \int_{-b}^{b} \left(1 - \frac{1}{b^2}y^2 \right) dy = a^2\pi \left(y - \frac{1}{3b^2}y^3 \right) \Big|_{-b}^{b} = \frac{4}{3}\pi a^2 b.$$

63. Verify the formula

$$\int_{x_1}^{x_2} (x - x_1)(x - x_2)\, dx = \frac{1}{6}(x_1 - x_2)^3 \qquad \boxed{3}$$

Then prove that the solid obtained by rotating the shaded region in Figure 19 about the x-axis has volume $V = \frac{\pi}{6}BH^2$, with B and H as in the figure. *Hint:* Let x_1 and x_2 be the roots of $f(x) = ax + b - (mx + c)^2$, where $x_1 < x_2$. Show that

$$V = \pi \int_{x_1}^{x_2} f(x)\, dx$$

and use Eq. (3).

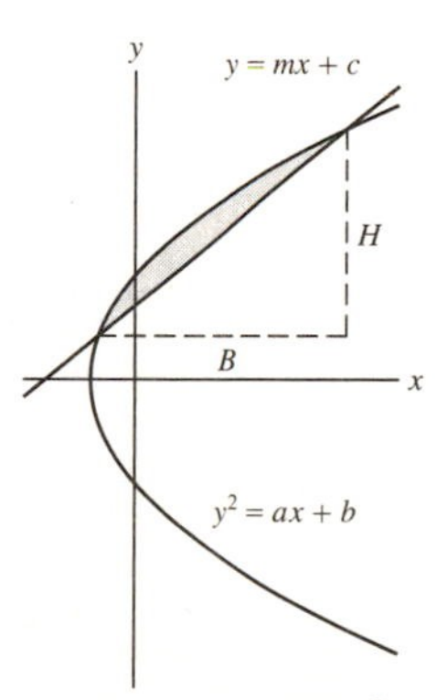

FIGURE 19 The line $y = mx + c$ intersects the parabola $y^2 = ax + b$ at two points above the x-axis.

SOLUTION First, we calculate

$$\int_{x_1}^{x_2} (x - x_1)(x - x_2)\, dx = \left(\frac{1}{3}x^3 - \frac{1}{2}(x_1 + x_2)x^2 + x_1x_2x \right) \Big|_{x_1}^{x_2} = \frac{1}{6}x_1^3 - \frac{1}{2}x_1^2x_2 + \frac{1}{2}x_1x_2^2 - \frac{1}{6}x_2^3$$

$$= \frac{1}{6}\left(x_1^3 - 3x_1^2x_2 + 3x_1x_2^2 - x_2^3 \right) = \frac{1}{6}(x_1 - x_2)^3.$$

Now, consider the region enclosed by the parabola $y^2 = ax + b$ and the line $y = mx + c$, and let x_1 and x_2 denote the x-coordinates of the points of intersection between the two curves with $x_1 < x_2$. Rotating the region about the y-axis produces a solid whose cross sections are washers with outer radius $R = \sqrt{ax + b}$ and inner radius $r = mx + c$. The volume of the solid of revolution is then

$$V = \pi \int_{x_1}^{x_2} \left(ax + b - (mx + c)^2 \right) dx$$

Because x_1 and x_2 are roots of the equation $ax + b - (mx + c)^2 = 0$ and $ax + b - (mx + c)^2$ is a quadratic polynomial in x with leading coefficient $-m^2$, it follows that $ax + b - (mx + c)^2 = -m^2(x - x_1)(x - x_2)$. Therefore,

$$V = -\pi m^2 \int_{x_1}^{x_2} (x - x_1)(x - x_2)\,dx = \frac{\pi}{6} m^2 (x_2 - x_1)^3,$$

where we have used Eq. (3). From the diagram, we see that

$$B = x_2 - x_1 \qquad \text{and} \qquad H = mB,$$

so

$$V = \frac{\pi}{6} m^2 B^3 = \frac{\pi}{6} B\,(mB)^2 = \frac{\pi}{6} BH^2.$$

6.4 The Method of Cylindrical Shells

Preliminary Questions

1. Consider the region $\mathcal{R}$ under the graph of the constant function $f(x) = h$ over the interval $[0, r]$. Give the height and the radius of the cylinder generated when $\mathcal{R}$ is rotated about:

(a) the x-axis **(b)** the y-axis

SOLUTION

(a) When the region is rotated about the x-axis, each shell will have radius h and height r.

(b) When the region is rotated about the y-axis, each shell will have radius r and height h.

2. Let V be the volume of a solid of revolution about the y-axis.

(a) Does the Shell Method for computing V lead to an integral with respect to x or y?

(b) Does the Disk or Washer Method for computing V lead to an integral with respect to x or y?

SOLUTION

(a) The Shell method requires slicing the solid parallel to the axis of rotation. In this case, that will mean slicing the solid in the vertical direction, so integration will be with respect to x.

(b) The Disk or Washer method requires slicing the solid perpendicular to the axis of rotation. In this case, that means slicing the solid in the horizontal direction, so integration will be with respect to y.

Exercises

In Exercises 1–6, sketch the solid obtained by rotating the region underneath the graph of the function over the given interval about the y-axis, and find its volume.

1. $f(x) = x^3$, $[0, 1]$

SOLUTION A sketch of the solid is shown below. Each shell has radius x and height x^3, so the volume of the solid is

$$2\pi \int_0^1 x \cdot x^3\,dx = 2\pi \int_0^1 x^4\,dx = 2\pi \left(\frac{1}{5}x^5\right)\Big|_0^1 = \frac{2}{5}\pi.$$

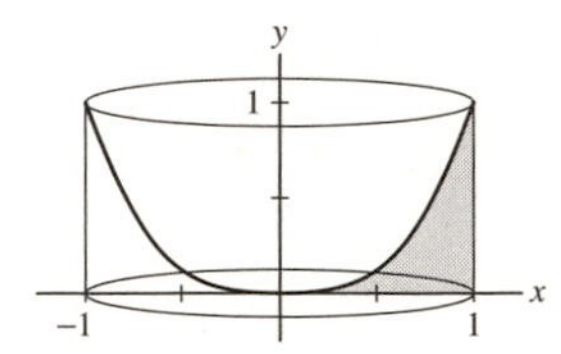

3. $f(x) = x^{-1}$, $[1, 3]$

SOLUTION A sketch of the solid is shown below. Each shell has radius x and height x^{-1}, so the volume of the solid is

$$2\pi \int_1^3 x(x^{-1})\,dx = 2\pi \int_1^3 1\,dx = 2\pi\,(x)\Big|_1^3 = 4\pi.$$

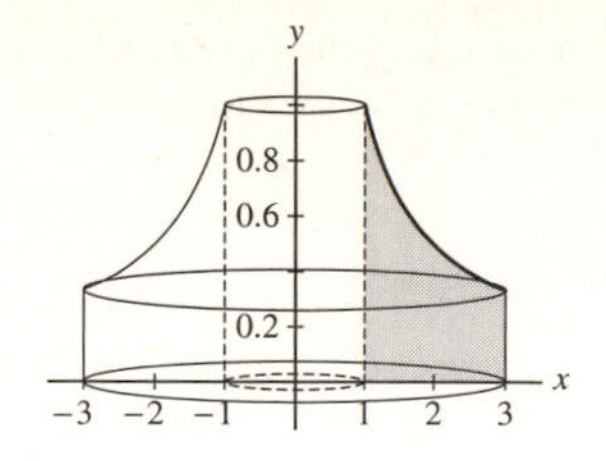

5. $f(x) = \sqrt{x^2 + 9}, \quad [0, 3]$

SOLUTION A sketch of the solid is shown below. Each shell has radius x and height $\sqrt{x^2 + 9}$, so the volume of the solid is

$$2\pi \int_0^3 x\sqrt{x^2 + 9}\, dx.$$

Let $u = x^2 + 9$. Then $du = 2x\, dx$ and

$$2\pi \int_0^3 x\sqrt{x^2 + 9}\, dx = \pi \int_9^{18} \sqrt{u}\, du = \pi \left(\frac{2}{3} u^{3/2} \right) \Big|_9^{18} = 18\pi(2\sqrt{2} - 1).$$

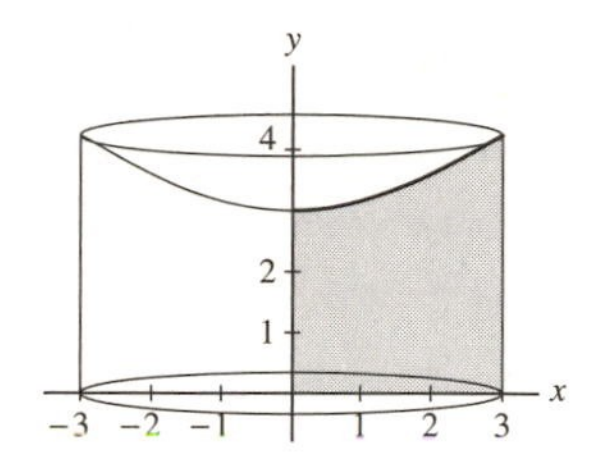

In Exercises 7–12, use the Shell Method to compute the volume obtained by rotating the region enclosed by the graphs as indicated, about the y-axis.

7. $y = 3x - 2, \quad y = 6 - x, \quad x = 0$

SOLUTION The region enclosed by $y = 3x - 2$, $y = 6 - x$ and $x = 0$ is shown below. When rotating this region about the y-axis, each shell has radius x and height $6 - x - (3x - 2) = 8 - 4x$. The volume of the resulting solid is

$$2\pi \int_0^2 x(8 - 4x)\, dx = 2\pi \int_0^2 (8x - 4x^2)\, dx = 2\pi \left(4x^2 - \frac{4}{3}x^3 \right) \Big|_0^2 = \frac{32}{3}\pi.$$

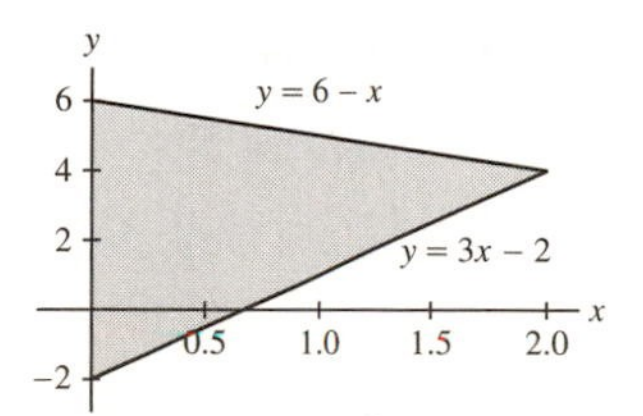

9. $y = x^2, \quad y = 8 - x^2, \quad x = 0, \quad \text{for } x \geq 0$

SOLUTION The region enclosed by $y = x^2$, $y = 8 - x^2$ and the y-axis is shown below. When rotating this region about the y-axis, each shell has radius x and height $8 - x^2 - x^2 = 8 - 2x^2$. The volume of the resulting solid is

$$2\pi \int_0^2 x(8 - 2x^2)\, dx = 2\pi \int_0^2 (8x - 2x^3)\, dx = 2\pi \left(4x^2 - \frac{1}{2}x^4 \right) \Big|_0^2 = 16\pi.$$

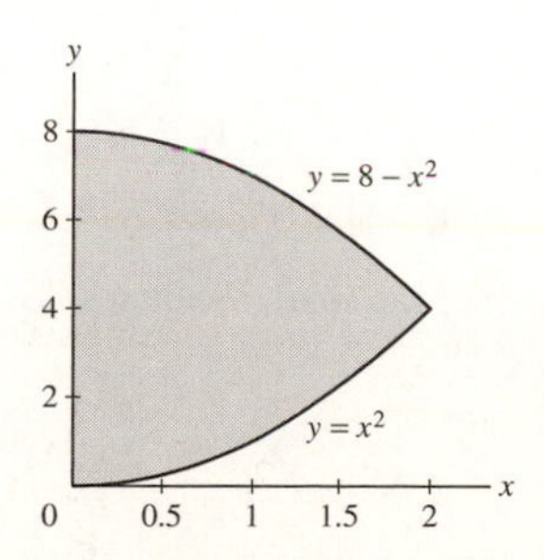

11. $y = (x^2+1)^{-2}, \quad y = 2-(x^2+1)^{-2}, \quad x = 2$

SOLUTION The region enclosed by $y = (x^2+1)^{-2}$, $y = 2-(x^2+1)^{-2}$ and $x = 2$ is shown below. When rotating this region about the y-axis, each shell has radius x and height $2-(x^2+1)^{-2}-(x^2+1)^{-2} = 2-2(x^2+1)^{-2}$. The volume of the resulting solid is

$$2\pi\int_0^2 x(2-2(x^2+1)^{-2})\,dx = 2\pi\int_0^2\left(2x-\frac{2x}{(x^2+1)^{-2}}\right)dx = 2\pi\left(x^2+\frac{1}{x^2+1}\right)\Big|_0^2 = \frac{32}{5}\pi.$$

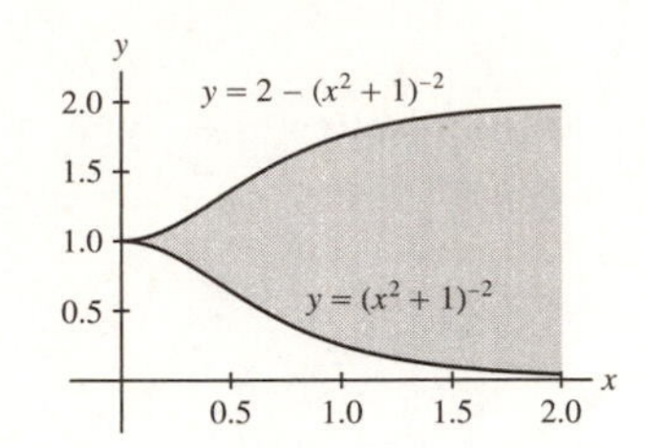

In Exercises 13 and 14, use a graphing utility to find the points of intersection of the curves numerically and then compute the volume of rotation of the enclosed region about the y-axis.

13. GU $y = \frac{1}{2}x^2, \quad y = \sin(x^2), \quad x \ge 0$

SOLUTION The region enclosed by $y = \frac{1}{2}x^2$ and $y = \sin x^2$ is shown below. When rotating this region about the y-axis, each shell has radius x and height $\sin x^2 - \frac{1}{2}x^2$. Using a computer algebra system, we find that the x-coordinate of the point of intersection on the right is $x = 1.376769504$. Thus, the volume of the resulting solid of revolution is

$$2\pi\int_0^{1.376769504} x\left(\sin x^2 - \frac{1}{2}x^2\right)dx = 1.321975576.$$

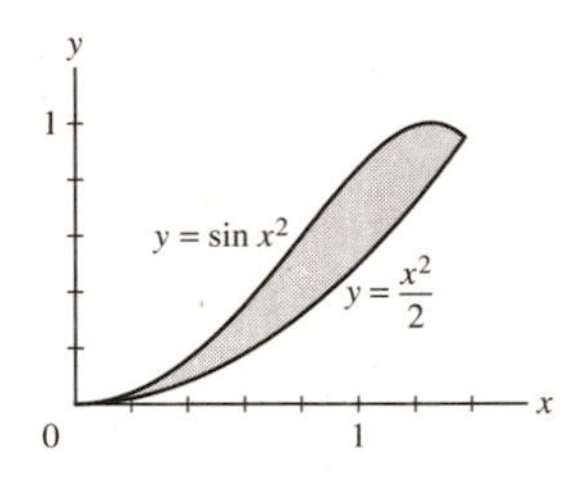

In Exercises 15–20, sketch the solid obtained by rotating the region underneath the graph of f(x) over the interval about the given axis, and calculate its volume using the Shell Method.

15. $f(x) = x^3, \quad [0,1], \quad$ about $x = 2$

SOLUTION A sketch of the solid is shown below. Each shell has radius $2-x$ and height x^3, so the volume of the solid is

$$2\pi\int_0^1 (2-x)\left(x^3\right)dx = 2\pi\int_0^1 (2x^3-x^4)\,dx = 2\pi\left(\frac{x^4}{2}-\frac{x^5}{5}\right)\Big|_0^1 = \frac{3\pi}{5}.$$

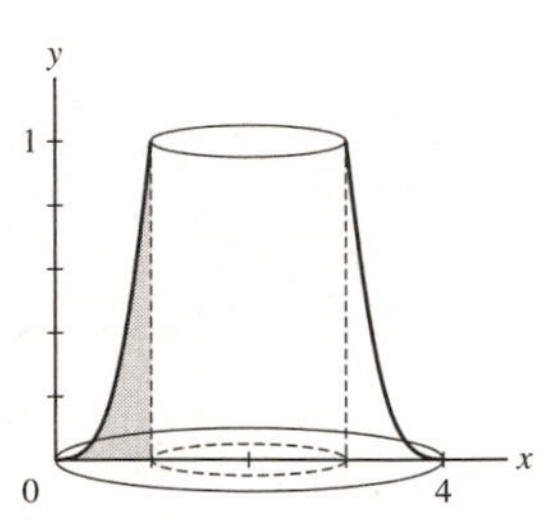

17. $f(x) = x^{-4}, \quad [-3,-1], \quad$ about $x = 4$

SOLUTION A sketch of the solid is shown below. Each shell has radius $4-x$ and height x^{-4}, so the volume of the solid is

$$2\pi\int_{-3}^{-1} (4-x)\left(x^{-4}\right)dx = 2\pi\int_{-3}^{-1} (4x^{-4}-x^{-3})\,dx = 2\pi\left(\frac{1}{2}x^{-2}-\frac{4}{3}x^{-3}\right)\Big|_{-3}^{-1} = \frac{280\pi}{81}.$$

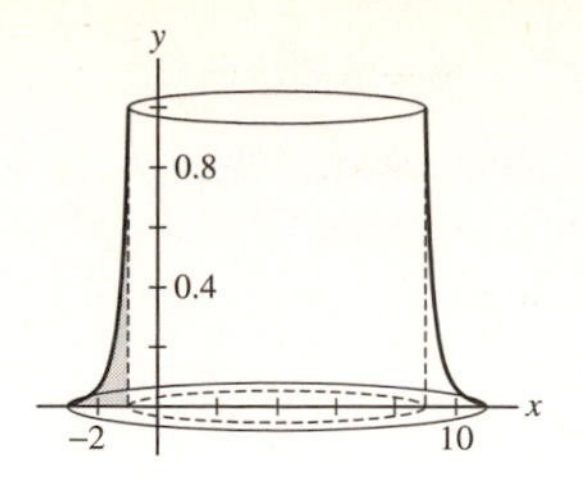

19. $f(x) = a - x$ with $a > 0$, $[0, a]$, about $x = -1$

SOLUTION A sketch of the solid is shown below. Each shell has radius $x - (-1) = x + 1$ and height $a - x$, so the volume of the solid is

$$\begin{aligned}2\pi \int_0^a (x+1)(a-x)\, dx &= 2\pi \int_0^a \left(a + (a-1)x - x^2\right) dx \\ &= 2\pi \left(ax + \frac{a-1}{2}x^2 - \frac{1}{3}x^3\right)\Big|_0^a \\ &= 2\pi \left(a^2 + \frac{a^2(a-1)}{2} - \frac{a^3}{3}\right) = \frac{a^2(a+3)}{3}\pi.\end{aligned}$$

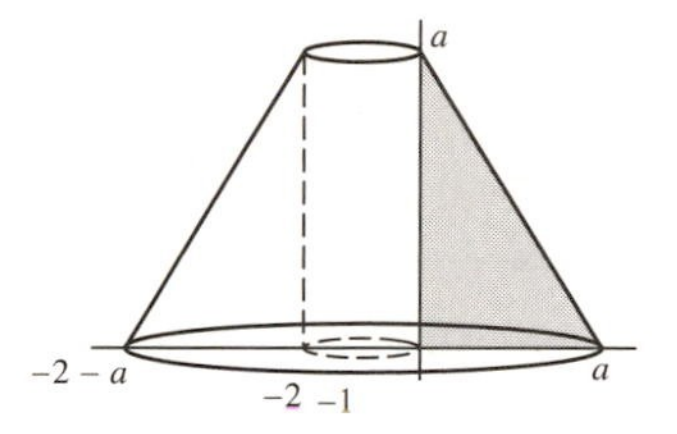

In Exercises 21–26, sketch the enclosed region and use the Shell Method to calculate the volume of rotation about the x-axis.

21. $x = y$, $y = 0$, $x = 1$

SOLUTION When the region shown below is rotated about the x-axis, each shell has radius y and height $1 - y$. The volume of the resulting solid is

$$2\pi \int_0^1 y(1-y)\, dy = 2\pi \int_0^1 (y - y^2)\, dy = 2\pi \left(\frac{1}{2}y^2 - \frac{1}{3}y^3\right)\Big|_0^1 = \frac{\pi}{3}.$$

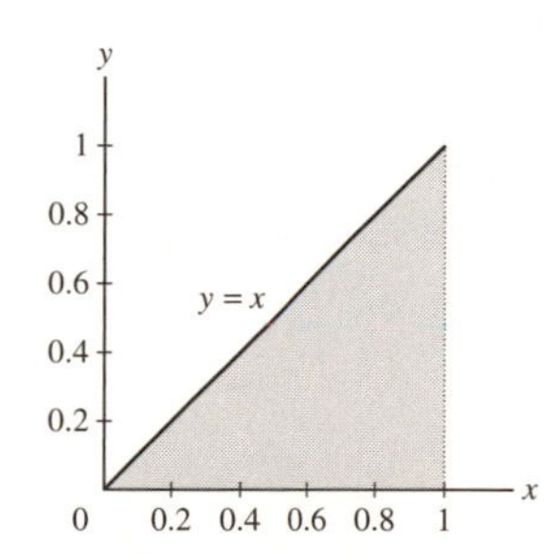

23. $x = y(4 - y)$, $y = 0$

SOLUTION When the region shown below is rotated about the x-axis, each shell has radius y and height $y(4 - y)$. The volume of the resulting solid is

$$2\pi \int_0^4 y^2(4-y)\, dy = 2\pi \int_0^4 (4y^2 - y^3)\, dy = 2\pi \left(\frac{4}{3}y^3 - \frac{1}{4}y^4\right)\Big|_0^4 = \frac{128\pi}{3}.$$

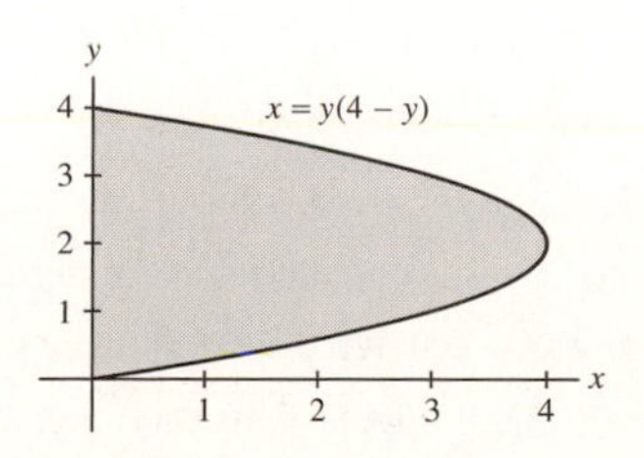

25. $y = 4 - x^2, \quad x = 0, \quad y = 0$

SOLUTION When the region shown below is rotated about the x-axis, each shell has radius y and height $\sqrt{4-y}$. The volume of the resulting solid is

$$2\pi \int_0^4 y\sqrt{4-y}\,dy.$$

Let $u = 4 - y$. Then $du = -dy$, $y = 4 - u$, and

$$2\pi \int_0^4 y\sqrt{4-y}\,dy = -2\pi \int_4^0 (4-u)\sqrt{u}\,du = 2\pi \int_0^4 \left(4\sqrt{u} - u^{3/2}\right) du$$

$$= 2\pi \left(\frac{8}{3}u^{3/2} - \frac{2}{5}u^{5/2}\right)\Big|_0^4 = \frac{256\pi}{15}.$$

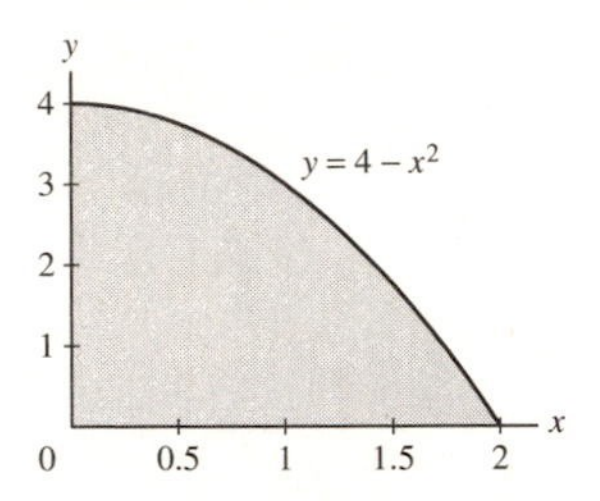

27. Use both the Shell and Disk Methods to calculate the volume obtained by rotating the region under the graph of $f(x) = 8 - x^3$ for $0 \le x \le 2$ about:

(a) the x-axis

(b) the y-axis

SOLUTION

(a) x-axis: Using the disk method, the cross sections are disks with radius $R = 8 - x^3$; hence the volume of the solid is

$$\pi \int_0^2 (8 - x^3)^2\,dx = \pi\left(64x - 4x^4 + \frac{1}{7}x^7\right)\Big|_0^2 = \frac{576\pi}{7}.$$

With the shell method, each shell has radius y and height $(8 - y)^{1/3}$. The volume of the solid is

$$2\pi \int_0^8 y\,(8-y)^{1/3}\,dy$$

Let $u = 8 - y$. Then $dy = -du$, $y = 8 - u$ and

$$2\pi \int_0^8 y\,(8-y)^{1/3}\,dy = 2\pi \int_0^8 (8-u)\cdot u^{1/3}\,du = 2\pi \int_0^8 (8u^{1/3} - u^{4/3})\,du$$

$$= 2\pi\left(6u^{4/3} - \frac{3}{7}u^{7/3}\right)\Big|_0^8 = \frac{576\pi}{7}.$$

(b) y-axis: With the shell method, each shell has radius x and height $8 - x^3$. The volume of the solid is

$$2\pi \int_0^2 x(8 - x^3)\,dx = 2\pi\left(4x^2 - \frac{1}{5}x^5\right)\Big|_0^2 = \frac{96\pi}{5}.$$

Using the disk method, the cross sections are disks with radius $R = (8 - y)^{1/3}$. The volume is then given by

$$\pi \int_0^8 (8-y)^{2/3}\,dy = -\frac{3\pi}{5}(8-y)^{5/3}\Big|_0^8 = \frac{96\pi}{5}.$$

29. The graph in Figure 11(A) can be described by both $y = f(x)$ and $x = h(y)$, where h is the inverse of f. Let V be the volume obtained by rotating the region under the graph about the y-axis.

(a) Describe the figures generated by rotating segments $\overline{AB}$ and $\overline{CB}$ about the y-axis.

(b) Set up integrals that compute V by the Shell and Disk Methods.

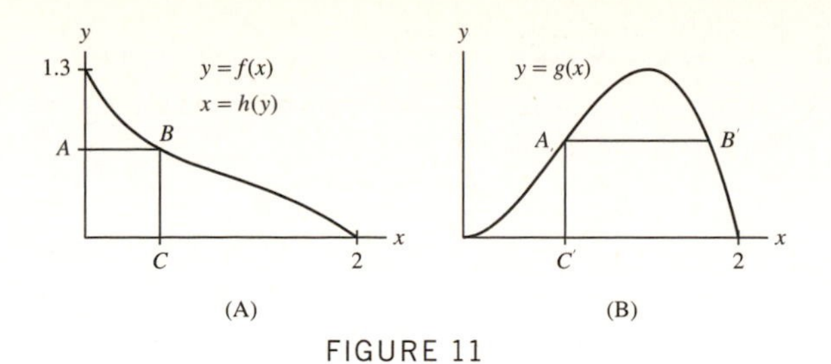

FIGURE 11

SOLUTION

(a) When rotated about the y-axis, the segment $\overline{AB}$ generates a disk with radius $R = h(y)$ and the segment $\overline{CB}$ generates a shell with radius x and height $f(x)$.

(b) Based on Figure 11(A) and the information from part (a), when using the Shell Method,

$$V = 2\pi \int_0^2 x f(x)\, dx;$$

when using the Disk Method,

$$V = \pi \int_0^{1.3} (h(y))^2\, dy.$$

31. Let R be the region under the graph of $y = 9 - x^2$ for $0 \le x \le 2$. Use the Shell Method to compute the volume of rotation of R about the x-axis as a sum of two integrals along the y-axis. *Hint:* The shells generated depend on whether $y \in [0, 5]$ or $y \in [5, 9]$.

SOLUTION The region R is sketched below. When rotating this region about the x-axis, we produce a solid with two different shell structures. For $0 \le y \le 5$, the shell has radius y and height 2; for $5 \le y \le 9$, the shell has radius y and height $\sqrt{9-y}$. The volume of the solid is therefore

$$V = 2\pi \int_0^5 2y\, dy + 2\pi \int_5^9 y\sqrt{9-y}\, dy$$

For the first integral, we calculate

$$2\pi \int_0^5 2y\, dy = 2\pi y^2 \Big|_0^5 = 50\pi.$$

For the second integral, we make the substitution $u = 9 - y$, $du = -dy$ and find

$$\begin{aligned} 2\pi \int_5^9 y\sqrt{9-y}\, dy &= -2\pi \int_4^0 (9-u)\sqrt{u}\, du \\ &= 2\pi \int_0^4 (9u^{1/2} - u^{3/2})\, du \\ &= 2\pi \left(6u^{3/2} - \frac{2}{5}u^{5/2} \right) \Big|_0^4 \\ &= 2\pi \left(48 - \frac{64}{5} \right) = \frac{352\pi}{5}. \end{aligned}$$

Thus, the total volume is

$$V = 50\pi + \frac{352\pi}{5} = \frac{602\pi}{5}.$$

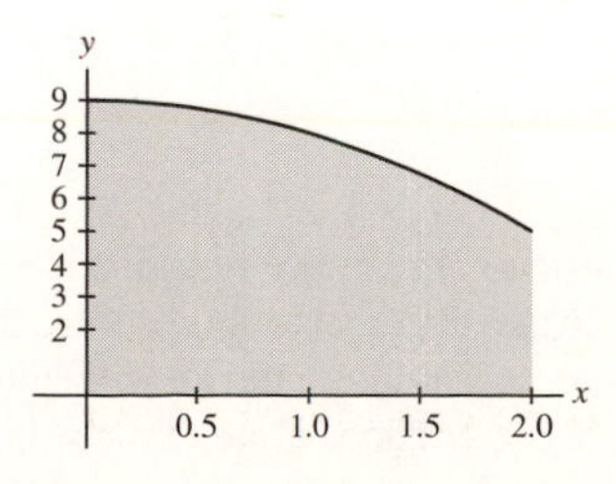

In Exercises 33–38, use the Shell Method to find the volume obtained by rotating region A in Figure 12 about the given axis.

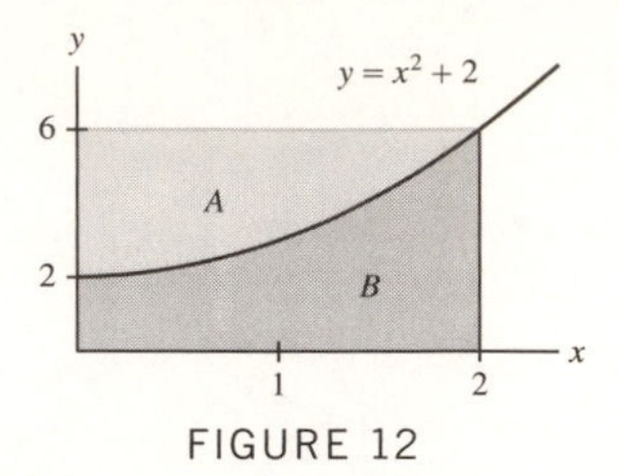

FIGURE 12

33. y-axis

SOLUTION When rotating region A about the y-axis, each shell has radius x and height $6-(x^2+2)=4-x^2$. The volume of the resulting solid is

$$2\pi\int_0^2 x(4-x^2)\,dx = 2\pi\int_0^2 (4x-x^3)\,dx = 2\pi\left(2x^2-\frac{1}{4}x^4\right)\Big|_0^2 = 8\pi.$$

35. $x=2$

SOLUTION When rotating region A about $x=2$, each shell has radius $2-x$ and height $6-(x^2+2)=4-x^2$. The volume of the resulting solid is

$$2\pi\int_0^2 (2-x)\left(4-x^2\right)\,dx = 2\pi\int_0^2 \left(8-2x^2-4x+x^3\right)\,dx = 2\pi\left(8x-\frac{2}{3}x^3-2x^2+\frac{1}{4}x^4\right)\Big|_0^2 = \frac{40\pi}{3}.$$

37. $y=-2$

SOLUTION When rotating region A about $y=-2$, each shell has radius $y-(-2)=y+2$ and height $\sqrt{y-2}$. The volume of the resulting solid is

$$2\pi\int_2^6 (y+2)\sqrt{y-2}\,dy$$

Let $u=y-2$. Then $du=dy$, $y+2=u+4$ and

$$2\pi\int_2^6 (y+2)\sqrt{y-2}\,dy = 2\pi\int_0^4 (u+4)\sqrt{u}\,du = 2\pi\left(\frac{2}{5}u^{5/2}+\frac{8}{3}u^{3/2}\right)\Big|_0^4 = \frac{1024\pi}{15}.$$

In Exercises 39–44, use the most convenient method (Disk or Shell Method) to find the volume obtained by rotating region B in Figure 12 about the given axis.

39. y-axis

SOLUTION Because a vertical slice of region B will produce a solid with a single cross section while a horizontal slice will produce a solid with two different cross sections, we will use a vertical slice. Now, because a vertical slice is parallel to the axis of rotation, we will use the Shell Method. Each shell has radius x and height x^2+2. The volume of the resulting solid is

$$2\pi\int_0^2 x(x^2+2)\,dx = 2\pi\int_0^2 (x^3+2x)\,dx = 2\pi\left(\frac{1}{4}x^4+x^2\right)\Big|_0^2 = 16\pi.$$

41. $x=2$

SOLUTION Because a vertical slice of region B will produce a solid with a single cross section while a horizontal slice will produce a solid with two different cross sections, we will use a vertical slice. Now, because a vertical slice is parallel to the axis of rotation, we will use the Shell Method. Each shell has radius $2-x$ and height x^2+2. The volume of the resulting solid is

$$2\pi\int_0^2 (2-x)\left(x^2+2\right)\,dx = 2\pi\int_0^2 \left(2x^2-x^3+4-2x\right)\,dx = 2\pi\left(\frac{2}{3}x^3-\frac{1}{4}x^4+4x-x^2\right)\Big|_0^2 = \frac{32\pi}{3}.$$

43. $y=-2$

SOLUTION Because a vertical slice of region B will produce a solid with a single cross section while a horizontal slice will produce a solid with two different cross sections, we will use a vertical slice. Now, because a vertical slice is perpendicular to the axis of rotation, we will use the Disk Method. Each disk has outer radius $R=x^2+2-(-2)=x^2+4$ and inner radius $r=0-(-2)=2$. The volume of the solid is then

$$\pi \int_0^2 \left((x^2+4)^2 - 2^2\right) dx = \pi \int_0^2 (x^4 + 8x^2 + 12)\, dx$$
$$= \pi \left(\frac{1}{5}x^5 + \frac{8}{3}x^3 + 12x\right)\Big|_0^2$$
$$= \pi \left(\frac{32}{5} + \frac{64}{3} + 24\right) = \frac{776\pi}{15}.$$

In Exercises 45–50, use the most convenient method (Disk or Shell Method) to find the given volume of rotation.

45. Region between $x = y(5-y)$ and $x = 0$, rotated about the y-axis

SOLUTION Examine the figure below, which shows the region bounded by $x = y(5-y)$ and $x = 0$. If the indicated region is sliced vertically, then the top of the slice lies along one branch of the parabola $x = y(5-y)$ and the bottom lies along the other branch. On the other hand, if the region is sliced horizontally, then the right endpoint of the slice always lies along the parabola and left endpoint always lies along the y-axis. Clearly, it will be easier to slice the region horizontally.

Now, suppose the region is rotated about the y-axis. Because a horizontal slice is perpendicular to the y-axis, we will calculate the volume of the resulting solid using the disk method. Each cross section is a disk of radius $R = y(5-y)$, so the volume is

$$\pi \int_0^5 y^2(5-y)^2\, dy = \pi \int_0^5 (25y^2 - 10y^3 + y^4)\, dy = \pi \left(\frac{25}{3}y^3 - \frac{5}{2}y^4 + \frac{1}{5}y^5\right)\Big|_0^5 = \frac{625\pi}{6}.$$

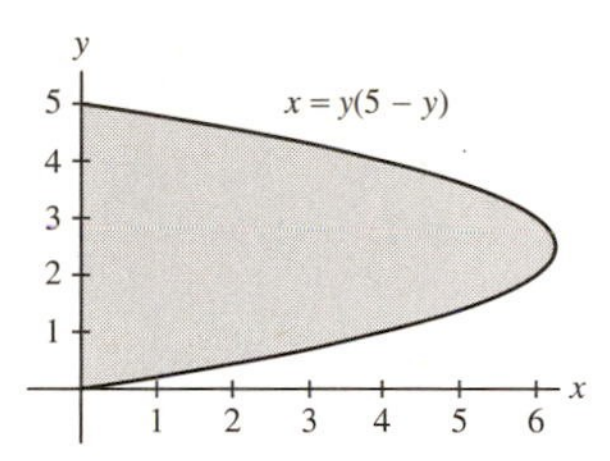

47. Region in Figure 13, rotated about the x-axis

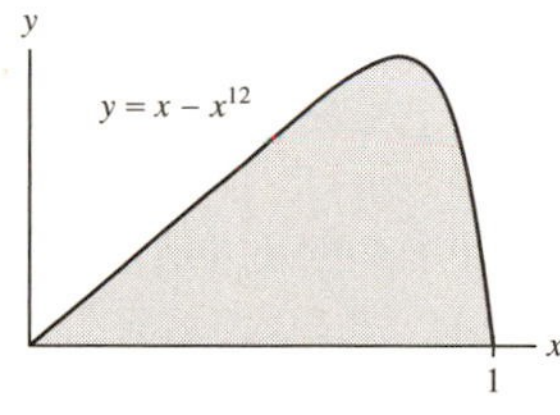

FIGURE 13

SOLUTION Examine Figure 13. If the indicated region is sliced vertically, then the top of the slice lies along the curve $y = x - x^{12}$ and the bottom lies along the curve $y = 0$ (the x-axis). On the other hand, if the region is sliced horizontally, the equation $y = x - x^{12}$ must be solved for x in order to determine the endpoint locations. Clearly, it will be easier to slice the region vertically.

Now, suppose the region in Figure 13 is rotated about the x-axis. Because a vertical slice is perpendicular to the x-axis, we will calculate the volume of the resulting solid using the disk method. Each cross section is a disk of radius $R = x - x^{12}$, so the volume is

$$\pi \int_0^1 \left(x - x^{12}\right)^2 dx = \pi \left(\frac{1}{3}x^3 - \frac{1}{7}x^{14} + \frac{1}{25}x^{25}\right)\Big|_0^1 = \frac{121\pi}{525}.$$

49. Region in Figure 14, rotated about $x = 4$

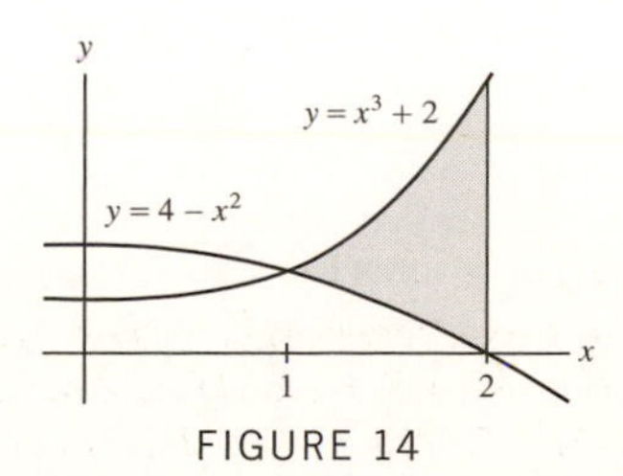

FIGURE 14

SOLUTION Examine Figure 14. If the indicated region is sliced vertically, then the top of the slice lies along the curve $y = x^3 + 2$ and the bottom lies along the curve $y = 4 - x^2$. On the other hand, the left end of a horizontal slice switches from $y = 4 - x^2$ to $y = x^3 + 2$ at $y = 3$. Here, vertical slices will be more convenient.

Now, suppose the region in Figure 14 is rotated about $x = 4$. Because a vertical slice is parallel to $x = 4$, we will calculate the volume of the resulting solid using the shell method. Each shell has radius $4 - x$ and height $x^3 + 2 - (4 - x^2) = x^3 + x^2 - 2$, so the volume is

$$2\pi \int_1^2 (4 - x)(x^3 + x^2 - 2)\, dx = 2\pi \left(-\frac{1}{5}x^5 + \frac{3}{4}x^4 + \frac{4}{3}x^3 + x^2 - 8x \right)\Bigg|_1^2 = \frac{563\pi}{30}.$$

In Exercises 51–54, use the Shell Method to find the given volume of rotation.

51. A sphere of radius r

SOLUTION A sphere of radius r can be generated by rotating the region under the semicircle $y = \sqrt{r^2 - x^2}$ about the x-axis. Each shell has radius y and height

$$\sqrt{r^2 - y^2} - \left(-\sqrt{r^2 - y^2} \right) = 2\sqrt{r^2 - y^2}.$$

Thus, the volume of the sphere is

$$2\pi \int_0^r 2y\sqrt{r^2 - y^2}\, dy.$$

Let $u = r^2 - y^2$. Then $du = -2y\, dy$ and

$$2\pi \int_0^r 2y\sqrt{r^2 - y^2}\, dy = 2\pi \int_0^{r^2} \sqrt{u}\, du = 2\pi \left(\frac{2}{3}u^{3/2} \right)\Bigg|_0^{r^2} = \frac{4}{3}\pi r^3.$$

53. The torus obtained by rotating the circle $(x - a)^2 + y^2 = b^2$ about the y-axis, where $a > b$ (compare with Exercise 53 in Section 5.3). *Hint:* Evaluate the integral by interpreting part of it as the area of a circle.

SOLUTION When rotating the region enclosed by the circle $(x - a)^2 + y^2 = b^2$ about the y-axis each shell has radius x and height

$$\sqrt{b^2 - (x - a)^2} - \left(-\sqrt{b^2 - (x - a)^2} \right) = 2\sqrt{b^2 - (x - a)^2}.$$

The volume of the resulting torus is then

$$2\pi \int_{a-b}^{a+b} 2x\sqrt{b^2 - (x - a)^2}\, dx.$$

Let $u = x - a$. Then $du = dx$, $x = u + a$ and

$$2\pi \int_{a-b}^{a+b} 2x\sqrt{b^2 - (x - a)^2}\, dx = 2\pi \int_{-b}^{b} 2(u + a)\sqrt{b^2 - u^2}\, du$$

$$= 4\pi \int_{-b}^{b} u\sqrt{b^2 - u^2}\, du + 4a\pi \int_{-b}^{b} \sqrt{b^2 - u^2}\, du.$$

Now,

$$\int_{-b}^{b} u\sqrt{b^2 - u^2}\, du = 0$$

because the integrand is an odd function and the integration interval is symmetric with respect to zero. Moreover, the other integral is one-half the area of a circle of radius b; thus,

$$\int_{-b}^{b} \sqrt{b^2 - u^2}\, du = \frac{1}{2}\pi b^2.$$

Finally, the volume of the torus is

$$4\pi(0) + 4a\pi \left(\frac{1}{2}\pi b^2 \right) = 2\pi^2 a b^2.$$

Further Insights and Challenges

55. The surface area of a sphere of radius r is $4\pi r^2$. Use this to derive the formula for the volume V of a sphere of radius R in a new way.

(a) Show that the volume of a thin spherical shell of inner radius r and thickness Δr is approximately $4\pi r^2 \Delta r$.

(b) Approximate V by decomposing the sphere of radius R into N thin spherical shells of thickness $\Delta r = R/N$.

(c) Show that the approximation is a Riemann sum that converges to an integral. Evaluate the integral.

SOLUTION

(a) The volume of a thin spherical shell of inner radius r and thickness Δx is given by the product of the surface area of the shell, $4\pi r^2$ and the thickness. Thus, we have $4\pi r^2 \Delta x$.

(b) The volume of the sphere is approximated by

$$R_N = 4\pi \left(\frac{R}{N}\right) \sum_{k=1}^{N} (x_k)^2$$

where $x_k = k\frac{R}{N}$.

(c) $V = 4\pi \lim_{N\to\infty} \left(\frac{R}{N}\right) \sum_{k=1}^{N} (x_k)^2 = 4\pi \int_0^R x^2\,dx = 4\pi \left(\frac{1}{3}x^3\right)\Big|_0^R = \frac{4}{3}\pi R^3.$

57. The bell-shaped curve $y = f(x)$ in Figure 16 satisfies $dy/dx = -xy$. Use the Shell Method and the substitution $u = f(x)$ to show that the solid obtained by rotating the region R about the y-axis has volume $V = 2\pi(1 - c)$, where $c = f(a)$. Observe that as $c \to 0$, the region R becomes infinite but the volume V approaches 2π.

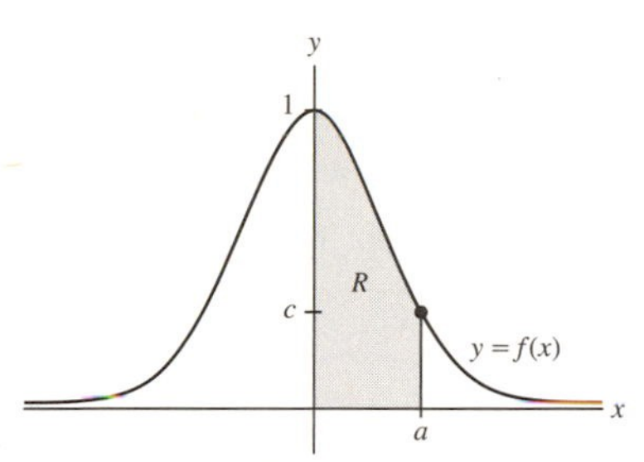

FIGURE 16 The bell-shaped curve.

SOLUTION Let $y = f(x)$ be the exponential function depicted in Figure 16. When rotating the region R about the y-axis, each shell in the resulting solid has radius x and height $f(x)$. The volume of the solid is then

$$V = 2\pi \int_0^a x f(x)\,dx.$$

Now, let $u = f(x)$. Then $du = f'(x)\,dx = -xf(x)\,dx$; hence, $xf(x)dx = -du$, and

$$V = 2\pi \int_1^c (-du) = 2\pi \int_c^1 du = 2\pi(1 - c).$$

6.5 Work and Energy

Preliminary Questions

1. Why is integration needed to compute the work performed in stretching a spring?

SOLUTION Recall that the force needed to extend or compress a spring depends on the amount by which the spring has already been extended or compressed from its equilibrium position. In other words, the force needed to move a spring is variable. Whenever the force is variable, work needs to be computed with an integral.

2. Why is integration needed to compute the work performed in pumping water out of a tank but not to compute the work performed in lifting up the tank?

SOLUTION To lift a tank through a vertical distance d, the force needed to move the tank remains constant; hence, no integral is needed to calculate the work done in lifting the tank. On the other hand, pumping water from a tank requires that different layers of the water be lifted through different distances, and, depending on the shape of the tank, may require different forces. Thus, pumping water from a tank requires that an integral be evaluated.

3. Which of the following represents the work required to stretch a spring (with spring constant k) a distance x beyond its equilibrium position: kx, $-kx$, $\frac{1}{2}mk^2$, $\frac{1}{2}kx^2$, or $\frac{1}{2}mx^2$?

SOLUTION The work required to stretch a spring with spring constant k a distance x beyond its equilibrium position is

$$\int_0^x ky\,dy = \frac{1}{2}ky^2\Big|_0^x = \frac{1}{2}kx^2.$$

Exercises

1. How much work is done raising a 4-kg mass to a height of 16 m above ground?

SOLUTION The force needed to lift a 4-kg object is a constant

$$(4\text{ kg})(9.8\text{ m/s}^2) = 39.2\text{ N}.$$

The work done in lifting the object to a height of 16 m is then

$$(39.2\text{ N})(16\text{ m}) = 627.2\text{ J}.$$

In Exercises 3–6, compute the work (in joules) required to stretch or compress a spring as indicated, assuming a spring constant of $k = 800$ N/m.

3. Stretching from equilibrium to 12 cm past equilibrium

SOLUTION The work required to stretch the spring 12 cm past equilibrium is

$$\int_0^{0.12} 800x\,dx = 400x^2\Big|_0^{0.12} = 5.76\text{ J}.$$

5. Stretching from 5 cm to 15 cm past equilibrium

SOLUTION The work required to stretch the spring from 5 cm to 15 cm past equilibrium is

$$\int_{0.05}^{0.15} 800x\,dx = 400x^2\Big|_{0.05}^{0.15} = 8\text{ J}.$$

7. If 5 J of work are needed to stretch a spring 10 cm beyond equilibrium, how much work is required to stretch it 15 cm beyond equilibrium?

SOLUTION First, we determine the value of the spring constant as follows:

$$\int_0^{0.1} kx\,dx = \frac{1}{2}kx^2\Big|_0^{0.1} = 0.005k = 5\text{ J}.$$

Thus, $k = 1000$ N/m. Next, we calculate the work required to stretch the spring 15 cm beyond equilibrium:

$$\int_0^{0.15} 1000x\,dx = 500x^2\Big|_0^{0.15} = 11.25\text{ J}.$$

9. A spring obeys a force law $F(x) = -kx^{1.1}$ with $k = 100$ N/m$^{1.1}$. Find the work required to stretch the spring 0.3 m past equilibrium.

SOLUTION The work required to stretch this spring 0.3 m past equilibrium is

$$\int_0^{0.3} 100x^{1.1}\,dx = \frac{100}{1.1}x^{2.1}\Big|_0^{0.3} \approx 7.25\text{ J}.$$

In Exercises 11–14, use the method of Examples 2 and 3 to calculate the work against gravity required to build the structure out of a lightweight material of density 600 kg/m^3.

11. Box of height 3 m and square base of side 2 m

SOLUTION The volume of one layer is $4\Delta y$ m^3 and so the weight of one layer is $23520\Delta y$ N. Thus, the work done against gravity to build the tower is

$$W = \int_0^3 23520y\,dy = 11760y^2\Big|_0^3 = 105840\text{ J}.$$

13. Right circular cone of height 4 m and base of radius 1.2 m

SOLUTION By similar triangles, the layer of the cone at a height y above the base has radius $r = 0.3(4 - y)$ meters. Thus, the volume of the small layer at this height is $0.09\pi(4 - y)^2\Delta y$ m^3, and the weight is $529.2\pi(4 - y)^2\Delta y$ N. Finally, the total work done against gravity to build the tower is

$$\int_0^4 529.2\pi(4 - y)^2 y\,dy = 11289.6\pi \text{ J} \approx 35467.3 \text{ J}.$$

15. Built around 2600 BCE, the Great Pyramid of Giza in Egypt (Figure 7) is 146 m high and has a square base of side 230 m. Find the work (against gravity) required to build the pyramid if the density of the stone is estimated at 2000 kg/m^3.

FIGURE 7 The Great Pyramid in Giza, Egypt.

SOLUTION From similar triangles, the area of one layer is

$$\left(230 - \frac{230}{146}y\right)^2 \text{ m}^2,$$

so the volume of each small layer is

$$\left(230 - \frac{230}{146}y\right)^2 \Delta y \text{ m}^3.$$

The weight of one layer is then

$$19600\left(230 - \frac{230}{146}y\right)^2 \Delta y \text{ N}.$$

Finally, the total work needed to build the pyramid was

$$\int_0^{146} 19600\left(230 - \frac{230}{146}y\right)^2 y\,dy \approx 1.84 \times 10^{12} \text{ J}.$$

In Exercises 17–22, calculate the work (in joules) required to pump all of the water out of a full tank. Distances are in meters, and the density of water is 1000 kg/m^3.

17. Rectangular tank in Figure 8; water exits from a small hole at the top.

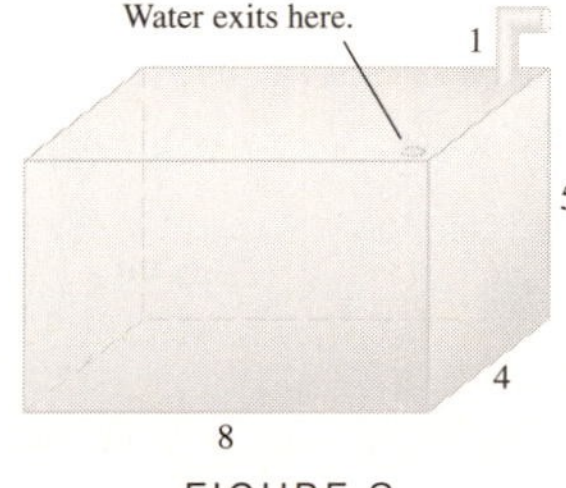

FIGURE 8

SOLUTION Place the origin on the top of the box, and let the positive y-axis point downward. The volume of one layer of water is $32\Delta y$ m^3, so the force needed to lift each layer is

$$(9.8)(1000)32\Delta y = 313600\Delta y \text{ N}.$$

Each layer must be lifted y meters, so the total work needed to empty the tank is

$$\int_0^5 313600y\,dy = 156800y^2\Big|_0^5 = 3.92 \times 10^6 \text{ J}.$$

19. Hemisphere in Figure 9; water exits through the spout.

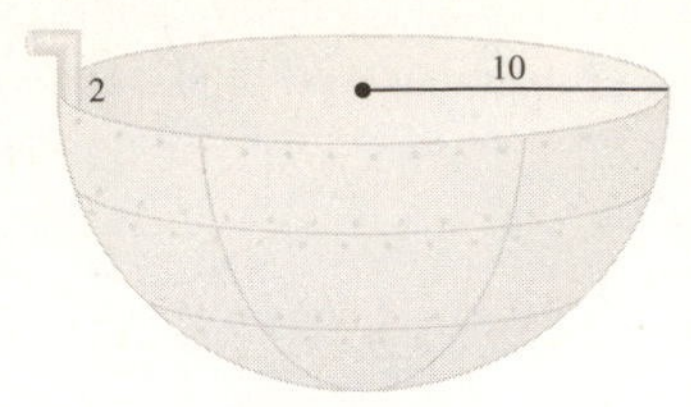

FIGURE 9

SOLUTION Place the origin at the center of the hemisphere, and let the positive y-axis point downward. The radius of a layer of water at depth y is $\sqrt{100 - y^2}$ m, so the volume of the layer is $\pi(100 - y^2)\Delta y$ m^3, and the force needed to lift the layer is $9800\pi(100 - y^2)\Delta y$ N. The layer must be lifted $y + 2$ meters, so the total work needed to empty the tank is

$$\int_0^{10} 9800\pi(100 - y^2)(y + 2)\, dy = \frac{112700000\pi}{3} \text{ J} \approx 1.18 \times 10^8 \text{ J}.$$

21. Horizontal cylinder in Figure 11; water exits from a small hole at the top. *Hint:* Evaluate the integral by interpreting part of it as the area of a circle.

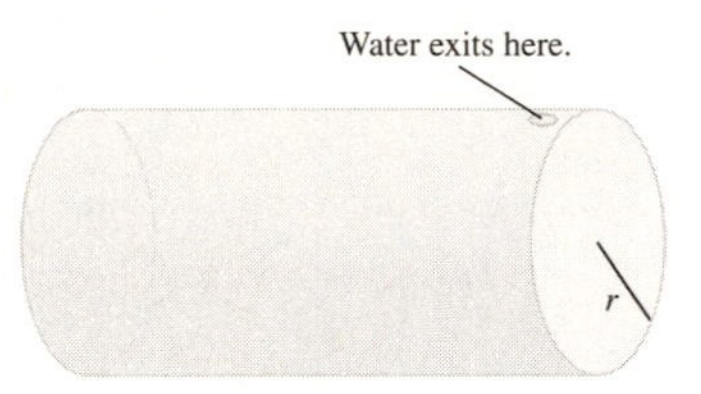

FIGURE 11

SOLUTION Place the origin along the axis of the cylinder. At location y, the layer of water is a rectangular slab of length ℓ, width $2\sqrt{r^2 - y^2}$ and thickness Δy. Thus, the volume of the layer is $2\ell\sqrt{r^2 - y^2}\Delta y$, and the force needed to lift the layer is $19{,}600\ell\sqrt{r^2 - y^2}\Delta y$. The layer must be lifted a distance $r - y$, so the total work needed to empty the tank is given by

$$\int_{-r}^{r} 19{,}600\ell\sqrt{r^2 - y^2}(r - y)\, dy = 19{,}600\ell r \int_{-r}^{r} \sqrt{r^2 - y^2}\, dy - 19{,}600\ell \int_{-r}^{r} y\sqrt{r^2 - y^2}\, dy.$$

Now,

$$\int_{-r}^{r} y\sqrt{r^2 - y^2}\, du = 0$$

because the integrand is an odd function and the integration interval is symmetric with respect to zero. Moreover, the other integral is one-half the area of a circle of radius r; thus,

$$\int_{-r}^{r} \sqrt{r^2 - y^2}\, dy = \frac{1}{2}\pi r^2.$$

Finally, the total work needed to empty the tank is

$$19{,}600\ell r\left(\frac{1}{2}\pi r^2\right) - 19{,}600\ell(0) = 9800\ell\pi r^3 \text{ J}.$$

23. Find the work W required to empty the tank in Figure 8 through the hole at the top if the tank is half full of water.

SOLUTION Place the origin on the top of the box, and let the positive y-axis point downward. Note that with this coordinate system, the bottom half of the box corresponds to y values from 2.5 to 5. The volume of one layer of water is $32\Delta y$ m^3, so the force needed to lift each layer is

$$(9.8)(1000)32\Delta y = 313{,}600\Delta y \text{ N}.$$

Each layer must be lifted y meters, so the total work needed to empty the tank is

$$\int_{2.5}^{5} 313{,}600y\, dy = 156{,}800y^2\Big|_{2.5}^{5} = 2.94 \times 10^6 \text{ J}.$$

25. Assume the tank in Figure 10 is full. Find the work required to pump out half of the water. *Hint:* First, determine the level H at which the water remaining in the tank is equal to one-half the total capacity of the tank.

SOLUTION Our first step is to determine the level H at which the water remaining in the tank is equal to one-half the total capacity of the tank. From Figure 10 and similar triangles, we see that the radius of the cone at level H is $H/2$ so the volume of water is

$$V = \frac{1}{3}\pi r^2 H = \frac{1}{3}\pi \left(\frac{H}{2}\right)^2 H = \frac{1}{12}\pi H^3.$$

The total capacity of the tank is $250\pi/3$ m^3, so the water level when the water remaining in the tank is equal to one-half the total capacity of the tank satisfies

$$\frac{1}{12}\pi H^3 = \frac{125}{3}\pi \quad \text{or} \quad H = \frac{10}{2^{1/3}} \text{ m.}$$

Place the origin at the vertex of the inverted cone, and let the positive y-axis point upward. Now, consider a layer of water at a height of y meters. From similar triangles, the area of the layer is

$$\pi \left(\frac{y}{2}\right)^2 \text{ m}^2,$$

so the volume is

$$\pi \left(\frac{y}{2}\right)^2 \Delta y \text{ m}^3.$$

Thus the weight of one layer is

$$9800\pi \left(\frac{y}{2}\right)^2 \Delta y \text{ N.}$$

The layer must be lifted $12 - y$ meters, so the total work needed to empty the half-full tank is

$$\int_{10/2^{1/3}}^{10} 9800\pi \left(\frac{y}{2}\right)^2 (12 - y)\, dy \approx 3.79 \times 10^6 \text{ J.}$$

27. Calculate the work required to lift a 10-m chain over the side of a building (Figure 13) Assume that the chain has a density of 8 kg/m. *Hint:* Break up the chain into N segments, estimate the work performed on a segment, and compute the limit as $N \to \infty$ as an integral.

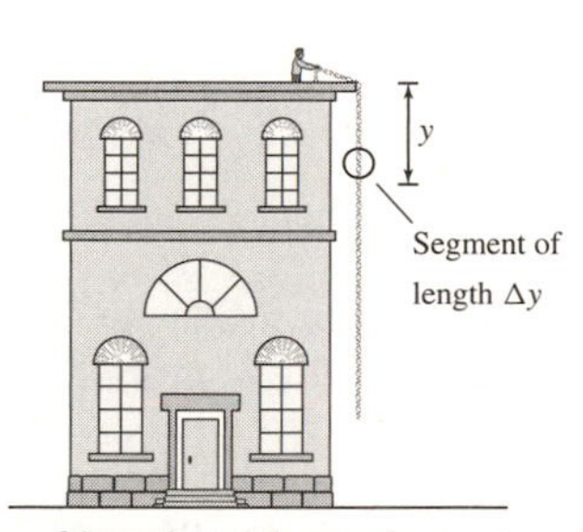

FIGURE 13 The small segment of the chain of length Δy located y meters from the top is lifted through a vertical distance y.

SOLUTION In this example, each part of the chain is lifted a different distance. Therefore, we divide the chain into N small segments of length $\Delta y = 10/N$. Suppose that the ith segment is located a distance y_i from the top of the building. This segment weighs $8(9.8)\Delta y$ kilograms and it must be lifted approximately y_i meters (not exactly y_i meters, because each point along the segment is a slightly different distance from the top). The work W_i done on this segment is approximately $W_i \approx 78.4 y_i \Delta y$ N. The total work W is the sum of the W_i and we have

$$W = \sum_{j=1}^{N} W_i \approx \sum_{j=1}^{N} 78.4 y_j\, \Delta y.$$

Passing to the limit as $N \to \infty$, we obtain

$$W = \int_0^{10} 78.4\, y\, dy = 39.2 y^2 \Big|_0^{10} = 3920 \text{ J.}$$

29. A 6-m chain has mass 18 kg. Find the work required to lift the chain over the side of a building.

SOLUTION First, note that the chain has a mass density of 3 kg/m. Now, consider a segment of the chain of length Δy located a distance y_j feet from the top of the building. The work needed to lift this segment of the chain to the top of the building is approximately

$$W_j \approx (3\Delta y)9.8y_j \text{ ft-lb}.$$

Summing over all segments of the chain and passing to the limit as $\Delta y \to 0$, it follows that the total work is

$$\int_0^6 29.4y\,dy = 14.7y^2\Big|_0^6 = 529.2 \text{ J}.$$

31. How much work is done lifting a 12-m chain that has mass density 3 kg/m (initially coiled on the ground) so that its top end is 10 m above the ground?

SOLUTION Consider a segment of the chain of length Δy that must be lifted y_j feet off the ground. The work needed to lift this segment of the chain is approximately

$$W_j \approx (3\Delta y)9.8y_j \text{ J}.$$

Summing over all segments of the chain and passing to the limit as $\Delta y \to 0$, it follows that the total work is

$$\int_0^{10} 29.4y\,dy = 14.7y^2\Big|_0^{10} = 1470 \text{ J}.$$

33. Calculate the work required to lift a 3-m chain over the side of a building if the chain has variable density of $\rho(x) = x^2 - 3x + 10$ kg/m for $0 \le x \le 3$.

SOLUTION Consider a segment of the chain of length Δx that must be lifted x_j feet. The work needed to lift this segment is approximately

$$W_j \approx \left(\rho(x_j)\Delta x\right) 9.8x_j \text{ J}.$$

Summing over all segments of the chain and passing to the limit as $\Delta x \to 0$, it follows that the total work is

$$\int_0^3 9.8\rho(x)x\,dx = 9.8\int_0^3 \left(x^3 - 3x^2 + 10x\right) dx$$

$$= 9.8\left(\frac{1}{4}x^4 - x^3 + 5x^2\right)\Big|_0^3 = 374.85 \text{ J}.$$

Exercises 35–37: The gravitational force between two objects of mass m and M, separated by a distance r, has magnitude GMm/r^2, where $G = 6.67 \times 10^{-11}\ \text{m}^3\text{kg}^{-1}\text{s}^{-1}$.

35. Show that if two objects of mass M and m are separated by a distance r_1, then the work required to increase the separation to a distance r_2 is equal to $W = GMm(r_1^{-1} - r_2^{-1})$.

SOLUTION The work required to increase the separation from a distance r_1 to a distance r_2 is

$$\int_{r_1}^{r_2} \frac{GMm}{r^2}\,dr = -\frac{GMm}{r}\Big|_{r_1}^{r_2} = GMm(r_1^{-1} - r_2^{-1}).$$

37. Use the result of Exercise 35 to compute the work required to move a 1500-kg satellite from an orbit 1000 to an orbit 1500 km above the surface of the earth.

SOLUTION The satellite will move from a distance $r_1 = R_e + 1{,}000{,}000$ to a distance $r_2 = R_e + 1{,}500{,}000$. Thus, from Exercise 35,

$$W = (6.67 \times 10^{-11})(5.98 \times 10^{24})(1500) \times \left(\frac{1}{6.37 \times 10^6 + 1{,}000{,}000} - \frac{1}{6.37 \times 10^6 + 1{,}500{,}000}\right)$$

$$\approx 5.16 \times 10^9 \text{ J}.$$

Further Insights and Challenges

39. Work-Energy Theorem An object of mass m moves from x_1 to x_2 during the time interval $[t_1, t_2]$ due to a force $F(x)$ acting in the direction of motion. Let $x(t)$, $v(t)$, and $a(t)$ be the position, velocity, and acceleration at time t. The object's kinetic energy is $\text{KE} = \frac{1}{2}mv^2$.

(a) Use the change-of-variables formula to show that the work performed is equal to

$$W = \int_{x_1}^{x_2} F(x)\,dx = \int_{t_1}^{t_2} F(x(t))v(t)\,dt$$

(b) Use Newton's Second Law, $F(x(t)) = ma(t)$, to show that

$$\frac{d}{dt}\left(\frac{1}{2}mv(t)^2\right) = F(x(t))v(t)$$

(c) Use the FTC to prove the Work-Energy Theorem: The change in kinetic energy during the time interval $[t_1, t_2]$ is equal to the work performed.

SOLUTION

(a) Let $x_1 = x(t_1)$ and $x_2 = x(t_2)$, then $x = x(t)$ gives $dx = v(t)\,dt$. By substitution we have

$$W = \int_{x_1}^{x_2} F(x)\,dx = \int_{t_1}^{t_2} F(x(t))v(t)\,dt.$$

(b) Knowing $F(x(t)) = m \cdot a(t)$, we have

$$\begin{aligned}\frac{d}{dt}\left(\frac{1}{2}m \cdot v(t)^2\right) &= m \cdot v(t)\,v'(t) && \text{(Chain Rule)}\\ &= m \cdot v(t)\,a(t)\\ &= v(t) \cdot F(x(t)) && \text{(Newton's 2nd law)}\end{aligned}$$

(c) From the FTC,

$$\frac{1}{2}m \cdot v(t)^2 = \int F(x(t))\,v(t)\,dt.$$

Since $KE = \frac{1}{2}m\,v^2$,

$$\Delta KE = KE(t_2) - KE(t_1) = \frac{1}{2}m\,v(t_2)^2 - \frac{1}{2}m\,v(t_1)^2 = \int_{t_1}^{t_2} F(x(t))\,v(t)\,dt.$$

(d)

$$\begin{aligned}W &= \int_{x_1}^{x_2} F(x)\,dx = \int_{t_1}^{t_2} F(x(t))\,v(t)\,dt && \text{(Part (a))}\\ &= KE(t_2) - KE(t_1)\\ &= \Delta KE && \text{(as required)}\end{aligned}$$

41. With what initial velocity v_0 must we fire a rocket so it attains a maximum height r above the earth? *Hint:* Use the results of Exercises 35 and 39. As the rocket reaches its maximum height, its KE decreases from $\frac{1}{2}mv_0^2$ to zero.

SOLUTION The work required to move the rocket a distance r from the surface of the earth is

$$W(r) = GM_em\left(\frac{1}{R_e} - \frac{1}{r + R_e}\right).$$

As the rocket climbs to a height r, its kinetic energy is reduced by the amount $W(r)$. The rocket reaches its maximum height when its kinetic energy is reduced to zero, that is, when

$$\frac{1}{2}mv_0^2 = GM_em\left(\frac{1}{R_e} - \frac{1}{r + R_e}\right).$$

Therefore, its initial velocity must be

$$v_0 = \sqrt{2GM_e\left(\frac{1}{R_e} - \frac{1}{r + R_e}\right)}.$$

43. Calculate **escape velocity,** the minimum initial velocity of an object to ensure that it will continue traveling into space and never fall back to earth (assuming that no force is applied after takeoff). *Hint:* Take the limit as $r \to \infty$ in Exercise 41.

SOLUTION The result of Exercise 41 leads to an interesting conclusion. The initial velocity v_0 required to reach a height r does not increase beyond all bounds as r tends to infinity; rather, it approaches a finite limit, called the escape velocity:

$$v_{\text{esc}} = \lim_{r\to\infty} \sqrt{2GM_e\left(\frac{1}{R_e} - \frac{1}{r+R_e}\right)} = \sqrt{\frac{2GM_e}{R_e}}$$

In other words, v_{esc} is large enough to insure that the rocket reaches a height r for every value of r! Therefore, a rocket fired with initial velocity v_{esc} never returns to earth. It continues traveling indefinitely into outer space.

Now, let's see how large escape velocity actually is:

$$v_{\text{esc}} = \left(\frac{2\cdot 6.67\times 10^{-11}\cdot 5.989\times 10^{24}}{6.37\times 10^{6}}\right)^{1/2} \approx 11{,}190 \text{ m/sec.}$$

Since one meter per second is equal to 2.236 miles per hour, escape velocity is approximately $11{,}190(2.236) = 25{,}020$ miles per hour.

CHAPTER REVIEW EXERCISES

1. Compute the area of the region in Figure 1(A) enclosed by $y = 2 - x^2$ and $y = -2$.

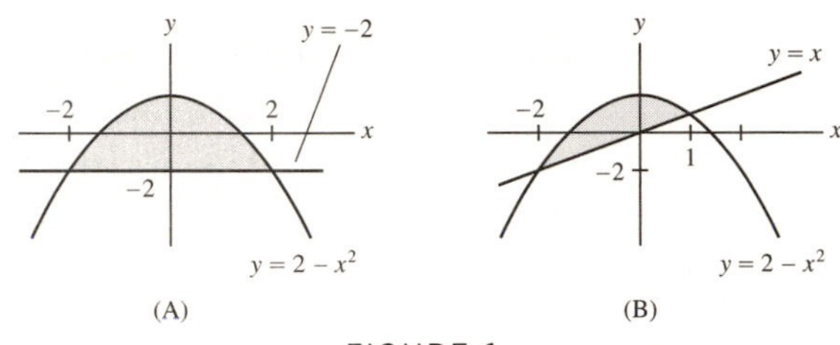

FIGURE 1

SOLUTION The graphs of $y = 2 - x^2$ and $y = -2$ intersect where $2 - x^2 = -2$, or $x = \pm 2$. Therefore, the enclosed area lies over the interval $[-2, 2]$. The region enclosed by the graphs lies below $y = 2 - x^2$ and above $y = -2$, so the area is

$$\int_{-2}^{2} \left((2-x^2) - (-2)\right) dx = \int_{-2}^{2} (4 - x^2)\, dx = \left(4x - \frac{1}{3}x^3\right)\Big|_{-2}^{2} = \frac{32}{3}.$$

In Exercises 3–12, find the area of the region enclosed by the graphs of the functions.

3. $y = x^3 - 2x^2 + x, \quad y = x^2 - x$

SOLUTION The region bounded by the graphs of $y = x^3 - 2x^2 + x$ and $y = x^2 - x$ over the interval $[0, 2]$ is shown below. For $x \in [0, 1]$, the graph of $y = x^3 - 2x^2 + x$ lies above the graph of $y = x^2 - x$, whereas, for $x \in [1, 2]$, the graph of $y = x^2 - x$ lies above the graph of $y = x^3 - 2x^2 + x$. The area of the region is therefore given by

$$\int_0^1 \left((x^3 - 2x^2 + x) - (x^2 - x)\right) dx + \int_1^2 \left((x^2 - x) - (x^3 - 2x^2 + x)\right) dx$$

$$= \left(\frac{1}{4}x^4 - x^3 + x^2\right)\Big|_0^1 + \left(x^3 - x^2 - \frac{1}{4}x^4\right)\Big|_1^2$$

$$= \frac{1}{4} - 1 + 1 + (8 - 4 - 4) - \left(1 - 1 - \frac{1}{4}\right) = \frac{1}{2}.$$

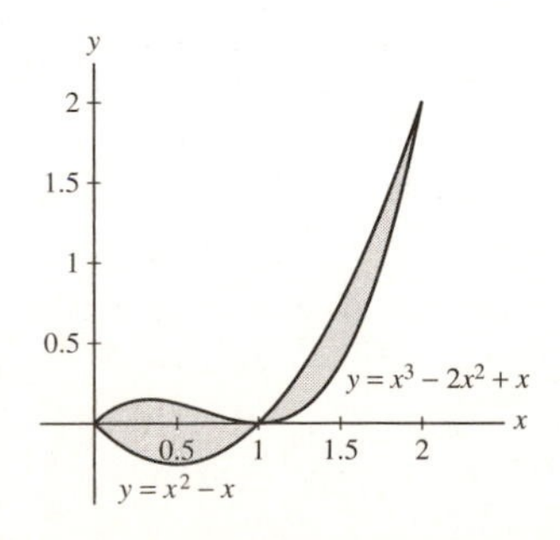

5. $x = 4y, \quad x = 24 - 8y, \quad y = 0$

SOLUTION The region bounded by the graphs $x = 4y$, $x = 24 - 8y$ and $y = 0$ is shown below. For each $0 \le y \le 2$, the graph of $x = 24 - 8y$ lies to the right of $x = 4y$. The area of the region is therefore

$$A = \int_0^2 (24 - 8y - 4y)\, dy = \int_0^2 (24 - 12y)\, dy$$

$$= (24y - 6y^2)\Big|_0^2 = 24.$$

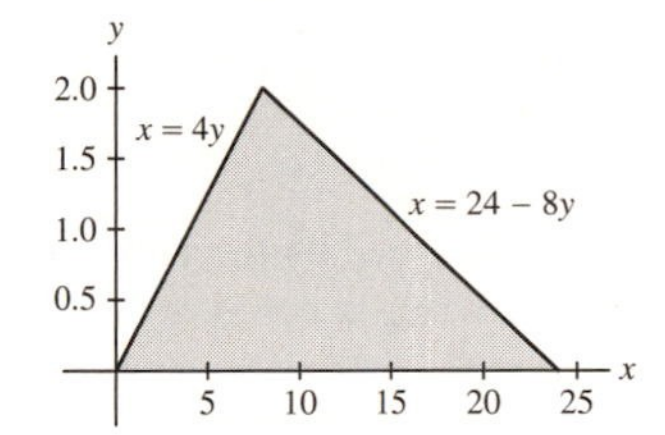

7. $y = 4 - x^2, \quad y = 3x, \quad y = 4$

SOLUTION The region bounded by the graphs of $y = 4 - x^2$, $y = 3x$ and $y = 4$ is shown below. For $x \in [0, 1]$, the graph of $y = 4$ lies above the graph of $y = 4 - x^2$, whereas, for $x \in [1, \frac{4}{3}]$, the graph of $y = 4$ lies above the graph of $y = 3x$. The area of the region is therefore given by

$$\int_0^1 (4 - (4 - x^2))\, dx + \int_1^{4/3} (4 - 3x)\, dx = \frac{1}{3}x^3\Big|_0^1 + \left(4x - \frac{3}{2}x^2\right)\Big|_1^{4/3} = \frac{1}{3} + \left(\frac{16}{3} - \frac{8}{3}\right) - \left(4 - \frac{3}{2}\right) = \frac{1}{2}.$$

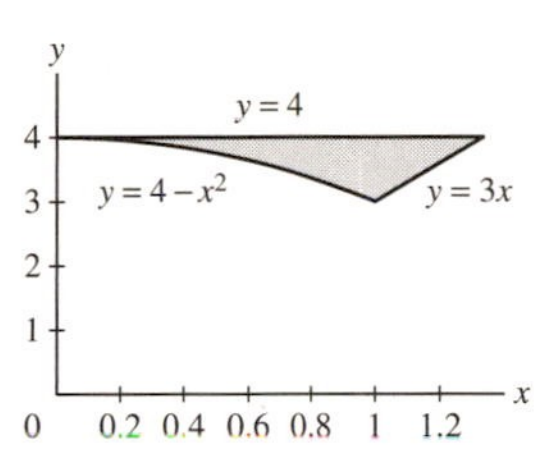

9. $y = \sin x, \quad y = \cos x, \quad 0 \le x \le \dfrac{5\pi}{4}$

SOLUTION The region bounded by the graphs of $y = \sin x$ and $y = \cos x$ over the interval $[0, \frac{5\pi}{4}]$ is shown below. For $x \in [0, \frac{\pi}{4}]$, the graph of $y = \cos x$ lies above the graph of $y = \sin x$, whereas, for $x \in [\frac{\pi}{4}, \frac{5\pi}{4}]$, the graph of $y = \sin x$ lies above the graph of $y = \cos x$. The area of the region is therefore given by

$$\int_0^{\pi/4} (\cos x - \sin x)\, dx + \int_{\pi/4}^{5\pi/4} (\sin x - \cos x)\, dx$$

$$= (\sin x + \cos x)\Big|_0^{\pi/4} + (-\cos x - \sin x)\Big|_{\pi/4}^{5\pi/4}$$

$$= \frac{\sqrt{2}}{2} + \frac{\sqrt{2}}{2} - (0 + 1) + \left(\frac{\sqrt{2}}{2} + \frac{\sqrt{2}}{2}\right) - \left(-\frac{\sqrt{2}}{2} - \frac{\sqrt{2}}{2}\right) = 3\sqrt{2} - 1.$$

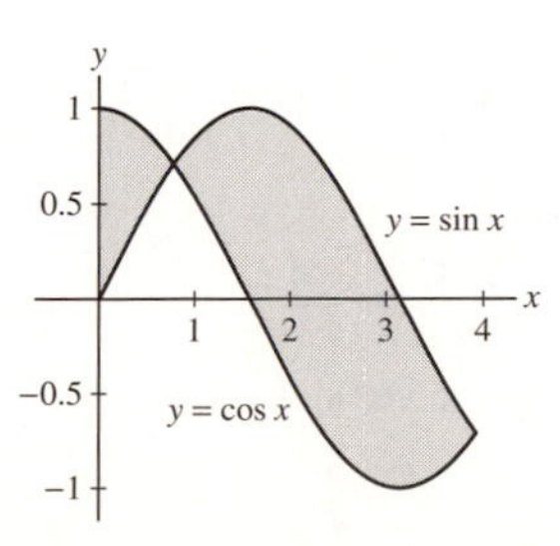

11. $y = e^x, \quad y = 1 - x, \quad x = 1$

SOLUTION The region bounded by the graphs of $y = e^x$, $y = 1 - x$ and $x = 1$ is shown below. As the graph of $y = e^x$ lies above the graph of $y = 1 - x$, the area of the region is given by

$$\int_0^1 (e^x - (1 - x))\, dx = \left(e^x - x + \frac{1}{2}x^2\right)\Big|_0^1 = \left(e - 1 + \frac{1}{2}\right) - 1 = e - \frac{3}{2}.$$

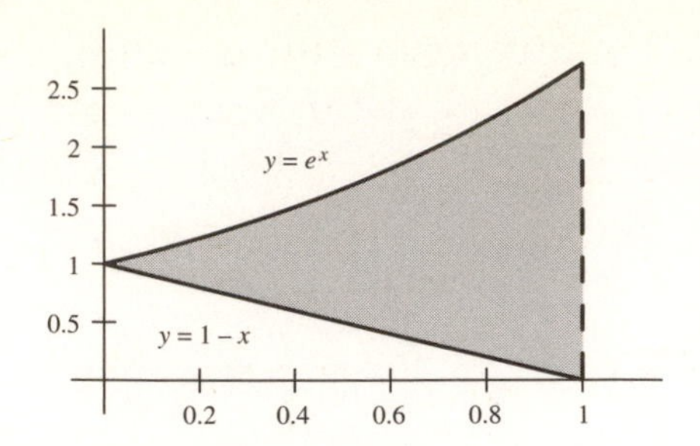

13. GU Use a graphing utility to locate the points of intersection of $y = e^{-x}$ and $y = 1 - x^2$ and find the area between the two curves (approximately).

SOLUTION The region bounded by the graphs of $y = e^{-x}$ and $y = 1 - x^2$ is shown below. One point of intersection clearly occurs at $x = 0$. Using a computer algebra system, we find that the other point of intersection occurs at $x = 0.7145563847$. As the graph of $y = 1 - x^2$ lies above the graph of $y = e^{-x}$, the area of the region is given by

$$\int_0^{0.7145563847} \left(1 - x^2 - e^{-x}\right) dx = 0.08235024596$$

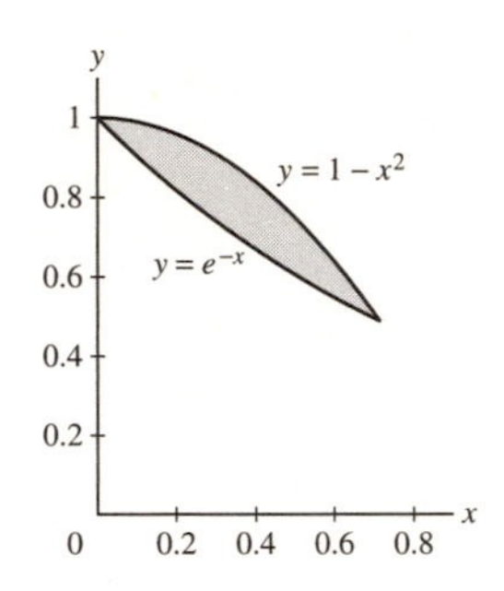

15. The base of a solid is the unit circle $x^2 + y^2 = 1$, and its cross sections perpendicular to the x-axis are rectangles of height 4. Find its volume.

SOLUTION Because the cross sections are rectangles of constant height 4, the figure is a cylinder of radius 1 and height 4. The volume is therefore $\pi r^2 h = 4\pi$.

17. Find the total mass of a rod of length 1.2 m with linear density $\rho(x) = (1 + 2x + \frac{2}{9}x^3)$ kg/m.

SOLUTION The total weight of the rod is

$$\int_0^{1.2} \rho(x)\, dx = \left(x + x^2 + \frac{1}{18}x^4\right)\Big|_0^{1.2} = 2.7552 \text{ kg}.$$

In Exercises 19–24, find the average value of the function over the interval.

19. $f(x) = x^3 - 2x + 2, \quad [-1, 2]$

SOLUTION The average value is

$$\frac{1}{2-(-1)} \int_{-1}^{2} \left(x^3 - 2x + 2\right) dx = \frac{1}{3}\left(\frac{1}{4}x^4 - x^2 + 2x\right)\Big|_{-1}^{2} = \frac{1}{3}\left[(4 - 4 + 4) - \left(\frac{1}{4} - 1 - 2\right)\right] = \frac{9}{4}.$$

21. $f(x) = x\cosh(x^2), \quad [0, 1]$

SOLUTION The average value is

$$\frac{1}{1-0} \int_0^1 x\cosh(x^2)\, dx.$$

To evaluate the integral, let $u = x^2$. Then $du = 2x\, dx$ and

$$\frac{1}{1-0} \int_0^1 x\cosh(x^2)\, dx = \frac{1}{2}\int_0^1 \cosh u\, du = \frac{1}{2}\sinh u\Big|_0^1 = \frac{1}{2}\sinh 1.$$

23. $f(x) = \sqrt{9 - x^2}$, $[0, 3]$ *Hint:* Use geometry to evaluate the integral.

SOLUTION The region below the graph of $y = \sqrt{9 - x^2}$ but above the x-axis over the interval $[0, 3]$ is one-quarter of a circle of radius 3; consequently,

$$\int_0^3 \sqrt{9 - x^2}\,dx = \frac{1}{4}\pi(3)^2 = \frac{9\pi}{4}.$$

The average value is then

$$\frac{1}{3-0}\int_0^3 \sqrt{9 - x^2}\,dx = \frac{1}{3}\left(\frac{9\pi}{4}\right) = \frac{3\pi}{4}.$$

25. Find $\int_2^5 g(t)\,dt$ if the average value of $g(t)$ on $[2, 5]$ is 9.

SOLUTION The average value of the function $g(t)$ on $[2, 5]$ is given by

$$\frac{1}{5-2}\int_2^5 g(t)\,dt = \frac{1}{3}\int_2^5 g(t)\,dt.$$

Therefore,

$$\int_2^5 g(t)\,dt = 3(\text{average value}) = 3(9) = 27.$$

27. Use the Washer Method to find the volume obtained by rotating the region in Figure 3 about the x-axis.

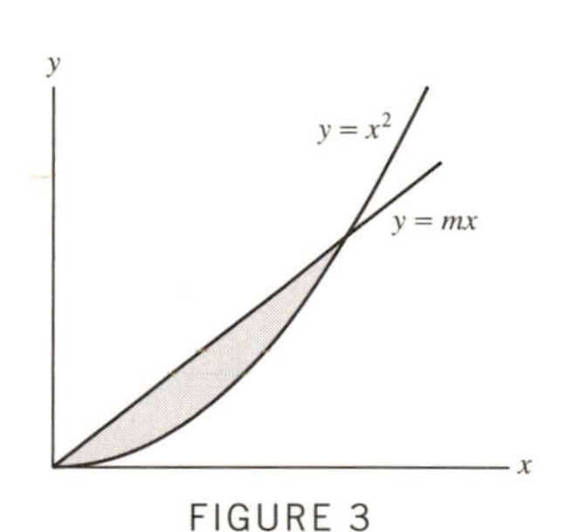

FIGURE 3

SOLUTION Setting $x^2 = mx$ yields $x(x - m) = 0$, so the two curves intersect at $(0, 0)$ and (m, m^2). To use the washer method, we must slice the solid perpendicular to the axis of rotation; as we are revolving about the y-axis, this implies a horizontal slice and integration in y. For each $y \in [0, m^2]$, the cross section is a washer with outer radius $R = \sqrt{y}$ and inner radius $r = \frac{y}{m}$. The volume of the solid is therefore given by

$$\pi\int_0^{m^2}\left((\sqrt{y})^2 - \left(\frac{y}{m}\right)^2\right)dy = \pi\left(\frac{1}{2}y^2 - \frac{y^3}{3m^2}\right)\Bigg|_0^{m^2} = \pi\left(\frac{m^4}{2} - \frac{m^4}{3}\right) = \frac{\pi}{6}m^4.$$

In Exercises 29–40, use any method to find the volume of the solid obtained by rotating the region enclosed by the curves about the given axis.

29. $y = x^2 + 2$, $y = x + 4$, x-axis

SOLUTION Let's choose to slice the region bounded by the graphs of $y = x^2 + 2$ and $y = x + 4$ (see the figure below) vertically. Because a vertical slice is perpendicular to the axis of rotation, we will use the washer method to calculate the volume of the solid of revolution. For each $x \in [-1, 2]$, the washer has outer radius $x + 4$ and inner radius $x^2 + 2$. The volume of the solid is therefore given by

$$\begin{aligned}\pi\int_{-1}^{2}((x+4)^2 - (x^2+2)^2)\,dx &= \pi\int_{-1}^{2}(-x^4 - 3x^2 + 8x + 12)\,dx\\ &= \pi\left(-\frac{1}{5}x^5 - x^3 + 4x^2 + 12x\right)\Bigg|_{-1}^{2}\\ &= \pi\left(\frac{128}{5} + \frac{34}{5}\right) = \frac{162\pi}{5}.\end{aligned}$$

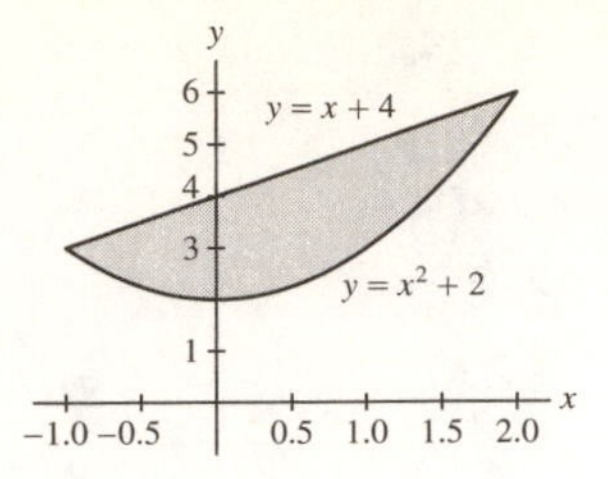

31. $x = y^2 - 3, \quad x = 2y, \quad \text{axis } y = 4$

SOLUTION Let's choose to slice the region bounded by the graphs of $x = y^2 - 3$ and $x = 2y$ (see the figure below) horizontally. Because a horizontal slice is parallel to the axis of rotation, we will use the shell method to calculate the volume of the solid of revolution. For each $y \in [-1, 3]$, the shell has radius $4 - y$ and height $2y - (y^2 - 3) = 3 + 2y - y^2$. The volume of the solid is therefore given by

$$2\pi \int_{-1}^{3} (4 - y)(3 + 2y - y^2)\, dy = 2\pi \int_{-1}^{3} (12 + 5y - 6y^2 + y^3)\, dy$$

$$= 2\pi \left(12y + \frac{5}{2}y^2 - 2y^3 + \frac{1}{4}y^4\right)\Big|_{-1}^{3}$$

$$= 2\pi \left(\frac{99}{4} + \frac{29}{4}\right) = 64\pi.$$

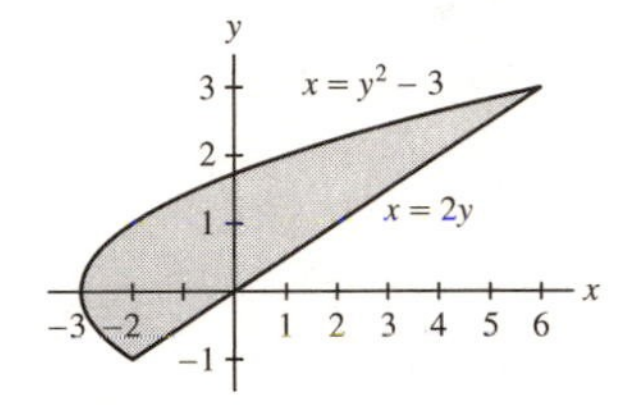

33. $y = x^2 - 1, \quad y = 2x - 1, \quad \text{axis } x = -2$

SOLUTION The region bounded by the graphs of $y = x^2 - 1$ and $y = 2x - 1$ is shown below. Let's choose to slice the region vertically. Because a vertical slice is parallel to the axis of rotation, we will use the shell method to calculate the volume of the solid of revolution. For each $x \in [0, 2]$, the shell has radius $x - (-2) = x + 2$ and height $(2x - 1) - (x^2 - 1) = 2x - x^2$. The volume of the solid is therefore given by

$$2\pi \int_{0}^{2} (x + 2)(2x - x^2)\, dx = 2\pi \left(2x^2 - \frac{1}{4}x^4\right)\Big|_{0}^{2} = 2\pi(8 - 4) = 8\pi.$$

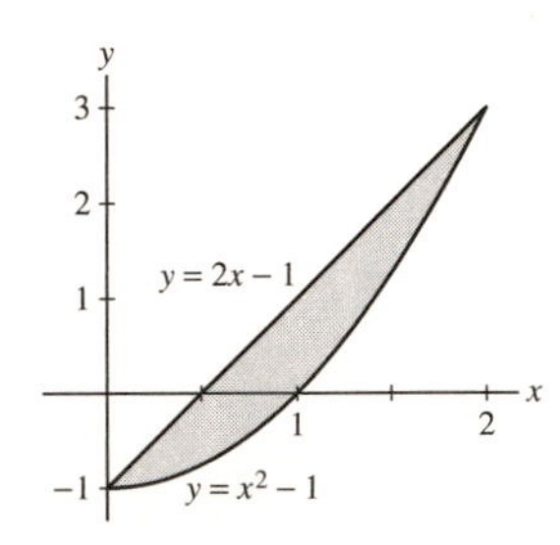

35. $y = -x^2 + 4x - 3, \quad y = 0, \quad \text{axis } y = -1$

SOLUTION The region bounded by the graph of $y = -x^2 + 4x - 3$ and the x-axis is shown below. Let's choose to slice the region vertically. Because a vertical slice is perpendicular to the axis of rotation, we will use the washer method to calculate the volume of the solid of revolution. For each $x \in [1, 3]$, the cross section is a washer with outer radius $R = -x^2 + 4x - 3 - (-1) = -x^2 + 4x - 2$ and inner radius $r = 0 - (-1) = 1$. The volume of the solid is therefore given by

$$\pi \int_{1}^{3} \left((-x^2 + 4x - 2)^2 - 1\right) dx = \pi \left(\frac{1}{5}x^5 - 2x^4 + \frac{20}{3}x^3 - 8x^2 + 3x\right)\Big|_{1}^{3}$$

$$= \pi \left[\left(\frac{243}{5} - 162 + 180 - 72 + 9\right) - \left(\frac{1}{5} - 2 + \frac{20}{3} - 8 + 3\right)\right] = \frac{56\pi}{15}.$$

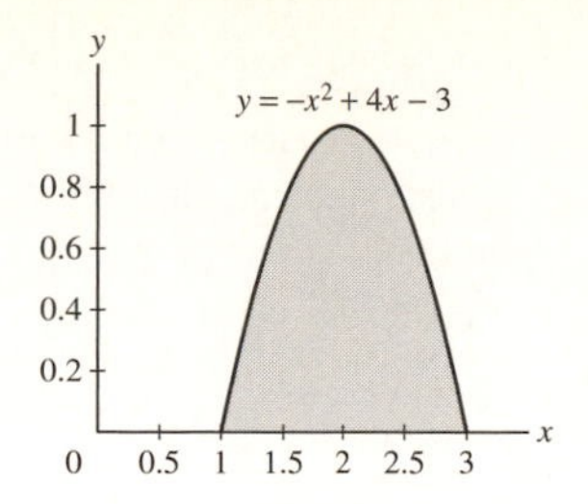

37. $x = 4y - y^3, \quad x = 0, \quad y \geq 0, \quad x\text{-axis}$

SOLUTION The region bounded by the graphs of $x = 4y - y^3$ and $x = 0$ for $y \geq 0$ is shown below. Let's choose to slice this region horizontally. Because a horizontal slice is parallel to the axis of rotation, we will use the shell method to calculate the volume of the solid of revolution. For each $y \in [0, 2]$, the shell has radius y and height $4y - y^3$. The volume of the solid is therefore given by

$$\begin{aligned} 2\pi \int_0^2 y(4y - y^3)\, dy &= 2\pi \int_0^2 (4y^2 - y^4)\, dy \\ &= 2\pi \left(\frac{4}{3}y^3 - \frac{1}{5}y^5 \right)\Big|_0^2 \\ &= 2\pi \left(\frac{32}{3} - \frac{32}{5} \right) = \frac{128\pi}{15}. \end{aligned}$$

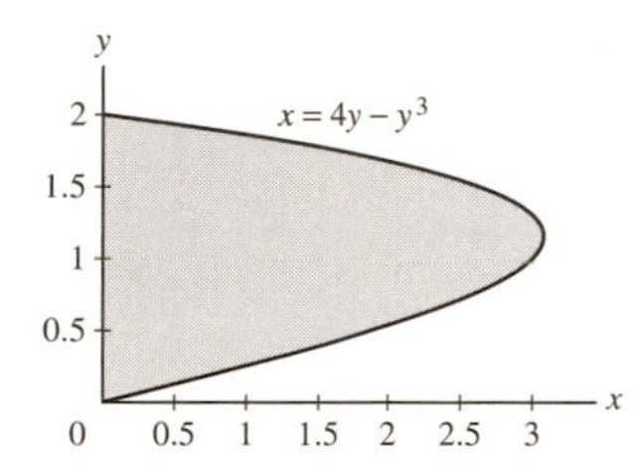

39. $y = e^{-x^2/2}, \quad y = -e^{-x^2/2}, \quad x = 0, \quad x = 1, \quad y\text{-axis}$

SOLUTION Let's choose to slice the region bounded by the graphs of $y = e^{-x^2/2}$ and $y = -e^{-x^2/2}$ (see the figure below) vertically. Because a vertical slice is parallel to the axis of rotation, we will use the shell method to calculate the volume of the solid of revolution. For each $x \in [0, 1]$, the shell has radius x and height $e^{-x^2/2} - (-e^{-x^2/2}) = 2e^{-x^2/2}$. The volume of the solid is therefore given by

$$\begin{aligned} 2\pi \int_0^1 2xe^{-x^2/2}\, dx &= -4\pi e^{-x^2/2}\Big|_0^1 \\ &= -4\pi(e^{-1/2} - 1) = 4\pi(1 - e^{-1/2}). \end{aligned}$$

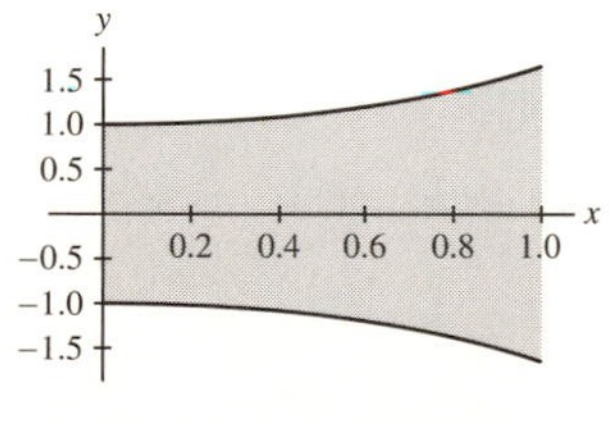

In Exercises 41–44, find the volume obtained by rotating the region about the given axis. The regions refer to the graph of the hyperbola $y^2 - x^2 = 1$ in Figure 4.

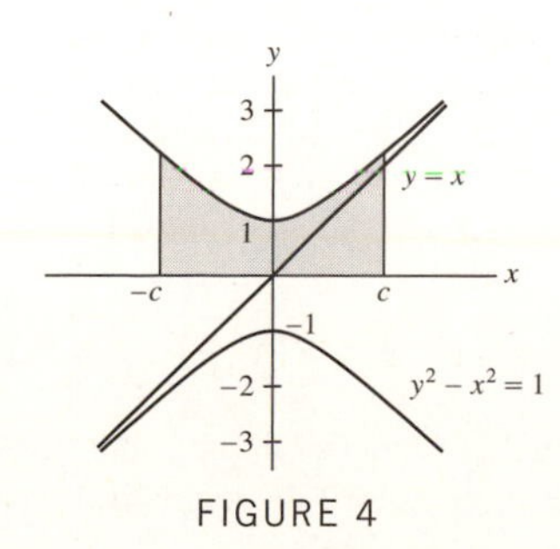

FIGURE 4

41. The shaded region between the upper branch of the hyperbola and the x-axis for $-c \le x \le c$, about the x-axis.

SOLUTION Let's choose to slice the region vertically. Because a vertical slice is perpendicular to the axis of rotation, we will use the washer method to calculate the volume of the solid of revolution. For each $x \in [-c, c]$, cross sections are circular disks with radius $R = \sqrt{1+x^2}$. The volume of the solid is therefore given by

$$\pi \int_{-c}^{c} (1+x^2)\,dx = \pi \left(x + \frac{1}{3}x^3\right)\Bigg|_{-c}^{c} = \pi \left[\left(c + \frac{c^3}{3}\right) - \left(-c - \frac{c^3}{3}\right)\right] = 2\pi \left(c + \frac{c^3}{3}\right).$$

43. The region between the upper branch of the hyperbola and the line $y = x$ for $0 \le x \le c$, about the x-axis.

SOLUTION Let's choose to slice the region vertically. Because a vertical slice is perpendicular to the axis of rotation, we will use the washer method to calculate the volume of the solid of revolution. For each $x \in [0, c]$, cross sections are washers with outer radius $R = \sqrt{1+x^2}$ and inner radius $r = x$. The volume of the solid is therefore given by

$$\pi \int_0^c \left((1+x^2) - x^2\right) dx = \pi x \Big|_0^c = c\pi.$$

45. Let R be the intersection of the circles of radius 1 centered at $(1, 0)$ and $(0, 1)$. Express as an integral (but do not evaluate): **(a)** the area of R and **(b)** the volume of revolution of R about the x-axis.

SOLUTION The region R is shown below.

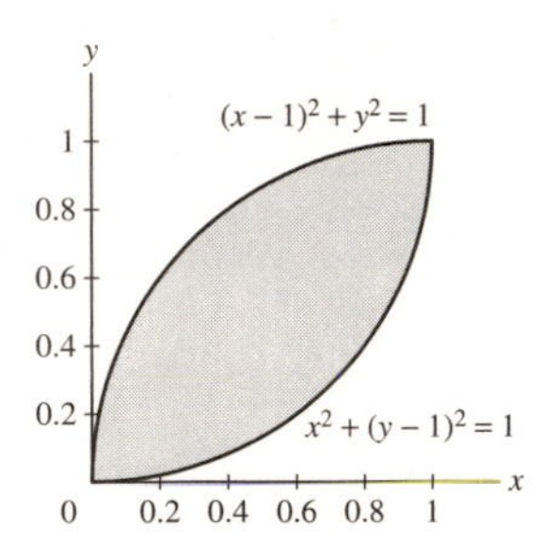

(a) A vertical slice of R has its top along the upper left arc of the circle $(x-1)^2 + y^2 = 1$ and its bottom along the lower right arc of the circle $x^2 + (y-1)^2 = 1$. The area of R is therefore given by

$$\int_0^1 \left(\sqrt{1-(x-1)^2} - (1 - \sqrt{1-x^2})\right) dx.$$

(b) If we revolve R about the x-axis and use the washer method, each cross section is a washer with outer radius $\sqrt{1-(x-1)^2}$ and inner radius $1 - \sqrt{1-x^2}$. The volume of the solid is therefore given by

$$\pi \int_0^1 \left[(1-(x-1)^2) - (1-\sqrt{1-x^2})^2\right] dx.$$

47. If 12 J of work are needed to stretch a spring 20 cm beyond equilibrium, how much work is required to compress it 6 cm beyond equilibrium?

SOLUTION First, we determine the value of the spring constant k as follows:

$$\frac{1}{2}k(0.2)^2 = 12 \quad \text{so} \quad k = 600 \text{ N/m}.$$

Now, the work needed to compress the spring 6 cm beyond equilibrium is

$$W = \int_0^{0.06} 600x\,dx = 300x^2\Big|_0^{0.06} = 1.08 \text{ J}.$$

49. If 18 ft-lb of work are needed to stretch a spring 1.5 ft beyond equilibrium, how far will the spring stretch if a 12-lb weight is attached to its end?

SOLUTION First, we determine the value of the spring constant as follows:

$$\frac{1}{2}k(1.5)^2 = 18 \quad \text{so} \quad k = 16 \text{ lb/ft}.$$

Now, if a 12-lb weight is attached to the end of the spring, balancing the forces acting on the weight, we have $12 = 16d$, which implies $d = 0.75$ ft. A 12-lb weight will therefore stretch the spring 9 inches.

In Exercises 51 and 52, water is pumped into a spherical tank of radius 2 m *from a source located* 1 m *below a hole at the bottom (Figure 5). The density of water is* 1000 kg/m^3.

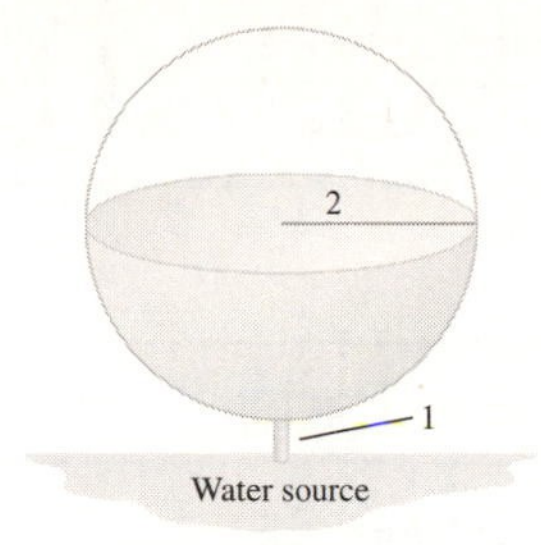

FIGURE 5

51. Calculate the work required to fill the tank.

SOLUTION Place the origin at the base of the sphere with the positive y-axis pointing upward. The equation for the great circle of the sphere is then $x^2 + (y-2)^2 = 4$. At location y, the horizontal cross section is a circle of radius $\sqrt{4-(y-2)^2} = \sqrt{4y-y^2}$; the volume of the layer is then $\pi(4y-y^2)\Delta y$ m^3, and the force needed to lift the layer is $1000(9.8)\pi(4y-y^2)\Delta y$ N. The layer of water must be lifted $y+1$ meters, so the work required to fill the tank is given by

$$\begin{aligned} 9800\pi \int_0^4 (y+1)(4y-y^2)\,dy &= 9800\pi \int_0^4 (3y^2+4y-y^3)\,dy \\ &= 9800\pi \left(y^3 + 2y^2 - \frac{1}{4}y^4 \right)\Bigg|_0^4 \\ &= 313{,}600\pi \approx 985{,}203.5 \text{ J.} \end{aligned}$$

53. A tank of mass 20 kg containing 100 kg of water (density 1000 kg/m^3) is raised vertically at a constant speed of 100 m/min for one minute, during which time it leaks water at a rate of 40 kg/min. Calculate the total work performed in raising the container.

SOLUTION Let t denote the elapsed time in minutes and let y denote the height of the container. Given that the speed of ascent is 100 m/min, $y = 100t$; moreover, the mass of water in the container is

$$100 - 40t = 100 - 0.4y\text{kg}.$$

The force needed to lift the container and its contents is then

$$9.8\,(20 + (100 - 0.4y)) = 1176 - 3.92y \text{ N},$$

and the work required to lift the container and its contents is

$$\int_0^{100} (1176 - 3.92y)\,dy = (1176y - 1.96y^2)\Bigg|_0^{100} = 98{,}000\text{J}.$$

7 TECHNIQUES OF INTEGRATION

7.1 Integration by Parts

Preliminary Questions

1. Which derivative rule is used to derive the Integration by Parts formula?

SOLUTION The Integration by Parts formula is derived from the Product Rule.

2. For each of the following integrals, state whether substitution or Integration by Parts should be used:

$$\int x\cos(x^2)\,dx, \qquad \int x\cos x\,dx, \qquad \int x^2e^x\,dx, \qquad \int xe^{x^2}\,dx$$

SOLUTION

(a) $\int x\cos(x^2)\,dx$: use the substitution $u = x^2$.

(b) $\int x\cos x\,dx$: use Integration by Parts.

(c) $\int x^2e^x\,dx$; use Integration by Parts.

(d) $\int xe^{x^2}\,dx$; use the substitution $u = x^2$.

3. Why is $u = \cos x$, $v' = x$ a poor choice for evaluating $\int x\cos x\,dx$?

SOLUTION Transforming $v' = x$ into $v = \frac{1}{2}x^2$ increases the power of x and makes the new integral harder than the original.

Exercises

In Exercises 1–6, evaluate the integral using the Integration by Parts formula with the given choice of u and v'.

1. $\int x\sin x\,dx; \quad u = x,\ v' = \sin x$

SOLUTION Using the given choice of u and v' results in

$$\begin{aligned} u &= x & v &= -\cos x \\ u' &= 1 & v' &= \sin x \end{aligned}$$

Using Integration by Parts,

$$\int x\sin x\,dx = x(-\cos x) - \int (1)(-\cos x)\,dx = -x\cos x + \int \cos x\,dx = -x\cos x + \sin x + C.$$

3. $\int (2x+9)e^x\,dx; \quad u = 2x+9,\ v' = e^x$

SOLUTION Using $u = 2x+9$ and $v' = e^x$ gives us

$$\begin{aligned} u &= 2x+9 & v &= e^x \\ u' &= 2 & v' &= e^x \end{aligned}$$

Integration by Parts gives us

$$\int (2x+9)e^x\,dx = (2x+9)e^x - \int 2e^x\,dx = (2x+9)e^x - 2e^x + C = e^x(2x+7) + C.$$

5. $\int x^3 \ln x\, dx; \quad u = \ln x,\ v' = x^3$

SOLUTION Using $u = \ln x$ and $v' = x^3$ gives us

$$u = \ln x \qquad v = \tfrac{1}{4}x^4$$
$$u' = \tfrac{1}{x} \qquad v' = x^3$$

Integration by Parts gives us

$$\int x^3 \ln x\, dx = (\ln x)\left(\frac{1}{4}x^4\right) - \int \left(\frac{1}{x}\right)\left(\frac{1}{4}x^4\right) dx$$
$$= \frac{1}{4}x^4 \ln x - \frac{1}{4}\int x^3\, dx = \frac{1}{4}x^4 \ln x - \frac{1}{16}x^4 + C = \frac{x^4}{16}(4\ln x - 1) + C.$$

In Exercises 7–36, evaluate using Integration by Parts.

7. $\int (4x - 3)e^{-x}\, dx$

SOLUTION Let $u = 4x - 3$ and $v' = e^{-x}$. Then we have

$$u = 4x - 3 \qquad v = -e^{-x}$$
$$u' = 4 \qquad v' = e^{-x}$$

Using Integration by Parts, we get

$$\int (4x - 3)e^{-x}\, dx = (4x - 3)(-e^{-x}) - \int (4)(-e^{-x})\, dx$$
$$= -e^{-x}(4x - 3) + 4\int e^{-x}\, dx = -e^{-x}(4x - 3) - 4e^{-x} + C = -e^{-x}(4x + 1) + C.$$

9. $\int x\, e^{5x+2}\, dx$

SOLUTION Let $u = x$ and $v' = e^{5x+2}$. Then we have

$$u = x \qquad v = \frac{1}{5}e^{5x+2}$$
$$u' = 1 \qquad v' = e^{5x+2}$$

Using Integration by Parts, we get

$$\int xe^{5x+2}\, dx = x\left(\frac{1}{5}e^{5x+2}\right) - \int (1)\left(\frac{1}{5}e^{5x+2}\right) dx = \frac{1}{5}xe^{5x+2} - \frac{1}{5}\int e^{5x+2}\, dx$$
$$= \frac{1}{5}xe^{5x+2} - \frac{1}{25}e^{5x+2} + C = \left(\frac{x}{5} - \frac{1}{25}\right)e^{5x+2} + C$$

11. $\int x \cos 2x\, dx$

SOLUTION Let $u = x$ and $v' = \cos 2x$. Then we have

$$u = x \qquad v = \tfrac{1}{2}\sin 2x$$
$$u' = 1 \qquad v' = \cos 2x$$

Using Integration by Parts, we get

$$\int x \cos 2x\, dx = x\left(\frac{1}{2}\sin 2x\right) - \int (1)\left(\frac{1}{2}\sin 2x\right) dx$$
$$= \frac{1}{2}x\sin 2x - \frac{1}{2}\int \sin 2x\, dx = \frac{1}{2}x\sin 2x + \frac{1}{4}\cos 2x + C.$$

13. $\int x^2 \sin x \, dx$

SOLUTION Let $u = x^2$ and $v' = \sin x$. Then we have

$$u = x^2 \qquad v = -\cos x$$
$$u' = 2x \qquad v' = \sin x$$

Using Integration by Parts, we get

$$\int x^2 \sin x \, dx = x^2(-\cos x) - \int 2x(-\cos x)\, dx = -x^2 \cos x + 2\int x \cos x \, dx.$$

We must apply Integration by Parts again to evaluate $\int x \cos x \, dx$. Taking $u = x$ and $v' = \cos x$, we get

$$\int x \cos x \, dx = x \sin x - \int \sin x \, dx = x \sin x + \cos x + C.$$

Plugging this into the original equation gives us

$$\int x^2 \sin x \, dx = -x^2 \cos x + 2(x \sin x + \cos x) + C = -x^2 \cos x + 2x \sin x + 2\cos x + C.$$

15. $\int e^{-x} \sin x \, dx$

SOLUTION Let $u = e^{-x}$ and $v' = \sin x$. Then we have

$$u = e^{-x} \qquad v = -\cos x$$
$$u' = -e^{-x} \qquad v' = \sin x$$

Using Integration by Parts, we get

$$\int e^{-x} \sin x \, dx = -e^{-x} \cos x - \int (-e^{-x})(-\cos x)\, dx = -e^{-x} \cos x - \int e^{-x} \cos x \, dx.$$

We must apply Integration by Parts again to evaluate $\int e^{-x} \cos x \, dx$. Using $u = e^{-x}$ and $v' = \cos x$, we get

$$\int e^{-x} \cos x \, dx = e^{-x} \sin x - \int (-e^{-x})(\sin x)\, dx = e^{-x} \sin x + \int e^{-x} \sin x \, dx.$$

Plugging this into the original equation, we get

$$\int e^{-x} \sin x \, dx = -e^{-x} \cos x - \left[e^{-x} \sin x + \int e^{-x} \sin x \, dx \right].$$

Solving this equation for $\int e^{-x} \sin x \, dx$ gives us

$$\int e^{-x} \sin x \, dx = -\frac{1}{2} e^{-x}(\sin x + \cos x) + C.$$

17. $\int e^{-5x} \sin x \, dx$

SOLUTION Let $u = \sin x$ and $v' = e^{-5x}$. Then we have

$$u = \sin x \qquad v = -\frac{1}{5} e^{-5x}$$
$$u' = \cos x \qquad v' = e^{-5x}$$

Using Integration by Parts, we get

$$\int e^{-5x} \sin x \, dx = -\frac{1}{5} e^{-5x} \sin x - \int \cos x \left(-\frac{1}{5} e^{-5x} \right) dx = -\frac{1}{5} e^{-5x} \sin x + \frac{1}{5} \int e^{-5x} \cos x \, dx$$

Apply Integration by Parts again to this integral, with $u = \cos x$ and $v' = e^{-5x}$ to get

$$\int e^{-5x} \cos x \, dx = -\frac{1}{5} e^{-5x} \cos x - \frac{1}{5} \int e^{-5x} \sin x \, dx$$

Plugging this into the original equation, we get

$$\int e^{-5x}\sin x\,dx = -\frac{1}{5}e^{-5x}\sin x + \frac{1}{5}\left(-\frac{1}{5}e^{-5x}\cos x - \frac{1}{5}\int e^{-5x}\sin x\,dx\right)$$
$$= -\frac{1}{5}e^{-5x}\sin x - \frac{1}{25}e^{-5x}\cos x - \frac{1}{25}\int e^{-5x}\sin x\,dx$$

Solving this equation for $\int e^{-5x}\sin x\,dx$ gives us

$$\int e^{-5x}\sin x\,dx = -\frac{5}{26}e^{-5x}\sin x - \frac{1}{26}e^{-5x}\cos x + C = -\frac{1}{26}e^{-5x}(5\sin x + \cos x) + C$$

19. $\int x\ln x\,dx$

SOLUTION Let $u = \ln x$ and $v' = x$. Then we have

$$u = \ln x \qquad v = \tfrac{1}{2}x^2$$
$$u' = \tfrac{1}{x} \qquad v' = x$$

Using Integration by Parts, we get

$$\int x\ln x\,dx = \frac{1}{2}x^2\ln x - \int\left(\frac{1}{x}\right)\left(\frac{1}{2}x^2\right)dx$$
$$= \frac{1}{2}x^2\ln x - \frac{1}{2}\int x\,dx = \frac{1}{2}x^2\ln x - \frac{1}{2}\left(\frac{x^2}{2}\right) + C = \frac{1}{4}x^2(2\ln x - 1) + C.$$

21. $\int x^2\ln x\,dx$

SOLUTION Let $u = \ln x$ and $v' = x^2$. Then we have

$$u = \ln x \qquad v = \tfrac{1}{3}x^3$$
$$u' = \tfrac{1}{x} \qquad v' = x^2$$

Using Integration by Parts, we get

$$\int x^2\ln x\,dx = \frac{1}{3}x^3\ln x - \int\frac{1}{x}\left(\frac{1}{3}x^3\right)dx = \frac{1}{3}x^3\ln x - \frac{1}{3}\int x^2\,dx$$
$$= \frac{1}{3}x^3\ln x - \frac{1}{3}\left(\frac{x^3}{3}\right) + C = \frac{x^3}{3}\left(\ln x - \frac{1}{3}\right) + C.$$

23. $\int(\ln x)^2\,dx$

SOLUTION Let $u = (\ln x)^2$ and $v' = 1$. Then we have

$$u = (\ln x)^2 \qquad v = x$$
$$u' = \frac{2}{x}\ln x \qquad v' = 1$$

Using Integration by Parts, we get

$$\int(\ln x)^2\,dx = (\ln x)^2(x) - \int\left(\frac{2}{x}\ln x\right)x\,dx = x(\ln x)^2 - 2\int\ln x\,dx.$$

We must apply Integration by Parts again to evaluate $\int\ln x\,dx$. Using $u = \ln x$ and $v' = 1$, we have

$$\int\ln x\,dx = x\ln x - \int\frac{1}{x}\cdot x\,dx = x\ln x - \int dx = x\ln x - x + C.$$

Plugging this into the original equation, we get

$$\int(\ln x)^2\,dx = x(\ln x)^2 - 2(x\ln x - x) + C = x\left[(\ln x)^2 - 2\ln x + 2\right] + C.$$

25. $\int x \sec^2 x \, dx$

SOLUTION Let $u = x$ and $v' = \sec^2 x$. Then we have

$$u = x \qquad v = \tan x$$
$$u' = 1 \qquad v' = \sec^2 x$$

Using Integration by Parts, we get

$$\int x \sec^2 x \, dx = x \tan x - \int (1) \tan x \, dx = x \tan x - \ln |\sec x| + C.$$

27. $\int \cos^{-1} x \, dx$

SOLUTION Let $u = \cos^{-1} x$ and $v' = 1$. Then we have

$$u = \cos^{-1} x \qquad v = x$$
$$u' = \frac{-1}{\sqrt{1-x^2}} \qquad v' = 1$$

Using Integration by Parts, we get

$$\int \cos^{-1} x \, dx = x \cos^{-1} x - \int \frac{-x}{\sqrt{1-x^2}} \, dx.$$

We can evaluate $\int \frac{-x}{\sqrt{1-x^2}} \, dx$ by making the substitution $w = 1 - x^2$. Then $dw = -2x \, dx$, and we have

$$\int \cos^{-1} x \, dx = x \cos^{-1} x - \frac{1}{2} \int \frac{-2x \, dx}{\sqrt{1-x^2}} = x \cos^{-1} x - \frac{1}{2} \int w^{-1/2} \, dw$$
$$= x \cos^{-1} x - \frac{1}{2}(2w^{1/2}) + C = x \cos^{-1} x - \sqrt{1-x^2} + C.$$

29. $\int \sec^{-1} x \, dx$

SOLUTION We are forced to choose $u = \sec^{-1} x$, $v' = 1$, so that $u' = \frac{1}{x\sqrt{x^2-1}}$ and $v = x$. Using Integration by parts, we get:

$$\int \sec^{-1} x \, dx = x \sec^{-1} x - \int \frac{x \, dx}{x\sqrt{x^2-1}} = x \sec^{-1} x - \int \frac{dx}{\sqrt{x^2-1}}.$$

Via the substitution $\sqrt{x^2-1} = \tan\theta$ (so that $x = \sec\theta$ and $dx = \sec\theta \tan\theta \, d\theta$), we get:

$$\int \sec^{-1} x \, dx = x \sec^{-1} x - \int \frac{\sec\theta \tan\theta \, d\theta}{\tan\theta} = x \sec^{-1} x - \int \sec\theta \, d\theta$$
$$= x \sec^{-1} x - \ln|\sec\theta + \tan\theta| + C = x \sec^{-1} x - \ln|x + \sqrt{x^2-1}| + C.$$

31. $\int 3^x \cos x \, dx$

SOLUTION Let $u = \cos x$ and $v' = 3^x$. Then we have

$$u = \cos x \qquad v = \frac{3^x}{\ln 3}$$
$$u' = -\sin x \qquad v' = 3^x$$

Using Integration by Parts, we get

$$\int 3^x \cos x \, dx = \frac{3^x}{\ln 3} \cos x + \frac{1}{\ln 3} \int 3^x \sin x \, dx$$

Apply Integration by Parts to the remaining integral, with $u = \sin x$ and $v' = 3^x$; then

$$\int 3^x \sin x \, dx = \frac{3^x}{\ln 3} \sin x - \frac{1}{\ln 3} \int 3^x \cos x \, dx$$

Plug this into the first equation to get

$$\int 3^x \cos x\,dx = \frac{3^x}{\ln 3}\cos x + \frac{1}{\ln 3}\left(\frac{3^x}{\ln 3}\sin x - \frac{1}{\ln 3}\int 3^x \cos x\,dx\right)$$

$$= \frac{3^x}{\ln 3}\cos x + \frac{3^x}{(\ln 3)^2}\sin x - \frac{1}{(\ln 3)^2}\int 3^x \cos x\,dx$$

Solving for $\int 3^x \cos x\,dx$ gives

$$\int 3^x \cos x\,dx = \frac{3^x \ln 3 \cos x}{1+(\ln 3)^2} + \frac{3^x \sin x}{1+(\ln 3)^2} + C = \frac{3^x}{1+(\ln 3)^2}(\ln 3 \cos x + \sin x) + C$$

33. $\int x^2 \cosh x\,dx$

SOLUTION Let $u = x^2$, $v' = \cosh x$. Then

$$u = x^2 \qquad v = \sinh x$$
$$u' = 2x \qquad v' = \cosh x$$

Integration by Parts gives us (along with Exercise 32)

$$\int x^2 \cosh x\,dx = x^2 \sinh x - 2\int x \sinh x,\,dx = x^2 \sinh x - 2x\cosh x + 2\sinh x + C$$

35. $\int \tanh^{-1} 4x\,dx$

SOLUTION Using $u = \tanh^{-1} 4x$ and $v' = 1$ gives us

$$u = \tanh^{-1} 4x \qquad v = x$$
$$u' = \frac{4}{1-16x^2} \qquad v' = 1$$

Integration by Parts gives us

$$\int \tanh^{-1} 4x\,dx = x\tanh^{-1} 4x - \int \left(\frac{4}{1-16x^2}\right) x\,dx.$$

For the integral on the right we'll use the substitution $w = 1 - 16x^2$, $dw = -32x\,dx$. Then we have

$$\int \tanh^{-1} 4x\,dx = x\tanh^{-1} 4x + \frac{1}{8}\int \frac{dw}{w} = x\tanh^{-1} 4x + \frac{1}{8}\ln|w| + C$$

$$= x\tanh^{-1} 4x + \frac{1}{8}\ln|1-16x^2| + C.$$

In Exercises 37 and 38, evaluate using substitution and then Integration by Parts.

37. $\int e^{\sqrt{x}}\,dx$ *Hint:* Let $u = x^{1/2}$

SOLUTION Let $w = x^{1/2}$. Then $dw = \frac{1}{2}x^{-1/2}dx$, or $dx = 2x^{1/2}\,dw = 2w\,dw$. Now,

$$\int e^{\sqrt{x}}\,dx = 2\int we^w\,dw.$$

Using Integration by Parts with $u = w$ and $v' = e^w$, we get

$$2\int we^w\,dw = 2(we^w - e^w) + C.$$

Substituting back, we find

$$\int e^{\sqrt{x}}\,dx = 2e^{\sqrt{x}}(\sqrt{x} - 1) + C.$$

In Exercises 39–48, evaluate using Integration by Parts, substitution, or both if necessary.

39. $\int x \cos 4x \, dx$

SOLUTION Let $u = x$ and $v' = \cos 4x$. Then we have

$$u = x \qquad v = \tfrac{1}{4}\sin 4x$$
$$u' = 1 \qquad v' = \cos 4x$$

Using Integration by Parts, we get

$$\int x \cos 4x \, dx = \frac{1}{4}x \sin 4x - \int (1)\frac{1}{4}\sin 4x \, dx = \frac{1}{4}x \sin 4x - \frac{1}{4}\left(-\frac{1}{4}\cos 4x\right) + C$$
$$= \frac{1}{4}x \sin 4x + \frac{1}{16}\cos 4x + C.$$

41. $\int \frac{x \, dx}{\sqrt{x+1}}$

SOLUTION Let $u = x + 1$. Then $du = dx$, $x = u - 1$, and

$$\int \frac{x \, dx}{\sqrt{x+1}} = \int \frac{(u-1)\, du}{\sqrt{u}} = \int \left(\frac{u}{\sqrt{u}} - \frac{1}{\sqrt{u}}\right) du = \int (u^{1/2} - u^{-1/2})\, du$$
$$= \frac{2}{3}u^{3/2} - 2u^{1/2} + C = \frac{2}{3}(x+1)^{3/2} - 2(x+1)^{1/2} + C.$$

43. $\int \cos x \ln(\sin x)\, dx$

SOLUTION Let $w = \sin x$. Then $dw = \cos x \, dx$, and

$$\int \cos x \, \ln(\sin x)\, dx = \int \ln w \, dw.$$

Now use Integration by Parts with $u = \ln w$ and $v' = 1$. Then $u' = 1/w$ and $v = w$, which gives us

$$\int \cos x \, \ln(\sin x)\, dx = \int \ln w \, dw = w \ln w - w + C = \sin x \ln(\sin x) - \sin x + C.$$

45. $\int \sqrt{x} e^{\sqrt{x}} \, dx$

SOLUTION Let $w = \sqrt{x}$. Then $dw = \frac{1}{2\sqrt{x}}\, dx$ and

$$\int \sqrt{x} e^{\sqrt{x}} \, dx = 2\int w^2 e^w \, dw.$$

Now, use Integration by Parts with $u = w^2$ and $v' = e^w$. This gives

$$\int \sqrt{x} e^{\sqrt{x}} \, dx = 2\int w^2 e^w \, dw = 2w^2 e^w - 4\int w e^w \, dw.$$

We need to use Integration by Parts again, this time with $u = w$ and $v' = e^w$. We find

$$\int w e^w \, dw = w e^w - \int e^w \, dw = w e^w - e^w + C;$$

finally,

$$\int \sqrt{x} e^{\sqrt{x}} \, dx = 2w^2 e^w - 4w e^w + 4e^w + C = 2x e^{\sqrt{x}} - 4\sqrt{x} e^{\sqrt{x}} + 4e^{\sqrt{x}} + C.$$

47. $\int \frac{\ln(\ln x)\ln x \, dx}{x}$

SOLUTION Let $w = \ln x$. Then $dw = dx/x$, and

$$\int \frac{\ln(\ln x)\ln x \, dx}{x} = \int w \ln w \, dw.$$

Now use Integration by Parts, with $u = \ln w$ and $v' = w$. Then,

$$u = \ln w \qquad v = \frac{1}{2}w^2$$
$$u' = w^{-1} \qquad v' = w$$

and

$$\int \frac{\ln(\ln x)\ln x\,dx}{x} = \frac{1}{2}w^2\ln w - \frac{1}{2}\int w\,dw = \frac{1}{2}w^2\ln w - \frac{1}{2}\left(\frac{w^2}{2}\right) + C$$
$$= \frac{1}{2}(\ln x)^2\ln(\ln x) - \frac{1}{4}(\ln x)^2 + C = \frac{1}{4}(\ln x)^2[2\ln(\ln x) - 1] + C.$$

In Exercises 49–54, compute the definite integral.

49. $\displaystyle\int_0^3 xe^{4x}\,dx$

SOLUTION Let $u = x$, $v' = e^{4x}$. Then $u' = 1$ and $v = \frac{1}{4}e^{4x}$. Using Integration by Parts,

$$\int_0^3 xe^{4x}\,dx = \left(\frac{1}{4}xe^{4x}\right)\Big|_0^3 - \frac{1}{4}\int_0^3 e^{4x}\,dx = \frac{3}{4}e^{12} - \frac{1}{16}e^{12} + \frac{1}{16} = \frac{11}{16}e^{12} + \frac{1}{16}$$

51. $\displaystyle\int_1^2 x\ln x\,dx$

SOLUTION Let $u = \ln x$ and $v' = x$. Then $u' = \frac{1}{x}$ and $v = \frac{1}{2}x^2$. Using Integration by Parts gives

$$\int_1^2 x\ln x\,dx = \left(\frac{1}{2}x^2\ln x\right)\Big|_1^2 - \frac{1}{2}\int_1^2 x\,dx = 2\ln 2 - \frac{1}{4}x^2\Big|_1^2 = 2\ln 2 - \frac{3}{4}$$

53. $\displaystyle\int_0^\pi e^x\sin x\,dx$

SOLUTION Let $u = \sin x$ and $v' = e^x$; then $u' = \cos x$ and $v = e^x$. Integration by Parts gives

$$\int_0^\pi e^x\sin x\,dx = e^x\sin x\Big|_0^\pi - \int_0^\pi e^x\cos x\,dx = -\int_0^\pi e^x\cos x\,dx$$

Apply integration by parts again to this integral, with $u = \cos x$ and $v' = e^x$; then $u' = -\sin x$ and $v = e^x$, so we get

$$\int_0^\pi e^x\sin x\,dx = -\left((e^x\cos x)\Big|_0^\pi + \int_0^\pi e^x\sin x\,dx\right) = e^\pi + 1 - \int_0^\pi e^x\sin x\,dx$$

Solving for $\displaystyle\int_0^\pi e^x\sin x\,dx$ gives

$$\int_0^\pi e^x\sin x\,dx = \frac{e^\pi + 1}{2}$$

55. Use Eq. (5) to evaluate $\displaystyle\int x^4e^x\,dx$.

SOLUTION

$$\int x^4e^x\,dx = x^4e^x - 4\int x^3e^x\,dx = x^4e^x - 4\left[x^3e^x - 3\int x^2e^x\,dx\right]$$
$$= x^4e^x - 4x^3e^x + 12\int x^2e^x\,dx = x^4e^x - 4x^3e^x + 12\left[x^2e^x - 2\int xe^x\,dx\right]$$
$$= x^4e^x - 4x^3e^x + 12x^2e^x - 24\int xe^x\,dx = x^4e^x - 4x^3e^x + 12x^2e^x - 24\left[xe^x - \int e^x\,dx\right]$$
$$= x^4e^x - 4x^3e^x + 12x^2e^x - 24\left[xe^x - e^x\right] + C.$$

Thus,

$$\int x^4e^x\,dx = e^x(x^4 - 4x^3 + 12x^2 - 24x + 24) + C.$$

57. Find a reduction formula for $\int x^n e^{-x}\,dx$ similar to Eq. (5).

SOLUTION Let $u = x^n$ and $v' = e^{-x}$. Then

$$u = x^n \qquad v = -e^{-x}$$
$$u' = nx^{n-1} \qquad v' = e^{-x}$$

Using Integration by Parts, we get

$$\int x^n e^{-x}\,dx = -x^n e^{-x} - \int nx^{n-1}(-e^{-x})\,dx = -x^n e^{-x} + n\int x^{n-1}e^{-x}\,dx.$$

In Exercises 59–66, indicate a good method for evaluating the integral (but do not evaluate). Your choices are algebraic manipulation, substitution (specify u and du), and Integration by Parts (specify u and v′). If it appears that the techniques you have learned thus far are not sufficient, state this.

59. $\int \sqrt{x}\ln x\,dx$

SOLUTION Use Integration by Parts, with $u = \ln x$ and $v' = \sqrt{x}$.

61. $\int \dfrac{x^3\,dx}{\sqrt{4-x^2}}$

SOLUTION Use substitution, followed by algebraic manipulation: Let $u = 4 - x^2$. Then $du = -2x\,dx$, $x^2 = 4 - u$, and

$$\int \frac{x^3}{\sqrt{4-x^2}}dx = -\frac{1}{2}\int \frac{(x^2)(-2x\,dx)}{\sqrt{u}} = -\frac{1}{2}\int \frac{(4-u)(du)}{\sqrt{u}} = -\frac{1}{2}\int\left(\frac{4}{\sqrt{u}} - \frac{u}{\sqrt{u}}\right)du.$$

63. $\int \dfrac{x+2}{x^2+4x+3}\,dx$

SOLUTION Use substitution. Let $u = x^2 + 4x + 3$; then $du = 2x + 4\,dx = 2(x+2)\,dx$, and

$$\int \frac{x+2}{x^2+4x+3}\,dx = \frac{1}{2}\int \frac{1}{u}\,du$$

65. $\int x\sin(3x+4)\,dx$

SOLUTION Use Integration by Parts, with $u = x$ and $v' = \sin(3x+4)$.

67. Evaluate $\int (\sin^{-1} x)^2\,dx$. *Hint:* Use Integration by Parts first and then substitution.

SOLUTION First use integration by parts with $v' = 1$ to get

$$\int (\sin^{-1} x)^2\,dx = x(\sin^{-1} x)^2 - 2\int \frac{x\sin^{-1} x\,dx}{\sqrt{1-x^2}}.$$

Now use substitution on the integral on the right, with $u = \sin^{-1} x$. Then $du = dx/\sqrt{1-x^2}$ and $x = \sin u$, and we get (using Integration by Parts again)

$$\int \frac{x\sin^{-1} x\,dx}{\sqrt{1-x^2}} = \int u\sin u\,du = -u\cos u + \sin u + C = -\sqrt{1-x^2}\,\sin^{-1} x + x + C.$$

where $\cos u = \sqrt{1-\sin^2 u} = \sqrt{1-x^2}$. So the final answer is

$$\int (\sin^{-1} x)^2\,dx = x(\sin^{-1} x)^2 + 2\sqrt{1-x^2}\,\sin^{-1} x - 2x + C.$$

69. Evaluate $\int x^7\cos(x^4)\,dx$.

SOLUTION First, let $w = x^4$. Then $dw = 4x^3\,dx$ and

$$\int x^7\cos(x^4)\,dx = \frac{1}{4}\int w\cos x\,dw.$$

Now, use Integration by Parts with $u = w$ and $v' = \cos w$. Then

$$\int x^7\cos(x^4)\,dx = \frac{1}{4}\left(w\sin w - \int \sin w\,dw\right) = \frac{1}{4}w\sin w + \frac{1}{4}\cos w + C = \frac{1}{4}x^4\sin(x^4) + \frac{1}{4}\cos(x^4) + C.$$

71. Find the volume of the solid obtained by revolving the region under $y = e^x$ for $0 \le x \le 2$ about the y-axis.

SOLUTION By the Method of Cylindrical Shells, the volume V of the solid is

$$V = \int_a^b (2\pi r)h\,dx = 2\pi \int_0^2 xe^x\,dx.$$

Using Integration by Parts with $u = x$ and $v' = e^x$, we find

$$V = 2\pi \ (xe^x - e^x)\Big|_0^2 = 2\pi\left[(2e^2 - e^2) - (0 - 1)\right] = 2\pi(e^2 + 1).$$

73. Recall that the *present value* (PV) of an investment that pays out income continuously at a rate $R(t)$ for T years is $\int_0^T R(t)e^{-rt}\,dt$, where r is the interest rate. Find the PV if $R(t) = 5000 + 100t$ \$/year, $r = 0.05$ and $T = 10$ years.

SOLUTION The present value is given by

$$PV = \int_0^T R(t)e^{-rt}\,dt = \int_0^{10} (5000 + 100t)e^{-rt}\,dt = 5000\int_0^{10} e^{-rt}\,dt + 100\int_0^{10} te^{-rt}\,dt.$$

Using Integration by Parts for the integral on the right, with $u = t$ and $v' = e^{-rt}$, we find

$$\begin{aligned}
PV &= 5000\left(-\frac{1}{r}e^{-rt}\right)\Big|_0^{10} + 100\left[\left(-\frac{t}{r}e^{-rt}\right)\Big|_0^{10} - \int_0^{10} \frac{-1}{r}e^{-rt}\,dt\right]\\
&= -\frac{5000}{r}e^{-rt}\Big|_0^{10} - \frac{100}{r}\left(te^{-rt} + \frac{1}{r}e^{-rt}\right)\Big|_0^{10}\\
&= -\frac{5000}{r}(e^{-10r} - 1) - \frac{100}{r}\left[\left(10e^{-10r} + \frac{1}{r}e^{-10r}\right) - \left(0 + \frac{1}{r}\right)\right]\\
&= e^{-10r}\left[-\frac{5000}{r} - \frac{1000}{r} - \frac{100}{r^2}\right] + \frac{5000}{r} + \frac{100}{r^2}\\
&= \frac{5000r + 100 - e^{-10r}(6000r + 100)}{r^2}.
\end{aligned}$$

75. Use Eq. (6) to calculate $\int (\ln x)^k\,dx$ for $k = 2, 3$.

SOLUTION

$$\int (\ln x)^2\,dx = x(\ln x)^2 - 2\int \ln x\,dx = x(\ln x)^2 - 2(x\ln x - x) + C = x(\ln x)^2 - 2x\ln x + 2x + C;$$

$$\begin{aligned}
\int (\ln x)^3\,dx &= x(\ln x)^3 - 3\int (\ln x)^2\,dx = x(\ln x)^3 - 3\left[x(\ln x)^2 - 2x\ln x + 2x\right] + C\\
&= x(\ln x)^3 - 3x(\ln x)^2 + 6x\ln x - 6x + C.
\end{aligned}$$

77. Prove that $\int xb^x\,dx = b^x\left(\frac{x}{\ln b} - \frac{1}{\ln^2 b}\right) + C.$

SOLUTION Let $u = x$ and $v' = b^x$. Then $u' = 1$ and $v = b^x/\ln b$. Using Integration by Parts, we get

$$\int x\,b^x\,dx = \frac{xb^x}{\ln b} - \frac{1}{\ln b}\int b^x\,dx = \frac{xb^x}{\ln b} - \frac{1}{\ln b}\cdot\frac{b^x}{\ln b} + C = b^x\left(\frac{x}{\ln b} - \frac{1}{(\ln b)^2}\right) + C.$$

Further Insights and Challenges

79. The Integration by Parts formula can be written

$$\int u(x)v(x)\,dx = u(x)V(x) - \int u'(x)V(x)\,dx \qquad \boxed{7}$$

where $V(x)$ satisfies $V'(x) = v(x)$.

(a) Show directly that the right-hand side of Eq. (7) does not change if $V(x)$ is replaced by $V(x) + C$, where C is a constant.

(b) Use $u = \tan^{-1} x$ and $v = x$ in Eq. (7) to calculate $\int x \tan^{-1} x\,dx$, but carry out the calculation twice: first with $V(x) = \frac{1}{2}x^2$ and then with $V(x) = \frac{1}{2}x^2 + \frac{1}{2}$. Which choice of $V(x)$ results in a simpler calculation?

SOLUTION

(a) Replacing $V(x)$ with $V(x) + C$ in the expression $u(x)V(x) - \int V(x)u'(x)\,dx$, we get

$$\begin{aligned} u(x)(V(x)+C) - \int (V(x)+C)u'(x)\,dx &= u(x)V(x) + u(x)C - \int V(x)u'(x)\,dx - C\int u'(x)\,dx \\ &= u(x)V(x) - \int V(x)u'(x)\,dx + C\left[u(x) - \int u'(x)\,dx\right] \\ &= u(x)V(x) - \int V(x)u'(x)\,dx + C\,[u(x) - u(x)] \\ &= u(x)V(x) - \int V(x)u'(x)\,dx. \end{aligned}$$

(b) If we evaluate $\int x \tan^{-1} x\,dx$ with $u = \tan^{-1} x$ and $v' = x$, and if we don't add a constant to v, Integration by Parts gives us

$$\int x \tan^{-1} x\,dx = \frac{x^2}{2}\tan^{-1} x - \frac{1}{2}\int \frac{x^2 dx}{x^2+1}.$$

The integral on the right requires algebraic manipulation in order to evaluate. But if we take $V(x) = \frac{1}{2}x^2 + \frac{1}{2}$ instead of $V(x) = \frac{1}{2}x^2$, then

$$\begin{aligned} \int x \tan^{-1} x\,dx &= \left(\frac{1}{2}x^2 + \frac{1}{2}\right)\tan^{-1} x - \frac{1}{2}\int \frac{x^2+1}{x^2+1}\,dx = \frac{1}{2}(x^2+1)\tan^{-1} x - \frac{1}{2}x + C \\ &= \frac{1}{2}(x^2\tan^{-1} x - x + \tan^{-1} x) + C. \end{aligned}$$

81. Assume that $f(0) = f(1) = 0$ and that f'' exists. Prove

$$\int_0^1 f''(x)f(x)\,dx = -\int_0^1 f'(x)^2\,dx \qquad \boxed{9}$$

Use this to prove that if $f(0) = f(1) = 0$ and $f''(x) = \lambda f(x)$ for some constant λ, then $\lambda < 0$. Can you think of a function satisfying these conditions for some λ?

SOLUTION Let $u = f(x)$ and $v' = f''(x)$. Using Integration by Parts, we get

$$\int_0^1 f''(x)f(x)\,dx = f(x)f'(x)\Big|_0^1 - \int_0^1 f'(x)^2\,dx = f(1)f'(1) - f(0)f'(0) - \int_0^1 f'(x)^2\,dx = -\int_0^1 f'(x)^2\,dx.$$

Now assume that $f''(x) = \lambda f(x)$ for some constant λ. Then

$$\int_0^1 f''(x)f(x)\,dx = \lambda\int_0^1 [f(x)]^2\,dx = -\int_0^1 f'(x)^2\,dx < 0.$$

Since $\int_0^1 [f(x)]^2\,dx > 0$, we must have $\lambda < 0$. An example of a function satisfying these properties for some λ is $f(x) = \sin \pi x$.

83. Let $I_n = \int x^n \cos(x^2)\,dx$ and $J_n = \int x^n \sin(x^2)\,dx$.

(a) Find a reduction formula that expresses I_n in terms of J_{n-2}. *Hint:* Write $x^n \cos(x^2)$ as $x^{n-1}(x\cos(x^2))$.

(b) Use the result of (a) to show that I_n can be evaluated explicitly if n is odd.

(c) Evaluate I_3.

SOLUTION

(a) Integration by Parts with $u = x^{n-1}$ and $v' = x\cos(x^2)\,dx$ yields

$$I_n = \frac{1}{2}x^{n-1}\sin(x^2) - \frac{n-1}{2}\int x^{n-2}\sin(x^2)\,dx = \frac{1}{2}x^{n-1}\sin(x^2) - \frac{n-1}{2}J_{n-2}.$$

(b) If n is odd, the reduction process will eventually lead to either

$$\int x\cos(x^2)\,dx \quad \text{or} \quad \int x\sin(x^2)\,dx,$$

both of which can be evaluated using the substitution $u = x^2$.

(c) Starting with the reduction formula from part (a), we find

$$I_3 = \frac{1}{2}x^2\sin(x^2) - \frac{2}{2}\int x\sin(x^2)\,dx = \frac{1}{2}x^2\sin(x^2) + \frac{1}{2}\cos(x^2) + C.$$

7.2 Trigonometric Integrals

Preliminary Questions

1. Describe the technique used to evaluate $\int \sin^5 x\,dx$.

SOLUTION Because the sine function is raised to an odd power, rewrite $\sin^5 x = \sin x \sin^4 x = \sin x(1-\cos^2 x)^2$ and then substitute $u = \cos x$.

2. Describe a way of evaluating $\int \sin^6 x\,dx$.

SOLUTION Repeatedly use the reduction formula for powers of $\sin x$.

3. Are reduction formulas needed to evaluate $\int \sin^7 x\cos^2 x\,dx$? Why or why not?

SOLUTION No, a reduction formula is not needed because the sine function is raised to an odd power.

4. Describe a way of evaluating $\int \sin^6 x\cos^2 x\,dx$.

SOLUTION Because both trigonometric functions are raised to even powers, write $\cos^2 x = 1 - \sin^2 x$ and then apply the reduction formula for powers of the sine function.

5. Which integral requires more work to evaluate?

$$\int \sin^{798} x\cos x\,dx \quad \text{or} \quad \int \sin^4 x\cos^4 x\,dx$$

Explain your answer.

SOLUTION The first integral can be evaluated using the substitution $u = \sin x$, whereas the second integral requires the use of reduction formulas. The second integral therefore requires more work to evaluate.

Exercises

In Exercises 1–6, use the method for odd powers to evaluate the integral.

1. $\int \cos^3 x\,dx$

SOLUTION Use the identity $\cos^2 x = 1 - \sin^2 x$ to rewrite the integrand:

$$\int \cos^3 x\,dx = \int \left(1-\sin^2 x\right)\cos x\,dx.$$

Now use the substitution $u = \sin x$, $du = \cos x\,dx$:

$$\int \cos^3 x\,dx = \int \left(1-u^2\right)du = u - \frac{1}{3}u^3 + C = \sin x - \frac{1}{3}\sin^3 x + C.$$

3. $\int \sin^3\theta\cos^2\theta\,d\theta$

SOLUTION Write $\sin^3\theta = \sin^2\theta\sin\theta = (1-\cos^2\theta)\sin\theta$. Then

$$\int \sin^3\theta\cos^2\theta\,d\theta = \int \left(1-\cos^2\theta\right)\cos^2\theta\sin\theta\,d\theta.$$

Now use the substitution $u = \cos\theta$, $du = -\sin\theta\,d\theta$:

$$\int \sin^3\theta\cos^2\theta\,d\theta = -\int \left(1-u^2\right)u^2\,du = -\int \left(u^2-u^4\right)du$$

$$= -\frac{1}{3}u^3 + \frac{1}{5}u^5 + C = -\frac{1}{3}\cos^3\theta + \frac{1}{5}\cos^5\theta + C.$$

5. $\int \sin^3 t \cos^3 t\, dt$

SOLUTION Write $\sin^3 t = (1-\cos^2 t)\sin t\, dt$. Then

$$\int \sin^3 t \cos^3 t\, dt = \int (1-\cos^2 t)\cos^3 t \sin t\, dt = \int \left(\cos^3 t - \cos^5 t\right)\sin t\, dt.$$

Now use the substitution $u = \cos t$, $du = -\sin t\, dt$:

$$\int \sin^3 t \cos^3 t\, dt = -\int \left(u^3 - u^5\right) du = -\frac{1}{4}u^4 + \frac{1}{6}u^6 + C = -\frac{1}{4}\cos^4 t + \frac{1}{6}\cos^6 t + C.$$

7. Find the area of the shaded region in Figure 1.

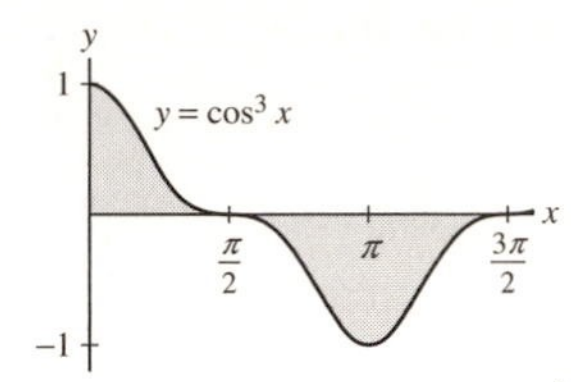

FIGURE 1 Graph of $y = \cos^3 x$.

SOLUTION First evaluate the indefinite integral by writing $\cos^3 x = (1-\sin^2 x)\cos x$, and using the substitution $u = \sin x$, $du = \cos x\, dx$:

$$\int \cos^3 x\, dx = \int \left(1-\sin^2 x\right)\cos x\, dx = \int \left(1-u^2\right) du = u - \frac{1}{3}u^3 + C = \sin x - \frac{1}{3}\sin^3 x + C.$$

The area is given by

$$A = \int_0^{\pi/2} \cos^3 x\, dx - \int_{\pi/2}^{3\pi/2} \cos^3 x\, dx = \left(\sin x - \frac{1}{3}\sin^3 x\right)\Big|_0^{\pi/2} - \left(\sin x - \frac{1}{3}\sin^3 x\right)\Big|_{\pi/2}^{3\pi/2}$$

$$= \left[\left(\sin\frac{\pi}{2} - \frac{1}{3}\sin^3\frac{\pi}{2}\right) - 0\right] - \left[\left(\sin\frac{3\pi}{2} - \frac{1}{3}\sin^3\frac{3\pi}{2}\right) - \left(\sin\frac{\pi}{2} - \frac{1}{3}\sin^3\frac{\pi}{2}\right)\right]$$

$$= 1 - \frac{1}{3}(1)^3 - (-1) + \frac{1}{3}(-1)^3 + 1 - \frac{1}{3}(1)^3 = 2.$$

In Exercises 9–12, evaluate the integral using methods employed in Examples 3 and 4.

9. $\int \cos^4 y\, dy$

SOLUTION Using the reduction formula for $\cos^m y$, we get

$$\int \cos^4 y\, dy = \frac{1}{4}\cos^3 y \sin y + \frac{3}{4}\int \cos^2 y\, dy = \frac{1}{4}\cos^3 y \sin y + \frac{3}{4}\left(\frac{1}{2}\cos y \sin y + \frac{1}{2}\int dy\right)$$

$$= \frac{1}{4}\cos^3 y \sin y + \frac{3}{8}\cos y \sin y + \frac{3}{8}y + C.$$

11. $\int \sin^4 x \cos^2 x\, dx$

SOLUTION Use the identity $\cos^2 x = 1 - \sin^2 x$ to write:

$$\int \sin^4 x \cos^2 x\, dx = \int \sin^4 x\left(1-\sin^2 x\right) dx = \int \sin^4 x\, dx - \int \sin^6 x\, dx.$$

Using the reduction formula for $\sin^m x$:

$$\int \sin^4 x \cos^2 x\, dx = \int \sin^4 x\, dx - \left[-\frac{1}{6}\sin^5 x \cos x + \frac{5}{6}\int \sin^4 x\, dx\right]$$

$$= \frac{1}{6}\sin^5 x \cos x + \frac{1}{6}\int \sin^4 x\, dx = \frac{1}{6}\sin^5 x \cos x + \frac{1}{6}\left(-\frac{1}{4}\sin^3 x \cos x + \frac{3}{4}\int \sin^2 x\, dx\right)$$

$$= \frac{1}{6}\sin^5 x \cos x - \frac{1}{24}\sin^3 x \cos x + \frac{1}{8}\int \sin^2 x\, dx$$

$$= \frac{1}{6}\sin^5 x\cos x - \frac{1}{24}\sin^3 x\cos x + \frac{1}{8}\left(-\frac{1}{2}\sin x\cos x + \frac{1}{2}\int dx\right)$$

$$= \frac{1}{6}\sin^5 x\cos x - \frac{1}{24}\sin^3 x\cos x - \frac{1}{16}\sin x\cos x + \frac{1}{16}x + C.$$

In Exercises 13 and 14, evaluate using Eq. (13).

13. $\int \sin^3 x\cos^2 x\,dx$

SOLUTION First rewrite $\sin^3 x = \sin x \cdot \sin^2 x = \sin x(1 - \cos^2 x)$, so that

$$\int \sin^3 x\cos^2 x\,dx = \int \sin x(1-\cos^2 x)\cos^2 x\,dx = \int \sin x(\cos^2 x - \cos^4 x)\,dx$$

Now make the substitution $u = \cos x$, $du = -\sin x\,dx$:

$$\int \sin x(\cos^2 x - \cos^4 x)\,dx = -\int u^2 - u^4\,du = \frac{1}{5}u^5 - \frac{1}{3}u^3 + C = \frac{1}{5}\cos^5 x - \frac{1}{3}\cos^3 x + C$$

In Exercises 15–18, evaluate the integral using the method described on page 409 and the reduction formulas on page 410 as necessary.

15. $\int \tan^3 x\sec x\,dx$

SOLUTION Use the identity $\tan^2 x = \sec^2 x - 1$ to rewrite $\tan^3 x\sec x = (\sec^2 x - 1)\sec x\tan x$. Then use the substitution $u = \sec x$, $du = \sec x\tan x\,dx$:

$$\int \tan^3 x\sec x\,dx = \int (\sec^2 x - 1)\sec x\tan x\,dx = \int u^2 - 1\,du = \frac{1}{3}u^3 - u + C = \frac{1}{3}\sec^3 x - \sec x + C$$

17. $\int \tan^2 x\sec^4 x\,dx$

SOLUTION First use the identity $\tan^2 x = \sec^2 x - 1$:

$$\int \tan^2 x\sec^4 x\,dx = \int (\sec^2 x - 1)\sec^4 x\,dx = \int \sec^6 x - \sec^4 x\,dx = \int \sec^6 x\,dx - \int \sec^4 x,dx$$

We evaluate the second integral using the reduction formula:

$$\int \sec^4 x\,dx = \frac{1}{3}\tan x\sec^2 x + \frac{2}{3}\int \sec^2 x\,dx$$

$$= \frac{1}{3}\tan x\sec^2 x + \frac{2}{3}\tan x$$

Then

$$\int \sec^6 x\,dx = \frac{1}{5}\tan x\sec^4 x + \frac{4}{5}\int \sec^4 x\,dx$$

$$= \frac{1}{5}\tan x\sec^4 x + \frac{4}{5}\left(\frac{1}{3}\tan x\sec^2 x + \frac{2}{3}\tan x\right)$$

$$= \frac{1}{5}\tan x\sec^4 x + \frac{4}{15}\tan x\sec^2 x + \frac{8}{15}\tan x$$

so that

$$\int \tan^2 x\sec^4 x\,dx = \int \sec^6 x\,dx - \int \sec^4 x\,dx$$

$$= \frac{1}{5}\tan x\sec^4 x - \frac{1}{15}\tan x\sec^2 x - \frac{2}{15}\tan x + C$$

In Exercises 19–22, evaluate using methods similar to those that apply to integral $\tan^m x\sec^n$.

19. $\int \cot^3 x\,dx$

SOLUTION Using the reduction formula for $\cot^m x$, we get

$$\int \cot^3 x\,dx = -\frac{1}{2}\cot^2 x - \int \cot x\,dx = -\frac{1}{2}\cot^2 x + \ln|\csc x| + C.$$

21. $\displaystyle\int \cot^5 x \csc^2 x\,dx$

SOLUTION Make the substitution $u = \cot x$, $du = -\csc^2 x\,dx$; then

$$\int \cot^5 x \csc^2 x\,dx = -\int u^5\,du = -\frac{1}{6}u^6 = -\frac{1}{6}\cot^6 x + C$$

In Exercises 23–46, evaluate the integral.

23. $\displaystyle\int \cos^5 x \sin x\,dx$

SOLUTION Use the substitution $u = \cos x$, $du = -\sin x\,dx$. Then

$$\int \cos^5 x \sin x\,dx = -\int u^5\,du = -\frac{1}{6}u^6 + C = -\frac{1}{6}\cos^6 x + C.$$

25. $\displaystyle\int \cos^4(3x+2)\,dx$

SOLUTION First use the substitution $u = 3x + 2$, $du = 3\,dx$ and then apply the reduction formula for $\cos^n x$:

$$\begin{aligned}\int \cos^4(3x+2)\,dx &= \frac{1}{3}\cos^4 u\,du = \frac{1}{3}\left(\frac{1}{4}\cos^3 u \sin u + \frac{3}{4}\int \cos^2 u\,du\right)\\
&= \frac{1}{12}\cos^3 u \sin u + \frac{1}{4}\left(\frac{u}{2} + \frac{\sin 2u}{4}\right) + C\\
&= \frac{1}{12}\cos^3(3x+2)\sin(3x+2) + \frac{1}{8}(3x+2) + \frac{1}{16}\sin(6x+4) + C\end{aligned}$$

27. $\displaystyle\int \cos^3(\pi\theta)\sin^4(\pi\theta)\,d\theta$

SOLUTION Use the substitution $u = \pi\theta$, $du = \pi\,d\theta$, and the identity $\cos^2 u = 1 - \sin^2 u$ to write

$$\int \cos^3(\pi\theta)\sin^4(\pi\theta)\,d\theta = \frac{1}{\pi}\int \cos^3 u \sin^4 u\,du = \frac{1}{\pi}\int\left(1 - \sin^2 u\right)\sin^4 u \cos u\,du.$$

Now use the substitution $w = \sin u$, $dw = \cos u\,du$:

$$\begin{aligned}\int \cos^3(\pi\theta)\sin^4(\pi\theta)\,d\theta &= \frac{1}{\pi}\int\left(1 - w^2\right)w^4\,dw = \frac{1}{\pi}\int\left(w^4 - w^6\right)dw = \frac{1}{5\pi}w^5 - \frac{1}{7\pi}w^7 + C\\
&= \frac{1}{5\pi}\sin^5(\pi\theta) - \frac{1}{7\pi}\sin^7(\pi\theta) + C.\end{aligned}$$

29. $\displaystyle\int \sin^4(3x)\,dx$

SOLUTION Use the substitution $u = 3x$, $du = 3\,dx$ and the reduction formula for $\sin^m x$:

$$\begin{aligned}\int \sin^4(3x)\,dx &= \frac{1}{3}\int \sin^4 u\,du = -\frac{1}{12}\sin^3 u \cos u + \frac{1}{4}\int \sin^2 u\,du\\
&= -\frac{1}{12}\sin^3 u \cos u + \frac{1}{4}\left(-\frac{1}{2}\sin u \cos u + \frac{1}{2}\int du\right)\\
&= -\frac{1}{12}\sin^3 u \cos u - \frac{1}{8}\sin u \cos u + \frac{1}{8}u + C\\
&= -\frac{1}{12}\sin^3(3x)\cos(3x) - \frac{1}{8}\sin(3x)\cos(3x) + \frac{3}{8}x + C.\end{aligned}$$

31. $\displaystyle\int \csc^2(3-2x)\,dx$

SOLUTION First make the substitution $u = 3 - 2x$, $du = -2\,dx$, so that

$$\int \csc^2(3-2x)\,dx = \frac{1}{2}\int(-\csc^2 u)\,du = \frac{1}{2}\cot u + C = \frac{1}{2}\cot(3-2x) + C$$

33. $\int \tan x \sec^2 x\, dx$

SOLUTION Use the substitution $u = \tan x$, $du = \sec^2 x\, dx$. Then

$$\int \tan x \sec^2 x\, dx = \int u\, du = \frac{1}{2}u^2 + C = \frac{1}{2}\tan^2 x + C.$$

35. $\int \tan^5 x \sec^4 x\, dx$

SOLUTION Use the identity $\tan^2 x = \sec^2 x - 1$ to write

$$\int \tan^5 x \sec^4 x\, dx = \int \left(\sec^2 x - 1\right)^2 \sec^3 x(\sec x \tan x\, dx).$$

Now use the substitution $u = \sec x$, $du = \sec x \tan x\, dx$:

$$\int \tan^5 x \sec^4 x\, dx = \int \left(u^2 - 1\right)^2 u^3\, du = \int \left(u^7 - 2u^5 + u^3\right) du$$
$$= \frac{1}{8}u^8 - \frac{1}{3}u^6 + \frac{1}{4}u^4 + C = \frac{1}{8}\sec^8 x - \frac{1}{3}\sec^6 x + \frac{1}{4}\sec^4 x + C.$$

37. $\int \tan^6 x \sec^4 x\, dx$

SOLUTION Use the identity $\sec^2 x = \tan^2 x + 1$ to write

$$\int \tan^6 x \sec^4 x\, dx = \int \tan^6 x \left(\tan^2 x + 1\right) \sec^2 x\, dx.$$

Now use the substitution $u = \tan x$, $du = \sec^2 x\, dx$:

$$\int \tan^6 x \sec^4 x\, dx = \int u^6\left(u^2 + 1\right) du = \int \left(u^8 + u^6\right) du = \frac{1}{9}u^9 + \frac{1}{7}u^7 + C = \frac{1}{9}\tan^9 x + \frac{1}{7}\tan^7 x + C.$$

39. $\int \cot^5 x \csc^5 x\, dx$

SOLUTION First use the identity $\cot^2 x = \csc^2 x - 1$ to rewrite the integral:

$$\int \cot^5 x \csc^5 x\, dx = \int (\csc^2 x - 1)^2 \csc^4 x(\cot x \csc x)\, dx = \int (\csc^8 x - 2\csc^6 x + \csc^4 x)(\cot x \csc x)\, dx$$

Now use the substitution $u = \csc x$ and $du = -\cot x \csc x\, dx$ to get

$$\int \cot^5 x \csc^5 x\, dx = -\int u^8 - 2u^6 + u^4\, du = -\frac{1}{9}u^9 + \frac{2}{7}u^7 - \frac{1}{5}u^5 + C$$
$$= -\frac{1}{9}\csc^9 x + \frac{2}{7}\csc^7 x - \frac{1}{5}\csc^5 x + C$$

41. $\int \sin 2x \cos 2x\, dx$

SOLUTION Use the substitution $u = \sin 2x$, $du = 2\cos 2x\, dx$:

$$\int \sin 2x \cos 2x\, dx = \frac{1}{2}\int \sin 2x(2\cos 2x\, dx) = \frac{1}{2}\int u\, du = \frac{1}{4}u^2 + C = \frac{1}{4}\sin^2 2x + C.$$

43. $\int t \cos^3(t^2)\, dt$

SOLUTION Use the substitution $u = t^2$, $du = 2t\, dt$, followed by the reduction formula for $\cos^m x$:

$$\int t\cos^3(t^2)\, dt = \frac{1}{2}\int \cos^3 u\, du = \frac{1}{6}\cos^2 u \sin u + \frac{1}{3}\int \cos u\, du$$
$$= \frac{1}{6}\cos^2 u \sin u + \frac{1}{3}\sin u + C = \frac{1}{6}\cos^2(t^2)\sin(t^2) + \frac{1}{3}\sin(t^2) + C.$$

45. $\int \cos^2(\sin t)\cos t\,dt$

SOLUTION Use the substitution $u = \sin t$, $du = \cos t\,dt$, followed by the reduction formula for $\cos^m x$:

$$\int \cos^2(\sin t)\cos t\,dt = \int \cos^2 u\,du = \frac{1}{2}\cos u \sin u + \frac{1}{2}\int du$$
$$= \frac{1}{2}\cos u \sin u + \frac{1}{2}u + C = \frac{1}{2}\cos(\sin t)\sin(\sin t) + \frac{1}{2}\sin t + C.$$

In Exercises 47–60, evaluate the definite integral.

47. $\int_0^{2\pi} \sin^2 x\,dx$

SOLUTION Use the formula for $\int \sin^2 x\,dx$:

$$\int_0^{2\pi} \sin^2 x\,dx = \left(\frac{x}{2} - \frac{\sin 2x}{4}\right)\Bigg|_0^{2\pi} = \left(\frac{2\pi}{2} - \frac{\sin 4\pi}{4}\right) - \left(\frac{0}{2} - \frac{\sin 0}{4}\right) = \pi.$$

49. $\int_0^{\pi/2} \sin^5 x\,dx$

SOLUTION Use the identity $\sin^2 x = 1 - \cos^2 x$ followed by the substitution $u = \cos x$, $du = -\sin x\,dx$ to get

$$\int_0^{\pi/2} \sin^5 x\,dx = \int_0^{\pi/2} (1 - \cos^2 x)^2 \sin x\,dx = \int_0^{\pi/2} (1 - 2\cos^2 x + \cos^4 x)\sin x\,dx$$
$$= -\int_1^0 (1 - 2u^2 + u^4)\,du = -\left(u - \frac{2}{3}u^3 + \frac{1}{5}u^5\right)\Bigg|_1^0 = 1 - \frac{2}{3} + \frac{1}{5} = \frac{8}{15}$$

51. $\int_0^{\pi/4} \frac{dx}{\cos x}$

SOLUTION Use the definition of $\sec x$ to simplify the integral:

$$\int_0^{\pi/4} \frac{dx}{\cos x} = \int_0^{\pi/4} \sec x\,dx = \ln|\sec x + \tan x|\Big|_0^{\pi/4} = \ln\left|\sqrt{2} + 1\right| - \ln|1 + 0| = \ln\left(\sqrt{2} + 1\right).$$

53. $\int_0^{\pi/3} \tan x\,dx$

SOLUTION Use the formula for $\int \tan x\,dx$:

$$\int_0^{\pi/3} \tan x\,dx = \ln|\sec x|\Big|_0^{\pi/3} = \ln 2 - \ln 1 = \ln 2.$$

55. $\int_{-\pi/4}^{\pi/4} \sec^4 x\,dx$

SOLUTION First use the reduction formula for $\sec^m x$ to evaluate the indefinite integral:

$$\int \sec^4 x\,dx = \frac{1}{3}\tan x \sec^2 x + \frac{2}{3}\int \sec^2 x\,dx = \frac{1}{3}\tan x \sec^2 x + \frac{2}{3}\tan x + C.$$

Now compute the definite integral:

$$\int_{-\pi/4}^{\pi/4} \sec^4 x\,dx = \left(\frac{1}{3}\tan x \sec^2 x + \frac{2}{3}\tan x\right)\Bigg|_{-\pi/4}^{\pi/4}$$
$$= \left[\frac{1}{3}(1)\left(\sqrt{2}\right)^2 + \frac{2}{3}(1)\right] - \left[\frac{1}{3}(-1)\left(\sqrt{2}\right)^2 + \frac{2}{3}(-1)\right] = \frac{4}{3} - \left(-\frac{4}{3}\right) = \frac{8}{3}.$$

57. $\int_0^{\pi} \sin 3x \cos 4x\,dx$

SOLUTION Use the formula for $\int \sin mx \cos nx\,dx$:

$$\int_0^{\pi} \sin 3x \cos 4x\,dx = \left(-\frac{\cos(3-4)x}{2(3-4)} - \frac{\cos(3+4)x}{2(3+4)}\right)\Bigg|_0^{\pi} = \left(-\frac{\cos(-x)}{-2} - \frac{\cos 7x}{14}\right)\Bigg|_0^{\pi}$$
$$= \left(\frac{1}{2}\cos x - \frac{1}{14}\cos 7x\right)\Bigg|_0^{\pi} = \left[\frac{1}{2}(-1) - \frac{1}{14}(-1)\right] - \left[\frac{1}{2}(1) - \frac{1}{14}(1)\right] = -\frac{6}{7}.$$

59. $\displaystyle\int_0^{\pi/6} \sin 2x \cos 4x\, dx$

SOLUTION Using the formula for $\displaystyle\int \sin mx \cos nx\, dx$, we have

$$\int_0^{\pi/6} \sin 2x \cos 4x\, dx = \left(-\frac{1}{-4}\cos(-2x) - \frac{1}{2\cdot 6}\cos(6x)\right)\Big|_0^{\pi/6} = \left(\frac{1}{4}\cos 2x - \frac{1}{12}\cos 6x\right)\Big|_0^{\pi/6}$$
$$= \left(\frac{1}{4}\cdot\frac{1}{2} - \frac{1}{12}\cdot(-1)\right) - \left(\frac{1}{4} - \frac{1}{12}\right) = \frac{1}{24}$$

Here we've used the fact that $\cos x$ is an even function: $\cos(-x) = \cos x$.

61. Use the identities for $\sin 2x$ and $\cos 2x$ on page 407 to verify that the following formulas are equivalent.

$$\int \sin^4 x\, dx = \frac{1}{32}(12x - 8\sin 2x + \sin 4x) + C$$
$$\int \sin^4 x\, dx = -\frac{1}{4}\sin^3 x \cos x - \frac{3}{8}\sin x \cos x + \frac{3}{8}x + C$$

SOLUTION First, observe

$$\sin 4x = 2\sin 2x \cos 2x = 2\sin 2x(1 - 2\sin^2 x)$$
$$= 2\sin 2x - 4\sin 2x \sin^2 x = 2\sin 2x - 8\sin^3 x \cos x.$$

Then

$$\frac{1}{32}(12x - 8\sin 2x + \sin 4x) + C = \frac{3}{8}x - \frac{3}{16}\sin 2x - \frac{1}{4}\sin^3 x \cos x + C$$
$$= \frac{3}{8}x - \frac{3}{8}\sin x \cos x - \frac{1}{4}\sin^3 x \cos x + C.$$

63. Find the volume of the solid obtained by revolving $y = \sin x$ for $0 \le x \le \pi$ about the x-axis.

SOLUTION Using the disk method, the volume is given by

$$V = \int_0^{\pi} \pi(\sin x)^2\, dx = \pi\int_0^{\pi} \sin^2 x\, dx = \pi\left(\frac{x}{2} - \frac{\sin 2x}{4}\right)\Big|_0^{\pi} = \pi\left[\left(\frac{\pi}{2} - 0\right) - (0)\right] = \frac{\pi^2}{2}.$$

In Exercises 65–68, use the following alternative method for evaluating the integral $J = \int \sin^m x \cos^n x\, dx$ when m and n are both even. Use the identities

$$\sin^2 x = \frac{1}{2}(1 - \cos 2x), \qquad \cos^2 x = \frac{1}{2}(1 + \cos 2x)$$

to write $J = \frac{1}{4}\int (1 - \cos 2x)^{m/2}(1 + \cos 2x)^{n/2}\, dx$, and expand the right-hand side as a sum of integrals involving smaller powers of sine and cosine in the variable $2x$.

65. $\displaystyle\int \sin^2 x \cos^2 x\, dx$

SOLUTION Using the identities $\sin^2 x = \frac{1}{2}(1 - \cos 2x)$ and $\cos^2 x = \frac{1}{2}(1 + \cos 2x)$, we have

$$J = \int \sin^2 x \cos^2 x\, dx = \frac{1}{4}\int (1 - \cos 2x)(1 + \cos 2x)\, dx$$
$$= \frac{1}{4}\int \left(1 - \cos^2 2x\right) dx = \frac{1}{4}\int dx - \frac{1}{4}\int \cos^2 2x\, dx.$$

Now use the substitution $u = 2x$, $du = 2\,dx$, and the formula for $\displaystyle\int \cos^2 u\, du$:

$$J = \frac{1}{4}x - \frac{1}{8}\int \cos^2 u\, du = \frac{1}{4}x - \frac{1}{8}\left(\frac{u}{2} + \frac{1}{2}\sin u \cos u\right) + C$$
$$= \frac{1}{4}x - \frac{1}{16}(2x) - \frac{1}{16}\sin 2x \cos 2x + C = \frac{1}{8}x - \frac{1}{16}\sin 2x \cos 2x + C.$$

67. $\int \sin^4 x \cos^2 x\,dx$

SOLUTION Using the identities $\sin^2 x = \frac{1}{2}(1-\cos 2x)$ and $\cos^2 x = \frac{1}{2}(1+\cos 2x)$, we have

$$J = \int \sin^4 x \cos^2 x\,dx = \frac{1}{8}\int (1-\cos 2x)^2(1+\cos 2x)\,dx$$
$$= \frac{1}{8}\int \left(1 - 2\cos 2x + \cos^2 2x\right)(1+\cos 2x)\,dx$$
$$= \frac{1}{8}\int \left(1 - \cos 2x - \cos^2 2x + \cos^3 2x\right)dx.$$

Now use the substitution $u = 2x$, $du = 2\,dx$, together with the reduction formula for $\cos^m x$:

$$J = \frac{1}{8}x - \frac{1}{16}\int \cos u\,du - \frac{1}{16}\int \cos^2 u\,du + \frac{1}{16}\int \cos^3 u\,du$$
$$= \frac{1}{8}x - \frac{1}{16}\sin u - \frac{1}{16}\left(\frac{u}{2} + \frac{1}{2}\sin u\cos u\right) + \frac{1}{16}\left(\frac{1}{3}\cos^2 u\sin u + \frac{2}{3}\int \cos u\,du\right)$$
$$= \frac{1}{8}x - \frac{1}{16}\sin 2x - \frac{1}{32}(2x) - \frac{1}{32}\sin 2x\cos 2x + \frac{1}{48}\cos^2 2x\sin 2x + \frac{1}{24}\sin 2x + C$$
$$= \frac{1}{16}x - \frac{1}{48}\sin 2x - \frac{1}{32}\sin 2x\cos 2x + \frac{1}{48}\cos^2 2x\sin 2x + C.$$

69. Prove the reduction formula

$$\int \tan^k x\,dx = \frac{\tan^{k-1} x}{k-1} - \int \tan^{k-2} x\,dx$$

Hint: $\tan^k x = (\sec^2 x - 1)\tan^{k-2} x$.

SOLUTION Use the identity $\tan^2 x = \sec^2 x - 1$ to write

$$\int \tan^k x\,dx = \int \tan^{k-2} x\left(\sec^2 x - 1\right)dx = \int \tan^{k-2} x\sec^2 x\,dx - \int \tan^{k-2} x\,dx.$$

Now use the substitution $u = \tan x$, $du = \sec^2 x\,dx$:

$$\int \tan^k x\,dx = \int u^{k-2}\,du - \int \tan^{k-2} x\,dx = \frac{1}{k-1}u^{k-1} - \int \tan^{k-2} x\,dx = \frac{\tan^{k-1} x}{k-1} - \int \tan^{k-2} x\,dx.$$

71. Let $I_m = \int_0^{\pi/2} \sin^m x\,dx$.

(a) Show that $I_0 = \frac{\pi}{2}$ and $I_1 = 1$.

(b) Prove that, for $m \geq 2$,

$$I_m = \frac{m-1}{m}I_{m-2}$$

(c) Use (a) and (b) to compute I_m for $m = 2, 3, 4, 5$.

SOLUTION

(a) We have

$$I_0 = \int_0^{\pi/2} \sin^0 x\,dx = \int_0^{\pi/2} 1\,dx = \frac{\pi}{2}$$
$$I_1 = \int_0^{\pi/2} \sin x\,dx = -\cos x\Big|_0^{\pi/2} = 1$$

(b) Using the reduction formula for $\sin^m x$, we get for $m \geq 2$

$$I_m = \int_0^{\pi/2} \sin^m x\,dx = -\frac{1}{m}\sin^{m-1} x\cos x\Big|_0^{\pi/2} + \frac{m-1}{m}\int_0^{\pi/2} \sin^{m-2} x\,dx$$
$$= -\frac{1}{m}\sin^{m-1}\left(\frac{\pi}{2}\right)\cos\left(\frac{\pi}{2}\right) + \frac{1}{m}\sin^{m-1}(0)\cos(0) + \frac{m-1}{m}I_{m-2}$$
$$= \frac{1}{m}(-1\cdot 0 + 0\cdot 1) + \frac{m-1}{m}I_{m-2}$$
$$= \frac{m-1}{m}I_{m-2}$$

(c)

$$I_2 = \frac{1}{2}I_0 = \frac{1}{2}\cdot\frac{\pi}{2} = \frac{\pi}{4}$$
$$I_3 = \frac{2}{3}I_1 = \frac{2}{3}$$
$$I_4 = \frac{3}{4}I_2 = \frac{3}{4}\cdot\frac{\pi}{4} = \frac{3}{16}\pi$$
$$I_5 = \frac{4}{5}I_3 = \frac{8}{15}$$

73. Evaluate $\int \sin x \ln(\sin x)\,dx$. *Hint:* Use Integration by Parts as a first step.

SOLUTION Start by using integration by parts with $u = \ln(\sin x)$ and $v' = \sin x$, so that $u' = \cot x$ and $v = -\cos x$. Then

$$I = \int \sin x \ln(\sin x)\,dx = -\cos x \ln(\sin x) + \int \cot x \cos x\,dx = -\cos x \ln(\sin x) + \int \frac{\cos^2 x}{\sin x}\,dx$$
$$= -\cos x \ln(\sin x) + \int \frac{1-\sin^2 x}{\sin x}\,dx = -\cos x \ln(\sin x) - \int \sin x\,dx + \int \csc x\,dx$$
$$= -\cos x \ln(\sin x) + \cos x + \int \csc x\,dx$$

Using the table, $\int \csc x\,dx = \ln|\csc x - \cot x| + C$, so finally

$$I = -\cos x \ln(\sin x) + \cos x + \ln|\csc x - \cot x| + C$$

75. Let m, n be integers with $m \neq \pm n$. Use Eqs. (23)–(25) to prove the so-called **orthogonality relations** that play a basic role in the theory of Fourier Series (Figure 2):

$$\int_0^{\pi} \sin mx \sin nx\,dx = 0$$
$$\int_0^{\pi} \cos mx \cos nx\,dx = 0$$
$$\int_0^{2\pi} \sin mx \cos nx\,dx = 0$$

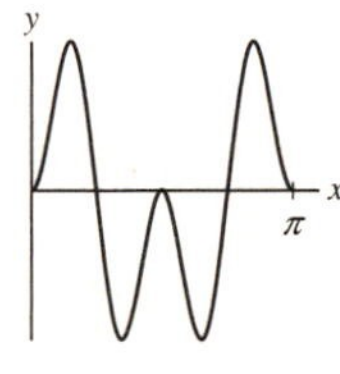

$y = \sin 2x \sin 4x$ $\qquad$ $y = \sin 3x \cos 4x$

FIGURE 2 The integrals are zero by the orthogonality relations.

SOLUTION If m, n are integers, then $m-n$ and $m+n$ are integers, and therefore $\sin(m-n)\pi = \sin(m+n)\pi = 0$, since $\sin k\pi = 0$ if k is an integer. Thus we have

$$\int_0^{\pi} \sin mx \sin nx\,dx = \left(\frac{\sin(m-n)x}{2(m-n)} - \frac{\sin(m+n)x}{2(m+n)}\right)\Bigg|_0^{\pi} = \left(\frac{\sin(m-n)\pi}{2(m-n)} - \frac{\sin(m+n)\pi}{2(m+n)}\right) - 0 = 0;$$
$$\int_0^{\pi} \cos mx \cos nx\,dx = \left(\frac{\sin(m-n)x}{2(m-n)} + \frac{\sin(m+n)x}{2(m+n)}\right)\Bigg|_0^{\pi} = \left(\frac{\sin(m-n)\pi}{2(m-n)} + \frac{\sin(m+n)\pi}{2(m+n)}\right) - 0 = 0.$$

If k is an integer, then $\cos 2k\pi = 1$. Using this fact, we have

$$\int_0^{2\pi} \sin mx \cos nx\,dx = \left(-\frac{\cos(m-n)x}{2(m-n)} - \frac{\cos(m+n)x}{2(m+n)}\right)\Bigg|_0^{2\pi}$$
$$= \left(-\frac{\cos(m-n)2\pi}{2(m-n)} - \frac{\cos(m+n)2\pi}{2(m+n)}\right) - \left(-\frac{1}{2(m-n)} - \frac{1}{2(m+n)}\right)$$
$$= \left(-\frac{1}{2(m-n)} - \frac{1}{2(m+n)}\right) - \left(-\frac{1}{2(m-n)} - \frac{1}{2(m+n)}\right) = 0.$$

Further Insights and Challenges

77. Use Integration by Parts to prove that (for $m \neq 1$)

$$\int \sec^m x\,dx = \frac{\tan x \sec^{m-2} x}{m-1} + \frac{m-2}{m-1}\int \sec^{m-2} x\,dx$$

SOLUTION Using Integration by Parts with $u = \sec^{m-2} x$ and $v' = \sec^2 x$, we have $v = \tan x$ and

$$u' = (m-2)\sec^{m-3} x(\sec x \tan x) = (m-2)\tan x \sec^{m-2} x.$$

Then,

$$\begin{aligned}\int \sec^m x\,dx &= \tan x \sec^{m-2} x - (m-2)\int \tan^2 x \sec^{m-2} x\,dx\\ &= \tan x \sec^{m-2} x - (m-2)\int \left(\sec^2 x - 1\right)\sec^{m-2} x\,dx\\ &= \tan x \sec^{m-2} x - (m-2)\int \sec^m x\,dx + (m-2)\int \sec^{m-2} x\,dx.\end{aligned}$$

Solving this equation for $\int \sec^m x\,dx$, we get

$$\begin{aligned}(m-1)\int \sec^m x\,dx &= \tan x \sec^{m-2} x + (m-2)\int \sec^{m-2} x\,dx\\ \int \sec^m x\,dx &= \frac{\tan x \sec^{m-2} x}{m-1} + \frac{m-2}{m-1}\int \sec^{m-2} x\,dx.\end{aligned}$$

79. This is a continuation of Exercise 78.

(a) Prove that $I_{2m+1} \le I_{2m} \le I_{2m-1}$. *Hint:* $\sin^{2m+1} x \le \sin^{2m} x \le \sin^{2m-1} x$ for $0 \le x \le \frac{\pi}{2}$.

(b) Show that $\dfrac{I_{2m-1}}{I_{2m+1}} = 1 + \dfrac{1}{2m}$.

(c) Show that $1 \le \dfrac{I_{2m}}{I_{2m+1}} \le 1 + \dfrac{1}{2m}$.

(d) Prove that $\displaystyle\lim_{m\to\infty} \frac{I_{2m}}{I_{2m+1}} = 1$.

(e) Finally, deduce the infinite product for $\frac{\pi}{2}$ discovered by English mathematician John Wallis (1616–1703):

$$\frac{\pi}{2} = \lim_{m\to\infty} \frac{2}{1}\cdot\frac{2}{3}\cdot\frac{4}{3}\cdot\frac{4}{5}\cdots\frac{2m\cdot 2m}{(2m-1)(2m+1)}$$

SOLUTION

(a) For $0 \le x \le \frac{\pi}{2}$, $0 \le \sin x \le 1$. Multiplying this last inequality by $\sin x$, we obtain

$$0 \le \sin^2 x \le \sin x.$$

Continuing to multiply this inequality by $\sin x$, we obtain, more generally,

$$\sin^{2m+1} x \le \sin^{2m} x \le \sin^{2m-1} x.$$

Integrating these functions over $[0, \frac{\pi}{2}]$, we get

$$\int_0^{\pi/2} \sin^{2m+1} x\,dx \le \int_0^{\pi/2} \sin^{2m} x\,dx \le \int_0^{\pi/2} \sin^{2m-1} x\,dx,$$

which is the same as

$$I_{2m+1} \le I_{2m} \le I_{2m-1}.$$

(b) Using the relation $I_m = ((m-1)/m)I_{m-2}$, we have

$$\frac{I_{2m-1}}{I_{2m+1}} = \frac{I_{2m-1}}{\left(\frac{2m}{2m+1}\right) I_{2m-1}} = \frac{2m+1}{2m} = \frac{2m}{2m} + \frac{1}{2m} = 1 + \frac{1}{2m}.$$

(c) First start with the inequality of part (a):

$$I_{2m+1} \le I_{2m} \le I_{2m-1}.$$

Divide through by I_{2m+1}:

$$1 \le \frac{I_{2m}}{I_{2m+1}} \le \frac{I_{2m-1}}{I_{2m+1}}.$$

Use the result from part (b):

$$1 \le \frac{I_{2m}}{I_{2m+1}} \le 1 + \frac{1}{2m}.$$

(d) Taking the limit of this inequality, and applying the Squeeze Theorem, we have

$$\lim_{m\to\infty} 1 \le \lim_{m\to\infty} \frac{I_{2m}}{I_{2m+1}} \le \lim_{m\to\infty} \left(1 + \frac{1}{2m}\right).$$

Because

$$\lim_{m\to\infty} 1 = 1 \quad \text{and} \quad \lim_{m\to\infty} \left(1 + \frac{1}{2m}\right) = 1,$$

we obtain

$$1 \le \lim_{m\to\infty} \frac{I_{2m}}{I_{2m+1}} \le 1.$$

Therefore

$$\lim_{m\to\infty} \frac{I_{2m}}{I_{2m+1}} = 1.$$

(e) Take the limit of both sides of the equation obtained in Exercise 78(d):

$$\lim_{m\to\infty} \frac{\pi}{2} = \lim_{m\to\infty} \frac{2\cdot 2}{1\cdot 3} \cdot \frac{4\cdot 4}{3\cdot 5} \cdots \frac{2m\cdot 2m}{(2m-1)(2m+1)} \frac{I_{2m}}{I_{2m+1}}$$

$$\frac{\pi}{2} = \left(\lim_{m\to\infty} \frac{2\cdot 2}{1\cdot 3} \cdot \frac{4\cdot 4}{3\cdot 5} \cdots \frac{2m\cdot 2m}{(2m-1)(2m+1)}\right)\left(\lim_{m\to\infty} \frac{I_{2m}}{I_{2m+1}}\right).$$

Finally, using the result from (d), we have

$$\frac{\pi}{2} = \lim_{m\to\infty} \frac{2\cdot 2}{1\cdot 3} \cdot \frac{4\cdot 4}{3\cdot 5} \cdots \frac{2m\cdot 2m}{(2m-1)(2m+1)}.$$

7.3 Trigonometric Substitution

Preliminary Questions

1. State the trigonometric substitution appropriate to the given integral:

(a) $\int \sqrt{9 - x^2}\,dx$

(b) $\int x^2(x^2 - 16)^{3/2}\,dx$

(c) $\int x^2(x^2 + 16)^{3/2}\,dx$

(d) $\int (x^2 - 5)^{-2}\,dx$

SOLUTION

(a) $x = 3\sin\theta$ **(b)** $x = 4\sec\theta$ **(c)** $x = 4\tan\theta$ **(d)** $x = \sqrt{5}\sec\theta$

2. Is trigonometric substitution needed to evaluate $\int x\sqrt{9 - x^2}\,dx$?

SOLUTION No. There is a factor of x in the integrand outside the radical and the derivative of $9 - x^2$ is $-2x$, so we may use the substitution $u = 9 - x^2$, $du = -2x\,dx$ to evaluate this integral.

3. Express $\sin 2\theta$ in terms of $x = \sin\theta$.

SOLUTION First note that if $\sin\theta = x$, then $\cos\theta = \sqrt{1 - \sin^2\theta} = \sqrt{1 - x^2}$. Thus,

$$\sin 2\theta = 2\sin\theta\cos\theta = 2x\sqrt{1 - x^2}.$$

4. Draw a triangle that would be used together with the substitution $x = 3\sec\theta$.

SOLUTION

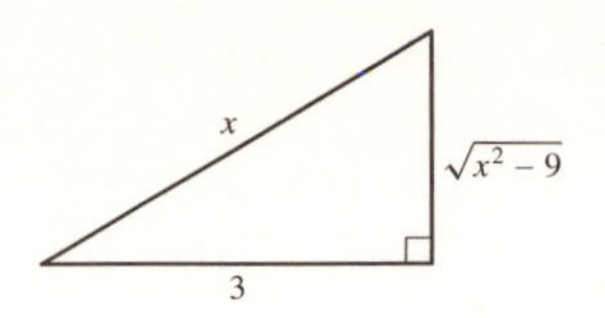

Exercises

In Exercises 1–4, evaluate the integral by following the steps given.

1. $I = \int \frac{dx}{\sqrt{9-x^2}}$

(a) Show that the substitution $x = 3\sin\theta$ transforms I into $\int d\theta$, and evaluate I in terms of θ.

(b) Evaluate I in terms of x.

SOLUTION

(a) Let $x = 3\sin\theta$. Then $dx = 3\cos\theta\, d\theta$, and

$$\sqrt{9-x^2} = \sqrt{9-9\sin^2\theta} = 3\sqrt{1-\sin^2\theta} = 3\sqrt{\cos^2\theta} = 3\cos\theta.$$

Thus,

$$I = \int \frac{dx}{\sqrt{9-x^2}} = \int \frac{3\cos\theta\, d\theta}{3\cos\theta} = \int d\theta = \theta + C.$$

(b) If $x = 3\sin\theta$, then $\theta = \sin^{-1}(\frac{x}{3})$. Thus,

$$I = \theta + C = \sin^{-1}\left(\frac{x}{3}\right) + C.$$

3. $I = \int \frac{dx}{\sqrt{4x^2+9}}$

(a) Show that the substitution $x = \frac{3}{2}\tan\theta$ transforms I into $\frac{1}{2}\int \sec\theta\, d\theta$.

(b) Evaluate I in terms of θ (refer to the table of integrals on page 410 in Section 7.2 if necessary).

(c) Express I in terms of x.

SOLUTION

(a) If $x = \frac{3}{2}\tan\theta$, then $dx = \frac{3}{2}\sec^2\theta\, d\theta$, and

$$\sqrt{4x^2+9} = \sqrt{4\cdot\left(\frac{3}{2}\tan\theta\right)^2+9} = \sqrt{9\tan^2\theta+9} = 3\sqrt{\sec^2\theta} = 3\sec\theta$$

Thus,

$$I = \int \frac{dx}{\sqrt{4x^2+9}} = \int \frac{\frac{3}{2}\sec^2\theta\, d\theta}{3\sec\theta} = \frac{1}{2}\int \sec\theta\, d\theta$$

(b)

$$I = \frac{1}{2}\int \sec\theta\, d\theta = \frac{1}{2}\ln|\sec\theta + \tan\theta| + C$$

(c) Since $x = \frac{3}{2}\tan\theta$, we construct a right triangle with $\tan\theta = \frac{2x}{3}$:

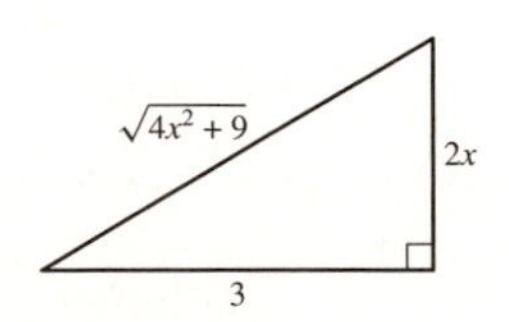

From this triangle, we see that $\sec\theta = \frac{1}{3}\sqrt{4x^2+9}$, and therefore

$$I = \frac{1}{2}\ln|\sec\theta + \tan\theta| + C = \frac{1}{2}\ln\left|\frac{1}{3}\sqrt{4x^2+9} + \frac{2x}{3}\right| + C$$

$$= \frac{1}{2}\ln\left|\frac{\sqrt{4x^2+9}+2x}{3}\right| + C = \frac{1}{2}\ln|\sqrt{4x^2+9}+2x| - \frac{1}{2}\ln 3 + C = \frac{1}{2}\ln|\sqrt{4x^2+9}+2x| + C$$

In Exercises 5–10, use the indicated substitution to evaluate the integral.

5. $\int \sqrt{16-5x^2}\,dx, \quad x = \frac{4}{\sqrt{5}}\sin\theta$

SOLUTION Let $x = \frac{4}{\sqrt{5}}\sin\theta$. Then $dx = \frac{4}{\sqrt{5}}\cos\theta\,d\theta$, and

$$I = \int \sqrt{16-5x^2}\,dx = \int \sqrt{16 - 5\left(\frac{4}{\sqrt{5}}\sin\theta\right)^2}\cdot\frac{4}{\sqrt{5}}\cos\theta\,d\theta = \frac{4}{\sqrt{5}}\int\sqrt{16-16\sin^2\theta}\cdot\cos\theta\,d\theta$$

$$= \frac{4}{\sqrt{5}}\cdot 4\int\cos\theta\cdot\cos\theta\,d\theta = \frac{16}{\sqrt{5}}\int\cos^2\theta\,d\theta$$

$$= \frac{16}{\sqrt{5}}\left(\frac{1}{2}\theta + \frac{1}{2}\sin\theta\cos\theta\right) + C = \frac{8}{\sqrt{5}}(\theta + \sin\theta\cos\theta) + C$$

Since $x = \frac{4}{\sqrt{5}}\sin\theta$, we construct a right triangle with $\sin\theta = \frac{x\sqrt{5}}{4}$:

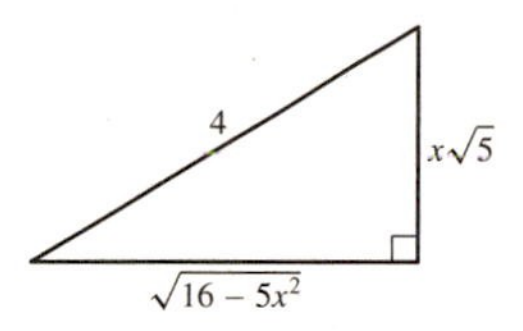

From this triangle we see that $\cos\theta = \frac{1}{4}\sqrt{16-5x^2}$, so we have

$$I = \frac{8}{\sqrt{5}}(\theta + \sin\theta\cos\theta) + C$$

$$= \frac{8}{\sqrt{5}}\left(\sin^{-1}\left(\frac{x\sqrt{5}}{4}\right) + \frac{x\sqrt{5}}{4}\cdot\frac{1}{4}\sqrt{16-5x^2}\right) + C$$

$$= \frac{8}{\sqrt{5}}\sin^{-1}\left(\frac{x\sqrt{5}}{4}\right) + \frac{1}{2}x\sqrt{16-5x^2} + C$$

7. $\int \frac{dx}{x\sqrt{x^2-9}}, \quad x = 3\sec\theta$

SOLUTION Let $x = 3\sec\theta$. Then $dx = 3\sec\theta\tan\theta\,d\theta$, and

$$\sqrt{x^2-9} = \sqrt{9\sec^2\theta - 9} = 3\sqrt{\sec^2\theta - 1} = 3\sqrt{\tan^2\theta} = 3\tan\theta.$$

Thus,

$$\int\frac{dx}{x\sqrt{x^2-9}} = \int\frac{(3\sec\theta\tan\theta\,d\theta)}{(3\sec\theta)(3\tan\theta)} = \frac{1}{3}\int d\theta = \frac{1}{3}\theta + C.$$

Since $x = 3\sec\theta$, $\theta = \sec^{-1}(\frac{x}{3})$, and

$$\int\frac{dx}{x\sqrt{x^2-9}} = \frac{1}{3}\sec^{-1}\left(\frac{x}{3}\right) + C.$$

9. $\int \frac{dx}{(x^2-4)^{3/2}}, \quad x = 2\sec\theta$

SOLUTION Let $x = 2\sec\theta$. Then $dx = 2\sec\theta\tan\theta\,d\theta$, and

$$x^2 - 4 = 4\sec^2\theta - 4 = 4(\sec^2\theta - 1) = 4\tan^2\theta.$$

This gives

$$I = \int \frac{dx}{(x^2-4)^{3/2}} = \int \frac{2\sec\theta\tan\theta\,d\theta}{(4\tan^2\theta)^{3/2}} = \int \frac{2\sec\theta\tan\theta\,d\theta}{8\tan^3\theta} = \frac{1}{4}\int \frac{\sec\theta\,d\theta}{\tan^2\theta} = \frac{1}{4}\int \frac{\cos\theta}{\sin^2\theta}\,d\theta.$$

Now use substitution with $u = \sin\theta$ and $du = \cos\theta\,d\theta$. Then

$$I = \frac{1}{4}\int u^{-2}\,du = -\frac{1}{4}u^{-1} + C = \frac{-1}{4\sin\theta} + C.$$

Since $x = 2\sec\theta$, we construct a right triangle with $\sec\theta = \frac{x}{2}$:

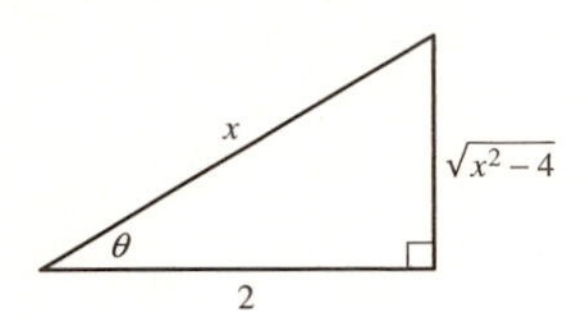

From this triangle we see that $\sin\theta = \sqrt{x^2-4}/x$, so therefore

$$I = \frac{-1}{4(\sqrt{x^2-4}/x)} + C = \frac{-x}{4\sqrt{x^2-4}} + C.$$

11. Evaluate $\int \frac{x\,dx}{\sqrt{x^2-4}}$ in two ways: using the direct substitution $u = x^2 - 4$ and by trigonometric substitution.

SOLUTION Let $u = x^2 - 4$. Then $du = 2x\,dx$, and

$$I_1 = \int \frac{x\,dx}{\sqrt{x^2-4}} = \frac{1}{2}\int \frac{du}{\sqrt{u}} = \frac{1}{2}\left(2u^{1/2}\right) + C = \sqrt{u} + C = \sqrt{x^2-4} + C.$$

To use trigonometric substitution, let $x = 2\sec\theta$. Then $dx = 2\sec\theta\tan\theta\,d\theta$, $x^2 - 4 = 4\sec^2\theta - 4 = 4\tan^2\theta$, and

$$I_1 = \int \frac{x\,dx}{\sqrt{x^2-4}} = \int \frac{2\sec\theta(2\sec\theta\tan\theta\,d\theta)}{2\tan\theta} = 2\int \sec^2\theta\,d\theta = 2\tan\theta + C.$$

Since $x = 2\sec\theta$, we construct a right triangle with $\sec\theta = \frac{x}{2}$:

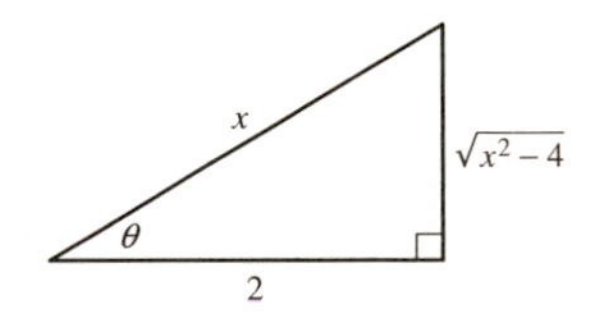

From this triangle we see that

$$I_1 = 2\left(\frac{\sqrt{x^2-4}}{2}\right) + C = \sqrt{x^2-4} + C.$$

13. Evaluate using the substitution $u = 1 - x^2$ or trigonometric substitution.

(a) $\int \frac{x}{\sqrt{1-x^2}}\,dx$

(b) $\int x^2\sqrt{1-x^2}\,dx$

(c) $\int x^3\sqrt{1-x^2}\,dx$

(d) $\int \frac{x^4}{\sqrt{1-x^2}}\,dx$

SOLUTION

(a) Let $u = 1 - x^2$. Then $du = -2x\,dx$, and we have

$$\int \frac{x}{\sqrt{1-x^2}}\,dx = -\frac{1}{2}\int \frac{-2x\,dx}{\sqrt{1-x^2}} = -\frac{1}{2}\int \frac{du}{u^{1/2}}.$$

(b) Let $x = \sin\theta$. Then $dx = \cos\theta\,d\theta$, $1 - x^2 = \cos^2\theta$, and so

$$\int x^2\sqrt{1-x^2}\,dx = \int \sin^2\theta(\cos\theta)\cos\theta\,d\theta = \int \sin^2\theta\cos^2\theta\,d\theta.$$

(c) Use the substitution $u = 1 - x^2$. Then $du = -2x\,dx$, $x^2 = 1 - u$, and so

$$\int x^3\sqrt{1-x^2}\,dx = -\frac{1}{2}\int x^2\sqrt{1-x^2}(-2x\,dx) = -\frac{1}{2}\int (1-u)u^{1/2}\,du.$$

(d) Let $x = \sin\theta$. Then $dx = \cos\theta\,d\theta$, $1 - x^2 = \cos^2\theta$, and so

$$\int \frac{x^4}{\sqrt{1-x^2}}\,dx = \int \frac{\sin^4\theta}{\cos\theta}\cos\theta\,d\theta = \int \sin^4\theta\,d\theta.$$

In Exercises 15–32, evaluate using trigonometric substitution. Refer to the table of trigonometric integrals as necessary.

15. $\displaystyle\int \frac{x^2\,dx}{\sqrt{9-x^2}}$

SOLUTION Let $x = 3\sin\theta$. Then $dx = 3\cos\theta\,d\theta$,

$$9 - x^2 = 9 - 9\sin^2\theta = 9(1-\sin^2\theta) = 9\cos^2\theta,$$

and

$$I = \int \frac{x^2\,dx}{\sqrt{9-x^2}} = \int \frac{9\sin^2\theta(3\cos\theta\,d\theta)}{3\cos\theta} = 9\int \sin^2\theta\,d\theta = 9\left[\frac{1}{2}\theta - \frac{1}{2}\sin\theta\cos\theta\right] + C.$$

Since $x = 3\sin\theta$, we construct a right triangle with $\sin\theta = \frac{x}{3}$:

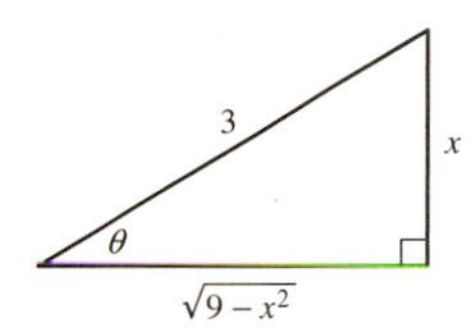

From this we see that $\cos\theta = \sqrt{9-x^2}/3$, and so

$$I = \frac{9}{2}\sin^{-1}\left(\frac{x}{3}\right) - \frac{9}{2}\left(\frac{x}{3}\right)\left(\frac{\sqrt{9-x^2}}{3}\right) + C = \frac{9}{2}\sin^{-1}\left(\frac{x}{3}\right) - \frac{1}{2}x\sqrt{9-x^2} + C.$$

17. $\displaystyle\int \frac{dx}{x\sqrt{x^2+16}}$

SOLUTION Use the substitution $x = 4\tan\theta$, so that $dx = 4\sec^2\theta\,d\theta$. Then

$$x\sqrt{x^2+16} = 4\tan\theta\sqrt{(4\tan\theta)^2+16} = 4\tan\theta\sqrt{16(\tan^2\theta+1)} = 16\tan\theta\sec\theta$$

so that

$$I = \int \frac{dx}{x\sqrt{x^2+16}} = \int \frac{4\sec^2\theta}{16\tan\theta\sec\theta}\,d\theta = \frac{1}{4}\int \frac{\sec\theta}{\tan\theta}\,d\theta = \frac{1}{4}\int \csc\theta\,d\theta = -\frac{1}{4}\ln|\csc x + \cot x| + C$$

Since $x = 4\tan\theta$, we construct a right triangle with $\tan\theta = \frac{x}{4}$:

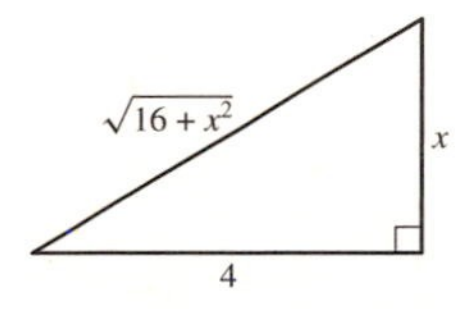

From this, we see that $\csc x = \frac{\sqrt{x^2+16}}{x}$ and $\cot x = \frac{4}{x}$, so that

$$I = -\frac{1}{4}\ln|\csc x + \cot x| + C = -\frac{1}{4}\ln\left|\frac{\sqrt{x^2+16}}{x} + \frac{4}{x}\right| + C = -\frac{1}{4}\ln\left|\frac{4+\sqrt{x^2+16}}{x}\right| + C$$

19. $\displaystyle\int \frac{dx}{\sqrt{x^2-9}}$

SOLUTION Let $x = 3\sec\theta$. Then $dx = 3\sec\theta\tan\theta\,d\theta$,

$$x^2 - 9 = 9\sec^2\theta - 9 = 9(\sec^2\theta - 1) = 9\tan^2\theta,$$

and

$$I = \int \frac{dx}{\sqrt{x^2 - 9}} = \int \frac{3\sec\theta\tan\theta\,d\theta}{3\tan\theta} = \int \sec\theta\,d\theta = \ln|\sec\theta + \tan\theta| + C.$$

Since $x = 3\sec\theta$, we construct a right triangle with $\sec\theta = \frac{x}{3}$:

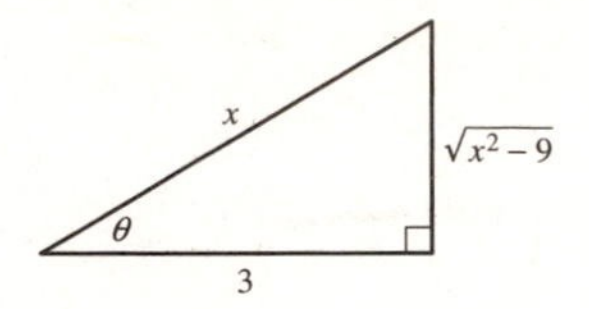

From this we see that $\tan\theta = \sqrt{x^2 - 9}/3$, and so

$$I = \ln\left|\frac{x}{3} + \frac{\sqrt{x^2 - 9}}{3}\right| + C_1 = \ln\left|x + \sqrt{x^2 - 9}\right| + \ln\left(\frac{1}{3}\right) + C_1 = \ln\left|x + \sqrt{x^2 - 9}\right| + C,$$

where $C = \ln\left(\frac{1}{3}\right) + C_1$.

21. $\displaystyle\int \frac{dy}{y^2\sqrt{5 - y^2}}$

SOLUTION Let $y = \sqrt{5}\sin\theta$. Then $dy = \sqrt{5}\cos\theta\,d\theta$,

$$5 - y^2 = 5 - 5\sin^2\theta = 5(1 - \sin^2\theta) = 5\cos^2\theta,$$

and

$$I = \int \frac{dy}{y^2\sqrt{5 - y^2}} = \int \frac{\sqrt{5}\cos\theta\,d\theta}{(5\sin^2\theta)(\sqrt{5}\cos\theta)} = \frac{1}{5}\int \frac{d\theta}{\sin^2\theta} = \frac{1}{5}\int \csc^2\theta\,d\theta = \frac{1}{5}(-\cot\theta) + C.$$

Since $y = \sqrt{5}\sin\theta$, we construct a right triangle with $\sin\theta = \frac{y}{\sqrt{5}}$:

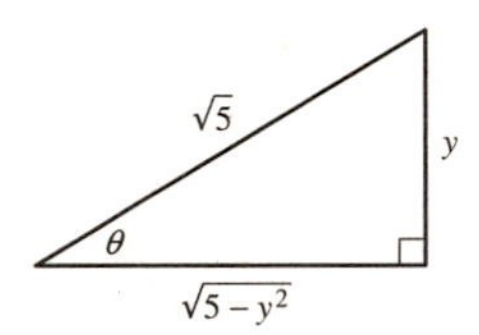

From this we see that $\cot\theta = \sqrt{5 - y^2}/y$, which gives us

$$I = \frac{1}{5}\left(\frac{-\sqrt{5 - y^2}}{y}\right) + C = -\frac{\sqrt{5 - y^2}}{5y} + C.$$

23. $\displaystyle\int \frac{dx}{\sqrt{25x^2 + 2}}$

SOLUTION Let $x = \frac{\sqrt{2}}{5}\tan\theta$. Then $dx = \frac{\sqrt{2}}{5}\sec^2\theta\,d\theta$, $25x^2 + 2 = 2\tan^2\theta + 2 = 2\sec^2\theta$, and

$$I = \int \frac{dx}{\sqrt{25x^2 + 2}} = \int \frac{\frac{\sqrt{2}}{5}\sec^2\theta\,d\theta}{\sqrt{2}\sec\theta} = \frac{1}{5}\int \sec\theta\,d\theta = \frac{1}{5}\ln|\sec\theta + \tan\theta| + C.$$

Since $x = \frac{\sqrt{2}}{5}\tan\theta$, we construct a right triangle with $\tan\theta = \frac{5x}{\sqrt{2}}$:

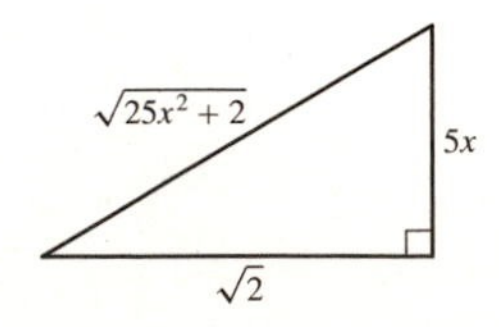

From this we see that $\sec\theta = \frac{1}{\sqrt{2}}\sqrt{25x^2+2}$, so that

$$I = \frac{1}{5}\ln|\sec\theta + \tan\theta| + C = \frac{1}{5}\ln\left|\frac{\sqrt{25x^2+2}}{\sqrt{2}} + \frac{5x}{\sqrt{2}}\right| + C$$

$$= \frac{1}{5}\ln\left|\frac{5x+\sqrt{25x^2+2}}{\sqrt{2}}\right| + C = \frac{1}{5}\ln\left|5x+\sqrt{25x^2+2}\right| - \frac{1}{5}\ln\sqrt{2} + C$$

$$= \frac{1}{5}\ln\left|5x+\sqrt{25x^2+2}\right| + C$$

25. $\displaystyle\int \frac{dz}{z^3\sqrt{z^2-4}}$

SOLUTION Let $z = 2\sec\theta$. Then $dz = 2\sec\theta\tan\theta\,d\theta$,

$$z^2 - 4 = 4\sec^2\theta - 4 = 4(\sec^2\theta - 1) = 4\tan^2\theta,$$

and

$$I = \int \frac{dz}{z^3\sqrt{z^2-4}} = \int \frac{2\sec\theta\tan\theta\,d\theta}{(8\sec^3\theta)(2\tan\theta)} = \frac{1}{8}\int\frac{d\theta}{\sec^2\theta} = \frac{1}{8}\int\cos^2\theta\,d\theta$$

$$= \frac{1}{8}\left[\frac{1}{2}\theta + \frac{1}{2}\sin\theta\cos\theta\right] + C = \frac{1}{16}\theta + \frac{1}{16}\sin\theta\cos\theta + C.$$

As explained in the text, this computation is valid if we choose θ in $[0, \pi/2)$ if $z > 2$ and in $[\pi, 3\pi/2)$ if $z \le -2$.

If $z \ge 2$, we may construct a right triangle with $\sec\theta = \frac{z}{2}$:

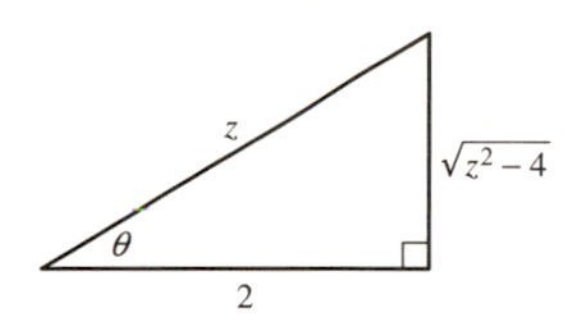

From this we see that $\sin\theta = \sqrt{z^2-4}/z$ and $\cos\theta = 2/z$. Then

$$I = \frac{1}{16}\sec^{-1}\left(\frac{z}{2}\right) + \frac{1}{16}\left(\frac{\sqrt{z^2-4}}{z}\right)\left(\frac{2}{z}\right) + C = \frac{1}{16}\sec^{-1}\left(\frac{z}{2}\right) + \frac{\sqrt{z^2-4}}{8z^2} + C.$$

However, if $z \le -2$, then $\sec^{-1}\left(\frac{z}{2}\right)$ lies in $\left(\frac{\pi}{2}, \pi\right]$ according to the definition of $\sec^{-1} x$ used in the text. But since θ is the angle in $\left[\pi, \frac{3\pi}{2}\right)$ satisfying $\sec\theta = z/2$, we find that $\theta = 2\pi - \sec^{-1}\left(\frac{z}{2}\right)$. Similarly,

$$\sin\theta = -\sqrt{z^2-4}/z \quad \text{and} \quad \cos\theta = -2/z$$

So for $z \le -2$,

$$I = -\frac{1}{16}\sec^{-1}\left(\frac{z}{2}\right) + \frac{\sqrt{z^2-4}}{8z^2} + C.$$

Note that although $\theta = 2\pi - \sec^{-1}\left(\frac{z}{2}\right)$, the 2π is not needed in the expression for I because it may be absorbed in the constant C.

27. $\displaystyle\int \frac{x^2\,dx}{(6x^2-49)^{1/2}}$

SOLUTION Let $x = \frac{7}{\sqrt{6}}\sec\theta$; then $dx = \frac{7}{\sqrt{6}}\sec\theta\tan\theta\,d\theta$, and

$$6x^2 - 49 = 6\left(\frac{7}{\sqrt{6}}\sec\theta\right)^2 - 49 = 49(\sec^2\theta - 1) = 49\tan^2\theta$$

so that

$$I = \int \frac{x^2\,dx}{(6x^2-49)^{1/2}} = \int \frac{\frac{49}{6}\sec^2\theta\left(\frac{7}{\sqrt{6}}\sec\theta\tan\theta\right)}{7\tan\theta}\,d\theta$$

$$= \frac{49}{6\sqrt{6}}\int\sec^3\theta\,d\theta = \frac{49}{6\sqrt{6}}\left(\frac{1}{2}\tan\theta\sec\theta + \frac{1}{2}\int\sec\theta\,d\theta\right)$$

$$= \frac{49}{12\sqrt{6}}(\tan\theta\sec\theta + \ln|\sec\theta + \tan\theta|) + C$$

Since $x = \frac{7}{\sqrt{6}} \sec\theta$, we construct a right triangle with $\sec\theta = \frac{x\sqrt{6}}{7}$:

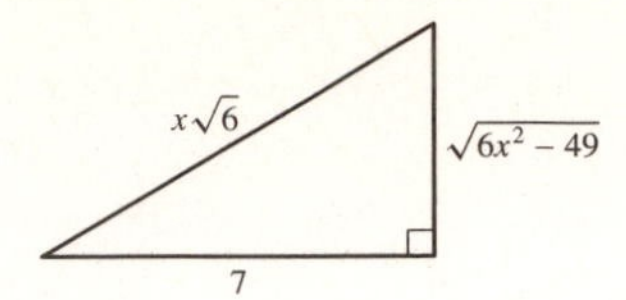

From this we see that $\tan\theta = \frac{1}{7}\sqrt{6x^2-49}$, so that

$$\begin{aligned} I &= \frac{49}{12\sqrt{6}}\left(\frac{x\sqrt{6}\sqrt{6x^2-49}}{49} + \ln\left|\frac{x\sqrt{6}+\sqrt{6x^2-49}}{7}\right|\right) + C \\ &= \frac{49}{12\sqrt{6}}\left(\frac{x\sqrt{6}\sqrt{6x^2-49}}{49} + \ln\left|x\sqrt{6}+\sqrt{6x^2-49}\right| - \ln 7\right) + C \\ &= \frac{1}{12\sqrt{6}}\left(x\sqrt{6}\sqrt{6x^2-49} + 49\ln\left|x\sqrt{6}+\sqrt{6x^2-49}\right|\right) + C \end{aligned}$$

29. $\displaystyle\int \frac{dt}{(t^2+9)^2}$

SOLUTION Let $t = 3\tan\theta$. Then $dt = 3\sec^2\theta\, d\theta$,

$$t^2+9 = 9\tan^2\theta + 9 = 9(\tan^2\theta+1) = 9\sec^2\theta,$$

and

$$I = \int \frac{dt}{(t^2+9)^2} = \int \frac{3\sec^2\theta\, d\theta}{81\sec^4\theta} = \frac{1}{27}\int \cos^2\theta\, d\theta = \frac{1}{27}\left[\frac{1}{2}\theta + \frac{1}{2}\sin\theta\cos\theta\right] + C.$$

Since $t = 3\tan\theta$, we construct a right triangle with $\tan\theta = \frac{t}{3}$:

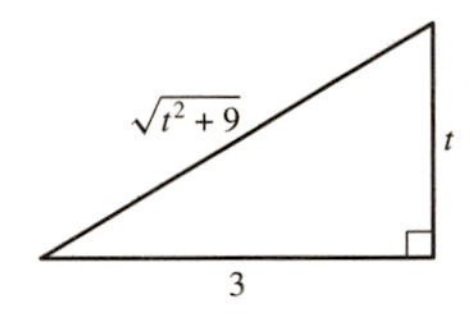

From this we see that $\sin\theta = t/\sqrt{t^2+9}$ and $\cos\theta = 3/\sqrt{t^2+9}$. Thus

$$I = \frac{1}{54}\tan^{-1}\left(\frac{t}{3}\right) + \frac{1}{54}\left(\frac{t}{\sqrt{t^2+9}}\right)\left(\frac{3}{\sqrt{t^2+9}}\right) + C = \frac{1}{54}\tan^{-1}\left(\frac{t}{3}\right) + \frac{t}{18(t^2+9)} + C.$$

31. $\displaystyle\int \frac{x^2\, dx}{(x^2-1)^{3/2}}$

SOLUTION Let $x = \sec\theta$. Then $dx = \sec\theta\tan\theta\, d\theta$, and $x^2 - 1 = \sec^2\theta - 1 = \tan^2\theta$. Thus

$$\begin{aligned} I &= \int \frac{x^2}{(x^2-1)^{3/2}}\, dx = \int \frac{\sec^2\theta}{(\tan^2\theta)^{3/2}} \sec\theta\tan\theta\, d\theta \\ &= \int \frac{\sec^2\theta\sec\theta\tan\theta}{\tan^3\theta}\, d\theta = \int \frac{\sec^3\theta}{\tan^2\theta}\, d\theta \\ &= \int \frac{\sec^2\theta}{\tan^2\theta}\sec\theta\, d\theta = \int \csc^2\theta\sec\theta\, d\theta = \int (1+\cot^2\theta)\sec\theta\, d\theta \\ &= \int \sec\theta + \cot\theta\csc\theta\, d\theta = \ln|\sec\theta + \tan\theta| - \csc\theta + C \end{aligned}$$

Since $x = \sec\theta$, we construct the following right triangle:

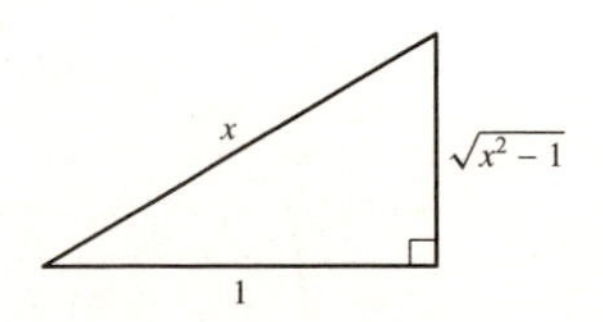

From this we see that $\tan\theta = \sqrt{x^2-1}$ and that $\csc\theta = \frac{x}{\sqrt{x^2-1}}$, so that

$$I = \ln\left|x + \sqrt{x^2-1}\right| - \frac{x}{\sqrt{x^2-1}} + C$$

33. Prove for $a > 0$:

$$\int \frac{dx}{x^2+a} = \frac{1}{\sqrt{a}}\tan^{-1}\frac{x}{\sqrt{a}} + C$$

SOLUTION Let $x = \sqrt{a}\,u$. Then, $x^2 = au^2$, $dx = \sqrt{a}\,du$, and

$$\int \frac{dx}{x^2+a} = \frac{1}{\sqrt{a}}\int \frac{du}{u^2+1} = \frac{1}{\sqrt{a}}\tan^{-1}u + C = \frac{1}{\sqrt{a}}\tan^{-1}\left(\frac{x}{\sqrt{a}}\right) + C.$$

35. Let $I = \displaystyle\int \frac{dx}{\sqrt{x^2-4x+8}}$.

(a) Complete the square to show that $x^2 - 4x + 8 = (x-2)^2 + 4$.

(b) Use the substitution $u = x - 2$ to show that $I = \displaystyle\int \frac{du}{\sqrt{u^2+2^2}}$. Evaluate the u-integral.

(c) Show that $I = \ln\left|\sqrt{(x-2)^2+4} + x - 2\right| + C$.

SOLUTION

(a) Completing the square, we get

$$x^2 - 4x + 8 = x^2 - 4x + 4 + 4 = (x-2)^2 + 4.$$

(b) Let $u = x - 2$. Then $du = dx$, and

$$I = \int \frac{dx}{\sqrt{x^2-4x+8}} = \int \frac{dx}{\sqrt{(x-2)^2+4}} = \int \frac{du}{\sqrt{u^2+4}}.$$

Now let $u = 2\tan\theta$. Then $du = 2\sec^2\theta\,d\theta$,

$$u^2 + 4 = 4\tan^2\theta + 4 = 4(\tan^2\theta + 1) = 4\sec^2\theta,$$

and

$$I = \int \frac{2\sec^2\theta\,d\theta}{2\sec\theta} = \int \sec\theta\,d\theta = \ln|\sec\theta + \tan\theta| + C.$$

Since $u = 2\tan\theta$, we construct a right triangle with $\tan\theta = \frac{u}{2}$:

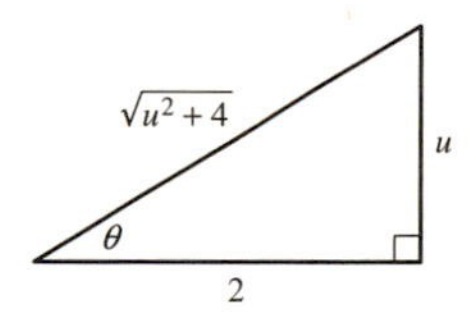

From this we see that $\sec\theta = \sqrt{u^2+4}/2$. Thus

$$I = \ln\left|\frac{\sqrt{u^2+4}}{2} + \frac{u}{2}\right| + C_1 = \ln\left|\sqrt{u^2+4} + u\right| + \left(\ln\frac{1}{2} + C_1\right) = \ln\left|\sqrt{u^2+4} + u\right| + C.$$

(c) Substitute back for x in the result of part (b):

$$I = \ln\left|\sqrt{(x-2)^2+4} + x - 2\right| + C.$$

In Exercises 37–42, evaluate the integral by completing the square and using trigonometric substitution.

37. $\displaystyle\int \frac{dx}{\sqrt{x^2+4x+13}}$

SOLUTION First complete the square:

$$x^2 + 4x + 13 = x^2 + 4x + 4 + 9 = (x+2)^2 + 9.$$

Let $u = x + 2$. Then $du = dx$, and

$$I = \int \frac{dx}{\sqrt{x^2+4x+13}} = \int \frac{dx}{\sqrt{(x+2)^2+9}} = \int \frac{du}{\sqrt{u^2+9}}.$$

Now let $u = 3\tan\theta$. Then $du = 3\sec^2\theta\, d\theta$,

$$u^2 + 9 = 9\tan^2\theta + 9 = 9(\tan^2\theta + 1) = 9\sec^2\theta,$$

and

$$I = \int \frac{3\sec^2\theta\, d\theta}{3\sec\theta} = \int \sec\theta\, d\theta = \ln|\sec\theta + \tan\theta| + C.$$

Since $u = 3\tan\theta$, we construct the following right triangle:

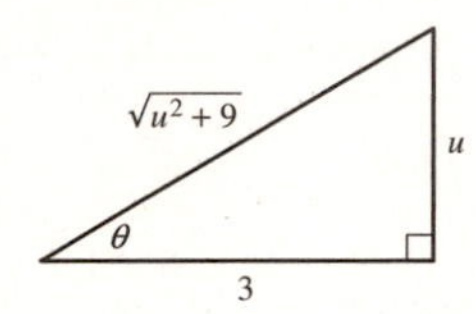

From this we see that $\sec\theta = \sqrt{u^2+9}/3$. Thus

$$\begin{aligned} I &= \ln\left|\frac{\sqrt{u^2+9}}{3} + \frac{u}{3}\right| + C_1 = \ln\left|\sqrt{u^2+9} + u\right| + \left(\ln\frac{1}{3} + C_1\right) \\ &= \ln\left|\sqrt{(x+2)^2+9} + x + 2\right| + C = \ln\left|\sqrt{x^2+4x+13} + x + 2\right| + C. \end{aligned}$$

39. $\displaystyle\int \frac{dx}{\sqrt{x+6x^2}}$

SOLUTION First complete the square:

$$6x^2 + x = \left(6x^2 + x + \frac{1}{24}\right) - \frac{1}{24} = \left(\sqrt{6}x + \frac{1}{2\sqrt{6}}\right)^2 - \frac{1}{24}$$

Let $u = \sqrt{6}x + \frac{1}{2\sqrt{6}}$ so that $du = \sqrt{6}\,dx$. Then

$$I = \int \frac{1}{\sqrt{x+6x^2}}\,dx = \int \frac{1}{\sqrt{\left(\sqrt{6}x + \frac{1}{2\sqrt{6}}\right)^2 - \frac{1}{24}}}\,dx = \frac{1}{\sqrt{6}}\int \frac{1}{\sqrt{u^2 - \frac{1}{24}}}\,du$$

Now let $u = \frac{1}{2\sqrt{6}}\sec\theta$. Then $du = \frac{1}{2\sqrt{6}}\sec\theta\tan\theta$, and

$$u^2 - \frac{1}{24} = \frac{1}{24}(\sec^2\theta - 1) = \frac{1}{24}\tan^2\theta$$

so that

$$I = \frac{1}{\sqrt{6}}\int \frac{1}{\frac{1}{2\sqrt{6}}\tan\theta}\frac{1}{2\sqrt{6}}\sec\theta\tan\theta\, d\theta = \frac{1}{\sqrt{6}}\int \sec\theta\, d\theta = \frac{1}{\sqrt{6}}\ln|\sec\theta + \tan\theta| + C$$

Since $u = \frac{1}{2\sqrt{6}}\sec\theta$, we construct the following right triangle:

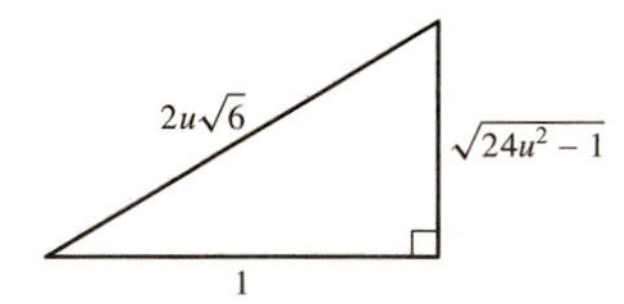

from which we see that $\tan\theta = \sqrt{24u^2-1}$ and $\sec\theta = 2u\sqrt{6}$. Thus

$$\begin{aligned} I &= \frac{1}{\sqrt{6}}\ln\left|2u\sqrt{6} + \sqrt{24u^2-1}\right| + C = \frac{1}{\sqrt{6}}\ln\left|2\sqrt{6}\left(\sqrt{6}x + \frac{1}{2\sqrt{6}}\right) + \sqrt{24\left(6x^2 + x + \frac{1}{24}\right) - 1}\right| + C \\ &= \frac{1}{\sqrt{6}}\ln\left|12x + 1 + \sqrt{144x^2 + 24x}\right| + C \end{aligned}$$

41. $\displaystyle\int \sqrt{x^2 - 4x + 3}\,dx$

SOLUTION First complete the square:

$$x^2 - 4x + 3 = x^2 - 4x + 4 - 1 = (x-2)^2 - 1.$$

Let $u = x - 2$. Then $du = dx$, and

$$I = \int \sqrt{x^2 - 4x + 3}\,dx = \int \sqrt{(x-2)^2 - 1}\,dx = \int \sqrt{u^2 - 1}\,du.$$

Now let $u = \sec\theta$. Then $du = \sec\theta\tan\theta\,d\theta$, $u^2 - 1 = \sec^2\theta - 1 = \tan^2\theta$, and

$$\begin{aligned} I &= \int \sqrt{\tan^2\theta}(\sec\theta\tan\theta\,d\theta) = \int \tan^2\theta\sec\theta\,d\theta = \int \left(\sec^2\theta - 1\right)\sec\theta\,d\theta \\ &= \int \sec^3\theta\,d\theta - \int \sec\theta\,d\theta = \left(\frac{\tan\theta\sec\theta}{2} + \frac{1}{2}\int \sec\theta\,d\theta\right) - \int \sec\theta\,d\theta \\ &= \frac{1}{2}\tan\theta\sec\theta - \frac{1}{2}\int \sec\theta\,d\theta = \frac{1}{2}\tan\theta\sec\theta - \frac{1}{2}\ln|\sec\theta + \tan\theta| + C. \end{aligned}$$

Since $u = \sec\theta$, we construct the following right triangle:

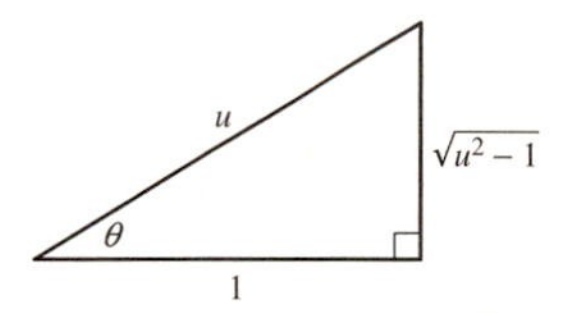

From this we see that $\tan\theta = \sqrt{u^2 - 1}$. Thus

$$\begin{aligned} I &= \frac{1}{2}u\sqrt{u^2-1} - \frac{1}{2}\ln\left|u + \sqrt{u^2-1}\right| + C = \frac{1}{2}(x-2)\sqrt{(x-2)^2-1} - \frac{1}{2}\ln\left|x - 2 + \sqrt{(x-2)^2-1}\right| + C \\ &= \frac{1}{2}(x-2)\sqrt{x^2-4x+3} - \frac{1}{2}\ln\left|x - 2 + \sqrt{x^2-4x+3}\right| + C. \end{aligned}$$

In Exercises 43–52, indicate a good method for evaluating the integral (but do not evaluate). Your choices are: substitution (specify u and du), Integration by Parts (specify u and v′), a trigonometric method, or trigonometric substitution (specify). If it appears that these techniques are not sufficient, state this.

43. $\displaystyle\int \frac{x\,dx}{\sqrt{12 - 6x - x^2}}$

SOLUTION Complete the square so the the denominator is $\sqrt{15 - (x+3)^2}$ and then use trigonometric substitution with $x + 3 = \sin\theta$.

45. $\displaystyle\int \sin^3 x\cos^3 x\,dx$

SOLUTION Use one of the following trigonometric methods: rewrite $\sin^3 x = (1 - \cos^2 x)\sin x$ and let $u = \cos x$, or rewrite $\cos^3 x = (1 - \sin^2 x)\cos x$ and let $u = \sin x$.

47. $\displaystyle\int \frac{dx}{\sqrt{9 - x^2}}$

SOLUTION Either use the substitution $x = 3u$ and then recognize the formula for the inverse sine:

$$\int \frac{du}{\sqrt{1-u^2}} = \sin^{-1} u + C,$$

or use trigonometric substitution, with $x = 3\sin\theta$.

49. $\displaystyle\int \sin^{3/2} x\,dx$

SOLUTION Not solvable by any method yet considered.

51. $\displaystyle\int \frac{dx}{(x+1)(x+2)^3}$

SOLUTION The techniques we have covered thus far are not sufficient to treat this integral. This integral requires a technique known as partial fractions.

In Exercises 53–56, evaluate using Integration by Parts as a first step.

53. $\displaystyle\int \sec^{-1} x\,dx$

SOLUTION Let $u = \sec^{-1} x$ and $v' = 1$. Then $v = x$, $u' = 1/x\sqrt{x^2-1}$, and

$$I = \int \sec^{-1} x\,dx = x\sec^{-1} x - \int \frac{x}{x\sqrt{x^2-1}}\,dx = x\sec^{-1} x - \int \frac{dx}{\sqrt{x^2-1}}.$$

To evaluate the integral on the right, let $x = \sec\theta$. Then $dx = \sec\theta\tan\theta\,d\theta$, $x^2 - 1 = \sec^2\theta - 1 = \tan^2\theta$, and

$$\int \frac{dx}{\sqrt{x^2-1}} = \int \frac{\sec\theta\tan\theta\,d\theta}{\tan\theta} = \int \sec\theta\,d\theta = \ln|\sec\theta + \tan\theta| + C = \ln\left|x + \sqrt{x^2-1}\right| + C.$$

Thus, the final answer is

$$I = x\sec^{-1}x - \ln\left|x + \sqrt{x^2-1}\right| + C.$$

55. $\displaystyle\int \ln(x^2+1)\,dx$

SOLUTION Start by using integration by parts, with $u = \ln(x^2+1)$ and $v' = 1$; then $u' = \frac{2x}{x^2+1}$ and $v = x$, so that

$$I = \int \ln(x^2+1)\,dx = x\ln(x^2+1) - 2\int \frac{x^2}{x^2+1}\,dx = x\ln(x^2+1) - 2\int\left(1 - \frac{1}{x^2+1}\right)dx$$

$$= x\ln(x^2+1) - 2x + 2\int \frac{1}{x^2+1}\,dx$$

To deal with the remaining integral, use the substitution $x = \tan\theta$, so that $dx = \sec^2\theta\,d\theta$ and

$$\int \frac{1}{x^2+1}\,dx = \int \frac{\sec^2\theta}{\tan^2\theta+1}\,d\theta = \int \frac{\sec^2\theta}{\sec^2\theta}\,d\theta = \int 1\,d\theta = \theta = \tan^{-1}x + C$$

so that finally

$$I = x\ln(x^2+1) - 2x + 2\tan^{-1}x + C$$

57. Find the average height of a point on the semicircle $y = \sqrt{1-x^2}$ for $-1 \le x \le 1$.

SOLUTION The average height is given by the formula

$$y_{\text{ave}} = \frac{1}{1-(-1)}\int_{-1}^{1}\sqrt{1-x^2}\,dx = \frac{1}{2}\int_{-1}^{1}\sqrt{1-x^2}\,dx$$

Let $x = \sin\theta$. Then $dx = \cos\theta\,d\theta$, $1 - x^2 = \cos^2\theta$, and

$$\int \sqrt{1-x^2}\,dx = \int (\cos\theta)(\cos\theta\,d\theta) = \int \cos^2\theta\,d\theta = \frac{1}{2}\theta + \frac{1}{2}\sin\theta\cos\theta + C.$$

Since $x = \sin\theta$, we construct the following right triangle:

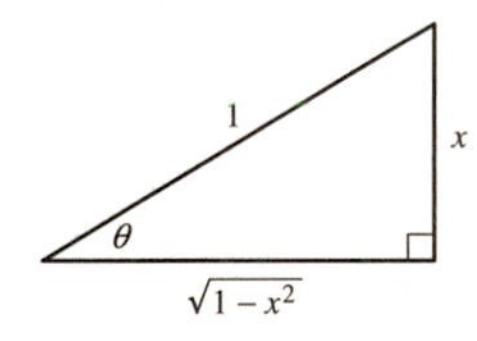

From this we see that $\cos\theta = \sqrt{1-x^2}$. Therefore,

$$y_{\text{ave}} = \frac{1}{2}\left(\frac{1}{2}\sin^{-1}x + \frac{1}{2}x\sqrt{1-x^2}\right)\Big|_{-1}^{1} = \frac{1}{2}\left[\left(\frac{1}{2}\pi + 0\right) - \left(-\frac{1}{2}\pi + 0\right)\right] = \frac{\pi}{4}.$$

59. Find the volume of the solid obtained by revolving the region between the graph of $y^2 - x^2 = 1$ and the line $y = 2$ about the line $y = 2$.

SOLUTION First solve the equation $y^2 - x^2 = 1$ for y:

$$y = \pm\sqrt{x^2+1}.$$

The region in question is bounded in part by the top half of this hyperbola, which is the equation

$$y = \sqrt{x^2+1}.$$

The limits of integration are obtained by finding the points of intersection of this equation with $y = 2$:

$$2 = \sqrt{x^2+1} \Rightarrow x = \pm\sqrt{3}.$$

The radius of each disk is given by $2 - \sqrt{x^2+1}$; the volume is therefore given by

$$V = \int_{-\sqrt{3}}^{\sqrt{3}} \pi r^2\,dx = 2\pi \int_0^{\sqrt{3}} \left(2 - \sqrt{x^2+1}\right)^2 dx = 2\pi \int_0^{\sqrt{3}} \left[4 - 4\sqrt{x^2+1} + (x^2+1)\right] dx$$

$$= 8\pi \int_0^{\sqrt{3}} dx - 8\pi \int_0^{\sqrt{3}} \sqrt{x^2+1}\,dx + 2\pi \int_0^{\sqrt{3}} (x^2+1)\,dx.$$

To evaluate the integral $\int \sqrt{x^2+1}\,dx$, let $x = \tan\theta$. Then $dx = \sec^2\theta\,d\theta$, $x^2+1 = \sec^2\theta$, and

$$\int \sqrt{x^2+1}\,dx = \int \sec^3\theta\,d\theta = \frac{1}{2}\tan\theta\sec\theta + \frac{1}{2}\int \sec\theta\,d\theta$$

$$= \frac{1}{2}\tan\theta\sec\theta + \frac{1}{2}\ln|\sec\theta + \tan\theta| + C = \frac{1}{2}x\sqrt{x^2+1} + \frac{1}{2}\ln\left|\sqrt{x^2+1} + x\right| + C.$$

Now we can compute the volume:

$$V = \left[8\pi x - 8\pi\left(\frac{1}{2}x\sqrt{x^2+1} + \frac{1}{2}\ln\left|\sqrt{x^2+1}+x\right|\right) + \frac{2}{3}\pi x^3 + 2\pi x\right]\Bigg|_0^{\sqrt{3}}$$

$$= \left(10\pi x + \frac{2}{3}\pi x^3 - 4\pi x\sqrt{x^2+1} - 4\pi\ln\left|\sqrt{x^2+1}+x\right|\right)\Bigg|_0^{\sqrt{3}}$$

$$= \left(10\pi\sqrt{3} + 2\pi\sqrt{3} - 8\pi\sqrt{3} - 4\pi\ln\left|2+\sqrt{3}\right|\right) - (0) = 4\pi\left[\sqrt{3} - \ln\left|2+\sqrt{3}\right|\right].$$

61. Compute $\int \frac{dx}{x^2-1}$ in two ways and verify that the answers agree: first via trigonometric substitution and then using the identity

$$\frac{1}{x^2-1} = \frac{1}{2}\left(\frac{1}{x-1} - \frac{1}{x+1}\right)$$

SOLUTION Using trigonometric substitution, let $x = \sec\theta$. Then $dx = \sec\theta\tan\theta d\theta$, $x^2 - 1 = \sec^2\theta - 1 = \tan^2\theta$, and

$$I = \int \frac{dx}{x^2-1} = \int \frac{\sec\theta\tan\theta\,d\theta}{\tan^2\theta} = \int \frac{\sec\theta}{\tan\theta}\,d\theta = \int \frac{d\theta}{\sin\theta} = \int \csc\theta\,d\theta = \ln|\csc\theta - \cot\theta| + C.$$

Since $x = \sec\theta$, we construct the following right triangle:

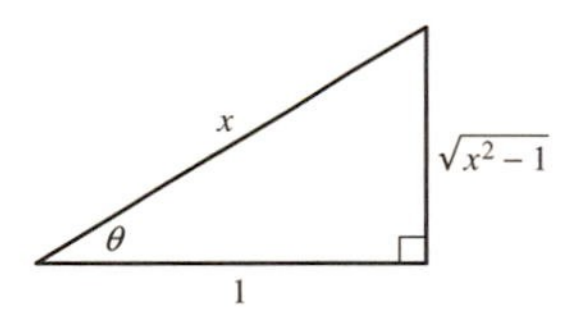

From this we see that $\csc\theta = x/\sqrt{x^2-1}$ and $\cot\theta = 1/\sqrt{x^2-1}$. This gives us

$$I = \ln\left|\frac{x}{\sqrt{x^2-1}} - \frac{1}{\sqrt{x^2-1}}\right| + C = \ln\left|\frac{x-1}{\sqrt{x^2-1}}\right| + C.$$

Using the given identity, we get

$$I = \int \frac{dx}{x^2-1} = \frac{1}{2}\int\left(\frac{1}{x-1} - \frac{1}{x+1}\right)dx = \frac{1}{2}\int\frac{dx}{x-1} - \frac{1}{2}\int\frac{dx}{x+1} = \frac{1}{2}\ln|x-1| - \frac{1}{2}\ln|x+1| + C.$$

To confirm that these answers agree, note that

$$\frac{1}{2}\ln|x-1| - \frac{1}{2}\ln|x+1| = \frac{1}{2}\ln\left|\frac{x-1}{x+1}\right| = \ln\sqrt{\left|\frac{x-1}{x+1}\right|} = \ln\left|\frac{\sqrt{x-1}}{\sqrt{x+1}}\cdot\frac{\sqrt{x-1}}{\sqrt{x-1}}\right| = \ln\left|\frac{x-1}{\sqrt{x^2-1}}\right|.$$

63. A charged wire creates an electric field at a point P located at a distance D from the wire (Figure 7). The component $E_\perp$ of the field perpendicular to the wire (in N/C) is

$$E_\perp = \int_{x_1}^{x_2} \frac{k\lambda D}{(x^2+D^2)^{3/2}}\,dx$$

where λ is the charge density (coulombs per meter), $k = 8.99 \times 10^9$ N·m²/C² (Coulomb constant), and x_1, x_2 are as in the figure. Suppose that $\lambda = 6 \times 10^{-4}$ C/m, and $D = 3$ m. Find $E_\perp$ if (a) $x_1 = 0$ and $x_2 = 30$ m, and (b) $x_1 = -15$ m and $x_2 = 15$ m.

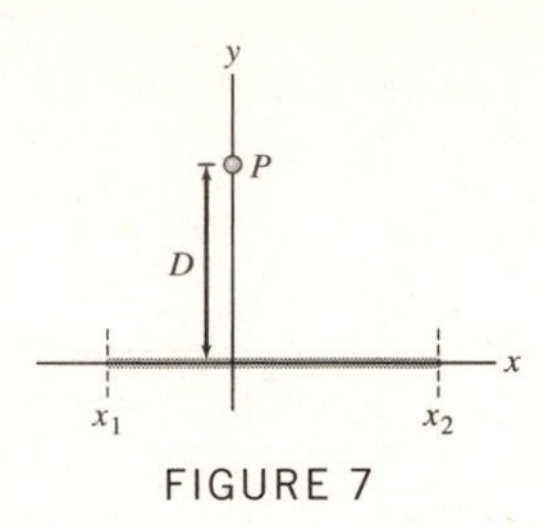

FIGURE 7

SOLUTION Let $x = D\tan\theta$. Then $dx = D\sec^2\theta\, d\theta$,

$$x^2 + D^2 = D^2\tan^2\theta + D^2 = D^2(\tan^2\theta + 1) = D^2\sec^2\theta,$$

and

$$\begin{aligned} E_\perp &= \int_{x_1}^{x_2} \frac{k\lambda D}{(x^2+D^2)^{3/2}}\,dx = k\lambda D \int_{x_1}^{x_2} \frac{D\sec^2\theta\,d\theta}{(D^2\sec^2\theta)^{3/2}} \\ &= \frac{k\lambda D^2}{D^3}\int_{x_1}^{x_2} \frac{\sec^2\theta\,d\theta}{\sec^3\theta} = \frac{k\lambda}{D}\int_{x_1}^{x_2}\cos\theta\,d\theta = \frac{k\lambda}{D}\sin\theta\bigg|_{x_1}^{x_2} \end{aligned}$$

Since $x = D\tan\theta$, we construct a right triangle with $\tan\theta = x/D$:

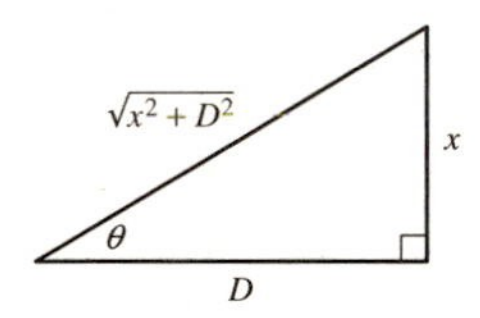

From this we see that $\sin\theta = x/\sqrt{x^2+D^2}$. Then

$$E_\perp = \frac{k\lambda}{D}\left(\frac{x}{\sqrt{x^2+D^2}}\right)\bigg|_{x_1}^{x_2}$$

(a) Plugging in the values for the constants k, λ, D, and evaluating the antiderivative for $x_1 = 0$ and $x_2 = 30$, we get

$$E_\perp = \frac{(8.99\times 10^9)(6\times 10^{-4})}{3}\left[\frac{30}{\sqrt{30^2+3^2}} - 0\right] \approx 1.789 \times 10^6\ \frac{\text{V}}{\text{m}}$$

(b) If $x_1 = -15$ m and $x_2 = 15$ m, we get

$$E_\perp = \frac{(8.99\times 10^9)(6\times 10^{-4})}{3}\left[\frac{15}{\sqrt{15^2+3^2}} - \frac{-15}{\sqrt{(-15)^2+3^2}}\right] \approx 3.526 \times 10^6\ \frac{\text{V}}{\text{m}}$$

Further Insights and Challenges

65. Prove the formula

$$\int \sqrt{1-x^2}\,dx = \frac{1}{2}\sin^{-1}x + \frac{1}{2}x\sqrt{1-x^2} + C$$

using geometry by interpreting the integral as the area of part of the unit circle.

SOLUTION The integral $\int_0^a \sqrt{1-x^2}\,dx$ is the area bounded by the unit circle, the x-axis, the y-axis, and the line $x = a$. This area can be divided into two regions as follows:

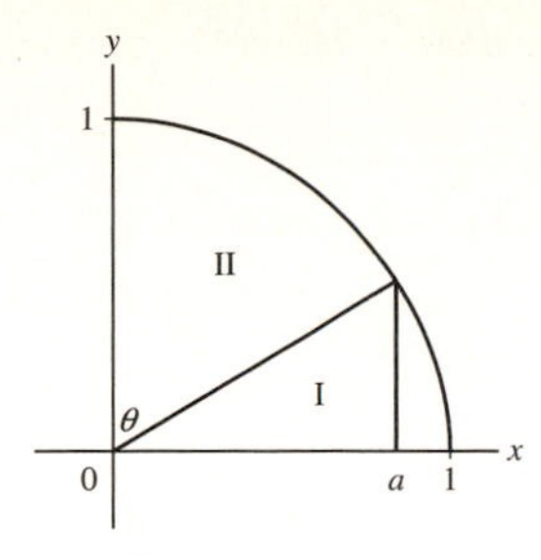

Region I is a triangle with base a and height $\sqrt{1-a^2}$. Region II is a sector of the unit circle with central angle $\theta = \frac{\pi}{2} - \cos^{-1} a = \sin^{-1} a$. Thus,

$$\int_0^a \sqrt{1-x^2}\,dx = \frac{1}{2}a\sqrt{1-a^2} + \frac{1}{2}\sin^{-1} a = \left(\frac{1}{2}x\sqrt{1-x^2} + \frac{1}{2}\sin^{-1} x\right)\Big|_0^a.$$

7.4 Integrals Involving Hyperbolic and Inverse Hyperbolic Functions

Preliminary Questions

1. Which hyperbolic substitution can be used to evaluate the following integrals?

(a) $\displaystyle\int \frac{dx}{\sqrt{x^2+1}}$ **(b)** $\displaystyle\int \frac{dx}{\sqrt{x^2+9}}$ **(c)** $\displaystyle\int \frac{dx}{\sqrt{9x^2+1}}$

SOLUTION The appropriate hyperbolic substitutions are

(a) $x = \sinh t$

(b) $x = 3\sinh t$

(c) $3x = \sinh t$

2. Which two of the hyperbolic integration formulas differ from their trigonometric counterparts by a minus sign?

SOLUTION The integration formulas for $\sinh x$ and $\tanh x$ differ from their trigonometric counterparts by a minus sign.

3. Which antiderivative of $y = (1-x^2)^{-1}$ should we use to evaluate the integral $\displaystyle\int_3^5 (1-x^2)^{-1}\,dx$?

SOLUTION Because the integration interval lies outside $-1 < x < 1$, the appropriate antiderivative of $y = (1-x^2)^{-1}$ is $\frac{1}{2}\ln\left|\frac{1+x}{1-x}\right|$.

Exercises

In Exercises 1–16, calculate the integral.

1. $\displaystyle\int \cosh(3x)\,dx$

SOLUTION $\displaystyle\int \cosh(3x)\,dx = \frac{1}{3}\sinh 3x + C.$

3. $\displaystyle\int x\sinh(x^2+1)\,dx$

SOLUTION $\displaystyle\int x\sinh(x^2+1)\,dx = \frac{1}{2}\cosh(x^2+1) + C.$

5. $\displaystyle\int \operatorname{sech}^2(1-2x)\,dx$

SOLUTION $\displaystyle\int \operatorname{sech}^2(1-2x)\,dx = -\frac{1}{2}\tanh(1-2x) + C.$

7. $\displaystyle\int \tanh x \operatorname{sech}^2 x\,dx$

SOLUTION Let $u = \tanh x$. Then $du = \operatorname{sech}^2 x\,dx$ nd

$$\int \tanh x \operatorname{sech}^2 x\,dx = \int u\,du = \frac{1}{2}u^2 + C = \frac{\tanh^2 x}{2} + C.$$

9. $\int \tanh x\,dx$

SOLUTION $\int \tanh x\,dx = \ln\cosh x + C.$

11. $\int \frac{\cosh x}{\sinh x}\,dx$

SOLUTION $\int \frac{\cosh x}{\sinh x}\,dx = \ln|\sinh x| + C.$

13. $\int \sinh^2(4x-9)\,dx$

SOLUTION $\int \sinh^2(4x-9)\,dx = \frac{1}{2}\int(\cosh(8x-18)-1)\,dx = \frac{1}{16}\sinh(8x-18) - \frac{1}{2}x + C.$

15. $\int \sinh^2 x\cosh^2 x\,dx$

SOLUTION

$$\int \sinh^2 x\cosh^2 x\,dx = \frac{1}{4}\int \sinh^2 2x\,dx = \frac{1}{8}\int(\cosh 4x - 1)\,dx = \frac{1}{32}\sinh 4x - \frac{1}{8}x + C.$$

In Exercises 17–30, calculate the integral in terms of the inverse hyperbolic functions.

17. $\int \frac{dx}{\sqrt{x^2-1}}$

SOLUTION $\int \frac{dx}{\sqrt{x^2-1}} = \cosh^{-1} x + C.$

19. $\int \frac{dx}{\sqrt{16+25x^2}}$

SOLUTION $\int \frac{dx}{\sqrt{16+25x^2}} = \frac{1}{5}\sinh^{-1}\left(\frac{5x}{4}\right) + C.$

21. $\int \sqrt{x^2-1}\,dx$

SOLUTION Let $x = \cosh t$. Then $dx = \sinh t\,dt$ and

$$\begin{aligned}\int \sqrt{x^2-1}\,dx &= \int \sinh^2 t\,dt = \frac{1}{2}\int(\cosh 2t - 1)\,dt = \frac{1}{4}\sinh 2t - \frac{1}{2}t + C\\ &= \frac{1}{2}\sinh t\cosh t - \frac{1}{2}t + C = \frac{1}{2}x\sqrt{x^2-1} - \frac{1}{2}\cosh^{-1} x + C.\end{aligned}$$

23. $\int_{-1/2}^{1/2} \frac{dx}{1-x^2}$

SOLUTION

$$\int_{-1/2}^{1/2} \frac{dx}{1-x^2} = \tanh^{-1} x\Big|_{-1/2}^{1/2} = \tanh^{-1}\left(\frac{1}{2}\right) - \tanh^{-1}\left(-\frac{1}{2}\right) = 2\tanh^{-1}\left(\frac{1}{2}\right).$$

25. $\int_0^1 \frac{dx}{\sqrt{1+x^2}}$

SOLUTION $\int_0^1 \frac{dx}{\sqrt{1+x^2}} = \sinh^{-1}\Big|_0^1 = \sinh^{-1}(1) - \sinh^{-1}(0) = \sinh^{-1} 1.$

27. $\int_{-3}^{-1} \frac{dx}{x\sqrt{x^2+16}}$

SOLUTION $\int_{-3}^{-1} \frac{dx}{x\sqrt{x^2+16}} = \frac{1}{4}\operatorname{csch}^{-1}\left(\frac{x}{4}\right)\Big|_{-3}^{-1} = \frac{1}{4}\left(\operatorname{csch}^{-1}\left(-\frac{1}{4}\right) - \operatorname{csch}^{-1}\left(-\frac{3}{4}\right)\right).$

29. $\int \frac{\sqrt{x^2-1}\,dx}{x^2}$

SOLUTION Let $x = \cosh t$. Then $dx = \sinh t\,dt$ and

$$\int \frac{\sqrt{x^2-1}\,dx}{x^2} = \int \frac{\sinh^2 t}{\cosh^2 t}\,dt = \int \tanh^2 t\,dt = \int (1 - \operatorname{sech}^2 t)\,dt$$

$$= t - \tanh t + C = \cosh^{-1} x - \frac{\sqrt{x^2-1}}{x} + C.$$

31. Verify the formulas

$$\sinh^{-1} x = \ln |x + \sqrt{x^2+1}|$$

$$\cosh^{-1} x = \ln |x + \sqrt{x^2-1}| \qquad (\text{for } x \geq 1)$$

SOLUTION Let $x = \sinh t$. Then

$$\cosh t = \sqrt{1 + \sinh^2 t} = \sqrt{1+x^2}.$$

Moreover, because

$$\sinh t + \cosh t = \frac{e^t - e^{-t}}{2} + \frac{e^t + e^{-t}}{2} = e^t,$$

it follows that

$$\sinh^{-1} x = t = \ln(\sinh t + \cosh t) = \ln(x + \sqrt{x^2+1}).$$

Now, Let $x = \cosh t$. Then

$$\sinh t = \sqrt{\cosh^2 t - 1} = \sqrt{x^2-1}.$$

and

$$\cosh^{-1} x = t = \ln(\sinh t + \cosh t) = \ln(x + \sqrt{x^2-1}).$$

Because $\cosh t \geq 1$ for all t, this last expression is only valid for $x = \cosh t \geq 1$.

33. Evaluate $\int \sqrt{x^2+16}\,dx$ using trigonometric substitution. Then use Exercise 31 to verify that your answer agrees with the answer in Example 3.

SOLUTION Let $x = 4\tan\theta$. Then $dx = 4\sec^2\theta\,d\theta$ and

$$\int \sqrt{x^2+16}\,dx = 16\int \sec^3\theta\,d\theta = 8\tan\theta\sec\theta + 8\int \sec\theta\,d\theta = 8\tan\theta\sec\theta + 8\ln|\sec\theta + \tan\theta| + C$$

$$= 8 \cdot \frac{x}{4} \cdot \frac{\sqrt{x^2+16}}{4} + 8\ln\left|\frac{\sqrt{x^2+16}}{4} + \frac{x}{4}\right| + C$$

$$= \frac{1}{2}x\sqrt{x^2+16} + 8\ln\left|\frac{x}{4} + \sqrt{\left(\frac{x}{4}\right)^2 + 1}\right| + C.$$

Using Exercise 31,

$$\ln\left|\frac{x}{4} + \sqrt{\left(\frac{x}{4}\right)^2 + 1}\right| = \sinh^{-1}\left(\frac{x}{4}\right),$$

so we can write the antiderivative as

$$\frac{1}{2}x\sqrt{x^2+16} + 8\sinh^{-1}\left(\frac{x}{4}\right) + C,$$

which agrees with the answer in Example 3.

35. Prove the reduction formula for $n \geq 2$:

$$\int \cosh^n x\,dx = \frac{1}{n}\cosh^{n-1} x \sinh x + \frac{n-1}{n}\int \cosh^{n-2} x\,dx$$

5

SOLUTION Using Integration by Parts with $u = \cosh^{n-1} x$ and $v' = \cosh x$, we have

$$\int \cosh^n x\,dx = \cosh^{n-1} x \sinh x - (n-1)\int \cosh^{n-2} x \sinh^2 x\,dx$$

$$= \cosh^{n-1} x \sinh x - (n-1)\int \cosh^n x\,dx + (n-1)\int \cosh^{n-2} x\,dx.$$

Adding $(n-1)\int \cosh^n x\,dx$ to both sides then yields

$$n\int \cosh^n x\,dx = \cosh^{n-1} x \sinh x + (n-1)\int \cosh^{n-2} x\,dx.$$

Finally,

$$\int \cosh^n x\,dx = \frac{1}{n}\cosh^{n-1} x \sinh x + \frac{n-1}{n}\int \cosh^{n-2} x\,dx.$$

In Exercises 37–40, evaluate the integral.

37. $\displaystyle\int \frac{\tanh^{-1} x\,dx}{x^2-1}$

SOLUTION Let $u = \tanh^{-1} x$. Then $du = \dfrac{1}{1-x^2}\,dx = -\dfrac{1}{x^2-1}\,dx$ and

$$\int \frac{\tanh^{-1} x}{x^2-1}\,dx = -\int u\,du = -\frac{1}{2}u^2 + C = -\frac{1}{2}\left(\tanh^{-1} x\right)^2 + C.$$

39. $\displaystyle\int \tanh^{-1} x\,dx$

SOLUTION Using Integration by Parts with $u = \tanh^{-1} x$ and $v' = 1$,

$$\int \tanh^{-1} x\,dx = x\tanh^{-1} x - \int \frac{x}{1-x^2}\,dx = x\tanh^{-1} x + \frac{1}{2}\ln|1-x^2| + C.$$

Further Insights and Challenges

41. Show that if $u = \tanh(x/2)$, then

$$\cosh x = \frac{1+u^2}{1-u^2}, \qquad \sinh x = \frac{2u}{1-u^2}, \qquad dx = \frac{2du}{1-u^2}$$

Hint: For the first relation, use the identities

$$\sinh^2\left(\frac{x}{2}\right) = \frac{1}{2}(\cosh x - 1), \qquad \cosh^2\left(\frac{x}{2}\right) = \frac{1}{2}(\cosh x + 1)$$

SOLUTION Let $u = \tanh(x/2)$. Then

$$u = \frac{\sinh(x/2)}{\cosh(x/2)} = \sqrt{\frac{\cosh x - 1}{\cosh x + 1}}.$$

Solving for $\cosh x$ yields

$$\cosh x = \frac{1+u^2}{1-u^2}.$$

Next,

$$\sinh x = \sqrt{\cosh^2 x - 1} = \sqrt{\frac{(1+u^2)^2 - (1-u^2)^2}{(1-u^2)^2}} = \frac{2u}{1-u^2}.$$

Finally, if $u = \tanh(x/2)$, then $x = 2\tanh^{-1} u$ and

$$dx = \frac{2\,du}{1-u^2}.$$

Exercises 42 and 43: evaluate using the substitution of Exercise 41.

43. $\int \frac{dx}{1+\cosh x}$

SOLUTION Let $u = \tanh(x/2)$. Then, by Exercise 41,

$$1 + \cosh x = 1 + \frac{1+u^2}{1-u^2} = \frac{2}{1-u^2} \quad \text{and} \quad dx = \frac{2\,du}{1-u^2},$$

so

$$\int \frac{dx}{1+\cosh x} = \int du = u + C = \tanh\frac{x}{2} + C.$$

Exercises 45–48 refer to the function $gd(y) = \tan^{-1}(\sinh y)$, called the ***gudermannian****. In a map of the earth constructed by Mercator projection, points located y radial units from the equator correspond to points on the globe of latitude $gd(y)$.*

45. Prove that $\frac{d}{dy} gd(y) = \operatorname{sech} y$.

SOLUTION Let $gd(y) = \tan^{-1}(\sinh y)$. Then

$$\frac{d}{dy} gd(y) = \frac{1}{1+\sinh^2 y}\cosh y = \frac{1}{\cosh y} = \operatorname{sech} y,$$

where we have used the identity $1 + \sinh^2 y = \cosh^2 y$.

47. Let $t(y) = \sinh^{-1}(\tan y)$. Show that $t(y)$ is the inverse of $gd(y)$ for $0 \le y < \pi/2$.

SOLUTION Let $x = gd(y) = \tan^{-1}(\sinh y)$. Solving for y yields $y = \sinh^{-1}(\tan x)$. Therefore,

$$gd^{-1}(y) = \sinh^{-1}(\tan y).$$

49. The relations $\cosh(it) = \cos t$ and $\sinh(it) = i \sin t$ were discussed in the Excursion. Use these relations to show that the identity $\cos^2 t + \sin^2 t = 1$ results from setting $x = it$ in the identity $\cosh^2 x - \sinh^2 x = 1$.

SOLUTION Let $x = it$. Then

$$\cosh^2 x = (\cosh(it))^2 = \cos^2 t$$

and

$$\sinh^2 x = (\sinh(it))^2 = i^2 \sin^2 t = -\sin^2 t.$$

Thus,

$$1 = \cosh^2(it) - \sinh^2(it) = \cos^2 t - (-\sin^2 t) = \cos^2 t + \sin^2 t,$$

as desired.

7.5 The Method of Partial Fractions

Preliminary Questions

1. Suppose that $\int f(x)\,dx = \ln x + \sqrt{x+1} + C$. Can $f(x)$ be a rational function? Explain.

SOLUTION No, $f(x)$ cannot be a rational function because the integral of a rational function cannot contain a term with a non-integer exponent such as $\sqrt{x+1}$.

2. Which of the following are *proper* rational functions?

(a) $\frac{x}{x-3}$ **(b)** $\frac{4}{9-x}$

(c) $\frac{x^2+12}{(x+2)(x+1)(x-3)}$ **(d)** $\frac{4x^3-7x}{(x-3)(2x+5)(9-x)}$

SOLUTION

(a) No, this is not a proper rational function because the degree of the numerator is not less than the degree of the denominator.

(b) Yes, this is a proper rational function.

(c) Yes, this is a proper rational function.

(d) No, this is not a proper rational function because the degree of the numerator is not less than the degree of the denominator.

3. Which of the following quadratic polynomials are irreducible? To check, complete the square if necessary.

(a) $x^2 + 5$

(b) $x^2 - 5$

(c) $x^2 + 4x + 6$

(d) $x^2 + 4x + 2$

SOLUTION

(a) Square is already completed; irreducible.

(b) Square is already completed; factors as $(x - \sqrt{5})(x + \sqrt{5})$.

(c) $x^2 + 4x + 6 = (x + 2)^2 + 2$; irreducible.

(d) $x^2 + 4x + 2 = (x + 2)^2 - 2$; factors as $(x + 2 - \sqrt{2})(x + 2 + \sqrt{2})$.

4. Let $P(x)/Q(x)$ be a proper rational function where $Q(x)$ factors as a product of distinct linear factors $(x - a_i)$. Then

$$\int \frac{P(x)\,dx}{Q(x)}$$

(choose the correct answer):

(a) is a sum of logarithmic terms $A_i \ln(x - a_i)$ for some constants A_i.

(b) may contain a term involving the arctangent.

SOLUTION The correct answer is **(a)**: the integral is a sum of logarithmic terms $A_i \ln(x - a_i)$ for some constants A_i.

Exercises

1. Match the rational functions (a)–(d) with the corresponding partial fraction decompositions (i)–(iv).

(a) $\dfrac{x^2 + 4x + 12}{(x + 2)(x^2 + 4)}$

(b) $\dfrac{2x^2 + 8x + 24}{(x + 2)^2(x^2 + 4)}$

(c) $\dfrac{x^2 - 4x + 8}{(x - 1)^2(x - 2)^2}$

(d) $\dfrac{x^4 - 4x + 8}{(x + 2)(x^2 + 4)}$

(i) $x - 2 + \dfrac{4}{x + 2} - \dfrac{4x - 4}{x^2 + 4}$

(ii) $\dfrac{-8}{x - 2} + \dfrac{4}{(x - 2)^2} + \dfrac{8}{x - 1} + \dfrac{5}{(x - 1)^2}$

(iii) $\dfrac{1}{x + 2} + \dfrac{2}{(x + 2)^2} + \dfrac{-x + 2}{x^2 + 4}$

(iv) $\dfrac{1}{x + 2} + \dfrac{4}{x^2 + 4}$

SOLUTION

(a) $\dfrac{x^2 + 4x + 12}{(x + 2)(x^2 + 4)} = \dfrac{1}{x + 2} + \dfrac{4}{x^2 + 4}.$

(b) $\dfrac{2x^2 + 8x + 24}{(x + 2)^2(x^2 + 4)} = \dfrac{1}{x + 2} + \dfrac{2}{(x + 2)^2} + \dfrac{-x + 2}{x^2 + 4}.$

(c) $\dfrac{x^2 - 4x + 8}{(x - 1)^2(x - 2)^2} = \dfrac{-8}{x - 2} + \dfrac{4}{(x - 2)^2} + \dfrac{8}{x - 1} + \dfrac{5}{(x - 1)^2}.$

(d) $\dfrac{x^4 - 4x + 8}{(x + 2)(x^2 + 4)} = x - 2 + \dfrac{4}{x + 2} - \dfrac{4x - 4}{x^2 + 4}.$

3. Clear denominators in the following partial fraction decomposition and determine the constant B (substitute a value of x or use the method of undetermined coefficients).

$$\frac{3x^2 + 11x + 12}{(x + 1)(x + 3)^2} = \frac{1}{x + 1} - \frac{B}{x + 3} - \frac{3}{(x + 3)^2}$$

SOLUTION Clearing denominators gives

$$3x^2 + 11x + 12 = (x + 3)^2 - B(x + 1)(x + 3) - 3(x + 1).$$

Setting $x = 0$ then yields

$$12 = 9 - B(1)(3) - 3(1) \qquad \text{or} \qquad B = -2.$$

To use the method of undetermined coefficients, expand the right-hand side and gather like terms:

$$3x^2 + 11x + 12 = (1 - B)x^2 + (3 - 4B)x + (6 - 3B).$$

Equating x^2-coefficients on both sides, we find

$$3 = 1 - B \qquad \text{or} \qquad B = -2.$$

In Exercises 5–8, evaluate using long division first to write $f(x)$ as the sum of a polynomial and a proper rational function.

5. $\displaystyle\int \frac{x\,dx}{3x - 4}$

SOLUTION Long division gives us

$$\frac{x}{3x - 4} = \frac{1}{3} + \frac{4/3}{3x - 4}$$

Therefore the integral is

$$\int \frac{x}{3x - 4}\,dx = \int \frac{1}{3} - \frac{4}{9x - 12}\,dx = \frac{1}{3}x - \frac{4}{9}\ln|9x - 12| + C$$

7. $\displaystyle\int \frac{(x^3 + 2x^2 + 1)\,dx}{x + 2}$

SOLUTION Long division gives us

$$\frac{x^3 + 2x^2 + 1}{x + 2} = x^2 + \frac{1}{x + 2}$$

Therefore the integral is

$$\int \frac{x^3 + 2x^2 + 1}{x + 2}\,dx = \int x^2 + \frac{1}{x + 2}\,dx = \frac{1}{3}x^3 + \ln|x + 2| + C$$

In Exercises 9–44, evaluate the integral.

9. $\displaystyle\int \frac{dx}{(x - 2)(x - 4)}$

SOLUTION The partial fraction decomposition has the form:

$$\frac{1}{(x - 2)(x - 4)} = \frac{A}{x - 2} + \frac{B}{x - 4}.$$

Clearing denominators gives us

$$1 = A(x - 4) + B(x - 2).$$

Setting $x = 2$ then yields

$$1 = A(2 - 4) + 0 \qquad \text{or} \qquad A = -\frac{1}{2},$$

while setting $x = 4$ yields

$$1 = 0 + B(4 - 2) \qquad \text{or} \qquad B = \frac{1}{2}.$$

The result is:

$$\frac{1}{(x - 2)(x - 4)} = \frac{-\frac{1}{2}}{x - 2} + \frac{\frac{1}{2}}{x - 4}.$$

Thus,

$$\int \frac{dx}{(x - 2)(x - 4)} = -\frac{1}{2}\int \frac{dx}{x - 2} + \frac{1}{2}\int \frac{dx}{x - 4} = -\frac{1}{2}\ln|x - 2| + \frac{1}{2}\ln|x - 4| + C.$$

11. $\displaystyle\int \frac{dx}{x(2x + 1)}$

SOLUTION The partial fraction decomposition has the form:

$$\frac{1}{x(2x + 1)} = \frac{A}{x} + \frac{B}{2x + 1}.$$

Clearing denominators gives us

$$1 = A(2x + 1) + Bx.$$

Setting $x = 0$ then yields

$$1 = A(1) + 0 \quad \text{or} \quad A = 1,$$

while setting $x = -\frac{1}{2}$ yields

$$1 = 0 + B\left(-\frac{1}{2}\right) \quad \text{or} \quad B = -2.$$

The result is:

$$\frac{1}{x(2x+1)} = \frac{1}{x} + \frac{-2}{2x+1}.$$

Thus,

$$\int \frac{dx}{x(2x+1)} = \int \frac{dx}{x} - \int \frac{2\,dx}{2x+1} = \ln|x| - \ln|2x+1| + C.$$

For the integral on the right, we have used the substitution $u = 2x + 1$, $du = 2\,dx$.

13. $\displaystyle\int \frac{x^2\,dx}{x^2+9}$

SOLUTION

$$\int \frac{x^2}{x^2+9}\,dx = \int 1 - \frac{9}{x^2+9}\,dx = x - 3\tan^{-1}\left(\frac{x}{3}\right) + C$$

15. $\displaystyle\int \frac{(x^2+3x-44)\,dx}{(x+3)(x+5)(3x-2)}$

SOLUTION The partial fraction decomposition has the form:

$$\frac{x^2+3x-44}{(x+3)(x+5)(3x-2)} = \frac{A}{x+3} + \frac{B}{x+5} + \frac{C}{3x-2}.$$

Clearing denominators gives us

$$x^2 + 3x - 44 = A(x+5)(3x-2) + B(x+3)(3x-2) + C(x+3)(x+5).$$

Setting $x = -3$ then yields

$$9 - 9 - 44 = A(2)(-11) + 0 + 0 \quad \text{or} \quad A = 2,$$

while setting $x = -5$ yields

$$25 - 15 - 44 = 0 + B(-2)(-17) + 0 \quad \text{or} \quad B = -1,$$

and setting $x = \frac{2}{3}$ yields

$$\frac{4}{9} + 2 - 44 = 0 + 0 + C\left(\frac{11}{3}\right)\left(\frac{17}{3}\right) \quad \text{or} \quad C = -2.$$

The result is:

$$\frac{x^2+3x-44}{(x+3)(x+5)(3x-2)} = \frac{2}{x+3} + \frac{-1}{x+5} + \frac{-2}{3x-2}.$$

Thus,

$$\int \frac{(x^2+3x-44)\,dx}{(x+3)(x+5)(3x-2)} = 2\int \frac{dx}{x+3} - \int \frac{dx}{x+5} - 2\int \frac{dx}{3x-2}$$
$$= 2\ln|x+3| - \ln|x+5| - \frac{2}{3}\ln|3x-2| + C.$$

To evaluate the last integral, we have made the substitution $u = 3x - 2$, $du = 3\,dx$.

17. $\int \frac{(x^2+11x)\,dx}{(x-1)(x+1)^2}$

SOLUTION The partial fraction decomposition has the form:

$$\frac{x^2+11x}{(x-1)(x+1)^2} = \frac{A}{x-1} + \frac{B}{x+1} + \frac{C}{(x+1)^2}.$$

Clearing denominators gives us

$$x^2+11x = A(x+1)^2 + B(x-1)(x+1) + C(x-1).$$

Setting $x=1$ then yields

$$12 = A(4)+0+0 \qquad \text{or} \qquad A=3,$$

while setting $x=-1$ yields

$$-10 = 0+0+C(-2) \qquad \text{or} \qquad C=5.$$

Plugging in these values results in

$$x^2+11x = 3(x+1)^2 + B(x-1)(x+1) + 5(x-1).$$

The constant B can be determined by plugging in for x any value other than 1 or -1. If we plug in $x=0$, we get

$$0 = 3+B(-1)(1)+5(-1) \qquad \text{or} \qquad B=-2.$$

The result is

$$\frac{x^2+11x}{(x-1)(x+1)^2} = \frac{3}{x-1} + \frac{-2}{x+1} + \frac{5}{(x+1)^2}.$$

Thus,

$$\int \frac{(x^2+11x)\,dx}{(x-1)(x+1)^2} = 3\int \frac{dx}{x-1} - 2\int \frac{dx}{x+1} + 5\int \frac{dx}{(x+1)^2} = 3\ln|x-1| - 2\ln|x+1| - \frac{5}{x+1} + C.$$

19. $\int \frac{dx}{(x-1)^2(x-2)^2}$

SOLUTION The partial fraction decomposition has the form:

$$\frac{1}{(x-1)^2(x-2)^2} = \frac{A}{x-1} + \frac{B}{(x-1)^2} + \frac{C}{x-2} + \frac{D}{(x-2)^2}.$$

Clearing denominators gives us

$$1 = A(x-1)(x-2)^2 + B(x-2)^2 + C(x-2)(x-1)^2 + D(x-1)^2.$$

Setting $x=1$ then yields

$$1 = B(1) \qquad \text{or} \qquad B=1,$$

while setting $x=2$ yields

$$1 = D(1) \qquad \text{or} \qquad D=1.$$

Plugging in these values gives us

$$1 = A(x-1)(x-2)^2 + (x-2)^2 + C(x-2)(x-1)^2 + (x-1)^2.$$

Setting $x=0$ now yields

$$1 = A(-1)(4)+4+C(-2)(1)+1 \qquad \text{or} \qquad -4 = -4A-2C,$$

while setting $x=3$ yields

$$1 = A(2)(1)+1+C(1)(4)+4 \qquad \text{or} \qquad -4 = 2A+4C.$$

Solving this system of two equations in two unknowns gives $A = 2$ and $C = -2$. The result is

$$\frac{1}{(x-1)^2(x-2)^2} = \frac{2}{x-1} + \frac{1}{(x-1)^2} + \frac{-2}{x-2} + \frac{1}{(x-2)^2}.$$

Thus,

$$\int \frac{dx}{(x-1)^2(x-2)^2} = 2\int \frac{dx}{x-1} + \int \frac{dx}{(x-1)^2} - 2\int \frac{dx}{x-2} + \int \frac{dx}{(x-2)^2}$$
$$= 2\ln|x-1| - \frac{1}{x-1} - 2\ln|x-2| - \frac{1}{x-2} + C.$$

21. $\displaystyle\int \frac{8\,dx}{x(x+2)^3}$

SOLUTION The partial fraction decomposition is

$$\frac{8}{x(x+2)^3} = \frac{A}{x} + \frac{B}{x+2} + \frac{C}{(x+2)^2} + \frac{D}{(x+2)^3}$$

Clearing fractions gives

$$8 = A(x+2)^3 + Bx(x+2)^2 + Cx(x+2) + Dx$$

Setting $x = 0$ gives $8 = 8A$ so $A = 1$; setting $x = -2$ gives $8 = -2D$ so that $D = -4$; the result is

$$8 = (x+2)^3 + Bx(x+2)^2 + Cx(x+2) - 4x$$

The coefficient of x^3 on the right-hand side must be zero, since it is zero on the left. We compute it to be $1 + B$, so that $B = -1$. Finally, we look at the coefficient of x^2 on the right-hand side; it must be zero as well. We compute it to be

$$3 \cdot 2 - 4 + C = C + 2$$

so that $C = -2$ and the partial fraction decomposition is

$$\frac{8}{x(x+2)^3} = \frac{1}{x} - \frac{1}{x+2} - \frac{2}{(x+2)^2} - \frac{4}{(x+2)^3}$$

and

$$\int \frac{8}{x(x+2)^3}\,dx = \int \frac{1}{x}\,dx - \frac{1}{x+2}\,dx - 2\int (x+2)^{-2}\,dx - 4\int (x+2)^{-3}\,dx$$
$$= \ln|x| - \ln|x+2| + 2(x+2)^{-1} + 2(x+2)^{-2} + C = \ln\left|\frac{x}{x+2}\right| + \frac{2}{x+2} + \frac{2}{(x+2)^2} + C$$

23. $\displaystyle\int \frac{dx}{2x^2-3}$

SOLUTION The partial fraction decomposition has the form

$$\frac{1}{2x^2-3} = \frac{1}{(\sqrt{2}x-\sqrt{3})(\sqrt{2}x+\sqrt{3})} = \frac{A}{\sqrt{2}x-\sqrt{3}} + \frac{B}{\sqrt{2}x+\sqrt{3}}.$$

Clearing denominators, we get

$$1 = A\left(\sqrt{2}x+\sqrt{3}\right) + B\left(\sqrt{2}x-\sqrt{3}\right).$$

Setting $x = \sqrt{3}/\sqrt{2}$ then yields

$$1 = A\left(\sqrt{3}+\sqrt{3}\right) + 0 \quad \text{or} \quad A = \frac{1}{2\sqrt{3}},$$

while setting $x = -\sqrt{3}/\sqrt{2}$ yields

$$1 = 0 + B\left(-\sqrt{3}-\sqrt{3}\right) \quad \text{or} \quad B = \frac{-1}{2\sqrt{3}}.$$

The result is

$$\frac{1}{2x^2-3} = \frac{1/2\sqrt{3}}{\sqrt{2}x-\sqrt{3}} - \frac{1/2\sqrt{3}}{\sqrt{2}x+\sqrt{3}}.$$

Thus,

$$\int \frac{dx}{2x^2-3} = \frac{1}{2\sqrt{3}} \int \frac{dx}{\sqrt{2}x-\sqrt{3}} - \frac{1}{2\sqrt{3}} \int \frac{dx}{\sqrt{2}x+\sqrt{3}}.$$

For the first integral, let $u = \sqrt{2}x - \sqrt{3}$, $du = \sqrt{2}\,dx$, and for the second, let $w = \sqrt{2}x + \sqrt{3}$, $dw = \sqrt{2}\,dx$. Then we have

$$\int \frac{dx}{2x^2-3} = \frac{1}{2\sqrt{3}(\sqrt{2})} \int \frac{du}{u} - \frac{1}{2\sqrt{3}(\sqrt{2})} \int \frac{dw}{w} = \frac{1}{2\sqrt{6}} \ln\left|\sqrt{2}x - \sqrt{3}\right| - \frac{1}{2\sqrt{6}} \ln\left|\sqrt{2}x + \sqrt{3}\right| + C.$$

25. $\displaystyle\int \frac{4x^2-20}{(2x+5)^3}\,dx$

SOLUTION The partial fraction decomposition is

$$\frac{4x^2-20}{(2x+5)^3} = \frac{A}{2x+5} + \frac{B}{(2x+5)^2} + \frac{C}{(2x+5)^3}$$

Clearing fractions gives

$$4x^2 - 20 = A(2x+5)^2 + B(2x+5) + C$$

Setting $x = -5/2$ gives $5 = C$ so that $C = 5$. The coefficient of x^2 on the left-hand side is 4, and on the right-hand side is $4A$, so that $A = 1$ and we have

$$4x^2 - 20 = (2x+5)^2 + B(2x+5) + 5$$

Considering the constant terms now gives $-20 = 25 + 5B + 5$ so that $B = -10$. Thus

$$\int \frac{4x^2-20}{(2x+5)^3} = \int \frac{1}{2x+5}\,dx - 10 \int \frac{1}{(2x+5)^2}\,dx + 5 \int \frac{1}{(2x+5)^3}\,dx$$

$$= \frac{1}{2} \ln|2x+5| + \frac{5}{2x+5} - \frac{5}{4(2x+5)^2} + C$$

27. $\displaystyle\int \frac{dx}{x(x-1)^3}$

SOLUTION The partial fraction decomposition has the form:

$$\frac{1}{x(x-1)^3} = \frac{A}{x} + \frac{B}{x-1} + \frac{C}{(x-1)^2} + \frac{D}{(x-1)^3}.$$

Clearing denominators, we get

$$1 = A(x-1)^3 + Bx(x-1)^2 + Cx(x-1) + Dx.$$

Setting $x = 0$ then yields

$$1 = A(-1) + 0 + 0 + 0 \quad \text{or} \quad A = -1,$$

while setting $x = 1$ yields

$$1 = 0 + 0 + 0 + D(1) \quad \text{or} \quad D = 1.$$

Plugging in $A = -1$ and $D = 1$ gives us

$$1 = -(x-1)^3 + Bx(x-1)^2 + Cx(x-1) + x.$$

Now, setting $x = 2$ yields

$$1 = -1 + 2B + 2C + 2 \quad \text{or} \quad 2B + 2C = 0,$$

and setting $x = 3$ yields

$$1 = -8 + 12B + 6C + 3 \quad \text{or} \quad 2B + C = 1.$$

Solving these two equations in two unknowns, we find $B = 1$ and $C = -1$. The result is

$$\frac{1}{x(x-1)^3} = \frac{-1}{x} + \frac{1}{x-1} + \frac{-1}{(x-1)^2} + \frac{1}{(x-1)^3}.$$

Thus,

$$\int \frac{dx}{x(x-1)^3} = -\int \frac{dx}{x} + \int \frac{dx}{x-1} - \int \frac{dx}{(x-1)^2} + \int \frac{dx}{(x-1)^3}$$
$$= -\ln|x| + \ln|x-1| + \frac{1}{x-1} - \frac{1}{2(x-1)^2} + C.$$

29. $\int \frac{(x^2 - x + 1)\,dx}{x^2 + x}$

SOLUTION First use long division to write

$$\frac{x^2 - x + 1}{x^2 + x} = 1 + \frac{-2x + 1}{x^2 + x} = 1 + \frac{-2x + 1}{x(x+1)}.$$

The partial fraction decomposition of the term on the right has the form:

$$\frac{-2x+1}{x(x+1)} = \frac{A}{x} + \frac{B}{x+1}.$$

Clearing denominators gives us

$$-2x + 1 = A(x+1) + Bx.$$

Setting $x = 0$ then yields

$$1 = A(1) + 0 \qquad \text{or} \qquad A = 1,$$

while setting $x = -1$ yields

$$3 = 0 + B(-1) \qquad \text{or} \qquad B = -3.$$

The result is

$$\frac{-2x+1}{x(x+1)} = \frac{1}{x} + \frac{-3}{x+1}.$$

Thus,

$$\int \frac{(x^2 - x + 1)\,dx}{x^2 + x} = \int dx + \int \frac{dx}{x} - 3\int \frac{dx}{x+1} = x + \ln|x| - 3\ln|x+1| + C.$$

31. $\int \frac{(3x^2 - 4x + 5)\,dx}{(x-1)(x^2+1)}$

SOLUTION The partial fraction decomposition has the form:

$$\frac{3x^2 - 4x + 5}{(x-1)(x^2+1)} = \frac{A}{x-1} + \frac{Bx + C}{x^2 + 1}.$$

Clearing denominators, we get

$$3x^2 - 4x + 5 = A(x^2 + 1) + (Bx + C)(x - 1).$$

Setting $x = 1$ then yields

$$3 - 4 + 5 = A(2) + 0 \qquad \text{or} \qquad A = 2.$$

This gives us

$$3x^2 - 4x + 5 = 2(x^2 + 1) + (Bx + C)(x - 1) = (B + 2)x^2 + (C - B)x + (2 - C).$$

Equating x^2-coefficients, we find

$$3 = B + 2 \qquad \text{or} \qquad B = 1;$$

while equating constant coefficients yields

$$5 = 2 - C \qquad \text{or} \qquad C = -3.$$

The result is

$$\frac{3x^2-4x+5}{(x-1)(x^2+1)}=\frac{2}{x-1}+\frac{x-3}{x^2+1}.$$

Thus,

$$\int\frac{(3x^2-4x+5)\,dx}{(x-1)(x^2+1)}=2\int\frac{dx}{x-1}+\int\frac{(x-3)\,dx}{x^2+1}=2\int\frac{dx}{x-1}+\int\frac{x\,dx}{x^2+1}-3\int\frac{dx}{x^2+1}.$$

For the second integral, use the substitution $u=x^2+1$, $du=2x\,dx$. The final answer is

$$\int\frac{(3x^2-4x+5)\,dx}{(x-1)(x^2+1)}=2\ln|x-1|+\frac{1}{2}\ln|x^2+1|-3\tan^{-1}x+C.$$

33. $\displaystyle\int\frac{dx}{x(x^2+25)}$

SOLUTION The partial fraction decomposition has the form:

$$\frac{1}{x(x^2+25)}=\frac{A}{x}+\frac{Bx+C}{x^2+25}.$$

Clearing denominators, we get

$$1=A(x^2+25)+(Bx+C)x.$$

Setting $x=0$ then yields

$$1=A(25)+0\qquad\text{or}\qquad A=\frac{1}{25}.$$

This gives us

$$1=\frac{1}{25}x^2+1+Bx^2+Cx=\left(B+\frac{1}{25}\right)x^2+Cx+1.$$

Equating x^2-coefficients, we find

$$0=B+\frac{1}{25}\qquad\text{or}\qquad B=-\frac{1}{25},$$

while equating x-coefficients yields $C=0$. The result is

$$\frac{1}{x(x^2+25)}=\frac{\frac{1}{25}}{x}+\frac{-\frac{1}{25}x}{x^2+25}.$$

Thus,

$$\int\frac{dx}{x(x^2+25)}=\frac{1}{25}\int\frac{dx}{x}-\frac{1}{25}\int\frac{x\,dx}{x^2+25}.$$

For the integral on the right, use $u=x^2+25$, $du=2x\,dx$. Then we have

$$\int\frac{dx}{x(x^2+25)}=\frac{1}{25}\ln|x|-\frac{1}{50}\ln|x^2+25|+C.$$

35. $\displaystyle\int\frac{(6x^2+2)\,dx}{x^2+2x-3}$

SOLUTION Long division gives

$$\frac{6x^2+2}{x^2+2x-3}=6-\frac{12x-20}{x^2+2x-3}=6-\frac{12x-20}{(x+3)(x-1)}$$

The partial fraction decomposition of the second term is

$$\frac{12x-20}{(x+3)(x-1)}=\frac{A}{x+3}+\frac{B}{x-1}$$

Clear fractions to get

$$12x-20=A(x-1)+B(x+3)$$

Set $x = 1$ to get $-8 = 4B$ so that $B = -2$. Set $x = -3$ to get $-56 = -4A$ so that $A = 14$, and we have

$$\int \frac{6x^2+2}{x^2+2x-3} = \int 6 - \frac{14}{x+3} + \frac{2}{x-1}\,dx = \int 6\,dx - 14\int \frac{1}{x+3}\,dx + 2\int \frac{1}{x-1}\,dx$$
$$= 6x - 14\ln|x+3| + 2\ln|x-1| + C$$

37. $\displaystyle\int \frac{10\,dx}{(x-1)^2(x^2+9)}$

SOLUTION The partial fraction decomposition has the form:

$$\frac{10}{(x-1)^2(x^2+9)} = \frac{A}{x-1} + \frac{B}{(x-1)^2} + \frac{Cx+D}{x^2+9}.$$

Clearing denominators, we get

$$10 = A(x-1)(x^2+9) + B(x^2+9) + (Cx+D)(x-1)^2.$$

Setting $x = 1$ then yields

$$10 = 0 + B(10) + 0 \qquad \text{or} \qquad B = 1.$$

Expanding the right-hand side, we have

$$10 = (A+C)x^3 + (1 - A - 2C + D)x^2 + (9A + C - 2D)x + (9 - 9A + D).$$

Equating coefficients of like powers of x then yields

$$A + C = 0$$
$$1 - A - 2C + D = 0$$
$$9A + C - 2D = 0$$
$$9 - 9A + D = 10$$

From the first equation, we have $C = -A$, and from the fourth equation we have $D = 1 + 9A$. Substituting these into the second equation, we get

$$1 - A - 2(-A) + (1 + 9A) = 0 \qquad \text{or} \qquad A = -\frac{1}{5}.$$

Finally, $C = \frac{1}{5}$ and $D = -\frac{4}{5}$. The result is

$$\frac{10}{(x-1)^2(x^2+9)} = \frac{-\frac{1}{5}}{x-1} + \frac{1}{(x-1)^2} + \frac{\frac{1}{5}x - \frac{4}{5}}{x^2+9}.$$

Thus,

$$\int \frac{10\,dx}{(x-1)^2(x^2+9)} = -\frac{1}{5}\int \frac{dx}{x-1} + \int \frac{dx}{(x-1)^2} + \frac{1}{5}\int \frac{x\,dx}{x^2+9} - \frac{4}{5}\int \frac{dx}{x^2+9}$$
$$= -\frac{1}{5}\ln|x-1| - \frac{1}{x-1} + \frac{1}{10}\ln|x^2+9| - \frac{4}{15}\tan^{-1}\left(\frac{x}{3}\right) + C.$$

39. $\displaystyle\int \frac{dx}{x(x^2+8)^2}$

SOLUTION The partial fraction decomposition has the form:

$$\frac{1}{x(x^2+8)^2} = \frac{A}{x} + \frac{Bx+C}{x^2+8} + \frac{Dx+E}{(x^2+8)^2}.$$

Clearing denominators, we get

$$1 = A(x^2+8)^2 + (Bx+C)x(x^2+8) + (Dx+E)x.$$

Expanding the right-hand side gives us

$$1 = (A+B)x^4 + Cx^3 + (16A + 8B + D)x^2 + (8C + E)x + 64A.$$

Equating coefficients of like powers of x yields

$$\begin{aligned} A + B &= 0 \\ C &= 0 \\ 16A + 8B + D &= 0 \\ 8C + E &= 0 \\ 64A &= 1 \end{aligned}$$

The solution to this system of equations is

$$A = \frac{1}{64}, \qquad B = -\frac{1}{64}, \qquad C = 0, \qquad D = -\frac{1}{8}, \qquad E = 0.$$

Therefore

$$\frac{1}{x(x^2+8)^2} = \frac{\frac{1}{64}}{x} + \frac{-\frac{1}{64}x}{x^2+8} + \frac{-\frac{1}{8}x}{(x^2+8)^2},$$

and

$$\int \frac{dx}{x(x^2+8)^2} = \frac{1}{64}\int \frac{dx}{x} - \frac{1}{64}\int \frac{x\,dx}{x^2+8} - \frac{1}{8}\int \frac{x\,dx}{(x^2+8)^2}.$$

For the second and third integrals, use the substitution $u = x^2 + 8$, $du = 2x\,dx$. Then we have

$$\int \frac{dx}{x(x^2+8)^2} = \frac{1}{64}\ln|x| - \frac{1}{128}\ln|x^2+8| + \frac{1}{16(x^2+8)} + C.$$

41. $\displaystyle\int \frac{dx}{(x+2)(x^2+4x+10)}$

SOLUTION The partial fraction decomposition has the form:

$$\frac{1}{(x+2)(x^2+4x+10)} = \frac{A}{x+2} + \frac{Bx+C}{x^2+4x+10}.$$

Clearing denominators, we get

$$1 = A(x^2+4x+10) + (Bx+C)(x+2).$$

Setting $x = -2$ then yields

$$1 = A(6) + 0 \qquad \text{or} \qquad A = \frac{1}{6}.$$

Expanding the right-hand side gives us

$$1 = \left(\frac{1}{6} + B\right)x^2 + \left(\frac{2}{3} + 2B + C\right)x + \left(\frac{5}{3} + 2C\right).$$

Equating x^2-coefficients yields

$$0 = \frac{1}{6} + B \qquad \text{or} \qquad B = -\frac{1}{6},$$

while equating constant coefficients yields

$$1 = \frac{5}{3} + 2C \qquad \text{or} \qquad C = -\frac{1}{3}.$$

The result is

$$\frac{1}{(x+2)(x^2+4x+10)} = \frac{\frac{1}{6}}{x+2} + \frac{-\frac{1}{6}x - \frac{1}{3}}{x^2+4x+10}.$$

Thus,

$$\int \frac{dx}{(x+2)(x^2+4x+10)} = \frac{1}{6}\int \frac{dx}{x+2} - \frac{1}{6}\int \frac{(x+2)\,dx}{x^2+4x+10}.$$

For the second integral, let $u = x^2 + 4x + 10$. Then $du = (2x+4)\,dx$, and

$$\int \frac{dx}{(x+2)(x^2+4x+10)} = \frac{1}{6}\ln|x+2| - \frac{1}{12}\int \frac{(2x+4)\,dx}{x^2+4x+10}$$

$$= \frac{1}{6}\ln|x+2| - \frac{1}{12}\ln|x^2+4x+10| + C.$$

43. $\displaystyle\int \frac{25\,dx}{x(x^2+2x+5)^2}$

SOLUTION The partial fraction decomposition has the form

$$\frac{25}{x(x^2+2x+5)^2} = \frac{A}{x} + \frac{Bx+C}{x^2+2x+5} + \frac{Dx+E}{(x^2+2x+5)^2}.$$

Clearing denominators yields:

$$25 = A(x^2+2x+5)^2 + x(Bx+C)(x^2+2x+5) + x(Dx+E)$$

$$= (Ax^4 + 4Ax^3 + 14Ax^2 + 20Ax + 25A) + (Bx^4 + Cx^3 + 2Bx^3 + 2Cx^2 + 5Bx^2 + 5Cx) + Dx^2 + Ex.$$

Equating constant terms yields

$$25A = 25 \quad \text{or} \quad A = 1,$$

while equating x^4-coefficients yields

$$A + B = 0 \quad \text{or} \quad B = -A = -1.$$

Equating x^3-coefficients yields

$$4A + C + 2B = 0 \quad \text{or} \quad C = -2,$$

and equating x^2-coefficients yields

$$14A + 2C + 5B + D = 0 \quad \text{or} \quad D = -5.$$

Finally, equating x-coefficients yields

$$20A + 5C + E = 0 \quad \text{or} \quad E = -10.$$

Thus,

$$\int \frac{25\,dx}{x(x^2+2x+5)^2} = \int \left(\frac{1}{x} - \frac{x+2}{x^2+2x+5} - 5\frac{x+2}{(x^2+2x+5)^2}\right) dx$$

$$= \ln|x| - \int \frac{x+2}{x^2+2x+5}\,dx - 5\int \frac{x+2}{(x^2+2x+5)^2}\,dx.$$

The two integrals on the right both require the substitution $u = x+1$, so that $x^2+2x+5 = (x+1)^2+4 = u^2+4$ and $du = dx$. This means:

$$\int \frac{25\,dx}{x(x^2+2x+5)^2} = \ln|x| - \int \frac{u+1}{u^2+4}\,du - 5\int \frac{u+1}{(u^2+4)^2}\,du$$

$$= \ln|x| - \int \frac{u}{u^2+4}\,du - \int \frac{1}{u^2+4}\,du - 5\int \frac{u}{(u^2+4)^2}\,du - 5\int \frac{1}{(u^2+4)^2}\,du.$$

For the first and third integrals, we make the substitution $w = u^2+4$, $dw = 2u\,du$. Then we have

$$\int \frac{25\,dx}{x(x^2+2x+5)^2} = \ln|x| - \frac{1}{2}\ln|u^2+4| - \frac{1}{2}\tan^{-1}\left(\frac{u}{2}\right) + \frac{5}{2(u^2+4)} - 5\int \frac{du}{(u^2+4)^2}$$

$$= \ln|x| - \frac{1}{2}\ln|x^2+2x+5| - \frac{1}{2}\tan^{-1}\left(\frac{x+1}{2}\right) + \frac{5}{2(x^2+2x+5)} - 5\int \frac{du}{(u^2+4)^2}.$$

For the remaining integral, we use the trigonometric substitution $2\tan w = u$, so that $u^2+4 = 4\tan^2 w + 4 = 4\sec^2 w$ and $du = 2\sec^2 w\,dw$. This means

$$\int \frac{1}{(u^2+4)^2}\,du = \frac{1}{8}\int \frac{1}{\sec^4 w}\sec^2 w\,dw = \frac{1}{8}\int \cos^2 w\,dw$$

$$= \frac{1}{8}\left(\frac{1}{4}\sin 2w + \frac{w}{2}\right) + C = \left(\frac{1}{16}\sin w \cos w + \frac{w}{16}\right) + C$$

$$= \frac{1}{16}\frac{u}{\sqrt{u^2+4}}\frac{2}{\sqrt{u^2+4}} + \frac{1}{16}\tan^{-1}\left(\frac{u}{2}\right) + C = \frac{1}{8}\frac{u}{u^2+4} + \frac{1}{16}\tan^{-1}\left(\frac{u}{2}\right) + C$$

$$= \frac{1}{8}\frac{x+1}{x^2+2x+5} + \frac{1}{16}\tan^{-1}\left(\frac{x+1}{2}\right).$$

Hence, the integral is

$$\int \frac{25\,dx}{x(x^2+2x+5)^2} = \ln|x| - \frac{1}{2}\ln|x^2+2x+5| - \frac{1}{2}\tan^{-1}\left(\frac{x+1}{2}\right)$$

$$+ \frac{5}{2(x^2+2x+5)} - \frac{5}{8}\frac{x+1}{x^2+2x+5} - \frac{5}{16}\tan^{-1}\left(\frac{x+1}{2}\right)$$

$$= \ln|x| + \frac{15-5x}{8(x^2+2x+5)} - \frac{13}{16}\tan^{-1}\left(\frac{x+1}{2}\right) - \frac{1}{2}\ln|x^2+2x+5| + C.$$

In Exercises 45–48, evaluate by using first substitution and then partial fractions if necessary.

45. $\displaystyle\int \frac{x\,dx}{x^4+1}$

SOLUTION Use the substitution $u = x^2$ so that $du = 2x\,dx$, and

$$\int \frac{x}{x^4+1}\,dx = \frac{1}{2}\int \frac{1}{u^2+1}\,du = \frac{1}{2}\tan^{-1}u = \frac{1}{2}\tan^{-1}(x^2)$$

47. $\displaystyle\int \frac{e^x\,dx}{e^{2x}-e^x}$

SOLUTION Use the substitution $u = e^x$. Then $du = e^x\,dx = u\,dx$ so that $dx = \frac{1}{u}\,du$. Then

$$\int \frac{e^x\,dx}{e^{2x}-e^x} = \int \frac{u\cdot\frac{1}{u}\,du}{u^2-u} = \int \frac{1}{u(u-1)}\,du$$

Using partial fractions, we have

$$\frac{1}{u(u-1)} = \frac{A}{u} + \frac{B}{u-1} = \frac{(A+B)u - A}{u(u-1)}$$

Upon equating coefficients in the numerators, we have $A + B = 0$, $A = -1$ so that $B = 1$. Then

$$\int \frac{e^x\,dx}{e^{2x}-e^x} = -\int \frac{1}{u}\,du + \int \frac{1}{u-1}\,du = \ln|u-1| - \ln|u| + C = \ln|e^x-1| - \ln e^x + C$$

49. Evaluate $\displaystyle\int \frac{\sqrt{x}\,dx}{x-1}$. *Hint:* Use the substitution $u = \sqrt{x}$ (sometimes called a **rationalizing substitution**).

SOLUTION Let $u = \sqrt{x}$. Then $du = (1/2\sqrt{x})\,dx = (1/2u)\,dx$. Thus

$$\int \frac{\sqrt{x}\,dx}{x-1} = \int \frac{u(2u\,du)}{u^2-1} = 2\int \frac{u^2\,du}{u^2-1} = 2\int \frac{(u^2-1+1)\,du}{u^2-1}$$

$$= 2\int \left(\frac{u^2-1}{u^2-1} + \frac{1}{u^2-1}\right)du = 2\int du + \int \frac{2\,du}{u^2-1} = 2u + \int \frac{2\,du}{u^2-1}.$$

The partial fraction decomposition of the remaining integral has the form:

$$\frac{2}{u^2-1} = \frac{2}{(u-1)(u+1)} = \frac{A}{u-1} + \frac{B}{u+1}.$$

Clearing denominators gives us

$$2 = A(u+1) + B(u-1).$$

Setting $u = 1$ yields $2 = A(2) + 0$ or $A = 1$, while setting $u = -1$ yields $2 = 0 + B(-2)$ or $B = -1$. The result is

$$\frac{2}{u^2-1} = \frac{1}{u-1} + \frac{-1}{u+1}.$$

Thus,

$$\int \frac{2\,du}{u^2-1} = \int \frac{du}{u-1} - \int \frac{du}{u+1} = \ln|u-1| - \ln|u+1| + C.$$

The final answer is

$$\int \frac{\sqrt{x}\,dx}{x-1} = 2u + \ln|u-1| - \ln|u+1| + C = 2\sqrt{x} + \ln|\sqrt{x}-1| - \ln|\sqrt{x}+1| + C.$$

51. Evaluate $\displaystyle\int \frac{dx}{x^2-1}$ in two ways: using partial fractions and using trigonometric substitution. Verify that the two answers agree.

SOLUTION The partial fraction decomposition has the form:

$$\frac{1}{x^2-1} = \frac{1}{(x-1)(x+1)} = \frac{A}{x-1} + \frac{B}{x+1}.$$

Clearing denominators gives us

$$1 = A(x+1) + B(x-1).$$

Setting $x = 1$, we get $1 = A(2)$ or $A = \frac{1}{2}$; while setting $x = -1$, we get $1 = B(-2)$ or $B = -\frac{1}{2}$. The result is

$$\frac{1}{x^2-1} = \frac{\frac{1}{2}}{x-1} + \frac{-\frac{1}{2}}{x+1}.$$

Thus,

$$\int \frac{dx}{x^2-1} = \frac{1}{2}\int \frac{dx}{x-1} - \frac{1}{2}\int \frac{dx}{x+1} = \frac{1}{2}\ln|x-1| - \frac{1}{2}\ln|x+1| + C.$$

Using trigonometric substitution, let $x = \sec\theta$. Then $dx = \tan\theta\sec\theta\,d\theta$, and $x^2 - 1 = \sec^2\theta - 1 = \tan^2\theta$. Thus

$$\int \frac{dx}{x^2-1} = \int \frac{\tan\theta\sec\theta\,d\theta}{\tan^2\theta} = \int \frac{\sec\theta\,d\theta}{\tan\theta} = \int \frac{\cos\theta\,d\theta}{\sin\theta\cos\theta}$$
$$= \int \csc\theta\,d\theta = \ln|\csc\theta - \cot\theta| + C.$$

Now we construct a right triangle with $\sec\theta = x$:

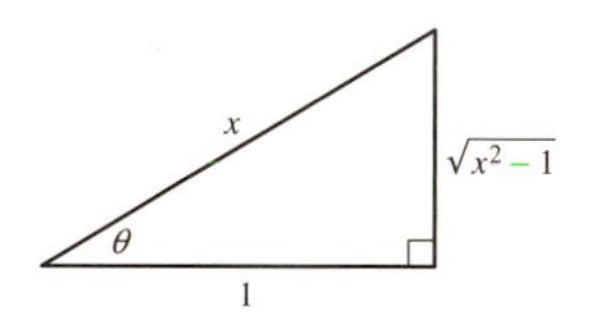

From this we see that $\csc\theta = x/\sqrt{x^2-1}$ and $\cot\theta = 1/\sqrt{x^2-1}$. Thus

$$\int \frac{dx}{x^2-1} = \ln\left|\frac{x}{\sqrt{x^2-1}} - \frac{1}{\sqrt{x^2-1}}\right| + C = \ln\left|\frac{x-1}{\sqrt{x^2-1}}\right| + C.$$

To check that these two answers agree, we write

$$\frac{1}{2}\ln|x-1| - \frac{1}{2}\ln|x+1| = \frac{1}{2}\left|\frac{x-1}{x+1}\right| = \ln\left|\sqrt{\frac{x-1}{x+1}}\right| = \ln\left|\frac{\sqrt{x-1}}{\sqrt{x+1}}\cdot\frac{\sqrt{x-1}}{\sqrt{x-1}}\right| = \ln\left|\frac{x-1}{\sqrt{x^2-1}}\right|.$$

In Exercises 53–66, evaluate the integral using the appropriate method or combination of methods covered thus far in the text.

53. $\displaystyle\int \frac{dx}{x^2\sqrt{4-x^2}}$

SOLUTION Use the trigonometric substitution $x = 2\sin\theta$. Then $dx = 2\cos\theta\,d\theta$,

$$4 - x^2 = 4 - 4\sin^2\theta = 4(1-\sin^2\theta) = 4\cos^2\theta,$$

and

$$\int \frac{dx}{x^2\sqrt{4-x^2}} = \int \frac{2\cos\theta\,d\theta}{(4\sin^2\theta)(2\cos\theta)} = \frac{1}{4}\int \csc^2\theta\,d\theta = -\frac{1}{4}\cot\theta + C.$$

Now construct a right triangle with $\sin\theta = x/2$:

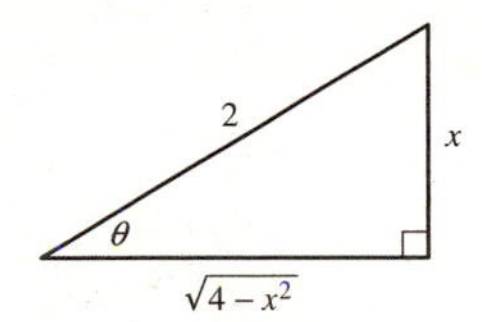

From this we see that $\cot\theta = \sqrt{4-x^2}/x$. Thus

$$\int \frac{dx}{x^2\sqrt{4-x^2}} = -\frac{1}{4}\left(\frac{\sqrt{4-x^2}}{x}\right) + C = -\frac{\sqrt{4-x^2}}{4x} + C.$$

55. $\displaystyle\int \cos^2 4x\,dx$

SOLUTION Use the substitution $u = 4x$, $du = 4\,dx$. Then we have

$$\int \cos^2(4x)\,dx = \frac{1}{4}\int \cos^2(4x)4\,dx = \frac{1}{4}\int \cos^2 u\,du = \frac{1}{4}\left[\frac{1}{2}u + \frac{1}{2}\sin u\cos u\right] + C$$
$$= \frac{1}{8}u + \frac{1}{8}\sin u\cos u + C = \frac{1}{2}x + \frac{1}{8}\sin 4x\cos 4x + C.$$

57. $\displaystyle\int \frac{dx}{(x^2+9)^2}$

SOLUTION Use the trigonometric substitution $x = 3\tan\theta$. Then $dx = 3\sec^2\theta\,d\theta$,

$$x^2 + 9 = 9\tan^2\theta + 9 = 9(\tan^2\theta + 1) = 9\sec^2\theta,$$

and

$$\int \frac{dx}{(x^2+9)^2} = \int \frac{3\sec^2\theta\,d\theta}{(9\sec^2\theta)^2} = \frac{3}{81}\int \frac{\sec^2\theta\,d\theta}{\sec^4\theta} = \frac{1}{27}\int \cos^2\theta\,d\theta = \frac{1}{27}\left(\frac{1}{2}\theta + \frac{1}{2}\sin\theta\cos\theta\right) + C.$$

Now construct a right triangle with $\tan\theta = x/3$:

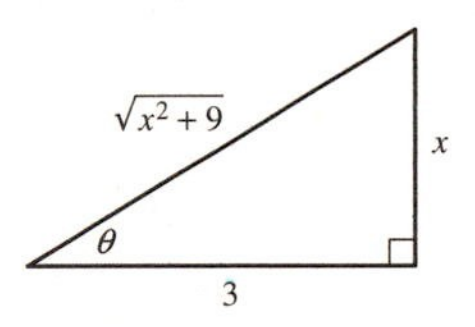

From this we see that $\sin\theta = x/\sqrt{x^2+9}$ and $\cos\theta = 3/\sqrt{x^2+9}$. Thus

$$\int \frac{dx}{\sqrt{x^2+9}^2} = \frac{1}{54}\tan^{-1}\left(\frac{x}{3}\right) + \frac{1}{54}\left(\frac{x}{\sqrt{x^2+9}}\right)\left(\frac{3}{\sqrt{x^2+9}}\right) + C = \frac{1}{54}\tan^{-1}\left(\frac{x}{3}\right) + \frac{x}{18(x^2+9)} + C.$$

59. $\displaystyle\int \tan^5 x\sec x\,dx$

SOLUTION Use the trigonometric identity $\tan^2 x = \sec^2 x - 1$ to write

$$\int \tan^5 x\sec x\,dx = \int \left(\sec^2 x - 1\right)^2 \tan x\sec x\,dx.$$

Now use the substitution $u = \sec x$, $du = \sec x\tan x\,dx$:

$$\int \tan^5 x\sec x\,dx = \int (u^2-1)^2\,du = \int \left(u^4 - 2u^2 + 1\right)du$$
$$= \frac{1}{5}u^5 - \frac{2}{3}u^3 + u + C = \frac{1}{5}\sec^5 x - \frac{2}{3}\sec^3 x + \sec x + C.$$

61. $\int \ln(x^4 - 1)\,dx$

SOLUTION Apply integration by parts with $u = \ln(x^4 - 1)$, $v' = 1$; then $u' = \frac{4x^3}{x^4-1}$ and $v = x$, so after simplification,

$$\begin{aligned}\int \ln(x^4-1)\,dx &= x\ln(x^4-1) - 4\int \frac{x^4}{x^4-1}\,dx = x\ln(x^4-1) - 4\int 1 + \frac{1}{x^4-1}\,dx \\ &= x\ln(x^4-1) - 4\int 1\,dx - 4\int \frac{1}{x^4-1}\,dx \\ &= x\ln(x^4-1) - 4x - 4\int \frac{1}{2}\left(\frac{1}{x^2-1} - \frac{1}{x^2+1}\right)dx \\ &= x\ln(x^4-1) - 4x - 2\int \frac{1}{x^2-1}\,dx + 2\int \frac{1}{x^2+1}\,dx \\ &= x\ln(x^4-1) - 4x + 2\tanh^{-1}x + 2\tan^{-1}x + C\end{aligned}$$

63. $\int \frac{x^2\,dx}{(x^2-1)^{3/2}}$

SOLUTION Use the trigonometric substitution $x = \sec\theta$. Then $dx = \sec\theta\tan\theta\,d\theta$,

$$x^2 - 1 = \sec^2\theta - 1 = \tan^2\theta,$$

and

$$\begin{aligned}\int \frac{x^2\,dx}{(x^2-1)^{3/2}} &= \int \frac{(\sec^2\theta)\sec\theta\tan\theta\,d\theta}{(\tan^2\theta)^{3/2}} = \int \frac{\sec^3\theta\,d\theta}{\tan^2\theta} = \int \frac{(\tan^2\theta+1)\sec\theta\,d\theta}{\tan^2\theta} \\ &= \int \frac{\tan^2\theta\sec\theta\,d\theta}{\tan^2\theta} + \int \frac{\sec\theta\,d\theta}{\tan^2\theta} = \int \sec\theta\,d\theta + \int \csc\theta\cot\theta\,d\theta \\ &= \ln|\sec\theta + \tan\theta| - \csc\theta + C.\end{aligned}$$

Now construct a right triangle with $\sec\theta = x$:

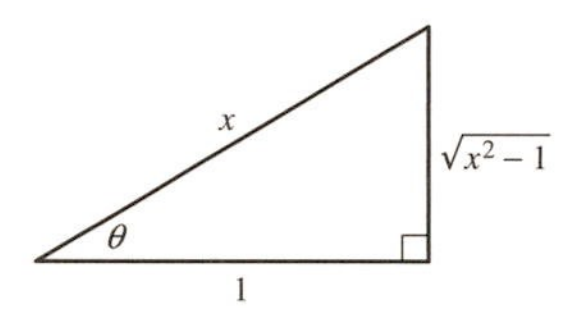

From this we see that $\tan\theta = \sqrt{x^2-1}$ and $\csc\theta = x/\sqrt{x^2-1}$. So the final answer is

$$\int \frac{x^2\,dx}{(x^2-1)^{3/2}} = \ln\left|x + \sqrt{x^2-1}\right| - \frac{x}{\sqrt{x^2-1}} + C.$$

65. $\int \frac{\sqrt{x}\,dx}{x^3+1}$

SOLUTION Use the substitution $u = x^{3/2}$, $du = \frac{3}{2}x^{1/2}\,dx$. Then $x^3 = (x^{3/2})^2 = u^2$, so we have

$$\int \frac{\sqrt{x}\,dx}{x^3+1} = \frac{2}{3}\int \frac{du}{u^2+1} = \frac{2}{3}\tan^{-1}u + C = \frac{2}{3}\tan^{-1}(x^{3/2}) + C.$$

67. Show that the substitution $\theta = 2\tan^{-1}t$ (Figure 2) yields the formulas

$$\cos\theta = \frac{1-t^2}{1+t^2}, \qquad \sin\theta = \frac{2t}{1+t^2}, \qquad d\theta = \frac{2\,dt}{1+t^2} \qquad \boxed{10}$$

This substitution transforms the integral of any rational function of $\cos\theta$ and $\sin\theta$ into an integral of a rational function of t (which can then be evaluated using partial fractions). Use it to evaluate $\int \frac{d\theta}{\cos\theta + \frac{3}{4}\sin\theta}$.

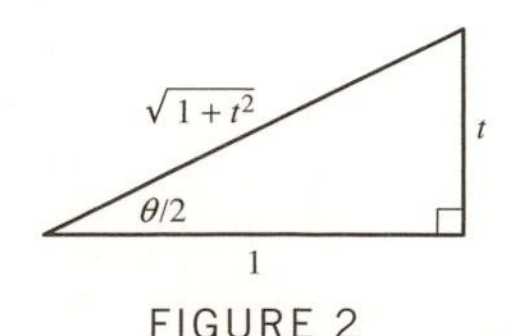

FIGURE 2

SOLUTION If $\theta = 2\tan^{-1} t$, then $d\theta = 2\,dt/(1+t^2)$. We also have that $\cos(\frac{\theta}{2}) = 1/\sqrt{1+t^2}$ and $\sin(\frac{\theta}{2}) = t/\sqrt{1+t^2}$. To find $\cos\theta$, we use the double angle identity $\cos\theta = 1 - 2\sin^2(\frac{\theta}{2})$. This gives us

$$\cos\theta = 1 - 2\left(\frac{t}{\sqrt{1+t^2}}\right)^2 = 1 - \frac{2t^2}{1+t^2} = \frac{1+t^2-2t^2}{1+t^2} = \frac{1-t^2}{1+t^2}.$$

To find $\sin\theta$, we use the double angle identity $\sin\theta = 2\sin(\frac{\theta}{2})\cos(\frac{\theta}{2})$. This gives us

$$\sin\theta = 2\left(\frac{t}{\sqrt{1+t^2}}\right)\left(\frac{1}{\sqrt{1+t^2}}\right) = \frac{2t}{1+t^2}.$$

With these formulas, we have

$$\int \frac{d\theta}{\cos\theta + (3/4)\sin\theta} = \int \frac{\frac{2\,dt}{1+t^2}}{\left(\frac{1-t^2}{1+t^2}\right) + \frac{3}{4}\left(\frac{2t}{1+t^2}\right)} = \int \frac{8\,dt}{4(1-t^2)+3(2t)} = \int \frac{8\,dt}{4+6t-4t^2} = \int \frac{4\,dt}{2+3t-2t^2}.$$

The partial fraction decomposition has the form

$$\frac{4}{2+3t-2t^2} = \frac{A}{2-t} + \frac{B}{1+2t}.$$

Clearing denominators gives us

$$4 = A(1+2t) + B(2-t).$$

Setting $t = 2$ then yields

$$4 = A(5) + 0 \qquad \text{or} \qquad A = \frac{4}{5},$$

while setting $t = -\frac{1}{2}$ yields

$$4 = 0 + B\left(\frac{5}{2}\right) \qquad \text{or} \qquad B = \frac{8}{5}.$$

The result is

$$\frac{4}{2+3t-2t^2} = \frac{\frac{4}{5}}{2-t} + \frac{\frac{8}{5}}{1+2t}.$$

Thus,

$$\int \frac{4}{2+3t-2t^2}\,dt = \frac{4}{5}\int \frac{dt}{2-t} + \frac{8}{5}\int \frac{dt}{1+2t} = -\frac{4}{5}\ln|2-t| + \frac{4}{5}\ln|1+2t| + C.$$

The original substitution was $\theta = 2\tan^{-1} t$, which means that $t = \tan(\frac{\theta}{2})$. The final answer is then

$$\int \frac{d\theta}{\cos\theta + \frac{3}{4}\sin\theta} = -\frac{4}{5}\ln\left|2 - \tan\left(\frac{\theta}{2}\right)\right| + \frac{4}{5}\ln\left|1 + 2\tan\left(\frac{\theta}{2}\right)\right| + C.$$

Further Insights and Challenges

69. Prove the general formula

$$\int \frac{dx}{(x-a)(x-b)} = \frac{1}{a-b}\ln\frac{x-a}{x-b} + C$$

where a, b are constants such that $a \neq b$.

SOLUTION The partial fraction decomposition has the form:

$$\frac{1}{(x-a)(x-b)} = \frac{A}{x-a} + \frac{B}{x-b}.$$

Clearing denominators, we get

$$1 = A(x-b) + B(x-a).$$

Setting $x = a$ then yields

$$1 = A(a - b) + 0 \quad \text{or} \quad A = \frac{1}{a - b},$$

while setting $x = b$ yields

$$1 = 0 + B(b - a) \quad \text{or} \quad B = \frac{1}{b - a}.$$

The result is

$$\frac{1}{(x - a)(x - b)} = \frac{\frac{1}{a-b}}{x - a} + \frac{\frac{1}{b-a}}{x - b}.$$

Thus,

$$\int \frac{dx}{(x - a)(x - b)} = \frac{1}{a - b} \int \frac{dx}{x - a} + \frac{1}{b - a} \int \frac{dx}{x - b} = \frac{1}{a - b} \ln|x - a| + \frac{1}{b - a} \ln|x - b| + C$$

$$= \frac{1}{a - b} \ln|x - a| - \frac{1}{a - b} \ln|x - b| + C = \frac{1}{a - b} \ln\left|\frac{x - a}{x - b}\right| + C.$$

71. Suppose that $Q(x) = (x - a)(x - b)$, where $a \neq b$, and let $P(x)/Q(x)$ be a proper rational function so that

$$\frac{P(x)}{Q(x)} = \frac{A}{(x - a)} + \frac{B}{(x - b)}$$

(a) Show that $A = \dfrac{P(a)}{Q'(a)}$ and $B = \dfrac{P(b)}{Q'(b)}$.

(b) Use this result to find the partial fraction decomposition for $P(x) = 3x - 2$ and $Q(x) = x^2 - 4x - 12$.

SOLUTION

(a) Clearing denominators gives us

$$P(x) = A(x - b) + B(x - a).$$

Setting $x = a$ then yields

$$P(a) = A(a - b) + 0 \quad \text{or} \quad A = \frac{P(a)}{a - b},$$

while setting $x = b$ yields

$$P(b) = 0 + B(b - a) \quad \text{or} \quad B = \frac{P(b)}{b - a}.$$

Now use the product rule to differentiate $Q(x)$:

$$Q'(x) = (x - a)(1) + (1)(x - b) = x - a + x - b = 2x - a - b;$$

therefore,

$$Q'(a) = 2a - a - b = a - b$$

$$Q'(b) = 2b - a - b = b - a$$

Substituting these into the above results, we find

$$A = \frac{P(a)}{Q'(a)} \quad \text{and} \quad B = \frac{P(b)}{Q'(b)}.$$

(b) The partial fraction decomposition has the form:

$$\frac{P(x)}{Q(x)} = \frac{3x - 2}{x^2 - 4x - 12} = \frac{3x - 2}{(x - 6)(x + 2)} = \frac{A}{x - 6} + \frac{B}{x + 2};$$

$$A = \frac{P(6)}{Q'(6)} = \frac{3(6) - 2}{2(6) - 4} = \frac{16}{8} = 2;$$

$$B = \frac{P(-2)}{Q'(-2)} = \frac{3(-2) - 2}{2(-2) - 4} = \frac{-8}{-8} = 1.$$

The result is

$$\frac{3x - 2}{x^2 - 4x - 12} = \frac{2}{x - 6} + \frac{1}{x + 2}.$$

7.6 Improper Integrals

Preliminary Questions

1. State whether the integral converges or diverges:

(a) $\int_1^\infty x^{-3}\,dx$ **(b)** $\int_0^1 x^{-3}\,dx$

(c) $\int_1^\infty x^{-2/3}\,dx$ **(d)** $\int_0^1 x^{-2/3}\,dx$

SOLUTION

(a) The integral is improper because one of the limits of integration is infinite. Because the power of x in the integrand is less than -1, this integral converges.

(b) The integral is improper because the integrand is undefined at $x = 0$. Because the power of x in the integrand is less than -1, this integral diverges.

(c) The integral is improper because one of the limits of integration is infinite. Because the power of x in the integrand is greater than -1, this integral diverges.

(d) The integral is improper because the integrand is undefined at $x = 0$. Because the power of x in the integrand is greater than -1, this integral converges.

2. Is $\int_0^{\pi/2} \cot x\,dx$ an improper integral? Explain.

SOLUTION Because the integrand $\cot x$ is undefined at $x = 0$, this is an improper integral.

3. Find a value of $b > 0$ that makes $\int_0^b \frac{1}{x^2-4}\,dx$ an improper integral.

SOLUTION Any value of b satisfying $|b| \geq 2$ will make this an improper integral.

4. Which comparison would show that $\int_0^\infty \frac{dx}{x+e^x}$ converges?

SOLUTION Note that, for $x > 0$,

$$\frac{1}{x+e^x} < \frac{1}{e^x} = e^{-x}.$$

Moreover

$$\int_0^\infty e^{-x}\,dx$$

converges. Therefore,

$$\int_0^\infty \frac{1}{x+e^x}\,dx$$

converges by the comparison test.

5. Explain why it is not possible to draw any conclusions about the convergence of $\int_1^\infty \frac{e^{-x}}{x}\,dx$ by comparing with the integral $\int_1^\infty \frac{dx}{x}$.

SOLUTION For $1 \leq x < \infty$,

$$\frac{e^{-x}}{x} < \frac{1}{x},$$

but

$$\int_1^\infty \frac{dx}{x}$$

diverges. Knowing that an integral is smaller than a divergent integral does not allow us to draw any conclusions using the comparison test.

Exercises

1. Which of the following integrals is improper? Explain your answer, but do not evaluate the integral.

(a) $\int_0^2 \frac{dx}{x^{1/3}}$ **(b)** $\int_1^\infty \frac{dx}{x^{0.2}}$ **(c)** $\int_{-1}^\infty e^{-x}\,dx$

(d) $\int_0^1 e^{-x}\,dx$ **(e)** $\int_0^{\pi/2} \sec x\,dx$ **(f)** $\int_0^\infty \sin x\,dx$

(g) $\int_0^1 \sin x\,dx$ **(h)** $\int_0^1 \frac{dx}{\sqrt{3-x^2}}$ **(i)** $\int_1^\infty \ln x\,dx$

(j) $\int_0^3 \ln x\,dx$

SOLUTION

(a) Improper. The function $x^{-1/3}$ is infinite at 0.
(b) Improper. Infinite interval of integration.
(c) Improper. Infinite interval of integration.
(d) Proper. The function e^{-x} is continuous on the finite interval $[0, 1]$.
(e) Improper. The function $\sec x$ is infinite at $\frac{\pi}{2}$.
(f) Improper. Infinite interval of integration.
(g) Proper. The function $\sin x$ is continuous on the finite interval $[0, 1]$.
(h) Proper. The function $1/\sqrt{3-x^2}$ is continuous on the finite interval $[0, 1]$.
(i) Improper. Infinite interval of integration.
(j) Improper. The function $\ln x$ is infinite at 0.

3. Prove that $\int_1^\infty x^{-2/3}\,dx$ diverges by showing that

$$\lim_{R\to\infty} \int_1^R x^{-2/3}\,dx = \infty$$

SOLUTION First compute the proper integral:

$$\int_1^R x^{-2/3}\,dx = 3x^{1/3}\Big|_1^R = 3R^{1/3} - 3 = 3\left(R^{1/3} - 1\right).$$

Then show divergence:

$$\int_1^\infty x^{-2/3}\,dx = \lim_{R\to\infty} \int_1^R x^{-2/3}\,dx = \lim_{R\to\infty} 3\left(R^{1/3} - 1\right) = \infty.$$

In Exercises 5–40, determine whether the improper integral converges and, if so, evaluate it.

5. $\int_1^\infty \frac{dx}{x^{19/20}}$

SOLUTION First evaluate the integral over the finite interval $[1, R]$ for $R > 1$:

$$\int_1^R \frac{dx}{x^{19/20}} = 20x^{1/20}\Big|_1^R = 20R^{1/20} - 20.$$

Now compute the limit as $R \to \infty$:

$$\int_1^\infty \frac{dx}{x^{19/20}} = \lim_{R\to\infty} \int_1^R \frac{dx}{x^{19/20}} = \lim_{R\to\infty} \left(20R^{1/20} - 20\right) = \infty.$$

The integral does not converge.

7. $\int_{-\infty}^4 e^{0.0001t}\,dt$

SOLUTION First evaluate the integral over the finite interval $[R, 4]$ for $R < 4$:

$$\int_R^4 e^{(0.0001)t}\,dt = \frac{e^{(0.0001)t}}{0.0001}\Bigg|_R^4 = 10{,}000\left(e^{0.0004} - e^{(0.0001)R}\right).$$

Now compute the limit as $R \to -\infty$:

$$\int_{-\infty}^{4} e^{(0.0001)t}\,dt = \lim_{R\to-\infty}\int_{R}^{4} e^{(0.0001)t}\,dt = \lim_{R\to-\infty} 10{,}000\left(e^{0.0004} - e^{(0.0001)R}\right)$$
$$= 10{,}000\left(e^{0.0004} - 0\right) = 10{,}000e^{0.0004}.$$

9. $\displaystyle\int_0^5 \frac{dx}{x^{20/19}}$

SOLUTION The function $x^{-20/19}$ is infinite at the endpoint 0, so we'll first evaluate the integral on the finite interval $[R, 5]$ for $0 < R < 5$:

$$\int_R^5 \frac{dx}{x^{20/19}} = -19x^{-1/19}\Big|_R^5 = -19\left(5^{-1/19} - R^{-1/19}\right) = 19\left(\frac{1}{R^{1/19}} - \frac{1}{5^{1/19}}\right).$$

Now compute the limit as $R \to 0^+$:

$$\int_0^5 \frac{dx}{x^{20/19}} = \lim_{R\to0^+}\int_R^5 \frac{dx}{x^{20/19}} = \lim_{R\to0^+} 19\left(\frac{1}{R^{1/19}} - \frac{1}{5^{1/19}}\right) = \infty;$$

thus, the integral does not converge.

11. $\displaystyle\int_0^4 \frac{dx}{\sqrt{4-x}}$

SOLUTION The function $1/\sqrt{4-x}$ is infinite at $x = 4$, so we'll first evaluate the integral on the interval $[0, R]$ for $0 < R < 4$:

$$\int_0^R \frac{dx}{\sqrt{4-x}} = -2\sqrt{4-x}\Big|_0^R = -2\sqrt{4-R} - (-2)\sqrt{4} = 4 - 2\sqrt{4-R}.$$

Now compute the limit as $R \to 4^-$:

$$\int_0^4 \frac{dx}{\sqrt{4-x}} = \lim_{R\to4^-}\int_0^R \frac{dx}{\sqrt{4-x}} = \lim_{R\to4^-}\left(4 - 2\sqrt{4-R}\right) = 4 - 0 = 4.$$

13. $\displaystyle\int_2^\infty x^{-3}\,dx$

SOLUTION First evaluate the integral on the finite interval $[2, R]$ for $2 < R$:

$$\int_2^R x^{-3}\,dx = \frac{x^{-2}}{-2}\Big|_2^R = \frac{-1}{2R^2} - \frac{-1}{2(2^2)} = \frac{1}{8} - \frac{1}{2R^2}.$$

Now compute the limit as $R \to \infty$:

$$\int_2^\infty x^{-3}\,dx = \lim_{R\to\infty}\int_2^R x^{-3}\,dx = \lim_{R\to\infty}\left(\frac{1}{8} - \frac{1}{2R^2}\right) = \frac{1}{8}.$$

15. $\displaystyle\int_{-3}^\infty \frac{dx}{(x+4)^{3/2}}$

SOLUTION First evaluate the integral on the finite interval $[-3, R]$ for $R > -3$:

$$\int_{-3}^R \frac{dx}{(x+4)^{3/2}} = -2(x+4)^{-1/2}\Big|_{-3}^R = \frac{-2}{\sqrt{R+4}} - \frac{-2}{\sqrt{1}} = 2 - \frac{2}{\sqrt{R+4}}.$$

Now compute the limit as $R \to \infty$:

$$\int_{-3}^\infty \frac{dx}{(x+4)^{3/2}} = \lim_{R\to\infty}\int_{-3}^R \frac{dx}{(x+4)^{3/2}} = \lim_{R\to\infty}\left(2 - \frac{2}{\sqrt{R+4}}\right) = 2 - 0 = 2.$$

17. $\displaystyle\int_0^1 \frac{dx}{x^{0.2}}$

SOLUTION The function $x^{-0.2}$ is infinite at $x = 0$, so we'll first evaluate the integral on the interval $[R, 1]$ for $0 < R < 1$:

$$\int_R^1 \frac{dx}{x^{0.2}} = \frac{x^{0.8}}{0.8}\Big|_R^1 = 1.25\left(1 - R^{0.8}\right).$$

Now compute the limit as $R \to 0^+$:

$$\int_0^1 \frac{dx}{x^{0.2}} = \lim_{R\to 0+} \int_R^1 \frac{dx}{x^{0.2}} = \lim_{R\to 0+} 1.25\left(1 - R^{0.8}\right) = 1.25(1-0) = 1.25.$$

19. $\int_4^\infty e^{-3x}\,dx$

SOLUTION First evaluate the integral on the finite interval $[4, R]$ for $R > 4$:

$$\int_4^R e^{-3x}\,dx = \left.\frac{e^{-3x}}{-3}\right|_4^R = -\frac{1}{3}\left(e^{-3R} - e^{-12}\right) = \frac{1}{3}\left(e^{-12} - e^{-3R}\right).$$

Now compute the limit as $R \to \infty$:

$$\int_4^\infty e^{-3x}\,dx = \lim_{R\to\infty} \int_4^R e^{-3x}\,dx = \lim_{R\to\infty} \frac{1}{3}\left(e^{-12} - e^{-3R}\right) = \frac{1}{3}\left(e^{-12} - 0\right) = \frac{1}{3e^{12}}.$$

21. $\int_{-\infty}^0 e^{3x}\,dx$

SOLUTION First evaluate the integral on the finite interval $[R, 0]$ for $R < 0$:

$$\int_R^0 e^{3x}\,dx = \left.\frac{e^{3x}}{3}\right|_R^0 = \frac{1}{3} - \frac{e^{3R}}{3}.$$

Now compute the limit as $R \to -\infty$:

$$\int_{-\infty}^0 e^{3x}\,dx = \lim_{R\to-\infty} \int_R^0 e^{3x}\,dx = \lim_{R\to-\infty}\left(\frac{1}{3} - \frac{e^{3R}}{3}\right) = \frac{1}{3} - 0 = \frac{1}{3}.$$

23. $\int_1^3 \frac{dx}{\sqrt{3-x}}$

SOLUTION The function $f(x) = 1/\sqrt{3-x}$ is infinite at $x = 3$, so we first evaluate the integral on the interval $[1, R]$ for $1 < R < 3$:

$$\int_1^R \frac{dx}{\sqrt{3-x}} = -2\sqrt{3-x}\Big|_1^R = -2\sqrt{3-R} + 2\sqrt{2}.$$

Now compute the limit as $R \to 3^-$:

$$\int_1^3 \frac{dx}{\sqrt{3-x}} = \lim_{R\to 3-} \int_1^R \frac{dx}{\sqrt{3-x}} = 0 + 2\sqrt{2} = 2\sqrt{2}.$$

25. $\int_0^\infty \frac{dx}{1+x}$

SOLUTION First evaluate the integral on the finite interval $[0, R]$ for $R > 0$:

$$\int_0^R \frac{dx}{1+x} = \ln|1+x|\Big|_0^R = \ln|1+R| - \ln 1 = \ln|1+R|.$$

Now compute the limit as $R \to \infty$:

$$\int_0^\infty \frac{dx}{1+x} = \lim_{R\to\infty} \int_0^R \frac{dx}{1+x} = \lim_{R\to\infty} \ln|1+R| = \infty;$$

thus, the integral does not converge.

27. $\int_0^\infty \frac{x\,dx}{(1+x^2)^2}$

SOLUTION First evaluate the indefinite integral, using the substitution $u = x^2$, $du = 2x\,dx$; then

$$\int \frac{x\,dx}{(1+x^2)^2} = \frac{1}{2}\int \frac{1}{(1+u)^2}\,du = -\frac{1}{2(u+1)} + C = -\frac{1}{2(x^2+1)} + C$$

Thus, for $R > 0$,

$$\int_0^R \frac{x\,dx}{(x^2+1)^2} = \left(-\frac{1}{2(x^2+1)}\right)\Bigg|_0^R = -\frac{1}{2(R^2+1)} + \frac{1}{2}$$

and thus in the limit

$$\int_0^\infty \frac{x\,dx}{(x^2+1)^2} = \lim_{R\to\infty}\int_0^R \frac{x\,dx}{(x^2+1)^2} = \frac{1}{2} - \lim_{R\to\infty}\frac{1}{2(R^2+1)} = \frac{1}{2}$$

29. $\displaystyle\int_0^\infty e^{-x}\cos x\,dx$

SOLUTION First evaluate the indefinite integral using Integration by Parts, with $u = e^{-x}$, $v' = \cos x$. Then $u' = -e^{-x}$, $v = \sin x$, and

$$\int e^{-x}\cos x\,dx = e^{-x}\sin x - \int \sin x(-e^{-x})\,dx = e^{-x}\sin x + \int e^{-x}\sin x\,dx.$$

Now use Integration by Parts again, with $u = e^{-x}$, $v' = \sin x$. Then $u' = -e^{-x}$, $v = -\cos x$, and

$$\int e^{-x}\cos x\,dx = e^{-x}\sin x + \left[-e^{-x}\cos x - \int e^{-x}\cos x\,dx\right].$$

Solving this equation for $\int e^{-x}\cos x\,dx$, we find

$$\int e^{-x}\cos x\,dx = \frac{1}{2}e^{-x}(\sin x - \cos x) + C.$$

Thus,

$$\int_0^R e^{-x}\cos x\,dx = \frac{1}{2}e^{-x}(\sin x - \cos x)\Big|_0^R = \frac{\sin R - \cos R}{2e^R} - \frac{\sin 0 - \cos 0}{2} = \frac{\sin R - \cos R}{2e^R} + \frac{1}{2},$$

and

$$\int_0^\infty e^{-x}\cos x\,dx = \lim_{R\to\infty}\left(\frac{\sin R - \cos R}{2e^R} + \frac{1}{2}\right) = 0 + \frac{1}{2} = \frac{1}{2}.$$

31. $\displaystyle\int_0^3 \frac{dx}{\sqrt{9-x^2}}$

SOLUTION The function $(9-x^2)^{-1/2}$ is infinite at $x = 3$, so we'll first evaluate the integral on the interval $[0, R]$ for $0 < R < 3$:

$$\int_0^R \frac{dx}{\sqrt{9-x^2}} = \sin^{-1}\frac{x}{3}\Big|_0^R = \sin^{-1}\frac{R}{3} - \sin^{-1}0 = \sin^{-1}\frac{R}{3}.$$

Thus,

$$\int_0^3 \frac{dx}{\sqrt{9-x^2}} = \lim_{R\to 3^-}\sin^{-1}\frac{R}{3} = \sin^{-1}1 = \frac{\pi}{2}.$$

33. $\displaystyle\int_1^\infty \frac{e^{\sqrt{x}}\,dx}{\sqrt{x}}$

SOLUTION Let $u = \sqrt{x}$, $du = \frac{1}{2}x^{-1/2}\,dx$. Then

$$\int \frac{e^{\sqrt{x}}\,dx}{\sqrt{x}} = 2\int e^{\sqrt{x}}\left(\frac{dx}{2\sqrt{x}}\right) = 2\int e^u\,du = 2e^u + C = 2e^{\sqrt{x}} + C,$$

and

$$\int_1^\infty \frac{e^{\sqrt{x}}\,dx}{\sqrt{x}} = \lim_{R\to\infty}\int_1^R \frac{e^{\sqrt{x}}\,dx}{\sqrt{x}} = \lim_{R\to\infty} 2e^{\sqrt{x}}\Big|_1^R = \lim_{R\to\infty}\left(2e^{\sqrt{R}} - 2e\right) = \infty.$$

The integral does not converge.

35. $\displaystyle\int_0^\infty \sin x\,dx$

SOLUTION First evaluate the integral on the finite interval $[0, R]$ for $R > 0$:

$$\int_0^R \sin x\,dx = -\cos x\Big|_0^R = -\cos R + \cos 0 = 1 - \cos R.$$

Thus,

$$\int_0^R \sin x\, dx = \lim_{R\to\infty}(1-\cos R) = 1 - \lim_{R\to\infty}\cos R.$$

This limit does not exist, since the value of $\cos R$ oscillates between 1 and -1 as R approaches infinity. Hence the integral does not converge.

37. $\displaystyle\int_0^1 \ln x\, dx$

SOLUTION The function $\ln x$ is infinite at $x = 0$, so we'll first evaluate the integral on $[R, 1]$ for $0 < R < 1$. Use Integration by Parts with $u = \ln x$ and $v' = 1$. Then $u' = 1/x$, $v = x$, and we have

$$\int_R^1 \ln x\, dx = x\ln x\Big|_R^1 - \int_R^1 dx = (x\ln x - x)\Big|_R^1 = (\ln 1 - 1) - (R\ln R - R) = R - 1 - R\ln R.$$

Thus,

$$\int_0^1 \ln x\, dx = \lim_{R\to 0+}(R - 1 - R\ln R) = -1 - \lim_{R\to 0+} R\ln R.$$

To compute the limit, rewrite the function as a quotient and apply L'Hôpital's Rule:

$$\int_0^1 \ln x\, dx = -1 - \lim_{R\to 0+}\frac{\ln R}{\frac{1}{R}} = -1 - \lim_{R\to 0+}\frac{\frac{1}{R}}{\frac{-1}{R^2}} = -1 - \lim_{R\to 0+}(-R) = -1 - 0 = -1.$$

39. $\displaystyle\int_0^1 \frac{\ln x}{x^2}\, dx$

SOLUTION Use Integration by Parts, with $u = \ln x$ and $v' = x^{-2}$. Then $u' = 1/x$, $v = -x^{-1}$, and

$$\int \frac{\ln x}{x^2}\, dx = -\frac{1}{x}\ln x + \int \frac{dx}{x^2} = -\frac{1}{x}\ln x - \frac{1}{x} + C.$$

The function is infinite at $x = 0$, so we'll first evaluate the integral on $[R, 1]$ for $0 < R < 1$:

$$\int_a^1 \frac{\ln x}{x^2}\, dx = \left(-\frac{1}{x}\ln x - \frac{1}{x}\right)\Big|_R^1 = \left(-\frac{1}{1}\ln 1 - \frac{1}{1}\right) - \left(-\frac{1}{R}\ln R - \frac{1}{R}\right) = \frac{1}{R}\ln R + \frac{1}{R} - 1.$$

Thus,

$$\int_0^1 \frac{\ln x}{x^2}\, dx = \lim_{R\to 0+}\frac{1}{R}\ln R + \frac{1}{R} - 1 = -1 + \lim_{R\to 0+}\frac{\ln R + 1}{R} = -\infty.$$

The integral does not converge.

41. Let $I = \displaystyle\int_4^\infty \frac{dx}{(x-2)(x-3)}$.

(a) Show that for $R > 4$,

$$\int_4^R \frac{dx}{(x-2)(x-3)} = \ln\left|\frac{R-3}{R-2}\right| - \ln\frac{1}{2}$$

(b) Then show that $I = \ln 2$.

SOLUTION

(a) The partial fraction decomposition takes the form

$$\frac{1}{(x-2)(x-3)} = \frac{A}{x-2} + \frac{B}{x-3}.$$

Clearing denominators gives us

$$1 = A(x-3) + B(x-2).$$

Setting $x = 2$ then yields $A = -1$, while setting $x = 3$ yields $B = 1$. Thus,

$$\int \frac{dx}{(x-2)(x-3)} = \int \frac{dx}{x-3} - \int \frac{dx}{x-2} = \ln|x-3| - \ln|x-2| + C = \ln\left|\frac{x-3}{x-2}\right| + C,$$

and, for $R > 4$,

$$\int_4^R \frac{dx}{(x-2)(x-3)} = \ln\left|\frac{x-3}{x-2}\right|\Big|_4^R = \ln\left|\frac{R-3}{R-2}\right| - \ln\frac{1}{2}.$$

(b) Using the result from part (a),

$$I = \lim_{R\to\infty}\left(\ln\left|\frac{R-3}{R-2}\right| - \ln\frac{1}{2}\right) = \ln 1 - \ln\frac{1}{2} = \ln 2.$$

43. Evaluate $I = \int_0^1 \frac{dx}{x(2x+5)}$ or state that it diverges.

SOLUTION The partial fraction decomposition takes the form

$$\frac{1}{x(2x+5)} = \frac{A}{x} + \frac{B}{2x+5}.$$

Clearing denominators gives us

$$1 = A(2x+5) + Bx.$$

Setting $x = 0$ then yields $A = \frac{1}{5}$, while setting $x = -\frac{5}{2}$ yields $B = -\frac{2}{5}$. Thus,

$$\int \frac{dx}{x(2x+5)} = \frac{1}{5}\int\frac{dx}{x} - \frac{2}{5}\int\frac{dx}{2x+5} = \frac{1}{5}\ln|x| - \frac{1}{5}\ln|2x+5| + C = \frac{1}{5}\ln\left|\frac{x}{2x+5}\right| + C,$$

and, for $0 < R < 1$,

$$\int_R^1 \frac{dx}{x(2x+5)} = \frac{1}{5}\ln\left|\frac{x}{2x+5}\right|\Big|_R^1 = \frac{1}{5}\ln\frac{1}{7} - \frac{1}{5}\ln\left|\frac{R}{2R+5}\right|.$$

Thus,

$$I = \lim_{R\to 0+}\left(\frac{1}{5}\ln\frac{1}{7} - \frac{1}{5}\ln\left|\frac{R}{2R+5}\right|\right) = \infty.$$

The integral does not converge.

In Exercises 45–48, determine whether the doubly infinite improper integral converges and, if so, evaluate it. Use definition (2).

45. $\int_{-\infty}^{\infty} \frac{x\,dx}{1+x^2}$

SOLUTION Using the substitution $u = x^2 + 1$, $du = 2x\,dx$, we obtain

$$\int \frac{x\,dx}{1+x^2} = \frac{1}{2}\ln(x^2+1) + C.$$

Thus,

$$\int_0^{\infty} \frac{x\,dx}{1+x^2} = \lim_{R\to\infty}\int_0^R \frac{x\,dx}{1+x^2} = \lim_{R\to\infty}\frac{1}{2}\ln(R^2+1) = \infty;$$

$$\int_{-\infty}^{0} \frac{x\,dx}{1+x^2} = \lim_{R\to-\infty}\int_R^0 \frac{x\,dx}{1+x^2} = \lim_{R\to-\infty}\frac{1}{2}\ln(R^2+1) = \infty;$$

It follows that

$$\int_{-\infty}^{\infty} \frac{x\,dx}{1+x^2}$$

diverges.

47. $\int_{-\infty}^{\infty} xe^{-x^2}\,dx$

SOLUTION First note that

$$\int xe^{-x^2}\,dx = -\frac{1}{2}e^{-x^2} + C.$$

Thus,

$$\int_0^\infty xe^{-x^2}\,dx = \lim_{R\to\infty}\int_0^R xe^{-x^2}\,dx = \lim_{R\to\infty}\left(\frac{1}{2}-\frac{1}{2}e^{-R^2}\right) = \frac{1}{2};$$

$$\int_{-\infty}^0 xe^{-x^2}\,dx = \lim_{R\to-\infty}\int_R^0 xe^{-x^2}\,dx = \lim_{R\to-\infty}\left(-\frac{1}{2}+\frac{1}{2}e^{-R^2}\right) = -\frac{1}{2};$$

and

$$\int_{-\infty}^\infty xe^{-x^2}\,dx = \frac{1}{2}-\frac{1}{2} = 0.$$

49. Define $J = \int_{-1}^1 \frac{dx}{x^{1/3}}$ as the sum of the two improper integrals $\int_{-1}^0 \frac{dx}{x^{1/3}} + \int_0^1 \frac{dx}{x^{1/3}}$. Show that J converges and that $J = 0$.

SOLUTION Note that since $x^{-1/3}$ is an odd function, one might expect this integral over a symmetric interval to be zero. To prove this, we start by evaluating the indefinite integral:

$$\int \frac{dx}{x^{1/3}} = \frac{3}{2}x^{2/3} + C$$

Then

$$\int_{-1}^0 \frac{dx}{x^{1/3}} = \lim_{R\to0-}\int_{-1}^R \frac{dx}{x^{1/3}} = \lim_{R\to0-}\frac{3}{2}x^{2/3}\Big|_{-1}^R = \lim_{R\to0-}\frac{3}{2}R^{2/3}-\frac{3}{2} = -\frac{3}{2}$$

$$\int_0^1 \frac{dx}{x^{1/3}} = \lim_{R\to0+}\int_R^1 \frac{dx}{x^{1/3}} = \lim_{R\to0+}\frac{3}{2}x^{2/3}\Big|_R^1 = \frac{3}{2} - \lim_{R\to0+}\frac{3}{2}R^{2/3} = \frac{3}{2}$$

so that

$$J = \int_{-1}^1 \frac{dx}{x^{1/3}} = \int_{-1}^0 \frac{dx}{x^{1/3}} + \int_0^1 \frac{dx}{x^{1/3}} = -\frac{3}{2}+\frac{3}{2} = 0$$

51. For which values of a does $\int_0^\infty e^{ax}\,dx$ converge?

SOLUTION First evaluate the integral on the finite interval $[0, R]$ for $R > 0$:

$$\int_0^R e^{ax}\,dx = \frac{1}{a}e^{ax}\Big|_0^R = \frac{1}{a}\left(e^{aR}-1\right).$$

Thus,

$$\int_0^\infty e^{ax}\,dx = \lim_{R\to\infty}\frac{1}{a}\left(e^{aR}-1\right).$$

If $a > 0$, then $e^{aR}\to\infty$ as $R\to\infty$. If $a < 0$, then $e^{aR}\to 0$ as $R\to\infty$, and

$$\int_0^\infty e^{ax}\,dx = \lim_{R\to\infty}\frac{1}{a}\left(e^{aR}-1\right) = -\frac{1}{a}.$$

The integral converges for $a < 0$.

53. Sketch the region under the graph of $f(x) = \frac{1}{1+x^2}$ for $-\infty < x < \infty$, and show that its area is π.

SOLUTION The graph is shown below.

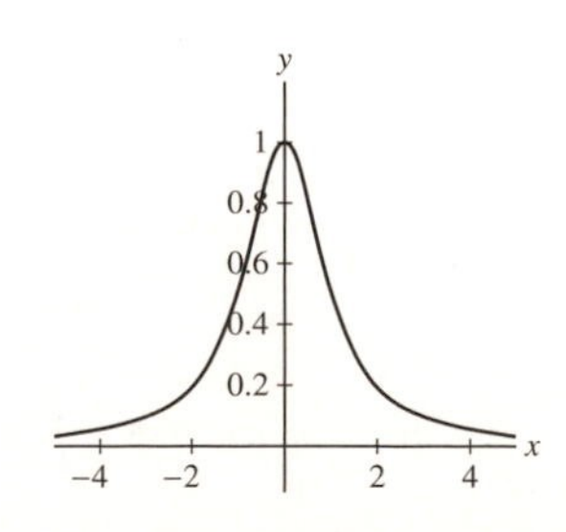

Since $(1+x^2)^{-1}$ is an even function, we can first compute the area under the graph for $x > 0$:

$$\int_0^R \frac{dx}{1+x^2} = \tan^{-1} x \Big|_0^R = \tan^{-1} R - \tan^{-1} 0 = \tan^{-1} R.$$

Thus,

$$\int_0^\infty \frac{dx}{1+x^2} = \lim_{R\to\infty} \tan^{-1} R = \frac{\pi}{2}.$$

By symmetry, we have

$$\int_{-\infty}^\infty \frac{dx}{1+x^2} = \int_{-\infty}^0 \frac{dx}{1+x^2} + \int_0^\infty \frac{dx}{1+x^2} = \frac{\pi}{2} + \frac{\pi}{2} = \pi.$$

55. Show that $\displaystyle\int_1^\infty \frac{dx}{x^3+4}$ converges by comparing with $\displaystyle\int_1^\infty x^{-3}\,dx$.

SOLUTION The integral $\displaystyle\int_1^\infty x^{-3}\,dx$ converges because $3 > 1$. Since $x^3+4 \ge x^3$, it follows that

$$\frac{1}{x^3+4} \le \frac{1}{x^3}.$$

Therefore, by the comparison test,

$$\int_1^\infty \frac{dx}{x^3+4} \text{ converges.}$$

57. Show that $0 \le e^{-x^2} \le e^{-x}$ for $x \ge 1$ (Figure 10). Use the Comparison Test to show that $\int_0^\infty e^{-x^2}\,dx$ converges. *Hint:* It suffices (why?) to make the comparison for $x \ge 1$ because

$$\int_0^\infty e^{-x^2}\,dx = \int_0^1 e^{-x^2}\,dx + \int_1^\infty e^{-x^2}\,dx$$

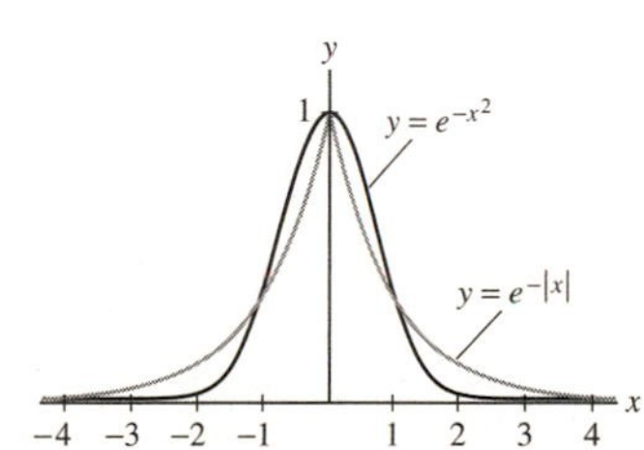

FIGURE 10 Comparison of $y = e^{-|x|}$ and $y = e^{-x^2}$.

SOLUTION For $x \ge 1$, $x^2 \ge x$, so $-x^2 \le -x$ and $e^{-x^2} \le e^{-x}$. Now

$$\int_1^\infty e^{-x}\,dx \text{ converges,} \quad \text{so} \quad \int_1^\infty e^{-x^2}\,dx \text{ converges}$$

by the comparison test. Finally, because e^{-x^2} is continuous on $[0, 1]$,

$$\int_0^\infty e^{-x^2}\,dx \text{ converges.}$$

We conclude that our integral converges by writing it as a sum:

$$\int_0^\infty e^{-x^2}\,dx = \int_0^1 e^{-x^2}\,dx + \int_1^\infty e^{-x^2}\,dx$$

59. Show that $\displaystyle\int_1^\infty \frac{1-\sin x}{x^2}\,dx$ converges.

SOLUTION Let $f(x) = \dfrac{1-\sin x}{x^2}$. Since $f(x) \le \dfrac{2}{x^2}$ and $\displaystyle\int_1^\infty 2x^{-2}\,dx = 2$, it follows that

$$\int_1^\infty \frac{1-\sin x}{x^2}\,dx \text{ converges}$$

by the comparison test.

In Exercises 61–74, use the Comparison Test to determine whether or not the integral converges.

61. $\displaystyle\int_1^\infty \frac{1}{\sqrt{x^5+2}}\,dx$

SOLUTION Since $\sqrt{x^5+2} \ge \sqrt{x^5} = x^{5/2}$, it follows that

$$\frac{1}{\sqrt{x^5+2}} \le \frac{1}{x^{5/2}}.$$

The integral $\int_1^\infty dx/x^{5/2}$ converges because $\frac{5}{2} > 1$. Therefore, by the comparison test:

$$\int_1^\infty \frac{dx}{\sqrt{x^5+2}} \text{ also converges.}$$

63. $\displaystyle\int_3^\infty \frac{dx}{\sqrt{x}-1}$

SOLUTION Since $\sqrt{x} \ge \sqrt{x}-1$, we have (for $x > 1$)

$$\frac{1}{\sqrt{x}} \le \frac{1}{\sqrt{x}-1}.$$

The integral $\int_1^\infty dx/\sqrt{x} = \int_1^\infty dx/x^{1/2}$ diverges because $\frac{1}{2} < 1$. Since the function $x^{-1/2}$ is continuous (and therefore finite) on $[1,3]$, we also know that $\int_3^\infty dx/x^{1/2}$ diverges. Therefore, by the comparison test,

$$\int_3^\infty \frac{dx}{\sqrt{x}-1} \text{ also diverges.}$$

65. $\displaystyle\int_1^\infty e^{-(x+x^{-1})}\,dx$

SOLUTION For all $x \ge 1$, $\frac{1}{x} > 0$ so $x + \frac{1}{x} \ge x$. Then

$$-(x+x^{-1}) \le -x \qquad \text{and} \qquad e^{-(x+x^{-1})} \le e^{-x}.$$

The integral $\int_1^\infty e^{-x}\,dx$ converges by direct computation:

$$\int_1^\infty e^{-x}\,dx = \lim_{R\to\infty}\int_1^R e^{-x}\,dx = \lim_{R\to\infty} -e^{-x}\Big|_1^R = \lim_{R\to\infty} -e^{-R} + e^{-1} = 0 + e^{-1} = e^{-1}.$$

Therefore, by the comparison test,

$$\int_1^\infty e^{-(x+x^{-1})} \text{ also converges.}$$

67. $\displaystyle\int_0^1 \frac{e^x}{x^2}\,dx$

SOLUTION For $0 < x < 1$, $e^x > 1$, and therefore

$$\frac{1}{x^2} < \frac{e^x}{x^2}.$$

The integral $\int_0^1 dx/x^2$ diverges since $2 > 1$. Therefore, by the comparison test,

$$\int_0^1 \frac{e^x}{x^2} \text{ also diverges.}$$

69. $\displaystyle\int_0^1 \frac{1}{x^4+\sqrt{x}}\,dx$

SOLUTION For $0 < x < 1$, $x^4 + \sqrt{x} \ge \sqrt{x}$, and

$$\frac{1}{x^4+\sqrt{x}} \le \frac{1}{\sqrt{x}}.$$

The integral $\int_0^1 (1/\sqrt{x})\,dx$ converges, since $p = \frac{1}{2} < 1$. Therefore, by the comparison test,

$$\int_0^1 \frac{dx}{x^4 + \sqrt{x}} \text{ also converges.}$$

71. $\displaystyle\int_1^\infty \frac{dx}{\sqrt{x^{1/3} + x^3}}$

SOLUTION For $x \geq 0$, $\sqrt{x^{1/3} + x^3} \geq \sqrt{x^3} = x^{3/2}$, so that

$$\frac{1}{\sqrt{x^{1/3} + x^3}} \leq \frac{1}{x^{3/2}}$$

The integral $\int_1^\infty x^{-3/2}\,dx$ converges since $p = 3/2 > 1$. Therefore, by the comparison test,

$$\int \frac{1}{\sqrt{x^{1/3} + x^3}}\,dx \quad \text{also converges.}$$

73. $\displaystyle\int_1^\infty \frac{dx}{(x + x^2)^{1/3}}$

SOLUTION For $x > 1$, $x < x^2$ so that $x + x^2 < 2x^2$; then

$$\int_1^\infty \frac{1}{(x + x^2)^{1/3}}\,dx \geq \int_1^\infty \frac{1}{(2x^2)^{1/3}}\,dx = \frac{1}{2^{1/3}} \int_1^\infty \frac{1}{x^{2/3}}\,dx$$

But $\int_1^\infty \frac{1}{x^{2/3}}\,dx$ diverges since $p = 2/3 < 1$. Therefore, by the comparison test,

$$\int_1^\infty \frac{1}{(x + x^2)^{1/3}}\,dx \quad \text{diverges as well.}$$

75. Define $J = \displaystyle\int_0^\infty \frac{dx}{x^{1/2}(x + 1)}$ as the sum of the two improper integrals

$$\int_0^1 \frac{dx}{x^{1/2}(x + 1)} + \int_1^\infty \frac{dx}{x^{1/2}(x + 1)}$$

Use the Comparison Test to show that J converges.

SOLUTION For the first integral, note that for $0 \leq x \leq 1$, we have $1 \leq 1 + x$, so that $x^{1/2}(x + 1) \geq x^{1/2}$. It follows that

$$\int_0^1 \frac{1}{x^{1/2}(x + 1)}\,dx \leq \int_0^1 \frac{1}{x^{1/2}}\,dx$$

which converges since $p = 1/2 < 1$. Thus the first integral converges by the comparison test. For the second integral, for $1 \leq x$, we have $x^{1/2}(x + 1) = x^{3/2} + x^{1/2} \geq x^{3/2}$, so that

$$\int_1^\infty \frac{1}{x^{1/2}(x + 1)}\,dx = \int_1^\infty \frac{1}{x^{3/2} + x^{1/2}}\,dx \leq \int_1^\infty \frac{1}{x^{3/2}}\,dx$$

which converges since $p = 3/2 > 1$. Thus the second integral converges as well by the comparison test, and therefore J, which is the sum of the two, converges.

77. An investment pays a dividend of \$250/year continuously forever. If the interest rate is 7%, what is the present value of the entire income stream generated by the investment?

SOLUTION The present value of the income stream after T years is

$$\int_0^T 250e^{-0.07t}\,dt = \left.\frac{250e^{-0.07t}}{-0.07}\right|_0^T = \frac{-250}{0.07}\left(e^{-0.07T} - 1\right) = \frac{250}{0.07}\left(1 - e^{-0.07T}\right).$$

Therefore the present value of the entire income stream is

$$\int_0^\infty 250e^{-0.07t} = \lim_{T\to\infty} \int_0^T 250e^{-0.07t} = \lim_{T\to\infty} \frac{250}{0.07}\left(1 - e^{-0.07T}\right) = \frac{250}{0.07}(1 - 0) = \frac{250}{0.07} = \$3571.43.$$

79. Compute the present value of an investment that generates income at a rate of $5000te^{0.01t}$ dollars per year forever, assuming an interest rate of 6%.

SOLUTION The present value of the income stream after T years is

$$\int_0^T \left(5000te^{0.01t}\right)e^{-0.06t}\,dt = 5000\int_0^T te^{-0.05t}\,dt$$

Compute the indefinite integral using Integration by Parts, with $u = t$ and $v' = e^{-0.05t}$. Then $u' = 1$, $v = (-1/0.05)e^{-0.05t}$, and

$$\int te^{-0.05t}\,dt = \frac{-t}{0.05}e^{-0.05t} + \frac{1}{0.05}\int e^{-0.05t}\,dt = -20te^{-0.05t} + \frac{20}{-0.05}e^{-0.05t} + C$$
$$= e^{-0.05t}(-20t - 400) + C.$$

Thus,

$$5000\int_0^T te^{-0.05t}\,dt = 5000e^{-0.05t}(-20t-400)\Big|_0^T = 5000e^{-0.05T}(-20T-400) - 5000(-400)$$
$$= 2{,}000{,}000 - 5000e^{-0.05T}(20T+400).$$

Use L'Hôpital's Rule to compute the limit:

$$\lim_{T\to\infty}\left(2{,}000{,}000 - \frac{5000(20T+400)}{e^{0.05T}}\right) = 2{,}000{,}000 - \lim_{T\to\infty}\frac{5000(20)}{0.05e^{0.05T}} = 2{,}000{,}000 - 0 = \$2{,}000{,}000.$$

81. The solid S obtained by rotating the region below the graph of $y = x^{-1}$ about the x-axis for $1 \le x < \infty$ is called **Gabriel's Horn** (Figure 11).

(a) Use the Disk Method (Section 6.3) to compute the volume of S. Note that the volume is finite even though S is an infinite region.

(b) It can be shown that the surface area of S is

$$A = 2\pi\int_1^\infty x^{-1}\sqrt{1+x^{-4}}\,dx$$

Show that A is infinite. If S were a container, you could fill its interior with a finite amount of paint, but you could not paint its surface with a finite amount of paint.

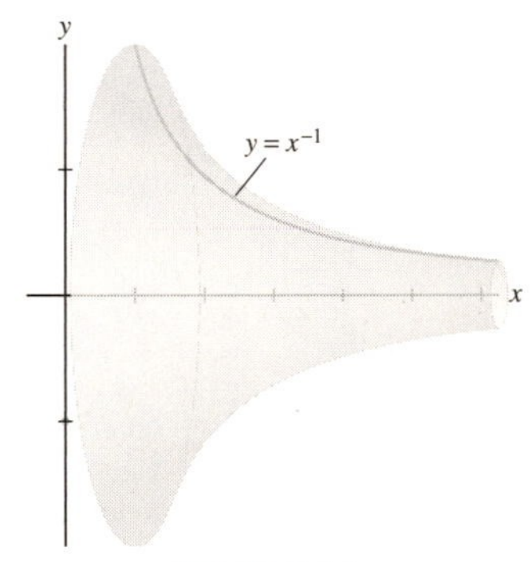

FIGURE 11

SOLUTION

(a) The volume is given by

$$V = \int_1^\infty \pi\left(\frac{1}{x}\right)^2 dx.$$

First compute the volume over a finite interval:

$$\int_1^R \pi\left(\frac{1}{x}\right)^2 dx = \pi\int_1^R x^{-2}\,dx = \pi\frac{x^{-1}}{-1}\Big|_1^R = \pi\left(\frac{-1}{R} - \frac{-1}{1}\right) = \pi\left(1 - \frac{1}{R}\right).$$

Thus,

$$V = \lim_{R\to\infty}\int_1^\infty \pi x^{-2}\,dx = \lim_{R\to\infty}\pi\left(1 - \frac{1}{R}\right) = \pi.$$

(b) For $x > 1$, we have

$$\frac{1}{x}\sqrt{1+\frac{1}{x^4}} = \frac{1}{x}\sqrt{\frac{x^4+1}{x^4}} = \frac{\sqrt{x^4+1}}{x^3} \geq \frac{\sqrt{x^4}}{x^3} = \frac{x^2}{x^3} = \frac{1}{x}.$$

The integral $\int_1^\infty \frac{1}{x}\,dx$ diverges, since $p = 1 \geq 1$. Therefore, by the comparison test,

$$\int_1^\infty \frac{1}{x}\sqrt{1+\frac{1}{x^4}}\,dx \text{ also diverges.}$$

Finally,

$$A = 2\pi \int_1^\infty \frac{1}{x}\sqrt{1+\frac{1}{x^4}}\,dx$$

diverges.

83. When a capacitor of capacitance C is charged by a source of voltage V, the power expended at time t is

$$P(t) = \frac{V^2}{R}(e^{-t/RC} - e^{-2t/RC})$$

where R is the resistance in the circuit. The total energy stored in the capacitor is

$$W = \int_0^\infty P(t)\,dt$$

Show that $W = \frac{1}{2}CV^2$.

SOLUTION The total energy contained after the capacitor is fully charged is

$$W = \frac{V^2}{R}\int_0^\infty \left(e^{-t/RC} - e^{-2t/RC}\right)dt.$$

The energy after a finite amount of time ($t = T$) is

$$\frac{V^2}{R}\int_0^T \left(e^{-t/RC} - e^{-2t/RC}\right)dt = \frac{V^2}{R}\left(-RCe^{-t/RC} + \frac{RC}{2}e^{-2t/RC}\right)\Bigg|_0^T$$

$$= V^2C\left[\left(-e^{-T/RC} + \frac{1}{2}e^{-2T/RC}\right) - \left(-1 + \frac{1}{2}\right)\right]$$

$$= CV^2\left(\frac{1}{2} - e^{-T/RC} + \frac{1}{2}e^{-2T/RC}\right).$$

Thus,

$$W = \lim_{T\to\infty} CV^2\left(\frac{1}{2} - e^{-T/RC} + \frac{1}{2}e^{-2T/RC}\right) = CV^2\left(\frac{1}{2} - 0 + 0\right) = \frac{1}{2}CV^2.$$

85. Conservation of Energy can be used to show that when a mass m oscillates at the end of a spring with spring constant k, the period of oscillation is

$$T = 4\sqrt{m}\int_0^{\sqrt{2E/k}} \frac{dx}{\sqrt{2E - kx^2}}$$

where E is the total energy of the mass. Show that this is an improper integral with value $T = 2\pi\sqrt{m/k}$.

SOLUTION The integrand is infinite at the upper limit of integration, $x = \sqrt{2E/k}$, so the integral is improper. Now, let

$$T(R) = 4\sqrt{m}\int_0^R \frac{dx}{\sqrt{2E - kx^2}} = 4\sqrt{m}\frac{1}{\sqrt{2E}}\int_0^R \frac{dx}{\sqrt{1 - (\frac{k}{2E})x^2}}$$

$$= 4\sqrt{\frac{m}{2E}}\sqrt{\frac{2E}{k}}\sin^{-1}\left(\sqrt{\frac{k}{2E}}R\right) = 4\sqrt{m/k}\sin^{-1}\left(\sqrt{\frac{k}{2E}}R\right).$$

Therefore

$$T = \lim_{R\to\sqrt{2E/k}} T(R) = 4\sqrt{\frac{m}{k}}\sin^{-1}(1) = 2\pi\sqrt{\frac{m}{k}}.$$

In Exercises 86–89, the **Laplace transform** *of a function $f(x)$ is the function $Lf(s)$ of the variable s defined by the improper integral (if it converges):*

$$Lf(s) = \int_0^\infty f(x)e^{-sx}\,dx$$

Laplace transforms are widely used in physics and engineering.

87. Show that if $f(x) = \sin\alpha x$, then $Lf(s) = \dfrac{\alpha}{s^2+\alpha^2}$.

SOLUTION If $f(x) = \sin\alpha x$, then the Laplace transform of $f(x)$ is

$$Lf(s) = \int_0^\infty e^{-sx}\sin\alpha x\,dx$$

First evaluate the indefinite integral using Integration by Parts, with $u = \sin\alpha x$ and $v' = e^{-sx}$. Then $u' = \alpha\cos\alpha x$, $v = -\frac{1}{s}e^{-sx}$, and

$$\int e^{-sx}\sin\alpha x\,dx = -\frac{1}{s}e^{-sx}\sin\alpha x + \frac{\alpha}{s}\int e^{-sx}\cos\alpha x\,dx.$$

Use Integration by Parts again, with $u = \cos\alpha x$, $v' = e^{-sx}$. Then $u' = -\alpha\sin\alpha x$, $v = -\frac{1}{s}e^{-sx}$, and

$$\int e^{-sx}\cos\alpha x\,dx = -\frac{1}{s}e^{-sx}\cos\alpha x - \frac{\alpha}{s}\int e^{-sx}\sin\alpha x\,dx.$$

Substituting this into the first equation and solving for $\int e^{-sx}\sin\alpha x\,dx$, we get

$$\int e^{-sx}\sin\alpha x\,dx = -\frac{1}{s}e^{-sx}\sin\alpha x - \frac{\alpha}{s^2}e^{-sx}\cos\alpha x - \frac{\alpha^2}{s^2}\int e^{-sx}\sin\alpha x\,dx$$

$$\int e^{-sx}\sin\alpha x\,dx = \frac{-e^{-sx}\left(\frac{1}{s}\sin\alpha x + \frac{\alpha}{s^2}\cos\alpha x\right)}{\left(1+\frac{\alpha^2}{s^2}\right)} = \frac{-e^{-sx}(s\sin\alpha x + \alpha\cos\alpha x)}{s^2+\alpha^2}$$

Thus,

$$\int_0^R e^{-sx}\sin\alpha x\,dx = \frac{1}{s^2+\alpha^2}\left[\frac{s\sin\alpha R + \alpha\cos\alpha R}{-e^{sR}} - \frac{0+\alpha}{-1}\right] = \frac{1}{s^2+\alpha^2}\left[\alpha - \frac{s\sin\alpha R + \alpha\cos\alpha R}{e^{sR}}\right].$$

Finally we take the limit, noting the fact that, for all values of R, $|s\sin\alpha R + \alpha\cos\alpha R| \le s + |\alpha|$

$$Lf(s) = \lim_{R\to\infty}\frac{1}{s^2+\alpha^2}\left[\alpha - \frac{s\sin\alpha R + \alpha\cos\alpha R}{e^{sR}}\right] = \frac{1}{s^2+\alpha^2}(\alpha - 0) = \frac{\alpha}{s^2+\alpha^2}.$$

89. Compute $Lf(s)$, where $f(x) = \cos\alpha x$ and $s > 0$.

SOLUTION If $f(x) = \cos\alpha x$, then the Laplace transform of $f(x)$ is

$$Lf(x) = \int_0^\infty e^{-sx}\cos\alpha x\,dx$$

First evaluate the indefinite integral using Integration by Parts, with $u = \cos\alpha x$ and $v' - e^{-sx}$. Then $u' = -\alpha\sin\alpha x$, $v = -\frac{1}{s}e^{-sx}$, and

$$\int e^{-sx}\cos\alpha x\,dx = -\frac{1}{s}e^{-sx}\cos\alpha x - \frac{\alpha}{s}\int e^{-sx}\sin\alpha x\,dx.$$

Use Integration by Parts again, with $u = \sin\alpha x\,dx$ and $v' = -e^{-sx}$. Then $u' = \alpha\cos\alpha x$, $v = -\frac{1}{s}e^{-sx}$, and

$$\int e^{-sx}\sin\alpha x\,dx = -\frac{1}{s}e^{-sx}\sin\alpha x + \frac{\alpha}{s}\int e^{-sx}\cos\alpha x\,dx.$$

Substituting this into the first equation and solving for $\int e^{-sx}\cos\alpha x\,dx$, we get

$$\int e^{-sx}\cos\alpha x\,dx = -\frac{1}{s}e^{-sx}\cos\alpha x - \frac{\alpha}{s}\left[-\frac{1}{s}e^{-sx}\sin\alpha x + \frac{\alpha}{s}\int e^{-sx}\cos\alpha\,dx\right]$$

$$= -\frac{1}{s}e^{-sx}\cos\alpha x + \frac{\alpha}{s^2}e^{-sx}\sin\alpha x - \frac{\alpha^2}{s^2}\int e^{-sx}\cos\alpha x\,dx$$

$$\int e^{-sx}\cos\alpha x\,dx = \frac{e^{-sx}\left(\frac{\alpha}{s^2}\sin\alpha x - \frac{1}{s}\cos\alpha x\right)}{1+\frac{\alpha^2}{s^2}} = \frac{e^{-sx}(\alpha\sin\alpha x - s\cos\alpha x)}{s^2+\alpha^2}$$

Thus,

$$\int_0^R e^{-sx}\cos\alpha x\,dx = \frac{1}{s^2+\alpha^2}\left[\frac{\alpha\sin\alpha R - s\cos\alpha R}{e^{sR}} - \frac{0-s}{1}\right].$$

Finally we take the limit, noting the fact that, for all values of R, $|\alpha\sin\alpha R - s\cos\alpha R| \le |\alpha| + s$

$$Lf(s) = \lim_{R\to\infty}\frac{1}{s^2+\alpha^2}\left[s + \frac{\alpha\sin\alpha R - s\cos\alpha R}{e^{sR}}\right] = \frac{1}{s^2+\alpha^2}(s+0) = \frac{s}{s^2+\alpha^2}.$$

91. Let $J_n = \int_0^\infty x^n e^{-\alpha x}\,dx$, where $n \ge 1$ is an integer and $\alpha > 0$. Prove that

$$J_n = \frac{n}{\alpha}J_{n-1}$$

and $J_0 = 1/\alpha$. Use this to compute J_4. Show that $J_n = n!/\alpha^{n+1}$.

SOLUTION Using Integration by Parts, with $u = x^n$ and $v' = e^{-\alpha x}$, we get $u' = nx^{n-1}$, $v = -\frac{1}{\alpha}e^{-\alpha x}$, and

$$\int x^n e^{-\alpha x}\,dx = -\frac{1}{\alpha}x^n e^{-\alpha x} + \frac{n}{\alpha}\int x^{n-1}e^{-\alpha x}\,dx.$$

Thus,

$$J_n = \int_0^\infty x^n e^{-\alpha x}\,dx = \lim_{R\to\infty}\left(-\frac{1}{\alpha}x^n e^{-\alpha x}\right)\Big|_0^R + \frac{n}{\alpha}\int_0^\infty x^{n-1}e^{-\alpha x}\,dx = \lim_{R\to\infty}\frac{-R^n}{\alpha e^{\alpha R}} + 0 + \frac{n}{\alpha}J_{n-1}.$$

Use L'Hôpital's Rule repeatedly to compute the limit:

$$\lim_{R\to\infty}\frac{-R_n}{\alpha e^{\alpha R}} = \lim_{R\to\infty}\frac{-nR^{n-1}}{\alpha^2 e^{\alpha R}} = \lim_{R\to\infty}\frac{-n(n-1)R^{n-2}}{\alpha^3 e^{\alpha R}} = \cdots = \lim_{R\to\infty}\frac{-n(n-1)(n-2)\cdots(3)(2)(1)}{\alpha^{n+1}e^{\alpha R}} = 0.$$

Finally,

$$J_n = 0 + \frac{n}{\alpha}J_{n-1} = \frac{n}{\alpha}J_{n-1}.$$

J_0 can be computed directly:

$$J_0 = \int_0^\infty e^{-\alpha x}\,dx = \lim_{R\to\infty}\int_0^R e^{-\alpha x}\,dx = \lim_{R\to\infty} -\frac{1}{\alpha}e^{-\alpha x}\Big|_0^R = \lim_{R\to\infty} -\frac{1}{\alpha}\left(e^{-\alpha R} - 1\right) = -\frac{1}{\alpha}(0-1) = \frac{1}{\alpha}.$$

With this starting point, we can work up to J_4:

$$J_1 = \frac{1}{\alpha}J_0 = \frac{1}{\alpha}\left(\frac{1}{\alpha}\right) = \frac{1}{\alpha^2};$$

$$J_2 = \frac{2}{\alpha}J_1 = \frac{2}{\alpha}\left(\frac{1}{\alpha^2}\right) = \frac{2}{\alpha^3} = \frac{2!}{\alpha^{2+1}};$$

$$J_3 = \frac{3}{\alpha}J_2 = \frac{3}{\alpha}\left(\frac{2}{\alpha^3}\right) = \frac{6}{\alpha^4} = \frac{3!}{\alpha^{3+1}};$$

$$J_4 = \frac{4}{\alpha}J_3 = \frac{4}{\alpha}\left(\frac{6}{\alpha^4}\right) = \frac{24}{\alpha^5} = \frac{4!}{\alpha^{4+1}}.$$

We can use induction to prove the formula for J_n. If

$$J_{n-1} = \frac{(n-1)!}{\alpha^n},$$

then we have

$$J_n = \frac{n}{\alpha} J_{n-1} = \frac{n}{\alpha} \cdot \frac{(n-1)!}{\alpha^n} = \frac{n!}{\alpha^{n+1}}.$$

93. According to **Planck's Radiation Law**, the amount of electromagnetic energy with frequency between ν and $\nu + \Delta\nu$ that is radiated by a so-called black body at temperature T is proportional to $F(\nu)\,\Delta\nu$, where

$$F(\nu) = \left(\frac{8\pi h}{c^3}\right) \frac{\nu^3}{e^{h\nu/kT} - 1}$$

where c, h, k are physical constants. Use Exercise 92 to show that the total radiated energy

$$E = \int_0^\infty F(\nu)\,d\nu$$

is finite. To derive his law, Planck introduced the quantum hypothesis in 1900, which marked the birth of quantum mechanics.

SOLUTION The total radiated energy E is given by

$$E = \int_0^\infty F(\nu)\,d\nu = \frac{8\pi h}{c^3} \int_0^\infty \frac{\nu^3}{e^{h\nu/kT} - 1}\,d\nu.$$

Let $\alpha = h/kT$. Then

$$E = \frac{8\pi h}{c^3} \int_0^\infty \frac{\nu^3}{e^{\alpha\nu} - 1}\,d\nu.$$

Because $\alpha > 0$ and $8\pi h/c^3$ is a constant, we know E is finite by Exercise 92.

Further Insights and Challenges

95. Let

$$F(x) = \int_2^x \frac{dt}{\ln t} \qquad \text{and} \qquad G(x) = \frac{x}{\ln x}$$

Verify that L'Hôpital's Rule applies to the limit $L = \lim_{x\to\infty} \frac{F(x)}{G(x)}$ and evaluate L.

SOLUTION Because $\ln t < t$ for $t > 2$, we have $\frac{1}{\ln t} > \frac{1}{t}$ for $t > 2$, and so

$$F(x) = \int_2^x \frac{dt}{\ln t} > \int_2^x \frac{dt}{t} = \ln x - \ln 2$$

Thus, $F(x) \to \infty$ as $x \to \infty$. Moreover, by L'Hôpital's Rule

$$\lim_{x\to\infty} G(x) = \lim_{x\to\infty} \frac{1}{1/x} = \lim_{x\to\infty} x = \infty.$$

Thus, $\lim_{x\to\infty} \frac{F(x)}{G(x)}$ is of the form ∞/∞, and L'Hôpital's Rule applies. Finally,

$$L = \lim_{x\to\infty} \frac{F(x)}{G(x)} = \lim_{x\to\infty} \frac{\frac{1}{\ln x}}{\frac{\ln x - 1}{(\ln x)^2}} = \lim_{x\to\infty} \frac{\ln x}{\ln x - 1} = \lim_{x\to\infty} \frac{1}{1 - (1/\ln x)} = 1.$$

In Exercises 96–98, an improper integral $I = \int_a^\infty f(x)\,dx$ is called ***absolutely convergent*** *if $\int_a^\infty |f(x)|\,dx$ converges. It can be shown that if I is absolutely convergent, then it is convergent.*

97. Show that $\int_1^\infty e^{-x^2} \cos x\,dx$ is absolutely convergent.

SOLUTION By the result of Exercise 57, we know that $\int_0^\infty e^{-x^2}\,dx$ is convergent. Then $\int_1^\infty e^{-x^2}\,dx$ is also convergent. Because $|\cos x| \le 1$ for all x, we have

$$\left|e^{-x^2} \cos x\right| = |\cos x| \left|e^{-x^2}\right| \le \left|e^{-x^2}\right| = e^{-x^2}.$$

Therefore, by the comparison test, we have

$$\int_1^{\infty} \left| e^{-x^2} \cos x \right| dx \text{ also converges.}$$

Since $\int_1^{\infty} e^{-x^2} \cos x \, dx$ converges absolutely, it itself converges.

99. The **gamma function**, which plays an important role in advanced applications, is defined for $n \geq 1$ by

$$\Gamma(n) = \int_0^{\infty} t^{n-1} e^{-t} \, dt$$

(a) Show that the integral defining $\Gamma(n)$ converges for $n \geq 1$ (it actually converges for all $n > 0$). *Hint:* Show that $t^{n-1}e^{-t} < t^{-2}$ for t sufficiently large.

(b) Show that $\Gamma(n+1) = n\Gamma(n)$ using Integration by Parts.

(c) Show that $\Gamma(n+1) = n!$ if $n \geq 1$ is an integer. *Hint:* Use (a) repeatedly. Thus, $\Gamma(n)$ provides a way of defining n-factorial when n is not an integer.

SOLUTION

(a) By repeated use of L'Hôpital's Rule, we can compute the following limit:

$$\lim_{t \to \infty} \frac{e^t}{t^{n+1}} = \lim_{t \to \infty} \frac{e^t}{(n+1)t^n} = \cdots = \lim_{t \to \infty} \frac{e^t}{(n+1)!} = \infty.$$

This implies that, for t sufficiently large, we have

$$e^t \geq t^{n+1};$$

therefore

$$\frac{e^t}{t^{n-1}} \geq \frac{t^{n+1}}{t^{n-1}} = t^2 \quad \text{or} \quad t^{n-1}e^{-t} \leq t^{-2}.$$

The integral $\int_1^{\infty} t^{-2} \, dt$ converges because $p = 2 > 1$. Therefore, by the comparison test,

$$\int_M^{\infty} t^{n-1} e^{-t} \, dt \text{ also converges,}$$

where M is the value above which the above comparisons hold. Finally, because the function $t^{n-1}e^{-t}$ is continuous for all t, we know that

$$\Gamma(n) = \int_0^{\infty} t^{n-1} e^{-t} \, dt \text{ converges} \quad \text{for all } n \geq 1.$$

(b) Using Integration by Parts, with $u = t^n$ and $v' - e^{-t}$, we have $u' = nt^{n-1}$, $v = -e^{-t}$, and

$$\Gamma(n+1) = \int_0^{\infty} t^n e^{-t} \, dt = -t^n e^{-t} \Big|_0^{\infty} + n \int_0^{\infty} t^{n-1} e^{-t} \, dt$$

$$= \lim_{R \to \infty} \left(\frac{-R^n}{e^R} - 0 \right) + n\Gamma(n) = 0 + n\Gamma(n) = n\Gamma(n).$$

Here, we've computed the limit as in part (a) with repeated use of L'Hôpital's Rule.

(c) By the result of part (b), we have

$$\Gamma(n+1) = n\Gamma(n) = n(n-1)\Gamma(n-1) = n(n-1)(n-2)\Gamma(n-2) = \cdots = n!\,\Gamma(1).$$

If $n = 1$, then

$$\Gamma(1) = \int_0^{\infty} e^{-t} \, dt = \lim_{R \to \infty} -e^{-t} \Big|_0^R = \lim_{R \to \infty} \left(1 - e^{-R}\right) = 1.$$

Thus

$$\Gamma(n+1) = n!\,(1) = n!$$

7.7 Probability and Integration

Preliminary Questions

1. The function $p(x) = \cos x$ satisfies $\int_{-\pi/2}^{\pi} p(x)\,dx = 1$. Is $p(x)$ a probability density function on $[-\pi/2, \pi]$?

SOLUTION Since $p(x) = \cos x < 0$ for some points in $(-\pi/2, \pi)$, $p(x)$ is not a probability density function.

2. Estimate $P(2 \le X \le 2.1)$ assuming that the probability density function of X satisfies $p(2) = 0.2$.

SOLUTION $P(2 \le X \le 2.1) \approx p(2) \cdot (2.1 - 2) = 0.02$.

3. Which exponential probability density has mean $\mu = \frac{1}{4}$?

SOLUTION $\frac{1}{1/4}e^{-x/(1/4)} = 4e^{-4x}$.

Exercises

In Exercises 1–6, find a constant C such that p(x) is a probability density function on the given interval, and compute the probability indicated.

1. $p(x) = \dfrac{C}{(x+1)^3}$ on $[0, \infty)$; $P(0 \le X \le 1)$

SOLUTION Compute the indefinite integral using the substitution $u = x + 1$, $du = dx$:

$$\int p(x)\,dx = \int \frac{C}{(x+1)^3}\,dx = -\frac{1}{2}C(x+1)^{-2} + K$$

For $p(x)$ to be a probability density function, we must have

$$1 = \int_0^{\infty} p(x)\,dx = -\frac{1}{2}C \lim_{R\to\infty} (x+1)^{-2}\Big|_0^R = \frac{1}{2}C - \frac{1}{2}C \lim_{R\to\infty} (R+1)^{-2} = \frac{1}{2}C$$

so that $C = 2$, and $p(x) = \frac{2}{(x+1)^3}$. Then using the indefinite integral above,

$$P(0 \le X \le 1) = \int_0^1 \frac{2}{(x+1)^3} = -\frac{1}{2} \cdot 2 \cdot (x+1)^{-2}\Big|_0^1 = -\frac{1}{4} + 1 = \frac{3}{4}$$

3. $p(x) = \dfrac{C}{\sqrt{1-x^2}}$ on $(-1, 1)$; $P\left(-\frac{1}{2} \le X \le \frac{1}{2}\right)$

SOLUTION Compute the indefinite integral:

$$\int p(x)\,dx = C\int \frac{1}{\sqrt{1-x^2}}\,dx = C\sin^{-1} x + K$$

valid for $-1 < x < 1$. For $p(x)$ to be a probability density function, we must have

$$1 = \int_{-1}^{1} p(x)\,dx = \int_{-1}^{0} p(x)\,dx + \int_0^1 p(x)\,dx = C\left(\lim_{R\to -1^+} \sin^{-1} x\Big|_R^0 + \lim_{R\to 1^-} \sin^{-1} x\Big|_0^R\right)$$

$$= C\left(\sin^{-1}(0) - \lim_{R\to -1^+} \sin^{-1}(R) + \lim_{R\to 1^-} \sin^{-1} R - \sin^{-1}(0)\right)$$

$$= C\left(-\sin^{-1}(-1) + \sin^{-1}(1)\right) = \pi C$$

so that $C = \frac{1}{\pi}$ and $p(x) = \frac{1}{\pi\sqrt{1-x^2}}$. Then using the indefinite integral above,

$$P\left(-\frac{1}{2} \le X \le \frac{1}{2}\right) = \int_{-1/2}^{1/2} p(x)\,dx = \frac{1}{\pi}\sin^{-1} x\Big|_{-1/2}^{1/2} = \frac{1}{\pi}\left(\frac{\pi}{6} - \frac{-\pi}{6}\right) = \frac{1}{3}$$

5. $p(x) = C\sqrt{1-x^2}$ on $(-1, 1)$; $P\left(-\frac{1}{2} \le X \le 1\right)$

SOLUTION Compute the indefinite integral using the substitution $x = \sin u$, so that $dx = \cos u\, du$:

$$\int p(x)\,dx = C\int \sqrt{1-x^2}\,dx = C\int \sqrt{1-\sin^2 u}\cos u\,du = C\int \cos^2 u\,du$$
$$= C\left(\frac{1}{2}u + \frac{1}{2}\cos u \sin u\right) + K$$

Since $x = \sin u$, we construct the following right triangle:

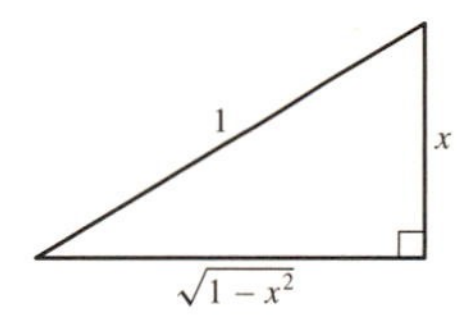

and we see that $\cos u = \sqrt{1-x^2}$, so that

$$\int p(x)\,dx = \frac{1}{2}C\left(\sin^{-1}x + x\sqrt{1-x^2}\right) + K$$

For $p(x)$ to be a probability density function, we must have

$$1 = \int_{-1}^{1} p(x)\,dx = \frac{1}{2}C\left(\sin^{-1}x + x\sqrt{1-x^2}\right)\Big|_{-1}^{1} = \frac{1}{2}C(\sin^{-1}1 - \sin^{-1}(-1)) = \frac{\pi}{2}C$$

so that $C = \frac{2}{\pi}$ and $p(x) = \frac{2}{\pi}\sqrt{1-x^2}$. Then using the indefinite integral above,

$$P\left(-\frac{1}{2} \le X \le 1\right) = \int_{-1/2}^{1} \frac{2}{\pi}\sqrt{1-x^2}\,dx = \frac{1}{\pi}\left(\sin^{-1}x + x\sqrt{1-x^2}\right)\Big|_{-1/2}^{1}$$
$$= \frac{1}{\pi}\left(\sin^{-1}1 + 0 - \sin^{-1}\left(-\frac{1}{2}\right) - \frac{-1}{2}\sqrt{1-\frac{1}{4}}\right)$$
$$= \frac{1}{\pi}\left(\frac{\pi}{2} - \frac{-\pi}{6} + \frac{\sqrt{3}}{4}\right) = \frac{2}{3} + \frac{\sqrt{3}}{4\pi} \approx 0.804$$

7. Verify that $p(x) = 3x^{-4}$ is a probability density function on $[1, \infty)$ and calculate its mean value.

SOLUTION We have

$$\int_{1}^{\infty} 3x^{-4}\,dx = \lim_{R\to\infty}\left(-x^{-3}\right)\Big|_{1}^{R} = \lim_{R\to\infty}\left(-\frac{1}{R^3}\right) + 1 = 1$$

so that $p(x)$ is a probability density function on $[1, \infty)$. Its mean value is

$$\int_{1}^{\infty} x \cdot 3x^{-4}\,dx = \int_{1}^{\infty} 3x^{-3}\,dx = -\frac{3}{2}x^{-2}\Big|_{1}^{R} = \lim_{R\to\infty}\left(-\frac{3}{2R^2}\right) + \frac{3}{2} = \frac{3}{2}$$

9. Verify that $p(t) = \frac{1}{50}e^{-t/50}$ satisfies the condition $\int_0^\infty p(t)\,dt = 1$.

SOLUTION Use the substitution $u = \frac{t}{50}$, so that $du = \frac{1}{50}\,dt$. Then

$$\int_{0}^{\infty} p(t)\,dt = \int_{0}^{\infty} \frac{1}{50}e^{-t/50}\,dt = \int_{0}^{\infty} e^{-u}\,du = \lim_{R\to\infty}(-e^{-u})\Big|_{0}^{R} = \lim_{R\to\infty} 1 - e^{-R} = 1$$

11. The life X (in hours) of a battery in constant use is a random variable with exponential density. What is the probability that the battery will last more than 12 hours if the average life is 8 hours?

SOLUTION If the average life is 8 hours, then the mean of the exponential distribution is 8, so that the distribution is

$$p(x) = \frac{1}{8}e^{-x/8}$$

The probability that the battery will last more than 12 hours is given by (using the substitution $u = x/8$, so that $du = 1/8\,dx$ and $x = 12$ corresponds to $u = 3/2$)

$$P(X \geq 12) = \int_{12}^{\infty} p(x)\,dx = \int_{12}^{\infty} \frac{1}{8}e^{-x/8}\,dx = \int_{3/2}^{\infty} e^{-u}\,du = \lim_{R\to\infty} (-e^{-u})\Big|_{3/2}^{R}$$

$$= e^{-3/2} - \lim_{R\to\infty} e^{-R} = e^{-3/2} \approx 0.223$$

13. The distance r between the electron and the nucleus in a hydrogen atom (in its lowest energy state) is a random variable with probability density $p(r) = 4a_0^{-3}r^2e^{-2r/a_0}$ for $r \geq 0$, where a_0 is the Bohr radius (Figure 7). Calculate the probability P that the electron is within one Bohr radius of the nucleus. The value of a_0 is approximately 5.29×10^{-11} m, but this value is not needed to compute P.

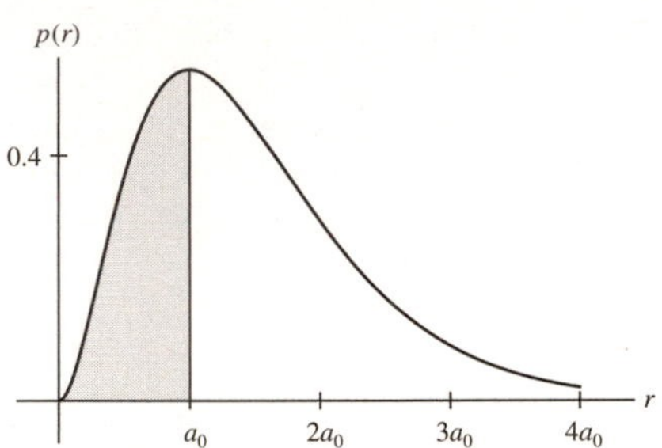

FIGURE 7 Probability density function $p(r) = 4a_0^{-3}r^2e^{-2r/a_0}$.

SOLUTION The probability P is the area of the shaded region in Figure 7. To calculate p, use the substitution $u = 2r/a_0$:

$$P = \int_0^{a_0} p(r)\,dr = \frac{4}{a_0^3}\int_0^{a_0} r^2e^{-2r/a_0}\,dr = \left(\frac{4}{a_0^3}\right)\left(\frac{a_0^3}{8}\right)\int_0^2 u^2e^{-u}\,du$$

The constant in front simplifies to $\frac{1}{2}$ and the formula in the margin gives us

$$P = \frac{1}{2}\int_0^2 u^2e^{-u}\,du = \frac{1}{2}\left(-(u^2+2u+2)e^{-u}\right)\Big|_0^2 = \frac{1}{2}\left(2 - 10e^{-2}\right) \approx 0.32$$

Thus, the electron within a distance a_0 of the nucleus with probability 0.32.

In Exercises 15–21, $F(z)$ denotes the cumulative normal distribution function. Refer to a calculator, computer algebra system, or online resource to obtain values of $F(z)$.

15. Express the area of region A in Figure 8 in terms of $F(z)$ and compute its value.

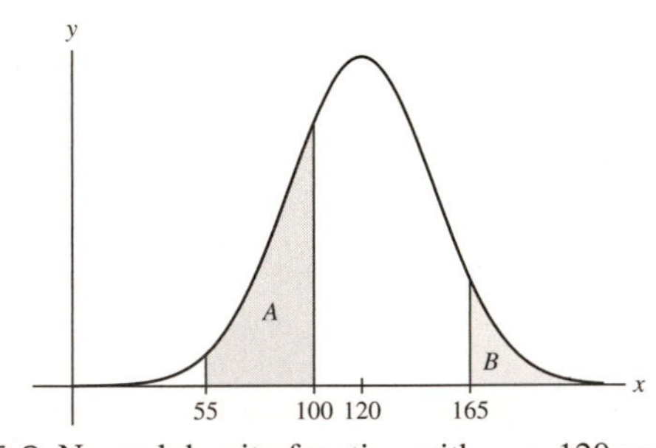

FIGURE 8 Normal density function with $\mu = 120$ and $\sigma = 30$.

SOLUTION The area of region A is $P(55 \leq X \leq 100)$. By Theorem 1, we have

$$P(55 \leq X \leq 100) = F\left(\frac{100-120}{30}\right) - F\left(\frac{55-120}{30}\right) = F\left(-\frac{2}{3}\right) - F\left(-\frac{13}{6}\right) \approx 0.237$$

17. Assume X has a standard normal distribution ($\mu = 0, \sigma = 1$). Find:

(a) $P(X \leq 1.2)$ **(b)** $P(X \geq -0.4)$

SOLUTION

(a) $P(X \leq 1.2) = F(1.2) \approx 0.8849$

(b) $P(X \geq -0.4) = 1 - P(X \leq -0.4) = 1 - F(-0.4) \approx 1 - 0.3446 \approx 0.6554$

19. Use a graph to show that $F(-z) = 1 - F(z)$ for all z. Then show that if $p(x)$ is a normal density function with mean μ and standard deviation σ, then for all $r \geq 0$,

$$P(\mu - r\sigma \leq X \leq \mu + r\sigma) = 2F(r) - 1$$

SOLUTION Consider the graph of the standard normal density function in Figure 5. This graph is symmetric around the y-axis, so that the area under the curve from z to ∞, which is $1 - F(z)$, is equal to the area under the curve from $-\infty$ to $-z$, which is $F(-z)$. Thus $1 - F(z) = F(-z)$. Now, if $p(x)$ is a normal density function with mean μ and standard deviation σ, then for $r \geq 0$ (so that the range $\mu - r\sigma \leq X \leq \mu + r\sigma$ is nonempty),

$$\begin{aligned} P(\mu - r\sigma \leq X \leq \mu + r\sigma) &= F\left(\frac{\mu + r\sigma - \mu}{\sigma}\right) - F\left(\frac{\mu - r\sigma - \mu}{\sigma}\right) \\ &= F(r) - F(-r) = F(r) - (1 - F(r)) = 2F(r) - 1 \end{aligned}$$

21. A bottling company produces bottles of fruit juice that are filled, on average, with 32 ounces of juice. Due to random fluctuations in the machinery, the actual volume of juice is normally distributed with a standard deviation of 0.4 ounce. Let P be the probability of a bottle having less than 31 ounces. Express P as an integral of an appropriate density function and compute its value numerically.

SOLUTION The associated cumulative distribution function is

$$f(z) = \frac{1}{0.4\sqrt{2\pi}} \int_{-\infty}^{z} e^{-(x-32)^2/(2\cdot 0.4^2)}\, dx$$

To compute the value numerically, we use the standard normal cumulative distribution function $F(z)$:

$$P(X \leq 31) = F\left(\frac{31 - 32}{0.4}\right) = F\left(-\frac{5}{2}\right) = \frac{1}{\sqrt{2\pi}} \int_{-\infty}^{-5/2} e^{-x^2/2}\, dx \approx 0.0062$$

In Exercises 23–26, calculate μ and σ, where σ is the ***standard deviation****, defined by*

$$\sigma^2 = \int_{-\infty}^{\infty} (x - \mu)^2\, p(x)\, dx$$

The smaller the value of σ, the more tightly clustered are the values of the random variable X about the mean μ.

23. $p(x) = \dfrac{5}{2x^{7/2}}$ on $[1, \infty)$

SOLUTION The mean is

$$\int_1^{\infty} xp(x)\, dx = \int_1^{\infty} \frac{5}{2} x^{-5/2}\, dx = -\frac{5}{3} x^{-3/2}\Big|_1^{\infty} = \frac{5}{3}$$

and

$$\begin{aligned} \sigma^2 &= \int_1^{\infty} (x - \mu)^2 p(x)\, dx = \int_1^{\infty} (x^2 - 2\mu x + \mu^2) \frac{5}{2} x^{-7/2}\, dx \\ &= \frac{5}{2} \int_1^{\infty} x^{-3/2} - 2\mu x^{-5/2} + \mu^2 x^{-7/2}\, dx = \frac{5}{2}\left(-2x^{-1/2} + \frac{4}{3}\mu x^{-3/2} - \frac{2}{5}\mu^2 x^{-5/2}\right)\Big|_1^{\infty} \\ &= \frac{5}{2}\left(2 - \frac{4}{3}\mu + \frac{2}{5}\mu^2\right) = \frac{5}{2}\left(2 - \frac{4}{3}\cdot\frac{5}{3} + \frac{2}{5}\cdot\frac{25}{9}\right) = \frac{20}{9} \end{aligned}$$

25. $p(x) = \dfrac{1}{3} e^{-x/3}$ on $[0, \infty)$

SOLUTION This is an exponential density function with mean $\mu = 3$. The standard deviation is

$$\begin{aligned} \sigma^2 &= \frac{1}{3} \int_0^{\infty} (x - 3)^2 e^{-x/3}\, dx = \frac{1}{3} \int_0^{\infty} \left(x^2 e^{-x/3} - 6xe^{-x/3} + 9e^{-x/3}\right) dx \\ &= \frac{1}{3} \int_0^{\infty} x^2 e^{-x/3}\, dx - 2 \int_0^{\infty} xe^{-x/3}\, dx + 3 \int_0^{\infty} e^{-x/3}\, dx \end{aligned}$$

We tackle the third integral first:

$$\int_0^\infty e^{-x/3}\,dx = -3e^{-x/3}\Big|_0^\infty = 3$$

For the second integral, use integration by parts with $u = x$, $v' = e^{-x/3}$ so that $u' = 1$ and $v = -3e^{-x/3}$. Then

$$\int_0^\infty xe^{-x/3}\,dx = -3xe^{-x/3}\Big|_0^\infty + 3\int_0^\infty e^{-x/3}\,dx = 0 + 3\cdot 3 = 9$$

Finally, the first integral is solved using integration by parts with $u = x^2$, $v' = e^{-x/3}$ so that $u' = 2x$ and $v = -3e^{-x/3}$; then

$$\int_0^\infty x^2e^{-x/3}\,dx = -3x^2e^{-x/3}\Big|_0^\infty + 6\int_0^\infty xe^{-x/3}\,dx = 0 + 6\cdot 9 = 54$$

and, finally,

$$\begin{aligned}\sigma^2 &= \frac{1}{3}\int_0^\infty x^2e^{-x/3}\,dx - 2\int_0^\infty xe^{-x/3}\,dx + 3\int_0^\infty e^{-x/3}\,dx \\ &= \frac{1}{3}\cdot 54 - 2\cdot 9 + 3\cdot 3 = 9\end{aligned}$$

Further Insights and Challenges

27. The time to decay of an atom in a radioactive substance is a random variable X. The law of radioactive decay states that if N atoms are present at time $t = 0$, then $Nf(t)$ atoms will be present at time t, where $f(t) = e^{-kt}$ ($k > 0$ is the decay constant). Explain the following statements:

(a) The fraction of atoms that decay in a small time interval $[t, t + \Delta t]$ is approximately $-f'(t)\Delta t$.

(b) The probability density function of X is $-f'(t)$.

(c) The average time to decay is $1/k$.

SOLUTION

(a) The number of atoms that decay in the interval $[t, t + \Delta t]$ is just $f(t) - f(t + \Delta t)$; the statement simply says that $f(t) - f(t + \Delta t) \approx -f'(t)\Delta t$, which is the same as saying that

$$f'(t) \approx \frac{f(t) - f(t + \Delta t)}{\Delta t} = \frac{f(t + \Delta t) - f(t)}{\Delta t}$$

which is true by the definition of the derivative. Intuitively, since $f'(t)$ is the instantaneous rate of decay, we would expect that over a short interval, the number of atoms decaying is proportional to both $f'(t)$ and the size of the interval.

(b) The probability density function is defined by the property in (a): the probability that X lies in a small interval $[t, t + \Delta t]$ is approximately $p(t)\Delta t$, so that $p(t) = -f'(t)$.

(c) The average time to decay is the mean of the distribution, which is

$$\mu = \int_0^\infty t\cdot(-f'(t))\,dt = -\int_0^\infty tf'(t)\,dt$$

We compute this integral using integration by parts, with $u = t$, $v' = f'(t)$. Then $u' = 1$, $v = f(t)$, and

$$\mu = -\int_0^\infty tf'(t)\,dt = -tf(t)\Big|_0^\infty + \int_0^\infty f(t)\,dt.$$

Since $f(t) = e^{-kt}$, we have

$$-tf(t)\big|_0^\infty = \lim_{R\to\infty} -te^{-kt}\Big|_0^R = \lim_{R\to\infty} -Re^{-Rt} + 0 = \lim_{R\to\infty}\frac{-R}{e^{Rt}} = \lim_{R\to\infty}\frac{-1}{Re^{Rt}} = 0.$$

Here we used L'Hôpital's Rule to compute the limit. Thus

$$\mu = \int_0^\infty f(t)\,dt = \int_0^\infty e^{-kt}\,dt.$$

Now,

$$\int_0^R e^{-kt}\,dt = -\frac{1}{k}e^{-kt}\Big|_0^R = -\frac{1}{k}\left(e^{-kR} - 1\right) = \frac{1}{k}\left(1 - e^{-kR}\right),$$

so

$$\mu = \lim_{R\to\infty} \frac{1}{k}\left(1 - e^{-kR}\right) = \frac{1}{k}(1-0) = \frac{1}{k}.$$

Because k has units of $(\text{time})^{-1}$, μ does in fact have the appropriate units of time.

7.8 Numerical Integration

Preliminary Questions

1. What are T_1 and T_2 for a function on $[0, 2]$ such that $f(0) = 3$, $f(1) = 4$, and $f(2) = 3$?

SOLUTION Using the given function values

$$T_1 = \frac{1}{2}(2)(3+3) = 6 \quad \text{and} \quad T_2 = \frac{1}{2}(1)(3+8+3) = 7.$$

2. For which graph in Figure 16 will T_N *overestimate* the integral? What about M_N?

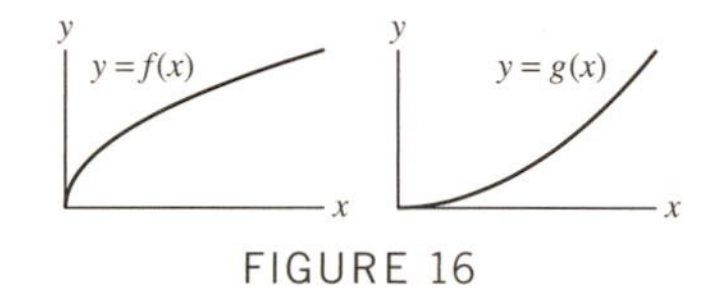

FIGURE 16

SOLUTION T_N overestimates the value of the integral when the integrand is concave up; thus, T_N will overestimate the integral of $y = g(x)$. On the other hand, M_N overestimates the value of the integral when the integrand is concave down; thus, M_N will overestimate the integral of $y = f(x)$.

3. How large is the error when the Trapezoidal Rule is applied to a linear function? Explain graphically.

SOLUTION The Trapezoidal Rule integrates linear functions exactly, so the error will be zero.

4. What is the maximum possible error if T_4 is used to approximate

$$\int_0^3 f(x)\,dx$$

where $|f''(x)| \le 2$ for all x.

SOLUTION The maximum possible error in T_4 is

$$\max|f''(x)|\frac{(b-a)^3}{12n^2} \le \frac{2(3-0)^3}{12(4)^2} = \frac{9}{32}.$$

5. What are the two graphical interpretations of the Midpoint Rule?

SOLUTION The two graphical interpretations of the Midpoint Rule are the sum of the areas of the midpoint rectangles and the sum of the areas of the tangential trapezoids.

Exercises

In Exercises 1–12, calculate T_N and M_N for the value of N indicated.

1. $\displaystyle\int_0^2 x^2\,dx, \quad N = 4$

SOLUTION Let $f(x) = x^2$. We divide $[0, 2]$ into 4 subintervals of width

$$\Delta x = \frac{2-0}{4} = \frac{1}{2}$$

with endpoints 0, 0.5, 1, 1.5, 2, and midpoints 0.25, 0.75, 1.25, 1.75. With this data, we get

$$T_4 = \frac{1}{2}\cdot\frac{1}{2}\left(0^2 + 2(0.5)^2 + 2(1)^2 + 2(1.5)^2 + 2^2\right) = 2.75; \text{ and}$$

$$M_4 = \frac{1}{2}\left(0.25^2 + 0.75^2 + 1.25^2 + 1.75^2\right) = 2.625.$$

3. $\int_1^4 x^3\,dx, \quad N = 6$

SOLUTION Let $f(x) = x^3$. We divide $[1, 4]$ into 6 subintervals of width

$$\Delta x = \frac{4-1}{6} = \frac{1}{2}$$

with endpoints 1, 1.5, 2, 2.5, 3, 3.5, 4, and midpoints 1.25, 1.75, 2.25, 2.75, 3.25, 3.75. With this data, we get

$$T_6 = \frac{1}{2}\left(\frac{1}{2}\right)\left(1^3 + 2(1.5)^3 + 2(2)^3 + 2(2.5)^3 + 2(3)^3 + 2(3.5)^3 + 4^3\right) = 64.6875; \text{ and}$$

$$M_6 = \frac{1}{2}\left(1.25^3 + 1.75^3 + 2.25^3 + 2.75^3 + 3.25^3 + 3.75^3\right) = 63.28125.$$

5. $\int_1^4 \frac{dx}{x}, \quad N = 6$

SOLUTION Let $f(x) = 1/x$. We divide $[1, 4]$ into 6 subintervals of width

$$\Delta x = \frac{4-1}{6} = \frac{1}{2}$$

with endpoints 1, 1.5, 2, 2.5, 3, 3.5, 4, and midpoints 1.25, 1.75, 2.25, 2.75, 3.25, 3.75. With this data, we get

$$T_6 = \frac{1}{2}\left(\frac{1}{2}\right)\left(\frac{1}{1} + \frac{2}{1.5} + \frac{2}{2} + \frac{2}{2.5} + \frac{2}{3} + \frac{2}{3.5} + \frac{1}{4}\right) \approx 1.40536; \text{ and}$$

$$M_6 = \frac{1}{2}\left(\frac{1}{1.25} + \frac{1}{1.75} + \frac{1}{2.25} + \frac{1}{2.75} + \frac{1}{3.25} + \frac{1}{3.75}\right) \approx 1.37693.$$

7. $\int_0^{\pi/2} \sqrt{\sin x}\,dx, \quad N = 6$

SOLUTION Let $f(x) = \sqrt{\sin x}$. We divide $[0, \pi/2]$ into 6 subintervals of width

$$\Delta x = \frac{\frac{\pi}{2} - 0}{6} = \frac{\pi}{12}$$

with endpoints

$$0, \frac{\pi}{12}, \frac{2\pi}{12}, \ldots, \frac{6\pi}{12} = \frac{\pi}{2},$$

and midpoints

$$\frac{\pi}{24}, \frac{3\pi}{24}, \ldots, \frac{11\pi}{24}.$$

With this data, we get

$$T_6 = \frac{1}{2}\left(\frac{\pi}{12}\right)\left(\sqrt{\sin(0)} + 2\sqrt{\sin(\pi/12)} + \cdots + \sqrt{\sin(6\pi/12)}\right) \approx 1.17029; \text{ and}$$

$$M_6 = \frac{\pi}{12}\left(\sqrt{\sin(\pi/24)} + \sqrt{\sin(3\pi/24)} + \cdots + \sqrt{\sin(11\pi/24)}\right) \approx 1.20630.$$

9. $\int_1^2 \ln x\,dx, \quad N = 5$

SOLUTION Let $f(x) = \ln x$. We divide $[1, 2]$ into 5 subintervals of width

$$\Delta x = \frac{2-1}{5} = \frac{1}{5} = 0.2$$

with endpoints 1, 1.2, 1.4, 1.6, 1.8, 2, and midpoints 1.1, 1.3, 1.5, 1.7, 1.9. With this data, we get

$$T_5 = \frac{1}{2}\left(\frac{1}{5}\right)\left(\ln 1 + 2\ln 1.2 + 2\ln 1.4 + 2\ln 1.6 + 2\ln 1.8 + \ln 2\right) \approx 0.384632; \text{ and}$$

$$M_5 = \frac{1}{5}\left(\ln 1.1 + \ln 1.3 + \ln 1.5 + \ln 1.7 + \ln 1.9\right) \approx 0.387124.$$

11. $\int_0^1 e^{-x^2}\,dx, \quad N = 5$

SOLUTION Let $f(x) = e^{-x^2}$. We divide $[0, 1]$ into 5 subintervals of width

$$\Delta x = \frac{1-0}{5} = \frac{1}{5} = 0.2$$

with endpoints

$$0, \frac{1}{5}, \frac{2}{5}, \frac{3}{5}, \frac{4}{5}, 1$$

and midpoints

$$\frac{1}{10}, \frac{3}{10}, \frac{5}{10}, \frac{7}{10}, \frac{9}{10}.$$

With this data, we get

$$T_5 = \frac{1}{2}\left(\frac{1}{5}\right)\left(e^{-0^2} + 2e^{-(1/5)^2} + 2e^{-(2/5)^2} + 2e^{-(3/5)^2} + 2e^{-(4/5)^2} + e^{-1^2}\right) \approx 0.74437; \text{ and}$$

$$M_5 = \frac{1}{5}\left(e^{-(1/10)^2} + e^{-(3/10)^2} + e^{-(5/10)^2} + e^{-(7/10)^2} + e^{-(9/10)^2}\right) \approx 0.74805.$$

In Exercises 13–22, calculate S_N given by Simpson's Rule for the value of N indicated.

13. $\int_0^4 \sqrt{x}\,dx, \quad N = 4$

SOLUTION Let $f(x) = \sqrt{x}$. We divide $[0, 4]$ into 4 subintervals of width

$$\Delta x = \frac{4-0}{4} = 1$$

with endpoints 0, 1, 2, 3, 4. With this data, we get

$$S_4 = \frac{1}{3}(1)\left(\sqrt{0} + 4\sqrt{1} + 2\sqrt{2} + 4\sqrt{3} + \sqrt{4}\right) \approx 5.25221.$$

15. $\int_0^3 \frac{dx}{x^4+1}, \quad N = 6$

SOLUTION Let $f(x) = 1/(x^4+1)$. We divide $[0, 3]$ into 6 subintervals of length

$$\Delta x = \frac{3-0}{6} = \frac{1}{2} = 0.5$$

with endpoints 0, 0.5, 1, 1.5, 2, 2.5, 3. With this data, we get

$$S_6 = \frac{1}{3}\left(\frac{1}{2}\right)\left[\frac{1}{0^4+1} + \frac{4}{0.5^4+1} + \frac{2}{1^4+1} + \frac{4}{1.5^4+1} + \frac{2}{2^4+1} + \frac{4}{2.5^4+1} + \frac{1}{3^4+1}\right] \approx 1.10903.$$

17. $\int_0^1 e^{-x^2}\,dx, \quad N = 4$

SOLUTION Let $f(x) = e^{-x^2}$. We divide $[0, 1]$ into 4 subintervals of length

$$\Delta x = \frac{1-0}{4} = \frac{1}{4}$$

with endpoints $0, \frac{1}{4}, \frac{2}{4}, \frac{3}{4}, \frac{4}{4} = 1$. With this data, we get

$$S_4 = \frac{1}{3}\left(\frac{1}{4}\right)\left[e^{-0^2} + 4e^{-(1/4)^2} + 2e^{-(2/4)^2} + 4e^{-(3/4)^2} + e^{-(1)^2}\right] \approx 0.746855.$$

19. $\int_1^4 \ln x\,dx, \quad N = 8$

SOLUTION Let $f(x) = \ln x$. We divide $[1, 4]$ into 8 subintervals of length

$$\Delta x = \frac{4-1}{8} = \frac{3}{8} = 0.375$$

with endpoints 1, 1.375, 1.75, 2.125, 2.5, 2.875, 3.25, 3.625, 4. With this data, we get

$$S_8 = \frac{1}{3}\left(\frac{3}{8}\right)\left[\ln 1 + 4\ln(1.375) + 2\ln(1.75) + \cdots + 4\ln(3.625) + \ln 4\right] \approx 2.54499.$$

21. $\displaystyle\int_0^{\pi/4} \tan\theta \, d\theta, \quad N = 10$

SOLUTION Let $f(\theta) = \tan\theta$. We divide $[0, \frac{\pi}{4}]$ into 10 subintervals of width

$$\Delta\theta = \frac{\frac{\pi}{4} - 0}{10} = \frac{\pi}{40}$$

with endpoints $0, \frac{\pi}{40}, \frac{2\pi}{40}, \frac{3\pi}{40}, \dots, \frac{10\pi}{40} = \frac{\pi}{4}$. With this data, we get

$$S_{10} = \frac{1}{3}\left(\frac{\pi}{40}\right)\left[\tan(0) + 4\tan\left(\frac{\pi}{40}\right) + 2\tan\left(\frac{2\pi}{40}\right) + \cdots + 4\tan\left(\frac{9\pi}{40}\right) + \tan\left(\frac{10\pi}{40}\right)\right] \approx 0.346576.$$

In Exercises 23–26, calculate the approximation to the volume of the solid obtained by rotating the graph around the given axis.

23. $y = \cos x$; $\left[0, \frac{\pi}{2}\right]$; x-axis; M_8

SOLUTION Using the disk method, the volume is given by

$$V = \int_0^{\pi/2} \pi r^2 \, dx = \pi \int_0^{\pi/2} (\cos x)^2 \, dx$$

which can be estimated as

$$\pi \int_0^{\pi/2} (\cos x)^2 \, dx \approx \pi[M_8].$$

Let $f(x) = \cos^2 x$. We divide $[0, \pi/2]$ into 8 subintervals of length

$$\Delta x = \frac{\frac{\pi}{2} - 0}{8} = \frac{\pi}{16}$$

with midpoints

$$\frac{\pi}{32}, \frac{3\pi}{32}, \frac{5\pi}{32}, \dots, \frac{15\pi}{32}.$$

With this data, we get

$$V \approx \pi[M_8] = \pi\left[\Delta x(y_1 + y_2 + \cdots + y_8)\right] = \frac{\pi^2}{16}\left[\cos^2\left(\frac{\pi}{32}\right) + \cos^2\left(\frac{3\pi}{32}\right) + \cdots + \cos^2\left(\frac{15\pi}{32}\right)\right] \approx 2.46740.$$

25. $y = e^{-x^2}$; $[0, 1]$; x-axis; T_8

SOLUTION Using the disk method, the volume is given by

$$V = \int_0^1 \pi r^2 \, dx = \pi \int_0^1 (e^{-x^2})^2 \, dx = \pi \int_0^1 e^{-2x^2} \, dx.$$

We can use the approximation

$$V = \pi \int_0^1 e^{-2x^2} \, dx \approx \pi[T_8],$$

where $f(x) = e^{-2x^2}$. Divide $[0, 1]$ into 8 subintervals of length

$$\Delta x = \frac{1 - 0}{8} = \frac{1}{8},$$

with endpoints

$$0, \frac{1}{8}, \frac{2}{8}, \dots, 1.$$

With this data, we get

$$V \approx \pi[T_8] = \pi\left[\frac{1}{2} \cdot \frac{1}{8}\left(e^{-2(0^2)} + 2e^{-2(1/8)^2} + \cdots + 2e^{-2(7/8)^2} + e^{-2(1)^2}\right)\right] \approx 1.87691.$$

27. An airplane's velocity is recorded at 5-min intervals during a 1-hour period with the following results, in miles per hour:

550, 575, 600, 580, 610, 640, 625,
595, 590, 620, 640, 640, 630

Use Simpson's Rule to estimate the distance traveled during the hour.

SOLUTION The distance traveled is equal to the integral $\int_0^1 v(t)\,dt$, where t is in hours. Since 5 minutes is $1/12$ of an hour, we have $\Delta t = 1/12$. Simpson's Rule gives us

$$S_{12} = \frac{1}{3} \cdot \frac{1}{12}\Big[550 + 4 \cdot 575 + 2 \cdot 600 + 4 \cdot 580 + 2 \cdot 610 + \cdots + 4 \cdot 640 + 630\Big] \approx 608.611.$$

The distance traveled during the hour is approximately 608.6 miles.

29. **Tsunami Arrival Times** Scientists estimate the arrival times of tsunamis (seismic ocean waves) based on the point of origin P and ocean depths. The speed s of a tsunami in miles per hour is approximately $s = \sqrt{15d}$, where d is the ocean depth in feet.

(a) Let $f(x)$ be the ocean depth x miles from P (in the direction of the coast). Argue using Riemann sums that the time T required for the tsunami to travel M miles toward the coast is

$$T = \int_0^M \frac{dx}{\sqrt{15 f(x)}}$$

(b) Use Simpson's Rule to estimate T if $M = 1000$ and the ocean depths (in feet), measured at 100-mile intervals starting from P, are

13,000, 11,500, 10,500, 9000, 8500,
7000, 6000, 4400, 3800, 3200, 2000

SOLUTION

(a) At a given distance from shore, say, x_i, the speed of the tsunami in mph is $s = \sqrt{15 f(x_i)}$. If we assume the speed s is constant over a small interval Δx, then the time to cover that interval at that speed is

$$t_i = \frac{\text{distance}}{\text{speed}} = \frac{\Delta x}{\sqrt{15 f(x_i)}}.$$

Now divide the interval $[0, M]$ into N subintervals of length Δx. The total time T is given by

$$T = \sum_{i=1}^{N} t_i = \sum_{i=1}^{N} \frac{\Delta x}{\sqrt{15 f(x_i)}}.$$

Taking the limit as $N \to \infty$, we get

$$T = \int_0^M \frac{dx}{\sqrt{15 f(x)}}.$$

(b) We have $\Delta x = 100$. Simpson's Rule gives us

$$S_{10} = \frac{1}{3} \cdot 100 \left[\frac{1}{\sqrt{15(13{,}000)}} + \frac{4}{\sqrt{15(11{,}500)}} + \cdots + \frac{1}{\sqrt{15(2000)}}\right] \approx 3.347.$$

It will take the tsunami about 3 hours and 21 minutes to reach shore.

31. Calculate T_6 for the integral $I = \displaystyle\int_0^2 x^3\,dx$.

(a) Is T_6 too large or too small? Explain graphically.

(b) Show that $K_2 = |f''(2)|$ may be used in the error bound and find a bound for the error.

(c) Evaluate I and check that the actual error is less than the bound computed in (b).

SOLUTION Let $f(x) = x^3$. Divide $[0, 2]$ into 6 subintervals of length $\Delta x = \frac{2-0}{6} = \frac{1}{3}$ with endpoints $0, \frac{1}{3}, \frac{2}{3}, \ldots, 2$. With this data, we get

$$T_6 = \frac{1}{2} \cdot \frac{1}{3}\left[0^3 + 2\left(\frac{1}{3}\right)^3 + 2\left(\frac{2}{3}\right)^3 + 2\left(\frac{3}{3}\right)^3 + 2\left(\frac{4}{3}\right)^3 + 2\left(\frac{5}{3}\right)^3 + (1)2^3\right] \approx 4.11111.$$

(a) Since x^3 is concave up on $[0, 2]$, T_6 is too large.

(b) We have $f'(x) = 3x^2$ and $f''(x) = 6x$. Since $|f''(x)| = |6x|$ is *increasing* on $[0, 2]$, its maximum value occurs at $x = 2$ and we may take $K_2 = |f''(2)| = 12$. Then

$$\text{Error}(T_6) \le \frac{K_2(b-a)^3}{12N^2} = \frac{12(2-0)^3}{12(6)^2} = \frac{2}{9} \approx 0.22222.$$

(c) The exact value is

$$\int_0^2 x^3\,dx = \frac{1}{4}x^4\Big|_0^2 = \frac{1}{4}(16-0) = 4.$$

We can use this to compute the actual error:

$$\text{Error}(T_6) = |T_6 - 4| \approx |4.11111 - 4| \approx 0.11111.$$

Since $0.11111 < 0.22222$, the actual error is indeed less than the maximum possible error.

In Exercises 33–36, state whether T_N or M_N underestimates or overestimates the integral and find a bound for the error (but do not calculate T_N or M_N).

33. $\displaystyle\int_1^4 \frac{1}{x}\,dx, \quad T_{10}$

SOLUTION Let $f(x) = \frac{1}{x}$. Then $f'(x) = \frac{-1}{x^2}$ and $f''(x) = \frac{2}{x^3} > 0$ on $[1, 4]$, so $f(x)$ is concave up, and T_{10} overestimates the integral. Since $|f''(x)| = |\frac{2}{x^3}|$ has its maximum value on $[1, 4]$ at $x = 1$, we can take $K_2 = \frac{2}{1^3} = 2$, and

$$\text{Error}(T_{10}) \le \frac{K_2(4-1)^3}{12N^2} = \frac{2(3)^3}{12(10)^2} = 0.045.$$

35. $\displaystyle\int_1^4 \ln x\,dx, \quad M_{10}$

SOLUTION Let $f(x) = \ln x$. Then $f'(x) = 1/x$ and

$$f''(x) = -\frac{1}{x^2} < 0$$

on $[1, 4]$, so $f(x)$ is concave down, and M_{10} overestimates the integral. Since $|f''(x)| = |-1/x^2|$ has its maximum value on $[1, 4]$ at $x = 1$, we can take $K_2 = |-1/1^2| = 1$, and

$$\text{Error}(M_{10}) \le \frac{K_2(4-1)^3}{24N^2} = \frac{(1)(3)^3}{24(10)^2} = 0.01125.$$

CAS *In Exercises 37–40, use the error bound to find a value of N for which Error(T_N) $\le 10^{-6}$. If you have a computer algebra system, calculate the corresponding approximation and confirm that the error satisfies the required bound.*

37. $\displaystyle\int_0^1 x^4\,dx$

SOLUTION Let $f(x) = x^4$. Then $f'(x) = 4x^3$ and $|f''(x)| = |12x^2|$, which has its maximum value on $[0, 1]$ at $x = 1$, so we can take $K_2 = |12(1)^2| = 12$. Then we have

$$\text{Error}(T_N) \le \frac{K_2(1-0)^3}{12N^2} = \frac{12}{12N^2} = \frac{1}{N^2}.$$

To ensure that the error is at most 10^{-6}, we must choose N such that

$$\frac{1}{N^2} \le \frac{1}{10^6}.$$

This gives $N^2 \ge 10^6$ or $N \ge 10^3$. Thus let $N = 1000$. The exact value of the integral is

$$\int_0^1 x^4\,dx = \frac{x^5}{5}\Big|_0^1 = \frac{1}{5} = 0.2.$$

Using a CAS, we find that

$$T_{1000} \approx 0.2000003333.$$

The actual error is approximately $|0.2000003333 - 0.2| \approx 3.333 \times 10^{-7}$, and is indeed less than 10^{-6}.

39. $\displaystyle\int_2^5 \frac{1}{x}\,dx$

SOLUTION Let $f(x) = 1/x$. Then $f'(x) = -1/x^2$ and $|f''(x)| = |2/x^3|$, which has its maximum value on $[2, 5]$ at $x = 2$, so we can take $K_2 = |2/2^3| = 1/4$. Then we have

$$\text{Error}(T_N) \le \frac{K_2(5-2)^3}{12N^2} = \frac{(1/4)3^3}{12N^2} = \frac{9}{16N^2}.$$

To ensure that the error is at most 10^{-6}, we must choose N such that

$$\frac{9}{16N^2} \le \frac{1}{10^6}.$$

This gives us

$$N^2 \ge \frac{9 \cdot 10^6}{16} \Rightarrow N \ge \sqrt{\frac{9 \cdot 10^6}{16}} = 750.$$

Thus let $N = 750$. The exact value of the integral is

$$\int_2^5 \frac{1}{x}\,dx = \ln 5 - \ln 2 \approx 0.9162907314.$$

Using a CAS, we find that

$$T_{750} \approx 0.9162910119.$$

The error is approximately

$$|0.9162907314 - 0.9162910119| \approx 2.805 \times 10^{-7}$$

and is indeed less than 10^{-6}.

41. Compute the error bound for the approximations T_{10} and M_{10} to $\int_0^3 (x^3+1)^{-1/2}\,dx$, using Figure 17 to determine a value of K_2. Then find a value of N such that the error in M_N is at most 10^{-6}.

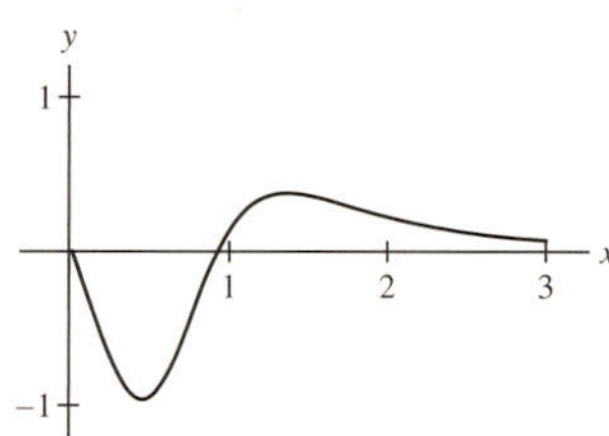

FIGURE 17 Graph of $f''(x)$, where $f(x) = (x^3+1)^{-1/2}$.

SOLUTION Clearly, in the range $0 \le x \le 3$, we have $|f''(x)| \le 1$, so we may choose $K_2 = 1$. Then

$$\text{Error}(T_{10}) \le \frac{K_2(3-0)^3}{12N^2} = \frac{27}{12 \cdot 10^2} = \frac{27}{1200} = 0.0225$$

$$\text{Error}(M_{10}) \le \frac{K_2(3-0)^3}{24N^2} = \frac{27}{24 \cdot 10^2} = \frac{27}{2400} = 0.01125$$

In order for the error in M_N to be at most 10^{-6}, we must have

$$\text{Error}(M_N) \le \frac{K_2(3-0)^3}{24N^2} = \frac{9}{8N^2} \le 10^{-6}$$

so that $8N^2 \ge 9 \times 10^6$ and $N^2 \ge 1{,}125{,}000$. Thus we must choose $N \ge \sqrt{1{,}125{,}000} \approx 1060.7$, so that $N = 1061$.

43. Calculate S_8 for $\int_1^5 \ln x\,dx$ and calculate the error bound. Then find a value of N such that S_N has an error of at most 10^{-6}.

SOLUTION Let $f(x) = \ln x$. We divide $[1, 5]$ into eight subintervals of length $\Delta x = (5-1)/8 = 0.5$, with endpoints $1, 1.5, 2, \ldots, 5$. With this data, we get

$$S_8 = \frac{1}{3} \cdot \frac{1}{2}\left[\ln 1 + 4\ln 1.5 + 2\ln 2 + \cdots + 4\ln 4.5 + \ln 5\right] \approx 4.046655.$$

To find the maximum possible error, we first take derivatives:

$$f'(x) = \frac{1}{x}, \quad f''(x) = -\frac{1}{x^2}, \quad f^{(3)}(x) = \frac{2}{x^3}, \quad f^{(4)}(x) = -\frac{6}{x^4}.$$

Since $|f^{(4)}(x)| = |-6x^{-4}| = 6x^{-4}$, assumes its maximum value on $[1, 5]$ at $x = 1$, we can set $K_4 = 6(1)^{-4} = 6$. Then we have

$$\text{Error}(S_8) \le \frac{K_4(5-1)^5}{180N^4} = \frac{6 \cdot 4^5}{180 \cdot 8^4} \approx 0.0083333.$$

To ensure that S_N has error at most 10^{-6}, we must find N such that

$$\frac{6 \cdot 4^5}{180N^4} \le \frac{1}{10^6}.$$

This gives us

$$N^4 \ge \frac{6 \cdot 4^5 \cdot 10^6}{180} \Rightarrow N \ge \left(\frac{6 \cdot 4^5 \cdot 10^6}{180}\right)^{1/4} \approx 76.435.$$

Thus let $N = 78$ (remember that N must be even when using Simpson's Rule).

45. CAS Use a computer algebra system to compute and graph $f^{(4)}(x)$ for $f(x) = \sqrt{1+x^4}$ and find a bound for the error in the approximation S_{40} to $\int_0^5 f(x)\,dx$.

SOLUTION From the graph of $f^{(4)}(x)$ shown below, we see that $|f^{(4)}(x)| \le 15$ on $[0, 5]$. Therefore we set $K_4 = 15$. Now we have

$$\text{Error}(S_{40}) \le \frac{15(5-0)^5}{180(40)^4} = \frac{5}{49152} \approx 1.017 \times 10^{-4}.$$

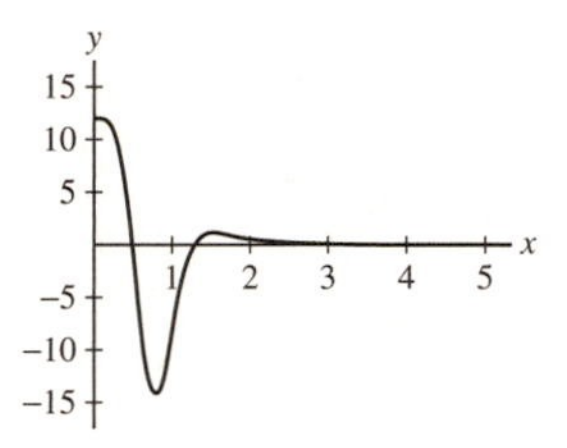

In Exercises 47–50, use the error bound to find a value of N for which Error(S_N) $\le 10^{-9}$.

47. $\int_1^6 x^{4/3}\,dx$

SOLUTION Let $f(x) = x^{4/3}$. We start by taking derivatives:

$$f'(x) = \frac{4}{3}x^{1/3}$$

$$f''(x) = \frac{4}{9}x^{-2/3}$$

$$f'''(x) = -\frac{8}{27}x^{-5/3}$$

$$f^{(4)}(x) = \frac{40}{81}x^{-8/3}$$

For $x \ge 1$, $f^{(4)}(x)$ is a decreasing function of x, so it takes its maximum value on $[1, 6]$ at $x = 1$. That maximum value is $\frac{40}{81}$, which is quite close to (but smaller than) $\frac{1}{2}$. For simplicity, we take $K_4 = \frac{1}{2}$. Then

$$\text{Error}(S_N) \le \frac{K_4(b-a)^5}{180N^4} = \frac{(6-1)^5}{2 \cdot 180 \cdot N^4} = \frac{5^5}{360N^4} = \frac{625}{72N^4} \le 10^{-9}$$

Thus $72N^4 \ge 625 \times 10^9$, so that

$$N \ge \left(\frac{625 \times 10^9}{72}\right)^{1/4} \approx 305.24$$

so we can take $N = 306$.

49. $\int_0^1 e^{x^2}\,dx$

SOLUTION Let $f(x) = e^{x^2}$. To find K_4, we first take derivatives:

$$f'(x) = 2xe^{x^2}$$

$$f''(x) = 4x^2e^{x^2} + 2e^{x^2}$$

$$f^{(3)}(x) = 8x^3e^{x^2} + 12xe^{x^2}$$

$$f^{(4)}(x) = 16x^4e^{x^2} + 48x^2e^{x^2} + 12e^{x^2}.$$

On the interval [0, 1], $|f^{(4)}(x)|$ assumes its maximum value at $x = 1$. Therefore we set

$$K_4 = |f^{(4)}(1)| = 16e + 48e + 12e = 76e.$$

Now we have

$$\text{Error}(S_N) \le \frac{K_4(1-0)^5}{180N^4} = \frac{76e}{180N^4}.$$

To ensure that S_N has error at most 10^{-9}, we must find N such that

$$\frac{76e}{180N^4} \le \frac{1}{10^9}.$$

This gives us

$$N^4 \ge \frac{76e \cdot 10^9}{180} \Rightarrow N \ge \left(\frac{76e \cdot 10^9}{180}\right)^{1/4} \approx 184.06.$$

Thus we let $N = 186$ (remember that N must be even when using Simpson's Rule).

51. CAS Show that $\int_0^1 \frac{dx}{1+x^2} = \frac{\pi}{4}$ [use Eq. (3) in Section 5.7].

(a) Use a computer algebra system to graph $f^{(4)}(x)$ for $f(x) = (1+x^2)^{-1}$ and find its maximum on [0, 1].

(b) Find a value of N such that S_N approximates the integral with an error of at most 10^{-6}. Calculate the corresponding approximation and confirm that you have computed $\frac{\pi}{4}$ to at least four places.

SOLUTION Recall from Section 3.9 that

$$\frac{d}{dx}\tan^{-1}(x) = \frac{1}{1+x^2}.$$

So then

$$\int_0^1 \frac{dx}{1+x^2} = \tan^{-1} x\Big|_0^1 = \tan^{-1}(1) - \tan^{-1}(0) = \frac{\pi}{4}.$$

(a) From the graph of $f^{(4)}(x)$ shown below, we can see that the maximum value of $|f^{(4)}(x)|$ on the interval [0, 1] is 24.

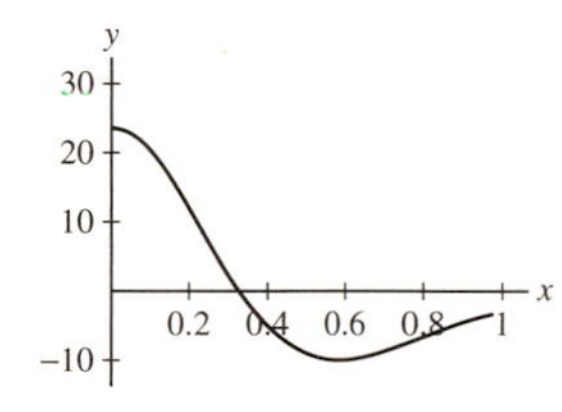

(b) From part (a), we set $K_4 = 24$. Then we have

$$\text{Error}(S_N) \le \frac{24(1-0)^5}{180N^4} = \frac{2}{15N^4}.$$

To ensure that S_N has error at most 10^{-6}, we must find N such that

$$\frac{2}{15N^4} \le \frac{1}{10^6}.$$

This gives us

$$N^4 \ge \frac{2 \cdot 10^6}{15} \Rightarrow N \ge \left(\frac{2 \cdot 10^6}{15}\right)^{1/4} \approx 19.1.$$

Thus let $N = 20$. To compute S_{20}, let $\Delta x = (1-0)/20 = 0.05$. The endpoints of [0, 1] are $0,\ 0.05,\ \ldots,\ 1$. With this data, we get

$$S_{20} = \frac{1}{3}\left(\frac{1}{20}\right)\left[\frac{1}{1+0^2} + \frac{4}{1+(0.05)^2} + \frac{2}{1+(0.1)^2} + \cdots + \frac{1}{1+1^2}\right] \approx 0.785398163242.$$

The actual error is

$$|0.785398163242 - \pi/4| = |0.785398163242 - 0.785398163397| = 1.55 \times 10^{-10}.$$

53. Let $f(x) = \sin(x^2)$ and $I = \int_0^1 f(x)\,dx$.

(a) Check that $f''(x) = 2\cos(x^2) - 4x^2\sin(x^2)$. Then show that $|f''(x)| \le 6$ for $x \in [0, 1]$. *Hint:* Note that $|2\cos(x^2)| \le 2$ and $|4x^2\sin(x^2)| \le 4$ for $x \in [0, 1]$.

(b) Show that $\text{Error}(M_N)$ is at most $\dfrac{1}{4N^2}$.

(c) Find an N such that $|I - M_N| \le 10^{-3}$.

SOLUTION

(a) Taking derivatives, we get

$$f'(x) = 2x\cos(x^2)$$
$$f''(x) = 2x(-\sin(x^2)\cdot 2x) + 2\cos(x^2) = 2\cos(x^2) - 4x^2\sin(x^2).$$

On the interval $[0, 1]$,

$$|f''(x)| = |2\cos(x^2) - 4x^2\sin(x^2)| \le |2\cos(x^2)| + |4x^2\sin(x^2)| \le 2 + 4 = 6.$$

(b) Using $K_2 = 6$, we get

$$\text{Error}(M_N) \le \frac{K_2(1-0)^3}{24N^2} = \frac{6}{24N^2} = \frac{1}{4N^2}.$$

(c) To ensure that M_N has error at most 10^{-3}, we must find N such that

$$\frac{1}{4N^2} \le \frac{1}{10^3}.$$

This gives us

$$N^2 \ge \frac{10^3}{4} = 250 \Rightarrow N \ge \sqrt{250} \approx 15.81.$$

Thus let $N = 16$.

55. CAS Observe that the error bound for T_N (which has 12 in the denominator) is twice as large as the error bound for M_N (which has 24 in the denominator). Compute the actual error in T_N for $\int_0^\pi \sin x\,dx$ for $N = 4, 8, 16, 32$, and 64 and compare with the calculations of Exercise 54. Does the actual error in T_N seem to be roughly twice as large as the error in M_N in this case?

SOLUTION The exact value of the integral is

$$\int_0^\pi \sin x\,dx = -\cos x\Big|_0^\pi = -(-1) - (1) = 2.$$

To compute T_4, we have $\Delta x = (\pi - 0)/4 = \pi/4$, and endpoints $0,\ \pi/4,\ 2\pi/4,\ 3\pi/4,\ \pi$. With this data, we get

$$T_4 = \frac{1}{2}\cdot\frac{\pi}{4}\left[\sin(0) + 2\sin\left(\frac{\pi}{4}\right) + 2\sin\left(\frac{2\pi}{4}\right) + 2\sin\left(\frac{3\pi}{4}\right) + \sin(\pi)\right] \approx 1.896119.$$

The values for T_8, T_{16}, T_{32}, and T_{64} are computed similarly:

$$T_8 = \frac{1}{2}\cdot\frac{\pi}{8}\left[\sin(0) + 2\sin\left(\frac{\pi}{8}\right) + 2\sin\left(\frac{2\pi}{8}\right) + \cdots + 2\sin\left(\frac{7\pi}{8}\right) + \sin(\pi)\right] \approx 1.974232;$$

$$T_{16} = \frac{1}{2}\cdot\frac{\pi}{16}\left[\sin(0) + 2\sin\left(\frac{\pi}{16}\right) + 2\sin\left(\frac{2\pi}{16}\right) + \cdots + 2\sin\left(\frac{15\pi}{16}\right) + \sin(\pi)\right] \approx 1.993570;$$

$$T_{32} = \frac{1}{2}\cdot\frac{\pi}{32}\left[\sin(0) + 2\sin\left(\frac{\pi}{32}\right) + 2\sin\left(\frac{2\pi}{32}\right) + \cdots + 2\sin\left(\frac{31\pi}{32}\right) + \sin(\pi)\right] \approx 1.998393;$$

$$T_{64} = \frac{1}{2}\cdot\frac{\pi}{64}\left[\sin(0) + 2\sin\left(\frac{\pi}{64}\right) + 2\sin\left(\frac{2\pi}{64}\right) + \cdots + 2\sin\left(\frac{63\pi}{64}\right) + \sin(\pi)\right] \approx 1.999598.$$

Now we can compute the actual errors for each N:

$$\text{Error}(T_4) = |2 - 1.896119| = 0.103881$$
$$\text{Error}(T_8) = |2 - 1.974232| = 0.025768$$
$$\text{Error}(T_{16}) = |2 - 1.993570| = 0.006430$$
$$\text{Error}(T_{32}) = |2 - 1.998393| = 0.001607$$
$$\text{Error}(T_{64}) = |2 - 1.999598| = 0.000402$$

Comparing these results with the calculations of Exercise 54, we see that the actual error in T_N is in fact about twice as large as the error in M_N.

57. Verify that S_2 yields the exact value of $\int_0^1 (x - x^3)\,dx$.

SOLUTION Let $f(x) = x - x^3$. Clearly $f^{(4)}(x) = 0$, so we may take $K_4 = 0$ in the error bound estimate for S_2. Then

$$\text{Error}(S_2) \le \frac{K_4(1-0)^5}{180 \cdot 2^4} = 0 \cdot \frac{1}{2880} = 0$$

so that S_2 yields the exact value of the integral.

Further Insights and Challenges

59. Show that if $f(x) = rx + s$ is a linear function (r, s constants), then $T_N = \int_a^b f(x)\,dx$ for all N and all endpoints a, b.

SOLUTION First, note that

$$\int_a^b (rx + s)\,dx = \frac{r(b^2 - a^2)}{2} + s(b - a).$$

Now,

$$T_N(rx+s) = \frac{b-a}{2N}\left[f(a) + 2\sum_{i=1}^{N-1} f(x_i) + f(b)\right] = \frac{r(b-a)}{2N}\left[a + 2\sum_{i=1}^{N-1} a + 2\frac{b-a}{N}\sum_{i=1}^{N-1} i + b\right] + s\frac{b-a}{2N}(2N)$$
$$= \frac{r(b-a)}{2N}\left[(2N-1)a + 2\frac{b-a}{N}\frac{(N-1)N}{2} + b\right] + s(b-a) = \frac{r(b^2-a^2)}{2} + s(b-a).$$

61. For N even, divide $[a, b]$ into N subintervals of width $\Delta x = \frac{b-a}{N}$. Set $x_j = a + j\,\Delta x$, $y_j = f(x_j)$, and

$$S_2^{2j} = \frac{b-a}{3N}\left(y_{2j} + 4y_{2j+1} + y_{2j+2}\right)$$

(a) Show that S_N is the sum of the approximations on the intervals $[x_{2j}, x_{2j+2}]$—that is, $S_N = S_2^0 + S_2^2 + \cdots + S_2^{N-2}$.

(b) By Exercise 60, $S_2^{2j} = \int_{x_{2j}}^{x_{2j+2}} f(x)\,dx$ if $f(x)$ is a quadratic polynomial. Use (a) to show that S_N is exact *for all* N if $f(x)$ is a quadratic polynomial.

SOLUTION

(a) This result follows because the even-numbered interior endpoints overlap:

$$\sum_{i=0}^{(N-2)/2} S_2^{2j} = \frac{b-a}{6}\left[(y_0 + 4y_1 + y_2) + (y_2 + 4y_3 + y_4) + \cdots\right]$$
$$= \frac{b-a}{6}\left[y_0 + 4y_1 + 2y_2 + 4y_3 + 2y_4 + \cdots + 4y_{N-1} + y_N\right] = S_N.$$

(b) If $f(x)$ is a quadratic polynomial, then by part (a) we have

$$S_N = S_2^0 + S_2^2 + \cdots + S_2^{N-2} = \int_{x_0}^{x_2} f(x)\,dx + \int_{x_2}^{x_4} f(x)\,dx + \cdots + \int_{x_{N-2}}^{x_N} f(x)\,dx = \int_a^b f(x)\,dx.$$

63. Use the error bound for S_N to obtain another proof that Simpson's Rule is exact for all cubic polynomials.

SOLUTION Let $f(x) = ax^3 + bx^2 + cx + d$, with $a \neq 0$, be any cubic polynomial. Then, $f^{(4)}(x) = 0$, so we can take $K_4 = 0$. This yields

$$\text{Error}(S_N) \leq \frac{0}{180N^4} = 0.$$

In other words, S_N is exact for all cubic polynomials for all N.

CHAPTER REVIEW EXERCISES

1. Match the integrals (a)–(e) with their antiderivatives (i)–(v) on the basis of the general form (do not evaluate the integrals).

(a) $\displaystyle\int \frac{x\,dx}{x^2-4}$

(b) $\displaystyle\int \frac{(2x+9)\,dx}{x^2+4}$

(c) $\displaystyle\int \sin^3 x \cos^2 x\,dx$

(d) $\displaystyle\int \frac{dx}{x\sqrt{16x^2-1}}$

(e) $\displaystyle\int \frac{16\,dx}{x(x-4)^2}$

(i) $\sec^{-1} 4x + C$

(ii) $\log|x| - \log|x-4| - \dfrac{4}{x-4} + C$

(iii) $\dfrac{1}{30}(3\cos^5 x - 3\cos^3 x \sin^2 x - 7\cos^3 x) + C$

(iv) $\dfrac{9}{2}\tan^{-1}\dfrac{x}{2} + \ln(x^2+4) + C$

(v) $\sqrt{x^2-4} + C$

SOLUTION

(a) $\displaystyle\int \frac{x\,dx}{\sqrt{x^2-4}}$

Since x is a constant multiple of the derivative of x^2-4, the substitution method implies that the integral is a constant multiple of $\int \frac{du}{\sqrt{u}}$ where $u = x^2-4$, that is a constant multiple of $\sqrt{u} = \sqrt{x^2-4}$. It corresponds to the function in **(v)**.

(b) $\displaystyle\int \frac{(2x+9)\,dx}{x^2+4}$

The part $\int \frac{2x}{x^2+4}\,dx$ corresponds to $\ln(x^2+4)$ in **(iv)** and the part $\int \frac{9}{x^2+4}\,dx$ corresponds to $\frac{9}{2}\tan^{-1}\frac{x}{2}$. Hence the integral corresponds to the function in **(iv)**.

(c) $\displaystyle\int \sin^3 x \cos^2 x\,dx$

The reduction formula for $\int \sin^m x \cos^n x\,dx$ shows that this integral is equal to a sum of constant multiples of products in the form $\cos^i x \sin^j x$ as in **(iii)**.

(d) $\displaystyle\int \frac{dx}{x\sqrt{16x^2-1}}$

Since $\int \frac{dx}{|x|\sqrt{x^2-1}} = \sec^{-1} x + C$, we expect the integral $\int \frac{dx}{x\sqrt{16x^2-1}}$ to be equal to the function in **(i)**.

(e) $\displaystyle\int \frac{16\,dx}{x(x-4)^2}$

The partial fraction decomposition of the integrand has the form:

$$\frac{A}{x} + \frac{B}{x-4} + \frac{C}{(x-4)^2}$$

The term $\frac{A}{x}$ contributes the function $A\ln|x|$ to the integral, the term $\frac{B}{x-4}$ contributes $B\ln|x-4|$ and the term $\frac{C}{(x-4)^2}$ contributes $-\frac{C}{x-4}$. Therefore, we expect the integral to be equal to the function in **(ii)**.

In Exercises 3–12, evaluate using the suggested method.

3. $\int \cos^3\theta \sin^8\theta \, d\theta$ [write $\cos^3\theta$ as $\cos\theta(1-\sin^2\theta)$]

SOLUTION We use the identity $\cos^2\theta = 1 - \sin^2\theta$ to rewrite the integral:

$$\int \cos^3\theta \sin^8\theta \, d\theta = \int \cos^2\theta \sin^8\theta \cos\theta \, d\theta = \int \left(1 - \sin^2\theta\right) \sin^8\theta \cos\theta \, d\theta.$$

Now, we use the substitution $u = \sin\theta$, $du = \cos\theta \, d\theta$:

$$\int \cos^3\theta \sin^8\theta \, d\theta = \int \left(1 - u^2\right) u^8 \, du = \int \left(u^8 - u^{10}\right) du = \frac{u^9}{9} - \frac{u^{11}}{11} + C = \frac{\sin^9\theta}{9} - \frac{\sin^{11}\theta}{11} + C.$$

5. $\int \sec^3\theta \tan^4\theta \, d\theta$ (trigonometric identity, reduction formula)

SOLUTION We use the identity $1 + \tan^2\theta = \sec^2\theta$ to write $\tan^4\theta = \left(\sec^2\theta - 1\right)^2$ and to rewrite the integral as

$$\int \sec^3\theta \tan^4\theta \, d\theta \int \sec^3\theta \left(1 - \sec^2\theta\right)^2 d\theta = \int \sec^3\theta \left(1 - 2\sec^2\theta + \sec^4\theta\right) d\theta$$

$$= \int \sec^7\theta \, d\theta - 2\int \sec^5\theta \, d\theta + \int \sec^3\theta \, d\theta.$$

Now we use the reduction formula

$$\int \sec^m\theta \, d\theta = \frac{\tan\theta \sec^{m-2}\theta}{m-1} + \frac{m-2}{m-1}\int \sec^{m-2}\theta \, d\theta.$$

We have

$$\int \sec^5\theta \, d\theta = \frac{\tan\theta \sec^3\theta}{4} + \frac{3}{4}\int \sec^3\theta \, d\theta + C,$$

and

$$\int \sec^7\theta \, d\theta = \frac{\tan\theta \sec^5\theta}{6} + \frac{5}{6}\int \sec^5\theta \, d\theta = \frac{\tan\theta \sec^5\theta}{6} + \frac{5}{6}\left(\frac{\tan\theta \sec^3\theta}{4} + \frac{3}{4}\int \sec^3\theta \, d\theta\right) + C$$

$$= \frac{\tan\theta \sec^5\theta}{6} + \frac{5}{24}\tan\theta \sec^3\theta + \frac{5}{8}\int \sec^3\theta \, d\theta + C.$$

Therefore,

$$\int \sec^3\theta \tan^4\theta \, d\theta = \left(\frac{\tan\theta \sec^5\theta}{6} + \frac{5}{24}\tan\theta \sec^3\theta + \frac{5}{8}\int \sec^3\theta \, d\theta\right)$$

$$- 2\left(\frac{\tan\theta \sec^3\theta}{4} + \frac{3}{4}\int \sec^3\theta \, d\theta\right) + \int \sec^3\theta \, d\theta$$

$$= \frac{\tan\theta \sec^5\theta}{6} - \frac{7\tan\theta \sec^3\theta}{24} + \frac{1}{8}\int \sec^3\theta \, d\theta.$$

We again use the reduction formula to compute

$$\int \sec^3\theta \, d\theta = \frac{\tan\theta \sec\theta}{2} + \frac{1}{2}\int \sec\theta \, d\theta = \frac{\tan\theta \sec\theta}{2} + \frac{1}{2}\ln|\sec\theta + \tan\theta| + C.$$

Finally,

$$\int \sec^3\theta \tan^4\theta \, d\theta = \frac{\tan\theta \sec^5\theta}{6} - \frac{7\tan\theta \sec^3\theta}{24} + \frac{\tan\theta \sec\theta}{16} + \frac{1}{16}\ln|\sec\theta + \tan\theta| + C.$$

7. $\int \frac{dx}{x(x^2-1)^{3/2}} \, dx$ (trigonometric substitution)

SOLUTION Substitute $x = \sec\theta$, $dx = \sec\theta \tan\theta \, d\theta$. Then,

$$\left(x^2 - 1\right)^{3/2} = \left(\sec^2\theta - 1\right)^{3/2} = \left(\tan^2\theta\right)^{3/2} = \tan^3\theta,$$

and

$$\int \frac{dx}{x(x^2-1)^{3/2}} = \int \frac{\sec\theta\tan\theta\, d\theta}{\sec\theta\tan^3\theta} = \int \frac{d\theta}{\tan^2\theta} = \int \cot^2\theta\, d\theta.$$

Using a reduction formula we find that:

$$\int \cot^2\theta\, d\theta = -\cot\theta - \theta + C$$

so

$$\int \frac{dx}{x(x^2-1)^{3/2}} = -\cot\theta - \theta + C.$$

We now must return to the original variable x. We use the relation $x = \sec\theta$ and the figure to obtain:

$$\int \frac{dx}{x(x^2-1)^{3/2}} = -\frac{1}{\sqrt{x^2-1}} - \sec^{-1}x + C.$$

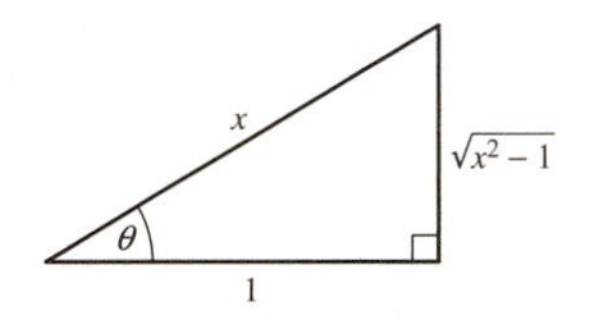

9. $\displaystyle\int \frac{dx}{x^{3/2}+x^{1/2}}$ (substitution)

SOLUTION Let $t = x^{1/2}$. Then $dt = \frac{1}{2}x^{-1/2}\,dx$ or $dx = 2x^{1/2}\,dt = 2t\,dt$. Therefore,

$$\int \frac{dx}{x^{3/2}+x^{1/2}} = \int \frac{2t\,dt}{t^3+t} = \int \frac{2\,dt}{t^2+1} = 2\tan^{-1}t + C = 2\tan^{-1}\sqrt{x} + C.$$

11. $\displaystyle\int x^{-2}\tan^{-1}x\,dx$ (Integration by Parts)

SOLUTION We use Integration by Parts with $u = \tan^{-1}x$ and $v' = x^{-2}$. Then $u' = \frac{1}{1+x^2}$, $v = -x^{-1}$ and

$$\int x^{-2}\tan^{-1}x\,dx = -\frac{\tan^{-1}x}{x} + \int \frac{dx}{x\left(1+x^2\right)}.$$

For the remaining integral, the partial fraction decomposition takes the form

$$\frac{1}{x(1+x^2)} = \frac{A}{x} + \frac{Bx+C}{1+x^2}.$$

Clearing denominators gives us

$$1 = A(1+x^2) + (Bx+C)x.$$

Setting $x = 0$ then yields $A = 1$. Next, equating the x^2-coefficients gives

$$0 = A + B \qquad \text{so} \qquad B = -1,$$

while equating x-coefficients gives $C = 0$. Hence,

$$\frac{1}{x\left(1+x^2\right)} = \frac{1}{x} - \frac{x}{1+x^2},$$

and

$$\int \frac{dx}{x(1+x^2)} = \int \frac{1}{x}\,dx - \int \frac{x\,dx}{1+x^2} = \ln|x| - \frac{1}{2}\ln\left(1+x^2\right) + C.$$

Therefore,

$$\int x^{-2}\tan^{-1}x\,dx = -\frac{\tan^{-1}x}{x} + \ln|x| - \frac{1}{2}\ln\left(1+x^2\right) + C.$$

In Exercises 13–64, evaluate using the appropriate method or combination of methods.

13. $\int_0^1 x^2 e^{4x}\,dx$

SOLUTION We evaluate the indefinite integral using Integration by Parts with $u = x^2$ and $v' = e^{4x}$. Then $u' = 2x$, $v = \frac{1}{4}e^{4x}$ and

$$\int x^2 e^{4x}\,dx = \frac{x^2}{4}e^{4x} - \frac{1}{2}\int xe^{4x}\,dx.$$

We compute the resulting integral using Integration by Parts again, this time with $u = x$ and $v' = e^{4x}$. Then $u' = 1$, $v = \frac{1}{4}e^{4x}$ and

$$\int xe^{4x}\,dx = x\cdot\frac{1}{4}e^{4x} - \int \frac{1}{4}e^{4x}\,dx = \frac{x}{4}e^{4x} - \frac{1}{16}e^{4x} + C.$$

Therefore,

$$\int x^2 e^{4x}\,dx = \frac{x^2}{4}e^{4x} - \frac{1}{2}\left(\frac{x}{4}e^{4x} - \frac{1}{16}e^{4x}\right) + C = \frac{e^{4x}}{32}\left(8x^2 - 4x + 1\right) + C.$$

Finally,

$$\int_0^1 x^2 e^{4x}\,dx = \left(\frac{e^{4x}}{32}\left(8x^2 - 4x + 1\right)\right)\Bigg|_0^1 = \frac{e^4}{32}(8 - 4 + 1) - \frac{1}{32}(1) = \frac{5e^4 - 1}{32}$$

15. $\int \cos^9 6\theta \sin^3 6\theta\,d\theta$

SOLUTION We use the identity $\sin^2 6\theta = 1 - \cos^2 6\theta$ to rewrite the integral:

$$\int \cos^9 6\theta \sin^3 6\theta\,d\theta = \int \cos^9 6\theta \sin^2 6\theta \sin 6\theta\,d\theta = \int \cos^9 6\theta\left(1 - \cos^2 6\theta\right)\sin 6\theta\,d\theta.$$

Now, we use the substitution $u = \cos 6\theta$, $du = -6\sin 6\theta\,d\theta$:

$$\int \cos^9 6\theta \sin^3 6\theta\,d\theta = \int u^9\left(1 - u^2\right)\left(-\frac{du}{6}\right) = -\frac{1}{6}\int\left(u^9 - u^{11}\right)du$$

$$= -\frac{1}{6}\left(\frac{u^{10}}{10} - \frac{u^{12}}{12}\right) + C = \frac{\cos^{12}6\theta}{72} - \frac{\cos^{10}6\theta}{60} + C.$$

17. $\int \frac{(6x+4)\,dx}{x^2 - 1}$

SOLUTION The partial fraction decomposition takes the form

$$\frac{6x+4}{(x-1)(x+1)} = \frac{A}{x-1} + \frac{B}{x+1}.$$

Clearing the denominators gives us

$$6x + 4 = A(x+1) + B(x-1).$$

Setting $x = 1$ then yields $A = 5$, while setting $x = -1$ yields $B = 1$. Hence,

$$\int \frac{(6x+4)dx}{x^2-1} = \int \frac{5}{x-1}\,dx + \int \frac{1}{x+1}\,dx = 5\ln|x-1| + \ln|x+1| + C.$$

19. $\int \frac{d\theta}{\cos^4\theta}$

SOLUTION We use the identity $1 + \tan^2\theta = \sec^2\theta$ to rewrite the integral:

$$\int \frac{d\theta}{\cos^4\theta} = \int \sec^4\theta\,d\theta = \int\left(1 + \tan^2\theta\right)\sec^2\theta\,d\theta.$$

Now, we substitute $u = \tan\theta$. Then, $du = \sec^2\theta\,d\theta$ and

$$\int \frac{d\theta}{\cos^4\theta} = \int\left(1 + u^2\right)du = u + \frac{u^3}{3} + C = \frac{\tan^3\theta}{3} + \tan\theta + C.$$

21. $\int_0^1 \ln(4-2x)\,dx$

SOLUTION Note that $\ln(4-2x) = \ln(2(2-x)) = \ln 2 + \ln(2-x)$. Use integration by parts to integrate $\ln(2-x)$, with $u = \ln(2-x)$, $v' = 1$, so that $u' = -\frac{1}{2-x}$ and $v = x$. Then

$$I = \int_0^1 \ln(4-2x)\,dx = \int_0^1 \ln 2\,dx + \int_0^1 \ln(2-x)\,dx = \ln 2 + (x\ln(2-x))\Big|_0^1 + \int_0^1 \frac{x}{2-x}\,dx$$

Now use long division on the remaining integral, and the substitution $u = 2-x$:

$$I = \ln 2 + (x\ln(2-x))\Big|_0^1 + \int_0^1 \left(-1 + \frac{2}{2-x}\right) dx$$

$$= \ln 2 + 1\ln 1 - \int_0^1 1\,dx + 2\int_0^1 \frac{1}{2-x}\,dx = \ln 2 - 1 - 2\int_2^1 \frac{1}{u}\,du$$

$$= \ln 2 - 1 - 2\ln u\Big|_2^1 = \ln 2 - 1 + 2\ln 2 = 3\ln 2 - 1$$

23. $\int \sin^5\theta\,d\theta$

SOLUTION We use the trigonometric identity $\sin^2\theta = 1 - \cos^2\theta$ to rewrite the integral:

$$\int \sin^5\theta\,d\theta = \int \sin^4\theta \sin\theta\,d\theta = \int \left(1-\cos^2\theta\right)^2 \sin\theta\,d\theta.$$

Now, we substitute $u = \cos\theta$. Then $du = -\sin\theta\,d\theta$ and

$$\int \sin^5\theta\,d\theta = \int \left(1-u^2\right)^2(-du) = -\int \left(1-2u^2+u^4\right) du$$

$$= -\left(u - \frac{2}{3}u^3 + \frac{u^5}{5}\right) + C = -\frac{\cos^5\theta}{5} + \frac{2\cos^3\theta}{3} - \cos\theta + C.$$

25. $\int_0^{\pi/4} \sin 3x \cos 5x\,dx$

SOLUTION First compute the indefinite integral, using the trigonometric identity:

$$\sin\alpha\cos\beta = \frac{1}{2}\left(\sin(\alpha+\beta) + \sin(\alpha-\beta)\right).$$

For $\alpha = 3x$ and $\beta = 5x$ we get:

$$\sin 3x \cos 5x = \frac{1}{2}\left(\sin 8x + \sin(-2x)\right) = \frac{1}{2}(\sin 8x - \sin 2x).$$

Hence,

$$\int \sin 3x \cos 5x\,dx = \frac{1}{2}\int \sin 8x\,dx - \frac{1}{2}\int \sin 2x\,dx = -\frac{1}{16}\cos 8x + \frac{1}{4}\cos 2x + C.$$

Then

$$\int_0^{\pi/4} \sin 3x \cos 5x\,dx = \left(\frac{1}{4}\cos 2x - \frac{1}{16}\cos 8x\right)\Big|_0^{\pi/4} = \frac{1}{4}\cos\frac{\pi}{2} - \frac{1}{16}\cos 2\pi - \frac{1}{4}\cos 0 + \frac{1}{16}\cos 0 = -\frac{1}{4}$$

27. $\int \sqrt{\tan x}\,\sec^2 x\,dx$

SOLUTION We substitute $u = \tan x$. Then $du = \sec^2 x\,dx$ and we obtain:

$$\int \sqrt{\tan x}\,\sec^2 x\,dx = \int \sqrt{u}\,du = \frac{2}{3}u^{3/2} + C = \frac{2}{3}(\tan x)^{3/2} + C.$$

29. $\int \sin^5\theta \cos^3\theta\,d\theta$

SOLUTION We use the identity $\cos^2\theta = 1 - \sin^2\theta$ to rewrite the integral:

$$\int \sin^5\theta \cos^3\theta\,d\theta = \int \sin^5\theta \cos^2\theta \cos\theta\,d\theta = \int \sin^5\theta\left(1-\sin^2\theta\right)\cos\theta\,d\theta.$$

Now, we use the substitution $u = \sin\theta$, $du = \cos\theta\, d\theta$:

$$\int \sin^5\theta \cos^3\theta\, d\theta = \int u^5\left(1 - u^2\right) du = \int \left(u^5 - u^7\right) du = \frac{u^6}{6} - \frac{u^8}{8} + C = \frac{\sin^6\theta}{6} - \frac{\sin^8\theta}{8} + C.$$

31. $\displaystyle\int \cot^2 x \csc^2 x\, dx$

SOLUTION Use the substitution $u = \cot x$, $du = -\csc^2 x\, dx$:

$$\int \cot^2 x \csc^2 x\, dx = -\int \cot^2 x \left(-\csc^2 x\, dx\right) = -\int u^2\, du = -\frac{1}{3}u^3 + C = -\frac{1}{3}\cot^3 x + C.$$

33. $\displaystyle\int_{\pi/4}^{\pi/2} \cot^2 x \csc^3 x\, dx$

SOLUTION To compute the indefinite integral, use the identity $\cot^2 x = \csc^2 x - 1$ to write

$$\int \cot^2 x \csc^3 x\, dx = \int \left(\csc^2 x - 1\right)\csc^3 x\, dx = \int \csc^5 x\, dx - \int \csc^3 x\, dx.$$

Now use the reduction formula for $\csc^m x$:

$$\begin{aligned}
\int \cot^2 x \csc^3 x\, dx &= \left(-\frac{1}{4}\cot x \csc^3 x + \frac{3}{4}\int \csc^3 x\, dx\right) - \int \csc^3 x\, dx \\
&= -\frac{1}{4}\cot x \csc^3 x - \frac{1}{4}\int \csc^3 x\, dx \\
&= -\frac{1}{4}\cot x \csc^3 x - \frac{1}{4}\left(-\frac{1}{2}\cot x \csc x + \frac{1}{2}\int \csc x\, dx\right) \\
&= -\frac{1}{4}\cot x \csc^3 x + \frac{1}{8}\cot x \csc x - \frac{1}{8}\ln|\csc x - \cot x| + C.
\end{aligned}$$

Then

$$\begin{aligned}
\int_{\pi/4}^{\pi/2} \cos^2 x \csc^3 x\, dx &= \left(-\frac{1}{4}\cot x \csc^3 x + \frac{1}{8}\cot x \csc x - \frac{1}{8}\ln|\csc x - \cot x|\right)\Bigg|_{\pi/4}^{\pi/2} \\
&= -\frac{1}{4}\cot\frac{\pi}{2}\csc^3\frac{\pi}{2} + \frac{1}{8}\cot\frac{\pi}{2}\csc\frac{\pi}{2} - \frac{1}{8}\ln\left|\csc\frac{\pi}{2} - \cot\frac{\pi}{2}\right| \\
&\quad + \frac{1}{4}\cot\frac{\pi}{4}\csc^3\frac{\pi}{4} - \frac{1}{8}\cot\frac{\pi}{4}\csc\frac{\pi}{4} + \frac{1}{8}\ln\left|\csc\frac{\pi}{4} - \cot\frac{\pi}{4}\right| \\
&= 0 + 0 - \frac{1}{8}\ln|1 - 0| + \frac{1}{4}\cdot 1\cdot(\sqrt{2})^3 - \frac{1}{8}\cdot 1\cdot\sqrt{2} + \frac{1}{8}\ln\left|\sqrt{2} - 1\right| \\
&= \frac{\sqrt{2}}{2} - \frac{\sqrt{2}}{8} + \frac{1}{8}\ln(\sqrt{2} - 1) = \frac{3}{8}\sqrt{2} + \frac{1}{8}\ln(\sqrt{2} - 1)
\end{aligned}$$

35. $\displaystyle\int \frac{dt}{(t-3)^2(t+4)}$

SOLUTION The partial fraction decomposition has the form

$$\frac{1}{(t-3)^2(t+4)} = \frac{A}{t+4} + \frac{B}{t-3} + \frac{C}{(t-3)^2}.$$

Clearing denominators gives us

$$1 = A(t-3)^2 + B(t-3)(t+4) + C(t+4).$$

Setting $t = 3$ then yields $C = \frac{1}{7}$, while setting $t = -4$ yields $A = \frac{1}{49}$. Lastly, setting $t = 0$ yields

$$1 = 9A - 12B + 4C \qquad \text{or} \qquad B = -\frac{1}{49}.$$

Hence,

$$\int \frac{dt}{(t-3)^2(t+4)} = \frac{1}{49}\int \frac{dt}{t+4} - \frac{1}{49}\int \frac{dt}{t-3} + \frac{1}{7}\int \frac{dt}{(t-3)^2}$$
$$= \frac{1}{49}\ln|t+4| - \frac{1}{49}\ln|t-3| + \frac{1}{7}\cdot\frac{-1}{t-3} + C = \frac{1}{49}\ln\left|\frac{t+4}{t-3}\right| - \frac{1}{7}\cdot\frac{1}{t-3} + C.$$

37. $\displaystyle\int \frac{dx}{x\sqrt{x^2-4}}$

SOLUTION Substitute $x = 2\sec\theta$, $dx = 2\sec\theta\tan\theta\,d\theta$. Then

$$\sqrt{x^2-4} = \sqrt{4\sec^2\theta - 4} = \sqrt{4\left(\sec^2\theta - 1\right)} = \sqrt{4\tan^2\theta} = 2\tan\theta,$$

and

$$\int \frac{dx}{x\sqrt{x^2-4}} = \int \frac{2\sec\theta\tan\theta\,d\theta}{2\sec\theta\cdot 2\tan\theta} = \frac{1}{2}\int d\theta = \frac{1}{2}\theta + C.$$

Now, return to the original variable x. Since $x = 2\sec\theta$, we have $\sec\theta = \frac{x}{2}$ or $\theta = \sec^{-1}\frac{x}{2}$. Thus,

$$\int \frac{dx}{x\sqrt{x^2-4}} = \frac{1}{2}\sec^{-1}\frac{x}{2} + C.$$

39. $\displaystyle\int \frac{dx}{x^{3/2}+ax^{1/2}}$

SOLUTION Let $u = x^{1/2}$ or $x = u^2$. Then $dx = 2u\,du$ and

$$\int \frac{dx}{x^{3/2}+ax^{1/2}} = \int \frac{2u\,du}{u^3+au} = 2\int \frac{du}{u^2+a}.$$

If $a > 0$, then

$$\int \frac{dx}{x^{3/2}+ax^{1/2}} = 2\int \frac{du}{u^2+a} = \frac{2}{\sqrt{a}}\tan^{-1}\left(\frac{u}{\sqrt{a}}\right) + C = \frac{2}{\sqrt{a}}\tan^{-1}\sqrt{\frac{x}{a}} + C.$$

If $a = 0$, then

$$\int \frac{dx}{x^{3/2}} = -\frac{2}{\sqrt{x}} + C.$$

Finally, if $a < 0$, then

$$\int \frac{du}{u^2+a} = \int \frac{du}{u^2 - \left(\sqrt{-a}\right)^2},$$

and the partial fraction decomposition takes the form

$$\frac{1}{u^2-\left(\sqrt{-a}\right)^2} = \frac{A}{u-\sqrt{-a}} + \frac{B}{u+\sqrt{-a}}.$$

Clearing denominators gives us

$$1 = A(u+\sqrt{-a}) + B(u-\sqrt{-a}).$$

Setting $u = \sqrt{-a}$ then yields $A = \frac{1}{2\sqrt{-a}}$, while setting $u = -\sqrt{-a}$ yields $B = -\frac{1}{2\sqrt{-a}}$. Hence,

$$\int \frac{dx}{x^{3/2}+ax^{1/2}} = 2\int \frac{du}{u^2+a} = \frac{1}{\sqrt{-a}}\int \frac{du}{u-\sqrt{-a}} - \frac{1}{\sqrt{-a}}\int \frac{du}{u+\sqrt{-a}}$$
$$= \frac{1}{\sqrt{-a}}\ln|u-\sqrt{-a}| - \frac{1}{\sqrt{-a}}\ln|u+\sqrt{-a}| + C$$
$$= \frac{1}{\sqrt{-a}}\ln\left|\frac{u-\sqrt{-a}}{u+\sqrt{-a}}\right| + C = \frac{1}{\sqrt{-a}}\ln\left|\frac{\sqrt{x}-\sqrt{-a}}{\sqrt{x}+\sqrt{-a}}\right| + C.$$

In summary,

$$\int \frac{dx}{x^{3/2}+ax^{1/2}} = \begin{cases} \frac{2}{\sqrt{a}}\tan^{-1}\sqrt{\frac{x}{a}}+C & a>0 \\ \frac{1}{\sqrt{-a}}\ln\left|\frac{\sqrt{x}-\sqrt{-a}}{\sqrt{x}+\sqrt{-a}}\right|+C & a<0 \\ -\frac{2}{\sqrt{x}}+C & a=0 \end{cases}$$

41. $\displaystyle\int \frac{(x^2-x)\,dx}{(x+2)^3}$

SOLUTION The partial fraction decomposition has the form

$$\frac{x^2-x}{(x+2)^3} = \frac{A}{x+2}+\frac{B}{(x+2)^2}+\frac{C}{(x+2)^3}.$$

Clearing denominators gives us

$$x^2-x = A(x+2)^2+B(x+2)+C.$$

Setting $x=-2$ then yields $C=6$. Equating x^2-coefficients gives us $A=1$, and equating x-coefficients yields $4A+B=-1$, or $B=-5$. Thus,

$$\int \frac{x^2-x}{(x+2)^3}\,dx = \int\frac{dx}{x+2}+\int\frac{-5\,dx}{(x+2)^2}+\int\frac{6\,dx}{(x+2)^3} = \ln|x+2|+\frac{5}{x+2}-\frac{3}{(x+2)^2}+C.$$

43. $\displaystyle\int \frac{16\,dx}{(x-2)^2(x^2+4)}$

SOLUTION The partial fraction decomposition has the form

$$\frac{16}{(x-2)^2\left(x^2+4\right)} = \frac{A}{x-2}+\frac{B}{(x-2)^2}+\frac{Cx+D}{x^2+4}.$$

Clearing denominators gives us

$$16 = A(x-2)\left(x^2+4\right)+B\left(x^2+4\right)+(Cx+D)(x-2)^2.$$

Setting $x=2$ then yields $B=2$. With $B=2$,

$$16 = A\left(x^3-2x^2+4x-8\right)+2\left(x^2+4\right)+Cx^3+(D-4C)x^2+(4C-4D)x+4D$$

$$16 = (A+C)x^3+(-2A+2+D-4C)\,x^2+(4A+4C-4D)x+(-8A+8+4D)$$

Equating coefficients of like powers of x now gives us the system of equations

$$A+C=0$$

$$-2A-4C+D+2=0$$

$$4A+4C-4D=0$$

$$-8A+4D+8=1$$

whose solution is

$$A=-1,\ C=1,\ D=0.$$

Thus,

$$\int\frac{dx}{(x-2)^2\left(x^2+4\right)} = -\int\frac{dx}{x-2}+2\int\frac{dx}{(x-2)^2}+\int\frac{x}{x^2+4}\,dx$$

$$= -\ln|x-2|-2\frac{1}{x-2}+\frac{1}{2}\ln\left(x^2+4\right)+C.$$

45. $\displaystyle\int \frac{dx}{x^2+8x+25}$

SOLUTION Complete the square to rewrite the denominator as

$$x^2+8x+25 = (x+4)^2+9.$$

Now, let $u = x + 4$, $du = dx$. Then,

$$\int \frac{dx}{x^2 + 8x + 25} = \int \frac{du}{u^2 + 9} = \frac{1}{3}\tan^{-1}\frac{u}{3} + C = \frac{1}{3}\tan^{-1}\left(\frac{x+4}{3}\right) + C.$$

47. $\int \frac{(x^2 - x)\,dx}{(x+2)^3}$

SOLUTION The partial fraction decomposition has the form

$$\frac{x^2 - x}{(x+2)^3} = \frac{A}{x+2} + \frac{B}{(x+2)^2} + \frac{C}{(x+2)^3}.$$

Clearing denominators gives us

$$x^2 - x = A(x+2)^2 + B(x+2) + C.$$

Setting $x = -2$ then yields $C = 6$. Equating x^2-coefficients gives us $A = 1$, and equating x-coefficients yields $4A + B = -1$, or $B = -5$. Thus,

$$\int \frac{x^2 - x}{(x+2)^3}\,dx = \int \frac{dx}{x+2} + \int \frac{-5\,dx}{(x+2)^2} + \int \frac{6\,dx}{(x+2)^3} = \ln|x+2| + \frac{5}{x+2} - \frac{3}{(x+2)^2} + C.$$

49. $\int \frac{dx}{x^4\sqrt{x^2+4}}$

SOLUTION Substitute $x = 2\tan\theta$, $dx = 2\sec^2\theta\,d\theta$. Then

$$\sqrt{x^2+4} = \sqrt{4\tan^2\theta + 4} = \sqrt{4\left(\tan^2\theta + 1\right)} = 2\sqrt{\sec^2\theta} = 2\sec\theta,$$

and

$$\int \frac{dx}{x^4\sqrt{x^2+4}} = \int \frac{2\sec^2\theta\,d\theta}{16\tan^4\theta \cdot 2\sec\theta} = \int \frac{\sec\theta\,d\theta}{16\tan^4\theta}.$$

We have

$$\frac{\sec\theta}{\tan^4\theta} = \frac{\cos^3\theta}{\sin^4\theta}.$$

Hence,

$$\int \frac{dx}{x^4\sqrt{x^2+4}} = \frac{1}{16}\int \frac{\cos^3\theta\,d\theta}{\sin^4\theta} = \frac{1}{16}\int \frac{\cos^2\theta\cos\theta\,d\theta}{\sin^4\theta} = \frac{1}{16}\int \frac{\left(1 - \sin^2\theta\right)\cos\theta\,d\theta}{\sin^4\theta}.$$

Now substitute $u = \sin\theta$ and $du = \cos\theta\,d\theta$ to obtain

$$\int \frac{dx}{x^4\sqrt{x^2+4}} = \frac{1}{16}\int \frac{1-u^2}{u^4}\,du = \frac{1}{16}\int \left(u^{-4} - u^{-2}\right)du = -\frac{1}{48u^3} + \frac{1}{16}\frac{1}{u} + C$$

$$= -\frac{1}{48}\cdot\frac{1}{\sin^3\theta} + \frac{1}{16}\frac{1}{\sin\theta} + C = -\frac{1}{48}\csc^3\theta + \frac{1}{16}\csc\theta + C.$$

Finally, return to the original to the original variable x using the relation $x = 2\tan\theta$ and the figure below.

$$\int \frac{dx}{x^4\sqrt{x^2+4}} = -\frac{1}{48}\left(\frac{\sqrt{x^2+4}}{x}\right)^3 + \frac{1}{16}\frac{\sqrt{x^2+4}}{x} + C = -\frac{\left(x^2+4\right)^{3/2}}{48x^3} + \frac{\sqrt{x^2+4}}{16x} + C.$$

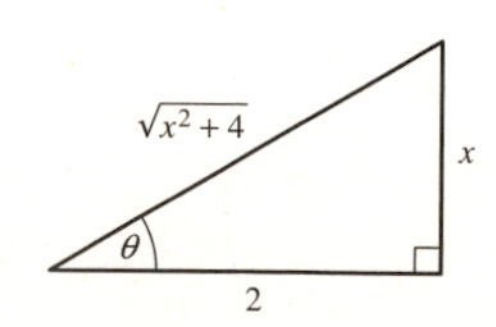

51. $\int (x+1)e^{4-3x}\,dx$

SOLUTION We compute the integral using Integration by Parts with $u = x+1$ and $v' = e^{4-3x}$. Then $u' = 1$, $v = -\frac{1}{3}e^{4-3x}$ and

$$\int (x+1)e^{4-3x}\,dx = -\frac{1}{3}(x+1)e^{4-3x} + \frac{1}{3}\int e^{4-3x}\,dx = -\frac{1}{3}(x+1)e^{4-3x} + \frac{1}{3}\cdot\left(-\frac{1}{3}\right)e^{4-3x} + C$$
$$= -\frac{1}{9}e^{4-3x}(3x+4) + C.$$

53. $\int x^3\cos(x^2)\,dx$

SOLUTION Substitute $t = x^2$, $dt = 2x\,dx$. Then

$$\int x^3\cos\left(x^2\right)dx = \frac{1}{2}\int t\cos t\,dt.$$

We compute the resulting integral using Integration by Parts with $u = t$ and $v' = \cos t$. Then $u' = 1$, $v = \sin t$ and

$$\int t\cos t\,dt = t\sin t - \int \sin t\,dt = t\sin t + \cos t + C.$$

Thus,

$$\int x^3\cos\left(x^2\right)dx = \frac{1}{2}x^2\sin x^2 + \frac{1}{2}\cos x^2 + C.$$

55. $\int x\tanh^{-1}x\,dx$

SOLUTION We use Integration by Parts with $u = \tanh^{-1}x$ and $v' = x$. Then $u' = \frac{1}{1-x^2}$, $v = \frac{x^2}{2}$ and

$$\int x\tanh^{-1}x\,dx = \frac{x^2}{2}\tanh^{-1}x - \frac{1}{2}\int\frac{x^2}{1-x^2}\,dx.$$

Now

$$\frac{x^2}{1-x^2} = \frac{x^2-1+1}{1-x^2} = -1 + \frac{1}{1-x^2},$$

and the partial fraction decomposition for the remaining fraction takes the form

$$\frac{1}{1-x^2} = \frac{A}{1-x} + \frac{B}{1+x}.$$

Clearing denominators gives us

$$1 = A(1+x) + B(1-x).$$

Setting $x = 1$ then yields $A = \frac{1}{2}$, while setting $x = -1$ yields $B = \frac{1}{2}$. Thus,

$$\int\frac{x^2}{1-x^2} = -\int dx + \frac{1}{2}\int\frac{1}{1-x}\,dx + \frac{1}{2}\int\frac{1}{1+x}\,dx$$
$$= -x - \frac{1}{2}\ln|1-x| + \frac{1}{2}\ln|1+x| + C = -x + \frac{1}{2}\ln\left|\frac{1+x}{1-x}\right| + C.$$

Therefore,

$$\int x\tanh^{-1}x\,dx = \frac{x^2}{2}\tanh^{-1}x - \frac{1}{2}\left(-x + \frac{1}{2}\ln\left|\frac{1+x}{1-x}\right|\right) + C = \frac{x^2}{2}\tanh^{-1}x + \frac{x}{2} - \frac{1}{4}\ln\left|\frac{1+x}{1-x}\right| + C.$$

57. $\int \ln(x^2+9)\,dx$

SOLUTION We compute the integral using Integration by Parts with $u = \ln\left(x^2+9\right)$ and $v' = 1$. Then $u' = \frac{2x}{x^2+9}$, $v = x$, and

$$\int \ln\left(x^2+9\right)dx = x\ln\left(x^2+9\right) - \int\frac{2x^2}{x^2+9}\,dx.$$

To compute this integral we write:

$$\frac{x^2}{x^2+9} = \frac{\left(x^2+9\right)-9}{x^2+9} = 1 - \frac{9}{x^2+9};$$

hence,

$$\int \frac{x^2}{x^2+9}\,dx = \int 1\,dx - 9\int \frac{dx}{x^2+9} = x - 3\tan^{-1}\frac{x}{3} + C.$$

Therefore,

$$\int \ln\left(x^2+9\right)dx = x\ln\left(x^2+9\right) - 2x + 6\tan^{-1}\left(\frac{x}{3}\right) + C.$$

59. $\int_0^1 \cosh 2t\,dt$

SOLUTION $\int_0^1 \cosh 2t\,dt = \frac{1}{2}\sinh 2t\Big|_0^1 = \frac{1}{2}\sinh 2.$

61. $\int \coth^2(1-4t)\,dt$

SOLUTION $\int \coth^2(1-4t)\,dt = \int \left(1+\operatorname{csch}^2(1-4t)\right)dt = t + \frac{1}{4}\coth(1-4t) + C.$

63. $\int_0^{3\sqrt{3}/2} \frac{dx}{\sqrt{9-x^2}}$

SOLUTION $\int_0^{3\sqrt{3}/2} \frac{dx}{\sqrt{9-x^2}} = \sin^{-1}\frac{x}{3}\Big|_0^{3\sqrt{3}/2} = \sin^{-1}\frac{\sqrt{3}}{2} = \frac{\pi}{3}.$

65. Use the substitution $u = \tanh t$ to evaluate $\int \frac{dt}{\cosh^2 t + \sinh^2 t}$.

SOLUTION Let $u = \tanh t$. Then $du = \operatorname{sech}^2 t\,dt$ and

$$\int \frac{dt}{\cosh^2 t + \sinh^2 t} = \int \frac{\operatorname{sech}^2 t}{1+\tanh^2 t}\,dt = \int \frac{du}{1+u^2} = \tan^{-1} u + C = \tan^{-1}(\tanh x) + C.$$

67. Let $I_n = \int \frac{x^n\,dx}{x^2+1}$.

(a) Prove that $I_n = \frac{x^{n-1}}{n-1} - I_{n-2}$.

(b) Use (a) to calculate I_n for $0 \le n \le 5$.

(c) Show that, in general,

$$I_{2n+1} = \frac{x^{2n}}{2n} - \frac{x^{2n-2}}{2n-2} + \cdots + (-1)^{n-1}\frac{x^2}{2} + (-1)^n\frac{1}{2}\ln(x^2+1) + C$$

$$I_{2n} = \frac{x^{2n-1}}{2n-1} - \frac{x^{2n-3}}{2n-3} + \cdots + (-1)^{n-1}x + (-1)^n\tan^{-1}x + C$$

SOLUTION

(a) $I_n = \int \frac{x^n}{x^2+1}\,dx = \int \frac{x^{n-2}(x^2+1-1)}{x^2+1}\,dx = \int x^{n-2}\,dx - \int \frac{x^{n-2}}{x^2+1}\,dx = \frac{x^{n-1}}{n-1} - I_{n-2}.$

(b) First compute I_0 and I_1 directly:

$$I_0 = \int \frac{x^0\,dx}{x^2+1} = \int \frac{dx}{x^2+1} = \tan^{-1}x + C \quad\text{and}\quad I_1 = \int \frac{x\,dx}{x^2+1} = \frac{1}{2}\ln\left(x^2+1\right) + C.$$

We now use the equality obtained in part (a) to compute I_2, I_3, I_4 and I_5:

$$I_2 = \frac{x^{2-1}}{2-1} - I_{2-2} = x - I_0 = x - \tan^{-1}x + C;$$

$$I_3 = \frac{x^{3-1}}{3-1} - I_{3-2} = \frac{x^2}{2} - I_1 = \frac{x^2}{2} - \frac{1}{2}\ln\left(x^2+1\right) + C;$$

$$I_4 = \frac{x^{4-1}}{4-1} - I_{4-2} = \frac{x^3}{3} - I_2 = \frac{x^3}{3} - \left(x - \tan^{-1}x\right) + C = \frac{x^3}{3} - x + \tan^{-1}x + C;$$

$$I_5 = \frac{x^{5-1}}{5-1} - I_{5-2} = \frac{x^4}{4} - I_3 = \frac{x^4}{4} - \left(\frac{x^2}{2} - \frac{1}{2}\ln\left(x^2+1\right)\right) + C = \frac{x^4}{4} - \frac{x^2}{2} + \frac{1}{2}\ln\left(x^2+1\right) + C.$$

(c) We prove the two identities using mathematical induction. We first prove that for $n \ge 1$:

$$I_{2n+1} = \frac{x^{2n}}{2n} - \frac{x^{2n-2}}{2n-2} + \cdots + (-1)^n \cdot \frac{1}{2}\ln\left(x^2+1\right) + C.$$

We verify the equality for $n = 1$. Setting $n = 1$, we find

$$I_3 = \frac{x^2}{2} + (-1)^1 \cdot \frac{1}{2}\ln\left(x^2+1\right) + C = \frac{x^2}{2} - \frac{1}{2}\ln\left(x^2+1\right) + C,$$

which agrees with the value obtained in part (b). We now assume that for $n = k$:

$$I_{2k+1} = \frac{x^{2k}}{2k} - \frac{x^{2k-2}}{2k-2} + \cdots + (-1)^k \cdot \frac{1}{2}\ln\left(x^2+1\right) + C.$$

We use this assumption to prove the equality for $n = k + 1$. By part (a) and the induction hypothesis

$$\begin{aligned} I_{2k+3} = \frac{x^{2k+2}}{2k+2} - I_{2k+1} &= \frac{x^{2k+2}}{2k+2} - \frac{x^{2k}}{2k} + \frac{x^{2k-2}}{2k-2} - \cdots - (-1)^k \cdot \frac{1}{2}\ln\left(x^2+1\right) + C \\ &= \frac{x^{2k+2}}{2k+2} - \frac{x^{2k}}{2k} + \cdots + (-1)^{k+1} \cdot \frac{1}{2}\ln\left(x^2+1\right) + C \end{aligned}$$

as required. We now prove the second identity for $n \ge 1$:

$$I_{2n} = \frac{x^{2n-1}}{2n-1} - \frac{x^{2n-3}}{2n-3} + \cdots + (-1)^n \tan^{-1}x + C.$$

We verify this equality for $n = 1$:

$$I_2 = x - \tan^{-1}x + C,$$

which agrees with the value obtained in part (b). We now assume that for $n = k$

$$I_{2k} = \frac{x^{2k-1}}{2k-1} - \frac{x^{2k-3}}{2k-3} + \cdots + (-1)^k \tan^{-1}x + C.$$

We use this assumption to prove the equality for $n = k + 1$. By part (a) and the induction hypothesis

$$\begin{aligned} I_{2k+2} = \frac{x^{2k+1}}{2k+1} - I_{2k} &= \frac{x^{2k+1}}{2k+1} - \frac{x^{2k-1}}{2k-1} + \frac{x^{2k-3}}{2k-3} - \cdots - (-1)^k \cdot \tan^{-1}x + C \\ &= \frac{x^{2k+1}}{2k+1} - \frac{x^{2k-1}}{2k-1} + \cdots + (-1)^{k+1} \cdot \tan^{-1}x + C \end{aligned}$$

as required.

69. Compute $p(X \le 1)$, where X is a continuous random variable with probability density $p(x) = \dfrac{1}{\pi(x^2+1)}$.

SOLUTION

$$P(X \le 1) = \int_{-\infty}^{1} p(x)\,dx = \frac{1}{\pi}\int_{-\infty}^{1} \frac{1}{x^2+1}\,dx = \frac{1}{\pi}\tan^{-1}x\Big|_{-\infty}^{1} = \frac{1}{\pi}\cdot\left(\frac{\pi}{4} - \frac{-\pi}{2}\right) = \frac{3}{4}$$

71. Find a constant C such that $p(x) = Cx^3e^{-x^2}$ is a probability density and compute $p(0 \le X \le 1)$.

SOLUTION We first find the indefinite integral of $p(x)$ using integration by parts, with $u = x^2$, $v' = xe^{-x^2}$, so that $u' = 2x$ and $v = -\frac{1}{2}e^{-x^2}$:

$$\int Cx^3e^{-x^2}\,dx = C\left(-\frac{1}{2}x^2e^{-x^2} + \int xe^{-x^2}\,dx\right) = C\left(-\frac{1}{2}x^2e^{-x^2} - \frac{1}{2}e^{-x^2}\right) = -\frac{C}{2}e^{-x^2}(x^2+1)$$

To determine the constant C, the value of the integral on the interval $[0, \infty)$ must be 1:

$$1 = \int_0^\infty Cx^3e^{-x^2}\,dx = -\frac{C}{2}e^{-x/2}(x^2+1)\Big|_0^\infty = -\frac{C}{2}\left(\lim_{R\to\infty}\frac{x^2+1}{e^{x/2}} - 1\right) = \frac{C}{2}$$

so that $C = 2$. Then

$$P(0 \le X \le 1) = \int_0^1 2x^3e^{-x^2}\,dx = -e^{-x^2}(x^2+1)\Big|_0^1 = 1 - 2e^{-1} \approx 0.13212$$

73. Calculate the following probabilities, assuming that X is normally distributed with mean $\mu = 40$ and $\sigma = 5$.

(a) $p(X \ge 45)$ **(b)** $p(0 \le X \le 40)$

SOLUTION Let F be the standard normal cumulative distribution function. Then by Theorem 1 in Section 7.7,

(a)

$$p(X \ge 45) = 1 - p(X \le 45) = 1 - F\left(\frac{45-40}{5}\right) = 1 - F(1) \approx 1 - 0.8413 \approx 0.1587$$

(b)

$$p(0 \le X \le 40) = p(X \le 40) - p(X \le 0) = F\left(\frac{40-40}{5}\right) - F\left(\frac{0-40}{5}\right)$$
$$= F(0) - F(-8) = \frac{1}{2} - F(-8) \approx \frac{1}{2} - 0 = \frac{1}{2}$$

Note that $p(X \le 40)$ is exactly $\frac{1}{2}$ since 40 is the mean. Also, since -8 is so far to the left in the standard normal distribution, the probability of its occurrence is quite small (approximately 8×10^{-11}).

In Exercises 75–84, determine whether the improper integral converges and, if so, evaluate it.

75. $\displaystyle\int_0^\infty \frac{dx}{(x+2)^2}$

SOLUTION

$$\int_0^\infty \frac{dx}{(x+2)^2} = \lim_{R\to\infty}\int_0^R \frac{dx}{(x+2)^2} = \lim_{R\to\infty} -\frac{1}{x+2}\Big|_0^R$$
$$= \lim_{R\to\infty}\left(-\frac{1}{R+2} + \frac{1}{0+2}\right) = \lim_{R\to\infty}\left(-\frac{1}{R+2} + \frac{1}{2}\right) = 0 + \frac{1}{2} = \frac{1}{2}.$$

77. $\displaystyle\int_0^4 \frac{dx}{x^{2/3}}$

SOLUTION

$$\int_0^4 \frac{dx}{x^{2/3}} = \lim_{R\to 0+}\int_R^4 \frac{dx}{x^{2/3}} = \lim_{R\to 0+} 3x^{1/3}\Big|_R^4 = \lim_{R\to 0+}\left(3\cdot 4^{1/3} - 3\cdot R^{1/3}\right) = 3\sqrt[3]{4}.$$

79. $\displaystyle\int_{-\infty}^0 \frac{dx}{x^2+1}$

SOLUTION

$$\int_{-\infty}^0 \frac{dx}{x^2+1} = \lim_{R\to-\infty}\int_R^0 \frac{dx}{x^2+1} = \lim_{R\to-\infty} \tan^{-1}x\Big|_R^0 = \lim_{R\to-\infty}\left(\tan^{-1}0 - \tan^{-1}R\right)$$
$$= \lim_{R\to-\infty}\left(-\tan^{-1}R\right) = -\left(-\frac{\pi}{2}\right) = \frac{\pi}{2}.$$

81. $\int_0^{\pi/2} \cot\theta\, d\theta$

SOLUTION

$$\int_0^{\pi/2} \cot\theta\, d\theta = \lim_{R\to 0+} \int_R^{\pi/2} \cot\theta\, d\theta = \lim_{R\to 0+} \ln|\sin\theta|\Big|_R^{\pi/2} = \lim_{R\to 0+} \left(\ln\left(\sin\frac{\pi}{2}\right) - \ln(\sin R)\right)$$

$$= \lim_{R\to 0+} (\ln 1 - \ln(\sin R)) = \lim_{R\to 0+} \ln\left(\frac{1}{\sin R}\right) = \infty.$$

We conclude that the improper integral diverges.

83. $\int_0^{\infty} (5+x)^{-1/3}\, dx$

SOLUTION

$$\int_0^{\infty} (5+x)^{-1/3}\, dx = \lim_{R\to\infty} \int_0^R (5+x)^{-1/3}\, dx = \lim_{R\to\infty} \frac{3}{2}(5+x)^{2/3}\Big|_0^R$$

$$= \lim_{R\to\infty} \left(\frac{3}{2}(5+R)^{2/3} - \frac{3}{2}5^{2/3}\right) = \infty.$$

We conclude that the improper integral diverges.

In Exercises 85–90, use the Comparison Test to determine whether the improper integral converges or diverges.

85. $\int_8^{\infty} \frac{dx}{x^2-4}$

SOLUTION For $x \geq 8$, $\frac{1}{2}x^2 \geq 4$, so that

$$-\frac{1}{2}x^2 \leq -4$$

$$\frac{1}{2}x^2 \leq x^2 - 4$$

and

$$\frac{1}{x^2-4} \leq \frac{2}{x^2}.$$

Now, $\int_1^{\infty} \frac{dx}{x^2}$ converges, so $\int_8^{\infty} \frac{2}{x^2}\, dx$ also converges. Therefore, by the comparison test,

$$\int_8^{\infty} \frac{dx}{x^2-4} \text{ converges.}$$

87. $\int_3^{\infty} \frac{dx}{x^4+\cos^2 x}$

SOLUTION For $x \geq 1$, we have

$$\frac{1}{x^4+\cos^2 x} \leq \frac{1}{x^4}.$$

Since $\int_1^{\infty} \frac{dx}{x^4}$ converges, the Comparison Test guarantees that $\int_1^{\infty} \frac{dx}{x^4+\cos^2 x}$ also converges. The integral $\int_1^{3} \frac{dx}{x^4+\cos^2 x}$ has a finite value (notice that $x^4+\cos^2 x \neq 0$) hence we conclude that the integral $\int_3^{\infty} \frac{dx}{x^4+\cos^2 x}$ also converges.

89. $\displaystyle\int_0^1 \frac{dx}{x^{1/3}+x^{2/3}}$

SOLUTION For $0 \le x \le 1$,

$$x^{1/3}+x^{2/3} \ge x^{1/3} \quad \text{so} \quad \frac{1}{x^{1/3}+x^{2/3}} \le \frac{1}{x^{1/3}}.$$

Now, $\displaystyle\int_0^1 x^{-1/3}\,dx$ converges. Therefore, by the Comparison Test, the improper integral $\displaystyle\int_0^1 \frac{dx}{x^{1/3}+x^{2/3}}$ also converges.

91. Calculate the volume of the infinite solid obtained by rotating the region under $y=(x^2+1)^{-2}$ for $0 \le x < \infty$ about the y-axis.

SOLUTION Using the Shell Method, the volume of the infinite solid obtained by rotating the region under the graph of $y=\left(x^2+1\right)^{-2}$ over the interval $[0,\infty)$ about the y-axis is

$$V = 2\pi \int_0^\infty \frac{x}{(x^2+1)^2}\,dx.$$

Now,

$$\int_0^\infty \frac{x}{(x^2+1)^2}\,dx = \lim_{R\to\infty}\int_0^R \frac{x\,dx}{(x^2+1)^2}$$

We substitute $t=x^2+1$, $dt=2x\,dx$. The new limits of integration are $t=1$ and $t=R^2+1$. Thus,

$$\int_0^R \frac{x\,dx}{(x^2+1)^2} = \int_1^{R^2+1} \frac{\frac{1}{2}\,dt}{t^2} = -\frac{1}{2t}\bigg|_1^{R^2+1} = \frac{1}{2}\left(1-\frac{1}{R^2+1}\right).$$

Taking the limit as $R\to\infty$ yields:

$$\int_0^\infty \frac{x\,dx}{(x^2+1)^2} = \lim_{R\to\infty}\frac{1}{2}\left(1-\frac{1}{R^2+1}\right) = \frac{1}{2}(1-0) = \frac{1}{2}.$$

Therefore,

$$V = 2\pi\cdot\frac{1}{2} = \pi.$$

93. Show that $\int_0^\infty x^n e^{-x^2}\,dx$ converges for all $n>0$. *Hint:* First observe that $x^n e^{-x^2} < x^n e^{-x}$ for $x>1$. Then show that $x^n e^{-x} < x^{-2}$ for x sufficiently large.

SOLUTION For $x>1$, $x^2>x$; hence $e^{x^2}>e^x$, and $0<e^{-x^2}<e^{-x}$. Therefore, for $x>1$ the following inequality holds:

$$x^{n+2}e^{-x^2} < x^{n+2}e^{-x}.$$

Now, using L'Hôpital's Rule $n+2$ times, we find

$$\lim_{x\to\infty} x^{n+2}e^{-x} = \lim_{x\to\infty}\frac{x^{n+2}}{e^x} = \lim_{x\to\infty}\frac{(n+2)x^{n+1}}{e^x} = \lim_{x\to\infty}\frac{(n+2)(n+1)x^n}{e^x}$$
$$= \cdots = \lim_{x\to\infty}\frac{(n+2)!}{e^x} = 0.$$

Therefore,

$$\lim_{x\to\infty} x^{n+2}e^{-x^2} = 0$$

by the Squeeze Theorem, and there exists a number $R>1$ such that, for all $x>R$:

$$x^{n+2}e^{-x^2} < 1 \quad \text{or} \quad x^n e^{-x^2} < x^{-2}.$$

Finally, write

$$\int_0^\infty x^n e^{-x^2}\,dx = \int_0^R x^n e^{-x^2}\,dx + \int_R^\infty x^n e^{-x^2}\,dx.$$

The first integral on the right-hand side has finite value since the integrand is a continuous function. The second integral converges since on the interval of integration, $x^n e^{-x^2} < x^{-2}$ and we know that $\int_R^\infty x^{-2}\,dx = \int_R^\infty \frac{dx}{x^2}$ converges. We conclude that the integral $\int_0^\infty x^n e^{-x^2}\,dx$ converges.

95. Compute the Laplace transform $Lf(s)$ of the function $f(x) = x^2 e^{\alpha x}$ for $s > \alpha$.

SOLUTION The Laplace transform is the following integral:

$$L\left(x^2 e^{\alpha x}\right)(s) = \int_0^\infty x^2 e^{\alpha x} e^{-sx}\,dx = \int_0^\infty x^2 e^{(\alpha-s)x}\,dx = \lim_{R\to\infty}\int_0^R x^2 e^{(\alpha-s)x}\,dx.$$

We compute the definite integral using Integration by Parts with $u = x^2$, $v' = e^{(\alpha-s)x}$. Then $u' = 2x$, $v = \frac{1}{\alpha-s}e^{(\alpha-s)x}$ and

$$\int_0^R x^2 e^{(\alpha-s)x}\,dx = \frac{1}{\alpha-s}x^2 e^{(\alpha-s)x}\Big|_{x=0}^R - \int_0^R 2x\cdot\frac{1}{\alpha-s}e^{(\alpha-s)x}\,dx$$
$$= \frac{1}{\alpha-s}R^2 e^{(\alpha-s)R} - \frac{2}{\alpha-s}\int_0^R x e^{(\alpha-s)x}\,dx.$$

We compute the resulting integral using Integration by Parts again, this time with $u = x$ and $v' = e^{(\alpha-s)x}$. Then $u' = 1$, $v = \frac{1}{\alpha-s}e^{(\alpha-s)x}$ and

$$\int_0^R x e^{(\alpha-s)x}\,dx = x\cdot\frac{1}{\alpha-s}e^{(\alpha-s)x}\Big|_{x=0}^R - \frac{1}{\alpha-s}\int_0^R e^{(\alpha-s)x}\,dx = \left(\frac{x}{\alpha-s}e^{(\alpha-s)x} - \frac{1}{(\alpha-s)^2}e^{(\alpha-s)x}\right)\Big|_{x=0}^R$$
$$= \frac{R}{\alpha-s}e^{(\alpha-s)R} - \frac{1}{(\alpha-s)^2}\left(e^{(\alpha-s)R} - e^0\right) = \frac{1}{(\alpha-s)^2} - \frac{1}{(\alpha-s)^2}e^{(\alpha-s)R} + \frac{R}{\alpha-s}e^{(\alpha-s)R}.$$

Thus,

$$\int_0^R x^2 e^{(\alpha-s)x}\,dx = \frac{1}{\alpha-s}R^2 e^{(\alpha-s)R} - \frac{2}{\alpha-s}\left(\frac{1}{(\alpha-s)^2} - \frac{1}{(\alpha-s)^2}e^{(\alpha-s)R} + \frac{R}{\alpha-s}e^{(\alpha-s)R}\right)$$
$$= \frac{1}{\alpha-s}R^2 e^{(\alpha-s)R} - \frac{2}{(\alpha-s)^3} + \frac{2}{(\alpha-s)^3}e^{(\alpha-s)R} - \frac{2R}{(\alpha-s)^2}e^{(\alpha-s)R},$$

and

$$L\left(x^2 e^{\alpha x}\right)(s) = \frac{2}{(s-\alpha)^3} - \frac{1}{s-\alpha}\lim_{R\to\infty} R^2 e^{-(s-\alpha)R} - \frac{2}{(s-\alpha)^3}\lim_{R\to\infty} e^{-(s-\alpha)R} - \frac{2}{(s-\alpha)^2}\lim_{R\to\infty} Re^{-(s-\alpha)R}.$$

Now, since $s > \alpha$, $\lim_{R\to\infty} e^{-(s-\alpha)R} = 0$. We use L'Hôpital's Rule to compute the other two limits:

$$\lim_{R\to\infty} Re^{-(s-\alpha)R} = \lim_{R\to\infty}\frac{R}{e^{(s-\alpha)R}} = \lim_{R\to\infty}\frac{1}{(s-\alpha)e^{(s-\alpha)R}} = 0;$$

$$\lim_{R\to\infty} R^2 e^{-(s-\alpha)R} = \lim_{R\to\infty}\frac{R^2}{e^{(s-\alpha)R}} = \lim_{R\to\infty}\frac{2R}{(s-\alpha)e^{(s-\alpha)R}} = \lim_{R\to\infty}\frac{2}{(s-\alpha)^2 e^{(s-\alpha)R}} = 0.$$

Finally,

$$L\left(x^2 e^{\alpha x}\right)(s) = \frac{2}{(s-\alpha)^3} - 0 - 0 - 0 = \frac{2}{(s-\alpha)^3}.$$

97. State whether the approximation M_N or T_N is larger or smaller than the integral.

(a) $\int_0^\pi \sin x\,dx$

(b) $\int_\pi^{2\pi} \sin x\,dx$

(c) $\int_1^8 \frac{dx}{x^2}$

(d) $\int_2^5 \ln x\,dx$

SOLUTION

(a) Because $f(x) = \sin x$ is concave down on the interval $[0, \pi]$,

$$T_N \le \int_0^\pi \sin x\,dx \le M_N;$$

that is, T_N is smaller and M_N is larger than the integral.

(b) On the interval $[\pi, 2\pi]$, the function $f(x) = \sin x$ is concave up, therefore

$$M_N \le \int_{\pi}^{2\pi} \sin x\, dx \le T_N;$$

that is, M_N is smaller and T_N is larger than the integral.

(c) The function $f(x) = \frac{1}{x^2}$ is concave up on the interval $[1, 8]$; therefore,

$$M_N \le \int_1^8 \frac{dx}{x^2} \le T_N;$$

that is, M_N is smaller and T_N is larger than the integral.

(d) The integrand $y = \ln x$ is concave down on the interval $[2, 5]$; hence,

$$T_N \le \int_2^5 \ln x\, dx \le M_N;$$

that is, T_N is smaller and M_N is larger than the integral.

In Exercises 99–104, compute the given approximation to the integral.

99. $\int_0^1 e^{-x^2}\, dx, \quad M_5$

SOLUTION Divide the interval $[0, 1]$ into 5 subintervals of length $\Delta x = \frac{1-0}{5} = \frac{1}{5}$, with midpoints $c_1 = \frac{1}{10}$, $c_2 = \frac{3}{10}$, $c_3 = \frac{1}{2}$, $c_4 = \frac{7}{10}$, and $c_5 = \frac{9}{10}$. Then

$$M_5 = \Delta x \left[f\left(\frac{1}{10}\right) + f\left(\frac{3}{10}\right) + f\left(\frac{1}{2}\right) + f\left(\frac{7}{10}\right) + f\left(\frac{9}{10}\right) \right]$$
$$= \frac{1}{5} \left[e^{-(1/10)^2} + e^{-(3/10)^2} + e^{-(1/2)^2} + e^{-(7/10)^2} + e^{-(9/10)^2} \right] = 0.748053.$$

101. $\int_{\pi/4}^{\pi/2} \sqrt{\sin\theta}\, d\theta, \quad M_4$

SOLUTION Divide the interval $\left[\frac{\pi}{4}, \frac{\pi}{2}\right]$ into 4 subintervals of length $\Delta x = \frac{\frac{\pi}{2} - \frac{\pi}{4}}{4} = \frac{\pi}{16}$ with midpoints $\frac{9\pi}{32}$, $\frac{11\pi}{32}$, $\frac{13\pi}{32}$, and $\frac{15\pi}{32}$. Then

$$M_4 = \Delta x \left(f\left(\frac{9\pi}{32}\right) + f\left(\frac{11\pi}{32}\right) + f\left(\frac{13\pi}{32}\right) + f\left(\frac{15\pi}{32}\right) \right)$$
$$= \frac{\pi}{16} \left(\sqrt{\sin\frac{9\pi}{32}} + \sqrt{\sin\frac{11\pi}{32}} + \sqrt{\sin\frac{13\pi}{32}} + \sqrt{\sin\frac{15\pi}{32}} \right) = 0.744978.$$

103. $\int_0^1 e^{-x^2}\, dx, \quad S_4$

SOLUTION Divide the interval $[0, 1]$ into 4 subintervals of length $\Delta x = \frac{1}{4}$ with endpoints $0, \frac{1}{4}, \frac{1}{2}, \frac{3}{4}, 1$. Then

$$S_6 = \frac{1}{3}\Delta x \left(f(0) + 4f\left(\frac{1}{4}\right) + 2f\left(\frac{1}{2}\right) + 4f\left(\frac{3}{4}\right) + f(1) \right)$$
$$= \frac{1}{3} \cdot \frac{1}{4} \left(e^{-0^2} + 4e^{-(1/4)^2} + 2e^{-(1/2)^2} + 4e^{-(3/4)^2} + e^{-1^2} \right) = 0.746855.$$

105. The following table gives the area $A(h)$ of a horizontal cross section of a pond at depth h. Use the Trapezoidal Rule to estimate the volume V of the pond (Figure 1).

h (ft)	$A(h)$ (acres)	h (ft)	$A(h)$ (acres)
0	2.8	10	0.8
2	2.4	12	0.6
4	1.8	14	0.2
6	1.5	16	0.1
8	1.2	18	0

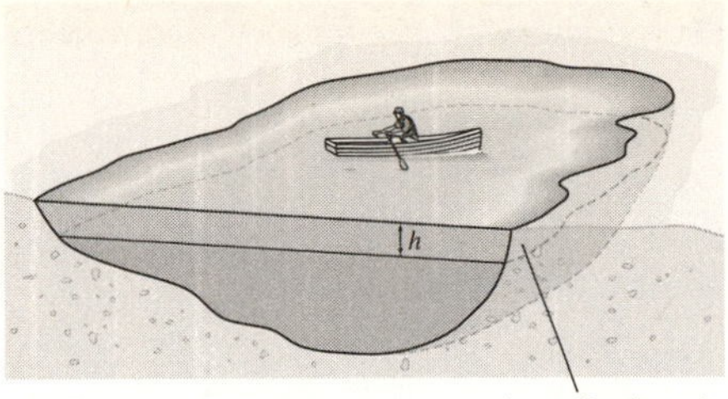

FIGURE 1

SOLUTION The volume of the pond is the following integral:

$$V = \int_0^{18} A(h)dh$$

We approximate the integral using the trapezoidal approximation T_9. The interval of depth [0, 18] is divided to 9 subintervals of length $\Delta x = 2$ with endpoints 0, 2, 4, 6, 8, 10, 12, 14, 16, 18. Thus,

$$\begin{aligned} V \approx T_9 &= \frac{1}{2} \cdot 2(2.8 + 2 \cdot 2.4 + 2 \cdot 1.8 + 2 \cdot 1.5 + 2 \cdot 1.2 + 2 \cdot 0.8 + 2 \cdot 0.6 + 2 \cdot 0.2 + 2 \cdot 0.1 + 0) \\ &= 20 \text{ acre} \cdot \text{ft} = 871{,}200 \text{ ft}^3, \end{aligned}$$

where we have used the fact that 1 acre = 43,560 ft^2.

107. Find a bound for the error $\left|M_{16} - \int_1^3 x^3\,dx\right|$.

SOLUTION The Error Bound for the Midpoint Rule states that

$$\left|M_N - \int_a^b f(x)\,dx\right| \le \frac{K_2(b-a)^3}{24N^2},$$

where K_2 is a number such that $|f''(x)| \le K_2$ for all $x \in [1, 3]$. Here $b - a = 3 - 1 = 2$ and $N = 16$. Therefore,

$$\left|M_{16} - \int_1^3 x^3\,dx\right| \le \frac{K_2 \cdot 2^3}{24 \cdot 16^2} = \frac{K_2}{768}.$$

To find K_2, we differentiate $f(x) = x^3$ twice:

$$f'(x) = 3x^2 \qquad \text{and} \qquad f''(x) = 6x.$$

On the interval [1, 3] we have $|f''(x)| = 6x \le 6 \cdot 3 = 18$; hence, we may take $K_2 = 18$. Thus,

$$\left|M_{16} - \int_1^3 x^3\,dx\right| \le \frac{18}{768} = \frac{3}{128} = 0.0234375.$$

109. Find a value of N such that

$$\left|M_N - \int_0^{\pi/4} \tan x\,dx\right| \le 10^{-4}$$

SOLUTION To use the Error Bound we must find the second derivative of $f(x) = \tan x$. We differentiate f twice to obtain:

$$\begin{aligned} f'(x) &= \sec^2 x \\ f''(x) &= 2\sec x \tan x = \frac{2\sin x}{\cos^2 x} \end{aligned}$$

For $0 \le x \le \frac{\pi}{4}$, we have $\sin x \le \sin\frac{\pi}{4} = \frac{1}{\sqrt{2}}$ and $\cos x \ge \frac{1}{\sqrt{2}}$ or $\cos^2 x \ge \frac{1}{2}$. Therefore, for $0 \le x \le \frac{\pi}{4}$ we have:

$$f''(x) = \frac{2\sin x}{\cos^2 x} \le \frac{2 \cdot \frac{1}{\sqrt{2}}}{\frac{1}{2}} = 2\sqrt{2}.$$

Using the Error Bound with $b = \frac{\pi}{4}$, $a = 0$ and $K_2 = 2\sqrt{2}$ we have:

$$\left| M_N - \int_0^{\pi/4} \tan x\, dx \right| \le \frac{2\sqrt{2} \cdot \left(\frac{\pi}{4} - 0\right)^3}{24N^2} = \frac{\pi^3\sqrt{2}}{768N^2}.$$

We must choose a value of N such that:

$$\frac{\pi^3\sqrt{2}}{768N^2} \le 10^{-4}$$

$$N^2 \ge \frac{10^4 \cdot \sqrt{2}\pi^3}{768}$$

$$N \ge 23.9$$

The smallest integer that is needed to obtain the required precision is $N = 24$.

8 FURTHER APPLICATIONS OF THE INTEGRAL AND TAYLOR POLYNOMIALS

8.1 Arc Length and Surface Area

Preliminary Questions

1. Which integral represents the length of the curve $y = \cos x$ between 0 and π?

$$\int_0^{\pi} \sqrt{1+\cos^2 x}\, dx, \qquad \int_0^{\pi} \sqrt{1+\sin^2 x}\, dx$$

SOLUTION Let $y = \cos x$. Then $y' = -\sin x$, and $1 + (y')^2 = 1 + \sin^2 x$. Thus, the length of the curve $y = \cos x$ between 0 and π is

$$\int_0^{\pi} \sqrt{1+\sin^2 x}\, dx.$$

2. Use the formula for arc length to show that for any constant C, the graphs $y = f(x)$ and $y = f(x) + C$ have the same length over every interval $[a, b]$. Explain geometrically.

SOLUTION The graph of $y = f(x) + C$ is a vertical translation of the graph of $y = f(x)$; hence, the two graphs should have the same arc length. We can explicitly establish this as follows:

$$\text{length of } y = f(x) + C = \int_a^b \sqrt{1 + \left[\frac{d}{dx}(f(x) + C)\right]^2}\, dx = \int_a^b \sqrt{1 + [f'(x)]^2}\, dx = \text{length of } y = f(x).$$

3. Use the formula for arc length to show that the length of a graph over $[1, 4]$ cannot be less than 3.

SOLUTION Note that $f'(x)^2 \geq 0$, so that $\sqrt{1 + [f'(x)]^2} \geq \sqrt{1} = 1$. Then the arc length of the graph of $f(x)$ on $[1, 4]$ is

$$\int_1^4 \sqrt{1 + [f'(x)]^2}\, dx \geq \int_1^4 1\, dx = 3$$

Exercises

1. Express the arc length of the curve $y = x^4$ between $x = 2$ and $x = 6$ as an integral (but do not evaluate).

SOLUTION Let $y = x^4$. Then $y' = 4x^3$ and

$$s = \int_2^6 \sqrt{1 + (4x^3)^2}\, dx = \int_2^6 \sqrt{1 + 16x^6}\, dx.$$

3. Find the arc length of $y = \frac{1}{12}x^3 + x^{-1}$ for $1 \leq x \leq 2$. *Hint:* Show that $1 + (y')^2 = \left(\frac{1}{4}x^2 + x^{-2}\right)^2$.

SOLUTION Let $y = \dfrac{1}{12}x^3 + x^{-1}$. Then $y' = \dfrac{x^2}{4} x^{-2}$, and

$$(y')^2 + 1 = \left(\frac{x^2}{4} - x^{-2}\right)^2 + 1 = \frac{x^4}{16} - \frac{1}{2} + x^{-4} + 1 = \frac{x^4}{16} + \frac{1}{2} + x^{-4} = \left(\frac{x^2}{4} + x^{-2}\right)^2.$$

Thus,

$$s = \int_1^2 \sqrt{1 + (y')^2}\, dx = \int_1^2 \sqrt{\left(\frac{x^2}{4} + \frac{1}{x^2}\right)^2}\, dx = \int_1^2 \left|\frac{x^2}{4} + \frac{1}{x^2}\right| dx$$

$$= \int_1^2 \left(\frac{x^2}{4} + \frac{1}{x^2}\right) dx \quad \text{since} \quad \frac{x^2}{4} + \frac{1}{x^2} > 0$$

$$= \left(\frac{x^3}{12} - \frac{1}{x}\right)\Big|_1^2 = \frac{13}{12}.$$

In Exercises 5–10, calculate the arc length over the given interval.

5. $y = 3x + 1, \quad [0, 3]$

SOLUTION Let $y = 3x + 1$. Then $y' = 3$, and $s = \int_0^3 \sqrt{1+9}\, dx = 3\sqrt{10}$.

7. $y = x^{3/2}, \quad [1, 2]$

SOLUTION Let $y = x^{3/2}$. Then $y' = \frac{3}{2}x^{1/2}$, and

$$s = \int_1^2 \sqrt{1 + \frac{9}{4}x}\, dx = \frac{8}{27}\left(1 + \frac{9}{4}x\right)^{3/2}\Big|_1^2 = \frac{8}{27}\left(\left(\frac{11}{2}\right)^{3/2} - \left(\frac{13}{4}\right)^{3/2}\right) = \frac{1}{27}\left(22\sqrt{22} - 13\sqrt{13}\right).$$

9. $y = \frac{1}{4}x^2 - \frac{1}{2}\ln x, \quad [1, 2e]$

SOLUTION Let $y = \frac{1}{4}x^2 - \frac{1}{2}\ln x$. Then

$$y' = \frac{x}{2} - \frac{1}{2x},$$

and

$$1 + (y')^2 = 1 + \left(\frac{x}{2} - \frac{1}{2x}\right)^2 = \frac{x^2}{4} + \frac{1}{2} + \frac{1}{4x^2} = \left(\frac{x}{2} + \frac{1}{2x}\right)^2.$$

Hence,

$$s = \int_1^{2e} \sqrt{1 + (y')^2}\, dx = \int_1^{2e} \sqrt{\left(\frac{x}{2} + \frac{1}{2x}\right)^2}\, dx = \int_1^{2e} \left|\frac{x}{2} + \frac{1}{2x}\right| dx$$

$$= \int_1^{2e} \left(\frac{x}{2} + \frac{1}{2x}\right) dx \quad \text{since} \quad \frac{x}{2} + \frac{1}{2x} > 0 \text{ on } [1, 2e]$$

$$= \left(\frac{x^2}{4} + \frac{1}{2}\ln x\right)\Big|_1^{2e} = e^2 + \frac{\ln 2}{2} + \frac{1}{4}.$$

In Exercises 11–14, approximate the arc length of the curve over the interval using the Trapezoidal Rule T_N, the Midpoint Rule M_N, or Simpson's Rule S_N as indicated.

11. $y = \frac{1}{4}x^4, \quad [1, 2], \quad T_5$

SOLUTION Let $y = \frac{1}{4}x^4$. Then

$$1 + (y')^2 = 1 + (x^3)^2 = 1 + x^6.$$

Therefore, the arc length over $[1, 2]$ is

$$\int_1^2 \sqrt{1 + x^6}\, dx.$$

Now, let $f(x) = \sqrt{1 + x^6}$. With $n = 5$,

$$\Delta x = \frac{2-1}{5} = \frac{1}{5} \quad \text{and} \quad \{x_i\}_{i=0}^5 = \left\{1, \frac{6}{5}, \frac{7}{5}, \frac{8}{5}, \frac{9}{5}, 2\right\}.$$

Using the Trapezoidal Rule,

$$\int_1^2 \sqrt{1 + x^6}\, dx \approx \frac{\Delta x}{2}\left[f(x_0) + 2\sum_{i=1}^4 f(x_i) + f(x_5)\right] = 3.957736.$$

The arc length is approximately 3.957736 units.

13. $y = x^{-1}$, $[1, 2]$, S_8

SOLUTION Let $y = x^{-1}$. Then $y' = -x^{-2}$ and

$$1 + (y')^2 = 1 + \frac{1}{x^4}.$$

Therefore, the arc length over $[1, 2]$ is

$$\int_1^2 \sqrt{1 + \frac{1}{x^4}}\, dx.$$

Now, let $f(x) = \sqrt{1 + \frac{1}{x^4}}$. With $n = 8$,

$$\Delta x = \frac{2-1}{8} = \frac{1}{8} \quad \text{and} \quad \{x_i\}_{i=0}^{8} = \left\{1, \frac{9}{8}, \frac{5}{4}, \frac{11}{8}, \frac{3}{2}, \frac{13}{8}, \frac{7}{4}, \frac{15}{8}, 2\right\}.$$

Using Simpson's Rule,

$$\int_1^2 \sqrt{1 + \frac{1}{x^4}}\, dx \approx \frac{\Delta x}{3}\left[f(x_0) + 4\sum_{i=1}^{4} f(x_{2i-1}) + 2\sum_{i=1}^{3} f(x_{2i}) + f(x_8)\right] = 1.132123.$$

The arc length is approximately 1.132123 units.

15. Calculate the length of the astroid $x^{2/3} + y^{2/3} = 1$ (Figure 11).

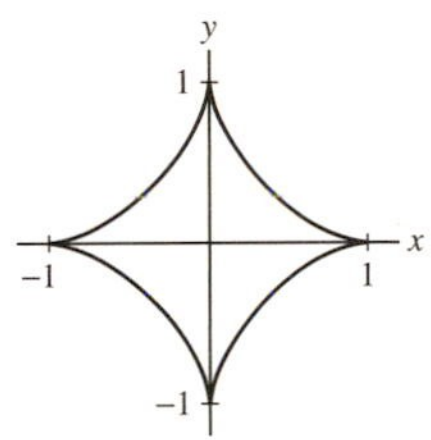

FIGURE 11 Graph of $x^{2/3} + y^{2/3} = 1$.

SOLUTION We will calculate the arc length of the portion of the asteroid in the first quadrant and then multiply by 4. By implicit differentiation

$$\frac{2}{3}x^{-1/3} + \frac{2}{3}y^{-1/3}y' = 0,$$

so

$$y' = -\frac{x^{-1/3}}{y^{-1/3}} = -\frac{y^{1/3}}{x^{1/3}}.$$

Thus

$$1 + (y')^2 = 1 + \frac{y^{2/3}}{x^{2/3}} = \frac{x^{2/3} + y^{2/3}}{x^{2/3}} = \frac{1}{x^{2/3}},$$

and

$$s = \int_0^1 \frac{1}{x^{1/3}}\, dx = \frac{3}{2}.$$

The total arc length is therefore $4 \cdot \frac{3}{2} = 6$.

17. Let $a, r > 0$. Show that the arc length of the curve $x^r + y^r = a^r$ for $0 \le x \le a$ is proportional to a.

SOLUTION Using implicit differentiation, we find $y' = -(x/y)^{r-1}$ and

$$1 + (y')^2 = 1 + (x/y)^{2r-2} = \frac{x^{2r-1} + y^{2r-2}}{y^{2r-2}} = \frac{x^{2r-2} + (a^r - x^r)^{2-2/r}}{(a^r - x^r)^{2-2/r}}.$$

The arc length is then

$$s = \int_0^a \sqrt{\frac{x^{2r-2} + (a^r - x^r)^{2-2/r}}{(a^r - x^r)^{2-2/r}}}\, dx.$$

Using the substitution $x = au$, we obtain

$$s = a\int_0^1 \sqrt{\frac{u^{2r-2} + (1-u^r)^{2-2/r}}{(1-u^r)^{2-2/r}}}\,du,$$

where the integral is independent of a.

19. Find the value of a such that the arc length of the *catenary* $y = \cosh x$ for $-a \le x \le a$ equals 10.

SOLUTION Let $y = \cosh x$. Then $y' = \sinh x$ and

$$1 + (y')^2 = 1 + \sinh^2 x = \cosh^2 x.$$

Thus,

$$s = \int_{-a}^{a} \cosh x\,dx = \sinh(a) - \sinh(-a) = 2\sinh a.$$

Setting this expression equal to 10 and solving for a yields $a = \sinh^{-1}(5) = \ln(5 + \sqrt{26})$.

21. Show that the circumference of the unit circle is equal to

$$2\int_{-1}^{1} \frac{dx}{\sqrt{1-x^2}} \quad \text{(an improper integral)}$$

Evaluate, thus verifying that the circumference is 2π.

SOLUTION Note the circumference of the unit circle is twice the arc length of the upper half of the curve defined by $x^2 + y^2 = 1$. Thus, let $y = \sqrt{1-x^2}$. Then

$$y' = -\frac{x}{\sqrt{1-x^2}} \quad \text{and} \quad 1 + (y')^2 = 1 + \frac{x^2}{1-x^2} = \frac{1}{1-x^2}.$$

Finally, the circumference of the unit circle is

$$2\int_{-1}^{1} \frac{dx}{\sqrt{1-x^2}} = 2\sin^{-1} x\Big|_{-1}^{1} = \pi - (-\pi) = 2\pi.$$

23. Calculate the arc length of $y = x^2$ over $[0, a]$. *Hint:* Use trigonometric substitution. Evaluate for $a = 1$.

SOLUTION Let $y = x^2$. Then $y' = 2x$ and

$$s = \int_0^a \sqrt{1 + 4x^2}\,dx.$$

Using the substitution $2x = \tan\theta$, $2\,dx = \sec^2\theta\,d\theta$, we find

$$s = \frac{1}{2}\int_{x=0}^{x=a} \sec^3\theta\,d\theta.$$

Next, using a reduction formula for the integral of $\sec^3\theta$, we see that

$$s = \left(\frac{1}{4}\sec\theta\tan\theta + \frac{1}{4}\ln|\sec\theta + \tan\theta|\right)\Big|_{x=0}^{x=a} = \left(\frac{1}{2}x\sqrt{1+4x^2} + \frac{1}{4}\ln|\sqrt{1+4x^2} + 2x|\right)\Big|_0^a$$

$$= \frac{a}{2}\sqrt{1+4a^2} + \frac{1}{4}\ln|\sqrt{1+4a^2} + 2a|$$

Thus, when $a = 1$,

$$s = \frac{1}{2}\sqrt{5} + \frac{1}{4}\ln(\sqrt{5} + 2) \approx 1.478943.$$

25. Find the arc length of $y = e^x$ over $[0, a]$. *Hint:* Try the substitution $u = \sqrt{1 + e^{2x}}$ followed by partial fractions.

SOLUTION Let $y = e^x$. Then $1 + (y')^2 = 1 + e^{2x}$, and the arc length over $[0, a]$ is

$$\int_0^a \sqrt{1 + e^{2x}}\,dx.$$

Now, let $u = \sqrt{1 + e^{2x}}$. Then

$$du = \frac{1}{2} \cdot \frac{2e^{2x}}{\sqrt{1+e^{2x}}}\,dx = \frac{u^2 - 1}{u}\,dx$$

and the arc length is

$$\begin{aligned}
\int_0^a \sqrt{1+e^{2x}}\,dx &= \int_{x=0}^{x=a} u \cdot \frac{u}{u^2-1}\,du = \int_{x=0}^{x=a} \frac{u^2}{u^2-1}\,du = \int_{x=0}^{x=a} \left(1 + \frac{1}{u^2-1}\right) du \\
&= \int_{x=0}^{x=a} \left(1 + \frac{1}{2}\frac{1}{u-1} - \frac{1}{2}\frac{1}{u+1}\right) du = \left(u + \frac{1}{2}\ln(u-1) - \frac{1}{2}\ln(u+1)\right)\Bigg|_{x=0}^{x=a} \\
&= \left[\sqrt{1+e^{2x}} + \frac{1}{2}\ln\left(\frac{\sqrt{1+e^{2x}}-1}{\sqrt{1+e^{2x}}+1}\right)\right]\Bigg|_0^a \\
&= \sqrt{1+e^{2a}} + \frac{1}{2}\ln\frac{\sqrt{1+e^{2a}}-1}{\sqrt{1+e^{2a}}+1} - \sqrt{2} + \frac{1}{2}\ln\frac{1+\sqrt{2}}{\sqrt{2}-1} \\
&= \sqrt{1+e^{2a}} + \frac{1}{2}\ln\frac{\sqrt{1+e^{2a}}-1}{\sqrt{1+e^{2a}}+1} - \sqrt{2} + \ln(1+\sqrt{2}).
\end{aligned}$$

27. Use Eq. (4) to compute the arc length of $y = \ln(\sin x)$ for $\frac{\pi}{4} \le x \le \frac{\pi}{2}$.

SOLUTION With $f(x) = \sin x$, Eq. (4) yields

$$s = \int_{\pi/4}^{\pi/2} \frac{\sqrt{\sin^2 x + \cos^2 x}}{\sin x}\,dx = \int_{\pi/4}^{\pi/2} \csc x\,dx = \ln(\csc x - \cot x)\Bigg|_{\pi/4}^{\pi/2}$$

$$= \ln 1 - \ln(\sqrt{2}-1) = \ln\frac{1}{\sqrt{2}-1} = \ln(\sqrt{2}+1).$$

29. Show that if $0 \le f'(x) \le 1$ for all x, then the arc length of $y = f(x)$ over $[a, b]$ is at most $\sqrt{2}(b-a)$. Show that for $f(x) = x$, the arc length equals $\sqrt{2}(b-a)$.

SOLUTION If $0 \le f'(x) \le 1$ for all x, then

$$s = \int_a^b \sqrt{1 + f'(x)^2}\,dx \le \int_a^b \sqrt{1+1}\,dx = \sqrt{2}(b-a).$$

If $f(x) = x$, then $f'(x) = 1$ and

$$s = \int_a^b \sqrt{1+1}\,dx = \sqrt{2}(b-a).$$

31. Approximate the arc length of one-quarter of the unit circle (which we know is $\frac{\pi}{2}$) by computing the length of the polygonal approximation with $N = 4$ segments (Figure 14).

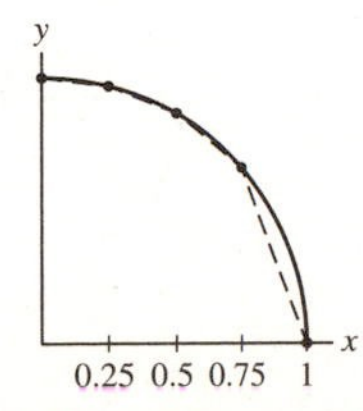

FIGURE 14 One-quarter of the unit circle

SOLUTION With $y = \sqrt{1-x^2}$, the five points along the curve are

$$P_0(0,1),\ P_1(1/4, \sqrt{15}/4),\ P_2(1/2, \sqrt{3}/2),\ P_3(3/4, \sqrt{7}/4),\ P_4(1,0)$$

Then

$$\overline{P_0P_1} = \sqrt{\frac{1}{16} + \left(\frac{4-\sqrt{15}}{4}\right)^2} \approx 0.252009$$

$$\overline{P_1P_2} = \sqrt{\frac{1}{16} + \left(\frac{2\sqrt{3}-\sqrt{15}}{4}\right)^2} \approx 0.270091$$

$$\overline{P_2P_3} = \sqrt{\frac{1}{16} + \left(\frac{2\sqrt{3}-\sqrt{7}}{4}\right)^2} \approx 0.323042$$

$$\overline{P_3P_4} = \sqrt{\frac{1}{16} + \frac{7}{16}} \approx 0.707108$$

and the total approximate distance is 1.552250 whereas $\pi/2 \approx 1.570796$.

In Exercises 33–40, compute the surface area of revolution about the x-axis over the interval.

33. $y = x$, $[0, 4]$

SOLUTION $1 + (y')^2 = 2$ so that

$$SA = 2\pi \int_0^4 x\sqrt{2}\,dx = 2\pi\sqrt{2}\frac{1}{2}x^2\Big|_0^4 = 16\pi\sqrt{2}$$

35. $y = x^3$, $[0, 2]$

SOLUTION $1 + (y')^2 = 1 + 9x^4$, so that

$$SA = 2\pi \int_0^2 x^3\sqrt{1+9x^4}\,dx = \frac{2\pi}{36}\int_0^2 36x^3\sqrt{1+9x^4}\,dx = \frac{\pi}{18}(1+9x^4)^{3/2}\Big|_0^2 = \frac{\pi}{18}\left(145^{3/2} - 1\right)$$

37. $y = (4 - x^{2/3})^{3/2}$, $[0, 8]$

SOLUTION Let $y = (4 - x^{2/3})^{3/2}$. Then

$$y' = -x^{-1/3}(4 - x^{2/3})^{1/2},$$

and

$$1 + (y')^2 = 1 + \frac{4 - x^{2/3}}{x^{2/3}} = \frac{4}{x^{2/3}}.$$

Therefore,

$$SA = 2\pi \int_0^8 (4 - x^{2/3})^{3/2}\left(\frac{2}{x^{1/3}}\right)dx.$$

Using the substitution $u = 4 - x^{2/3}$, $du = -\frac{2}{3}x^{-1/3}\,dx$, we find

$$SA = 2\pi \int_4^0 u^{3/2}(-3)\,du = 6\pi \int_0^4 u^{3/2}\,du = \frac{12}{5}\pi u^{5/2}\Big|_0^4 = \frac{384\pi}{5}.$$

39. $y = \frac{1}{4}x^2 - \frac{1}{2}\ln x$, $[1, e]$

SOLUTION We have $y' = \frac{x}{2} - \frac{1}{2x}$, and

$$1 + (y')^2 = 1 + \left(\frac{x}{2} - \frac{1}{2x}\right)^2 = 1 + \frac{x^2}{4} - \frac{1}{2} + \frac{1}{4x^2} = \frac{x^2}{4} + \frac{1}{2} + \frac{1}{4x^2} = \left(\frac{x}{2} + \frac{1}{2x}\right)^2.$$

Thus,

$$SA = 2\pi \int_1^e \left(\frac{x^2}{4} - \frac{\ln x}{2}\right)\left(\frac{x}{2} + \frac{1}{2x}\right)dx = 2\pi \int_1^e \frac{x^3}{8} + \frac{x}{8} - \frac{x\ln x}{4} - \frac{\ln x}{4x}\,dx$$

$$= 2\pi\left(\frac{x^4}{32} + \frac{x^2}{16} - \frac{x^2\ln x}{8} + \frac{x^2}{16} - \frac{(\ln x)^2}{8}\right)\Big|_1^e$$

$$= 2\pi\left(\frac{e^4}{32} + \frac{e^2}{16} - \frac{e^2}{8} + \frac{e^2}{16} - \frac{1}{8} - \left(\frac{1}{32} + \frac{1}{16} + 0 + \frac{1}{16} - 0\right)\right)$$

$$= 2\pi\left(\frac{e^4}{32} - \frac{1}{8} - \frac{1}{32} - \frac{1}{16} - \frac{1}{16}\right)$$

$$= \frac{\pi}{16}(e^4 - 9)$$

CAS *In Exercises 41–44, use a computer algebra system to find the approximate surface area of the solid generated by rotating the curve about the x-axis.*

41. $y = x^{-1}$, $[1, 3]$

SOLUTION

$$SA = 2\pi\int_1^3 \frac{1}{x}\sqrt{1 + \left(-\frac{1}{x^2}\right)^2}\,dx = 2\pi\int_1^3 \frac{1}{x}\sqrt{1 + \frac{1}{x^4}}\,dx \approx 7.603062807$$

using Maple.

43. $y = e^{-x^2/2}$, $[0, 2]$

SOLUTION

$$SA = 2\pi\int_0^2 e^{-x^2/2}\sqrt{1 + (-xe^{-x^2/2})^2}\,dx = 2\pi\int_0^2 e^{-x^2/2}\sqrt{1 + x^2e^{-x^2}}\,dx \approx 8.222695606$$

using Maple.

45. Find the area of the surface obtained by rotating $y = \cosh x$ over $[-\ln 2, \ln 2]$ around the x-axis.

SOLUTION Let $y = \cosh x$. Then $y' = \sinh x$, and

$$\sqrt{1 + (y')^2} = \sqrt{1 + \sinh^2 x} = \sqrt{\cosh^2 x} = \cosh x.$$

Therefore,

$$SA = 2\pi\int_{-\ln 2}^{\ln 2} \cosh^2 x\,dx = \pi\int_{-\ln 2}^{\ln 2} (1 + \cosh 2x)\,dx = \pi\left(x + \frac{1}{2}\sinh 2x\right)\Big|_{-\ln 2}^{\ln 2}$$

$$= \pi\left(\ln 2 + \frac{1}{2}\sinh(2\ln 2) + \ln 2 - \frac{1}{2}\sinh(-2\ln 2)\right) = 2\pi\ln 2 + \pi\sinh(2\ln 2).$$

We can simplify this answer by recognizing that

$$\sinh(2\ln 2) = \frac{e^{2\ln 2} - e^{-2\ln 2}}{2} = \frac{4 - \frac{1}{4}}{2} = \frac{15}{8}.$$

Thus,

$$SA = 2\pi\ln 2 + \frac{15\pi}{8}.$$

47. Find the surface area of the torus obtained by rotating the circle $x^2 + (y - b)^2 = a^2$ about the x-axis (Figure 17).

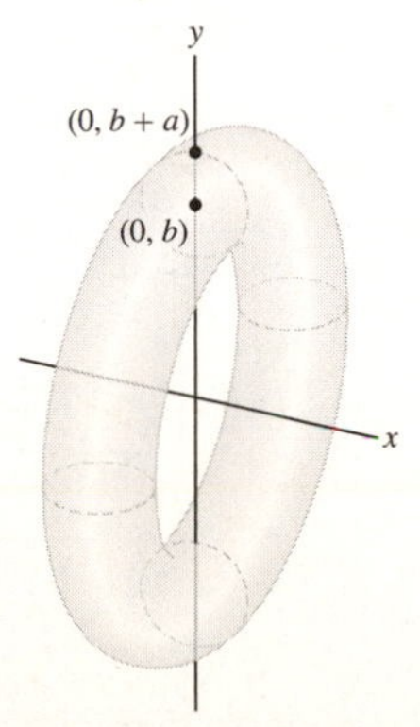

FIGURE 17 Torus obtained by rotating a circle about the x-axis.

SOLUTION $y = b + \sqrt{a^2 - x^2}$ gives the top half of the circle and $y = b - \sqrt{a^2 - x^2}$ gives the bottom half. Note that in each case,

$$1 + (y')^2 = 1 + \frac{x^2}{a^2 - x^2} = \frac{a^2}{a^2 - x^2}.$$

Rotating the two halves of the circle around the x-axis then yields

$$\begin{aligned} SA &= 2\pi \int_{-a}^{a} (b + \sqrt{a^2 - x^2}) \frac{a}{\sqrt{a^2 - x^2}}\, dx + 2\pi \int_{-a}^{a} (b - \sqrt{a^2 - x^2}) \frac{a}{\sqrt{a^2 - x^2}}\, dx \\ &= 2\pi \int_{-a}^{a} 2b \frac{a}{\sqrt{a^2 - x^2}}\, dx = 4\pi ba \int_{-a}^{a} \frac{1}{\sqrt{a^2 - x^2}}\, dx \\ &= 4\pi ba \cdot \sin^{-1}\left(\frac{x}{a}\right)\Big|_{-a}^{a} = 4\pi ba \left(\frac{\pi}{2} - \left(-\frac{\pi}{2}\right)\right) = 4\pi^2 ba. \end{aligned}$$

Further Insights and Challenges

49. Find the surface area of the ellipsoid obtained by rotating the ellipse $\left(\frac{x}{a}\right)^2 + \left(\frac{y}{b}\right)^2 = 1$ about the x-axis.

SOLUTION Taking advantage of symmetry, we can find the surface area of the ellipsoid by doubling the surface area obtained by rotating the portion of the ellipse in the first quadrant about the x-axis. The equation for the portion of the ellipse in the first quadrant is

$$y = \frac{b}{a}\sqrt{a^2 - x^2}.$$

Thus,

$$1 + (y')^2 = 1 + \frac{b^2 x^2}{a^2(a^2 - x^2)} = \frac{a^4 + (b^2 - a^2)x^2}{a^2(a^2 - x^2)},$$

and

$$SA = 4\pi \int_0^a \frac{b}{a}\sqrt{a^2 - x^2} \frac{\sqrt{a^4 + (b^2 - a^2)x^2}}{a\sqrt{a^2 - x^2}}\, dx = 4\pi b \int_0^a \sqrt{1 + \left(\frac{b^2 - a^2}{a^4}\right) x^2}\, dx.$$

We now consider two cases. If $b^2 > a^2$, then we make the substitution

$$\frac{\sqrt{b^2 - a^2}}{a^2} x = \tan\theta, \quad dx = \frac{a^2}{\sqrt{b^2 - a^2}} \sec^2\theta\, d\theta,$$

and find that

$$\begin{aligned} SA &= 4\pi b \frac{a^2}{\sqrt{b^2 - a^2}} \int_{x=0}^{x=a} \sec^3\theta\, d\theta = 2\pi b \frac{a^2}{\sqrt{b^2 - a^2}} (\sec\theta \tan\theta + \ln|\sec\theta + \tan\theta|)\Big|_{x=0}^{x=a} \\ &= \left(2\pi bx\sqrt{1 + \left(\frac{b^2 - a^2}{a^4}\right)x^2} + 2\pi b \frac{a^2}{\sqrt{b^2 - a^2}} \ln\left|\sqrt{1 + \left(\frac{b^2 - a^2}{a^4}\right)x^2} + \frac{\sqrt{b^2 - a^2}}{a^2}x\right|\right)\Big|_0^a \\ &= 2\pi b^2 + 2\pi b \frac{a^2}{\sqrt{b^2 - a^2}} \ln\left(\frac{b}{a} + \frac{\sqrt{b^2 - a^2}}{a}\right). \end{aligned}$$

On the other hand, if $a^2 > b^2$, then we make the substitution

$$\frac{\sqrt{a^2 - b^2}}{a^2} x = \sin\theta, \quad dx = \frac{a^2}{\sqrt{a^2 - b^2}} \cos\theta\, d\theta,$$

and find that

$$\begin{aligned} SA &= 4\pi b \frac{a^2}{\sqrt{a^2 - b^2}} \int_{x=0}^{x=a} \cos^2\theta\, d\theta = 2\pi b \frac{a^2}{\sqrt{a^2 - b^2}} (\theta + \sin\theta\cos\theta)\Big|_{x=0}^{x=a} \\ &= \left[2\pi bx\sqrt{1 - \left(\frac{a^2 - b^2}{a^4}\right)x^2} + 2\pi b \frac{a^2}{\sqrt{a^2 - b^2}} \sin^{-1}\left(\frac{\sqrt{a^2 - b^2}}{a^2}x\right)\right]\Big|_0^a \\ &= 2\pi b^2 + 2\pi b \frac{a^2}{\sqrt{a^2 - b^2}} \sin^{-1}\left(\frac{\sqrt{a^2 - b^2}}{a}\right). \end{aligned}$$

Observe that in both cases, as a approaches b, the value of the surface area of the ellipsoid approaches $4\pi b^2$, the surface area of a sphere of radius b.

51. CAS Let L be the arc length of the upper half of the ellipse with equation

$$y = \frac{b}{a}\sqrt{a^2 - x^2}$$

(Figure 18) and let $\eta = \sqrt{1 - (b^2/a^2)}$. Use substitution to show that

$$L = a\int_{-\pi/2}^{\pi/2} \sqrt{1 - \eta^2 \sin^2\theta}\, d\theta$$

Use a computer algebra system to approximate L for $a = 2$, $b = 1$.

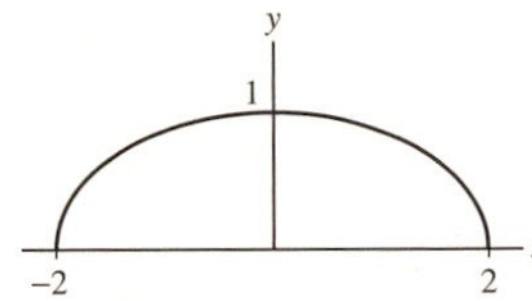

FIGURE 18 Graph of the ellipse $y = \frac{1}{2}\sqrt{4 - x^2}$.

SOLUTION Let $y = \frac{b}{a}\sqrt{a^2 - x^2}$. Then

$$1 + (y')^2 = \frac{b^2x^2 + a^2(a^2 - x^2)}{a^2(a^2 - x^2)}$$

and

$$s = \int_{-a}^{a} \sqrt{\frac{b^2x^2 + a^2(a^2 - x^2)}{a^2(a^2 - x^2)}}\, dx.$$

With the substitution $x = a\sin t$, $dx = a\cos t\, dt$, $a^2 - x^2 = a^2\cos^2 t$ and

$$s = a\int_{-\pi/2}^{\pi/2} \cos t\sqrt{\frac{a^2b^2\sin^2 t + a^2a^2\cos^2 t}{a^2(a^2\cos^2 t)}}\, dt = a\int_{\pi/2}^{\pi/2} \sqrt{\frac{b^2\sin^2 t}{a^2} + \cos^2 t}\, dt$$

Because

$$\eta = \sqrt{1 - \frac{b^2}{a^2}}, \quad \eta^2 = 1 - \frac{b^2}{a^2}$$

we then have

$$1 - \eta^2\sin^2 t = 1 - \left(1 - \frac{b^2}{a^2}\right)\sin^2 t = 1 - \sin^2 t + \frac{b^2}{a^2}\sin^2 t = \cos^2 t + \frac{b^2}{a^2}\sin^2 t$$

which is the same as the expression under the square root above. Substituting, we get

$$s = a\int_{-\pi/2}^{\pi/2} \sqrt{1 - \eta^2\sin^2 t}\, dt$$

When $a = 2$ and $b = 1$, $\eta^2 = \frac{3}{4}$. Using a computer algebra system to approximate the value of the definite integral, we find $s \approx 4.84422$.

53. Suppose that the observer in Exercise 52 moves off to infinity—that is, $d \to \infty$. What do you expect the limiting value of the observed area to be? Check your guess by calculating the limit using the formula for the area in the previous exercise.

SOLUTION We would assume the observed surface area would approach $2\pi R^2$ which is the surface area of a hemisphere of radius R. To verify this, observe:

$$\lim_{d\to\infty} SA = \lim_{d\to\infty} \frac{2\pi R^2 d}{R + d} = \lim_{d\to\infty} \frac{2\pi R^2}{1} = 2\pi R^2.$$

55. Let $f(x)$ be an increasing function on $[a, b]$ and let $g(x)$ be its inverse. Argue on the basis of arc length that the following equality holds:

$$\int_a^b \sqrt{1 + f'(x)^2}\, dx = \int_{f(a)}^{f(b)} \sqrt{1 + g'(y)^2}\, dy \qquad \boxed{5}$$

Then use the substitution $u = f(x)$ to prove Eq. (5).

SOLUTION Since the graphs of $f(x)$ and $g(x)$ are symmetric with respect to the line $y = x$, the arc length of the curves will be equal on the respective domains. Since the domain of g is the range of f, on $f(a)$ to $f(b)$, $g(x)$ will have the same arc length as $f(x)$ on a to b. If $g(x) = f^{-1}(x)$ and $u = f(x)$, then $x = g(u)$ and $du = f'(x)\,dx$. But

$$g'(u) = \frac{1}{f'(g(u))} = \frac{1}{f'(x)} \Rightarrow f'(x) = \frac{1}{g'(u)}$$

Now substituting $u = f(x)$,

$$s = \int_a^b \sqrt{1 + f'(x)^2}\,dx = \int_{f(a)}^{f(b)} \sqrt{1 + \left(\frac{1}{g'(u)}\right)^2}\, g'(u)\,du = \int_{f(a)}^{f(b)} \sqrt{g'(u)^2 + 1}\,du$$

8.2 Fluid Pressure and Force

Preliminary Questions

1. How is pressure defined?

SOLUTION Pressure is defined as force per unit area.

2. Fluid pressure is proportional to depth. What is the factor of proportionality?

SOLUTION The factor of proportionality is the weight density of the fluid, $w = \rho g$, where ρ is the mass density of the fluid.

3. When fluid force acts on the side of a submerged object, in which direction does it act?

SOLUTION Fluid force acts in the direction perpendicular to the side of the submerged object.

4. Why is fluid pressure on a surface calculated using thin horizontal strips rather than thin vertical strips?

SOLUTION Pressure depends only on depth and does not change horizontally at a given depth.

5. If a thin plate is submerged horizontally, then the fluid force on one side of the plate is equal to pressure times area. Is this true if the plate is submerged vertically?

SOLUTION When a plate is submerged vertically, the pressure is not constant along the plate, so the fluid force is not equal to the pressure times the area.

Exercises

1. A box of height 6 m and square base of side 3 m is submerged in a pool of water. The top of the box is 2 m below the surface of the water.

(a) Calculate the fluid force on the top and bottom of the box.

(b) Write a Riemann sum that approximates the fluid force on a side of the box by dividing the side into N horizontal strips of thickness $\Delta y = 6/N$.

(c) To which integral does the Riemann sum converge?

(d) Compute the fluid force on a side of the box.

SOLUTION

(a) At a depth of 2 m, the pressure on the top of the box is $\rho g h = 10^3 \cdot 9.8 \cdot 2 = 19{,}600$ Pa. The top has area 9 m^2, and the pressure is constant, so the force on the top of the box is $19{,}600 \cdot 9 = 176{,}400N$. At a depth of 8 m, the pressure on the bottom of the box is $\rho g h = 10^3 \cdot 9.8 \cdot 8 = 78{,}400$ Pa, so the force on the bottom of the box is $78{,}400 \cdot 9 = 705{,}600N$.

(b) Let y_j denote the depth of the jth strip, for $j = 1, 2, 3, \ldots, N$; the pressure at this depth is $10^3 \cdot 9.8 \cdot y_j = 9800y_j$ Pa. The strip has thickness Δy m and length 3 m, so has area $3\Delta y$ m^2. Thus the force on the strip is $29{,}400y_j\Delta y$ N. Sum over all the strips to conclude that the force on one side of the box is approximately

$$F \approx \sum_{j=1}^{N} 29{,}400 y_j \Delta y.$$

(c) As $N \to \infty$, the Riemann sum in part (b) converges to the definite integral $29{,}400 \int_2^8 y\,dy$.

(d) Using the result from part (c), the fluid force on one side of the box is

$$29{,}400 \int_2^8 y\,dy = 14{,}700y^2 \Big|_2^8 = 882{,}000\ N$$

3. Repeat Exercise 2, but assume that the top of the triangle is located 3 m below the surface of the water.

SOLUTION

(a) Examine the figure below. By similar triangles, $\dfrac{y-3}{2} = \dfrac{f(y)}{1}$ so $f(y) = \dfrac{y-3}{2}$.

(b) The pressure at a depth of y feet is $\rho g y$ lb/ Pa, and the area of the strip is approximately $f(y)\,\Delta y = \frac{1}{2}(y-3)\Delta y$ m^2. Therefore, the fluid force on this strip is approximately

$$\rho g y\left(\frac{1}{2}(y-3)\Delta y\right) = \frac{1}{2}\rho g y(y-3)\Delta y \text{ N.}$$

(c) $F \approx \sum_{j=1}^{N} \rho g \frac{y_j^2 - 3y_j}{2}\Delta y$. As $N \to \infty$, the Riemann sum converges to the definite integral

$$\frac{\rho g}{2}\int_3^5 (y^2 - 3y)\,dy.$$

(d) Using the result of part (c),

$$F = \frac{\rho g}{2}\int_3^5 (y^2-3y)\,dy = \frac{\rho g}{2}\left(\frac{y^3}{3} - \frac{3y^2}{2}\right)\Bigg|_3^5 = \frac{9800}{2}\left[\left(\frac{125}{3} - \frac{75}{2}\right) - \left(9 - \frac{27}{2}\right)\right] = \frac{127{,}400}{3} \text{ N.}$$

5. Let F be the fluid force on a side of a semicircular plate of radius r meters, submerged vertically in water so that its diameter is level with the water's surface (Figure 9).

(a) Show that the width of the plate at depth y is $2\sqrt{r^2 - y^2}$.

(b) Calculate F as a function of r using Eq. (2).

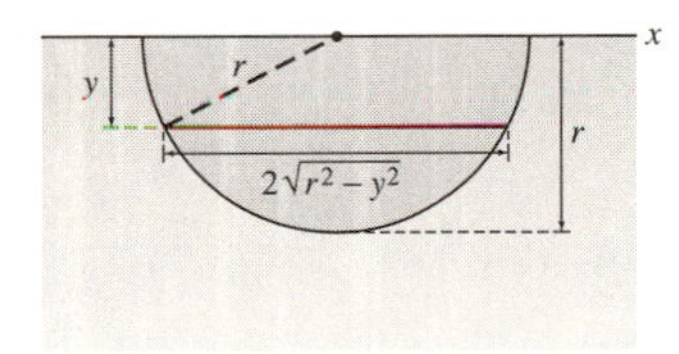

FIGURE 9

SOLUTION

(a) Place the origin at the center of the semicircle and point the positive y-axis downward. The equation for the edge of the semicircular plate is then $x^2 + y^2 = r^2$. At a depth of y, the plate extends from the point $(-\sqrt{r^2-y^2}, y)$ on the left to the point $(\sqrt{r^2-y^2}, y)$ on the right. The width of the plate at depth y is then

$$\sqrt{r^2-y^2} - \left(-\sqrt{r^2-y^2}\right) = 2\sqrt{r^2-y^2}.$$

(b) With $w = 9800$ N/m^3,

$$F = 2w\int_0^r y\sqrt{r^2-y^2}\,dy = -\frac{19{,}600}{3}(r^2-y^2)^{3/2}\Big|_0^r = \frac{19{,}600r^3}{3} \text{ N.}$$

7. A semicircular plate of radius r meters, oriented as in Figure 9, is submerged in water so that its diameter is located at a depth of m meters. Calculate the fluid force on one side of the plate in terms of m and r.

SOLUTION Place the origin at the center of the semicircular plate with the positive y-axis pointing downward. The water surface is then at $y = -m$. Moreover, at location y, the width of the plate is $2\sqrt{r^2-y^2}$ and the depth is $y + m$. Thus,

$$F = 2\rho g\int_0^r (y+m)\sqrt{r^2-y^2}\,dy.$$

Now,

$$\int_0^r y\sqrt{r^2 - y^2}\, dy = -\frac{1}{3}(r^2 - y^2)^{3/2}\Big|_0^r = \frac{1}{3}r^3.$$

Geometrically,

$$\int_0^r \sqrt{r^2 - y^2}\, dy$$

represents the area of one quarter of a circle of radius r, and thus has the value $\frac{\pi r^2}{4}$. Bringing these results together, we find that

$$F = 2\rho g\left(\frac{1}{3}r^3 + \frac{\pi}{4}r^2\right) = \frac{19{,}600}{3}r^3 + 4900\pi r^2 \text{ N}.$$

9. Figure 10 shows the wall of a dam on a water reservoir. Use the Trapezoidal Rule and the width and depth measurements in the figure to estimate the fluid force on the wall.

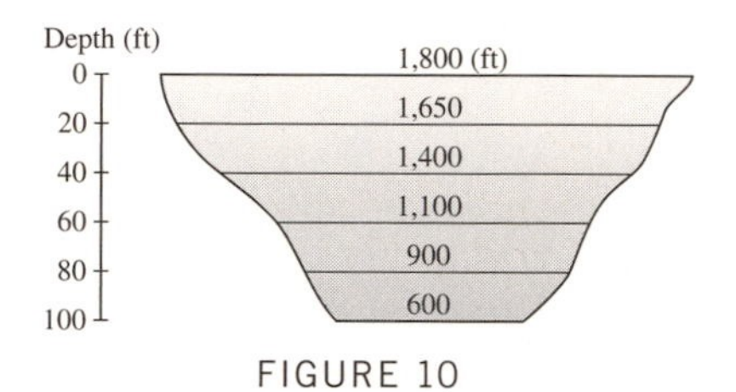

FIGURE 10

SOLUTION Let $f(y)$ denote the width of the dam wall at depth y feet. Then the force on the dam wall is

$$F = w\int_0^{100} yf(y)\, dy.$$

Using the Trapezoidal Rule and the width and depth measurements in the figure,

$$F \approx w\frac{20}{2}[0 \cdot f(0) + 2 \cdot 20 \cdot f(20) + 2 \cdot 40 \cdot f(40) + 2 \cdot 60 \cdot f(60) + 2 \cdot 80 \cdot f(80) + 100 \cdot f(100)]$$
$$= 10w(0 + 66{,}000 + 112{,}000 + 132{,}000 + 144{,}000 + 60{,}000) = 321{,}250{,}000 \text{ lb}.$$

11. Calculate the fluid force on a side of the plate in Figure 11(B), submerged in a fluid of mass density $\rho = 800$ kg/m^3.

SOLUTION Because the fluid has a mass density of $\rho = 800$ kg/m^3,

$$w = (800)(9.8) = 7840 \text{ N/m}^3.$$

For depths up to 2 meters, the width of the plate at depth y is y; for depths from 2 meters to 6 meters, the width of the plate is a constant 2 meters. Thus,

$$F = w\int_0^2 y(y)\, dy + w\int_2^6 2y\, dy = w\frac{y^3}{3}\Big|_0^2 + wy^2\Big|_2^6 = \frac{8w}{3} + 32w = \frac{104w}{3} = \frac{815{,}360}{3} \text{ N}.$$

13. Let R be the plate in the shape of the region under $y = \sin x$ for $0 \le x \le \frac{\pi}{2}$ in Figure 13(A). Find the fluid force on a side of R if it is rotated counterclockwise by 90° and submerged in a fluid of density 1100 kg/m^3 with its top edge level with the surface of the fluid as in (B).

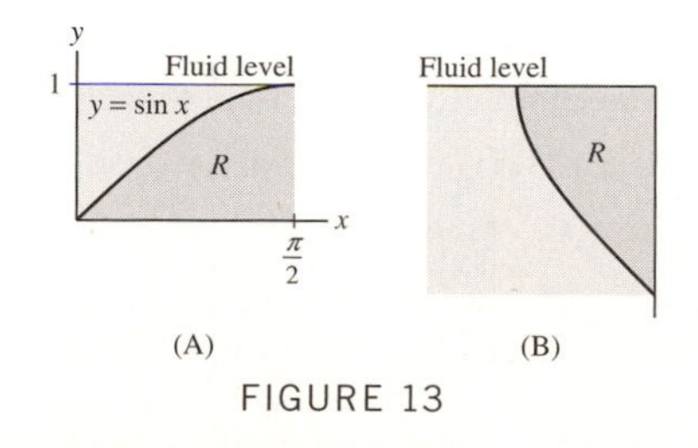

FIGURE 13

SOLUTION Place the origin at the bottom corner of the plate with the positive y-axis pointing upward. The fluid surface is then at height $y = \frac{\pi}{2}$, and the horizontal strip of the plate at height y is at a depth of $\frac{\pi}{2} - y$ and has a width of $\sin y$. Now, using integration by parts we find

$$F = \rho g \int_0^{\pi/2} \left(\frac{\pi}{2} - y\right) \sin y \, dy = \rho g \left[-\left(\frac{\pi}{2} - y\right) \cos y - \sin y \right]\Big|_0^{\pi/2} = \rho g \left(\frac{\pi}{2} - 1\right)$$
$$= 1100 \cdot 9.8 \left(\frac{\pi}{2} - 1\right) \approx 6153.184 \text{ N}.$$

15. Calculate the fluid force on one side of a plate in the shape of region A shown Figure 14. The water surface is at $y = 1$, and the fluid has density $\rho = 900$ kg/m^3.

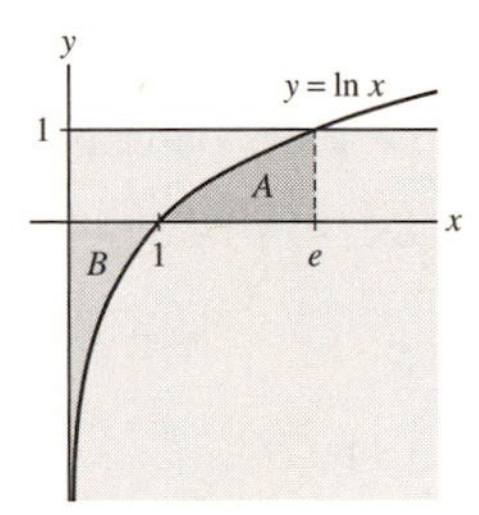

FIGURE 14

SOLUTION Because the fluid surface is at height $y = 1$, the horizontal strip at height y is at a depth of $1 - y$. Moreover, this strip has a width of $e - e^y$. Thus,

$$F = \rho g \int_0^1 (1 - y)(e - e^y) \, dy = e\rho g \int_0^1 (1 - y) \, dy - \rho g \int_0^1 (1 - y)e^y \, dy.$$

Now,

$$\int_0^1 (1 - y) \, dy = \left(y - \frac{1}{2}y^2\right)\Big|_0^1 = \frac{1}{2},$$

and using integration by parts

$$\int_0^1 (1 - y)e^y \, dy = ((1 - y)e^y + e^y)\Big|_0^1 = e - 2.$$

Combining these results, we find that

$$F = \rho g \left(\frac{1}{2}e - (e - 2)\right) = \rho g \left(2 - \frac{1}{2}e\right) = 900 \cdot 9.8 \left(2 - \frac{1}{2}e\right) \approx 5652.37 \text{ N}.$$

17. Figure 15(A) shows a ramp inclined at 30° leading into a swimming pool. Calculate the fluid force on the ramp.

SOLUTION A horizontal strip at depth y has length 6 and width

$$\frac{\Delta y}{\sin 30°} = 2\Delta y.$$

Thus,

$$F = 2\rho g \int_0^4 6y \, dy = 96\rho g.$$

If distances are in feet, then $\rho g = w = 62.5$ lb/ft^3 and $F = 6000$ lb; if distances are in meters, then $\rho g = 9800$ N/m^3 and $F = 940{,}800$ N.

19. The massive Three Gorges Dam on China's Yangtze River has height 185 m (Figure 16). Calculate the force on the dam, assuming that the dam is a trapezoid of base 2000 m and upper edge 3000 m, inclined at an angle of 55° to the horizontal (Figure 17).

FIGURE 16 Three Gorges Dam on the Yangtze River

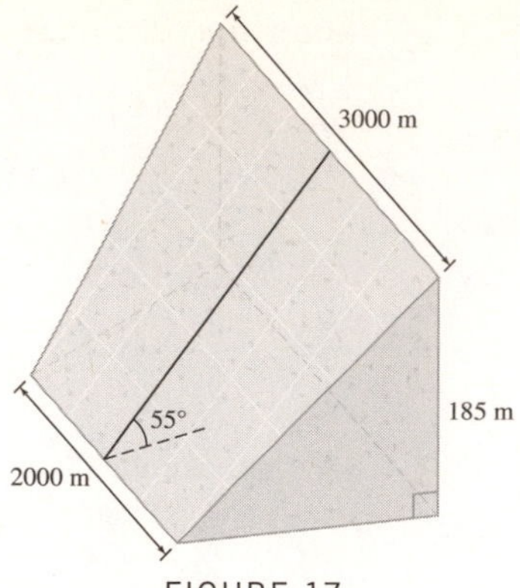

FIGURE 17

SOLUTION Let $y = 0$ be at the bottom of the dam, so that the top of the dam is at $y = 185$. Then the width of the dam at height y is $2000 + \frac{1000y}{185}$. The dam is inclined at an angle of 55° to the horizontal, so the height of a horizontal strip is

$$\frac{\Delta y}{\sin 55°} \approx 1.221\Delta y$$

so that the area of such a strip is

$$1.221\left(2000 + \frac{1000y}{185}\right)\Delta y$$

Then

$$F = \rho g \int_0^{185} 1.221y\left(2000 + \frac{1000y}{185}\right) dy = \rho g \int_0^{185} 2442y + 6.6y^2\, dy = \rho g(1221y^2 + 2.2y^3)\Big|_0^{185}$$

$$= 55{,}718{,}300\rho g = 55{,}718{,}300 \cdot 9800 = 5.460393400 \times 10^{11}\text{ N}.$$

21. The trough in Figure 18 is filled with corn syrup, whose weight density is 90 lb/ft^3. Calculate the force on the front side of the trough.

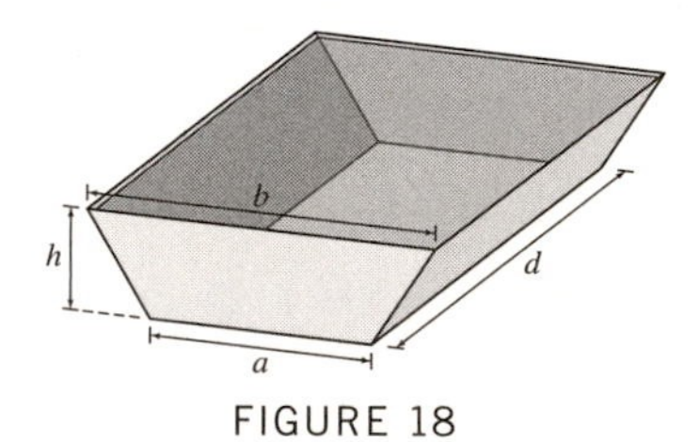

FIGURE 18

SOLUTION Place the origin along the top edge of the trough with the positive y-axis pointing downward. The width of the front side of the trough varies linearly from b when $y = 0$ to a when $y = h$; thus, the width of the front side of the trough at depth y feet is given by

$$b + \frac{a-b}{h}y.$$

Now,

$$F = w\int_0^h y\left(b + \frac{a-b}{h}y\right) dy = w\left(\frac{1}{2}by^2 + \frac{a-b}{3h}y^3\right)\Big|_0^h = w\left(\frac{b}{6} + \frac{a}{3}\right)h^2 = (15b + 30a)h^2\text{ lb}.$$

Further Insights and Challenges

23. The end of the trough in Figure 19 is an equilateral triangle of side 3. Assume that the trough is filled with water to height H. Calculate the fluid force on each side of the trough as a function of H and the length l of the trough.

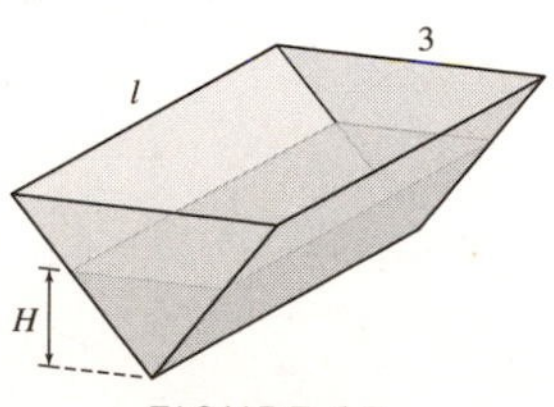

FIGURE 19

SOLUTION Place the origin at the lower vertex of the trough and orient the positive y-axis pointing upward. First, consider the faces at the front and back ends of the trough. A horizontal strip at height y has a length of $\frac{2y}{\sqrt{3}}$ and is at a depth of $H - y$. Thus,

$$F = w\int_0^H (H-y)\frac{2y}{\sqrt{3}}\,dy = w\left(\frac{H}{\sqrt{3}}y^2 - \frac{2}{3\sqrt{3}}y^3\right)\Big|_0^H = \frac{\sqrt{3}}{9}wH^3.$$

For the slanted sides, we note that each side makes an angle of 60° with the horizontal. If we let ℓ denote the length of the trough, then

$$F = \frac{2w\ell}{\sqrt{3}}\int_0^H (H-y)\,dy = \frac{\sqrt{3}}{3}\ell w H^2.$$

25. Prove that the force on the side of a rectangular plate of area A submerged vertically in a fluid is equal to $p_0 A$, where p_0 is the fluid pressure at the center point of the rectangle.

SOLUTION Let ℓ denote the length of the vertical side of the rectangle, x denote the length of the horizontal side of the rectangle, and suppose the top edge of the rectangle is at depth $y = m$. The pressure at the center of the rectangle is then

$$p_0 = w\left(m + \frac{\ell}{2}\right),$$

and the force on the side of the rectangular plate is

$$F = \int_m^{\ell+m} wxy\,dy = \frac{wx}{2}\left[(\ell+m)^2 - m^2\right] = \frac{wx\ell}{2}(\ell + 2m) = Aw\left(\frac{\ell}{2} + m\right) = Ap_0.$$

8.3 Center of Mass

Preliminary Questions

1. What are the x- and y-moments of a lamina whose center of mass is located at the origin?

SOLUTION Because the center of mass is located at the origin, it follows that $M_x = M_y = 0$.

2. A thin plate has mass 3. What is the x-moment of the plate if its center of mass has coordinates $(2, 7)$?

SOLUTION The x-moment of the plate is the product of the mass of the plate and the y-coordinate of the center of mass. Thus, $M_x = 3(7) = 21$.

3. The center of mass of a lamina of total mass 5 has coordinates $(2, 1)$. What are the lamina's x- and y-moments?

SOLUTION The x-moment of the plate is the product of the mass of the plate and the y-coordinate of the center of mass, whereas the y-moment is the product of the mass of the plate and the x-coordinate of the center of mass. Thus, $M_x = 5(1) = 5$, and $M_y = 5(2) = 10$.

4. Explain how the Symmetry Principle is used to conclude that the centroid of a rectangle is the center of the rectangle.

SOLUTION Because a rectangle is symmetric with respect to both the vertical line and the horizontal line through the center of the rectangle, the Symmetry Principle guarantees that the centroid of the rectangle must lie along both of these lines. The only point in common to both lines of symmetry is the center of the rectangle, so the centroid of the rectangle must be the center of the rectangle.

Exercises

1. Four particles are located at points $(1, 1), (1, 2), (4, 0), (3, 1)$.

(a) Find the moments M_x and M_y and the center of mass of the system, assuming that the particles have equal mass m.

(b) Find the center of mass of the system, assuming the particles have masses 3, 2, 5, and 7, respectively.

SOLUTION

(a) Because each particle has mass m,

$$M_x = m(1) + m(2) + m(0) + m(1) = 4m;$$

$$M_y = m(1) + m(1) + m(4) + m(3) = 9m;$$

and the total mass of the system is $4m$. Thus, the coordinates of the center of mass are

$$\left(\frac{M_y}{M}, \frac{M_x}{M}\right) = \left(\frac{9m}{4m}, \frac{4m}{4m}\right) = \left(\frac{9}{4}, 1\right).$$

(b) With the indicated masses of the particles,

$$M_x = 3(1) + 2(2) + 5(0) + 7(1) = 14;$$

$$M_y = 3(1) + 2(1) + 5(4) + 7(3) = 46;$$

and the total mass of the system is 17. Thus, the coordinates of the center of mass are

$$\left(\frac{M_y}{M}, \frac{M_x}{M}\right) = \left(\frac{46}{17}, \frac{14}{17}\right).$$

3. Point masses of equal size are placed at the vertices of the triangle with coordinates $(a, 0)$, $(b, 0)$, and $(0, c)$. Show that the center of mass of the system of masses has coordinates $\left(\frac{1}{3}(a+b), \frac{1}{3}c\right)$.

SOLUTION Let each particle have mass m. The total mass of the system is then $3m$. and the moments are

$$M_x = 0(m) + 0(m) + c(m) = cm; \text{ and}$$

$$M_y = a(m) + b(m) + 0(m) = (a+b)m.$$

Thus, the coordinates of the center of mass are

$$\left(\frac{M_y}{M}, \frac{M_x}{M}\right) = \left(\frac{(a+b)m}{3m}, \frac{cm}{3m}\right) = \left(\frac{a+b}{3}, \frac{c}{3}\right).$$

5. Sketch the lamina S of constant density $\rho = 3$ g/cm^2 occupying the region beneath the graph of $y = x^2$ for $0 \le x \le 3$.
(a) Use Eqs. (1) and (2) to compute M_x and M_y.
(b) Find the area and the center of mass of S.

SOLUTION A sketch of the lamina is shown below

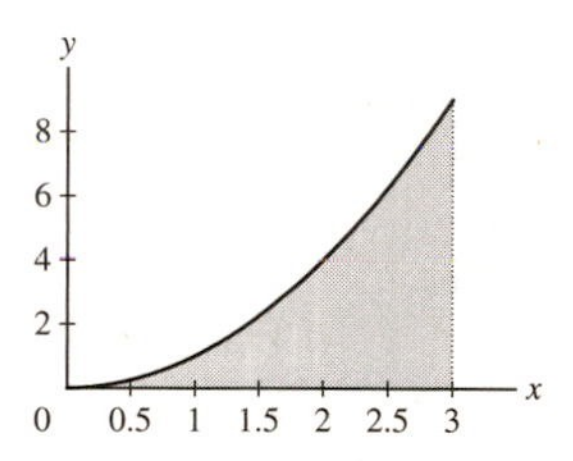

(a) Using Eq. (2),

$$M_x = 3\int_0^9 y(3 - \sqrt{y})\,dy = \left(\frac{9y^2}{2} - \frac{6}{5}y^{5/2}\right)\Bigg|_0^9 = \frac{729}{10}.$$

Using Eq. (1),

$$M_y = 3\int_0^3 x(x^2)\,dx = \frac{3x^4}{4}\Bigg|_0^3 = \frac{243}{4}.$$

(b) The area of the lamina is

$$A = \int_0^3 x^2\,dx = \frac{x^3}{3}\Bigg|_0^3 = 9 \text{ cm}^2.$$

With a constant density of $\rho = 3$ g/cm^2, the mass of the lamina is $M = 27$ grams, and the coordinates of the center of mass are

$$\left(\frac{M_y}{M}, \frac{M_x}{M}\right) = \left(\frac{243/4}{27}, \frac{729/10}{27}\right) = \left(\frac{9}{4}, \frac{27}{10}\right).$$

7. Find the moments and center of mass of the lamina of uniform density ρ occupying the region underneath $y = x^3$ for $0 \le x \le 2$.

SOLUTION With uniform density ρ,

$$M_x = \frac{1}{2}\rho\int_0^2 (x^3)^2\,dx = \frac{64\rho}{7} \quad \text{and} \quad M_y = \rho\int_0^2 x(x^3)\,dx = \frac{32\rho}{5}.$$

The mass of the lamina is

$$M = \rho\int_0^2 x^3\,dx = 4\rho,$$

so the coordinates of the center of mass are

$$\left(\frac{M_y}{M}, \frac{M_x}{M}\right) = \left(\frac{8}{5}, \frac{16}{7}\right).$$

9. Let T be the triangular lamina in Figure 17.
(a) Show that the horizontal cut at height y has length $4 - \frac{2}{3}y$ and use Eq. (2) to compute M_x (with $\rho = 1$).
(b) Use the Symmetry Principle to show that $M_y = 0$ and find the center of mass.

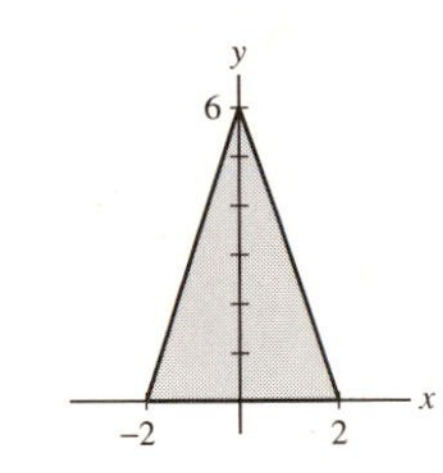

FIGURE 17 Isosceles triangle.

SOLUTION
(a) The equation of the line from $(2, 0)$ to $(0, 6)$ is $y = -3x + 6$, so

$$x = 2 - \frac{1}{3}y.$$

The length of the horizontal cut at height y is then

$$2\left(2 - \frac{1}{3}y\right) = 4 - \frac{2}{3}y,$$

and

$$M_x = \int_0^6 y\left(4 - \frac{2}{3}y\right)\, dy = 24.$$

(b) Because the triangular lamina is symmetric with respect to the y-axis, $x_{cm} = 0$, which implies that $M_y = 0$. The total mass of the lamina is

$$M = 2\int_0^2 (-3x + 6)\, dx = 12,$$

so $y_{cm} = 24/12$. Finally, the coordinates of the center of mass are $(0, 2)$.

In Exercises 10–17, find the centroid of the region lying underneath the graph of the function over the given interval.

11. $f(x) = \sqrt{x}$, $[1, 4]$

SOLUTION The moments of the region are

$$M_x = \frac{1}{2}\int_1^4 x\, dx = \frac{15}{4} \quad \text{and} \quad M_y = \int_1^4 x\sqrt{x}\, dx = \frac{62}{5}.$$

The area of the region is

$$A = \int_1^4 \sqrt{x}\, dx = \frac{14}{3},$$

so the coordinates of the centroid are

$$\left(\frac{M_y}{A}, \frac{M_x}{A}\right) = \left(\frac{93}{35}, \frac{45}{56}\right).$$

13. $f(x) = 9 - x^2$, $[0, 3]$

SOLUTION The moments of the region are

$$M_x = \frac{1}{2}\int_0^3 (9 - x^2)^2\, dx = \frac{324}{5} \quad \text{and} \quad M_y = \int_0^3 x(9 - x^2)\, dx = \frac{81}{4}.$$

The area of the region is

$$A = \int_0^3 (9 - x^2)\, dx = 18,$$

so the coordinates of the centroid are

$$\left(\frac{M_y}{A}, \frac{M_x}{A}\right) = \left(\frac{9}{8}, \frac{18}{5}\right).$$

15. $f(x) = e^{-x}$, $[0, 4]$

SOLUTION The moments of the region are

$$M_x = \frac{1}{2}\int_0^4 e^{-2x}\,dx = \frac{1}{4}\left(1 - e^{-8}\right) \quad \text{and} \quad M_y = \int_0^4 xe^{-x}\,dx = -e^{-x}(x+1)\Big|_0^4 = 1 - 5e^{-4}.$$

The area of the region is

$$A = \int_0^4 e^{-x}\,dx = 1 - e^{-4},$$

so the coordinates of the centroid are

$$\left(\frac{M_y}{A}, \frac{M_x}{A}\right) = \left(\frac{1-5e^{-4}}{1-e^{-4}}, \frac{1-e^{-8}}{4(1-e^{-4})}\right).$$

17. $f(x) = \sin x$, $[0, \pi]$

SOLUTION The moments of the region are

$$M_x = \frac{1}{2}\int_0^\pi \sin^2 x\,dx = \frac{1}{4}(x - \sin x\cos x)\Big|_0^\pi = \frac{\pi}{4}; \text{ and}$$

$$M_y = \int_0^\pi x\sin x\,dx = (-x\cos x + \sin x)\Big|_0^\pi = \pi.$$

The area of the region is

$$A = \int_0^\pi \sin x\,dx = 2,$$

so the coordinates of the centroid are

$$\left(\frac{M_y}{A}, \frac{M_x}{A}\right) = \left(\frac{\pi}{2}, \frac{\pi}{8}\right).$$

19. Sketch the region between $y = x + 4$ and $y = 2 - x$ for $0 \le x \le 2$. Using symmetry, explain why the centroid of the region lies on the line $y = 3$. Verify this by computing the moments and the centroid.

SOLUTION A sketch of the region is shown below.

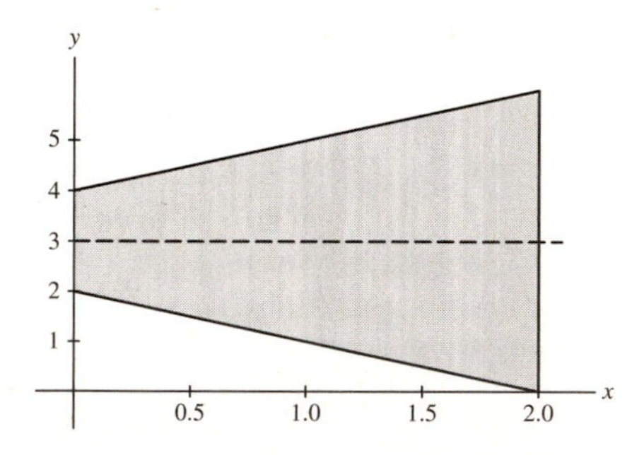

The region is clearly symmetric about the line $y = 3$, so we expect the centroid of the region to lie along this line. We find

$$M_x = \frac{1}{2}\int_0^2 \left((x+4)^2 - (2-x)^2\right)\,dx = 24;$$

$$M_y = \int_0^2 x\left((x+4) - (2-x)\right)\,dx = \frac{28}{3}; \text{ and}$$

$$A = \int_0^2 \left((x+4) - (2-x)\right)\,dx = 8.$$

Thus, the coordinates of the centroid are $\left(\frac{7}{6}, 3\right)$.

In Exercises 20–25, find the centroid of the region lying between the graphs of the functions over the given interval.

21. $y = x^2, \quad y = \sqrt{x}, \quad [0, 1]$

SOLUTION The moments of the region are

$$M_x = \frac{1}{2}\int_0^1 (x - x^4)\,dx = \frac{3}{20} \quad \text{and} \quad M_y = \int_0^1 x(\sqrt{x} - x^2)\,dx = \frac{3}{20}.$$

The area of the region is

$$A = \int_0^1 (\sqrt{x} - x^2)\,dx = \frac{1}{3},$$

so the coordinates of the centroid are

$$\left(\frac{9}{20}, \frac{9}{20}\right).$$

Note: This makes sense, since the functions are inverses of each other. This makes the region symmetric with respect to the line $y = x$. Thus, by the symmetry principle, the center of mass must lie on that line.

23. $y = e^x, \quad y = 1, \quad [0, 1]$

SOLUTION The moments of the region are

$$M_x = \frac{1}{2}\int_0^1 (e^{2x} - 1)\,dx = \frac{e^2 - 3}{4} \quad \text{and} \quad M_y = \int_0^1 x(e^x - 1)\,dx = \left(xe^x - e^x - \frac{1}{2}x^2\right)\bigg|_0^1 = \frac{1}{2}.$$

The area of the region is

$$A = \int_0^1 (e^x - 1)\,dx = e - 2,$$

so the coordinates of the centroid are

$$\left(\frac{1}{2(e-2)}, \frac{e^2 - 3}{4(e-2)}\right).$$

25. $y = \sin x, \quad y = \cos x, \quad [0, \pi/4]$

SOLUTION The moments of the region are

$$M_x = \frac{1}{2}\int_0^{\pi/4} (\cos^2 x - \sin^2 x)\,dx = \frac{1}{2}\int_0^{\pi/4} \cos 2x\,dx = \frac{1}{4}; \text{ and}$$

$$M_y = \int_0^{\pi/4} x(\cos x - \sin x)\,dx = [(x-1)\sin x + (x+1)\cos x]\bigg|_0^{\pi/4} = \frac{\pi\sqrt{2}}{4} - 1.$$

The area of the region is

$$A = \int_0^{\pi/4} (\cos x - \sin x)\,dx = \sqrt{2} - 1,$$

so the coordinates of the centroid are

$$\left(\frac{\pi\sqrt{2} - 4}{4(\sqrt{2} - 1)}, \frac{1}{4(\sqrt{2} - 1)}\right).$$

27. Sketch the region enclosed by $y = 0$, $y = (x+1)^3$, and $y = (1-x)^3$, and find its centroid.

SOLUTION A sketch of the region is shown below.

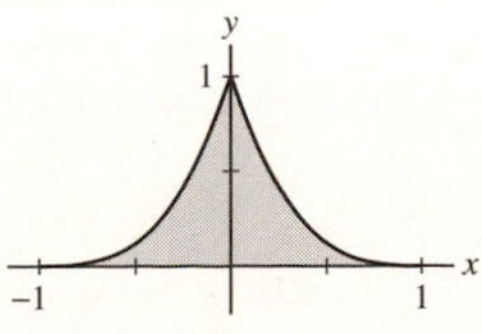

The moments of the region are

$$M_x = \frac{1}{2}\left(\int_{-1}^{0} (x+1)^6\,dx + \int_{0}^{1} (1-x)^6\,dx\right) = \frac{1}{7};\ \text{and}$$

$$M_y = 0 \text{ by the Symmetry Principle.}$$

The area of the region is

$$A = \int_{-1}^{0} (x+1)^3\,dx + \int_{0}^{1} (1-x)^3\,dx = \frac{1}{2},$$

so the coordinates of the centroid are $\left(0, \frac{2}{7}\right)$.

In Exercises 28–32, find the centroid of the region.

29. Top half of the ellipse $\left(\frac{x}{a}\right)^2 + \left(\frac{y}{b}\right)^2 = 1$ for arbitrary $a, b > 0$

SOLUTION The equation of the top half of the ellipse is

$$y = \sqrt{b^2 - \frac{b^2x^2}{a^2}}$$

Thus,

$$M_x = \frac{1}{2}\int_{-a}^{a} \left(\sqrt{b^2 - \frac{b^2x^2}{a^2}}\right)^2 dx = \frac{2ab^2}{3}.$$

By the Symmetry Principle, $M_y = 0$. The area of the region is one-half the area of an ellipse with axes of length a and b; i.e., $\frac{1}{2}\pi ab$. Finally, the coordinates of the centroid are

$$\left(0, \frac{4b}{3\pi}\right).$$

31. Quarter of the unit circle lying in the first quadrant

SOLUTION By the Symmetry Principle, the center of mass must lie on the line $y = x$ in the first quadrant. Therefore we need only calculate one of the moments of the region. With $y = \sqrt{1-x^2}$, we find

$$M_y = \int_{0}^{1} x\sqrt{1-x^2}\,dx = \frac{1}{3}.$$

The area of the region is one-quarter of the area of a unit circle; i.e., $\frac{1}{4}\pi$. Thus, the coordinates of the centroid are

$$\left(\frac{4}{3\pi}, \frac{4}{3\pi}\right).$$

33. Find the centroid of the shaded region of the semicircle of radius r in Figure 18. What is the centroid when $r =$ and $h = \frac{1}{2}$? *Hint:* Use geometry rather than integration to show that the *area* of the region is $r^2 \sin^{-1}(\sqrt{1 - h^2/r^2}) - h\sqrt{r^2 - h^2})$.

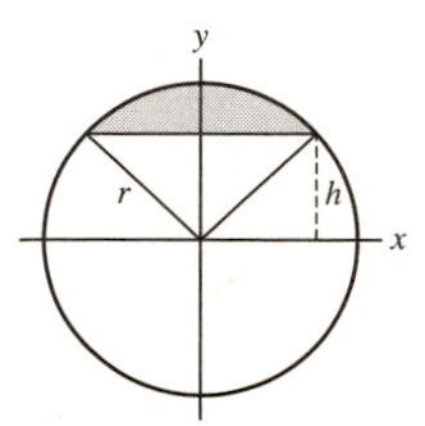

FIGURE 18

SOLUTION From the symmetry of the region, it is obvious that the centroid lies along the y-axis. To determine th y-coordinate of the centroid, we must calculate the moment about the x-axis and the area of the region. Now, the lengt of the horizontal cut of the semicircle at height y is

$$\sqrt{r^2 - y^2} - \left(-\sqrt{r^2 - y^2}\right) = 2\sqrt{r^2 - y^2}.$$

Therefore, taking $\rho = 1$, we find

$$M_x = 2\int_h^r y\sqrt{r^2 - y^2}\,dy = \frac{2}{3}(r^2 - h^2)^{3/2}.$$

Observe that the region is comprised of a sector of the circle with the triangle between the two radii removed. The angle of the sector is 2θ, where $\theta = \sin^{-1}\sqrt{1 - h^2/r^2}$, so the area of the sector is $\frac{1}{2}r^2(2\theta) = r^2\sin^{-1}\sqrt{1 - h^2/r^2}$. The triangle has base $2\sqrt{r^2 - h^2}$ and height h, so the area is $h\sqrt{r^2 - h^2}$. Therefore,

$$y_{CM} = \frac{M_x}{A} = \frac{\frac{2}{3}(r^2 - h^2)^{3/2}}{r^2\sin^{-1}\sqrt{1 - h^2/r^2} - h\sqrt{r^2 - h^2}}.$$

When $r = 1$ and $h = 1/2$, we find

$$y_{CM} = \frac{\frac{2}{3}(3/4)^{3/2}}{\sin^{-1}\frac{\sqrt{3}}{2} - \frac{\sqrt{3}}{4}} = \frac{3\sqrt{3}}{4\pi - 3\sqrt{3}}.$$

In Exercises 35–37, use the additivity of moments to find the COM of the region.

35. Isosceles triangle of height 2 on top of a rectangle of base 4 and height 3 (Figure 19)

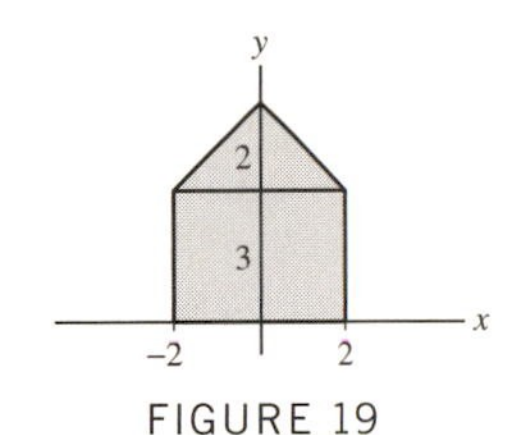

FIGURE 19

SOLUTION The region is symmetric with respect to the y-axis, so $M_y = 0$ by the Symmetry Principle. The moment about the x-axis for the rectangle is

$$M_x^{\text{rect}} = \frac{1}{2}\int_{-2}^{2} 3^2\,dx = 18,$$

whereas the moment about the x-axis for the triangle is

$$M_x^{\text{triangle}} = \int_3^5 y(10 - 2y)\,dy = \frac{44}{3}.$$

The total moment about the x-axis is then

$$M_x = M_x^{\text{rect}} + M_x^{\text{triangle}} = 18 + \frac{44}{3} = \frac{98}{3}.$$

Because the area of the region is $12 + 4 = 16$, the coordinates of the center of mass are

$$\left(0, \frac{49}{24}\right).$$

37. Three-quarters of the unit circle (remove the part in the fourth quadrant)

SOLUTION By the Symmetry Principle, the center of mass must lie on the line $y = -x$. Let region 1 be the semicircle above the x-axis and region 2 be the quarter circle in the third quadrant. Because region 1 is symmetric with respect to the y-axis, $M_y^1 = 0$ by the Symmetry Principle. Furthermore

$$M_y^2 = \int_{-1}^{0} x\sqrt{1 - x^2}\,dx = -\frac{1}{3}.$$

Thus, $M_y = M_y^1 + M_y^2 = 0 + (-\frac{1}{3}) = -\frac{1}{3}$. The area of the region is $3\pi/4$, so the coordinates of the centroid are

$$\left(-\frac{4}{9\pi}, \frac{4}{9\pi}\right).$$

39. Find the COM of the laminas in Figure 22 obtained by removing squares of side 2 from a square of side 8.

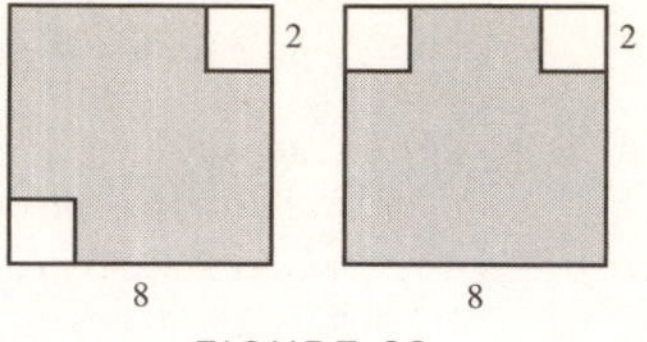

FIGURE 22

SOLUTION Start with the square on the left. Place the square so that the bottom left corner is at $(0, 0)$. By the Symmetry Principle, the center of mass must lie on the lines $y = x$ and $y = 8 - x$. The only point in common to these two lines is $(4, 4)$, so the center of mass is $(4, 4)$.

Now consider the square on the right. Place the square as above. By the symmetry principle, $x_{cm} = 4$. Now, let $s1$ denote the square in the upper left, $s2$ denote the square in the upper right, and B denote the entire square. Then

$$M_x^{s1} = \frac{1}{2}\int_0^2 (8^2 - 6^2)\,dx = 28;$$

$$M_x^{s2} = \frac{1}{2}\int_6^8 (8^2 - 6^2)\,dx = 28; \text{ and}$$

$$M_x^{B} = \frac{1}{2}\int_0^8 8^2\,dx = 256.$$

By the additivity of moments, $M_x = 256 - 28 - 28 = 200$. Finally, the area of the region is $A = 64 - 4 - 4 = 56$, so the coordinates of the center of mass are

$$\left(4, \frac{200}{56}\right) = \left(4, \frac{25}{7}\right).$$

Further Insights and Challenges

41. Let P be the COM of a system of two weights with masses m_1 and m_2 separated by a distance d. Prove Archimedes' Law of the (weightless) Lever: P is the point on a line between the two weights such that $m_1L_1 = m_2L_2$, where L_j is the distance from mass j to P.

SOLUTION Place the lever along the x-axis with mass m_1 at the origin. Then $M_y = m_2d$ and the x-coordinate of the center of mass, P, is

$$\frac{m_2d}{m_1 + m_2}.$$

Thus,

$$L_1 = \frac{m_2d}{m_1 + m_2}, \quad L_2 = d - \frac{m_2d}{m_1 + m_2} = \frac{m_1d}{m_1 + m_2},$$

and

$$L_1m_1 = m_1\frac{m_2d}{m_1 + m_2} = m_2\frac{m_1d}{m_1 + m_2} = L_2m_2.$$

43. **Symmetry Principle** Let $\mathcal{R}$ be the region under the graph of $f(x)$ over the interval $[-a, a]$, where $f(x) \geq 0$. Assume that $\mathcal{R}$ is symmetric with respect to the y-axis.

(a) Explain why $f(x)$ is even—that is, why $f(x) = f(-x)$.

(b) Show that $xf(x)$ is an *odd* function.

(c) Use (b) to prove that $M_y = 0$.

(d) Prove that the COM of $\mathcal{R}$ lies on the y-axis (a similar argument applies to symmetry with respect to the x-axis).

SOLUTION

(a) By the definition of symmetry with respect to the y-axis, $f(x) = f(-x)$, so f is even.

(b) Let $g(x) = xf(x)$ where f is even. Then

$$g(-x) = -xf(-x) = -xf(x) = -g(x),$$

and thus g is odd.

(c) $M_y = \rho \int_{-a}^{a} xf(x)\,dx = 0$ since $xf(x)$ is an odd function.

(d) By part (c), $x_{cm} = \frac{M_y}{M} = \frac{0}{M} = 0$ so the center of mass lies along the y-axis.

45. Let R be a lamina of uniform density submerged in a fluid of density w (Figure 23). Prove the following law: The fluid force on one side of R is equal to the area of R times the fluid pressure on the centroid. *Hint:* Let $g(y)$ be the horizontal width of R at depth y. Express both the fluid pressure [Eq. (2) in Section 8.2] and y-coordinate of the centroid in terms of $g(y)$.

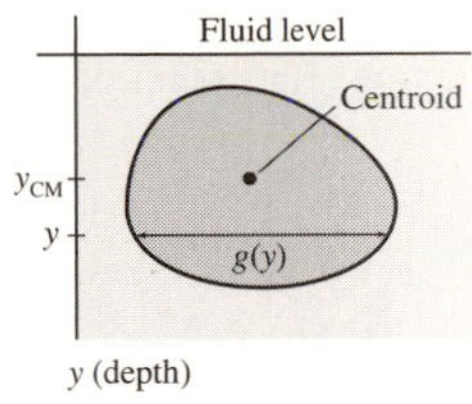

FIGURE 23

SOLUTION Let ρ denote the uniform density of the submerged lamina. Then

$$M_x = \rho \int_a^b yg(y)\,dy,$$

and the mass of the lamina is

$$M = \rho \int_a^b g(y)\,dy = \rho A,$$

where A is the area of the lamina. Thus, the y-coordinate of the centroid is

$$y_{cm} = \frac{\rho \int_a^b yg(y)\,dy}{\rho A} = \frac{\int_a^b yg(y)\,dy}{A}.$$

Now, the fluid force on the lamina is

$$F = w \int_a^b yg(y)\,dy = w\frac{\int_a^b yg(y)\,dy}{A}A = wy_{cm}A.$$

In other words, the fluid force on the lamina is equal to the fluid pressure at the centroid of the lamina times the area of the lamina.

8.4 Taylor Polynomials

Preliminary Questions

1. What is $T_3(x)$ centered at $a = 3$ for a function $f(x)$ such that $f(3) = 9$, $f'(3) = 8$, $f''(3) = 4$, and $f'''(3) = 12$?

SOLUTION In general, with $a = 3$,

$$T_3(x) = f(3) + f'(3)(x-3) + \frac{f''(3)}{2}(x-3)^2 + \frac{f'''(3)}{6}(x-3)^3.$$

Using the information provided, we find

$$T_3(x) = 9 + 8(x-3) + 2(x-3)^2 + 2(x-3)^3.$$

2. The dashed graphs in Figure 9 are Taylor polynomials for a function $f(x)$. Which of the two is a Maclaurin polynomial?

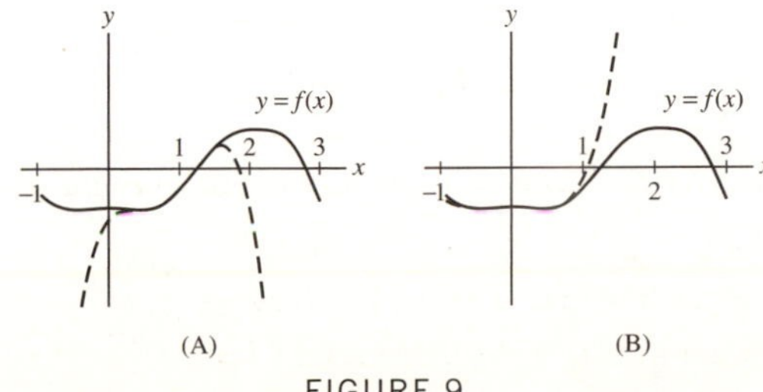

FIGURE 9

SOLUTION A Maclaurin polynomial always gives the value of $f(0)$ exactly. This is true for the Taylor polynomial sketched in (B); thus, this is the Maclaurin polynomial.

3. For which value of x does the Maclaurin polynomial $T_n(x)$ satisfy $T_n(x) = f(x)$, no matter what $f(x)$ is?

SOLUTION A Maclaurin polynomial always gives the value of $f(0)$ exactly.

4. Let $T_n(x)$ be the Maclaurin polynomial of a function $f(x)$ satisfying $|f^{(4)}(x)| \le 1$ for all x. Which of the following statements follow from the error bound?

(a) $|T_4(2) - f(2)| \le \frac{2}{3}$

(b) $|T_3(2) - f(2)| \le \frac{2}{3}$

(c) $|T_3(2) - f(2)| \le \frac{1}{3}$

SOLUTION For a function $f(x)$ satisfying $|f^{(4)}(x)| \le 1$ for all x,

$$|T_3(2) - f(2)| \le \frac{1}{24}|f^{(4)}(x)|2^4 \le \frac{16}{24} < \frac{2}{3}.$$

Thus, **(b)** is the correct answer.

Exercises

In Exercises 1–14, calculate the Taylor polynomials $T_2(x)$ and $T_3(x)$ centered at $x = a$ for the given function and value of a.

1. $f(x) = \sin x, \quad a = 0$

SOLUTION First, we calculate and evaluate the needed derivatives:

$$\begin{aligned} f(x) &= \sin x & f(a) &= 0 \\ f'(x) &= \cos x & f'(a) &= 1 \\ f''(x) &= -\sin x & f''(a) &= 0 \\ f'''(x) &= -\cos x & f'''(a) &= -1 \end{aligned}$$

Now,

$$T_2(x) = f(a) + f'(a)(x-a) + \frac{f''(a)}{2}(x-a)^2 = 0 + 1(x-0) + \frac{0}{2}(x-0)^2 = x; \text{ and}$$

$$\begin{aligned} T_3(x) &= f(a) + f'(a)(x-a) + \frac{f''(a)}{2}(x-a)^2 + \frac{f'''(a)}{6}(x-a)^3 \\ &= 0 + 1(x-0) + \frac{0}{2}(x-0)^2 + \frac{-1}{6}(x-0)^3 = x - \frac{1}{6}x^3. \end{aligned}$$

3. $f(x) = \dfrac{1}{1+x}, \quad a = 2$

SOLUTION First, we calculate and evaluate the needed derivatives:

$$\begin{aligned} f(x) &= \frac{1}{1+x} & f(a) &= \frac{1}{3} \\ f'(x) &= \frac{-1}{(1+x)^2} & f'(a) &= -\frac{1}{9} \\ f''(x) &= \frac{2}{(1+x)^3} & f''(a) &= \frac{2}{27} \\ f'''(x) &= \frac{-6}{(1+x)^4} & f'''(a) &= -\frac{2}{27} \end{aligned}$$

Now,

$$\begin{aligned} T_2(x) &= f(a) + f'(a)(x-a) + \frac{f''(a)}{2!}(x-a)^2 = \frac{1}{3} - \frac{1}{9}(x-2) + \frac{2/27}{2!}(x-2)^2 \\ &= \frac{1}{3} - \frac{1}{9}(x-2) + \frac{1}{27}(x-2)^2 \\ T_3(x) &= f(a) + f'(a)(x-a) + \frac{f''(a)}{2!}(x-a)^2 + \frac{f'''(a)}{3!}(x-a)^3 \\ &= \frac{1}{3} - \frac{1}{9}(x-2) + \frac{2/27}{2!}(x-2)^2 - \frac{2/27}{3!}(x-2)^3 = \frac{1}{3} - \frac{1}{9}(x-2) + \frac{1}{27}(x-2)^2 - \frac{1}{81}(x-2)^3 \end{aligned}$$

5. $f(x) = x^4 - 2x, \quad a = 3$

SOLUTION First calculate and evaluate the needed derivatives:

$$\begin{aligned} f(x) &= x^4 - 2x & f(a) &= 75 \\ f'(x) &= 4x^3 - 2 & f'(a) &= 106 \\ f''(x) &= 12x^2 & f''(a) &= 108 \\ f'''(x) &= 24x & f'''(a) &= 72 \end{aligned}$$

Now,

$$\begin{aligned} T_2(x) &= f(a) + f'(a)(x-a) + \frac{f''(a)}{2}(x-a)^2 = 75 + 106(x-3) + \frac{108}{2}(x-3)^2 \\ &= 75 + 106(x-3) + 54(x-3)^2 \\ T_3(x) &= f(a) + f'(a)(x-a) + \frac{f''(a)}{2}(x-a)^2 + \frac{f'''(a)}{3!}(x-a)^3 \\ &= 75 + 106(x-3) + \frac{108}{2}(x-3)^2 + \frac{72}{3!}(x-3)^3 \\ &= 75 + 106(x-3) + 54(x-3)^2 + 12(x-3)^3 \end{aligned}$$

7. $f(x) = \tan x, \quad a = 0$

SOLUTION First, we calculate and evaluate the needed derivatives:

$$\begin{aligned} f(x) &= \tan x & f(a) &= 0 \\ f'(x) &= \sec^2 x & f'(a) &= 1 \\ f''(x) &= 2\sec^2 x \tan x & f''(a) &= 0 \\ f'''(x) &= 2\sec^4 x + 4\sec^2 x \tan^2 x & f'''(a) &= 2 \end{aligned}$$

Now,

$$\begin{aligned} T_2(x) &= f(a) + f'(a)(x-a) + \frac{f''(a)}{2}(x-a)^2 = 0 + 1(x-0) + \frac{0}{2}(x-0)^2 = x; \text{ and} \\ T_3(x) &= f(a) + f'(a)(x-a) + \frac{f''(a)}{2}(x-a)^2 + \frac{f'''(a)}{6}(x-a)^3 \\ &= 0 + 1(x-0) + \frac{0}{2}(x-0)^2 + \frac{2}{6}(x-0)^3 = x + \frac{1}{3}x^3. \end{aligned}$$

9. $f(x) = e^{-x} + e^{-2x}, \quad a = 0$

SOLUTION First, we calculate and evaluate the needed derivatives:

$$\begin{aligned} f(x) &= e^{-x} + e^{-2x} & f(a) &= 2 \\ f'(x) &= -e^{-x} - 2e^{-2x} & f'(a) &= -3 \\ f''(x) &= e^{-x} + 4e^{-2x} & f''(a) &= 5 \\ f'''(x) &= -e^{-x} - 8e^{-2x} & f'''(a) &= -9 \end{aligned}$$

Now,

$$\begin{aligned} T_2(x) &= f(a) + f'(a)(x-a) + \frac{f''(a)}{2}(x-a)^2 \\ &= 2 + (-3)(x-0) + \frac{5}{2}(x-0)^2 = 2 - 3x + \frac{5}{2}x^2; \text{ and} \\ T_3(x) &= f(a) + f'(a)(x-a) + \frac{f''(a)}{2}(x-a)^2 + \frac{f'''(a)}{6}(x-a)^3 \\ &= 2 + (-3)(x-0) + \frac{5}{2}(x-0)^2 + \frac{-9}{6}(x-0)^3 = 2 - 3x + \frac{5}{2}x^2 - \frac{3}{2}x^3. \end{aligned}$$

11. $f(x) = x^2 e^{-x}, \quad a = 1$

SOLUTION First, we calculate and evaluate the needed derivatives:

$$\begin{aligned}
f(x) &= x^2 e^{-x} & f(a) &= 1/e \\
f'(x) &= (2x - x^2)e^{-x} & f'(a) &= 1/e \\
f''(x) &= (x^2 - 4x + 2)e^{-x} & f''(a) &= -1/e \\
f'''(x) &= (-x^2 + 6x - 6)e^{-x} & f'''(a) &= -1/e
\end{aligned}$$

Now,

$$\begin{aligned}
T_2(x) &= f(a) + f'(a)(x-a) + \frac{f''(a)}{2}(x-a)^2 \\
&= \frac{1}{e} + \frac{1}{e}(x-1) + \frac{-1/e}{2}(x-1)^2 = \frac{1}{e} + \frac{1}{e}(x-1) - \frac{1}{2e}(x-1)^2; \text{ and} \\
T_3(x) &= f(a) + f'(a)(x-a) + \frac{f''(a)}{2}(x-a)^2 + \frac{f'''(a)}{6}(x-a)^3 \\
&= \frac{1}{e} + \frac{1}{e}(x-1) + \frac{-1/e}{2}(x-1)^2 + \left(\frac{-1/e}{6}\right)(x-1)^3 \\
&= \frac{1}{e} + \frac{1}{e}(x-1) - \frac{1}{2e}(x-1)^2 - \frac{1}{6e}(x-1)^3.
\end{aligned}$$

13. $f(x) = \dfrac{\ln x}{x}, \quad a = 1$

SOLUTION First calculate and evaluate the needed derivatives:

$$\begin{aligned}
f(x) &= \frac{\ln x}{x} & f(a) &= 0 \\
f'(x) &= \frac{1 - \ln x}{x^2} & f(a) &= 1 \\
f''(x) &= \frac{-3 + 2\ln x}{x^3} & f(a) &= -3 \\
f'''(x) &= \frac{11 - 6\ln x}{x^4} & f(a) &= 11
\end{aligned}$$

so that

$$\begin{aligned}
T_2(x) &= f(a) + f'(a)(x-a) + \frac{f''(a)}{2!}(x-a)^2 = 0 + 1(x-1) + \frac{-3}{2!}(x-1)^2 \\
&= (x-1) - \frac{3}{2}(x-1)^2 \\
T_3(x) &= f(a) + f'(a)(x-a) + \frac{f''(a)}{2!}(x-a)^2 + \frac{f'''(a)}{3!}(x-a)^3 \\
&= 0 + 1(x-1) + \frac{-3}{2!}(x-1)^2 + \frac{11}{3!}(x-1)^3 \\
&= (x-1) - \frac{3}{2}(x-1)^2 + \frac{11}{6}(x-1)^3
\end{aligned}$$

15. Show that the nth Maclaurin polynomial for e^x is

$$T_n(x) = 1 + \frac{x}{1!} + \frac{x^2}{2!} + \cdots + \frac{x^n}{n!}$$

SOLUTION With $f(x) = e^x$, it follows that $f^{(n)}(x) = e^x$ and $f^{(n)}(0) = 1$ for all n. Thus,

$$T_n(x) = 1 + 1(x-0) + \frac{1}{2}(x-0)^2 + \cdots + \frac{1}{n!}(x-0)^n = 1 + x + \frac{x^2}{2} + \cdots + \frac{x^n}{n!}.$$

17. Show that the Maclaurin polynomials for $\sin x$ are

$$T_{2n+1}(x) = T_{2n+2}(x) = x - \frac{x^3}{3!} + \frac{x^5}{5!} - \cdots + (-1)^n \frac{x^{2n+1}}{(2n+1)!}$$

SOLUTION Let $f(x) = \sin x$. Then

$$\begin{aligned} f(x) &= \sin x & f(0) &= 0 \\ f'(x) &= \cos x & f'(0) &= 1 \\ f''(x) &= -\sin x & f''(0) &= 0 \\ f'''(x) &= -\cos x & f'''(0) &= -1 \\ f^{(4)}(x) &= \sin x & f^{(4)}(0) &= 0 \\ f^{(5)}(x) &= \cos x & f^{(5)}(0) &= 1 \\ &\vdots & &\vdots \end{aligned}$$

Consequently,

$$T_{2n+1}(x) = x - \frac{x^3}{3!} + \frac{x^5}{5!} + \cdots + (-1)^n \frac{x^{2n+1}}{(2n+1)!}$$

and

$$T_{2n+2}(x) = x - \frac{x^3}{3!} + \frac{x^5}{5!} + \cdots + (-1)^n \frac{x^{2n+1}}{(2n+1)!} + 0 = T_{2n+1}(x).$$

In Exercises 19–24, find $T_n(x)$ at $x = a$ for all n.

19. $f(x) = \dfrac{1}{1+x}, \quad a = 0$

SOLUTION We have

$$\frac{1}{1+x} = (\ln(1+x))'$$

so that from Exercise 18, letting $g(x) = \ln(1+x)$,

$$f^{(n)}(x) = g^{(n+1)}(x) = (-1)^n n!(x+1)^{-1-n} \quad \text{and} \quad f^{(n)}(0) = (-1)^n n!$$

Then

$$\begin{aligned} T_n(x) &= f(0) + f'(0)x + \frac{f''(0)}{2!}x^2 + \cdots + \frac{f^{(n)}(0)}{n!}x^n \\ &= 1 - x + \frac{2!}{2!}x^2 - \frac{3!}{3!}x^3 + \cdots + (-1)^n \frac{n!}{n!}x^n \\ &= 1 - x + x^2 - x^3 + \cdots + (-1)^n x^n \end{aligned}$$

21. $f(x) = e^x, \quad a = 1$

SOLUTION Let $f(x) = e^x$. Then $f^{(n)}(x) = e^x$ and $f^{(n)}(1) = e$ for all n. Therefore,

$$T_n(x) = e + e(x-1) + \frac{e}{2!}(x-1)^2 + \cdots + \frac{e}{n!}(x-1)^n.$$

23. $f(x) = \cos x, \quad a = \dfrac{\pi}{4}$

SOLUTION Let $f(x) = \cos x$. Then

$$\begin{aligned} f(x) &= \cos x & f(\pi/4) &= \frac{1}{\sqrt{2}} \\ f'(x) &= -\sin x & f'(\pi/4) &= -\frac{1}{\sqrt{2}} \\ f''(x) &= -\cos x & f''(\pi/4) &= -\frac{1}{\sqrt{2}} \\ f'''(x) &= \sin x & f'''(\pi/4) &= \frac{1}{\sqrt{2}} \end{aligned}$$

This pattern of four values repeats indefinitely. Thus,

$$f^{(n)}(\pi/4) = \begin{cases} (-1)^{(n+1)/2}\dfrac{1}{\sqrt{2}}, & n \text{ odd} \\ (-1)^{n/2}\dfrac{1}{\sqrt{2}}, & n \text{ even} \end{cases}$$

and

$$T_n(x) = \frac{1}{\sqrt{2}} - \frac{1}{\sqrt{2}}\left(x - \frac{\pi}{4}\right) - \frac{1}{2\sqrt{2}}\left(x - \frac{\pi}{4}\right)^2 + \frac{1}{6\sqrt{2}}\left(x - \frac{\pi}{4}\right)^3 \cdots.$$

In general, the coefficient of $(x - \pi/4)^n$ is

$$\pm\frac{1}{(\sqrt{2})n!}$$

with the pattern of signs $+, -, -, +, +, -, -, \ldots$.

In Exercises 25–28, find $T_2(x)$ and use a calculator to compute the error $|f(x) - T_2(x)|$ for the given values of a and x.

25. $y = e^x, \quad a = 0, \quad x = -0.5$

SOLUTION Let $f(x) = e^x$. Then $f'(x) = e^x$, $f''(x) = e^x$, $f(a) = 1$, $f'(a) = 1$ and $f''(a) = 1$. Therefore

$$T_2(x) = 1 + 1(x - 0) + \frac{1}{2}(x - 0)^2 = 1 + x + \frac{1}{2}x^2,$$

and

$$T_2(-0.5) = 1 + (-0.5) + \frac{1}{2}(-0.5)^2 = 0.625.$$

Using a calculator, we find

$$f(-0.5) = \frac{1}{\sqrt{e}} = 0.606531,$$

so

$$|T_2(-0.5) - f(-0.5)| = 0.0185.$$

27. $y = x^{-2/3}, \quad a = 1, \quad x = 1.2$

SOLUTION Let $f(x) = x^{-2/3}$. Then $f'(x) = -\frac{2}{3}x^{-5/3}$, $f''(x) = \frac{10}{9}x^{-8/3}$, $f(1) = 1$, $f'(1) = -\frac{2}{3}$, and $f''(1) = \frac{10}{9}$. Thus

$$T_2(x) = 1 - \frac{2}{3}(x - 1) + \frac{10}{2 \cdot 9}(x - 1)^2 = 1 - \frac{2}{3}(x - 1) + \frac{5}{9}(x - 1)^2$$

and

$$T_2(1.2) = 1 - \frac{2}{3}(0.2) + \frac{5}{9}(0.2)^2 = \frac{8}{9} \approx 0.88889$$

Using a calculator, $f(1.2) = (1.2)^{-2/3} \approx 0.88555$ so that

$$|T_2(1.2) - f(1.2)| \approx 0.00334$$

29. GU Compute $T_3(x)$ for $f(x) = \sqrt{x}$ centered at $a = 1$. Then use a plot of the error $|f(x) - T_3(x)|$ to find a value $c > 1$ such that the error on the interval $[1, c]$ is at most 0.25.

SOLUTION We have

$$\begin{aligned} f(x) &= x^{1/2} & f(1) &= 1 \\ f'(x) &= \frac{1}{2}x^{-1/2} & f'(1) &= \frac{1}{2} \\ f''(x) &= -\frac{1}{4}x^{-3/2} & f''(1) &= -\frac{1}{4} \\ f'''(x) &= \frac{3}{8}x^{-5/2} & f'''(1) &= \frac{3}{8} \end{aligned}$$

Therefore

$$T_3(x) = 1 + \frac{1}{2}(x - 1) - \frac{1}{4 \cdot 2!}(x - 1)^2 + \frac{3}{8 \cdot 3!}(x - 1)^3 = 1 + \frac{1}{2}(x - 1) - \frac{1}{8}(x - 1)^2 + \frac{1}{16}(x - 1)^3$$

A plot of $|f(x) - T_3(x)|$ is below.

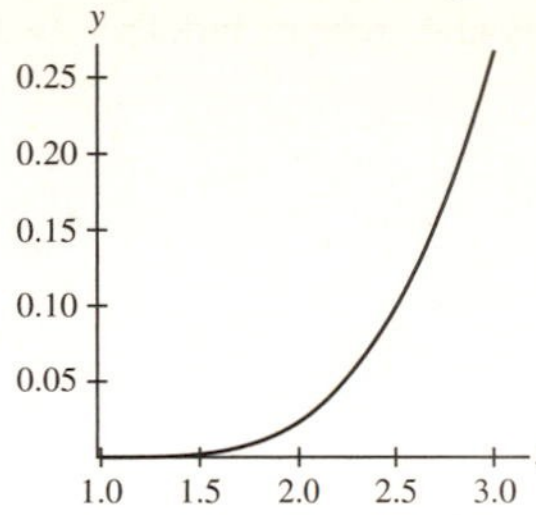

It appears that for $x \in [1, 2.9]$ that the error does not exceed 0.25. The error at $x = 3$ appears to just exceed 0.25.

31. Let $T_3(x)$ be the Maclaurin polynomial of $f(x) = e^x$. Use the error bound to find the maximum possible value of $|f(1.1) - T_3(1.1)|$. Show that we can take $K = e^{1.1}$.

SOLUTION Since $f(x) = e^x$, we have $f^{(n)}(x) = e^x$ for all n; since e^x is increasing, the maximum value of e^x on the interval $[0, 1.1]$ is $K = e^{1.1}$. Then by the error bound,

$$\left|e^{1.1} - T_3(1.1)\right| \le K\frac{(1.1-0)^4}{4!} = \frac{e^{1.1}1.1^4}{24} \approx 0.183$$

In Exercises 33–36, compute the Taylor polynomial indicated and use the error bound to find the maximum possible size of the error. Verify your result with a calculator.

33. $f(x) = \cos x, \quad a = 0; \quad |\cos 0.25 - T_5(0.25)|$

SOLUTION The Maclaurin series for $\cos x$ is

$$1 - \frac{x^2}{2!} + \frac{x^4}{4!} - \frac{x^6}{6!} + \dots$$

so that

$$T_5(x) = 1 - \frac{x^2}{2} + \frac{x^4}{24}$$

$$T_5(0.25) \approx 0.9689127604$$

In addition, $f^{(6)}(x) = -\cos x$ so that $|f^{(6)}(x)| \le 1$ and we may take $K = 1$ in the error bound formula. Then

$$|\cos 0.25 - T_5(0.25)| \le K\frac{0.25^6}{6!} = \frac{1}{2^{12}\cdot 6!} \approx 3.390842014 \cdot 10^{-7}$$

(The true value is $\cos 0.25 \approx 0.9689124217$ and the difference is in fact $\approx 3.387 \cdot 10^{-7}$.)

35. $f(x) = x^{-1/2}, \quad a = 4; \quad |f(4.3) - T_3(4.3)|$

SOLUTION We have

$$f(x) = x^{-1/2} \qquad f(4) = \frac{1}{2}$$

$$f'(x) = -\frac{1}{2}x^{-3/2} \qquad f'(4) = -\frac{1}{16}$$

$$f''(x) = \frac{3}{4}x^{-5/2} \qquad f''(4) = \frac{3}{128}$$

$$f'''(x) = -\frac{15}{8}x^{-7/2} \qquad f'''(4) = -\frac{15}{1024}$$

$$f^{(4)}(x) = \frac{105}{16}x^{-9/2}$$

so that

$$T_3(x) = \frac{1}{2} - \frac{1}{16}(x-4) + \frac{3}{256}(x-4)^2 - \frac{5}{2048}(x-4)^3$$

Using the error bound formula,

$$|f(4.3) - T_3(4.3)| \le K\frac{|4.3-4|^4}{4!} = \frac{27K}{80{,}000}$$

where K is a number such that $|f^{(4)}(x)| \le K$ for x between 4 and 4.3. Now, $f^{(4)}(x)$ is a decreasing function for $x > 1$, so it takes its maximum value on [4, 4.3] at $x = 4$; there, its value is

$$K = \frac{105}{16}4^{-9/2} = \frac{105}{8192}$$

so that

$$|f(4.3) - T_3(4.3)| \le \frac{27\frac{105}{8192}}{80{,}000} = \frac{27 \cdot 105}{8192 \cdot 80{,}000} \approx 4.3258667 \cdot 10^{-6}$$

37. Calculate the Maclaurin polynomial $T_3(x)$ for $f(x) = \tan^{-1} x$. Compute $T_3(\frac{1}{2})$ and use the error bound to find a bound for the error $\left|\tan^{-1}\frac{1}{2} - T_3(\frac{1}{2})\right|$. Refer to the graph in Figure 10 to find an acceptable value of K. Verify your result by computing $\left|\tan^{-1}\frac{1}{2} - T_3(\frac{1}{2})\right|$ using a calculator.

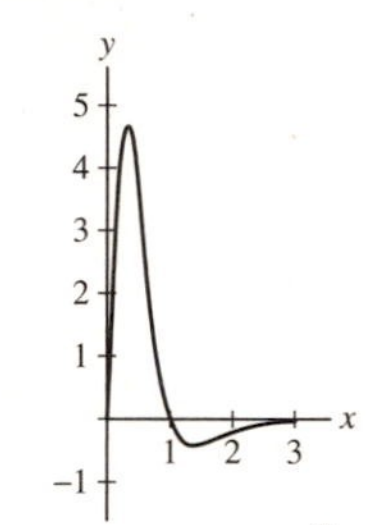

FIGURE 10 Graph of $f^{(4)}(x) = \dfrac{-24x(x^2-1)}{(x^2+1)^4}$, where $f(x) = \tan^{-1} x$.

SOLUTION Let $f(x) = \tan^{-1} x$. Then

$$f(x) = \tan^{-1} x \qquad f(0) = 0$$

$$f'(x) = \frac{1}{1+x^2} \qquad f'(0) = 1$$

$$f''(x) = \frac{-2x}{(1+x^2)^2} \qquad f''(0) = 0$$

$$f'''(x) = \frac{(1+x^2)^2(-2) - (-2x)(2)(1+x^2)(2x)}{(1+x^2)^4} \qquad f'''(0) = -2$$

and

$$T_3(x) = 0 + 1(x-0) + \frac{0}{2}(x-0)^2 + \frac{-2}{6}(x-0)^3 = x - \frac{x^3}{3}.$$

Since $f^{(4)}(x) \le 5$ for $x \ge 0$, we may take $K = 5$ in the error bound; then,

$$\left|\tan^{-1}\left(\frac{1}{2}\right) - T_3\left(\frac{1}{2}\right)\right| \le \frac{5(1/2)^4}{4!} = \frac{5}{384}.$$

39. GU Let $T_2(x)$ be the Taylor polynomial at $a = 0.5$ for $f(x) = \cos(x^2)$. Use the error bound to find the maximum possible value of $|f(0.6) - T_2(0.6)|$. Plot $f^{(3)}(x)$ to find an acceptable value of K.

SOLUTION We have

$$f(x) = \cos(x^2) \qquad f(0.5) = \cos(0.25) \approx 0.9689$$

$$f'(x) = -2x\sin(x^2) \qquad f'(0.5) = -\sin(0.25) \approx -0.2474039593$$

$$f''(x) = -4x^2\cos(x^2) - 2\sin(x^2) \qquad f''(0.5) = -\cos(0.25) - 2\sin(0.25) \approx -1.463720340$$

$$f^{(3)}(x) = 8x^3\sin(x^2) - 12x\cos(x^2)$$

so that

$$T_2(x) = 0.9689 - 0.2472039593(x - 0.5) - 0.73186017(x - 0.5)^2$$

and $T_2(0.6) \approx 0.9368534237$. A graph of $f^{(3)}(x)$ for x near 0.5 is below.

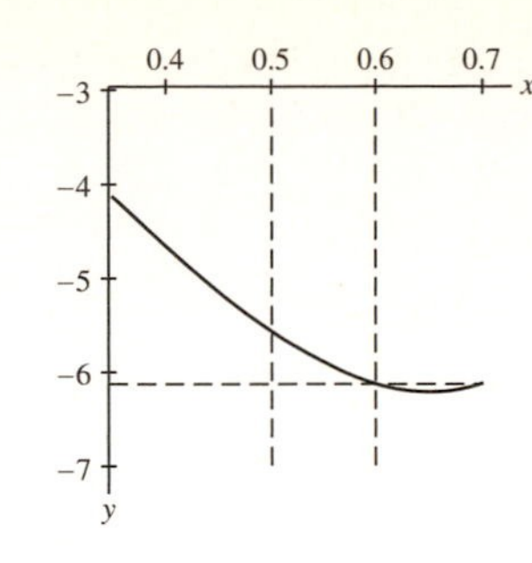

Clearly the maximum value of $|f^{(3)}(x)|$ on $[0.5, 0.6]$ is bounded by 7 (near $x = 0.5$), so we may take $K = 7$; then

$$|f(0.6) - T_2(0.6)| \leq K\frac{|0.6 - 0.5|^3}{3!} = \frac{7}{6000} \approx 0.0011666667$$

In Exercises 41–44, use the error bound to find a value of n for which the given inequality is satisfied. Then verify your result using a calculator.

41. $|\cos 0.1 - T_n(0.1)| \leq 10^{-7}, \quad a = 0$

SOLUTION Using the error bound with $K = 1$ (every derivative of $f(x) = \cos x$ is $\pm \sin x$ or $\pm \cos x$, so $|f^{(n)}(x)| \leq 1$ for all n), we have

$$|T_n(0.1) - \cos 0.1| \leq \frac{(0.1)^{n+1}}{(n+1)!}.$$

With $n = 3$,

$$\frac{(0.1)^4}{4!} \approx 4.17 \times 10^{-6} > 10^{-7},$$

but with $n = 4$,

$$\frac{(0.1)^5}{5!} \approx 8.33 \times 10^{-8} < 10^{-7},$$

so we choose $n = 4$. Now,

$$T_4(x) = 1 - \frac{1}{2}x^2 + \frac{1}{24}x^4,$$

so

$$T_4(0.1) = 1 - \frac{1}{2}(0.1)^2 + \frac{1}{24}(0.1)^4 = 0.995004166.$$

Using a calculator, $\cos 0.1 = 0.995004165$, so

$$|T_4(0.1) - \cos 0.1| = 1.387 \times 10^{-8} < 10^{-7}.$$

43. $|\sqrt{1.3} - T_n(1.3)| \leq 10^{-6}, \quad a = 1$

SOLUTION Using the Error Bound, we have

$$|\sqrt{1.3} - T_n(1.3)| \leq K\frac{|1.3 - 1|^{n+1}}{(n+1)!} = K\frac{|0.3|^{n+1}}{(n+1)!},$$

where K is a number such that $|f^{(n+1)}(x)| \leq K$ for x between 1 and 1.3. For $f(x) = \sqrt{x}$, $|f^{(n)}(x)|$ is decreasing for $x > 1$, hence the maximum value of $|f^{(n+1)}(x)|$ occurs at $x = 1$. We may therefore take

$$K = |f^{(n+1)}(1)| = \frac{1 \cdot 3 \cdot 5 \cdots (2n+1)}{2^{n+1}}$$
$$= \frac{1 \cdot 3 \cdot 5 \cdots (2n+1)}{2^{n+1}} \cdot \frac{2 \cdot 4 \cdot 6 \cdots (2n+2)}{2 \cdot 4 \cdot 6 \cdots (2n+2)} = \frac{(2n+2)!}{(n+1)!2^{2n+2}}.$$

Then

$$|\sqrt{1.3} - T_n(1.3)| \leq \frac{(2n+2)!}{(n+1)!2^{2n+2}} \cdot \frac{|0.3|^{n+1}}{(n+1)!} = \frac{(2n+2)!}{[(n+1)!]^2}(0.075)^{n+1}.$$

With $n = 9$

$$\frac{(20)!}{[(10)!]^2}(0.075)^{10} = 1.040 \times 10^{-6} > 10^{-6},$$

but with $n = 10$

$$\frac{(22)!}{[(11)!]^2}(0.075)^{11} = 2.979 \times 10^{-7} < 10^{-6}.$$

Hence, $n = 10$ will guarantee the desired accuracy. Using technology to compute and evaluate $T_{10}(1.3)$ gives

$$T_{10}(1.3) \approx 1.140175414, \qquad \sqrt{1.3} \approx 1.140175425$$

and

$$|\sqrt{1.3} - T_{10}(1.3)| \approx 1.1 \times 10^{-8} < 10^{-6}$$

45. Let $f(x) = e^{-x}$ and $T_3(x) = 1 - x + \frac{x^2}{2} - \frac{x^3}{6}$. Use the error bound to show that for all $x \ge 0$,

$$|f(x) - T_3(x)| \le \frac{x^4}{24}$$

If you have a GU, illustrate this inequality by plotting $f(x) - T_3(x)$ and $x^4/24$ together over [0, 1].

SOLUTION Note that $f^{(n)}(x) = \pm e^{-x}$, so that $|f^{(n)}(x)| = f(x)$. Now, $f(x)$ is a decreasing function for $x \ge 0$, so that for any $c > 0$, $|f^{(n)}(x)|$ takes its maximum value at $x = 0$; this value is $e^0 = 1$. Thus we may take $K = 1$ in the error bound equation. Thus for any x,

$$|f(x) - T_3(x)| \le K\frac{|x - 0|^4}{4!} = \frac{x^4}{24}$$

A plot of $f(x) - T_3(x)$ and $\frac{x^4}{24}$ is shown below.

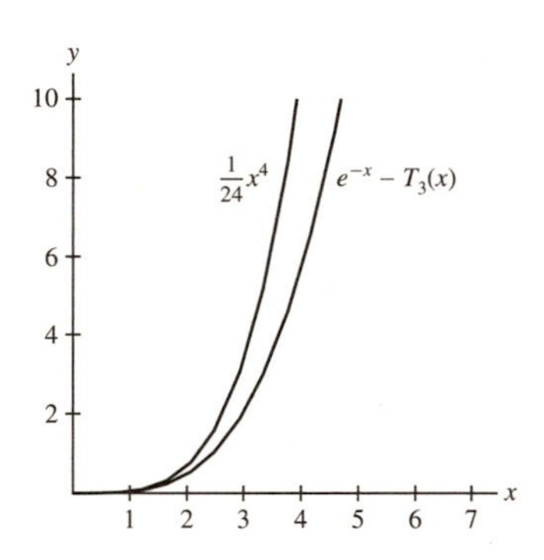

47. Let $T_n(x)$ be the Taylor polynomial for $f(x) = \ln x$ at $a = 1$, and let $c > 1$. Show that

$$|\ln c - T_n(c)| \le \frac{|c - 1|^{n+1}}{n + 1}$$

Then find a value of n such that $|\ln 1.5 - T_n(1.5)| \le 10^{-2}$.

SOLUTION With $f(x) = \ln x$, we have

$$f'(x) = x^{-1}, \ f''(x) = -x^{-2}, \ f'''(x) = 2x^{-3}, \ f^{(4)}(x) = -6x^{-4},$$

and, in general,

$$f^{(k+1)}(x) = (-1)^k k!\, x^{-k-1}.$$

Notice that $|f^{(k+1)}(x)| = k!|x|^{-k-1}$ is a decreasing function for $x > 0$. Therefore, the maximum value of $|f^{(k+1)}(x)|$ on $[1, c]$ is $|f^{(k+1)}(1)|$. Using the Error Bound, we have

$$|\ln c - T_n(c)| \le K\frac{|c - 1|^{n+1}}{(n + 1)!},$$

where K is a number such that $|f^{(n+1)}(x)| \leq K$ for x between 1 and c. From part (a), we know that we may take $K = |f^{(n+1)}(1)| = n!$. Then

$$|\ln c - T_n(c)| \leq n! \frac{|c-1|^{n+1}}{(n+1)!} = \frac{|c-1|^{n+1}}{n+1}.$$

Evaluating at $c = 1.5$ gives

$$|\ln 1.5 - T_n(1.5)| \leq \frac{|1.5-1|^{n+1}}{n+1} = \frac{(0.5)^{n+1}}{n+1}.$$

With $n = 3$,

$$\frac{(0.5)^4}{4} = 0.015625 > 10^{-2}.$$

but with $n = 4$,

$$\frac{(0.5)^5}{5} = 0.00625 < 10^{-2}.$$

Hence, $n = 4$ will guarantee the desired accuracy.

49. Verify that the third Maclaurin polynomial for $f(x) = e^x \sin x$ is equal to the product of the third Maclaurin polynomials of e^x and $\sin x$ (after discarding terms of degree greater than 3 in the product).

SOLUTION Let $f(x) = e^x \sin x$. Then

$$\begin{aligned} f(x) &= e^x \sin x & f(0) &= 0 \\ f'(x) &= e^x(\cos x + \sin x) & f'(0) &= 1 \\ f''(x) &= 2e^x \cos x & f''(0) &= 2 \\ f'''(x) &= 2e^x(\cos x - \sin x) & f'''(0) &= 2 \end{aligned}$$

and

$$T_3(x) = 0 + (1)x + \frac{2}{2!}x^2 + \frac{2}{3!}x^3 = x + x^2 + \frac{x^3}{3}.$$

Now, the third Maclaurin polynomial for e^x is $1 + x + \frac{x^2}{2} + \frac{x^3}{6}$, and the third Maclaurin polynomial for $\sin x$ is $x - \frac{x^3}{6}$. Multiplying these two polynomials, and then discarding terms of degree greater than 3, yields

$$e^x \sin x \approx x + x^2 + \frac{x^3}{3},$$

which agrees with the Maclaurin polynomial obtained from the definition.

51. Find the Maclaurin polynomials $T_n(x)$ for $f(x) = \cos(x^2)$. You may use the fact that $T_n(x)$ is equal to the sum of the terms up to degree n obtained by substituting x^2 for x in the nth Maclaurin polynomial of $\cos x$.

SOLUTION The Maclaurin polynomials for $\cos x$ are of the form

$$T_{2n}(x) = 1 - \frac{x^2}{2} + \frac{x^4}{4!} + \cdots + (-1)^n \frac{x^{2n}}{(2n)!}.$$

Accordingly, the Maclaurin polynomials for $\cos(x^2)$ are of the form

$$T_{4n}(x) = 1 - \frac{x^4}{2} + \frac{x^8}{4!} + \cdots + (-1)^n \frac{x^{4n}}{(2n)!}.$$

53. Let $f(x) = 3x^3 + 2x^2 - x - 4$. Calculate $T_j(x)$ for $j = 1, 2, 3, 4, 5$ at both $a = 0$ and $a = 1$. Show that $T_3(x) = f(x)$ in both cases.

SOLUTION Let $f(x) = 3x^3 + 2x^2 - x - 4$. Then

$$\begin{aligned} f(x) &= 3x^3 + 2x^2 - x - 4 & f(0) &= -4 & f(1) &= 0 \\ f'(x) &= 9x^2 + 4x - 1 & f'(0) &= -1 & f'(1) &= 12 \\ f''(x) &= 18x + 4 & f''(0) &= 4 & f''(1) &= 22 \\ f'''(x) &= 18 & f'''(0) &= 18 & f'''(1) &= 18 \\ f^{(4)}(x) &= 0 & f^{(4)}(0) &= 0 & f^{(4)}(1) &= 0 \\ f^{(5)}(x) &= 0 & f^{(5)}(0) &= 0 & f^{(5)}(1) &= 0 \end{aligned}$$

At $a = 0$,

$$T_1(x) = -4 - x;$$
$$T_2(x) = -4 - x + 2x^2;$$
$$T_3(x) = -4 - x + 2x^2 + 3x^3 = f(x);$$
$$T_4(x) = T_3(x); \text{ and}$$
$$T_5(x) = T_3(x).$$

At $a = 1$,

$$T_1(x) = 12(x - 1);$$
$$T_2(x) = 12(x - 1) + 11(x - 1)^2;$$
$$T_3(x) = 12(x - 1) + 11(x - 1)^2 + 3(x - 1)^3 = -4 - x + 2x^2 + 3x^3 = f(x);$$
$$T_4(x) = T_3(x); \text{ and}$$
$$T_5(x) = T_3(x).$$

55. Let $s(t)$ be the distance of a truck to an intersection. At time $t = 0$, the truck is 60 meters from the intersection, travels away from it with a velocity of 24 m/s, and begins to slow down with an acceleration of $a = -3$ m/s^2. Determine the second Maclaurin polynomial of $s(t)$, and use it to estimate the truck's distance from the intersection after 4 s.

SOLUTION Place the origin at the intersection, so that $s(0) = 60$ (the truck is traveling away from the intersection). The second Maclaurin polynomial of $s(t)$ is

$$T_2(t) = s(0) + s'(0)t + \frac{s''(0)}{2}t^2$$

The conditions of the problem tell us that $s(0) = 60$, $s'(0) = 24$, and $s''(0) = -3$. Thus

$$T_2(t) = 60 + 24t - \frac{3}{2}t^2$$

so that after 4 seconds,

$$T_2(4) = 60 + 24 \cdot 4 - \frac{3}{2} \cdot 4^2 = 132 \text{ m}$$

The truck is 132 m past the intersection.

57. A narrow, negatively charged ring of radius R exerts a force on a positively charged particle P located at distance x above the center of the ring of magnitude

$$F(x) = -\frac{kx}{(x^2 + R^2)^{3/2}}$$

where $k > 0$ is a constant (Figure 12).

(a) Compute the third-degree Maclaurin polynomial for $F(x)$.

(b) Show that $F \approx -(k/R^3)x$ to second order. This shows that when x is small, $F(x)$ behaves like a restoring force similar to the force exerted by a spring.

(c) Show that $F(x) \approx -k/x^2$ when x is large by showing that

$$\lim_{x \to \infty} \frac{F(x)}{-k/x^2} = 1$$

Thus, $F(x)$ behaves like an inverse square law, and the charged ring looks like a point charge from far away.

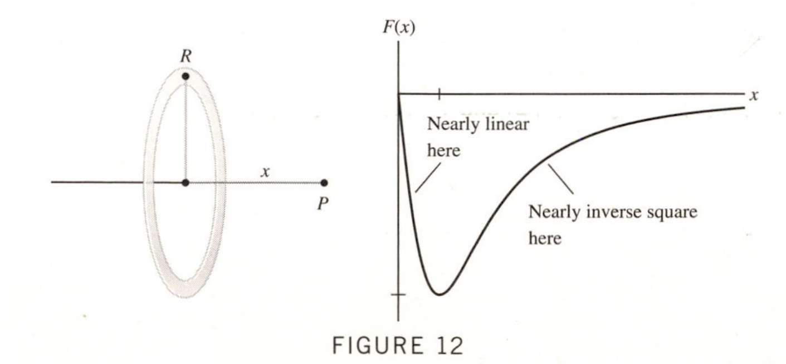

FIGURE 12

SOLUTION

(a) Start by computing and evaluating the necessary derivatives:

$$F(x) = -\frac{kx}{(x^2+R^2)^{3/2}} \qquad F(0) = 0$$

$$F'(x) = \frac{k(2x^2-R^2)}{(x^2+R^2)^{5/2}} \qquad F'(0) = -\frac{k}{R^3}$$

$$F''(x) = \frac{3kx(3R^2-2x^2)}{(x^2+R^2)^{7/2}} \qquad F''(0) = 0$$

$$F'''(x) = \frac{3k(8x^4-24x^2R^2+3R^4)}{(x^2+R^2)^{9/2}} \qquad F'''(0) = \frac{9k}{R^5}$$

so that

$$T_3(x) = F(0) + F'(0)x + \frac{F''(0)}{2!}x^2 + \frac{F'''(0)}{3!}x^3 = -\frac{k}{R^3}x + \frac{3k}{2R^5}x^3$$

(b) To degree 2, $F(x) \approx T_3(x) \approx -\frac{k}{R^3}x$ as we may ignore the x^3 term of $T_3(x)$.

(c) We have

$$\lim_{x\to\infty}\frac{F(x)}{-k/x^2} = \lim_{x\to\infty}\left(-\frac{x^2}{k}\cdot\frac{-kx}{(x^2+R^2)^{3/2}}\right) = \lim_{x\to\infty}\frac{x^3}{(x^2+R^2)^{3/2}}$$

$$= \lim_{x\to\infty}\frac{1}{x^{-3}(x^2+R^2)^{3/2}} = \lim_{x\to\infty}\frac{1}{(1+R^2/x^2)^{3/2}}$$

$$= 1$$

Thus as x grows large, $F(x)$ looks like an inverse square function.

59. Referring to Figure 14, let a be the length of the chord $\overline{AC}$ of angle θ of the unit circle. Derive the following approximation for the excess of the arc over the chord.

$$\theta - a \approx \frac{\theta^3}{24}$$

Hint: Show that $\theta - a = \theta - 2\sin(\theta/2)$ and use the third Maclaurin polynomial as an approximation.

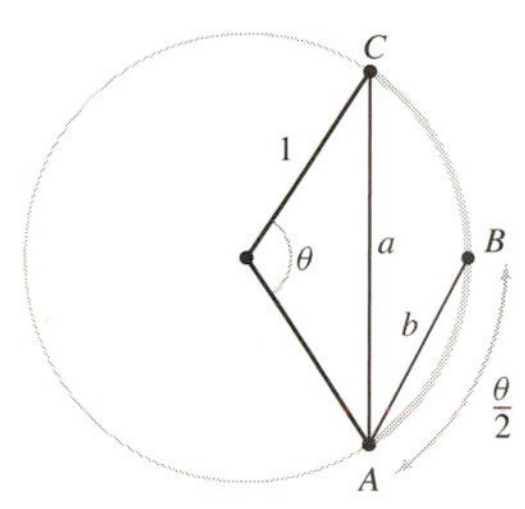

FIGURE 14 Unit circle.

SOLUTION Draw a line from the center O of the circle to B, and label the point of intersection of this line with AC as D. Then $CD = \frac{a}{2}$, and the angle COB is $\frac{\theta}{2}$. Since $CO = 1$, we have

$$\sin\frac{\theta}{2} = \frac{a}{2}$$

so that $a = 2\sin(\theta/2)$. Thus $\theta - a = \theta - 2\sin(\theta/2)$. Now, the third Maclaurin polynomial for $f(\theta) = \sin(\theta/2)$ can be computed as follows: $f(0) = 0$, $f'(x) = \frac{1}{2}\cos(\theta/2)$ so that $f'(0) = \frac{1}{2}$. $f''(x) = -\frac{1}{4}\sin(\theta/2)$ and $f''(0) = 0$. Finally, $f'''(x) = -\frac{1}{8}\cos(\theta/2)$ and $f'''(0) = -\frac{1}{8}$. Thus

$$T_3(\theta) = f(0) + f'(0)\theta + \frac{f''(0)}{2!}\theta^2 + \frac{f'''(0)}{3!}\theta^3 = \frac{1}{2}\theta - \frac{1}{48}\theta^3$$

Finally,

$$\theta - a = \theta - 2\sin\frac{\theta}{2} \approx \theta - 2T_3(\theta) = \theta - \left(\theta - \frac{1}{24}\theta^3\right) = \frac{\theta^3}{24}$$

Further Insights and Challenges

61. Show that the nth Maclaurin polynomial of $f(x) = \arcsin x$ for n odd is

$$T_n(x) = x + \frac{1}{2}\frac{x^3}{3} + \frac{1 \cdot 3}{2 \cdot 4}\frac{x^5}{5} + \cdots + \frac{1 \cdot 3 \cdot 5 \cdots (n-2)}{2 \cdot 4 \cdot 6 \cdots (n-1)}\frac{x^n}{n}$$

SOLUTION Let $f(x) = \sin^{-1} x$. Then

$$\begin{aligned}
f(x) &= \sin^{-1} x & f(0) &= 0 \\
f'(x) &= \frac{1}{\sqrt{1-x^2}} & f'(0) &= 1 \\
f''(x) &= -\frac{1}{2}(1-x^2)^{-3/2}(-2x) & f''(0) &= 0 \\
f'''(x) &= \frac{2x^2+1}{(1-x^2)^{5/2}} & f'''(0) &= 1 \\
f^{(4)}(x) &= \frac{-3x(2x^2+3)}{(1-x^2)^{7/2}} & f^{(4)}(0) &= 0 \\
f^{(5)}(x) &= \frac{24x^4+72x^2+9}{(1-x^2)^{9/2}} & f^{(5)}(0) &= 9 \\
&\vdots & &\vdots \\
& & f^{(7)}(0) &= 225
\end{aligned}$$

and

$$T_7(x) = x + \frac{x^3}{3!} + \frac{9x^5}{5!} + \frac{225x^7}{7!} = x + \frac{1}{2}\frac{x^3}{3} + \frac{1}{2}\frac{3}{4}\frac{x^5}{5} + \frac{1}{2}\frac{3}{4}\frac{5}{6}\frac{x^7}{7}.$$

Thus, we can infer that

$$T_n(x) = x + \frac{1}{2} \cdot \frac{x^3}{3} + \frac{1}{2}\frac{3}{4}\frac{x^5}{5} + \frac{1}{2}\frac{3}{4}\frac{5}{6}\frac{x^7}{7} + \cdots + \frac{1}{2}\frac{3}{4}\cdots\frac{n-2}{n-1}\frac{x^n}{n}.$$

63. Use Exercise 62 to show that for $x \geq 0$ and all n,

$$e^x \geq 1 + x + \frac{x^2}{2!} + \cdots + \frac{x^n}{n!}$$

Sketch the graphs of e^x, $T_1(x)$, and $T_2(x)$ on the same coordinate axes. Does this inequality remain true for $x < 0$?

SOLUTION Let $f(x) = e^x$. Then $f^{(n)}(x) = e^x$ for all n. Because $e^x > 0$ for all x, it follows from Exercise 62 that $f(x) \geq T_n(x)$ for all $x \geq 0$ and for all n. For $f(x) = e^x$,

$$T_n(x) = 1 + x + \frac{x^2}{2!} + \cdots + \frac{x^n}{n!},$$

thus,

$$e^x \geq 1 + x + \frac{x^2}{2!} + \cdots + \frac{x^n}{n!}.$$

From the figure below, we see that the inequality does not remain true for $x < 0$, as $T_2(x) \geq e^x$ for $x < 0$.

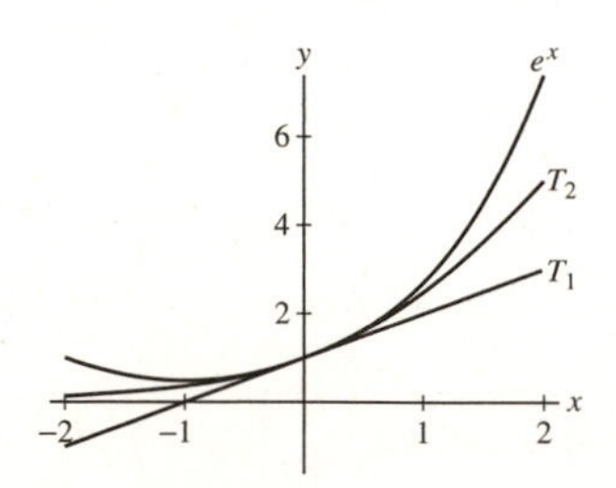

In Exercises 65–69, we estimate integrals using Taylor polynomials. Exercise 66 is used to estimate the error.

65. Find the fourth Maclaurin polynomial $T_4(x)$ for $f(x) = e^{-x^2}$, and calculate $I = \int_0^{1/2} T_4(x)\,dx$ as an estimate $\int_0^{1/2} e^{-x^2}\,dx$. A CAS yields the value $I \approx 0.461281$. How large is the error in your approximation? *Hint:* $T_4(x)$ is obtained by substituting $-x^2$ in the second Maclaurin polynomial for e^x.

SOLUTION Following the hint, since the second Maclaurin polynomial for e^x is

$$1 + x + \frac{x^2}{2}$$

we substitute $-x^2$ for x to get the fourth Maclaurin polynomial for e^{x^2}:

$$T_4(x) = 1 - x^2 + \frac{x^4}{2}$$

Then

$$\int_0^{1/2} e^{-x^2}\,dx \approx \int_0^{1/2} T_4(x)\,dx = \left(x - \frac{1}{3}x^3 + \frac{1}{10}x^5\right)\Bigg|_0^{1/2} = \frac{443}{960} \approx 0.4614583333$$

Using a CAS, we have $\int_0^{1/2} e^{-x^2}\,dx \approx 0.4612810064$, so the error is about 1.77×10^{-4}.

67. Let $T_4(x)$ be the fourth Maclaurin polynomial for $\cos x$.

(a) Show that $|\cos x - T_4(x)| \le \left(\frac{1}{2}\right)^6/6!$ for all $x \in \left[0, \frac{1}{2}\right]$. *Hint:* $T_4(x) = T_5(x)$.

(b) Evaluate $\int_0^{1/2} T_4(x)\,dx$ as an approximation to $\int_0^{1/2} \cos x\,dx$. Use Exercise 66 to find a bound for the size of the error.

SOLUTION

(a) Let $f(x) = \cos x$. Then

$$T_4(x) = 1 - \frac{x^2}{2} + \frac{x^4}{24}.$$

Moreover, with $a = 0$, $T_4(x) = T_5(x)$ and

$$|\cos x - T_4(x)| \le K\frac{|x|^6}{6!},$$

where K is a number such that $|f^{(6)}(u)| \le K$ for u between 0 and x. Now $|f^{(6)}(u)| = |\cos u| \le 1$, so we may take $K = 1$. Finally, with the restriction $x \in [0, \frac{1}{2}]$,

$$|\cos x - T_4(x)| \le \frac{(1/2)^6}{6!} \approx 0.000022.$$

(b)

$$\int_0^{1/2} \left(1 - \frac{x^2}{2} + \frac{x^4}{24}\right) dx = \frac{1841}{3840} \approx 0.479427.$$

By (a) and Exercise 66, the error associated with this approximation is less than or equal to

$$\frac{(1/2)^6}{6!}\left(\frac{1}{2} - 0\right) = \frac{1}{92{,}160} \approx 1.1 \times 10^{-5}.$$

Note that $\displaystyle\int_0^{1/2} \cos x\,dx \approx 0.4794255$, so the actual error is roughly 1.5×10^{-6}.

69. (a) Compute the sixth Maclaurin polynomial $T_6(x)$ for $\sin(x^2)$ by substituting x^2 in $P(x) = x - x^3/6$, the third Maclaurin polynomial for $\sin x$.

(b) Show that $|\sin(x^2) - T_6(x)| \le \dfrac{|x|^{10}}{5!}$.

Hint: Substitute x^2 for x in the error bound for $|\sin x - P(x)|$, noting that $P(x)$ is also the fourth Maclaurin polynomial for $\sin x$.

(c) Use $T_6(x)$ to approximate $\displaystyle\int_0^{1/2} \sin(x^2)\,dx$ and find a bound for the error.

SOLUTION Let $s(x) = \sin x$ and $f(x) = \sin(x^2)$. Then

(a) The third Maclaurin polynomial for $\sin x$ is

$$S_3(x) = x - \frac{x^3}{6}$$

so, substituting x^2 for x, we see that the sixth Maclaurin polynomial for $\sin(x^2)$ is

$$T_6(x) = x^2 - \frac{x^6}{6}$$

(b) Since all derivatives of $s(x)$ are either $\pm \cos x$ or $\pm \sin x$, they are bounded in magnitude by 1, so we may take $K = 1$ in the Error Bound for $\sin x$. Since the third Maclaurin polynomial $S_3(x)$ for $\sin x$ is also the fourth Maclaurin polynomial $S_4(x)$, we have

$$|\sin x - S_3(x)| = |\sin x - S_4(x)| \le K\frac{|x|^5}{5!} = \frac{|x|^5}{5!}$$

Now substitute x^2 for x in the above inequality and note from part (a) that $S_3(x^2) = T_6(x)$ to get

$$|\sin(x^2) - S_3(x^2)| = |\sin(x^2) - T_6(x)| \le \frac{|x^2|^5}{5!} = \frac{|x|^{10}}{5!}$$

(c)

$$\int_0^{1/2} \sin(x^2)\,dx \approx \int_0^{1/2} T_6(x)\,dx = \left(\frac{1}{3}x^3 - \frac{1}{42}x^7\right)\Big|_0^{1/2} \approx 0.04148065476$$

From part (b), the error is bounded by

$$\frac{x^{10}}{5!} = \frac{(1/2)^{10}}{120} = \frac{1}{1024 \cdot 120} \approx 8.138020833 \times 10^{-6}$$

The true value of the integral is approximately 0.04148102420, which is consistent with the computed error bound.

71. Let a be any number and let

$$P(x) = a_n x^n + a_{n-1}x^{n-1} + \cdots + a_1 x + a_0$$

be a polynomial of degree n or less.

(a) Show that if $P^{(j)}(a) = 0$ for $j = 0, 1, \ldots, n$, then $P(x) = 0$, that is, $a_j = 0$ for all j. *Hint:* Use induction, noting that if the statement is true for degree $n - 1$, then $P'(x) = 0$.

(b) Prove that $T_n(x)$ is the only polynomial of degree n or less that agrees with $f(x)$ at $x = a$ to order n. *Hint:* If $Q(x)$ is another such polynomial, apply (a) to $P(x) = T_n(x) - Q(x)$.

SOLUTION

(a) Note first that if $n = 0$, i.e. if $P(x) = a_0$ is a constant, then the statement holds: if $P^{(0)}(a) = P(a) = 0$, then $a_0 = 0$ so that $P(x) = 0$. Next, assume the statement holds for all polynomials of degree $n - 1$ or less, and let $P(x)$ be a polynomial of degree at most n with $P^{(j)}(a) = 0$ for $j = 0, 1, \ldots, n$. If $P(x)$ has degree less than n, then we know $P(x) = 0$ by induction, so assume the degree of $P(x)$ is exactly n. Then

$$P(x) = a_n x^n + a_{n-1}x^{n-1} + \cdots + a_1 x + a_0$$

where $a_n \ne 0$; also,

$$P'(x) = na_n x^{n-1} + (n-1)a_{n-1}x^{n-2} + \cdots + a_1$$

Note that $P^{(j+1)}(a) = (P')^{(j)}(a)$ for $j = 0, 1, \ldots, n - 1$. But then

$$0 = P^{(j+1)}(a) = (P')^{(j)}(a) \quad \text{for all } j = 0, 1, \ldots, n-1$$

Since $P'(x)$ has degree at most $n - 1$, it follows by induction that $P'(x) = 0$. Thus $a_n = a_{n-1} = \cdots = a_1 = 0$ so that $P(x) = a_0$. But $P(a) = 0$ so that $a_0 = 0$ as well and thus $P(x) = 0$.

(b) Suppose $Q(x)$ is a polynomial of degree at most n that agrees with $f(x)$ at $x = a$ up to order n. Let $P(x) = T_n(x) - Q(x)$. Note that $P(x)$ is a polynomial of degree at most n since both $T_n(x)$ and $Q(x)$ are. Since both $T_n(x)$ and $Q(x)$ agree with $f(x)$ at $x = a$ to order n, we have

$$T_n^{(j)}(a) = f^{(j)}(a) = Q^{(j)}(a), \quad j = 0, 1, 2, \ldots, n$$

Thus

$$P^{(j)}(a) = T_n^{(j)}(a) - Q^{(j)}(a) = 0 \quad \text{for } j = 0, 1, 2, \ldots, n$$

But then by part (a), $P(x) = 0$ so that $T_n(x) = Q(x)$.

CHAPTER REVIEW EXERCISES

In Exercises 1–4, calculate the arc length over the given interval.

1. $y = \frac{x^5}{10} + \frac{x^{-3}}{6}$, $[1, 2]$

SOLUTION Let $y = \frac{x^5}{10} + \frac{x^{-3}}{6}$. Then

$$1 + (y')^2 = 1 + \left(\frac{x^4}{2} - \frac{x^{-4}}{2}\right)^2 = 1 + \frac{x^8}{4} - \frac{1}{2} + \frac{x^{-8}}{4}$$

$$= \frac{x^8}{4} + \frac{1}{2} + \frac{x^{-8}}{4} = \left(\frac{x^4}{2} + \frac{x^{-4}}{2}\right)^2.$$

Because $\frac{1}{2}(x^4 + x^{-4}) > 0$ on $[1, 2]$, the arc length is

$$s = \int_1^2 \sqrt{1 + (y')^2}\, dx = \int_1^2 \left(\frac{x^4}{2} + \frac{x^{-4}}{2}\right) dx = \left(\frac{x^5}{10} - \frac{x^{-3}}{6}\right)\Bigg|_1^2 = \frac{779}{240}.$$

3. $y = 4x - 2$, $[-2, 2]$

SOLUTION Let $y = 4x - 2$. Then

$$\sqrt{1 + (y')^2} = \sqrt{1 + 4^2} = \sqrt{17}.$$

Hence,

$$s = \int_{-2}^2 \sqrt{17}\, dx = 4\sqrt{17}.$$

5. Show that the arc length of $y = 2\sqrt{x}$ over $[0, a]$ is equal to $\sqrt{a(a+1)} + \ln(\sqrt{a} + \sqrt{a+1})$. *Hint:* Apply the substitution $x = \tan^2\theta$ to the arc length integral.

SOLUTION Let $y = 2\sqrt{x}$. Then $y' = \frac{1}{\sqrt{x}}$, and

$$\sqrt{1 + (y')^2} = \sqrt{1 + \frac{1}{x}} = \sqrt{\frac{x+1}{x}} = \frac{1}{\sqrt{x}}\sqrt{x+1}.$$

Thus,

$$s = \int_0^a \frac{1}{\sqrt{x}}\sqrt{1+x}\, dx.$$

We make the substitution $x = \tan^2\theta$, $dx = 2\tan\theta\sec^2\theta\, d\theta$. Then

$$s = \int_{x=0}^{x=a} \frac{1}{\tan\theta}\sec\theta \cdot 2\tan\theta\sec^2\theta\, d\theta = 2\int_{x=0}^{x=a} \sec^3\theta\, d\theta.$$

We use a reduction formula to obtain

$$s = 2\left(\frac{\tan\theta\sec\theta}{2} + \frac{1}{2}\ln|\sec\theta + \tan\theta|\right)\Bigg|_{x=0}^{x=a} = \left(\sqrt{x}\sqrt{1+x} + \ln|\sqrt{1+x} + \sqrt{x}|\right)\Big|_0^a$$

$$= \sqrt{a}\sqrt{1+a} + \ln|\sqrt{1+a} + \sqrt{a}| = \sqrt{a(a+1)} + \ln\left(\sqrt{a} + \sqrt{a+1}\right).$$

In Exercises 7–10, calculate the surface area of the solid obtained by rotating the curve over the given interval about the x-axis.

7. $y = x + 1$, $[0, 4]$

SOLUTION Let $y = x + 1$. Then $y' = 1$, and

$$y\sqrt{1 + y'^2} = (x+1)\sqrt{1+1} = \sqrt{2}(x+1).$$

Thus,

$$SA = 2\pi \int_0^4 \sqrt{2}(x+1)\,dx = 2\sqrt{2}\pi\left(\frac{x^2}{2}+x\right)\Big|_0^4 = 24\sqrt{2}\pi.$$

9. $y = \frac{2}{3}x^{3/2} - \frac{1}{2}x^{1/2}, \quad [1, 2]$

SOLUTION Let $y = \frac{2}{3}x^{3/2} - \frac{1}{2}x^{1/2}$. Then

$$y' = \sqrt{x} - \frac{1}{4\sqrt{x}},$$

and

$$1+(y')^2 = 1+\left(\sqrt{x}-\frac{1}{4\sqrt{x}}\right)^2 = 1+\left(x-\frac{1}{2}+\frac{1}{16x}\right) = x+\frac{1}{2}+\frac{1}{16x} = \left(\sqrt{x}+\frac{1}{4\sqrt{x}}\right)^2.$$

Because $\sqrt{x}+\frac{1}{\sqrt{x}} \geq 0$, the surface area is

$$2\pi\int_a^b y\sqrt{1+(y')^2}\,dx = 2\pi\int_1^2\left(\frac{2}{3}x^{3/2}-\frac{\sqrt{x}}{2}\right)\left(\sqrt{x}+\frac{1}{4\sqrt{x}}\right)dx$$

$$= 2\pi\int_1^2\left(\frac{2}{3}x^2+\frac{1}{6}x-\frac{1}{2}x-\frac{1}{8}\right)dx = 2\pi\left(\frac{2x^3}{9}-\frac{x^2}{6}-\frac{1}{8}x\right)\Big|_1^2 = \frac{67}{36}\pi.$$

11. Compute the total surface area of the coin obtained by rotating the region in Figure 1 about the x-axis. The top and bottom parts of the region are semicircles with a radius of 1 mm.

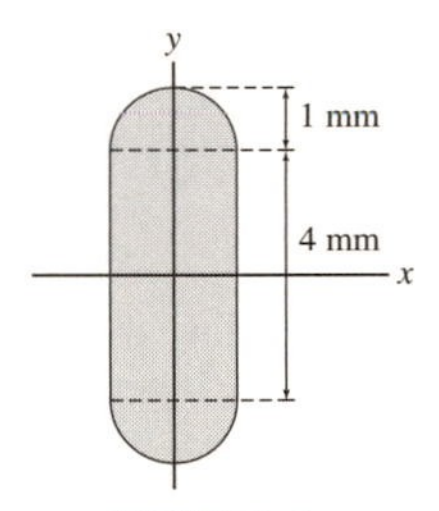

FIGURE 1

SOLUTION The generating half circle of the edge is $y = 2+\sqrt{1-x^2}$. Then,

$$y' = \frac{-2x}{2\sqrt{1-x^2}} = \frac{-x}{\sqrt{1-x^2}},$$

and

$$1+(y')^2 = 1+\frac{x^2}{1-x^2} = \frac{1}{1-x^2}.$$

The surface area of the edge of the coin is

$$2\pi\int_{-1}^1 y\sqrt{1+(y')^2}dx = 2\pi\int_{-1}^1\left(2+\sqrt{1-x^2}\right)\frac{1}{\sqrt{1-x^2}}dx$$

$$= 2\pi\left(2\int_{-1}^1\frac{dx}{\sqrt{1-x^2}}+\int_{-1}^1\frac{\sqrt{1-x^2}}{\sqrt{1-x^2}}dx\right)$$

$$= 2\pi\left(2\arcsin x|_{-1}^1+\int_{-1}^1 dx\right)$$

$$= 2\pi(2\pi+2) = 4\pi^2+4\pi.$$

We now add the surface area of the two sides of the disk, which are circles of radius 2. Hence the surface area of the coin is:

$$\left(4\pi^2+4\pi\right)+2\pi\cdot 2^2 = 4\pi^2+12\pi.$$

13. Calculate the fluid force on the side of a right triangle of height 3 m and base 2 m submerged in water vertically, with its upper vertex located at a depth of 4 m.

SOLUTION We need to find an expression for the horizontal width $f(y)$ at depth y.

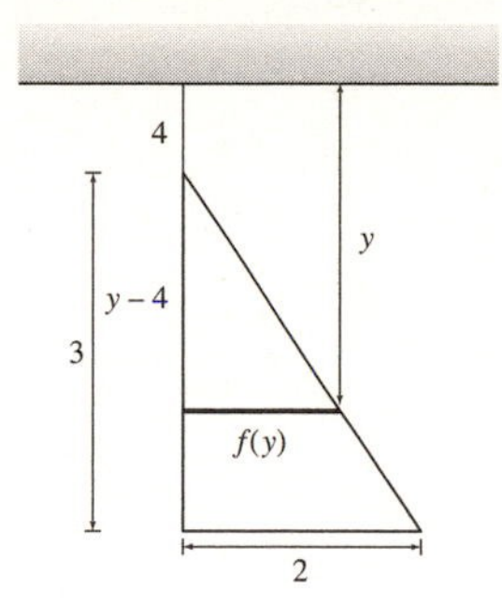

By similar triangles we have:

$$\frac{f(y)}{y-4} = \frac{2}{3} \quad \text{so} \quad f(y) = \frac{2(y-4)}{3}.$$

Hence, the force on the side of the triangle is

$$F = \rho g \int_4^7 y f(y)\, dy = \frac{2\rho g}{3} \int_4^7 \left(y^2 - 4y\right) dy = \frac{2\rho g}{3} \left(\frac{y^3}{3} - 2y^2\right)\Bigg|_4^7 = 18\rho g.$$

For water, $\rho = 10^3$; $g = 9.8$, so $F = 18 \cdot 9800 = 176{,}400$ N.

15. Figure 3 shows an object whose face is an equilateral triangle with 5-m sides. The object is 2 m thick and is submerged in water with its vertex 3 m below the water surface. Calculate the fluid force on both a triangular face and a slanted rectangular edge of the object.

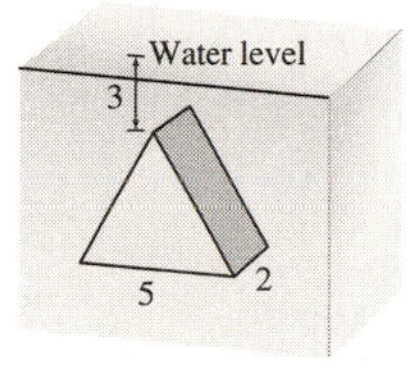

FIGURE 3

SOLUTION Start with each triangular face of the object. Place the origin at the upper vertex of the triangle, with the positive y-axis pointing downward. Note that because the equilateral triangle has sides of length 5 feet, the height of the triangle is $\frac{5\sqrt{3}}{2}$ feet. Moreover, the width of the triangle at location y is $\frac{2y}{\sqrt{3}}$. Thus,

$$F = \frac{2\rho g}{\sqrt{3}} \int_0^{5\sqrt{3}/2} (y+3)y\, dy = \frac{2\rho g}{\sqrt{3}} \left(\frac{1}{3}y^3 + \frac{3}{2}y^2\right)\Bigg|_0^{5\sqrt{3}/2} = \frac{\rho g}{4}(125 + 75\sqrt{3}) \approx 624{,}514 \text{ N}.$$

Now, consider the slanted rectangular edges of the object. Each edge is a constant 2 feet wide and makes an angle of 60° with the horizontal. Therefore,

$$F = \frac{\rho g}{\sin 60^\circ} \int_0^{5\sqrt{3}/2} 2(y+3)\, dy = \frac{2\rho g}{\sqrt{3}} \left(y^2 + 6y\right)\Bigg|_0^{5\sqrt{3}/2} = \rho g \left(\frac{25\sqrt{3}}{2} + 30\right) \approx 506{,}176 \text{ N}.$$

The force on the bottom face can be computed without calculus:

$$F = \left(3 + \frac{5\sqrt{3}}{2}\right)(2)(5)\rho g \approx 718{,}352 \text{ N}.$$

17. Calculate the moments and COM of the lamina occupying the region under $y = x(4-x)$ for $0 \le x \le 4$, assuming a density of $\rho = 1200$ kg/m^3.

SOLUTION Because the lamina is symmetric with respect to the vertical line $x = 2$, by the symmetry principle, we know that $x_{\text{cm}} = 2$. Now,

$$M_x = \frac{\rho}{2} \int_0^4 f(x)^2\, dx = \frac{1200}{2} \int_0^4 x^2(4-x)^2\, dx = \frac{1200}{2} \left(\frac{16}{3}x^3 - 2x^4 + \frac{1}{5}x^5\right)\Bigg|_0^4 = 20{,}480.$$

Moreover, the mass of the lamina is

$$M = \rho \int_0^4 f(x)\,dx = 1200 \int_0^4 x(4-x)\,dx = 1200 \left(2x^2 - \frac{1}{3}x^3\right)\Big|_0^4 = 12{,}800.$$

Thus, the coordinates of the center of mass are

$$\left(2, \frac{20{,}480}{12{,}800}\right) = \left(2, \frac{8}{5}\right).$$

19. Find the centroid of the region between the semicircle $y = \sqrt{1-x^2}$ and the top half of the ellipse $y = \frac{1}{2}\sqrt{1-x^2}$ (Figure 2).

SOLUTION Since the region is symmetric with respect to the y-axis, the centroid lies on the y-axis. To find y_{cm} we calculate

$$M_x = \frac{1}{2}\int_{-1}^{1}\left[\left(\sqrt{1-x^2}\right)^2 - \left(\frac{\sqrt{1-x^2}}{2}\right)^2\right] dx$$

$$= \frac{1}{2}\int_{-1}^{1} \frac{3}{4}\left(1-x^2\right) dx = \frac{3}{8}\left(x - \frac{1}{3}x^3\right)\Big|_{-1}^{1} = \frac{1}{2}.$$

The area of the lamina is $\frac{\pi}{2} - \frac{\pi}{4} = \frac{\pi}{4}$, so the coordinates of the centroid are

$$\left(0, \frac{1/2}{\pi/4}\right) = \left(0, \frac{2}{\pi}\right).$$

In Exercises 21–26, find the Taylor polynomial at $x = a$ for the given function.

21. $f(x) = x^3, \quad T_3(x), \quad a = 1$

SOLUTION We start by computing the first three derivatives of $f(x) = x^3$:

$$f'(x) = 3x^2$$
$$f''(x) = 6x$$
$$f'''(x) = 6$$

Evaluating the function and its derivatives at $x = 1$, we find

$$f(1) = 1,\ f'(1) = 3,\ f''(1) = 6,\ f'''(1) = 6.$$

Therefore,

$$T_3(x) = f(1) + f'(1)(x-1) + \frac{f''(1)}{2!}(x-2)^2 + \frac{f'''(1)}{3!}(x-1)^3$$

$$= 1 + 3(x-1) + \frac{6}{2!}(x-2)^2 + \frac{6}{3!}(x-1)^3$$

$$= 1 + 3(x-1) + 3(x-2)^2 + (x-1)^3.$$

23. $f(x) = x\ln(x), \quad T_4(x), \quad a = 1$

SOLUTION We start by computing the first four derivatives of $f(x) = x\ln x$:

$$f'(x) = \ln x + x \cdot \frac{1}{x} = \ln x + 1$$
$$f''(x) = \frac{1}{x}$$
$$f'''(x) = -\frac{1}{x^2}$$
$$f^{(4)}(x) = \frac{2}{x^3}$$

Evaluating the function and its derivatives at $x = 1$, we find

$$f(1) = 0,\ f'(1) = 1,\ f''(1) = 1,\ f'''(1) = -1,\ f^{(4)}(1) = 2.$$

Therefore,

$$\begin{aligned}T_4(x) &= f(1) + f'(1)(x-1) + \frac{f''(1)}{2!}(x-1)^2 + \frac{f'''(1)}{3!}(x-1)^3 + \frac{f^{(4)}(1)}{4!}(x-1)^4\\ &= 0 + 1(x-1) + \frac{1}{2!}(x-1)^2 - \frac{1}{3!}(x-1)^3 + \frac{2}{4!}(x-1)^4\\ &= (x-1) + \frac{1}{2}(x-1)^2 - \frac{1}{6}(x-1)^3 + \frac{1}{12}(x-1)^4.\end{aligned}$$

25. $f(x) = xe^{-x^2}$, $T_4(x)$, $a = 0$

SOLUTION We start by computing the first four derivatives of $f(x) = xe^{-x^2}$:

$$\begin{aligned}f'(x) &= e^{-x^2} + x\cdot(-2x)e^{-x^2} = (1-2x^2)e^{-x^2}\\ f''(x) &= -4xe^{-x^2} + (1-2x^2)\cdot(-2x)e^{-x^2} = (4x^3-6x)e^{-x^2}\\ f'''(x) &= (12x^2-6)e^{-x^2} + (4x^3-6x)\cdot(-2x)e^{-x^2} = (-8x^4+24x^2-6)e^{-x^2}\\ f^{(4)}(x) &= (-32x^3+48x)e^{-x^2} + (-8x^4+24x^2-6)\cdot(-2x)e^{-x^2} = (16x^5-80x^3+60x)e^{-x^2}\end{aligned}$$

Evaluating the function and its derivatives at $x = 0$, we find

$$f(0) = 0,\ f'(0) = 1,\ f''(0) = 0,\ f'''(0) = -6,\ f^{(4)}(0) = 0.$$

Therefore,

$$\begin{aligned}T_4(x) &= f(0) + f'(0)x + \frac{f''(0)}{2!}x^2 + \frac{f'''(0)}{3!}x^3 + \frac{f^{(4)}(0)}{4!}x^4\\ &= 0 + x + 0\cdot x^2 - \frac{6}{3!}x^3 + 0\cdot x^4 = x - x^3.\end{aligned}$$

27. Find the nth Maclaurin polynomial for $f(x) = e^{3x}$.

SOLUTION We differentiate the function $f(x) = e^{3x}$ repeatedly, looking for a pattern:

$$\begin{aligned}f'(x) &= 3e^{3x} = 3^1e^{3x}\\ f''(x) &= 3\cdot 3e^{3x} = 3^2e^{3x}\\ f'''(x) &= 3\cdot 3^2e^{3x} = 3^3e^{3x}\end{aligned}$$

Thus, for general n, $f^{(n)}(x) = 3^ne^{3x}$ and $f^{(n)}(0) = 3^n$. Substituting into the formula for the nth Taylor polynomial, we obtain:

$$\begin{aligned}T_n(x) &= 1 + \frac{3x}{1!} + \frac{3^2x^2}{2!} + \frac{3^3x^3}{3!} + \frac{3^4x^4}{4!} + \cdots + \frac{3^nx^n}{n!}\\ &= 1 + 3x + \frac{1}{2!}(3x)^2 + \frac{1}{3!}(3x)^3 + \cdots + \frac{1}{n!}(3x)^n.\end{aligned}$$

29. Use the third Taylor polynomial of $f(x) = \tan^{-1}x$ at $a = 1$ to approximate $f(1.1)$. Use a calculator to determine the error.

SOLUTION We start by computing the first three derivatives of $f(x) = \tan^{-1}x$:

$$\begin{aligned}f'(x) &= \frac{1}{1+x^2}\\ f''(x) &= -\frac{2x}{(1+x^2)^2}\\ f'''(x) &= \frac{-2\left(1+x^2\right)^2 + 2x\cdot 2\left(1+x^2\right)\cdot 2x}{(1+x^2)^4} = \frac{2\left(3x^2-1\right)}{(1+x^2)^3}\end{aligned}$$

Evaluating the function and its derivatives at $x = 1$, we find

$$f(1) = \frac{\pi}{4},\ f'(1) = \frac{1}{2},\ f''(1) = -\frac{1}{2},\ f'''(1) = \frac{1}{2}.$$

Therefore,

$$T_3(x) = f(1) + f'(1)(x-1) + \frac{f''(1)}{2!}(x-1)^2 + \frac{f'''(1)}{3!}(x-1)^3$$
$$= \frac{\pi}{4} + \frac{1}{2}(x-1) - \frac{1}{4}(x-1)^2 + \frac{1}{12}(x-1)^3.$$

Setting $x = 1.1$ yields:

$$T_3(1.1) = \frac{\pi}{4} + \frac{1}{2}(0.1) - \frac{1}{4}(0.1)^2 + \frac{1}{12}(0.1)^3 = 0.832981496.$$

Using a calculator, we find $\tan^{-1} 1.1 = 0.832981266$. The error in the Taylor polynomial approximation is

$$\left|T_3(1.1) - \tan^{-1} 1.1\right| = |0.832981496 - 0.832981266| = 2.301 \times 10^{-7}.$$

31. Find n such that $|e - T_n(1)| < 10^{-8}$, where $T_n(x)$ is the nth Maclaurin polynomial for $f(x) = e^x$.

SOLUTION Using the Error Bound, we have

$$|e - T_n(1)| \le K\frac{|1-0|^{n+1}}{(n+1)!} = \frac{K}{(n+1)!}$$

where K is a number such that $\left|f^{(n+1)}(x)\right| = e^x \le K$ for all $0 \le x \le 1$. Since e^x is increasing, the maximum value on the interval $0 \le x \le 1$ is attained at the endpoint $x = 1$. Thus, for $0 \le u \le 1$, $e^u \le e^1 < 2.8$. Hence we may take $K = 2.8$ to obtain:

$$|e - T_n(1)| \le \frac{2.8}{(n+1)!}$$

We now choose n such that

$$\frac{2.8}{(n+1)!} < 10^{-8}$$
$$\frac{(n+1)!}{2.8} > 10^8$$
$$(n+1)! > 2.8 \times 10^8$$

For $n = 10$, $(n+1)! = 3.99 \times 10^7 < 2.8 \times 10^8$ and for $n = 11$, $(n+1)! = 4.79 \times 10^8 > 2.8 \times 10^8$. Hence, to make the error less than 10^{-8}, $n = 11$ is sufficient; that is,

$$|e - T_{11}(1)| < 10^{-8}.$$

33. Verify that $T_n(x) = 1 + x + x^2 + \cdots + x^n$ is the nth Maclaurin polynomial of $f(x) = 1/(1-x)$. Show using substitution that the nth Maclaurin polynomial for $f(x) = 1/(1 - x/4)$ is

$$T_n(x) = 1 + \frac{1}{4}x + \frac{1}{4^2}x^2 + \cdots + \frac{1}{4^n}x^n$$

What is the nth Maclaurin polynomial for $g(x) = \dfrac{1}{1+x}$?

SOLUTION Let $f(x) = (1-x)^{-1}$. Then, $f'(x) = (1-x)^{-2}$, $f''(x) = 2(1-x)^{-3}$, $f'''(x) = 3!(1-x)^{-4}$, and, in general, $f^{(n)}(x) = n!(1-x)^{-(n+1)}$. Therefore, $f^{(n)}(0) = n!$ and

$$T_n(x) = 1 + \frac{1!}{1!}x + \frac{2!}{2!}x^2 + \cdots + \frac{n!}{n!}x^n = 1 + x + x^2 + \cdots + x^n.$$

Upon substituting $x/4$ for x, we find that the nth Maclaurin polynomial for $f(x) = \dfrac{1}{1 - x/4}$ is

$$T_n(x) = 1 + \frac{1}{4}x + \frac{1}{4^2}x^2 + \cdots + \frac{1}{4^n}x^n.$$

Substituting $-x$ for x, the nth Maclaurin polynomial for $g(x) = \dfrac{1}{1+x}$ is

$$T_n(x) = 1 - x + x^2 - x^3 + - \cdots + (-x)^n.$$

35. Let $T_n(x)$ be the nth Maclaurin polynomial for the function $f(x) = \sin x + \sinh x$.

(a) Show that $T_5(x) = T_6(x) = T_7(x) = T_8(x)$.

(b) Show that $|f^n(x)| \le 1 + \cosh x$ for all n. *Hint:* Note that $|\sinh x| \le |\cosh x|$ for all x.

(c) Show that $|T_8(x) - f(x)| \le \frac{2.6}{9!}|x|^9$ for $-1 \le x \le 1$.

SOLUTION

(a) Let $f(x) = \sin x + \sinh x$. Then

$$f'(x) = \cos x + \cosh x$$
$$f''(x) = -\sin x + \sinh x$$
$$f'''(x) = -\cos x + \cosh x$$
$$f^{(4)}(x) = \sin x + \sinh x.$$

From this point onward, the pattern of derivatives repeats indefinitely. Thus

$$f(0) = f^{(4)}(0) = f^{(8)}(0) = \sin 0 + \sinh 0 = 0$$
$$f'(0) = f^{(5)}(0) = \cos 0 + \cosh 0 = 2$$
$$f''(0) = f^{(6)}(0) = -\sin 0 + \sinh 0 = 0$$
$$f'''(0) = f^{(7)}(0) = -\cos 0 + \cosh 0 = 0.$$

Consequently,

$$T_5(x) = f'(0)x + \frac{f^{(5)}(0)}{5!}x^5 = 2x + \frac{1}{60}x^5,$$

and, because $f^{(6)}(0) = f^{(7)}(0) = f^{(8)}(0) = 0$, it follows that

$$T_6(x) = T_7(x) = T_8(x) = T_5(x) = 2x + \frac{1}{60}x^5.$$

(b) First note that $|\sin x| \le 1$ and $|\cos x| \le 1$ for all x. Moreover,

$$|\sinh x| = \left|\frac{e^x - e^{-x}}{2}\right| \le \frac{e^x + e^{-x}}{2} = \cosh x.$$

Now, recall from part (a), that all derivatives of $f(x)$ contain two terms: the first is $\pm \sin x$ or $\pm \cos x$, while the second is either $\sinh x$ or $\cosh x$. In absolute value, the trigonometric term is always less than or equal to 1, while the hyperbolic term is always less than or equal to $\cosh x$. Thus, for all n,

$$f^{(n)}(x) \le 1 + \cosh x.$$

(c) Using the Error Bound, we have

$$|T_8(x) - f(x)| \le \frac{K|x-0|^9}{9!} = \frac{K|x|^9}{9!},$$

where K is a number such that $\left|f^{(9)}(u)\right| \le K$ for all u between 0 and x. By part (b), we know that

$$f^{(9)}(u) \le 1 + \cosh u.$$

Now, $\cosh u$ is an even function that is increasing on $(0, \infty)$. The maximum value for u between 0 and x is therefore $\cosh x$. Moreover, for $-1 \le x \le 1$, $\cosh x \le \cosh 1 \approx 1.543 < 1.6$. Hence, we may take $K = 1 + 1.6 = 2.6$, and

$$|T_8(x) - f(x)| \le \frac{2.6}{9!}|x|^9.$$

9 INTRODUCTION TO DIFFERENTIAL EQUATIONS

9.1 Solving Differential Equations

Preliminary Questions

1. Determine the order of the following differential equations:

(a) $x^5 y' = 1$

(b) $(y')^3 + x = 1$

(c) $y''' + x^4 y' = 2$

(d) $\sin(y'') + x = y$

SOLUTION

(a) The highest order derivative that appears in this equation is a first derivative, so this is a first order equation.

(b) The highest order derivative that appears in this equation is a first derivative, so this is a first order equation.

(c) The highest order derivative that appears in this equation is a third derivative, so this is a third order equation.

(d) The highest order derivative that appears in this equation is a second derivative, so this is a second order equation.

2. Is $y'' = \sin x$ a linear differential equation?

SOLUTION Yes.

3. Give an example of a nonlinear differential equation of the form $y' = f(y)$.

SOLUTION One possibility is $y' = y^2$.

4. Can a nonlinear differential equation be separable? If so, give an example.

SOLUTION Yes. An example is $y' = y^2$.

5. Give an example of a linear, nonseparable differential equation.

SOLUTION One example is $y' + y = x$.

Exercises

1. Which of the following differential equations are first-order?

(a) $y' = x^2$

(b) $y'' = y^2$

(c) $(y')^3 + yy' = \sin x$

(d) $x^2 y' - e^x y = \sin y$

(e) $y'' + 3y' = \dfrac{y}{x}$

(f) $yy' + x + y = 0$

SOLUTION

(a) The highest order derivative that appears in this equation is a first derivative, so this is a first order equation.

(b) The highest order derivative that appears in this equation is a second derivative, so this is not a first order equation.

(c) The highest order derivative that appears in this equation is a first derivative, so this is a first order equation.

(d) The highest order derivative that appears in this equation is a first derivative, so this is a first order equation.

(e) The highest order derivative that appears in this equation is a second derivative, so this is not a first order equation.

(f) The highest order derivative that appears in this equation is a first derivative, so this is a first order equation.

In Exercises 3–8, verify that the given function is a solution of the differential equation.

3. $y' - 8x = 0, \quad y = 4x^2$

SOLUTION Let $y = 4x^2$. Then $y' = 8x$ and

$$y' - 8x = 8x - 8x = 0.$$

5. $y' + 4xy = 0, \quad y = 25e^{-2x^2}$

SOLUTION Let $y = 25e^{-2x^2}$. Then $y' = -100xe^{-2x^2}$, and

$$y' + 4xy = -100xe^{-2x^2} + 4x(25e^{-2x^2}) = 0.$$

7. $y'' - 2xy' + 8y = 0, \quad y = 4x^4 - 12x^2 + 3$

SOLUTION Let $y = 4x^4 - 12x^2 + 3$. Then $y' = 16x^3 - 24x$, $y'' = 48x^2 - 24$, and

$$\begin{aligned} y'' - 2xy' + 8y &= (48x^2 - 24) - 2x(16x^3 - 24x) + 8(4x^4 - 12x^2 + 3) \\ &= 48x^2 - 24 - 32x^4 + 48x^2 + 32x^4 - 96x^2 + 24 = 0. \end{aligned}$$

9. Which of the following equations are separable? Write those that are separable in the form $y' = f(x)g(y)$ (but do not solve).

(a) $xy' - 9y^2 = 0$

(b) $\sqrt{4 - x^2}\,y' = e^{3y} \sin x$

(c) $y' = x^2 + y^2$

(d) $y' = 9 - y^2$

SOLUTION

(a) $xy' - 9y^2 = 0$ is separable:

$$\begin{aligned} xy' - 9y^2 &= 0 \\ xy' &= 9y^2 \\ y' &= \frac{9}{x}y^2 \end{aligned}$$

(b) $\sqrt{4 - x^2}\,y' = e^{3y} \sin x$ is separable:

$$\begin{aligned} \sqrt{4 - x^2}\,y' &= e^{3y} \sin x \\ y' &= e^{3y} \frac{\sin x}{\sqrt{4 - x^2}}. \end{aligned}$$

(c) $y' = x^2 + y^2$ is not separable; y' is already isolated, but is not equal to a product $f(x)g(y)$.

(d) $y' = 9 - y^2$ is separable: $y' = (1)(9 - y^2)$.

11. Consider the differential equation $y^3 y' - 9x^2 = 0$.

(a) Write it as $y^3\,dy = 9x^2\,dx$.

(b) Integrate both sides to obtain $\frac{1}{4}y^4 = 3x^3 + C$.

(c) Verify that $y = (12x^3 + C)^{1/4}$ is the general solution.

(d) Find the particular solution satisfying $y(1) = 2$.

SOLUTION Solving $y^3 y' - 9x^2 = 0$ for y' gives $y' = 9x^2 y^{-3}$.

(a) Separating variables in the equation above yields

$$y^3\,dy = 9x^2\,dx$$

(b) Integrating both sides gives

$$\frac{y^4}{4} = 3x^3 + C$$

(c) Simplify the equation above to get $y^4 = 12x^3 + C$, or $y = (12x^3 + C)^{1/4}$.

(d) Solve $2 = (12 \cdot 1^3 + C)^{1/4}$ to get $16 = 12 + C$, or $C = 4$. Thus the particular solution is $y = (12x^3 + 4)^{1/4}$.

In Exercises 13–28, use separation of variables to find the general solution.

13. $y' + 4xy^2 = 0$

SOLUTION Rewrite

$$y' + 4xy^2 = 0 \quad \text{as} \quad \frac{dy}{dx} = -4xy^2 \quad \text{and then as} \quad y^{-2}\,dy = -4x\,dx$$

Integrating both sides of this equation gives

$$\begin{aligned} \int y^{-2}\,dy &= -4\int x\,dx \\ -y^{-1} &= -2x^2 + C \\ y^{-1} &= 2x^2 + C \end{aligned}$$

Solving for y gives

$$y = \frac{1}{2x^2 + C}$$

where C is an arbitrary constant.

15. $\dfrac{dy}{dt} - 20t^4e^{-y} = 0$

SOLUTION Rewrite

$$\frac{dy}{dt} - 20t^4e^{-y} = 0 \quad \text{as} \quad \frac{dy}{dt} = 20t^4e^{-y} \quad \text{and then as} \quad e^y\,dy = 20t^4\,dt$$

Integrating both sides of this equation gives

$$\int e^y\,dy = \int 20t^4\,dt$$
$$e^y = 4t^5 + C$$

Solve for y to get $y = \ln(4t^5 + C)$, where C is an arbitrary constant.

17. $2y' + 5y = 4$

SOLUTION Rewrite

$$2y' + 5y = 4 \quad \text{as} \quad y' = 2 - \frac{5}{2}y \quad \text{and then as} \quad (4 - 5y)^{-1}\,dy = \frac{1}{2}\,dx$$

Integrating both sides and solving for y gives

$$\int \frac{dy}{4 - 5y} = \frac{1}{2}\int 1\,dx$$
$$-\frac{1}{5}\ln|4 - 5y| = \frac{1}{2}x + C_1$$
$$\ln|4 - 5y| = C_2 - \frac{5}{2}x$$
$$4 - 5y = C_3e^{-5x/2}$$
$$5y = 4 - C_3e^{-5x/2}$$
$$y = Ce^{-5x/2} + \frac{4}{5}$$

where C is an arbitrary constant.

19. $\sqrt{1 - x^2}\,y' = xy$

SOLUTION Rewrite

$$\sqrt{1 - x^2}\,\frac{dy}{dx} = xy \qquad \text{as} \qquad \frac{dy}{y} = \frac{x}{\sqrt{1 - x^2}}\,dx.$$

Integrating both sides of this equation yields

$$\int \frac{dy}{y} = \int \frac{x}{\sqrt{1 - x^2}}\,dx$$
$$\ln|y| = -\sqrt{1 - x^2} + C.$$

Solving for y, we find

$$|y| = e^{-\sqrt{1-x^2}+C} = e^Ce^{-\sqrt{1-x^2}}$$
$$y = \pm e^Ce^{-\sqrt{1-x^2}} = Ae^{-\sqrt{1-x^2}},$$

where A is an arbitrary constant.

21. $yy' = x$

SOLUTION Rewrite

$$y\frac{dy}{dx} = x \qquad \text{as} \qquad y\,dy = x\,dx.$$

Integrating both sides of this equation yields

$$\int y\,dy = \int x\,dx$$
$$\frac{1}{2}y^2 = \frac{1}{2}x^2 + C.$$

Solving for y, we find

$$y^2 = x^2 + 2C$$
$$y = \pm\sqrt{x^2 + A},$$

where $A = 2C$ is an arbitrary constant.

23. $\dfrac{dx}{dt} = (t+1)(x^2+1)$

SOLUTION Rewrite

$$\frac{dx}{dt} = (t+1)(x^2+1) \qquad \text{as} \qquad \frac{1}{x^2+1}\,dx = (t+1)\,dt.$$

Integrating both sides of this equation yields

$$\int \frac{1}{x^2+1}\,dx = \int (t+1)\,dt$$
$$\tan^{-1} x = \frac{1}{2}t^2 + t + C.$$

Solving for x, we find

$$x = \tan\left(\frac{1}{2}t^2 + t + C\right).$$

where $A = \tan C$ is an arbitrary constant.

25. $y' = x \sec y$

SOLUTION Rewrite

$$\frac{dy}{dx} = x \sec y \qquad \text{as} \qquad \cos y\,dy = x\,dx.$$

Integrating both sides of this equation yields

$$\int \cos y\,dy = \int x\,dx$$
$$\sin y = \frac{1}{2}x^2 + C.$$

Solving for y, we find

$$y = \sin^{-1}\left(\frac{1}{2}x^2 + C\right),$$

where C is an arbitrary constant.

27. $\dfrac{dy}{dt} = y \tan t$

SOLUTION Rewrite

$$\frac{dy}{dt} = y \tan t \qquad \text{as} \qquad \frac{1}{y}\,dy = \tan t\,dt.$$

Integrating both sides of this equation yields

$$\int \frac{1}{y}\,dy = \int \tan t\,dt$$
$$\ln|y| = \ln|\sec t| + C.$$

Solving for y, we find

$$|y| = e^{\ln|\sec t| + C} = e^C\,|\sec t|$$
$$y = \pm e^C \sec t = A \sec t,$$

where $A = \pm e^C$ is an arbitrary constant.

In Exercises 29–42, solve the initial value problem.

29. $y' + 2y = 0, \quad y(\ln 5) = 3$

SOLUTION First, we find the general solution of the differential equation. Rewrite

$$\frac{dy}{dx} + 2y = 0 \qquad \text{as} \qquad \frac{1}{y}\,dy = -2\,dx,$$

and then integrate to obtain

$$\ln|y| = -2x + C.$$

Thus,

$$y = Ae^{-2x},$$

where $A = \pm e^C$ is an arbitrary constant. The initial condition $y(\ln 5) = 3$ allows us to determine the value of A.

$$3 = Ae^{-2(\ln 5)}; \quad 3 = A\frac{1}{25}; \quad \text{so} \quad 75 = A.$$

Finally,

$$y = 75e^{-2x}.$$

31. $yy' = xe^{-y^2}, \quad y(0) = -2$

SOLUTION First, we find the general solution of the differential equation. Rewrite

$$y\frac{dy}{dx} = xe^{-y^2} \qquad \text{as} \qquad ye^{y^2}\,dy = x\,dx,$$

and then integrate to obtain

$$\frac{1}{2}e^{y^2} = \frac{1}{2}x^2 + C.$$

Thus,

$$y = \pm\sqrt{\ln(x^2 + A)},$$

where $A = 2C$ is an arbitrary constant. The initial condition $y(0) = -2$ allows us to determine the value of A. Since $y(0) < 0$, we have $y = -\sqrt{\ln(x^2 + A)}$, and

$$-2 = -\sqrt{\ln(A)}; \quad 4 = \ln(A); \quad \text{so} \quad e^4 = A.$$

Finally,

$$y = -\sqrt{\ln(x^2 + e^4)}.$$

33. $y' = (x - 1)(y - 2), \quad y(2) = 4$

SOLUTION First, we find the general solution of the differential equation. Rewrite

$$\frac{dy}{dx} = (x - 1)(y - 2) \qquad \text{as} \qquad \frac{1}{y - 2}\,dy = (x - 1)\,dx,$$

and then integrate to obtain

$$\ln|y - 2| = \frac{1}{2}x^2 - x + C.$$

Thus,

$$y = Ae^{(1/2)x^2 - x} + 2,$$

where $A = \pm e^C$ is an arbitrary constant. The initial condition $y(2) = 4$ allows us to determine the value of A.

$$4 = Ae^0 + 2 \quad \text{so} \quad A = 2.$$

Finally,

$$y = 2e^{(1/2)x^2 - x} + 2.$$

35. $y' = x(y^2+1), \quad y(0) = 0$

SOLUTION First, find the general solution of the differential equation. Rewrite

$$\frac{dy}{dx} = x(y^2+1) \qquad \text{as} \qquad \frac{1}{y^2+1}\,dy = x\,dx$$

and integrate to obtain

$$\tan^{-1} y = \frac{1}{2}x^2 + C$$

so that

$$y = \tan\left(\frac{1}{2}x^2 + C\right)$$

where C is an arbitrary constant. The initial condition $y(0) = 0$ allows us to determine the value of C: $0 = \tan(C)$, so $C = 0$. Finally,

$$y = \tan\left(\frac{1}{2}x^2\right)$$

37. $\frac{dy}{dt} = ye^{-t}, \quad y(0) = 1$

SOLUTION First, we find the general solution of the differential equation. Rewrite

$$\frac{dy}{dt} = ye^{-t} \qquad \text{as} \qquad \frac{1}{y}\,dy = e^{-t}\,dt,$$

and then integrate to obtain

$$\ln|y| = -e^{-t} + C.$$

Thus,

$$y = Ae^{-e^{-t}},$$

where $A = \pm e^C$ is an arbitrary constant. The initial condition $y(0) = 1$ allows us to determine the value of A.

$$1 = Ae^{-1} \quad \text{so} \quad A = e.$$

Finally,

$$y = (e)e^{-e^{-t}} = e^{1-e^{-t}}.$$

39. $t^2\frac{dy}{dt} - t = 1 + y + ty, \quad y(1) = 0$

SOLUTION First, we find the general solution of the differential equation. Rewrite

$$t^2\frac{dy}{dt} = 1 + t + y + ty = (1+t)(1+y)$$

as

$$\frac{1}{1+y}\,dy = \frac{1+t}{t^2}\,dt,$$

and then integrate to obtain

$$\ln|1+y| = -t^{-1} + \ln|t| + C.$$

Thus,

$$y = A\frac{t}{e^{1/t}} - 1,$$

where $A = \pm e^C$ is an arbitrary constant. The initial condition $y(1) = 0$ allows us to determine the value of A.

$$0 = A\left(\frac{1}{e}\right) - 1 \quad \text{so} \quad A = e.$$

Finally,

$$y = \frac{et}{e^{1/t}} - 1.$$

41. $y' = \tan y, \quad y(\ln 2) = \frac{\pi}{2}$

SOLUTION First, we find the general solution of the differential equation. Rewrite

$$\frac{dy}{dx} = \tan y \quad \text{as} \quad \frac{dy}{\tan y} = dx,$$

and then integrate to obtain

$$\ln|\sin y| = x + C.$$

Thus,

$$y = \sin^{-1}(Ae^x),$$

where $A = \pm e^C$ is an arbitrary constant. The initial condition $y(\ln 2) = \frac{\pi}{2}$ allows us to determine the value of A.

$$\frac{\pi}{2} = \sin^{-1}(2A); \quad 1 = 2A \quad \text{so} \quad A = \frac{1}{2}.$$

Finally,

$$y = \sin^{-1}\left(\frac{1}{2}e^x\right).$$

43. Find all values of a such that $y = x^a$ is a solution of

$$y'' - 12x^{-2}y = 0$$

SOLUTION Let $y = x^a$. Then

$$y' = ax^{a-1} \quad \text{and} \quad y'' = a(a-1)x^{a-2}.$$

Substituting into the differential equation, we find

$$y'' - 12x^{-2}y = a(a-1)x^{a-2} - 12x^{a-2} = x^{a-2}(a^2 - a - 12).$$

Thus, $y'' - 12x^{-2}y = 0$ if and only if

$$a^2 - a - 12 = (a-4)(a+3) = 0.$$

Hence, $y = x^a$ is a solution of the differential equation $y'' - 12x^{-2}y = 0$ provided $a = 4$ or $a = -3$.

In Exercises 45 and 46, let $y(t)$ be a solution of $(\cos y + 1)\frac{dy}{dt} = 2t$ such that $y(2) = 0$.

45. Show that $\sin y + y = t^2 + C$. We cannot solve for y as a function of t, but, assuming that $y(2) = 0$, find the values of t at which $y(t) = \pi$.

SOLUTION Rewrite

$$(\cos y + 1)\frac{dy}{dt} = 2t \quad \text{as} \quad (\cos y + 1)\, dy = 2t\, dt$$

and integrate to obtain

$$\sin y + y = t^2 + C$$

where C is an arbitrary constant. Since $y(2) = 0$, we have $\sin 0 + 0 = 4 + C$ so that $C = -4$ and the particular solution we seek is $\sin y + y = t^2 - 4$. To find values of t at which $y(t) = \pi$, we must solve $\sin \pi + \pi = t^2 - 4$, or $t^2 - 4 = \pi$; thus $t = \pm\sqrt{\pi + 4}$.

In Exercises 47–52, use Eq. (4) and Torricelli's Law [Eq. (5)].

47. Water leaks through a hole of area $0.002\ \text{m}^2$ at the bottom of a cylindrical tank that is filled with water and has height 3 m and a base of area $10\ \text{m}^2$. How long does it take (a) for half of the water to leak out and (b) for the tank to empty?

SOLUTION Because the tank has a constant cross-sectional area of $10\ \text{m}^2$ and the hole has an area of $0.002\ \text{m}^2$, the differential equation for the height of the water in the tank is

$$\frac{dy}{dt} = \frac{0.002v}{10} = 0.0002v.$$

By Torricelli's Law,

$$v = -\sqrt{2gy} = -\sqrt{19.6y},$$

using $g = 9.8\ \text{m/s}^2$. Thus,

$$\frac{dy}{dt} = -0.0002\sqrt{19.6y} = -0.0002\sqrt{19.6} \cdot \sqrt{y}.$$

Separating variables and then integrating yields

$$y^{-1/2}\,dy = -0.0002\sqrt{19.6}\,dt$$
$$2y^{1/2} = -0.0002\sqrt{19.6}t + C$$

Solving for y, we find

$$y(t) = \left(C - 0.0001\sqrt{19.6}t\right)^2.$$

Since the tank is originally full, we have the initial condition $y(0) = 10$, whence $\sqrt{10} = C$. Therefore,

$$y(t) = \left(\sqrt{10} - 0.0001\sqrt{19.6}t\right)^2.$$

When half of the water is out of the tank, $y = 1.5$, so we solve:

$$1.5 = \left(\sqrt{10} - 0.0001\sqrt{19.6}t\right)^2$$

for t, finding

$$t = \frac{1}{0.0002\sqrt{19.6}}(2\sqrt{10} - \sqrt{6}) \approx 4376.44 \text{ sec}.$$

When all of the water is out of the tank, $y = 0$, so

$$\sqrt{10} - 0.0001\sqrt{19.6}t = 0 \quad \text{and} \quad t = \frac{\sqrt{10}}{0.0001\sqrt{19.6}} \approx 7142.86 \text{ sec}.$$

49. The tank in Figure 7(B) is a cylinder of radius 4 m and height 15 m. Assume that the tank is half-filled with water and that water leaks through a hole in the bottom of area $B = 0.001\ \text{m}^2$. Determine the water level $y(t)$ and the time t_e when the tank is empty.

SOLUTION When the water is at height y over the bottom, the top cross section is a rectangle with length 15 m, and with width x satisfying the equation:

$$(x/2)^2 + (y-4)^2 = 16.$$

Thus, $x = 2\sqrt{8y - y^2}$, and

$$A(y) = 15x = 30\sqrt{8y - y^2}.$$

With $B = 0.001\ \text{m}^2$ and $v = -\sqrt{2gy} = -\sqrt{19.6}\sqrt{y}$, it follows that

$$\frac{dy}{dt} = -\frac{0.001\sqrt{19.6}\sqrt{y}}{30\sqrt{8y - y^2}} = -\frac{0.001\sqrt{19.6}}{30\sqrt{8-y}}.$$

Separating variables and integrating then yields:

$$\sqrt{8-y}\,dy = -\frac{0.001\sqrt{19.6}}{30}\,dt = -\frac{0.0001\sqrt{19.6}}{3}\,dt$$
$$-\frac{2}{3}(8-y)^{3/2} = -\frac{0.0001\sqrt{19.6}}{3}t + C$$

When $t = 0$, $y = 4$, so $C = -\frac{2}{3}4^{3/2} = -\frac{16}{3}$, and

$$-\frac{2}{3}(8-y)^{3/2} = -\frac{0.0001\sqrt{19.6}}{3}t - \frac{16}{3}$$
$$y(t) = 8 - \left(\frac{0.0001\sqrt{19.6}}{2}t + 8\right)^{2/3}.$$

The tank is empty when $y = 0$. Thus, t_e satisfies the equation

$$8 - \left(\frac{0.0001\sqrt{19.6}}{2}t + 8\right)^{2/3} = 0.$$

It follows that

$$t_e = \frac{2(8^{3/2} - 8)}{0.0001\sqrt{19.6}} \approx 66{,}079.9 \text{ seconds.}$$

51. A tank has the shape of the parabola $y = ax^2$ (where a is a constant) revolved around the y-axis. Water drains from a hole of area B m^2 at the bottom of the tank.

(a) Show that the water level at time t is

$$y(t) = \left(y_0^{3/2} - \frac{3aB\sqrt{2g}}{2\pi}t\right)^{2/3}$$

where y_0 is the water level at time $t = 0$.

(b) Show that if the total volume of water in the tank has volume V at time $t = 0$, then $y_0 = \sqrt{2aV/\pi}$. *Hint:* Compute the volume of the tank as a volume of rotation.

(c) Show that the tank is empty at time

$$t_e = \left(\frac{2}{3B\sqrt{g}}\right)\left(\frac{2\pi V^3}{a}\right)^{1/4}$$

We see that for fixed initial water volume V, the time t_e is proportional to $a^{-1/4}$. A large value of a corresponds to a tall thin tank. Such a tank drains more quickly than a short wide tank of the same initial volume.

SOLUTION

(a) When the water is at height y, the surface of the water is a circle of radius $\sqrt{y/a}$, so that the cross-sectional area is $A(y) = \pi y/a$. With $v = -\sqrt{2gy} = -\sqrt{2g}\sqrt{y}$, we have

$$\frac{dy}{dt} = -\frac{B\sqrt{2g}\sqrt{y}}{A} = -\frac{aB\sqrt{2g}\sqrt{y}}{\pi y} = -\frac{aB\sqrt{2g}}{\pi}y^{-1/2}$$

Separating variables and integrating gives

$$\sqrt{y}\,dy = -\frac{aB\sqrt{2g}}{\pi}\,dt$$

$$\frac{2}{3}y^{3/2} = -\frac{aB\sqrt{2g}}{\pi}t + C_1$$

$$y^{3/2} = -\frac{3aB\sqrt{2g}}{2\pi}t + C$$

Since $y(0) = y_0$, we have $C = y_0^{3/2}$; solving for y gives

$$y = \left(y_0^{3/2} - \frac{3aB\sqrt{2g}}{2\pi}t\right)^{2/3}$$

(b) The volume of the tank can be computed as a volume of rotation. Using the disk method and applying it to the function $x = \sqrt{y/a}$, we have

$$V = \int_0^{y_0} \pi\sqrt{\frac{y}{a}}^2\,dy = \frac{\pi}{a}\int_0^{y_0} y\,dy = \frac{\pi}{2a}y^2\Big|_0^{y_0} = \frac{\pi}{2a}y_0^2$$

Solving for y_0 gives

$$y_0 = \sqrt{2aV/\pi}$$

(c) The tank is empty when $y = 0$; this occurs when

$$y_0^{3/2} - \frac{3aB\sqrt{2g}}{2\pi}t = 0$$

From part (b), we have

$$y_0^{3/2} = \sqrt{2aV/\pi}^{3/2} = ((2aV/\pi)^{1/2})^{3/2} = (2aV/\pi)^{3/4}$$

so that

$$t_e = \frac{2\pi y_0^{3/2}}{3aB\sqrt{2g}} = \frac{2\pi \sqrt[4]{8a^3V^3}}{3\pi^{3/4}B\sqrt[4]{a^4}\sqrt[4]{4}\sqrt{g}} = \frac{2\pi^{1/4}\sqrt[4]{2V^3a^{-1}}}{3B\sqrt{g}} = \left(\frac{2}{3B\sqrt{g}}\right)\left(\frac{2\pi V^3}{a}\right)^{1/4}$$

53. Figure 8 shows a circuit consisting of a resistor of R ohms, a capacitor of C farads, and a battery of voltage V. When the circuit is completed, the amount of charge $q(t)$ (in coulombs) on the plates of the capacitor varies according to the differential equation (t in seconds)

$$R\frac{dq}{dt} + \frac{1}{C}q = V$$

where R, C, and V are constants.

(a) Solve for $q(t)$, assuming that $q(0) = 0$.

(b) Show that $\lim_{t\to\infty} q(t) = CV$.

(c) Show that the capacitor charges to approximately 63% of its final value CV after a time period of length $\tau = RC$ (τ is called the time constant of the capacitor).

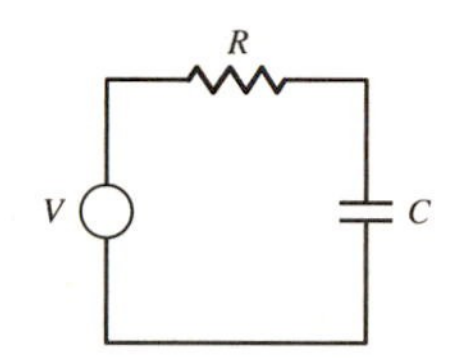

FIGURE 8 An RC circuit.

SOLUTION

(a) Upon rearranging the terms of the differential equation, we have

$$\frac{dq}{dt} = -\frac{q - CV}{RC}.$$

Separating the variables and integrating both sides, we obtain

$$\frac{dq}{q - CV} = -\frac{dt}{RC}$$

$$\int \frac{dq}{q - CV} = -\int \frac{dt}{RC}$$

and

$$\ln|q - CV| = -\frac{t}{RC} + k,$$

where k is an arbitrary constant. Solving for $q(t)$ yields

$$q(t) = CV + Ke^{-\frac{1}{RC}t},$$

where $K = \pm e^k$. We use the initial condition $q(0) = 0$ to solve for K:

$$0 = CV + K \quad \Rightarrow \quad K = -CV$$

so that the particular solution is

$$q(t) = CV(1 - e^{-\frac{1}{RC}t})$$

(b) Using the result from part (a), we calculate

$$\lim_{t\to\infty} q(t) = \lim_{t\to\infty} CV(1 - e^{-\frac{1}{RC}t}) = CV(1 - \lim_{t\to\infty} 1 - e^{-\frac{1}{RC}t}) = CV.$$

(c) We have

$$q(\tau) = q(RC) = CV(1 - e^{-\frac{1}{RC}RC}) = CV(1 - e^{-1}) \approx 0.632CV.$$

55. According to one hypothesis, the growth rate dV/dt of a cell's volume V is proportional to its surface area A. Since V has cubic units such as cm^3 and A has square units such as cm^2, we may assume roughly that $A \propto V^{2/3}$, and hence $dV/dt = kV^{2/3}$ for some constant k. If this hypothesis is correct, which dependence of volume on time would we expect to see (again, roughly speaking) in the laboratory?

(a) Linear **(b)** Quadratic **(c)** Cubic

SOLUTION Rewrite

$$\frac{dV}{dt} = kV^{2/3} \quad \text{as} \quad V^{-2/3}\,dv = k\,dt,$$

and then integrate both sides to obtain

$$3V^{1/3} = kt + C$$
$$V = (kt/3 + C)^3.$$

Thus, we expect to see V increasing roughly like the cube of time.

57. In general, $(fg)'$ is not equal to $f'g'$, but let $f(x) = e^{3x}$ and find a function $g(x)$ such that $(fg)' = f'g'$. Do the same for $f(x) = x$.

SOLUTION If $(fg)' = f'g'$, we have

$$f'(x)g(x) + g'(x)f(x) = f'(x)g'(x)$$
$$g'(x)(f(x) - f'(x)) = -g(x)f'(x)$$
$$\frac{g'(x)}{g(x)} = \frac{f'(x)}{f'(x) - f(x)}$$

Now, let $f(x) = e^{3x}$. Then $f'(x) = 3e^{3x}$ and

$$\frac{g'(x)}{g(x)} = \frac{3e^{3x}}{3e^{3x} - e^{3x}} = \frac{3}{2}.$$

Integrating and solving for $g(x)$, we find

$$\frac{dg}{g} = \frac{3}{2}\,dx$$
$$\ln|g| = \frac{3}{2}x + C$$
$$g(x) = Ae^{(3/2)x},$$

where $A = \pm e^C$ is an arbitrary constant.

If $f(x) = x$, then $f'(x) = 1$, and

$$\frac{g'(x)}{g(x)} = \frac{1}{1-x}.$$

Thus,

$$\frac{dg}{g} = \frac{1}{1-x}\,dx$$
$$\ln|g| = -\ln|1-x| + C$$
$$g(x) = \frac{A}{1-x},$$

where $A = \pm e^C$ is an arbitrary constant.

59. Show that the differential equations $y' = 3y/x$ and $y' = -x/3y$ define **orthogonal families** of curves; that is, the graphs of solutions to the first equation intersect the graphs of the solutions to the second equation in right angles (Figure 10). Find these curves explicitly.

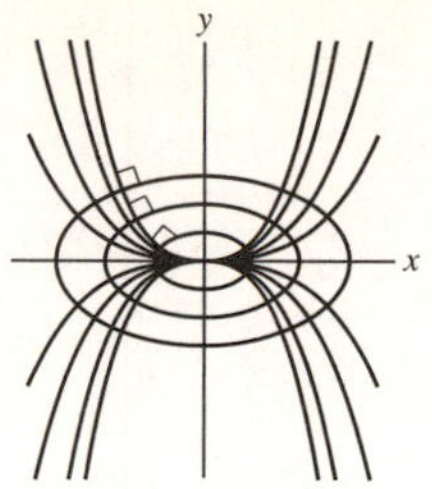

FIGURE 10 Two orthogonal families of curves.

SOLUTION Let y_1 be a solution to $y' = \frac{3y}{x}$ and let y_2 be a solution to $y' = -\frac{x}{3y}$. Suppose these two curves intersect at a point (x_0, y_0). The line tangent to the curve $y_1(x)$ at (x_0, y_0) has a slope of $\frac{3y_0}{x_0}$ and the line tangent to the curve $y_2(x)$ has a slope of $-\frac{x_0}{3y_0}$. The slopes are negative reciprocals of one another; hence the tangent lines are perpendicular.

Separation of variables and integration applied to $y' = \frac{3y}{x}$ gives

$$\frac{dy}{y} = 3\frac{dx}{x}$$

$$\ln|y| = 3\ln|x| + C$$

$$y = Ax^3$$

On the other hand, separation of variables and integration applied to $y' = -\frac{x}{3y}$ gives

$$3y\,dy = -x\,dx$$

$$3y^2/2 = -x^2/2 + C$$

$$y = \pm\sqrt{C - x^2/3}$$

61. A 50-kg model rocket lifts off by expelling fuel downward at a rate of $k = 4.75$ kg/s for 10 s. The fuel leaves the end of the rocket with an exhaust velocity of $b = -100$ m/s. Let $m(t)$ be the mass of the rocket at time t. From the law of conservation of momentum, we find the following differential equation for the rocket's velocity $v(t)$ (in meters per second):

$$m(t)v'(t) = -9.8m(t) + b\frac{dm}{dt}$$

(a) Show that $m(t) = 50 - 4.75t$ kg.

(b) Solve for $v(t)$ and compute the rocket's velocity at rocket burnout (after 10 s).

SOLUTION

(a) For $0 \le t \le 10$, the rocket is expelling fuel at a constant rate of 4.75 kg/s, giving $m'(t) = -4.75$. Hence, $m(t) = -4.75t + C$. Initially, the rocket has a mass of 50 kg, so $C = 50$. Therefore, $m(t) = 50 - 4.75t$.

(b) With $m(t) = 50 - 4.75t$ and $\frac{dm}{dt} = -4.75$, the equation for v becomes

$$\frac{dv}{dt} = -9.8 + \frac{b\frac{dm}{dt}}{50 - 4.75t} = -9.8 + \frac{(-100)(-4.75)}{50 - 4.75t}$$

and therefore

$$v(t) = -9.8t + 100\int \frac{4.75\,dt}{50 - 4.75t} = -9.8t - 100\ln(50 - 4.75t) + C$$

Because $v(0) = 0$, we find $C = 100\ln 50$ and

$$v(t) = -9.8t - 100\ln(50 - 4.75t) + 100\ln(50).$$

After 10 seconds the velocity is:

$$v(10) = -98 - 100\ln(2.5) + 100\ln(50) \approx 201.573 \text{ m/s}.$$

63. If a bucket of water spins about a vertical axis with constant angular velocity ω (in radians per second), the water climbs up the side of the bucket until it reaches an equilibrium position (Figure 11). Two forces act on a particle located at a distance x from the vertical axis: the gravitational force $-mg$ acting downward and the force of the bucket on the particle (transmitted indirectly through the liquid) in the direction perpendicular to the surface of the water. These two forces must combine to supply a centripetal force $m\omega^2 x$, and this occurs if the diagonal of the rectangle in Figure 11 is normal to the water's surface (that is, perpendicular to the tangent line). Prove that if $y = f(x)$ is the equation of the curve obtained by taking a vertical cross section through the axis, then $-1/y' = -g/(\omega^2 x)$. Show that $y = f(x)$ is a parabola.

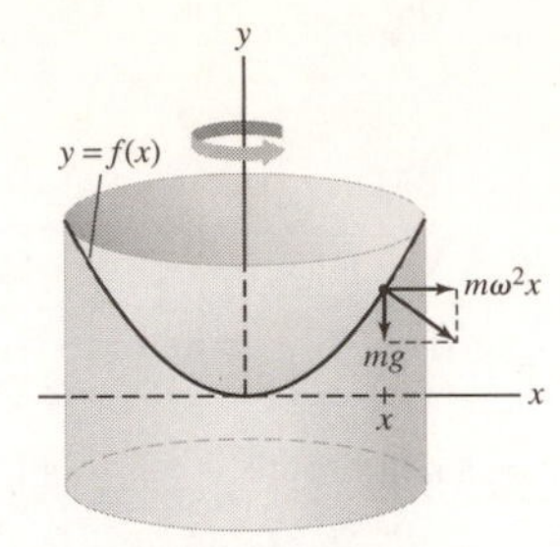

FIGURE 11

SOLUTION At any point along the surface of the water, the slope of the tangent line is given by the value of y' at that point; hence, the slope of the line perpendicular to the surface of the water is given by $-1/y'$. The slope of the resultant force generated by the gravitational force and the centrifugal force is

$$\frac{-mg}{m\omega^2 x} = -\frac{g}{\omega^2 x}.$$

Therefore, the curve obtained by taking a vertical cross-section of the water surface is determined by the equation

$$-\frac{1}{y'} = -\frac{g}{\omega^2 x} \quad \text{or} \quad y' = \frac{\omega^2}{g}x.$$

Performing one integration yields

$$y = f(x) = \frac{\omega^2}{2g}x^2 + C,$$

where C is a constant of integration. Thus, $y = f(x)$ is a parabola.

Further Insights and Challenges

65. A basic theorem states that a *linear* differential equation of order n has a general solution that depends on n arbitrary constants. There are, however, nonlinear exceptions.

(a) Show that $(y')^2 + y^2 = 0$ is a first-order equation with only one solution $y = 0$.

(b) Show that $(y')^2 + y^2 + 1 = 0$ is a first-order equation with no solutions.

SOLUTION

(a) $(y')^2 + y^2 \geq 0$ and equals zero if and only if $y' = 0$ and $y = 0$

(b) $(y')^2 + y^2 + 1 \geq 1 > 0$ for all y' and y, so $(y')^2 + y^2 + 1 = 0$ has no solution

67. A spherical tank of radius R is half-filled with water. Suppose that water leaks through a hole in the bottom of area B. Let $y(t)$ be the water level at time t (seconds).

(a) Show that $\dfrac{dy}{dt} = \dfrac{-\sqrt{2g}B\sqrt{y}}{\pi(2Ry - y^2)}$.

(b) Show that for some constant C,

$$\frac{2\pi}{15B\sqrt{2g}}\left(10Ry^{3/2} - 3y^{5/2}\right) = C - t$$

(c) Use the initial condition $y(0) = R$ to compute C, and show that $C = t_e$, the time at which the tank is empty.

(d) Show that t_e is proportional to $R^{5/2}$ and inversely proportional to B.

SOLUTION

(a) At height y above the bottom of the tank, the cross section is a circle of radius

$$r = \sqrt{R^2 - (R - y)^2} = \sqrt{2Ry - y^2}.$$

The cross-sectional area function is then $A(y) = \pi(2Ry - y^2)$. The differential equation for the height of the water in the tank is then

$$\frac{dy}{dt} = -\frac{\sqrt{2g}B\sqrt{y}}{\pi(2Ry - y^2)}$$

by Torricelli's law.

(b) Rewrite the differential equation as

$$\frac{\pi}{\sqrt{2g}B}\left(2Ry^{1/2} - y^{3/2}\right)dy = -\,dt,$$

and then integrate both sides to obtain

$$\frac{2\pi}{\sqrt{2g}B}\left(\frac{2}{3}Ry^{3/2} - \frac{1}{5}y^{5/2}\right) = C - t,$$

where C is an arbitrary constant. Simplifying gives

$$\frac{2\pi}{15B\sqrt{2g}}(10Ry^{3/2} - 3y^{5/2}) = C - t \qquad (*)$$

(c) From Equation (*) we see that $y = 0$ when $t = C$. It follows that $C = t_e$, the time at which the tank is empty. Moreover, the initial condition $y(0) = R$ allows us to determine the value of C:

$$\frac{2\pi}{15B\sqrt{2g}}(10R^{5/2} - 3R^{5/2}) = \frac{14\pi}{15B\sqrt{2g}}R^{5/2} = C$$

(d) From part (c),

$$t_e = \frac{14\pi}{15\sqrt{2g}} \cdot \frac{R^{5/2}}{B},$$

from which it is clear that t_e is proportional to $R^{5/2}$ and inversely proportional to B.

9.2 Models Involving $y' = k(y - b)$

Preliminary Questions

1. Write down a solution to $y' = 4(y - 5)$ that tends to $-\infty$ as $t \to \infty$.

SOLUTION The general solution is $y(t) = 5 + Ce^{4t}$ for any constant C; thus the solution tends to $-\infty$ as $t \to \infty$ whenever $C < 0$. One specific example is $y(t) = 5 - e^{4t}$.

2. Does $y' = -4(y - 5)$ have a solution that tends to ∞ as $t \to \infty$?

SOLUTION The general solution is $y(t) = 5 + Ce^{-4t}$ for any constant C. As $t \to \infty$, $y(t) \to 5$. Thus, there is no solution of $y' = -4(y - 5)$ that tends to ∞ as $t \to \infty$.

3. True or false? If $k > 0$, then all solutions of $y' = -k(y - b)$ approach the same limit as $t \to \infty$.

SOLUTION True. The general solution of $y' = -k(y - b)$ is $y(t) = b + Ce^{-kt}$ for any constant C. If $k > 0$, then $y(t) \to b$ as $t \to \infty$.

4. As an object cools, its rate of cooling slows. Explain how this follows from Newton's Law of Cooling.

SOLUTION Newton's Law of Cooling states that $y' = -k(y - T_0)$ where $y(t)$ is the temperature and T_0 is the ambient temperature. Thus as $y(t)$ gets closer to T_0, $y'(t)$, the rate of cooling, gets smaller and the rate of cooling slows.

Exercises

1. Find the general solution of $y' = 2(y - 10)$. Then find the two solutions satisfying $y(0) = 25$ and $y(0) = 5$, and sketch their graphs.

SOLUTION The general solution of $y' = 2(y - 10)$ is $y(t) = 10 + Ce^{2t}$ for any constant C. If $y(0) = 25$, then $10 + C = 25$, or $C = 15$; therefore, $y(t) = 10 + 15e^{2t}$. On the other hand, if $y(0) = 5$, then $10 + C = 5$, or $C = -5$; therefore, $y(t) = 10 - 5e^{2t}$. Graphs of these two functions are given below.

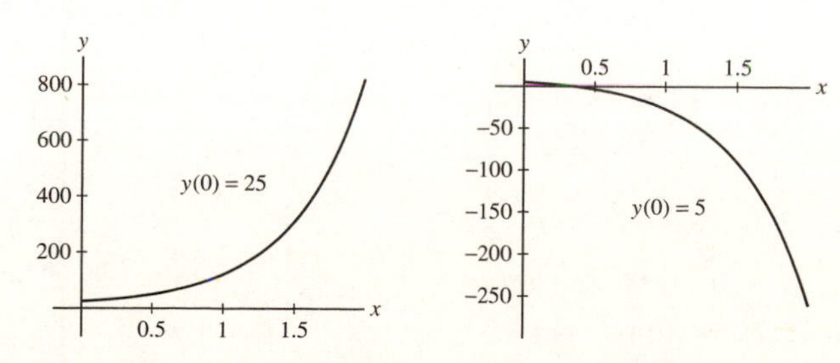

3. Solve $y' = 4y + 24$ subject to $y(0) = 5$.

SOLUTION Rewrite

$$y' = 4y + 24 \quad \text{as} \quad \frac{1}{4y + 24}\,dy = 1\,dt$$

Integrating gives

$$\frac{1}{4}\ln|4y + 24| = t + C$$
$$\ln|4y + 24| = 4t + C$$
$$4y + 24 = \pm e^{4t+C}$$
$$y = Ae^{4t} - 6$$

where $A = \pm e^C/4$ is any constant. Since $y(0) = 5$ we have $5 = A - 6$ so that $A = 11$, and the solution is $y = 11e^{4t} - 6$.

In Exercises 5–12, use Newton's Law of Cooling.

5. A hot anvil with cooling constant $k = 0.02\ \text{s}^{-1}$ is submerged in a large pool of water whose temperature is 10°C. Let $y(t)$ be the anvil's temperature t seconds later.

(a) What is the differential equation satisfied by $y(t)$?

(b) Find a formula for $y(t)$, assuming the object's initial temperature is 100°C.

(c) How long does it take the object to cool down to 20°?

SOLUTION

(a) By Newton's Law of Cooling, the differential equation is

$$y' = -0.02(y - 10)$$

(b) Separating variables gives

$$\frac{1}{y - 10}\,dy = -0.02\,dt$$

Integrate to get

$$\ln|y - 10| = -0.02t + C$$
$$y - 10 = \pm e^{-0.02t+C}$$
$$y = 10 + Ae^{-0.02t}$$

where $A = \pm e^C$ is a constant. Since the initial temperature is 100°C, we have $y(0) = 100 = 10 + A$ so that $A = 90$, and $y = 10 + 90e^{-0.02t}$.

(c) We must find the value of t such that $y(t) = 20$, so we need to solve $20 = 10 + 90e^{-0.02t}$. Thus

$$10 = 90e^{-0.02t} \quad \Rightarrow \quad \frac{1}{9} = e^{-0.02t} \quad \Rightarrow \quad -\ln 9 = -0.02t \quad \Rightarrow \quad t = 50\ln 9 \approx 109.86\ \text{s}$$

7. At 10:30 AM, detectives discover a dead body in a room and measure its temperature at 26°C. One hour later, the body's temperature had dropped to 24.8°C. Determine the time of death (when the body temperature was a normal 37°C), assuming that the temperature in the room was held constant at 20°C.

SOLUTION Let $t = 0$ be the time when the person died, and let t_0 denote 10:30AM. The differential equation satisfied by the body temperature, $y(t)$, is

$$y' = -k(y - 20)$$

by Newton's Law of Cooling. Separating variables gives $\frac{1}{y - 20}\,dy = -k\,dt$. Integrate to get

$$\ln|y - 20| = -kt + C$$
$$y - 20 = \pm e^{-kt+C}$$
$$y = 20 + Ae^{-kt}$$

where $A = \pm e^C$ is a constant. Since normal body temperature is 37°C, we have $y(0) = 37 = 20 + A$ so that $A = 17$. To determine k, note that

$$26 = 20 + 17e^{-kt_0} \qquad \text{and} \qquad 24.8 = 20 + 17e^{-k(t_0+1)}$$

$$kt_0 = -\ln\frac{6}{17} \qquad\qquad kt_0 + k = -\ln\frac{4.8}{17}$$

Subtracting these equations gives

$$k = \ln\frac{6}{17} - \ln\frac{4.8}{17} = \ln\frac{6}{4.8} \approx 0.223$$

We thus have

$$y = 20 + 17e^{-0.223t}$$

as the equation for the body temperature at time t. Since $y(t_0) = 26$, we have

$$26 = 20 + 17e^{-0.223t} \quad \Rightarrow \quad e^{-0.223t} = \frac{6}{17} \quad \Rightarrow \quad t = -\frac{1}{0.223}\ln\frac{6}{17} \approx 4.667 \text{ h}$$

so that the time of death was approximately 4 hours and 40 minutes ago.

9. A cold metal bar at -30°C is submerged in a pool maintained at a temperature of 40°C. Half a minute later, the temperature of the bar is 20°C. How long will it take for the bar to attain a temperature of 30°C?

SOLUTION With $T_0 = 40$°C, the temperature of the bar is given by $F(t) = 40 + Ce^{-kt}$ for some constants C and k. From the initial condition, $F(0) = 40 + C = -30$, so $C = -70$. After 30 seconds, $F(30) = 40 - 70e^{-30k} = 20$, so

$$k = -\frac{1}{30}\ln\left(\frac{20}{70}\right) \approx 0.0418 \text{ seconds}^{-1}.$$

To attain a temperature of 30°C we must solve $40 - 70e^{-0.0418t} = 30$ for t. This yields

$$t = \frac{\ln\left(\frac{10}{70}\right)}{-0.0418} \approx 46.55 \text{ seconds}.$$

11. GU Objects A and B are placed in a warm bath at temperature $T_0 = 40$°C. Object A has initial temperature -20°C and cooling constant $k = 0.004\ \text{s}^{-1}$. Object B has initial temperature 0°C and cooling constant $k = 0.002\ \text{s}^{-1}$. Plot the temperatures of A and B for $0 \le t \le 1000$. After how many seconds will the objects have the same temperature?

SOLUTION With $T_0 = 40$°C, the temperature of A and B are given by

$$A(t) = 40 + C_A e^{-0.004t} \qquad B(t) = 40 + C_B e^{-0.002t}$$

Since $A(0) = -20$ and $B(0) = 0$, we have

$$A(t) = 40 - 60e^{-0.004t} \qquad B(t) = 40 - 40e^{-0.002t}$$

The two objects will have the same temperature whenever $A(t) = B(t)$, so we must solve

$$40 - 60e^{-0.004t} = 40 - 40e^{-0.002t} \quad \Rightarrow \quad 3e^{-0.004t} = 2e^{-0.002t}$$

Take logs to get

$$-0.004t + \ln 3 = -0.002t + \ln 2 \quad \Rightarrow \quad t = \frac{\ln 3 - \ln 2}{0.002} \approx 202.7 \text{ s}$$

or about 3 minutes 22 seconds.

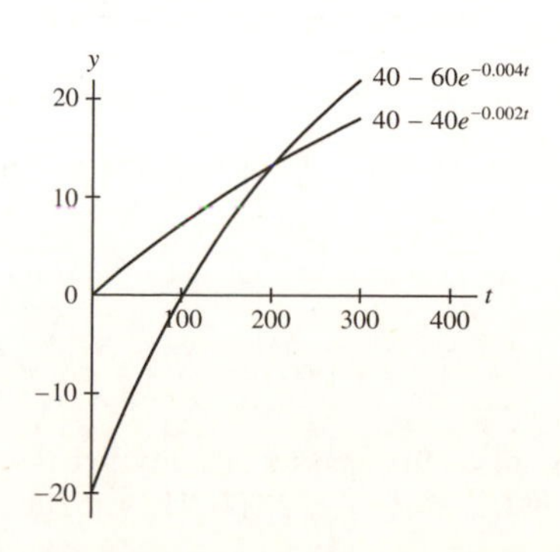

In Exercises 13–16, use Eq. (3) as a model for free-fall with air resistance.

13. A 60-kg skydiver jumps out of an airplane. What is her terminal velocity, in meters per second, assuming that $k = 10$ kg/s for free-fall (no parachute)?

SOLUTION The free-fall terminal velocity is

$$\frac{-gm}{k} = \frac{-9.8(60)}{10} = -58.8 \text{ m/s}.$$

15. A 80-kg skydiver jumps out of an airplane (with zero initial velocity). Assume that $k = 12$ kg/s with a closed parachute and $k = 70$ kg/s with an open parachute. What is the skydiver's velocity at $t = 25$ s if the parachute opens after 20 s of free fall?

SOLUTION We first compute the skydiver's velocity after 20 s of free fall, then use that as the initial velocity to calculate her velocity after an additional 5 s of restrained fall. We have $m = 80$ and $g = 9.8$; for free fall, $k = 12$, so

$$\frac{k}{m} = \frac{12}{80} = 0.15, \qquad \frac{-mg}{k} = \frac{-80 \cdot 9.8}{12} \approx -65.33$$

The general solution is thus $v(t) = -65.33 + Ce^{-0.15t}$. Since $v(0) = 0$, we have $C = 65.33$, so that

$$v(t) = -65.33(1 - e^{-0.15t})$$

After 20 s of free fall, the diver's velocity is thus

$$v(20) = -65.33(1 - e^{-0.15 \cdot 20}) \approx -62.08 \text{ m/s}$$

Once the parachute opens, $k = 70$, so

$$\frac{k}{m} = \frac{70}{80} = 0.875, \qquad \frac{mg}{k} = \frac{80 \cdot 9.8}{70} = 11.2$$

so that the general solution for the restrained fall model is $v_r(t) = -11.2 + Ce^{-0.875t}$. Here $v_r(0) = -62.08$, so that $C = 11.2 - 62.08 = -50.88$ and $v_r(t) = -11.20 - 50.88e^{-0.875t}$. After 5 additional seconds, the diver's velocity is therefore

$$v_r(5) = -11.20 - 50.88e^{-0.875 \cdot 5} \approx -11.84 \text{ m/s}$$

17. A continuous annuity with withdrawal rate $N = \$5000$/year and interest rate $r = 5\%$ is funded by an initial deposit of $P_0 = \$50{,}000$.

(a) What is the balance in the annuity after 10 years?

(b) When will the annuity run out of funds?

SOLUTION

(a) From Equation , the value of the annuity is given by

$$P(t) = \frac{5000}{0.05} + Ce^{0.05t} = 100{,}000 + Ce^{0.05t}$$

for some constant C. Since $P(0) = 50{,}000$, we have $C = -50{,}000$ and $P(t) = 100{,}000 - 50{,}000e^{0.05t}$. After ten years, then, the balance in the annuity is

$$P(10) = 100{,}000 - 50{,}000e^{0.05 \cdot 10} = 100{,}000 - 50{,}000e^{0.5} \approx \$17{,}563.94$$

(b) The annuity will run out of funds when $P(t) = 0$:

$$0 = 100{,}000 - 50{,}000e^{0.05t} \quad \Rightarrow \quad e^{0.05t} = 2 \quad \Rightarrow \quad t = \frac{\ln 2}{0.05} \approx 13.86$$

The annuity will run out of funds after approximately 13 years 10 months.

19. Find the minimum initial deposit P_0 that will allow an annuity to pay out \$6000/year indefinitely if it earns interest at a rate of 5%.

SOLUTION Let $P(t)$ denote the balance of the annuity at time t measured in years. Then

$$P(t) = \frac{N}{r} + Ce^{rt} = \frac{6000}{0.05} + Ce^{0.05t} = 120{,}000 + Ce^{0.05t}$$

for some constant C. To fund the annuity indefinitely, we must have $C \geq 0$. If the initial deposit is P_0, then P_0 $120{,}000 + C$ and $C = P_0 - 120{,}000$. Thus, to fund the annuity indefinitely, we must have $P_0 \geq \$120{,}000$.

21. An initial deposit of 100,000 euros are placed in an annuity with a French bank. What is the minimum interest rate the annuity must earn to allow withdrawals at a rate of 8000 euros/year to continue indefinitely?

SOLUTION Let $P(t)$ denote the balance of the annuity at time t measured in years. Then

$$P(t) = \frac{N}{r} + Ce^{rt} = \frac{8000}{r} + Ce^{rt}$$

for some constant C. To fund the annuity indefinitely, we need $C \geq 0$. If the initial deposit is 100,000 euros, then $100{,}000 = \frac{8000}{r} + C$ and $C = 100{,}000 - \frac{8000}{r}$. Thus, to fund the annuity indefinitely, we need $100{,}000 - \frac{8000}{r} \geq 0$, or $r \geq 0.08$. The bank must pay at least 8%.

23. Sam borrows \$10,000 from a bank at an interest rate of 9% and pays back the loan continuously at a rate of N dollars per year. Let $P(t)$ denote the amount still owed at time t.

(a) Explain why $P(t)$ satisfies the differential equation

$$y' = 0.09y - N$$

(b) How long will it take Sam to pay back the loan if $N = \$1200$?

(c) Will the loan ever be paid back if $N = \$800$?

SOLUTION

(a)

$$\begin{aligned}\text{Rate of Change of Loan} &= \text{(Amount still owed)(Interest rate)} - \text{(Payback rate)}\\ &= P(t) \cdot r - N = r\left(P - \frac{N}{r}\right).\end{aligned}$$

Therefore, if $y = P(t)$,

$$y' = r\left(y - \frac{N}{r}\right) = ry - N$$

(b) From the differential equation derived in part (a), we know that $P(t) = \frac{N}{r} + Ce^{rt} = 13{,}333.33 + Ce^{0.09t}$. Since \$10,000 was initially borrowed, $P(0) = 13{,}333.33 + C = 10{,}000$, and $C = -3333.33$. The loan is paid off when $P(t) = 13{,}333.33 - 3333.33e^{0.09t} = 0$. This yields

$$t = \frac{1}{0.09}\ln\left(\frac{13{,}333.33}{3333.33}\right) \approx 15.4 \text{ years.}$$

(c) If the annual rate of payment is \$800, then $P(t) = 800/0.09 + Ce^{0.09t} = 8888.89 + Ce^{0.09t}$. With $P(0) = 8888.89 + C = 10{,}000$, it follows that $C = 1111.11$. Since $C > 0$ and $e^{0.09t} \to \infty$ as $t \to \infty$, $P(t) \to \infty$, and the loan will never be paid back.

25. Let $N(t)$ be the fraction of the population who have heard a given piece of news t hours after its initial release. According to one model, the rate $N'(t)$ at which the news spreads is equal to k times the fraction of the population that has not yet heard the news, for some constant $k > 0$.

(a) Determine the differential equation satisfied by $N(t)$.

(b) Find the solution of this differential equation with the initial condition $N(0) = 0$ in terms of k.

(c) Suppose that half of the population is aware of an earthquake 8 hours after it occurs. Use the model to calculate k and estimate the percentage that will know about the earthquake 12 hours after it occurs.

SOLUTION

(a) $N'(t) = k(1 - N(t)) = -k(N(t) - 1)$.

(b) The general solution of the differential equation from part (a) is $N(t) = 1 + Ce^{-kt}$. The initial condition determines the value of C: $N(0) = 1 + C = 0$ so $C = -1$. Thus, $N(t) = 1 - e^{-kt}$.

(c) Knowing that $N(8) = 1 - e^{-8k} = \frac{1}{2}$, we find that

$$k = -\frac{1}{8}\ln\left(\frac{1}{2}\right) \approx 0.0866 \text{ hours}^{-1}.$$

With the value of k determined, we estimate that

$$N(12) = 1 - e^{-0.0866(12)} \approx 0.6463 = 64.63\%$$

of the population will know about the earthquake after 12 hours.

Further Insights and Challenges

27. Show that the cooling constant of an object can be determined from two temperature readings $y(t_1)$ and $y(t_2)$ at times $t_1 \neq t_2$ by the formula

$$k = \frac{1}{t_1 - t_2} \ln\left(\frac{y(t_2) - T_0}{y(t_1) - T_0}\right)$$

SOLUTION We know that $y(t_1) = T_0 + Ce^{-kt_1}$ and $y(t_2) = T_0 + Ce^{-kt_2}$. Thus, $y(t_1) - T_0 = Ce^{-kt_1}$ and $y(t_2) - T_0 = Ce^{-kt_2}$. Dividing the latter equation by the former yields

$$e^{-kt_2+kt_1} = \frac{y(t_2) - T_0}{y(t_1) - T_0},$$

so that

$$k(t_1 - t_2) = \ln\left(\frac{y(t_2) - T_0}{y(t_1) - T_0}\right) \quad \text{and} \quad k = \frac{1}{t_1 - t_2} \ln\left(\frac{y(t_2) - T_0}{y(t_1) - T_0}\right).$$

29. Air Resistance A projectile of mass $m = 1$ travels straight up from ground level with initial velocity v_0. Suppose that the velocity v satisfies $v' = -g - kv$.

(a) Find a formula for $v(t)$.

(b) Show that the projectile's height $h(t)$ is given by

$$h(t) = C(1 - e^{-kt}) - \frac{g}{k}t$$

where $C = k^{-2}(g + kv_0)$.

(c) Show that the projectile reaches its maximum height at time $t_{max} = k^{-1} \ln(1 + kv_0/g)$.

(d) In the absence of air resistance, the maximum height is reached at time $t = v_0/g$. In view of this, explain why we should expect that

$$\lim_{k \to 0} \frac{\ln(1 + \frac{kv_0}{g})}{k} = \frac{v_0}{g} \tag{8}$$

(e) Verify Eq. (8). *Hint:* Use Theorem 2 in Section 5.8 to show that $\lim_{k \to 0} \left(1 + \frac{kv_0}{g}\right)^{1/k} = e^{v_0/g}$ or use L'Hôpital's Rule.

SOLUTION

(a) Since $v' = -g - kv = -k\left(v - \frac{-g}{k}\right)$ it follows that $v(t) = \frac{-g}{k} + Be^{-kt}$ for some constant B. The initial condition $v(0) = v_0$ determines B: $v_0 = -\frac{g}{k} + B$, so $B = v_0 + \frac{g}{k}$. Thus,

$$v(t) = -\frac{g}{k} + \left(v_0 + \frac{g}{k}\right)e^{-kt}.$$

(b) $v(t) = h'(t)$ so

$$h(t) = \int \left(-\frac{g}{k} + \left(v_0 + \frac{g}{k}\right)e^{-kt}\right) dt = -\frac{g}{k}t - \frac{1}{k}\left(v_0 + \frac{g}{k}\right)e^{-kt} + D.$$

The initial condition $h(0) = 0$ determines

$$D = \frac{1}{k}\left(v_0 + \frac{g}{k}\right) = \frac{1}{k^2}(v_0 k + g).$$

Let $C = \frac{1}{k^2}(v_0 k + g)$. Then

$$h(t) = C(1 - e^{-kt}) - \frac{g}{k}t.$$

(c) The projectile reaches its maximum height when $v(t) = 0$. This occurs when

$$-\frac{g}{k} + \left(v_0 + \frac{g}{k}\right)e^{-kt} = 0,$$

or

$$t = \frac{1}{-k} \ln\left(\frac{g}{kv_0 + g}\right) = \frac{1}{k} \ln\left(1 + \frac{kv_0}{g}\right).$$

(d) Recall that k is the proportionality constant for the force due to air resistance. Thus, as $k \to 0$, the effect of air resistance disappears. We should therefore expect that, as $k \to 0$, the time at which the maximum height is achieved from part (c) should approach v_0/g. In other words, we should expect

$$\lim_{k \to 0} \frac{1}{k} \ln\left(1 + \frac{kv_0}{g}\right) = \frac{v_0}{g}.$$

(e) Recall that

$$e^x = \lim_{n\to\infty}\left(1+\frac{x}{n}\right)^n.$$

If we substitute $x = v_0/g$ and $k = 1/n$, we find

$$e^{v_0/g} = \lim_{k\to 0}\left(1+\frac{v_0 k}{g}\right)^{1/k}.$$

Then

$$\lim_{k\to 0}\frac{1}{k}\ln\left(1+\frac{kv_0}{g}\right) = \lim_{k\to 0}\ln\left(1+\frac{v_0 k}{g}\right)^{1/k} = \ln\left(\lim_{k\to 0}\left(1+\frac{v_0 k}{g}\right)^{1/k}\right) = \ln(e^{v_0/g}) = \frac{v_0}{g}.$$

9.3 Graphical and Numerical Methods

Preliminary Questions

1. What is the slope of the segment in the slope field for $\dot{y} = ty + 1$ at the point (2, 3)?

SOLUTION The slope of the segment in the slope field for $\dot{y} = ty + 1$ at the point (2, 3) is $(2)(3) + 1 = 7$.

2. What is the equation of the isocline of slope $c = 1$ for $\dot{y} = y^2 - t$?

SOLUTION The isocline of slope $c = 1$ has equation $y^2 - t = 1$, or $y = \pm\sqrt{1+t}$.

3. For which of the following differential equations are the slopes at points on a vertical line $t = C$ all equal?

(a) $\dot{y} = \ln y$ **(b)** $\dot{y} = \ln t$

SOLUTION Only for the equation in part (b). The slope at a point is simply the value of $\dot{y}$ at that point, so for part (a), the slope depends on y, while for part (b), the slope depends only on t.

4. Let $y(t)$ be the solution to $\dot{y} = F(t, y)$ with $y(1) = 3$. How many iterations of Euler's Method are required to approximate $y(3)$ if the time step is $h = 0.1$?

SOLUTION The initial condition is specified at $t = 1$ and we want to obtain an approximation to the value of the solution at $t = 3$. With a time step of $h = 0.1$,

$$\frac{3-1}{0.1} = 20$$

iterations of Euler's method are required.

Exercises

1. Figure 8 shows the slope field for $\dot{y} = \sin y \sin t$. Sketch the graphs of the solutions with initial conditions $y(0) = 1$ and $y(0) = -1$. Show that $y(t) = 0$ is a solution and add its graph to the plot.

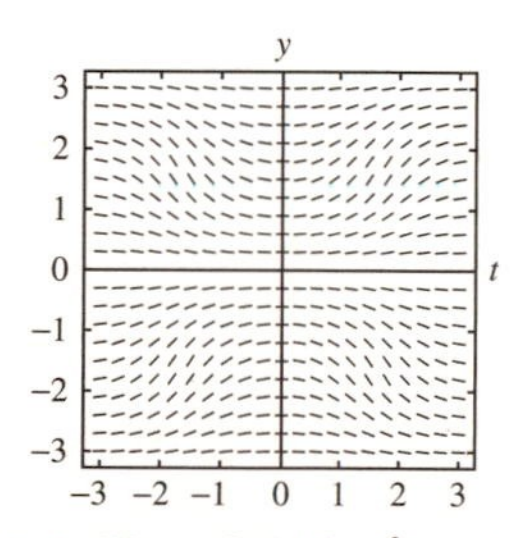

FIGURE 8 Slope field for $\dot{y} = \sin y \sin t$.

SOLUTION The sketches of the solutions appear below.

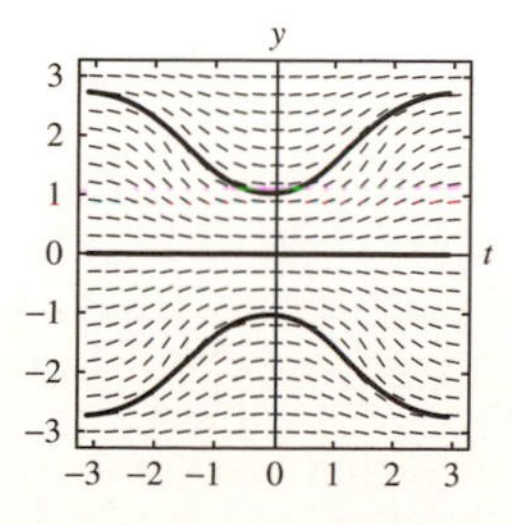

If $y(t) = 0$, then $y' = 0$; moreover, $\sin 0 \sin t = 0$. Thus, $y(t) = 0$ is a solution of $\dot{y} = \sin y \sin t$.

3. Show that $f(t) = \frac{1}{2}\left(t - \frac{1}{2}\right)$ is a solution to $\dot{y} = t - 2y$. Sketch the four solutions with $y(0) = \pm 0.5, \pm 1$ on the slope field in Figure 10. The slope field suggests that every solution approaches $f(t)$ as $t \to \infty$. Confirm this by showing that $y = f(t) + Ce^{-2t}$ is the general solution.

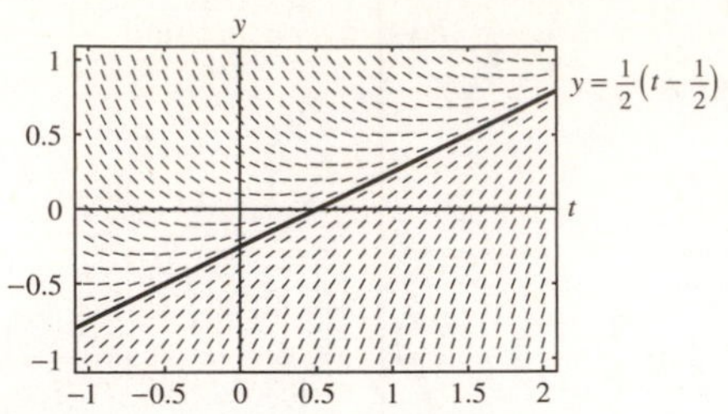

FIGURE 10 Slope field for $\dot{y} = t - 2y$.

SOLUTION Let $y = f(t) = \frac{1}{2}(t - \frac{1}{2})$. Then $\dot{y} = \frac{1}{2}$ and

$$\dot{y} + 2y = \frac{1}{2} + t - \frac{1}{2} = t,$$

so $f(t) = \frac{1}{2}(t - \frac{1}{2})$ is a solution to $\dot{y} = t - 2y$. The slope field with the four required solutions is shown below.

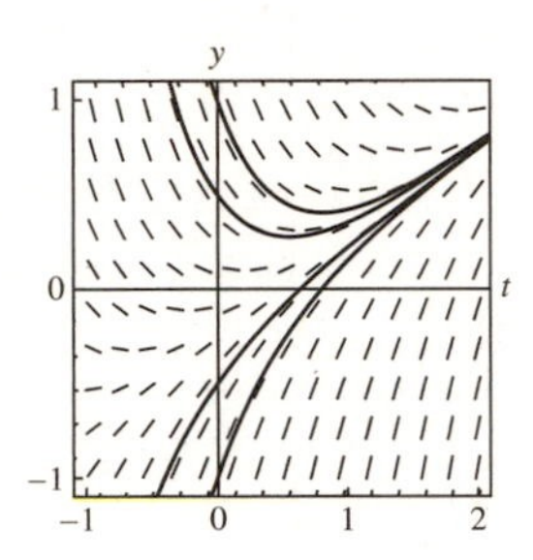

Now, let $y = f(t) + Ce^{-2t} = \frac{1}{2}(t - \frac{1}{2}) + Ce^{-2t}$. Then

$$\dot{y} = \frac{1}{2} - 2Ce^{-2t},$$

and

$$\dot{y} + 2y = \frac{1}{2} - 2Ce^{-2t} + \left(t - \frac{1}{2}\right) + 2Ce^{-2t} = t.$$

Thus, $y = f(t) + Ce^{-2t}$ is the general solution to the equation $\dot{y} = t - 2y$.

5. Consider the differential equation $\dot{y} = t - y$.

(a) Sketch the slope field of the differential equation $\dot{y} = t - y$ in the range $-1 \le t \le 3$, $-1 \le y \le 3$. As an aid, observe that the isocline of slope c is the line $t - y = c$, so the segments have slope c at points on the line $y = t - c$.

(b) Show that $y = t - 1 + Ce^{-t}$ is a solution for all C. Since $\lim_{t\to\infty} e^{-t} = 0$, these solutions approach the particular solution $y = t - 1$ as $t \to \infty$. Explain how this behavior is reflected in your slope field.

SOLUTION

(a) Here is a sketch of the slope field:

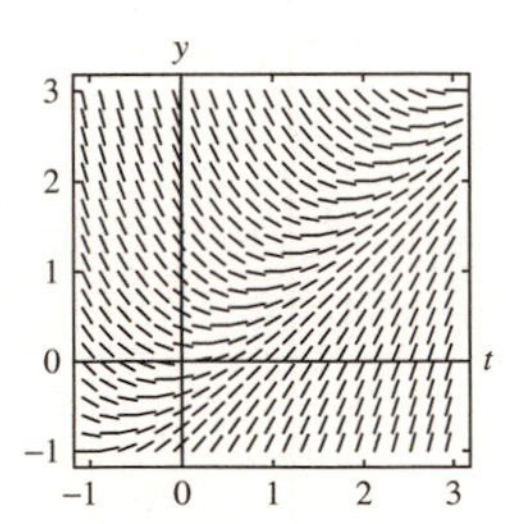

(b) Let $y = t - 1 + Ce^{-t}$. Then $\dot{y} = 1 - C^{-t}$, and

$$t - y = t - (t - 1 + Ce^{-t}) = 1 - Ce^{-t}.$$

Thus, $y = t - 1 + Ce^{-t}$ is a solution of $\dot{y} = t - y$. On the slope field, we can see that the isoclines of 1 all lie along the line $y = t - 1$. Whenever $y > t - 1$, $\dot{y} = t - y < 1$, so the solution curve will converge downward towards the line $y = t - 1$. On the other hand, if $y < t - 1$, $\dot{y} = t - y > 1$, so the solution curve will converge upward towards $y = t - 1$. In either case, the solution is approaching $t - 1$.

7. Show that the isoclines of $\dot{y} = t$ are vertical lines. Sketch the slope field for $-2 \le t \le 2$, $-2 \le y \le 2$ and plot the integral curves passing through $(0, -1)$ and $(0, 1)$.

SOLUTION The isocline of slope c for the differential equation $\dot{y} = t$ has equation $t = c$, which is the equation of a vertical line. The slope field and the required solution curves are shown below.

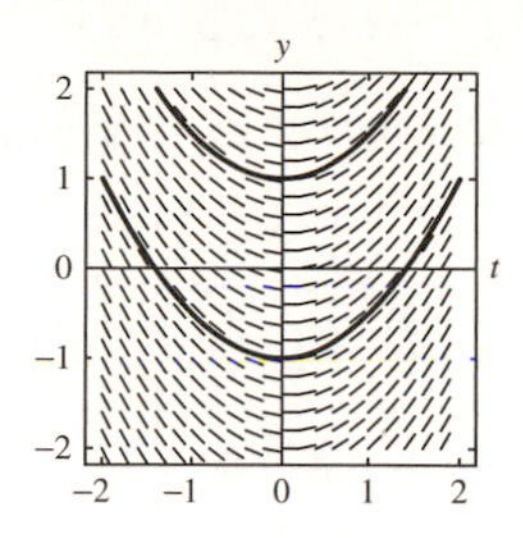

9. Match each differential equation with its slope field in Figures 12(A)–(F).

(i) $\dot{y} = -1$

(ii) $\dot{y} = \dfrac{y}{t}$

(iii) $\dot{y} = t^2 y$

(iv) $\dot{y} = ty^2$

(v) $\dot{y} = t^2 + y^2$

(vi) $\dot{y} = t$

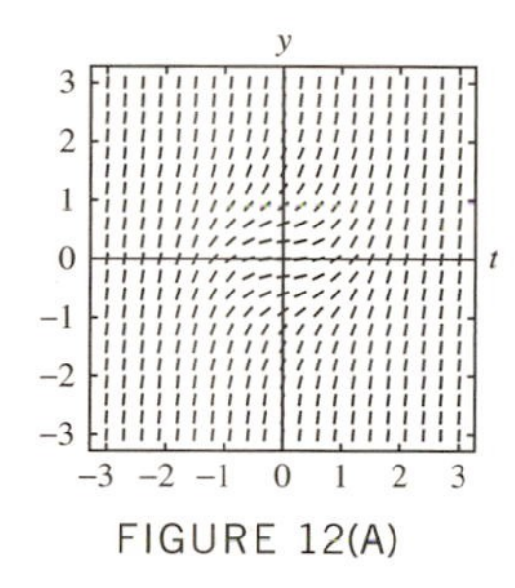
FIGURE 12(A)

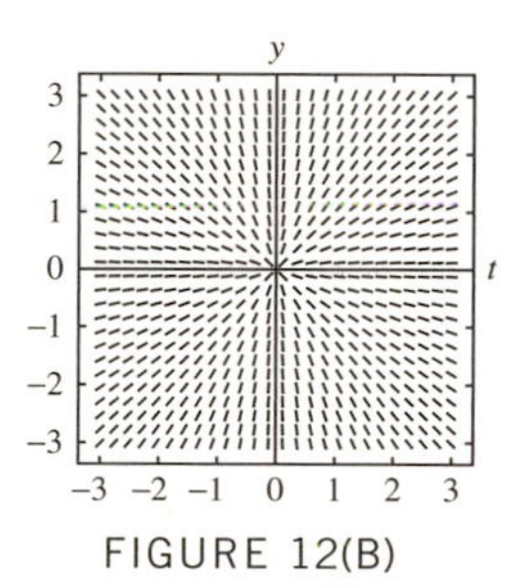
FIGURE 12(B)

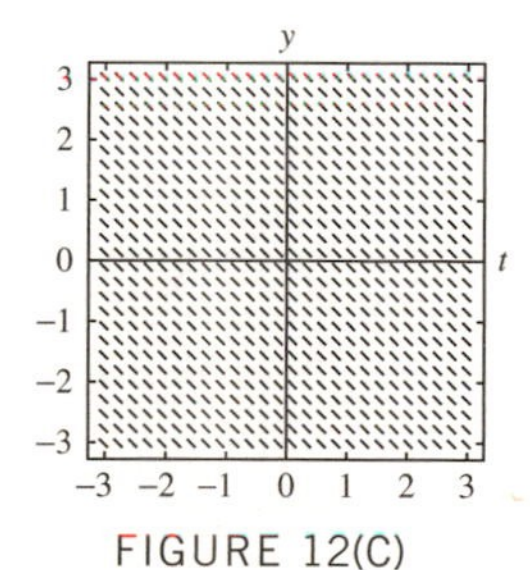
FIGURE 12(C)

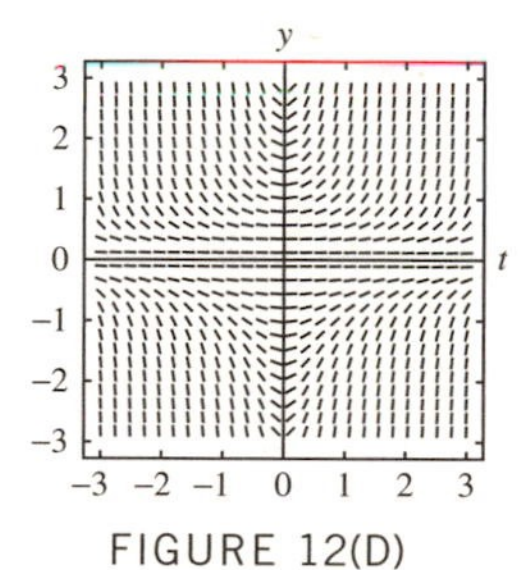
FIGURE 12(D)

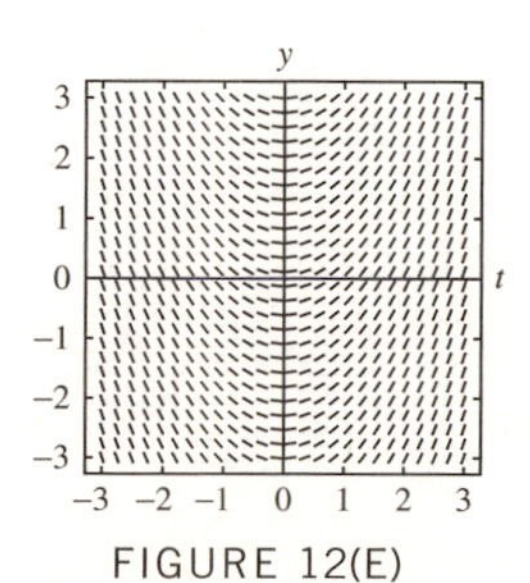
FIGURE 12(E)

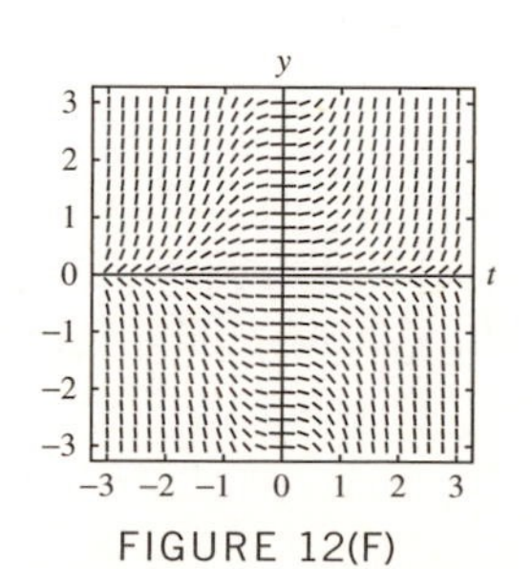
FIGURE 12(F)

SOLUTION

(i) Every segment in the slope field for $\dot{y} = -1$ will have slope -1; this matches Figure 12(C).

(ii) The segments in the slope field for $\dot{y} = \dfrac{y}{t}$ will have positive slope in the first and third quadrants and negative slopes in the second and fourth quadrant; this matches Figure 12(B).

(iii) The segments in the slope field for $\dot{y} = t^2 y$ will have positive slope in the upper half of the plane and negative slopes in the lower half of the plane; this matches Figure 12(F).

(iv) The segments in the slope field for $\dot{y} = ty^2$ will have positive slope on the right side of the plane and negative slopes on the left side of the plane; this matches Figure 12(D).

(v) Every segment in the slope field for $\dot{y} = t^2 + y^2$, except at the origin, will have positive slope; this matches Figure 12(A).

(vi) The isoclines for $\dot{y} = t$ are vertical lines; this matches Figure 12(E).

11. **(a)** Sketch the slope field of $\dot{y} = t/y$ in the region $-2 \le t \le 2$, $-2 \le y \le 2$.

(b) Check that $y = \pm\sqrt{t^2 + C}$ is the general solution.

(c) Sketch the solutions on the slope field with initial conditions $y(0) = 1$ and $y(0) = -1$.

SOLUTION

(a) The slope field is shown below:

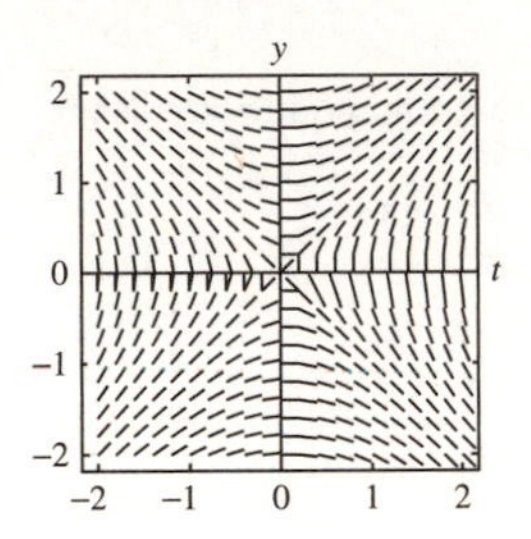

(b) Rewrite

$$\frac{dy}{dt} = \frac{t}{y} \quad \text{as} \quad y\,dy = t\,dt,$$

and then integrate both sides to obtain

$$\frac{1}{2}y^2 = \frac{1}{2}t^2 + C.$$

Solving for y, we find that the general solution is

$$y = \pm\sqrt{t^2 + C}.$$

(c) The sketches of the two solutions are shown below:

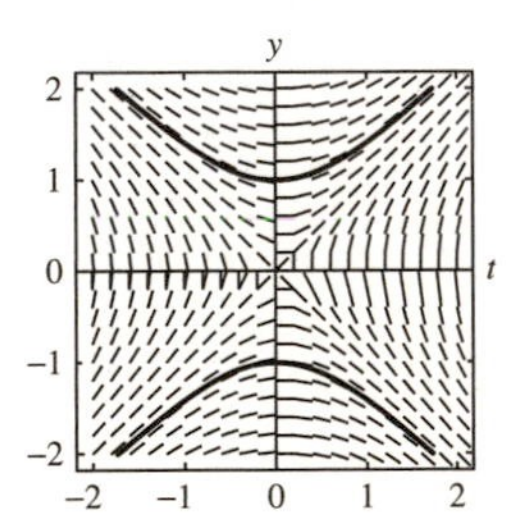

13. Let $F(t, y) = t^2 - y$ and let $y(t)$ be the solution of $\dot{y} = F(t, y)$ satisfying $y(2) = 3$. Let $h = 0.1$ be the time step in Euler's Method, and set $y_0 = y(2) = 3$.

(a) Calculate $y_1 = y_0 + hF(2, 3)$.

(b) Calculate $y_2 = y_1 + hF(2.1, y_1)$.

(c) Calculate $y_3 = y_2 + hF(2.2, y_2)$ and continue computing y_4, y_5, and y_6.

(d) Find approximations to $y(2.2)$ and $y(2.5)$.

SOLUTION

(a) With $y_0 = 3$, $t_0 = 2$, $h = 0.1$, and $F(t, y) = t^2 - y$, we find

$$y_1 = y_0 + hF(t_0, y_0) = 3 + 0.1(1) = 3.1.$$

(b) With $y_1 = 3.1$, $t_1 = 2.1$, $h = 0.1$, and $F(t, y) = t^2 - y$, we find

$$y_2 = y_1 + hF(t_1, y_1) = 3.1 + 0.1(4.41 - 3.1) = 3.231.$$

(c) Continuing as in the previous two parts, we find

$$y_3 = y_2 + hF(t_2, y_2) = 3.3919;$$
$$y_4 = y_3 + hF(t_3, y_3) = 3.58171;$$
$$y_5 = y_4 + hF(t_4, y_4) = 3.799539;$$
$$y_6 = y_5 + hF(t_5, y_5) = 4.0445851.$$

(d) $y(2.2) \approx y_2 = 3.231$, and $y(2.5) \approx y_5 = 3.799539$.

In Exercises 15–20, use Euler's Method to approximate the given value of $y(t)$ with the time step h indicated.

15. $y(0.5)$; $\dot{y} = y + t$, $y(0) = 1$, $h = 0.1$

SOLUTION With $y_0 = 1$, $t_0 = 0$, $h = 0.1$, and $F(t, y) = y + t$, we compute

n	t_n	y_n
0	0	1
1	0.1	$y_0 + hF(t_0, y_0) = 1.1$
2	0.2	$y_1 + hF(t_1, y_1) = 1.22$
3	0.3	$y_2 + hF(t_2, y_2) = 1.362$
4	0.4	$y_3 + hF(t_3, y_3) = 1.5282$
5	0.5	$y_4 + hF(t_4, y_4) = 1.72102$

17. $y(3.3)$; $\dot{y} = t^2 - y$, $y(3) = 1$, $h = 0.05$

SOLUTION With $y_0 = 1$, $t_0 = 3$, $h = 0.05$, and $F(t, y) = t^2 - y$, we compute

n	t_n	y_n
0	3	1
1	3.05	$y_0 + hF(t_0, y_0) = 1.4$
2	3.1	$y_1 + hF(t_1, y_1) = 1.795125$
3	3.15	$y_2 + hF(t_2, y_2) = 2.185869$
4	3.2	$y_3 + hF(t_3, y_3) = 2.572700$
5	3.25	$y_4 + hF(t_4, y_4) = 2.956065$
6	3.3	$y_5 + hF(t_5, y_5) = 3.336387$

19. $y(2)$; $\dot{y} = t \sin y$, $y(1) = 2$, $h = 0.2$

SOLUTION Let $F(t, y) = t \sin y$. With $t_0 = 1$, $y_0 = 2$ and $h = 0.2$, we compute

n	t_n	y_n
0	1	2
1	1.2	$y_0 + hF(t_0, y_0) = 2.181859$
2	1.4	$y_1 + hF(t_1, y_1) = 2.378429$
3	1.6	$y_2 + hF(t_2, y_2) = 2.571968$
4	1.8	$y_3 + hF(t_3, y_3) = 2.744549$
5	2.0	$y_4 + hF(t_4, y_4) = 2.883759$

Further Insights and Challenges

21. If $f(t)$ is continuous on $[a, b]$, then the solution to $\dot{y} = f(t)$ with initial condition $y(a) = 0$ is $y(t) = \int_a^t f(u)\,du$. Show that Euler's Method with time step $h = (b - a)/N$ for N steps yields the Nth left-endpoint approximation to $y(b) = \int_a^b f(u)\,du$.

SOLUTION For a differential equation of the form $\dot{y} = f(t)$, the equation for Euler's method reduces to

$$y_k = y_{k-1} + hf(t_{k-1}).$$

With a step size of $h = (b - a)/N$, $y(b) =\approx y_N$. Starting from $y_0 = 0$, we compute

$$y_1 = y_0 + hf(t_0) = hf(t_0)$$
$$y_2 = y_1 + hf(t_1) = h[f(t_0) + f(t_1)]$$
$$y_3 = y_2 + hf(t_2) = h[f(t_0) + f(t_1) + f(t_2)]$$
$$\vdots$$
$$y_N = y_{N_1} + hf(t_{N-1}) = h\left[f(t_0) + f(t_1) + f(t_2) + \ldots + f(t_{N-1})\right] = h\sum_{k=0}^{N-1} f(t_k)$$

Observe this last expression is exactly the Nth left-endpoint approximation to $y(b) = \int_a^b f(u)\,du$.

Exercises 22–27: ***Euler's Midpoint Method*** *is a variation on Euler's Method that is significantly more accurate in general. For time step h and initial value $y_0 = y(t_0)$, the values y_k are defined successively by*

$$y_k = y_{k-1} + hm_{k-1}$$

where $m_{k-1} = F\left(t_{k-1} + \frac{h}{2}, y_{k-1} + \frac{h}{2}F(t_{k-1}, y_{k-1})\right)$.

In Exercises 23–26, use Euler's Midpoint Method with the time step indicated to approximate the given value of $y(t)$.

23. $y(0.5)$; $\dot{y} = y + t$, $y(0) = 1$, $h = 0.1$

SOLUTION With $t_0 = 0$, $y_0 = 1$, $F(t, y) = y + t$, and $h = 0.1$ we compute

n	t_n	y_n
0	0	1
1	0.1	$y_0 + hF(t_0 + h/2, y_0 + (h/2)F(t_0, y_0)) = 1.11$
2	0.2	$y_1 + hF(t_1 + h/2, y_1 + (h/2)F(t_1, y_1)) = 1.242050$
3	0.3	$y_2 + hF(t_2 + h/2, y_2 + (h/2)F(t_2, y_2)) = 1.398465$
4	0.4	$y_3 + hF(t_3 + h/2, y_3 + (h/2)F(t_3, y_3)) = 1.581804$
5	0.5	$y_4 + hF(t_4 + h/2, y_4 + (h/2)F(t_4, y_4)) = 1.794894$

25. $y(0.25)$; $\dot{y} = \cos(y + t)$, $y(0) = 1$, $h = 0.05$

SOLUTION With $t_0 = 0$, $y_0 = 1$, $F(t, y) = \cos(y + t)$, and $h = 0.05$ we compute

n	t_n	y_n
0	0	1
1	0.05	$y_0 + hF(t_0 + h/2, y_0 + (h/2)F(t_0, y_0)) = 1.025375$
2	0.10	$y_1 + hF(t_1 + h/2, y_1 + (h/2)F(t_1, y_1)) = 1.047507$
3	0.15	$y_2 + hF(t_2 + h/2, y_2 + (h/2)F(t_2, y_2)) = 1.066425$
4	0.20	$y_3 + hF(t_3 + h/2, y_3 + (h/2)F(t_3, y_3)) = 1.082186$
5	0.25	$y_4 + hF(t_4 + h/2, y_4 + (h/2)F(t_4, y_4)) = 1.094871$

27. Assume that $f(t)$ is continuous on $[a, b]$. Show that Euler's Midpoint Method applied to $\dot{y} = f(t)$ with initial condition $y(a) = 0$ and time step $h = (b - a)/N$ for N steps yields the Nth midpoint approximation to

$$y(b) = \int_a^b f(u)\,du$$

SOLUTION For a differential equation of the form $\dot{y} = f(t)$, the equations for Euler's midpoint method reduce to

$$m_{k-1} = f\left(t_{k-1} + \frac{h}{2}\right) \quad \text{and} \quad y_k = y_{k-1} + hf\left(t_{k-1} + \frac{h}{2}\right).$$

With a step size of $h = (b-a)/N$, $y(b) =\approx y_N$. Starting from $y_0 = 0$, we compute

$$y_1 = y_0 + hf\left(t_0 + \frac{h}{2}\right) = hf\left(t_0 + \frac{h}{2}\right)$$
$$y_2 = y_1 + hf\left(t_1 + \frac{h}{2}\right) = h\left[f\left(t_0 + \frac{h}{2}\right) + f\left(t_1 + \frac{h}{2}\right)\right]$$
$$y_3 = y_2 + hf\left(t_2 + \frac{h}{2}\right) = h\left[f\left(t_0 + \frac{h}{2}\right) + f\left(t_1 + \frac{h}{2}\right) + f\left(t_2 + \frac{h}{2}\right)\right]$$
$$\vdots$$
$$y_N = y_{N_1} + hf\left(t_{N-1} + \frac{h}{2}\right) = h\left[f\left(t_0 + \frac{h}{2}\right) + f\left(t_1 + \frac{h}{2}\right) + f\left(t_2 + \frac{h}{2}\right) + \ldots + f\left(t_{N-1} + \frac{h}{2}\right)\right]$$
$$= h\sum_{k=0}^{N-1} f\left(t_k + \frac{h}{2}\right)$$

Observe this last expression is exactly the Nth midpoint approximation to $y(b) = \int_a^b f(u)\,du$.

9.4 The Logistic Equation

Preliminary Questions

1. Which of the following differential equations is a logistic differential equation?

(a) $\dot{y} = 2y(1 - y^2)$

(b) $\dot{y} = 2y\left(1 - \frac{y}{3}\right)$

(c) $\dot{y} = 2y\left(1 - \frac{t}{4}\right)$

(d) $\dot{y} = 2y(1 - 3y)$

SOLUTION The differential equations in **(b)** and **(d)** are logistic equations. The equation in **(a)** is not a logistic equation because of the y^2 term inside the parentheses on the right-hand side; the equation in **(c)** is not a logistic equation because of the presence of the independent variable on the right-hand side.

2. Is the logistic equation a linear differential equation?

SOLUTION No, the logistic equation is not linear.

$$\dot{y} = ky\left(1 - \frac{y}{A}\right) \quad \text{can be rewritten} \quad \dot{y} = ky - \frac{k}{A}y^2$$

and we see that a term involving y^2 occurs.

3. Is the logistic equation separable?

SOLUTION Yes, the logistic equation is a separable differential equation.

Exercises

1. Find the general solution of the logistic equation

$$\dot{y} = 3y\left(1 - \frac{y}{5}\right)$$

Then find the particular solution satisfying $y(0) = 2$.

SOLUTION $\dot{y} = 3y(1 - y/5)$ is a logistic equation with $k = 3$ and $A = 5$; therefore, the general solution is

$$y = \frac{5}{1 - e^{-3t}/C}.$$

The initial condition $y(0) = 2$ allows us to determine the value of C:

$$2 = \frac{5}{1 - 1/C}; \quad 1 - \frac{1}{C} = \frac{5}{2}; \quad \text{so} \quad C = -\frac{2}{3}.$$

The particular solution is then

$$y = \frac{5}{1 + \frac{3}{2}e^{-3t}} = \frac{10}{2 + 3e^{-3t}}.$$

3. Let $y(t)$ be a solution of $\dot{y} = 0.5y(1 - 0.5y)$ such that $y(0) = 4$. Determine $\lim_{t\to\infty} y(t)$ without finding $y(t)$ explicitly.

SOLUTION This is a logistic equation with $k = \frac{1}{2}$ and $A = 2$, so the carrying capacity is 2. Thus the required limit is 2.

5. A population of squirrels lives in a forest with a carrying capacity of 2000. Assume logistic growth with growth constant $k = 0.6\ \text{yr}^{-1}$.

(a) Find a formula for the squirrel population $P(t)$, assuming an initial population of 500 squirrels.

(b) How long will it take for the squirrel population to double?

SOLUTION

(a) Since $k = 0.6$ and the carrying capacity is $A = 2000$, the population $P(t)$ of the squirrels satisfies the differential equation

$$P'(t) = 0.6P(t)(1 - P(t)/2000),$$

with general solution

$$P(t) = \frac{2000}{1 - e^{-0.6t}/C}.$$

The initial condition $P(0) = 500$ allows us to determine the value of C:

$$500 = \frac{2000}{1 - 1/C}; \quad 1 - \frac{1}{C} = 4; \quad \text{so} \quad C = -\frac{1}{3}.$$

The formula for the population is then

$$P(t) = \frac{2000}{1 + 3e^{-0.6t}}.$$

(b) The squirrel population will have doubled at the time t where $P(t) = 1000$. This gives

$$1000 = \frac{2000}{1 + 3e^{-0.6t}}; \quad 1 + 3e^{-0.6t} = 2; \quad \text{so} \quad t = \frac{5}{3}\ln 3 \approx 1.83.$$

It therefore takes approximately 1.83 years for the squirrel population to double.

7. Sunset Lake is stocked with 2000 rainbow trout, and after 1 year the population has grown to 4500. Assuming logistic growth with a carrying capacity of 20,000, find the growth constant k (specify the units) and determine when the population will increase to 10,000.

SOLUTION Since $A = 20{,}000$, the trout population $P(t)$ satisfies the logistic equation

$$P'(t) = kP(t)(1 - P(t)/20{,}000),$$

with general solution

$$P(t) = \frac{20{,}000}{1 - e^{-kt}/C}.$$

The initial condition $P(0) = 2000$ allows us to determine the value of C:

$$2000 = \frac{20{,}000}{1 - 1/C}; \quad 1 - \frac{1}{C} = 10; \quad \text{so} \quad C = -\frac{1}{9}.$$

After one year, we know the population has grown to 4500. Let's measure time in years. Then

$$\begin{aligned} 4500 &= \frac{20{,}000}{1 + 9e^{-k}} \\ 1 + 9e^{-k} &= \frac{40}{9} \\ e^{-k} &= \frac{31}{81} \\ k &= \ln\frac{81}{31} \approx 0.9605\ \text{years}^{-1}. \end{aligned}$$

The population will increase to 10,000 at time t where $P(t) = 10{,}000$. This gives

$$\begin{aligned} 10{,}000 &= \frac{20{,}000}{1 + 9e^{-0.9605t}} \\ 1 + 9e^{-0.9605t} &= 2 \\ e^{-0.9605t} &= \frac{1}{9} \\ t &= \frac{1}{0.9605}\ln 9 \approx 2.29 \text{ years.} \end{aligned}$$

9. A rumor spreads through a school with 1000 students. At 8 AM, 80 students have heard the rumor, and by noon, half the school has heard it. Using the logistic model of Exercise 8, determine when 90% of the students will have heard the rumor.

SOLUTION Let $y(t)$ be the proportion of students that have heard the rumor at a time t hours after 8 AM. In the logistic model of Exercise 8, we have a capacity of $A = 1$ (100% of students) and an unknown growth factor of k. Hence,

$$y(t) = \frac{1}{1 - e^{-kt}/C}.$$

The initial condition $y(0) = 0.08$ allows us to determine the value of C:

$$\frac{2}{25} = \frac{1}{1 - 1/C}; \quad 1 - \frac{1}{C} = \frac{25}{2}; \quad \text{so} \quad C = -\frac{2}{23}.$$

so that

$$y(t) = \frac{2}{2 + 23e^{-kt}}.$$

The condition $y(4) = 0.5$ now allows us to determine the value of k:

$$\frac{1}{2} = \frac{2}{2 + 23e^{-4k}}; \quad 2 + 23e^{-4k} = 4; \quad \text{so} \quad k = \frac{1}{4}\ln\frac{23}{2} \approx 0.6106 \text{ hours}^{-1}.$$

90% of the students have heard the rumor when $y(t) = 0.9$. Thus

$$\begin{aligned} \frac{9}{10} &= \frac{2}{2 + 23e^{-0.6106t}} \\ 2 + 23e^{-0.6106t} &= \frac{20}{9} \\ t &= \frac{1}{0.6106}\ln\frac{207}{2} \approx 7.6 \text{ hours.} \end{aligned}$$

Thus, 90% of the students have heard the rumor after 7.6 hours, or at 3:36 PM.

11. Let $k = 1$ and $A = 1$ in the logistic equation.

(a) Find the solutions satisfying $y_1(0) = 10$ and $y_2(0) = -1$.

(b) Find the time t when $y_1(t) = 5$.

(c) When does $y_2(t)$ become infinite?

SOLUTION The general solution of the logistic equation with $k = 1$ and $A = 1$ is

$$y(t) = \frac{1}{1 - e^{-t}/C}.$$

(a) Given $y_1(0) = 10$, we find $C = \frac{10}{9}$, and

$$y_1(t) = \frac{1}{1 - \frac{10}{9}e^{-t}} = \frac{10}{10 - 9e^{-t}}.$$

On the other hand, given $y_2(0) = -1$, we find $C = \frac{1}{2}$, and

$$y_2(t) = \frac{1}{1 - 2e^{-t}}.$$

(b) From part (a), we have

$$y_1(t) = \frac{10}{10 - 9e^{-t}}.$$

Thus, $y_1(t) = 5$ when

$$5 = \frac{10}{10 - 9e^{-t}}; \quad 10 - 9e^{-t} = 2; \quad \text{so} \quad t = \ln \frac{9}{8}.$$

(c) From part (a), we have

$$y_2(t) = \frac{1}{1 - 2e^{-t}}.$$

Thus, $y_2(t)$ becomes infinite when

$$1 - 2e^{-t} = 0 \quad \text{or} \quad t = \ln 2.$$

13. GU In the model of Exercise 12, let $A(t)$ be the area at time t (hours) of a growing tissue culture with initial size $A(0) = 1 \text{ cm}^2$, assuming that the maximum area is $M = 16 \text{ cm}^2$ and the growth constant is $k = 0.1$.
(a) Find a formula for $A(t)$. *Note:* The initial condition is satisfied for two values of the constant C. Choose the value of C for which $A(t)$ is increasing.
(b) Determine the area of the culture at $t = 10$ hours.
(c) GU Graph the solution using a graphing utility.

SOLUTION

(a) From the values for M and k we have

$$A(t) = 16\left(\frac{Ce^{t/40} - 1}{Ce^{t/40} + 1}\right)^2$$

and the initial condition then gives us

$$A(0) = 1 = 16\left(\frac{Ce^{0/40} - 1}{Ce^{0/40} + 1}\right)^2$$

so, simplifying,

$$1 = 16\left(\frac{C - 1}{C + 1}\right)^2 \quad \Rightarrow \quad C^2 + 2C + 1 = 16C^2 - 32C + 16 \quad \Rightarrow \quad 15C^2 - 34C + 15 = 0$$

and thus $C = \frac{5}{3}$ or $C = \frac{3}{5}$. The derivative of $A(t)$ is

$$A'(t) = \frac{16Ce^{t/40}}{(Ce^{t/40} + 1)^3} \cdot (Ce^{t/40} - 1)$$

For $C = 3/5$, $A'(t)$ can be negative, while for $C = 5/3$, it is always positive. So let $C = 5/3$.
(b) From part (a), we have

$$A(t) = 16\left(\frac{\frac{5}{3}e^{t/40} - 1}{\frac{5}{3}e^{t/40} + 1}\right)^2$$

and $A(10) \approx 2.11$.
(c)

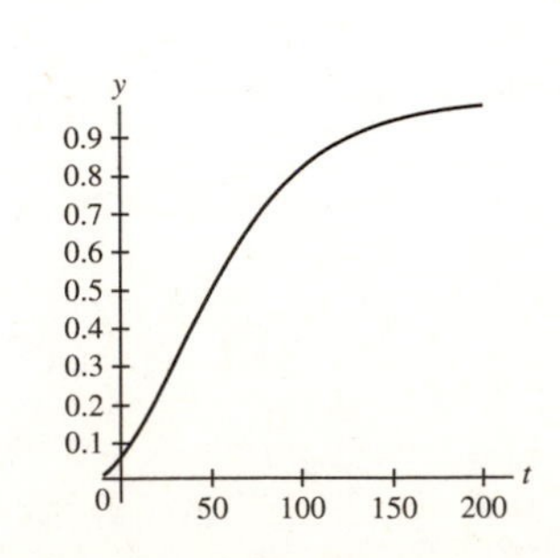

15. In 1751, Benjamin Franklin predicted that the U.S. population $P(t)$ would increase with growth constant $k = 0.028$ year^{-1}. According to the census, the U.S. population was 5 million in 1800 and 76 million in 1900. Assuming logistic growth with $k = 0.028$, find the predicted carrying capacity for the U.S. population. *Hint:* Use Eqs. (3) and (4) to show that

$$\frac{P(t)}{P(t) - A} = \frac{P_0}{P_0 - A} e^{kt}$$

SOLUTION Assuming the population grows according to the logistic equation,

$$\frac{P(t)}{P(t) - A} = Ce^{kt}.$$

But

$$C = \frac{P_0}{P_0 - A},$$

so

$$\frac{P(t)}{P(t) - A} = \frac{P_0}{P_0 - A} e^{kt}.$$

Now, let $t = 0$ correspond to the year 1800. Then the year 1900 corresponds to $t = 100$, and with $k = 0.028$, we have

$$\frac{76}{76 - A} = \frac{5}{5 - A} e^{(0.028)(100)}.$$

Solving for A, we find

$$A = \frac{5(e^{2.8} - 1)}{\frac{5}{76}e^{2.8} - 1} \approx 943.07.$$

Thus, the predicted carrying capacity for the U.S. population is approximately 943 million.

Further Insights and Challenges

In Exercises 17 and 18, let $y(t)$ be a solution of the logistic equation

$$\frac{dy}{dt} = ky\left(1 - \frac{y}{A}\right) \qquad \boxed{9}$$

where $A > 0$ and $k > 0$.

17. (a) Differentiate Eq. (9) with respect to t and use the Chain Rule to show that

$$\frac{d^2y}{dt^2} = k^2y\left(1 - \frac{y}{A}\right)\left(1 - \frac{2y}{A}\right)$$

(b) Show that $y(t)$ is concave up if $0 < y < A/2$ and concave down if $A/2 < y < A$.

(c) Show that if $0 < y(0) < A/2$, then $y(t)$ has a point of inflection at $y = A/2$ (Figure 6).

(d) Assume that $0 < y(0) < A/2$. Find the time t when $y(t)$ reaches the inflection point.

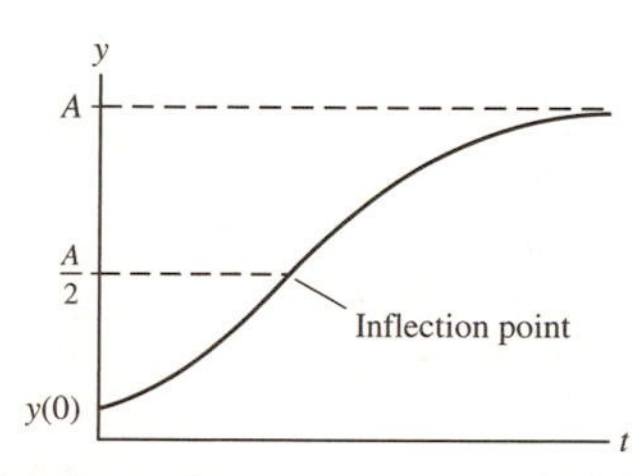

FIGURE 6 An inflection point occurs at $y = A/2$ in the logistic curve.

SOLUTION

(a) The derivative of Eq. (9) with respect to t is

$$y'' = ky' - \frac{2kyy'}{A} = ky'\left(1 - \frac{2y}{A}\right) = k\left(1 - \frac{y}{A}\right)ky\left(1 - \frac{2y}{A}\right) = k^2y\left(1 - \frac{y}{A}\right)\left(1 - \frac{2y}{A}\right).$$

(b) If $0 < y < A/2$, $1 - \frac{y}{A}$ and $1 - \frac{2y}{A}$ are both positive, so $y'' > 0$. Therefore, y is concave up. If $A/2 < y < A$, $1 - \frac{y}{A} > 0$, but $1 - \frac{2y}{A} < 0$, so $y'' < 0$, so y is concave down.

(c) If $y_0 < A$, y grows and $\lim_{t\to\infty} y(t) = A$. If $0 < y < A/2$, y is concave up at first. Once y passes $A/2$, y becomes concave down, so y has an inflection point at $y = A/2$.

(d) The general solution to Eq. (9) is

$$y = \frac{A}{1 - e^{-kt}/C};$$

thus, $y = A/2$ when

$$\frac{A}{2} = \frac{A}{1 - e^{-kt}/C}$$

$$1 - e^{-kt}/C = 2$$

$$t = -\frac{1}{k}\ln(-C)$$

Now, $C = y_0/(y_0 - A)$, so

$$t = -\frac{1}{k}\ln\frac{y_0}{A - y_0} = \frac{1}{k}\ln\frac{A - y_0}{y_0}.$$

9.5 First-Order Linear Equations

Preliminary Questions

1. Which of the following are first-order linear equations?

(a) $y' + x^2 y = 1$

(b) $y' + xy^2 = 1$

(c) $x^5 y' + y = e^x$

(d) $x^5 y' + y = e^y$

SOLUTION The equations in **(a)** and **(c)** are first-order linear differential equations. The equation in **(b)** is not linear because of the y^2 factor in the second term on the left-hand side of the equation; the equation in **(d)** is not linear because of the e^y term on the right-hand side of the equation.

2. If $\alpha(x)$ is an integrating factor for $y' + A(x)y = B(x)$, then $\alpha'(x)$ is equal to (choose the correct answer):

(a) $B(x)$

(b) $\alpha(x)A(x)$

(c) $\alpha(x)A'(x)$

(d) $\alpha(x)B(x)$

SOLUTION The correct answer is **(b)**: $\alpha(x)A(x)$.

Exercises

1. Consider $y' + x^{-1}y = x^3$.

(a) Verify that $\alpha(x) = x$ is an integrating factor.

(b) Show that when multiplied by $\alpha(x)$, the differential equation can be written $(xy)' = x^4$.

(c) Conclude that xy is an antiderivative of x^4 and use this information to find the general solution.

(d) Find the particular solution satisfying $y(1) = 0$.

SOLUTION

(a) The equation is of the form

$$y' + A(x)y = B(x)$$

for $A(x) = x^{-1}$ and $B(x) = x^3$. By Theorem 1, $\alpha(x)$ is defined by

$$\alpha(x) = e^{\int A(x)\,dx} = e^{\ln x} = x.$$

(b) When multiplied by $\alpha(x)$, the equation becomes:

$$xy' + y = x^4.$$

Now, $xy' + y = xy' + (x)'y = (xy)'$, so

$$(xy)' = x^4.$$

(c) Since $(xy)' = x^4$, $(xy) = \frac{x^5}{5} + C$ and

$$y = \frac{x^4}{5} + \frac{C}{x}$$

(d) If $y(1) = 0$, we find

$$0 = \frac{1}{5} + C \quad \text{so} \quad -\frac{1}{5} = C.$$

The solution, therefore, is

$$y = \frac{x^4}{5} - \frac{1}{5x}.$$

3. Let $\alpha(x) = e^{x^2}$. Verify the identity

$$(\alpha(x)y)' = \alpha(x)(y' + 2xy)$$

and explain how it is used to find the general solution of

$$y' + 2xy = x$$

SOLUTION Let $\alpha(x) = e^{x^2}$. Then

$$(\alpha(x)y)' = (e^{x^2}y)' = 2xe^{x^2}y + e^{x^2}y' = e^{x^2}\left(2xy + y'\right) = \alpha(x)\left(y' + 2xy\right).$$

If we now multiply both sides of the differential equation $y' + 2xy = x$ by $\alpha(x)$, we obtain

$$\alpha(x)(y' + 2xy) = x\alpha(x) = xe^{x^2}.$$

But $\alpha(x)(y' + 2xy) = (\alpha(x)y)'$, so by integration we find

$$\alpha(x)y = \int xe^{x^2}\,dx = \frac{1}{2}e^{x^2} + C.$$

Finally,

$$y(x) = \frac{1}{2} + Ce^{-x^2}.$$

In Exercises 5–18, find the general solution of the first-order linear differential equation.

5. $xy' + y = x$

SOLUTION Rewrite the equation as

$$y' + \frac{1}{x}y = 1,$$

which is in standard linear form with $A(x) = \frac{1}{x}$ and $B(x) = 1$. By Theorem 1, the integrating factor is

$$\alpha(x) = e^{\int A(x)\,dx} = e^{\ln x} = x.$$

When multiplied by the integrating factor, the rewritten differential equation becomes

$$xy' + y = x \qquad \text{or} \qquad (xy)' = x.$$

Integration of both sides now yields

$$xy = \frac{1}{2}x^2 + C.$$

Finally,

$$y(x) = \frac{1}{2}x + \frac{C}{x}.$$

7. $3xy' - y = x^{-1}$

SOLUTION Rewrite the equation as

$$y' - \frac{1}{3x}y = \frac{1}{3x^2},$$

which is in standard form with $A(x) = -\frac{1}{3}x^{-1}$ and $B(x) = \frac{1}{3}x^{-2}$. By Theorem 1, the integrating factor is

$$\alpha(x) = e^{\int A(x)\,dx} = e^{-(1/3)\ln x} = x^{-1/3}.$$

When multiplied by the integrating factor, the rewritten differential equation becomes

$$x^{-1/3}y' - \frac{1}{3}x^{-4/3} = \frac{1}{3}x^{-7/3} \qquad \text{or} \qquad (x^{-1/3}y)' = \frac{1}{3}x^{-7/3}.$$

Integration of both sides now yields

$$x^{-1/3}y = -\frac{1}{4}x^{-4/3} + C.$$

Finally,

$$y(x) = -\frac{1}{4}x^{-1} + Cx^{1/3}.$$

9. $y' + 3x^{-1}y = x + x^{-1}$

SOLUTION This equation is in standard form with $A(x) = 3x^{-1}$ and $B(x) = x + x^{-1}$. By Theorem 1, the integrating factor is

$$\alpha(x) = e^{\int 3x^{-1}} = e^{3\ln x} = x^3.$$

When multiplied by the integrating factor, the original differential equation becomes

$$x^3y' + 3x^2y = x^4 + x^2 \qquad \text{or} \qquad (x^3y)' = x^4 + x^3.$$

Integration of both sides now yields

$$x^3y = \frac{1}{5}x^5 + \frac{1}{3}x^3 + C.$$

Finally,

$$y(x) = \frac{1}{5}x^2 + \frac{1}{3} + Cx^{-3}.$$

11. $xy' = y - x$

SOLUTION Rewrite the equation as

$$y' - \frac{1}{x}y = -1,$$

which is in standard form with $A(x) = -\frac{1}{x}$ and $B(x) = -1$. By Theorem 1, the integrating factor is

$$\alpha(x) = e^{\int -(1/x)\,dx} = e^{-\ln x} = x^{-1}.$$

When multiplied by the integrating factor, the rewritten differential equation becomes

$$\frac{1}{x}y' - \frac{1}{x^2}y = -\frac{1}{x} \qquad \text{or} \qquad \left(\frac{1}{x}y\right)' = -\frac{1}{x}.$$

Integration on both sides now yields

$$\frac{1}{x}y = -\ln x + C.$$

Finally,

$$y(x) = -x\ln x + Cx.$$

13. $y' + y = e^x$

SOLUTION This equation is in standard form with $A(x) = 1$ and $B(x) = e^x$. By Theorem 1, the integrating factor is

$$\alpha(x) = e^{\int 1\,dx} = e^x.$$

When multiplied by the integrating factor, the original differential equation becomes

$$e^x y' + e^x y = e^{2x} \qquad \text{or} \qquad (e^x y)' = e^{2x}.$$

Integration on both sides now yields

$$e^x y = \frac{1}{2}e^{2x} + C.$$

Finally,

$$y(x) = \frac{1}{2}e^x + Ce^{-x}.$$

15. $y' + (\tan x)y = \cos x$

SOLUTION This equation is in standard form with $A(x) = \tan x$ and $B(x) = \cos x$. By Theorem 1, the integrating factor is

$$\alpha(x) = e^{\int \tan x\,dx} = e^{\ln \sec x} = \sec x.$$

When multiplied by the integrating factor, the original differential equation becomes

$$\sec x y' + \sec x \tan x y = 1 \qquad \text{or} \qquad (y \sec x)' = 1.$$

Integration on both sides now yields

$$y \sec x = x + C.$$

Finally,

$$y(x) = x \cos x + C \cos x.$$

17. $y' - (\ln x)y = x^x$

SOLUTION This equation is in standard form with $A(x) = -\ln x$ and $B(x) = x^x$. By Theorem 1, the integrating factor is

$$\alpha(x) = e^{\int -\ln x\,dx} = e^{x - x\ln x} = \frac{e^x}{x^x}.$$

When multiplied by the integrating factor, the original differential equation becomes

$$x^{-x}e^x y' - (\ln x)x^{-x}e^x y = e^x \qquad \text{or} \qquad (x^{-x}e^x y)' = e^x.$$

Integration on both sides now yields

$$x^{-x}e^x y = e^x + C.$$

Finally,

$$y(x) = x^x + Cx^x e^{-x}.$$

In Exercises 19–26, solve the initial value problem.

19. $y' + 3y = e^{2x}, \quad y(0) = -1$

SOLUTION First, we find the general solution of the differential equation. This linear equation is in standard form with $A(x) = 3$ and $B(x) = e^{2x}$. By Theorem 1, the integrating factor is

$$\alpha(x) = e^{3x}.$$

When multiplied by the integrating factor, the original differential equation becomes

$$(e^{3x}y)' = e^{5x}.$$

Integration on both sides now yields

$$(e^{3x}y) = \frac{1}{5}e^{5x} + C;$$

hence,

$$y(x) = \frac{1}{5}e^{2x} + Ce^{-3x}.$$

The initial condition $y(0) = -1$ allows us to determine the value of C:

$$-1 = \frac{1}{5} + C \qquad \text{so} \qquad C = -\frac{6}{5}.$$

The solution to the initial value problem is therefore

$$y(x) = \frac{1}{5}e^{2x} - \frac{6}{5}e^{-3x}.$$

21. $y' + \dfrac{1}{x+1}y = x^{-2}, \quad y(1) = 2$

SOLUTION First, we find the general solution of the differential equation. This linear equation is in standard form with $A(x) = \frac{1}{x+1}$ and $B(x) = x^{-2}$. By Theorem 1, the integrating factor is

$$\alpha(x) = e^{\int 1/(x+1)\,dx} = e^{\ln(x+1)} = x + 1.$$

When multiplied by the integrating factor, the original differential equation becomes

$$((x+1)y)' = x^{-1} + x^{-2}.$$

Integration on both sides now yields

$$(x+1)y = \ln x - x^{-1} + C;$$

hence,

$$y(x) = \frac{1}{x+1}\left(C + \ln x - \frac{1}{x}\right).$$

The initial condition $y(1) = 2$ allows us to determine the value of C:

$$2 = \frac{1}{2}(C - 1) \qquad \text{so} \qquad C = 5.$$

The solution to the initial value problem is therefore

$$y(x) = \frac{1}{x+1}\left(5 + \ln x - \frac{1}{x}\right).$$

23. $(\sin x)y' = (\cos x)y + 1, \quad y\left(\frac{\pi}{4}\right) = 0$

SOLUTION First, we find the general solution of the differential equation. Rewrite the equation as

$$y' - (\cot x)y = \csc x,$$

which is in standard form with $A(x) = -\cot x$ and $B(x) = \csc x$. By Theorem 1, the integrating factor is

$$\alpha(x) = e^{\int -\cot x\,dx} = e^{-\ln \sin x} = \csc x.$$

When multiplied by the integrating factor, the rewritten differential equation becomes

$$(\csc x y)' = \csc^2 x.$$

Integration on both sides now yields

$$(\csc x)y = -\cot x + C;$$

hence,

$$y(x) = -\cos x + C\sin x.$$

The initial condition $y(\pi/4) = 0$ allows us to determine the value of C:

$$0 = -\frac{\sqrt{2}}{2} + C\frac{\sqrt{2}}{2} \qquad \text{so} \qquad C = 1.$$

The solution to the initial value problem is therefore

$$y(x) = -\cos x + \sin x.$$

25. $y' + (\tanh x)y = 1, \quad y(0) = 3$

SOLUTION First, we find the general solution of the differential equation. This equation is in standard form with $A(x) = \tanh x$ and $B(x) = 1$. By Theorem 1, the integrating factor is

$$\alpha(x) = e^{\int \tanh x\,dx} = e^{\ln \cosh x} = \cosh x.$$

When multiplied by the integrating factor, the original differential equation becomes

$$(\cosh x y)' = \cosh x.$$

Integration on both sides now yields

$$(\cosh x y) = \sinh x + C;$$

hence,

$$y(x) = \tanh x + C \operatorname{sech} x.$$

The initial condition $y(0) = 3$ allows us to determine the value of C:

$$3 = C.$$

The solution to the initial value problem is therefore

$$y(x) = \tanh x + 3 \operatorname{sech} x.$$

27. Find the general solution of $y' + ny = e^{mx}$ for all m, n. *Note:* The case $m = -n$ must be treated separately.

SOLUTION For any m, n, Theorem 1 gives us the formula for $\alpha(x)$:

$$\alpha(x) = e^{\int n\,dx} = e^{nx}.$$

When multiplied by the integrating factor, the original differential equation becomes

$$(e^{nx}y)' = e^{(m+n)x}.$$

If $m \neq -n$, integration on both sides yields

$$e^{nx}y = \frac{1}{m+n}e^{(m+n)x} + C,$$

so

$$y(x) = \frac{1}{m+n}e^{mx} + Ce^{-nx}.$$

However, if $m = -n$, then $m + n = 0$ and the equation reduces to

$$(e^{nx}y)' = 1,$$

so integration yields

$$e^{nx}y = x + C \qquad \text{or} \qquad y(x) = (x + C)e^{-nx}.$$

In Exercises 29–32, a 1000 L *tank contains* 500 L *of water with a salt concentration of* 10 g/L. *Water with a salt concentration of* 50 g/L *flows into the tank at a rate of* 80 L/min. *The fluid mixes instantaneously and is pumped out at a specified rate* R_{out}. *Let* $y(t)$ *denote the quantity of salt in the tank at time* t.

29. Assume that $R_{\text{out}} = 40$ L/min.

(a) Set up and solve the differential equation for $y(t)$.

(b) What is the salt concentration when the tank overflows?

SOLUTION Because water flows into the tank at the rate of 80 L/min but flows out at the rate of $R_{\text{out}} = 40$ L/min, there is a net inflow of 40 L/min. Therefore, at any time t, there are $500 + 40t$ liters of water in the tank.

(a) The net flow of salt into the tank at time t is

$$\frac{dy}{dt} = \text{salt rate in} - \text{salt rate out} = \left(80\frac{\text{L}}{\text{min}}\right)\left(50\frac{\text{g}}{\text{L}}\right) - \left(40\frac{\text{L}}{\text{min}}\right)\left(\frac{y\text{ g}}{500 + 40t\text{ L}}\right) = 4000 - 40 \cdot \frac{y}{500 + 40t}$$

Rewriting this linear equation in standard form, we have

$$\frac{dy}{dt} + \frac{4}{50+4t}y = 4000,$$

so $A(t) = \frac{4}{50+4t}$ and $B(t) = 4000$. By Theorem 1, the integrating factor is

$$\alpha(t) = e^{\int 4(50+4t)^{-1}\,dt} = e^{\ln(50+4t)} = 50 + 4t.$$

When multiplied by the integrating factor, the rewritten differential equation becomes

$$((50+4t)y)' = 4000(50+4t).$$

Integration on both sides now yields

$$(50+4t)y = 200{,}000t + 16{,}000t^2 + C;$$

hence,

$$y(t) = \frac{200{,}000t + 8000t^2 + C}{50+4t}.$$

The initial condition $y(0) = 10$ allows us to determine the value of C:

$$10 = \frac{C}{50} \qquad \text{so} \qquad C = 500.$$

The solution to the initial value problem is therefore

$$y(t) = \frac{200{,}000t + 8000t^2 + 500}{50+4t} = \frac{250 + 4000t^2 + 100{,}000t}{25+2t}.$$

(b) The tank overflows when $t = 25/2 = 12.5$. The amount of salt in the tank at that time is

$$y(12.5) = 37{,}505 \text{ g},$$

so the concentration of salt is

$$\frac{37{,}505 \text{ g}}{1000 \text{ L}} = 37.505 \text{ g/L}.$$

31. Find the limiting salt concentration as $t \to \infty$ assuming that $R_{\text{out}} = 80$ L/min.

SOLUTION The total volume of water is now constant at 500 liters, so the net flow of salt at time t is

$$\frac{dy}{dt} = \text{salt rate in} - \text{salt rate out} = \left(80\frac{\text{L}}{\text{min}}\right)\left(50\frac{\text{g}}{\text{L}}\right) - \left(80\frac{\text{L}}{\text{min}}\right)\left(\frac{y\text{ g}}{500\text{ L}}\right) = 4000 - \frac{8}{50}y$$

Rewriting this equation in standard form gives

$$\frac{dy}{dt} + \frac{8}{50}y = 4000$$

so that the integrating factor is

$$e^{\int (8/50)\,dt} = e^{0.16t}$$

Multiplying both sides by the integrating factor gives

$$(e^{0.16t}y)' = 4000e^{0.16t}$$

Integrate both sides to get

$$e^{0.16t}y = 25{,}000e^{0.16t} + C \quad \text{so that} \quad y = 25{,}000 + Ce^{-0.16t}$$

As $t \to \infty$, the exponential term tends to zero, so that the amount of salt tends to 25,000g, or 50 g/L. (Note that this is precisely what would be expected naïvely, since the salt concentration flowing in is also 50 g/L).

33. Water flows into a tank at the variable rate of $R_{\text{in}} = 20/(1+t)$ gal/min and out at the constant rate $R_{\text{out}} = 5$ gal/min. Let $V(t)$ be the volume of water in the tank at time t.

(a) Set up a differential equation for $V(t)$ and solve it with the initial condition $V(0) = 100$.

(b) Find the maximum value of V.

(c) CAS Plot $V(t)$ and estimate the time t when the tank is empty.

SOLUTION

(a) The rate of change of the volume of water in the tank is given by

$$\frac{dV}{dt} = R_{\text{in}} - R_{\text{out}} = \frac{20}{1+t} - 5.$$

Because the right-hand side depends only on the independent variable t, we integrate to obtain

$$V(t) = 20\ln(1+t) - 5t + C.$$

The initial condition $V(0) = 100$ allows us to determine the value of C:

$$100 = 20\ln 1 - 0 + C \qquad \text{so} \qquad C = 100.$$

Therefore

$$V(t) = 20\ln(1+t) - 5t + 100.$$

(b) Using the result from part (a),

$$\frac{dV}{dt} = \frac{20}{1+t} - 5 = 0$$

when $t = 3$. Because $\frac{dV}{dt} > 0$ for $t < 3$ and $\frac{dV}{dt} < 0$ for $t > 3$, it follows that

$$V(3) = 20\ln 4 - 15 + 100 \approx 112.726 \text{ gal}$$

is the maximum volume.

(c) $V(t)$ is plotted in the figure below at the left. On the right, we zoom in near the location where the curve crosses the t-axis. From this graph, we estimate that the tank is empty after roughly 34.25 minutes.

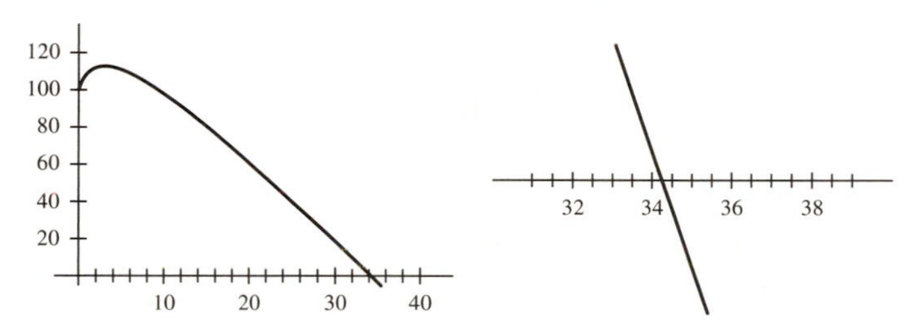

In Exercises 35–38, consider a series circuit (Figure 4) consisting of a resistor of R ohms, an inductor of L henries, and a variable voltage source of V(t) volts (time t in seconds). The current through the circuit I(t) (in amperes) satisfies the differential equation

$$\frac{dI}{dt} + \frac{R}{L}I = \frac{1}{L}V(t) \qquad \boxed{10}$$

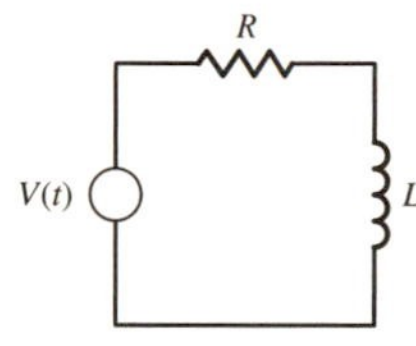

FIGURE 4 RL circuit.

35. Solve Eq. (10) with initial condition $I(0) = 0$, assuming that $R = 100\ \Omega$, $L = 5$ H, and $V(t)$ is constant with $V(t) = 10$ volts.

SOLUTION If $R = 100$, $V(t) = 10$, and $L = 5$, the differential equation becomes

$$\frac{dI}{dt} + 20I = 2,$$

which is a linear equation in standard form with $A(t) = 20$ and $B(t) = 2$. The integrating factor is $\alpha(t) = e^{20t}$, and when multiplied by the integrating factor, the differential equation becomes

$$(e^{20t}I)' = 2e^{20t}.$$

Integration of both sides now yields

$$e^{20t}I = \frac{1}{10}e^{20t} + C;$$

hence,

$$I(t) = \frac{1}{10} + Ce^{-20t}.$$

The initial condition $I(0) = 0$ allows us to determine the value of C:

$$0 = \frac{1}{10} + C \qquad \text{so} \qquad C = -\frac{1}{10}.$$

Finally,

$$I(t) = \frac{1}{10}\left(1 - e^{-20t}\right).$$

37. Assume that $V(t) = V$ is constant and $I(0) = 0$.

(a) Solve for $I(t)$.

(b) Show that $\lim_{t\to\infty} I(t) = V/R$ and that $I(t)$ reaches approximately 63% of its limiting value after L/R seconds.

(c) How long does it take for $I(t)$ to reach 90% of its limiting value if $R = 500\ \Omega$, $L = 4$ H, and $V = 20$ volts?

SOLUTION

(a) The equation

$$\frac{dI}{dt} + \frac{R}{L}I = \frac{1}{L}V$$

is a linear equation in standard form with $A(t) = \frac{R}{L}$ and $B(t) = \frac{1}{L}V(t)$. By Theorem 1, the integrating factor is

$$\alpha(t) = e^{\int (R/L)\,dt} = e^{(R/L)\,t}.$$

When multiplied by the integrating factor, the original differential equation becomes

$$(e^{(R/L)\,t}I)' = e^{(R/L)\,t}\frac{V}{L}.$$

Integration on both sides now yields

$$(e^{(R/L)\,t}I) = \frac{V}{R}e^{(R/L)\,t} + C;$$

hence,

$$I(t) = \frac{V}{R} + Ce^{-(R/L)\,t}.$$

The initial condition $I(0) = 0$ allows us to determine the value of C:

$$0 = \frac{V}{R} + C \qquad \text{so} \qquad C = -\frac{V}{R}.$$

Therefore the current is given by

$$I(t) = \frac{V}{R}\left(1 - e^{-(R/L)\,t}\right).$$

(b) As $t \to \infty$, $e^{-(R/L)\,t} \to 0$, so $I(t) \to \frac{V}{R}$. Moreover, when $t = (L/R)$ seconds, we have

$$I\left(\frac{L}{R}\right) = \frac{V}{R}\left(1 - e^{-(R/L)\,(L/R)}\right) = \frac{V}{R}\left(1 - e^{-1}\right) \approx 0.632\frac{V}{R}.$$

(c) Using the results from part (a) and part (b), $I(t)$ reaches 90% of its limiting value when

$$\frac{9}{10} = 1 - e^{-(R/L)\,t},$$

or when

$$t = \frac{L}{R}\ln 10.$$

With $L = 4$ and $R = 500$, this takes approximately 0.0184 seconds.

39. Tank 1 in Figure 5 is filled with V_1 liters of water containing blue dye at an initial concentration of c_0 g/L. Water flows into the tank at a rate of R L/min, is mixed instantaneously with the dye solution, and flows out through the bottom at the same rate R. Let $c_1(t)$ be the dye concentration in the tank at time t.

(a) Explain why c_1 satisfies the differential equation $\dfrac{dc_1}{dt} = -\dfrac{R}{V_1}c_1$.

(b) Solve for $c_1(t)$ with $V_1 = 300$ L, $R = 50$, and $c_0 = 10$ g/L.

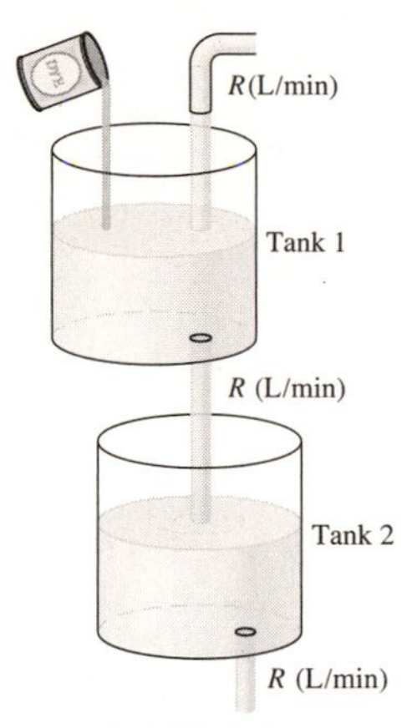

FIGURE 5

SOLUTION

(a) Let $g_1(t)$ be the number of grams of dye in the tank at time t. Then $g_1(t) = V_1c_1(t)$ and $g_1'(t) = V_1c_1'(t)$. Now,

$$g_1'(t) = \text{grams of dye in} - \text{grams of dye out} = 0 - \frac{g(t)}{V_1}\text{ g/L} \cdot R\text{ L/min} = -\frac{R}{V_1}g(t)$$

Substituting gives

$$V_1c_1'(t) = -\frac{R}{V_1}c_1(t)V_1 \quad \text{and simplifying yields} \quad c_1'(t) = -\frac{R}{V_1}c_1(t)$$

(b) In standard form, the equation is

$$c_1'(t) + \frac{R}{V_1}c_1(t) = 0$$

so that $A(t) = \dfrac{R}{V_1}$ and $B(t) = 0$. The integrating factor is $e^{(R/V_1)t}$; multiplying through gives

$$(e^{(R/V_1)t}c_1(t))' = 0 \quad \text{so, integrating,} \quad e^{(R/V_1)t}c_1(t) = C$$

and thus $c_1(t) = Ce^{-(R/V_1)t}$. With $R = 50$ and $V_1 = 300$ we have $c_1(t) = Ce^{-t/6}$; the initial condition $c_1(0) = c_0 = 10$ gives $C = 10$. Finally,

$$c_1(t) = 10e^{-t/6}$$

41. Let a, b, r be constants. Show that

$$y = Ce^{-kt} + a + bk\left(\frac{k\sin rt - r\cos rt}{k^2 + r^2}\right)$$

is a general solution of

$$\frac{dy}{dt} = -k\Big(y - a - b\sin rt\Big)$$

SOLUTION This is a linear differential equation; in standard form, it is

$$\frac{dy}{dt} + ky = k(a + b\sin rt)$$

The integrating factor is then e^{kt}; multiplying through gives

$$(e^{kt}y)' = kae^{kt} + kbe^{kt}\sin rt \qquad (*)$$

The first term on the right-hand side has integral ae^{kt}. To integrate the second term, use integration by parts twice; this result in an equation of the form

$$\int kbe^{kt}\sin rt = F(t) + A\int kbe^{kt}\sin rt$$

for some function $F(t)$ and constant A. Solving for the integral gives

$$\int kbe^{kt}\sin rt = kbe^{kt}\frac{k\sin rt - r\cos rt}{k^2+r^2}$$

so that integrating equation (*) gives

$$e^{kt}y = ae^{kt} + kbe^{kt}\frac{k\sin rt - r\cos rt}{k^2+r^2} + C$$

Divide through by e^{kt} to get

$$y = a + bk\left(\frac{k\sin rt - r\cos rt}{k^2+r^2}\right) + Ce^{-kt}$$

Further Insights and Challenges

43. Let $\alpha(x)$ be an integrating factor for $y' + A(x)y = B(x)$. The differential equation $y' + A(x)y = 0$ is called the associated **homogeneous equation**.

(a) Show that $1/\alpha(x)$ is a solution of the associated homogeneous equation.

(b) Show that if $y = f(x)$ is a particular solution of $y' + A(x)y = B(x)$, then $f(x) + C/\alpha(x)$ is also a solution for any constant C.

SOLUTION

(a) Remember that $\alpha'(x) = A(x)\alpha(x)$. Now, let $y(x) = (\alpha(x))^{-1}$. Then

$$y' + A(x)y = -\frac{1}{(\alpha(x))^2}\alpha'(x) + \frac{A(x)}{\alpha(x)} = -\frac{1}{(\alpha(x))^2}A(x)\alpha(x) + \frac{A(x)}{\alpha(x)} = 0.$$

(b) Suppose $f(x)$ satisfies $f'(x) + A(x)f(x) = B(x)$. Now, let $y(x) = f(x) + C/\alpha(x)$, where C is an arbitrary constant. Then

$$\begin{aligned} y' + A(x)y &= f'(x) - \frac{C}{(\alpha(x))^2}\alpha'(x) + A(x)f(x) + \frac{CA(x)}{\alpha(x)} \\ &= \left(f'(x) + A(x)f(x)\right) + \frac{C}{\alpha(x)}\left(A(x) - \frac{\alpha'(x)}{\alpha(x)}\right) = B(x) + 0 = B(x). \end{aligned}$$

45. Transient Currents Suppose the circuit described by Eq. (10) is driven by a sinusoidal voltage source $V(t) = V\sin\omega t$ (where V and ω are constant).

(a) Show that

$$I(t) = \frac{V}{R^2 + L^2\omega^2}(R\sin\omega t - L\omega\cos\omega t) + Ce^{-(R/L)t}$$

(b) Let $Z = \sqrt{R^2 + L^2\omega^2}$. Choose θ so that $Z\cos\theta = R$ and $Z\sin\theta = L\omega$. Use the addition formula for the sine function to show that

$$I(t) = \frac{V}{Z}\sin(\omega t - \theta) + Ce^{-(R/L)t}$$

This shows that the current in the circuit varies sinusoidally apart from a DC term (called the **transient current** in electronics) that decreases exponentially.

SOLUTION

(a) With $V(t) = V\sin\omega t$, the equation

$$\frac{dI}{dt} + \frac{R}{L}I = \frac{1}{L}V(t)$$

becomes

$$\frac{dI}{dt} + \frac{R}{L}I = \frac{V}{L}\sin\omega t.$$

This is a linear equation in standard form with $A(t) = \frac{R}{L}$ and $B(t) = \frac{V}{L}\sin\omega t$. By Theorem 1, the integrating factor is

$$\alpha(t) = \int e^{\int A(t)\,dt} = e^{(R/L)t}.$$

When multiplied by the integrating factor, the equation becomes

$$(e^{(R/L)t}I)' = \frac{V}{L}e^{(R/L)t}\sin\omega t.$$

Integration on both sides (integration by parts is needed for the integral on the right-hand side) now yields

$$(e^{(R/L)t}I) = \frac{V}{R^2+L^2\omega^2}e^{(R/L)t}(R\sin\omega t - L\omega\cos\omega t) + C;$$

hence,

$$I(t) = \frac{V}{R^2+L^2\omega^2}(R\sin\omega t - L\omega\cos\omega t) + Ce^{-(R/L)t}.$$

(b) Let $Z = \sqrt{R^2+L^2\omega^2}$, and choose θ so that $Z\cos\theta = R$ and $Z\sin\theta = L\omega$. Then

$$\begin{aligned}\frac{V}{R^2+L^2\omega^2}(R\sin\omega t - L\omega\cos\omega t) &= \frac{V}{Z^2}(Z\cos\theta\sin\omega t - Z\sin\theta\cos\omega t)\\ &= \frac{V}{Z}(\cos\theta\sin\omega t - \sin\theta\cos\omega t) = \frac{V}{Z}\sin(\omega t - \theta).\end{aligned}$$

Thus,

$$I(t) = \frac{V}{Z}\sin(\omega t - \theta) + Ce^{-(R/L)t}.$$

CHAPTER REVIEW EXERCISES

1. Which of the following differential equations are linear? Determine the order of each equation.

(a) $y' = y^5 - 3x^4y$ **(b)** $y' = x^5 - 3x^4y$

(c) $y = y''' - 3x\sqrt{y}$ **(d)** $\sin x \cdot y'' = y - 1$

SOLUTION

(a) y^5 is a nonlinear term involving the dependent variable, so this is not a linear equation; the highest order derivative that appears in the equation is a first derivative, so this is a first-order equation.

(b) This is linear equation; the highest order derivative that appears in the equation is a first derivative, so this is a first-order equation.

(c) $\sqrt{y}$ is a nonlinear term involving the dependent variable, so this is not a linear equation; the highest order derivative that appears in the equation is a third derivative, so this is a third-order equation.

(d) This is linear equation; the highest order derivative that appears in the equation is a second derivative, so this is a second-order equation.

In Exercises 3–6, solve using separation of variables.

3. $\frac{dy}{dt} = t^2y^{-3}$

SOLUTION Rewrite the equation as

$$y^3\,dy = t^2\,dt.$$

Upon integrating both sides of this equation, we obtain:

$$\int y^3\,dy = \int t^2\,dt$$

$$\frac{y^4}{4} = \frac{t^3}{3} + C.$$

Thus,

$$y = \pm\left(\frac{4}{3}t^3 + C\right)^{1/4},$$

where C is an arbitrary constant.

5. $x\dfrac{dy}{dx} - y = 1$

SOLUTION Rewrite the equation as

$$\frac{dy}{1+y} = \frac{dx}{x}.$$

upon integrating both sides of this equation, we obtain

$$\int \frac{dy}{1+y} = \int \frac{dx}{x}$$
$$\ln|1+y| = \ln|x| + C.$$

Thus,

$$y = -1 + Ax,$$

where $A = \pm e^C$ is an arbitrary constant.

In Exercises 7–10, solve the initial value problem using separation of variables.

7. $y' = \cos^2 x, \quad y(0) = \dfrac{\pi}{4}$

SOLUTION First, we find the general solution of the differential equation. Because the variables are already separated, we integrate both sides to obtain

$$y = \int \cos^2 x\, dx = \int \left(\frac{1}{2} + \frac{1}{2}\cos 2x\right) dx = \frac{x}{2} + \frac{\sin 2x}{4} + C.$$

The initial condition $y(0) = \frac{\pi}{4}$ allows us to determine $C = \frac{\pi}{4}$. Thus, the solution is:

$$y(x) = \frac{x}{2} + \frac{\sin 2x}{4} + \frac{\pi}{4}.$$

9. $y' = xy^2, \quad y(1) = 2$

SOLUTION First, we find the general solution of the differential equation. Rewrite

$$\frac{dy}{dx} = xy^2 \quad \text{as} \quad \frac{dy}{y^2} = x\,dx.$$

Upon integrating both sides of this equation, we find

$$\int \frac{dy}{y^2} = \int x\,dx$$
$$-\frac{1}{y} = \frac{1}{2}x^2 + C.$$

Thus,

$$y = -\frac{1}{\frac{1}{2}x^2 + C}.$$

The initial condition $y(1) = 2$ allows us to determine the value of C:

$$2 = -\frac{1}{\frac{1}{2}\cdot 1^2 + C} = -\frac{2}{1+2C}$$
$$1 + 2C = -1$$
$$C = -1$$

Hence, the solution to the initial value problem is

$$y = -\frac{1}{\frac{1}{2}x^2 - 1} = -\frac{2}{x^2 - 2}.$$

11. Figure 1 shows the slope field for $\dot{y} = \sin y + ty$. Sketch the graphs of the solutions with the initial conditions $y(0) = 1$, $y(0) = 0$, and $y(0) = -1$.

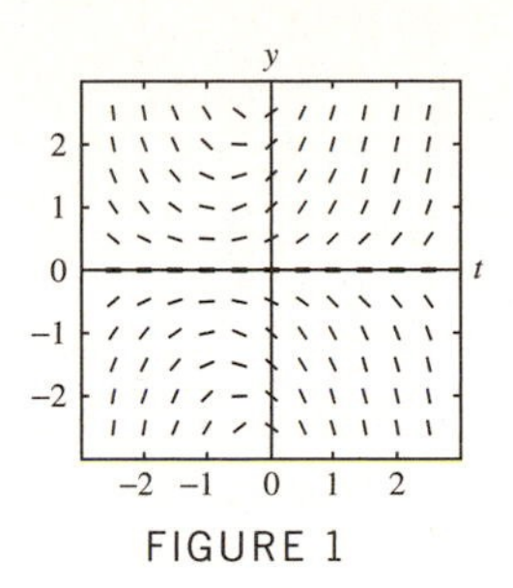

FIGURE 1

SOLUTION

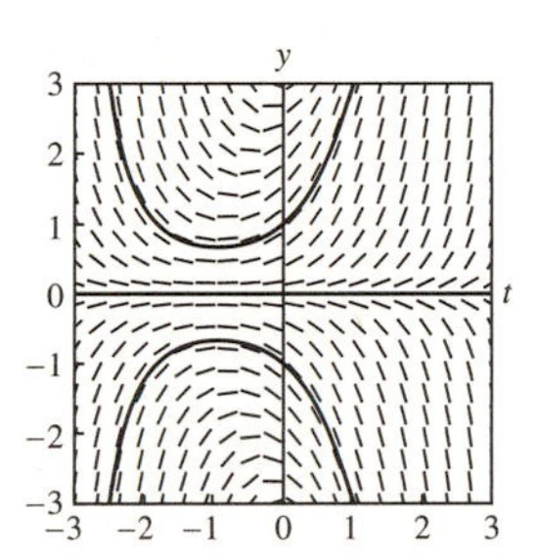

13. Let $y(t)$ be the solution to the differential equation with slope field as shown in Figure 2, satisfying $y(0) = 0$. Sketch the graph of $y(t)$. Then use your answer to Exercise 12 to solve for $y(t)$.

SOLUTION As explained in the previous exercise, the slope field in Figure 2 corresponds to the equation $\dot{y} = 1 + y^2$. The graph of the solution satisfying $y(0) = 0$ is:

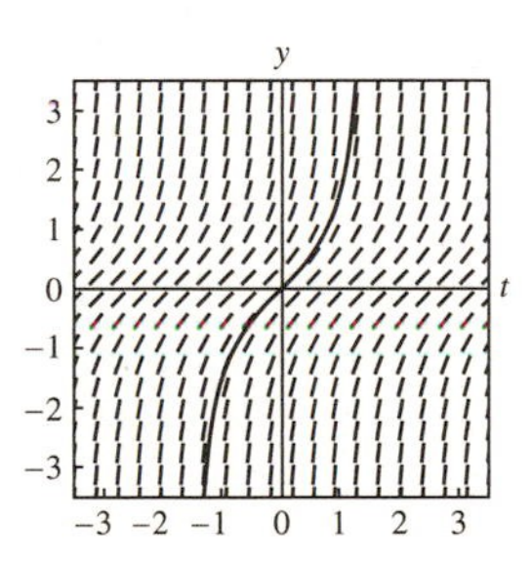

To solve the initial value problem $\dot{y} = 1 + y^2$, $y(0) = 0$, we first find the general solution of the differential equation. Separating variables yields:

$$\frac{dy}{1 + y^2} = dt.$$

Upon integrating both sides of this equation, we find

$$\tan^{-1} y = t + C \qquad \text{or} \qquad y = \tan(t + C).$$

The initial condition gives $C = 0$, so the solution is $y = \tan x$.

15. Let $y(t)$ be the solution of $(x^3 + 1)\dot{y} = y$ satisfying $y(0) = 1$. Compute approximations to $y(0.1)$, $y(0.2)$, and $y(0.3)$ using Euler's Method with time step $h = 0.1$.

SOLUTION Rewriting the equation as $\dot{y} = \frac{y}{x^3+1}$ we have $F(x, y) = \frac{y}{x^3+1}$. Using Euler's Method with $x_0 = 0$, $y_0 = 1$ and $h = 0.1$, we calculate

$$y(0.1) \approx y_1 = y_0 + hF(x_0, y_0) = 1 + 0.1 \cdot \frac{1}{0^3 + 1} = 1.1$$

$$y(0.2) \approx y_2 = y_1 + hF(x_1, y_1) = 1.209890$$

$$y(0.3) \approx y_3 = y_2 + hF(x_2, y_2) = 1.329919$$

In Exercises 16–19, solve using the method of integrating factors.

17. $\frac{dy}{dx} = \frac{y}{x} + x, \quad y(1) = 3$

SOLUTION First, we find the general solution of the differential equation. Rewrite the equation as

$$y' - \frac{1}{x}y = x,$$

which is in standard form with $A(x) = -\frac{1}{x}$ and $B(x) = x$. The integrating factor is

$$\alpha(x) = e^{\int -\frac{1}{x}\,dx} = e^{-\ln x} = \frac{1}{x}.$$

When multiplied by the integrating factor, the rewritten differential equation becomes

$$\left(\frac{1}{x}y\right)' = 1.$$

Integration on both sides now yields

$$\frac{1}{x}y = x + C;$$

hence,

$$y(x) = x^2 + Cx.$$

The initial condition $y(1) = 3$ allows us to determine the value of C:

$$3 = 1 + C \qquad \text{so} \qquad C = 2.$$

The solution to the initial value problem is then

$$y = x^2 + 2x.$$

19. $y' + 2y = 1 + e^{-x}, \quad y(0) = -4$

SOLUTION The equation is already in standard form with $A(x) = 2$ and $B(x) = 1 + e^{-x}$. The integrating factor is

$$\alpha(x) = e^{\int 2\,dx} = e^{2x}.$$

When multiplied by the integrating factor, the original differential equation becomes

$$(e^{2x}y)' = e^{2x} + e^{x}.$$

Integration on both sides now yields

$$e^{2x}y = \frac{1}{2}e^{2x} + e^{x} + C;$$

hence,

$$y(x) = \frac{1}{2} + e^{-x} + Ce^{-2x}.$$

The initial condition $y(0) = -4$ allows us to determine the value of C:

$$-4 = \frac{1}{2} + 1 + C \qquad \text{so} \qquad C = -\frac{11}{2}.$$

The solution to the initial value problem is then

$$y(x) = \frac{1}{2} + e^{-x} - \frac{11}{2}e^{-2x}.$$

In Exercises 20–27, solve using the appropriate method.

21. $y' + (\tan x)y = \cos^2 x, \quad y(\pi) = 2$

SOLUTION First, we find the general solution of the differential equation. As this is a first order linear equation with $A(x) = \tan x$ and $B(x) = \cos^2 x$, we compute the integrating factor

$$\alpha(x) = e^{\int A(x)\,dx} = e^{\int \tan x\,dx} = e^{-\ln \cos x} = \frac{1}{\cos x}.$$

When multiplied by the integrating factor, the original differential equation becomes

$$\left(\frac{1}{\cos x} y\right)' = \cos x.$$

Integration on both sides now yields

$$\frac{1}{\cos x} y = \sin x + C;$$

hence,

$$y(x) = \sin x \cos x + C \cos x = \frac{1}{2}\sin 2x + C\cos x.$$

The initial condition $y(\pi) = 2$ allows us to determine the value of C:

$$2 = 0 + C(-1) \qquad \text{so} \qquad C = -2.$$

The solution to the initial value problem is then

$$y = \frac{1}{2}\sin 2x - 2\cos x.$$

23. $(y-1)y' = t, \quad y(1) = -3$

SOLUTION First, we find the general solution of the differential equation. This is a separable equation that we rewrite as

$$(y-1)\,dy = t\,dt.$$

Upon integrating both sides of this equation, we find

$$\int (y-1)\,dy = \int t\,dt$$

$$\frac{y^2}{2} - y = \frac{1}{2}t^2 + C$$

$$y^2 - 2y + 1 = t^2 + C$$

$$(y-1)^2 = t^2 + C$$

$$y(t) = \pm\sqrt{t^2 + C} + 1$$

To satisfy the initial condition $y(1) = -3$ we must choose the negative square root; moreover,

$$-3 = -\sqrt{1 + C} + 1 \qquad \text{so} \qquad C = 15.$$

The solution to the initial value problem is then

$$y(t) = -\sqrt{t^2 + 15} + 1$$

25. $\dfrac{dw}{dx} = k\dfrac{1 + w^2}{x}, \quad w(1) = 1$

SOLUTION First, we find the general solution of the differential equation. This is a separable equation that we rewrite as

$$\frac{dw}{1 + w^2} = \frac{k}{x}\,dx.$$

Upon integrating both sides of this equation, we find

$$\int \frac{dw}{1 + w^2} = \int \frac{k}{x}\,dx$$

$$\tan^{-1} w = k \ln x + C$$

$$w(x) = \tan(k \ln x + C).$$

Because the initial condition is specified at $x = 1$, we are interested in the solution for $x > 0$; we can therefore omit the absolute value within the natural logarithm function. The initial condition $w(1) = 1$ allows us to determine the value of C:

$$1 = \tan(k \ln 1 + C) \quad \text{so} \quad C = \tan^{-1} 1 = \frac{\pi}{4}.$$

The solution to the initial value problem is then

$$w = \tan\left(k \ln x + \frac{\pi}{4}\right).$$

27. $y' + \dfrac{y}{x} = \sin x$

SOLUTION This is a first order linear equation with $A(x) = \frac{1}{x}$ and $B(x) = \sin x$. The integrating factor is

$$\alpha(x) = e^{\int A(x)\,dx} = e^{\ln x} = x.$$

When multiplied by the integrating factor, the original differential equation becomes

$$(xy)' = x \sin x.$$

Integration on both sides (integration by parts is needed for the integral on the right-hand side) now yields

$$xy = -x \cos x + \sin x + C;$$

hence,

$$y(x) = -\cos x + \frac{\sin x}{x} + \frac{C}{x}.$$

29. Find the solutions to $y' = -2y + 8$ satisfying $y(0) = 3$ and $y(0) = 4$, and sketch their graphs.

SOLUTION First, rewrite the differential equation as $y' = -2(y - 4)$; from here we see that the general solution is

$$y(t) = 4 + Ce^{-2t},$$

for some constant C. If $y(0) = 3$, then

$$3 = 4 + Ce^0 \quad \text{and} \quad C = -1.$$

Thus, $y(t) = 4 - e^{-2t}$. If $y(0) = 4$, then

$$4 = 4 + Ce^0 \quad \text{and} \quad C = 0;$$

hence, $y(t) = 4$. The graphs of the two solutions are shown below.

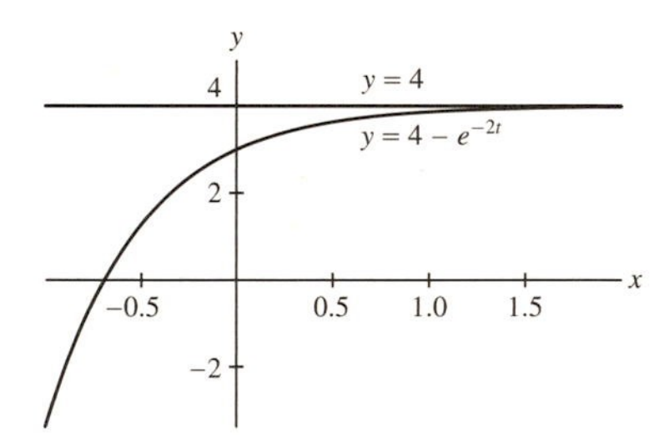

31. What is the limit $\lim_{t\to\infty} y(t)$ if $y(t)$ is a solution of:

(a) $\dfrac{dy}{dt} = -4(y - 12)$?

(b) $\dfrac{dy}{dt} = 4(y - 12)$?

(c) $\dfrac{dy}{dt} = -4y - 12$?

SOLUTION

(a) The general solution of $\dfrac{dy}{dt} = -4(y - 12)$ is $y(t) = 12 + Ce^{-4t}$, where C is an arbitrary constant. Regardless of the value of C,

$$\lim_{t\to\infty} y(t) = \lim_{t\to\infty} (12 + Ce^{-4t}) = 12.$$

(b) The general solution of $\frac{dy}{dt} = 4(y-12)$ is $y(t) = 12 + Ce^{4t}$, where C is an arbitrary constant. Here, the limit depends on the value of C. Specifically,

$$\lim_{t\to\infty} y(t) = \lim_{t\to\infty} (12 + Ce^{4t}) = \begin{cases} \infty, & C > 0 \\ 12, & C = 0 \\ -\infty, & C < 0 \end{cases}$$

(c) The general solution of $\frac{dy}{dt} = -4y - 12 = -4(y+3)$ is $y(t) = -3 + Ce^{-4t}$, where C is an arbitrary constant. Regardless of the value of C,

$$\lim_{t\to\infty} y(t) = \lim_{t\to\infty} (-3 + Ce^{-4t}) = -3.$$

In Exercises 32–35, let $P(t)$ denote the balance at time t (years) of an annuity that earns 5% interest continuously compounded and pays out \$20,000/year continuously.

33. Determine $P(5)$ if $P(0) = \$200{,}000$.

SOLUTION In the previous exercise we concluded that $P(t)$ satisfies the equation $P' = 0.05(P - 400{,}000)$. The general solution of this differential equation is

$$P(t) = 400{,}000 + Ce^{0.05t}.$$

Given $P(0) = 200{,}000$, it follows that

$$200{,}000 = 400{,}000 + Ce^{0.05\cdot 0} = 400{,}000 + C$$

or

$$C = -200{,}000.$$

Thus,

$$P(t) = 400{,}000 - 200{,}000e^{0.05t},$$

and

$$P(5) = 400{,}000 - 200{,}000e^{0.05(5)} \approx \$143{,}194.90.$$

35. What is the minimum initial balance that will allow the annuity to make payments indefinitely?

SOLUTION In Exercise 33, we found that the balance at time t is

$$P(t) = 400{,}000 + Ce^{0.05t}.$$

If initial balance is P_0 then

$$P_0 = P(0) = 400{,}000 + Ce^{0.05\cdot 0} = 400{,}000 + C$$

or

$$C = P_0 - 400{,}000.$$

Thus,

$$P(t) = 400{,}000 + (P_0 - 400{,}000)\, e^{0.05t}.$$

If $P_0 \ge 400{,}000$, then $P(t)$ is always positive. Therefore, the minimum initial balance that allows the annuity to make payments indefinitely is $P_0 = \$400{,}000$.

37. Let A and B be constants. Prove that if $A > 0$, then all solutions of $\frac{dy}{dt} + Ay = B$ approach the same limit as $t \to \infty$.

SOLUTION This is a linear first-order equation in standard form with integrating factor

$$\alpha(t) = e^{\int A\,dt} = e^{At}.$$

When multiplied by the integrating factor, the original differential equation becomes

$$(e^{At}y)' = Be^{At}.$$

Integration on both sides now yields

$$e^{At}y = \frac{B}{A}e^{At} + C;$$

hence,

$$y(t) = \frac{B}{A} + Ce^{-At}.$$

Because $A > 0$,

$$\lim_{t\to\infty} y(t) = \lim_{t\to\infty}\left(\frac{B}{A} + Ce^{-At}\right) = \frac{B}{A}.$$

We conclude that if $A > 0$, all solutions approach the limit $\frac{B}{A}$ as $t \to \infty$.

39. The trough in Figure 3 (dimensions in centimeters) is filled with water. At time $t = 0$ (in seconds), water begins leaking through a hole at the bottom of area 4 cm^2. Let $y(t)$ be the water height at time t. Find a differential equation for $y(t)$ and solve it to determine when the water level decreases to 60 cm.

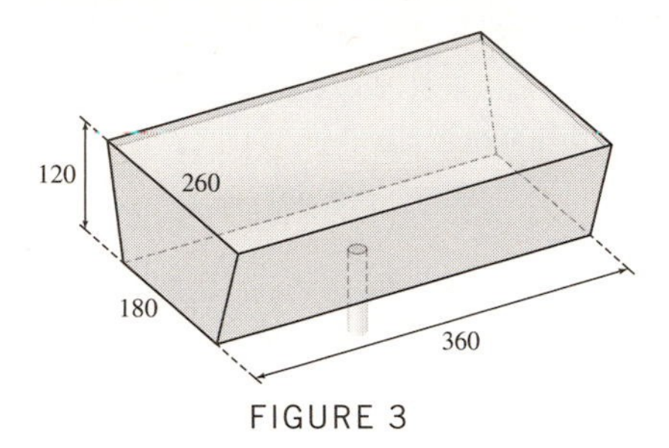

FIGURE 3

SOLUTION $y(t)$ obeys the differential equation:

$$\frac{dy}{dt} = \frac{Bv(y)}{A(y)},$$

where $v(y)$ denotes the velocity of the water flowing through the hole when the trough is filled to height y, B denotes the area of the hole and $A(y)$ denotes the area of the horizontal cross section of the trough at height y. Since measurements are all in centimeters, we will work in centimeters. We have

$$g = 9.8 \text{ m/s}^2 = 980 \text{ cm/s}^2$$

By Torricelli's Law, $v(y) = -\sqrt{2\cdot 980}\sqrt{y} = -14\sqrt{10}\sqrt{y}$ m/s. The area of the hole is $B = 4$ cm^2. The horizontal cross section of the trough at height y is a rectangle of length 360 and width $w(y)$. As $w(y)$ varies linearly from 180 when $y = 0$ to 260 when $y = 120$, it follows that

$$w(y) = 180 + \frac{80y}{120} = 180 + \frac{2}{3}y$$

so that the area of the horizontal cross-section at height y is

$$A(y) = 360w(y) = 64800 + 240y = 240(y + 270)$$

The differential equation for $y(t)$ then becomes

$$\frac{dy}{dt} = \frac{Bv(y)}{A(y)} = \frac{-4\cdot 14\sqrt{10}\sqrt{y}}{240(y+270)} = \frac{-7\sqrt{10}}{30}\cdot\frac{\sqrt{y}}{y+270}$$

This equation is separable, so

$$\frac{y+270}{\sqrt{y}}\,dy = \frac{-7\sqrt{10}}{30}\,dt$$

$$(y^{1/2} + 270y^{-1/2})\,dy = \frac{-7\sqrt{10}}{30}\,dt$$

$$\int (y^{1/2} + 270y^{-1/2})\,dy = \frac{-7\sqrt{10}}{30}\int 1\,dt$$

$$\frac{2}{3}y^{3/2} + 540y^{1/2} = -\frac{7\sqrt{10}}{30}t + C$$

$$y^{3/2} + 810y^{1/2} = -\frac{7\sqrt{10}}{20}t + C$$

The initial condition $y(0) = 120$ allows us to determine the value of C:

$$120^{3/2} + 810 \cdot 120^{1/2} = 0 + C \quad \text{so} \quad C = 930\sqrt{120} = 1860\sqrt{30}$$

Thus the height of the water is given implicitly by the equation

$$y^{3/2} + 810y^{1/2} = -\frac{7\sqrt{10}}{20}t + 1860\sqrt{30}$$

We want to find t such that $y(t) = 60$:

$$60^{3/2} + 810 \cdot 60^{1/2} = -\frac{7\sqrt{10}}{20}t + 1860\sqrt{30}$$

$$1740\sqrt{15} = -\frac{7\sqrt{10}}{20}t + 1860\sqrt{30}$$

$$t = \frac{120}{7}\sqrt{10}(31\sqrt{30} - 29\sqrt{15}) \approx 3115.88 \text{ s}$$

The height of the water in the tank is 60 cm after approximately 3116 seconds, or 51 minutes 56 seconds.

41. Let $y(t)$ be the solution of $\dot{y} = 0.3y(2 - y)$ with $y(0) = 1$. Determine $\lim_{t\to\infty} y(t)$ without solving for y explicitly.

SOLUTION We write the given equation in the form

$$\dot{y} = 0.6y\left(1 - \frac{y}{2}\right).$$

This is a logistic equation with $A = 2$ and $k = 0.6$. Because the initial condition $y(0) = y_0 = 1$ satisfies $0 < y_0 < A$, the solution is increasing and approaches A as $t \to \infty$. That is, $\lim_{t\to\infty} y(t) = 2$.

43. A lake has a carrying capacity of 1000 fish. Assume that the fish population grows logistically with growth constant $k = 0.2$ day^{-1}. How many days will it take for the population to reach 900 fish if the initial population is 20 fish?

SOLUTION Let $y(t)$ represent the fish population. Because the population grows logistically with $k = 0.2$ and $A = 1000$,

$$y(t) = \frac{1000}{1 - e^{-0.2t}/C}.$$

The initial condition $y(0) = 20$ allows us to determine the value of C:

$$20 = \frac{1000}{1 - \frac{1}{C}}; \quad 1 - \frac{1}{C} = 50; \quad \text{so} \quad C = -\frac{1}{49}.$$

Hence,

$$y(t) = \frac{1000}{1 + 49e^{-0.2t}}.$$

The population will reach 900 fish when

$$\frac{1000}{1 + 49e^{-0.2t}} = 900.$$

Solving for t, we find

$$t = 5 \ln 441 \approx 30.44 \text{ days}.$$

45. Show that $y = \sin(\tan^{-1} x + C)$ is the general solution of $y' = \sqrt{1 - y^2}/(1 + x^2)$. Then use the addition formula for the sine function to show that the general solution may be written

$$y = \frac{(\cos C)x + \sin C}{\sqrt{1 + x^2}}$$

SOLUTION Rewrite

$$\frac{dy}{dx} = \frac{\sqrt{1 - y^2}}{1 + x^2} \quad \text{as} \quad \frac{dy}{\sqrt{1 - y^2}} = \frac{dx}{1 + x^2}.$$

Upon integrating both sides of this equation, we find

$$\int \frac{dy}{\sqrt{1 - y^2}} = \int \frac{dx}{1 + x^2}$$

$$\sin^{-1} y = \tan^{-1} x + C$$

Thus,

$$y(x) = \sin\left(\tan^{-1} x + C\right).$$

To express the solution in the required form, we use the addition formula

$$\sin(\alpha + \beta) = \sin\alpha \cos\beta + \sin\beta \cos\alpha$$

This yields

$$y(x) = \sin\left(\tan^{-1} x\right) \cos C + \sin C \cos\left(\tan^{-1} x\right).$$

Using the figure below, we see that

$$\sin\left(\tan^{-1} x\right) = \frac{x}{\sqrt{1+x^2}}; \text{ and}$$

$$\cos\left(\tan^{-1} x\right) = \frac{1}{\sqrt{1+x^2}}.$$

Finally,

$$y = \frac{x\cos C}{\sqrt{1+x^2}} + \frac{\sin C}{\sqrt{1+x^2}} = \frac{(\cos C)x + \sin C}{\sqrt{1+x^2}}.$$

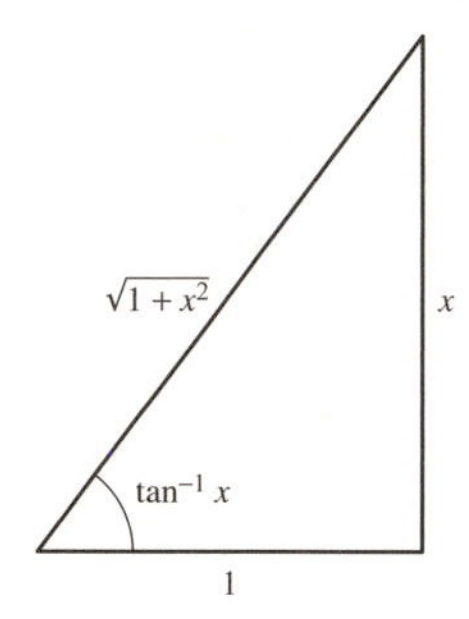

47. At $t = 0$, a tank of volume 300 L is filled with 100 L of water containing salt at a concentration of 8 g/L. Fresh water flows in at a rate of 40 L/min, mixes instantaneously, and exits at the same rate. Let $c_1(t)$ be the salt concentration at time t.

(a) Find a differential equation satisfied by $c_1(t)$ *Hint:* Find the differential equation for the quantity of salt $y(t)$, and observe that $c_1(t) = y(t)/100$.

(b) Find the salt concentration $c_1(t)$ in the tank as a function of time.

SOLUTION

(a) Let $y(t)$ be the amount of salt in the tank at time t; then $c_1(t) = y(t)/100$. The rate of change of the amount of salt in the tank is

$$\frac{dy}{dt} = \text{salt rate in} - \text{salt rate out} = \left(40\frac{\text{L}}{\text{min}}\right)\left(0\frac{\text{g}}{\text{L}}\right) - \left(40\frac{\text{L}}{\text{min}}\right)\left(\frac{y}{100}\cdot\frac{\text{g}}{\text{L}}\right)$$

$$= -\frac{2}{5}y$$

Now, $c_1'(t) = y'(t)/100$ and $c(t) = y(t)/100$, so that c_1 satisfies the same differential equation:

$$\frac{dc_1}{dt} = -\frac{2}{5}c_1$$

(b) This is a linear differential equation. Putting it in standard form gives

$$\frac{dc_1}{dt} + \frac{2}{5}c_1 = 0$$

The integrating factor is $e^{2t/5}$; multiplying both sides by the integrating factor gives

$$(e^{2t/5}c_1)' = 0$$

Integrate and multiply through by $e^{-2t/5}$ to get

$$c_1(t) = Ce^{-2t/5}$$

The initial condition tells us that $y(0) = Ce^{-2\cdot 0/5} = C = 8$, so that finally,

$$c_1(t) = 8e^{-2t/5}$$

10 INFINITE SERIES

10.1 Sequences

Preliminary Questions

1. What is a_4 for the sequence $a_n = n^2 - n$?

SOLUTION Substituting $n = 4$ in the expression for a_n gives

$$a_4 = 4^2 - 4 = 12.$$

2. Which of the following sequences converge to zero?

(a) $\dfrac{n^2}{n^2+1}$ **(b)** 2^n **(c)** $\left(\dfrac{-1}{2}\right)^n$

SOLUTION

(a) This sequence does not converge to zero:

$$\lim_{n\to\infty} \frac{n^2}{n^2+1} = \lim_{x\to\infty} \frac{x^2}{x^2+1} = \lim_{x\to\infty} \frac{1}{1+\frac{1}{x^2}} = \frac{1}{1+0} = 1.$$

(b) This sequence does not converge to zero: this is a geometric sequence with $r = 2 > 1$; hence, the sequence diverges to ∞.

(c) Recall that if $|a_n|$ converges to 0, then a_n must also converge to zero. Here,

$$\left|\left(-\frac{1}{2}\right)^n\right| = \left(\frac{1}{2}\right)^n,$$

which is a geometric sequence with $0 < r < 1$; hence, $(\frac{1}{2})^n$ converges to zero. It therefore follows that $(-\frac{1}{2})^n$ converges to zero.

3. Let a_n be the nth decimal approximation to $\sqrt{2}$. That is, $a_1 = 1$, $a_2 = 1.4$, $a_3 = 1.41$, etc. What is $\lim_{n\to\infty} a_n$?

SOLUTION $\lim_{n\to\infty} a_n = \sqrt{2}$.

4. Which of the following sequences is defined recursively?

(a) $a_n = \sqrt{4+n}$ **(b)** $b_n = \sqrt{4+b_{n-1}}$

SOLUTION

(a) a_n can be computed directly, since it depends on n only and not on preceding terms. Therefore a_n is defined explicitly and not recursively.

(b) b_n is computed in terms of the preceding term b_{n-1}, hence the sequence $\{b_n\}$ is defined recursively.

5. Theorem 5 says that every convergent sequence is bounded. Determine if the following statements are true or false and if false, give a counterexample.

(a) If $\{a_n\}$ is bounded, then it converges.

(b) If $\{a_n\}$ is not bounded, then it diverges.

(c) If $\{a_n\}$ diverges, then it is not bounded.

SOLUTION

(a) This statement is false. The sequence $a_n = \cos \pi n$ is bounded since $-1 \le \cos \pi n \le 1$ for all n, but it does not converge: since $a_n = \cos n\pi = (-1)^n$, the terms assume the two values 1 and -1 alternately, hence they do not approach one value.

(b) By Theorem 5, a converging sequence must be bounded. Therefore, if a sequence is not bounded, it certainly does not converge.

(c) The statement is false. The sequence $a_n = (-1)^n$ is bounded, but it does not approach one limit.

Exercises

1. Match each sequence with its general term:

$a_1, a_2, a_3, a_4, \ldots$	General term
(a) $\frac{1}{2}, \frac{2}{3}, \frac{3}{4}, \frac{4}{5}, \ldots$	(i) $\cos \pi n$
(b) $-1, 1, -1, 1, \ldots$	(ii) $\dfrac{n!}{2^n}$
(c) $1, -1, 1, -1, \ldots$	(iii) $(-1)^{n+1}$
(d) $\frac{1}{2}, \frac{2}{4}, \frac{6}{8}, \frac{24}{16} \cdots$	(iv) $\dfrac{n}{n+1}$

SOLUTION

(a) The numerator of each term is the same as the index of the term, and the denominator is one more than the numerator; hence $a_n = \frac{n}{n+1}$, $n = 1, 2, 3, \ldots$.

(b) The terms of this sequence are alternating between -1 and 1 so that the positive terms are in the even places. Since $\cos \pi n = 1$ for even n and $\cos \pi n = -1$ for odd n, we have $a_n = \cos \pi n$, $n = 1, 2, \ldots$.

(c) The terms a_n are 1 for odd n and -1 for even n. Hence, $a_n = (-1)^{n+1}$, $n = 1, 2, \ldots$

(d) The numerator of each term is $n!$, and the denominator is 2^n; hence, $a_n = \frac{n!}{2^n}$, $n = 1, 2, 3, \ldots$.

In Exercises 3–12, calculate the first four terms of the sequence, starting with $n = 1$.

3. $c_n = \dfrac{3^n}{n!}$

SOLUTION Setting $n = 1, 2, 3, 4$ in the formula for c_n gives

$$c_1 = \frac{3^1}{1!} = \frac{3}{1} = 3, \qquad c_2 = \frac{3^2}{2!} = \frac{9}{2},$$

$$c_3 = \frac{3^3}{3!} = \frac{27}{6} = \frac{9}{2}, \qquad c_4 = \frac{3^4}{4!} = \frac{81}{24} = \frac{27}{8}.$$

5. $a_1 = 2, \quad a_{n+1} = 2a_n^2 - 3$

SOLUTION For $n = 1, 2, 3$ we have:

$$a_2 = a_{1+1} = 2a_1^2 - 3 = 2 \cdot 4 - 3 = 5;$$

$$a_3 = a_{2+1} = 2a_2^2 - 3 = 2 \cdot 25 - 3 = 47;$$

$$a_4 = a_{3+1} = 2a_3^2 - 3 = 2 \cdot 2209 - 3 = 4415.$$

The first four terms of $\{a_n\}$ are 2, 5, 47, 4415.

7. $b_n = 5 + \cos \pi n$

SOLUTION For $n = 1, 2, 3, 4$ we have

$$b_1 = 5 + \cos \pi = 4;$$

$$b_2 = 5 + \cos 2\pi = 6;$$

$$b_3 = 5 + \cos 3\pi = 4;$$

$$b_4 = 5 + \cos 4\pi = 6.$$

The first four terms of $\{b_n\}$ are 4, 6, 4, 6.

9. $c_n = 1 + \dfrac{1}{2} + \dfrac{1}{3} + \cdots + \dfrac{1}{n}$

SOLUTION

$$c_1 = 1;$$

$$c_2 = 1 + \frac{1}{2} = \frac{3}{2};$$

$$c_3 = 1 + \frac{1}{2} + \frac{1}{3} = \frac{3}{2} + \frac{1}{3} = \frac{11}{6};$$

$$c_4 = 1 + \frac{1}{2} + \frac{1}{3} + \frac{1}{4} = \frac{11}{6} + \frac{1}{4} = \frac{25}{12}.$$

11. $b_1 = 2, \quad b_2 = 3, \quad b_n = 2b_{n-1} + b_{n-2}$

SOLUTION We need to find b_3 and b_4. Setting $n = 3$ and $n = 4$ and using the given values for b_1 and b_2 we obtain:

$$b_3 = 2b_{3-1} + b_{3-2} = 2b_2 + b_1 = 2 \cdot 3 + 2 = 8;$$

$$b_4 = 2b_{4-1} + b_{4-2} = 2b_3 + b_2 = 2 \cdot 8 + 3 = 19.$$

The first four terms of the sequence $\{b_n\}$ are 2, 3, 8, 19.

13. Find a formula for the nth term of each sequence.

(a) $\frac{1}{1}, \frac{-1}{8}, \frac{1}{27}, \ldots$ **(b)** $\frac{2}{6}, \frac{3}{7}, \frac{4}{8}, \ldots$

SOLUTION

(a) The denominators are the third powers of the positive integers starting with $n = 1$. Also, the sign of the terms is alternating with the sign of the first term being positive. Thus,

$$a_1 = \frac{1}{1^3} = \frac{(-1)^{1+1}}{1^3}; \quad a_2 = -\frac{1}{2^3} = \frac{(-1)^{2+1}}{2^3}; \quad a_3 = \frac{1}{3^3} = \frac{(-1)^{3+1}}{3^3}.$$

This rule leads to the following formula for the nth term:

$$a_n = \frac{(-1)^{n+1}}{n^3}.$$

(b) Assuming a starting index of $n = 1$, we see that each numerator is one more than the index and the denominator is four more than the numerator. Thus, the general term a_n is

$$a_n = \frac{n+1}{n+5}.$$

In Exercises 15–26, use Theorem 1 to determine the limit of the sequence or state that the sequence diverges.

15. $a_n = 12$

SOLUTION We have $a_n = f(n)$ where $f(x) = 12$; thus,

$$\lim_{n\to\infty} a_n = \lim_{x\to\infty} f(x) = \lim_{x\to\infty} 12 = 12.$$

17. $b_n = \dfrac{5n-1}{12n+9}$

SOLUTION We have $b_n = f(n)$ where $f(x) = \dfrac{5x-1}{12x+9}$; thus,

$$\lim_{n\to\infty} \frac{5n-1}{12n+9} = \lim_{x\to\infty} \frac{5x-1}{12x+9} = \frac{5}{12}.$$

19. $c_n = -2^{-n}$

SOLUTION We have $c_n = f(n)$ where $f(x) = -2^{-x}$; thus,

$$\lim_{n\to\infty} \left(-2^{-n}\right) = \lim_{x\to\infty} -2^{-x} = \lim_{x\to\infty} -\frac{1}{2^x} = 0.$$

21. $c_n = 9^n$

SOLUTION We have $c_n = f(n)$ where $f(x) = 9^x$; thus,

$$\lim_{n\to\infty} 9^n = \lim_{x\to\infty} 9^x = \infty$$

Thus, the sequence 9^n diverges.

23. $a_n = \dfrac{n}{\sqrt{n^2+1}}$

SOLUTION We have $a_n = f(n)$ where $f(x) = \dfrac{x}{\sqrt{x^2+1}}$; thus,

$$\lim_{n\to\infty} \frac{n}{\sqrt{n^2+1}} = \lim_{x\to\infty} \frac{x}{\sqrt{x^2+1}} = \lim_{x\to\infty} \frac{\frac{x}{x}}{\frac{\sqrt{x^2+1}}{x}} = \lim_{x\to\infty} \frac{1}{\sqrt{\frac{x^2+1}{x^2}}} = \lim_{x\to\infty} \frac{1}{\sqrt{1+\frac{1}{x^2}}} = \frac{1}{\sqrt{1+0}} = 1.$$

25. $a_n = \ln\left(\dfrac{12n+2}{-9+4n}\right)$

SOLUTION We have $a_n = f(n)$ where $f(x) = \ln\left(\dfrac{12x+2}{-9+4x}\right)$; thus,

$$\lim_{n\to\infty} \ln\left(\frac{12n+2}{-9+4n}\right) = \lim_{x\to\infty} \ln\left(\frac{12x+2}{-9+4x}\right) = \ln \lim_{x\to\infty}\left(\frac{12x+2}{-9+4x}\right) = \ln 3$$

In Exercises 27–30, use Theorem 4 to determine the limit of the sequence.

27. $a_n = \sqrt{4 + \dfrac{1}{n}}$

SOLUTION We have

$$\lim_{n\to\infty} 4 + \frac{1}{n} = \lim_{x\to\infty} 4 + \frac{1}{x} = 4$$

Since $\sqrt{x}$ is a continuous function for $x > 0$, Theorem 4 tells us that

$$\lim_{n\to\infty} \sqrt{4 + \frac{1}{n}} = \sqrt{\lim_{n\to\infty} 4 + \frac{1}{n}} = \sqrt{4} = 2$$

29. $a_n = \cos^{-1}\left(\dfrac{n^3}{2n^3+1}\right)$

SOLUTION We have

$$\lim_{n\to\infty} \frac{n^3}{2n^3+1} = \frac{1}{2}$$

Since $\cos^{-1}(x)$ is continuous for all x, Theorem 4 tells us that

$$\lim_{n\to\infty} \cos^{-1}\left(\frac{n^3}{2n^3+1}\right) = \cos^{-1}\left(\lim_{n\to\infty} \frac{n^3}{2n^3+1}\right) = \cos^{-1}(1/2) = \frac{\pi}{3}$$

31. Let $a_n = \dfrac{n}{n+1}$. Find a number M such that:

(a) $|a_n - 1| \le 0.001$ for $n \ge M$.

(b) $|a_n - 1| \le 0.00001$ for $n \ge M$.

Then use the limit definition to prove that $\lim_{n\to\infty} a_n = 1$.

SOLUTION

(a) We have

$$|a_n - 1| = \left|\frac{n}{n+1} - 1\right| = \left|\frac{n-(n+1)}{n+1}\right| = \left|\frac{-1}{n+1}\right| = \frac{1}{n+1}.$$

Therefore $|a_n - 1| \le 0.001$ provided $\frac{1}{n+1} \le 0.001$, that is, $n \ge 999$. It follows that we can take $M = 999$.

(b) By part (a), $|a_n - 1| \le 0.00001$ provided $\frac{1}{n+1} \le 0.00001$, that is, $n \ge 99999$. It follows that we can take $M = 99999$.

We now prove formally that $\lim_{n\to\infty} a_n = 1$. Using part (a), we know that

$$|a_n - 1| = \frac{1}{n+1} < \epsilon,$$

provided $n > \frac{1}{\epsilon} - 1$. Thus, Let $\epsilon > 0$ and take $M = \frac{1}{\epsilon} - 1$. Then, for $n > M$, we have

$$|a_n - 1| = \frac{1}{n+1} < \frac{1}{M+1} = \epsilon.$$

33. Use the limit definition to prove that $\lim_{n\to\infty} n^{-2} = 0$.

SOLUTION We see that

$$|n^{-2} - 0| = \left|\frac{1}{n^2}\right| = \frac{1}{n^2} < \epsilon$$

provided

$$n > \frac{1}{\sqrt{\epsilon}}.$$

Thus, let $\epsilon > 0$ and take $M = \frac{1}{\sqrt{\epsilon}}$. Then, for $n > M$, we have

$$|n^{-2} - 0| = \left|\frac{1}{n^2}\right| = \frac{1}{n^2} < \frac{1}{M^2} = \epsilon.$$

In Exercises 35–62, use the appropriate limit laws and theorems to determine the limit of the sequence or show that it diverges.

35. $a_n = 10 + \left(-\frac{1}{9}\right)^n$

SOLUTION By the Limit Laws for Sequences we have:

$$\lim_{n\to\infty}\left(10 + \left(-\frac{1}{9}\right)^n\right) = \lim_{n\to\infty} 10 + \lim_{n\to\infty}\left(-\frac{1}{9}\right)^n = 10 + \lim_{n\to\infty}\left(-\frac{1}{9}\right)^n.$$

Now,

$$-\left(\frac{1}{9}\right)^n \leq \left(-\frac{1}{9}\right)^n \leq \left(\frac{1}{9}\right)^n.$$

Because

$$\lim_{n\to\infty}\left(\frac{1}{9}\right)^n = 0,$$

by the Limit Laws for Sequences,

$$\lim_{n\to\infty} -\left(\frac{1}{9}\right)^n = -\lim_{n\to\infty}\left(\frac{1}{9}\right)^n = 0.$$

Thus, we have

$$\lim_{n\to\infty}\left(-\frac{1}{9}\right)^n = 0,$$

and

$$\lim_{n\to\infty}\left(10 + \left(-\frac{1}{9}\right)^n\right) = 10 + 0 = 10.$$

37. $c_n = 1.01^n$

SOLUTION Since $c_n = f(n)$ where $f(x) = 1.01^x$, we have

$$\lim_{n\to\infty} 1.01^n = \lim_{x\to\infty} 1.01^x = \infty$$

so that the sequence diverges.

39. $a_n = 2^{1/n}$

SOLUTION Because 2^x is a continuous function,

$$\lim_{n\to\infty} 2^{1/n} = \lim_{x\to\infty} 2^{1/x} = 2^{\lim_{x\to\infty}(1/x)} = 2^0 = 1.$$

41. $c_n = \frac{9^n}{n!}$

SOLUTION For $n \geq 9$, write

$$c_n = \frac{9^n}{n!} = \underbrace{\frac{9}{1}\cdot\frac{9}{2}\cdots\frac{9}{9}}_{\text{call this } C}\cdot\underbrace{\frac{9}{10}\cdot\frac{9}{11}\cdots\frac{9}{n-1}\cdot\frac{9}{n}}_{\text{Each factor is less than 1}}$$

Then clearly

$$0 \leq \frac{9^n}{n!} \leq C\frac{9}{n}$$

since each factor after the first nine is < 1. The squeeze theorem tells us that

$$\lim_{n\to\infty} 0 \le \lim_{n\to\infty} \frac{9^n}{n!} \le \lim_{n\to\infty} C\frac{9}{n} = C \lim_{n\to\infty} \frac{9}{n} = C \cdot 0 = 0$$

so that $\lim_{n\to\infty} c_n = 0$ as well.

43. $a_n = \dfrac{3n^2 + n + 2}{2n^2 - 3}$

SOLUTION

$$\lim_{n\to\infty} \frac{3n^2 + n + 2}{2n^2 - 3} = \lim_{x\to\infty} \frac{3x^2 + x + 2}{2x^2 - 3} = \frac{3}{2}.$$

45. $a_n = \dfrac{\cos n}{n}$

SOLUTION Since $-1 \le \cos n \le 1$ the following holds:

$$-\frac{1}{n} \le \frac{\cos n}{n} \le \frac{1}{n}.$$

We now apply the Squeeze Theorem for Sequences and the limits

$$\lim_{n\to\infty} -\frac{1}{n} = \lim_{n\to\infty} \frac{1}{n} = 0$$

to conclude that $\lim_{n\to\infty} \frac{\cos n}{n} = 0$.

47. $d_n = \ln 5^n - \ln n!$

SOLUTION Note that

$$d_n = \ln \frac{5^n}{n!}$$

so that

$$e^{d_n} = \frac{5^n}{n!} \quad \text{so} \quad \lim_{n\to\infty} e^{d_n} = \lim_{n\to\infty} \frac{5^n}{n!} = 0$$

by the method of Exercise 41. If d_n converged, we could, since $f(x) = e^x$ is continuous, then write

$$\lim_{n\to\infty} e^{d_n} = e^{\lim_{n\to\infty} d_n} = 0$$

which is impossible. Thus $\{d_n\}$ diverges.

49. $a_n = \left(2 + \dfrac{4}{n^2}\right)^{1/3}$

SOLUTION Let $a_n = \left(2 + \frac{4}{n^2}\right)^{1/3}$. Taking the natural logarithm of both sides of this expression yields

$$\ln a_n = \ln \left(2 + \frac{4}{n^2}\right)^{1/3} = \frac{1}{3} \ln \left(2 + \frac{4}{n^2}\right).$$

Thus,

$$\begin{aligned}\lim_{n\to\infty} \ln a_n &= \lim_{n\to\infty} \frac{1}{3} \ln \left(2 + \frac{4}{n^2}\right)^{1/3} = \frac{1}{3} \lim_{x\to\infty} \ln \left(2 + \frac{4}{x^2}\right) = \frac{1}{3} \ln \left(\lim_{x\to\infty} \left(2 + \frac{4}{x^2}\right)\right) \\ &= \frac{1}{3} \ln (2 + 0) = \frac{1}{3} \ln 2 = \ln 2^{1/3}.\end{aligned}$$

Because $f(x) = e^x$ is a continuous function, it follows that

$$\lim_{n\to\infty} a_n = \lim_{n\to\infty} e^{\ln a_n} = e^{\lim_{n\to\infty} (\ln a_n)} = e^{\ln 2^{1/3}} = 2^{1/3}.$$

51. $c_n = \ln \left(\dfrac{2n + 1}{3n + 4}\right)$

SOLUTION Because $f(x) = \ln x$ is a continuous function, it follows that

$$\lim_{n\to\infty} c_n = \lim_{x\to\infty} \ln \left(\frac{2x + 1}{3x + 4}\right) = \ln \left(\lim_{x\to\infty} \frac{2x + 1}{3x + 4}\right) = \ln \frac{2}{3}.$$

53. $y_n = \dfrac{e^n}{2^n}$

SOLUTION $\frac{e^n}{2^n} = \left(\frac{e}{2}\right)^n$ and $\frac{e}{2} > 1$. By the Limit of Geometric Sequences,we conclude that $\lim_{n\to\infty} \left(\frac{e}{2}\right)^n = \infty$. Thus, the given sequence diverges.

55. $y_n = \dfrac{e^n + (-3)^n}{5^n}$

SOLUTION

$$\lim_{n\to\infty} \frac{e^n + (-3)^n}{5^n} = \lim_{n\to\infty} \left(\frac{e}{5}\right)^n + \lim_{n\to\infty} \left(\frac{-3}{5}\right)^n$$

assuming both limits on the right-hand side exist. But by the Limit of Geometric Sequences, since

$$-1 < \frac{-3}{5} < 0 < \frac{e}{5} < 1$$

both limits on the right-hand side are 0, so that y_n converges to 0.

57. $a_n = n \sin \dfrac{\pi}{n}$

SOLUTION By the Theorem on Sequences Defined by a Function, we have

$$\lim_{n\to\infty} n \sin \frac{\pi}{n} = \lim_{x\to\infty} x \sin \frac{\pi}{x}.$$

Now,

$$\lim_{x\to\infty} x \sin \frac{\pi}{x} = \lim_{x\to\infty} \frac{\sin \frac{\pi}{x}}{\frac{1}{x}} = \lim_{x\to\infty} \frac{\left(\cos \frac{\pi}{x}\right)\left(-\frac{\pi}{x^2}\right)}{-\frac{1}{x^2}} = \lim_{x\to\infty} \left(\pi \cos \frac{\pi}{x}\right)$$

$$= \pi \lim_{x\to\infty} \cos \frac{\pi}{x} = \pi \cos 0 = \pi \cdot 1 = \pi.$$

Thus,

$$\lim_{n\to\infty} n \sin \frac{\pi}{n} = \pi.$$

59. $b_n = \dfrac{3 - 4^n}{2 + 7 \cdot 4^n}$

SOLUTION Divide the numerator and denominator by 4^n to obtain

$$a_n = \frac{3 - 4^n}{2 + 7 \cdot 4^n} = \frac{\frac{3}{4^n} - \frac{4^n}{4^n}}{\frac{2}{4^n} + \frac{7 \cdot 4^n}{4^n}} = \frac{\frac{3}{4^n} - 1}{\frac{2}{4^n} + 7}.$$

Thus,

$$\lim_{n\to\infty} a_n = \lim_{x\to\infty} \frac{\frac{3}{4^x} - 1}{\frac{2}{4^x} + 7} = \frac{\lim_{x\to\infty} \left(\frac{3}{4^x} - 1\right)}{\lim_{x\to\infty} \left(\frac{2}{4^x} + 7\right)} = \frac{3 \lim_{x\to\infty} \frac{1}{4^x} - \lim_{x\to\infty} 1}{2 \lim_{x\to\infty} \frac{1}{4^x} - \lim_{x\to\infty} 7} = \frac{3 \cdot 0 - 1}{2 \cdot 0 + 7} = -\frac{1}{7}.$$

61. $a_n = \left(1 + \dfrac{1}{n}\right)^n$

SOLUTION Taking the natural logarithm of both sides of this expression yields

$$\ln a_n = \ln \left(1 + \frac{1}{n}\right)^n = n \ln \left(1 + \frac{1}{n}\right) = \frac{\ln \left(1 + \frac{1}{n}\right)}{\frac{1}{n}}.$$

Thus,

$$\lim_{n\to\infty} (\ln a_n) = \lim_{x\to\infty} \frac{\ln \left(1 + \frac{1}{x}\right)}{\frac{1}{x}} = \lim_{x\to\infty} \frac{\frac{d}{dx}\left(\ln \left(1 + \frac{1}{x}\right)\right)}{\frac{d}{dx}\left(\frac{1}{x}\right)} = \lim_{x\to\infty} \frac{\frac{1}{1+\frac{1}{x}} \cdot \left(-\frac{1}{x^2}\right)}{-\frac{1}{x^2}} = \lim_{x\to\infty} \frac{1}{1 + \frac{1}{x}} = \frac{1}{1+0} = 1.$$

Because $f(x) = e^x$ is a continuous function, it follows that

$$\lim_{n\to\infty} a_n = \lim_{n\to\infty} e^{\ln a_n} = e^{\lim_{n\to\infty} (\ln a_n)} = e^1 = e.$$

In Exercises 63–66, find the limit of the sequence using L'Hôpital's Rule.

63. $a_n = \dfrac{(\ln n)^2}{n}$

SOLUTION

$$\lim_{n\to\infty} \frac{(\ln n)^2}{n} = \lim_{x\to\infty} \frac{(\ln x)^2}{x} = \lim_{x\to\infty} \frac{\frac{d}{dx}(\ln x)^2}{\frac{d}{dx}x} = \lim_{x\to\infty} \frac{\frac{2\ln x}{x}}{1} = \lim_{x\to\infty} \frac{2\ln x}{x}$$

$$= \lim_{x\to\infty} \frac{\frac{d}{dx}2\ln x}{\frac{d}{dx}x} = \lim_{x\to\infty} \frac{\frac{2}{x}}{1} = \lim_{x\to\infty} \frac{2}{x} = 0$$

65. $c_n = n\left(\sqrt{n^2+1} - n\right)$

SOLUTION

$$\lim_{n\to\infty} n\left(\sqrt{n^2+1} - n\right) = \lim_{x\to\infty} x\left(\sqrt{x^2+1} - x\right) = \lim_{x\to\infty} \frac{x\left(\sqrt{x^2+1} - x\right)\left(\sqrt{x^2+1} + x\right)}{\sqrt{x^2+1} + x}$$

$$= \lim_{x\to\infty} \frac{x}{\sqrt{x^2+1} + x} = \lim_{x\to\infty} \frac{\frac{d}{dx}x}{\frac{d}{dx}\sqrt{x^2+1} + x} = \lim_{x\to\infty} \frac{1}{1 + \frac{x}{\sqrt{x^2+1}}}$$

$$= \lim_{x\to\infty} \frac{1}{1 + \sqrt{\frac{x^2}{x^2+1}}} = \lim_{x\to\infty} \frac{1}{1 + \sqrt{\frac{1}{1+(1/x^2)}}} = \frac{1}{2}$$

In Exercises 67–70, use the Squeeze Theorem to evaluate $\lim_{n\to\infty} a_n$ *by verifying the given inequality.*

67. $a_n = \dfrac{1}{\sqrt{n^4+n^8}}, \quad \dfrac{1}{\sqrt{2}n^4} \le a_n \le \dfrac{1}{\sqrt{2}n^2}$

SOLUTION For all $n > 1$ we have $n^4 < n^8$, so the quotient $\frac{1}{\sqrt{n^4+n^8}}$ is smaller than $\frac{1}{\sqrt{n^4+n^4}}$ and larger than $\frac{1}{\sqrt{n^8+n^8}}$. That is,

$$a_n < \frac{1}{\sqrt{n^4+n^4}} = \frac{1}{\sqrt{n^4\cdot 2}} = \frac{1}{\sqrt{2}n^2}; \text{ and}$$

$$a_n > \frac{1}{\sqrt{n^8+n^8}} = \frac{1}{\sqrt{2n^8}} = \frac{1}{\sqrt{2}n^4}.$$

Now, since $\lim_{n\to\infty} \frac{1}{\sqrt{2}n^4} = \lim_{n\to\infty} \frac{1}{\sqrt{2}n^2} = 0$, the Squeeze Theorem for Sequences implies that $\lim_{n\to\infty} a_n = 0$.

69. $a_n = (2^n + 3^n)^{1/n}, \quad 3 \le a_n \le (2\cdot 3^n)^{1/n} = 2^{1/n}\cdot 3$

SOLUTION Clearly $2^n + 3^n \ge 3^n$ for all $n \ge 1$. Therefore:

$$(2^n + 3^n)^{1/n} \ge (3^n)^{1/n} = 3.$$

Also $2^n + 3^n \le 3^n + 3^n = 2\cdot 3^n$, so

$$(2^n + 3^n)^{1/n} \le (2\cdot 3^n)^{1/n} = 2^{1/n}\cdot 3.$$

Thus,

$$3 \le (2^n + 3^n)^{1/n} \le 2^{1/n}\cdot 3.$$

Because

$$\lim_{n\to\infty} 2^{1/n}\cdot 3 = 3\lim_{n\to\infty} 2^{1/n} = 3\cdot 1 = 3$$

and $\lim_{n\to\infty} 3 = 3$, the Squeeze Theorem for Sequences guarantees

$$\lim_{n\to\infty} (2^n + 3^n)^{1/n} = 3.$$

71. Which of the following statements is equivalent to the assertion $\lim_{n\to\infty} a_n = L$? Explain.

(a) For every $\epsilon > 0$, the interval $(L-\epsilon, L+\epsilon)$ contains at least one element of the sequence $\{a_n\}$.

(b) For every $\epsilon > 0$, the interval $(L-\epsilon, L+\epsilon)$ contains all but at most finitely many elements of the sequence $\{a_n\}$.

SOLUTION Statement (b) is equivalent to Definition 1 of the limit, since the assertion "$|a_n - L| < \epsilon$ for all $n > M$" means that $L - \epsilon < a_n < L + \epsilon$ for all $n > M$; that is, the interval $(L-\epsilon, L+\epsilon)$ contains all the elements a_n except (maybe) the finite number of elements $a_1, a_2, \ldots, a_M$.

Statement (a) is not equivalent to the assertion $\lim_{n\to\infty} a_n = L$. We show this, by considering the following sequence:

$$a_n = \begin{cases} \frac{1}{n} & \text{for odd } n \\ 1 + \frac{1}{n} & \text{for even } n \end{cases}$$

Clearly for every $\epsilon > 0$, the interval $(-\epsilon, \epsilon) = (L - \epsilon, L + \epsilon)$ for $L = 0$ contains at least one element of $\{a_n\}$, but the sequence diverges (rather than converges to $L = 0$). Since the terms in the odd places converge to 0 and the terms in the even places converge to 1. Hence, a_n does not approach one limit.

73. Show that $a_n = \dfrac{3n^2}{n^2+2}$ is increasing. Find an upper bound.

SOLUTION Let $f(x) = \frac{3x^2}{x^2+2}$. Then

$$f'(x) = \frac{6x(x^2+2) - 3x^2 \cdot 2x}{(x^2+2)^2} = \frac{12x}{(x^2+2)^2}.$$

$f'(x) > 0$ for $x > 0$, hence f is increasing on this interval. It follows that $a_n = f(n)$ is also increasing. We now show that $M = 3$ is an upper bound for a_n, by writing:

$$a_n = \frac{3n^2}{n^2+2} \le \frac{3n^2+6}{n^2+2} = \frac{3(n^2+2)}{n^2+2} = 3.$$

That is, $a_n \le 3$ for all n.

75. Give an example of a divergent sequence $\{a_n\}$ such that $\lim_{n\to\infty} |a_n|$ converges.

SOLUTION Let $a_n = (-1)^n$. The sequence $\{a_n\}$ diverges because the terms alternate between $+1$ and -1; however, the sequence $\{|a_n|\}$ converges because it is a constant sequence, all of whose terms are equal to 1.

77. Using the limit definition, prove that if $\{a_n\}$ converges and $\{b_n\}$ diverges, then $\{a_n + b_n\}$ diverges.

SOLUTION We will prove this result by contradiction. Suppose $\lim_{n\to\infty} a_n = L_1$ and that $\{a_n + b_n\}$ converges to a limit L_2. Now, let $\epsilon > 0$. Because $\{a_n\}$ converges to L_1 and $\{a_n + b_n\}$ converges to L_2, it follows that there exist numbers M_1 and M_2 such that:

$$|a_n - L_1| < \frac{\epsilon}{2} \qquad \text{for all } n > M_1,$$
$$|(a_n + b_n) - L_2| < \frac{\epsilon}{2} \qquad \text{for all } n > M_2.$$

Thus, for $n > M = \max\{M_1, M_2\}$,

$$|a_n - L_1| < \frac{\epsilon}{2} \quad \text{and} \quad |(a_n + b_n) - L_2| < \frac{\epsilon}{2}.$$

By the triangle inequality,

$$|b_n - (L_2 - L_1)| = |a_n + b_n - a_n - (L_2 - L_1)| = |(-a_n + L_1) + (a_n + b_n - L_2)|$$
$$\le |L_1 - a_n| + |a_n + b_n - L_2|.$$

Thus, for $n > M$,

$$|b_n - (L_2 - L_1)| < \frac{\epsilon}{2} + \frac{\epsilon}{2} = \epsilon;$$

that is, $\{b_n\}$ converges to $L_2 - L_1$, in contradiction to the given data. Thus, $\{a_n + b_n\}$ must diverge.

79. Theorem 1 states that if $\lim_{x\to\infty} f(x) = L$, then the sequence $a_n = f(n)$ converges and $\lim_{n\to\infty} a_n = L$. Show that the *converse* is false. In other words, find a function $f(x)$ such that $a_n = f(n)$ converges but $\lim_{x\to\infty} f(x)$ does not exist.

SOLUTION Let $f(x) = \sin \pi x$ and $a_n = \sin \pi n$. Then $a_n = f(n)$. Since $\sin \pi x$ is oscillating between -1 and 1 the limit $\lim_{x\to\infty} f(x)$ does not exist. However, the sequence $\{a_n\}$ is the constant sequence in which $a_n = \sin \pi n = 0$ for all n, hence it converges to zero.

81. Let $b_n = a_{n+1}$. Use the limit definition to prove that if $\{a_n\}$ converges, then $\{b_n\}$ also converges and $\lim_{n\to\infty} a_n = \lim_{n\to\infty} b_n$.

SOLUTION Suppose $\{a_n\}$ converges to L. Let $b_n = a_{n+1}$, and let $\epsilon > 0$. Because $\{a_n\}$ converges to L, there exists an M' such that $|a_n - L| < \epsilon$ for $n > M'$. Now, let $M = M' - 1$. Then, whenever $n > M$, $n + 1 > M + 1 = M'$. Thus, for $n > M$,

$$|b_n - L| = |a_{n+1} - L| < \epsilon.$$

Hence, $\{b_n\}$ converges to L.

83. Proceed as in Example 12 to show that the sequence $\sqrt{3}, \sqrt{3\sqrt{3}}, \sqrt{3\sqrt{3\sqrt{3}}}, \dots$ is increasing and bounded above by $M = 3$. Then prove that the limit exists and find its value.

SOLUTION This sequence is defined recursively by the formula:

$$a_{n+1} = \sqrt{3a_n}, \qquad a_1 = \sqrt{3}.$$

Consider the following inequalities:

$$\begin{aligned} a_2 &= \sqrt{3a_1} = \sqrt{3\sqrt{3}} > \sqrt{3} = a_1 &\Rightarrow\quad & a_2 > a_1; \\ a_3 &= \sqrt{3a_2} > \sqrt{3a_1} = a_2 &\Rightarrow\quad & a_3 > a_2; \\ a_4 &= \sqrt{3a_3} > \sqrt{3a_2} = a_3 &\Rightarrow\quad & a_4 > a_3. \end{aligned}$$

In general, if we assume that $a_k > a_{k-1}$, then

$$a_{k+1} = \sqrt{3a_k} > \sqrt{3a_{k-1}} = a_k.$$

Hence, by mathematical induction, $a_{n+1} > a_n$ for all n; that is, the sequence $\{a_n\}$ is increasing.

Because $a_{n+1} = \sqrt{3a_n}$, it follows that $a_n \geq 0$ for all n. Now, $a_1 = \sqrt{3} < 3$. If $a_k \leq 3$, then

$$a_{k+1} = \sqrt{3a_k} \leq \sqrt{3 \cdot 3} = 3.$$

Thus, by mathematical induction, $a_n \leq 3$ for all n.

Since $\{a_n\}$ is increasing and bounded, it follows by the Theorem on Bounded Monotonic Sequences that this sequence is converging. Denote the limit by $L = \lim_{n\to\infty} a_n$. Using Exercise 81, it follows that

$$L = \lim_{n\to\infty} a_{n+1} = \lim_{n\to\infty} \sqrt{3a_n} = \sqrt{3 \lim_{n\to\infty} a_n} = \sqrt{3L}.$$

Thus, $L^2 = 3L$, so $L = 0$ or $L = 3$. Because the sequence is increasing, we have $a_n \geq a_1 = \sqrt{3}$ for all n. Hence, the limit also satisfies $L \geq \sqrt{3}$. We conclude that the appropriate solution is $L = 3$; that is, $\lim_{n\to\infty} a_n = 3$.

Further Insights and Challenges

85. Show that $\lim_{n\to\infty} \sqrt[n]{n!} = \infty$. *Hint:* Verify that $n! \geq (n/2)^{n/2}$ by observing that half of the factors of $n!$ are greater than or equal to $n/2$.

SOLUTION We show that $n! \geq \left(\frac{n}{2}\right)^{n/2}$. For $n \geq 4$ even, we have:

$$n! = \underbrace{1 \cdot \dots \cdot \frac{n}{2}}_{\frac{n}{2}\text{ factors}} \cdot \underbrace{\left(\frac{n}{2}+1\right) \cdot \dots \cdot n}_{\frac{n}{2}\text{ factors}} \geq \underbrace{\left(\frac{n}{2}+1\right) \cdot \dots \cdot n}_{\frac{n}{2}\text{ factors}}.$$

Since each one of the $\frac{n}{2}$ factors is greater than $\frac{n}{2}$, we have:

$$n! \geq \underbrace{\left(\frac{n}{2}+1\right) \cdot \dots \cdot n}_{\frac{n}{2}\text{ factors}} \geq \underbrace{\frac{n}{2} \cdot \dots \cdot \frac{n}{2}}_{\frac{n}{2}\text{ factors}} = \left(\frac{n}{2}\right)^{n/2}.$$

For $n \geq 3$ odd, we have:

$$n! = \underbrace{1 \cdot \dots \cdot \frac{n-1}{2}}_{\frac{n-1}{2}\text{ factors}} \cdot \underbrace{\frac{n+1}{2} \cdot \dots \cdot n}_{\frac{n+1}{2}\text{ factors}} \geq \underbrace{\frac{n+1}{2} \cdot \dots \cdot n.}_{\frac{n+1}{2}\text{ factors}}$$

Since each one of the $\frac{n+1}{2}$ factors is greater than $\frac{n}{2}$, we have:

$$n! \geq \underbrace{\frac{n+1}{2} \cdot \cdots \cdot n}_{\frac{n+1}{2} \text{ factors}} \geq \underbrace{\frac{n}{2} \cdot \cdots \cdot \frac{n}{2}}_{\frac{n+1}{2} \text{ factors}} = \left(\frac{n}{2}\right)^{(n+1)/2} = \left(\frac{n}{2}\right)^{n/2} \sqrt{\frac{n}{2}} \geq \left(\frac{n}{2}\right)^{n/2}.$$

In either case we have $n! \geq \left(\frac{n}{2}\right)^{n/2}$. Thus,

$$\sqrt[n]{n!} \geq \sqrt{\frac{n}{2}}.$$

Since $\lim_{n\to\infty} \sqrt{\frac{n}{2}} = \infty$, it follows that $\lim_{n\to\infty} \sqrt[n]{n!} = \infty$. Thus, the sequence $a_n = \sqrt[n]{n!}$ diverges.

87. Given positive numbers $a_1 < b_1$, define two sequences recursively by

$$a_{n+1} = \sqrt{a_n b_n}, \qquad b_{n+1} = \frac{a_n + b_n}{2}$$

(a) Show that $a_n \leq b_n$ for all n (Figure 13).
(b) Show that $\{a_n\}$ is increasing and $\{b_n\}$ is decreasing.
(c) Show that $b_{n+1} - a_{n+1} \leq \dfrac{b_n - a_n}{2}$.
(d) Prove that both $\{a_n\}$ and $\{b_n\}$ converge and have the same limit. This limit, denoted $\text{AGM}(a_1, b_1)$, is called the **arithmetic-geometric mean** of a_1 and b_1.
(e) Estimate $\text{AGM}(1, \sqrt{2})$ to three decimal places.

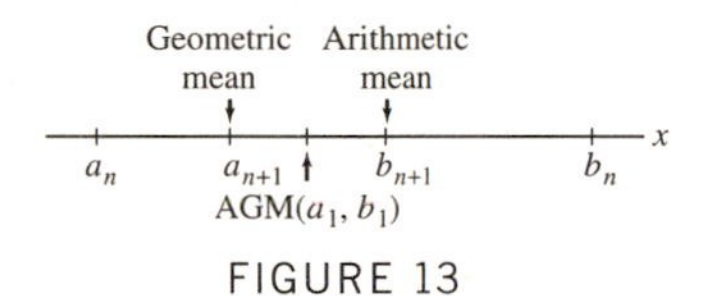

FIGURE 13

SOLUTION

(a) Examine the following:

$$b_{n+1} - a_{n+1} = \frac{a_n + b_n}{2} - \sqrt{a_n b_n} = \frac{a_n + b_n - 2\sqrt{a_n b_n}}{2} = \frac{\left(\sqrt{a_n}\right)^2 - 2\sqrt{a_n}\sqrt{b_n} + \left(\sqrt{b_n}\right)^2}{2}$$
$$= \frac{\left(\sqrt{a_n} - \sqrt{b_n}\right)^2}{2} \geq 0.$$

We conclude that $b_{n+1} \geq a_{n+1}$ for all $n > 1$. By the given information $b_1 > a_1$; hence, $b_n \geq a_n$ for all n.
(b) By part (a), $b_n \geq a_n$ for all n, so

$$a_{n+1} = \sqrt{a_n b_n} \geq \sqrt{a_n \cdot a_n} = \sqrt{a_n^2} = a_n$$

for all n. Hence, the sequence $\{a_n\}$ is increasing. Moreover, since $a_n \leq b_n$ for all n,

$$b_{n+1} = \frac{a_n + b_n}{2} \leq \frac{b_n + b_n}{2} = \frac{2b_n}{2} = b_n$$

for all n; that is, the sequence $\{b_n\}$ is decreasing.
(c) Since $\{a_n\}$ is increasing, $a_{n+1} \geq a_n$. Thus,

$$b_{n+1} - a_{n+1} \leq b_{n+1} - a_n = \frac{a_n + b_n}{2} - a_n = \frac{a_n + b_n - 2a_n}{2} = \frac{b_n - a_n}{2}.$$

Now, by part (a), $a_n \leq b_n$ for all n. By part (b), $\{b_n\}$ is decreasing. Hence $b_n \leq b_1$ for all n. Combining the two inequalities we conclude that $a_n \leq b_1$ for all n. That is, the sequence $\{a_n\}$ is increasing and bounded ($0 \leq a_n \leq b_1$). By the Theorem on Bounded Monotonic Sequences we conclude that $\{a_n\}$ converges. Similarly, since $\{a_n\}$ is increasing, $a_n \geq a_1$ for all n. We combine this inequality with $b_n \geq a_n$ to conclude that $b_n \geq a_1$ for all n. Thus, $\{b_n\}$ is decreasing and bounded ($a_1 \leq b_n \leq b_1$); hence this sequence converges.

To show that $\{a_n\}$ and $\{b_n\}$ converge to the same limit, note that

$$b_n - a_n \leq \frac{b_{n-1} - a_{n-1}}{2} \leq \frac{b_{n-2} - a_{n-2}}{2^2} \leq \cdots \leq \frac{b_1 - a_1}{2^{n-1}}.$$

Thus,

$$\lim_{n\to\infty} (b_n - a_n) = (b_1 - a_1) \lim_{n\to\infty} \frac{1}{2^{n-1}} = 0.$$

(d) We have

$$a_{n+1} = \sqrt{a_n b_n}, \quad a_1 = 1; \quad b_{n+1} = \frac{a_n + b_n}{2}, \quad b_1 = \sqrt{2}$$

Computing the values of a_n and b_n until the first three decimal digits are equal in successive terms, we obtain:

$$a_2 = \sqrt{a_1 b_1} = \sqrt{1 \cdot \sqrt{2}} = 1.1892$$

$$b_2 = \frac{a_1 + b_1}{2} = \frac{1 + \sqrt{2}}{2} = 1.2071$$

$$a_3 = \sqrt{a_2 b_2} = \sqrt{1.1892 \cdot 1.2071} = 1.1981$$

$$b_3 = \frac{a_2 + b_2}{2} = \frac{1.1892 \cdot 1.2071}{2} = 1.1981$$

$$a_4 = \sqrt{a_3 b_3} = 1.1981$$

$$b_4 = \frac{a_3 + b_3}{2} = 1.1981$$

Thus,

$$AGM\left(1, \sqrt{2}\right) \approx 1.198.$$

89. Let $a_n = H_n - \ln n$, where H_n is the nth harmonic number

$$H_n = 1 + \frac{1}{2} + \frac{1}{3} + \cdots + \frac{1}{n}$$

(a) Show that $a_n \geq 0$ for $n \geq 1$. *Hint:* Show that $H_n \geq \int_1^{n+1} \frac{dx}{x}$.

(b) Show that $\{a_n\}$ is decreasing by interpreting $a_n - a_{n+1}$ as an area.

(c) Prove that $\lim_{n\to\infty} a_n$ exists.

This limit, denoted γ, is known as *Euler's Constant*. It appears in many areas of mathematics, including analysis and number theory, and has been calculated to more than 100 million decimal places, but it is still not known whether γ is an irrational number. The first 10 digits are $\gamma \approx 0.5772156649$.

SOLUTION

(a) Since the function $y = \frac{1}{x}$ is decreasing, the left endpoint approximation to the integral $\int_1^{n+1} \frac{dx}{x}$ is greater than this integral; that is,

$$1 \cdot 1 + \frac{1}{2} \cdot 1 + \frac{1}{3} \cdot 1 + \cdots + \frac{1}{n} \cdot 1 \geq \int_1^{n+1} \frac{dx}{x}$$

or

$$H_n \geq \int_1^{n+1} \frac{dx}{x}.$$

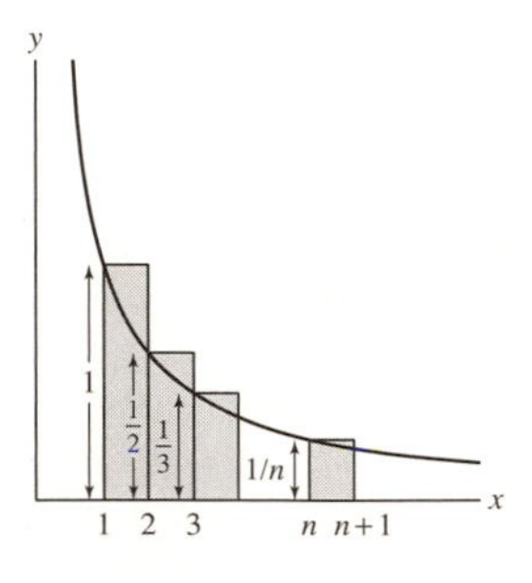

Moreover, since the function $y = \frac{1}{x}$ is positive for $x > 0$, we have:

$$\int_1^{n+1} \frac{dx}{x} \geq \int_1^{n} \frac{dx}{x}.$$

Thus,

$$H_n \geq \int_1^n \frac{dx}{x} = \ln x\Big|_1^n = \ln n - \ln 1 = \ln n,$$

and

$$a_n = H_n - \ln n \geq 0 \qquad \text{for all } n \geq 1.$$

(b) To show that $\{a_n\}$ is decreasing, we consider the difference $a_n - a_{n+1}$:

$$\begin{aligned} a_n - a_{n+1} &= H_n - \ln n - \big(H_{n+1} - \ln(n+1)\big) = H_n - H_{n+1} + \ln(n+1) - \ln n \\ &= 1 + \frac{1}{2} + \cdots + \frac{1}{n} - \left(1 + \frac{1}{2} + \cdots + \frac{1}{n} + \frac{1}{n+1}\right) + \ln(n+1) - \ln n \\ &= -\frac{1}{n+1} + \ln(n+1) - \ln n. \end{aligned}$$

Now, $\ln(n+1) - \ln n = \int_n^{n+1} \frac{dx}{x}$, whereas $\frac{1}{n+1}$ is the right endpoint approximation to the integral $\int_n^{n+1} \frac{dx}{x}$. Recalling $y = \frac{1}{x}$ is decreasing, it follows that

$$\int_n^{n+1} \frac{dx}{x} \geq \frac{1}{n+1}$$

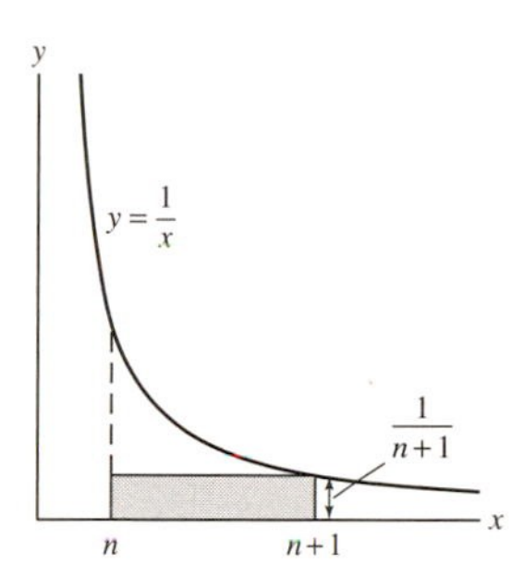

so

$$a_n - a_{n+1} \geq 0.$$

(c) By parts (a) and (b), $\{a_n\}$ is decreasing and 0 is a lower bound for this sequence. Hence $0 \leq a_n \leq a_1$ for all n. A monotonic and bounded sequence is convergent, so $\lim_{n\to\infty} a_n$ exists.

10.2 Summing an Infinite Series

Preliminary Questions

1. What role do partial sums play in defining the sum of an infinite series?

SOLUTION The sum of an infinite series is defined as the limit of the sequence of partial sums. If the limit of this sequence does not exist, the series is said to diverge.

2. What is the sum of the following infinite series?

$$\frac{1}{4} + \frac{1}{8} + \frac{1}{16} + \frac{1}{32} + \frac{1}{64} + \cdots$$

SOLUTION This is a geometric series with $c = \frac{1}{4}$ and $r = \frac{1}{2}$. The sum of the series is therefore

$$\frac{\frac{1}{4}}{1 - \frac{1}{2}} = \frac{\frac{1}{4}}{\frac{1}{2}} = \frac{1}{2}.$$

3. What happens if you apply the formula for the sum of a geometric series to the following series? Is the formula valid?

$$1 + 3 + 3^2 + 3^3 + 3^4 + \cdots$$

SOLUTION This is a geometric series with $c = 1$ and $r = 3$. Applying the formula for the sum of a geometric series then gives

$$\sum_{n=0}^{\infty} 3^n = \frac{1}{1-3} = -\frac{1}{2}.$$

Clearly, this is not valid: a series with all positive terms cannot have a negative sum. The formula is not valid in this case because a geometric series with $r = 3$ diverges.

4. Arvind asserts that $\sum_{n=1}^{\infty} \frac{1}{n^2} = 0$ because $\frac{1}{n^2}$ tends to zero. Is this valid reasoning?

SOLUTION Arvind's reasoning is not valid. Though the terms in the series do tend to zero, the general term in the sequence of partial sums,

$$S_n = 1 + \frac{1}{2^2} + \frac{1}{3^2} + \cdots + \frac{1}{n^2},$$

is clearly larger than 1. The sum of the series therefore cannot be zero.

5. Colleen claims that $\sum_{n=1}^{\infty} \frac{1}{\sqrt{n}}$ converges because

$$\lim_{n\to\infty} \frac{1}{\sqrt{n}} = 0$$

Is this valid reasoning?

SOLUTION Colleen's reasoning is not valid. Although the general term of a convergent series must tend to zero, a series whose general term tends to zero need not converge. In the case of $\sum_{n=1}^{\infty} \frac{1}{\sqrt{n}}$, the series diverges even though its general term tends to zero.

6. Find an N such that $S_N > 25$ for the series $\sum_{n=1}^{\infty} 2$.

SOLUTION The Nth partial sum of the series is:

$$S_N = \sum_{n=1}^{N} 2 = \underbrace{2 + \cdots + 2}_{N} = 2N.$$

7. Does there exist an N such that $S_N > 25$ for the series $\sum_{n=1}^{\infty} 2^{-n}$? Explain.

SOLUTION The series $\sum_{n=1}^{\infty} 2^{-n}$ is a convergent geometric series with the common ratio $r = \frac{1}{2}$. The sum of the series is:

$$S = \frac{\frac{1}{2}}{1 - \frac{1}{2}} = 1.$$

Notice that the sequence of partial sums $\{S_N\}$ is increasing and converges to 1; therefore $S_N \le 1$ for all N. Thus, there does not exist an N such that $S_N > 25$.

8. Give an example of a divergent infinite series whose general term tends to zero.

SOLUTION Consider the series $\sum_{n=1}^{\infty} \frac{1}{n^{\frac{9}{10}}}$. The general term tends to zero, since $\lim_{n\to\infty} \frac{1}{n^{\frac{9}{10}}} = 0$. However, the Nth partial sum satisfies the following inequality:

$$S_N = \frac{1}{1^{\frac{9}{10}}} + \frac{1}{2^{\frac{9}{10}}} + \cdots + \frac{1}{N^{\frac{9}{10}}} \ge \frac{N}{N^{\frac{9}{10}}} = N^{1-\frac{9}{10}} = N^{\frac{1}{10}}.$$

That is, $S_N \ge N^{\frac{1}{10}}$ for all N. Since $\lim_{N\to\infty} N^{\frac{1}{10}} = \infty$, the sequence of partial sums S_n diverges; hence, the series $\sum_{n=1}^{\infty} \frac{1}{n^{\frac{9}{10}}}$ diverges.

Exercises

1. Find a formula for the general term a_n (not the partial sum) of the infinite series.

(a) $\frac{1}{3}+\frac{1}{9}+\frac{1}{27}+\frac{1}{81}+\cdots$

(b) $\frac{1}{1}+\frac{5}{2}+\frac{25}{4}+\frac{125}{8}+\cdots$

(c) $\frac{1}{1}-\frac{2^2}{2\cdot 1}+\frac{3^3}{3\cdot 2\cdot 1}-\frac{4^4}{4\cdot 3\cdot 2\cdot 1}+\cdots$

(d) $\frac{2}{1^2+1}+\frac{1}{2^2+1}+\frac{2}{3^2+1}+\frac{1}{4^2+1}+\cdots$

SOLUTION

(a) The denominators of the terms are powers of 3, starting with the first power. Hence, the general term is:

$$a_n = \frac{1}{3^n}.$$

(b) The numerators are powers of 5, and the denominators are the same powers of 2. The first term is $a_1 = 1$ so,

$$a_n = \left(\frac{5}{2}\right)^{n-1}.$$

(c) The general term of this series is,

$$a_n = (-1)^{n+1}\frac{n^n}{n!}.$$

(d) Notice that the numerators of a_n equal 2 for odd values of n and 1 for even values of n. Thus,

$$a_n = \begin{cases} \dfrac{2}{n^2+1} & \text{odd } n \\ \dfrac{1}{n^2+1} & \text{even } n \end{cases}$$

The formula can also be rewritten as follows:

$$a_n = \frac{1+\frac{(-1)^{n+1}+1}{2}}{n^2+1}.$$

In Exercises 3–6, compute the partial sums S_2, S_4, and S_6.

3. $1+\frac{1}{2^2}+\frac{1}{3^2}+\frac{1}{4^2}+\cdots$

SOLUTION

$$S_2 = 1+\frac{1}{2^2} = \frac{5}{4};$$

$$S_4 = 1+\frac{1}{2^2}+\frac{1}{3^2}+\frac{1}{4^2} = \frac{205}{144};$$

$$S_6 = 1+\frac{1}{2^2}+\frac{1}{3^2}+\frac{1}{4^2}+\frac{1}{5^2}+\frac{1}{6^2} = \frac{5369}{3600}.$$

5. $\frac{1}{1\cdot 2}+\frac{1}{2\cdot 3}+\frac{1}{3\cdot 4}+\cdots$

SOLUTION

$$S_2 = \frac{1}{1\cdot 2}+\frac{1}{2\cdot 3} = \frac{1}{2}+\frac{1}{6} = \frac{4}{6} = \frac{2}{3};$$

$$S_4 = S_2+a_3+a_4 = \frac{2}{3}+\frac{1}{3\cdot 4}+\frac{1}{4\cdot 5} = \frac{2}{3}+\frac{1}{12}+\frac{1}{20} = \frac{4}{5};$$

$$S_6 = S_4+a_5+a_6 = \frac{4}{5}+\frac{1}{5\cdot 6}+\frac{1}{6\cdot 7} = \frac{4}{5}+\frac{1}{30}+\frac{1}{42} = \frac{6}{7}.$$

7. The series $S = 1 + \left(\frac{1}{5}\right) + \left(\frac{1}{5}\right)^2 + \left(\frac{1}{5}\right)^3 + \cdots$ converges to $\frac{5}{4}$. Calculate S_N for $N = 1, 2, \ldots$ until you find an S_N that approximates $\frac{5}{4}$ with an error less than 0.0001.

SOLUTION

$$S_1 = 1$$
$$S_2 = 1 + \frac{1}{5} = \frac{6}{5} = 1.2$$
$$S_3 = 1 + \frac{1}{5} + \frac{1}{25} = \frac{31}{25} = 1.24$$
$$S_3 = 1 + \frac{1}{5} + \frac{1}{25} + \frac{1}{125} = \frac{156}{125} = 1.248$$
$$S_4 = 1 + \frac{1}{5} + \frac{1}{25} + \frac{1}{125} + \frac{1}{625} = \frac{781}{625} = 1.2496$$
$$S_5 = 1 + \frac{1}{5} + \frac{1}{25} + \frac{1}{125} + \frac{1}{625} + \frac{1}{3125} = \frac{3906}{3125} = 1.24992$$

Note that

$$1.25 - S_5 = 1.25 - 1.24992 = 0.00008 < 0.0001$$

In Exercises 9 and 10, use a computer algebra system to compute S_{10}, S_{100}, S_{500}, and S_{1000} for the series. Do these values suggest convergence to the given value?

9. CAS

$$\frac{\pi - 3}{4} = \frac{1}{2 \cdot 3 \cdot 4} - \frac{1}{4 \cdot 5 \cdot 6} + \frac{1}{6 \cdot 7 \cdot 8} - \frac{1}{8 \cdot 9 \cdot 10} + \cdots$$

SOLUTION Write

$$a_n = \frac{(-1)^{n+1}}{2n \cdot (2n+1) \cdot (2n+2)}$$

Then

$$S_N = \sum_{i=1}^{N} a_n$$

Computing, we find

$$\frac{\pi - 3}{4} \approx 0.0353981635$$
$$S_{10} \approx 0.03535167962$$
$$S_{100} \approx 0.03539810274$$
$$S_{500} \approx 0.03539816290$$
$$S_{1000} \approx 0.03539816334$$

It appears that $S_N \to \frac{\pi-3}{4}$.

11. Calculate S_3, S_4, and S_5 and then find the sum of the telescoping series

$$S = \sum_{n=1}^{\infty} \left(\frac{1}{n+1} - \frac{1}{n+2} \right)$$

SOLUTION

$$S_3 = \left(\frac{1}{2} - \frac{1}{3}\right) + \left(\frac{1}{3} - \frac{1}{4}\right) + \left(\frac{1}{4} - \frac{1}{5}\right) = \frac{1}{2} - \frac{1}{5} = \frac{3}{10};$$
$$S_4 = S_3 + \left(\frac{1}{5} - \frac{1}{6}\right) = \frac{1}{2} - \frac{1}{6} = \frac{1}{3};$$
$$S_5 = S_4 + \left(\frac{1}{6} - \frac{1}{7}\right) = \frac{1}{2} - \frac{1}{7} = \frac{5}{14}.$$

The general term in the sequence of partial sums is

$$S_N = \left(\frac{1}{2} - \frac{1}{3}\right) + \left(\frac{1}{3} - \frac{1}{4}\right) + \left(\frac{1}{4} - \frac{1}{5}\right) + \cdots + \left(\frac{1}{N+1} - \frac{1}{N+2}\right) = \frac{1}{2} - \frac{1}{N+2};$$

thus,

$$S = \lim_{N\to\infty} S_N = \lim_{N\to\infty} \left(\frac{1}{2} - \frac{1}{N+2}\right) = \frac{1}{2}.$$

The sum of the telescoping series is therefore $\frac{1}{2}$.

13. Calculate S_3, S_4, and S_5 and then find the sum $S = \sum_{n=1}^{\infty} \frac{1}{4n^2 - 1}$ using the identity

$$\frac{1}{4n^2 - 1} = \frac{1}{2}\left(\frac{1}{2n-1} - \frac{1}{2n+1}\right)$$

SOLUTION

$$S_3 = \frac{1}{2}\left(\frac{1}{1} - \frac{1}{3}\right) + \frac{1}{2}\left(\frac{1}{3} - \frac{1}{5}\right) + \frac{1}{2}\left(\frac{1}{5} - \frac{1}{7}\right) = \frac{1}{2}\left(1 - \frac{1}{7}\right) = \frac{3}{7};$$

$$S_4 = S_3 + \frac{1}{2}\left(\frac{1}{7} - \frac{1}{9}\right) = \frac{1}{2}\left(1 - \frac{1}{9}\right) = \frac{4}{9};$$

$$S_5 = S_4 + \frac{1}{2}\left(\frac{1}{9} - \frac{1}{11}\right) = \frac{1}{2}\left(1 - \frac{1}{11}\right) = \frac{5}{11}.$$

The general term in the sequence of partial sums is

$$S_N = \frac{1}{2}\left(\frac{1}{1} - \frac{1}{3}\right) + \frac{1}{2}\left(\frac{1}{3} - \frac{1}{5}\right) + \frac{1}{2}\left(\frac{1}{5} - \frac{1}{7}\right) + \cdots + \frac{1}{2}\left(\frac{1}{2N-1} - \frac{1}{2N+1}\right) = \frac{1}{2}\left(1 - \frac{1}{2N+1}\right);$$

thus,

$$S = \lim_{N\to\infty} S_N = \lim_{N\to\infty} \frac{1}{2}\left(1 - \frac{1}{2N+1}\right) = \frac{1}{2}.$$

15. Find the sum of $\frac{1}{1\cdot 3} + \frac{1}{3\cdot 5} + \frac{1}{5\cdot 7} + \cdots$.

SOLUTION We may write this sum as

$$\sum_{n=1}^{\infty} \frac{1}{(2n-1)(2n+1)} = \sum_{n=1}^{\infty} \frac{1}{2}\left(\frac{1}{2n-1} - \frac{1}{2n+1}\right).$$

The general term in the sequence of partial sums is

$$S_N = \frac{1}{2}\left(\frac{1}{1} - \frac{1}{3}\right) + \frac{1}{2}\left(\frac{1}{3} - \frac{1}{5}\right) + \frac{1}{2}\left(\frac{1}{5} - \frac{1}{7}\right) + \cdots + \frac{1}{2}\left(\frac{1}{2N-1} - \frac{1}{2N+1}\right) = \frac{1}{2}\left(1 - \frac{1}{2N+1}\right);$$

thus,

$$\lim_{N\to\infty} S_N = \lim_{N\to\infty} \frac{1}{2}\left(1 - \frac{1}{2N+1}\right) = \frac{1}{2},$$

and

$$\sum_{n=1}^{\infty} \frac{1}{(2n-1)(2n+1)} = \frac{1}{2}.$$

In Exercises 17–22, use Theorem 3 to prove that the following series diverge.

17. $\sum_{n=1}^{\infty} \frac{n}{10n + 12}$

SOLUTION The general term, $\frac{n}{10n+12}$, has limit

$$\lim_{n\to\infty} \frac{n}{10n+12} = \lim_{n\to\infty} \frac{1}{10 + (12/n)} = \frac{1}{10}$$

Since the general term does not tend to zero, the series diverges.

19. $\frac{0}{1} - \frac{1}{2} + \frac{2}{3} - \frac{3}{4} + \cdots$

SOLUTION The general term $a_n = (-1)^{n-1}\frac{n-1}{n}$ does not tend to zero. In fact, because $\lim_{n\to\infty}\frac{n-1}{n} = 1$, $\lim_{n\to\infty} a_n$ does not exist. By Theorem 3, we conclude that the given series diverges.

21. $\cos\frac{1}{2} + \cos\frac{1}{3} + \cos\frac{1}{4} + \cdots$

SOLUTION The general term $a_n = \cos\frac{1}{n+1}$ tends to 1, not zero. By Theorem 3, we conclude that the given series diverges.

In Exercises 23–36, use the formula for the sum of a geometric series to find the sum or state that the series diverges.

23. $\frac{1}{1} + \frac{1}{8} + \frac{1}{8^2} + \cdots$

SOLUTION This is a geometric series with $c = 1$ and $r = \frac{1}{8}$, so its sum is

$$\frac{1}{1 - \frac{1}{8}} = \frac{1}{7/8} = \frac{8}{7}$$

25. $\sum_{n=3}^{\infty} \left(\frac{3}{11}\right)^{-n}$

SOLUTION Rewrite this series as

$$\sum_{n=3}^{\infty} \left(\frac{11}{3}\right)^n$$

This is a geometric series with $r = \frac{11}{3} > 1$, so it is divergent.

27. $\sum_{n=-4}^{\infty} \left(-\frac{4}{9}\right)^n$

SOLUTION This is a geometric series with $c = 1$ and $r = -\frac{4}{9}$, starting at $n = -4$. Its sum is thus

$$\frac{cr^{-4}}{1-r} = \frac{c}{r^4 - r^5} = \frac{1}{\frac{4^4}{9^4} + \frac{4^5}{9^5}} = \frac{9^5}{9 \cdot 4^4 + 4^5} = \frac{59{,}049}{3328}$$

29. $\sum_{n=1}^{\infty} e^{-n}$

SOLUTION Rewrite the series as

$$\sum_{n=1}^{\infty} \left(\frac{1}{e}\right)^n$$

to recognize it as a geometric series with $c = \frac{1}{e}$ and $r = \frac{1}{e}$. Thus,

$$\sum_{n=1}^{\infty} e^{-n} = \frac{\frac{1}{e}}{1 - \frac{1}{e}} = \frac{1}{e-1}.$$

31. $\sum_{n=0}^{\infty} \frac{8 + 2^n}{5^n}$

SOLUTION Rewrite the series as

$$\sum_{n=0}^{\infty} \frac{8}{5^n} + \sum_{n=0}^{\infty} \frac{2^n}{5^n} = \sum_{n=0}^{\infty} 8 \cdot \left(\frac{1}{5}\right)^n + \sum_{n=0}^{\infty} \left(\frac{2}{5}\right)^n,$$

which is a sum of two geometric series. The first series has $c = 8\left(\frac{1}{5}\right)^0 = 8$ and $r = \frac{1}{5}$; the second has $c = \left(\frac{2}{5}\right)^0 = 1$ and $r = \frac{2}{5}$. Thus,

$$\sum_{n=0}^{\infty} 8 \cdot \left(\frac{1}{5}\right)^n = \frac{8}{1 - \frac{1}{5}} = \frac{8}{\frac{4}{5}} = 10,$$

$$\sum_{n=0}^{\infty} \left(\frac{2}{5}\right)^n = \frac{1}{1 - \frac{2}{5}} = \frac{1}{\frac{3}{5}} = \frac{5}{3},$$

and

$$\sum_{n=0}^{\infty} \frac{8+2^n}{5^n} = 10 + \frac{5}{3} = \frac{35}{3}.$$

33. $5 - \frac{5}{4} + \frac{5}{4^2} - \frac{5}{4^3} + \cdots$

SOLUTION This is a geometric series with $c = 5$ and $r = -\frac{1}{4}$. Thus,

$$\sum_{n=0}^{\infty} 5 \cdot \left(-\frac{1}{4}\right)^n = \frac{5}{1-\left(-\frac{1}{4}\right)} = \frac{5}{1+\frac{1}{4}} = \frac{5}{\frac{5}{4}} = 4.$$

35. $\frac{7}{8} - \frac{49}{64} + \frac{343}{512} - \frac{2401}{4096} + \cdots$

SOLUTION This is a geometric series with $c = \frac{7}{8}$ and $r = -\frac{7}{8}$. Thus,

$$\sum_{n=0}^{\infty} \frac{7}{8} \cdot \left(-\frac{7}{8}\right)^n = \frac{\frac{7}{8}}{1-\left(-\frac{7}{8}\right)} = \frac{\frac{7}{8}}{\frac{15}{8}} = \frac{7}{15}.$$

37. Which of the following are *not* geometric series?

(a) $\displaystyle\sum_{n=0}^{\infty} \frac{7^n}{29^n}$

(b) $\displaystyle\sum_{n=3}^{\infty} \frac{1}{n^4}$

(c) $\displaystyle\sum_{n=0}^{\infty} \frac{n^2}{2^n}$

(d) $\displaystyle\sum_{n=5}^{\infty} \pi^{-n}$

SOLUTION

(a) $\displaystyle\sum_{n=0}^{\infty} \frac{7^n}{29^n} = \sum_{n=0}^{\infty} \left(\frac{7}{29}\right)^n$: this is a geometric series with common ratio $r = \frac{7}{29}$.

(b) The ratio between two successive terms is

$$\frac{a_{n+1}}{a_n} = \frac{\frac{1}{(n+1)^4}}{\frac{1}{n^4}} = \frac{n^4}{(n+1)^4} = \left(\frac{n}{n+1}\right)^4.$$

This ratio is not constant since it depends on n. Hence, the series $\displaystyle\sum_{n=3}^{\infty} \frac{1}{n^4}$ is not a geometric series.

(c) The ratio between two successive terms is

$$\frac{a_{n+1}}{a_n} = \frac{\frac{(n+1)^2}{2^{n+1}}}{\frac{n^2}{2^n}} = \frac{(n+1)^2}{n^2} \cdot \frac{2^n}{2^{n+1}} = \left(1 + \frac{1}{n}\right)^2 \cdot \frac{1}{2}.$$

This ratio is not constant since it depends on n. Hence, the series $\displaystyle\sum_{n=0}^{\infty} \frac{n^2}{2^n}$ is not a geometric series.

(d) $\displaystyle\sum_{n=5}^{\infty} \pi^{-n} = \sum_{n=5}^{\infty} \left(\frac{1}{\pi}\right)^n$: this is a geometric series with common ratio $r = \frac{1}{\pi}$.

39. Prove that if $\displaystyle\sum_{n=1}^{\infty} a_n$ converges and $\displaystyle\sum_{n=1}^{\infty} b_n$ diverges, then $\displaystyle\sum_{n=1}^{\infty} (a_n + b_n)$ diverges. *Hint:* If not, derive a contradiction by writing

$$\sum_{n=1}^{\infty} b_n = \sum_{n=1}^{\infty} (a_n + b_n) - \sum_{n=1}^{\infty} a_n$$

SOLUTION Suppose to the contrary that $\sum_{n=1}^{\infty} a_n$ converges, $\sum_{n=1}^{\infty} b_n$ diverges, but $\sum_{n=1}^{\infty}(a_n + b_n)$ converges. Then by the Linearity of Infinite Series, we have

$$\sum_{n=1}^{\infty} b_n = \sum_{n=1}^{\infty}(a_n + b_n) - \sum_{n=1}^{\infty} a_n$$

so that $\sum_{n=1}^{\infty} b_n$ converges, a contradiction.

41. Give a counterexample to show that each of the following statements is false.

(a) If the general term a_n tends to zero, then $\sum_{n=1}^{\infty} a_n = 0$.

(b) The Nth partial sum of the infinite series defined by $\{a_n\}$ is a_N.

(c) If a_n tends to zero, then $\sum_{n=1}^{\infty} a_n$ converges.

(d) If a_n tends to L, then $\sum_{n=1}^{\infty} a_n = L$.

SOLUTION

(a) Let $a_n = 2^{-n}$. Then $\lim_{n\to\infty} a_n = 0$, but a_n is a geometric series with $c = 2^0 = 1$ and $r = 1/2$, so its sum is $\dfrac{1}{1-(1/2)} = 2$.

(b) Let $a_n = 1$. Then the nth partial sum is $a_1 + a_2 + \cdots + a_n = n$ while $a_n = 1$.

(c) Let $a_n = \dfrac{1}{\sqrt{n}}$. An example in the text shows that while a_n tends to zero, the sum $\sum_{n=1}^{\infty} a_n$ does not converge.

(d) Let $a_n = 1$. Then clearly a_n tends to $L = 1$, while the series $\sum_{n=1}^{\infty} a_n$ obviously diverges.

43. Compute the total area of the (infinitely many) triangles in Figure 4.

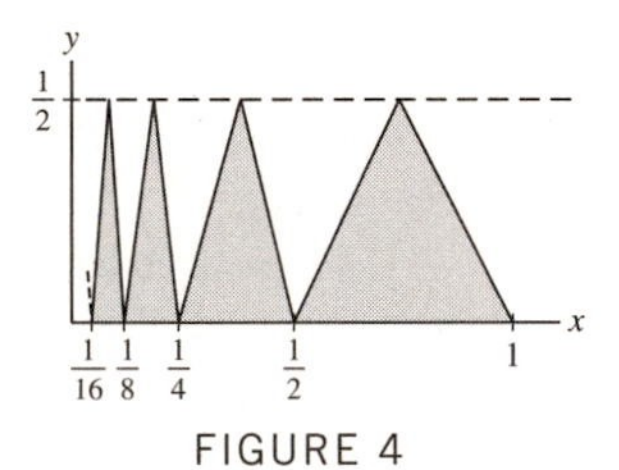

FIGURE 4

SOLUTION The area of a triangle with base B and height H is $A = \frac{1}{2}BH$. Because all of the triangles in Figure 4 have height $\frac{1}{2}$, the area of each triangle equals one-quarter of the base. Now, for $n \geq 0$, the nth triangle has a base which extends from $x = \frac{1}{2^{n+1}}$ to $x = \frac{1}{2^n}$. Thus,

$$B = \frac{1}{2^n} - \frac{1}{2^{n+1}} = \frac{1}{2^{n+1}} \quad \text{and} \quad A = \frac{1}{4}B = \frac{1}{2^{n+3}}.$$

The total area of the triangles is then given by the geometric series

$$\sum_{n=0}^{\infty} \frac{1}{2^{n+3}} = \sum_{n=0}^{\infty} \frac{1}{8}\left(\frac{1}{2}\right)^n = \frac{\frac{1}{8}}{1-\frac{1}{2}} = \frac{1}{4}.$$

45. Find the total length of the infinite zigzag path in Figure 5 (each zag occurs at an angle of $\frac{\pi}{4}$).

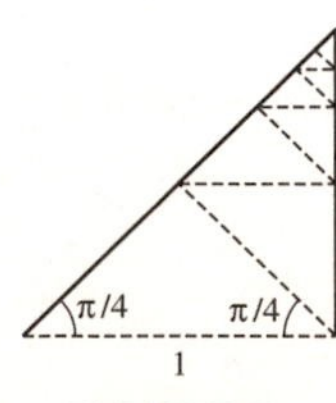

FIGURE 5

SOLUTION Because the angle at the lower left in Figure 5 has measure $\frac{\pi}{4}$ and each zag in the path occurs at an angle of $\frac{\pi}{4}$, every triangle in the figure is an isosceles right triangle. Accordingly, the length of each new segment in the path is $\frac{1}{\sqrt{2}}$ times the length of the previous segment. Since the first segment has length 1, the total length of the path is

$$\sum_{n=0}^{\infty}\left(\frac{1}{\sqrt{2}}\right)^n = \frac{1}{1-\frac{1}{\sqrt{2}}} = \frac{\sqrt{2}}{\sqrt{2}-1} = 2+\sqrt{2}.$$

47. Show that if a is a positive integer, then

$$\sum_{n=1}^{\infty}\frac{1}{n(n+a)} = \frac{1}{a}\left(1+\frac{1}{2}+\cdots+\frac{1}{a}\right)$$

SOLUTION By partial fraction decomposition

$$\frac{1}{n\,(n+a)} = \frac{A}{n}+\frac{B}{n+a};$$

clearing the denominators gives

$$1 = A(n+a)+Bn.$$

Setting $n=0$ then yields $A=\frac{1}{a}$, while setting $n=-a$ yields $B=-\frac{1}{a}$. Thus,

$$\frac{1}{n\,(n+a)} = \frac{\frac{1}{a}}{n}-\frac{\frac{1}{a}}{n+a} = \frac{1}{a}\left(\frac{1}{n}-\frac{1}{n+a}\right),$$

and

$$\sum_{n=1}^{\infty}\frac{1}{n(n+a)} = \sum_{n=1}^{\infty}\frac{1}{a}\left(\frac{1}{n}-\frac{1}{n+a}\right).$$

For $N>a$, the Nth partial sum is

$$S_N = \frac{1}{a}\left(1+\frac{1}{2}+\frac{1}{3}+\cdots+\frac{1}{a}\right)-\frac{1}{a}\left(\frac{1}{N+1}+\frac{1}{N+2}+\frac{1}{N+3}+\cdots+\frac{1}{N+a}\right).$$

Thus,

$$\sum_{n=1}^{\infty}\frac{1}{n(n+a)} = \lim_{N\to\infty}S_N = \frac{1}{a}\left(1+\frac{1}{2}+\frac{1}{3}+\cdots+\frac{1}{a}\right).$$

49. Let $\{b_n\}$ be a sequence and let $a_n = b_n - b_{n-1}$. Show that $\sum_{n=1}^{\infty} a_n$ converges if and only if $\lim_{n\to\infty} b_n$ exists.

SOLUTION Let $a_n = b_n - b_{n-1}$. The general term in the sequence of partial sums for the series $\sum_{n=1}^{\infty} a_n$ is then

$$S_N = (b_1-b_0)+(b_2-b_1)+(b_3-b_2)+\cdots+(b_N-b_{N-1}) = b_N - b_0.$$

Now, if $\lim_{N\to\infty} b_N$ exists, then so does $\lim_{N\to\infty} S_N$ and $\sum_{n=1}^{\infty} a_n$ converges. On the other hand, if $\sum_{n=1}^{\infty} a_n$ converges, then $\lim_{N\to\infty} S_N$ exists, which implies that $\lim_{N\to\infty} b_N$ also exists. Thus, $\sum_{n=1}^{\infty} a_n$ converges if and only if $\lim_{n\to\infty} b_n$ exists.

Further Insights and Challenges

Exercises 51–53 use the formula

$$1+r+r^2+\cdots+r^{N-1} = \frac{1-r^N}{1-r} \qquad \boxed{7}$$

51. Professor George Andrews of Pennsylvania State University observed that we can use Eq. (7) to calculate the derivative of $f(x)=x^N$ (for $N\geq 0$). Assume that $a\neq 0$ and let $x=ra$. Show that

$$f'(a) = \lim_{x\to a}\frac{x^N-a^N}{x-a} = a^{N-1}\lim_{r\to 1}\frac{r^N-1}{r-1}$$

and evaluate the limit.

SOLUTION According to the definition of derivative of $f(x)$ at $x = a$

$$f'(a) = \lim_{x \to a} \frac{x^N - a^N}{x - a}.$$

Now, let $x = ra$. Then $x \to a$ if and only if $r \to 1$, and

$$f'(a) = \lim_{x \to a} \frac{x^N - a^N}{x - a} = \lim_{r \to 1} \frac{(ra)^N - a^N}{ra - a} = \lim_{r \to 1} \frac{a^N \left(r^N - 1\right)}{a(r-1)} = a^{N-1} \lim_{r \to 1} \frac{r^N - 1}{r - 1}.$$

By Eq. (7) for a geometric sum,

$$\frac{1 - r^N}{1 - r} = \frac{r^N - 1}{r - 1} = 1 + r + r^2 + \cdots + r^{N-1},$$

so

$$\lim_{r \to 1} \frac{r^N - 1}{r - 1} = \lim_{r \to 1} \left(1 + r + r^2 + \cdots + r^{N-1}\right) = 1 + 1 + 1^2 + \cdots + 1^{N-1} = N.$$

Therefore, $f'(a) = a^{N-1} \cdot N = Na^{N-1}$

53. Verify the Gregory–Leibniz formula as follows.

(a) Set $r = -x^2$ in Eq. (7) and rearrange to show that

$$\frac{1}{1 + x^2} = 1 - x^2 + x^4 - \cdots + (-1)^{N-1} x^{2N-2} + \frac{(-1)^N x^{2N}}{1 + x^2}$$

(b) Show, by integrating over [0, 1], that

$$\frac{\pi}{4} = 1 - \frac{1}{3} + \frac{1}{5} - \frac{1}{7} + \cdots + \frac{(-1)^{N-1}}{2N - 1} + (-1)^N \int_0^1 \frac{x^{2N}\, dx}{1 + x^2}$$

(c) Use the Comparison Theorem for integrals to prove that

$$0 \le \int_0^1 \frac{x^{2N}\, dx}{1 + x^2} \le \frac{1}{2N + 1}$$

Hint: Observe that the integrand is $\le x^{2N}$.

(d) Prove that

$$\frac{\pi}{4} = 1 - \frac{1}{3} + \frac{1}{5} - \frac{1}{7} + \frac{1}{9} - \cdots$$

Hint: Use (b) and (c) to show that the partial sums S_N of satisfy $\left|S_N - \frac{\pi}{4}\right| \le \frac{1}{2N+1}$, and thereby conclude that $\lim_{N \to \infty} S_N = \frac{\pi}{4}$.

SOLUTION

(a) Start with Eq. (7), and substitute $-x^2$ for r:

$$1 + r + r^2 + \cdots + r^{N-1} = \frac{1 - r^N}{1 - r}$$

$$1 - x^2 + x^4 + \cdots + (-1)^{N-1} x^{2N-2} = \frac{1 - (-1)^N x^{2N}}{1 - (-x^2)}$$

$$1 - x^2 + x^4 + \cdots + (-1)^{N-1} x^{2N-2} = \frac{1}{1 + x^2} - \frac{(-1)^N x^{2N}}{1 + x^2}$$

$$\frac{1}{1 + x^2} = 1 - x^2 + x^4 + \cdots + (-1)^{N-1} x^{2N-2} + \frac{(-1)^N x^{2N}}{1 + x^2}$$

(b) The integrals of both sides must be equal. Now,

$$\int_0^1 \frac{1}{1 + x^2}\, dx = \tan^{-1} x \Big|_0^1 = \tan^{-1} 1 - \tan^{-1} 0 = \frac{\pi}{4}$$

while

$$\int_0^1 \left(1 - x^2 + x^4 + \cdots + (-1)^{N-1} x^{2N-2} + \frac{(-1)^N x^{2N}}{1 + x^2}\right) dx$$

$$= \left(x - \frac{1}{3}x^3 + \frac{1}{5}x^5 + \cdots + (-1)^{N-1}\frac{1}{2N-1}x^{2N-1}\right) + (-1)^N \int_0^1 \frac{x^{2N}\,dx}{1+x^2}$$

$$= 1 - \frac{1}{3} + \frac{1}{5} + \cdots + (-1)^{N-1}\frac{1}{2N-1} + (-1)^N \int_0^1 \frac{x^{2N}\,dx}{1+x^2}$$

(c) Note that for $x \in [0, 1]$, we have $1 + x^2 \geq 1$, so that

$$0 \leq \frac{x^{2N}}{1+x^2} \leq x^{2N}$$

By the Comparison Theorem for integrals, we then see that

$$0 \leq \int_0^1 \frac{x^{2N}\,dx}{1+x^2} \leq \int_0^1 x^{2N}\,dx = \frac{1}{2N+1}x^{2N+1}\bigg|_0^1 = \frac{1}{2N+1}$$

(d) Write

$$a_n = (-1)^n \frac{1}{2n-1}, \quad n \geq 1$$

and let S_N be the partial sums. Then

$$\left|S_N - \frac{\pi}{4}\right| = \left|(-1)^N \int_0^1 \frac{x^{2N}\,dx}{1+x^2}\right| = \int_0^1 \frac{x^{2N}\,dx}{1+x^2} \leq \frac{1}{2N+1}$$

Thus $\lim_{N\to\infty} S_N = \frac{\pi}{4}$ so that

$$\frac{\pi}{4} = 1 - \frac{1}{3} + \frac{1}{5} - \frac{1}{7} + \frac{1}{9} - \ldots$$

55. The **Koch snowflake** (described in 1904 by Swedish mathematician Helge von Koch) is an infinitely jagged "fractal" curve obtained as a limit of polygonal curves (it is continuous but has no tangent line at any point). Begin with an equilateral triangle (stage 0) and produce stage 1 by replacing each edge with four edges of one-third the length, arranged as in Figure 8. Continue the process: At the nth stage, replace each edge with four edges of one-third the length.
(a) Show that the perimeter P_n of the polygon at the nth stage satisfies $P_n = \frac{4}{3}P_{n-1}$. Prove that $\lim_{n\to\infty} P_n = \infty$. The snowflake has infinite length.
(b) Let A_0 be the area of the original equilateral triangle. Show that $(3)4^{n-1}$ new triangles are added at the nth stage, each with area $A_0/9^n$ (for $n \geq 1$). Show that the total area of the Koch snowflake is $\frac{8}{5}A_0$.

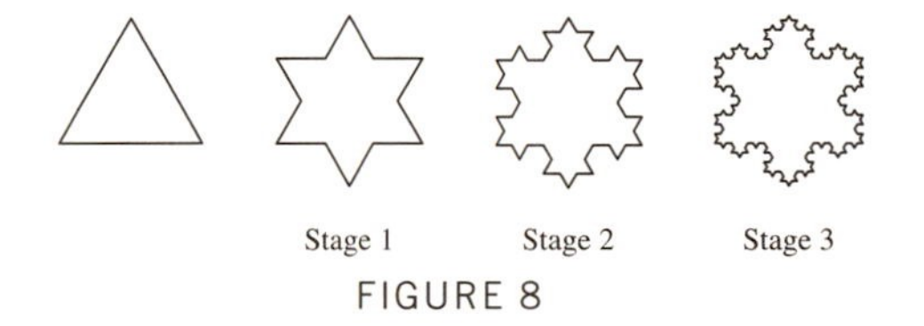

FIGURE 8

SOLUTION
(a) Each edge of the polygon at the $(n-1)$st stage is replaced by four edges of one-third the length; hence the perimeter of the polygon at the nth stage is $\frac{4}{3}$ times the perimeter of the polygon at the $(n-1)$th stage. That is, $P_n = \frac{4}{3}P_{n-1}$. Thus,

$$P_1 = \frac{4}{3}P_0; \quad P_2 = \frac{4}{3}P_1 = \left(\frac{4}{3}\right)^2 P_0, \quad P_3 = \frac{4}{3}P_2 = \left(\frac{4}{3}\right)^3 P_0,$$

and, in general, $P_n = \left(\frac{4}{3}\right)^n P_0$. As $n \to \infty$, it follows that

$$\lim_{n\to\infty} P_n = P_0 \lim_{n\to\infty} \left(\frac{4}{3}\right)^n = \infty.$$

(b) When each edge is replaced by four edges of one-third the length, one new triangle is created. At the $(n-1)$st stage, there are $3 \cdot 4^{n-1}$ edges in the snowflake, so $3 \cdot 4^{n-1}$ new triangles are generated at the nth stage. Because the area of an equilateral triangle is proportional to the square of its side length and the side length for each new triangle is one-third the side length of triangles from the previous stage, it follows that the area of the triangles added at each stage is reduced by a factor of $\frac{1}{9}$ from the area of the triangles added at the previous stage. Thus, each triangle added at the nth stage has an area of $A_0/9^n$. This means that the nth stage contributes

$$3 \cdot 4^{n-1} \cdot \frac{A_0}{9^n} = \frac{3}{4}A_0\left(\frac{4}{9}\right)^n$$

to the area of the snowflake. The total area is therefore

$$A = A_0 + \frac{3}{4}A_0 \sum_{n=1}^{\infty} \left(\frac{4}{9}\right)^n = A_0 + \frac{3}{4}A_0 \frac{\frac{4}{9}}{1-\frac{4}{9}} = A_0 + \frac{3}{4}A_0 \cdot \frac{4}{5} = \frac{8}{5}A_0.$$

10.3 Convergence of Series with Positive Terms

Preliminary Questions

1. Let $S = \sum_{n=1}^{\infty} a_n$. If the partial sums S_N are increasing, then (choose the correct conclusion):

(a) $\{a_n\}$ is an increasing sequence.

(b) $\{a_n\}$ is a positive sequence.

SOLUTION The correct response is **(b)**. Recall that $S_N = a_1 + a_2 + a_3 + \cdots + a_N$; thus, $S_N - S_{N-1} = a_N$. If S_N is increasing, then $S_N - S_{N-1} \geq 0$. It then follows that $a_N \geq 0$; that is, $\{a_n\}$ is a positive sequence.

2. What are the hypotheses of the Integral Test?

SOLUTION The hypotheses for the Integral Test are: A function $f(x)$ such that $a_n = f(n)$ must be positive, decreasing, and continuous for $x \geq 1$.

3. Which test would you use to determine whether $\sum_{n=1}^{\infty} n^{-3.2}$ converges?

SOLUTION Because $n^{-3.2} = \frac{1}{n^{3.2}}$, we see that the indicated series is a p-series with $p = 3.2 > 1$. Therefore, the series converges.

4. Which test would you use to determine whether $\sum_{n=1}^{\infty} \frac{1}{2^n + \sqrt{n}}$ converges?

SOLUTION Because

$$\frac{1}{2^n + \sqrt{n}} < \frac{1}{2^n} = \left(\frac{1}{2}\right)^n,$$

and

$$\sum_{n=1}^{\infty} \left(\frac{1}{2}\right)^n$$

is a convergent geometric series, the comparison test would be an appropriate choice to establish that the given series converges.

5. Ralph hopes to investigate the convergence of $\sum_{n=1}^{\infty} \frac{e^{-n}}{n}$ by comparing it with $\sum_{n=1}^{\infty} \frac{1}{n}$. Is Ralph on the right track?

SOLUTION No, Ralph is not on the right track. For $n \geq 1$,

$$\frac{e^{-n}}{n} < \frac{1}{n};$$

however, $\sum_{n=1}^{\infty} \frac{1}{n}$ is a divergent series. The Comparison Test therefore does not allow us to draw a conclusion about the convergence or divergence of the series $\sum_{n=1}^{\infty} \frac{e^{-n}}{n}$.

Exercises

In Exercises 1–14, use the Integral Test to determine whether the infinite series is convergent.

1. $\sum_{n=1}^{\infty} \frac{1}{n^4}$

SOLUTION Let $f(x) = \frac{1}{x^4}$. This function is continuous, positive and decreasing on the interval $x \geq 1$, so the Integral Test applies. Moreover,

$$\int_1^\infty \frac{dx}{x^4} = \lim_{R\to\infty} \int_1^R x^{-4}\,dx = -\frac{1}{3}\lim_{R\to\infty}\left(\frac{1}{R^3}-1\right) = \frac{1}{3}.$$

The integral converges; hence, the series $\sum_{n=1}^{\infty} \frac{1}{n^4}$ also converges.

3. $\sum_{n=1}^{\infty} n^{-1/3}$

SOLUTION Let $f(x) = x^{-\frac{1}{3}} = \frac{1}{\sqrt[3]{x}}$. This function is continuous, positive and decreasing on the interval $x \geq 1$, so the Integral Test applies. Moreover,

$$\int_1^\infty x^{-1/3}\,dx = \lim_{R\to\infty} \int_1^R x^{-1/3}\,dx = \frac{3}{2}\lim_{R\to\infty}\left(R^{2/3}-1\right) = \infty.$$

The integral diverges; hence, the series $\sum_{n=1}^{\infty} n^{-1/3}$ also diverges.

5. $\sum_{n=25}^{\infty} \frac{n^2}{(n^3+9)^{5/2}}$

SOLUTION Let $f(x) = \frac{x^2}{(x^3+9)^{5/2}}$. This function is positive and continuous for $x \geq 25$. Moreover, because

$$f'(x) = \frac{2x(x^3+9)^{5/2} - x^2 \cdot \frac{5}{2}(x^3+9)^{3/2} \cdot 3x^2}{(x^3+9)^5} = \frac{x(36-11x^3)}{2(x^3+9)^{7/2}},$$

we see that $f'(x) < 0$ for $x \geq 25$, so f is decreasing on the interval $x \geq 25$. The Integral Test therefore applies. To evaluate the improper integral, we use the substitution $u = x^3 + 9$, $du = 3x^2 dx$. We then find

$$\int_{25}^\infty \frac{x^2}{(x^3+9)^{5/2}}\,dx = \lim_{R\to\infty} \int_{25}^R \frac{x^2}{(x^3+9)^{5/2}}\,dx = \frac{1}{3}\lim_{R\to\infty}\int_{15634}^{R^3+9} \frac{du}{u^{5/2}}$$

$$= -\frac{2}{9}\lim_{R\to\infty}\left(\frac{1}{(R^3+9)^{3/2}} - \frac{1}{15634^{3/2}}\right) = \frac{2}{9 \cdot 15634^{3/2}}.$$

The integral converges; hence, the series $\sum_{n=25}^{\infty} \frac{n^2}{(n^3+9)^{5/2}}$ also converges.

7. $\sum_{n=1}^{\infty} \frac{1}{n^2+1}$

SOLUTION Let $f(x) = \frac{1}{x^2+1}$. This function is positive, decreasing and continuous on the interval $x \geq 1$, hence the Integral Test applies. Moreover,

$$\int_1^\infty \frac{dx}{x^2+1} = \lim_{R\to\infty}\int_1^R \frac{dx}{x^2+1} = \lim_{R\to\infty}\left(\tan^{-1} R - \frac{\pi}{4}\right) = \frac{\pi}{2} - \frac{\pi}{4} = \frac{\pi}{4}.$$

The integral converges; hence, the series $\sum_{n=1}^{\infty} \frac{1}{n^2+1}$ also converges.

9. $\sum_{n=1}^{\infty} \frac{1}{n(n+1)}$

SOLUTION Let $f(x) = \frac{1}{x(x+1)}$. This function is positive, continuous and decreasing on the interval $x \geq 1$, so the Integral Test applies. We compute the improper integral using partial fractions:

$$\int_1^\infty \frac{dx}{x(x+1)} = \lim_{R\to\infty}\int_1^R \left(\frac{1}{x} - \frac{1}{x+1}\right)dx = \lim_{R\to\infty} \ln \frac{x}{x+1}\Big|_1^R = \lim_{R\to\infty}\left(\ln\frac{R}{R+1} - \ln\frac{1}{2}\right) = \ln 1 - \ln\frac{1}{2} = \ln 2.$$

The integral converges; hence, the series $\sum_{n=1}^{\infty} \frac{1}{n(n+1)}$ converges.

11. $\sum_{n=2}^{\infty} \frac{1}{n(\ln n)^2}$

SOLUTION Let $f(x) = \frac{1}{x(\ln x)^2}$. This function is positive and continuous for $x \geq 2$. Moreover,

$$f'(x) = -\frac{1}{x^2(\ln x)^4}\left(1 \cdot (\ln x)^2 + x \cdot 2(\ln x) \cdot \frac{1}{x}\right) = -\frac{1}{x^2(\ln x)^4}\left((\ln x)^2 + 2\ln x\right).$$

Since $\ln x > 0$ for $x > 1$, $f'(x)$ is negative for $x > 1$; hence, f is decreasing for $x \geq 2$. To compute the improper integral, we make the substitution $u = \ln x$, $du = \frac{1}{x}\,dx$. We obtain:

$$\int_2^{\infty} \frac{1}{x(\ln x)^2}\,dx = \lim_{R\to\infty} \int_2^R \frac{1}{x(\ln x)^2}\,dx = \lim_{R\to\infty} \int_{\ln 2}^{\ln R} \frac{du}{u^2}$$
$$= -\lim_{R\to\infty}\left(\frac{1}{\ln R} - \frac{1}{\ln 2}\right) = \frac{1}{\ln 2}.$$

The integral converges; hence, the series $\sum_{n=2}^{\infty} \frac{1}{n(\ln n)^2}$ also converges.

13. $\sum_{n=1}^{\infty} \frac{1}{2^{\ln n}}$

SOLUTION Note that

$$2^{\ln n} = (e^{\ln 2})^{\ln n} = (e^{\ln n})^{\ln 2} = n^{\ln 2}.$$

Thus,

$$\sum_{n=1}^{\infty} \frac{1}{2^{\ln n}} = \sum_{n=1}^{\infty} \frac{1}{n^{\ln 2}}.$$

Now, let $f(x) = \frac{1}{x^{\ln 2}}$. This function is positive, continuous and decreasing on the interval $x \geq 1$; therefore, the Integral Test applies. Moreover,

$$\int_1^{\infty} \frac{dx}{x^{\ln 2}} = \lim_{R\to\infty} \int_1^R \frac{dx}{x^{\ln 2}} = \frac{1}{1-\ln 2} \lim_{R\to\infty} (R^{1-\ln 2} - 1) = \infty,$$

because $1 - \ln 2 > 0$. The integral diverges; hence, the series $\sum_{n=1}^{\infty} \frac{1}{2^{\ln n}}$ also diverges.

15. Show that $\sum_{n=1}^{\infty} \frac{1}{n^3 + 8n}$ converges by using the Comparison Test with $\sum_{n=1}^{\infty} n^{-3}$.

SOLUTION We compare the series with the p-series $\sum_{n=1}^{\infty} n^{-3}$. For $n \geq 1$,

$$\frac{1}{n^3 + 8n} \leq \frac{1}{n^3}.$$

Since $\sum_{n=1}^{\infty} \frac{1}{n^3}$ converges (it is a p-series with $p = 3 > 1$), the series $\sum_{n=1}^{\infty} \frac{1}{n^3 + 8n}$ also converges by the Comparison Test.

17. Let $S = \sum_{n=1}^{\infty} \frac{1}{n + \sqrt{n}}$. Verify that for $n \geq 1$,

$$\frac{1}{n + \sqrt{n}} \leq \frac{1}{n}, \qquad \frac{1}{n + \sqrt{n}} \leq \frac{1}{\sqrt{n}}$$

Can either inequality be used to show that S diverges? Show that $\frac{1}{n + \sqrt{n}} \geq \frac{1}{2n}$ and conclude that S diverges.

SOLUTION For $n \geq 1$, $n + \sqrt{n} \geq n$ and $n + \sqrt{n} \geq \sqrt{n}$. Taking the reciprocal of each of these inequalities yields

$$\frac{1}{n + \sqrt{n}} \leq \frac{1}{n} \quad \text{and} \quad \frac{1}{n + \sqrt{n}} \leq \frac{1}{\sqrt{n}}.$$

These inequalities indicate that the series $\sum_{n=1}^{\infty} \frac{1}{n+\sqrt{n}}$ is smaller than both $\sum_{n=1}^{\infty} \frac{1}{n}$ and $\sum_{n=1}^{\infty} \frac{1}{\sqrt{n}}$; however, $\sum_{n=1}^{\infty} \frac{1}{n}$ and $\sum_{n=1}^{\infty} \frac{1}{\sqrt{n}}$ both diverge so neither inequality allows us to show that S diverges.

On the other hand, for $n \geq 1$, $n \geq \sqrt{n}$, so $2n \geq n + \sqrt{n}$ and

$$\frac{1}{n+\sqrt{n}} \geq \frac{1}{2n}.$$

The series $\sum_{n=1}^{\infty} \frac{1}{2n} = 2\sum_{n=1}^{\infty} \frac{1}{n}$ diverges, since the harmonic series diverges. The Comparison Test then lets us conclude that the larger series $\sum_{n=1}^{\infty} \frac{1}{n+\sqrt{n}}$ also diverges.

In Exercises 19–30, use the Comparison Test to determine whether the infinite series is convergent.

19. $\sum_{n=1}^{\infty} \frac{1}{n2^n}$

SOLUTION We compare with the geometric series $\sum_{n=1}^{\infty} \left(\frac{1}{2}\right)^n$. For $n \geq 1$,

$$\frac{1}{n2^n} \leq \frac{1}{2^n} = \left(\frac{1}{2}\right)^n.$$

Since $\sum_{n=1}^{\infty} \left(\frac{1}{2}\right)^n$ converges (it is a geometric series with $r = \frac{1}{2}$), we conclude by the Comparison Test that $\sum_{n=1}^{\infty} \frac{1}{n2^n}$ also converges.

21. $\sum_{n=1}^{\infty} \frac{1}{n^{1/3}+2^n}$

SOLUTION For $n \geq 1$,

$$\frac{1}{n^{1/3}+2^n} \leq \frac{1}{2^n}$$

The series $\sum_{n=1}^{\infty} \frac{1}{2^n}$ is a geometric series with $r = \frac{1}{2}$, so it converges. By the Comparison test, so does $\sum_{n=1}^{\infty} \frac{1}{n^{1/3}+2^n}$.

23. $\sum_{m=1}^{\infty} \frac{4}{m!+4^m}$

SOLUTION For $m \geq 1$,

$$\frac{4}{m!+4^m} \leq \frac{4}{4^m} = \left(\frac{1}{4}\right)^{m-1}.$$

The series $\sum_{m=1}^{\infty} \left(\frac{1}{4}\right)^{m-1}$ is a geometric series with $r = \frac{1}{4}$, so it converges. By the Comparison Test we can therefore conclude that the series $\sum_{m=1}^{\infty} \frac{4}{m!+4^m}$ also converges.

25. $\sum_{k=1}^{\infty} \frac{\sin^2 k}{k^2}$

SOLUTION For $k \geq 1$, $0 \leq \sin^2 k \leq 1$, so

$$0 \leq \frac{\sin^2 k}{k^2} \leq \frac{1}{k^2}.$$

The series $\sum_{k=1}^{\infty} \frac{1}{k^2}$ is a p-series with $p = 2 > 1$, so it converges. By the Comparison Test we can therefore conclude that the series $\sum_{k=1}^{\infty} \frac{\sin^2 k}{k^2}$ also converges.

27. $\displaystyle\sum_{n=1}^{\infty} \frac{2}{3^n + 3^{-n}}$

SOLUTION Since $3^{-n} > 0$ for all n,

$$\frac{2}{3^n + 3^{-n}} \le \frac{2}{3^n} = 2\left(\frac{1}{3}\right)^n.$$

The series $\displaystyle\sum_{n=1}^{\infty} 2\left(\frac{1}{3}\right)^n$ is a geometric series with $r = \frac{1}{3}$, so it converges. By the Comparison Theorem we can therefore conclude that the series $\displaystyle\sum_{n=1}^{\infty} \frac{2}{3^n + 3^{-n}}$ also converges.

29. $\displaystyle\sum_{n=1}^{\infty} \frac{1}{(n+1)!}$

SOLUTION Note that for $n \ge 2$,

$$(n+1)! = 1 \cdot \underbrace{2 \cdot 3 \cdots n \cdot (n+1)}_{n \text{ factors}} \le 2^n$$

so that

$$\sum_{n=1}^{\infty} \frac{1}{(n+1)!} = 1 + \sum_{n=2}^{\infty} \frac{1}{(n+1)!} \le 1 + \sum_{n=2}^{\infty} \frac{1}{2^n}$$

But $\sum_{n=2}^{\infty} \frac{1}{2^n}$ is a geometric series with ratio $r = \frac{1}{2}$, so it converges. By the comparison test, $\displaystyle\sum_{n=1}^{\infty} \frac{1}{(n+1)!}$ converges as well.

Exercise 31–36: For all $a > 0$ and $b > 1$, the inequalities

$$\ln n \le n^a, \qquad n^a < b^n$$

are true for n sufficiently large (this can be proved using L'Hopital's Rule). Use this, together with the Comparison Theorem, to determine whether the series converges or diverges.

31. $\displaystyle\sum_{n=1}^{\infty} \frac{\ln n}{n^3}$

SOLUTION For n sufficiently large (say $n = k$, although in this case $n = 1$ suffices), we have $\ln n \le n$, so that

$$\sum_{n=k}^{\infty} \frac{\ln n}{n^3} \le \sum_{n=k}^{\infty} \frac{n}{n^3} = \sum_{n=k}^{\infty} \frac{1}{n^2}$$

This is a p-series with $p = 2 > 1$, so it converges. Thus $\sum_{n=k}^{\infty} \frac{\ln n}{n^3}$ also converges; adding back in the finite number of terms for $1 \le n \le k$ does not affect this result.

33. $\displaystyle\sum_{n=1}^{\infty} \frac{(\ln n)^{100}}{n^{1.1}}$

SOLUTION Choose N so that $\ln n \le n^{0.0005}$ for $n \ge N$. Then also for $n > N$, $(\ln n)^{100} \le (n^{0.0005})^{100} = n^{0.05}$. Then

$$\sum_{n=N}^{\infty} \frac{(\ln n)^{100}}{n^{1.1}} \le \sum_{n=N}^{\infty} \frac{n^{0.05}}{n^{1.1}} = \sum_{n=N}^{\infty} \frac{1}{n^{1.05}}$$

But $\displaystyle\sum_{n=N}^{\infty} \frac{1}{n^{1.05}}$ is a p-series with $p = 1.05 > 1$, so is convergent. It follows that $\sum_{n=N}^{\infty} \frac{(\ln n)^100}{n^{1.1}}$ is also convergent; adding back in the finite number of terms for $n = 1, 2, \ldots, N-1$ shows that $\displaystyle\sum_{n=1}^{\infty} \frac{(\ln n)^{100}}{n^{1.1}}$ converges as well.

35. $\sum_{n=1}^{\infty} \frac{n}{3^n}$

SOLUTION Choose N such that $n \le 2^n$ for $n \ge N$. Then

$$\sum_{n=N}^{\infty} \frac{n}{3^n} \le \sum_{n=N}^{\infty} \left(\frac{2}{3}\right)^n$$

The latter sum is a geometric series with $r = \frac{2}{3} < 1$, so it converges. Thus the series on the left converges as well. Adding back in the finite number of terms for $n < N$ shows that $\sum_{n=1}^{\infty} \frac{n}{3^n}$ converges.

37. Show that $\sum_{n=1}^{\infty} \sin \frac{1}{n^2}$ converges. *Hint:* Use $\sin x \le x$ for $x \ge 0$.

SOLUTION For $n \ge 1$,

$$0 \le \frac{1}{n^2} \le 1 < \pi;$$

therefore, $\sin \frac{1}{n^2} > 0$ for $n \ge 1$. Moreover, for $n \ge 1$,

$$\sin \frac{1}{n^2} \le \frac{1}{n^2}.$$

The series $\sum_{n=1}^{\infty} \frac{1}{n^2}$ is a p-series with $p = 2 > 1$, so it converges. By the Comparison Test we can therefore conclude that the series $\sum_{n=1}^{\infty} \sin \frac{1}{n^2}$ also converges.

In Exercises 39–48, use the Limit Comparison Test to prove convergence or divergence of the infinite series.

39. $\sum_{n=2}^{\infty} \frac{n^2}{n^4 - 1}$

SOLUTION Let $a_n = \frac{n^2}{n^4 - 1}$. For large n, $\frac{n^2}{n^4 - 1} \approx \frac{n^2}{n^4} = \frac{1}{n^2}$, so we apply the Limit Comparison Test with $b_n = \frac{1}{n^2}$. We find

$$L = \lim_{n\to\infty} \frac{a_n}{b_n} = \lim_{n\to\infty} \frac{\frac{n^2}{n^4-1}}{\frac{1}{n^2}} = \lim_{n\to\infty} \frac{n^4}{n^4 - 1} = 1.$$

The series $\sum_{n=1}^{\infty} \frac{1}{n^2}$ is a p-series with $p = 2 > 1$, so it converges; hence, $\sum_{n=2}^{\infty} \frac{1}{n^2}$ also converges. Because L exists, by the Limit Comparison Test we can conclude that the series $\sum_{n=2}^{\infty} \frac{n^2}{n^4 - 1}$ converges.

41. $\sum_{n=2}^{\infty} \frac{n}{\sqrt{n^3 + 1}}$

SOLUTION Let $a_n = \frac{n}{\sqrt{n^3 + 1}}$. For large n, $\frac{n}{\sqrt{n^3 + 1}} \approx \frac{n}{\sqrt{n^3}} = \frac{1}{\sqrt{n}}$, so we apply the Limit Comparison test with $b_n = \frac{1}{\sqrt{n}}$. We find

$$L = \lim_{n\to\infty} \frac{a_n}{b_n} = \lim_{n\to\infty} \frac{\frac{n}{\sqrt{n^3+1}}}{\frac{1}{\sqrt{n}}} = \lim_{n\to\infty} \frac{\sqrt{n^3}}{\sqrt{n^3 + 1}} = 1.$$

The series $\sum_{n=1}^{\infty} \frac{1}{\sqrt{n}}$ is a p-series with $p = \frac{1}{2} < 1$, so it diverges; hence, $\sum_{n=2}^{\infty} \frac{1}{\sqrt{n}}$ also diverges. Because $L > 0$, by the Limit Comparison Test we can conclude that the series $\sum_{n=2}^{\infty} \frac{n}{\sqrt{n^3 + 1}}$ diverges.

43. $\sum_{n=3}^{\infty} \frac{3n+5}{n(n-1)(n-2)}$

SOLUTION Let $a_n = \frac{3n+5}{n(n-1)(n-2)}$. For large n, $\frac{3n+5}{n(n-1)(n-2)} \approx \frac{3n}{n^3} = \frac{3}{n^2}$, so we apply the Limit Comparison Test with $b_n = \frac{1}{n^2}$. We find

$$L = \lim_{n\to\infty} \frac{a_n}{b_n} = \lim_{n\to\infty} \frac{\frac{3n+5}{n(n+1)(n+2)}}{\frac{1}{n^2}} = \lim_{n\to\infty} \frac{3n^3+5n^2}{n(n+1)(n+2)} = 3.$$

The series $\sum_{n=1}^{\infty} \frac{1}{n^2}$ is a p-series with $p = 2 > 1$, so it converges; hence, the series $\sum_{n=3}^{\infty} \frac{1}{n^2}$ also converges. Because L exists, by the Limit Comparison Test we can conclude that the series $\sum_{n=3}^{\infty} \frac{3n+5}{n(n-1)(n-2)}$ converges.

45. $\sum_{n=1}^{\infty} \frac{1}{\sqrt{n} + \ln n}$

SOLUTION Let

$$a_n = \frac{1}{\sqrt{n} + \ln n}$$

For large n, $\sqrt{n} + \ln n \approx \sqrt{n}$, so apply the Comparison Test with $b_n = \frac{1}{\sqrt{n}}$. We find

$$L = \lim_{n\to\infty} \frac{a_n}{b_n} = \lim_{n\to\infty} \frac{1}{\sqrt{n} + \ln n} \cdot \frac{\sqrt{n}}{1} = \lim_{n\to\infty} \frac{1}{1 + \frac{\ln n}{\sqrt{n}}} = 1$$

The series $\sum_{n=1}^{\infty} \frac{1}{\sqrt{n}}$ is a p-series with $p = \frac{1}{2} < 1$, so it diverges. Because L exists, the Limit Comparison Test tells us the the original series also diverges.

47. $\sum_{n=1}^{\infty} \left(1 - \cos \frac{1}{n}\right)$ *Hint:* Compare with $\sum_{n=1}^{\infty} n^{-2}$.

SOLUTION Let $a_n = 1 - \cos \frac{1}{n}$, and apply the Limit Comparison Test with $b_n = \frac{1}{n^2}$. We find

$$L = \lim_{n\to\infty} \frac{a_n}{b_n} = \lim_{n\to\infty} \frac{1 - \cos \frac{1}{n}}{\frac{1}{n^2}} = \lim_{x\to\infty} \frac{1 - \cos \frac{1}{x}}{\frac{1}{x^2}} = \lim_{x\to\infty} \frac{-\frac{1}{x^2} \sin \frac{1}{x}}{-\frac{2}{x^3}} = \frac{1}{2} \lim_{x\to\infty} \frac{\sin \frac{1}{x}}{\frac{1}{x}}.$$

As $x \to \infty$, $u = \frac{1}{x} \to 0$, so

$$L = \frac{1}{2} \lim_{x\to\infty} \frac{\sin \frac{1}{x}}{\frac{1}{x}} = \frac{1}{2} \lim_{u\to 0} \frac{\sin u}{u} = \frac{1}{2}.$$

The series $\sum_{n=1}^{\infty} \frac{1}{n^2}$ is a p-series with $p = 2 > 1$, so it converges. Because L exists, by the Limit Comparison Test we can conclude that the series $\sum_{n=1}^{\infty} \left(1 - \cos \frac{1}{n}\right)$ also converges.

In Exercises 49–78, determine convergence or divergence using any method covered so far.

49. $\sum_{n=4}^{\infty} \frac{1}{n^2 - 9}$

SOLUTION Apply the Limit Comparison Test with $a_n = \frac{1}{n^2 - 9}$ and $b_n = \frac{1}{n^2}$:

$$L = \lim_{n\to\infty} \frac{a_n}{b_n} = \lim_{n\to\infty} \frac{\frac{1}{n^2-9}}{\frac{1}{n^2}} = \lim_{n\to\infty} \frac{n^2}{n^2 - 9} = 1.$$

Since the p-series $\sum_{n=1}^{\infty} \frac{1}{n^2}$ converges, the series $\sum_{n=4}^{\infty} \frac{1}{n^2}$ also converges. Because L exists, by the Limit Comparison Test we can conclude that the series $\sum_{n=4}^{\infty} \frac{1}{n^2-9}$ converges.

51. $\sum_{n=1}^{\infty} \frac{\sqrt{n}}{4n+9}$

SOLUTION Apply the Limit Comparison Test with $a_n = \frac{\sqrt{n}}{4n+9}$ and $b_n = \frac{1}{\sqrt{n}}$:

$$L = \lim_{n\to\infty} \frac{a_n}{b_n} = \lim_{n\to\infty} \frac{\frac{\sqrt{n}}{4n+9}}{\frac{1}{\sqrt{n}}} = \lim_{n\to\infty} \frac{n}{4n+9} = \frac{1}{4}.$$

The series $\sum_{n=1}^{\infty} \frac{1}{\sqrt{n}}$ is a divergent p-series. Because $L > 0$, by the Limit Comparison Test we can conclude that the series $\sum_{n=1}^{\infty} \frac{\sqrt{n}}{4n+9}$ also diverges.

53. $\sum_{n=1}^{\infty} \frac{n^2-n}{n^5+n}$

SOLUTION First rewrite $a_n = \frac{n^2-n}{n^5+n} = \frac{n(n-1)}{n(n^4+1)} = \frac{n-1}{n^4+1}$ and observe

$$\frac{n-1}{n^4+1} < \frac{n}{n^4} = \frac{1}{n^3}$$

for $n \geq 1$. The series $\sum_{n=1}^{\infty} \frac{1}{n^3}$ is a convergent p-series, so by the Comparison Test we can conclude that the series $\sum_{n=1}^{\infty} \frac{n^2-n}{n^5+n}$ also converges.

55. $\sum_{n=5}^{\infty} (4/5)^{-n}$

SOLUTION

$$\sum_{n=5}^{\infty} \left(\frac{4}{5}\right)^{-n} = \sum_{n=5}^{\infty} \left(\frac{5}{4}\right)^{n}$$

which is a geometric series starting at $n = 5$ with ratio $r = \frac{5}{4} > 1$. Thus the series diverges.

57. $\sum_{n=2}^{\infty} \frac{1}{n^{3/2} \ln n}$

SOLUTION For $n \geq 3$, $\ln n > 1$, so $n^{3/2} \ln n > n^{3/2}$ and

$$\frac{1}{n^{3/2} \ln n} < \frac{1}{n^{3/2}}.$$

The series $\sum_{n=1}^{\infty} \frac{1}{n^{3/2}}$ is a convergent p-series, so the series $\sum_{n=3}^{\infty} \frac{1}{n^{3/2}}$ also converges. By the Comparison Test we can therefore conclude that the series $\sum_{n=3}^{\infty} \frac{1}{n^{3/2} \ln n}$ converges. Hence, the series $\sum_{n=2}^{\infty} \frac{1}{n^{3/2} \ln n}$ also converges.

59. $\sum_{k=1}^{\infty} 4^{1/k}$

SOLUTION

$$\lim_{k\to\infty} a_k = \lim_{k\to\infty} 4^{1/k} = 4^0 = 1 \neq 0;$$

therefore, the series $\sum_{k=1}^{\infty} 4^{1/k}$ diverges by the Divergence Test.

61. $\displaystyle\sum_{n=2}^{\infty} \frac{1}{(\ln n)^4}$

SOLUTION By the comment preceding Exercise 31, we can choose N so that for $n \geq N$, we have $\ln n < n^{1/8}$, so that $(\ln n)^4 < n^{1/2}$. Then

$$\sum_{n=N}^{\infty} \frac{1}{(\ln n)^4} > \sum_{n=N}^{\infty} \frac{1}{n^{1/2}}$$

which is a divergent p-series. Thus the series on the left diverges as well, and adding back in the finite number of terms for $n < N$ does not affect the result. Thus $\displaystyle\sum_{n=2}^{\infty} \frac{1}{(\ln n)^4}$ diverges.

63. $\displaystyle\sum_{n=1}^{\infty} \frac{1}{n \ln n - n}$

SOLUTION For $n \geq 2$, $n \ln n - n \leq n \ln n$; therefore,

$$\frac{1}{n \ln n - n} \geq \frac{1}{n \ln n}.$$

Now, let $f(x) = \dfrac{1}{x \ln x}$. For $x \geq 2$, this function is continuous, positive and decreasing, so the Integral Test applies. Using the substitution $u = \ln x$, $du = \frac{1}{x}\,dx$, we find

$$\int_2^{\infty} \frac{dx}{x \ln x} = \lim_{R\to\infty} \int_2^R \frac{dx}{x \ln x} = \lim_{R\to\infty} \int_{\ln 2}^{\ln R} \frac{du}{u} = \lim_{R\to\infty} (\ln(\ln R) - \ln(\ln 2)) = \infty.$$

The integral diverges; hence, the series $\displaystyle\sum_{n=2}^{\infty} \frac{1}{n \ln n}$ also diverges. By the Comparison Test we can therefore conclude that the series $\displaystyle\sum_{n=2}^{\infty} \frac{1}{n \ln n - n}$ diverges.

65. $\displaystyle\sum_{n=1}^{\infty} \frac{1}{n^n}$

SOLUTION For $n \geq 2$, $n^n \geq 2^n$; therefore,

$$\frac{1}{n^n} \leq \frac{1}{2^n} = \left(\frac{1}{2}\right)^n.$$

The series $\displaystyle\sum_{n=1}^{\infty} \left(\frac{1}{2}\right)^n$ is a convergent geometric series, so $\displaystyle\sum_{n=2}^{\infty} \left(\frac{1}{2}\right)^n$ also converges. By the Comparison Test we can therefore conclude that the series $\displaystyle\sum_{n=2}^{\infty} \frac{1}{n^n}$ converges. Hence, the series $\displaystyle\sum_{n=1}^{\infty} \frac{1}{n^n}$ converges.

67. $\displaystyle\sum_{n=1}^{\infty} \frac{1+(-1)^n}{n}$

SOLUTION Let

$$a_n = \frac{1+(-1)^n}{n}$$

Then

$$a_n = \begin{cases} 0 & n \text{ odd} \\ \frac{2}{2k} = \frac{1}{k} & n = 2k \text{ even} \end{cases}$$

Therefore, $\{a_n\}$ consists of 0s in the odd places and the harmonic series in the even places, so $\sum_{i=1}^{\infty} a_n$ is just the sum of the harmonic series, which diverges. Thus $\sum_{i=1}^{\infty} a_n$ diverges as well.

69. $\sum_{n=1}^{\infty} \sin \frac{1}{n}$

SOLUTION Apply the Limit Comparison Test with $a_n = \sin \frac{1}{n}$ and $b_n = \frac{1}{n}$:

$$L = \lim_{n\to\infty} \frac{\sin \frac{1}{n}}{\frac{1}{n}} = \lim_{u\to 0} \frac{\sin u}{u} = 1,$$

where $u = \frac{1}{n}$. The harmonic series diverges. Because $L > 0$, by the Limit Comparison Test we can conclude that the series $\sum_{n=1}^{\infty} \sin \frac{1}{n}$ also diverges.

71. $\sum_{n=1}^{\infty} \frac{2n+1}{4^n}$

SOLUTION For $n \geq 3$, $2n + 1 < 2^n$, so

$$\frac{2n+1}{4^n} < \frac{2^n}{4^n} = \left(\frac{1}{2}\right)^n.$$

The series $\sum_{n=1}^{\infty} \left(\frac{1}{2}\right)^n$ is a convergent geometric series, so $\sum_{n=3}^{\infty} \left(\frac{1}{2}\right)^n$ also converges. By the Comparison Test we can therefore conclude that the series $\sum_{n=3}^{\infty} \frac{2n+1}{4^n}$ converges. Finally, the series $\sum_{n=1}^{\infty} \frac{2n+1}{4^n}$ converges.

73. $\sum_{n=4}^{\infty} \frac{\ln n}{n^2 - 3n}$

SOLUTION By the comment preceding Exercise 31, we can choose $N \geq 4$ so that for $n \geq N$, $\ln n < n^{1/2}$. Then

$$\sum_{n=N}^{\infty} \frac{\ln n}{n^2 - 3n} \leq \sum_{n=N}^{\infty} \frac{n^{1/2}}{n^2 - 3n} = \sum_{n=N}^{\infty} \frac{1}{n^{3/2} - 3n^{1/2}}$$

To evaluate convergence of the latter series, let $a_n = \frac{1}{n^{3/2} - 3n^{1/2}}$ and $b_n = \frac{1}{n^{3/2}}$, and apply the Limit Comparison Test:

$$L = \lim_{n\to\infty} \frac{a_n}{b_n} = \lim_{n\to\infty} \frac{1}{n^{3/2} - 3n^{1/2}} \cdot n^{3/2} = \lim_{n\to\infty} \frac{1}{1 - 3n^{-1}} = 0$$

Thus $\sum a_n$ converges if $\sum b_n$ does. But $\sum b_n$ is a convergent p-series. Thus $\sum a_n$ converges and, by the comparison test, so does the original series. Adding back in the finite number of terms for $n < N$ does not affect convergence.

75. $\sum_{n=2}^{\infty} \frac{1}{n^{1/2} \ln n}$

SOLUTION By the comment preceding Exercise 31, we can choose $N \geq 2$ so that for $n \geq N$, $\ln n < n^{1/4}$. Then

$$\sum_{n=N}^{\infty} \frac{1}{n^{1/2} \ln n} > \sum_{n=N}^{\infty} \frac{1}{n^{3/4}}$$

which is a divergent p-series. Thus the original series diverges as well - as usual, adding back in the finite number of terms for $n < N$ does not affect convergence.

77. $\sum_{n=1}^{\infty} \frac{4n^2 + 15n}{3n^4 - 5n^2 - 17}$

SOLUTION Apply the Limit Comparison Test with

$$a_n = \frac{4n^2 + 15n}{3n^4 - 5n^2 - 17}, \qquad b_n = \frac{4n^2}{3n^4} = \frac{4}{3n^2}$$

We have

$$L = \lim_{n\to\infty} \frac{a_n}{b_n} = \lim_{n\to\infty} \frac{4n^2+15n}{3n^4-5n^2-17} \cdot \frac{3n^2}{4} = \lim_{n\to\infty} \frac{12n^4+45n^3}{12n^4-20n^2-68} = \lim_{n\to\infty} \frac{12+45/n}{12-20/n^2-68/n^4} = 1$$

Now, $\sum_{n=1}^{\infty} b_n$ is a p-series with $p = 2 > 1$, so converges. Since $L = 1$, we see that $\sum_{n=1}^{\infty} \frac{4n^2+15n}{3n^4-5n^2-17}$ converges as well.

79. For which a does $\sum_{n=2}^{\infty} \frac{1}{n(\ln n)^a}$ converge?

SOLUTION First consider the case $a > 0$ but $a \neq 1$. Let $f(x) = \frac{1}{x(\ln x)^a}$. This function is continuous, positive and decreasing for $x \geq 2$, so the Integral Test applies. Now,

$$\int_2^{\infty} \frac{dx}{x(\ln x)^a} = \lim_{R\to\infty} \int_2^R \frac{dx}{x(\ln x)^a} = \lim_{R\to\infty} \int_{\ln 2}^{\ln R} \frac{du}{u^a} = \frac{1}{1-a} \lim_{R\to\infty} \left(\frac{1}{(\ln R)^{a-1}} - \frac{1}{(\ln 2)^{a-1}} \right).$$

Because

$$\lim_{R\to\infty} \frac{1}{(\ln R)^{a-1}} = \begin{cases} \infty, & 0 < a < 1 \\ 0, & a > 1 \end{cases}$$

we conclude the integral diverges when $0 < a < 1$ and converges when $a > 1$. Therefore

$$\sum_{n=2}^{\infty} \frac{1}{n(\ln n)^a} \text{ converges for } a > 1 \text{ and diverges for } 0 < a < 1.$$

Next, consider the case $a = 1$. The series becomes $\sum_{n=2}^{\infty} \frac{1}{n \ln n}$. Let $f(x) = \frac{1}{x \ln x}$. For $x \geq 2$, this function is continuous, positive and decreasing, so the Integral Test applies. Using the substitution $u = \ln x$, $du = \frac{1}{x}\,dx$, we find

$$\int_2^{\infty} \frac{dx}{x \ln x} = \lim_{R\to\infty} \int_2^R \frac{dx}{x \ln x} = \lim_{R\to\infty} \int_{\ln 2}^{\ln R} \frac{du}{u} = \lim_{R\to\infty} (\ln(\ln R) - \ln(\ln 2)) = \infty.$$

The integral diverges; hence, the series also diverges.

Finally, consider the case $a < 0$. Let $b = -a > 0$ so the series becomes $\sum_{n=2}^{\infty} \frac{(\ln n)^b}{n}$. Since $\ln n > 1$ for all $n \geq 3$, it follows that

$$(\ln n)^b > 1 \quad \text{so} \quad \frac{(\ln n)^b}{n} > \frac{1}{n}.$$

The series $\sum_{n=3}^{\infty} \frac{1}{n}$ diverges, so by the Comparison Test we can conclude that $\sum_{n=3}^{\infty} \frac{(\ln n)^b}{n}$ also diverges. Consequently, $\sum_{n=2}^{\infty} \frac{(\ln n)^b}{n}$ diverges. Thus,

$$\sum_{n=2}^{\infty} \frac{1}{n(\ln n)^a} \text{ diverges for } a < 0.$$

To summarize:

$$\sum_{n=2}^{\infty} \frac{1}{n(\ln n)^a} \text{ converges if } a > 1 \text{ and diverges if } a \leq 1.$$

Approximating Infinite Sums *In Exercises 81–83, let $a_n = f(n)$, where $f(x)$ is a continuous, decreasing function such that $f(x) \geq 0$ and $\int_1^{\infty} f(x)\,dx$ converges.*

81. Show that

$$\int_1^{\infty} f(x)\,dx \leq \sum_{n=1}^{\infty} a_n \leq a_1 + \int_1^{\infty} f(x)\,dx$$

3

SOLUTION From the proof of the Integral Test, we know that

$$a_2 + a_3 + a_4 + \cdots + a_N \le \int_1^N f(x)\,dx \le \int_1^\infty f(x)\,dx;$$

that is,

$$S_N - a_1 \le \int_1^\infty f(x)\,dx \quad \text{or} \quad S_N \le a_1 + \int_1^\infty f(x)\,dx.$$

Also from the proof of the Integral test, we know that

$$\int_1^N f(x)\,dx \le a_1 + a_2 + a_3 + \cdots + a_{N-1} = S_N - a_N \le S_N.$$

Thus,

$$\int_1^N f(x)\,dx \le S_N \le a_1 + \int_1^\infty f(x)\,dx.$$

Taking the limit as $N \to \infty$ yields Eq. (3), as desired.

83. Let $S = \sum_{n=1}^{\infty} a_n$. Arguing as in Exercise 81, show that

$$\sum_{n=1}^{M} a_n + \int_{M+1}^\infty f(x)\,dx \le S \le \sum_{n=1}^{M+1} a_n + \int_{M+1}^\infty f(x)\,dx \qquad \boxed{4}$$

Conclude that

$$0 \le S - \left(\sum_{n=1}^{M} a_n + \int_{M+1}^\infty f(x)\,dx\right) \le a_{M+1} \qquad \boxed{5}$$

This provides a method for approximating S with an error of at most a_{M+1}.

SOLUTION Following the proof of the Integral Test and the argument in Exercise 81, but starting with $n = M + 1$ rather than $n = 1$, we obtain

$$\int_{M+1}^\infty f(x)\,dx \le \sum_{n=M+1}^{\infty} a_n \le a_{M+1} + \int_{M+1}^\infty f(x)\,dx.$$

Adding $\sum_{n=1}^{M} a_n$ to each part of this inequality yields

$$\sum_{n=1}^{M} a_n + \int_{M+1}^\infty f(x)\,dx \le \sum_{n=1}^{\infty} a_n = S \le \sum_{n=1}^{M+1} a_n + \int_{M+1}^\infty f(x)\,dx.$$

Subtracting $\sum_{n=1}^{M} a_n + \int_{M+1}^\infty f(x)\,dx$ from each part of this last inequality then gives us

$$0 \le S - \left(\sum_{n=1}^{M} a_n + \int_{M+1}^\infty f(x)\,dx\right) \le a_{M+1}.$$

85. CAS Apply Eq. (4) with $M = 40{,}000$ to show that

$$1.644934066 \le \sum_{n=1}^{\infty} \frac{1}{n^2} \le 1.644934068$$

Is this consistent with Euler's result, according to which this infinite series has sum $\pi^2/6$?

SOLUTION Using Eq. (4) with $f(x) = \frac{1}{x^2}$, $a_n = \frac{1}{n^2}$ and $M = 40{,}000$, we find

$$S_{40{,}000} + \int_{40{,}001}^\infty \frac{dx}{x^2} \le \sum_{n=1}^{\infty} \frac{1}{n^2} \le S_{40{,}001} + \int_{40{,}001}^\infty \frac{dx}{x^2}.$$

Now,

$$S_{40,000} = 1.6449090672;$$

$$S_{40,001} = S_{40,000} + \frac{1}{40,001} = 1.6449090678;$$

and

$$\int_{40,001}^{\infty} \frac{dx}{x^2} = \lim_{R\to\infty} \int_{40,001}^{R} \frac{dx}{x^2} = -\lim_{R\to\infty} \left(\frac{1}{R} - \frac{1}{40,001}\right) = \frac{1}{40,001} = 0.0000249994.$$

Thus,

$$1.6449090672 + 0.0000249994 \le \sum_{n=1}^{\infty} \frac{1}{n^2} \le 1.6449090678 + 0.0000249994,$$

or

$$1.6449340665 \le \sum_{n=1}^{\infty} \frac{1}{n^2} \le 1.6449340672.$$

Since $\frac{\pi^2}{6} \approx 1.6449340668$, our approximation is consistent with Euler's result.

87. CAS Using a CAS and Eq. (5), determine the value of $\sum_{n=1}^{\infty} n^{-5}$ to within an error less than 10^{-4}.

SOLUTION Using Eq. (5) with $f(x) = x^{-5}$ and $a_n = n^{-5}$, we have

$$0 \le \sum_{n=1}^{\infty} n^{-5} - \left(\sum_{n=1}^{M+1} n^{-5} + \int_{M+1}^{\infty} x^{-5}\, dx\right) \le (M+1)^{-5}.$$

To guarantee an error less than 10^{-4}, we need $(M+1)^{-5} \le 10^{-4}$. This yields $M \ge 10^{4/5} - 1 \approx 5.3$, so we choose $M = 6$. Now,

$$\sum_{n=1}^{7} n^{-5} = 1.0368498887,$$

and

$$\int_{7}^{\infty} x^{-5}\, dx = \lim_{R\to\infty} \int_{7}^{R} x^{-5}\, dx = -\frac{1}{4} \lim_{R\to\infty} \left(R^{-4} - 7^{-4}\right) = \frac{1}{4 \cdot 7^4} = 0.0001041233.$$

Thus,

$$\sum_{n=1}^{\infty} n^{-5} \approx \sum_{n=1}^{7} n^{-5} + \int_{7}^{\infty} x^{-5}\, dx = 1.0368498887 + 0.0001041233 = 1.0369540120.$$

89. The following argument proves the divergence of the harmonic series $S = \sum_{n=1}^{\infty} 1/n$ without using the Integral Test.

Let

$$S_1 = 1 + \frac{1}{3} + \frac{1}{5} + \cdots, \qquad S_2 = \frac{1}{2} + \frac{1}{4} + \frac{1}{6} + \cdots$$

Show that if S converges, then

(a) S_1 and S_2 also converge and $S = S_1 + S_2$.

(b) $S_1 > S_2$ and $S_2 = \frac{1}{2}S$.

Observe that (b) contradicts (a), and conclude that S diverges.

SOLUTION Assume throughout that S converges; we will derive a contradiction. Write

$$a_n = \frac{1}{n}, \qquad b_n = \frac{1}{2n-1}, \qquad c_n = \frac{1}{2n}$$

for the nth terms in the series S, S_1, and S_2. Since $2n-1 \geq n$ for $n \geq 1$, we have $b_n < a_n$. Since $S = \sum a_n$ converges, so does $S_1 = \sum b_n$ by the Comparison Test. Also, $c_n = \frac{1}{2}a_n$, so again by the Comparison Test, the convergence of S implies the convergence of $S_2 = \sum c_n$. Now, define two sequences

$$b'_n = \begin{cases} b_{(n+1)/2} & n \text{ odd} \\ 0 & n \text{ even} \end{cases}$$

$$c'_n = \begin{cases} 0 & n \text{ odd} \\ c_{n/2} & n \text{ even} \end{cases}$$

That is, b'_n and c'_n look like b_n and c_n, but have zeros inserted in the "missing" places compared to a_n. Then $a_n = b'_n + c'_n$; also $S_1 = \sum b_n = \sum b'_n$ and $S_2 = \sum c_n = \sum c'_n$. Finally, since S, S_1, and S_2 all converge, we have

$$S = \sum_{n=1}^{\infty} a_n = \sum_{n=1}^{\infty} (b'_n + c'_n) = \sum_{n=1}^{\infty} b'_n + \sum_{n=1}^{\infty} c'_n = \sum_{n=1}^{\infty} b_n + \sum_{n=1}^{\infty} c_n = S_1 + S_2$$

Now, $b_n > c_n$ for every n, so that $S_1 > S_2$. Also, we showed above that $c_n = \frac{1}{2}a_n$, so that $2S_2 = S$. Putting all this together gives

$$S = S_1 + S_2 > S_2 + S_2 = 2S_2 = S$$

so that $S > S$, a contradiction. Thus S must diverge.

Further Insights and Challenges

91. Kummer's Acceleration Method Suppose we wish to approximate $S = \sum_{n=1}^{\infty} 1/n^2$. There is a similar telescoping series whose value can be computed exactly (Example 1 in Section 10.2):

$$\sum_{n=1}^{\infty} \frac{1}{n(n+1)} = 1$$

(a) Verify that

$$S = \sum_{n=1}^{\infty} \frac{1}{n(n+1)} + \sum_{n=1}^{\infty} \left(\frac{1}{n^2} - \frac{1}{n(n+1)} \right)$$

Thus for M large,

$$S \approx 1 + \sum_{n=1}^{M} \frac{1}{n^2(n+1)} \qquad \boxed{6}$$

(b) Explain what has been gained. Why is Eq. (6) a better approximation to S than is $\sum_{n=1}^{M} 1/n^2$?

(c) CAS Compute

$$\sum_{n=1}^{1000} \frac{1}{n^2}, \qquad 1 + \sum_{n=1}^{100} \frac{1}{n^2(n+1)}$$

Which is a better approximation to S, whose exact value is $\pi^2/6$?

SOLUTION

(a) Because the series $\sum_{n=1}^{\infty} \frac{1}{n^2}$ and $\sum_{n=1}^{\infty} \frac{1}{n(n+1)}$ both converge,

$$\sum_{n=1}^{\infty} \frac{1}{n(n+1)} + \sum_{n=1}^{\infty} \left(\frac{1}{n^2} - \frac{1}{n(n+1)} \right) = \sum_{n=1}^{\infty} \frac{1}{n(n+1)} + \sum_{n=1}^{\infty} \frac{1}{n^2} - \sum_{n=1}^{\infty} \frac{1}{n(n+1)} = \sum_{n=1}^{\infty} \frac{1}{n^2} = S.$$

Now,

$$\frac{1}{n^2} - \frac{1}{n(n+1)} = \frac{n+1}{n^2(n+1)} - \frac{n}{n^2(n+1)} = \frac{1}{n^2(n+1)},$$

so, for M large,

$$S \approx 1 + \sum_{n=1}^{M} \frac{1}{n^2(n+1)}.$$

(b) The series $\sum_{n=1}^{\infty} \frac{1}{n^2(n+1)}$ converges more rapidly than $\sum_{n=1}^{\infty} \frac{1}{n^2}$ since the degree of n in the denominator is larger.

(c) Using a computer algebra system, we find

$$\sum_{n=1}^{1000} \frac{1}{n^2} = 1.6439345667 \quad \text{and} \quad 1 + \sum_{n=1}^{100} \frac{1}{n^2(n+1)} = 1.6448848903.$$

The second sum is more accurate because it is closer to the exact solution $\frac{\pi^2}{6} \approx 1.6449340668$.

10.4 Absolute and Conditional Convergence

Preliminary Questions

1. Give an example of a series such that $\sum a_n$ converges but $\sum |a_n|$ diverges.

SOLUTION The series $\sum \frac{(-1)^n}{\sqrt[3]{n}}$ converges by the Leibniz Test, but the positive series $\sum \frac{1}{\sqrt[3]{n}}$ is a divergent p-series.

2. Which of the following statements is equivalent to Theorem 1?

(a) If $\sum_{n=0}^{\infty} |a_n|$ diverges, then $\sum_{n=0}^{\infty} a_n$ also diverges.

(b) If $\sum_{n=0}^{\infty} a_n$ diverges, then $\sum_{n=0}^{\infty} |a_n|$ also diverges.

(c) If $\sum_{n=0}^{\infty} a_n$ converges, then $\sum_{n=0}^{\infty} |a_n|$ also converges.

SOLUTION The correct answer is **(b)**: If $\sum_{n=0}^{\infty} a_n$ diverges, then $\sum_{n=0}^{\infty} |a_n|$ also diverges. Take $a_n = (-1)^n \frac{1}{n}$ to see that statements **(a)** and **(c)** are not true in general.

3. Lathika argues that $\sum_{n=1}^{\infty} (-1)^n \sqrt{n}$ is an alternating series and therefore converges. Is Lathika right?

SOLUTION No. Although $\sum_{n=1}^{\infty} (-1)^n \sqrt{n}$ is an alternating series, the terms $a_n = \sqrt{n}$ do not form a decreasing sequence that tends to zero. In fact, $a_n = \sqrt{n}$ is an increasing sequence that tends to ∞, so $\sum_{n=1}^{\infty} (-1)^n \sqrt{n}$ diverges by the Divergence Test.

4. Suppose that a_n is positive, decreasing, and tends to 0, and let $S = \sum_{n=1}^{\infty} (-1)^{n-1} a_n$. What can we say about $|S - S_{100}|$ if $a_{101} = 10^{-3}$? Is S larger or smaller than S_{100}?

SOLUTION From the text, we know that $|S - S_{100}| < a_{101} = 10^{-3}$. Also, the Leibniz test tells us that $S_{2N} < S < S_{2N+1}$ for any $N \geq 1$, so that $S_{100} < S$.

Exercises

1. Show that

$$\sum_{n=0}^{\infty} \frac{(-1)^n}{2^n}$$

converges absolutely.

SOLUTION The positive series $\sum_{n=0}^{\infty} \frac{1}{2^n}$ is a geometric series with $r = \frac{1}{2}$. Thus, the positive series converges, and the given series converges absolutely.

In Exercises 3–10, determine whether the series converges absolutely, conditionally, or not at all.

3. $\displaystyle\sum_{n=1}^{\infty} \frac{(-1)^{n-1}}{n^{1/3}}$

SOLUTION The sequence $a_n = \frac{1}{n^{1/3}}$ is positive, decreasing, and tends to zero; hence, the series $\displaystyle\sum_{n=1}^{\infty} \frac{(-1)^{n-1}}{n^{1/3}}$ converges by the Leibniz Test. However, the positive series $\displaystyle\sum_{n=1}^{\infty} \frac{1}{n^{1/3}}$ is a divergent p-series, so the original series converges conditionally.

5. $\displaystyle\sum_{n=0}^{\infty} \frac{(-1)^{n-1}}{(1.1)^n}$

SOLUTION The positive series $\displaystyle\sum_{n=0}^{\infty} \left(\frac{1}{1.1}\right)^n$ is a convergent geometric series; thus, the original series converges absolutely.

7. $\displaystyle\sum_{n=2}^{\infty} \frac{(-1)^n}{n \ln n}$

SOLUTION Let $a_n = \frac{1}{n \ln n}$. Then a_n forms a decreasing sequence (note that n and $\ln n$ are both increasing functions of n) that tends to zero; hence, the series $\displaystyle\sum_{n=2}^{\infty} \frac{(-1)^n}{n \ln n}$ converges by the Leibniz Test. However, the positive series $\displaystyle\sum_{n=2}^{\infty} \frac{1}{n \ln n}$ diverges, so the original series converges conditionally.

9. $\displaystyle\sum_{n=2}^{\infty} \frac{\cos n\pi}{(\ln n)^2}$

SOLUTION Since $\cos n\pi$ alternates between $+1$ and -1,

$$\sum_{n=2}^{\infty} \frac{\cos n\pi}{(lnn)^2} = \sum_{n=2}^{\infty} \frac{(-1)^n}{(lnn)^2}$$

This is an alternating series whose general term decreases to zero, so it converges. The associated positive series,

$$\sum_{n=2}^{\infty} \frac{1}{(\ln n)^2}$$

is a divergent series, so the original series converges conditionally.

11. Let $S = \displaystyle\sum_{n=1}^{\infty} (-1)^{n+1} \frac{1}{n^3}$.

(a) Calculate S_n for $1 \le n \le 10$.

(b) Use Eq. (2) to show that $0.9 \le S \le 0.902$.

SOLUTION

(a)

$$S_1 = 1 \qquad S_6 = S_5 - \frac{1}{6^3} = 0.899782407$$

$$S_2 = 1 - \frac{1}{2^3} = \frac{7}{8} = 0.875 \qquad S_7 = S_6 + \frac{1}{7^3} = 0.902697859$$

$$S_3 = S_2 + \frac{1}{3^3} = 0.912037037 \qquad S_8 = S_7 - \frac{1}{8^3} = 0.900744734$$

$$S_4 = S_3 - \frac{1}{4^3} = 0.896412037 \qquad S_9 = S_8 + \frac{1}{9^3} = 0.902116476$$

$$S_5 = S_4 + \frac{1}{5^3} = 0.904412037 \qquad S_{10} = S_9 - \frac{1}{10^3} = 0.901116476$$

(b) By Eq. (2),

$$|S_{10} - S| \leq a_{11} = \frac{1}{11^3},$$

so

$$S_{10} - \frac{1}{11^3} \leq S \leq S_{10} + \frac{1}{11^3},$$

or

$$0.900365161 \leq S \leq 0.901867791.$$

13. Approximate $\sum_{n=1}^{\infty} \frac{(-1)^{n+1}}{n^4}$ to three decimal places.

SOLUTION Let $S = \sum_{n=1}^{\infty} \frac{(-1)^{n+1}}{n^4}$, so that $a_n = \frac{1}{n^4}$. By Eq. (2),

$$|S_N - S| \leq a_{N+1} = \frac{1}{(N+1)^4}.$$

To guarantee accuracy to three decimal places, we must choose N so that

$$\frac{1}{(N+1)^4} < 5 \times 10^{-4} \quad \text{or} \quad N > \sqrt[4]{2000} - 1 \approx 5.7.$$

The smallest value that satisfies the required inequality is then $N = 6$. Thus,

$$S \approx S_6 = 1 - \frac{1}{2^4} + \frac{1}{3^4} - \frac{1}{4^4} + \frac{1}{5^4} - \frac{1}{6^4} = 0.946767824.$$

In Exercises 15 and 16, find a value of N such that S_N approximates the series with an error of at most 10^{-5}. If you have a CAS, compute this value of S_N.

15. $\sum_{n=1}^{\infty} \frac{(-1)^{n+1}}{n(n+2)(n+3)}$

SOLUTION Let $S = \sum_{n=1}^{\infty} \frac{(-1)^{n+1}}{n\,(n+2)\,(n+3)}$, so that $a_n = \frac{1}{n\,(n+2)\,(n+3)}$. By Eq. (2),

$$|S_N - S| \leq a_{N+1} = \frac{1}{(N+1)(N+3)(N+4)}.$$

We must choose N so that

$$\frac{1}{(N+1)(N+3)(N+4)} \leq 10^{-5} \quad \text{or} \quad (N+1)(N+3)(N+4) \geq 10^5.$$

For $N = 43$, the product on the left hand side is 95,128, while for $N = 44$ the product is 101,520; hence, the smallest value of N which satisfies the required inequality is $N = 44$. Thus,

$$S \approx S_{44} = \sum_{n=1}^{44} \frac{(-1)^{n+1}}{n(n+2)(n+3)} = 0.0656746.$$

In Exercises 17–32, determine convergence or divergence by any method.

17. $\sum_{n=0}^{\infty} 7^{-n}$

SOLUTION This is a (positive) geometric series with $r = \frac{1}{7} < 1$, so it converges.

19. $\displaystyle\sum_{n=1}^{\infty} \frac{1}{5^n - 3^n}$

SOLUTION Use the Limit Comparison Test with $\dfrac{1}{5^n}$:

$$L = \lim_{n\to\infty} \frac{1/(5^n - 3^n)}{1/5^n} = \lim_{n\to\infty} \frac{5^n}{5^n - 3^n} = \lim_{n\to\infty} \frac{1}{1-(3/5)^n} = 1$$

But $\sum_{n=1}^{\infty} \dfrac{1}{5^n}$ is a convergent geometric series. Since $L = 1$, the Limit Comparison Test tells us that the original series converges as well.

21. $\displaystyle\sum_{n=1}^{\infty} \frac{1}{3n^4 + 12n}$

SOLUTION Use the Limit Comparison Test with $\dfrac{1}{3n^4}$:

$$L = \lim_{n\to\infty} \frac{(1/(3n^4 + 12n)}{1/3n^4} = \lim_{n\to\infty} \frac{3n^4}{3n^4 + 12n} = \lim_{n\to\infty} \frac{1}{1 + 4n^{-3}} = 1$$

But $\sum_{n=1}^{\infty} \dfrac{1}{3n^4} = \frac{1}{3} \sum_{n=1}^{\infty} \frac{1}{n^4}$ is a convergent p-series. Since $L = 1$, the Limit Comparison Test tells us that the original series converges as well.

23. $\displaystyle\sum_{n=1}^{\infty} \frac{1}{\sqrt{n^2 + 1}}$

SOLUTION Apply the Limit Comparison Test and compare the series with the divergent harmonic series:

$$L = \lim_{n\to\infty} \frac{\frac{1}{\sqrt{n^2+1}}}{\frac{1}{n}} = \lim_{n\to\infty} \frac{n}{\sqrt{n^2+1}} = 1.$$

Because $L > 0$, we conclude that the series $\displaystyle\sum_{n=1}^{\infty} \frac{1}{\sqrt{n^2+1}}$ diverges.

25. $\displaystyle\sum_{n=1}^{\infty} \frac{3^n + (-2)^n}{5^n}$

SOLUTION The series

$$\sum_{n=1}^{\infty} \frac{3^n}{5^n} = \sum_{n=1}^{\infty} \left(\frac{3}{5}\right)^n$$

is a convergent geometric series, as is the series

$$\sum_{n=1}^{\infty} \frac{(-1)^n\, 2^n}{5^n} = \sum_{n=1}^{\infty} \left(-\frac{2}{5}\right)^n.$$

Hence,

$$\sum_{n=1}^{\infty} \frac{3^n + (-1)^n 2^n}{5^n} = \sum_{n=1}^{\infty} \left(\frac{3}{5}\right)^n + \sum_{n=1}^{\infty} \left(-\frac{2}{5}\right)^n$$

also converges.

27. $\displaystyle\sum_{n=1}^{\infty} (-1)^n n^2 e^{-n^3/3}$

SOLUTION Consider the associated positive series $\displaystyle\sum_{n=1}^{\infty} n^2 e^{-n^3/3}$. This series can be seen to converge by the Integral Test:

$$\int_1^{\infty} x^2 e^{-x^3/3}\, dx = \lim_{R\to\infty} \int_1^R x^2 e^{-x^3/3}\, dx = -\lim_{R\to\infty} e^{-x^3/3}\Big|_1^R = e^{-1/3} + \lim_{R\to\infty} e^{-R^3/3} = e^{-1/3}.$$

The integral converges, so the original series converges absolutely.

29. $\displaystyle\sum_{n=2}^{\infty} \frac{(-1)^n}{n^{1/2}(\ln n)^2}$

SOLUTION This is an alternating series with $a_n = \dfrac{1}{n^{1/2}(\ln n)^2}$. Because a_n is a decreasing sequence which converges to zero, the series $\displaystyle\sum_{n=2}^{\infty} \frac{(-1)^n}{n^{1/2}(\ln n)^2}$ converges by the Leibniz Test. (Note that the series converges only conditionally, not absolutely; the associated positive series is eventually greater than $\dfrac{1}{n^{3/4}}$, which is a divergent p-series).

31. $\displaystyle\sum_{n=1}^{\infty} \frac{\ln n}{n^{1.05}}$

SOLUTION Choose N so that for $n \geq N$ we have $\ln n \leq n^{0.01}$. Then

$$\sum_{n=N}^{\infty} \frac{\ln n}{n^{1.05}} \leq \sum_{n=N}^{\infty} \frac{n^{0.01}}{n^{1.05}} = \sum_{n=N}^{\infty} \frac{1}{n^{1.04}}$$

This is a convergent p-series, so by the Comparison Test, the original series converges as well.

33. Show that

$$S = \frac{1}{2} - \frac{1}{2} + \frac{1}{3} - \frac{1}{3} + \frac{1}{4} - \frac{1}{4} + \cdots$$

converges by computing the partial sums. Does it converge absolutely?

SOLUTION The sequence of partial sums is

$$S_1 = \frac{1}{2}$$

$$S_2 = S_1 - \frac{1}{2} = 0$$

$$S_3 = S_2 + \frac{1}{3} = \frac{1}{3}$$

$$S_4 = S_3 - \frac{1}{3} = 0$$

and, in general,

$$S_N = \begin{cases} \dfrac{1}{N}, & \text{for odd } N \\ 0, & \text{for even } N \end{cases}$$

Thus, $\lim_{N\to\infty} S_N = 0$, and the series converges to 0. The positive series is

$$\frac{1}{2} + \frac{1}{2} + \frac{1}{3} + \frac{1}{3} + \frac{1}{4} + \frac{1}{4} + \cdots = 2\sum_{n=2}^{\infty} \frac{1}{n};$$

which diverges. Therefore, the original series converges conditionally, not absolutely.

35. **Assumptions Matter** Show by counterexample that the Leibniz Test does not remain true if the sequence a_n tends to zero but is not assumed nonincreasing. *Hint:* Consider

$$R = \frac{1}{2} - \frac{1}{4} + \frac{1}{3} - \frac{1}{8} + \frac{1}{4} - \frac{1}{16} + \cdots + \left(\frac{1}{n} - \frac{1}{2^n}\right) + \cdots$$

SOLUTION Let

$$R = \frac{1}{2} - \frac{1}{4} + \frac{1}{3} - \frac{1}{8} + \frac{1}{4} - \frac{1}{16} + \cdots + \left(\frac{1}{n+1} - \frac{1}{2^{n+1}}\right) + \cdots$$

This is an alternating series with

$$a_n = \begin{cases} \dfrac{1}{k+1}, & n = 2k-1 \\ \dfrac{1}{2^{k+1}}, & n = 2k \end{cases}$$

Note that $a_n \to 0$ as $n \to \infty$, but the sequence $\{a_n\}$ is not decreasing. We will now establish that R diverges.

For sake of contradiction, suppose that R converges. The geometric series

$$\sum_{n=1}^{\infty} \frac{1}{2^{n+1}}$$

converges, so the sum of R and this geometric series must also converge; however,

$$R + \sum_{n=1}^{\infty} \frac{1}{2^{n+1}} = \sum_{n=2}^{\infty} \frac{1}{n},$$

which diverges because the harmonic series diverges. Thus, the series R must diverge.

37. Prove that if $\sum a_n$ converges absolutely, then $\sum a_n^2$ also converges. Then give an example where $\sum a_n$ is only conditionally convergent and $\sum a_n^2$ diverges.

SOLUTION Suppose the series $\sum a_n$ converges absolutely. Because $\sum |a_n|$ converges, we know that

$$\lim_{n\to\infty} |a_n| = 0.$$

Therefore, there exists a positive integer N such that $|a_n| < 1$ for all $n \geq N$. It then follows that for $n \geq N$,

$$0 \leq a_n^2 = |a_n|^2 = |a_n| \cdot |a_n| < |a_n| \cdot 1 = |a_n|.$$

By the Comparison Test we can then conclude that $\sum a_n^2$ also converges.

Consider the series $\sum_{n=1}^{\infty} \frac{(-1)^n}{\sqrt{n}}$. This series converges by the Leibniz Test, but the corresponding positive series is a divergent p-series; that is, $\sum_{n=1}^{\infty} \frac{(-1)^n}{\sqrt{n}}$ is conditionally convergent. Now, $\sum_{n=1}^{\infty} a_n^2$ is the divergent harmonic series $\sum_{n=1}^{\infty} \frac{1}{n}$. Thus, $\sum a_n^2$ need not converge if $\sum a_n$ is only conditionally convergent.

Further Insights and Challenges

39. Use Exercise 38 to show that the following series converges:

$$S = \frac{1}{\ln 2} + \frac{1}{\ln 3} - \frac{2}{\ln 4} + \frac{1}{\ln 5} + \frac{1}{\ln 6} - \frac{2}{\ln 7} + \cdots$$

SOLUTION The given series has the structure of the generic series from Exercise 38 with $a_n = \frac{1}{\ln(n+1)}$. Because a_n is a positive, decreasing sequence with $\lim_{n\to\infty} a_n = 0$, we can conclude from Exercise 38 that the given series converges.

41. Show that the following series diverges:

$$S = 1 + \frac{1}{2} + \frac{1}{3} - \frac{2}{4} + \frac{1}{5} + \frac{1}{6} + \frac{1}{7} - \frac{2}{8} + \cdots$$

Hint: Use the result of Exercise 40 to write S as the sum of a convergent series and a divergent series.

SOLUTION Let

$$R = 1 + \frac{1}{2} + \frac{1}{3} - \frac{3}{4} + \frac{1}{5} + \frac{1}{6} + \frac{1}{7} - \frac{3}{8} + \cdots$$

and

$$S = 1 + \frac{1}{2} + \frac{1}{3} - \frac{2}{4} + \frac{1}{5} + \frac{1}{6} + \frac{1}{7} - \frac{2}{8} + \cdots$$

For sake of contradiction, suppose the series S converges. From Exercise 40, we know that the series R converges. Thus, the series $S - R$ must converge; however,

$$S - R = \frac{1}{4} + \frac{1}{8} + \frac{1}{12} + \cdots = \frac{1}{4}\sum_{k=1}^{\infty} \frac{1}{k},$$

which diverges because the harmonic series diverges. Thus, the series S must diverge.

43. We say that $\{b_n\}$ is a rearrangement of $\{a_n\}$ if $\{b_n\}$ has the same terms as $\{a_n\}$ but occurring in a different order. Show that if $\{b_n\}$ is a rearrangement of $\{a_n\}$ and $S = \sum_{n=1}^{\infty} a_n$ converges absolutely, then $T = \sum_{n=1}^{\infty} b_n$ also converges absolutely. (This result does not hold if S is only conditionally convergent.) *Hint:* Prove that the partial sums $\sum_{n=1}^{N} |b_n|$ are bounded. It can be shown further that $S = T$.

SOLUTION Suppose the series $S = \sum_{n=1}^{\infty} a_n$ converges absolutely and denote the corresponding positive series by

$$S^+ = \sum_{n=1}^{\infty} |a_n|.$$

Further, let $T_N = \sum_{n=1}^{N} |b_n|$ denote the Nth partial sum of the series $\sum_{n=1}^{\infty} |b_n|$. Because $\{b_n\}$ is a rearrangement of $\{a_n\}$, we know that

$$0 \le T_N \le \sum_{n=1}^{\infty} |a_n| = S^+;$$

that is, the sequence $\{T_N\}$ is bounded. Moreover,

$$T_{N+1} = \sum_{n=1}^{N+1} |b_n| = T_N + |b_{N+1}| \ge T_N;$$

that is, $\{T_N\}$ is increasing. It follows that $\{T_N\}$ converges, so the series $\sum_{n=1}^{\infty} |b_n|$ converges, which means the series $\sum_{n=1}^{\infty} b_n$ converges absolutely.

10.5 The Ratio and Root Tests

Preliminary Questions

1. In the Ratio Test, is ρ equal to $\lim_{n\to\infty} \left|\frac{a_{n+1}}{a_n}\right|$ or $\lim_{n\to\infty} \left|\frac{a_n}{a_{n+1}}\right|$?

SOLUTION In the Ratio Test ρ is the limit $\lim_{n\to\infty} \left|\frac{a_{n+1}}{a_n}\right|$.

2. Is the Ratio Test conclusive for $\sum_{n=1}^{\infty} \frac{1}{2^n}$? Is it conclusive for $\sum_{n=1}^{\infty} \frac{1}{n}$?

SOLUTION The general term of $\sum_{n=1}^{\infty} \frac{1}{2^n}$ is $a_n = \frac{1}{2^n}$; thus,

$$\left|\frac{a_{n+1}}{a_n}\right| = \frac{1}{2^{n+1}} \cdot \frac{2^n}{1} = \frac{1}{2},$$

and

$$\rho = \lim_{n\to\infty} \left|\frac{a_{n+1}}{a_n}\right| = \frac{1}{2} < 1.$$

Consequently, the Ratio Test guarantees that the series $\sum_{n=1}^{\infty} \frac{1}{2^n}$ converges.

The general term of $\sum_{n=1}^{\infty} \frac{1}{n}$ is $a_n = \frac{1}{n}$; thus,

$$\left|\frac{a_{n+1}}{a_n}\right| = \frac{1}{n+1} \cdot \frac{n}{1} = \frac{n}{n+1},$$

and

$$\rho = \lim_{n\to\infty}\left|\frac{a_{n+1}}{a_n}\right| = \lim_{n\to\infty}\frac{n}{n+1} = 1.$$

The Ratio Test is therefore inconclusive for the series $\sum_{n=1}^{\infty}\frac{1}{n}$.

3. Can the Ratio Test be used to show convergence if the series is only conditionally convergent?

SOLUTION No. The Ratio Test can only establish absolute convergence and divergence, not conditional convergence.

Exercises

In Exercises 1–20, apply the Ratio Test to determine convergence or divergence, or state that the Ratio Test is inconclusive.

1. $\sum_{n=1}^{\infty}\frac{1}{5^n}$

SOLUTION With $a_n = \frac{1}{5^n}$,

$$\left|\frac{a_{n+1}}{a_n}\right| = \frac{1}{5^{n+1}}\cdot\frac{5^n}{1} = \frac{1}{5} \quad\text{and}\quad \rho = \lim_{n\to\infty}\left|\frac{a_{n+1}}{a_n}\right| = \frac{1}{5} < 1.$$

Therefore, the series $\sum_{n=1}^{\infty}\frac{1}{5^n}$ converges by the Ratio Test.

3. $\sum_{n=1}^{\infty}\frac{1}{n^n}$

SOLUTION With $a_n = \frac{1}{n^n}$,

$$\left|\frac{a_{n+1}}{a_n}\right| = \frac{1}{(n+1)^{n+1}}\cdot\frac{n^n}{1} = \frac{1}{n+1}\left(\frac{n}{n+1}\right)^n = \frac{1}{n+1}\left(1+\frac{1}{n}\right)^{-n},$$

and

$$\rho = \lim_{n\to\infty}\left|\frac{a_{n+1}}{a_n}\right| = 0\cdot\frac{1}{e} = 0 < 1.$$

Therefore, the series $\sum_{n=1}^{\infty}\frac{1}{n^n}$ converges by the Ratio Test.

5. $\sum_{n=1}^{\infty}\frac{n}{n^2+1}$

SOLUTION With $a_n = \frac{n}{n^2+1}$,

$$\left|\frac{a_{n+1}}{a_n}\right| = \frac{n+1}{(n+1)^2+1}\cdot\frac{n^2+1}{n} = \frac{n+1}{n}\cdot\frac{n^2+1}{n^2+2n+2},$$

and

$$\rho = \lim_{n\to\infty}\left|\frac{a_{n+1}}{a_n}\right| = 1\cdot 1 = 1.$$

Therefore, for the series $\sum_{n=1}^{\infty}\frac{n}{n^2+1}$, the Ratio Test is inconclusive.

We can show that this series diverges by using the Limit Comparison Test and comparing with the divergent harmonic series.

7. $\sum_{n=1}^{\infty}\frac{2^n}{n^{100}}$

SOLUTION With $a_n = \frac{2^n}{n^{100}}$,

$$\left|\frac{a_{n+1}}{a_n}\right| = \frac{2^{n+1}}{(n+1)^{100}}\cdot\frac{n^{100}}{2^n} = 2\left(\frac{n}{n+1}\right)^{100} \quad\text{and}\quad \rho = \lim_{n\to\infty}\left|\frac{a_{n+1}}{a_n}\right| = 2\cdot 1^{100} = 2 > 1.$$

Therefore, the series $\sum_{n=1}^{\infty}\frac{2^n}{n^{100}}$ diverges by the Ratio Test.

9. $\displaystyle\sum_{n=1}^{\infty} \frac{10^n}{2^{n^2}}$

SOLUTION With $a_n = \frac{10^n}{2^{n^2}}$,

$$\left|\frac{a_{n+1}}{a_n}\right| = \frac{10^{n+1}}{2^{(n+1)^2}} \cdot \frac{2^{n^2}}{10^n} = 10 \cdot \frac{1}{2^{2n+1}} \quad \text{and} \quad \rho = \lim_{n\to\infty}\left|\frac{a_{n+1}}{a_n}\right| = 10 \cdot 0 = 0 < 1.$$

Therefore, the series $\displaystyle\sum_{n=1}^{\infty} \frac{10^n}{2^{n^2}}$ converges by the Ratio Test.

11. $\displaystyle\sum_{n=1}^{\infty} \frac{e^n}{n^n}$

SOLUTION With $a_n = \frac{e^n}{n^n}$,

$$\left|\frac{a_{n+1}}{a_n}\right| = \frac{e^{n+1}}{(n+1)^{n+1}} \cdot \frac{n^n}{e^n} = \frac{e}{n+1}\left(\frac{n}{n+1}\right)^n = \frac{e}{n+1}\left(1+\frac{1}{n}\right)^{-n},$$

and

$$\rho = \lim_{n\to\infty}\left|\frac{a_{n+1}}{a_n}\right| = 0 \cdot \frac{1}{e} = 0 < 1.$$

Therefore, the series $\displaystyle\sum_{n=1}^{\infty} \frac{e^n}{n^n}$ converges by the Ratio Test.

13. $\displaystyle\sum_{n=0}^{\infty} \frac{n!}{6^n}$

SOLUTION With $a_n = \frac{n!}{6^n}$,

$$\left|\frac{a_{n+1}}{a_n}\right| = \frac{(n+1)!}{6^{n+1}} \cdot \frac{6^n}{n!} = \frac{n+1}{6} \quad \text{and} \quad \rho = \lim_{n\to\infty}\left|\frac{a_{n+1}}{a_n}\right| = \infty > 1.$$

Therefore, the series $\displaystyle\sum_{n=0}^{\infty} \frac{n!}{6^n}$ diverges by the Ratio Test.

15. $\displaystyle\sum_{n=2}^{\infty} \frac{1}{n \ln n}$

SOLUTION With $a_n = \frac{1}{n \ln n}$,

$$\left|\frac{a_{n+1}}{a_n}\right| = \frac{1}{(n+1)\ln(n+1)} \cdot \frac{n \ln n}{1} = \frac{n}{n+1}\frac{\ln n}{\ln(n+1)},$$

and

$$\rho = \lim_{n\to\infty}\left|\frac{a_{n+1}}{a_n}\right| = 1 \cdot \lim_{n\to\infty}\frac{\ln n}{\ln(n+1)}.$$

Now,

$$\lim_{n\to\infty}\frac{\ln n}{\ln(n+1)} = \lim_{x\to\infty}\frac{\ln x}{\ln(x+1)} = \lim_{x\to\infty}\frac{1/(x+1)}{1/x} = \lim_{x\to\infty}\frac{x}{x+1} = 1.$$

Thus, $\rho = 1$, and the Ratio Test is inconclusive for the series $\displaystyle\sum_{n=2}^{\infty} \frac{1}{n \ln n}$.

Using the Integral Test, we can show that the series $\displaystyle\sum_{n=2}^{\infty} \frac{1}{n \ln n}$ diverges.

17. $\displaystyle\sum_{n=1}^{\infty} \frac{n^2}{(2n+1)!}$

SOLUTION With $a_n = \frac{n^2}{(2n+1)!}$,

$$\left|\frac{a_{n+1}}{a_n}\right| = \frac{(n+1)^2}{(2n+3)!} \cdot \frac{(2n+1)!}{n^2} = \left(\frac{n+1}{n}\right)^2 \frac{1}{(2n+3)(2n+2)},$$

and

$$\rho = \lim_{n\to\infty}\left|\frac{a_{n+1}}{a_n}\right| = 1^2 \cdot 0 = 0 < 1.$$

Therefore, the series $\sum_{n=1}^{\infty} \frac{n^2}{(2n+1)!}$ converges by the Ratio Test.

19. $\sum_{n=2}^{\infty} \frac{1}{2^n+1}$

SOLUTION With $a_n = \frac{1}{2^n+1}$,

$$\left|\frac{a_{n+1}}{a_n}\right| = \frac{1}{2^{n+1}+1} \cdot \frac{2^n+1}{1} = \frac{1+2^{-n}}{2+2^{-n}}$$

and

$$\rho = \lim_{n\to\infty}\left|\frac{a_{n+1}}{a_n}\right| = \frac{1}{2} < 1$$

Therefore, the series $\sum_{n=2}^{\infty} \frac{1}{2^n+1}$ converges by the Ratio Test.

21. Show that $\sum_{n=1}^{\infty} n^k\, 3^{-n}$ converges for all exponents k.

SOLUTION With $a_n = n^k 3^{-n}$,

$$\left|\frac{a_{n+1}}{a_n}\right| = \frac{(n+1)^k 3^{-(n+1)}}{n^k 3^{-n}} = \frac{1}{3}\left(1+\frac{1}{n}\right)^k,$$

and, for all k,

$$\rho = \lim_{n\to\infty}\left|\frac{a_{n+1}}{a_n}\right| = \frac{1}{3}\cdot 1 = \frac{1}{3} < 1.$$

Therefore, the series $\sum_{n=1}^{\infty} n^k\, 3^{-n}$ converges for all exponents k by the Ratio Test.

23. Show that $\sum_{n=1}^{\infty} 2^n x^n$ converges if $|x| < \frac{1}{2}$.

SOLUTION With $a_n = 2^n x^n$,

$$\left|\frac{a_{n+1}}{a_n}\right| = \frac{2^{n+1}|x|^{n+1}}{2^n|x|^n} = 2|x| \quad \text{and} \quad \rho = \lim_{n\to\infty}\left|\frac{a_{n+1}}{a_n}\right| = 2|x|.$$

Therefore, $\rho < 1$ and the series $\sum_{n=1}^{\infty} 2^n x^n$ converges by the Ratio Test provided $|x| < \frac{1}{2}$.

25. Show that $\sum_{n=1}^{\infty} \frac{r^n}{n}$ converges if $|r| < 1$.

SOLUTION With $a_n = \frac{r^n}{n}$,

$$\left|\frac{a_{n+1}}{a_n}\right| = \frac{|r|^{n+1}}{n+1} \cdot \frac{n}{|r|^n} = |r|\frac{n}{n+1} \quad \text{and} \quad \rho = \lim_{n\to\infty}\left|\frac{a_{n+1}}{a_n}\right| = 1 \cdot |r| = |r|.$$

Therefore, by the Ratio Test, the series $\sum_{n=1}^{\infty} \frac{r^n}{n}$ converges provided $|r| < 1$.

27. Show that $\sum_{n=1}^{\infty} \frac{n!}{n^n}$ converges. *Hint:* Use $\lim_{n\to\infty} \left(1+\frac{1}{n}\right)^n = e$.

SOLUTION With $a_n = \frac{n!}{n^n}$,

$$\left|\frac{a_{n+1}}{a_n}\right| = \frac{(n+1)!}{(n+1)^{n+1}} \cdot \frac{n^n}{n!} = \left(\frac{n}{n+1}\right)^n = \left(1+\frac{1}{n}\right)^{-n},$$

and

$$\rho = \lim_{n\to\infty}\left|\frac{a_{n+1}}{a_n}\right| = \frac{1}{e} < 1.$$

Therefore, the series $\sum_{n=1}^{\infty} \frac{n!}{n^n}$ converges by the Ratio Test.

In Exercises 28–33, assume that $|a_{n+1}/a_n|$ converges to $\rho = \frac{1}{3}$. What can you say about the convergence of the given series?

29. $\sum_{n=1}^{\infty} n^3 a_n$

SOLUTION Let $b_n = n^3 a_n$. Then

$$\rho = \lim_{n\to\infty}\left|\frac{b_{n+1}}{b_n}\right| = \lim_{n\to\infty}\left(\frac{n+1}{n}\right)^3 \left|\frac{a_{n+1}}{a_n}\right| = 1^3 \cdot \frac{1}{3} = \frac{1}{3} < 1.$$

Therefore, the series $\sum_{n=1}^{\infty} n^3 a_n$ converges by the Ratio Test.

31. $\sum_{n=1}^{\infty} 3^n a_n$

SOLUTION Let $b_n = 3^n a_n$. Then

$$\rho = \lim_{n\to\infty}\left|\frac{b_{n+1}}{b_n}\right| = \lim_{n\to\infty} \frac{3^{n+1}}{3^n}\left|\frac{a_{n+1}}{a_n}\right| = 3 \cdot \frac{1}{3} = 1.$$

Therefore, the Ratio Test is inconclusive for the series $\sum_{n=1}^{\infty} 3^n a_n$.

33. $\sum_{n=1}^{\infty} a_n^2$

SOLUTION Let $b_n = a_n^2$. Then

$$\rho = \lim_{n\to\infty}\left|\frac{b_{n+1}}{b_n}\right| = \lim_{n\to\infty}\left|\frac{a_{n+1}}{a_n}\right|^2 = \left(\frac{1}{3}\right)^2 = \frac{1}{9} < 1.$$

Therefore, the series $\sum_{n=1}^{\infty} a_n^2$ converges by the Ratio Test.

35. Is the Ratio Test conclusive for the p-series $\sum_{n=1}^{\infty} \frac{1}{n^p}$?

SOLUTION With $a_n = \frac{1}{n^p}$,

$$\left|\frac{a_{n+1}}{a_n}\right| = \frac{1}{(n+1)^p} \cdot \frac{n^p}{1} = \left(\frac{n}{n+1}\right)^p \quad \text{and} \quad \rho = \lim_{n\to\infty}\left|\frac{a_{n+1}}{a_n}\right| = 1^p = 1.$$

Therefore, the Ratio Test is inconclusive for the p-series $\sum_{n=1}^{\infty} \frac{1}{n^p}$.

In Exercises 36–41, use the Root Test to determine convergence or divergence (or state that the test is inconclusive).

37. $\displaystyle\sum_{n=1}^{\infty} \frac{1}{n^n}$

SOLUTION With $a_n = \frac{1}{n^n}$,

$$\sqrt[n]{a_n} = \sqrt[n]{\frac{1}{n^n}} = \frac{1}{n} \quad \text{and} \quad \lim_{n\to\infty} \sqrt[n]{a_n} = 0 < 1.$$

Therefore, the series $\displaystyle\sum_{n=1}^{\infty} \frac{1}{n^n}$ converges by the Root Test.

39. $\displaystyle\sum_{k=0}^{\infty} \left(\frac{k}{3k+1}\right)^k$

SOLUTION With $a_k = \left(\frac{k}{3k+1}\right)^k$,

$$\sqrt[k]{a_k} = \sqrt[k]{\left(\frac{k}{3k+1}\right)^k} = \frac{k}{3k+1} \quad \text{and} \quad \lim_{k\to\infty} \sqrt[k]{a_k} = \frac{1}{3} < 1.$$

Therefore, the series $\displaystyle\sum_{k=0}^{\infty} \left(\frac{k}{3k+1}\right)^k$ converges by the Root Test.

41. $\displaystyle\sum_{n=4}^{\infty} \left(1 + \frac{1}{n}\right)^{-n^2}$

SOLUTION With $a_k = \left(1 + \frac{1}{n}\right)^{-n^2}$,

$$\sqrt[n]{a_n} = \sqrt[n]{\left(1 + \frac{1}{n}\right)^{-n^2}} = \left(1 + \frac{1}{n}\right)^{-n} \quad \text{and} \quad \lim_{n\to\infty} \sqrt[n]{a_n} = e^{-1} < 1.$$

Therefore, the series $\displaystyle\sum_{n=4}^{\infty} \left(1 + \frac{1}{n}\right)^{-n^2}$ converges by the Root Test.

In Exercises 43–56, determine convergence or divergence using any method covered in the text so far.

43. $\displaystyle\sum_{n=1}^{\infty} \frac{2^n + 4^n}{7^n}$

SOLUTION Because the series

$$\sum_{n=1}^{\infty} \frac{2^n}{7^n} = \sum_{n=1}^{\infty} \left(\frac{2}{7}\right)^n \quad \text{and} \quad \sum_{n=1}^{\infty} \frac{4^n}{7^n} = \sum_{n=1}^{\infty} \left(\frac{4}{7}\right)^n$$

are both convergent geometric series, it follows that

$$\sum_{n=1}^{\infty} \frac{2^n + 4^n}{7^n} = \sum_{n=1}^{\infty} \left(\frac{2}{7}\right)^n + \sum_{n=1}^{\infty} \left(\frac{4}{7}\right)^n$$

also converges.

45. $\displaystyle\sum_{n=1}^{\infty} \frac{n^3}{5^n}$

SOLUTION The presence of the exponential term suggests applying the Ratio Test. With $a_n = \frac{n^3}{5^n}$,

$$\left|\frac{a_{n+1}}{a_n}\right| = \frac{(n+1)^3}{5^{n+1}} \cdot \frac{5^n}{n^3} = \frac{1}{5}\left(1 + \frac{1}{n}\right)^3 \quad \text{and} \quad \rho = \lim_{n\to\infty} \left|\frac{a_{n+1}}{a_n}\right| = \frac{1}{5} \cdot 1^3 = \frac{1}{5} < 1.$$

Therefore, the series $\displaystyle\sum_{n=1}^{\infty} \frac{n^3}{5^n}$ converges by the Ratio Test.

47. $\displaystyle\sum_{n=2}^{\infty} \frac{1}{\sqrt{n^3 - n^2}}$

SOLUTION This series is similar to a p-series; because

$$\frac{1}{\sqrt{n^3 - n^2}} \approx \frac{1}{\sqrt{n^3}} = \frac{1}{n^{3/2}}$$

for large n, we will apply the Limit Comparison Test comparing with the p-series with $p = \frac{3}{2}$. Now,

$$L = \lim_{n\to\infty} \frac{\frac{1}{\sqrt{n^3-n^2}}}{\frac{1}{n^{3/2}}} = \lim_{n\to\infty} \sqrt{\frac{n^3}{n^3 - n^2}} = 1.$$

The p-series with $p = \frac{3}{2}$ converges and L exists; therefore, the series $\displaystyle\sum_{n=2}^{\infty} \frac{1}{\sqrt{n^3 - n^2}}$ also converges.

49. $\displaystyle\sum_{n=1}^{\infty} n^{-0.8}$

SOLUTION

$$\sum_{n=1}^{\infty} n^{-0.8} = \sum_{n=1}^{\infty} \frac{1}{n^{0.8}}$$

so that this is a divergent p-series.

51. $\displaystyle\sum_{n=1}^{\infty} 4^{-2n+1}$

SOLUTION Observe

$$\sum_{n=1}^{\infty} 4^{-2n+1} = \sum_{n=1}^{\infty} 4 \cdot (4^{-2})^n = \sum_{n=1}^{\infty} 4\left(\frac{1}{16}\right)^n$$

is a geometric series with $r = \frac{1}{16}$; therefore, this series converges.

53. $\displaystyle\sum_{n=1}^{\infty} \sin \frac{1}{n^2}$

SOLUTION Here, we will apply the Limit Comparison Test, comparing with the p-series with $p = 2$. Now,

$$L = \lim_{n\to\infty} \frac{\sin \frac{1}{n^2}}{\frac{1}{n^2}} = \lim_{u\to 0} \frac{\sin u}{u} = 1,$$

where $u = \frac{1}{n^2}$. The p-series with $p = 2$ converges and L exists; therefore, the series $\displaystyle\sum_{n=1}^{\infty} \sin \frac{1}{n^2}$ also converges.

55. $\displaystyle\sum_{n=1}^{\infty} \frac{(-2)^n}{\sqrt{n}}$

SOLUTION Because

$$\lim_{n\to\infty} \frac{2^n}{\sqrt{n}} = \lim_{x\to\infty} \frac{2^x}{\sqrt{x}} = \lim_{x\to\infty} \frac{2^x \ln 2}{\frac{1}{2\sqrt{x}}} = \lim_{x\to\infty} 2^{x+1}\sqrt{x} \ln 2 = \infty \neq 0,$$

the general term in the series $\displaystyle\sum_{n=1}^{\infty} \frac{(-2)^n}{\sqrt{n}}$ does not tend toward zero; therefore, the series diverges by the Divergence Test.

Further Insights and Challenges

57. **Proof of the Root Test** Let $S = \displaystyle\sum_{n=0}^{\infty} a_n$ be a positive series, and assume that $L = \displaystyle\lim_{n\to\infty} \sqrt[n]{a_n}$ exists.

(a) Show that S converges if $L < 1$. *Hint:* Choose R with $L < R < 1$ and show that $a_n \leq R^n$ for n sufficiently large Then compare with the geometric series $\sum R^n$.

(b) Show that S diverges if $L > 1$.

SOLUTION Suppose $\lim_{n\to\infty} \sqrt[n]{a_n} = L$ exists.

(a) If $L < 1$, let $\epsilon = \dfrac{1-L}{2}$. By the definition of a limit, there is a positive integer N such that

$$-\epsilon \le \sqrt[n]{a_n} - L \le \epsilon$$

for $n \ge N$. From this, we conclude that

$$0 \le \sqrt[n]{a_n} \le L + \epsilon$$

for $n \ge N$. Now, let $R = L + \epsilon$. Then

$$R = L + \frac{1-L}{2} = \frac{L+1}{2} < \frac{1+1}{2} = 1,$$

and

$$0 \le \sqrt[n]{a_n} \le R \quad \text{or} \quad 0 \le a_n \le R^n$$

for $n \ge N$. Because $0 \le R < 1$, the series $\sum_{n=N}^{\infty} R^n$ is a convergent geometric series, so the series $\sum_{n=N}^{\infty} a_n$ converges by the Comparison Test. Therefore, the series $\sum_{n=0}^{\infty} a_n$ also converges.

(b) If $L > 1$, let $\epsilon = \dfrac{L-1}{2}$. By the definition of a limit, there is a positive integer N such that

$$-\epsilon \le \sqrt[n]{a_n} - L \le \epsilon$$

for $n \ge N$. From this, we conclude that

$$L - \epsilon \le \sqrt[n]{a_n}$$

for $n \ge N$. Now, let $R = L - \epsilon$. Then

$$R = L - \frac{L-1}{2} = \frac{L+1}{2} > \frac{1+1}{2} = 1,$$

and

$$R \le \sqrt[n]{a_n} \quad \text{or} \quad R^n \le a_n$$

for $n \ge N$. Because $R > 1$, the series $\sum_{n=N}^{\infty} R^n$ is a divergent geometric series, so the series $\sum_{n=N}^{\infty} a_n$ diverges by the Comparison Test. Therefore, the series $\sum_{n=0}^{\infty} a_n$ also diverges.

59. Let $S = \sum_{n=1}^{\infty} \dfrac{c^n n!}{n^n}$, where c is a constant.

(a) Prove that S converges absolutely if $|c| < e$ and diverges if $|c| > e$.

(b) It is known that $\lim_{n\to\infty} \dfrac{e^n n!}{n^{n+1/2}} = \sqrt{2\pi}$. Verify this numerically.

(c) Use the Limit Comparison Test to prove that S diverges for $c = e$.

SOLUTION

(a) With $a_n = \frac{c^n n!}{n^n}$,

$$\left|\frac{a_{n+1}}{a_n}\right| = \frac{|c|^{n+1}(n+1)!}{(n+1)^{n+1}} \cdot \frac{n^n}{|c|^n n!} = |c|\left(\frac{n}{n+1}\right)^n = |c|\left(1+\frac{1}{n}\right)^{-n},$$

and

$$\rho = \lim_{n\to\infty}\left|\frac{a_{n+1}}{a_n}\right| = |c|e^{-1}.$$

Thus, by the Ratio Test, the series $\sum_{n=1}^{\infty} \dfrac{c^n n!}{n^n}$ converges when $|c|e^{-1} < 1$, or when $|c| < e$. The series diverges when $|c| > e$.

(b) The table below lists the value of $\frac{e^n n!}{n^{n+1/2}}$ for several increasing values of n. Since $\sqrt{2\pi} = 2.506628275$, the numerical evidence verifies that

$$\lim_{n\to\infty} \frac{e^n n!}{n^{n+1/2}} = \sqrt{2\pi}.$$

n	100	1000	10000	100000
$\frac{e^n n!}{n^{n+1/2}}$	2.508717995	2.506837169	2.506649163	2.506630363

(c) With $c = e$, the series S becomes $\sum_{n=1}^{\infty} \frac{e^n n!}{n^n}$. Using the result from part (b),

$$L = \lim_{n\to\infty} \frac{\frac{e^n n!}{n^n}}{\sqrt{n}} = \lim_{n\to\infty} \frac{e^n n!}{n^{n+1/2}} = \sqrt{2\pi}.$$

Because the series $\sum_{n=1}^{\infty} \sqrt{n}$ diverges by the Divergence Test and $L > 0$, we conclude that $\sum_{n=1}^{\infty} \frac{e^n n!}{n^n}$ diverges by the Limit Comparison Test.

10.6 Power Series

Preliminary Questions

1. Suppose that $\sum a_n x^n$ converges for $x = 5$. Must it also converge for $x = 4$? What about $x = -3$?

SOLUTION The power series $\sum a_n x^n$ is centered at $x = 0$. Because the series converges for $x = 5$, the radius of convergence must be at least 5 and the series converges absolutely at least for the interval $|x| < 5$. Both $x = 4$ and $x = -3$ are inside this interval, so the series converges for $x = 4$ and for $x = -3$.

2. Suppose that $\sum a_n (x - 6)^n$ converges for $x = 10$. At which of the points (a)–(d) must it also converge?

(a) $x = 8$ **(b)** $x = 11$ **(c)** $x = 3$ **(d)** $x = 0$

SOLUTION The given power series is centered at $x = 6$. Because the series converges for $x = 10$, the radius of convergence must be at least $|10 - 6| = 4$ and the series converges absolutely at least for the interval $|x - 6| < 4$, or $2 < x < 10$.

(a) $x = 8$ is inside the interval $2 < x < 10$, so the series converges for $x = 8$.

(b) $x = 11$ is not inside the interval $2 < x < 10$, so the series may or may not converge for $x = 11$.

(c) $x = 3$ is inside the interval $2 < x < 10$, so the series converges for $x = 2$.

(d) $x = 0$ is not inside the interval $2 < x < 10$, so the series may or may not converge for $x = 0$.

3. What is the radius of convergence of $F(3x)$ if $F(x)$ is a power series with radius of convergence $R = 12$?

SOLUTION If the power series $F(x)$ has radius of convergence $R = 12$, then the power series $F(3x)$ has radius of convergence $R = \frac{12}{3} = 4$.

4. The power series $F(x) = \sum_{n=1}^{\infty} nx^n$ has radius of convergence $R = 1$. What is the power series expansion of $F'(x)$ and what is its radius of convergence?

SOLUTION We obtain the power series expansion for $F'(x)$ by differentiating the power series expansion for $F(x)$ term-by-term. Thus,

$$F'(x) = \sum_{n=1}^{\infty} n^2 x^{n-1}.$$

The radius of convergence for this series is $R = 1$, the same as the radius of convergence for the series expansion for $F(x)$.

Exercises

1. Use the Ratio Test to determine the radius of convergence R of $\sum_{n=0}^{\infty} \frac{x^n}{2^n}$. Does it converge at the endpoints $x = \pm R$?

SOLUTION With $a_n = \frac{x^n}{2^n}$,

$$\left|\frac{a_{n+1}}{a_n}\right| = \frac{|x|^{n+1}}{2^{n+1}} \cdot \frac{2^n}{|x|^n} = \frac{|x|}{2} \quad \text{and} \quad \rho = \lim_{n\to\infty} \left|\frac{a_{n+1}}{a_n}\right| = \frac{|x|}{2}.$$

By the Ratio Test, the series converges when $\rho = \frac{|x|}{2} < 1$, or $|x| < 2$, and diverges when $\rho = \frac{|x|}{2} > 1$, or $|x| > 2$. The radius of convergence is therefore $R = 2$. For $x = -2$, the left endpoint, the series becomes $\sum_{n=0}^{\infty}(-1)^n$, which is divergent. For $x = 2$, the right endpoint, the series becomes $\sum_{n=0}^{\infty} 1$, which is also divergent. Thus the series diverges at both endpoints.

3. Show that the power series (a)–(c) have the same radius of convergence. Then show that (a) diverges at both endpoints, (b) converges at one endpoint but diverges at the other, and (c) converges at both endpoints.

(a) $\sum_{n=1}^{\infty} \frac{x^n}{3^n}$ **(b)** $\sum_{n=1}^{\infty} \frac{x^n}{n3^n}$ **(c)** $\sum_{n=1}^{\infty} \frac{x^n}{n^2 3^n}$

SOLUTION

(a) With $a_n = \frac{x^n}{3^n}$,

$$\rho = \lim_{n\to\infty} \left|\frac{a_{n+1}}{a_n}\right| = \lim_{n\to\infty} \left|\frac{x^{n+1}}{3^{n+1}} \cdot \frac{3^n}{x^n}\right| = \lim_{n\to\infty} \left|\frac{x}{3}\right| = \left|\frac{x}{3}\right|$$

Then $\rho < 1$ if $|x| < 3$, so that the radius of convergence is $R = 3$. For the endpoint $x = 3$, the series becomes

$$\sum_{n=1}^{\infty} \frac{3^n}{3^n} = \sum_{n=1}^{\infty} 1,$$

which diverges by the Divergence Test. For the endpoint $x = -3$, the series becomes

$$\sum_{n=1}^{\infty} \frac{(-3)^n}{3^n} = \sum_{n=1}^{\infty} (-1)^n,$$

which also diverges by the Divergence Test.

(b) With $a_n = \frac{x^n}{n3^n}$,

$$\rho = \lim_{n\to\infty} \left|\frac{a_{n+1}}{a_n}\right| = \lim_{n\to\infty} \left|\frac{x^{n+1}}{(n+1)3^{n+1}} \cdot \frac{n3^n}{x^n}\right| = \lim_{n\to\infty} \left|\frac{x}{3}\left(\frac{n}{n+1}\right)\right| = \left|\frac{x}{3}\right|.$$

Then $\rho < 1$ when $|x| < 3$, so that the radius of convergence is $R = 3$. For the endpoint $x = 3$, the series becomes

$$\sum_{n=1}^{\infty} \frac{3^n}{n3^n} = \sum_{n=1}^{\infty} \frac{1}{n},$$

which is the divergent harmonic series. For the endpoint $x = -3$, the series becomes

$$\sum_{n=1}^{\infty} \frac{(-3)^n}{n3^n} = \sum_{n=1}^{\infty} \frac{(-1)^n}{n},$$

which converges by the Leibniz Test.

(c) With $a_n = \frac{x^n}{n^2 3^n}$,

$$\rho = \lim_{n\to\infty} \left|\frac{a_{n+1}}{a_n}\right| = \lim_{n\to\infty} \left|\frac{x^{n+1}}{(n+1)^2 3^{n+1}} \cdot \frac{n^2 3^n}{x^n}\right| = \lim_{n\to\infty} \left|\frac{x}{3}\left(\frac{n}{n+1}\right)^2\right| = \left|\frac{x}{3}\right|$$

Then $\rho < 1$ when $|x| < 3$, so that the radius of convergence is $R = 3$. For the endpoint $x = 3$, the series becomes

$$\sum_{n=1}^{\infty} \frac{3^n}{n^2 3^n} = \sum_{n=1}^{\infty} \frac{1}{n^2},$$

which is a convergent p-series. For the endpoint $x = -3$, the series becomes

$$\sum_{n=1}^{\infty} \frac{(-3)^n}{n^2 3^n} = \sum_{n=1}^{\infty} \frac{(-1)^n}{n^2},$$

which converges by the Leibniz Test.

5. Show that $\sum_{n=0}^{\infty} n^n x^n$ diverges for all $x \neq 0$.

SOLUTION With $a_n = n^n x^n$, and assuming $x \neq 0$,

$$\rho = \lim_{n\to\infty} \left|\frac{a_{n+1}}{a_n}\right| = \lim_{n\to\infty} \left|\frac{(n+1)^{n+1} x^{n+1}}{n^n x^n}\right| = \lim_{n\to\infty} \left|x\left(1 + \frac{1}{n}\right)^n (n+1)\right| = \infty$$

$\rho < 1$ only if $x = 0$, so that the radius of convergence is therefore $R = 0$. In other words, the power series converges only for $x = 0$.

7. Use the Ratio Test to show that $\sum_{n=0}^{\infty} \frac{x^{2n}}{3^n}$ has radius of convergence $R = \sqrt{3}$.

SOLUTION With $a_n = \frac{x^{2n}}{3^n}$,

$$\rho = \lim_{n\to\infty} \left|\frac{a_{n+1}}{a_n}\right| = \lim_{n\to\infty} \left|\frac{x^{2(n+1)}}{3^{n+1}} \cdot \frac{3^n}{x^{2n}}\right| = \lim_{n\to\infty} \left|\frac{x^2}{3}\right| = \left|\frac{x^2}{3}\right|$$

Then $\rho < 1$ when $|x^2| < 3$, or $x = \sqrt{3}$, so the radius of convergence is $R = \sqrt{3}$.

In Exercises 9–34, find the interval of convergence.

9. $\sum_{n=0}^{\infty} nx^n$

SOLUTION With $a_n = nx^n$,

$$\rho = \lim_{n\to\infty} \left|\frac{a_{n+1}}{a_n}\right| = \lim_{n\to\infty} \left|\frac{(n+1)x^{n+1}}{nx^n}\right| = \lim_{n\to\infty} \left|x\frac{n+1}{n}\right| = |x|$$

Then $\rho < 1$ when $|x| < 1$, so that the radius of convergence is $R = 1$, and the series converges absolutely on the interval $|x| < 1$, or $-1 < x < 1$. For the endpoint $x = 1$, the series becomes $\sum_{n=0}^{\infty} n$, which diverges by the Divergence Test. For the endpoint $x = -1$, the series becomes $\sum_{n=1}^{\infty} (-1)^n n$, which also diverges by the Divergence Test. Thus, the series $\sum_{n=0}^{\infty} nx^n$ converges for $-1 < x < 1$ and diverges elsewhere.

11. $\sum_{n=1}^{\infty} (-1)^n \frac{x^{2n+1}}{2^n n}$

SOLUTION With $a_n = (-1)^n \frac{x^{2n+1}}{2^n n}$,

$$\rho = \lim_{n\to\infty} \left|\frac{x^{2(n+1)+1}}{2^{n+1}(n+1)} \cdot \frac{2^n n}{x^{2n+1}}\right| = \lim_{n\to\infty} \left|\frac{x^2}{2} \cdot \frac{n}{n+1}\right| = \left|\frac{x^2}{2}\right|$$

Then $\rho < 1$ when $|x| < \sqrt{2}$, so the radius of convergence is $R = \sqrt{2}$, and the series converges absolutely on the interval $-\sqrt{2} < x < \sqrt{2}$. For the endpoint $x = -\sqrt{2}$, the series becomes $\sum_{n=1}^{\infty} (-1)^n \frac{-\sqrt{2}}{n} = \sum_{n=1}^{\infty} (-1)^{n+1} \frac{\sqrt{2}}{n}$, which converges by the Leibniz test. For the endpoint $x = \sqrt{2}$, the series becomes $\sum_{n=1}^{\infty} (-1)^n \frac{\sqrt{2}}{n}$ which also converges by the Leibniz test. Thus the series $\sum_{n=1}^{\infty} (-1)^n \frac{x^{2n+1}}{2^n n}$ converges for $-\sqrt{2} \le x \le \sqrt{2}$ and diverges elsewhere.

13. $\displaystyle\sum_{n=4}^{\infty} \frac{x^n}{n^5}$

SOLUTION With $a_n = \frac{x^n}{n^5}$,

$$\rho = \lim_{n\to\infty} \left|\frac{a_{n+1}}{a_n}\right| = \lim_{n\to\infty} \left|\frac{x^{n+1}}{(n+1)^5} \cdot \frac{n^5}{x^n}\right| = \lim_{n\to\infty} \left|x\left(\frac{n}{n+1}\right)^5\right| = |x|$$

Then $\rho < 1$ when $|x| < 1$, so the radius of convergence is $R = 1$, and the series converges absolutely on the interval $|x| < 1$, or $-1 < x < 1$. For the endpoint $x = 1$, the series becomes $\displaystyle\sum_{n=1}^{\infty} \frac{1}{n^5}$, which is a convergent p-series. For the endpoint $x = -1$, the series becomes $\displaystyle\sum_{n=1}^{\infty} \frac{(-1)^n}{n^5}$, which converges by the Leibniz Test. Thus, the series $\displaystyle\sum_{n=4}^{\infty} \frac{x^n}{n^5}$ converges for $-1 \le x \le 1$ and diverges elsewhere.

15. $\displaystyle\sum_{n=0}^{\infty} \frac{x^n}{(n!)^2}$

SOLUTION With $a_n = \frac{x^n}{(n!)^2}$,

$$\rho = \lim_{n\to\infty} \left|\frac{a_{n+1}}{a_n}\right| = \lim_{n\to\infty} \left|\frac{x^{n+1}}{((n+1)!)^2} \cdot \frac{(n!)^2}{x^n}\right| = \lim_{n\to\infty} \left|x\left(\frac{1}{n+1}\right)^2\right| = 0$$

$\rho < 1$ for all x, so the radius of convergence is $R = \infty$, and the series converges absolutely for all x.

17. $\displaystyle\sum_{n=0}^{\infty} \frac{(2n)!}{(n!)^3} x^n$

SOLUTION With $a_n = \frac{(2n)!x^n}{(n!)^3}$, and assuming $x \neq 0$,

$$\rho = \lim_{n\to\infty} \left|\frac{a_{n+1}}{a_n}\right| = \lim_{n\to\infty} \left|\frac{(2(n+1))!x^{n+1}}{((n+1)!)^3} \cdot \frac{(n!)^3}{(2n)!x^n}\right| = \lim_{n\to\infty} \left|x\frac{(2n+2)(2n+1)}{(n+1)^3}\right|$$

$$= \lim_{n\to\infty} \left|x\frac{4n^2+6n+2}{n^3+3n^2+3n+1}\right| = \lim_{n\to\infty} \left|x\frac{4n^{-1}+6n^{-1}+2n^{-3}}{1+3n^{-1}+3n^{-2}+n^{-3}}\right| = 0$$

Then $\rho < 1$ for all x, so the radius of convergence is $R = \infty$, and the series converges absolutely for all x.

19. $\displaystyle\sum_{n=0}^{\infty} \frac{(-1)^n x^n}{\sqrt{n^2+1}}$

SOLUTION With $a_n = \frac{(-1)^n x^n}{\sqrt{n^2+1}}$,

$$\rho = \lim_{n\to\infty} \left|\frac{a_{n+1}}{a_n}\right| = \lim_{n\to\infty} \left|\frac{(-1)^{n+1}x^{n+1}}{\sqrt{n^2+2n+2}} \cdot \frac{\sqrt{n^2+1}}{(-1)^n x^n}\right|$$

$$= \lim_{n\to\infty} \left|x\frac{\sqrt{n^2+1}}{\sqrt{n^2+2n+2}}\right| = \lim_{n\to\infty} \left|x\sqrt{\frac{n^2+1}{n^2+2n+2}}\right| = \lim_{n\to\infty} \left|x\sqrt{\frac{1+1/n^2}{1+2/n+2/n^2}}\right|$$

$$= |x|$$

Then $\rho < 1$ when $|x| < 1$, so the radius of convergence is $R = 1$, and the series converges absolutely on the interval $-1 < x < 1$. For the endpoint $x = 1$, the series becomes $\displaystyle\sum_{n=1}^{\infty} \frac{(-1)^n}{\sqrt{n^2+1}}$, which converges by the Leibniz Test. For the endpoint $x = -1$, the series becomes $\displaystyle\sum_{n=1}^{\infty} \frac{1}{\sqrt{n^2+1}}$, which diverges by the Limit Comparison Test comparing with the divergent harmonic series. Thus, the series $\displaystyle\sum_{n=0}^{\infty} \frac{(-1)^n x^n}{\sqrt{n^2+1}}$ converges for $-1 < x \le 1$ and diverges elsewhere.

21. $\sum_{n=15}^{\infty} \frac{x^{2n+1}}{3n+1}$

SOLUTION With $a_n = \frac{x^{2n+1}}{3n+1}$,

$$\rho = \lim_{n\to\infty} \left|\frac{a_{n+1}}{a_n}\right| = \lim_{n\to\infty} \left|\frac{x^{2n+3}}{3n+4} \cdot \frac{3n+1}{x^{2n+1}}\right| = \lim_{n\to\infty} \left|x^2 \frac{3n+1}{3n+4}\right| = |x^2|$$

Then $\rho < 1$ when $|x^2| < 1$, so the radius of convergence is $R = 1$, and the series converges absolutely for $-1 < x < 1$. For the endpoint $x = 1$, the series becomes $\sum_{n=15}^{\infty} \frac{1}{3n+1}$, which diverges by the Limit Comparison Test comparing with the divergent harmonic series. For the endpoint $x = -1$, the series becomes $\sum_{n=15}^{\infty} \frac{-1}{3n+1}$, which also diverges by the Limit Comparison Test comparing with the divergent harmonic series. Thus, the series $\sum_{n=15}^{\infty} \frac{x^{2n+1}}{3n+1}$ converges for $-1 < x < 1$ and diverges elsewhere.

23. $\sum_{n=2}^{\infty} \frac{x^n}{\ln n}$

SOLUTION With $a_n = \frac{x^n}{\ln n}$,

$$\rho = \lim_{n\to\infty} \left|\frac{a_{n+1}}{a_n}\right| = \lim_{n\to\infty} \left|\frac{x^{n+1}}{\ln(n+1)} \cdot \frac{\ln n}{x^n}\right| = \lim_{n\to\infty} \left|x \frac{\ln(n+1)}{\ln n}\right| = \lim_{n\to\infty} \left|x \frac{1/(n+1)}{1/n}\right| = \lim_{n\to\infty} \left|x \frac{n}{n+1}\right| = |x|$$

using L'Hôpital's rule. Then $\rho < 1$ when $|x| < 1$, so the radius of convergence is 1, and the series converges absolutely on the interval $|x| < 1$, or $-1 < x < 1$. For the endpoint $x = 1$, the series becomes $\sum_{n=2}^{\infty} \frac{1}{\ln n}$. Because $\frac{1}{\ln n} > \frac{1}{n}$ and $\sum_{n=2}^{\infty} \frac{1}{n}$ is the divergent harmonic series, the endpoint series diverges by the Comparison Test. For the endpoint $x = -1$, the series becomes $\sum_{n=2}^{\infty} \frac{(-1)^n}{\ln n}$, which converges by the Leibniz Test. Thus, the series $\sum_{n=2}^{\infty} \frac{x^n}{\ln n}$ converges for $-1 \le x < 1$ and diverges elsewhere.

25. $\sum_{n=1}^{\infty} n(x-3)^n$

SOLUTION With $a_n = n(x-3)^n$,

$$\rho = \lim_{n\to\infty} \left|\frac{a_{n+1}}{a_n}\right| = \lim_{n\to\infty} \left|\frac{(n+1)(x-3)^{n+1}}{n(x-3)^n}\right| = \lim_{n\to\infty} \left|(x-3) \cdot \frac{n+1}{n}\right| = |x-3|$$

Then $\rho < 1$ when $|x-3| < 1$, so the radius of convergence is 1, and the series converges absolutely on the interval $|x-3| < 1$, or $2 < x < 4$. For the endpoint $x = 4$, the series becomes $\sum_{n=1}^{\infty} n$, which diverges by the Divergence Test. For the endpoint $x = 2$, the series becomes $\sum_{n=1}^{\infty} (-1)^n n$, which also diverges by the Divergence Test. Thus, the series $\sum_{n=1}^{\infty} n(x-3)^n$ converges for $2 < x < 4$ and diverges elsewhere.

27. $\sum_{n=1}^{\infty} (-1)^n n^5 (x-7)^n$

SOLUTION With $a_n = (-1)^n n^5 (x-7)^n$,

$$\rho = \lim_{n\to\infty} \left|\frac{a_{n+1}}{a_n}\right| = \lim_{n\to\infty} \left|\frac{(-1)^{n+1}(n+1)^5(x-7)^{n+1}}{(-1)^n n^5 (x-7)^n}\right| = \lim_{n\to\infty} \left|(x-7) \cdot \frac{(n+1)^5}{n^5}\right|$$

$$= \lim_{n\to\infty} \left|(x-7) \cdot \frac{n^5 + \ldots}{n^5}\right| = |x-7|$$

Then $\rho < 1$ when $|x-7| < 1$, so the radius of convergence is 1, and the series converges absolutely on the interval $|x-7| < 1$, or $6 < x < 8$. For the endpoint $x = 6$, the series becomes $\sum_{n=1}^{\infty}(-1)^{2n}n^5 = \sum_{n=1}^{\infty} n^5$, which diverges by the Divergence Test. For the endpoint $x = 8$, the series becomes $\sum_{n=1}^{\infty}(-1)^n n^5$, which also diverges by the Divergence Test. Thus, the series $\sum_{n=1}^{\infty}(-1)^n n^5(x-7)^n$ converges for $6 < x < 8$ and diverges elsewhere.

29. $\sum_{n=1}^{\infty} \frac{2^n}{3n}(x+3)^n$

SOLUTION With $a_n = \frac{2^n(x+3)^n}{3n}$,

$$\rho = \lim_{n\to\infty}\left|\frac{a_{n+1}}{a_n}\right| = \lim_{n\to\infty}\left|\frac{2^{n+1}(x+3)^{n+1}}{3(n+1)} \cdot \frac{3n}{2^n(x+3)^n}\right| = \lim_{n\to\infty}\left|2(x+3)\cdot\frac{3n}{3n+3}\right|$$
$$= \lim_{n\to\infty}\left|2(x+3)\cdot\frac{1}{1+1/n}\right| = |2(x+3)|$$

Then $\rho < 1$ when $|2(x+3)| < 1$, so when $|x+3| < \frac{1}{2}$. Thus the radius of convergence is $\frac{1}{2}$, and the series converges absolutely on the interval $|x+3| < \frac{1}{2}$, or $-\frac{7}{2} < x < -\frac{5}{2}$. For the endpoint $x = -\frac{5}{2}$, the series becomes $\sum_{n=1}^{\infty}\frac{1}{3n}$, which diverges because it is a multiple of the divergent harmonic series. For the endpoint $x = -\frac{7}{2}$, the series becomes $\sum_{n=1}^{\infty}\frac{(-1)^n}{3n}$, which converges by the Leibniz Test. Thus, the series $\sum_{n=1}^{\infty}\frac{2^n}{3n}(x+3)^n$ converges for $-\frac{7}{2} \le x < -\frac{5}{2}$ and diverges elsewhere.

31. $\sum_{n=0}^{\infty} \frac{(-5)^n}{n!}(x+10)^n$

SOLUTION With $a_n = \frac{(-5)^n}{n!}(x+10)^n$,

$$\rho = \lim_{n\to\infty}\left|\frac{a_{n+1}}{a_n}\right| = \lim_{n\to\infty}\left|\frac{(-5)^{n+1}(x+10)^{n+1}}{(n+1)!} \cdot \frac{n!}{(-5)^n(x+10)^n}\right| = \lim_{n\to\infty}\left|5(x+10)\frac{1}{n}\right| = 0$$

Thus $\rho < 1$ for all x, so the radius of convergence is infinite, and $\sum_{n=0}^{\infty}\frac{(-5)^n}{n!}(x+10)^n$ converges for all x.

33. $\sum_{n=12}^{\infty} e^n(x-2)^n$

SOLUTION With $a_n = e^n(x-2)^n$,

$$\rho = \lim_{n\to\infty}\left|\frac{a_{n+1}}{a_n}\right| = \lim_{n\to\infty}\left|\frac{e^{n+1}(x-2)^{n+1}}{e^n(x-2)^n}\right| = \lim_{n\to\infty}|e(x-2)| = |e(x-2)|$$

Thus $\rho < 1$ when $|e(x-2)| < 1$, so when $|x-2| < e^{-1}$. Thus the radius of convergence is e^{-1}, and the series converges absolutely on the interval $|x-2| < e^{-1}$, or $2-e^{-1} < x < 2+e^{-1}$. For the endpoint $x = 2+e^{-1}$, the series becomes $\sum_{n=1}^{\infty} 1$, which diverges by the Divergence Test. For the endpoint $x = 2-e^{-1}$, the series becomes $\sum_{n=1}^{\infty}(-1)^n$, which also diverges by the Divergence Test. Thus, the series $\sum_{n=12}^{\infty} e^n(x-2)^n$ converges for $2-e^{-1} < x < 2+e^{-1}$ and diverges elsewhere.

In Exercises 35–40, use Eq. (2) to expand the function in a power series with center $c = 0$ and determine the interval of convergence.

35. $f(x) = \dfrac{1}{1-3x}$

SOLUTION Substituting $3x$ for x in Eq. (2), we obtain

$$\frac{1}{1-3x} = \sum_{n=0}^{\infty}(3x)^n = \sum_{n=0}^{\infty} 3^n x^n.$$

This series is valid for $|3x| < 1$, or $|x| < \frac{1}{3}$.

37. $f(x) = \dfrac{1}{3-x}$

SOLUTION First write

$$\frac{1}{3-x} = \frac{1}{3} \cdot \frac{1}{1-\frac{x}{3}}.$$

Substituting $\frac{x}{3}$ for x in Eq. (2), we obtain

$$\frac{1}{1-\frac{x}{3}} = \sum_{n=0}^{\infty}\left(\frac{x}{3}\right)^n = \sum_{n=0}^{\infty}\frac{x^n}{3^n};$$

Thus,

$$\frac{1}{3-x} = \frac{1}{3}\sum_{n=0}^{\infty}\frac{x^n}{3^n} = \sum_{n=0}^{\infty}\frac{x^n}{3^{n+1}}.$$

This series is valid for $|x/3| < 1$, or $|x| < 3$.

39. $f(x) = \dfrac{1}{1+x^2}$

SOLUTION Substituting $-x^2$ for x in Eq. (2), we obtain

$$\frac{1}{1+x^2} = \sum_{n=0}^{\infty}(-x^2)^n = \sum_{n=0}^{\infty}(-1)^n x^{2n}$$

This series is valid for $|x| < 1$.

41. Use the equalities

$$\frac{1}{1-x} = \frac{1}{-3-(x-4)} = \frac{-\frac{1}{3}}{1+\left(\frac{x-4}{3}\right)}$$

to show that for $|x-4| < 3$,

$$\frac{1}{1-x} = \sum_{n=0}^{\infty}(-1)^{n+1}\frac{(x-4)^n}{3^{n+1}}$$

SOLUTION Substituting $-\frac{x-4}{3}$ for x in Eq. (2), we obtain

$$\frac{1}{1+\left(\frac{x-4}{3}\right)} = \sum_{n=0}^{\infty}\left(-\frac{x-4}{3}\right)^n = \sum_{n=0}^{\infty}(-1)^n\frac{(x-4)^n}{3^n}.$$

Thus,

$$\frac{1}{1-x} = -\frac{1}{3}\sum_{n=0}^{\infty}(-1)^n\frac{(x-4)^n}{3^n} = \sum_{n=0}^{\infty}(-1)^{n+1}\frac{(x-4)^n}{3^{n+1}}.$$

This series is valid for $\left|-\frac{x-4}{3}\right| < 1$, or $|x-4| < 3$.

43. Use the method of Exercise 41 to expand $1/(4-x)$ in a power series with center $c = 5$. Determine the interval of convergence.

SOLUTION First write

$$\frac{1}{4-x} = \frac{1}{-1-(x-5)} = -\frac{1}{1+(x-5)}.$$

Substituting $-(x-5)$ for x in Eq. (2), we obtain

$$\frac{1}{1+(x-5)} = \sum_{n=0}^{\infty}(-(x-5))^n = \sum_{n=0}^{\infty}(-1)^n(x-5)^n.$$

Thus,

$$\frac{1}{4-x} = -\sum_{n=0}^{\infty}(-1)^n(x-5)^n = \sum_{n=0}^{\infty}(-1)^{n+1}(x-5)^n.$$

This series is valid for $|-(x-5)| < 1$, or $|x-5| < 1$.

45. Apply integration to the expansion

$$\frac{1}{1+x} = \sum_{n=0}^{\infty}(-1)^n x^n = 1 - x + x^2 - x^3 + \cdots$$

to prove that for $-1 < x < 1$,

$$\ln(1+x) = \sum_{n=1}^{\infty}\frac{(-1)^{n-1}x^n}{n} = x - \frac{x^2}{2} + \frac{x^3}{3} - \frac{x^4}{4} + \cdots$$

SOLUTION To obtain the first expansion, substitute $-x$ for x in Eq. (2):

$$\frac{1}{1+x} = \sum_{n=0}^{\infty}(-x)^n = \sum_{n=0}^{\infty}(-1)^n x^n.$$

This expansion is valid for $|-x| < 1$, or $-1 < x < 1$.

Upon integrating both sides of the above equation, we find

$$\ln(1+x) = \int \frac{dx}{1+x} = \int\left(\sum_{n=0}^{\infty}(-1)^n x^n\right)dx.$$

Integrating the series term-by-term then yields

$$\ln(1+x) = C + \sum_{n=0}^{\infty}(-1)^n\frac{x^{n+1}}{n+1}.$$

To determine the constant C, set $x = 0$. Then $0 = \ln(1+0) = C$. Finally,

$$\ln(1+x) = \sum_{n=0}^{\infty}(-1)^n\frac{x^{n+1}}{n+1} = \sum_{n=1}^{\infty}(-1)^{n-1}\frac{x^n}{n}.$$

47. Let $F(x) = (x+1)\ln(1+x) - x$.

(a) Apply integration to the result of Exercise 45 to prove that for $-1 < x < 1$,

$$F(x) = \sum_{n=1}^{\infty}(-1)^{n+1}\frac{x^{n+1}}{n(n+1)}$$

(b) Evaluate at $x = \frac{1}{2}$ to prove

$$\frac{3}{2}\ln\frac{3}{2} - \frac{1}{2} = \frac{1}{1\cdot 2\cdot 2^2} - \frac{1}{2\cdot 3\cdot 2^3} + \frac{1}{3\cdot 4\cdot 2^4} - \frac{1}{4\cdot 5\cdot 2^5} + \cdots$$

(c) Use a calculator to verify that the partial sum S_4 approximates the left-hand side with an error no greater than the term a_5 of the series.

SOLUTION

(a) Note that

$$\int \ln(x+1)\,dx = (x+1)\ln(x+1) - x + C$$

Then integrating both sides of the result of Exercise 45 gives

$$(x+1)\ln(x+1) - x = \int \ln(x+1)\,dx = \int \sum_{n=1}^{\infty} \frac{(-1)^{n-1}x^n}{n}\,dx$$

For $-1 < x < 1$, which is the interval of convergence of the series in Exercise 45, therefore, we can integrate term by term to get

$$(x+1)\ln(x+1) - x = \sum_{n=1}^{\infty} \frac{(-1)^{n-1}}{n} \int x^n\,dx = \sum_{n=1}^{\infty} \frac{(-1)^{n-1}}{n} \cdot \frac{x^{n+1}}{n+1} + C = \sum_{n=1}^{\infty} (-1)^{n+1} \frac{x^{n+1}}{n(n+1)} + C$$

(noting that $(-1)^{n-1} = (-1)^{n+1}$). To determine C, evaluate both sides at $x = 0$ to get

$$0 = \ln 1 - 0 = 0 + C$$

so that $C = 0$ and we get finally

$$(x+1)\ln(x+1) - x = \sum_{n=1}^{\infty} (-1)^{n+1} \frac{x^{n+1}}{n(n+1)}$$

(b) Evaluating the result of part(a) at $x = \frac{1}{2}$ gives

$$\frac{3}{2}\ln\frac{3}{2} - \frac{1}{2} = \sum_{n=1}^{\infty} (-1)^{n+1} \frac{1}{n(n+1)2^{n+1}}$$

$$= \frac{1}{1\cdot 2\cdot 2^2} - \frac{1}{2\cdot 3\cdot 2^3} + \frac{1}{3\cdot 4\cdot 2^4} - \frac{1}{4\cdot 5\cdot 2^5} + \dots$$

(c)

$$S_4 = \frac{1}{1\cdot 2\cdot 2^2} - \frac{1}{2\cdot 3\cdot 2^3} + \frac{1}{3\cdot 4\cdot 2^4} - \frac{1}{4\cdot 5\cdot 2^5} = 0.1078125$$

$$a_5 = \frac{1}{5\cdot 6\cdot 2^6} \approx 0.0005208$$

$$\frac{3}{2}\ln\frac{3}{2} - \frac{1}{2} \approx 0.10819766$$

and

$$\left| S_4 - \frac{3}{2}\ln\frac{3}{2} - \frac{1}{2} \right| \approx 0.0003852 < a_5$$

49. Use the result of Example 7 to show that

$$F(x) = \frac{x^2}{1\cdot 2} - \frac{x^4}{3\cdot 4} + \frac{x^6}{5\cdot 6} - \frac{x^8}{7\cdot 8} + \cdots$$

is an antiderivative of $f(x) = \tan^{-1} x$ satisfying $F(0) = 0$. What is the radius of convergence of this power series?

SOLUTION For $-1 < x < 1$, which is the interval of convergence for the power series for arctangent, we can integrate term-by-term, so integrate that power series to get

$$F(x) = \int \tan^{-1} x\,dx = \sum_{n=0}^{\infty} \int \frac{(-1)^n x^{2n+1}}{2n+1}\,dx = \sum_{n=0}^{\infty} (-1)^n \frac{x^{2n+2}}{(2n+1)(2n+2)}$$

$$= \frac{x^2}{1\cdot 2} - \frac{x^4}{3\cdot 4} + \frac{x^6}{5\cdot 6} - \frac{x^8}{7\cdot 8} + \cdots + C$$

If we assume $F(0) = 0$, then we have $C = 0$. The radius of convergence of this power series is the same as that of the original power series, which is 1.

51. Evaluate $\sum_{n=1}^{\infty} \frac{n}{2^n}$. *Hint:* Use differentiation to show that

$$(1-x)^{-2} = \sum_{n=1}^{\infty} nx^{n-1} \quad (\text{for } |x| < 1)$$

SOLUTION Differentiate both sides of Eq. (2) to obtain

$$\frac{1}{(1-x)^2} = \sum_{n=1}^{\infty} nx^{n-1}.$$

Setting $x = \frac{1}{2}$ then yields

$$\sum_{n=1}^{\infty} \frac{n}{2^{n-1}} = \frac{1}{\left(1-\frac{1}{2}\right)^2} = 4.$$

Divide this equation by 2 to obtain

$$\sum_{n=1}^{\infty} \frac{n}{2^n} = 2.$$

53. Show that the following series converges absolutely for $|x| < 1$ and compute its sum:

$$F(x) = 1 - x - x^2 + x^3 - x^4 - x^5 + x^6 - x^7 - x^8 + \cdots$$

Hint: Write $F(x)$ as a sum of three geometric series with common ratio x^3.

SOLUTION Because the coefficients in the power series are all ± 1, we find

$$r = \lim_{n\to\infty} \left| \frac{a_{n+1}}{a_n} \right| = 1.$$

The radius of convergence is therefore $R = r^{-1} = 1$, and the series converges absolutely for $|x| < 1$.

By Exercise 43 of Section 10.4, any rearrangement of the terms of an absolutely convergent series yields another absolutely convergent series with the same sum as the original series. Following the hint, we now rearrange the terms of $F(x)$ as the sum of three geometric series:

$$F(x) = \left(1 + x^3 + x^6 + \cdots\right) - \left(x + x^4 + x^7 + \cdots\right) - \left(x^2 + x^5 + x^8 + \cdots\right)$$

$$= \sum_{n=0}^{\infty} (x^3)^n - \sum_{n=0}^{\infty} x(x^3)^n - \sum_{n=0}^{\infty} x^2(x^3)^n = \frac{1}{1-x^3} - \frac{x}{1-x^3} - \frac{x^2}{1-x^3} = \frac{1-x-x^2}{1-x^3}.$$

55. Find all values of x such that $\sum_{n=1}^{\infty} \frac{x^{n^2}}{n!}$ converges.

SOLUTION With $a_n = \frac{x^{n^2}}{n!}$,

$$\left| \frac{a_{n+1}}{a_n} \right| = \frac{|x|^{(n+1)^2}}{(n+1)!} \cdot \frac{n!}{|x|^{n^2}} = \frac{|x|^{2n+1}}{n+1}.$$

if $|x| \le 1$, then

$$\lim_{n\to\infty} \frac{|x|^{2n+1}}{n+1} = 0,$$

and the series converges absolutely. On the other hand, if $|x| > 1$, then

$$\lim_{n\to\infty} \frac{|x|^{2n+1}}{n+1} = \infty,$$

and the series diverges. Thus, $\sum_{n=1}^{\infty} \frac{x^{n^2}}{n!}$ converges for $-1 \le x \le 1$ and diverges elsewhere.

57. Find a power series $P(x) = \sum_{n=0}^{\infty} a_n x^n$ satisfying the differential equation $y' = -y$ with initial condition $y(0) = 1$. Then use Theorem 1 of Section 5.8 to conclude that $P(x) = e^{-x}$.

SOLUTION Let $P(x) = \sum_{n=0}^{\infty} a_n x^n$ and note that $P(0) = a_0$; thus, to satisfy the initial condition $P(0) = 1$, we must take $a_0 = 1$. Now,

$$P'(x) = \sum_{n=1}^{\infty} n a_n x^{n-1},$$

so

$$P'(x) + P(x) = \sum_{n=1}^{\infty} na_n x^{n-1} + \sum_{n=0}^{\infty} a_n x^n = \sum_{n=0}^{\infty} \left[(n+1)a_{n+1} + a_n\right] x^n.$$

In order for this series to be equal to zero, the coefficient of x^n must be equal to zero for each n; thus

$$(n+1)a_{n+1} + a_n = 0 \quad \text{or} \quad a_{n+1} = -\frac{a_n}{n+1}.$$

Starting from $a_0 = 1$, we then calculate

$$a_1 = -\frac{a_0}{1} = -1;$$

$$a_2 = -\frac{a_1}{2} = \frac{1}{2};$$

$$a_3 = -\frac{a_2}{3} = -\frac{1}{6} = -\frac{1}{3!};$$

and, in general,

$$a_n = (-1)^n \frac{1}{n!}.$$

Hence,

$$P(x) = \sum_{n=0}^{\infty} (-1)^n \frac{x^n}{n!}.$$

The solution to the initial value problem $y' = -y$, $y(0) = 1$ is $y = e^{-x}$. Because this solution is unique, it follows that

$$P(x) = \sum_{n=0}^{\infty} (-1)^n \frac{x^n}{n!} = e^{-x}.$$

59. Use the power series for $y = e^x$ to show that

$$\frac{1}{e} = \frac{1}{2!} - \frac{1}{3!} + \frac{1}{4!} - \cdots$$

Use your knowledge of alternating series to find an N such that the partial sum S_N approximates e^{-1} to within an error of at most 10^{-3}. Confirm this using a calculator to compute both S_N and e^{-1}.

SOLUTION Recall that the series for e^x is

$$\sum_{n=0}^{\infty} \frac{x^n}{n!} = 1 + x + \frac{x^2}{2!} + \frac{x^3}{3!} + \frac{x^4}{4!} + \cdots.$$

Setting $x = -1$ yields

$$e^{-1} = 1 - 1 + \frac{1}{2!} - \frac{1}{3!} + \frac{1}{4!} - + \cdots = \frac{1}{2!} - \frac{1}{3!} + \frac{1}{4!} - + \cdots.$$

This is an alternating series with $a_n = \frac{1}{(n+1)!}$. The error in approximating e^{-1} with the partial sum S_N is therefore bounded by

$$|S_N - e^{-1}| \le a_{N+1} = \frac{1}{(N+2)!}.$$

To make the error at most 10^{-3}, we must choose N such that

$$\frac{1}{(N+2)!} \le 10^{-3} \quad \text{or} \quad (N+2)! \ge 1000.$$

For $N = 4$, $(N+2)! = 6! = 720 < 1000$, but for $N = 5$, $(N+2)! = 7! = 5040$; hence, $N = 5$ is the smallest value that satisfies the error bound. The corresponding approximation is

$$S_5 = \frac{1}{2!} - \frac{1}{3!} + \frac{1}{4!} - \frac{1}{5!} + \frac{1}{6!} = 0.368055555$$

Now, $e^{-1} = 0.367879441$, so

$$|S_5 - e^{-1}| = 1.761 \times 10^{-4} < 10^{-3}.$$

61. Find a power series $P(x)$ satisfying the differential equation

$$y'' - xy' + y = 0$$

9

with initial condition $y(0) = 1$, $y'(0) = 0$. What is the radius of convergence of the power series?

SOLUTION Let $P(x) = \sum_{n=0}^{\infty} a_n x^n$. Then

$$P'(x) = \sum_{n=1}^{\infty} n a_n x^{n-1} \quad \text{and} \quad P''(x) = \sum_{n=2}^{\infty} n(n-1)a_n x^{n-2}.$$

Note that $P(0) = a_0$ and $P'(0) = a_1$; in order to satisfy the initial conditions $P(0) = 1$, $P'(0) = 0$, we must have $a_0 = 1$ and $a_1 = 0$. Now,

$$\begin{aligned} P''(x) - xP'(x) + P(x) &= \sum_{n=2}^{\infty} n(n-1)a_n x^{n-2} - \sum_{n=1}^{\infty} n a_n x^n + \sum_{n=0}^{\infty} a_n x^n \\ &= \sum_{n=0}^{\infty} (n+2)(n+1)a_{n+2} x^n - \sum_{n=1}^{\infty} n a_n x^n + \sum_{n=0}^{\infty} a_n x^n \\ &= 2a_2 + a_0 + \sum_{n=1}^{\infty} \left[(n+2)(n+1)a_{n+2} - n a_n + a_n\right] x^n. \end{aligned}$$

In order for this series to be equal to zero, the coefficient of x^n must be equal to zero for each n; thus, $2a_2 + a_0 = 0$ and $(n+2)(n+1)a_{n+2} - (n-1)a_n = 0$, or

$$a_2 = -\frac{1}{2}a_0 \quad \text{and} \quad a_{n+2} = \frac{n-1}{(n+2)(n+1)}a_n.$$

Starting from $a_1 = 0$, we calculate

$$\begin{aligned} a_3 &= \frac{1-1}{(3)(2)}a_1 = 0; \\ a_5 &= \frac{2}{(5)(4)}a_3 = 0; \\ a_7 &= \frac{4}{(7)(6)}a_5 = 0; \end{aligned}$$

and, in general, all of the odd coefficients are zero. As for the even coefficients, we have $a_0 = 1$, $a_2 = -\frac{1}{2}$,

$$\begin{aligned} a_4 &= \frac{1}{(4)(3)}a_2 = -\frac{1}{4!}; \\ a_6 &= \frac{3}{(6)(5)}a_4 = -\frac{3}{6!}; \\ a_8 &= \frac{5}{(8)(7)}a_6 = -\frac{15}{8!} \end{aligned}$$

and so on. Thus,

$$P(x) = 1 - \frac{1}{2}x^2 - \frac{1}{4!}x^4 - \frac{3}{6!}x^6 - \frac{15}{8!}x^8 - \cdots$$

To determine the radius of convergence, treat this as a series in the variable x^2, and observe that

$$r = \lim_{k\to\infty} \left|\frac{a_{2k+2}}{a_{2k}}\right| = \lim_{k\to\infty} \frac{2k-1}{(2k+2)(2k+1)} = 0.$$

Thus, the radius of convergence is $R = r^{-1} = \infty$.

63. Prove that

$$J_2(x) = \sum_{k=0}^{\infty} \frac{(-1)^k}{2^{2k+2}\, k!\,(k+3)!} x^{2k+2}$$

is a solution of the Bessel differential equation of order 2:

$$x^2 y'' + xy' + (x^2 - 4)y = 0$$

SOLUTION Let $J_2(x) = \sum_{k=0}^{\infty} \frac{(-1)^k}{2^{2k+2}\,k!\,(k+2)!} x^{2k+2}$. Then

$$J_2'(x) = \sum_{k=0}^{\infty} \frac{(-1)^k (k+1)}{2^{2k+1}\,k!\,(k+2)!} x^{2k+1}$$

$$J_2''(x) = \sum_{k=0}^{\infty} \frac{(-1)^k (k+1)(2k+1)}{2^{2k+1}\,k!\,(k+2)!} x^{2k}$$

and

$$\begin{aligned} x^2 J_2''(x) + x J_2'(x) + (x^2 - 4) J_2(x) &= \sum_{k=0}^{\infty} \frac{(-1)^k (k+1)(2k+1)}{2^{2k+1}\,k!\,(k+2)!} x^{2k+2} + \sum_{k=0}^{\infty} \frac{(-1)^k (k+1)}{2^{2k+1}\,k!\,(k+2)!} x^{2k+2} \\ &\quad - \sum_{k=0}^{\infty} \frac{(-1)^k}{2^{2k+2}\,k!\,(k+2)!} x^{2k+4} - \sum_{k=0}^{\infty} \frac{(-1)^k}{2^{2k}\,k!\,(k+2)!} x^{2k+2} \\ &= \sum_{k=0}^{\infty} \frac{(-1)^k k(k+2)}{2^{2k} k!(k+2)!} x^{2k+2} + \sum_{k=1}^{\infty} \frac{(-1)^{k-1}}{2^{2k}\,(k-1)!\,(k+1)!} x^{2k+2} \\ &= \sum_{k=1}^{\infty} \frac{(-1)^k}{2^{2k}(k-1)!(k+1)!} x^{2k+2} - \sum_{k=1}^{\infty} \frac{(-1)^k}{2^{2k}(k-1)!(k+1)!} x^{2k+2} = 0. \end{aligned}$$

Further Insights and Challenges

65. Suppose that the coefficients of $F(x) = \sum_{n=0}^{\infty} a_n x^n$ are *periodic*; that is, for some whole number $M > 0$, we have $a_{M+n} = a_n$. Prove that $F(x)$ converges absolutely for $|x| < 1$ and that

$$F(x) = \frac{a_0 + a_1 x + \cdots + a_{M-1} x^{M-1}}{1 - x^M}$$

Hint: Use the hint for Exercise 53.

SOLUTION Suppose the coefficients of $F(x)$ are periodic, with $a_{M+n} = a_n$ for some whole number M and all n. The $F(x)$ can be written as the sum of M geometric series:

$$\begin{aligned} F(x) &= a_0\left(1 + x^M + x^{2M} + \cdots\right) + a_1\left(x + x^{M+1} + x^{2M+1} + \cdots\right) + \\ &= a_2\left(x^2 + x^{M+2} + x^{2M+2} + \cdots\right) + \cdots + a_{M-1}\left(x^{M-1} + x^{2M-1} + x^{3M-1} + \cdots\right) \\ &= \frac{a_0}{1 - x^M} + \frac{a_1 x}{1 - x^M} + \frac{a_2 x^2}{1 - x^M} + \cdots + \frac{a_{M-1} x^{M-1}}{1 - x^M} = \frac{a_0 + a_1 x + a_2 x^2 + \cdots + a_{M-1} x^{M-1}}{1 - x^M}. \end{aligned}$$

As each geometric series converges absolutely for $|x| < 1$, it follows that $F(x)$ also converges absolutely for $|x| < 1$.

10.7 Taylor Series

Preliminary Questions

1. Determine $f(0)$ and $f'''(0)$ for a function $f(x)$ with Maclaurin series

$$T(x) = 3 + 2x + 12x^2 + 5x^3 + \cdots$$

SOLUTION The Maclaurin series for a function f has the form

$$f(0) + \frac{f'(0)}{1!} x + \frac{f''(0)}{2!} x^2 + \frac{f'''(0)}{3!} x^3 + \cdots$$

Matching this general expression with the given series, we find $f(0) = 3$ and $\frac{f'''(0)}{3!} = 5$. From this latter equation, it follows that $f'''(0) = 30$.

2. Determine $f(-2)$ and $f^{(4)}(-2)$ for a function with Taylor series

$$T(x) = 3(x+2) + (x+2)^2 - 4(x+2)^3 + 2(x+2)^4 + \cdots$$

SOLUTION The Taylor series for a function f centered at $x = -2$ has the form

$$f(-2) + \frac{f'(-2)}{1!}(x+2) + \frac{f''(-2)}{2!}(x+2)^2 + \frac{f'''(-2)}{3!}(x+2)^3 + \frac{f^{(4)}(-2)}{4!}(x+2)^4 + \cdots$$

Matching this general expression with the given series, we find $f(-2) = 0$ and $\dfrac{f^{(4)}(-2)}{4!} = 2$. From this latter equation, it follows that $f^{(4)}(-2) = 48$.

3. What is the easiest way to find the Maclaurin series for the function $f(x) = \sin(x^2)$?

SOLUTION The easiest way to find the Maclaurin series for $\sin\left(x^2\right)$ is to substitute x^2 for x in the Maclaurin series for $\sin x$.

4. Find the Taylor series for $f(x)$ centered at $c = 3$ if $f(3) = 4$ and $f'(x)$ has a Taylor expansion

$$f'(x) = \sum_{n=1}^{\infty} \frac{(x-3)^n}{n}$$

SOLUTION Integrating the series for $f'(x)$ term-by-term gives

$$f(x) = C + \sum_{n=1}^{\infty} \frac{(x-3)^{n+1}}{n(n+1)}.$$

Substituting $x = 3$ then yields

$$f(3) = C = 4;$$

so

$$f(x) = 4 + \sum_{n=1}^{\infty} \frac{(x-3)^{n+1}}{n(n+1)}.$$

5. Let $T(x)$ be the Maclaurin series of $f(x)$. Which of the following guarantees that $f(2) = T(2)$?

(a) $T(x)$ converges for $x = 2$.

(b) The remainder $R_k(2)$ approaches a limit as $k \to \infty$.

(c) The remainder $R_k(2)$ approaches zero as $k \to \infty$.

SOLUTION The correct response is **(c)**: $f(2) = T(2)$ if and only if the remainder $R_k(2)$ approaches zero as $k \to \infty$.

Exercises

1. Write out the first four terms of the Maclaurin series of $f(x)$ if

$$f(0) = 2, \quad f'(0) = 3, \quad f''(0) = 4, \quad f'''(0) = 12$$

SOLUTION The first four terms of the Maclaurin series of $f(x)$ are

$$f(0) + f'(0)x + \frac{f''(0)}{2!}x^2 + \frac{f'''(0)}{3!}x^3 = 2 + 3x + \frac{4}{2}x^2 + \frac{12}{6}x^3 = 2 + 3x + 2x^2 + 2x^3.$$

In Exercises 3–18, find the Maclaurin series and find the interval on which the expansion is valid.

3. $f(x) = \dfrac{1}{1-2x}$

SOLUTION Substituting $2x$ for x in the Maclaurin series for $\frac{1}{1-x}$ gives

$$\frac{1}{1-2x} = \sum_{n=0}^{\infty} (2x)^n = \sum_{n=0}^{\infty} 2^n x^n.$$

This series is valid for $|2x| < 1$, or $|x| < \frac{1}{2}$.

5. $f(x) = \cos 3x$

SOLUTION Substituting $3x$ for x in the Maclaurin series for $\cos x$ gives

$$\cos 3x = \sum_{n=0}^{\infty}(-1)^n \frac{(3x)^{2n}}{(2n)!} = \sum_{n=0}^{\infty}(-1)^n \frac{9^n x^{2n}}{(2n)!}.$$

This series is valid for all x.

7. $f(x) = \sin(x^2)$

SOLUTION Substituting x^2 for x in the Maclaurin series for $\sin x$ gives

$$\sin x^2 = \sum_{n=0}^{\infty}(-1)^n \frac{(x^2)^{2n+1}}{(2n+1)!} = \sum_{n=0}^{\infty}(-1)^n \frac{x^{4n+2}}{(2n+1)!}.$$

This series is valid for all x.

9. $f(x) = \ln(1 - x^2)$

SOLUTION Substituting $-x^2$ for x in the Maclaurin series for $\ln(1 + x)$ gives

$$\ln(1 - x^2) = \sum_{n=1}^{\infty} \frac{(-1)^{n-1}(-x^2)^n}{n} = \sum_{n=1}^{\infty} \frac{(-1)^{2n-1}x^{2n}}{n} = -\sum_{n=1}^{\infty} \frac{x^{2n}}{n}.$$

This series is valid for $|x| < 1$.

11. $f(x) = \tan^{-1}(x^2)$

SOLUTION Substituting x^2 for x in the Maclaurin series for $\tan^{-1} x$ gives

$$\tan^{-1}(x^2) = \sum_{n=0}^{\infty}(-1)^n \frac{(x^2)^{2n+1}}{2n+1} = \sum_{n=0}^{\infty}(-1)^n \frac{x^{4n+2}}{2n+1}.$$

This series is valid for $|x| \le 1$.

13. $f(x) = e^{x-2}$

SOLUTION $e^{x-2} = e^{-2}e^x$; thus,

$$e^{x-2} = e^{-2}\sum_{n=0}^{\infty} \frac{x^n}{n!} = \sum_{n=0}^{\infty} \frac{x^n}{e^2 n!}.$$

This series is valid for all x.

15. $f(x) = \ln(1 - 5x)$

SOLUTION Substituting $-5x$ for x in the Maclaurin series for $\ln(1 + x)$ gives

$$\ln(1 - 5x) = \sum_{n=1}^{\infty} \frac{(-1)^{n-1}(-5x)^n}{n} = \sum_{n=1}^{\infty} \frac{(-1)^{2n-1}5^n x^n}{n} = -\sum_{n=1}^{\infty} \frac{5^n x^n}{n}.$$

This series is valid for $|5x| < 1$, or $|x| < \frac{1}{5}$, and for $x = -\frac{1}{5}$.

17. $f(x) = \sinh x$

SOLUTION Recall that

$$\sinh x = \frac{1}{2}(e^x - e^{-x}).$$

Therefore,

$$\sinh x = \frac{1}{2}\left(\sum_{n=0}^{\infty} \frac{x^n}{n!} - \sum_{n=0}^{\infty} \frac{(-x)^n}{n!}\right) = \sum_{n=0}^{\infty} \frac{x^n}{2(n!)}\left(1 - (-1)^n\right).$$

Now,

$$1 - (-1)^n = \begin{cases} 0, & n \text{ even} \\ 2, & n \text{ odd} \end{cases}$$

so

$$\sinh x = \sum_{k=0}^{\infty} 2\frac{x^{2k+1}}{2(2k+1)!} = \sum_{k=0}^{\infty} \frac{x^{2k+1}}{(2k+1)!}.$$

This series is valid for all x.

In Exercises 19–28, find the terms through degree four of the Maclaurin series of $f(x)$. Use multiplication and substitution as necessary.

19. $f(x) = e^x \sin x$

SOLUTION Multiply the fourth-order Taylor Polynomials for e^x and $\sin x$:

$$\left(1 + x + \frac{x^2}{2} + \frac{x^3}{6} + \frac{x^4}{24}\right)\left(x - \frac{x^3}{6}\right)$$

$$= x + x^2 - \frac{x^3}{6} + \frac{x^3}{2} - \frac{x^4}{6} + \frac{x^4}{6} + \text{ higher-order terms}$$

$$= x + x^2 + \frac{x^3}{3} + \text{ higher-order terms.}$$

The terms through degree four in the Maclaurin series for $f(x) = e^x \sin x$ are therefore

$$x + x^2 + \frac{x^3}{3}.$$

21. $f(x) = \dfrac{\sin x}{1 - x}$

SOLUTION Multiply the fourth order Taylor Polynomials for $\sin x$ and $\dfrac{1}{1-x}$:

$$\left(x - \frac{x^3}{6}\right)\left(1 + x + x^2 + x^3 + x^4\right)$$

$$= x + x^2 - \frac{x^3}{6} + x^3 + x^4 - \frac{x^4}{6} + \text{ higher-order terms}$$

$$= x + x^2 + \frac{5x^3}{6} + \frac{5x^4}{6} + \text{ higher-order terms.}$$

The terms through order four of the Maclaurin series for $f(x) = \dfrac{\sin x}{1-x}$ are therefore

$$x + x^2 + \frac{5x^3}{6} + \frac{5x^4}{6}.$$

23. $f(x) = (1 + x)^{1/4}$

SOLUTION The first five generalized binomial coefficients for $a = \frac{1}{4}$ are

$$1, \quad \frac{1}{4}, \quad \frac{\frac{1}{4}\left(\frac{-3}{4}\right)}{2!} = -\frac{3}{32}, \quad \frac{\frac{1}{4}\left(\frac{-3}{4}\right)\left(\frac{-7}{4}\right)}{3!} = \frac{7}{128}, \quad \frac{\frac{1}{4}\left(\frac{-3}{4}\right)\left(\frac{-7}{4}\right)\left(\frac{-11}{4}\right)}{4!} = \frac{-77}{2048}$$

Therefore, the first four terms in the binomial series for $(1 + x)^{1/4}$ are

$$1 + \frac{1}{4}x - \frac{3}{32}x^2 + \frac{7}{128}x^3 - \frac{77}{2048}x^4$$

25. $f(x) = e^x \tan^{-1} x$

SOLUTION Using the Maclaurin series for e^x and $\tan^{-1} x$, we find

$$e^x \tan^{-1} x = \left(1 + x + \frac{x^2}{2} + \frac{x^3}{6} + \cdots\right)\left(x - \frac{x^3}{3} + \cdots\right) = x + x^2 - \frac{x^3}{3} + \frac{x^3}{2} + \frac{x^4}{6} - \frac{x^4}{3} + \cdots$$

$$= x + x^2 + \frac{1}{6}x^3 - \frac{1}{6}x^4 + \cdots.$$

27. $f(x) = e^{\sin x}$

SOLUTION Substituting $\sin x$ for x in the Maclaurin series for e^x and then using the Maclaurin series for $\sin x$, we find

$$\begin{aligned} e^{\sin x} &= 1 + \sin x + \frac{\sin^2 x}{2} + \frac{\sin^3 x}{6} + \frac{\sin^4 x}{24} + \cdots \\ &= 1 + \left(x - \frac{x^3}{6} + \cdots\right) + \frac{1}{2}\left(x - \frac{x^3}{6} + \cdots\right)^2 + \frac{1}{6}(x - \cdots)^3 + \frac{1}{24}(x - \cdots)^4 \\ &= 1 + x + \frac{1}{2}x^2 - \frac{1}{6}x^3 + \frac{1}{6}x^3 - \frac{1}{6}x^4 + \frac{1}{24}x^4 + \cdots \\ &= 1 + x + \frac{1}{2}x^2 - \frac{1}{8}x^4 + \cdots. \end{aligned}$$

In Exercises 29–38, find the Taylor series centered at c and find the interval on which the expansion is valid.

29. $f(x) = \frac{1}{x}, \quad c = 1$

SOLUTION Write

$$\frac{1}{x} = \frac{1}{1 + (x-1)},$$

and then substitute $-(x-1)$ for x in the Maclaurin series for $\frac{1}{1-x}$ to obtain

$$\frac{1}{x} = \sum_{n=0}^{\infty} [-(x-1)]^n = \sum_{n=0}^{\infty} (-1)^n (x-1)^n.$$

This series is valid for $|x-1| < 1$.

31. $f(x) = \frac{1}{1-x}, \quad c = 5$

SOLUTION Write

$$\frac{1}{1-x} = \frac{1}{-4-(x-5)} = -\frac{1}{4} \cdot \frac{1}{1 + \frac{x-5}{4}}.$$

Substituting $-\frac{x-5}{4}$ for x in the Maclaurin series for $\frac{1}{1-x}$ yields

$$\frac{1}{1 + \frac{x-5}{4}} = \sum_{n=0}^{\infty} \left(-\frac{x-5}{4}\right)^n = \sum_{n=0}^{\infty} (-1)^n \frac{(x-5)^n}{4^n}.$$

Thus,

$$\frac{1}{1-x} = -\frac{1}{4} \sum_{n=0}^{\infty} (-1)^n \frac{(x-5)^n}{4^n} = \sum_{n=0}^{\infty} (-1)^{n+1} \frac{(x-5)^n}{4^{n+1}}.$$

This series is valid for $\left|\frac{x-5}{4}\right| < 1$, or $|x-5| < 4$.

33. $f(x) = x^4 + 3x - 1, \quad c = 2$

SOLUTION To determine the Taylor series with center $c = 2$, we compute

$$f'(x) = 4x^3 + 3, \quad f''(x) = 12x^2, \quad f'''(x) = 24x,$$

and $f^{(4)}(x) = 24$. All derivatives of order five and higher are zero. Now,

$$f(2) = 21, \quad f'(2) = 35, \quad f''(2) = 48, \quad f'''(2) = 48,$$

and $f^{(4)}(2) = 24$. Therefore, the Taylor series is

$$21 + 35(x-2) + \frac{48}{2}(x-2)^2 + \frac{48}{6}(x-2)^3 + \frac{24}{24}(x-2)^4,$$

or

$$21 + 35(x-2) + 24(x-2)^2 + 8(x-2)^3 + (x-2)^4.$$

35. $f(x) = \dfrac{1}{x^2}, \quad c = 4$

SOLUTION We will first find the Taylor series for $\frac{1}{x}$ and then differentiate to obtain the series for $\frac{1}{x^2}$. Write

$$\frac{1}{x} = \frac{1}{4 + (x - 4)} = \frac{1}{4} \cdot \frac{1}{1 + \frac{x-4}{4}}.$$

Now substitute $-\frac{x-4}{4}$ for x in the Maclaurin series for $\frac{1}{1-x}$ to obtain

$$\frac{1}{x} = \frac{1}{4} \sum_{n=}^{\infty} \left(-\frac{x-4}{4}\right)^n = \sum_{n=0}^{\infty} (-1)^n \frac{(x-4)^n}{4^{n+1}}.$$

Differentiating term-by-term yields

$$-\frac{1}{x^2} = \sum_{n=1}^{\infty} (-1)^n n \frac{(x-4)^{n-1}}{4^{n+1}},$$

so that

$$\frac{1}{x^2} = \sum_{n=1}^{\infty} (-1)^{n-1} n \frac{(x-4)^{n-1}}{4^{n+1}} = \sum_{n=0}^{\infty} (-1)^n (n+1) \frac{(x-4)^n}{4^{n+2}}.$$

This series is valid for $\left|\frac{x-4}{4}\right| < 1$, or $|x - 4| < 4$.

37. $f(x) = \dfrac{1}{1 - x^2}, \quad c = 3$

SOLUTION By partial fraction decomposition

$$\frac{1}{1 - x^2} = \frac{\frac{1}{2}}{1 - x} + \frac{\frac{1}{2}}{1 + x},$$

so

$$\frac{1}{1 - x^2} = \frac{\frac{1}{2}}{-2 - (x - 3)} + \frac{\frac{1}{2}}{4 + (x - 3)} = -\frac{1}{4} \cdot \frac{1}{1 + \frac{x-3}{2}} + \frac{1}{8} \cdot \frac{1}{1 + \frac{x-3}{4}}.$$

Substituting $-\frac{x-3}{2}$ for x in the Maclaurin series for $\frac{1}{1-x}$ gives

$$\frac{1}{1 + \frac{x-3}{2}} = \sum_{n=0}^{\infty} \left(-\frac{x-3}{2}\right)^n = \sum_{n=0}^{\infty} \frac{(-1)^n}{2^n} (x-3)^n,$$

while substituting $-\frac{x-3}{4}$ for x in the same series gives

$$\frac{1}{1 + \frac{x-3}{4}} = \sum_{n=0}^{\infty} \left(-\frac{x-3}{4}\right)^n = \sum_{n=0}^{\infty} \frac{(-1)^n}{4^n} (x-3)^n.$$

Thus,

$$\frac{1}{1 - x^2} = -\frac{1}{4} \sum_{n=0}^{\infty} \frac{(-1)^n}{2^n} (x-3)^n + \frac{1}{8} \sum_{n=0}^{\infty} \frac{(-1)^n}{4^n} (x-3)^n = \sum_{n=0}^{\infty} \frac{(-1)^{n+1}}{2^{n+2}} (x-3)^n + \sum_{n=0}^{\infty} \frac{(-1)^n}{2^{2n+3}} (x-3)^n$$

$$= \sum_{n=0}^{\infty} \left(\frac{(-1)^{n+1}}{2^{n+2}} + \frac{(-1)^n}{2^{2n+3}}\right) (x-3)^n = \sum_{n=0}^{\infty} \frac{(-1)^{n+1}(2^{n+1} - 1)}{2^{2n+3}} (x-3)^n.$$

This series is valid for $|x - 3| < 2$.

39. Use the identity $\cos^2 x = \frac{1}{2}(1 + \cos 2x)$ to find the Maclaurin series for $\cos^2 x$.

SOLUTION The Maclaurin series for $\cos 2x$ is

$$\sum_{n=0}^{\infty} (-1)^n \frac{(2x)^{2n}}{(2n)!} = \sum_{n=0}^{\infty} (-1)^n \frac{2^{2n} x^{2n}}{(2n)!}$$

so the Maclaurin series for $\cos^2 x = \frac{1}{2}(1 + \cos 2x)$ is

$$\frac{1 + \left(1 + \sum_{n=1}^{\infty} (-1)^n \frac{2^{2n} x^{2n}}{(2n)!}\right)}{2} = 1 + \sum_{n=1}^{\infty} (-1)^n \frac{2^{2n-1} x^{2n}}{(2n)!}$$

41. Use the Maclaurin series for $\ln(1 + x)$ and $\ln(1 - x)$ to show that

$$\frac{1}{2} \ln\left(\frac{1 + x}{1 - x}\right) = x + \frac{x^3}{3} + \frac{x^5}{5} + \cdots$$

for $|x| < 1$. What can you conclude by comparing this result with that of Exercise 40?

SOLUTION Using the Maclaurin series for $\ln(1 + x)$ and $\ln(1 - x)$, we have for $|x| < 1$

$$\begin{aligned} \ln(1 + x) - \ln(1 - x) &= \sum_{n=1}^{\infty} \frac{(-1)^{n-1}}{n} x^n - \sum_{n=1}^{\infty} \frac{(-1)^{n-1}}{n} (-x)^n \\ &= \sum_{n=1}^{\infty} \frac{(-1)^{n-1}}{n} x^n + \sum_{n=1}^{\infty} \frac{x^n}{n} = \sum_{n=1}^{\infty} \frac{1 + (-1)^{n-1}}{n} x^n. \end{aligned}$$

Since $1 + (-1)^{n-1} = 0$ for even n and $1 + (-1)^{n-1} = 2$ for odd n,

$$\ln(1 + x) - \ln(1 - x) = \sum_{k=0}^{\infty} \frac{2}{2k + 1} x^{2k+1}.$$

Thus,

$$\frac{1}{2} \ln\left(\frac{1 + x}{1 - x}\right) = \frac{1}{2}(\ln(1 + x) - \ln(1 - x)) = \frac{1}{2} \sum_{k=0}^{\infty} \frac{2}{2k + 1} x^{2k+1} = \sum_{k=0}^{\infty} \frac{x^{2k+1}}{2k + 1}.$$

Observe that this is the same series we found in Exercise 40; therefore,

$$\frac{1}{2} \ln\left(\frac{1 + x}{1 - x}\right) = \tanh^{-1} x.$$

43. Show, by integrating the Maclaurin series for $f(x) = \dfrac{1}{\sqrt{1 - x^2}}$, that for $|x| < 1$,

$$\sin^{-1} x = x + \sum_{n=1}^{\infty} \frac{1 \cdot 3 \cdot 5 \cdots (2n - 1)}{2 \cdot 4 \cdot 6 \cdots (2n)} \frac{x^{2n+1}}{2n + 1}$$

SOLUTION From Example 10, we know that for $|x| < 1$

$$\frac{1}{\sqrt{1 - x^2}} = \sum_{n=0}^{\infty} \frac{1 \cdot 3 \cdot 5 \cdots (2n - 1)}{2 \cdot 4 \cdot 6 \cdots (2n)} x^{2n} = 1 + \sum_{n=1}^{\infty} \frac{1 \cdot 3 \cdot 5 \cdots (2n - 1)}{2 \cdot 4 \cdot 6 \cdots (2n)} x^{2n},$$

so, for $|x| < 1$,

$$\sin^{-1} x = \int \frac{dx}{\sqrt{1 - x^2}} = C + x + \sum_{n=1}^{\infty} \frac{1 \cdot 3 \cdot 5 \cdots (2n - 1)}{2 \cdot 4 \cdot 6 \cdots (2n)} \frac{x^{2n+1}}{2n + 1}.$$

Since $\sin^{-1} 0 = 0$, we find that $C = 0$. Thus,

$$\sin^{-1} x = x + \sum_{n=1}^{\infty} \frac{1 \cdot 3 \cdot 5 \cdots (2n - 1)}{2 \cdot 4 \cdot 6 \cdots (2n)} \frac{x^{2n+1}}{2n + 1}.$$

45. How many terms of the Maclaurin series of $f(x) = \ln(1 + x)$ are needed to compute $\ln 1.2$ to within an error of at most 0.0001? Make the computation and compare the result with the calculator value.

SOLUTION Substitute $x = 0.2$ into the Maclaurin series for $\ln(1 + x)$ to obtain:

$$\ln 1.2 = \sum_{n=1}^{\infty} (-1)^{n-1} \frac{(0.2)^n}{n} = \sum_{n=1}^{\infty} (-1)^{n-1} \frac{1}{5^n n}.$$

This is an alternating series with $a_n = \dfrac{1}{n \cdot 5^n}$. Using the error bound for alternating series

$$|\ln 1.2 - S_N| \le a_{N+1} = \frac{1}{(N+1)5^{N+1}},$$

so we must choose N so that

$$\frac{1}{(N+1)5^{N+1}} < 0.0001 \quad \text{or} \quad (N+1)5^{N+1} > 10{,}000.$$

For $N = 3$, $(N+1)5^{N+1} = 4 \cdot 5^4 = 2500 < 10,000$, and for $N = 4$, $(N+1)5^{N+1} = 5 \cdot 5^5 = 15,625 > 10,000$; thus, the smallest acceptable value for N is $N = 4$. The corresponding approximation is:

$$S_4 = \sum_{n=1}^{4} \frac{(-1)^{n-1}}{5^n \cdot n} = \frac{1}{5} - \frac{1}{5^2 \cdot 2} + \frac{1}{5^3 \cdot 3} - \frac{1}{5^4 \cdot 4} = 0.182266666.$$

Now, $\ln 1.2 = 0.182321556$, so

$$|\ln 1.2 - S_4| = 5.489 \times 10^{-5} < 0.0001.$$

47. Use the Maclaurin expansion for e^{-t^2} to express the function $F(x) = \int_0^x e^{-t^2}\,dt$ as an alternating power series in x (Figure 4).

(a) How many terms of the Maclaurin series are needed to approximate the integral for $x = 1$ to within an error of at most 0.001?

(b) CAS Carry out the computation and check your answer using a computer algebra system.

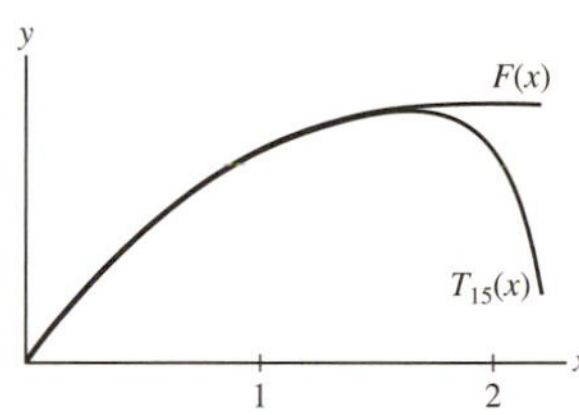

FIGURE 4 The Maclaurin polynomial $T_{15}(x)$ for $F(t) = \displaystyle\int_0^x e^{-t^2}\,dt$.

SOLUTION Substituting $-t^2$ for t in the Maclaurin series for e^t yields

$$e^{-t^2} = \sum_{n=0}^{\infty} \frac{(-t^2)^n}{n!} = \sum_{n=0}^{\infty} (-1)^n \frac{t^{2n}}{n!};$$

thus,

$$\int_0^x e^{-t^2}\,dt = \sum_{n=0}^{\infty} (-1)^n \frac{x^{2n+1}}{n!(2n+1)}.$$

(a) For $x = 1$,

$$\int_0^1 e^{-t^2}\,dt = \sum_{n=0}^{\infty} (-1)^n \frac{1}{n!(2n+1)}.$$

This is an alternating series with $a_n = \frac{1}{n!(2n+1)}$; therefore, the error incurred by using S_N to approximate the value of the definite integral is bounded by

$$\left| \int_0^1 e^{-t^2}\,dt - S_N \right| \le a_{N+1} = \frac{1}{(N+1)!(2N+3)}.$$

To guarantee the error is at most 0.001, we must choose N so that

$$\frac{1}{(N+1)!(2N+3)} < 0.001 \quad \text{or} \quad (N+1)!(2N+3) > 1000.$$

For $N = 3$, $(N+1)!(2N+3) = 4! \cdot 9 = 216 < 1000$ and for $N = 4$, $(N+1)!(2N+3) = 5! \cdot 11 = 1320 > 1000$; thus, the smallest acceptable value for N is $N = 4$. The corresponding approximation is

$$S_4 = \sum_{n=0}^{4} \frac{(-1)^n}{n!(2n+1)} = 1 - \frac{1}{3} + \frac{1}{2! \cdot 5} - \frac{1}{3! \cdot 7} + \frac{1}{4! \cdot 9} = 0.747486772.$$

(b) Using a computer algebra system, we find

$$\int_0^1 e^{-t^2}\,dt = 0.746824133;$$

therefore

$$\left|\int_0^1 e^{-t^2}\,dt - S_4\right| = 6.626 \times 10^{-4} < 10^{-3}.$$

In Exercises 49–52, express the definite integral as an infinite series and find its value to within an error of at most 10^{-4}.

49. $\int_0^1 \cos(x^2)\,dx$

SOLUTION Substituting x^2 for x in the Maclaurin series for $\cos x$ yields

$$\cos(x^2) = \sum_{n=0}^{\infty}(-1)^n \frac{(x^2)^{2n}}{(2n)!} = \sum_{n=0}^{\infty}(-1)^n \frac{x^{4n}}{(2n)!};$$

therefore,

$$\int_0^1 \cos(x^2)\,dx = \sum_{n=0}^{\infty}(-1)^n \left.\frac{x^{4n+1}}{(2n)!(4n+1)}\right|_0^1 = \sum_{n=0}^{\infty}\frac{(-1)^n}{(2n)!(4n+1)}.$$

This is an alternating series with $a_n = \frac{1}{(2n)!(4n+1)}$; therefore, the error incurred by using S_N to approximate the value of the definite integral is bounded by

$$\left|\int_0^1 \cos(x^2)\,dx - S_N\right| \le a_{N+1} = \frac{1}{(2N+2)!(4N+5)}.$$

To guarantee the error is at most 0.0001, we must choose N so that

$$\frac{1}{(2N+2)!(4N+5)} < 0.0001 \quad \text{or} \quad (2N+2)!(4N+5) > 10{,}000.$$

For $N = 2$, $(2N+2)!(4N+5) = 6! \cdot 13 = 9360 < 10{,}000$ and for $N = 3$, $(2N+2)!(4N+5) = 8! \cdot 17 = 685{,}440 > 10{,}000$; thus, the smallest acceptable value for N is $N = 3$. The corresponding approximation is

$$S_3 = \sum_{n=0}^{3}\frac{(-1)^n}{(2n)!(4n+1)} = 1 - \frac{1}{5\cdot 2!} + \frac{1}{9\cdot 4!} - \frac{1}{13\cdot 6!} = 0.904522792.$$

51. $\int_0^1 e^{-x^3}\,dx$

SOLUTION Substituting $-x^3$ for x in the Maclaurin series for e^x yields

$$e^{-x^3} = \sum_{n=0}^{\infty}\frac{(-x^3)^n}{n!} = \sum_{n=0}^{\infty}(-1)^n \frac{x^{3n}}{n!};$$

therefore,

$$\int_0^1 e^{-x^3}\,dx = \sum_{n=0}^{\infty}(-1)^n \left.\frac{x^{3n+1}}{n!(3n+1)}\right|_0^1 = \sum_{n=0}^{\infty}\frac{(-1)^n}{n!(3n+1)}.$$

This is an alternating series with $a_n = \frac{1}{n!(3n+1)}$; therefore, the error incurred by using S_N to approximate the value of the definite integral is bounded by

$$\left|\int_0^1 e^{-x^3}\,dx - S_N\right| \le a_{N+1} = \frac{1}{(N+1)!(3N+4)}.$$

To guarantee the error is at most 0.0001, we must choose N so that

$$\frac{1}{(N+1)!(3N+4)} < 0.0001 \quad \text{or} \quad (N+1)!(3N+4) > 10{,}000.$$

For $N = 4$, $(N+1)!(3N+4) = 5! \cdot 16 = 1920 < 10{,}000$ and for $N = 5$, $(N+1)!(3N+4) = 6! \cdot 19 = 13{,}680 > 10{,}000$; thus, the smallest acceptable value for N is $N = 5$. The corresponding approximation is

$$S_5 = \sum_{n=0}^{5} \frac{(-1)^n}{n!(3n+1)} = 0.807446200.$$

In Exercises 53–56, express the integral as an infinite series.

53. $\displaystyle\int_0^x \frac{1-\cos(t)}{t}\,dt$, for all x

SOLUTION The Maclaurin series for $\cos t$ is

$$\cos t = \sum_{n=0}^{\infty} (-1)^n \frac{t^{2n}}{(2n)!} = 1 + \sum_{n=1}^{\infty} (-1)^n \frac{t^{2n}}{(2n)!},$$

so

$$1 - \cos t = -\sum_{n=1}^{\infty} (-1)^n \frac{t^{2n}}{(2n)!} = \sum_{n=1}^{\infty} (-1)^{n+1} \frac{t^{2n}}{(2n)!},$$

and

$$\frac{1-\cos t}{t} = \frac{1}{t}\sum_{n=1}^{\infty} (-1)^{n+1} \frac{t^{2n}}{(2n)!} = \sum_{n=1}^{\infty} (-1)^{n+1} \frac{t^{2n-1}}{(2n)!}.$$

Thus,

$$\int_0^x \frac{1-\cos(t)}{t}\,dt = \sum_{n=1}^{\infty} (-1)^{n+1} \frac{t^{2n}}{(2n)!2n}\Bigg|_0^x = \sum_{n=1}^{\infty} (-1)^{n+1} \frac{x^{2n}}{(2n)!2n}.$$

55. $\displaystyle\int_0^x \ln(1+t^2)\,dt$, for $|x| < 1$

SOLUTION Substituting t^2 for t in the Maclaurin series for $\ln(1+t)$ yields

$$\ln(1+t^2) = \sum_{n=1}^{\infty} (-1)^{n-1} \frac{(t^2)^n}{n} = \sum_{n=1}^{\infty} (-1)^n \frac{t^{2n}}{n}.$$

Thus,

$$\int_0^x \ln(1+t^2)\,dt = \sum_{n=1}^{\infty} (-1)^n \frac{t^{2n+1}}{n(2n+1)}\Bigg|_0^x = \sum_{n=1}^{\infty} (-1)^n \frac{x^{2n+1}}{n(2n+1)}.$$

57. Which function has Maclaurin series $\displaystyle\sum_{n=0}^{\infty} (-1)^n 2^n x^n$?

SOLUTION We recognize that

$$\sum_{n=0}^{\infty} (-1)^n 2^n x^n = \sum_{n=0}^{\infty} (-2x)^n$$

is the Maclaurin series for $\frac{1}{1-x}$ with x replaced by $-2x$. Therefore,

$$\sum_{n=0}^{\infty} (-1)^n 2^n x^n = \frac{1}{1-(-2x)} = \frac{1}{1+2x}.$$

In Exercises 59–62, use Theorem 2 to prove that the $f(x)$ is represented by its Maclaurin series for all x.

59. $f(x) = \sin\left(\frac{x}{2}\right) + \cos\left(\frac{x}{3}\right)$,

SOLUTION All derivatives of $f(x)$ consist of sin or cos applied to each of $x/2$ and $x/3$ and added together, so each summand is bounded by 1. Thus $\left|f^{(n)}(x)\right| \le 2$ for all n and x. By Theorem 2, $f(x)$ is represented by its Taylor series for every x.

61. $f(x) = \sinh x$

SOLUTION By definition, $\sinh x = \frac{1}{2}(e^x - e^{-x})$, so if both e^x and e^{-x} are represented by their Taylor series centered at c, then so is $\sinh x$. But the previous exercise shows that e^{-x} is so represented, and the text shows that e^x is.

In Exercises 63–66, find the functions with the following Maclaurin series (refer to Table 1 on page 599).

63. $1 + x^3 + \dfrac{x^6}{2!} + \dfrac{x^9}{3!} + \dfrac{x^{12}}{4!} + \cdots$

SOLUTION We recognize

$$1 + x^3 + \frac{x^6}{2!} + \frac{x^9}{3!} + \frac{x^{12}}{4!} + \cdots = \sum_{n=0}^{\infty} \frac{x^{3n}}{n!} = \sum_{n=0}^{\infty} \frac{(x^3)^n}{n!}$$

as the Maclaurin series for e^x with x replaced by x^3. Therefore,

$$1 + x^3 + \frac{x^6}{2!} + \frac{x^9}{3!} + \frac{x^{12}}{4!} + \cdots = e^{x^3}.$$

65. $1 - \dfrac{5^3x^3}{3!} + \dfrac{5^5x^5}{5!} - \dfrac{5^7x^7}{7!} + \cdots$

SOLUTION Note

$$1 - \frac{5^3x^3}{3!} + \frac{5^5x^5}{5!} - \frac{5^7x^7}{7!} + \cdots = 1 - 5x + \left(5x - \frac{5^3x^3}{3!} + \frac{5^5x^5}{5!} - \frac{5^7x^7}{7!} + \cdots\right)$$

$$= 1 - 5x + \sum_{n=0}^{\infty} (-1)^n \frac{(5x)^{2n+1}}{(2n+1)!}.$$

The series is the Maclaurin series for $\sin x$ with x replaced by $5x$, so

$$1 - \frac{5^3x^3}{3!} + \frac{5^5x^5}{5!} - \frac{5^7x^7}{7!} + \cdots = 1 - 5x + \sin(5x).$$

In Exercises 67 and 68, let

$$f(x) = \frac{1}{(1-x)(1-2x)}$$

67. Find the Maclaurin series of $f(x)$ using the identity

$$f(x) = \frac{2}{1-2x} - \frac{1}{1-x}$$

SOLUTION Substituting $2x$ for x in the Maclaurin series for $\dfrac{1}{1-x}$ gives

$$\frac{1}{1-2x} = \sum_{n=0}^{\infty} (2x)^n = \sum_{n=0}^{\infty} 2^n x^n$$

which is valid for $|2x| < 1$, or $|x| < \frac{1}{2}$. Because the Maclaurin series for $\dfrac{1}{1-x}$ is valid for $|x| < 1$, the two series together are valid for $|x| < \frac{1}{2}$. Thus, for $|x| < \frac{1}{2}$,

$$\frac{1}{(1-2x)(1-x)} = \frac{2}{1-2x} - \frac{1}{1-x} = 2\sum_{n=0}^{\infty} 2^n x^n - \sum_{n=0}^{\infty} x^n$$

$$= \sum_{n=0}^{\infty} 2^{n+1}x^n - \sum_{n=0}^{\infty} x^n = \sum_{n=0}^{\infty} \left(2^{n+1} - 1\right) x^n.$$

69. When a voltage V is applied to a series circuit consisting of a resistor R and an inductor L, the current at time t is

$$I(t) = \left(\frac{V}{R}\right)\left(1 - e^{-Rt/L}\right)$$

Expand $I(t)$ in a Maclaurin series. Show that $I(t) \approx \dfrac{Vt}{L}$ for small t.

SOLUTION Substituting $-\frac{Rt}{L}$ for t in the Maclaurin series for e^t gives

$$e^{-Rt/L} = \sum_{n=0}^{\infty} \frac{\left(-\frac{Rt}{L}\right)^n}{n!} = \sum_{n=0}^{\infty} \frac{(-1)^n}{n!}\left(\frac{R}{L}\right)^n t^n = 1 + \sum_{n=1}^{\infty} \frac{(-1)^n}{n!}\left(\frac{R}{L}\right)^n t^n$$

Thus,

$$1 - e^{-Rt/L} = 1 - \left(1 + \sum_{n=1}^{\infty} \frac{(-1)^n}{n!}\left(\frac{R}{L}\right)^n t^n\right) = \sum_{n=1}^{\infty} \frac{(-1)^{n+1}}{n!}\left(\frac{Rt}{L}\right)^n,$$

and

$$I(t) = \frac{V}{R}\sum_{n=1}^{\infty} \frac{(-1)^{n+1}}{n!}\left(\frac{Rt}{L}\right)^n = \frac{Vt}{L} + \frac{V}{R}\sum_{n=2}^{\infty} \frac{(-1)^{n+1}}{n!}\left(\frac{Rt}{L}\right)^n.$$

If t is small, then we can approximate $I(t)$ by the first (linear) term, and ignore terms with higher powers of t; then we find

$$V(t) \approx \frac{Vt}{L}.$$

71. Find the Maclaurin series for $f(x) = \cos(x^3)$ and use it to determine $f^{(6)}(0)$.

SOLUTION The Maclaurin series for $\cos x$ is

$$\cos x = \sum_{n=0}^{\infty} (-1)^n \frac{x^{2n}}{(2n)!}$$

Substituting x^3 for x gives

$$\cos(x^3) = \sum_{n=0}^{\infty} (-1)^n \frac{x^{6n}}{(2n)!}$$

Now, the coefficient of x^6 in this series is

$$-\frac{1}{2!} = -\frac{1}{2} = \frac{f^{(6)}(0)}{6!}$$

so

$$f^{(6)}(0) = -\frac{6!}{2} = -360$$

73. Use substitution to find the first three terms of the Maclaurin series for $f(x) = e^{x^{20}}$. How does the result show that $f^{(k)}(0) = 0$ for $1 \le k \le 19$?

SOLUTION Substituting x^{20} for x in the Maclaurin series for e^x yields

$$e^{x^{20}} = \sum_{n=0}^{\infty} \frac{(x^{20})^n}{n!} = \sum_{n=0}^{\infty} \frac{x^{20n}}{n!};$$

the first three terms in the series are then

$$1 + x^{20} + \frac{1}{2}x^{40}.$$

Recall that the coefficient of x^k in the Maclaurin series for f is $\frac{f^{(k)}(0)}{k!}$. For $1 \le k \le 19$, the coefficient of x^k in the Maclaurin series for $f(x) = e^{x^{20}}$ is zero; it therefore follows that

$$\frac{f^{(k)}(0)}{k!} = 0 \quad \text{or} \quad f^{(k)}(0) = 0$$

for $1 \le k \le 19$.

75. Does the Maclaurin series for $f(x) = (1+x)^{3/4}$ converge to $f(x)$ at $x = 2$? Give numerical evidence to support your answer.

SOLUTION The Taylor series for $f(x) = (1+x)^{3/4}$ converges to $f(x)$ for $|x| < 1$; because $x = 2$ is not contained on this interval, the series does not converge to $f(x)$ at $x = 2$. The graph below displays

$$S_N = \sum_{n=0}^{N} \binom{\frac{3}{4}}{n} 2^n$$

for $0 \le N \le 14$. The divergent nature of the sequence of partial sums is clear.

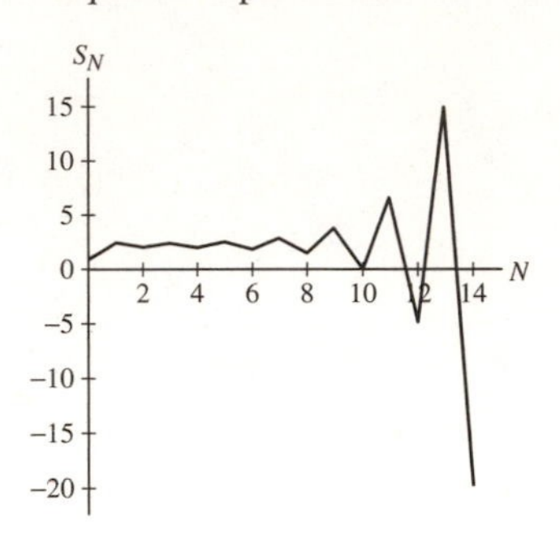

77. GU Let $f(x) = \sqrt{1+x}$.

(a) Use a graphing calculator to compare the graph of f with the graphs of the first five Taylor polynomials for f. What do they suggest about the interval of convergence of the Taylor series?

(b) Investigate numerically whether or not the Taylor expansion for f is valid for $x = 1$ and $x = -1$.

SOLUTION

(a) The five first terms of the Binomial series with $a = \frac{1}{2}$ are

$$\sqrt{1+x} = 1 + \frac{1}{2}x + \frac{\frac{1}{2}\left(\frac{1}{2}-1\right)}{2!}x^2 + \frac{\frac{1}{2}\left(\frac{1}{2}-1\right)\left(\frac{1}{2}-2\right)}{3!}x^3 + \frac{\frac{1}{2}\left(\frac{1}{2}-1\right)\left(\frac{1}{2}-2\right)\left(\frac{1}{2}-3\right)}{4!}x^4 + \cdots$$

$$= 1 + \frac{1}{2}x - \frac{1}{8}x^2 + \frac{9}{4}x^3 - \frac{45}{2}x^4 + \cdots$$

Therefore, the first five Taylor polynomials are

$$T_0(x) = 1;$$

$$T_1(x) = 1 + \frac{1}{2}x;$$

$$T_2(x) = 1 + \frac{1}{2}x - \frac{1}{8}x^2;$$

$$T_3(x) = 1 + \frac{1}{2}x - \frac{1}{8}x^2 + \frac{1}{8}x^3;$$

$$T_4(x) = 1 + \frac{1}{2}x - \frac{1}{8}x^2 + \frac{1}{8}x^3 - \frac{5}{128}x^4.$$

The figure displays the graphs of these Taylor polynomials, along with the graph of the function $f(x) = \sqrt{1+x}$, which is shown in red.

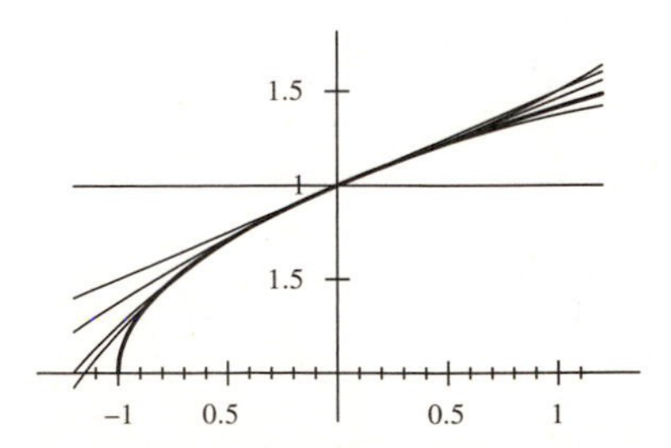

The graphs suggest that the interval of convergence for the Taylor series is $-1 < x < 1$.

(b) Using a computer algebra system to calculate $S_N = \sum_{n=0}^{N} \binom{\frac{1}{2}}{n} x^n$ for $x = 1$ we find

$$S_{10} = 1.409931183, \quad S_{100} = 1.414073048, \quad S_{1000} = 1.414209104,$$

which appears to be converging to $\sqrt{2}$ as expected. At $x = -1$ we calculate $S_N = \sum_{n=0}^{N} \binom{\frac{1}{2}}{n} \cdot (-1)^n$, and find

$$S_{10} = 0.176197052, \quad S_{100} = 0.056348479, \quad S_{1000} = 0.017839011,$$

which appears to be converging to zero, though slowly.

79. Use Example 11 and the approximation $\sin x \approx x$ to show that the period T of a pendulum released at an angle θ has the following second-order approximation:

$$T \approx 2\pi\sqrt{\frac{L}{g}}\left(1 + \frac{\theta^2}{16}\right)$$

SOLUTION The period T of a pendulum of length L released from an angle θ is

$$T = 4\sqrt{\frac{L}{g}}E(k),$$

where $g \approx 9.8$ m/s^2 is the acceleration due to gravity, $E(k)$ is the elliptic function of the first kind and $k = \sin\frac{\theta}{2}$. From Example 11, we know that

$$E(k) = \frac{\pi}{2}\sum_{n=0}^{\infty}\left(\frac{1 \cdot 3 \cdot 5 \cdots (2n-1)}{2 \cdot 4 \cdot 6 \cdots (2n)}\right)^2 k^{2n}.$$

Using the approximation $\sin x \approx x$, we have

$$k = \sin\frac{\theta}{2} \approx \frac{\theta}{2};$$

moreover, using the first two terms of the series for $E(k)$, we find

$$E(k) \approx \frac{\pi}{2}\left[1 + \left(\frac{1}{2}\right)^2\left(\frac{\theta}{2}\right)^2\right] = \frac{\pi}{2}\left(1 + \frac{\theta^2}{16}\right).$$

Therefore,

$$T = 4\sqrt{\frac{L}{g}}E(k) \approx 2\pi\sqrt{\frac{L}{g}}\left(1 + \frac{\theta^2}{16}\right).$$

In Exercises 80–83, find the Maclaurin series of the function and use it to calculate the limit.

81. $\lim_{x\to 0} \frac{\sin x - x + \frac{x^3}{6}}{x^5}$

SOLUTION Using the Maclaurin series for $\sin x$, we find

$$\sin x = \sum_{n=0}^{\infty}(-1)^n\frac{x^{2n+1}}{(2n+1)!} = x - \frac{x^3}{6} + \frac{x^5}{120} + \sum_{n=3}^{\infty}(-1)^n\frac{x^{2n+1}}{(2n+1)!}.$$

Thus,

$$\sin x - x + \frac{x^3}{6} = \frac{x^5}{120} + \sum_{n=3}^{\infty}(-1)^n\frac{x^{2n+1}}{(2n+1)!}$$

and

$$\frac{\sin x - x + \frac{x^3}{6}}{x^5} = \frac{1}{120} + \sum_{n=3}^{\infty}(-1)^n\frac{x^{2n-4}}{(2n+1)!}$$

Note that the radius of convergence for this series is infinite, and recall from the previous section that a convergent power series is continuous within its radius of convergence. Thus to calculate the limit of this power series as $x \to 0$ it suffices to evaluate it at $x = 0$:

$$\lim_{x\to 0}\frac{\sin x - x + \frac{x^3}{6}}{x^5} = \lim_{x\to 0}\left(\frac{1}{120} + \sum_{n=3}^{\infty}(-1)^n\frac{x^{2n-4}}{(2n+1)!}\right) = \frac{1}{120} + 0 = \frac{1}{120}$$

83. $\lim_{x\to 0}\left(\frac{\sin(x^2)}{x^4}-\frac{\cos x}{x^2}\right)$

SOLUTION We start with

$$\sin x=\sum_{n=0}^{\infty}(-1)^n\frac{x^{2n+1}}{(2n+1)!}\qquad \cos x=\sum_{n=0}^{\infty}(-1)^n\frac{x^{2n}}{(2n)!}$$

so that

$$\frac{\sin(x^2)}{x^4}=\sum_{n=0}^{\infty}(-1)^n\frac{x^{4n+2}}{(2n+1)!x^4}=\sum_{n=0}^{\infty}(-1)^n\frac{x^{4n-2}}{(2n+1)!}$$

$$\frac{\cos x}{x^2}=\sum_{n=0}^{\infty}(-1)^n\frac{x^{2n-2}}{(2n)!}$$

Expanding the first few terms gives

$$\frac{\sin(x^2)}{x^4}=\frac{1}{x^2}-\sum_{n=1}^{\infty}(-1)^n\frac{x^{4n-2}}{(2n+1)!}$$

$$\frac{\cos x}{x^2}=\frac{1}{x^2}-\frac{1}{2}+\sum_{n=2}^{\infty}(-1)^n\frac{x^{2n-2}}{(2n)!}$$

so that

$$\frac{\sin(x^2)}{x^4}-\frac{\cos x}{x^2}=\frac{1}{2}-\sum_{n=1}^{\infty}(-1)^n\frac{x^{4n-2}}{(2n+1)!}-\sum_{n=2}^{\infty}(-1)^n\frac{x^{2n-2}}{(2n)!}$$

Note that all terms under the summation signs have positive powers of x. Now, the radius of convergence of the series for both sin and cos is infinite, so the radius of convergence of this series is infinite. Recall from the previous section that a convergent power series is continuous within its radius of convergence. Thus to calculate the limit of this power series as $x\to 0$ it suffices to evaluate it at $x=0$:

$$\lim_{x\to 0}\left(\frac{\sin(x^2)}{x^4}-\frac{\cos x}{x^2}\right)=\lim_{x\to 0}\left(\frac{1}{2}-\sum_{n=1}^{\infty}(-1)^n\frac{x^{4n-2}}{(2n+1)!}-\sum_{n=2}^{\infty}(-1)^n\frac{x^{2n-2}}{(2n)!}\right)=\frac{1}{2}+0=\frac{1}{2}$$

Further Insights and Challenges

85. Let $g(t)=\frac{1}{1+t^2}-\frac{t}{1+t^2}$.

(a) Show that $\int_0^1 g(t)\,dt=\frac{\pi}{4}-\frac{1}{2}\ln 2$.

(b) Show that $g(t)=1-t-t^2+t^3-t^4-t^5-t^6+\dots$

(c) Evaluate $S=1-\frac{1}{2}-\frac{1}{3}+\frac{1}{4}-\frac{1}{5}-\frac{1}{6}-\frac{1}{7}+\dots$

SOLUTION

(a)

$$\int_0^1 g(t)\,dt=\left(\tan^{-1}t-\frac{1}{2}\ln(t^2+1)\right)\Big|_0^1=\tan^{-1}1-\frac{1}{2}\ln 2=\frac{\pi}{4}-\frac{1}{2}\ln 2$$

(b) Start with the Taylor series for $\frac{1}{1+t}$:

$$\frac{1}{1+t}=\sum_{n=0}^{\infty}(-1)^n t^n$$

and substitute t^2 for t to get

$$\frac{1}{1+t^2}=\sum_{n=0}^{\infty}(-1)^n t^{2n}=1-t^2+t^4-t^6+\dots$$

so that

$$\frac{t}{1+t^2} = \sum_{n=0}^{\infty}(-1)^n t^{2n+1} = t - t^3 + t^5 - t^7 + \dots$$

Finally,

$$g(t) = \frac{1}{1+t^2} - \frac{t}{1+t^2} = 1 - t - t^2 + t^3 + t^4 - t^5 - t^6 + t^7 + \dots$$

(c) We have

$$\int g(t)\,dt = \int (1 - t - t^2 + t^3 + t^4 - t^5 - \dots)\,dt = t - \frac{1}{2}t^2 - \frac{1}{3}t^3 + \frac{1}{4}t^4 + \frac{1}{5}t^5 - \frac{1}{6}t^6 - \dots + C$$

The radius of convergence of the series for $g(t)$ is 1, so the radius of convergence of this series is also 1. However, this series converges at the right endpoint, $t = 1$, since

$$\left(1 - \frac{1}{2}\right) - \left(\frac{1}{3} - \frac{1}{4}\right) + \left(\frac{1}{5} - \frac{1}{6}\right) - \dots$$

is an alternating series with general term decreasing to zero. Thus by part (a),

$$1 - \frac{1}{2} - \frac{1}{3} + \frac{1}{4} + \frac{1}{5} - \frac{1}{6} - \dots = \frac{\pi}{4} - \frac{1}{2}\ln 2$$

In Exercises 86 and 87, we investigate the convergence of the binomial series

$$T_a(x) = \sum_{n=0}^{\infty} \binom{a}{n} x^n$$

87. By Exercise 86, $T_a(x)$ converges for $|x| < 1$, but we do not yet know whether $T_a(x) = (1+x)^a$.

(a) Verify the identity

$$a\binom{a}{n} = n\binom{a}{n} + (n+1)\binom{a}{n+1}$$

(b) Use (a) to show that $y = T_a(x)$ satisfies the differential equation $(1+x)y' = ay$ with initial condition $y(0) = 1$.

(c) Prove that $T_a(x) = (1+x)^a$ for $|x| < 1$ by showing that the derivative of the ratio $\dfrac{T_a(x)}{(1+x)^a}$ is zero.

SOLUTION

(a)

$$\begin{aligned} n\binom{a}{n} + (n+1)\binom{a}{n+1} &= n \cdot \frac{a(a-1)\cdots(a-n+1)}{n!} + (n+1)\cdot\frac{a(a-1)\cdots(a-n+1)(a-n)}{(n+1)!} \\ &= \frac{a(a-1)\cdots(a-n+1)}{(n-1)!} + \frac{a(a-1)\cdots(a-n+1)(a-n)}{n!} \\ &= \frac{a(a-1)\cdots(a-n+1)(n+(a-n))}{n!} = a\cdot\binom{a}{n} \end{aligned}$$

(b) Differentiating $T_a(x)$ term-by-term yields

$$T_a'(x) = \sum_{n=1}^{\infty} n\binom{a}{n} x^{n-1}.$$

Thus,

$$\begin{aligned} (1+x)T_a'(x) &= \sum_{n=1}^{\infty} n\binom{a}{n}x^{n-1} + \sum_{n=1}^{\infty} n\binom{a}{n}x^n = \sum_{n=0}^{\infty}(n+1)\binom{a}{n+1}x^n + \sum_{n=0}^{\infty} n\binom{a}{n}x^n \\ &= \sum_{n=0}^{\infty}\left[(n+1)\binom{a}{n+1} + n\binom{a}{n}\right]x^n = a\sum_{n=0}^{\infty}\binom{a}{n}x^n = aT_a(x). \end{aligned}$$

Moreover,

$$T_a(0) = \binom{a}{0} = 1.$$

(c)

$$\frac{d}{dx}\left(\frac{T_a(x)}{(1+x)^a}\right) = \frac{(1+x)^a T_a'(x) - a(1+x)^{a-1}T_a(x)}{(1+x)^{2a}} = \frac{(1+x)T_a'(x) - aT_a(x)}{(1+x)^{a+1}} = 0.$$

Thus,

$$\frac{T_a(x)}{(1+x)^a} = C,$$

for some constant C. For $x = 0$,

$$\frac{T_a(0)}{(1+0)^a} = \frac{1}{1} = 1, \text{ so } C = 1.$$

Finally, $T_a(x) = (1+x)^a$.

89. Assume that $a < b$ and let L be the arc length (circumference) of the ellipse $\left(\frac{x}{a}\right)^2 + \left(\frac{y}{b}\right)^2 = 1$ shown in Figure 5. There is no explicit formula for L, but it is known that $L = 4bG(k)$, with $G(k)$ as in Exercise 88 and $k = \sqrt{1 - a^2/b^2}$. Use the first three terms of the expansion of Exercise 88 to estimate L when $a = 4$ and $b = 5$.

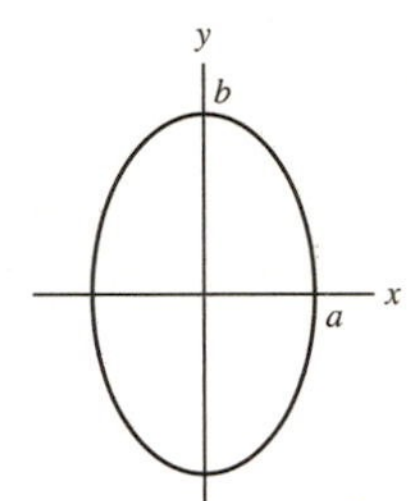

FIGURE 5 The ellipse $\left(\frac{x}{a}\right)^2 + \left(\frac{y}{b}\right)^2 = 1$.

SOLUTION With $a = 4$ and $b = 5$,

$$k = \sqrt{1 - \frac{4^2}{5^2}} = \frac{3}{5},$$

and the arc length of the ellipse $\left(\frac{x}{4}\right)^2 + \left(\frac{y}{5}\right)^2 = 1$ is

$$L = 20G\left(\frac{3}{5}\right) = 20\left(\frac{\pi}{2} - \frac{\pi}{2}\sum_{n=1}^{\infty}\left(\frac{1\cdot 3\cdots(2n-1)}{2\cdot 4\cdots(2n)}\right)^2 \frac{\left(\frac{3}{5}\right)^{2n}}{2n-1}\right).$$

Using the first three terms in the series for $G(k)$ gives

$$L \approx 10\pi - 10\pi\left(\left(\frac{1}{2}\right)^2 \cdot \frac{(3/5)^2}{1} + \left(\frac{1\cdot 3}{2\cdot 4}\right)^2 \cdot \frac{(3/5)^4}{3}\right) = 10\pi\left(1 - \frac{9}{100} - \frac{243}{40{,}000}\right) = \frac{36{,}157\pi}{4000} \approx 28.398.$$

91. Irrationality of e Prove that e is an irrational number using the following argument by contradiction. Suppose that $e = M/N$, where M, N are nonzero integers.

(a) Show that $M!\,e^{-1}$ is a whole number.

(b) Use the power series for e^x at $x = -1$ to show that there is an integer B such that $M!\,e^{-1}$ equals

$$B + (-1)^{M+1}\left(\frac{1}{M+1} - \frac{1}{(M+1)(M+2)} + \cdots\right)$$

(c) Use your knowledge of alternating series with decreasing terms to conclude that $0 < |M!\,e^{-1} - B| < 1$ and observe that this contradicts (a). Hence, e is not equal to M/N.

SOLUTION Suppose that $e = M/N$, where M, N are nonzero integers.

(a) With $e = M/N$,

$$M!e^{-1} = M!\frac{N}{M} = (M-1)!N,$$

which is a whole number.

(b) Substituting $x = -1$ into the Maclaurin series for e^x and multiplying the resulting series by $M!$ yields

$$M!e^{-1} = M!\left(1 - 1 + \frac{1}{2!} - \frac{1}{3!} + \cdots + \frac{(-1)^k}{k!} + \cdots\right).$$

For all $k \le M$, $\frac{M!}{k!}$ is a whole number, so

$$M!\left(1 - 1 + \frac{1}{2!} - \frac{1}{3!} + \cdots + \frac{(-1)^k}{M!}\right)$$

is an integer. Denote this integer by B. Thus,

$$M!e^{-1} = B + M!\left(\frac{(-1)^{M+1}}{(M+1)!} + \frac{(-1)^{M+2}}{(M+2)!} + \cdots\right) = B + (-1)^{M+1}\left(\frac{1}{M+1} - \frac{1}{(M+1)(M+2)} + \cdots\right).$$

(c) The series for $M!e^{-1}$ obtained in part (b) is an alternating series with $a_n = \frac{M!}{n!}$. Using the error bound for an alternating series and noting that $B = S_M$, we have

$$\left|M!e^{-1} - B\right| \le a_{M+1} = \frac{1}{M+1} < 1.$$

This inequality implies that $M!e^{-1} - B$ is not a whole number; however, B is a whole number so $M!e^{-1}$ cannot be a whole number. We get a contradiction to the result in part (a), which proves that the original assumption that e is a rational number is false.

CHAPTER REVIEW EXERCISES

1. Let $a_n = \dfrac{n-3}{n!}$ and $b_n = a_{n+3}$. Calculate the first three terms in each sequence.

(a) a_n^2

(b) b_n

(c) $a_n b_n$

(d) $2a_{n+1} - 3a_n$

SOLUTION

(a)

$$a_1^2 = \left(\frac{1-3}{1!}\right)^2 = (-2)^2 = 4;$$

$$a_2^2 = \left(\frac{2-3}{2!}\right)^2 = \left(-\frac{1}{2}\right)^2 = \frac{1}{4};$$

$$a_3^2 = \left(\frac{3-3}{3!}\right)^2 = 0.$$

(b)

$$b_1 = a_4 = \frac{4-3}{4!} = \frac{1}{24};$$

$$b_2 = a_5 = \frac{5-3}{5!} = \frac{1}{60};$$

$$b_3 = a_6 = \frac{6-3}{6!} = \frac{1}{240}.$$

(c) Using the formula for a_n and the values in (b) we obtain:

$$a_1b_1 = \frac{1-3}{1!} \cdot \frac{1}{24} = -\frac{1}{12};$$

$$a_2b_2 = \frac{2-3}{2!} \cdot \frac{1}{60} = -\frac{1}{120};$$

$$a_3b_3 = \frac{3-3}{3!} \cdot \frac{1}{240} = 0.$$

(d)

$$2a_2 - 3a_1 = 2\left(-\frac{1}{2}\right) - 3(-2) = 5;$$

$$2a_3 - 3a_2 = 2 \cdot 0 - 3\left(-\frac{1}{2}\right) = \frac{3}{2};$$

$$2a_4 - 3a_3 = 2 \cdot \frac{1}{24} - 3 \cdot 0 = \frac{1}{12}.$$

In Exercises 3–8, compute the limit (or state that it does not exist) assuming that $\lim_{n\to\infty} a_n = 2$.

3. $\lim_{n\to\infty} (5a_n - 2a_n^2)$

SOLUTION

$$\lim_{n\to\infty} \left(5a_n - 2a_n^2\right) = 5\lim_{n\to\infty} a_n - 2\lim_{n\to\infty} a_n^2 = 5\lim_{n\to\infty} a_n - 2\left(\lim_{n\to\infty} a_n\right)^2 = 5 \cdot 2 - 2 \cdot 2^2 = 2.$$

5. $\lim_{n\to\infty} e^{a_n}$

SOLUTION The function $f(x) = e^x$ is continuous, hence:

$$\lim_{n\to\infty} e^{a_n} = e^{\lim_{n\to\infty} a_n} = e^2.$$

7. $\lim_{n\to\infty} (-1)^n a_n$

SOLUTION Because $\lim_{n\to\infty} a_n \neq 0$, it follows that $\lim_{n\to\infty} (-1)^n a_n$ does not exist.

In Exercises 9–22, determine the limit of the sequence or show that the sequence diverges.

9. $a_n = \sqrt{n+5} - \sqrt{n+2}$

SOLUTION First rewrite a_n as follows:

$$a_n = \frac{\left(\sqrt{n+5} - \sqrt{n+2}\right)\left(\sqrt{n+5} + \sqrt{n+2}\right)}{\sqrt{n+5} + \sqrt{n+2}} = \frac{(n+5) - (n+2)}{\sqrt{n+5} + \sqrt{n+2}} = \frac{3}{\sqrt{n+5} + \sqrt{n+2}}.$$

Thus,

$$\lim_{n\to\infty} a_n = \lim_{n\to\infty} \frac{3}{\sqrt{n+5} + \sqrt{n+2}} = 0.$$

11. $a_n = 2^{1/n^2}$

SOLUTION The function $f(x) = 2^x$ is continuous, so

$$\lim_{n\to\infty} a_n = \lim_{n\to\infty} 2^{1/n^2} = 2^{\lim_{n\to\infty}(1/n^2)} = 2^0 = 1.$$

13. $b_m = 1 + (-1)^m$

SOLUTION Because $1 + (-1)^m$ is equal to 0 for m odd and is equal to 2 for m even, the sequence $\{b_m\}$ does not approach one limit; hence this sequence diverges.

15. $b_n = \tan^{-1}\left(\frac{n+2}{n+5}\right)$

SOLUTION The function $\tan^{-1}x$ is continuous, so

$$\lim_{n\to\infty} b_n = \lim_{n\to\infty} \tan^{-1}\left(\frac{n+2}{n+5}\right) = \tan^{-1}\left(\lim_{n\to\infty} \frac{n+2}{n+5}\right) = \tan^{-1} 1 = \frac{\pi}{4}.$$

17. $b_n = \sqrt{n^2+n} - \sqrt{n^2+1}$

SOLUTION Rewrite b_n as

$$b_n = \frac{\left(\sqrt{n^2+n} - \sqrt{n^2+1}\right)\left(\sqrt{n^2+n} + \sqrt{n^2+1}\right)}{\sqrt{n^2+n} + \sqrt{n^2+1}} = \frac{\left(n^2+n\right) - \left(n^2+1\right)}{\sqrt{n^2+n} + \sqrt{n^2+1}} = \frac{n-1}{\sqrt{n^2+n} + \sqrt{n^2+1}}.$$

Then

$$\lim_{n\to\infty} b_n = \lim_{n\to\infty} \frac{\frac{n}{n} - \frac{1}{n}}{\sqrt{\frac{n^2}{n^2} + \frac{n}{n^2}} + \sqrt{\frac{n^2}{n^2} + \frac{1}{n^2}}} = \lim_{n\to\infty} \frac{1 - \frac{1}{n}}{\sqrt{1 + \frac{1}{n}} + \sqrt{1 + \frac{1}{n^2}}} = \frac{1 - 0}{\sqrt{1+0} + \sqrt{1+0}} = \frac{1}{2}.$$

19. $b_m = \left(1 + \frac{1}{m}\right)^{3m}$

SOLUTION $\lim_{m\to\infty} b_m = \lim_{m\to\infty} \left(1 + \frac{1}{m}\right)^m = e.$

21. $b_n = n\big(\ln(n+1) - \ln n\big)$

SOLUTION Write

$$b_n = n \ln\left(\frac{n+1}{n}\right) = \frac{\ln\left(1 + \frac{1}{n}\right)}{\frac{1}{n}}.$$

Using L'Hôpital's Rule, we find

$$\lim_{n\to\infty} b_n = \lim_{n\to\infty} \frac{\ln\left(1 + \frac{1}{n}\right)}{\frac{1}{n}} = \lim_{x\to\infty} \frac{\ln\left(1 + \frac{1}{x}\right)}{\frac{1}{x}} = \lim_{x\to\infty} \frac{\left(1 + \frac{1}{x}\right)^{-1} \cdot \left(-\frac{1}{x^2}\right)}{-\frac{1}{x^2}} = \lim_{x\to\infty} \left(1 + \frac{1}{x}\right)^{-1} = 1.$$

23. Use the Squeeze Theorem to show that $\lim_{n\to\infty} \frac{\arctan(n^2)}{\sqrt{n}} = 0.$

SOLUTION For all x,

$$-\frac{\pi}{2} < \arctan x < \frac{\pi}{2},$$

so

$$-\frac{\pi/2}{\sqrt{n}} < \frac{\arctan(n^2)}{\sqrt{n}} < \frac{\pi/2}{\sqrt{n}},$$

for all n. Because

$$\lim_{n\to\infty} \left(-\frac{\pi/2}{\sqrt{n}}\right) = \lim_{n\to\infty} \frac{\pi/2}{\sqrt{n}} = 0,$$

it follows by the Squeeze Theorem that

$$\lim_{n\to\infty} \frac{\arctan(n^2)}{\sqrt{n}} = 0.$$

25. Calculate $\lim_{n\to\infty} \frac{a_{n+1}}{a_n}$, where $a_n = \frac{1}{2}3^n - \frac{1}{3}2^n$.

SOLUTION Because

$$\frac{1}{2}3^n - \frac{1}{3}2^n \geq \frac{1}{2}3^n - \frac{1}{3}3^n = \frac{3^n}{6}$$

and

$$\lim_{n\to\infty} \frac{3^n}{6} = \infty,$$

we conclude that $\lim_{n\to\infty} a_n = \infty$, so L'Hôpital's rule may be used:

$$\lim_{n\to\infty} \frac{a_{n+1}}{a_n} = \lim_{n\to\infty} \frac{\frac{1}{2}3^{n+1} - \frac{1}{3}2^{n+1}}{\frac{1}{2}3^n - \frac{1}{3}2^n} = \lim_{n\to\infty} \frac{3^{n+2} - 2^{n+2}}{3^{n+1} - 2^{n+1}} = \lim_{n\to\infty} \frac{3 - 2\left(\frac{2}{3}\right)^{n+1}}{1 - \left(\frac{2}{3}\right)^{n+1}} = \frac{3 - 0}{1 - 0} = 3.$$

27. Calculate the partial sums S_4 and S_7 of the series $\sum_{n=1}^{\infty} \frac{n-2}{n^2 + 2n}$.

SOLUTION

$$S_4 = -\frac{1}{3} + 0 + \frac{1}{15} + \frac{2}{24} = -\frac{11}{60} = -0.183333;$$

$$S_7 = -\frac{1}{3} + 0 + \frac{1}{15} + \frac{2}{24} + \frac{3}{35} + \frac{4}{48} + \frac{5}{63} = \frac{287}{4410} = 0.065079.$$

29. Find the sum $\frac{4}{9}+\frac{8}{27}+\frac{16}{81}+\frac{32}{243}+\cdots$.

SOLUTION This is a geometric series with common ratio $r=\frac{2}{3}$. Therefore,

$$\frac{4}{9}+\frac{8}{27}+\frac{16}{81}+\frac{32}{243}+\cdots=\frac{\frac{4}{9}}{1-\frac{2}{3}}=\frac{4}{3}.$$

31. Find the sum $\sum_{n=-1}^{\infty}\frac{2^{n+3}}{3^n}$.

SOLUTION Note

$$\sum_{n=-1}^{\infty}\frac{2^{n+3}}{3^n}=2^3\sum_{n=-1}^{\infty}\frac{2^n}{3^n}=8\sum_{n=-1}^{\infty}\left(\frac{2}{3}\right)^n;$$

therefore,

$$\sum_{n=-1}^{\infty}\frac{2^{n+3}}{3^n}=8\cdot\frac{3}{2}\cdot\frac{1}{1-\frac{2}{3}}=36.$$

33. Give an example of divergent series $\sum_{n=1}^{\infty}a_n$ and $\sum_{n=1}^{\infty}b_n$ such that $\sum_{n=1}^{\infty}(a_n+b_n)=1$.

SOLUTION Let $a_n=\left(\frac{1}{2}\right)^n+1$, $b_n=-1$. The corresponding series diverge by the Divergence Test; however,

$$\sum_{n=1}^{\infty}(a_n+b_n)=\sum_{n=1}^{\infty}\left(\frac{1}{2}\right)^n=\frac{\frac{1}{2}}{1-\frac{1}{2}}=1.$$

35. Evaluate $S=\sum_{n=3}^{\infty}\frac{1}{n(n+3)}$.

SOLUTION Note that

$$\frac{1}{n(n+3)}=\frac{1}{3}\left(\frac{1}{n}-\frac{1}{n+3}\right)$$

so that

$$\begin{aligned}\sum_{n=3}^{N}\frac{1}{n(n+3)}&=\frac{1}{3}\sum_{n=3}^{N}\left(\frac{1}{n}-\frac{1}{n+3}\right)\\&=\frac{1}{3}\left(\left(\frac{1}{3}-\frac{1}{6}\right)+\left(\frac{1}{4}-\frac{1}{7}\right)+\left(\frac{1}{5}-\frac{1}{8}\right)\right.\\&\qquad\left.\left(\frac{1}{6}-\frac{1}{9}\right)+\cdots+\left(\frac{1}{N-1}-\frac{1}{N+2}\right)+\left(\frac{1}{N}-\frac{1}{N+3}\right)\right)\\&=\frac{1}{3}\left(\frac{1}{3}+\frac{1}{4}+\frac{1}{5}-\frac{1}{N+1}-\frac{1}{N+2}-\frac{1}{N+3}\right)\end{aligned}$$

Thus

$$\begin{aligned}\sum_{n=3}^{\infty}\frac{1}{n(n+3)}&=\frac{1}{3}\lim_{N\to\infty}\sum_{n=3}^{N}\left(\frac{1}{n}-\frac{1}{n+3}\right)\\&=\frac{1}{3}\left(\frac{1}{3}+\frac{1}{4}+\frac{1}{5}-\frac{1}{N+1}-\frac{1}{N+2}-\frac{1}{N+3}\right)=\frac{1}{3}\left(\frac{1}{3}+\frac{1}{4}+\frac{1}{5}\right)=\frac{47}{180}\end{aligned}$$

In Exercises 37–40, use the Integral Test to determine whether the infinite series converges.

37. $\sum_{n=1}^{\infty}\frac{n^2}{n^3+1}$

SOLUTION Let $f(x)=\frac{x^2}{x^3+1}$. This function is continuous and positive for $x\geq 1$. Because

$$f'(x)=\frac{(x^3+1)(2x)-x^2(3x^2)}{(x^3+1)^2}=\frac{x(2-x^3)}{(x^3+1)^2},$$

we see that $f'(x) < 0$ and f is decreasing on the interval $x \geq 2$. Therefore, the Integral Test applies on the interval $x \geq 2$. Now,

$$\int_2^\infty \frac{x^2}{x^3+1}\,dx = \lim_{R\to\infty}\int_2^R \frac{x^2}{x^3+1}\,dx = \frac{1}{3}\lim_{R\to\infty}\left(\ln(R^3+1) - \ln 9\right) = \infty.$$

The integral diverges; hence, the series $\sum_{n=2}^{\infty} \frac{n^2}{n^3+1}$ diverges, as does the series $\sum_{n=1}^{\infty} \frac{n^2}{n^3+1}$.

39. $\sum_{n=1}^{\infty} \frac{1}{(n+2)(\ln(n+2))^3}$

SOLUTION Let $f(x) = \frac{1}{(x+2)\ln^3(x+2)}$. Using the substitution $u = \ln(x+2)$, so that $du = \frac{1}{x+2}\,dx$, we have

$$\int_0^\infty f(x)\,dx = \int_{\ln 2}^\infty \frac{1}{u^3}\,du = \lim_{R\to\infty}\int_{\ln 2}^\infty \frac{1}{u^3}\,du = \lim_{R\to\infty}\left(-\frac{1}{2u^2}\bigg|_{\ln 2}^R\right)$$

$$= \lim_{R\to\infty}\left(\frac{1}{2(\ln 2)^2} - \frac{1}{2(\ln R)^2}\right) = \frac{1}{2(\ln 2)^2}$$

Since the integral of $f(x)$ converges, so does the series.

In Exercises 41–48, use the Comparison or Limit Comparison Test to determine whether the infinite series converges.

41. $\sum_{n=1}^{\infty} \frac{1}{(n+1)^2}$

SOLUTION For all $n \geq 1$,

$$0 < \frac{1}{n+1} < \frac{1}{n} \quad \text{so} \quad \frac{1}{(n+1)^2} < \frac{1}{n^2}.$$

The series $\sum_{n=1}^{\infty} \frac{1}{n^2}$ is a convergent p-series, so the series $\sum_{n=1}^{\infty} \frac{1}{(n+1)^2}$ converges by the Comparison Test.

43. $\sum_{n=2}^{\infty} \frac{n^2+1}{n^{3.5}-2}$

SOLUTION Apply the Limit Comparison Test with $a_n = \frac{n^2+1}{n^{3.5}-2}$ and $b_n = \frac{1}{n^{1.5}}$. Now,

$$L = \lim_{n\to\infty} \frac{\frac{n^2+1}{n^{3.5}-2}}{\frac{1}{n^{1.5}}} = \lim_{n\to\infty} \frac{n^{3.5}+n^{1.5}}{n^{3.5}-2} = 1.$$

Because L exists and $\sum_{n=1}^{\infty} \frac{1}{n^{1.5}}$ is a convergent p-series, we conclude by the Limit Comparison Test that the series $\sum_{n=2}^{\infty} \frac{n^2+1}{n^{3.5}-2}$ also converges.

45. $\sum_{n=2}^{\infty} \frac{n}{\sqrt{n^5+5}}$

SOLUTION For all $n \geq 2$,

$$\frac{n}{\sqrt{n^5+5}} < \frac{n}{n^{5/2}} = \frac{1}{n^{3/2}}.$$

The series $\sum_{n=2}^{\infty} \frac{1}{n^{3/2}}$ is a convergent p-series, so the series $\sum_{n=2}^{\infty} \frac{n}{\sqrt{n^5+5}}$ converges by the Comparison Test.

47. $\sum_{n=1}^{\infty} \frac{n^{10}+10^n}{n^{11}+11^n}$

SOLUTION Apply the Limit Comparison Test with $a_n = \frac{n^{10}+10^n}{n^{11}+11^n}$ and $b_n = \left(\frac{10}{11}\right)^n$. Then,

$$L = \lim_{n\to\infty} \frac{a_n}{b_n} = \lim_{n\to\infty} \frac{\frac{n^{10}+10^n}{n^{11}+11^n}}{\left(\frac{10}{11}\right)^n} = \lim_{n\to\infty} \frac{\frac{n^{10}+10^n}{10^n}}{\frac{n^{11}+11^n}{11^n}} = \lim_{n\to\infty} \frac{\frac{n^{10}}{10^n}+1}{\frac{n^{11}}{11^n}+1} = 1.$$

The series $\sum_{n=1}^{\infty}\left(\frac{10}{11}\right)^n$ is a convergent geometric series; because L exists, we may therefore conclude by the Limit Comparison Test that the series $\sum_{n=1}^{\infty}\frac{n^{10}+10^n}{n^{11}+11^n}$ also converges.

49. Determine the convergence of $\sum_{n=1}^{\infty}\frac{2^n+n}{3^n-2}$ using the Limit Comparison Test with $b_n=\left(\frac{2}{3}\right)^n$.

SOLUTION With $a_n=\frac{2^n+n}{3^n-2}$, we have

$$L=\lim_{n\to\infty}\frac{a_n}{b_n}=\lim_{n\to\infty}\frac{2^n+n}{3^n-2}\cdot\frac{3^n}{2^n}=\lim_{n\to\infty}\frac{6^n+n3^n}{6^n-2^{n+1}}=\lim_{n\to\infty}\frac{1+n\left(\frac{1}{2}\right)^n}{1-2\left(\frac{1}{3}\right)^n}=1$$

Since $L=1$, the two series either both converge or both diverge. Since $\sum_{n=1}^{\infty}\left(\frac{2}{3}\right)^n$ is a convergent geometric series, the Limit Comparison Test tells us that $\sum_{n=1}^{\infty}\frac{2^n+n}{3^n-2}$ also converges.

51. Let $a_n=1-\sqrt{1-\frac{1}{n}}$. Show that $\lim_{n\to\infty}a_n=0$ and that $\sum_{n=1}^{\infty}a_n$ diverges. *Hint:* Show that $a_n\geq\frac{1}{2n}$.

SOLUTION

$$1-\sqrt{1-\frac{1}{n}}=1-\sqrt{\frac{n-1}{n}}=\frac{\sqrt{n}-\sqrt{n-1}}{\sqrt{n}}=\frac{n-(n-1)}{\sqrt{n}(\sqrt{n}+\sqrt{n-1})}=\frac{1}{n+\sqrt{n^2-n}}$$

$$\geq\frac{1}{n+\sqrt{n^2}}=\frac{1}{2n}.$$

The series $\sum_{n=2}^{\infty}\frac{1}{2n}$ diverges, so the series $\sum_{n=2}^{\infty}\left(1-\sqrt{1-\frac{1}{n}}\right)$ also diverges by the Comparison Test.

53. Let $S=\sum_{n=1}^{\infty}\frac{n}{(n^2+1)^2}$.

(a) Show that S converges.

(b) CAS Use Eq. (4) in Exercise 83 of Section 10.3 with $M=99$ to approximate S. What is the maximum size of the error?

SOLUTION

(a) For $n\geq 1$,

$$\frac{n}{(n^2+1)^2}<\frac{n}{(n^2)^2}=\frac{1}{n^3}.$$

The series $\sum_{n=1}^{\infty}\frac{1}{n^3}$ is a convergent p-series, so the series $\sum_{n=1}^{\infty}\frac{n}{(n^2+1)^2}$ also converges by the Comparison Test.

(b) With $a_n=\frac{n}{(n^2+1)^2}$, $f(x)=\frac{x}{(x^2+1)^2}$ and $M=99$, Eq. (4) in Exercise 83 of Section 10.3 becomes

$$\sum_{n=1}^{99}\frac{n}{(n^2+1)^2}+\int_{100}^{\infty}\frac{x}{(x^2+1)^2}\,dx\leq S\leq\sum_{n=1}^{100}\frac{n}{(n^2+1)^2}+\int_{100}^{\infty}\frac{x}{(x^2+1)^2}\,dx,$$

or

$$0\leq S-\left(\sum_{n=1}^{99}\frac{n}{(n^2+1)^2}+\int_{100}^{\infty}\frac{x}{(x^2+1)^2}\,dx\right)\leq\frac{100}{(100^2+1)^2}.$$

Now,

$$\sum_{n=1}^{99}\frac{n}{(n^2+1)^2}=0.397066274;\text{ and}$$

$$\int_{100}^{\infty}\frac{x}{(x^2+1)^2}\,dx=\lim_{R\to\infty}\int_{100}^{R}\frac{x}{(x^2+1)^2}\,dx=\frac{1}{2}\lim_{R\to\infty}\left(-\frac{1}{R^2+1}+\frac{1}{100^2+1}\right)$$

$$=\frac{1}{20002}=0.000049995;$$

thus,

$$S \approx 0.397066274 + 0.000049995 = 0.397116269.$$

The bound on the error in this approximation is

$$\frac{100}{(100^2+1)^2} = 9.998 \times 10^{-7}.$$

In Exercises 54–57, determine whether the series converges absolutely. If it does not, determine whether it converges conditionally.

55. $\displaystyle\sum_{n=1}^{\infty} \frac{(-1)^n}{n^{1.1}\ln(n+1)}$

SOLUTION Consider the corresponding positive series $\displaystyle\sum_{n=1}^{\infty} \frac{1}{n^{1.1}\ln(n+1)}$. Because

$$\frac{1}{n^{1.1}\ln(n+1)} < \frac{1}{n^{1.1}}$$

and $\displaystyle\sum_{n=1}^{\infty} \frac{1}{n^{1.1}}$ is a convergent p-series, we can conclude by the Comparison Test that $\displaystyle\sum_{n=1}^{\infty} \frac{(-1)^n}{n^{1.1}\ln(n+1)}$ also converges. Thus, $\displaystyle\sum_{n=1}^{\infty} \frac{(-1)^n}{n^{1.1}\ln(n+1)}$ converges absolutely.

57. $\displaystyle\sum_{n=1}^{\infty} \frac{\cos\left(\frac{\pi}{4}+2\pi n\right)}{\sqrt{n}}$

SOLUTION $\cos\left(\frac{\pi}{4}+2\pi n\right) = \cos\frac{\pi}{4} = \frac{\sqrt{2}}{2}$, so

$$\sum_{n=1}^{\infty} \frac{\cos\left(\frac{\pi}{4}+2\pi n\right)}{\sqrt{n}} = \frac{\sqrt{2}}{2}\sum_{n=1}^{\infty}\frac{1}{\sqrt{n}}.$$

This is a divergent p-series, so the series $\displaystyle\sum_{n=1}^{\infty} \frac{\cos\left(\frac{\pi}{4}+2\pi n\right)}{\sqrt{n}}$ diverges.

59. Catalan's constant is defined by $K = \displaystyle\sum_{k=0}^{\infty} \frac{(-1)^k}{(2k+1)^2}$.

(a) How many terms of the series are needed to calculate K with an error of less than 10^{-6}?

(b) CAS Carry out the calculation.

SOLUTION Using the error bound for an alternating series, we have

$$|S_N - K| \le \frac{1}{(2(N+1)+1)^2} = \frac{1}{(2N+3)^2}.$$

For accuracy to three decimal places, we must choose N so that

$$\frac{1}{(2N+3)^2} < 5 \times 10^{-3} \quad \text{or} \quad (2N+3)^2 > 2000.$$

Solving for N yields

$$N > \frac{1}{2}\left(\sqrt{2000}-3\right) \approx 20.9.$$

Thus,

$$K \approx \sum_{k=0}^{21} \frac{(-1)^k}{(2k+1)^2} = 0.915707728.$$

61. Let $\sum_{n=1}^{\infty} a_n$ be an absolutely convergent series. Determine whether the following series are convergent or divergent:

(a) $\sum_{n=1}^{\infty} \left(a_n + \frac{1}{n^2}\right)$

(b) $\sum_{n=1}^{\infty} (-1)^n a_n$

(c) $\sum_{n=1}^{\infty} \frac{1}{1+a_n^2}$

(d) $\sum_{n=1}^{\infty} \frac{|a_n|}{n}$

SOLUTION Because $\sum_{n=1}^{\infty} a_n$ converges absolutely, we know that $\sum_{n=1}^{\infty} a_n$ converges and that $\sum_{n=1}^{\infty} |a_n|$ converges.

(a) Because we know that $\sum_{n=1}^{\infty} a_n$ converges and the series $\sum_{n=1}^{\infty} \frac{1}{n^2}$ is a convergent p-series, the sum of these two series, $\sum_{n=1}^{\infty} \left(a_n + \frac{1}{n^2}\right)$ also converges.

(b) We have,

$$\sum_{n=1}^{\infty} \left|(-1)^n a_n\right| = \sum_{n=1}^{\infty} |a_n|$$

Because $\sum_{n=1}^{\infty} |a_n|$ converges, it follows that $\sum_{n=1}^{\infty} (-1)^n a_n$ converges absolutely, which implies that $\sum_{n=1}^{\infty} (-1)^n a_n$ converges.

(c) Because $\sum_{n=1}^{\infty} a_n$ converges, $\lim_{n\to\infty} a_n = 0$. Therefore,

$$\lim_{n\to\infty} \frac{1}{1+a_n^2} = \frac{1}{1+0^2} = 1 \neq 0,$$

and the series $\sum_{n=1}^{\infty} \frac{1}{1+a_n^2}$ diverges by the Divergence Test.

(d) $\frac{|a_n|}{n} \le |a_n|$ and the series $\sum_{n=1}^{\infty} |a_n|$ converges, so the series $\sum_{n=1}^{\infty} \frac{|a_n|}{n}$ also converges by the Comparison Test.

In Exercises 63–70, apply the Ratio Test to determine convergence or divergence, or state that the Ratio Test is inconclusive.

63. $\sum_{n=1}^{\infty} \frac{n^5}{5^n}$

SOLUTION With $a_n = \frac{n^5}{5^n}$,

$$\left|\frac{a_{n+1}}{a_n}\right| = \frac{(n+1)^5}{5^{n+1}} \cdot \frac{5^n}{n^5} = \frac{1}{5}\left(1+\frac{1}{n}\right)^5,$$

and

$$\rho = \lim_{n\to\infty} \left|\frac{a_{n+1}}{a_n}\right| = \frac{1}{5} \lim_{n\to\infty} \left(1+\frac{1}{n}\right)^5 = \frac{1}{5}\cdot 1 = \frac{1}{5}.$$

Because $\rho < 1$, the series converges by the Ratio Test.

65. $\sum_{n=1}^{\infty} \frac{1}{n2^n + n^3}$

SOLUTION With $a_n = \frac{1}{n2^n+n^3}$,

$$\left|\frac{a_{n+1}}{a_n}\right| = \frac{n2^n + n^3}{(n+1)2^{n+1} + (n+1)^3} = \frac{n2^n\left(1+\frac{n^2}{2^n}\right)}{(n+1)2^{n+1}\left(1+\frac{(n+1)^2}{2^{n+1}}\right)} = \frac{1}{2}\cdot\frac{n}{n+1}\cdot\frac{1+\frac{n^2}{2^n}}{1+\frac{(n+1)^2}{2^{n+1}}},$$

and

$$\rho = \lim_{n\to\infty} \left|\frac{a_{n+1}}{a_n}\right| = \frac{1}{2}\cdot 1\cdot 1 = \frac{1}{2}.$$

Because $\rho < 1$, the series converges by the Ratio Test.

67. $\sum_{n=1}^{\infty} \frac{2^{n^2}}{n!}$

SOLUTION With $a_n = \frac{2^{n^2}}{n!}$,

$$\left|\frac{a_{n+1}}{a_n}\right| = \frac{2^{(n+1)^2}}{(n+1)!} \cdot \frac{n!}{2^{n^2}} = \frac{2^{2n+1}}{n+1} \quad \text{and} \quad \rho = \lim_{n\to\infty} \left|\frac{a_{n+1}}{a_n}\right| = \infty.$$

Because $\rho > 1$, the series diverges by the Ratio Test.

69. $\sum_{n=1}^{\infty} \left(\frac{n}{2}\right)^n \frac{1}{n!}$

SOLUTION With $a_n = \left(\frac{n}{2}\right)^n \frac{1}{n!}$,

$$\left|\frac{a_{n+1}}{a_n}\right| = \left(\frac{n+1}{2}\right)^{n+1} \frac{1}{(n+1)!} \cdot \left(\frac{2}{n}\right)^n n! = \frac{1}{2}\left(\frac{n+1}{n}\right)^n = \frac{1}{2}\left(1+\frac{1}{n}\right)^n,$$

and

$$\rho = \lim_{n\to\infty} \left|\frac{a_{n+1}}{a_n}\right| = \frac{1}{2}e.$$

Because $\rho = \frac{e}{2} > 1$, the series diverges by the Ratio Test.

In Exercises 71–74, apply the Root Test to determine convergence or divergence, or state that the Root Test is inconclusive.

71. $\sum_{n=1}^{\infty} \frac{1}{4^n}$

SOLUTION With $a_n = \frac{1}{4^n}$,

$$L = \lim_{n\to\infty} \sqrt[n]{a_n} = \lim_{n\to\infty} \sqrt[n]{\frac{1}{4^n}} = \frac{1}{4}.$$

Because $L < 1$, the series converges by the Root Test.

73. $\sum_{n=1}^{\infty} \left(\frac{3}{4n}\right)^n$

SOLUTION With $a_n = \left(\frac{3}{4n}\right)^n$,

$$L = \lim_{n\to\infty} \sqrt[n]{a_n} = \lim_{n\to\infty} \sqrt[n]{\left(\frac{3}{4n}\right)^n} = \lim_{n\to\infty} \frac{3}{4n} = 0.$$

Because $L < 1$, the series converges by the Root Test.

In Exercises 75–92, determine convergence or divergence using any method covered in the text.

75. $\sum_{n=1}^{\infty} \left(\frac{2}{3}\right)^n$

SOLUTION This is a geometric series with ratio $r = \frac{2}{3} < 1$; hence, the series converges.

77. $\sum_{n=1}^{\infty} e^{-0.02n}$

SOLUTION This is a geometric series with common ratio $r = \frac{1}{e^{0.02}} \approx 0.98 < 1$; hence, the series converges.

79. $\sum_{n=1}^{\infty} \frac{(-1)^{n-1}}{\sqrt{n}+\sqrt{n+1}}$

SOLUTION In this alternating series, $a_n = \frac{1}{\sqrt{n}+\sqrt{n+1}}$. The sequence $\{a_n\}$ is decreasing, and

$$\lim_{n\to\infty} a_n = 0;$$

therefore the series converges by the Leibniz Test.

81. $\displaystyle\sum_{n=2}^{\infty} \frac{(-1)^n}{\ln n}$

SOLUTION The sequence $a_n = \frac{1}{\ln n}$ is decreasing for $n \geq 10$ and

$$\lim_{n\to\infty} a_n = 0;$$

therefore, the series converges by the Leibniz Test.

83. $\displaystyle\sum_{n=1}^{\infty} \frac{1}{n\sqrt{n} + \ln n}$

SOLUTION For $n \geq 1$,

$$\frac{1}{n\sqrt{n} + \ln n} \leq \frac{1}{n\sqrt{n}} = \frac{1}{n^{3/2}}.$$

The series $\displaystyle\sum_{n=1}^{\infty} \frac{1}{n^{3/2}}$ is a convergent p-series, so the series $\displaystyle\sum_{n=1}^{\infty} \frac{1}{n\sqrt{n} + \ln n}$ converges by the Comparison Test.

85. $\displaystyle\sum_{n=1}^{\infty} \left(\frac{1}{\sqrt{n}} - \frac{1}{\sqrt{n+1}}\right)$

SOLUTION This series telescopes:

$$\sum_{n=1}^{\infty} \left(\frac{1}{\sqrt{n}} - \frac{1}{\sqrt{n+1}}\right) = \left(1 - \frac{1}{\sqrt{2}}\right) + \left(\frac{1}{\sqrt{2}} - \frac{1}{\sqrt{3}}\right) + \left(\frac{1}{\sqrt{3}} - \frac{1}{\sqrt{4}}\right) + \dots$$

so that the nth partial sum S_n is

$$S_n = \left(1 - \frac{1}{\sqrt{2}}\right) + \left(\frac{1}{\sqrt{2}} - \frac{1}{\sqrt{3}}\right) + \left(\frac{1}{\sqrt{3}} - \frac{1}{\sqrt{4}}\right) + \dots + \left(\frac{1}{\sqrt{n}} - \frac{1}{\sqrt{n+1}}\right) = 1 - \frac{1}{\sqrt{n+1}}$$

and then

$$\sum_{n=1}^{\infty} \left(\frac{1}{\sqrt{n}} - \frac{1}{\sqrt{n+1}}\right) = \lim_{n\to\infty} S_n = 1 - \lim_{n\to\infty} \frac{1}{\sqrt{n+1}} = 1$$

87. $\displaystyle\sum_{n=1}^{\infty} \frac{1}{n + \sqrt{n}}$

SOLUTION For $n \geq 1$, $\sqrt{n} \leq n$, so that

$$\sum_{n=1}^{\infty} \frac{1}{n + \sqrt{n}} \geq \sum_{n=1}^{\infty} \frac{1}{2n}$$

which diverges since it is a constant multiple of the harmonic series. Thus $\displaystyle\sum_{n=1}^{\infty} \frac{1}{n + \sqrt{n}}$ diverges as well, by the Comparison Test.

89. $\displaystyle\sum_{n=2}^{\infty} \frac{1}{n^{\ln n}}$

SOLUTION For $n \geq N$ large enough, $\ln n \geq 2$ so that

$$\sum_{n=N}^{\infty} \frac{1}{n^{\ln n}} \leq \sum_{n=N}^{\infty} \frac{1}{n^2}$$

which is a convergent p-series. Thus by the Comparison Test, $\displaystyle\sum_{n=N}^{\infty} \frac{1}{n^{\ln n}}$ also converges; adding back in the terms for $n < N$ does not affect convergence.

91. $\displaystyle\sum_{n=1}^{\infty} \sin^2 \frac{\pi}{n}$

SOLUTION For all $x > 0$, $\sin x < x$. Therefore, $\sin^2 x < x^2$, and for $x = \frac{\pi}{n}$,

$$\sin^2 \frac{\pi}{n} < \frac{\pi^2}{n^2} = \pi^2 \cdot \frac{1}{n^2}.$$

The series $\displaystyle\sum_{n=1}^{\infty} \frac{1}{n^2}$ is a convergent p-series, so the series $\displaystyle\sum_{n=1}^{\infty} \sin^2 \frac{\pi}{n}$ also converges by the Comparison Test.

In Exercises 93–98, find the interval of convergence of the power series.

93. $\displaystyle\sum_{n=0}^{\infty} \frac{2^n x^n}{n!}$

SOLUTION With $a_n = \frac{2^n x^n}{n!}$,

$$\rho = \lim_{n\to\infty} \left|\frac{a_{n+1}}{a_n}\right| = \lim_{n\to\infty} \left|\frac{2^{n+1}x^{n+1}}{(n+1)!} \cdot \frac{n!}{2^n x^n}\right| = \lim_{n\to\infty} \left|x \cdot \frac{2}{n}\right| = 0$$

Then $\rho < 1$ for all x, so that the radius of convergence is $R = \infty$, and the series converges for all x.

95. $\displaystyle\sum_{n=0}^{\infty} \frac{n^6}{n^8+1}(x-3)^n$

SOLUTION With $a_n = \frac{n^6(x-3)^n}{n^8+1}$,

$$\begin{aligned}\rho = \lim_{n\to\infty} \left|\frac{a_{n+1}}{a_n}\right| &= \lim_{n\to\infty} \left|\frac{(n+1)^6(x-3)^{n+1}}{(n+1)^8-1} \cdot \frac{n^8+1}{n^6(x-3)^n}\right| \\ &= \lim_{n\to\infty} \left|(x-3) \cdot \frac{(n+1)^6(n^8+1)}{n^6((n+1)^8+1)}\right| \\ &= \lim_{n\to\infty} \left|(x-3) \cdot \frac{n^{14} + \text{terms of lower degree}}{n^{14} + \text{terms of lower degree}}\right| = |x-3|\end{aligned}$$

Then $\rho < 1$ when $|x-3| < 1$, so the radius of convergence is 1, and the series converges absolutely for $|x-3| < 1$, or $2 < x < 4$. For the endpoint $x = 4$, the series becomes $\displaystyle\sum_{n=0}^{\infty} \frac{n^6}{n^8+1}$, which converges by the Comparison Test comparing with the convergent p-series $\displaystyle\sum_{n=1}^{\infty} \frac{1}{n^2}$. For the endpoint $x = 2$, the series becomes $\displaystyle\sum_{n=0}^{\infty} \frac{n^6(-1)^n}{n^8+1}$, which converges by the Leibniz Test. The series $\displaystyle\sum_{n=0}^{\infty} \frac{n^6(x-3)^n}{n^8+1}$ therefore converges for $2 \le x \le 4$.

97. $\displaystyle\sum_{n=0}^{\infty} (nx)^n$

SOLUTION With $a_n = n^n x^n$, and assuming $x \neq 0$,

$$\rho = \lim_{n\to\infty} \left|\frac{a_{n+1}}{a_n}\right| = \lim_{n\to\infty} \left|\frac{(n+1)^{n+1}x^{n+1}}{n^n x^n}\right| = \lim_{n\to\infty} \left|x(n+1) \cdot \left(\frac{n+1}{n}\right)^n\right| = \infty$$

since $\left(\frac{n+1}{n}\right)^n = \left(1 + \frac{1}{n}\right)^n$ converges to e and the $(n+1)$ term diverges to ∞. Thus $\rho < 1$ only when $x = 0$, so the series converges only for $x = 0$.

99. Expand $f(x) = \dfrac{2}{4-3x}$ as a power series centered at $c = 0$. Determine the values of x for which the series converges.

SOLUTION Write

$$\frac{2}{4-3x} = \frac{1}{2}\frac{1}{1-\frac{3}{4}x}.$$

Substituting $\frac{3}{4}x$ for x in the Maclaurin series for $\frac{1}{1-x}$, we obtain

$$\frac{1}{1-\frac{3}{4}x} = \sum_{n=0}^{\infty} \left(\frac{3}{4}\right)^n x^n.$$

This series converges for $\left|\frac{3}{4}x\right| < 1$, or $|x| < \frac{4}{3}$. Hence, for $|x| < \frac{4}{3}$,

$$\frac{2}{4-3x} = \frac{1}{2}\sum_{n=0}^{\infty} \left(\frac{3}{4}\right)^n x^n.$$

101. Let $F(x) = \sum_{k=0}^{\infty} \frac{x^{2k}}{2^k \cdot k!}$.

(a) Show that $F(x)$ has infinite radius of convergence.

(b) Show that $y = F(x)$ is a solution of

$$y'' = xy' + y, \qquad y(0) = 1, \qquad y'(0) = 0$$

(c) CAS Plot the partial sums S_N for $N = 1, 3, 5, 7$ on the same set of axes.

SOLUTION

(a) With $a_k = \frac{x^{2k}}{2^k \cdot k!}$,

$$\left|\frac{a_{k+1}}{a_k}\right| = \frac{|x|^{2k+2}}{2^{k+1}\cdot(k+1)!} \cdot \frac{2^k \cdot k!}{|x|^{2k}} = \frac{x^2}{2(k+1)},$$

and

$$\rho = \lim_{k\to\infty}\left|\frac{a_{k+1}}{a_k}\right| = x^2 \cdot 0 = 0.$$

Because $\rho < 1$ for all x, we conclude that the series converges for all x; that is, $R = \infty$.

(b) Let

$$y = F(x) = \sum_{k=0}^{\infty} \frac{x^{2k}}{2^k \cdot k!}.$$

Then

$$y' = \sum_{k=1}^{\infty} \frac{2kx^{2k-1}}{2^k k!} = \sum_{k=1}^{\infty} \frac{x^{2k-1}}{2^{k-1}(k-1)!},$$

$$y'' = \sum_{k=1}^{\infty} \frac{(2k-1)x^{2k-2}}{2^{k-1}(k-1)!},$$

and

$$xy' + y = x\sum_{k=1}^{\infty} \frac{x^{2k-1}}{2^{k-1}(k-1)!} + \sum_{k=0}^{\infty} \frac{x^{2k}}{2^k k!} = \sum_{k=1}^{\infty} \frac{x^{2k}}{2^{k-1}(k-1)!} + 1 + \sum_{k=1}^{\infty} \frac{x^{2k}}{2^k k!}$$

$$= 1 + \sum_{k=1}^{\infty} \frac{(2k+1)x^{2k}}{2^k k!} = \sum_{k=0}^{\infty} \frac{(2k+1)x^{2k}}{2^k k!} = \sum_{k=1}^{\infty} \frac{(2k-1)x^{2k-2}}{2^{k-1}(k-1)!} = y''.$$

Moreover,

$$y(0) = 1 + \sum_{k=1}^{\infty} \frac{0^{2k}}{2^k k!} = 1 \quad \text{and} \quad y'(0) = \sum_{k=1}^{\infty} \frac{0^{2k-1}}{2^{k-1}(k-1)!} = 0.$$

Thus, $\sum_{k=0}^{\infty} \frac{x^{2k}}{2^k k!}$ is the solution to the equation $y'' = xy' + y$ satisfying $y(0) = 1$, $y'(0) = 0$.

(c) The partial sums S_1, S_3, S_5 and S_7 are plotted in the figure below.

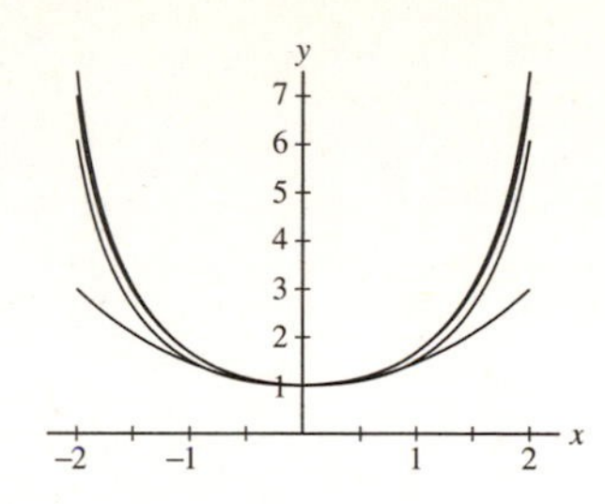

In Exercises 103–112, find the Taylor series centered at c.

103. $f(x) = e^{4x}, \quad c = 0$

SOLUTION Substituting $4x$ for x in the Maclaurin series for e^x yields

$$e^{4x} = \sum_{n=0}^{\infty} \frac{(4x)^n}{n!} = \sum_{n=0}^{\infty} \frac{4^n}{n!} x^n.$$

105. $f(x) = x^4, \quad c = 2$

SOLUTION We have

$$f'(x) = 4x^3 \quad f''(x) = 12x^2 \quad f'''(x) = 24x \quad f^{(4)}(x) = 24$$

and all higher derivatives are zero, so that

$$f(2) = 2^4 = 16 \quad f'(2) = 4 \cdot 2^3 = 32 \quad f''(2) = 12 \cdot 2^2 = 48 \quad f'''(2) = 24 \cdot 2 = 48 \quad f^{(4)}(2) = 24$$

Thus the Taylor series centered at $c = 2$ is

$$\begin{aligned}\sum_{n=0}^{4} \frac{f^{(n)}(2)}{n!}(x-2)^n &= 16 + \frac{32}{1!}(x-2) + \frac{48}{2!}(x-2)^2 + \frac{48}{3!}(x-2)^3 + \frac{24}{4!}(x-2)^4 \\ &= 16 + 32(x-2) + 24(x-2)^2 + 8(x-2)^3 + (x-2)^4\end{aligned}$$

107. $f(x) = \sin x, \quad c = \pi$

SOLUTION We have

$$f^{(4n)}(x) = \sin x \quad f^{(4n+1)}(x) = \cos x \quad f^{(4n+2)}(x) = -\sin x \quad f^{(4n+3)}(x) = -\cos x$$

so that

$$f^{(4n)}(\pi) = \sin \pi = 0 \quad f^{(4n+1)}(\pi) = \cos \pi = -1 \quad f^{(4n+2)}(\pi) = -\sin \pi = 0 \quad f^{(4n+3)}(\pi) = -\cos \pi = 1$$

Then the Taylor series centered at $c = \pi$ is

$$\begin{aligned}\sum_{n=0}^{\infty} \frac{f^{(n)}(\pi)}{n!}(x-\pi)^n &= \frac{-1}{1!}(x-\pi) + \frac{1}{3!}(x-\pi)^3 + \frac{-1}{5!}(x-\pi)^5 + \frac{1}{7!}(x-\pi)^7 - \dots \\ &= -(x-\pi) + \frac{1}{6}(x-\pi)^3 - \frac{1}{120}(x-\pi)^5 + \frac{1}{5040}(x-\pi)^7 - \dots\end{aligned}$$

109. $f(x) = \dfrac{1}{1-2x}, \quad c = -2$

SOLUTION Write

$$\frac{1}{1-2x} = \frac{1}{5-2(x+2)} = \frac{1}{5}\frac{1}{1-\frac{2}{5}(x+2)}.$$

Substituting $\frac{2}{5}(x+2)$ for x in the Maclaurin series for $\frac{1}{1-x}$ yields

$$\frac{1}{1-\frac{2}{5}(x+2)} = \sum_{n=0}^{\infty} \frac{2^n}{5^n}(x+2)^n;$$

hence,

$$\frac{1}{1-2x} = \frac{1}{5}\sum_{n=0}^{\infty}\frac{2^n}{5^n}(x+2)^n = \sum_{n=0}^{\infty}\frac{2^n}{5^{n+1}}(x+2)^n.$$

111. $f(x) = \ln\frac{x}{2}, \quad c = 2$

SOLUTION Write

$$\ln\frac{x}{2} = \ln\left(\frac{(x-2)+2}{2}\right) = \ln\left(1 + \frac{x-2}{2}\right).$$

Substituting $\frac{x-2}{2}$ for x in the Maclaurin series for $\ln(1+x)$ yields

$$\ln\frac{x}{2} = \sum_{n=1}^{\infty}\frac{(-1)^{n+1}\left(\frac{x-2}{2}\right)^n}{n} = \sum_{n=1}^{\infty}\frac{(-1)^{n+1}(x-2)^n}{n\cdot 2^n}.$$

This series is valid for $|x-2| < 2$.

In Exercises 113–116, find the first three terms of the Maclaurin series of $f(x)$ and use it to calculate $f^{(3)}(0)$.

113. $f(x) = (x^2 - x)e^{x^2}$

SOLUTION Substitute x^2 for x in the Maclaurin series for e^x to get

$$e^{x^2} = 1 + x^2 + \frac{1}{2}x^4 + \frac{1}{6}x^6 + \dots$$

so that the Maclaurin series for $f(x)$ is

$$(x^2 - x)e^{x^2} = x^2 + x^4 + \frac{1}{2}x^6 + \dots - x - x^3 - \frac{1}{2}x^5 - \dots = -x + x^2 - x^3 + x^4 + \dots$$

The coefficient of x^3 is

$$\frac{f'''(0)}{3!} = -1$$

so that $f'''(0) = -6$.

115. $f(x) = \dfrac{1}{1+\tan x}$

SOLUTION Substitute $-\tan x$ in the Maclaurin series for $\frac{1}{1-x}$ to get

$$\frac{1}{1+\tan x} = 1 - \tan x + (\tan x)^2 - (\tan x)^3 + \dots$$

We have not yet encountered the Maclaurin series for $\tan x$. We need only the terms up through x^3, so compute

$$\tan'(x) = \sec^2 x \quad \tan''(x) = 2(\tan x)\sec^2 x \quad \tan'''(x) = 2(1+\tan^2 x)\sec^2 x + 4(\tan^2 x)\sec^2 x$$

so that

$$\tan'(0) = 1 \quad \tan''(0) = 0 \quad \tan'''(0) = 2$$

Then the Maclaurin series for $\tan x$ is

$$\tan x = \tan 0 + \frac{\tan'(0)}{1!}x + \frac{\tan''(0)}{2!}x^2 + \frac{\tan'''(0)}{3!}x^3 + \dots = x + \frac{1}{3}x^3 + \dots$$

Substitute these into the series above to get

$$\begin{aligned}\frac{1}{1+\tan x} &= 1 - \left(x + \frac{1}{3}x^3\right) + \left(x + \frac{1}{3}x^3\right)^2 - \left(x + \frac{1}{3}x^3\right)^3 + \dots \\ &= 1 - x - \frac{1}{3}x^3 + x^2 - x^3 + \text{higher degree terms} \\ &= 1 - x + x^2 - \frac{4}{3}x^3 + \text{higher degree terms}\end{aligned}$$

The coefficient of x^3 is

$$\frac{f'''(0)}{3!} = -\frac{4}{3}$$

so that

$$f'''(0) = -6 \cdot \frac{4}{3} = -8$$

117. Calculate $\frac{\pi}{2} - \frac{\pi^3}{2^3 3!} + \frac{\pi^5}{2^5 5!} - \frac{\pi^7}{2^7 7!} + \cdots$.

SOLUTION We recognize that

$$\frac{\pi}{2} - \frac{\pi^3}{2^3 3!} + \frac{\pi^5}{2^5 5!} - \frac{\pi^7}{2^7 7!} + \cdots = \sum_{n=0}^{\infty} (-1)^n \frac{(\pi/2)^{2n+1}}{(2n+1)!}$$

is the Maclaurin series for $\sin x$ with x replaced by $\pi/2$. Therefore,

$$\frac{\pi}{2} - \frac{\pi^3}{2^3 3!} + \frac{\pi^5}{2^5 5!} - \frac{\pi^7}{2^7 7!} + \cdots = \sin \frac{\pi}{2} = 1.$$

11 PARAMETRIC EQUATIONS, POLAR COORDINATES, AND CONIC SECTIONS

11.1 Parametric Equations

Preliminary Questions

1. Describe the shape of the curve $x = 3\cos t$, $y = 3\sin t$.

SOLUTION For all t,

$$x^2 + y^2 = (3\cos t)^2 + (3\sin t)^2 = 9(\cos^2 t + \sin^2 t) = 9 \cdot 1 = 9,$$

therefore the curve is on the circle $x^2 + y^2 = 9$. Also, each point on the circle $x^2 + y^2 = 9$ can be represented in the form $(3\cos t, 3\sin t)$ for some value of t. We conclude that the curve $x = 3\cos t$, $y = 3\sin t$ is the circle of radius 3 centered at the origin.

2. How does $x = 4 + 3\cos t$, $y = 5 + 3\sin t$ differ from the curve in the previous question?

SOLUTION In this case we have

$$(x-4)^2 + (y-5)^2 = (3\cos t)^2 + (3\sin t)^2 = 9(\cos^2 t + \sin^2 t) = 9 \cdot 1 = 9$$

Therefore, the given equations parametrize the circle of radius 3 centered at the point $(4, 5)$.

3. What is the maximum height of a particle whose path has parametric equations $x = t^9$, $y = 4 - t^2$?

SOLUTION The particle's height is $y = 4 - t^2$. To find the maximum height we set the derivative equal to zero and solve:

$$\frac{dy}{dt} = \frac{d}{dt}(4 - t^2) = -2t = 0 \quad \text{or} \quad t = 0$$

The maximum height is $y(0) = 4 - 0^2 = 4$.

4. Can the parametric curve $(t, \sin t)$ be represented as a graph $y = f(x)$? What about $(\sin t, t)$?

SOLUTION In the parametric curve $(t, \sin t)$ we have $x = t$ and $y = \sin t$, therefore, $y = \sin x$. That is, the curve can be represented as a graph of a function. In the parametric curve $(\sin t, t)$ we have $x = \sin t$, $y = t$, therefore $x = \sin y$. This equation does not define y as a function of x, therefore the parametric curve $(\sin t, t)$ cannot be represented as a graph of a function $y = f(x)$.

5. Match the derivatives with a verbal description:

(a) $\dfrac{dx}{dt}$ **(b)** $\dfrac{dy}{dt}$ **(c)** $\dfrac{dy}{dx}$

(i) Slope of the tangent line to the curve

(ii) Vertical rate of change with respect to time

(iii) Horizontal rate of change with respect to time

SOLUTION

(a) The derivative $\dfrac{dx}{dt}$ is the horizontal rate of change with respect to time.

(b) The derivative $\dfrac{dy}{dt}$ is the vertical rate of change with respect to time.

(c) The derivative $\dfrac{dy}{dx}$ is the slope of the tangent line to the curve.

Hence, (a) $\leftrightarrow$ (iii), (b) $\leftrightarrow$ (ii), (c) $\leftrightarrow$ (i)

Exercises

1. Find the coordinates at times $t = 0, 2, 4$ of a particle following the path $x = 1 + t^3$, $y = 9 - 3t^2$.

SOLUTION Substituting $t = 0$, $t = 2$, and $t = 4$ into $x = 1 + t^3$, $y = 9 - 3t^2$ gives the coordinates of the particle at these times respectively. That is,

$$\begin{array}{lll} (t=0) & x = 1 + 0^3 = 1,\ y = 9 - 3 \cdot 0^2 = 9 & \Rightarrow (1, 9) \\ (t=2) & x = 1 + 2^3 = 9,\ y = 9 - 3 \cdot 2^2 = -3 & \Rightarrow (9, -3) \\ (t=4) & x = 1 + 4^3 = 65,\ y = 9 - 3 \cdot 4^2 = -39 & \Rightarrow (65, -39). \end{array}$$

3. Show that the path traced by the bullet in Example 3 is a parabola by eliminating the parameter.

SOLUTION The path traced by the bullet is given by the following parametric equations:

$$x = 200t,\ y = 400t - 16t^2$$

We eliminate the parameter. Since $x = 200t$, we have $t = \frac{x}{200}$. Substituting into the equation for y we obtain:

$$y = 400t - 16t^2 = 400 \cdot \frac{x}{200} - 16\left(\frac{x}{200}\right)^2 = 2x - \frac{x^2}{2500}$$

The equation $y = -\frac{x^2}{2500} + 2x$ is the equation of a parabola.

5. Graph the parametric curves. Include arrows indicating the direction of motion.

(a) (t, t), $-\infty < t < \infty$

(b) $(\sin t, \sin t)$, $0 \le t \le 2\pi$

(c) (e^t, e^t), $-\infty < t < \infty$

(d) (t^3, t^3), $-1 \le t \le 1$

SOLUTION

(a) For the trajectory $c(t) = (t, t)$, $-\infty < t < \infty$ we have $y = x$. Also the two coordinates tend to ∞ and $-\infty$ as $t \to \infty$ and $t \to -\infty$ respectively. The graph is shown next:

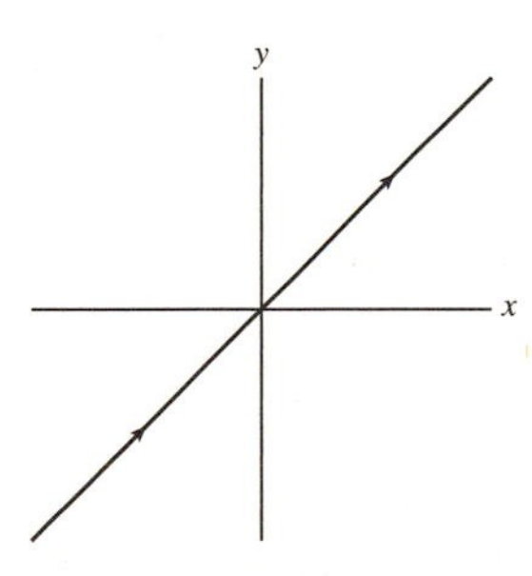

(b) For the curve $c(t) = (\sin t, \sin t)$, $0 \le t \le 2\pi$, we have $y = x$. $\sin t$ is increasing for $0 \le t \le \frac{\pi}{2}$, decreasing for $\frac{\pi}{2} \le t \le \frac{3\pi}{2}$ and increasing again for $\frac{3\pi}{2} \le t \le 2\pi$. Hence the particle moves from $c(0) = (0, 0)$ to $c(\frac{\pi}{2}) = (1, 1)$, then moves back to $c(\frac{3\pi}{2}) = (-1, -1)$ and then returns to $c(2\pi) = (0, 0)$. We obtain the following trajectory:

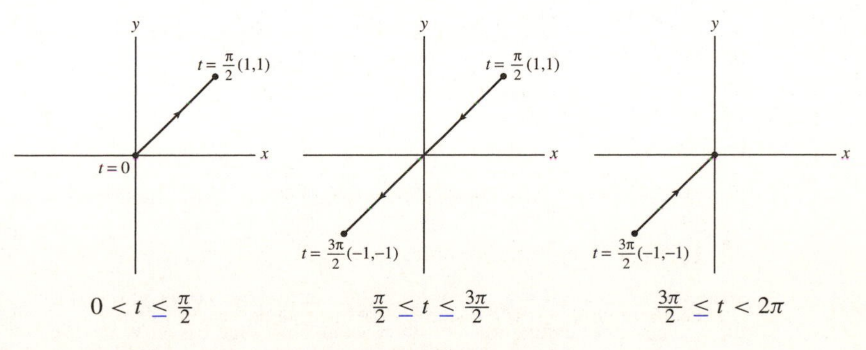

These three parts of the trajectory are shown together in the next figure:

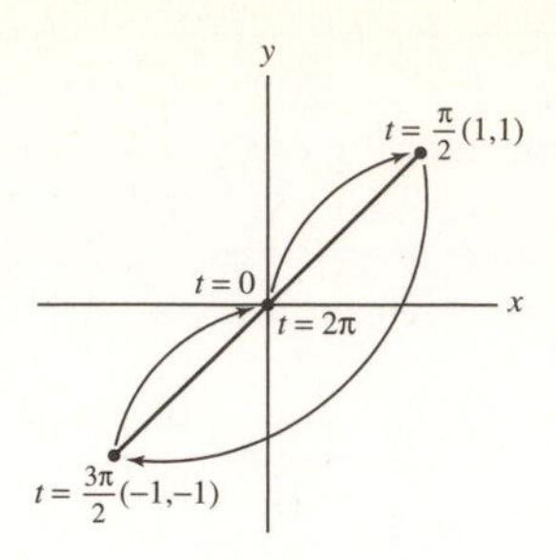

(c) For the trajectory $c(t) = (e^t, e^t)$, $-\infty < t < \infty$, we have $y = x$. However since $\lim_{t\to-\infty} e^t = 0$ and $\lim_{t\to\infty} e^t = \infty$, the trajectory is the part of the line $y = x$, $0 < x$.

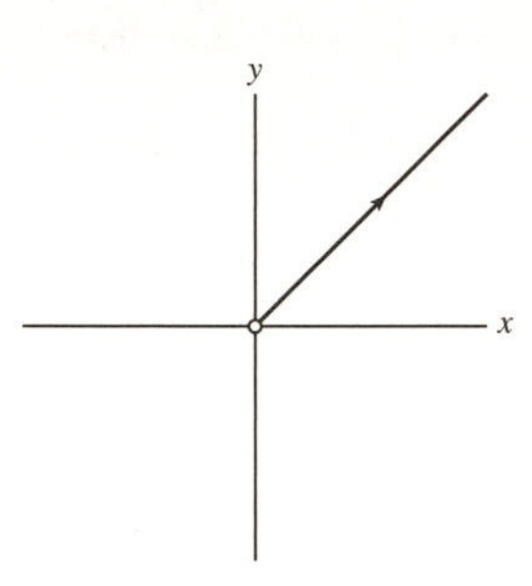

(d) For the trajectory $c(t) = (t^3, t^3)$, $-1 \le t \le 1$, we have again $y = x$. Since the function t^3 is increasing the particle moves in one direction starting at $((-1)^3, (-1)^3) = (-1, -1)$ and ending at $(1^3, 1^3) = (1, 1)$. The trajectory is shown next:

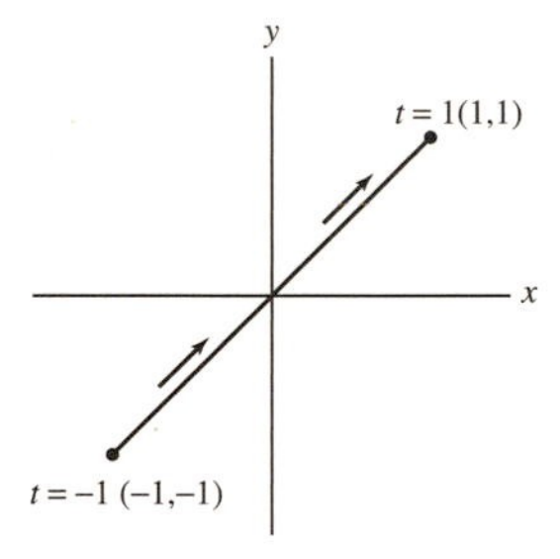

In Exercises 7–14, express in the form $y = f(x)$ by eliminating the parameter.

7. $x = t + 3, \quad y = 4t$

SOLUTION We eliminate the parameter. Since $x = t + 3$, we have $t = x - 3$. Substituting into $y = 4t$ we obtain

$$y = 4t = 4(x - 3) \Rightarrow y = 4x - 12$$

9. $x = t, \quad y = \tan^{-1}(t^3 + e^t)$

SOLUTION Replacing t by x in the equation for y we obtain $y = \tan^{-1}(x^3 + e^x)$.

11. $x = e^{-2t}, \quad y = 6e^{4t}$

SOLUTION We eliminate the parameter. Since $x = e^{-2t}$, we have $-2t = \ln x$ or $t = -\frac{1}{2}\ln x$. Substituting in $y = 6e^{4t}$ we get

$$y = 6e^{4t} = 6e^{4\cdot(-\frac{1}{2}\ln x)} = 6e^{-2\ln x} = 6e^{\ln x^{-2}} = 6x^{-2} \Rightarrow y = \frac{6}{x^2}, \quad x > 0.$$

13. $x = \ln t, \quad y = 2 - t$

SOLUTION Since $x = \ln t$ we have $t = e^x$. Substituting in $y = 2 - t$ we obtain $y = 2 - e^x$.

In Exercises 15–18, graph the curve and draw an arrow specifying the direction corresponding to motion.

15. $x = \frac{1}{2}t, \quad y = 2t^2$

SOLUTION Let $c(t) = (x(t), y(t)) = (\frac{1}{2}t, 2t^2)$. Then $c(-t) = (-x(t), y(t))$ so the curve is symmetric with respect to the y-axis. Also, the function $\frac{1}{2}t$ is increasing. Hence there is only one direction of motion on the curve. The corresponding function is the parabola $y = 2 \cdot (2x)^2 = 8x^2$. We obtain the following trajectory:

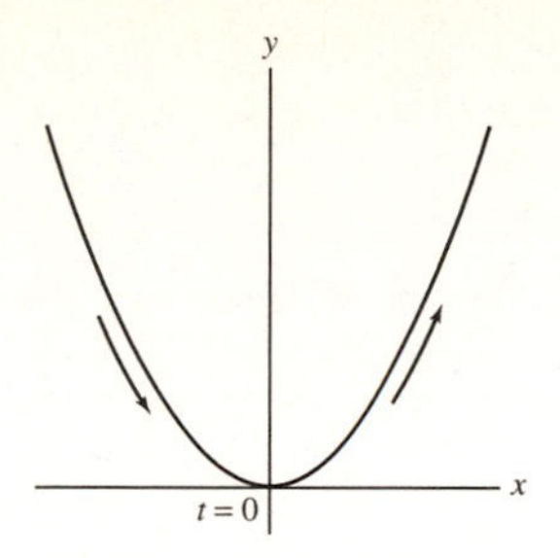

17. $x = \pi t, \quad y = \sin t$

SOLUTION We find the function by eliminating t. Since $x = \pi t$, we have $t = \frac{x}{\pi}$. Substituting $t = \frac{x}{\pi}$ into $y = \sin t$ we get $y = \sin \frac{x}{\pi}$. We obtain the following curve:

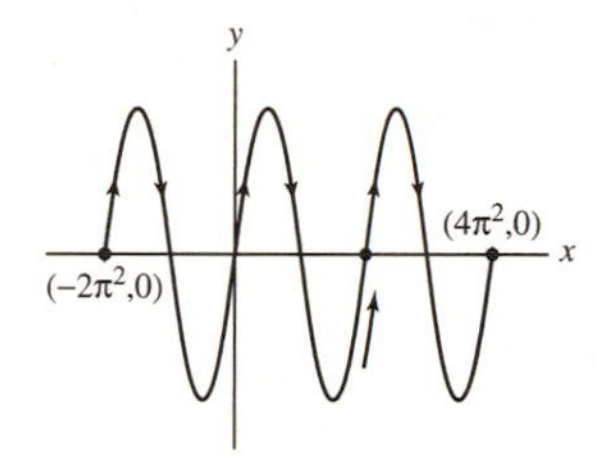

19. Match the parametrizations (a)–(d) below with their plots in Figure 14, and draw an arrow indicating the direction of motion.

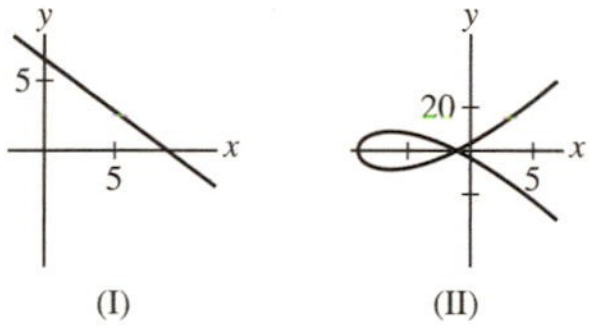

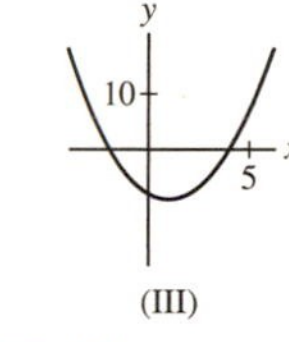

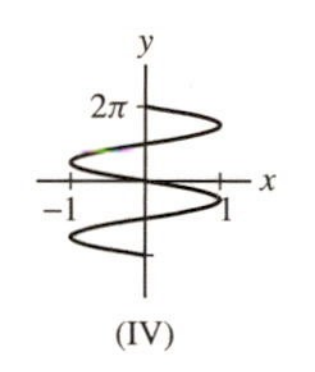

FIGURE 14

(a) $c(t) = (\sin t, -t)$

(b) $c(t) = (t^2 - 9, 8t - t^3)$

(c) $c(t) = (1 - t, t^2 - 9)$

(d) $c(t) = (4t + 2, 5 - 3t)$

SOLUTION

(a) In the curve $c(t) = (\sin t, -t)$ the x-coordinate is varying between -1 and 1 so this curve corresponds to plot IV. As t increases, the y-coordinate $y = -t$ is decreasing so the direction of motion is downward.

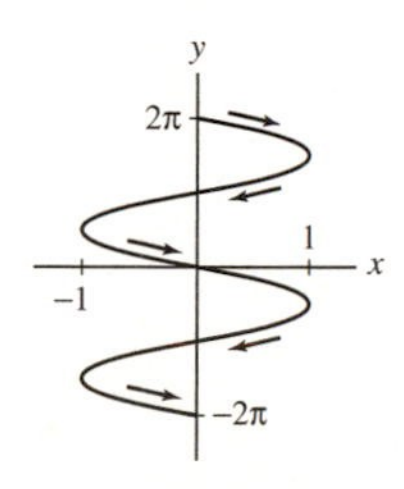

(IV) $c(t) = (\sin t, -t)$

(b) The curve $c(t) = (t^2 - 9, -t^3 - 8)$ intersects the x-axis where $y = -t^3 - 8 = 0$, or $t = -2$. The x-intercept is $(-5, 0)$. The y-intercepts are obtained where $x = t^2 - 9 = 0$, or $t = \pm 3$. The y-intercepts are $(0, -35)$ and $(0, 19)$. As t increases from $-\infty$ to 0, x and y decrease, and as t increases from 0 to ∞, x increases and y decreases. We obtain the following trajectory:

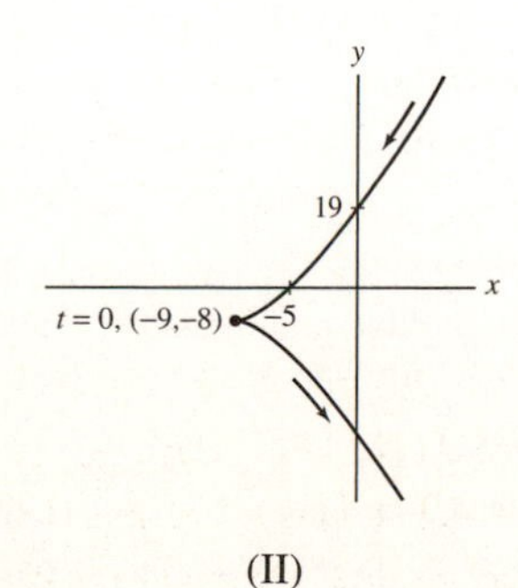

(II)

(c) The curve $c(t) = (1 - t, t^2 - 9)$ intersects the y-axis where $x = 1 - t = 0$, or $t = 1$. The y-intercept is $(0, -8)$. The x-intercepts are obtained where $t^2 - 9 = 0$ or $t = \pm 3$. These are the points $(-2, 0)$ and $(4, 0)$. Setting $t = 1 - x$ we get

$$y = t^2 - 9 = (1 - x)^2 - 9 = x^2 - 2x - 8.$$

As t increases the x coordinate decreases and we obtain the following trajectory:

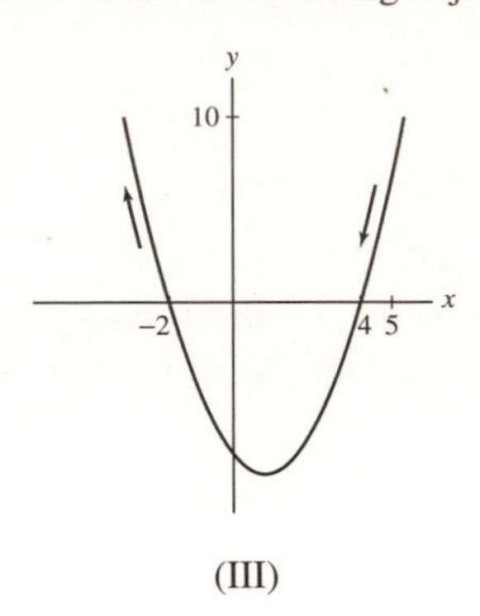

(III)

(d) The curve $c(t) = (4t + 2, 5 - 3t)$ is a straight line, since eliminating t in $x = 4t + 2$ and substituting in $y = 5 - 3t$ gives $y = 5 - 3 \cdot \frac{x-2}{4} = -\frac{3}{4}x + \frac{13}{2}$ which is the equation of a line. As t increases, the x coordinate $x = 4t + 2$ increases and the y-coordinate $y = 5 - 3t$ decreases. We obtain the following trajectory:

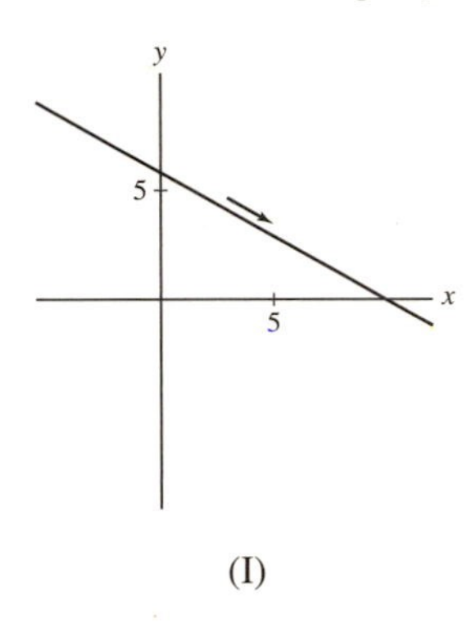

(I)

21. Find an interval of t-values such that $c(t) = (\cos t, \sin t)$ traces the lower half of the unit circle.

SOLUTION For $t = \pi$, we have $c(\pi) = (-1, 0)$. As t increases from π to 2π, the x-coordinate of $c(t)$ increases from -1 to 1, and the y-coordinate decreases from 0 to -1 (at $t = 3\pi/2$) and then returns to 0. Thus, for t in $[\pi, 2\pi]$, the equation traces the lower part of the circle.

In Exercises 23–38, find parametric equations for the given curve.

23. $y = 9 - 4x$

SOLUTION This is a line through $P = (0, 9)$ with slope $m = -4$. Using the parametric representation of a line, as given in Example 3, we obtain $c(t) = (t, 9 - 4t)$.

25. $4x - y^2 = 5$

SOLUTION We define the parameter $t = y$. Then, $x = \dfrac{5 + y^2}{4} = \dfrac{5 + t^2}{4}$, giving us the parametrization $c(t) = \left(\dfrac{5 + t^2}{4}, t\right)$.

27. $(x + 9)^2 + (y - 4)^2 = 49$

SOLUTION This is a circle of radius 7 centered at $(-9, 4)$. Using the parametric representation of a circle we get $c(t) = (-9 + 7\cos t, 4 + 7\sin t)$.

29. Line of slope 8 through $(-4, 9)$

SOLUTION Using the parametric representation of a line given in Example 3, we get the parametrization $c(t) = (-4 + t, 9 + 8t)$.

31. Line through $(3, 1)$ and $(-5, 4)$

SOLUTION We use the two-point parametrization of a line with $P = (a, b) = (3, 1)$ and $Q = (c, d) = (-5, 4)$. Then $c(t) = (3 - 8t, 1 + 3t)$ for $-\infty < t < \infty$.

33. Segment joining $(1, 1)$ and $(2, 3)$

SOLUTION We use the two-point parametrization of a line with $P = (a, b) = (1, 1)$ and $Q = (c, d) = (2, 3)$. Then $c(t) = (1 + t, 1 + 2t)$; since we want only the segment joining the two points, we want $0 \le t \le 1$.

35. Circle of radius 4 with center $(3, 9)$

SOLUTION Substituting $(a, b) = (3, 9)$ and $R = 4$ in the parametric equation of the circle we get $c(t) = (3 + 4\cos t, 9 + 4\sin t)$.

37. $y = x^2$, translated so that the minimum occurs at $(-4, -8)$

SOLUTION We may parametrize $y = x^2$ by (t, t^2) for $-\infty < t < \infty$. The minimum of $y = x^2$ occurs at $(0, 0)$, so the desired curve is translated by $(-4, -8)$ from $y = x^2$. Thus a parametrization of the desired curve is $c(t) = (-4 + t, -8 + t^2)$.

In Exercises 39–42, find a parametrization $c(t)$ of the curve satisfying the given condition.

39. $y = 3x - 4, \quad c(0) = (2, 2)$

SOLUTION Let $x(t) = t + a$ and $y(t) = 3x - 4 = 3(t + a) - 4$. We want $x(0) = 2$, thus we must use $a = 2$. Our line is $c(t) = (x(t), y(t)) = (t + 2, 3(t + 2) - 4) = (t + 2, 3t + 2)$.

41. $y = x^2, \quad c(0) = (3, 9)$

SOLUTION Let $x(t) = t + a$ and $y(t) = x^2 = (t + a)^2$. We want $x(0) = 3$, thus we must use $a = 3$. Our curve is $c(t) = (x(t), y(t)) = (t + 3, (t + 3)^2) = (t + 3, t^2 + 6t + 9)$.

43. Describe $c(t) = (\sec t, \tan t)$ for $0 \le t < \frac{\pi}{2}$ in the form $y = f(x)$. Specify the domain of x.

SOLUTION The function $x = \sec t$ has period 2π and $y = \tan t$ has period π. The graphs of these functions in the interval $-\pi \le t \le \pi$, are shown below:

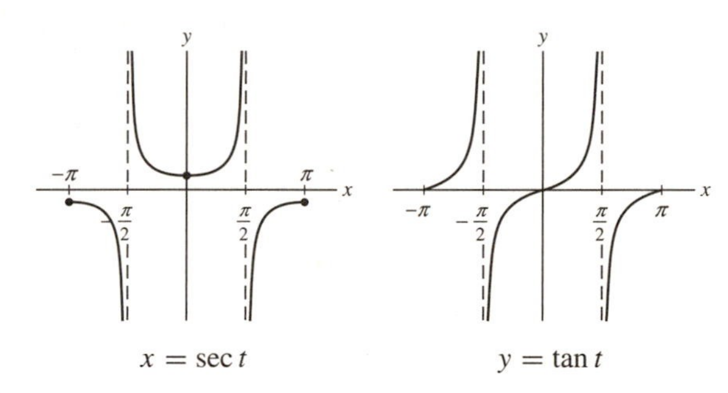

$$x = \sec t \Rightarrow x^2 = \sec^2 t$$

$$y = \tan t \Rightarrow y^2 = \tan^2 t = \frac{\sin^2 t}{\cos^2 t} = \frac{1 - \cos^2 t}{\cos^2 t} = \sec^2 t - 1 = x^2 - 1$$

Hence the graph of the curve is the hyperbola $x^2 - y^2 = 1$. The function $x = \sec t$ is an even function while $y = \tan t$ is odd. Also x has period 2π and y has period π. It follows that the intervals $-\pi \le t < -\frac{\pi}{2}$, $\frac{-\pi}{2} < t < \frac{\pi}{2}$ and $\frac{\pi}{2} < t < \pi$ trace the curve exactly once. The corresponding curve is shown next:

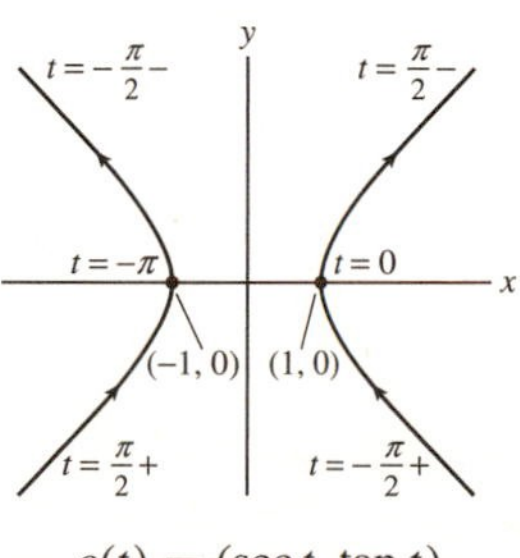

$c(t) = (\sec t, \tan t)$

45. The graphs of $x(t)$ and $y(t)$ as functions of t are shown in Figure 15(A). Which of (I)–(III) is the plot of $c(t) = (x(t), y(t))$? Explain.

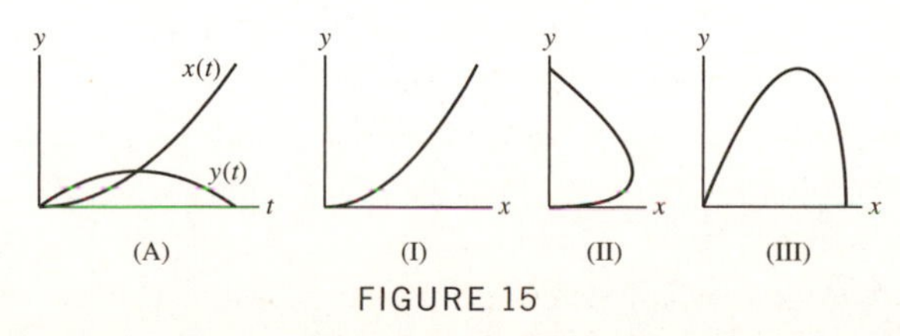

FIGURE 15

SOLUTION As seen in Figure 15(A), the x-coordinate is an increasing function of t, while $y(t)$ is first increasing and then decreasing. In Figure I, x and y are both increasing or both decreasing (depending on the direction on the curve). In Figure II, x does not maintain one tendency, rather, it is decreasing and increasing for certain values of t. The plot $c(t) = (x(t), y(t))$ is plot III.

47. Sketch $c(t) = (t^3 - 4t, t^2)$ following the steps in Example 7.

SOLUTION We note that $x(t) = t^3 - 4t$ is odd and $y(t) = t^2$ is even, hence $c(-t) = (x(-t), y(-t)) = (-x(t), y(t))$. It follows that $c(-t)$ is the reflection of $c(t)$ across y-axis. That is, $c(-t)$ and $c(t)$ are symmetric with respect to the y-axis; thus, it suffices to graph the curve for $t \geq 0$. For $t = 0$, we have $c(0) = (0, 0)$ and the y-coordinate $y(t) = t^2$ tends to ∞ as $t \to \infty$. To analyze the x-coordinate, we graph $x(t) = t^3 - 4t$ for $t \geq 0$:

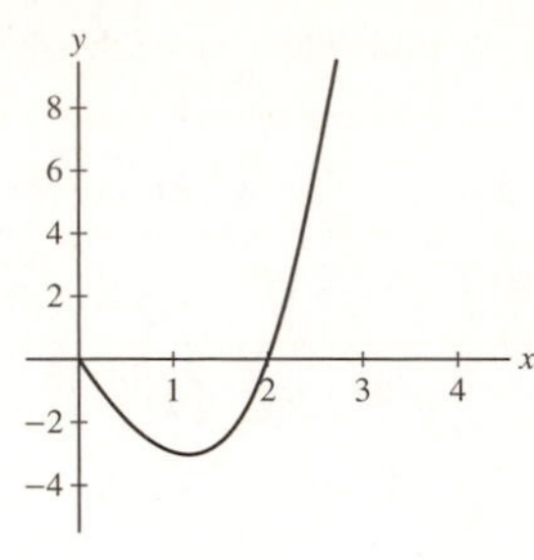

$x = t^3 - 4t$

We see that $x(t) < 0$ and decreasing for $0 < t < 2/\sqrt{3}$, $x(t) < 0$ and increasing for $2/\sqrt{3} < t < 2$ and $x(t) > 0$ and increasing for $t > 2$. Also $x(t)$ tends to ∞ as $t \to \infty$. Therefore, starting at the origin, the curve first directs to the left of the y-axis, then at $t = 2/\sqrt{3}$ it turns to the right, always keeping an upward direction. The part of the path for $t \leq 0$ is obtained by reflecting across the y-axis. We also use the points $c(0) = (0, 0)$, $c(1) = (-3, 1)$, $c(2) = (0, 4)$ to obtain the following graph for $c(t)$:

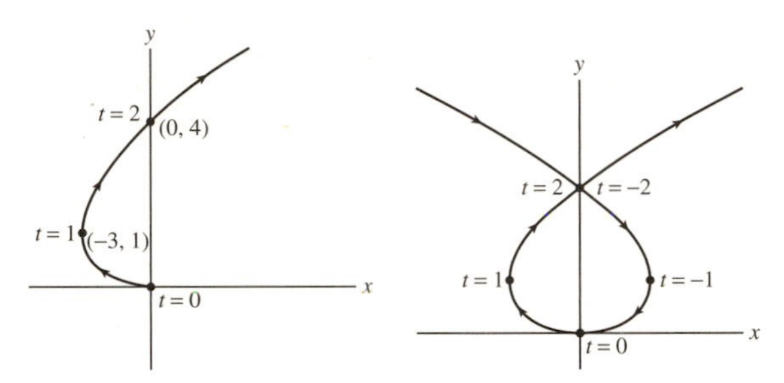

Graph of $c(t)$ for $t \geq 0$. Graph of $c(t)$ for all t.

In Exercises 49–52, use Eq. (7) to find dy/dx at the given point.

49. $(t^3, t^2 - 1)$, $t = -4$

SOLUTION By Eq. (7) we have

$$\frac{dy}{dx} = \frac{y'(t)}{x'(t)} = \frac{(t^2 - 1)'}{(t^3)'} = \frac{2t}{3t^2} = \frac{2}{3t}$$

Substituting $t = -4$ we get

$$\frac{dy}{dx} = \left.\frac{2}{3t}\right|_{t=-4} = \frac{2}{3 \cdot (-4)} = -\frac{1}{6}.$$

51. $(s^{-1} - 3s, s^3)$, $s = -1$

SOLUTION Using Eq. (7) we get

$$\frac{dy}{dx} = \frac{y'(s)}{x'(s)} = \frac{(s^3)'}{(s^{-1} - 3s)'} = \frac{3s^2}{-s^{-2} - 3} = \frac{3s^4}{-1 - 3s^2}$$

Substituting $s = -1$ we obtain

$$\frac{dy}{dx} = \left.\frac{3s^4}{-1 - 3s^2}\right|_{s=-1} = \frac{3 \cdot (-1)^4}{-1 - 3 \cdot (-1)^2} = -\frac{3}{4}.$$

In Exercises 53–56, find an equation $y = f(x)$ for the parametric curve and compute dy/dx in two ways: using Eq. (7) and by differentiating $f(x)$.

53. $c(t) = (2t + 1, 1 - 9t)$

SOLUTION Since $x = 2t + 1$, we have $t = \frac{x - 1}{2}$. Substituting in $y = 1 - 9t$ we have

$$y = 1 - 9\left(\frac{x - 1}{2}\right) = -\frac{9}{2}x + \frac{11}{2}$$

Differentiating $y = -\frac{9}{2}x + \frac{11}{2}$ gives $\frac{dy}{dx} = -\frac{9}{2}$. We now find $\frac{dy}{dx}$ using Eq. (7):

$$\frac{dy}{dx} = \frac{y'(t)}{x'(t)} = \frac{(1-9t)'}{(2t+1)'} = -\frac{9}{2}$$

55. $x = s^3, \quad y = s^6 + s^{-3}$

SOLUTION We find y as a function of x:

$$y = s^6 + s^{-3} = \left(s^3\right)^2 + \left(s^3\right)^{-1} = x^2 + x^{-1}.$$

We now differentiate $y = x^2 + x^{-1}$. This gives

$$\frac{dy}{dx} = 2x - x^{-2}.$$

Alternatively, we can use Eq. (7) to obtain the following derivative:

$$\frac{dy}{dx} = \frac{y'(s)}{x'(s)} = \frac{\left(s^6 + s^{-3}\right)'}{\left(s^3\right)'} = \frac{6s^5 - 3s^{-4}}{3s^2} = 2s^3 - s^{-6}.$$

Hence, since $x = s^3$,

$$\frac{dy}{dx} = 2x - x^{-2}.$$

57. Find the points on the curve $c(t) = (3t^2 - 2t, t^3 - 6t)$ where the tangent line has slope 3.

SOLUTION We solve

$$\frac{dy}{dx} = \frac{3t^2 - 6}{6t - 2} = 3$$

or $3t^2 - 6 = 18t - 6$, or $t^2 - 6t = 0$, so the slope is 3 at $t = 0, 6$ and the points are $(0, 0)$ and $(96, 180)$

In Exercises 59–62, let $c(t) = (t^2 - 9, t^2 - 8t)$ (see Figure 17).

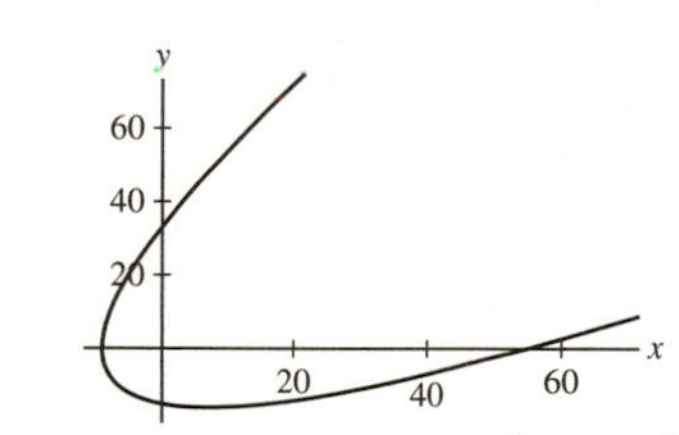

FIGURE 17 Plot of $c(t) = (t^2 - 9, t^2 - 8t)$.

59. Draw an arrow indicating the direction of motion, and determine the interval of t-values corresponding to the portion of the curve in each of the four quadrants.

SOLUTION We plot the functions $x(t) = t^2 - 9$ and $y(t) = t^2 - 8t$:

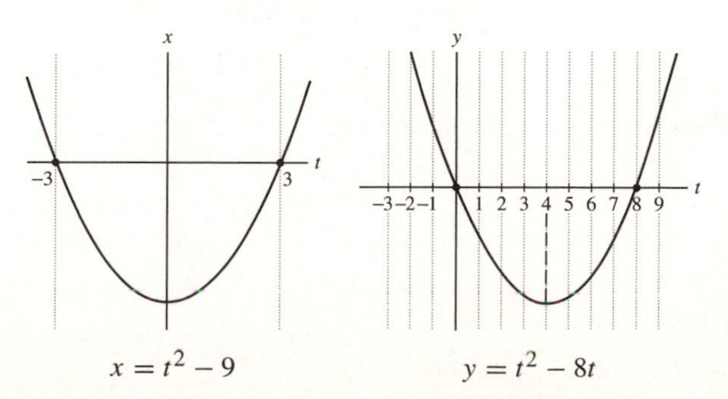

$x = t^2 - 9$ $\qquad$ $y = t^2 - 8t$

We note carefully where each of these graphs are positive or negative, increasing or decreasing. In particular, $x(t)$ is decreasing for $t < 0$, increasing for $t > 0$, positive for $|t| > 3$, and negative for $|t| < 3$. Likewise, $y(t)$ is decreasing for $t < 4$, increasing for $t > 4$, positive for $t > 8$ or $t < 0$, and negative for $0 < t < 8$. We now draw arrows on the path following the decreasing/increasing behavior of the coordinates as indicated above. We obtain:

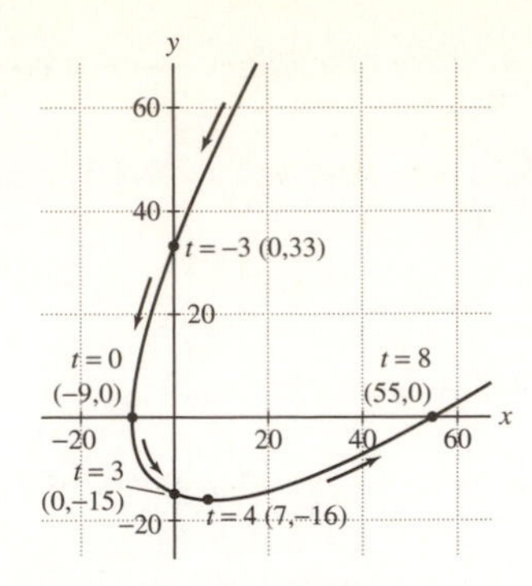

This plot also shows that:

- The graph is in the first quadrant for $t < -3$ or $t > 8$.
- The graph is in the second quadrant for $-3 < t < 0$.
- The graph is in the third quadrant for $0 < t < 3$.
- The graph is in the fourth quadrant for $3 < t < 8$.

61. Find the points where the tangent has slope $\frac{1}{2}$.

SOLUTION The slope of the tangent at t is

$$\frac{dy}{dx} = \frac{\left(t^2 - 8t\right)'}{\left(t^2 - 9\right)'} = \frac{2t - 8}{2t} = 1 - \frac{4}{t}$$

The point where the tangent has slope $\frac{1}{2}$ corresponds to the value of t that satisfies

$$\frac{dy}{dx} = 1 - \frac{4}{t} = \frac{1}{2} \Rightarrow \frac{4}{t} = \frac{1}{2} \Rightarrow t = 8.$$

We substitute $t = 8$ in $x(t) = t^2 - 9$ and $y(t) = t^2 - 8t$ to obtain the following point:

$$\begin{aligned} x(8) &= 8^2 - 9 = 55 \\ y(8) &= 8^2 - 8 \cdot 8 = 0 \end{aligned} \quad \Rightarrow \quad (55, 0)$$

63. Let A and B be the points where the ray of angle θ intersects the two concentric circles of radii $r < R$ centered at the origin (Figure 18). Let P be the point of intersection of the horizontal line through A and the vertical line through B. Express the coordinates of P as a function of θ and describe the curve traced by P for $0 \le \theta \le 2\pi$.

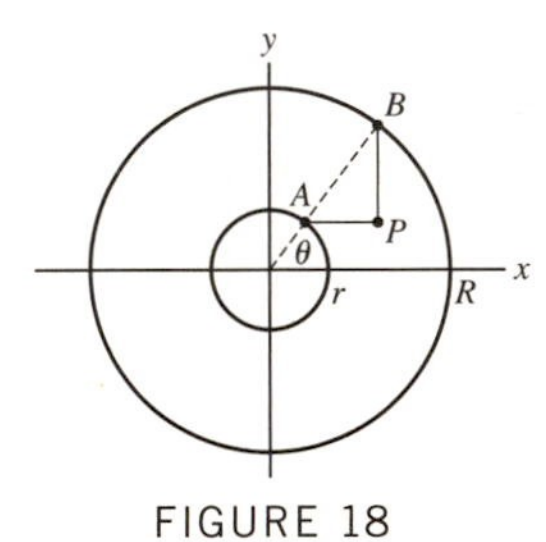

FIGURE 18

SOLUTION We use the parametric representation of a circle to determine the coordinates of the points A and B. That is

$$A = (r\cos\theta, r\sin\theta), \quad B = (R\cos\theta, R\sin\theta)$$

The coordinates of P are therefore

$$P = (R\cos\theta, r\sin\theta)$$

In order to identify the curve traced by P, we notice that the x and y coordinates of P satisfy $\frac{x}{R} = \cos\theta$ and $\frac{y}{r} = \sin\theta$. Hence

$$\left(\frac{x}{R}\right)^2 + \left(\frac{y}{r}\right)^2 = \cos^2\theta + \sin^2\theta = 1.$$

The equation

$$\left(\frac{x}{R}\right)^2 + \left(\frac{y}{r}\right)^2 = 1$$

is the equation of ellipse. Hence, the coordinates of P, $(R\cos\theta, r\sin\theta)$ describe an ellipse for $0 \le \theta \le 2\pi$.

In Exercises 65–68, refer to the Bézier curve defined by Eqs. (8) and (9).

65. Show that the Bézier curve with control points

$$P_0 = (1, 4), \quad P_1 = (3, 12), \quad P_2 = (6, 15), \quad P_3 = (7, 4)$$

has parametrization

$$c(t) = (1 + 6t + 3t^2 - 3t^3, 4 + 24t - 15t^2 - 9t^3)$$

Verify that the slope at $t = 0$ is equal to the slope of the segment $\overline{P_0 P_1}$.

SOLUTION For the given Bézier curve we have $a_0 = 1, a_1 = 3, a_2 = 6, a_3 = 7$, and $b_0 = 4, b_1 = 12, b_2 = 15, b_3 = 4$. Substituting these values in Eq. (8)–(9) and simplifying gives

$$\begin{aligned} x(t) &= (1-t)^3 + 9t(1-t)^2 + 18t^2(1-t) + 7t^3 \\ &= 1 - 3t + 3t^2 - t^3 + 9t(1 - 2t + t^2) + 18t^2 - 18t^3 + 7t^3 \\ &= 1 - 3t + 3t^2 - t^3 + 9t - 18t^2 + 9t^3 + 18t^2 - 18t^3 + 7t^3 \\ &= -3t^3 + 3t^2 + 6t + 1 \\ y(t) &= 4(1-t)^3 + 36t(1-t)^2 + 45t^2(1-t) + 4t^3 \\ &= 4(1 - 3t + 3t^2 - t^3) + 36t(1 - 2t + t^2) + 45t^2 - 45t^3 + 4t^3 \\ &= 4 - 12t + 12t^2 - 4t^3 + 36t - 72t^2 + 36t^3 + 45t^2 - 45t^3 + 4t^3 \\ &= 4 + 24t - 15t^2 - 9t^3 \end{aligned}$$

Then

$$c(t) = (1 + 6t + 3t^2 - 3t^3, 4 + 24t - 15t^2 - 9t^3), \quad 0 \le t \le 1.$$

We find the slope at $t = 0$. Using the formula for slope of the tangent line we get

$$\frac{dy}{dx} = \frac{(4 + 24t - 15t^2 - 9t^3)'}{(1 + 6t + 3t^2 - 3t^3)'} = \frac{24 - 30t - 27t^2}{6 + 6t - 9t^2} \Rightarrow \left.\frac{dy}{dx}\right|_{t=0} = \frac{24}{6} = 4.$$

The slope of the segment $\overline{P_0 P_1}$ is the slope of the line determined by the points $P_0 = (1, 4)$ and $P_1 = (3, 12)$. That is, $\frac{12-4}{3-1} = \frac{8}{2} = 4$. We see that the slope of the tangent line at $t = 0$ is equal to the slope of the segment $\overline{P_0 P_1}$, as expected.

67. CAS Find and plot the Bézier curve $c(t)$ passing through the control points

$$P_0 = (3, 2), \quad P_1 = (0, 2), \quad P_2 = (5, 4), \quad P_3 = (2, 4)$$

SOLUTION Setting $a_0 = 3$, $a_1 = 0$, $a_2 = 5$, $a_3 = 2$, and $b_0 = 2$, $b_1 = 2$, $b_2 = 4$, $b_3 = 4$ into Eq. (8)–(9) and simplifying gives

$$\begin{aligned} x(t) &= 3(1-t)^3 + 0 + 15t^2(1-t) + 2t^3 \\ &= 3(1 - 3t + 3t^2 - t^3) + 15t^2 - 15t^3 + 2t^3 = 3 - 9t + 24t^2 - 16t^3 \\ y(t) &= 2(1-t)^3 + 6t(1-t)^2 + 12t^2(1-t) + 4t^3 \\ &= 2(1 - 3t + 3t^2 - t^3) + 6t(1 - 2t + t^2) + 12t^2 - 12t^3 + 4t^3 \\ &= 2 - 6t + 6t^2 - 2t^3 + 6t - 12t^2 + 6t^3 + 12t^2 - 12t^3 + 4t^3 = 2 + 6t^2 - 4t^3 \end{aligned}$$

We obtain the following equation

$$c(t) = (3 - 9t + 24t^2 - 16t^3, 2 + 6t^2 - 4t^3), \quad 0 \le t \le 1.$$

The graph of the Bézier curve is shown in the following figure:

69. A bullet fired from a gun follows the trajectory

$$x = at, \qquad y = bt - 16t^2 \quad (a, b > 0)$$

Show that the bullet leaves the gun at an angle $\theta = \tan^{-1}\left(\frac{b}{a}\right)$ and lands at a distance $ab/16$ from the origin.

SOLUTION The height of the bullet equals the value of the y-coordinate. When the bullet leaves the gun, $y(t) = t(b - 16t) = 0$. The solutions to this equation are $t = 0$ and $t = \frac{b}{16}$, with $t = 0$ corresponding to the moment the bullet leaves the gun. We find the slope m of the tangent line at $t = 0$:

$$\frac{dy}{dx} = \frac{y'(t)}{x'(t)} = \frac{b - 32t}{a} \Rightarrow m = \left.\frac{b - 32t}{a}\right|_{t=0} = \frac{b}{a}$$

It follows that $\tan\theta = \frac{b}{a}$ or $\theta = \tan^{-1}\left(\frac{b}{a}\right)$. The bullet lands at $t = \frac{b}{16}$. We find the distance of the bullet from the origin at this time, by substituting $t = \frac{b}{16}$ in $x(t) = at$. This gives

$$x\left(\frac{b}{16}\right) = \frac{ab}{16}$$

71. CAS Plot the astroid $x = \cos^3\theta$, $y = \sin^3\theta$ and find the equation of the tangent line at $\theta = \frac{\pi}{3}$.

SOLUTION The graph of the astroid $x = \cos^3\theta$, $y = \sin^3\theta$ is shown in the following figure:

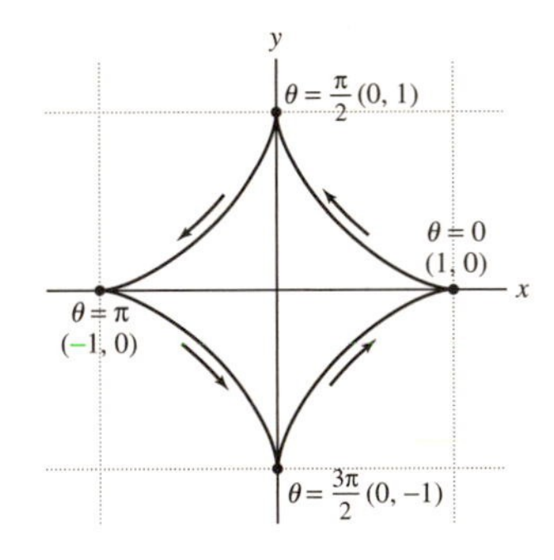

The slope of the tangent line at $\theta = \frac{\pi}{3}$ is

$$m = \left.\frac{dy}{dx}\right|_{\theta=\pi/3} = \left.\frac{(\sin^3\theta)'}{(\cos^3\theta)'}\right|_{\theta=\pi/3} = \left.\frac{3\sin^2\theta\cos\theta}{3\cos^2\theta(-\sin\theta)}\right|_{\theta=\pi/3} = -\left.\frac{\sin\theta}{\cos\theta}\right|_{\theta=\pi/3} = -\tan\theta\Big|_{\pi/3} = -\sqrt{3}$$

We find the point of tangency:

$$\left(x\left(\frac{\pi}{3}\right), y\left(\frac{\pi}{3}\right)\right) = \left(\cos^3\frac{\pi}{3}, \sin^3\frac{\pi}{3}\right) = \left(\frac{1}{8}, \frac{3\sqrt{3}}{8}\right)$$

The equation of the tangent line at $\theta = \frac{\pi}{3}$ is, thus,

$$y - \frac{3\sqrt{3}}{8} = -\sqrt{3}\left(x - \frac{1}{8}\right) \Rightarrow y = -\sqrt{3}x + \frac{\sqrt{3}}{2}$$

73. Find the points with horizontal tangent line on the cycloid with parametric equation (5).

SOLUTION The parametric equations of the cycloid are

$$x = t - \sin t, \quad y = 1 - \cos t$$

We find the slope of the tangent line at t:

$$\frac{dy}{dx} = \frac{(1-\cos t)'}{(t-\sin t)'} = \frac{\sin t}{1-\cos t}$$

The tangent line is horizontal where it has slope zero. That is,

$$\frac{dy}{dx} = \frac{\sin t}{1-\cos t} = 0 \quad \Rightarrow \quad \begin{matrix} \sin t = 0 \\ \cos t \neq 1 \end{matrix} \quad \Rightarrow \quad t = (2k-1)\pi, \quad k = 0, \pm 1, \pm 2, \ldots$$

We find the coordinates of the points with horizontal tangent line, by substituting $t = (2k-1)\pi$ in $x(t)$ and $y(t)$. This gives

$$x = (2k-1)\pi - \sin(2k-1)\pi = (2k-1)\pi$$
$$y = 1 - \cos((2k-1)\pi) = 1 - (-1) = 2$$

The required points are

$$((2k-1)\pi, 2), \quad k = 0, \pm 1, \pm 2, \ldots$$

75. A *curtate cycloid* (Figure 21) is the curve traced by a point at a distance h from the center of a circle of radius R rolling along the x-axis where $h < R$. Show that this curve has parametric equations $x = Rt - h\sin t$, $y = R - h\cos t$.

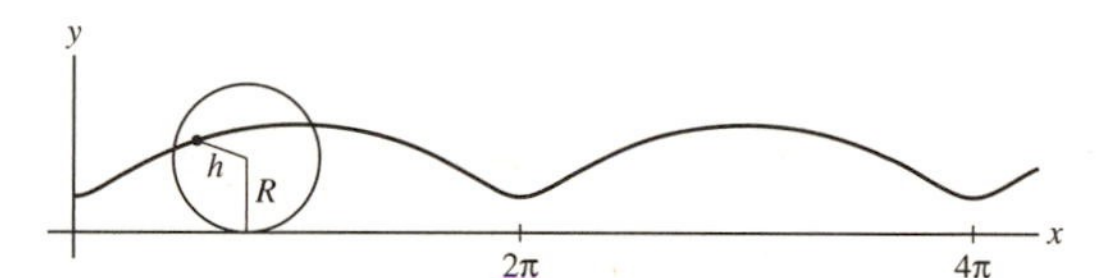

FIGURE 21 Curtate cycloid.

SOLUTION Let P be a point at a distance h from the center C of the circle. Assume that at $t = 0$, the line of CP is passing through the origin. When the circle rolls a distance Rt along the x-axis, the length of the arc $\widehat{SQ}$ (see figure) is also Rt and the angle $\angle SCQ$ has radian measure t. We compute the coordinates x and y of P.

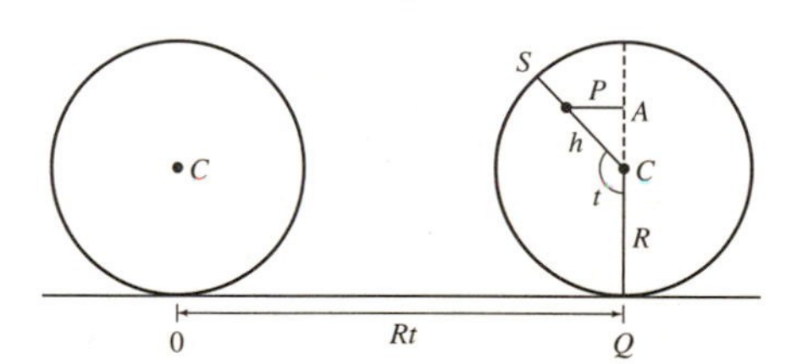

$$x = Rt - \overline{PA} = Rt - h\sin(\pi - t) = Rt - h\sin t$$
$$y = R + \overline{AC} = R + h\cos(\pi - t) = R - h\cos t$$

We obtain the following parametrization:

$$x = Rt - h\sin t, \quad y = R - h\cos t.$$

77. Show that the line of slope t through $(-1, 0)$ intersects the unit circle in the point with coordinates

$$x = \frac{1-t^2}{t^2+1}, \qquad y = \frac{2t}{t^2+1} \qquad \boxed{10}$$

Conclude that these equations parametrize the unit circle with the point $(-1, 0)$ excluded (Figure 22). Show further that $t = y/(x+1)$.

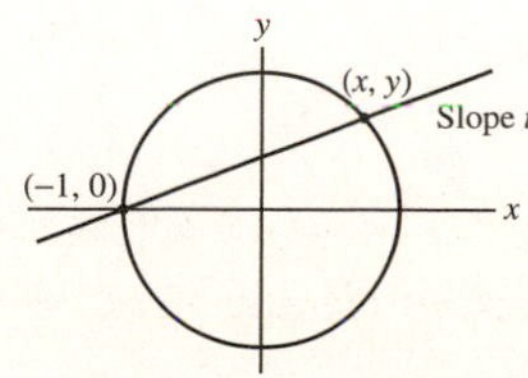

FIGURE 22 Unit circle.

SOLUTION The equation of the line of slope t through $(-1, 0)$ is $y = t(x + 1)$. The equation of the unit circle is $x^2 + y^2 = 1$. Hence, the line intersects the unit circle at the points (x, y) that satisfy the equations:

$$y = t(x + 1) \quad (1)$$
$$x^2 + y^2 = 1 \quad (2)$$

Substituting y from equation (1) into equation (2) and solving for x we obtain

$$x^2 + t^2(x + 1)^2 = 1$$
$$x^2 + t^2x^2 + 2t^2x + t^2 = 1$$
$$(1 + t^2)x^2 + 2t^2x + (t^2 - 1) = 0$$

This gives

$$x_{1,2} = \frac{-2t^2 \pm \sqrt{4t^4 - 4(t^2 + 1)(t^2 - 1)}}{2(1 + t^2)} = \frac{-2t^2 \pm 2}{2(1 + t^2)} = \frac{\pm 1 - t^2}{1 + t^2}$$

So $x_1 = -1$ and $x_2 = \frac{1 - t^2}{t^2 + 1}$. The solution $x = -1$ corresponds to the point $(-1, 0)$. We are interested in the second point of intersection that is varying as t varies. Hence the appropriate solution is

$$x = \frac{1 - t^2}{t^2 + 1}$$

We find the y-coordinate by substituting x in equation (1). This gives

$$y = t(x + 1) = t\left(\frac{1 - t^2}{t^2 + 1} + 1\right) = t \cdot \frac{1 - t^2 + t^2 + 1}{t^2 + 1} = \frac{2t}{t^2 + 1}$$

We conclude that the line and the unit circle intersect, besides at $(-1, 0)$, at the point with the following coordinates:

$$x = \frac{1 - t^2}{t^2 + 1}, \quad y = \frac{2t}{t^2 + 1} \quad (3)$$

Since these points determine all the points on the unit circle except for $(-1, 0)$ and no other points, the equations in (3) parametrize the unit circle with the point $(-1, 0)$ excluded.

We show that $t = \frac{y}{x + 1}$. Using (3) we have

$$\frac{y}{x + 1} = \frac{\frac{2t}{t^2+1}}{\frac{1-t^2}{t^2+1} + 1} = \frac{\frac{2t}{t^2+1}}{\frac{1-t^2+t^2+1}{t^2+1}} = \frac{\frac{2t}{t^2+1}}{\frac{2}{t^2+1}} = \frac{2t}{2} = t.$$

79. Use the results of Exercise 78 to show that the asymptote of the folium is the line $x + y = -a$. *Hint:* Show that $\lim_{t \to -1} (x + y) = -a$.

SOLUTION We must show that as $x \to \infty$ or $x \to -\infty$ the graph of the folium is getting arbitrarily close to the line $x + y = -a$, and the derivative $\frac{dy}{dx}$ is approaching the slope -1 of the line.

In Exercise 78 we showed that $x \to \infty$ when $t \to (-1^-)$ and $x \to -\infty$ when $t \to (-1^+)$. We first show that the graph is approaching the line $x + y = -a$ as $x \to \infty$ or $x \to -\infty$, by showing that $\lim_{t \to -1-} x + y = \lim_{t \to -1+} x + y = -a$.

For $x(t) = \frac{3at}{1 + t^3}$, $y(t) = \frac{3at^2}{1 + t^3}$, $a > 0$, calculated in Exercise 78, we obtain using L'Hôpital's Rule:

$$\lim_{t \to -1-} (x + y) = \lim_{t \to -1-} \frac{3at + 3at^2}{1 + t^3} = \lim_{t \to -1-} \frac{3a + 6at}{3t^2} = \frac{3a - 6a}{3} = -a$$
$$\lim_{t \to -1+} (x + y) = \lim_{t \to -1+} \frac{3at + 3at^2}{1 + t^3} = \lim_{t \to -1+} \frac{3a + 6at}{3t^2} = \frac{3a - 6a}{3} = -a$$

We now show that $\frac{dy}{dx}$ is approaching -1 as $t \to -1-$ and as $t \to -1+$. We use $\frac{dy}{dx} = \frac{6at - 3at^4}{3a - 6at^3}$ computed in Exercise 78 to obtain

$$\lim_{t\to-1-}\frac{dy}{dx}=\lim_{t\to-1-}\frac{6at-3at^4}{3a-6at^3}=\frac{-9a}{9a}=-1$$

$$\lim_{t\to-1+}\frac{dy}{dx}=\lim_{t\to-1+}\frac{6at-3at^4}{3a-6at^3}=\frac{-9a}{9a}=-1$$

We conclude that the line $x+y=-a$ is an asymptote of the folium as $x\to\infty$ and as $x\to-\infty$.

81. Second Derivative for a Parametrized Curve Given a parametrized curve $c(t)=(x(t),y(t))$, show that

$$\frac{d}{dt}\left(\frac{dy}{dx}\right)=\frac{x'(t)y''(t)-y'(t)x''(t)}{x'(t)^2}$$

Use this to prove the formula

$$\frac{d^2y}{dx^2}=\frac{x'(t)y''(t)-y'(t)x''(t)}{x'(t)^3} \qquad \boxed{11}$$

SOLUTION By the formula for the slope of the tangent line we have

$$\frac{dy}{dx}=\frac{y'(t)}{x'(t)}$$

Differentiating with respect to t, using the Quotient Rule, gives

$$\frac{d}{dt}\left(\frac{dy}{dx}\right)=\frac{d}{dt}\left(\frac{y'(t)}{x'(t)}\right)=\frac{x'(t)y''(t)-y'(t)x''(t)}{x'(t)^2}$$

By the Chain Rule we have

$$\frac{d^2y}{dx^2}=\frac{d}{dx}\left(\frac{dy}{dx}\right)=\frac{d}{dt}\left(\frac{dy}{dx}\right)\cdot\frac{dt}{dx}$$

Substituting into the above equation $\left(\text{and using } \frac{dt}{dx}=\frac{1}{dx/dt}=\frac{1}{x'(t)}\right)$ gives

$$\frac{d^2y}{dx^2}=\frac{x'(t)y''(t)-y'(t)x''(t)}{x'(t)^2}\cdot\frac{1}{x'(t)}=\frac{x'(t)y''(t)-y'(t)x''(t)}{x'(t)^3}$$

In Exercises 83–86, use Eq. (11) to find d^2y/dx^2.

83. $x=t^3+t^2,\quad y=7t^2-4,\quad t=2$

SOLUTION We find the first and second derivatives of $x(t)$ and $y(t)$:

$$\begin{aligned} x'(t)&=3t^2+2t \Rightarrow x'(2)=3\cdot 2^2+2\cdot 2=16\\ x''(t)&=6t+2 \quad\Rightarrow x''(2)=6\cdot 2+2=14\\ y'(t)&=14t \qquad\Rightarrow y'(2)=14\cdot 2=28\\ y''(t)&=14 \qquad\;\Rightarrow y''(2)=14 \end{aligned}$$

Using Eq. (11) we get

$$\left.\frac{d^2y}{dx^2}\right|_{t=2}=\left.\frac{x'(t)y''(t)-y'(t)x''(t)}{x'(t)^3}\right|_{t=2}=\frac{16\cdot 14-28\cdot 14}{16^3}=\frac{-21}{512}$$

85. $x=8t+9,\quad y=1-4t,\quad t=-3$

SOLUTION We compute the first and second derivatives of $x(t)$ and $y(t)$:

$$\begin{aligned} x'(t)&=8 \quad\Rightarrow x'(-3)=8\\ x''(t)&=0 \quad\Rightarrow x''(-3)=0\\ y'(t)&=-4 \Rightarrow y'(-3)=-4\\ y''(t)&=0 \quad\Rightarrow y''(-3)=0 \end{aligned}$$

Using Eq. (11) we get

$$\left.\frac{d^2y}{dx^2}\right|_{t=-3} = \frac{x'(-3)y''(-3) - y'(-3)x''(-3)}{x'(-3)^3} = \frac{8 \cdot 0 - (-4) \cdot 0}{8^3} = 0$$

87. Use Eq. (11) to find the t-intervals on which $c(t) = (t^2, t^3 - 4t)$ is concave up.

SOLUTION The curve is concave up where $\frac{d^2y}{dx^2} > 0$. Thus,

$$\frac{x'(t)y''(t) - y'(t)x''(t)}{x'(t)^3} > 0 \tag{1}$$

We compute the first and second derivatives:

$$x'(t) = 2t, \qquad x''(t) = 2$$
$$y'(t) = 3t^2 - 4, \quad y''(t) = 6t$$

Substituting in (1) and solving for t gives

$$\frac{12t^2 - (6t^2 - 8)}{8t^3} = \frac{6t^2 + 8}{8t^3}$$

Since $6t^2 + 8 > 0$ for all t, the quotient is positive if $8t^3 > 0$. We conclude that the curve is concave up for $t > 0$.

89. Area Under a Parametrized Curve Let $c(t) = (x(t), y(t))$, where $y(t) > 0$ and $x'(t) > 0$ (Figure 24). Show that the area A under $c(t)$ for $t_0 \le t \le t_1$ is

$$A = \int_{t_0}^{t_1} y(t)x'(t)\,dt \tag{12}$$

Hint: Because it is increasing, the function $x(t)$ has an inverse $t = g(x)$ and $c(t)$ is the graph of $y = y(g(x))$. Apply the change-of-variables formula to $A = \int_{x(t_0)}^{x(t_1)} y(g(x))\,dx$.

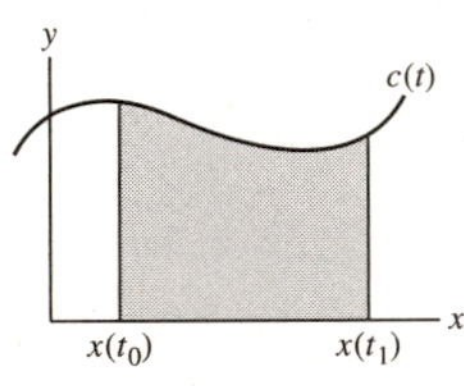

FIGURE 24

SOLUTION Let $x_0 = x(t_0)$ and $x_1 = x(t_1)$. We are given that $x'(t) > 0$, hence $x = x(t)$ is an increasing function of t, so it has an inverse function $t = g(x)$. The area A is given by $\int_{x_0}^{x_1} y(g(x))\,dx$. Recall that y is a function of t and $t = g(x)$, so the height y at any point x is given by $y = y(g(x))$. We find the new limits of integration. Since $x_0 = x(t_0)$ and $x_1 = x(t_1)$, the limits for t are t_0 and t_1, respectively. Also since $x'(t) = \frac{dx}{dt}$, we have $dx = x'(t)dt$. Performing this substitution gives

$$A = \int_{x_0}^{x_1} y(g(x))\,dx = \int_{t_0}^{t_1} y(g(x))x'(t)\,dt.$$

Since $g(x) = t$, we have $A = \int_{t_0}^{t_1} y(t)x'(t)\,dt$.

91. What does Eq. (12) say if $c(t) = (t, f(t))$?

SOLUTION In the parametrization $x(t) = t$, $y(t) = f(t)$ we have $x'(t) = 1$, $t_0 = x(t_0)$, $t_1 = x(t_1)$. Hence Eq. (12) becomes

$$A = \int_{t_0}^{t_1} y(t)x'(t)\,dt = \int_{x(t_0)}^{x(t_1)} f(t)\,dt$$

We see that in this parametrization Eq. (12) is the familiar formula for the area under the graph of a positive function.

93. Galileo tried unsuccessfully to find the area under a cycloid. Around 1630, Gilles de Roberval proved that the area under one arch of the cycloid $c(t) = (Rt - R\sin t, R - R\cos t)$ generated by a circle of radius R is equal to three times the area of the circle (Figure 25). Verify Roberval's result using Eq. (12).

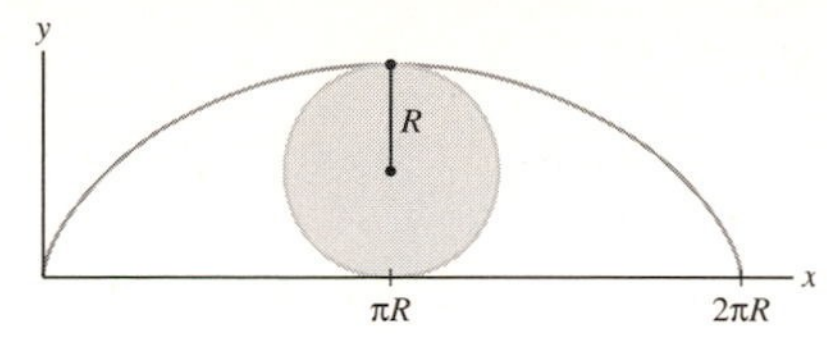

FIGURE 25 The area of one arch of the cycloid equals three times the area of the generating circle.

SOLUTION This reduces to

$$\int_0^{2\pi} (R - R\cos t)(Rt - R\sin t)'\,dt = \int_0^{2\pi} R^2(1 - \cos t)^2\,dt = 3\pi R^2.$$

Further Insights and Challenges

95. Derive the formula for the slope of the tangent line to a parametric curve $c(t) = (x(t), y(t))$ using a method different from that presented in the text. Assume that $x'(t_0)$ and $y'(t_0)$ exist and that $x'(t_0) \neq 0$. Show that

$$\lim_{h\to 0} \frac{y(t_0 + h) - y(t_0)}{x(t_0 + h) - x(t_0)} = \frac{y'(t_0)}{x'(t_0)}$$

Then explain why this limit is equal to the slope dy/dx. Draw a diagram showing that the ratio in the limit is the slope of a secant line.

SOLUTION Since $y'(t_0)$ and $x'(t_0)$ exist, we have the following limits:

$$\lim_{h\to 0} \frac{y(t_0 + h) - y(t_0)}{h} = y'(t_0), \quad \lim_{h\to 0} \frac{x(t_0 + h) - x(t_0)}{h} = x'(t_0) \tag{1}$$

We use Basic Limit Laws, the limits in (1) and the given data $x'(t_0) \neq 0$, to write

$$\lim_{h\to 0} \frac{y(t_0 + h) - y(t_0)}{x(t_0 + h) - x(t_0)} = \lim_{h\to 0} \frac{\frac{y(t_0+h)-y(t_0)}{h}}{\frac{x(t_0+h)-x(t_0)}{h}} = \frac{\lim_{h\to 0} \frac{y(t_0+h)-y(t_0)}{h}}{\lim_{h\to 0} \frac{x(t_0+h)-x(t_0)}{h}} = \frac{y'(t_0)}{x'(t_0)}$$

Notice that the quotient $\frac{y(t_0 + h) - y(t_0)}{x(t_0 + h) - x(t_0)}$ is the slope of the secant line determined by the points $P = (x(t_0), y(t_0))$ and $Q = (x(t_0 + h), y(t_0 + h))$. Hence, the limit of the quotient as $h \to 0$ is the slope of the tangent line at P, that is the derivative $\frac{dy}{dx}$.

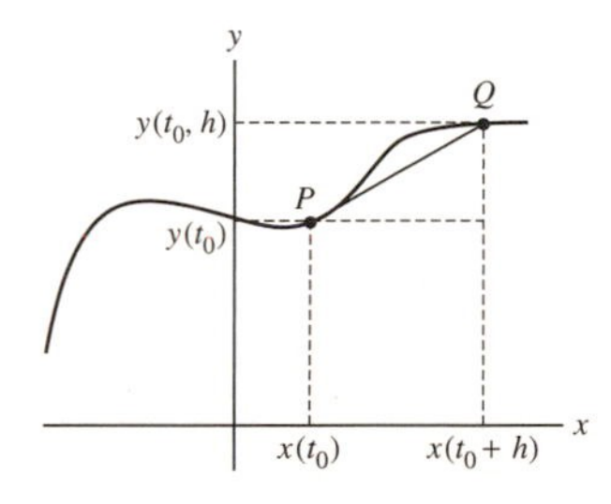

97. In Exercise 54 of Section 9.1 (LT Exercise 54 of Section 10.1), we described the tractrix by the differential equation

$$\frac{dy}{dx} = -\frac{y}{\sqrt{\ell^2 - y^2}}$$

Show that the curve $c(t)$ identified as the tractrix in Exercise 96 satisfies this differential equation. Note that the derivative on the left is taken with respect to x, not t.

SOLUTION Note that $dx/dt = 1 - \operatorname{sech}^2(t/\ell) = \tanh^2(t/\ell)$ and $dy/dt = -\operatorname{sech}(t/\ell)\tanh(t/\ell)$. Thus,

$$\frac{dy}{dx} = \frac{dy/dt}{dx/dt} = \frac{-\operatorname{sech}(t/\ell)}{\tanh(t/\ell)} = \frac{-y/\ell}{\sqrt{1 - y^2/\ell^2}}$$

Multiplying top and bottom by ℓ/ℓ gives

$$\frac{dy}{dx} = \frac{-y}{\sqrt{\ell^2 - y^2}}$$

In Exercises 98 and 99, refer to Figure 28.

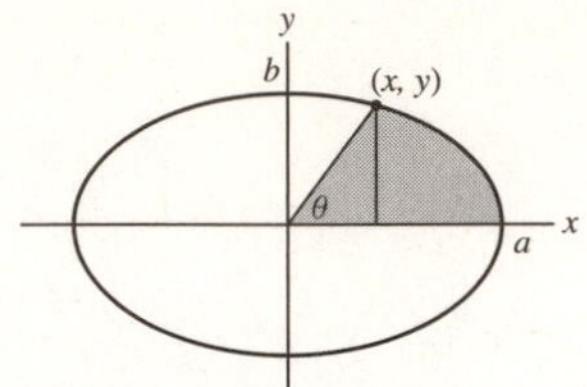

FIGURE 28 The parameter θ on the ellipse $\left(\frac{x}{a}\right)^2 + \left(\frac{y}{b}\right)^2 = 1$.

99. Show that the parametrization of the ellipse by the angle θ is

$$x = \frac{ab\cos\theta}{\sqrt{a^2\sin^2\theta + b^2\cos^2\theta}}$$

$$y = \frac{ab\sin\theta}{\sqrt{a^2\sin^2\theta + b^2\cos^2\theta}}$$

SOLUTION We consider the ellipse

$$\frac{x^2}{a^2} + \frac{y^2}{b^2} = 1.$$

For the angle θ we have $\tan\theta = \frac{y}{x}$, hence,

$$y = x\tan\theta \tag{1}$$

Substituting in the equation of the ellipse and solving for x we obtain

$$\frac{x^2}{a^2} + \frac{x^2\tan^2\theta}{b^2} = 1$$

$$b^2x^2 + a^2x^2\tan^2\theta = a^2b^2$$

$$(a^2\tan^2\theta + b^2)x^2 = a^2b^2$$

$$x^2 = \frac{a^2b^2}{a^2\tan^2\theta + b^2} = \frac{a^2b^2\cos^2\theta}{a^2\sin^2\theta + b^2\cos^2\theta}$$

We now take the square root. Since the sign of the x-coordinate is the same as the sign of $\cos\theta$, we take the positive root, obtaining

$$x = \frac{ab\cos\theta}{\sqrt{a^2\sin^2\theta + b^2\cos^2\theta}} \tag{2}$$

Hence by (1), the y-coordinate is

$$y = x\tan\theta = \frac{ab\cos\theta\tan\theta}{\sqrt{a^2\sin^2\theta + b^2\cos^2\theta}} = \frac{ab\sin\theta}{\sqrt{a^2\sin^2\theta + b^2\cos^2\theta}} \tag{3}$$

Equalities (2) and (3) give the following parametrization for the ellipse:

$$c_1(\theta) = \left(\frac{ab\cos\theta}{\sqrt{a^2\sin^2\theta + b^2\cos^2\theta}}, \frac{ab\sin\theta}{\sqrt{a^2\sin^2\theta + b^2\cos^2\theta}}\right)$$

11.2 Arc Length and Speed

Preliminary Questions

1. What is the definition of arc length?

SOLUTION A curve can be approximated by a polygonal path obtained by connecting points

$$p_0 = c(t_0),\ p_1 = c(t_1), \ldots, p_N = c(t_N)$$

on the path with segments. One gets an approximation by summing the lengths of the segments. The definition of arc length is the limit of that approximation when increasing the number of points so that the lengths of the segments approach zero. In doing so, we obtain the following theorem for the arc length:

$$S = \int_a^b \sqrt{x'(t)^2 + y'(t)^2}\, dt,$$

which is the length of the curve $c(t) = (x(t), y(t))$ for $a \le t \le b$.

2. What is the interpretation of $\sqrt{x'(t)^2 + y'(t)^2}$ for a particle following the trajectory $(x(t), y(t))$?

SOLUTION The expression $\sqrt{x'(t)^2 + y'(t)^2}$ denotes the speed at time t of a particle following the trajectory $(x(t), y(t))$.

3. A particle travels along a path from (0, 0) to (3, 4). What is the displacement? Can the distance traveled be determined from the information given?

SOLUTION The net displacement is the distance between the initial point (0, 0) and the endpoint (3, 4). That is

$$\sqrt{(3-0)^2 + (4-0)^2} = \sqrt{25} = 5.$$

The distance traveled can be determined only if the trajectory $c(t) = (x(t), y(t))$ of the particle is known.

4. A particle traverses the parabola $y = x^2$ with constant speed 3 cm/s. What is the distance traveled during the first minute? *Hint:* No computation is necessary.

SOLUTION Since the speed is constant, the distance traveled is the following product: $L = st = 3 \cdot 60 = 180$ cm.

Exercises

In Exercises 1–10, use Eq. (3) to find the length of the path over the given interval.

1. $(3t+1, 9-4t), \quad 0 \le t \le 2$

SOLUTION Since $x = 3t + 1$ and $y = 9 - 4t$ we have $x' = 3$ and $y' = -4$. Hence, the length of the path is

$$S = \int_0^2 \sqrt{3^2 + (-4)^2}\, dt = 5\int_0^2 dt = 10.$$

3. $(2t^2, 3t^2 - 1), \quad 0 \le t \le 4$

SOLUTION Since $x = 2t^2$ and $y = 3t^2 - 1$, we have $x' = 4t$ and $y' = 6t$. By the formula for the arc length we get

$$S = \int_0^4 \sqrt{x'(t)^2 + y'(t)^2}\, dt = \int_0^4 \sqrt{16t^2 + 36t^2}\, dt = \sqrt{52}\int_0^4 t\, dt = \sqrt{52} \cdot \frac{t^2}{2}\Big|_0^4 = 16\sqrt{13}$$

5. $(3t^2, 4t^3), \quad 1 \le t \le 4$

SOLUTION We have $x = 3t^2$ and $y = 4t^3$. Hence $x' = 6t$ and $y' = 12t^2$. By the formula for the arc length we get

$$S = \int_1^4 \sqrt{x'(t)^2 + y'(t)^2}\, dt = \int_1^4 \sqrt{36t^2 + 144t^4}\, dt = 6\int_1^4 \sqrt{1 + 4t^2}\, t\, dt.$$

Using the substitution $u = 1 + 4t^2$, $du = 8t\, dt$ we obtain

$$S = \frac{6}{8}\int_5^{65} \sqrt{u}\, du = \frac{3}{4} \cdot \frac{2}{3} u^{3/2}\Big|_5^{65} = \frac{1}{2}(65^{3/2} - 5^{3/2}) \approx 256.43$$

7. $(\sin 3t, \cos 3t), \quad 0 \le t \le \pi$

SOLUTION We have $x = \sin 3t$, $y = \cos 3t$, hence $x' = 3\cos 3t$ and $y' = -3\sin 3t$. By the formula for the arc length we obtain:

$$S = \int_0^\pi \sqrt{x'(t)^2 + y'(t)^2}\, dt = \int_0^\pi \sqrt{9\cos^2 3t + 9\sin^2 3t}\, dt = \int_0^\pi \sqrt{9}\, dt = 3\pi$$

In Exercises 9 and 10, use the identity

$$\frac{1 - \cos t}{2} = \sin^2 \frac{t}{2}$$

9. $(2\cos t - \cos 2t, 2\sin t - \sin 2t), \quad 0 \le t \le \frac{\pi}{2}$

SOLUTION We have $x = 2\cos t - \cos 2t$, $y = 2\sin t - \sin 2t$. Thus, $x' = -2\sin t + 2\sin 2t$ and $y' = 2\cos t - 2\cos 2t$. We get

$$\begin{aligned} x'(t)^2 + y'(t)^2 &= (-2\sin t + 2\sin 2t)^2 + (2\cos t - 2\cos 2t)^2 \\ &= 4\sin^2 t - 8\sin t \sin 2t + 4\sin^2 2t + 4\cos^2 t - 8\cos t\cos 2t + 4\cos^2 2t \\ &= 4(\sin^2 t + \cos^2 t) + 4(\sin^2 2t + \cos^2 2t) - 8(\sin t \sin 2t + \cos t \cos 2t) \\ &= 4 + 4 - 8\cos(2t - t) = 8 - 8\cos t = 8(1 - \cos t) \end{aligned}$$

We now use the formula for the arc length to obtain

$$\begin{aligned} S &= \int_0^{\pi/2} \sqrt{x'(t)^2 + y'(t)^2} = \int_0^{\pi/2} \sqrt{8(1-\cos t)}\,dt = \int_0^{\pi/2} \sqrt{16\sin^2 \frac{t}{2}}\,dt = 4\int_0^{\pi/2} \sin \frac{t}{2}\,dt \\ &= -8\cos \frac{t}{2}\Big|_0^{\pi/2} = -8\left(\cos\frac{\pi}{4} - \cos 0\right) = -8\left(\frac{\sqrt{2}}{2} - 1\right) \approx 2.34 \end{aligned}$$

11. Show that one arch of a cycloid generated by a circle of radius R has length $8R$.

SOLUTION Recall from earlier that the cycloid generated by a circle of radius R has parametric equations $x = Rt - R\sin t$, $y = R - R\cos t$. Hence, $x' = R - R\cos t$, $y' = R\sin t$. Using the identity $\sin^2 \frac{t}{2} = \frac{1-\cos t}{2}$, we get

$$\begin{aligned} x'(t)^2 + y'(t)^2 &= R^2(1-\cos t)^2 + R^2\sin^2 t = R^2(1 - 2\cos t + \cos^2 t + \sin^2 t) \\ &= R^2(1 - 2\cos t + 1) = 2R^2(1-\cos t) = 4R^2\sin^2\frac{t}{2} \end{aligned}$$

One arch of the cycloid is traced as t varies from 0 to 2π. Hence, using the formula for the arc length we obtain:

$$\begin{aligned} S &= \int_0^{2\pi} \sqrt{x'(t)^2 + y'(t)^2}\,dt = \int_0^{2\pi} \sqrt{4R^2\sin^2\frac{t}{2}}\,dt = 2R\int_0^{2\pi} \sin\frac{t}{2}\,dt = 4R\int_0^{\pi} \sin u\,du \\ &= -4R\cos u\Big|_0^{\pi} = -4R(\cos\pi - \cos 0) = 8R \end{aligned}$$

13. Find the length of the tractrix (see Figure 6)

$$c(t) = (t - \tanh(t), \operatorname{sech}(t)), \qquad 0 \le t \le A$$

SOLUTION Since $x = t - \tanh(t)$ and $y = \operatorname{sech}(t)$ we have $x' = 1 - \operatorname{sech}^2(t)$ and $y' = -\operatorname{sech}(t)\tanh(t)$. Hence,

$$\begin{aligned} x'(t)^2 + y'(t)^2 &= \left(1 - \operatorname{sech}^2(t)\right)^2 + \operatorname{sech}^2(t)\tanh^2(t) \\ &= 1 - 2\operatorname{sech}^2(t) + \operatorname{sech}^4(t) + \operatorname{sech}^2(t)\tanh^2(t) \\ &= 1 - 2\operatorname{sech}^2(t) + \operatorname{sech}^2(t)(\operatorname{sech}^2(t) + \tanh^2(t)) \\ &= 1 - 2\operatorname{sech}^2(t) + \operatorname{sech}^2(t) = 1 - \operatorname{sech}^2(t) = \tanh^2(t) \end{aligned}$$

Hence, using the formula for the arc length we get:

$$\begin{aligned} S &= \int_0^A \sqrt{x'(t)^2 + y'(t)^2}\,dt = \int_0^A \sqrt{\tanh^2(t)}\,dt = \int_0^A \tanh(t)\,dt = \ln(\cosh(t))\Big|_0^A \\ &= \ln(\cosh(A)) - \ln(\cosh(0)) = \ln(\cosh(A)) - \ln 1 = \ln(\cosh(A)) \end{aligned}$$

In Exercises 15–18, determine the speed s at time t (assume units of meters and seconds).

15. $(t^3, t^2), \quad t = 2$

SOLUTION We have $x(t) = t^3$, $y(t) = t^2$ hence $x'(t) = 3t^2$, $y'(t) = 2t$. The speed of the particle at time t is thus $\frac{ds}{dt} = \sqrt{x'(t)^2 + y'(t)^2} = \sqrt{9t^4 + 4t^2} = t\sqrt{9t^2 + 4}$. At time $t = 2$ the speed is

$$\frac{ds}{dt}\Big|_{t=2} = 2\sqrt{9 \cdot 2^2 + 4} = 2\sqrt{40} = 4\sqrt{10} \approx 12.65 \text{ m/s}.$$

17. $(5t+1, 4t-3), \quad t=9$

SOLUTION Since $x = 5t+1$, $y = 4t-3$, we have $x' = 5$ and $y' = 4$. The speed of the particle at time t is

$$\frac{ds}{dt} = \sqrt{x'(t) + y'(t)} = \sqrt{5^2+4^2} = \sqrt{41} \approx 6.4 \text{ m/s}.$$

We conclude that the particle has constant speed of 6.4 m/s.

19. Find the minimum speed of a particle with trajectory $c(t) = (t^3 - 4t, t^2 + 1)$ for $t \geq 0$. *Hint:* It is easier to find the minimum of the square of the speed.

SOLUTION We first find the speed of the particle. We have $x(t) = t^3 - 4t$, $y(t) = t^2 + 1$, hence $x'(t) = 3t^2 - 4$ and $y'(t) = 2t$. The speed is thus

$$\frac{ds}{dt} = \sqrt{(3t^2-4)^2 + (2t)^2} = \sqrt{9t^4 - 24t^2 + 16 + 4t^2} = \sqrt{9t^4 - 20t^2 + 16}.$$

The square root function is an increasing function, hence the minimum speed occurs at the value of t where the function $f(t) = 9t^4 - 20t^2 + 16$ has minimum value. Since $\lim_{t\to\infty} f(t) = \infty$, f has a minimum value on the interval $0 \leq t < \infty$, and it occurs at a critical point or at the endpoint $t = 0$. We find the critical point of f on $t \geq 0$:

$$f'(t) = 36t^3 - 40t = 4t(9t^2 - 10) = 0 \Rightarrow t = 0, t = \sqrt{\frac{10}{9}}.$$

We compute the values of f at these points:

$$f(0) = 9 \cdot 0^4 - 20 \cdot 0^2 + 16 = 16$$

$$f\left(\sqrt{\frac{10}{9}}\right) = 9\left(\sqrt{\frac{10}{9}}\right)^4 - 20\left(\sqrt{\frac{10}{9}}\right)^2 + 16 = \frac{44}{9} \approx 4.89$$

We conclude that the minimum value of f on $t \geq 0$ is 4.89. The minimum speed is therefore

$$\left(\frac{ds}{dt}\right)_{\min} \approx \sqrt{4.89} \approx 2.21.$$

21. Find the speed of the cycloid $c(t) = (4t - 4\sin t, 4 - 4\cos t)$ at points where the tangent line is horizontal.

SOLUTION We first find the points where the tangent line is horizontal. The slope of the tangent line is the following quotient:

$$\frac{dy}{dx} = \frac{dy/dt}{dx/dt} = \frac{4\sin t}{4 - 4\cos t} = \frac{\sin t}{1 - \cos t}.$$

To find the points where the tangent line is horizontal we solve the following equation for $t \geq 0$:

$$\frac{dy}{dx} = 0, \quad \frac{\sin t}{1 - \cos t} = 0 \Rightarrow \sin t = 0 \quad \text{and} \quad \cos t \neq 1.$$

Now, $\sin t = 0$ and $t \geq 0$ at the points $t = \pi k$, $k = 0, 1, 2, \ldots$. Since $\cos \pi k = (-1)^k$, the points where $\cos t \neq 1$ are $t = \pi k$ for k odd. The points where the tangent line is horizontal are, therefore:

$$t = \pi(2k-1), \quad k = 1, 2, 3, \ldots$$

The speed at time t is given by the following expression:

$$\begin{aligned}\frac{ds}{dt} &= \sqrt{x'(t)^2 + y'(t)^2} = \sqrt{(4 - 4\cos t)^2 + (4\sin t)^2} \\ &= \sqrt{16 - 32\cos t + 16\cos^2 t + 16\sin^2 t} = \sqrt{16 - 32\cos t + 16} \\ &= \sqrt{32(1-\cos t)} = \sqrt{32 \cdot 2\sin^2\frac{t}{2}} = 8\left|\sin\frac{t}{2}\right|\end{aligned}$$

That is, the speed of the cycloid at time t is

$$\frac{ds}{dt} = 8\left|\sin\frac{t}{2}\right|.$$

We now substitute

$$t = \pi(2k-1), \quad k = 1, 2, 3, \ldots$$

to obtain

$$\frac{ds}{dt} = 8\left|\sin\frac{\pi(2k-1)}{2}\right| = 8|(-1)^{k+1}| = 8$$

CAS *In Exercises 23–26, plot the curve and use the Midpoint Rule with $N = 10, 20, 30,$ and 50 to approximate its length.*

23. $c(t) = (\cos t, e^{\sin t})$ for $0 \le t \le 2\pi$

SOLUTION The curve of $c(t) = (\cos t, e^{\sin t})$ for $0 \le t \le 2\pi$ is shown in the figure below:

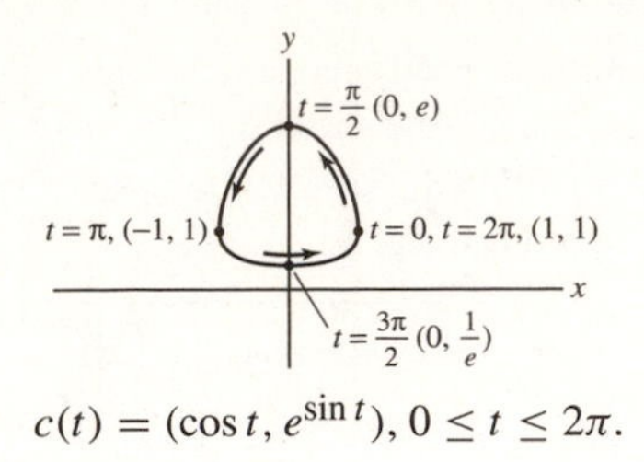

$c(t) = (\cos t, e^{\sin t}), 0 \le t \le 2\pi.$

The length of the curve is given by the following integral:

$$S = \int_0^{2\pi} \sqrt{x'(t)^2 + y'(t)^2}\, dt = \int_0^{2\pi} \sqrt{(-\sin t)^2 + (\cos t\, e^{\sin t})^2}\, dt.$$

That is, $S = \int_0^{2\pi} \sqrt{\sin^2 t + \cos^2 t\, e^{2\sin t}}\, dt$. We approximate the integral using the Mid-Point Rule with $N = 10, 20, 30, 50$. For $f(t) = \sqrt{\sin^2 t + \cos^2 t\, e^{2\sin t}}$ we obtain

$$(N = 10): \quad \Delta x = \frac{2\pi}{10} = \frac{\pi}{5}, c_i = \left(i - \frac{1}{2}\right) \cdot \frac{\pi}{5}$$

$$M_{10} = \frac{\pi}{5} \sum_{i=1}^{10} f(c_i) = 6.903734$$

$$(N = 20): \quad \Delta x = \frac{2\pi}{20} = \frac{\pi}{10}, c_i = \left(i - \frac{1}{2}\right) \cdot \frac{\pi}{10}$$

$$M_{20} = \frac{\pi}{10} \sum_{i=1}^{20} f(c_i) = 6.915035$$

$$(N = 30): \quad \Delta x = \frac{2\pi}{30} = \frac{\pi}{15}, c_i = \left(i - \frac{1}{2}\right) \cdot \frac{\pi}{15}$$

$$M_{30} = \frac{\pi}{15} \sum_{i=1}^{30} f(c_i) = 6.914949$$

$$(N = 50): \quad \Delta x = \frac{2\pi}{50} = \frac{\pi}{25}, c_i = \left(i - \frac{1}{2}\right) \cdot \frac{\pi}{25}$$

$$M_{50} = \frac{\pi}{25} \sum_{i=1}^{50} f(c_i) = 6.914951$$

25. The ellipse $\left(\frac{x}{5}\right)^2 + \left(\frac{y}{3}\right)^2 = 1$

SOLUTION We use the parametrization given in Example 4, section 12.1, that is, $c(t) = (5\cos t, 3\sin t), 0 \le t \le 2\pi$. The curve is shown in the figure below:

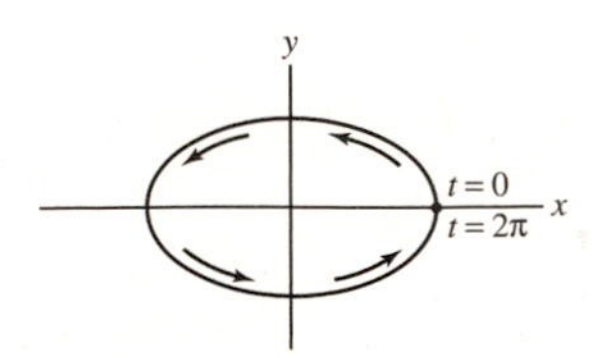

$c(t) = (5\cos t, 3\sin t), 0 \le t \le 2\pi.$

The length of the curve is given by the following integral:

$$S = \int_0^{2\pi} \sqrt{x'(t)^2 + y'(t)^2}\, dt = \int_0^{2\pi} \sqrt{(-5\sin t)^2 + (3\cos t)^2}\, dt$$

$$= \int_0^{2\pi} \sqrt{25\sin^2 t + 9\cos^2 t}\, dt = \int_0^{2\pi} \sqrt{9(\sin^2 t + \cos^2 t) + 16\sin^2 t}\, dt = \int_0^{2\pi} \sqrt{9 + 16\sin^2 t}\, dt.$$

That is,

$$S = \int_0^{2\pi} \sqrt{9 + 16\sin^2 t}\, dt.$$

We approximate the integral using the Mid-Point Rule with $N = 10, 20, 30, 50$, for $f(t) = \sqrt{9 + 16\sin^2 t}$. We obtain

$$(N = 10): \quad \Delta x = \frac{2\pi}{10} = \frac{\pi}{5}, \; c_i = \left(i - \frac{1}{2}\right) \cdot \frac{\pi}{5}$$

$$M_{10} = \frac{\pi}{5} \sum_{i=1}^{10} f(c_i) = 25.528309$$

$$(N = 20): \quad \Delta x = \frac{2\pi}{20} = \frac{\pi}{10}, \; c_i = \left(i - \frac{1}{2}\right) \cdot \frac{\pi}{10}$$

$$M_{20} = \frac{\pi}{10} \sum_{i=1}^{20} f(c_i) = 25.526999$$

$$(N = 30): \quad \Delta x = \frac{2\pi}{30} = \frac{\pi}{15}, \; c_i = \left(i - \frac{1}{2}\right) \cdot \frac{\pi}{15}$$

$$M_{30} = \frac{\pi}{15} \sum_{i=1}^{30} f(c_i) = 25.526999$$

$$(N = 50): \quad \Delta x = \frac{2\pi}{50} = \frac{\pi}{25}, \; c_i = \left(i - \frac{1}{2}\right) \cdot \frac{\pi}{25}$$

$$M_{50} = \frac{\pi}{25} \sum_{i=1}^{50} f(c_i) = 25.526999$$

27. If you unwind thread from a stationary circular spool, keeping the thread taut at all times, then the endpoint traces a curve C called the **involute** of the circle (Figure 9). Observe that $\overline{PQ}$ has length $R\theta$. Show that C is parametrized by

$$c(\theta) = \big(R(\cos\theta + \theta\sin\theta),\, R(\sin\theta - \theta\cos\theta)\big)$$

Then find the length of the involute for $0 \le \theta \le 2\pi$.

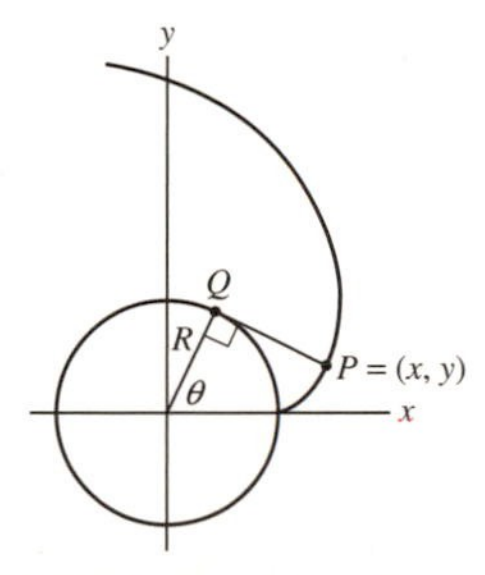

FIGURE 9 Involute of a circle.

SOLUTION Suppose that the arc $\widehat{QT}$ corresponding to the angle θ is unwound. Then the length of the segment $\overline{QP}$ equals the length of this arc. That is, $\overline{QP} = R\theta$. With the help of the figure we can see that

$$x = \overline{OA} + \overline{AB} = \overline{OA} + \overline{EP} = R\cos\theta + \overline{QP}\sin\theta = R\cos\theta + R\theta\sin\theta = R(\cos\theta + \theta\sin\theta).$$

Furthermore,

$$y = \overline{QA} - \overline{QE} = R\sin\theta - \overline{QP}\cos\theta = R\sin\theta - R\theta\cos\theta = R(\sin\theta - \theta\cos\theta)$$

The coordinates of P with respect to the parameter θ form the following parametrization of the curve:

$$c(\theta) = (R(\cos\theta + \theta\sin\theta),\, R(\sin\theta - \theta\cos\theta)), \qquad 0 \le \theta \le 2\pi.$$

We find the length of the involute for $0 \le \theta \le 2\pi$, using the formula for the arc length:

$$S = \int_0^{2\pi} \sqrt{x'(\theta)^2 + y'(\theta)^2}\, d\theta.$$

We compute the integrand:

$$x'(\theta) = \frac{d}{d\theta}(R(\cos\theta + \theta\sin\theta)) = R(-\sin\theta + \sin\theta + \theta\cos\theta) = R\theta\cos\theta$$

$$y'(\theta) = \frac{d}{d\theta}(R(\sin\theta - \theta\cos\theta)) = R(\cos\theta - (\cos\theta - \theta\sin\theta)) = R\theta\sin\theta$$

$$\sqrt{x'(\theta)^2 + y'(\theta)^2} = \sqrt{(R\theta\cos\theta)^2 + (R\theta\sin\theta)^2} = \sqrt{R^2\theta^2(\cos^2\theta + \sin^2\theta)} = \sqrt{R^2\theta^2} = R\theta$$

We now compute the arc length:

$$S = \int_0^{2\pi} R\theta\, d\theta = \left.\frac{R\theta^2}{2}\right|_0^{2\pi} = \frac{R\cdot(2\pi)^2}{2} = 2\pi^2 R.$$

In Exercises 29–32, use Eq. (4) to compute the surface area of the given surface.

29. The cone generated by revolving $c(t) = (t, mt)$ about the x-axis for $0 \le t \le A$

SOLUTION Substituting $y(t) = mt$, $y'(t) = m$, $x'(t) = 1$, $a = 0$, and $b = 0$ in the formula for the surface area, we get

$$S = 2\pi \int_0^A mt\sqrt{1+m^2}\, dt = 2\pi\sqrt{1+m^2}m \int_0^A t\, dt = 2\pi m\sqrt{1+m^2}\cdot\left.\frac{t^2}{2}\right|_0^A = m\sqrt{1+m^2}\pi A^2$$

31. The surface generated by revolving one arch of the cycloid $c(t) = (t - \sin t, 1 - \cos t)$ about the x-axis

SOLUTION One arch of the cycloid is traced as t varies from 0 to 2π. Since $x(t) = t - \sin t$ and $y(t) = 1 - \cos t$, we have $x'(t) = 1 - \cos t$ and $y'(t) = \sin t$. Hence, using the identity $1 - \cos t = 2\sin^2\frac{t}{2}$, we get

$$x'(t)^2 + y'(t)^2 = (1-\cos t)^2 + \sin^2 t = 1 - 2\cos t + \cos^2 t + \sin^2 t = 2 - 2\cos t = 4\sin^2\frac{t}{2}$$

By the formula for the surface area we obtain:

$$S = 2\pi\int_0^{2\pi} y(t)\sqrt{x'(t)^2 + y'(t)^2}\, dt = 2\pi\int_0^{2\pi}(1-\cos t)\cdot 2\sin\frac{t}{2}\, dt$$

$$= 2\pi\int_0^{2\pi} 2\sin^2\frac{t}{2}\cdot 2\sin\frac{t}{2}\, dt = 8\pi\int_0^{2\pi}\sin^3\frac{t}{2}\, dt = 16\pi\int_0^{\pi}\sin^3 u\, du$$

We use a reduction formula to compute this integral, obtaining

$$S = 16\pi\left[\frac{1}{3}\cos^3 u - \cos u\right]\Bigg|_0^{\pi} = 16\pi\left[\frac{4}{3}\right] = \frac{64\pi}{3}$$

Further Insights and Challenges

33. CAS Let $b(t)$ be the "Butterfly Curve":

$$x(t) = \sin t\left(e^{\cos t} - 2\cos 4t - \sin\left(\frac{t}{12}\right)^5\right)$$

$$y(t) = \cos t\left(e^{\cos t} - 2\cos 4t - \sin\left(\frac{t}{12}\right)^5\right)$$

(a) Use a computer algebra system to plot $b(t)$ and the speed $s'(t)$ for $0 \le t \le 12\pi$.

(b) Approximate the length $b(t)$ for $0 \le t \le 10\pi$.

SOLUTION

(a) Let $f(t) = e^{\cos t} - 2\cos 4t - \sin\left(\frac{t}{12}\right)^5$, then

$$x(t) = \sin t f(t)$$

$$y(t) = \cos t f(t)$$

and so

$$(x'(t))^2 + (y'(t))^2 = [\sin t f'(t) + \cos t f(t)]^2 + [\cos t f'(t) - \sin t f(t)]^2$$

Using the identity $\sin^2 t + \cos^2 t = 1$, we get

$$(x'(t))^2 + (y'(t))^2 = (f'(t))^2 + (f(t))^2.$$

Thus, $s'(t)$ is the following:

$$\sqrt{\left[e^{\cos t} - 2\cos 4t - \sin\left(\frac{t}{12}\right)^5\right]^2 + \left[-\sin t e^{\cos t} + 8\sin 4t - \frac{5}{12}\left(\frac{t}{12}\right)^4 \cos\left(\frac{t}{12}\right)^5\right]^2}.$$

The following figures show the curves of $b(t)$ and the speed $s'(t)$ for $0 \le t \le 10\pi$:

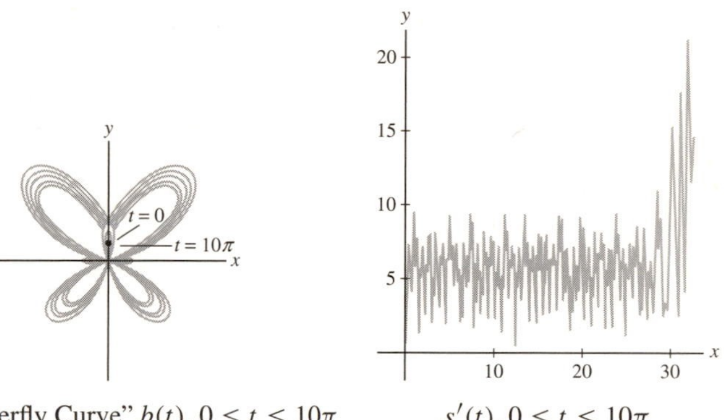

The "Butterfly Curve" $b(t)$, $0 \le t \le 10\pi$ $\quad$ $s'(t)$, $0 \le t \le 10\pi$

Looking at the graph, we see it would be difficult to compute the length using numeric integration; due to the high frequency oscillations, very small steps would be needed.

(b) The length of $b(t)$ for $0 \le t \le 10\pi$ is given by the integral: $L = \int_0^{10\pi} s'(t)\,dt$ where $s'(t)$ is given in part (a). We approximate the length using the Midpoint Rule with $N = 30$. The numerical methods in Mathematica approximate the answer by 211.952. Using the Midpoint Rule with $N = 50$, we get 204.48; with $N = 500$, we get 211.6; and with $N = 5000$, we get 212.09.

35. A satellite orbiting at a distance R from the center of the earth follows the circular path $x = R\cos\omega t$, $y = R\sin\omega t$.

(a) Show that the period T (the time of one revolution) is $T = 2\pi/\omega$.

(b) According to Newton's laws of motion and gravity,

$$x''(t) = -Gm_e\frac{x}{R^3}, \qquad y''(t) = -Gm_e\frac{y}{R^3}$$

where G is the universal gravitational constant and m_e is the mass of the earth. Prove that $R^3/T^2 = Gm_e/4\pi^2$. Thus, R^3/T^2 has the same value for all orbits (a special case of Kepler's Third Law).

SOLUTION

(a) As shown in Example 4, the circular path has constant speed of $\frac{ds}{dt} = \omega R$. Since the length of one revolution is $2\pi R$, the period T is

$$T = \frac{2\pi R}{\omega R} = \frac{2\pi}{\omega}.$$

(b) Differentiating $x = R\cos\omega t$ twice with respect to t gives

$$x'(t) = -R\omega\sin\omega t$$
$$x''(t) = -R\omega^2\cos\omega t$$

Substituting $x(t)$ and $x''(t)$ in the equation $x''(t) = -Gm_e\frac{x}{R^3}$ and simplifying, we obtain

$$-R\omega^2\cos\omega t = -Gm_e \cdot \frac{R\cos\omega t}{R^3}$$

$$-R\omega^2 = -\frac{Gm_e}{R^2} \Rightarrow R^3 = \frac{Gm_e}{\omega^2}$$

By part (a), $T = \frac{2\pi}{\omega}$. Hence, $\omega = \frac{2\pi}{T}$. Substituting yields

$$R^3 = \frac{Gm_e}{\frac{4\pi^2}{T^2}} = \frac{T^2 Gm_e}{4\pi^2} \Rightarrow \frac{R^3}{T^2} = \frac{Gm_e}{4\pi^2}$$

11.3 Polar Coordinates

Preliminary Questions

1. Points P and Q with the same radial coordinate (choose the correct answer):

(a) Lie on the same circle with the center at the origin.

(b) Lie on the same ray based at the origin.

SOLUTION Two points with the same radial coordinate are equidistant from the origin, therefore they lie on the same circle centered at the origin. The angular coordinate defines a ray based at the origin. Therefore, if the two points have the same angular coordinate, they lie on the same ray based at the origin.

2. Give two polar representations for the point $(x, y) = (0, 1)$, one with negative r and one with positive r.

SOLUTION The point $(0, 1)$ is on the y-axis, distant one unit from the origin, hence the polar representation with positive r is $(r, \theta) = \left(1, \frac{\pi}{2}\right)$. The point $(r, \theta) = \left(-1, \frac{\pi}{2}\right)$ is the reflection of $(r, \theta) = \left(1, \frac{\pi}{2}\right)$ through the origin, hence we must add π to return to the original point.

We obtain the following polar representation of $(0, 1)$ with negative r:

$$(r, \theta) = \left(-1, \frac{\pi}{2} + \pi\right) = \left(-1, \frac{3\pi}{2}\right).$$

3. Describe each of the following curves:

(a) $r = 2$ **(b)** $r^2 = 2$ **(c)** $r \cos\theta = 2$

SOLUTION

(a) Converting to rectangular coordinates we get

$$\sqrt{x^2 + y^2} = 2 \quad \text{or} \quad x^2 + y^2 = 2^2.$$

This is the equation of the circle of radius 2 centered at the origin.

(b) We convert to rectangular coordinates, obtaining $x^2 + y^2 = 2$. This is the equation of the circle of radius $\sqrt{2}$, centered at the origin.

(c) We convert to rectangular coordinates. Since $x = r\cos\theta$ we obtain the following equation: $x = 2$. This is the equation of the vertical line through the point $(2, 0)$.

4. If $f(-\theta) = f(\theta)$, then the curve $r = f(\theta)$ is symmetric with respect to the (choose the correct answer):

(a) x-axis **(b)** y-axis **(c)** origin

SOLUTION The equality $f(-\theta) = f(\theta)$ for all θ implies that whenever a point (r, θ) is on the curve, also the point $(r, -\theta)$ is on the curve. Since the point $(r, -\theta)$ is the reflection of (r, θ) with respect to the x-axis, we conclude that the curve is symmetric with respect to the x-axis.

Exercises

1. Find polar coordinates for each of the seven points plotted in Figure 16.

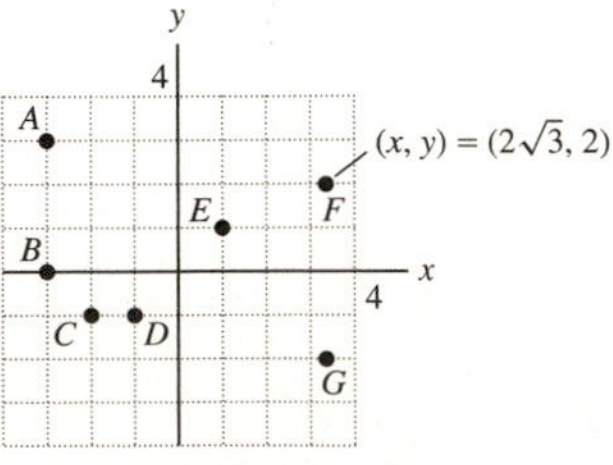

FIGURE 16

SOLUTION We mark the points as shown in the figure.

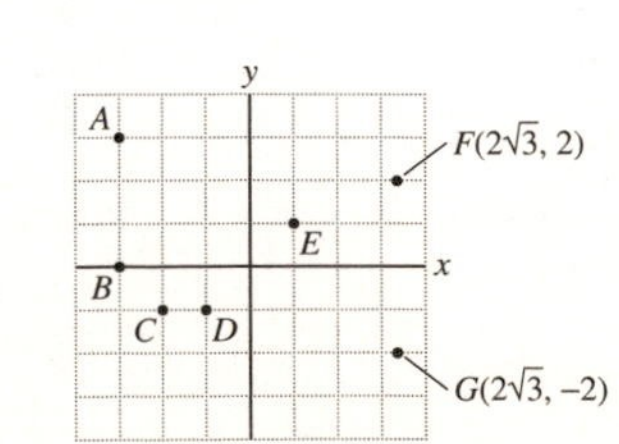

Using the data given in the figure for the x and y coordinates and the quadrants in which the point are located, we obtain:

(A), with rectangular coordinates $(-3, 4)$: $\begin{array}{l} r = \sqrt{(-3)^2 + 3^2} = \sqrt{18} \\ \theta = \pi - \frac{\pi}{4} = \frac{3\pi}{4} \end{array} \Rightarrow (r, \theta) = \left(3\sqrt{2}, \frac{3\pi}{4}\right)$

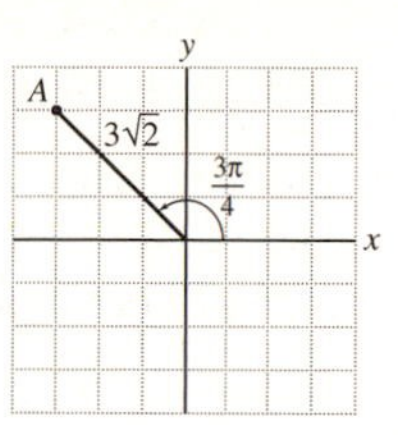

(B), with rectangular coordinates $(-3, 0)$: $\begin{array}{l} r = 3 \\ \theta = \pi \end{array} \Rightarrow (r, \theta) = (3, \pi)$

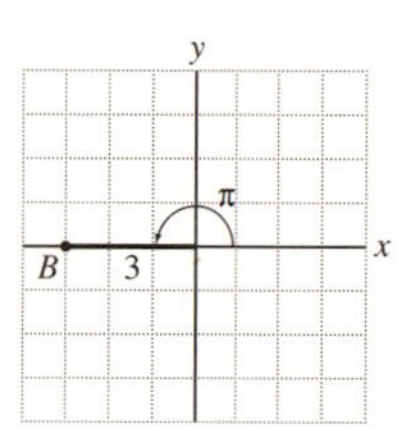

(C), with rectangular coordinates $(-2, -1)$:

$$\begin{array}{l} r = \sqrt{2^2 + 1^2} = \sqrt{5} \approx 2.2 \\ \theta = \tan^{-1}\left(\frac{-1}{-2}\right) = \tan^{-1}\left(\frac{1}{2}\right) = \pi + 0.46 \approx 3.6 \end{array} \Rightarrow (r, \theta) \approx \left(\sqrt{5}, 3.6\right)$$

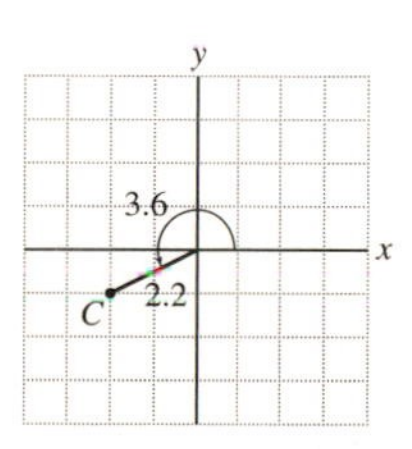

(D), with rectangular coordinates $(-1, -1)$: $\begin{array}{l} r = \sqrt{1^2 + 1^2} = \sqrt{2} \approx 1.4 \\ \theta = \pi + \frac{\pi}{4} = \frac{5\pi}{4} \end{array} \Rightarrow (r, \theta) \approx \left(\sqrt{2}, \frac{5\pi}{4}\right)$

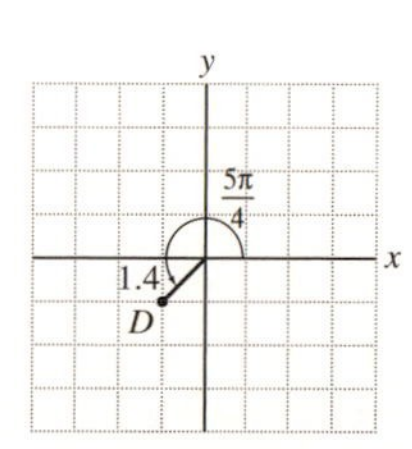

(E), with rectangular coordinates $(1, 1)$: $\begin{array}{l} r = \sqrt{1^2 + 1^2} = \sqrt{2} \approx 1.4 \\ \theta = \tan^{-1}\left(\frac{1}{1}\right) = \frac{\pi}{4} \end{array} \Rightarrow (r, \theta) \approx \left(\sqrt{2}, \frac{\pi}{4}\right)$

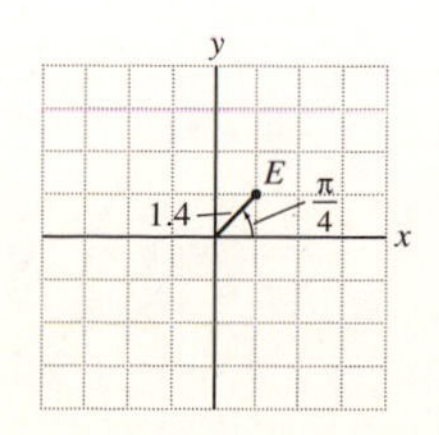

(F), with rectangular coordinates $(2\sqrt{3}, 2)$: $\begin{array}{l} r = \sqrt{\left(2\sqrt{3}\right)^2 + 2^2} = \sqrt{16} = 4 \\ \theta = \tan^{-1}\left(\frac{2}{2\sqrt{3}}\right) = \tan^{-1}\left(\frac{1}{\sqrt{3}}\right) = \frac{\pi}{6} \end{array}$ $\Rightarrow (r, \theta) = \left(4, \frac{\pi}{6}\right)$

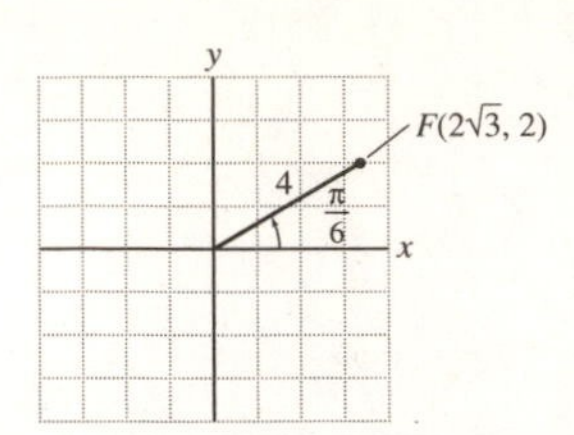

(G), with rectangular coordinates $(2\sqrt{3}, -2)$: G is the reflection of F about the x axis, hence the two points have equal radial coordinates, and the angular coordinate of G is obtained from the angular coordinate of F: $\theta = 2\pi - \frac{\pi}{6} = \frac{11\pi}{6}$. Hence, the polar coordinates of G are $\left(4, \frac{11\pi}{6}\right)$.

3. Convert from rectangular to polar coordinates.

(a) $(1, 0)$ **(b)** $(3, \sqrt{3})$ **(c)** $(-2, 2)$ **(d)** $(-1, \sqrt{3})$

SOLUTION

(a) The point $(1, 0)$ is on the positive x axis distanced one unit from the origin. Hence, $r = 1$ and $\theta = 0$. Thus, $(r, \theta) = (1, 0)$.

(b) The point $\left(3, \sqrt{3}\right)$ is in the first quadrant so $\theta = \tan^{-1}\left(\frac{\sqrt{3}}{3}\right) = \frac{\pi}{6}$. Also, $r = \sqrt{3^2 + \left(\sqrt{3}\right)^2} = \sqrt{12}$. Hence, $(r, \theta) = \left(\sqrt{12}, \frac{\pi}{6}\right)$.

(c) The point $(-2, 2)$ is in the second quadrant. Hence,

$$\theta = \tan^{-1}\left(\frac{2}{-2}\right) = \tan^{-1}(-1) = \pi - \frac{\pi}{4} = \frac{3\pi}{4}.$$

Also, $r = \sqrt{(-2)^2 + 2^2} = \sqrt{8}$. Hence, $(r, \theta) = \left(\sqrt{8}, \frac{3\pi}{4}\right)$.

(d) The point $\left(-1, \sqrt{3}\right)$ is in the second quadrant, hence,

$$\theta = \tan^{-1}\left(\frac{\sqrt{3}}{-1}\right) = \tan^{-1}\left(-\sqrt{3}\right) = \pi - \frac{\pi}{3} = \frac{2\pi}{3}.$$

Also, $r = \sqrt{(-1)^2 + \left(\sqrt{3}\right)^2} = \sqrt{4} = 2$. Hence, $(r, \theta) = \left(2, \frac{2\pi}{3}\right)$.

5. Convert from polar to rectangular coordinates:

(a) $\left(3, \frac{\pi}{6}\right)$ **(b)** $\left(6, \frac{3\pi}{4}\right)$ **(c)** $\left(0, \frac{\pi}{5}\right)$ **(d)** $\left(5, -\frac{\pi}{2}\right)$

SOLUTION

(a) Since $r = 3$ and $\theta = \frac{\pi}{6}$, we have:

$$\begin{array}{l} x = r\cos\theta = 3\cos\frac{\pi}{6} = 3 \cdot \frac{\sqrt{3}}{2} \approx 2.6 \\ y = r\sin\theta = 3\sin\frac{\pi}{6} = 3 \cdot \frac{1}{2} = 1.5 \end{array} \quad \Rightarrow \quad (x, y) \approx (2.6, 1.5).$$

(b) For $\left(6, \frac{3\pi}{4}\right)$ we have $r = 6$ and $\theta = \frac{3\pi}{4}$. Hence,

$$\begin{array}{l} x = r\cos\theta = 6\cos\frac{3\pi}{4} \approx -4.24 \\ y = r\sin\theta = 6\sin\frac{3\pi}{4} \approx 4.24 \end{array} \quad \Rightarrow \quad (x, y) \approx (-4.24, 4.24).$$

(c) For $\left(0, \frac{\pi}{5}\right)$, we have $r = 0$, so that the rectangular coordinates are $(x, y) = (0, 0)$.

(d) Since $r = 5$ and $\theta = -\frac{\pi}{2}$ we have

$$\begin{array}{l} x = r\cos\theta = 5\cos\left(-\frac{\pi}{2}\right) = 5 \cdot 0 = 0 \\ y = r\sin\theta = 5\sin\left(-\frac{\pi}{2}\right) = 5 \cdot (-1) = -5 \end{array} \quad \Rightarrow \quad (x, y) = (0, -5)$$

7. Describe each shaded sector in Figure 17 by inequalities in r and θ.

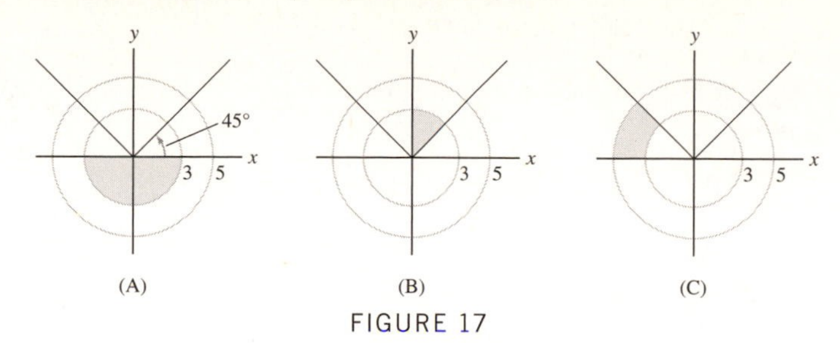

FIGURE 17

SOLUTION

(a) In the sector shown below r is varying between 0 and 3 and θ is varying between π and 2π. Hence the following inequalities describe the sector:

$$0 \le r \le 3$$
$$\pi \le \theta \le 2\pi$$

(b) In the sector shown below r is varying between 0 and 3 and θ is varying between $\frac{\pi}{4}$ and $\frac{\pi}{2}$. Hence, the inequalities for the sector are:

$$0 \le r \le 3$$
$$\frac{\pi}{4} \le \theta \le \frac{\pi}{2}$$

(c) In the sector shown below r is varying between 3 and 5 and θ is varying between $\frac{3\pi}{4}$ and π. Hence, the inequalities are:

$$3 \le r \le 5$$
$$\frac{3\pi}{4} \le \theta \le \pi$$

9. What is the slope of the line $\theta = \frac{3\pi}{5}$?

SOLUTION This line makes an angle $\theta_0 = \frac{3\pi}{5}$ with the positive x-axis, hence the slope of the line is $m = \tan \frac{3\pi}{5} \approx -3.1$.

In Exercises 11–16, convert to an equation in rectangular coordinates.

11. $r = 7$

SOLUTION $r = 7$ describes the points having distance 7 from the origin, that is, the circle with radius 7 centered at the origin. The equation of the circle in rectangular coordinates is

$$x^2 + y^2 = 7^2 = 49.$$

13. $r = 2 \sin \theta$

SOLUTION We multiply the equation by r and substitute $r^2 = x^2 + y^2$, $r \sin \theta = y$. This gives

$$r^2 = 2r \sin \theta$$
$$x^2 + y^2 = 2y$$

Moving the $2y$ and completing the square yield: $x^2 + y^2 - 2y = 0$ and $x^2 + (y - 1)^2 = 1$. Thus, $r = 2 \sin \theta$ is the equation of a circle of radius 1 centered at $(0, 1)$.

15. $r = \dfrac{1}{\cos \theta - \sin \theta}$

SOLUTION We multiply the equation by $\cos \theta - \sin \theta$ and substitute $y = r \sin \theta$, $x = r \cos \theta$. This gives

$$r(\cos \theta - \sin \theta) = 1$$
$$r \cos \theta - r \sin \theta = 1$$

$x - y = 1 \Rightarrow y = x - 1$. Thus,

$$r = \frac{1}{\cos \theta - \sin \theta}$$

is the equation of the line $y = x - 1$.

In Exercises 17–20, convert to an equation in polar coordinates.

17. $x^2 + y^2 = 5$

SOLUTION We make the substitution $x^2 + y^2 = r^2$ to obtain; $r^2 = 5$ or $r = \sqrt{5}$.

19. $y = x^2$

SOLUTION Substituting $y = r\sin\theta$ and $x = r\cos\theta$ yields

$$r\sin\theta = r^2\cos^2\theta.$$

Then, dividing by $r\cos^2\theta$ we obtain,

$$\frac{\sin\theta}{\cos^2\theta} = r \qquad \text{so} \qquad r = \tan\theta\sec\theta$$

21. Match each equation with its description.

(a) $r = 2$	**(i)** Vertical line
(b) $\theta = 2$	**(ii)** Horizontal line
(c) $r = 2\sec\theta$	**(iii)** Circle
(d) $r = 2\csc\theta$	**(iv)** Line through origin

SOLUTION

(a) $r = 2$ describes the points 2 units from the origin. Hence, it is the equation of a circle.

(b) $\theta = 2$ describes the points P so that $\overline{OP}$ makes an angle of $\theta_0 = 2$ with the positive x-axis. Hence, it is the equation of a line through the origin.

(c) This is $r\cos\theta = 2$, which is $x = 2$, a vertical line.

(d) Converting to rectangular coordinates, we get $r = 2\csc\theta$, so $r\sin\theta = 2$ and $y = 2$. This is the equation of a horizontal line.

23. Suppose that $P = (x, y)$ has polar coordinates (r, θ). Find the polar coordinates for the points:

(a) $(x, -y)$ **(b)** $(-x, -y)$ **(c)** $(-x, y)$ **(d)** (y, x)

SOLUTION

(a) $(x, -y)$ is the symmetric point of (x, y) with respect to the x-axis, hence the two points have the same radial coordinate, and the angular coordinate of $(x, -y)$ is $2\pi - \theta$. Hence, $(x, -y) = (r, 2\pi - \theta)$.

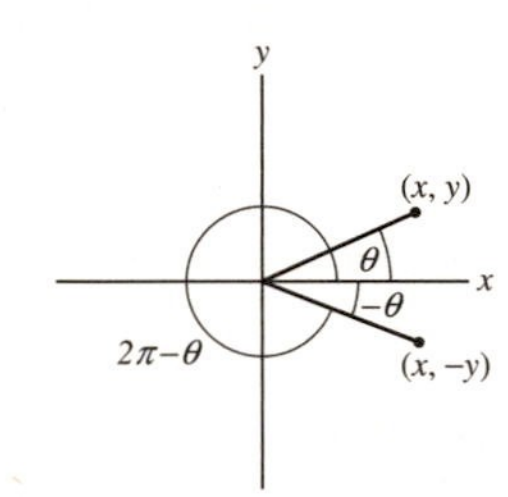

(b) $(-x, -y)$ is the symmetric point of (x, y) with respect to the origin. Hence, $(-x, -y) = (r, \theta + \pi)$.

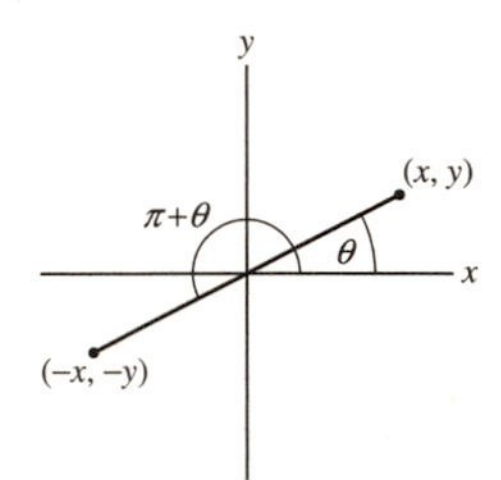

(c) $(-x, y)$ is the symmetric point of (x, y) with respect to the y-axis. Hence the two points have the same radial coordinates and the angular coordinate of $(-x, y)$ is $\pi - \theta$. Hence, $(-x, y) = (r, \pi - \theta)$.

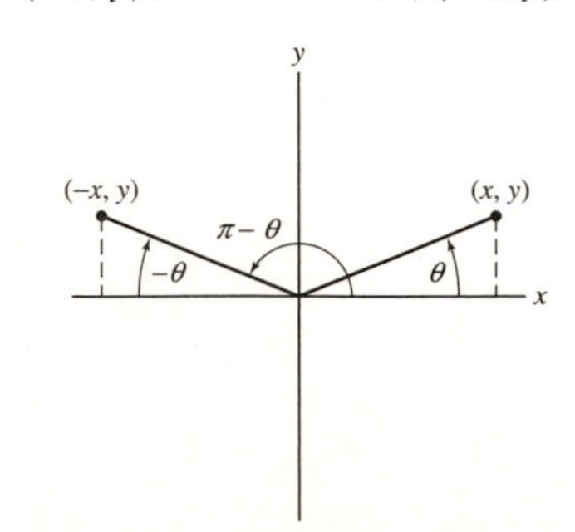

(d) Let (r_1, θ_1) denote the polar coordinates of (y, x). Hence,

$$r_1 = \sqrt{y^2 + x^2} = \sqrt{x^2 + y^2} = r$$

$$\tan\theta_1 = \frac{x}{y} = \frac{1}{y/x} = \frac{1}{\tan\theta} = \cot\theta = \tan\left(\frac{\pi}{2} - \theta\right)$$

Since the points (x, y) and (y, x) are in the same quadrant, the solution for θ_1 is $\theta_1 = \frac{\pi}{2} - \theta$. We obtain the following polar coordinates: $(y, x) = \left(r, \frac{\pi}{2} - \theta\right)$.

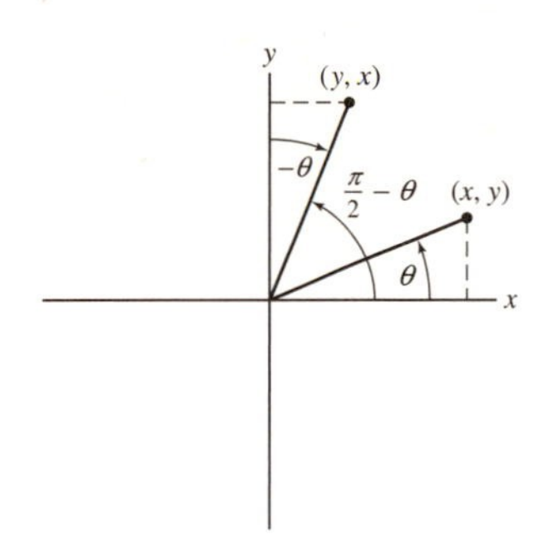

25. What are the polar equations of the lines parallel to the line $r\cos\left(\theta - \frac{\pi}{3}\right) = 1$?

SOLUTION The line $r\cos\left(\theta - \frac{\pi}{3}\right) = 1$, or $r = \sec\left(\theta - \frac{\pi}{3}\right)$, is perpendicular to the ray $\theta = \frac{\pi}{3}$ and at distance $d = 1$ from the origin. Hence, the lines parallel to this line are also perpendicular to the ray $\theta = \frac{\pi}{3}$, so the polar equations of these lines are $r = d\sec\left(\theta - \frac{\pi}{3}\right)$ or $r\cos\left(\theta - \frac{\pi}{3}\right) = d$.

27. Sketch the curve $r = \frac{1}{2}\theta$ (the spiral of Archimedes) for θ between 0 and 2π by plotting the points for $\theta = 0, \frac{\pi}{4}, \frac{\pi}{2}, \ldots, 2\pi$.

SOLUTION We first plot the following points (r, θ) on the spiral:

$$O = (0, 0)\,,\ A = \left(\frac{\pi}{8}, \frac{\pi}{4}\right),\ B = \left(\frac{\pi}{4}, \frac{\pi}{2}\right),\ C = \left(\frac{3\pi}{8}, \frac{3\pi}{4}\right),\ D = \left(\frac{\pi}{2}, \pi\right),$$

$$E = \left(\frac{5\pi}{8}, \frac{5\pi}{4}\right),\ F = \left(\frac{3\pi}{4}, \frac{3\pi}{2}\right),\ G = \left(\frac{7\pi}{8}, \frac{7\pi}{4}\right),\ H = (\pi, 2\pi)\,.$$

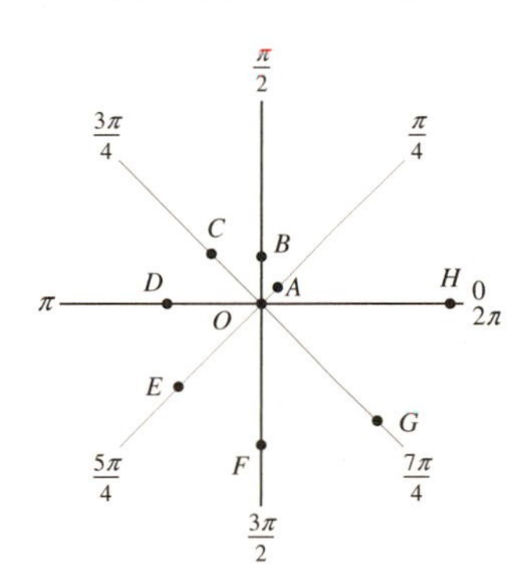

Since $r(0) = \frac{0}{2} = 0$, the graph begins at the origin and moves toward the points A, B, C, D, E, F, G and H as θ varies from $\theta = 0$ to the other values stated above. Connecting the points in this direction we obtain the following graph for $0 \le \theta \le 2\pi$:

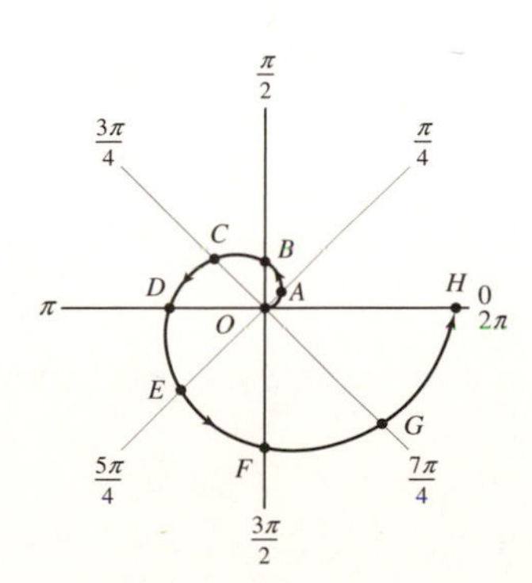

29. Sketch the cardioid curve $r = 1 + \cos\theta$.

SOLUTION Since $\cos\theta$ is period with period 2π, the entire curve will be traced out as θ varies from 0 to 2π. Additionally, since $\cos(2\pi - \theta) = \cos(\theta)$, we can sketch the curve for θ between 0 and π and reflect the result through the x axis to obtain the whole curve. Use the values $\theta = 0, \frac{\pi}{6}, \frac{\pi}{4}, \frac{\pi}{3}, \frac{\pi}{2}, \frac{2\pi}{3}, \frac{3\pi}{4}, \frac{5\pi}{6}$, and π:

θ	r	point
0	$1 + \cos 0 = 2$	$(2, 0)$
$\frac{\pi}{6}$	$1 + \cos\frac{\pi}{6} = \frac{2+\sqrt{3}}{2}$	$\left(\frac{2+\sqrt{3}}{2}, \frac{\pi}{6}\right)$
$\frac{\pi}{4}$	$1 + \cos\frac{\pi}{4} = \frac{2+\sqrt{2}}{2}$	$\left(\frac{2+\sqrt{2}}{2}, \frac{\pi}{4}\right)$
$\frac{\pi}{3}$	$1 + \cos\frac{\pi}{3} = \frac{3}{2}$	$\left(\frac{3}{2}, \frac{\pi}{3}\right)$
$\frac{\pi}{2}$	$1 + \cos\frac{\pi}{2} = 1$	$\left(1, \frac{\pi}{2}\right)$
$\frac{2\pi}{3}$	$1 + \cos\frac{2\pi}{3} = \frac{1}{2}$	$\left(\frac{1}{2}, \frac{2\pi}{3}\right)$
$\frac{3\pi}{4}$	$1 + \cos\frac{3\pi}{4} = \frac{2-\sqrt{2}}{2}$	$\left(\frac{2-\sqrt{2}}{2}, \frac{3\pi}{4}\right)$
$\frac{5\pi}{6}$	$1 + \cos\frac{5\pi}{6} = \frac{2-\sqrt{3}}{2}$	$\left(\frac{2-\sqrt{3}}{2}, \frac{5\pi}{6}\right)$

$\theta = 0$ corresponds to the point $(2, 0)$, and the graph moves clockwise as θ increases from 0 to π. Thus the graph is

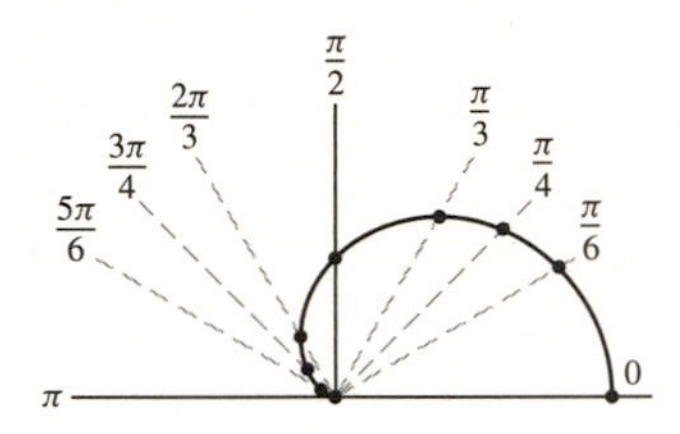

Reflecting through the x axis gives the other half of the curve:

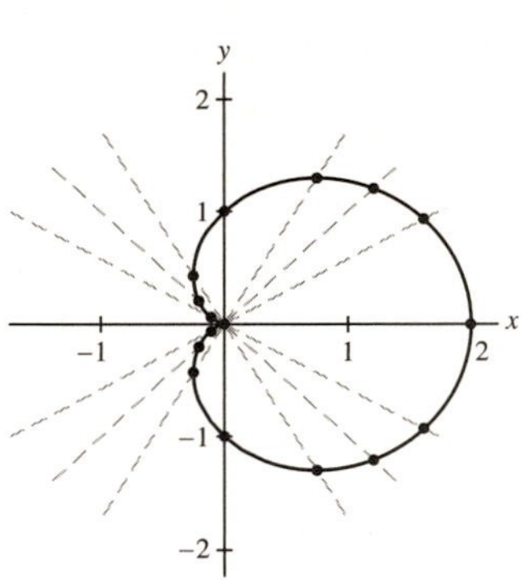

31. Figure 20 displays the graphs of $r = \sin 2\theta$ in rectangular coordinates and in polar coordinates, where it is a "rose with four petals." Identify:

(a) The points in (B) corresponding to points A–I in (A).

(b) The parts of the curve in (B) corresponding to the angle intervals $\left[0, \frac{\pi}{2}\right]$, $\left[\frac{\pi}{2}, \pi\right]$, $\left[\pi, \frac{3\pi}{2}\right]$, and $\left[\frac{3\pi}{2}, 2\pi\right]$.

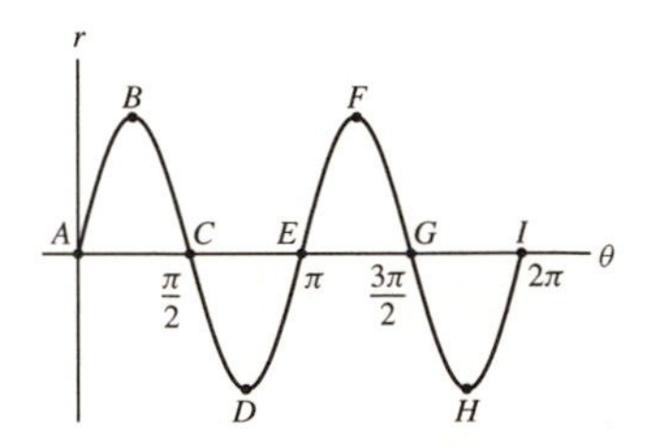

(A) Graph of r as a function of θ, where $r = \sin 2\theta$

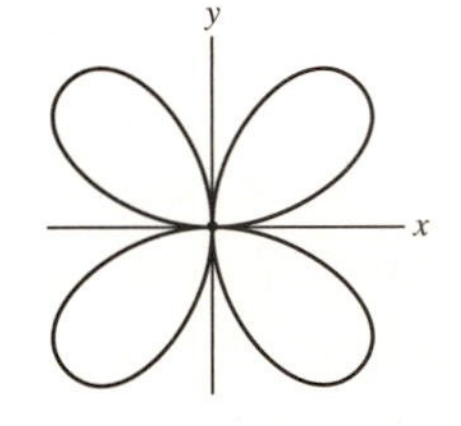

(B) Graph of $r = \sin 2\theta$ in polar coordinates

FIGURE 20

SOLUTION

(a) The graph (A) gives the following polar coordinates of the labeled points:

$$A: \quad \theta = 0, \quad r = 0$$

$$B: \quad \theta = \frac{\pi}{4}, \quad r = \sin\frac{2\pi}{4} = 1$$

$$C: \ \theta = \frac{\pi}{2}, \quad r = 0$$

$$D: \ \theta = \frac{3\pi}{4}, \quad r = \sin \frac{2 \cdot 3\pi}{4} = -1$$

$$E: \ \theta = \pi, \quad r = 0$$

$$F: \ \theta = \frac{5\pi}{4}, \quad r = 1$$

$$G: \ \theta = \frac{3\pi}{2}, \quad r = 0$$

$$H: \ \theta = \frac{7\pi}{4}, \quad r = -1$$

$$I: \ \theta = 2\pi, \quad r = 0.$$

Since the maximal value of $|r|$ is 1, the points with $r = 1$ or $r = -1$ are the furthest points from the origin. The corresponding quadrant is determined by the value of θ and the sign of r. If $r_0 < 0$, the point (r_0, θ_0) is on the ray $\theta = -\theta_0$. These considerations lead to the following identification of the points in the xy plane. Notice that A, C, G, E, and I are the same point.

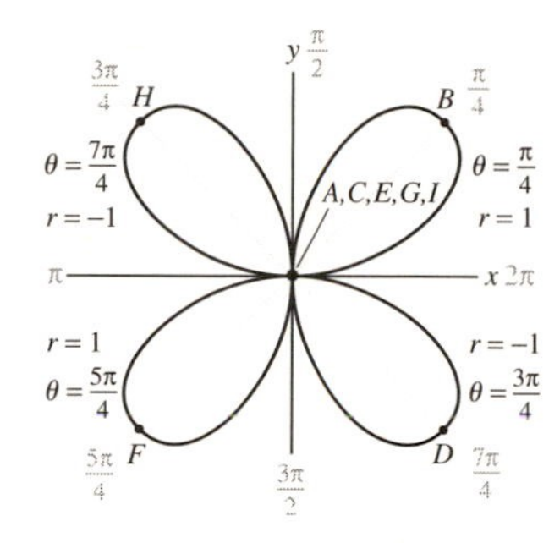

(b) We use the graph (A) to find the sign of $r = \sin 2\theta$: $0 \le \theta \le \frac{\pi}{2} \Rightarrow r \ge 0 \Rightarrow (r, \theta)$ is in the first quadrant. $\frac{\pi}{2} \le \theta \le \pi \Rightarrow r \le 0 \Rightarrow (r, \theta)$ is in the fourth quadrant. $\pi \le \theta \le \frac{3\pi}{2} \Rightarrow r \ge 0 \Rightarrow (r, \theta)$ is in the third quadrant. $\frac{3\pi}{2} \le \theta \le 2\pi \Rightarrow r \le 0 \Rightarrow (r, \theta)$ is in the second quadrant. That is,

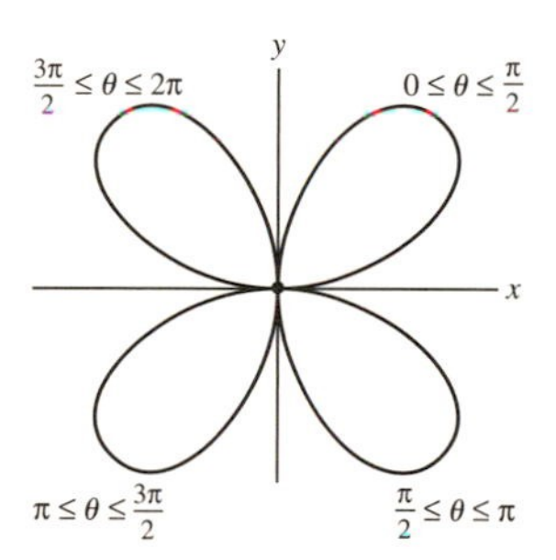

33. CAS Plot the **cissoid** $r = 2 \sin \theta \tan \theta$ and show that its equation in rectangular coordinates is

$$y^2 = \frac{x^3}{2 - x}$$

SOLUTION Using a CAS we obtain the following curve of the cissoid:

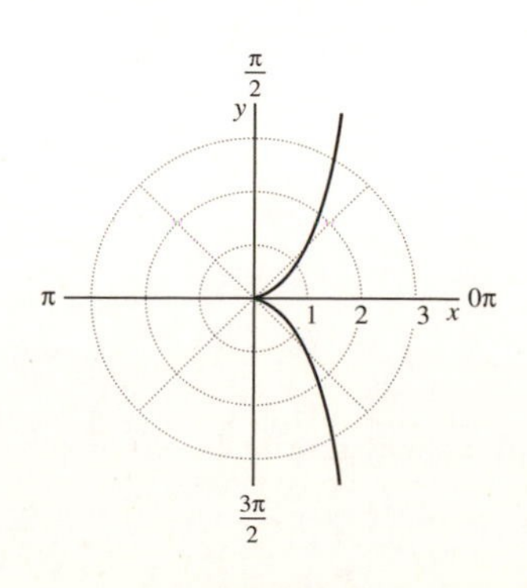

We substitute $\sin\theta = \frac{y}{r}$ and $\tan\theta = \frac{y}{x}$ in $r = 2\sin\theta\tan\theta$ to obtain

$$r = 2\frac{y}{r}\cdot\frac{y}{x}.$$

Multiplying by rx, setting $r^2 = x^2 + y^2$ and simplifying, yields

$$\begin{aligned} r^2 x &= 2y^2 \\ (x^2+y^2)x &= 2y^2 \\ x^3 + y^2 x &= 2y^2 \\ y^2(2-x) &= x^3 \end{aligned}$$

so

$$y^2 = \frac{x^3}{2-x}$$

35. Show that

$$r = a\cos\theta + b\sin\theta$$

is the equation of a circle passing through the origin. Express the radius and center (in rectangular coordinates) in terms of a and b.

SOLUTION We multiply the equation by r and then make the substitution $x = r\cos\theta$, $y = r\sin\theta$, and $r^2 = x^2 + y^2$. This gives

$$\begin{aligned} r^2 &= ar\cos\theta + br\sin\theta \\ x^2 + y^2 &= ax + by \end{aligned}$$

Transferring sides and completing the square yields

$$\begin{aligned} x^2 - ax + y^2 - by &= 0 \\ \left(x^2 - 2\cdot\frac{a}{2}x + \left(\frac{a}{2}\right)^2\right) + \left(y^2 - 2\cdot\frac{b}{2}y + \left(\frac{b}{2}\right)^2\right) &= \left(\frac{a}{2}\right)^2 + \left(\frac{b}{2}\right)^2 \\ \left(x - \frac{a}{2}\right)^2 + \left(y - \frac{b}{2}\right)^2 &= \frac{a^2+b^2}{4} \end{aligned}$$

This is the equation of the circle with radius $\frac{\sqrt{a^2+b^2}}{2}$ centered at the point $\left(\frac{a}{2}, \frac{b}{2}\right)$. By plugging in $x = 0$ and $y = 0$ it is clear that the circle passes through the origin.

37. Use the identity $\cos 2\theta = \cos^2\theta - \sin^2\theta$ to find a polar equation of the hyperbola $x^2 - y^2 = 1$.

SOLUTION We substitute $x = r\cos\theta$, $y = r\sin\theta$ in $x^2 - y^2 = 1$ to obtain

$$\begin{aligned} r^2\cos^2\theta - r^2\sin^2\theta &= 1 \\ r^2(\cos^2\theta - \sin^2\theta) &= 1 \end{aligned}$$

Using the identity $\cos 2\theta = \cos^2\theta - \sin^2\theta$ we obtain the following equation of the hyperbola:

$$r^2\cos 2\theta = 1 \quad \text{or} \quad r^2 = \sec 2\theta.$$

39. Show that $\cos 3\theta = \cos^3\theta - 3\cos\theta\sin^2\theta$ and use this identity to find an equation in rectangular coordinates for the curve $r = \cos 3\theta$.

SOLUTION We use the identities $\cos(\alpha+\beta) = \cos\alpha\cos\beta - \sin\alpha\sin\beta$, $\cos 2\alpha = \cos^2\alpha - \sin^2\alpha$, and $\sin 2\alpha = 2\sin\alpha\cos\alpha$ to write

$$\begin{aligned} \cos 3\theta = \cos(2\theta + \theta) &= \cos 2\theta\cos\theta - \sin 2\theta\sin\theta \\ &= (\cos^2\theta - \sin^2\theta)\cos\theta - 2\sin\theta\cos\theta\sin\theta \\ &= \cos^3\theta - \sin^2\theta\cos\theta - 2\sin^2\theta\cos\theta \\ &= \cos^3\theta - 3\sin^2\theta\cos\theta \end{aligned}$$

Using this identity we may rewrite the equation $r = \cos 3\theta$ as follows:

$$r = \cos^3\theta - 3\sin^2\theta\cos\theta \qquad (1)$$

Since $x = r\cos\theta$ and $y = r\sin\theta$, we have $\cos\theta = \frac{x}{r}$ and $\sin\theta = \frac{y}{r}$. Substituting into (1) gives:

$$r = \left(\frac{x}{r}\right)^3 - 3\left(\frac{y}{r}\right)^2\left(\frac{x}{r}\right)$$

$$r = \frac{x^3}{r^3} - \frac{3y^2x}{r^3}$$

We now multiply by r^3 and make the substitution $r^2 = x^2 + y^2$ to obtain the following equation for the curve:

$$r^4 = x^3 - 3y^2x$$

$$\left(x^2 + y^2\right)^2 = x^3 - 3y^2x$$

In Exercises 41–44, find an equation in polar coordinates of the line $\mathcal{L}$ with the given description.

41. The point on $\mathcal{L}$ closest to the origin has polar coordinates $\left(2, \frac{\pi}{9}\right)$.

SOLUTION In Example 5, it is shown that the polar equation of the line where (r, α) is the point on the line closest to the origin is $r = d\sec(\theta - \alpha)$. Setting $(d, \alpha) = \left(2, \frac{\pi}{9}\right)$ we obtain the following equation of the line:

$$r = 2\sec\left(\theta - \frac{\pi}{9}\right).$$

43. $\mathcal{L}$ is tangent to the circle $r = 2\sqrt{10}$ at the point with rectangular coordinates $(-2, -6)$.

SOLUTION

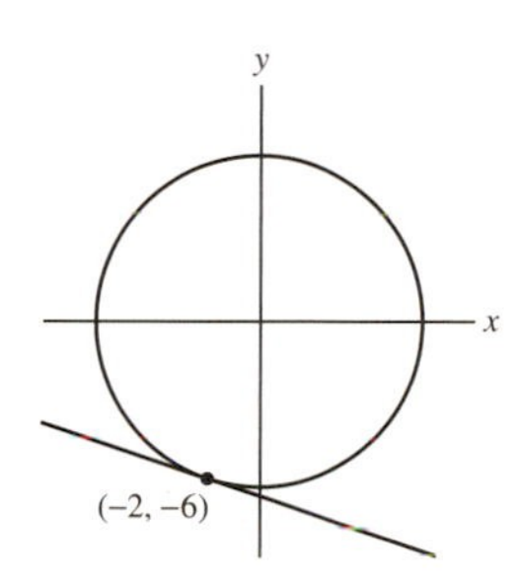

Since $\mathcal{L}$ is tangent to the circle at the point $(-2, -6)$, this is the point on $\mathcal{L}$ closest to the center of the circle which is at the origin. Therefore, we may use the polar coordinates (d, α) of this point in the equation of the line:

$$r = d\sec(\theta - \alpha) \tag{1}$$

We thus must convert the coordinates $(-2, -6)$ to polar coordinates. This point is in the third quadrant so $\pi < \alpha < \frac{3\pi}{2}$. We get

$$d = \sqrt{(-2)^2 + (-6)^2} = \sqrt{40} = 2\sqrt{10}$$

$$\alpha = \tan^{-1}\left(\frac{-6}{-2}\right) = \tan^{-1} 3 \approx \pi + 1.25 \approx 4.39$$

Substituting in (1) yields the following equation of the line:

$$r = 2\sqrt{10}\sec(\theta - 4.39).$$

45. Show that every line that does not pass through the origin has a polar equation of the form

$$r = \frac{b}{\sin\theta - a\cos\theta}$$

where $b \neq 0$.

SOLUTION Write the equation of the line in rectangular coordinates as $y = ax + b$. Since the line does not pass through the origin, we have $b \neq 0$. Substitute for y and x to convert to polar coordinates, and simplify:

$$y = ax + b$$

$$r\sin\theta = ar\cos\theta + b$$

$$r(\sin\theta - a\cos\theta) = b$$

$$r = \frac{b}{\sin\theta - a\cos\theta}$$

47. For $a > 0$, a **lemniscate curve** is the set of points P such that the product of the distances from P to $(a, 0)$ and $(-a, 0)$ is a^2. Show that the equation of the lemniscate is

$$(x^2 + y^2)^2 = 2a^2(x^2 - y^2)$$

Then find the equation in polar coordinates. To obtain the simplest form of the equation, use the identity $\cos 2\theta = \cos^2\theta - \sin^2\theta$. Plot the lemniscate for $a = 2$ if you have a computer algebra system.

SOLUTION We compute the distances d_1 and d_2 of $P(x, y)$ from the points $(a, 0)$ and $(-a, 0)$ respectively. We obtain:

$$d_1 = \sqrt{(x-a)^2 + (y-0)^2} = \sqrt{(x-a)^2 + y^2}$$
$$d_2 = \sqrt{(x+a)^2 + (y-0)^2} = \sqrt{(x+a)^2 + y^2}$$

For the points $P(x, y)$ on the lemniscate we have $d_1 d_2 = a^2$. That is,

$$\begin{aligned} a^2 &= \sqrt{(x-a)^2 + y^2}\sqrt{(x+a)^2 + y^2} = \sqrt{\left[(x-a)^2 + y^2\right]\left[(x+a)^2 + y^2\right]} \\ &= \sqrt{(x-a)^2(x+a)^2 + y^2(x-a)^2 + y^2(x+a)^2 + y^4} \\ &= \sqrt{(x^2-a^2)^2 + y^2\left[(x-a)^2 + (x+a)^2\right] + y^4} \\ &= \sqrt{x^4 - 2a^2x^2 + a^4 + y^2\left(x^2 - 2xa + a^2 + x^2 + 2xa + a^2\right) + y^4} \\ &= \sqrt{x^4 - 2a^2x^2 + a^4 + 2y^2x^2 + 2y^2a^2 + y^4} \\ &= \sqrt{x^4 + 2x^2y^2 + y^4 + 2a^2(y^2 - x^2) + a^4} \\ &= \sqrt{(x^2 + y^2)^2 + 2a^2(y^2 - x^2) + a^4}. \end{aligned}$$

Squaring both sides and simplifying yields

$$a^4 = (x^2 + y^2)^2 + 2a^2(y^2 - x^2) + a^4$$
$$0 = (x^2 + y^2)^2 + 2a^2(y^2 - x^2)$$

so

$$(x^2 + y^2)^2 = 2a^2(x^2 - y^2)$$

We now find the equation in polar coordinates. We substitute $x = r\cos\theta$, $y = r\sin\theta$ and $x^2 + y^2 = r^2$ into the equation of the lemniscate. This gives

$$(r^2)^2 = 2a^2(r^2\cos^2\theta - r^2\sin^2\theta) = 2a^2r^2(\cos^2\theta - \sin^2\theta) = 2a^2r^2\cos 2\theta$$
$$r^4 = 2a^2r^2\cos 2\theta$$

$r = 0$ is a solution, hence the origin is on the curve. For $r \neq 0$ we divide the equation by r^2 to obtain $r^2 = 2a^2\cos 2\theta$. This curve also includes the origin ($r = 0$ is obtained for $\theta = \frac{\pi}{4}$ for example), hence this is the polar equation of the lemniscate. Setting $a = 2$ we get $r^2 = 8\cos 2\theta$.

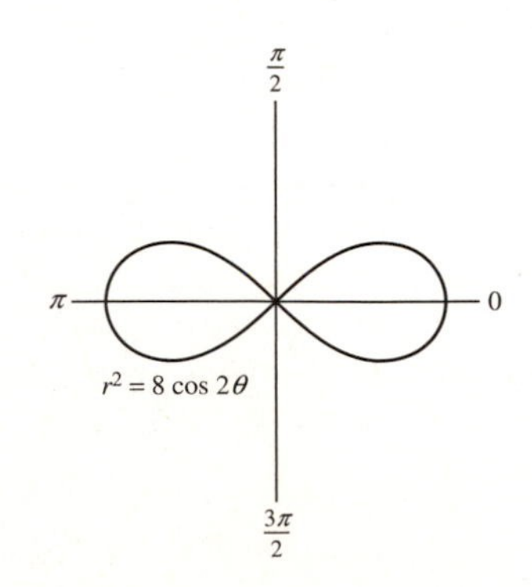

49. The Derivative in Polar Coordinates Show that a polar curve $r = f(\theta)$ has parametric equations

$$x = f(\theta)\cos\theta, \qquad y = f(\theta)\sin\theta$$

Then apply Theorem 2 of Section 11.1 to prove

$$\frac{dy}{dx} = \frac{f(\theta)\cos\theta + f'(\theta)\sin\theta}{-f(\theta)\sin\theta + f'(\theta)\cos\theta} \qquad \boxed{2}$$

where $f'(\theta) = df/d\theta$.

SOLUTION Multiplying both sides of the given equation by $\cos\theta$ yields $r\cos\theta = f(\theta)\cos\theta$; multiplying both sides by $\sin\theta$ yields $r\sin\theta = f(\theta)\sin\theta$. The left-hand sides of these two equations are the x and y coordinates in rectangular coordinates, so for any θ we have $x = f(\theta)\cos\theta$ and $y = f(\theta)\sin\theta$, showing that the parametric equations are as claimed. Now, by the formula for the derivative we have

$$\frac{dy}{dx} = \frac{y'(\theta)}{x'(\theta)} \qquad (1)$$

We differentiate the functions $x = f(\theta)\cos\theta$ and $y = f(\theta)\sin\theta$ using the Product Rule for differentiation. This gives

$$y'(\theta) = f'(\theta)\sin\theta + f(\theta)\cos\theta$$

$$x'(\theta) = f'(\theta)\cos\theta - f(\theta)\sin\theta$$

Substituting in (1) gives

$$\frac{dy}{dx} = \frac{f'(\theta)\sin\theta + f(\theta)\cos\theta}{f'(\theta)\cos\theta - f(\theta)\sin\theta} = \frac{f(\theta)\cos\theta + f'(\theta)\sin\theta}{-f(\theta)\sin\theta + f'(\theta)\cos\theta}.$$

51. Use Eq. (2) to find the slope of the tangent line to $r = \theta$ at $\theta = \frac{\pi}{2}$ and $\theta = \pi$.

SOLUTION In the given curve we have $r = f(\theta) = \theta$. Using Eq. (2) we obtain the following derivative, which is the slope of the tangent line at (r, θ).

$$\frac{dy}{dx} = \frac{f(\theta)\cos\theta + f'(\theta)\sin\theta}{-f(\theta)\sin\theta + f'(\theta)\cos\theta} = \frac{\theta\cos\theta + 1\cdot\sin\theta}{-\theta\sin\theta + 1\cdot\cos\theta} \qquad (1)$$

The slope, m, of the tangent line at $\theta = \frac{\pi}{2}$ and $\theta = \pi$ is obtained by substituting these values in (1). We get ($\theta = \frac{\pi}{2}$):

$$m = \frac{\frac{\pi}{2}\cos\frac{\pi}{2} + \sin\frac{\pi}{2}}{-\frac{\pi}{2}\sin\frac{\pi}{2} + \cos\frac{\pi}{2}} = \frac{\frac{\pi}{2}\cdot 0 + 1}{-\frac{\pi}{2}\cdot 1 + 0} = \frac{1}{-\frac{\pi}{2}} = -\frac{2}{\pi}.$$

($\theta = \pi$):

$$m = \frac{\pi\cos\pi + \sin\pi}{-\pi\sin\pi + \cos\pi} = \frac{-\pi}{-1} = \pi.$$

53. Find the polar coordinates of the points on the lemniscate $r^2 = \cos 2t$ in Figure 23 where the tangent line is horizontal.

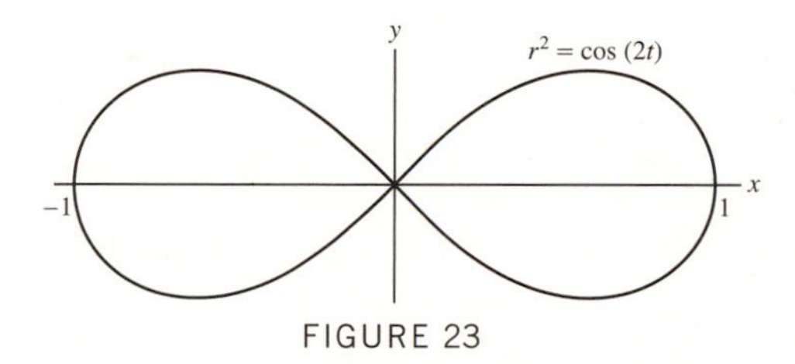

FIGURE 23

SOLUTION This curve is defined for $-\frac{\pi}{2} \le 2t \le \frac{\pi}{2}$ (where $\cos 2t \ge 0$), so for $-\frac{\pi}{4} \le t \le \frac{\pi}{4}$. For each θ in that range, there are two values of r satisfying the equation ($\pm\sqrt{\cos 2t}$). By symmetry, we need only calculate the coordinates of the points corresponding to the positive square root (i.e. to the right of the y axis). Then the equation becomes $r = \sqrt{\cos 2t}$. Now, by Eq. (2), with $f(t) = \sqrt{\cos(2t)}$ and $f'(t) = -\sin(2t)(\cos(2t))^{-1/2}$, we have

$$\frac{dy}{dx} = \frac{f(t)\cos t + f'(t)\sin t}{-f(t)\sin t + f'(t)\cos t} = \frac{\cos t\sqrt{\cos(2t)} - \sin(2t)\sin t(\cos(2t))^{-1/2}}{-\sin t\sqrt{\cos(2t)} - \sin(2t)\cos t(\cos(2t))^{-1/2}}$$

The tangent line is horizontal when this derivative is zero, which occurs when the numerator of the fraction is zero and the denominator is not. Multiply top and bottom of the fraction by $\sqrt{\cos(2t)}$, and use the identities $\cos 2t = \cos^2 t - \sin^2 t$, $\sin 2t = 2\sin t \cos t$ to get

$$-\frac{\cos t \cos 2t - \sin t \sin 2t}{\sin t \cos 2t + \cos t \sin 2t} = -\frac{\cos t(\cos^2 t - 3\sin^2 t)}{\sin t \cos 2t + \cos t \sin 2t}$$

The numerator is zero when $\cos t = 0$, so when $t = \frac{\pi}{2}$ or $t = \frac{3\pi}{2}$, or when $\tan t = \pm\frac{1}{\sqrt{3}}$, so when $t = \pm\frac{\pi}{6}$ or $t = \pm\frac{5\pi}{6}$. Of these possibilities, only $t = \pm\frac{\pi}{6}$ lie in the range $-\frac{\pi}{4} \le t \le \frac{\pi}{4}$. Note that the denominator is nonzero for $t = \pm\frac{\pi}{6}$, so these are the two values of t for which the tangent line is horizontal. The corresponding values of r are solutions to

$$r^2 = \cos\left(2 \cdot \frac{\pi}{6}\right) = \cos\left(\frac{\pi}{3}\right) = \frac{1}{2}$$

$$r^2 = \cos\left(2 \cdot \frac{-\pi}{6}\right) = \cos\left(-\frac{\pi}{3}\right) = \frac{1}{2}$$

Finally, the four points are $(r, t) =$

$$\left(\frac{1}{\sqrt{2}}, \frac{\pi}{6}\right), \quad \left(-\frac{1}{\sqrt{2}}, \frac{\pi}{6}\right), \quad \left(\frac{1}{\sqrt{2}}, -\frac{pi}{6}\right), \quad \left(-\frac{1}{\sqrt{2}}, -\frac{\pi}{6}\right)$$

If desired, we can change the second and fourth points by adding π to the angle and making r positive, to get

$$\left(\frac{1}{\sqrt{2}}, \frac{\pi}{6}\right), \quad \left(\frac{1}{\sqrt{2}}, \frac{7\pi}{6}\right), \quad \left(\frac{1}{\sqrt{2}}, -\frac{pi}{6}\right), \quad \left(\frac{1}{\sqrt{2}}, \frac{5\pi}{6}\right)$$

55. Use Eq. (2) to show that for $r = \sin\theta + \cos\theta$,

$$\frac{dy}{dx} = \frac{\cos 2\theta + \sin 2\theta}{\cos 2\theta - \sin 2\theta}$$

Then calculate the slopes of the tangent lines at points A, B, C in Figure 19.

SOLUTION In Exercise 49 we proved that for a polar curve $r = f(\theta)$ the following formula holds:

$$\frac{dy}{dx} = \frac{f(\theta)\cos\theta + f'(\theta)\sin\theta}{-f(\theta)\sin\theta + f'(\theta)\cos\theta} \tag{1}$$

For the given circle we have $r = f(\theta) = \sin\theta + \cos\theta$, hence $f'(\theta) = \cos\theta - \sin\theta$. Substituting in (1) we have

$$\frac{dy}{dx} = \frac{(\sin\theta + \cos\theta)\cos\theta + (\cos\theta - \sin\theta)\sin\theta}{-(\sin\theta + \cos\theta)\sin\theta + (\cos\theta - \sin\theta)\cos\theta} = \frac{\sin\theta\cos\theta + \cos^2\theta + \cos\theta\sin\theta - \sin^2\theta}{-\sin^2\theta - \cos\theta\sin\theta + \cos^2\theta - \sin\theta\cos\theta}$$

$$= \frac{\cos^2\theta - \sin^2\theta + 2\sin\theta\cos\theta}{\cos^2\theta - \sin^2\theta - 2\sin\theta\cos\theta}$$

We use the identities $\cos^2\theta - \sin^2\theta = \cos 2\theta$ and $2\sin\theta\cos\theta = \sin 2\theta$ to obtain

$$\frac{dy}{dx} = \frac{\cos 2\theta + \sin 2\theta}{\cos 2\theta - \sin 2\theta} \tag{2}$$

The derivative $\frac{dy}{dx}$ is the slope of the tangent line at (r, θ). The slopes of the tangent lines at the points with polar coordinates $A = \left(1, \frac{\pi}{2}\right)$ $B = \left(0, \frac{3\pi}{4}\right)$ $C = (1, 0)$ are computed by substituting the values of θ in (2). This gives

$$\left.\frac{dy}{dx}\right|_A = \frac{\cos\left(2 \cdot \frac{\pi}{2}\right) + \sin\left(2 \cdot \frac{\pi}{2}\right)}{\cos\left(2 \cdot \frac{\pi}{2}\right) - \sin\left(2 \cdot \frac{\pi}{2}\right)} = \frac{\cos\pi + \sin\pi}{\cos\pi - \sin\pi} = \frac{-1+0}{-1-0} = 1$$

$$\left.\frac{dy}{dx}\right|_B = \frac{\cos\left(2 \cdot \frac{3\pi}{4}\right) + \sin\left(2 \cdot \frac{3\pi}{4}\right)}{\cos\left(2 \cdot \frac{3\pi}{4}\right) - \sin\left(2 \cdot \frac{3\pi}{4}\right)} = \frac{\cos\frac{3\pi}{2} + \sin\frac{3\pi}{2}}{\cos\frac{3\pi}{2} - \sin\frac{3\pi}{2}} = \frac{0-1}{0+1} = -1$$

$$\left.\frac{dy}{dx}\right|_C = \frac{\cos(2 \cdot 0) + \sin(2 \cdot 0)}{\cos(2 \cdot 0) - \sin(2 \cdot 0)} = \frac{\cos 0 + \sin 0}{\cos 0 - \sin 0} = \frac{1+0}{1-0} = 1$$

Further Insights and Challenges

57. GU Use a graphing utility to convince yourself that the polar equations $r = f_1(\theta) = 2\cos\theta - 1$ and $r = f_2(\theta) = 2\cos\theta + 1$ have the same graph. Then explain why. *Hint:* Show that the points $(f_1(\theta+\pi), \theta+\pi)$ and $(f_2(\theta), \theta)$ coincide.

SOLUTION The graphs of $r = 2\cos\theta - 1$ and $r = 2\cos\theta + 1$ in the xy-plane coincide as shown in the graph obtained using a CAS.

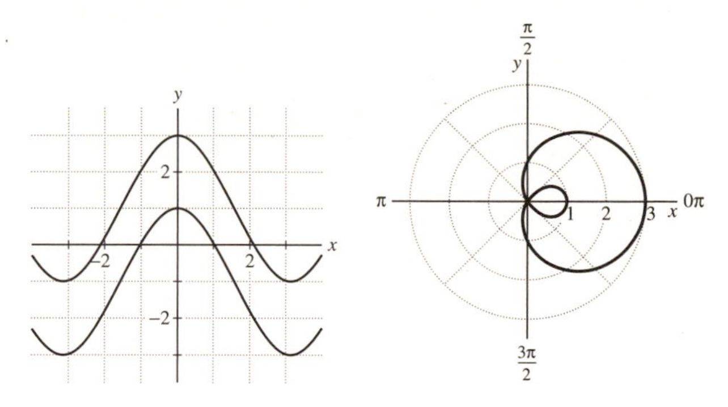

Recall that (r, θ) and $(-r, \theta+\pi)$ represent the same point. Replacing θ by $\theta+\pi$ and r by $(-r)$ in $r = 2\cos\theta - 1$ we obtain

$$-r = 2\cos(\theta+\pi) - 1$$
$$-r = -2\cos\theta - 1$$
$$r = 2\cos\theta + 1$$

Thus, the two equations define the same graph. (One could also convert both equations to rectangular coordinates and note that they come out identical.)

11.4 Area and Arc Length in Polar Coordinates

Preliminary Questions

1. Polar coordinates are suited to finding the area (choose one):

(a) Under a curve between $x = a$ and $x = b$.

(b) Bounded by a curve and two rays through the origin.

SOLUTION Polar coordinates are best suited to finding the area bounded by a curve and two rays through the origin. The formula for the area in polar coordinates gives the area of this region.

2. Is the formula for area in polar coordinates valid if $f(\theta)$ takes negative values?

SOLUTION The formula for the area

$$\frac{1}{2}\int_{\alpha}^{\beta} f(\theta)^2\, d\theta$$

always gives the actual (positive) area, even if $f(\theta)$ takes on negative values.

3. The horizontal line $y = 1$ has polar equation $r = \csc\theta$. Which area is represented by the integral $\frac{1}{2}\int_{\pi/6}^{\pi/2} \csc^2\theta\, d\theta$ (Figure 12)?

(a) $\square ABCD$ **(b)** $\triangle ABC$ **(c)** $\triangle ACD$

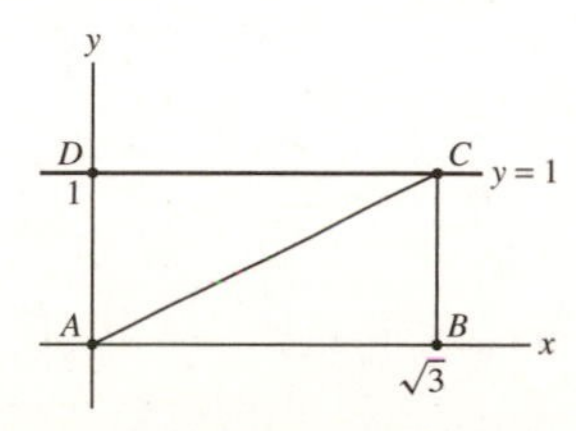

FIGURE 12

SOLUTION This integral represents an area taken from $\theta = \pi/6$ to $\theta = \pi/2$, which can only be the triangle $\triangle ACD$, as seen in part (c).

Exercises

1. Sketch the area bounded by the circle $r = 5$ and the rays $\theta = \frac{\pi}{2}$ and $\theta = \pi$, and compute its area as an integral in polar coordinates.

SOLUTION The region bounded by the circle $r = 5$ and the rays $\theta = \frac{\pi}{2}$ and $\theta = \pi$ is the shaded region in the figure. The area of the region is given by the following integral:

$$\frac{1}{2}\int_{\pi/2}^{\pi} r^2\,d\theta = \frac{1}{2}\int_{\pi/2}^{\pi} 5^2\,d\theta = \frac{25}{2}\left(\pi - \frac{\pi}{2}\right) = \frac{25\pi}{4}$$

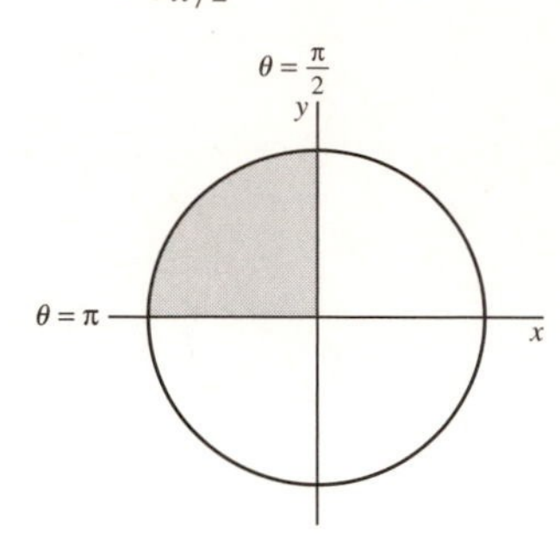

3. Calculate the area of the circle $r = 4\sin\theta$ as an integral in polar coordinates (see Figure 4). Be careful to choose the correct limits of integration.

SOLUTION The equation $r = 4\sin\theta$ defines a circle of radius 2 tangent to the x-axis at the origin as shown in the figure:

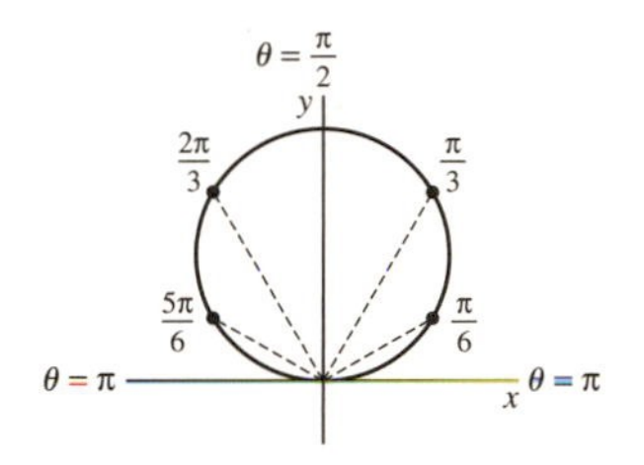

The circle is traced as θ varies from 0 to π. We use the area in polar coordinates and the identity

$$\sin^2\theta = \frac{1}{2}(1 - \cos 2\theta)$$

to obtain the following area:

$$A = \frac{1}{2}\int_0^{\pi} r^2\,d\theta = \frac{1}{2}\int_0^{\pi}(4\sin\theta)^2\,d\theta = 8\int_0^{\pi}\sin^2\theta\,d\theta = 4\int_0^{\pi}(1 - \cos 2\theta)\,d\theta = 4\left[\theta - \frac{\sin 2\theta}{2}\right]_0^{\pi}$$
$$= 4\left(\left(\pi - \frac{\sin 2\pi}{2}\right) - 0\right) = 4\pi.$$

5. Find the area of the shaded region in Figure 14. Note that θ varies from 0 to $\frac{\pi}{2}$.

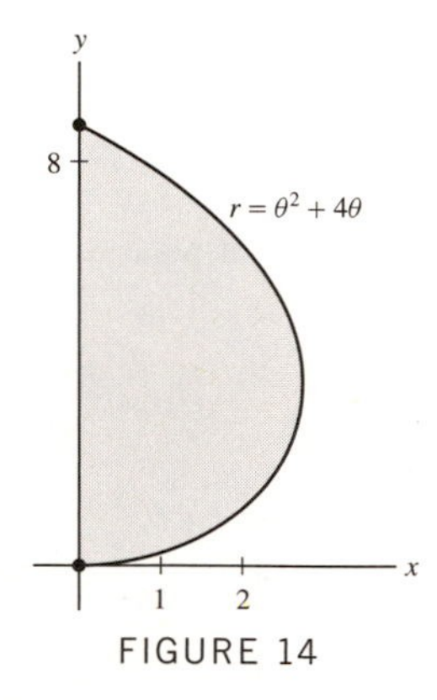

FIGURE 14

SOLUTION Since θ varies from 0 to $\frac{\pi}{2}$, the area is

$$\frac{1}{2}\int_0^{\pi/2} r^2\,d\theta = \frac{1}{2}\int_0^{\pi/2}(\theta^2 + 4\theta)^2\,d\theta = \frac{1}{2}\int_0^{\pi/2}\theta^4 + 8\theta^3 + 16\theta^2\,d\theta$$
$$= \frac{1}{2}\left(\frac{1}{5}\theta^5 + 2\theta^4 + \frac{16}{3}\theta^3\right)\Bigg|_0^{\pi/2} = \frac{\pi^5}{320} + \frac{\pi^4}{16} + \frac{\pi^2}{3}$$

7. Find the total area enclosed by the cardioid in Figure 16.

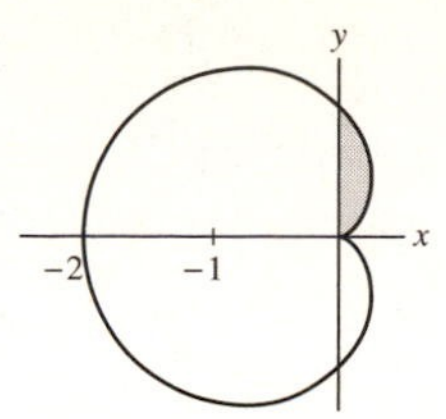

FIGURE 16 The cardioid $r = 1 - \cos\theta$.

SOLUTION We graph $r = 1 - \cos\theta$ in r and θ (cartesian, not polar, this time):

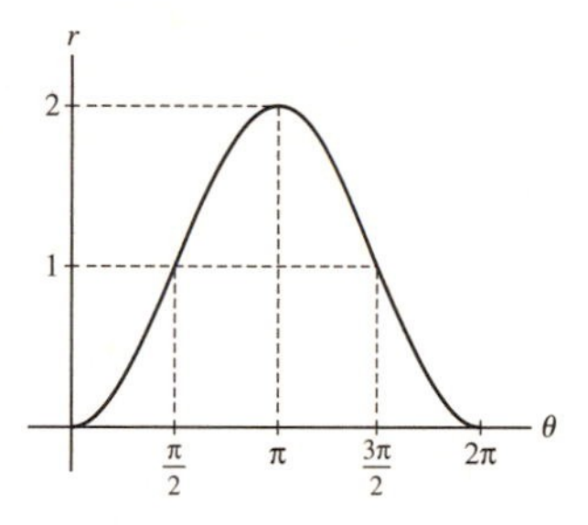

We see that as θ varies from 0 to π, the radius r increases from 0 to 2, so we get the upper half of the cardioid (the lower half is obtained as θ varies from π to 2π and consequently r decreases from 2 to 0). Since the cardioid is symmetric with respect to the x-axis we may compute the upper area and double the result. Using

$$\cos^2\theta = \frac{\cos 2\theta + 1}{2}$$

we get

$$A = 2 \cdot \frac{1}{2}\int_0^{\pi} r^2\,d\theta = \int_0^{\pi}(1-\cos\theta)^2\,d\theta = \int_0^{\pi}\left(1 - 2\cos\theta + \cos^2\theta\right)d\theta$$

$$= \int_0^{\pi}\left(1 - 2\cos\theta + \frac{\cos 2\theta + 1}{2}\right)d\theta = \int_0^{\pi}\left(\frac{3}{2} - 2\cos\theta + \frac{1}{2}\cos 2\theta\right)d\theta$$

$$= \frac{3}{2}\theta - 2\sin\theta + \frac{1}{4}\sin 2\theta\Big|_0^{\pi} = \frac{3\pi}{2}$$

The total area enclosed by the cardioid is $A = \frac{3\pi}{2}$.

9. Find the area of one leaf of the "four-petaled rose" $r = \sin 2\theta$ (Figure 17). Then prove that the total area of the rose is equal to one-half the area of the circumscribed circle.

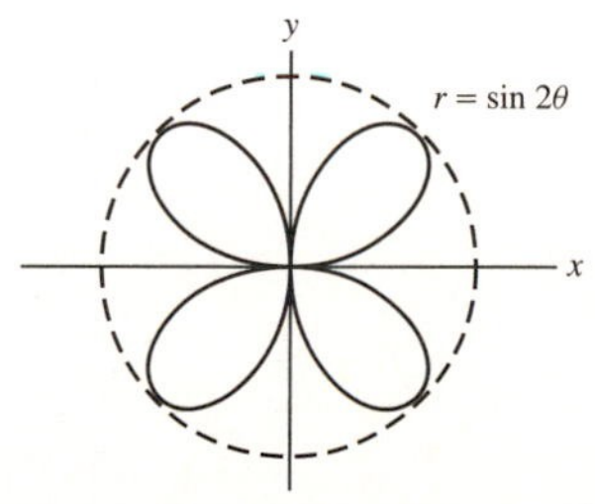

FIGURE 17 Four-petaled rose $r = \sin 2\theta$.

SOLUTION We consider the graph of $r = \sin 2\theta$ in cartesian and in polar coordinates:

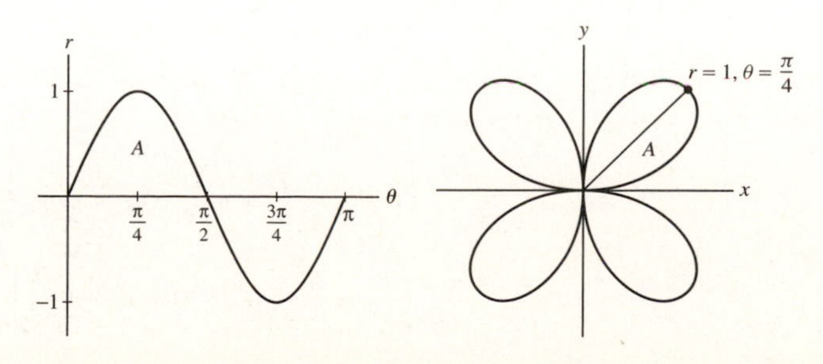

We see that as θ varies from 0 to $\frac{\pi}{4}$ the radius r is increasing from 0 to 1, and when θ varies from $\frac{\pi}{4}$ to $\frac{\pi}{2}$, r is decreasing back to zero. Hence, the leaf in the first quadrant is traced as θ varies from 0 to $\frac{\pi}{2}$. The area of the leaf (the four leaves have equal areas) is thus

$$A = \frac{1}{2}\int_0^{\pi/2} r^2\,d\theta = \frac{1}{2}\int_0^{\pi/2} \sin^2 2\theta\,d\theta.$$

Using the identity

$$\sin^2 2\theta = \frac{1-\cos 4\theta}{2}$$

we get

$$A = \frac{1}{2}\int_0^{\pi/2}\left(\frac{1}{2}-\frac{\cos 4\theta}{2}\right)d\theta = \frac{1}{2}\left(\frac{\theta}{2}-\frac{\sin 4\theta}{8}\right)\Big|_0^{\pi/2} = \frac{1}{2}\left(\left(\frac{\pi}{4}-\frac{\sin 2\pi}{8}\right)-0\right) = \frac{\pi}{8}$$

The area of one leaf is $A = \frac{\pi}{8} \approx 0.39$. It follows that the area of the entire rose is $\frac{\pi}{2}$. Since the "radius" of the rose (the point where $\theta = \frac{\pi}{4}$) is 1, and the circumscribed circle is tangent there, the circumscribed circle has radius 1 and thus area π. Hence the area of the rose is half that of the circumscribed circle.

11. Sketch the spiral $r = \theta$ for $0 \le \theta \le 2\pi$ and find the area bounded by the curve and the first quadrant.

SOLUTION The spiral $r = \theta$ for $0 \le \theta \le 2\pi$ is shown in the following figure in the xy-plane:

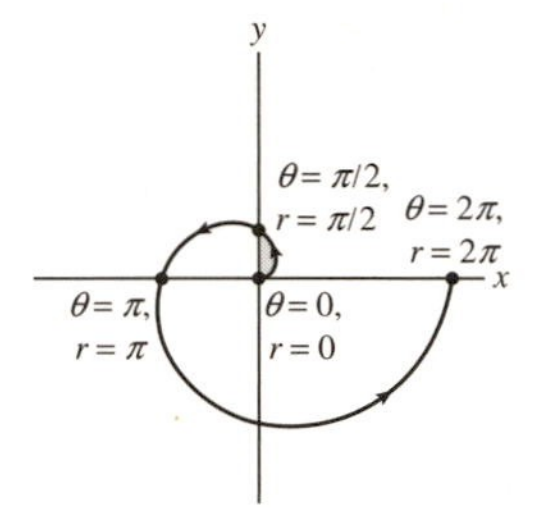

The spiral $r = \theta$

We must compute the area of the shaded region. This region is traced as θ varies from 0 to $\frac{\pi}{2}$. Using the formula for the area in polar coordinates we get

$$A = \frac{1}{2}\int_0^{\pi/2} r^2\,d\theta = \frac{1}{2}\int_0^{\pi/2} \theta^2\,d\theta = \frac{1}{2}\frac{\theta^3}{3}\Big|_0^{\pi/2} = \frac{1}{6}\left(\frac{\pi}{2}\right)^3 = \frac{\pi^3}{48}$$

13. Find the area of region A in Figure 19.

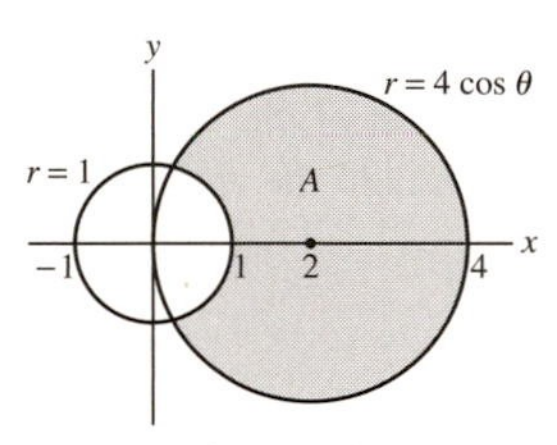

FIGURE 19

SOLUTION We first find the values of θ at the points of intersection of the two circles, by solving the following equation for $-\frac{\pi}{2} \le x \le \frac{\pi}{2}$:

$$4\cos\theta = 1 \Rightarrow \cos\theta = \frac{1}{4} \Rightarrow \theta_1 = \cos^{-1}\left(\frac{1}{4}\right)$$

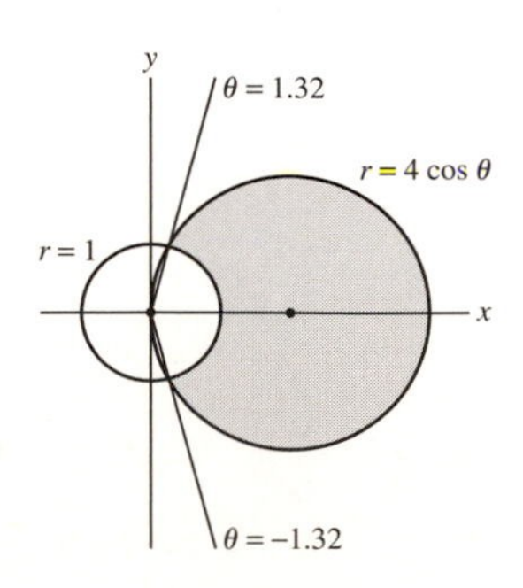

We now compute the area using the formula for the area between two curves:

$$A = \frac{1}{2}\int_{-\theta_1}^{\theta_1} \left((4\cos\theta)^2 - 1^2\right) d\theta = \frac{1}{2}\int_{-\theta_1}^{\theta_1} \left(16\cos^2\theta - 1\right) d\theta$$

Using the identity $\cos^2\theta = \frac{\cos 2\theta + 1}{2}$ we get

$$A = \frac{1}{2}\int_{-\theta_1}^{\theta_1} \left(\frac{16(\cos 2\theta + 1)}{2} - 1\right) d\theta = \frac{1}{2}\int_{-\theta_1}^{\theta_1} (8\cos 2\theta + 7)\, d\theta = \frac{1}{2}(4\sin 2\theta + 7\theta)\Big|_{-\theta_1}^{\theta_1}$$

$$= 4\sin 2\theta_1 + 7\theta_1 = 8\sin\theta_1\cos\theta_1 + 7\theta_1 = 8\sqrt{1-\cos^2\theta_1}\cos\theta_1 + 7\theta_1$$

Using the fact that $\cos\theta_1 = \frac{1}{4}$ we get

$$A = \frac{\sqrt{15}}{2} + 7\cos^{-1}\left(\frac{1}{4}\right) \approx 11.163$$

15. Find the area of the inner loop of the limaçon with polar equation $r = 2\cos\theta - 1$ (Figure 21).

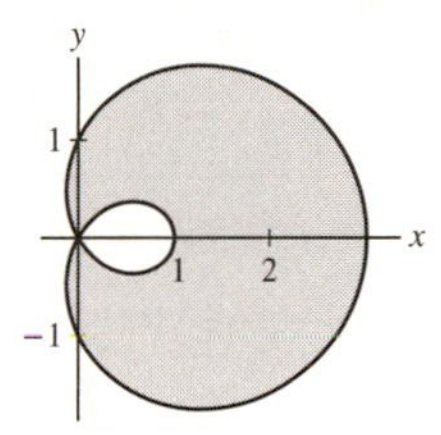

FIGURE 21 The limaçon $r = 2\cos\theta - 1$.

SOLUTION We consider the graph of $r = 2\cos\theta - 1$ in cartesian and in polar, for $-\frac{\pi}{2} \le x \le \frac{\pi}{2}$:

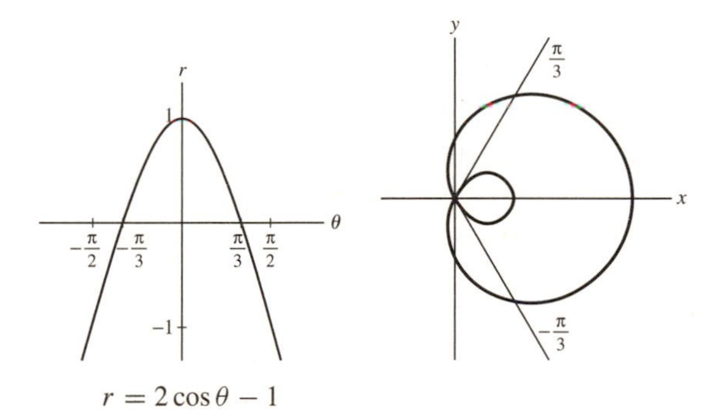

$r = 2\cos\theta - 1$

As θ varies from $-\frac{\pi}{3}$ to 0, r increases from 0 to 1. As θ varies from 0 to $\frac{\pi}{3}$, r decreases from 1 back to 0. Hence, the inner loop of the limaçon is traced as θ varies from $-\frac{\pi}{3}$ to $\frac{\pi}{3}$. The area of the shaded region is thus

$$A = \frac{1}{2}\int_{-\pi/3}^{\pi/3} r^2\, d\theta = \frac{1}{2}\int_{-\pi/3}^{\pi/3} (2\cos\theta - 1)^2\, d\theta = \frac{1}{2}\int_{-\pi/3}^{\pi/3} \left(4\cos^2\theta - 4\cos\theta + 1\right) d\theta$$

$$= \frac{1}{2}\int_{-\pi/3}^{\pi/3} (2(\cos 2\theta + 1) - 4\cos\theta + 1)\, d\theta = \frac{1}{2}\int_{-\pi/3}^{\pi/3} (2\cos 2\theta - 4\cos\theta + 3)\, d\theta$$

$$= \frac{1}{2}(\sin 2\theta - 4\sin\theta + 3\theta)\Big|_{-\pi/3}^{\pi/3} = \frac{1}{2}\left(\left(\sin\frac{2\pi}{3} - 4\sin\frac{\pi}{3} + \pi\right) - \left(\sin\left(-\frac{2\pi}{3}\right) - 4\sin\left(-\frac{\pi}{3}\right) - \pi\right)\right)$$

$$= \frac{\sqrt{3}}{2} - \frac{4\sqrt{3}}{2} + \pi = \pi - \frac{3\sqrt{3}}{2} \approx 0.54$$

17. Find the area of the part of the circle $r = \sin\theta + \cos\theta$ in the fourth quadrant (see Exercise 26 in Section 11.3).

SOLUTION The value of θ corresponding to the point B is the solution of $r = \sin\theta + \cos\theta = 0$ for $-\pi \le \theta \le \pi$.

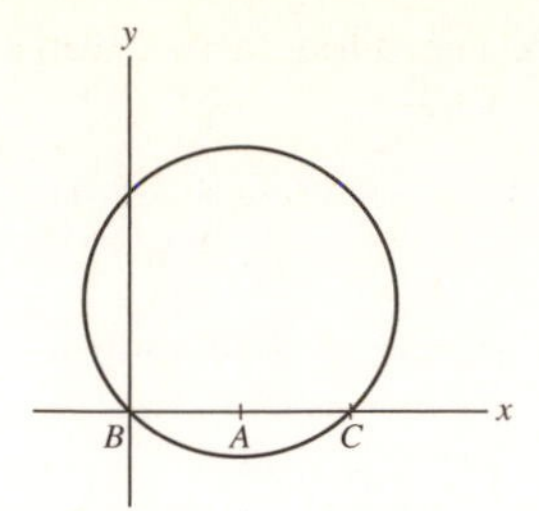

That is,

$$\sin\theta + \cos\theta = 0 \Rightarrow \sin\theta = -\cos\theta \Rightarrow \tan\theta = -1 \Rightarrow \theta = -\frac{\pi}{4}$$

At the point C, we have $\theta = 0$. The part of the circle in the fourth quadrant is traced if θ varies between $-\frac{\pi}{4}$ and 0. This leads to the following area:

$$A = \frac{1}{2}\int_{-\pi/4}^{0} r^2\,d\theta = \frac{1}{2}\int_{-\pi/4}^{0} (\sin\theta + \cos\theta)^2\,d\theta = \frac{1}{2}\int_{-\pi/4}^{0} \left(\sin^2\theta + 2\sin\theta\cos\theta + \cos^2\theta\right)\,d\theta$$

Using the identities $\sin^2\theta + \cos^2\theta = 1$ and $2\sin\theta\cos\theta = \sin 2\theta$ we get:

$$A = \frac{1}{2}\int_{-\pi/4}^{0} (1 + \sin 2\theta)\,d\theta = \frac{1}{2}\left(\theta - \frac{\cos 2\theta}{2}\right)\Big|_{-\pi/4}^{0}$$

$$= \frac{1}{2}\left(\left(0 - \frac{1}{2}\right) - \left(-\frac{\pi}{4} - \frac{\cos\left(\frac{-\pi}{2}\right)}{2}\right)\right) = \frac{1}{2}\left(\frac{\pi}{4} - \frac{1}{2}\right) = \frac{\pi}{8} - \frac{1}{4} \approx 0.14.$$

19. Find the area between the two curves in Figure 22(A).

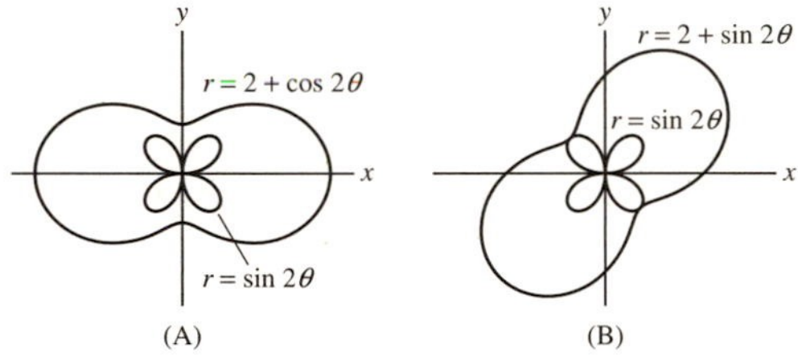

FIGURE 22

SOLUTION We compute the area A between the two curves as the difference between the area A_1 of the region enclosed in the outer curve $r = 2 + \cos 2\theta$ and the area A_2 of the region enclosed in the inner curve $r = \sin 2\theta$. That is,

$$A = A_1 - A_2.$$

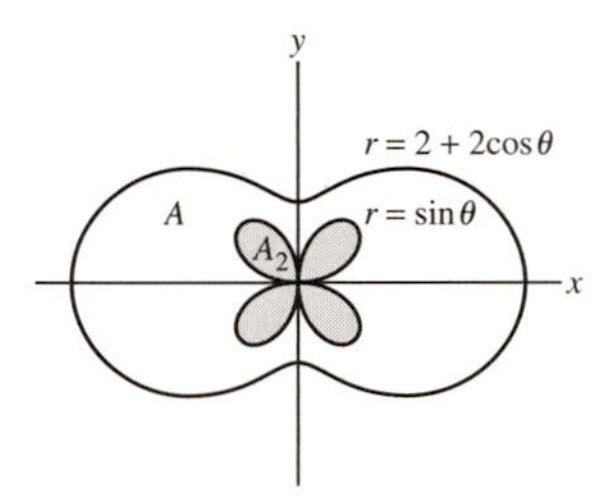

In Exercise 9 we showed that $A_2 = \frac{\pi}{2}$, hence,

$$A = A_1 - \frac{\pi}{2} \qquad (1)$$

We compute the area A_1.

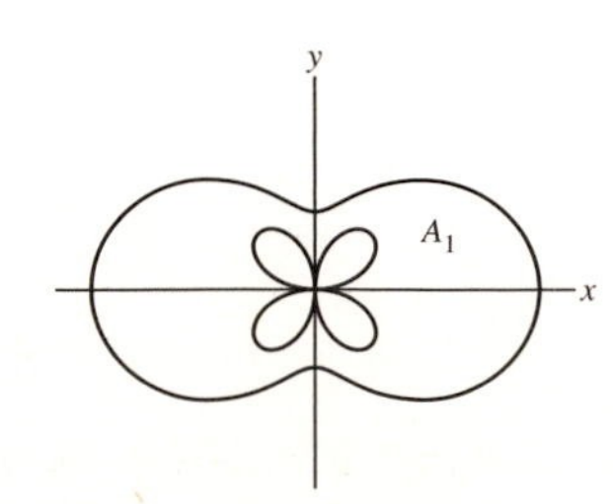

Using symmetry, the area is four times the area enclosed in the first quadrant. That is,

$$A_1 = 4 \cdot \frac{1}{2} \int_0^{\pi/2} r^2\, d\theta = 2 \int_0^{\pi/2} (2 + \cos 2\theta)^2\, d\theta = 2 \int_0^{\pi/2} \left(4 + 4\cos 2\theta + \cos^2 2\theta\right) d\theta$$

Using the identity $\cos^2 2\theta = \frac{1}{2}\cos 4\theta + \frac{1}{2}$ we get

$$A_1 = 2 \int_0^{\pi/2} \left(4 + 4\cos 2\theta + \frac{1}{2}\cos 4\theta + \frac{1}{2}\right) d\theta = 2 \int_0^{\pi/2} \left(\frac{9}{2} + \frac{1}{2}\cos 4\theta + 4\cos 2\theta\right) d\theta$$

$$= 2\left(\frac{9\theta}{2} + \frac{\sin 4\theta}{8} + 2\sin 2\theta\right)\Bigg|_0^{\pi/2} = 2\left(\left(\frac{9\pi}{4} + \frac{\sin 2\pi}{8} + 2\sin \pi\right) - 0\right) = \frac{9\pi}{2} \tag{2}$$

Combining (1) and (2) we obtain

$$A = \frac{9\pi}{2} - \frac{\pi}{2} = 4\pi.$$

21. Find the area inside both curves in Figure 23.

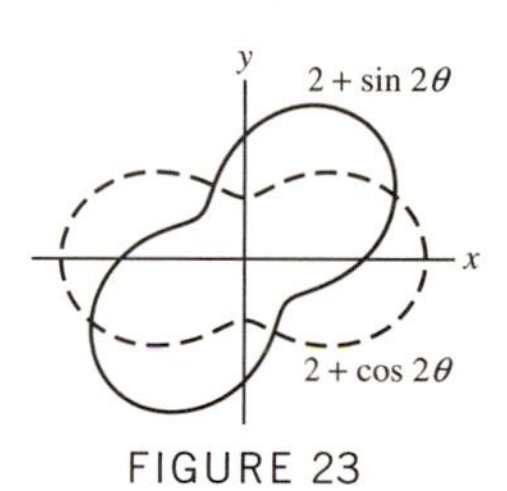

FIGURE 23

SOLUTION The area we need to find is the area of the shaded region in the figure.

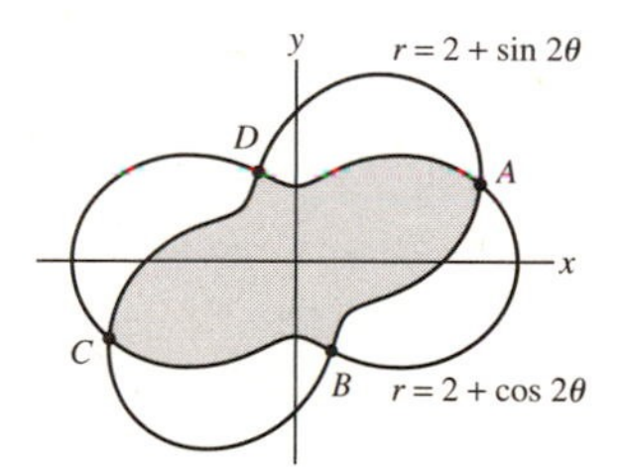

We first find the values of θ at the points of intersection A, B, C, and D of the two curves, by solving the following equation for $-\pi \le \theta \le \pi$:

$$2 + \cos 2\theta = 2 + \sin 2\theta$$

$$\cos 2\theta = \sin 2\theta$$

$$\tan 2\theta = 1 \Rightarrow 2\theta = \frac{\pi}{4} + \pi k \Rightarrow \theta = \frac{\pi}{8} + \frac{\pi k}{2}$$

The solutions for $-\pi \le \theta \le \pi$ are

$$A:\quad \theta = \frac{\pi}{8}.$$

$$B:\quad \theta = -\frac{3\pi}{8}.$$

$$C:\quad \theta = -\frac{7\pi}{8}.$$

$$D:\quad \theta = \frac{5\pi}{8}.$$

Using symmetry, we compute the shaded area in the figure below and multiply it by 4:

$$A = 4 \cdot A_1 = 4 \cdot \frac{1}{2} \cdot \int_{\pi/8}^{5\pi/8} (2 + \cos 2\theta)^2 \, d\theta = 2 \int_{\pi/8}^{5\pi/8} \left(4 + 4\cos 2\theta + \cos^2 2\theta\right) d\theta$$

$$= 2 \int_{\pi/8}^{5\pi/8} \left(4 + 4\cos 2\theta + \frac{1 + \cos 4\theta}{2}\right) d\theta = \int_{\pi/8}^{5\pi/8} (9 + 8\cos 2\theta + \cos 4\theta) \, d\theta$$

$$= 9\theta + 4\sin 2\theta + \frac{\sin 4\theta}{4}\Big|_{\pi/8}^{5\pi/8} = 9\left(\frac{5\pi}{8} - \frac{\pi}{8}\right) + 4\left(\sin\frac{5\pi}{4} - \sin\frac{\pi}{4}\right) + \frac{1}{4}\left(\sin\frac{5\pi}{2} - \sin\frac{\pi}{2}\right) = \frac{9\pi}{2} - 4\sqrt{2}$$

23. Calculate the total length of the circle $r = 4\sin\theta$ as an integral in polar coordinates.

SOLUTION We use the formula for the arc length:

$$S = \int_{\alpha}^{\beta} \sqrt{f(\theta)^2 + f'(\theta)^2} \, d\theta \tag{1}$$

In this case, $f(\theta) = 4\sin\theta$ and $f'(\theta) = 4\cos\theta$, hence

$$\sqrt{f(\theta)^2 + f'(\theta)^2} = \sqrt{(4\sin\theta)^2 + (4\cos\theta)^2} = \sqrt{16} = 4$$

The circle is traced as θ is varied from 0 to π. Substituting $\alpha = 0$, $\beta = \pi$ in (1) yields $S = \int_0^{\pi} 4 \, d\theta = 4\pi$.

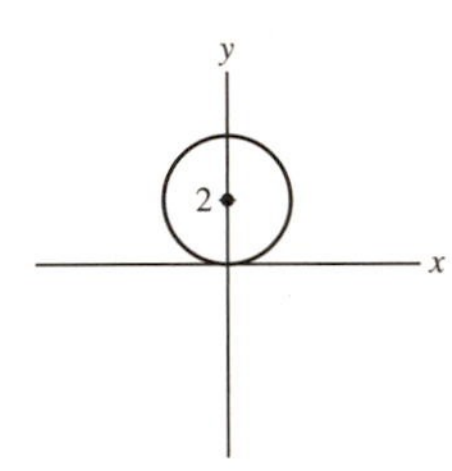

The circle $r = 4\sin\theta$

In Exercises 25–30, compute the length of the polar curve.

25. The length of $r = \theta^2$ for $0 \le \theta \le \pi$

SOLUTION We use the formula for the arc length. In this case $f(\theta) = \theta^2$, $f'(\theta) = 2\theta$, so we obtain

$$S = \int_0^{\pi} \sqrt{\left(\theta^2\right)^2 + (2\theta)^2} \, d\theta = \int_0^{\pi} \sqrt{\theta^4 + 4\theta^2} \, d\theta = \int_0^{\pi} \theta\sqrt{\theta^2 + 4} \, d\theta$$

We compute the integral using the substitution $u = \theta^2 + 4$, $du = 2\theta \, d\theta$. This gives

$$S = \frac{1}{2}\int_4^{\pi^2+4} \sqrt{u} \, du = \frac{1}{2} \cdot \frac{2}{3} u^{3/2}\Big|_4^{\pi^2+4} = \frac{1}{3}\left(\left(\pi^2 + 4\right)^{3/2} - 4^{3/2}\right) = \frac{1}{3}\left(\left(\pi^2 + 4\right)^{3/2} - 8\right) \approx 14.55$$

27. The equiangular spiral $r = e^{\theta}$ for $0 \le \theta \le 2\pi$

SOLUTION Since $f(\theta) = e^{\theta}$, by the formula for the arc length we have:

$$L = \int_0^{2\pi} \sqrt{f'(\theta)^2 + f(\theta)} \, d\theta + \int_0^{2\pi} \sqrt{\left(e^{\theta}\right)^2 + \left(e^{\theta}\right)^2} \, d\theta = \int_0^{2\pi} \sqrt{2e^{2\theta}} \, d\theta$$

$$= \sqrt{2}\int_0^{2\pi} e^{\theta} \, d\theta = \sqrt{2}e^{\theta}\Big|_0^{2\pi} = \sqrt{2}\left(e^{2\pi} - e^0\right) = \sqrt{2}\left(e^{2\pi} - 1\right) \approx 755.9$$

29. The cardioid $r = 1 - \cos\theta$ in Figure 16

SOLUTION In the equation of the cardioid, $f(\theta) = 1 - \cos\theta$. Using the formula for arc length in polar coordinates we have:

$$L = \int_{\alpha}^{\beta} \sqrt{f(\theta)^2 + f'(\theta)^2}\, d\theta \tag{1}$$

We compute the integrand:

$$\sqrt{f(\theta)^2 + f'(\theta)^2} = \sqrt{(1-\cos\theta)^2 + (\sin\theta)^2} = \sqrt{1 - 2\cos\theta + \cos^2\theta + \sin^2\theta} = \sqrt{2(1-\cos\theta)}$$

We identify the interval of θ. Since $-1 \le \cos\theta \le 1$, every $0 \le \theta \le 2\pi$ corresponds to a nonnegative value of r. Hence, θ varies from 0 to 2π. By (1) we obtain

$$L = \int_0^{2\pi} \sqrt{2(1-\cos\theta)}\, d\theta$$

Now, $1 - \cos\theta = 2\sin^2(\theta/2)$, and on the interval $0 \le \theta \le \pi$, $\sin(\theta/2)$ is nonnegative, so that $\sqrt{2(1-\cos\theta)} = \sqrt{4\sin^2(\theta/2)} = 2\sin(\theta/2)$ there. The graph is symmetric, so it suffices to compute the integral for $0 \le \theta \le \pi$, and we have

$$L = 2\int_0^{\pi} \sqrt{2(1-\cos\theta)}\, d\theta = 2\int_0^{\pi} 2\sin(\theta/2)\, d\theta = 8\sin\frac{\theta}{2}\Big|_0^{\pi} = 8$$

In Exercises 31 and 32, express the length of the curve as an integral but do not evaluate it.

31. $r = (2-\cos\theta)^{-1}, \quad 0 \le \theta \le 2\pi$

SOLUTION We have $f(\theta) = (2-\cos\theta)^{-1}$, $f'(\theta) = -(2-\cos\theta)^{-2}\sin\theta$, hence,

$$\sqrt{f^2(\theta) + f'(\theta)^2} = \sqrt{(2-\cos\theta)^{-2} + (2-\cos\theta)^{-4}\sin^2\theta} = \sqrt{(2-\cos\theta)^{-4}\left((2-\cos\theta)^2 + \sin^2\theta\right)}$$

$$= (2-\cos\theta)^{-2}\sqrt{4 - 4\cos\theta + \cos^2\theta + \sin^2\theta} = (2-\cos\theta)^{-2}\sqrt{5 - 4\cos\theta}$$

Using the integral for the arc length we get

$$L = \int_0^{2\pi} \sqrt{5-4\cos\theta}\,(2-\cos\theta)^{-2}\, d\theta.$$

In Exercises 33–36, use a computer algebra system to calculate the total length to two decimal places.

33. CAS The three-petal rose $r = \cos 3\theta$ in Figure 20

SOLUTION We have $f(\theta) = \cos 3\theta$, $f'(\theta) = -3\sin 3\theta$, so that

$$\sqrt{f(\theta)^2 + f'(\theta)^2} = \sqrt{\cos^2 3\theta + 9\sin^2 3\theta} = \sqrt{\cos^2 3\theta + \sin^2 3\theta + 8\sin^2 3\theta} = \sqrt{1 + 8\sin^2 3\theta}$$

Note that the curve is traversed completely for $0 \le \theta \le \pi$. Using the arc length formula and evaluating with Maple gives

$$L = \int_0^{\pi} \sqrt{f(\theta)^2 + f'(\theta)^2}\, d\theta = \int_0^{\pi} \sqrt{1 + 8\sin^2 3\theta}\, d\theta \approx 6.682446608$$

35. CAS The curve $r = \theta\sin\theta$ in Figure 24 for $0 \le \theta \le 4\pi$

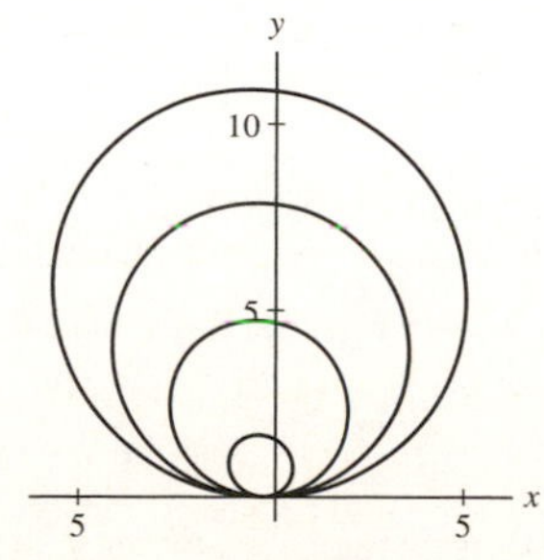

FIGURE 24 $r = \theta\sin\theta$ for $0 \le \theta \le 4\pi$.

SOLUTION We have $f(\theta) = \theta \sin\theta$, $f'(\theta) = \sin\theta + \theta\cos\theta$, so that

$$\sqrt{f(\theta)^2 + f'(\theta)^2} = \sqrt{\theta^2 \sin^2\theta + (\sin\theta + \theta\cos\theta)^2} = \sqrt{\theta^2\sin^2\theta + \sin^2\theta + 2\theta\sin\theta\cos\theta + \theta^2\cos^2\theta}$$
$$= \sqrt{\theta^2 + \sin^2\theta + \theta\sin 2\theta}$$

using the identities $\sin^2\theta + \cos^2\theta = 1$ and $2\sin\theta\cos\theta = \sin 2\theta$. Thus by the arc length formula and evaluating with Maple, we have

$$L = \int_0^{4\pi} \sqrt{f(\theta)^2 + f'(\theta)^2}\,d\theta = \int_0^{4\pi} \sqrt{\theta^2 + \sin^2\theta + \theta\sin 2\theta}\,d\theta \approx 79.56423976$$

Further Insights and Challenges

37. Suppose that the polar coordinates of a moving particle at time t are $(r(t), \theta(t))$. Prove that the particle's speed is equal to $\sqrt{(dr/dt)^2 + r^2(d\theta/dt)^2}$.

SOLUTION The speed of the particle in rectangular coordinates is:

$$\frac{ds}{dt} = \sqrt{x'(t)^2 + y'(t)^2} \tag{1}$$

We need to express the speed in polar coordinates. The x and y coordinates of the moving particles as functions of t are

$$x(t) = r(t)\cos\theta(t), \quad y(t) = r(t)\sin\theta(t)$$

We differentiate $x(t)$ and $y(t)$, using the Product Rule for differentiation. We obtain (omitting the independent variable t)

$$x' = r'\cos\theta - r(\sin\theta)\theta'$$
$$y' = r'\sin\theta - r(\cos\theta)\theta'$$

Hence,

$$x'^2 + y'^2 = (r'\cos\theta - r\theta'\sin\theta)^2 + (r'\sin\theta + r\theta'\cos\theta)^2$$
$$= r'^2\cos^2\theta - 2r'r\theta'\cos\theta\sin\theta + r^2\theta'^2\sin^2\theta + r'^2\sin^2\theta + 2r'r\theta'\sin^2\theta\cos\theta + r^2\theta'^2\cos^2\theta$$
$$= r'^2\left(\cos^2\theta + \sin^2\theta\right) + r^2\theta'^2\left(\sin^2\theta + \cos^2\theta\right) = r'^2 + r^2\theta'^2 \tag{2}$$

Substituting (2) into (1) we get

$$\frac{ds}{dt} = \sqrt{r'^2 + r^2\theta'^2} = \sqrt{\left(\frac{dr}{dt}\right)^2 + r^2\left(\frac{d\theta}{dt}\right)^2}$$

11.5 Conic Sections

Preliminary Questions

1. Which of the following equations defines an ellipse? Which does not define a conic section?

(a) $4x^2 - 9y^2 = 12$

(b) $-4x + 9y^2 = 0$

(c) $4y^2 + 9x^2 = 12$

(d) $4x^3 + 9y^3 = 12$

SOLUTION

(a) This is the equation of the hyperbola $\left(\frac{x}{\sqrt{3}}\right)^2 - \left(\frac{y}{\frac{2}{\sqrt{3}}}\right)^2 = 1$, which is a conic section.

(b) The equation $-4x + 9y^2 = 0$ can be rewritten as $x = \frac{9}{4}y^2$, which defines a parabola. This is a conic section.

(c) The equation $4y^2 + 9x^2 = 12$ can be rewritten in the form $\left(\frac{y}{\sqrt{3}}\right)^2 + \left(\frac{x}{\frac{2}{\sqrt{3}}}\right)^2 = 1$, hence it is the equation of ar ellipse, which is a conic section.

(d) This is not the equation of a conic section, since it is not an equation of degree two in x and y.

2. For which conic sections do the vertices lie between the foci?

SOLUTION If the vertices lie between the foci, the conic section is a hyperbola.

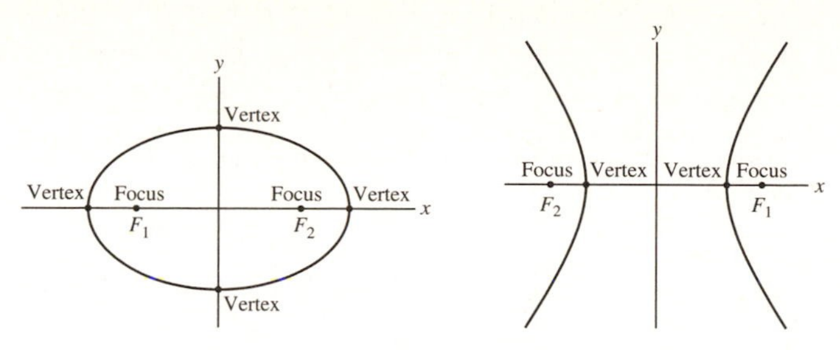

ellipse: foci between vertices   hyperbola: vertices between foci

3. What are the foci of

$$\left(\frac{x}{a}\right)^2 + \left(\frac{y}{b}\right)^2 = 1 \quad \text{if } a < b?$$

SOLUTION If $a < b$ the foci of the ellipse $\left(\frac{x}{a}\right)^2 + \left(\frac{y}{b}\right)^2 = 1$ are at the points $(0, c)$ and $(0, -c)$ on the y-axis, where $c = \sqrt{b^2 - a^2}$.

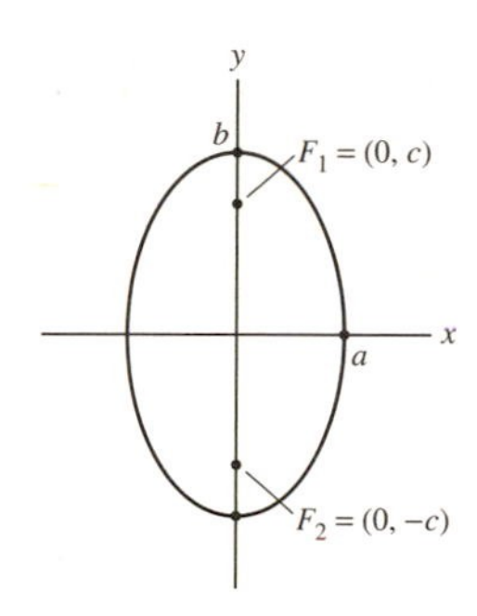

$\left(\frac{x}{a}\right)^2 + \left(\frac{y}{b}\right)^2 = 1;\ a < b$

4. What is the geometric interpretation of b/a in the equation of a hyperbola in standard position?

SOLUTION The vertices, i.e., the points where the focal axis intersects the hyperbola, are at the points $(a, 0)$ and $(-a, 0)$. The values $\pm\frac{b}{a}$ are the slopes of the two asymptotes of the hyperbola.

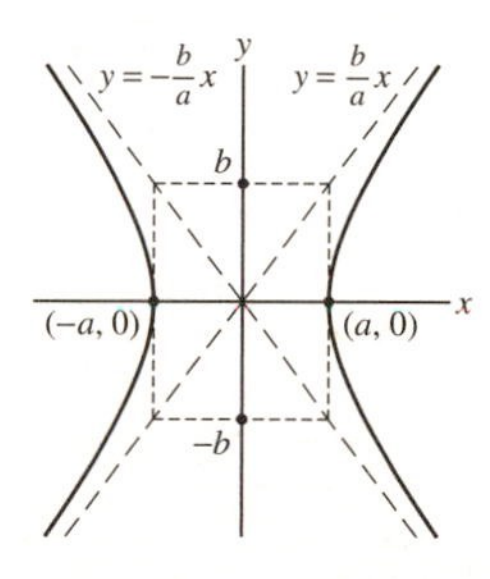

Hyperbola in standard position

Exercises

In Exercises 1–6, find the vertices and foci of the conic section.

1. $\left(\frac{x}{9}\right)^2 + \left(\frac{y}{4}\right)^2 = 1$

SOLUTION This is an ellipse in standard position with $a = 9$ and $b = 4$. Hence, $c = \sqrt{9^2 - 4^2} = \sqrt{65} \approx 8.06$. The foci are at $F_1 = (-8.06, 0)$ and $F_2 = (8.06, 0)$, and the vertices are $(9, 0)$, $(-9, 0)$, $(0, 4)$, $(0, -4)$.

3. $\left(\frac{x}{4}\right)^2 - \left(\frac{y}{9}\right)^2 = 1$

SOLUTION This is a hyperbola in standard position with $a = 4$ and $b = 9$. Hence, $c = \sqrt{a^2 + b^2} = \sqrt{97} \approx 9.85$. The foci are at $(\pm\sqrt{97}, 0)$ and the vertices are $(\pm 2, 0)$.

5. $\left(\frac{x-3}{7}\right)^2 - \left(\frac{y+1}{4}\right)^2 = 1$

SOLUTION We first consider the hyperbola $\left(\frac{x}{7}\right)^2 - \left(\frac{y}{4}\right)^2 = 1$. For this hyperbola, $a = 7$, $b = 4$ and $c = \sqrt{7^2 + 4^2} \approx 8.06$. Hence, the foci are at $(8.06, 0)$ and $(-8.06, 0)$ and the vertices are at $(7, 0)$ and $(-7, 0)$. Since the given hyperbola is obtained by translating the center of the hyperbola $\left(\frac{x}{7}\right)^2 - \left(\frac{y}{4}\right)^2 = 1$ to the point $(3, -1)$, the foci are at $F_1 = (8.06 + 3, 0 - 1) = (11.06, -1)$ and $F_2 = (-8.06 + 3, 0 - 1) = (-5.06, -1)$ and the vertices are $A = (7 + 3, 0 - 1) = (10, -1)$ and $A' = (-7 + 3, 0 - 1) = (-4, -1)$.

In Exercises 7–10, find the equation of the ellipse obtained by translating (as indicated) the ellipse

$$\left(\frac{x-8}{6}\right)^2 + \left(\frac{y+4}{3}\right)^2 = 1$$

7. Translated with center at the origin

SOLUTION Recall that the equation

$$\frac{(x-h)^2}{a^2} + \frac{(y-k)^2}{b^2} = 1$$

describes an ellipse with center (h, k). Thus, for our ellipse to be located at the origin, it must have equation

$$\frac{x^2}{6^2} + \frac{y^2}{3^2} = 1$$

9. Translated to the right six units

SOLUTION Recall that the equation

$$\frac{(x-h)^2}{a^2} + \frac{(y-k)^2}{b^2} = 1$$

describes an ellipse with center (h, k). The original ellipse has center at $(8, -4)$, so we want an ellipse with center $(14, -4)$. Thus its equation is

$$\frac{(x-14)^2}{6^2} + \frac{(y+4)^2}{3^2} = 1$$

In Exercises 11–14, find the equation of the given ellipse.

11. Vertices $(\pm 5, 0)$ and $(0, \pm 7)$

SOLUTION Since both sets of vertices are symmetric around the origin, the center of the ellipse is at $(0, 0)$. We have $a = 5$ and $b = 7$, so the equation of the ellipse is

$$\left(\frac{x}{5}\right)^2 + \left(\frac{y}{7}\right)^2 = 1$$

13. Foci $(0, \pm 10)$ and eccentricity $e = \frac{3}{5}$

SOLUTION Since the foci are on the y axis, this ellipse has a vertical major axis with center $(0, 0)$, so its equation is

$$\left(\frac{x}{b}\right)^2 + \left(\frac{y}{a}\right)^2 = 1$$

We have $a = \frac{c}{e} = \frac{10}{3/5} = \frac{50}{3}$ and

$$b = \sqrt{a^2 - c^2} = \sqrt{\frac{2500}{9} - 100} = \frac{1}{3}\sqrt{2500 - 900} = \frac{40}{3}$$

Thus the equation of the ellipse is

$$\left(\frac{x}{40/3}\right)^2 + \left(\frac{y}{50/3}\right)^2 = 1$$

In Exercises 15–20, find the equation of the given hyperbola.

15. Vertices $(\pm 3, 0)$ and foci $(\pm 5, 0)$

SOLUTION The equation is $\left(\frac{x}{a}\right)^2 - \left(\frac{y}{b}\right)^2 = 1$. The vertices are $(\pm a, 0)$ with $a = 3$ and the foci $(\pm c, 0)$ with $c = 5$. We use the relation $c = \sqrt{a^2 + b^2}$ to find b:

$$b = \sqrt{c^2 - a^2} = \sqrt{5^2 - 3^2} = \sqrt{16} = 4$$

Therefore, the equation of the hyperbola is

$$\left(\frac{x}{3}\right)^2 - \left(\frac{y}{4}\right)^2 = 1.$$

17. Foci $(\pm 4, 0)$ and eccentricity $e = 2$

SOLUTION We have $c = 4$ and $e = 2$; from $c = ae$ we get $a = 2$, and then

$$b = \sqrt{c^2 - a^2} = \sqrt{4^2 - 2^2} = 2\sqrt{3}$$

The hyperbola has center at $(0, 0)$ and horizontal axis, so its equation is

$$\left(\frac{x}{2}\right)^2 - \left(\frac{y}{2\sqrt{3}}\right)^2 = 1$$

19. Vertices $(-3, 0)$, $(7, 0)$ and eccentricity $e = 3$

SOLUTION The center is at $\frac{-3+7}{2} = 2$ with a horizontal focal axis, so the equation is

$$\left(\frac{x-2}{a}\right)^2 - \left(\frac{y}{b}\right)^2 = 1.$$

Then $a = 7 - 2 = 5$, and $c = ae = 5 \cdot 3 = 15$. Finally,

$$b = \sqrt{c^2 - a^2} = \sqrt{15^2 - 5^2} = 10\sqrt{2}$$

so that the equation of the hyperbola is

$$\left(\frac{x-2}{5}\right)^2 - \left(\frac{y}{10\sqrt{2}}\right)^2 = 1$$

In Exercises 21–28, find the equation of the parabola with the given properties.

21. Vertex $(0, 0)$, focus $\left(\frac{1}{12}, 0\right)$

SOLUTION Since the focus is on the x-axis rather than the y-axis, and the vertex is $(0, 0)$, the equation is $x = \frac{1}{4c}y^2$. The focus is $(0, c)$ with $c = \frac{1}{12}$, so the equation is

$$x = \frac{1}{4 \cdot \frac{1}{12}}y^2 = 3y^2$$

23. Vertex $(0, 0)$, directrix $y = -5$

SOLUTION The equation is $y = \frac{1}{4c}x^2$. The directrix is $y = -c$ with $c = 5$, hence $y = \frac{1}{20}x^2$.

25. Focus $(0, 4)$, directrix $y = -4$

SOLUTION The focus is $(0, c)$ with $c = 4$ and the directrix is $y = -c$ with $c = 4$, hence the equation of the parabola is

$$y = \frac{1}{4c}x^2 = \frac{x^2}{16}.$$

27. Focus $(2, 0)$, directrix $x = -2$

SOLUTION The focus is on the x-axis rather than on the y-axis and the directrix is a vertical line rather than horizontal as in the parabola in standard position. Therefore, the equation of the parabola is obtained by interchanging x and y in $y = \frac{1}{4c}x^2$. Also, by the given information $c = 2$. Hence, $x = \frac{1}{4c}y^2 = \frac{1}{4 \cdot 2}y^2$ or $x = \frac{y^2}{8}$.

In Exercises 29–38, find the vertices, foci, center (if an ellipse or a hyperbola), and asymptotes (if a hyperbola).

29. $x^2 + 4y^2 = 16$

SOLUTION We first divide the equation by 16 to convert it to the equation in standard form:

$$\frac{x^2}{16} + \frac{4y^2}{16} = 1 \Rightarrow \frac{x^2}{16} + \frac{y^2}{4} = 1 \Rightarrow \left(\frac{x}{4}\right)^2 + \left(\frac{y}{2}\right)^2 = 1$$

For this ellipse, $a = 4$ and $b = 2$ hence $c = \sqrt{4^2 - 2^2} = \sqrt{12} \approx 3.5$. Since $a > b$ we have:

- The vertices are at $(\pm 4, 0)$, $(0, \pm 2)$.
- The foci are $F_1 = (-3.5, 0)$ and $F_2 = (3.5, 0)$.
- The focal axis is the x-axis and the conjugate axis is the y-axis.
- The ellipse is centered at the origin.

31. $\left(\frac{x-3}{4}\right)^2 - \left(\frac{y+5}{7}\right)^2 = 1$

SOLUTION For this hyperbola $a = 4$ and $b = 7$ so $c = \sqrt{4^2+7^2} = \sqrt{65} \approx 8.06$. For the standard hyperbola $\left(\frac{x}{4}\right)^2 - \left(\frac{y}{7}\right)^2 = 1$, we have

- The vertices are $A = (4, 0)$ and $A' = (-4, 0)$.
- The foci are $F = (\sqrt{65}, 0)$ and $F' = (-\sqrt{65}, 0)$.
- The focal axis is the x-axis $y = 0$, and the conjugate axis is the y-axis $x = 0$.
- The center is at the midpoint of $\overline{FF'}$; that is, at the origin.
- The asymptotes $y = \pm\frac{b}{a}x$ are $y = \pm\frac{7}{4}x$.

The given hyperbola is a translation of the standard hyperbola, 3 units to the right and 5 units downward. Hence the following holds:

- The vertices are at $A = (7, -5)$ and $A' = (-1, -5)$.
- The foci are at $F = (3 + \sqrt{65}, -5)$ and $F' = (3 - \sqrt{65}, -5)$.
- The focal axis is $y = -5$ and the conjugate axis is $x = 3$.
- The center is at $(3, -5)$.
- The asymptotes are $y + 5 = \pm\frac{7}{4}(x - 3)$.

33. $4x^2 - 3y^2 + 8x + 30y = 215$

SOLUTION Since there is no cross term, we complete the square of the terms involving x and y separately:

$$4x^2 - 3y^2 + 8x + 30y = 4\left(x^2 + 2x\right) - 3\left(y^2 - 10y\right) = 4(x+1)^2 - 4 - 3(y-5)^2 + 75 = 215$$

Hence,

$$4(x+1)^2 - 3(y-5)^2 = 144$$

$$\frac{4(x+1)^2}{144} - \frac{3(y-5)^2}{144} = 1$$

$$\left(\frac{x+1}{6}\right)^2 - \left(\frac{y-5}{\sqrt{48}}\right)^2 = 1$$

This is the equation of the hyperbola obtained by translating the hyperbola $\left(\frac{x}{6}\right)^2 - \left(\frac{y}{\sqrt{48}}\right)^2 = 1$ one unit to the left and five units upwards. Since $a = 6$, $b = \sqrt{48}$, we have $c = \sqrt{36+48} = \sqrt{84} \sim 9.2$. We obtain the following table:

	Standard position	Translated hyperbola
vertices	$(6, 0), (-6, 0)$	$(5, 5), (-7, 5)$
foci	$(\pm 9.2, 0)$	$(8.2, 5), (-10.2, 5)$
focal axis	The x-axis	$y = 5$
conjugate axis	The y-axis	$x = -1$
center	The origin	$(-1, 5)$
asymptotes	$y = \pm 1.15x$	$y = -1.15x + 3.85$ $y = 1.15x + 6.15$

35. $y = 4(x-4)^2$

SOLUTION By Exercise 34, the parabola $y = 4x^2$ has the vertex at the origin, the focus at $\left(0, \frac{1}{16}\right)$ and its axis is the y-axis. Our parabola is a translation of the standard parabola four units to the right. Hence its vertex is at $(4, 0)$, the focus is at $\left(4, \frac{1}{16}\right)$ and its axis is the vertical line $x = 4$.

37. $4x^2 + 25y^2 - 8x - 10y = 20$

SOLUTION Since there are no cross terms this conic section is obtained by translating a conic section in standard position. To identify the conic section we complete the square of the terms involving x and y separately:

$$4x^2+25y^2-8x-10y=4\left(x^2-2x\right)+25\left(y^2-\frac{2}{5}y\right)$$
$$=4(x-1)^2-4+25\left(y-\frac{1}{5}\right)^2-1$$
$$=4(x-1)^2+25\left(y-\frac{1}{5}\right)^2-5=20$$

Hence,

$$4(x-1)^2+25\left(y-\frac{1}{5}\right)^2=25$$
$$\frac{4}{25}(x-1)^2+\left(y-\frac{1}{5}\right)^2=1$$
$$\left(\frac{x-1}{\frac{5}{2}}\right)^2+\left(y-\frac{1}{5}\right)^2=1$$

This is the equation of the ellipse obtained by translating the ellipse in standard position $\left(\frac{x}{\frac{5}{2}}\right)^2+y^2=1$ one unit to the right and $\frac{1}{5}$ unit upward. Since $a=\frac{5}{2}$, $b=1$ we have $c=\sqrt{\left(\frac{5}{2}\right)^2-1}\approx 2.3$, so we obtain the following table:

	Standard position	Translated ellipse
Vertices	$\left(\pm\frac{5}{2},0\right)$, $(0,\pm1)$	$\left(1\pm\frac{5}{2},\frac{1}{5}\right)$, $\left(1,\frac{1}{5}\pm1\right)$
Foci	$(-2.3,0)$, $(2.3,0)$	$\left(-1.3,\frac{1}{5}\right)$, $\left(3.3,\frac{1}{5}\right)$
Focal axis	The x-axis	$y=\frac{1}{5}$
Conjugate axis	The y-axis	$x=1$
Center	The origin	$\left(1,\frac{1}{5}\right)$

In Exercises 39–42, use the Discriminant Test to determine the type of the conic section (in each case, the equation is nondegenerate). Plot the curve if you have a computer algebra system.

39. $4x^2+5xy+7y^2=24$

SOLUTION Here, $D=25-4\cdot4\cdot7=-87$, so the conic section is an ellipse.

41. $2x^2-8xy+3y^2-4=0$

SOLUTION Here, $D=64-4\cdot2\cdot3=40$, giving us a hyperbola.

43. Show that the "conic" $x^2+3y^2-6x+12y+23=0$ has no points.

SOLUTION Complete the square in each variable separately:

$$-23=x^2-6x+3y^2+12y=(x^2-6x+9)+(3y^2+12y+12)-9-12=(x-3)^2+3(y+2)^2-21$$

Collecting constants and reversing sides gives

$$(x-3)^2+3(y+2)^2=-2$$

which has no solutions since the left-hand side is a sum of two squares so is always nonnegative.

45. Show that $\frac{b}{a}=\sqrt{1-e^2}$ for a standard ellipse of eccentricity e.

SOLUTION By the definition of eccentricity:

$$e=\frac{c}{a} \qquad (1)$$

For the ellipse in standard position, $c=\sqrt{a^2-b^2}$. Substituting into (1) and simplifying yields

$$e=\frac{\sqrt{a^2-b^2}}{a}=\sqrt{\frac{a^2-b^2}{a^2}}=\sqrt{1-\left(\frac{b}{a}\right)^2}$$

We square the two sides and solve for $\frac{b}{a}$:

$$e^2=1-\left(\frac{b}{a}\right)^2\Rightarrow\left(\frac{b}{a}\right)^2=1-e^2\Rightarrow\frac{b}{a}=\sqrt{1-e^2}$$

47. Explain why the dots in Figure 23 lie on a parabola. Where are the focus and directrix located?

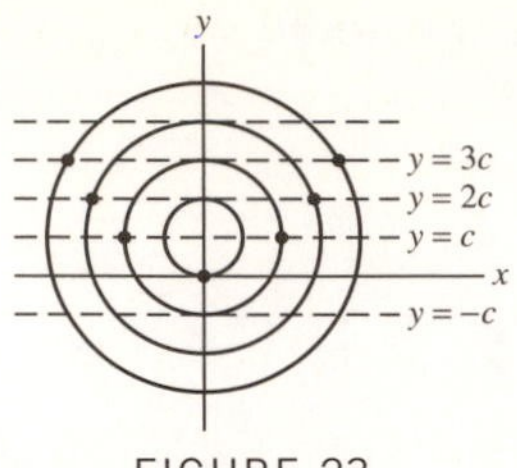

FIGURE 23

SOLUTION All the circles are centered at $(0, c)$ and the kth circle has radius kc. Hence the indicated point P_k on the kth circle has a distance kc from the point $F = (0, c)$. The point P_k also has distance kc from the line $y = -c$. That is, the indicated point on each circle is equidistant from the point $F = (0, c)$ and the line $y = -c$, hence it lies on the parabola with focus at $F = (0, c)$ and directrix $y = -c$.

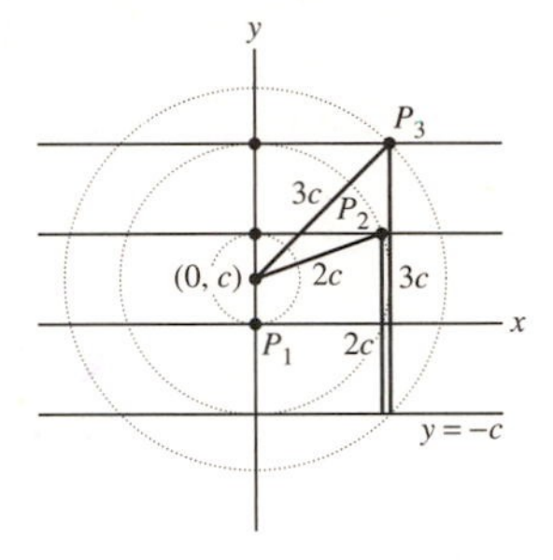

49. A **latus rectum** of a conic section is a chord through a focus parallel to the directrix. Find the area bounded by the parabola $y = x^2/(4c)$ and its latus rectum (refer to Figure 8).

SOLUTION The directrix is $y = -c$, and the focus is $(0, c)$. The chord through the focus parallel to $y = -c$ is clearly $y = c$; this line intersects the parabola when $c = x^2/(4c)$ or $4c^2 = x^2$, so when $x = \pm 2c$. The desired area is then

$$\int_{-2c}^{2c} c - \frac{1}{4c}x^2\,dx = \left(cx - \frac{1}{12c}x^3\right)\Big|_{-2c}^{2c}$$
$$= 2c^2 - \frac{8c^3}{12c} - \left(-2c^2 - \frac{(-2c)^3}{12c}\right) = 4c^2 - \frac{4}{3}c^2 = \frac{8}{3}c^2$$

In Exercises 51–54, find the polar equation of the conic with the given eccentricity and directrix, and focus at the origin.

51. $e = \frac{1}{2}$, $x = 3$

SOLUTION Substituting $e = \frac{1}{2}$ and $d = 3$ in the polar equation of a conic section we obtain

$$r = \frac{ed}{1 + e\cos\theta} = \frac{\frac{1}{2}\cdot 3}{1 + \frac{1}{2}\cos\theta} = \frac{3}{2 + \cos\theta} \Rightarrow r = \frac{3}{2 + \cos\theta}$$

53. $e = 1$, $x = 4$

SOLUTION We substitute $e = 1$ and $d = 4$ in the polar equation of a conic section to obtain

$$r = \frac{ed}{1 + e\cos\theta} = \frac{1\cdot 4}{1 + 1\cdot\cos\theta} = \frac{4}{1 + \cos\theta} \Rightarrow r = \frac{4}{1 + \cos\theta}$$

In Exercises 55–58, identify the type of conic, the eccentricity, and the equation of the directrix.

55. $r = \dfrac{8}{1 + 4\cos\theta}$

SOLUTION Matching with the polar equation $r = \frac{ed}{1+e\cos\theta}$ we get $ed = 8$ and $e = 4$ yielding $d = 2$. Since $e > 1$, the conic section is a hyperbola, having eccentricity $e = 4$ and directrix $x = 2$ (referring to the focus-directrix definition (11)).

57. $r = \dfrac{8}{4 + 3\cos\theta}$

SOLUTION We first rewrite the equation in the form $r = \frac{ed}{1+e\cos\theta}$, obtaining

$$r = \frac{2}{1 + \frac{3}{4}\cos\theta}$$

Hence, $ed = 2$ and $e = \frac{3}{4}$ yielding $d = \frac{8}{3}$. Since $e < 1$, the conic section is an ellipse, having eccentricity $e = \frac{3}{4}$ and directrix $x = \frac{8}{3}$.

59. Find a polar equation for the hyperbola with focus at the origin, directrix $x = -2$, and eccentricity $e = 1.2$.

SOLUTION We substitute $d = -2$ and $e = 1.2$ in the polar equation $r = \frac{ed}{1+e\cos\theta}$ and use Exercise 40 to obtain

$$r = \frac{1.2 \cdot (-2)}{1 + 1.2\cos\theta} = \frac{-2.4}{1 + 1.2\cos\theta} = \frac{-12}{5 + 6\cos\theta} = \frac{12}{5 - 6\cos\theta}$$

61. Find an equation in rectangular coordinates of the conic

$$r = \frac{16}{5 + 3\cos\theta}$$

Hint: Use the results of Exercise 60.

SOLUTION Put this equation in the form of the referenced exercise:

$$\frac{16}{5 + 3\cos\theta} = \frac{\frac{16}{5}}{1 + \frac{3}{5}\cos\theta} = \frac{\frac{16}{3} \cdot \frac{3}{5}}{1 + \frac{3}{5}\cos\theta}$$

so that $e = \frac{3}{5}$ and $d = \frac{16}{3}$. Then the center of the ellipse has x-coordinate

$$-\frac{de^2}{1 - e^2} = -\frac{\frac{16}{3} \cdot \frac{9}{25}}{1 - \frac{9}{25}} = -\frac{16}{3} \cdot \frac{9}{25} \cdot \frac{25}{16} = -3$$

and y-coordinate 0, and A' has x-coordinate

$$-\frac{de}{1 - e} = -\frac{\frac{16}{3} \cdot \frac{3}{5}}{1 - \frac{3}{5}} = -\frac{16}{3} \cdot \frac{3}{5} \cdot \frac{5}{2} = -8$$

and y-coordinate 0, so $a = -3 - (-8) = 5$, and the equation is

$$\left(\frac{x+3}{5}\right)^2 + \left(\frac{y}{b}\right)^2 = 1$$

To find b, set $\theta = \frac{\pi}{2}$; then $r = \frac{16}{5}$. But the point corresponding to $\theta = \frac{\pi}{2}$ lies on the y-axis, so has coordinates $\left(0, \frac{16}{5}\right)$. This point is on the ellipse, so that

$$\left(\frac{0+3}{5}\right)^2 + \left(\frac{\frac{16}{5}}{b}\right)^2 = 1 \quad \Rightarrow \quad \frac{256}{25 \cdot b^2} = \frac{16}{25} \quad \Rightarrow \quad \frac{256}{b^2} = 16 \quad \Rightarrow \quad b = 4$$

and the equation is

$$\left(\frac{x+3}{5}\right)^2 + \left(\frac{y}{4}\right)^2 = 1$$

63. Kepler's First Law states that planetary orbits are ellipses with the sun at one focus. The orbit of Pluto has eccentricity $e \approx 0.25$. Its **perihelion** (closest distance to the sun) is approximately 2.7 billion miles. Find the **aphelion** (farthest distance from the sun).

SOLUTION We define an xy-coordinate system so that the orbit is an ellipse in standard position, as shown in the figure.

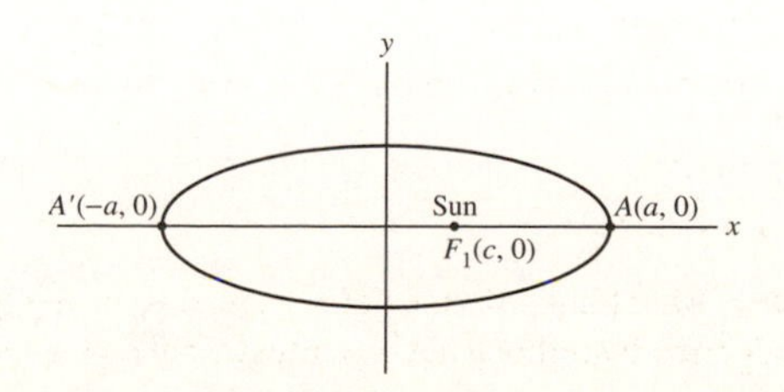

The aphelion is the length of $\overline{A'F_1}$, that is $a + c$. By the given data, we have

$$0.25 = e = \frac{c}{a} \Rightarrow c = 0.25a$$
$$a - c = 2.7 \Rightarrow c = a - 2.7$$

Equating the two expressions for c we get

$$0.25a = a - 2.7$$
$$0.75a = 2.7 \Rightarrow a = \frac{2.7}{0.75} = 3.6, \; c = 3.6 - 2.7 = 0.9$$

The aphelion is thus

$$\overline{A'F_0} = a + c = 3.6 + 0.9 = 4.5 \text{ billion miles.}$$

Further Insights and Challenges

65. Verify Theorem 2.

SOLUTION Let $F_1 = (c, 0)$ and $F_2 = (-c, 0)$ and let $P(x, y)$ be an arbitrary point on the hyperbola. Then for some constant a,

$$\overline{PF_1} - \overline{PF_2} = \pm 2a$$

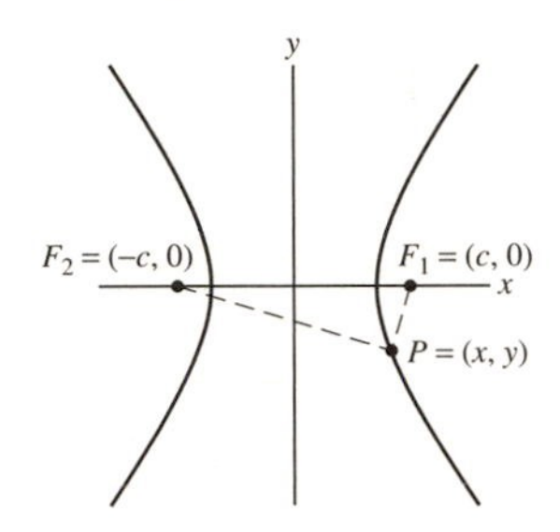

Using the distance formula we write this as

$$\sqrt{(x-c)^2 + y^2} - \sqrt{(x+c)^2 + y^2} = \pm 2a.$$

Moving the second term to the right and squaring both sides gives

$$\sqrt{(x-c)^2 + y^2} = \sqrt{(x+c)^2 + y^2} \pm 2a$$
$$(x-c)^2 + y^2 = (x+c)^2 + y^2 \pm 4a\sqrt{(x+c)^2 + y^2} + 4a^2$$
$$(x-c)^2 - (x+c)^2 - 4a^2 = \pm 4a\sqrt{(x+c)^2 + y^2}$$
$$xc + a^2 = \pm a\sqrt{(x+c)^2 + y^2}$$

We square and simplify to obtain

$$x^2c^2 + 2xca^2 + a^4 = a^2\left((x+c)^2 + y^2\right)$$
$$= a^2x^2 + 2a^2xc + a^2c^2 + a^2y^2$$
$$\left(c^2 - a^2\right)x^2 - a^2y^2 = a^2\left(c^2 - a^2\right)$$
$$\frac{x^2}{a^2} - \frac{y^2}{c^2 - a^2} = 1$$

For $b = \sqrt{c^2 - a^2}$ (or $c = \sqrt{a^2 + b^2}$) we get

$$\frac{x^2}{a^2} - \frac{y^2}{b^2} = 1 \Rightarrow \left(\frac{x}{a}\right)^2 - \left(\frac{y}{b}\right)^2 = 1.$$

67. Verify that if $e > 1$, then Eq. (11) defines a hyperbola of eccentricity e, with its focus at the origin and directrix at $x = d$.

SOLUTION The points $P = (r, \theta)$ on the hyperbola satisfy $\overline{PF} = e\overline{PD}$, $e > 1$. Referring to the figure we see that

$$\overline{PF} = r, \; \overline{PD} = d - r\cos\theta \tag{1}$$

Hence

$$r = e(d - r\cos\theta)$$

$$r = ed - er\cos\theta$$

$$r(1 + e\cos\theta) = ed \Rightarrow r = \frac{ed}{1 + e\cos\theta}$$

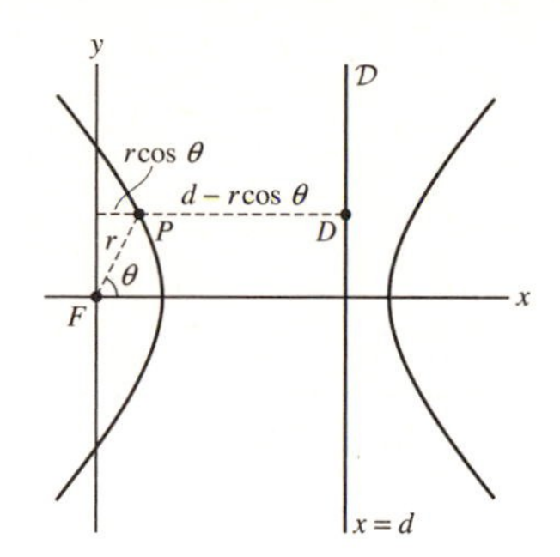

Remark: Equality (1) holds also for $\theta > \frac{\pi}{2}$. For example, in the following figure, we have

$$\overline{PD} = d + r\cos(\pi - \theta) = d - r\cos\theta$$

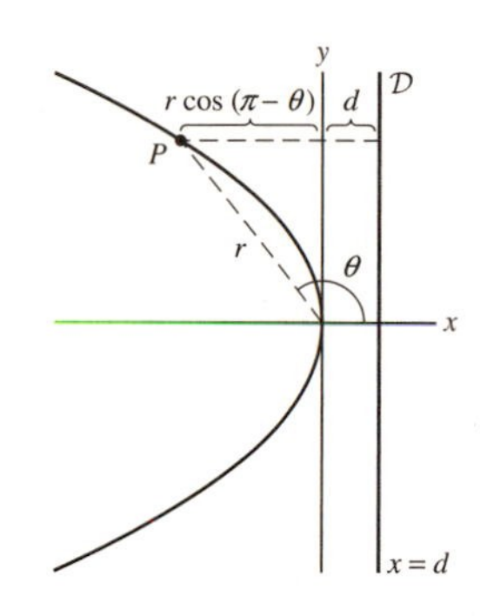

Reflective Property of the Ellipse *In Exercises 68–70, we prove that the focal radii at a point on an ellipse make equal angles with the tangent line* $\mathcal{L}$. *Let* $P = (x_0, y_0)$ *be a point on the ellipse in Figure 25 with foci* $F_1 = (-c, 0)$ *and* $F_2 = (c, 0)$, *and eccentricity* $e = c/a$.

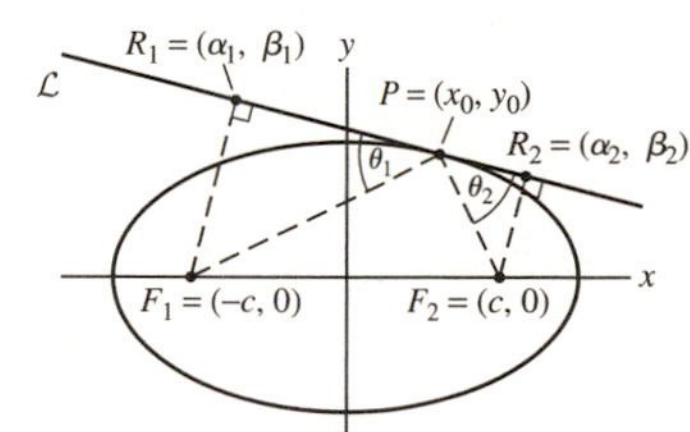

FIGURE 25 The ellipse $\left(\frac{x}{a}\right)^2 + \left(\frac{y}{b}\right)^2 = 1$.

69. Points R_1 and R_2 in Figure 25 are defined so that $\overline{F_1R_1}$ and $\overline{F_2R_2}$ are perpendicular to the tangent line.

(a) Show, with A and B as in Exercise 68, that

$$\frac{\alpha_1 + c}{\beta_1} = \frac{\alpha_2 - c}{\beta_2} = \frac{A}{B}$$

(b) Use (a) and the distance formula to show that

$$\frac{F_1R_1}{F_2R_2} = \frac{\beta_1}{\beta_2}$$

(c) Use (a) and the equation of the tangent line in Exercise 68 to show that

$$\beta_1 = \frac{B(1 + Ac)}{A^2 + B^2}, \qquad \beta_2 = \frac{B(1 - Ac)}{A^2 + B^2}$$

SOLUTION

(a) Since $R_1 = (\alpha_1, \beta_1)$ and $R_2 = (\alpha_2, \beta_2)$ lie on the tangent line at P, that is on the line $Ax + By = 1$, we have

$$A\alpha_1 + B\beta_1 = 1 \quad \text{and} \quad A\alpha_2 + B\beta_2 = 1$$

The slope of the line R_1F_1 is $\frac{\beta_1}{\alpha_1+c}$ and it is perpendicular to the tangent line having slope $-\frac{A}{B}$. Similarly, the slope of the line R_2F_2 is $\frac{\beta_2}{\alpha_2-c}$ and it is also perpendicular to the tangent line. Hence,

$$\frac{\alpha_1+c}{\beta_1}=\frac{A}{B} \quad\text{and}\quad \frac{\alpha_2-c}{\beta_2}=\frac{A}{B}.$$

(b) Using the distance formula, we have

$$\overline{R_1F_1}^2=(\alpha_1+c)^2+\beta_1^2$$

Thus,

$$\overline{R_1F_1}^2=\beta_1^2\left(\left(\frac{\alpha_1+c}{\beta_1}\right)^2+1\right) \tag{1}$$

By part (a), $\frac{\alpha_1+c}{\beta_1}=\frac{A}{B}$. Substituting in (1) gives

$$\overline{R_1F_1}^2=\beta_1^2\left(\frac{A^2}{B^2}+1\right) \tag{2}$$

Likewise,

$$\overline{R_2F_2}^2=(\alpha_2-c)^2+\beta_2{}^2=\beta_2{}^2\left(\left(\frac{\alpha_2-c}{\beta_2}\right)^2+1\right) \tag{3}$$

but since $\frac{\alpha_2-c}{\beta_2}=\frac{A}{B}$, substituting in (3) gives

$$\overline{R_2F_2}^2=\beta_2^2\left(\frac{A^2}{B^2}+1\right). \tag{4}$$

Dividing, we find that

$$\frac{\overline{R_1F_1}^2}{\overline{R_2F_2}^2}=\frac{\beta_1^2}{\beta_2^2} \quad\text{so}\quad \frac{\overline{R_1F_1}}{\overline{R_2F_2}}=\frac{\beta_1}{\beta_2},$$

as desired.

(c) In part (a) we showed that

$$\begin{cases} A\alpha_1+B\beta_1=1 \\ \dfrac{\beta_1}{\alpha_1+c}=\dfrac{B}{A} \end{cases}$$

Eliminating α_1 and solving for β_1 gives

$$\beta_1=\frac{B(1+Ac)}{A^2+B^2}. \tag{5}$$

Similarly, we have

$$\begin{cases} A\alpha_2+B\beta_2=1 \\ \dfrac{\beta_2}{\alpha_2-c}=\dfrac{B}{A} \end{cases}$$

Eliminating α_2 and solving for β_2 yields

$$\beta_2=\frac{B\,(1-Ac)}{A^2+B^2} \tag{6}$$

71. Here is another proof of the Reflective Property.

(a) Figure 25 suggests that $\mathcal{L}$ is the unique line that intersects the ellipse only in the point P. Assuming this, prove that $QF_1+QF_2>PF_1+PF_2$ for all points Q on the tangent line other than P.

(b) Use the Principle of Least Distance (Example 6 in Section 4.7) to prove that $\theta_1=\theta_2$.

SOLUTION

(a) Consider a point $Q \neq P$ on the line $\mathcal{L}$ (see figure). Since $\mathcal{L}$ intersects the ellipse in only one point, the remainder of the line lies outside the ellipse, so that QR does not have zero length, and F_2QR is a triangle. Thus

$$QF_1 + QF_2 = QR + RF_1 + QF_2 = RF_1 + (QR + QF_2) > RF_1 + RF_2$$

since the sum of lengths of two sides of a triangle exceeds the length of the third side. But since point R lies on the ellipse, $RF_2 + RF_2 = PF_1 + PF_2$, and we are done.

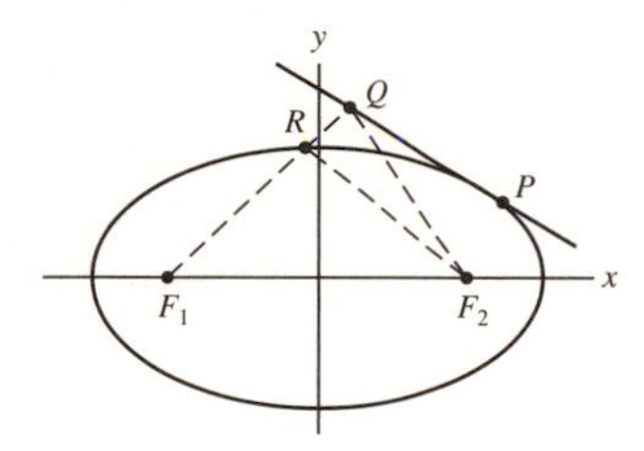

(b) Consider a beam of light traveling from F_1 to F_2 by reflection off of the line $\mathcal{L}$. By the principle of least distance, the light takes the shortest path, which by part (a) is the path through P. By Example 6 in Section 4.7, this shortest path has the property that the angle of incidence (θ_1) is equal to the angle of reflection (θ_2).

73. Show that $y = x^2/4c$ is the equation of a parabola with directrix $y = -c$, focus $(0, c)$, and the vertex at the origin, as stated in Theorem 3.

SOLUTION The points $P = (x, y)$ on the parabola are equidistant from $F = (0, c)$ and the line $y = -c$.

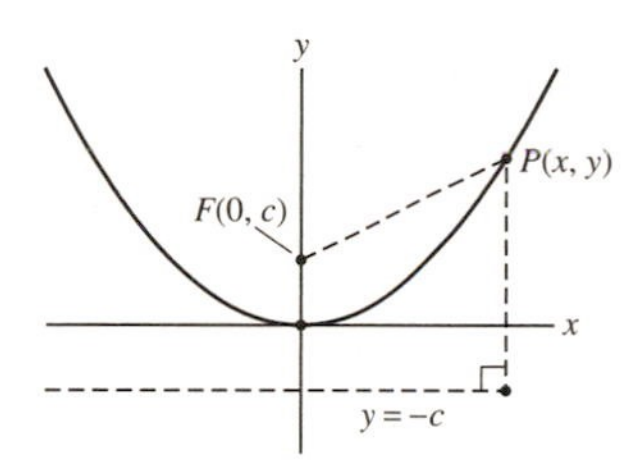

That is, by the distance formula, we have

$$\overline{PF} = \overline{PD}$$

$$\sqrt{x^2 + (y - c)^2} = |y + c|$$

Squaring and simplifying yields

$$x^2 + (y - c)^2 = (y + c)^2$$

$$x^2 + y^2 - 2yc + c^2 = y^2 + 2yc + c^2$$

$$x^2 - 2yc = 2yc$$

$$x^2 = 4yc \Rightarrow y = \frac{x^2}{4c}$$

Thus, we showed that the points that are equidistant from the focus $F = (0, c)$ and the directrix $y = -c$ satisfy the equation $y = \frac{x^2}{4c}$.

75. Derive Eqs. (13) and (14) in the text as follows. Write the coordinates of P with respect to the rotated axes in Figure 21 in polar form $x' = r\cos\alpha$, $y' = r\sin\alpha$. Explain why P has polar coordinates $(r, \alpha + \theta)$ with respect to the standard x and y-axes and derive Eqs. (13) and (14) using the addition formulas for cosine and sine.

SOLUTION If the polar coordinates of P with respect to the rotated axes are (r, α), then the line from the origin to P has length r and makes an angle of α with the rotated x-axis (the x'-axis). Since the x'-axis makes an angle of θ with the x-axis, it follows that the line from the origin to P makes an angle of $\alpha + \theta$ with the x-axis, so that the polar coordinates of P with respect to the standard axes are $(r, \alpha + \theta)$. Write (x', y') for the rectangular coordinates of P with respect to the rotated axes and (x, y) for the rectangular coordinates of P with respect to the standard axes. Then

$$x = r\cos(\alpha + \theta) = (r\cos\alpha)\cos\theta - (r\sin\alpha)\sin\theta = x'\cos\theta - y'\sin\theta$$

$$y = r\sin(\alpha + \theta) = r\sin\alpha\cos\theta + r\cos\alpha\sin\theta = (r\cos\alpha)\sin\theta + (r\sin\alpha)\cos\theta = x'\sin\theta + y'\cos\theta$$

CHAPTER REVIEW EXERCISES

1. Which of the following curves pass through the point $(1, 4)$?

(a) $c(t) = (t^2, t + 3)$ **(b)** $c(t) = (t^2, t - 3)$

(c) $c(t) = (t^2, 3 - t)$ **(d)** $c(t) = (t - 3, t^2)$

SOLUTION To check whether it passes through the point $(1, 4)$, we solve the equations $c(t) = (1, 4)$ for the given curves.

(a) Comparing the second coordinate of the curve and the point yields:

$$t + 3 = 4$$
$$t = 1$$

We substitute $t = 1$ in the first coordinate, to obtain

$$t^2 = 1^2 = 1$$

Hence the curve passes through $(1, 4)$.

(b) Comparing the second coordinate of the curve and the point yields:

$$t - 3 = 4$$
$$t = 7$$

We substitute $t = 7$ in the first coordinate to obtain

$$t^2 = 7^2 = 49 \neq 1$$

Hence the curve does not pass through $(1, 4)$.

(c) Comparing the second coordinate of the curve and the point yields

$$3 - t = 4$$
$$t = -1$$

We substitute $t = -1$ in the first coordinate, to obtain

$$t^2 = (-1)^2 = 1$$

Hence the curve passes through $(1, 4)$.

(d) Comparing the first coordinate of the curve and the point yields

$$t - 3 = 1$$
$$t = 4$$

We substitute $t = 4$ in the second coordinate, to obtain:

$$t^2 = 4^2 = 16 \neq 4$$

Hence the curve does not pass through $(1, 4)$.

3. Find parametric equations for the circle of radius 2 with center $(1, 1)$. Use the equations to find the points of intersectior of the circle with the x- and y-axes.

SOLUTION Using the standard technique for parametric equations of curves, we obtain

$$c(t) = (1 + 2\cos t, 1 + 2\sin t)$$

We compare the x coordinate of $c(t)$ to 0:

$$1 + 2\cos t = 0$$
$$\cos t = -\frac{1}{2}$$
$$t = \pm\frac{2\pi}{3}$$

Substituting in the y coordinate yields

$$1 + 2\sin\left(\pm\frac{2\pi}{3}\right) = 1 \pm 2\frac{\sqrt{3}}{2} = 1 \pm \sqrt{3}$$

Hence, the intersection points with the y-axis are $(0, 1 \pm \sqrt{3})$. We compare the y coordinate of $c(t)$ to 0:

$$1 + 2\sin t = 0$$
$$\sin t = -\frac{1}{2}$$
$$t = -\frac{\pi}{6} \quad \text{or} \quad \frac{7}{6}\pi$$

Substituting in the x coordinates yields

$$1 + 2\cos\left(-\frac{\pi}{6}\right) = 1 + 2\frac{\sqrt{3}}{2} = 1 + \sqrt{3}$$
$$1 + 2\cos\left(\frac{7}{6}\pi\right) = 1 - 2\cos\left(\frac{\pi}{6}\right) = 1 - 2\frac{\sqrt{3}}{2} = 1 - \sqrt{3}$$

Hence, the intersection points with the x-axis are $(1 \pm \sqrt{3}, 0)$.

5. Find a parametrization $c(\theta)$ of the unit circle such that $c(0) = (-1, 0)$.

SOLUTION The unit circle has the parametrization

$$c(t) = (\cos t, \sin t)$$

This parametrization does not satisfy $c(0) = (-1, 0)$. We replace the parameter t by a parameter θ so that $t = \theta + \alpha$, to obtain another parametrization for the circle:

$$c^*(\theta) = (\cos(\theta + \alpha), \sin(\theta + \alpha)) \qquad (1)$$

We need that $c^*(0) = (1, 0)$, that is,

$$c^*(0) = (\cos\alpha, \sin\alpha) = (-1, 0)$$

Hence

$$\begin{aligned} \cos\alpha &= -1 \\ \sin\alpha &= 0 \end{aligned} \quad \Rightarrow \quad \alpha = \pi$$

Substituting in (1) we obtain the following parametrization:

$$c^*(\theta) = (\cos(\theta + \pi), \sin(\theta + \pi))$$

7. Find a path $c(t)$ that traces the line $y = 2x + 1$ from $(1, 3)$ to $(3, 7)$ for $0 \le t \le 1$.

SOLUTION Solution 1: By one of the examples in section 12.1, the line through $P = (1, 3)$ with slope 2 has the parametrization

$$c(t) = (1 + t, 3 + 2t)$$

But this parametrization does not satisfy $c(1) = (3, 7)$. We replace the parameter t by a parameter s so that $t = \alpha s + \beta$. We get

$$c^*(s) = (1 + \alpha s + \beta, 3 + 2(\alpha s + \beta)) = (\alpha s + \beta + 1, 2\alpha s + 2\beta + 3)$$

We need that $c^*(0) = (1, 3)$ and $c^*(1) = (3, 7)$. Hence,

$$c^*(0) = (1 + \beta, 3 + 2\beta) = (1, 3)$$
$$c^*(1) = (\alpha + \beta + 1, 2\alpha + 2\beta + 3) = (3, 7)$$

We obtain the equations

$$\begin{aligned} 1 + \beta &= 1 \\ 3 + 2\beta &= 3 \\ \alpha + \beta + 1 &= 3 \\ 2\alpha + 2\beta + 3 &= 7 \end{aligned} \quad \Rightarrow \quad \beta = 0, \alpha = 2$$

Substituting in (1) gives

$$c^*(s) = (2s + 1, 4s + 3)$$

Solution 2: The segment from $(1,3)$ to $(3,7)$ has the following vector parametrization:

$$(1-t)\langle 1,3\rangle + t\langle 3,7\rangle = \langle 1-t+3t, 3(1-t)+7t\rangle = \langle 1+2t, 3+4t\rangle$$

The parametrization is thus

$$c(t) = (1+2t, 3+4t)$$

In Exercises 9–12, express the parametric curve in the form $y = f(x)$.

9. $c(t) = (4t-3, 10-t)$

SOLUTION We use the given equation to express t in terms of x.

$$x = 4t - 3$$
$$4t = x + 3$$
$$t = \frac{x+3}{4}$$

Substituting in the equation of y yields

$$y = 10 - t = 10 - \frac{x+3}{4} = -\frac{x}{4} + \frac{37}{4}$$

That is,

$$y = -\frac{x}{4} + \frac{37}{4}$$

11. $c(t) = \left(3 - \frac{2}{t}, t^3 + \frac{1}{t}\right)$

SOLUTION We use the given equation to express t in terms of x:

$$x = 3 - \frac{2}{t}$$
$$\frac{2}{t} = 3 - x$$
$$t = \frac{2}{3-x}$$

Substituting in the equation of y yields

$$y = \left(\frac{2}{3-x}\right)^3 + \frac{1}{2/(3-x)} = \frac{8}{(3-x)^3} + \frac{3-x}{2}$$

In Exercises 13–16, calculate dy/dx at the point indicated.

13. $c(t) = (t^3 + t, t^2 - 1), \quad t = 3$

SOLUTION The parametric equations are $x = t^3 + t$ and $y = t^2 - 1$. We use the theorem on the slope of the tangent line to find $\frac{dy}{dx}$:

$$\frac{dy}{dx} = \frac{\frac{dy}{dt}}{\frac{dx}{dt}} = \frac{2t}{3t^2+1}$$

We now substitute $t = 3$ to obtain

$$\left.\frac{dy}{dx}\right|_{t=3} = \frac{2\cdot 3}{3\cdot 3^2 + 1} = \frac{3}{14}$$

15. $c(t) = (e^t - 1, \sin t), \quad t = 20$

SOLUTION We use the theorem for the slope of the tangent line to find $\frac{dy}{dx}$:

$$\frac{dy}{dx} = \frac{\frac{dy}{dt}}{\frac{dx}{dt}} = \frac{(\sin t)'}{(e^t-1)'} = \frac{\cos t}{e^t}$$

We now substitute $t = 20$:

$$\left.\frac{dy}{dx}\right|_{t=0} = \frac{\cos 20}{e^{20}}$$

17. CAS Find the point on the cycloid $c(t) = (t - \sin t, 1 - \cos t)$ where the tangent line has slope $\frac{1}{2}$.

SOLUTION Since $x = t - \sin t$ and $y = 1 - \cos t$, the theorem on the slope of the tangent line gives

$$\frac{dy}{dx} = \frac{\frac{dy}{dt}}{\frac{dx}{dt}} = \frac{\sin t}{1 - \cos t}$$

The points where the tangent line has slope $\frac{1}{2}$ are those where $\frac{dy}{dx} = \frac{1}{2}$. We solve for t:

$$\frac{dy}{dx} = \frac{1}{2}$$

$$\frac{\sin t}{1 - \cos t} = \frac{1}{2} \tag{1}$$

$$2 \sin t = 1 - \cos t$$

We let $u = \sin t$. Then $\cos t = \pm\sqrt{1 - \sin^2 t} = \pm\sqrt{1 - u^2}$. Hence

$$2u = 1 \pm \sqrt{1 - u^2}$$

We transfer sides and square to obtain

$$\pm\sqrt{1 - u^2} = 2u - 1$$

$$1 - u^2 = 4u^2 - 4u + 1$$

$$5u^2 - 4u = u(5u - 4) = 0$$

$$u = 0, \ u = \frac{4}{5}$$

We find t by the relation $u = \sin t$:

$$u = 0: \quad \sin t = 0 \Rightarrow t = 0, t = \pi$$

$$u = \frac{4}{5}: \quad \sin t = \frac{4}{5} \Rightarrow t \approx 0.93, t \approx 2.21$$

These correspond to the points $(0, 1)$, $(\pi, 2)$, $(0.13, 0.40)$, and $(1.41, 1.60)$, respectively, for $0 < t < 2\pi$.

19. Find the equation of the Bézier curve with control points

$$P_0 = (-1, -1), \quad P_1 = (-1, 1), \quad P_2 = (1, 1), \quad P_3(1, -1)$$

SOLUTION We substitute the given points in the appropriate formulas in the text to find the parametric equations of the Bézier curve. We obtain

$$x(t) = -(1 - t)^3 - 3t(1 - t)^2 + t^2(1 - t) + t^3$$
$$= -(1 - 3t + 3t^2 - t^3) - (3t - 6t^2 + 3t^3) + (t^2 - t^3) + t^3$$
$$= (-2t^3 + 4t^2 - 1)$$
$$y(t) = -(1 - t)^3 + 3t(1 - t)^2 + t^2(1 - t) - t^3$$
$$= -(1 - 3t + 3t^2 - t^3) + (3t - 6t^2 + 3t^3) + (t^2 - t^3) - t^3$$
$$= (2t^3 - 8t^2 + 6t - 1)$$

21. Find the speed (as a function of t) of a particle whose position at time t seconds is $c(t) = (\sin t + t, \cos t + t)$. What is the particle's maximal speed?

SOLUTION We use the parametric definition to find the speed. We obtain

$$\frac{ds}{dt} = \sqrt{((\sin t + t)')^2 + ((\cos t + t)')^2} = \sqrt{(\cos t + 1)^2 + (1 - \sin t)^2}$$

$$= \sqrt{\cos^2 t + 2\cos t + 1 + 1 - 2\sin t + \sin^2 t} = \sqrt{3 + 2(\cos t - \sin t)}$$

We now differentiate the speed function to find its maximum:

$$\frac{d^2s}{dt^2} = \left(\sqrt{3+2(\cos t - \sin t)}\right)' = \frac{-\sin t - \cos t}{\sqrt{3+2(\cos t - \sin t)}}$$

We equate the derivative to zero, to obtain the maximum point:

$$\frac{d^2s}{dt^2} = 0$$

$$\frac{-\sin t - \cos t}{\sqrt{3+2(\cos t - \sin t)}} = 0$$

$$-\sin t - \cos t = 0$$

$$-\sin t = \cos t$$

$$\sin(-t) = \cos(-t)$$

$$-t = \frac{\pi}{4} + \pi k$$

$$t = -\frac{\pi}{4} + \pi k$$

Substituting t in the function of speed we obtain the value of the maximal speed:

$$\sqrt{3+2\left(\cos -\frac{\pi}{4} - \sin -\frac{\pi}{4}\right)} = \sqrt{3+2\left(\frac{\sqrt{2}}{2} - \left(-\frac{\sqrt{2}}{2}\right)\right)} = \sqrt{3+2\sqrt{2}}$$

In Exercises 23 and 24, let $c(t) = (e^{-t}\cos t, e^{-t}\sin t)$.

23. Show that $c(t)$ for $0 \le t < \infty$ has finite length and calculate its value.

SOLUTION We use the formula for arc length, to obtain:

$$\begin{aligned} s &= \int_0^\infty \sqrt{((e^{-t}\cos t)')^2 + ((e^{-t}\sin t)')^2 dt} \\ &= \int_0^\infty \sqrt{(-e^{-t}\cos t - e^{-t}\sin t)^2 + (-e^{-t}\sin t + e^{-t}\cos t)^2 dt} \\ &= \int_0^\infty \sqrt{e^{-2t}(\cos t + \sin t)^2 + e^{-2t}(\cos t - \sin t)^2 dt} \\ &= \int_0^\infty e^{-t}\sqrt{\cos^2 t + 2\sin t\cos t + \sin^2 t + \cos^2 t - 2\sin t\cos t + \sin^2 t dt} \\ &= \int_0^\infty e^{-t}\sqrt{2}dt = \sqrt{2}(-e^{-t})\Big|_0^\infty = -\sqrt{2}\left(\lim_{t\to\infty} e^{-t} - e^0\right) \\ &= -\sqrt{2}(0-1) = \sqrt{2} \end{aligned}$$

25. CAS Plot $c(t) = (\sin 2t, 2\cos t)$ for $0 \le t \le \pi$. Express the length of the curve as a definite integral, an approximate it using a computer algebra system.

SOLUTION We use a CAS to plot the curve. The resulting graph is shown here.

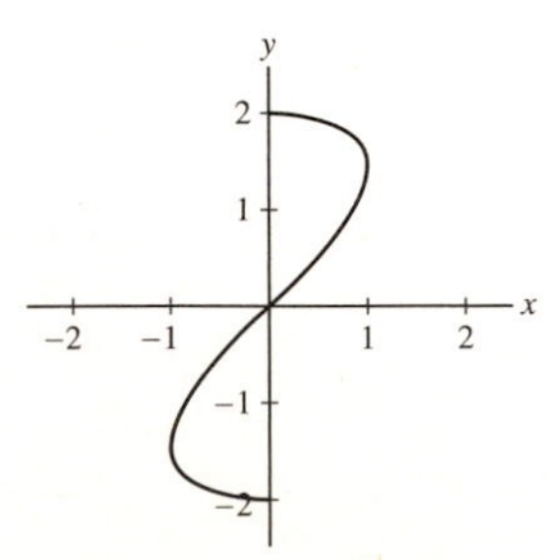

Plot of the curve $(\sin 2t, 2\cos t)$

To calculate the arc length we use the formula for the arc length to obtain

$$s = \int_0^{\pi} \sqrt{(2\cos 2t)^2 + (-2\sin t)^2}\, dt = 2\int_0^{\pi} \sqrt{\cos^2 2t + \sin^2 t}\, dt$$

We use a CAS to obtain $s = 6.0972$.

27. Convert the points $(r, \theta) = \left(1, \frac{\pi}{6}\right), \left(3, \frac{5\pi}{4}\right)$ from polar to rectangular coordinates.

SOLUTION We convert the points from polar coordinates to cartesian coordinates. For the first point we have

$$x = r\cos\theta = 1 \cdot \cos\frac{\pi}{6} = \frac{\sqrt{3}}{2}$$

$$y = r\sin\theta = 1 \cdot \sin\frac{\pi}{6} = \frac{1}{2}$$

For the second point we have

$$x = r\cos\theta = 3\cos\frac{5\pi}{4} = -\frac{3\sqrt{2}}{2}$$

$$y = r\sin\theta = 3\sin\frac{5\pi}{4} = -\frac{3\sqrt{2}}{2}$$

29. Write $r = \dfrac{2\cos\theta}{\cos\theta - \sin\theta}$ as an equation in rectangular coordinates.

SOLUTION We use the formula for converting from polar coordinates to cartesian coordinates to substitute x and y for r and θ:

$$r = \frac{2\cos\theta}{\cos\theta - \sin\theta}$$

$$\sqrt{x^2 + y^2} = \frac{2r\cos\theta}{r\cos\theta - r\sin\theta}$$

$$\sqrt{x^2 + y^2} = \frac{2x}{x - y}$$

31. GU Convert the equation

$$9(x^2 + y^2) = (x^2 + y^2 - 2y)^2$$

to polar coordinates, and plot it with a graphing utility.

SOLUTION We use the formula for converting from cartesian coordinates to polar coordinates to substitute r and θ for x and y:

$$9(x^2 + y^2) = (x^2 + y^2 - 2y)^2$$

$$9r^2 = (r^2 - 2r\sin\theta)^2$$

$$3r = r^2 - 2r\sin\theta$$

$$3 = r - 2\sin\theta$$

$$r = 3 + 2\sin\theta$$

The plot of $r = 3 + 2\sin\theta$ is shown here:

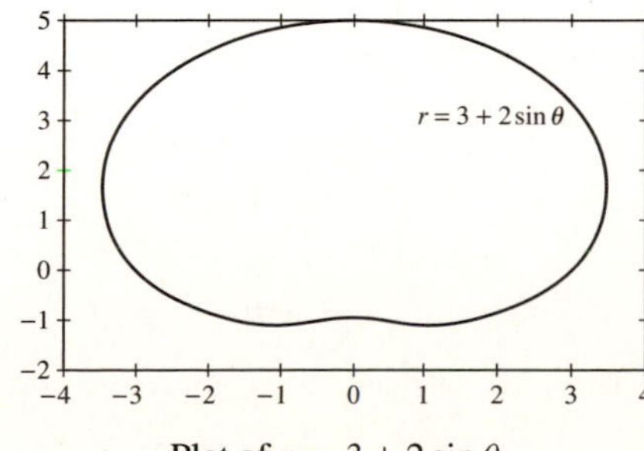

Plot of $r = 3 + 2\sin\theta$

33. Calculate the area of one petal of $r = \sin 4\theta$ (see Figure 1).

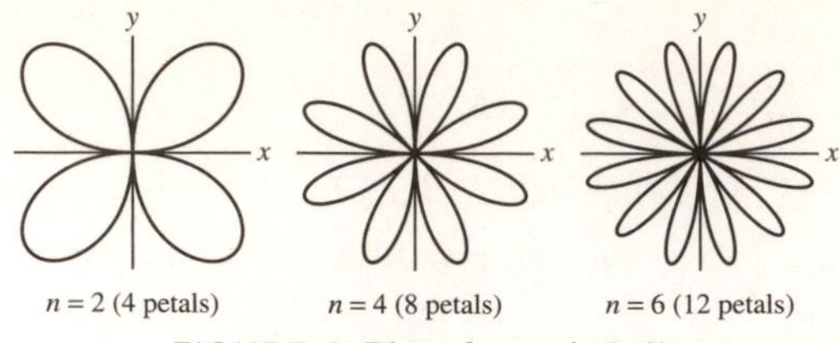

FIGURE 1 Plot of $r = \sin(n\theta)$.

SOLUTION We use a CAS to generate the plot, as shown here.

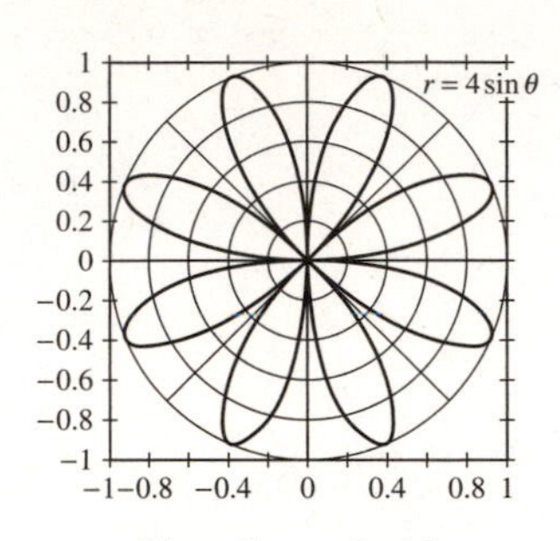

Plot of $r = \sin 4\theta$

We can see that one leaf lies between the rays $\theta = 0$ and $\theta = \frac{\theta}{4}$. We now use the formula for area in polar coordinates to obtain

$$A = \frac{1}{2}\int_0^{\pi/4} \sin^2 4\theta \, d\theta = \frac{1}{4}\int_0^{\pi/4} (1 - \cos 8\theta)\, d\theta = \frac{1}{4}\left(\theta - \frac{\sin 8\theta}{8}\Big|_0^{\pi/4}\right)$$

$$= \frac{\pi}{16} - \frac{1}{32}(\sin 2\pi - \sin 0) = \frac{\pi}{16}$$

35. Calculate the total area enclosed by the curve $r^2 = \cos\theta e^{\sin\theta}$ (Figure 2).

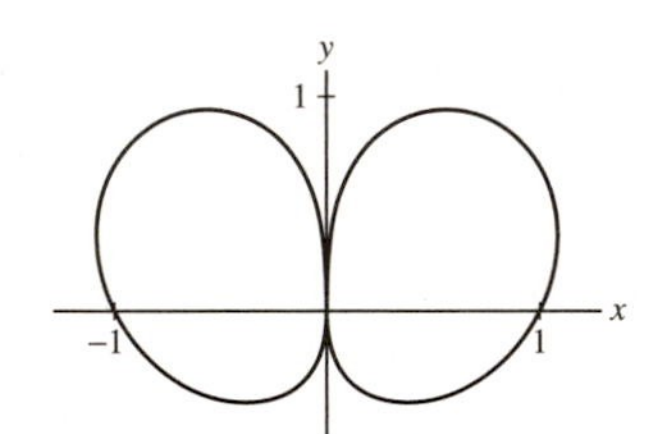

FIGURE 2 Graph of $r^2 = \cos\theta e^{\sin\theta}$.

SOLUTION Note that this is defined only for θ between $-\pi/2$ and $\pi/2$. We use the formula for area in polar coordinates to obtain:

$$A = \frac{1}{2}\int_{-\pi/2}^{\pi/2} r^2 \, d\theta = \frac{1}{2}\int_{-\pi/2}^{\pi/2} \cos\theta e^{\sin\theta}\, d\theta$$

We evaluate the integral by making the substitution $x = \sin\theta \; dx = \cos\theta \, d\theta$:

$$A = \frac{1}{2}\int_{-\pi/2}^{\pi/2} \cos\theta e^{\sin\theta}\, d\theta = \frac{1}{2}e^x\Big|_{-1}^{1} = \frac{1}{2}\left(e - e^{-1}\right)$$

37. Find the area enclosed by the cardioid $r = a(1 + \cos\theta)$, where $a > 0$.

SOLUTION The graph of $r = a\,(1 + \cos\theta)$ in the $r\theta$-plane for $0 \le \theta \le 2\pi$ and the cardioid in the xy-plane are shown in the following figures:

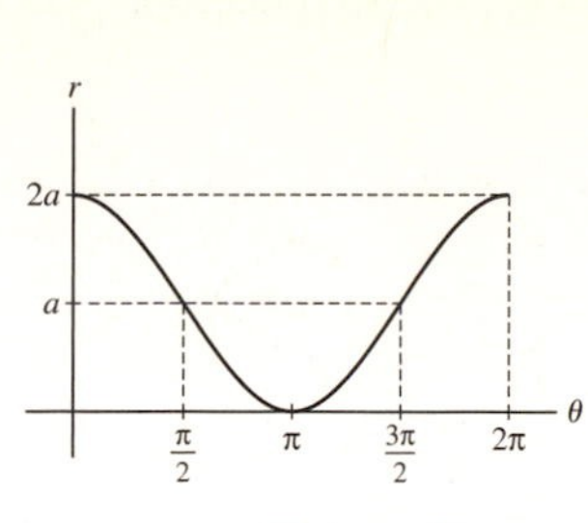

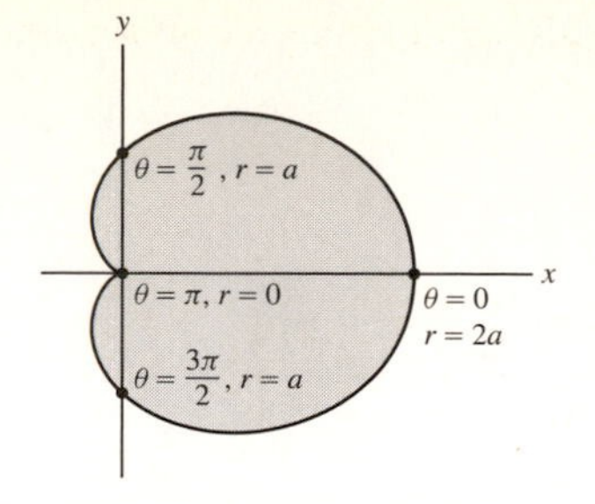

$r = a(1 + \cos\theta)$ The cardioid $r = a(1 + \cos\theta)$, $a > 0$

As θ varies from 0 to π the radius r decreases from $2a$ to 0, and this gives the upper part of the cardioid.

The lower part is traced as θ varies from π to 2π and consequently r increases from 0 back to $2a$. We compute the area enclosed by the upper part of the cardioid and the x-axis, using the following integral (we use the identity $\cos^2\theta = \frac{1}{2} + \frac{1}{2}\cos 2\theta$):

$$\frac{1}{2}\int_0^{\pi} r^2\,d\theta = \frac{1}{2}\int_0^{\pi} a^2(1+\cos\theta)^2\,d\theta = \frac{a^2}{2}\int_0^{\pi}\left(1 + 2\cos\theta + \cos^2\theta\right)d\theta$$

$$= \frac{a^2}{2}\int_0^{\pi}\left(1 + 2\cos\theta + \frac{1}{2} + \frac{1}{2}\cos 2\theta\right)d\theta = \frac{a^2}{2}\int_0^{\pi}\left(\frac{3}{2} + 2\cos\theta + \frac{1}{2}\cos 2\theta\right)d\theta$$

$$= \frac{a^2}{2}\left[\frac{3\theta}{2} + 2\sin\theta + \frac{1}{4}\sin 2\theta\right]\Bigg|_0^{\pi} = \frac{a^2}{2}\left[\frac{3\pi}{2} + 2\sin\pi + \frac{1}{4}\sin 2\pi - 0\right] = \frac{3\pi a^2}{4}$$

Using symmetry, the total area A enclosed by the cardioid is

$$A = 2\cdot\frac{3\pi a^2}{4} = \frac{3\pi a^2}{2}$$

39. CAS Figure 5 shows the graph of $r = e^{0.5\theta}\sin\theta$ for $0 \le \theta \le 2\pi$. Use a computer algebra system to approximate the difference in length between the outer and inner loops.

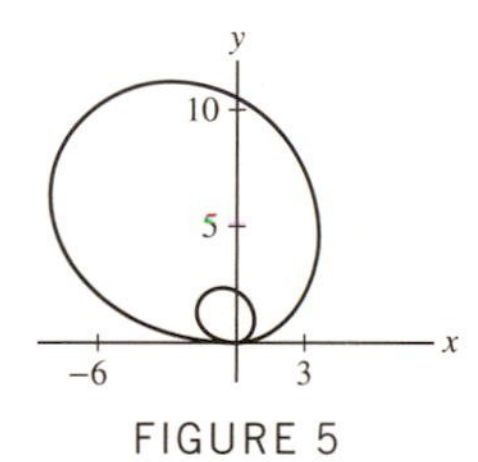

FIGURE 5

SOLUTION We note that the inner loop is the curve for $\theta \in [0,\pi]$, and the outer loop is the curve for $\theta \in [\pi, 2\pi]$. We express the length of these loops using the formula for the arc length. The length of the inner loop is

$$s_1 = \int_0^{\pi}\sqrt{(e^{0.5\theta}\sin\theta)^2 + ((e^{0.5\theta}\sin\theta)')^2 d\theta} = \int_0^{\pi}\sqrt{e^{\theta}\sin^2\theta + \left(\frac{e^{0.5\theta}\sin\theta}{2} + e^{0.5\theta}\cos\theta\right)^2}\,d\theta$$

and the length of the outer loop is

$$s_2 = \int_{\pi}^{2\pi}\sqrt{e^{\theta}\sin^2\theta + \left(\frac{e^{0.5\theta}\sin\theta}{2} + e^{0.5\theta}\cos\theta\right)^2}\,d\theta$$

We now use the CAS to calculate the arc length of each of the loops. We obtain that the length of the inner loop is 7.5087 and the length of the outer loop is 36.121, hence the outer one is 4.81 times longer than the inner one.

In Exercises 41–44, identify the conic section. Find the vertices and foci.

41. $\left(\frac{x}{3}\right)^2 + \left(\frac{y}{2}\right)^2 = 1$

SOLUTION This is an ellipse in standard position. Its foci are $(\pm\sqrt{3^2 - 2^2}, 0) = (\pm\sqrt{5}, 0)$ and its vertices are $(\pm 3, 0)$, $(0, \pm 2)$.

43. $\left(2x + \frac{1}{2}y\right)^2 = 4 - (x - y)^2$

SOLUTION We simplify the equation:

$$\left(2x + \frac{1}{2}y\right)^2 = 4 - (x - y)^2$$
$$4x^2 + 2xy + \frac{1}{4}y^2 = 4 - x^2 + 2xy - y^2$$
$$5x^2 + \frac{5}{4}y^2 = 4$$
$$\frac{5x^2}{4} + \frac{5y^2}{16} = 1$$
$$\left(\frac{x}{\frac{2}{\sqrt{5}}}\right)^2 + \left(\frac{y}{\frac{4}{\sqrt{5}}}\right)^2 = 1$$

This is an ellipse in standard position, with foci $\left(0, \pm\sqrt{\left(\frac{4}{\sqrt{5}}\right)^2 - \left(\frac{2}{\sqrt{5}}\right)^2}\right) = \left(0, \pm\sqrt{\frac{12}{5}}\right)$ and vertices $\left(\pm\frac{2}{\sqrt{5}}, 0\right)$, $\left(0, \pm\frac{4}{\sqrt{5}}\right)$.

In Exercises 45–50, find the equation of the conic section indicated.

45. Ellipse with vertices $(\pm 8, 0)$ and foci $(\pm\sqrt{3}, 0)$

SOLUTION Since the foci of the desired ellipse are on the x-axis, we conclude that $a > b$. We are given that the points $(\pm 8, 0)$ are vertices of the ellipse, and since they are on the x-axis, $a = 8$. We are given that the foci are $(\pm\sqrt{3}, 0)$ and we have shown that $a > b$, hence we have that $\sqrt{a^2 - b^2} = \sqrt{3}$. Solving for b yields

$$\sqrt{a^2 - b^2} = \sqrt{3}$$
$$a^2 - b^2 = 3$$
$$8^2 - b^2 = 3$$
$$b^2 = 61$$
$$b = \sqrt{61}$$

Next we use a and b to construct the equation of the ellipse:

$$\left(\frac{x}{8}\right)^2 + \left(\frac{y}{\sqrt{61}}\right)^2 = 1.$$

47. Hyperbola with vertices $(\pm 8, 0)$, asymptotes $y = \pm\frac{3}{4}x$

SOLUTION Since the asymptotes of the hyperbola are $y = \pm\frac{3}{4}x$, and the equation of the asymptotes for a genera hyperbola in standard position is $y = \pm\frac{b}{a}x$, we conclude that $\frac{b}{a} = \frac{3}{4}$. We are given that the vertices are $(\pm 8, 0)$, thus $a = 8$. We substitute and solve for b:

$$\frac{b}{a} = \frac{3}{4}$$
$$\frac{b}{8} = \frac{3}{4}$$
$$b = 6$$

Next we use a and b to construct the equation of the hyperbola:

$$\left(\frac{x}{8}\right)^2 - \left(\frac{y}{6}\right)^2 = 1.$$

49. Parabola with focus $(8, 0)$, directrix $x = -8$

SOLUTION This is similar to the usual equation of a parabola, but we must use y as x, and x as y, to obtain

$$x = \frac{1}{32}y^2.$$

51. Find the asymptotes of the hyperbola $3x^2 + 6x - y^2 - 10y = 1$.

SOLUTION We complete the squares and simplify:

$$3x^2 + 6x - y^2 - 10y = 1$$
$$3(x^2 + 2x) - (y^2 + 10y) = 1$$
$$3(x^2 + 2x + 1 - 1) - (y^2 + 10y + 25 - 25) = 1$$
$$3(x + 1)^2 - 3 - (y + 5)^2 + 25 = 1$$
$$3(x + 1)^2 - (y + 5)^2 = -21$$
$$\left(\frac{y+5}{\sqrt{21}}\right)^2 - \left(\frac{x+1}{\sqrt{7}}\right)^2 = 1$$

We obtained a hyperbola with focal axis that is parallel to the y-axis, and is shifted -5 units on the y-axis, and -1 units in the x-axis. Therefore, the asymptotes are

$$x + 1 = \pm\frac{\sqrt{7}}{\sqrt{21}}(y + 5) \quad \text{or} \quad y + 5 = \pm\sqrt{3}(x + 1).$$

53. Show that the relation $\frac{dy}{dx} = (e^2 - 1)\frac{x}{y}$ holds on a standard ellipse or hyperbola of eccentricity e.

SOLUTION We differentiate the equations of the standard ellipse and the hyperbola with respect to x:

Ellipse:	Hyperbola:
$\frac{x^2}{a^2} + \frac{y^2}{b^2} = 1$	$\frac{x^2}{a^2} - \frac{y^2}{b^2} = 1$
$\frac{2x}{a^2} + \frac{2y}{b^2}\frac{dy}{dx} = 0$	$\frac{2x}{a^2} - \frac{2y}{b^2}\frac{dy}{dx} = 0$
$\frac{dy}{dx} = -\frac{b^2}{a^2}\frac{x}{y}$	$\frac{dy}{dx} = \frac{b^2}{a^2}\frac{x}{y}$

The eccentricity of the ellipse is $e = \frac{\sqrt{a^2-b^2}}{a}$, hence $e^2a^2 = a^2 - b^2$ or $e^2 = 1 - \frac{b^2}{a^2}$ yielding $\frac{b^2}{a^2} = 1 - e^2$.

The eccentricity of the hyperbola is $e = \frac{\sqrt{a^2+b^2}}{a}$, hence $e^2a^2 = a^2 + b^2$ or $e^2 = 1 + \frac{b^2}{a^2}$, giving $\frac{b^2}{a^2} = e^2 - 1$. Combining with the expressions for $\frac{dy}{dx}$ we get:

Ellipse:	Hyperbola:
$\frac{dy}{dx} = -(1 - e^2)\frac{x}{y} = (e^2 - 1)\frac{x}{y}$	$\frac{dy}{dx} = (e^2 - 1)\frac{x}{y}$

We, thus, proved that the relation $\frac{dy}{dx} = (e^2 - 1)\frac{x}{y}$ holds on a standard ellipse or hyperbola of eccentricity e.

55. Refer to Figure 25 in Section 11.5. Prove that the product of the perpendicular distances F_1R_1 and F_2R_2 from the foci to a tangent line of an ellipse is equal to the square b^2 of the semiminor axes.

SOLUTION We first consider the ellipse in standard position:

$$\frac{x^2}{a^2} + \frac{y^2}{b^2} = 1$$

The equation of the tangent line at $P = (x_0, y_0)$ is

$$\frac{x_0x}{a^2} + \frac{y_0y}{b^2} = 1$$

or

$$b^2x_0x + a^2y_0y - a^2b^2 = 0$$

The distances of the foci $F_1 = (c, 0)$ and $F_2 = (-c, 0)$ from the tangent line are

$$\overline{F_1R_1} = \frac{|b^2x_0c - a^2b^2|}{\sqrt{b^4x_0^2 + a^4y_0^2}}; \quad \overline{F_2R_2} = \frac{|b^2x_0c + a^2b^2|}{\sqrt{b^4x_0^2 + a^4y_0^2}}$$

We compute the product of the distances:

$$\overline{F_1R_1} \cdot \overline{F_2R_2} = \left| \frac{\left(b^2x_0c - a^2b^2\right)\left(b^2x_0c + a^2b^2\right)}{b^4x_0^2 + a^4y_0^2} \right| = \left| \frac{b^4x_0^2c^2 - a^4b^4}{b^4x_0^2 + a^4y_0^2} \right| \tag{1}$$

The point $P = (x_0, y_0)$ lies on the ellipse, hence:

$$\frac{x_0^2}{a^2} + \frac{y_0^2}{b^2} = 1 \Rightarrow a^4y_0^2 = a^4b^2 - a^2b^2x_0^2$$

We substitute in (1) to obtain (notice that $b^2 - a^2 = -c^2$)

$$\begin{aligned}\overline{F_1R_1} \cdot \overline{F_2R_2} &= \frac{|b^4x_0^2c^2 - a^4b^4|}{|b^4x_0^2 + a^4b^2 - a^2b^2x_0^2|} = \frac{|b^4x_0^2c^2 - a^4b^4|}{|b^2(b^2 - a^2)x_0^2 + a^4b^2|} \\ &= \frac{|b^4x_0^2c^2 - a^4b^4|}{|-b^2x_0^2c^2 + a^4b^2|} = \frac{|b^2(x_0^2c^2 - a^4)|}{|-(x_0^2c^2 - a^4)|} = |-b^2| = b^2\end{aligned}$$

The product $\overline{F_1R_1} \cdot \overline{F_2R_2}$ remains unchanged if we translate the standard ellipse.